KIRK-OTHMER ENCYCLOPEDIA OF

CHEMICAL TECHNOLOGY

Fifth Edition

VOLUME 7

KIRK-OTHMER ENCYCLOPEDIA OF CHEMICAL TECHNOLOGY, FIFTH EDITION
EDITORIAL STAFF

KIRK-OTHMER ENCYCLOPEDIA OF

CHEMICAL TECHNOLOGY

Fifth Edition

VOLUME 7

Kirk-Othmer Encyclopedia of Chemical Technology
is available Online in full color and with additional content at
http://www3.interscience.wiley.com/cgi-bin/mrwhome/104554789/HOME.

WILEY-INTERSCIENCE

A John Wiley & Sons, Inc., Publication

Library of Congress Cataloging-in-Publication Data:

Kirk-Othmer encyclopedia of chemical technology. – 5th ed.
    p. cm.
Editor-in-chief, Arza Seidel.
"A Wiley-Interscience publication."
Includes index.
   ISBN 0-471-48494-6 (set) – ISBN 0-471-48516-0 (v. 7)
   1. Chemistry, Technical–Encyclopedias. I. Title: Encyclopedia of chemical technology. II. Kroschwitz, Jacqueline I.
  TP9.K54 2004
  660′.03–dc22   2003021960

Printed in the United States of America

10 9 8 7 6 5 4 3 2 1

CONTENTS

CONTRIBUTORS

Eric J. Amis, *NIST, Polymers Division, Gaithersburg, MD,* Combinatorial Chemistry

Colin Anderson, *International Coatings, Ltd., Gateshead, United Kingdom,* Coatings, Antifoulings

José M. Asua, *The University of the Basque Country, San Sebastian, Spain,* Copolymers

Allan Bradbury, *Kraft Foods Corporation, Roehrmoos, Germany,* Coffee

Robert F. Brady, Jr., *U.S. Naval Research Laboratory, Washington, DC,* Coatings, Marine

Ronald N. Caron, *Olin Corporation, New Haven, CT,* Copper Alloys, Wrought

Robert A. Charvat, *Charvat and Associates, Inc., Cleveland, OH,* Colorants for Plastics

K. J. Coeling, *Nordson Corporation, Westlake, OH,* Coating Processes, Spray

Edward D. Cohen, *Consultant, Fountain Hills, AZ,* Coating Processes

Theodore Cruz, *Kraft Foods Corporation, Tarrytown, NY,* Coffee

Larry Dominey, *OM Group, Inc. (retired), Chagrin, OH,* Cobalt and Cobalt Alloys

William H. Dresher, *WHD Consulting, Tucson, AZ,* Copper

Richard W. Drisko, *U.S. Naval Civil Engineering Laboratory, Port Hueneme, CA,* Coatings, Marine

Richard A. Eppler, *Consultant, Cheshire, CT,* Colorants for Ceramics

Peter J. S. Foot, *Kingston University, Kingston, Surrey, United Kingdom,* Conducting Polymers

Katharina M. Fromm, *University of Geneva, Geneva, Switzerland,* Coordination Compounds

Edgar B. Gutoff, *Consultant, Brookline, MA,* Coating Processes

John D. Hewes, *Honeywell International, Inc., Washington, DC,* Combinatorial Chemistry

F. Galen Hodge, *Haynes International, Inc. (retired), Kokoma, IN,* Cobalt and Cobalt Alloys

W. R., Johns, *Chemcept Limited, Reading, United Kingdom,* Computer-Aided Chemical Engineering

Debra Kaiser, *NIST, Ceramics Division, Gaithersburg, MD,* Combinatorial Chemistry

Alan B. Kaiser, *Mac Diarmid Institute for Advanced Materials* and *Victoria University of Wellington, Wellington, New Zealand,* Conducting Polymers

Alamgir Karim, *National Institute of Standards and Technology, Gaithersburg, MD,* Combinatorial Chemistry

I. Fred Koenigsberg, *White & Case LLP, New York, NY,* Copyrights

K. J. A. Kundig, *Consultant, Randolph, NJ,* Copper

José R. Leiza, *The University of the Basque Country, San Sebastian, Spain,* Copolymers

Dayal T. Meshri, *Advance Research Chemicals, Inc., Catoosa, OK,* Cobalt Compounds, Copper Compounds

Patrick Moran, *U.S. Naval Academy, Annapolis, MD,* Corrosion and Corrosion Control

Kurt Nassau, *Consultant, Lebanon, NJ,* Color

Paul Natishan, *Naval Research Laboratory, Washington, DC,* Corrosion and Corrosion Control

Simon Penson, *Kraft Foods Corporation, Warwick, United Kingdom,* Coffee

V. Pisupati, *Pennsylvania State University, University Park, PA,* Combustion Science and Technology

H. Wayne Richardson, *CP Chemicals, Inc., Sumter, SC,* Cobalt Compounds, Copper Compounds

Douglas S. Richart, *D.S. Richart Associates, Reading, PA,* Coating Processes, Powder

Martin M. Rieger, *M & A Rieger, Associates, Morris Plains, NJ,* Cosmetics

Peter W. Robinson, *Olin Corporation, Glen Carbon, IL,* Copper Alloys, Wrought

Alan W. Scaroni, *Pennsylvania State University, University Park, PA,* Combustion Science and Technology

Laurier L. Schramm, *Saskatchewan Research Council, Saskatoon, Saskatchewan, Canada,* Colloids

Reza Sharifi Sarma, *Pennsylvania State University, University Park, PA,* Combustion Science and Technology

Gerald S. Wasserman, *Kraft Foods Corporation, Tarrytown, NY,* Coffee

Zeno W. Wicks Jr., *Consultant, Louisville, KY,* Coatings, Coatings for Corrosion Control, Organic

Richard M. Wilkins, *Newcastle University, Newcastle upon Tyne, United Kingdom,* Controlled Release Technology, Agricultural

CONVERSION FACTORS, ABBREVIATIONS, AND UNIT SYMBOLS

SI Units (Adopted 1960)

The International System of Units (abbreviated SI), is implemented throughout the world. This measurement system is a modernized version of the MKSA (meter, kilogram, second, ampere) system, and its details are published and controlled by an international treaty organization (The International Bureau of Weights and Measures) (1).

SI units are divided into three classes:

BASE UNITS

length	meter[†] (m)
mass	kilogram (kg)
time	second (s)
electric current	ampere (A)
thermodynamic temperature[‡]	kelvin (K)
amount of substance	mole (mol)
luminous intensity	candela (cd)

SUPPLEMENTARY UNITS

plane angle	radian (rad)
solid angle	steradian (sr)

DERIVED UNITS AND OTHER ACCEPTABLE UNITS

These units are formed by combining base units, suplementary units, and other derived units (2–4). Those derived units having special names and symbols are marked with an asterisk in the list below.

[†]The spellings "metre" and "litre" are preferred by ASTM; however, "-er" is used in the *Encyclopedia*.

[‡]Wide use is made of Celsius temperature (t) defined by

$$t = T - T_0$$

where T is the thermodynamic temperature, expressed in kelvin, and $T_0 = 273.15$ K by definition. A temperature interval may be expressed in degrees Celsius as well as in kelvin.

Quantity	Unit	Symbol	Acceptable equivalent
*absorbed dose	gray	Gy	J/Kg
acceleration	meter per second squared	m/s^2	
*activity (of a radionuclide)	becquerel	Bq	1/s
area	square kilometer	km^2	
	square hectometer	hm^2	ha (hectare)
	square meter	m^2	
concentration (of amount of substance)	mole per cubic meter	mol/m^3	
current density	ampere per square meter	A/m^2	
density, mass density	kilogram per cubic meter	kg/m^3	g/L; mg/cm^3
dipole moment (quantity)	coulomb meter	$C \cdot m$	
*dose equivalent	sievert	Sv	J/kg
*electric capacitance	farad	F	C/V
*electric charge, quantity of electricity	coulomb	C	$A \cdot s$
electric charge density	coulomb per cubic meter	C/m^3	
*electric conductance	siemens	S	A/V
electric field strength	volt per meter	V/m	
electric flux density	coulomb per square meter	C/m^2	
*electric potential, potential difference, electromotive force	volt	V	W/A
*electric resistance	ohm	Ω	V/A
*energy, work, quantity of heat	megajoule	MJ	
	kilojoule	kJ	
	joule	J	$N \cdot m$
	electronvolt[†]	eV[†]	
	kilowatt-hour[†]	$kW \cdot h$[†]	
energy density	joule per cubic meter	J/m^3	
*force	kilonewton	kN	
	newton	N	$kg \cdot m/s^2$

[†]This non-SI unit is recognized by the CIPM as having to be retained because of practical importance or use in specialized fields (1).

Quantity	Unit	Symbol	Acceptable equivalent
*frequency	megahertz	MHz	
	hertz	Hz	1/s
heat capacity, entropy	joule per kelvin	J/K	
heat capacity (specific), specific entropy	joule per kilogram kelvin	$J/(kg \cdot K)$	
heat-transfer coefficient	watt per square meter kelvin	$W/(m^2 \cdot K)$	
*illuminance	lux	lx	lm/m^2
*inductance	henry	H	Wb/A
linear density	kilogram per meter	kg/m	
luminance	candela per square meter	cd/m^2	
*luminous flux	lumen	lm	$cd \cdot sr$
magnetic field strength	ampere per meter	A/m	
*magnetic flux	weber	Wb	$V \cdot s$
*magnetic flux density	tesla	T	Wb/m^2
molar energy	joule per mole	J/mol	
molar entropy, molar heat capacity	joule per mole kelvin	$J/(mol \cdot K)$	
moment of force, torque	newton meter	$N \cdot m$	
momentum	kilogram meter per second	$kg \cdot m/s$	
permeability	henry per meter	H/m	
permittivity	farad per meter	F/m	
*power, heat flow rate, radiant flux	kilowatt	kW	
	watt	W	J/s
power density, heat flux density, irradiance	watt per square meter	W/m^2	
*pressure, stress	megapascal	MPa	
	kilopascal	kPa	
	pascal	Pa	N/m^2
sound level	decibel	dB	
specific energy	joule per kilogram	J/kg	
specific volume	cubic meter per kilogram	m^3/kg	
surface tension	newton per meter	N/m	
thermal conductivity	watt per meter kelvin	$W/(m \cdot K)$	
velocity	meter per second	m/s	
	kilometer per hour	km/h	
viscosity, dynamic	pascal second	$Pa \cdot s$	
	millipascal second	$mPa \cdot s$	
viscosity, kinematic	square meter per second	m^2/s	
	square millimeter per second	mm^2/s	

Quantity	Unit	Symbol	Acceptable equivalent
volume	cubic meter	m^3	
	cubic diameter	dm^3	L (liter) (5)
	cubic centimeter	cm^3	mL
wave number	1 per meter	m^{-1}	
	1 per centimeter	cm^{-1}	

In addition, there are 16 prefixes used to indicate order of magnitude, as follows

Multiplication factor	Prefix	symbol	Note
10^{18}	exa	E	
10^{15}	peta	P	
10^{12}	tera	T	
10^{9}	giga	G	
10^{6}	mega	M	
10^{3}	kilo	k	
10^{2}	hecto	h[a]	[a]Although hecto, deka, deci, and
10	deka	da[a]	centi are SI prefixes, their use
10^{-1}	deci	d[a]	should be avoided except for SI
10^{-2}	centi	c[a]	unit-multiples for area and
10^{-3}	milli	m	volume and nontechnical use of
10^{-6}	micro	μ	centimeter, as for body and
10^{-9}	nano	n	clothing measurement.
10^{-12}	pico	p	
10^{-15}	femto	f	
10^{-18}	atto	a	

For a complete description of SI and its use the reader is referred to ASTM E380 (4) and the article UNITS AND CONVERSION FACTORS which appears in Vol. 24.

A representative list of conversion factors from non-SI to SI units is presented herewith. Factors are given to four significant figures. Exact relationships are followed by a dagger. A more complete list is given in the latest editions of ASTM E380 (4) and ANSI Z210.1 (6).

Conversion Factors to SI Units

To convert from	To	Multiply by
acre	square meter (m^2)	4.047×10^3
angstrom	meter (m)	$1.0 \times 10^{-10\dagger}$
are	square meter (m^2)	$1.0 \times 10^{2\dagger}$
astronomical unit	meter (m)	1.496×10^{11}

$\dagger$Exact.

To convert from	To	Multiply by
atmosphere, standard	pascal (Pa)	1.013×10^5
bar	pascal (Pa)	$1.0 \times 10^{5\dagger}$
barn	square meter (m^2)	$1.0 \times 10^{-28\dagger}$
barrel (42 U.S. liquid gallons)	cubic meter (m^3)	0.1590
Bohr magneton (μ_B)	J/T	9.274×10^{-24}
Btu (International Table)	joule (J)	1.055×10^3
Btu (mean)	joule (J)	1.056×10^3
Btu (thermochemical)	joule (J)	1.054×10^3
bushel	cubic meter(m^3)	3.524×10^{-2}
calorie (International Table)	joule (J)	4.187
calorie (mean)	joule (J)	4.190
calorie (thermochemical)	joule (J)	$4.184^\dagger$
centipoise	pascal second (Pa $\cdot$ s)	$1.0 \times 10^{-3\dagger}$
centistokes	square millimeter per second (mm^2/s)	$1.0^\dagger$
cfm (cubic foot per minute)	cubic meter per second (m^{3}s)	4.72×10^{-4}
cubic inch	cubic meter (m^3)	1.639×10^{-5}
cubic foot	cubic meter (m^3)	2.832×10^{-2}
cubic yard	cubic meter (m^3)	0.7646
curie	becquerel (Bq)	$3.70 \times 10^{10\dagger}$
debye	coulomb meter (C $\cdot$ m)	3.336×10^{-30}
degree (angle)	radian (rad)	1.745×10^{-2}
denier (international)	kilogram per meter (kg/m)	1.111×10^{-7}
	tex‡	0.1111
dram (apothecaries')	kilogram (kg)	3.888×10^{-3}
dram (avoirdupois)	kilogram (kg)	1.772×10^{-3}
dram (U.S. fluid)	cubic meter (m^3)	3.697×10^{-6}
dyne	newton (N)	$1.0 \times 10^{-5\dagger}$
dyne/cm	newton per meter (N/m)	$1.0 \times 10^{-3\dagger}$
electronvolt	joule (J)	1.602×10^{-19}
erg	joule (J)	$1.0 \times 10^{-7\dagger}$
fathom	meter (m)	1.829
fluid ounce (U.S.)	cubic meter (m^3)	2.957×10^{-5}
foot	meter (m)	$0.3048^\dagger$
footcandle	lux (lx)	10.76
furlong	meter (m)	2.012×10^{-2}
gal	meter per second squared (m/s^2)	$1.0 \times 10^{-2\dagger}$
gallon (U.S. dry)	cubic meter (m^3)	4.405×10^{-3}
gallon (U.S. liquid)	cubic meter (m^3)	3.785×10^{-3}
gallon per minute (gpm)	cubic meter per second (m^3/s)	6.309×10^{-5}
	cubic meter per hour (m^3/h)	0.2271

†Exact.
‡See footnote on p. x.

To convert from	To	Multiply by
gauss	tesla (T)	1.0×10^{-4}
gilbert	ampere (A)	0.7958
gill (U.S.)	cubic meter (m^3)	1.183×10^{-4}
grade	radian	1.571×10^{-2}
grain	kilogram (kg)	6.480×10^{-5}
gram force per denier	newton per tex (N/tex)	8.826×10^{-2}
hectare	square meter (m^2)	$1.0 \times 10^{4\dagger}$
horsepower (550 ft · lbf/s)	watt (W)	7.457×10^{2}
horsepower (boiler)	watt (W)	9.810×10^{3}
horsepower (electric)	watt (W)	$7.46 \times 10^{2\dagger}$
hundredweight (long)	kilogram (kg)	50.80
hundredweight (short)	kilogram (kg)	45.36
inch	meter (m)	$2.54 \times 10^{-2\dagger}$
inch of mercury (32°F)	pascal (Pa)	3.386×10^{3}
inch of water (39.2°F)	pascal (Pa)	2.491×10^{2}
kilogram-force	newton (N)	9.807
kilowatt hour	megajoule (MJ)	$3.6^{\dagger}$
kip	newton (N)	4.448×10^{3}
knot (international)	meter per second (m/S)	0.5144
lambert	candela per square meter (cd/m^3)	3.183×10^{3}
league (British nautical)	meter (m)	5.559×10^{3}
league (statute)	meter (m)	4.828×10^{3}
light year	meter (m)	9.461×10^{15}
liter (for fluids only)	cubic meter (m^3)	$1.0 \times 10^{-3\dagger}$
maxwell	weber (Wb)	$1.0 \times 10^{-8\dagger}$
micron	meter (m)	$1.0 \times 10^{-6\dagger}$
mil	meter (m)	$2.54 \times 10^{-5\dagger}$
mile (statue)	meter (m)	1.609×10^{3}
mile (U.S. nautical)	meter (m)	$1.852 \times 10^{3\dagger}$
mile per hour	meter per second (m/s)	0.4470
millibar	pascal (Pa)	1.0×10^{2}
millimeter of mercury (0°C)	pascal (Pa)	$1.333 \times 10^{2\dagger}$
minute (angular)	radian	2.909×10^{-4}
myriagram	kilogram (Kg)	10
myriameter	kilometer (Km)	10
oersted	ampere per meter (A/m)	79.58
ounce (avoirdupois)	kilogram (kg)	2.835×10^{-2}
ounce (troy)	kilogram (kg)	3.110×10^{-2}
ounce (U.S. fluid)	cubic meter (m^3)	2.957×10^{-5}
ounce-force	newton (N)	0.2780
peck (U.S.)	cubic meter (m^3)	8.810×10^{-3}
pennyweight	kilogram (kg)	1.555×10^{-3}
pint (U.S. dry)	cubic meter (m^3)	5.506×10^{-4}

†Exact.

To convert from	To	Multiply by
pint (U.S. liquid)	cubic meter (m^3)	4.732×10^{-4}
poise (absolute viscosity)	pascal second (Pa·s)	$0.10^\dagger$
pound (avoirdupois)	kilogram (kg)	0.4536
pound (troy)	kilogram (kg)	0.3732
poundal	newton (N)	0.1383
pound-force	newton (N)	4.448
pound force per square inch (psi)	pascal (Pa)	6.895×10^3
quart (U.S. dry)	cubic meter (m^3)	1.101×10^{-3}
quart (U.S. liquid)	cubic meter (m^3)	9.464×10^{-4}
quintal	kilogram (kg)	$1.0 \times 10^{-2\dagger}$
rad	gray (Gy)	$1.0 \times 10^{-2\dagger}$
rod	meter (m)	5.029
roentgen	coulomb per kilogram (C/kg)	2.58×10^{-4}
second (angle)	radian (rad)	$4.848 \times 10^{-6\dagger}$
section	square meter (m^2)	2.590×10^6
slug	kilogram (kg)	14.59
spherical candle power	lumen (lm)	12.57
square inch	square meter (m^2)	6.452×10^{-4}
square foot	square meter (m^2)	9.290×10^{-2}
square mile	square meter (m^2)	2.590×10^6
square yard	square meter (m^2)	0.8361
stere	cubic meter (m^3)	$1.0^\dagger$
stokes (kinematic viscosity)	square meter per second (m^2/s)	$1.0 \times 10^{-4\dagger}$
tex	kilogram per meter (kg/m)	$1.0 \times 10^{-6\dagger}$
ton (long, 2240 pounds)	kilogram (kg)	1.016×10^3
ton (metric) (tonne)	kilogram (kg)	$1.0 \times 10^{3\dagger}$
ton (short, 2000 pounds)	kilogram (kg)	9.072×10^2
torr	pascal (Pa)	1.333×10^2
unit pole	weber (Wb)	1.257×10^{-7}
yard	meter (m)	$0.9144^\dagger$

†Exact.

Abbreviations and Unit Symbols

Following is a list of common abbreviations and unit symbols used in the
Encyclopedia. In general they agree with those listed in *American National
Standard Abbreviations for Use on Drawings and in Text* (*ANSI Y1.1*) (6) and
American National Standard Letter Symbols for Units in Science and Technology
(*ANSI Y10*) (6). Also included is a list of acronyms for a number of private and

government organizations as well as common industrial solvents, polymers, and other chemicals.

   Rules for Writing Unit Symbols (4):

1. Unit symbols are printed in upright letters (roman) regardless of the type style used in the surrounding text.
2. Unit symbols are unaltered in the plural.
3. Unit symbols are not followed by a period except when used at the end of a sentence.
4. Letter unit symbols are generally printed lower-case (for example, cd for candela) unless the unit name has been derived from a proper name, in which case the first letter of the symbol is capitalized (W, Pa). Prefixes and unit symbols retain their prescribed form regardless of the surrounding typography.
5. In the complete expression for a quantity, a space should be left between the numerical value and the unit symbol. For example, write 2.37 lm, *not* 2.37 lm, and 35 mm, *not* 35 mm. When the quantity is used in an adjectival sense, a hyphen is often used, for example, 35-mm film. *Exception:* No space is left between the numerical value and the symbols of degree, minute, and second of plane angle, degree Celsius, and the percent sign.
6. No space is used between the prefix and unit symbol (for example, kg).
7. Symbols, not abbreviations, should be used for units. For example, use "A," not "amp," for ampere.
8. When multiplying unit symbols, use a raised dot:

$$N \cdot m \text{ for newton meter}$$

In the case of $W \cdot h$, the dot may be omitted, thus:

$$Wh$$

An exception to this practice is made for computer printouts, automatic typewriter work, etc, where the raised dot is not possible, and a dot on the line may be used.

9. When dividing unit symbols, use one of the following forms:

$$\text{m/s} \quad or \quad \text{m} \cdot \text{s}^{-1} \quad or \quad \frac{\text{m}}{\text{s}}$$

In no case should more than one slash be used in the same expression unless parentheses are inserted to avoid ambiguity. For example, write:

$$J/(\text{mol} \cdot K) \quad or \quad J \cdot \text{mol}^{-1} \cdot K^{-1} \quad or \quad (J/\text{mol})/K$$

but *not*

$$J/\text{mol}/K$$

10. Do not mix symbols and unit names in the same expression. Write:

$$\text{joules per kilogram} \quad or \quad \text{J/kg} \quad or \quad \text{J} \cdot \text{kg}^{-1}$$

but *not*

$$\text{joules/kilogram} \quad nor \quad \text{Joules/kg} \quad nor \quad \text{Joules} \cdot \text{kg}^{-1}$$

ABBREVIATIONS AND UNITS

A	ampere	AOAC	Association of Official Analytical Chemists
A	anion (eg, HA)		
A	mass number	AOCS	American Oil Chemists' Society
a	atto (prefix for 10^{-18})		
AATCC	American Association of Textile Chemists and Colorists	APHA	American Public Health Association
		API	American Petroleum Institute
ABS	acrylonitrile–butadiene–styrene		
		aq	aqueous
abs	absolute	Ar	aryl
ac	alternating current, *n.*	*ar-*	aromatic
a-c	alternating current, *adj.*	*as-*	Asymmetric(al)
ac-	alicyclic	ASHRAE	American Society of Heating, Refrigerating, and Air Conditioning Engineers
acac	acetylacetonate		
ACGIH	American Conference of Governmental Industrial Hygienists		
		ASM	American Society for Metals
ACS	American Chemical Society		
		ASME	American Society of Mechanical Engineers
AGA	American Gas Association		
Ah	ampere hour	ASTM	American Society for Testing and Materials
AIChE	American Institute of Chemical Engineers		
		at no.	atomic number
AIME	American Institute of Mining, metallurgical, and Petroleum Engineers	at wt	atomic weight
		av(g)	average
		AWS	American Welding Society
		b	bonding orbital
AIP	American Institute of Physics	bbl	barrel
		bcc	body-centered cubic
AISI	American Iron and Steel Institute	BCT	body-centered tetragonal
		Bé	Baumé
alc	alcohol(ic)	BET	Brunauer-Emmett-Teller (adsorption equation)
Alk	alkyl		
alk	alkaline (not alkali)	bid	twice daily
amt	amount	Boc	*t*-butyloxycarbonyl
amu	atomic mass unit	BOD	biochemical (biological) oxygen demand
ANSI	American National Standards Institute		
		bp	boiling point
AO	atomic orbital	Bq	becquerel

C	coulomb	dil	dilute
°C	degree Celsius	DIN	Deutsche Industrie
C-	denoting attachment to		Normen
	carbon	*dl-*; DL-	racemic
c	centi (prefix for 10^{-2})	DMA	dimethylacetamide
c	critical	DMF	dimethylformamide
ca	circa (Approximately)	DMG	dimethyl glyoxime
cd	candela; current density;	DMSO	dimethyl sulfoxide
	circular dichroism	DOD	Department of Defense
CFR	Code of Federal	DOE	Department of Energy
	Regulations	DOT	Department of
cgs	centimeter-gram-second		Transportation
CI	Color Index	DP	degree of polymerization
cis-	isomer in which	dp	dew point
	substituted groups are	DPH	diamond pyramid
	on some side of double		hardness
	bond between C atoms	dstl(d)	distill(ed)
cl	carload	dta	differential thermal
cm	centimeter		analysis
cmil	circular mil	*(E)-*	entgegen; opposed
cmpd	compound	ϵ	dielectric constant
CNS	central nervous system		(unitless number)
CoA	coenzyme A	*e*	electron
COD	chemical oxygen demand	ECU	electrochemical unit
coml	commerical(ly)	ed.	edited, edition, editor
cp	chemically pure	ED	effective dose
cph	close-packed hexagonal	EDTA	ethylenediaminetetra-
CPSC	Consumer Product Safety		acetic acid
	Commission	emf	electromotive force
cryst	crystalline	emu	electromagnetic unit
cub	cubic	en	ethylene diamine
D	debye	eng	engineering
D-	denoting configurational	EPA	Environmental Protection
	relationship		Agency
d	differential operator	epr	electron paramagnetic
d	day; deci (prefix for 10^{-1})		resonance
d	density	eq.	equation
d-	*dextro-*, dextrorotatory	esca	electron spectroscopy for
da	deka (prefix for 10^{-1})		chemical analysis
dB	decibel	esp	especially
dc	direct current, *n.*	esr	electron-spin resonance
d-c	direct current, *adj.*	est(d)	estimate(d)
dec	decompose	estn	estimation
detd	determined	esu	electrostatic unit
detn	determination	exp	experiment, experimental
Di	didymium, a mixture of all	ext(d)	extract(ed)
	lanthanons	F	farad (capacitance)
dia	diameter	*F*	fraday (96,487 C)

f	femto (prefix for 10^{-15})	hyd	hydrated, hydrous
FAO	Food and Agriculture Organization (United Nations)	hyg	hygroscopic
		Hz	hertz
		^{i}(eg, Pr^i)	iso (eg, isopropyl)
fcc	face-centered cubic	i-	inactive (eg, i-methionine)
FDA	Food and Drug Administration	IACS	international Annealed Copper Standard
FEA	Federal Energy Administration	ibp	initial boiling point
		IC	integrated circuit
FHSA	Federal Hazardous Substances Act	ICC	Interstate Commerce Commission
fob	free on board	ICT	International Critical Table
fp	freezing point		
FPC	Federal Power Commission	ID	inside diameter; infective dose
FRB	Federal Reserve Board		
frz	freezing	ip	intraperitoneal
G	giga (prefix for 10^9)	IPS	iron pipe size
G	gravitational constant $= 6.67 \times 10^{11} \text{N} \cdot \text{m}^2/\text{kg}^2$	ir	infrared
		IRLG	Interagency Regulatory Liaison Group
g	gram		
(g)	gas, only as in $H_2O(g)$	ISO	International Organization Standardization
g	gravitational acceleration		
gc	gas chromatography		
gem-	geminal	ITS-90	International Temperature Scale (NIST)
glc	gas–liquid chromatography		
g-mol wt; gmw	gram-molecular weight	IU	International Unit
		IUPAC	International Union of Pure and Applied Chemistry
GNP	gross national product		
gpc	gel-permeation chromatography		
		IV	iodine value
GRAS	Generally Recognized as Safe	iv	intravenous
		J	joule
grd	ground	K	kelvin
Gy	gray	k	kilo (prefix for 10^3)
H	henry	kg	kilogram
h	hour; hecto (prefix for 10^2)	L	denoting configurational relationship
ha	hectare		
HB	Brinell hardness number	L	liter (for fluids only) (5)
Hb	hemoglobin	l-	$levo$-, levorotatory
hcp	hexagonal close-packed	(l)	liquid, only as in $NH_3(l)$
hex	hexagonal	LC_{50}	conc lethal to 50% of the animals tested
HK	Knoop hardness number		
hplc	high performance liquid chromatography	LCAO	linear combination of atomic orbitals
HRC	Rockwell hardness (C scale)	lc	liquid chromatography
		LCD	liquid crystal display
HV	Vickers hardness number	lcl	less than carload lots

LD_{50}	dose lethal to 50% of the animals tested	N	newton (force)
		N	normal (concentration); neutron number
LED	light-emitting diode		
liq	liquid	$N\text{-}$	denoting attachment to nitrogen
lm	lumen		
ln	logarithm (natural)	n (as n_{D}^{20})	index of refraction (for 20°C and sodium light)
LNG	liquefied natural gas		
log	logarithm (common)		
LOI	limiting oxygen index	n (as Bu^n),	normal (straight-chain structure)
LPG	liquefied petroleum gas	$n\text{-}$	
ltl	less than truckload lots	n	neutron
lx	lux	n	nano (prefix for 10^9)
M	mega (prefix for 10^6); metal (as in MA)	na	not available
		NAS	National Academy of Sciences
M	molar; actual mass		
$\overline{M}_w$	weight-average mol wt	NASA	National Aeronautics and Space Administration
$\overline{M}_n$	number-average mol wt		
m	meter; milli (prefix for 10^{-3})	nat	natural
		ndt	nondestructive testing
m	molal	neg	negative
$m\text{-}$	meta	NF	*National Formulary*
max	maximum	NIH	National Institutes of Health
MCA	Chemical Manufacturers' Association (was Manufacturing Chemists Association)		
		NIOSH	National Institute of Occupational Safety and Health
MEK	methyl ethyl ketone	NIST	National Institute of Standards and Technology (formerly National Bureau of Standards)
meq	milliequivalent		
mfd	manufactured		
mfg	manufacturing		
mfr	manufacturer		
MIBC	methyl isobutyl carbinol	nmr	nuclear magnetic resonance
MIBK	methyl isobutyl ketone		
MIC	minimum inhibiting concentration	NND	New and Nonofficial Drugs (AMA)
min	minute; minimum	no.	number
mL	milliliter	NOI-(BN)	not otherwise indexed (by name)
MLD	minimum lethal dose		
MO	molecular orbital	NOS	not otherwise specified
mo	month	nqr	nuclear quadruple resonance
mol	mole		
mol wt	molecular weight	NRC	Nuclear Regulatory Commission; National Research Council
mp	melting point		
MR	molar refraction		
ms	mass spectrometry	NRI	New Ring Index
MSDS	material safety data sheet	NSF	National Science Foundation
mxt	mixture		
μ	micro (prefix for 10^{-6})	NTA	nitrilotriacetic acid

NTP	normal temperature and pressure (25°C and 101.3 kPa or 1 atm)	pwd	powder
		py	pyridine
		qv	quod vide (which see)
NTSB	National Transportation Safety Board	R	univalent hydrocarbon radical
O-	denoting attachment to oxygen	(R)-	rectus (clockwise configuration)
o-	ortho	r	precision of data
OD	outside diameter	rad	radian; radius
OPEC	Organization of Petroleum Exporting Countries	RCRA	Resource Conservation and Recovery Act
o-phen	o-phenanthridine	rds	rate-determining step
OSHA	Occupational Safety and Health Administration	ref.	reference
		rf	radio frequency, n.
owf	on weight of fiber	r-f	radio frequency, adj.
Ω	ohm	rh	relative humidity
P	peta (prefix for 10^{15})	RI	Ring Index
p	pico (prefix for 10^{-12}	rms	root-mean square
p-	para	rpm	rotations per minute
p	proton	rps	revolutions per second
p.	page	RT	room temperature
Pa	Pascal (pressure)	RTECS	Registry of Toxic Effects of Chemical Substances
PEL	personal exposure limit based on an 8-h exposure	^{s}(eg, Bu^s); sec-	secondary (eg, secondary butyl)
pd	potential difference	S	siemens
pH	negative logarithm of the effective hydrogen ion concentration	(S)-	sinister (counterclockwise configuration)
		S-	denoting attachment to sulfur
phr	parts per hundred of resin (rubber)	s-	symmetric(al)
p-i-n	positive-intrinsic-negative	S	second
pmr	proton magnetic resonance	(s)	solid, only as in $H_2O(s)$
p-n	positive-negative	SAE	Society of Automotive Engineers
po	per os (oral)		
POP	polyoxypropylene	SAN	styrene-acrylonitrile
pos	positive	sat(d)	saturate(d)
pp.	pages	satn	saturation
ppb	parts per billion (10^9)	SBS	styrene–butadiene–styrene
ppm	parts per milion (10^6)	sc	subcutaneous
ppmv	parts per million by volume	SCF	self-consistent field; standard cubic feet
ppmwt	parts per million by weight		
PPO	poly(phenyl oxide)	Sch	Schultz number
ppt(d)	precipitate(d)	sem	scanning electron microscope(y)
pptn	precipitation		
Pr (no.)	foreign prototype (number)	SFs	Saybolt Furol seconds
pt	point; part	sl sol	slightly soluble
PVC	poly(vinyl chloride)	sol	soluble

soln	solution	*trans-*	isomer in which substituted groups are on opposite sides of double bond between C atoms
soly	solubility		
sp	specific; species		
sp gr	specific gravity		
sr	steradian		
std	standard	TSCA	Toxic Substances Control Act
STP	standard temperature and pressure (0°C and 101.3 kPa)		
		TWA	time-weighted average
		Twad	Twaddell
sub	sublime(s)	UL	Underwriters' Laboratory
SUs	Saybolt Universal seconds	USDA	United States Department of Agriculture
syn	synthetic		
t (eg, Bu^t), t-, *tert*-	tertiary (eg, tertiary butyl)	USP	*United States Pharmacopeia*
T	tera (prefix for 10^{12}); tesla (magnetic flux density)	uv	ultraviolet
		V	volt (emf)
t	metric to (tonne)	var	variable
t	temperature	*vic-*	vicinal
TAPPI	Technical Association of the Pulp and Paper Industry	vol	volume (not volatile)
		vs	versus
		v sol	very soluble
TCC	Tagliabue closed cup	W	watt
tex	tex (linear density)	Wb	weber
T_g	glass-transition temperature	Wh	watt hour
		WHO	World Health Organization (United Nations)
tga	thermogravimetric analysis		
		wk	week
THF	tetrahydrofuran	yr	year
tlc	thin layer chromatography	(Z)-	zusammen; together; atomic number
TLV	threshold limit value		

Non-SI (Unacceptable and Obsolete) Units		Use
Å	angstrom	nm
at	atmosphere, technical	Pa
atm	atmosphere, standard	Pa
b	barn	cm^2
bar†	bar	Pa
bbl	barrel	m^3
bhp	brake horsepower	W
Btu	British thermal unit	J
bu	bushel	m^3; L
cal	calorie	J
cfm	cubic foot per minute	m^3/s
Ci	curie	Bq
cSt	centistokes	mm^2/s
c/s	cycle per second	Hz
cu	cubic	exponential form

†Do not use bar (10^5 Pa) or millibar (10^2 Pa) because they are not SI units, and are accepted internationally only in special fields because of existing usage.

Non-SI (Unacceptable and Obsolete) Units		Use
D	debye	$C \cdot m$
den	denier	tex
dr	dram	kg
dyn	dyne	N
dyn/cm	dyne per centimeter	mN/m
erg	erg	J
eu	entropy unit	J/K
°F	degree Fahrenheit	°C; K
fc	footcandle	lx
fl	footlambert	lx
fl oz	fluid ounce	m^3; L
ft	foot	m
ft $\cdot$ lbf	foot pound-force	J
gf den	gram-force per denier	N/tex
G	gauss	T
Gal	gal	m/s^2
gal	gallon	m^3; L
Gb	gilbert	A
gpm	gallon per minute	(m^3/s); (m^3/h)
gr	grain	kg
hp	horsepower	W
ihp	indicated horsepower	W
in.	inch	m
in. Hg	inch of mercury	Pa
in. H_2O	inch of water	Pa
in.-lbf	inch pound-force	J
kcal	kilo-calorie	J
kgf	kilogram-force	N
kilo	for kilogram	kg
L	lambert	lx
lb	pound	kg
lbf	pound-force	N
mho	mho	S
mi	mile	m
MM	million	M
mm Hg	millimeter of mercury	Pa
$m\mu$	millimicron	nm
mph	miles per hour	km/h
μ	micron	μm
Oe	oersted	A/m
oz	ounce	kg
ozf	ounce-force	N
η	poise	$Pa \cdot s$
P	poise	$Pa \cdot s$
ph	phot	lx
psi	pounds-force per square inch	Pa
psia	pounds-force per square inch absolute	Pa
psig	pounds-force per square inch gage	Pa
qt	quart	m^3; L
°R	degree Rankine	K
rd	rad	Gy
sb	stilb	lx
SCF	standard cubic foot	m^3
sq	square	exponential form
thm	therm	J
yd	yard	m

BIBLIOGRAPHY

1. The International Bureau of Weights and Measures, BIPM (Parc Saint-Cloud, France) is described in Ref. 4. This bureau operates under the exclusive supervision of the International Committee for Weights and Measures (CIPM).
2. *Metric Editorial Guide (ANMC-78-1)*, latest ed., American National Metric Council, 900 Mix Avenue, Suite 1 Hamden CT 06514-5106, 1981.
3. *SI Units and Recommendations for the Use of Their Multiples and of Certain Other Units (ISO 1000-1992)*, American National Standards Institute, 25 W 43rd St., New York, 10036, 1992.
4. Based on IEEE/ASTM-SI-10 *Standard for use of the International System of Units (SI): The Modern Metric System* (Replaces ASTM380 and ANSI/IEEE Std 268-1992), ASTM International, West Conshohocken, PA., 2002. See also www.astm.org
5. *Fed. Reg.*, Dec. 10, 1976 (41 FR 36414).
6. For ANSI address, see Ref. 3. See also www.ansi.org

C

Continued

COATING PROCESSES

1. Introduction

Coating process technology is in widespread use because there are few single materials that are suitable for the intended final use without treating the surface (1–7) to meet all the functional needs and requirements of the product. The modification is accomplished by applying a coating or series of coatings to the material—the substrate—to improve its performance and make it more suitable for use, or to give it different characteristics. The coating process is defined here as replacing the air at a substrate with a new material—the coating.

Typical coatings are the paints and the diverse surface coatings used to protect houses, bridges, appliances, automobiles, etc. These coatings protect the surface from corrosion and degradation, and may provide other functional advantages such as making the materials waterproof or flameproof, and improving the appearance. Adhesives are applied to paper or plastic to produce labels and tapes for a variety of uses. A thin layer of adhesive is coated onto paper to produce self-sticking note pads. Glass windows are coated with a variety of materials to make them stronger and to control light penetration into the structure. High energy lithium batteries contain coated structures. Plastic food wrap has layers to reduce oxygen penetration and retain moisture for the product to retain its freshness. Packaging materials for electronic products are coated with antistatic compounds to protect the sensitive components. Other important coated products are photographic films for medical, industrial, graphic arts, and consumer use; optical and magnetic media for audio and visual use data storage; printing plates; and glossy paper for magazines.

Several industries are based on coating process technology. Printing itself is a coating process, and much of the paper used in the printing industry had

1

previously been coated to improve its gloss, strength, and ink acceptability. Lithographic printing plates for printing presses are photosensitive coatings on aluminum. Photographic film, itself a coated product, is used to expose the plates, set the type, and prepare the printed pictures. The entertainment industry uses magnetic tape, silver halide film, and coated optical disks to record the material for distribution to the consumer. The electronics industry uses coated products such as photoresist films to fabricate circuit boards and to add functionality to the circuit boards. The computer industry stores information on coated magnetic structures such as hard drives and floppy disks.

2. Coating Machines

The basic steps in continuously producing a coated structure are

1. preparing the coating solution or dispersion
2. unwinding the roll of substrate
3. transporting it through the coater
4. applying the coating from a solvent, or as a liquid to be cross-linked, or from the vapor
5. drying or solidifying the coating
6. winding the final coated roll
7. converting the product to the final size and shape needed

Other operations that are often used are

1. surface treatment of the substrate to improve adhesion
2. cleaning of the substrate prior to coating, to reduce contamination
3. lamination, where a protective cover sheet is added to the coating structure.

Different types of machinery are used to produce coated products. Depending on the substrate, they can be web coaters, sheet coaters, and coaters for non-flat applications. Web coaters, the most prevalent, coat onto continuous webs of material. Magnetic tapes, window films, wallpaper; barrier coatings for plastic films, and many printed goods are all produced using this process. A typical web coater with all the process steps is shown in Figure 1.

These machines are commercially available in sizes from pilot coaters using narrow webs, 6–24 in. wide, and running at low speeds, 10–50 ft/min, to production machines using wide webs, over 5 ft wide and coating at 500–5000 ft/min. A typical pilot coater is shown in Figure 2.

Sheet coaters are available to coat individual sheets. Many printing operations and all copying machines are sheet-fed. Sheet coaters are also used as laboratory coaters to develop new products where many different solutions need to be coated and only small volumes of sample are available. These methods use a variety of devices such as draw-down blades, dies, or wire-wound rods to spread a uniform layer of solution across the web. Most applicators can provide

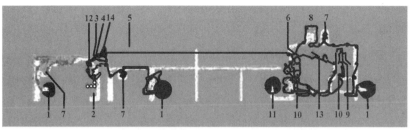

1. Unwinder   6. Dry laminating station   11. Rewinder
2. Coating unit   7. Corona station   12. Wet laminating station
3. Infrared lamp (short wave)   8. Edge guide unit   13. Uv curing unit
4. Infrared lamp (medium wave)   9. Remoisturising station   14. Curtain coater
5. Air floatation dryer   10. Cooling station

Fig. 1. Coating line showing components. Courtesy of Polytype America Corp.

wet coverages of 0.2–50 mil (5–1270 μm). Spray coaters may also be used to coat sheets. The coated webs or sheets are then dried by ambient air or in an oven. Laboratory automated sheet coaters are available. These give better reproducibility and control but are more expensive than hand applicators. They can be coupled to feed directly into dryers, with the temperature and residence time controlled, as shown in Figure 3.

In nonweb applications the coating is applied to a specific part at the end of the fabrication process. The part is usually three-dimensional and of varying shape. Automobiles, appliances, and steel structures all have the coating applied to the individual items as they are being built. It should be noted that many smaller steel items are made from prepainted sheet steel.

Fig. 2. Pilot coater. Courtesy of Texmax, Inc.

Fig. 3.   Laboratory bench-top coater. Courtesy of Werner Mathis.

3. Coating Processes

The application of a liquid to a traveling web or substrate is accomplished by one of the many coating methods. Widely used commercial coating methods are reverse roll, wire-wound or Mayer rod, direct and offset gravure, slot die, blade, hot melt, curtain, knife over roll, extrusion, air knife, spray, rotary screen, multilayer slide, coextrusion, meniscus, comma and microgravure coaters (based on analysis of methods reported by coaters in various trade sources). Each of these has many designs and hardware arrangements leading to many specific coating configurations. Powder coating is covered in the article Coating Methods, Powder Technology.

The choice of the method depends on the nature of the support to be coated, the rheology of the coating fluid, the solvent, the wet-coating weight or coverage desired, the needed coating uniformity, the desired coating width and speed; the number of layers to be coated simultaneously, cost considerations, environmental considerations, and whether the coating is to be continuous or intermittent.

The method should be chosen based on the specific requirements. Often a method is selected based on the availability of a specific coating applicator, even though it may not be the best choice. All applicators can apply a coating at some conditions. Much time and money can be wasted by trying to make the product by a process that is not suitable. The coating window may be too narrow at the conditions selected, or it may be impossible to ever obtain a quality coating. The successful process will provide defect-free film over a wide range of conditions. A process that works well at low speeds in the laboratory may not be appropriate for a manufacturing plant coating at high speeds, and conversely, a high speed coating process may not be appropriate for laboratory trials.

The first step in the selection process is to establish the requirements for the product to be coated. These requirements are then matched with the capabilities of the process and the best methods are evaluated experimentally to determine the one to use. Some of the basic characteristics of the principal coating processes are listed in Table 1.

Table 1. **Summary of Coating Methods**[a]

Process	Viscosity, mPa · s (= cP)	No. of layers	Wet thickness, μm	Coating accuracy, %	Max. speed, m/min
Self-metered					
Rod	20–1000	1	5–50	10	250
Dip	20–1000	1	5–100	10	150
Forward roll	20–1000	1	10–200	8	150
Reverse roll	100–50,000	1	5–400	5	400
Air knife	5–500	1	2–40	5	500
Knife over roll	100–50,000	1	25–750	10	150
Blade	500–40,000	1	25–750	7	1500
Premetered					
Slot	5–20,000	1–3	15–250	2	500
Extrusion	50,000– 5,000,000	1–3	15–750	5	700
Slide	5–500	1–18	15–250	2	300
Curtain	5–500	1–18	2–500	2	300
Hybrid					
Gravure, direct	1–5000	1	1–25	2	700
Gravure, offset	100–50,000	1	5–400	5	300
Microgravure	1–4000	1	1–40	2	100

[a] Values given are only guidelines.

The processes are grouped according to the principle used to control the coverage or coating weight of the coating and its resulting uniformity. We have three groupings, but there are no generally accepted definitions of the terminology we use: self-metered, premetered and hybrid. Self-metered processes are those in which the coverage is a function of the liquid properties and the system geometry, the web speed, the roll speeds, and any doctoring device. Examples are dip coating in which viscosity and web-speed control coverage, and blade and air knife coating in which excess is applied and then removed. Premetered processes deliver a set flow rate of solution per unit width to the applicator and all the material is transferred to the web. If a smooth coating is obtained then the coverage is fixed. Hybrid processes use features of both self- and premetered coating to achieve coating weight control. Gravure is an example of this method. The cell transfer determines coverage but doctoring is used to remove excess fluid from the gravure cylinder.

In most coating operations a single layer is coated. When more than one layer must be applied one can make multiple passes, or use tandem coaters where the next layer is applied at another coating station immediately following the dryer section for the previous layer, or a multilayer coating station can be used. Slot, extrusion, slide, and curtain coaters are used to apply multiple layers simultaneously. Slide and curtain coaters can apply an unlimited number of layers simultaneously, whereas slot coaters are limited by the complexity of the die internals and extrusion coaters by the ability of the combining adapter, ahead of the extrusion die, to handle many layers.

The precision or uniformity of the coating is very important for some products such as photographic or magnetic coatings. Some processes are better suited for precise control of coverage. When properly designed, slot, slide, curtain, gravure, and reverse-roll coaters are able to maintain coverage uniformity to within 2%. In many of the other coating processes the coverage control may be only 10%. Table 1 lists generally accepted attainable control.

The substrates or support coated on include paper and paper board, cellophane, poly(ethylene terephthalate), poly(ethylene naphthalate), polyethylene, polypropylene, polystyrene, poly(vinyl chloride), poly(vinyl fluoride), poly(vinylidene fluoride), polyimide, metal foils, woven and nonwoven fabrics, fibers, and metal coils. The surfaces of these supports can be impervious as in plastic films, or there may be a pore structure such as in paper. Primer coatings may be applied to seal these pores to give a uniform surface for the coating and to improve adhesion. The surfaces can also be modified with surface treatments such as flame treatment, plasma treatment, or corona discharge. These treatments increase the surface energy, thereby improving wetability and adhesion.

The web coating process can be used for intermittent coatings, such as in the printing process and to form coated batteries, as well as for the more common continuous coatings such as photographic films. In general, there is an ideal coater arrangement for any given product. However, most coating machines produce many different products and coating thickness and the machine is therefore usually a compromise made for the several applications.

3.1. Limits of Coatability.   In any coating process there is a maximum coating speed above which coating does not occur. At higher speeds air is entrained, resulting in many bubbles in the coating, or in ribs and finally rivulets, or in wet and dry patches. In slot coating below a critical speed, to coat thinner often means to coat slower. Above the critical speed the minimum thickness depends only on the gap. Above some higher speed a coating cannot be made (5). Using bead vacuum, thinner coatings can be obtained. Similar effects were found in slide coating, except the critical speed is never reached (6). The maximum coating speed in slide coating for thick coatings, where no bead vacuum or electrostatic assist is used, is identical to the velocity of air entrainment for a tape plunging into a pool of coating fluid. Lower viscosity liquids can be coated faster and thinner. Polymer solutions can be coated at higher speeds than Newtonian liquids. These phenomena have been explained in terms of a balance of forces acting on the coating bead, ie, the coating liquid in the region where it first makes contact with the web (7). Stabilizing forces are mainly bead vacuum or electrostatic assist, if used. The destabilizing forces are primarily the drag force on the coating liquid and the momentum of the air film carried along by the web. Thus, with no bead vacuum or electrostatic assist, there is a net destabilizing force which is balanced by the cohesive strength of the liquid. Limits of coatability occur in all coating operations but under different conditions in each process. A good description of the window of coatability in slot coating can be found in Reference 8.

The air entrainment velocity for plunging tapes does not depend on the wettability of the surface, but does increase with surface roughness (9,10). Presumably the rough surface lets air that would otherwise be entrained escape in the valleys between the peaks that are covered with coating liquid (11). In the converting industry, which involves coatings on rough and porous paper surfaces, much higher (up to 25 m/s) coating speeds can be attained than in photographic coatings on smooth plastic films. Although the wettability itself plays little or no role in coatability, it does play an extremely important role in coating. On a nonwetting surface, immediately after coating, the fluid will dewet and ball up into distinct islands.

3.2. Knife and Blade Coatings. Knife and blade coatings are in many ways similar. In both cases the knife or the blade doctors off excess coating that has been put on the web. Knives are usually held perpendicular to the web, whereas blades are usually tilted so that they form an acute angle with the incoming web. Typically blades are thin, only 0.2–0.5 mm thick, and can be rigid or flexible (of spring steel). Knives are thicker and are always rigid. Blades, being thinner, wear faster and have to be changed relatively often, perhaps two to four times a day. Blades are always pressed against the web, which is supported by a backing roll either made of chrome-plated steel or rubber-covered. Knives may be pressed against web on a rubber-covered backing roll; they may be pressed against the unsupported web which is held taut by the tension in the web; or they may be held at a fixed gap from the web that is supported by a backing roll.

The ends of the knives can be square, beveled or rounded. If the end is square and parallel to the web, if the upstream face is perpendicular to the web, and if there is a fixed gap between the end of the knife and the web, then the wet coverage is exactly one-half the gap. On the other hand, if there is a low angle in a converging section of the knife or of the blade, leading up to a tight gap as there is for many knives and for all bent blades, then strong hydrodynamic forces build up and tend to lift the knife or blade away from the web. This forces more fluid under the knife or blade, so that the coated thickness is greater than half the gap.

In all cases of knife and blade coatings, except in knife coating at a fixed gap, a rigid member and a flexible member are pressed together. The flexible member can deflect to allow for nonuniformities in the web. In knife coating and beveled blade coating, the knife or blade is rigid, and the unsupported web or the web on a rubber-covered roll is flexible. In bent blade coating the blade is flexible and the web on the roll is rigid or relatively rigid as in the case of a rubber-covered roll. Knife coating against unsupported web is more difficult to control than the other knife- and blade-coating techniques because, here, the web tension is a very important variable.

The simplest and least expensive, but still effective, coating method is knife coating, either against a backing roll or on unsupported web. Coating against a backing roll is more accurate, as it is independent of web tension. The knife, held perpendicular to the web, acts as a doctor blade and removes excess coating liquid. The coating can be applied by any convenient method, such as with an applicator roll, or by pumping the fluid into a pool formed by the web, the knife, and two end dams. The control of the coverage is by proper positioning of the knife. The unsupported knife shown in Figure 4a is used for coating open fabric webs where coating penetration is desired or cannot be prevented. A full width endless belt can be used to support a weak web and pull it through the knife area without tearing so as to overcome the drag of the knife.

The knife over roll coater (Fig. 4b) is probably the most common of the knife coaters. It is simple and compact. The driven back-up roll may be precision-made and chrome-plated, having a controlled gap between the web

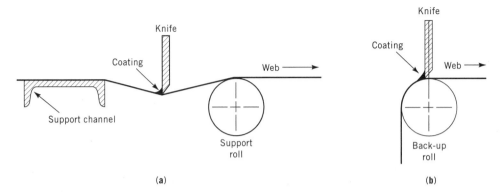

Fig. 4. (**a**) Unsupported knife; (**b**) knife over roll.

and the knife. The backing roll may also be rubber-covered, the knife pressing against the web. Here the coating weight is determined by the pressure against the knife. Higher pressures give lower coating weights.

Knife coaters can apply high coverages, up to 2.5-mm wet, and can handle high viscosities, up to 100,000 mPa·s(= cP). They tend to level rough surfaces rather than give a uniform coverage, a characteristic that can be desirable or not depending on the needs of the finished coating. Streaks and scratches are hard to avoid, especially using high viscosity liquids.

Blade Coating.   Flexible blade coaters can be used either with a downward moving web, as shown in Figure 5**a**, or with an upward moving web, as shown in Figure 5**b**. As with knife coaters there are many ways of feeding the metering blade. A puddle behind the blade is shown in Figure 5**a**, a forward turning applicator roll in Figure 5**b**, and a slot applicator or die fountain in Figure 5**c**. Jet fountains, where the coating liquid spurts out to the web 25–50 mm away, are occasionally used.

Blade coaters are commonly used on pigmented coatings. They have the unique feature of troweling in the low areas in a paper web, thus producing a coated surface that has excellent smoothness and printing qualities. The backing roll is usually covered with resilient material and is driven at the same speed as the web to stabilize the web and draw it past the blade. A replaceable blade is rigidly clamped at one end and the unsupported end is forced against the substrate. The wet coverage is adjusted by varying blade thickness, the blade angle, and the force pushing the blade against the substrate. The force on the blade can be obtained by various means: a rubber tube between the blade and a rigid member can be inflated with varying air pressures or the blade holder can be rotated so as to apply a greater or lesser force at the tip, while keeping constant the angle the blade tip makes with the web.

As the force on the blade increases, and as the force is concentrated at the blade tip, the wet coverage decreases rapidly. However, further increases in force bend the blade and a larger area of the blade presses the liquid against the web. Increasing the loading of the blade then causes the tip to lift up and the coverage now increases. At further increases in load the coverage again decreases.

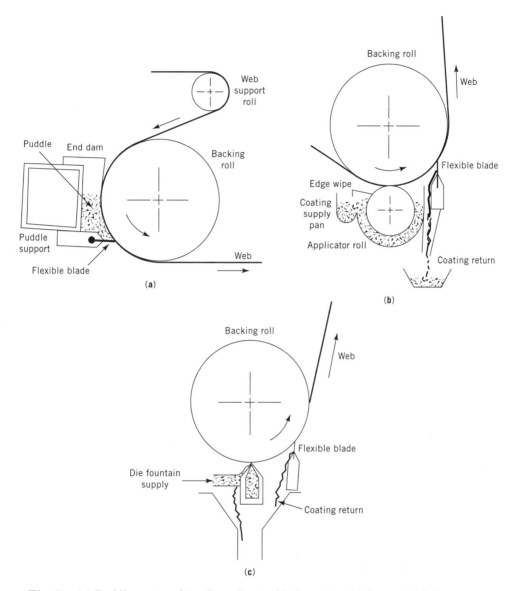

Fig. 5. (**a**) Puddle coater; (**b**) roll applicator blade coater; (**c**) fountain blade coater.

The beveled blade coater uses a rigid blade held at an angle of 40–55° to the web. The end of the blade is parallel to the substrate and pressed against it. If initially the end of the blade is not parallel to the web, it soon is as a result of abrasion by the pigmented fluid. When the loading on the blade increases, the wet coverage decreases. With the same force but using a thicker blade, the pressure, or force per unit area, on the coating fluid between the blade and the web decreases and the coverage increases.

In the rod–blade coater unit, a rod is mounted at the end of the blade. This coater behaves more like the beveled blade coater than a flexible blade coater.

Two-Blade Coaters.   In order to coat both sides of a web simultaneously, two flexible blade coaters can be used back-to-back, ie, with both blades pressing against each other and the web between them. The web usually travels vertically upward. Different coatings can be applied on each side of the web. The blades tend to be thinner and more flexible than the standard blades and the angle to the web is lower. The web has to have sufficient tensile strength to be pulled through the nip.

Simultaneous coatings can also be made with one flexible blade against the web on the roll, where the web moves downward. On one side the coating fluid is supplied to a puddle in front of the blade, and on the other side the fluid is carried into the nip by the roll. Edge dams between the web and the blade and between the web and the roll keep the fluid contained. The roll may rotate faster than the web. Figure 6 shows a version where the fluid on the roll side is supplied by a transfer roll.

Air-Knife Coater.   The air-knife coater is a versatile coating process in use for a wide range of products. A coating pan and roll are used to apply the coating solution and then an air knife is positioned after the pan to regulate the final wet-coating weight by applying a focused jet of air to the web. The excess solution is collected in an overflow pan and can be either recirculated and used again or scrapped. The air knife can function either in the precision or in the squeegee mode. These give very different types of coating and performance characteristics, although the same name is used for both processes.

In the precision mode, the air knife uses low pressures and doctors off some of the coating to control the coating weight and to level the surface to give a uniform coating of reasonable quality. The coating weight is a function of web speed, viscosity of solution, surface tension, and air-knife pressure. This precision mode has been used to coat photographic films where the air velocities are 13–130 m/min and air pressures are 50–2500 Pa (0.2–10 in. of water) to give 1- to 200-μm wet thickness.

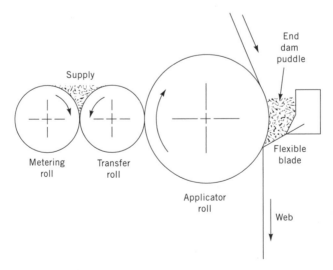

Fig. 6.   Billblade with transfer rolls.

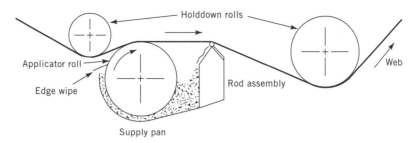

Fig. 7.   Wire-wound rod coater.

In the squeegee mode, the air knife operates at much higher pressures and coating speeds than in the precision mode and effectively doctors off the majority of the coating. This process is used for porous supports, such as paper, where the coating is absorbed into the voids. After the air knife, which effectively functions as a leveling device, the coating solids remain in the voids and in a thin surface layer.

The advantages of the air-knife process are low initial cost, versatility for coating a variety of webs and solutions, ease of changing and maintaining the coating, and the good coating quality. The disadvantages are the noise and contamination problems created by the air stream and the resulting spray, solution viscosity limitations, a somewhat restricted coating weight range, and the high cost to operate the air blowers.

3.3. Wire-Wound Rod Coating.   The wire-wound rod coater shown in Figure 7, called a Mayer rod, meters off excess applied coating solution. The rod is often rotated to increase its life by causing even wear and to prevent particles from getting caught under the rod and causing streaks. Normal rotation is in the reverse direction to the web travel. The wire size controls the coating weight. As the rod has an undulating surface because of the wire, one would expect the coating to have a similar unevenness. However, the down-web lines that form are frequently spaced at other than the wire diameter and are due to ribbing. If the solution is not self-leveling, a smoothing rod may be used to smooth out the surface. Rod coaters are best used with low viscosity liquids.

From the geometry of the wire-wound rod, one would expect the wet thickness to be 10.7% of the wire diameter. It is frequently less. It has been shown that with increasing load on the rod the wet thickness decreases to about 7% of the wire diameter (11).

Rod coaters are commonly used for low solids, low viscosity coatings such as those used to coat adhesives and barrier layers on poly(vinylidene chloride) carbon paper and silicone release papers. Coating weights range from 1.5–10 g/m^2, and speeds are as high as 300 m/min. The wire-wound rod can be held against unsupported web, as shown in Figure 6, or against a backing roll. When used against unsupported web the web tension affects the coverage. Coating rods are compact, simple, and inexpensive, but wear rapidly when used with abrasive fluids.

3.4. Roll Coating.   *Meniscus or Bead-Roll Coater.*   One of the simplest and most widely used coaters is the meniscus or bead-roll coater (Fig. 8). In this process the web passes over a backup roll that is just above the liquid level in a pan. A meniscus or coating bead is formed between the web and the

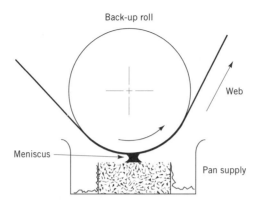

Fig. 8.   Pan-type Meniscus coater.

coating solution by raising the pan, and the solution transfers to the web. The coverage is determined by the viscosity of the solution and the coating speed. The pan design is very critical and a variety of configurations are available. The coating speed is very slow, only about 10 m/min on low viscosity liquids. High quality optical coatings may be produced. These coaters have also been used for adhesives.

Kiss Coaters.   In kiss coating fluid is transferred from a rotating applicator roll, called a kiss roll, to unsupported web. There are many types of kiss coaters. The kiss roll can turn in the direction of the web or in the reverse direction, but usually operates in the web direction. Kiss coaters are tension sensitive and are often used to apply excess coating prior to a metering device.

Forward-Roll Coaters.   In forward-roll coating the web passes between two rolls rotating in the same direction, the applicator roll and the backing roll. The applicator roll drags fluid into the nip, as shown in Figure 9 (12). The fluid exiting the nip splits in two, with some adhering to the web and some to the applicator roll. One might expect that if both rolls are rotating at the same surface speeds, then the fluid between them should move at that same speed and the flow rate per unit width through the nip, q, would equal the product of the surface speed, U, and the gap in the nip, G. Actually it is more than this because of the buildup of pressure as the fluid approaches the nip, which produces a greater flow. The dimensionless flow rate γ is defined as the ratio of the actual flow rate per unit width to GU, the "expected" value, or

$$\gamma = \frac{q}{GU}$$

where U is now the average surface speed of the two rolls. The dimensionless flow rate is approximately equal to 1.3 to 0.5, depending upon conditions.

Each roll carries away some of the flow. The ratio of film thickness on the two rolls, t_2/t_1, depends on the speed ratio, and for Newtonian fluids is

$$t_2/t_1 = (U_2/U_1)^{0.65}$$

where the subscript 1 corresponds to the web and the subscript 2 to the applicator roll. The total flow through the gap per unit width, q, is equal to the sum of

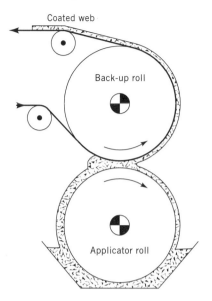

Fig. 9.   Two-roll forward-roll coater (12).

the flow on the web, t_1U_1, and that on the exit side of the applicator roll, t_2U_2. With shear-thinning fluids, when the roll speeds differ the split is more symmetrical than the equation indicates.

In forward-roll coating it is fairly common to have an instability called *ribbing*, where the coating thickness varies sinusoidally across the web and the coating looks as if a giant comb were dragged down the wet coating. Ribbing occurs when the *capillary number*, Ca, the ratio of viscous to surface forces, exceeds a certain value, depending on the gap-to-diameter ratio. The capillary number is

$$Ca = \eta U / \sigma$$

where U is the average surface speed of the two rolls, η is the viscosity of the coating fluid, and σ is the surface tension.

It is very difficult to avoid ribbing in forward-roll coating. When the fluid is not self-leveling, a smoothing bar is often used to smooth out the ribs. It has been found that a fine wire or thread stretched across the gap exit and touching the liquid eliminates ribbing (10).

Reverse-Roll Coating.   Reverse-roll coating is an extremely versatile coating method and can give a very uniform, defect-free coating (12–1200 μm thick) at a very wide range of coating speeds, using coating fluids with viscosities ranging from low to extremely high. In reverse-roll coating, the coating fluid is applied to the applicator roll by any of a number of techniques, such as having the applicator roll rotate in a pan of fluid, using a fountain roll or a fountain or slot die. The excess fluid is then metered off by a reverse-turning metering roll and the remaining fluid is completely transferred to the web traveling in the reverse direction. Two of the many possible configurations are shown in Figure 10.

All the flow remaining on the applicator roll after it rotates past the metering roll is transferred to the web; therefore it is important to know what this flow

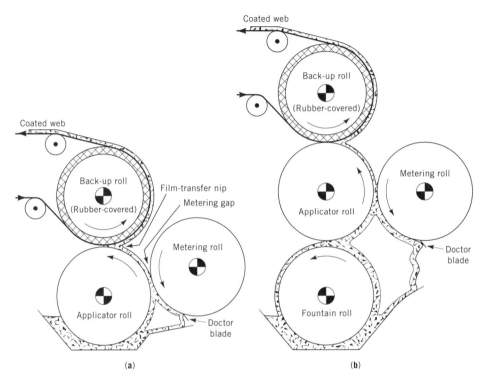

Fig. 10.   Pan-fed reverse-roll coaters: (**a**) three roll; (**b**) four roll (12).

is. The thickness of the metered coating on the applicator roll, t_a, is found to be a function of the gap, of the ratio of the speed of the metering roll to that of the applicator roll, and of the capillary number based on the applicator roll speed (Fig. 11).

In reverse-roll coating, as in forward-roll coating, instabilities can form. However, it is possible to obtain defect-free coatings at high coating speeds. Sometimes increasing the speed can lead to a smooth coating when a ribbing condition is present.

Another defect, called cascade or seashore, can form in reverse-roll coating. This defect is caused by the entrapment of air under certain conditions and can appear in the metered flow on the applicator roll. An operability diagram, showing the region of stable flow as well as the regions where these defects form, is given in Figure 12 for two gaps. The region where stable coatings can be made is at high capillary numbers, ie, at high speeds. There is also a stable region at very low speeds, but low speeds are not usually desirable. The principal advantage of reverse-roll coating is that conditions can be adjusted to give a stable, defect-free coating at high coating speeds. Using precision bearings, reverse-roll coaters can lay down as uniform a coating as any coating process, about ±2%.

3.5. Gravure Coating.   Gravure coating is an accurate way of coating thin (1- to 25-μm wet coverage) layers of low [10–5000 mPa·s (=cP)] viscosity liquids. The coating liquid is picked up by a patterned chrome-plated roll, the excess doctored off and the liquid transferred from the filled cells to the web.

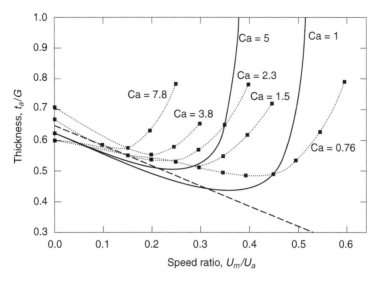

Fig. 11. Reverse-roll, metered film thickness on the applicator roll divided by gap, t_a/G, as a function of the ratio of the metering roll speed U_m to applicator roll speed U_a for various capillary numbers based on U_a. (——) represents theoretical values; (· · ·) experimental ones; and (– – –) is the lubrication model (12).

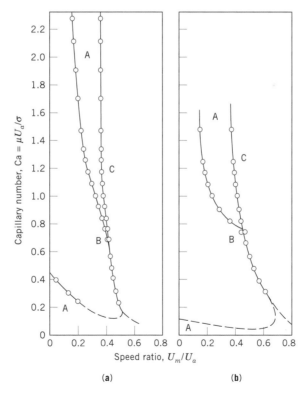

Fig. 12. Operability diagram for reverse-roll coating, where A represents a stable coatings area; B, ribbing; and C, cascade; for (**a**) a gap $G = 750\ \mu m$ and (**b**) a gap $G = 250\ \mu m$ (12).

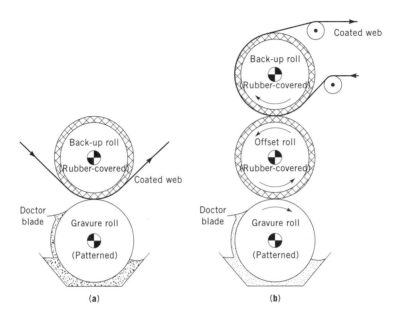

Fig. 13. Gravure coaters (**a**) direct; (**b**) offset (12).

Figure 13 illustrates two types of gravure coaters. In direct gravure the liquid is transferred directly from the gravure roll to the web. In offset gravure the liquid in the cells is first transferred to a rubber-covered offset roll before the final transfer to the web. In reverse gravure the gravure roll or the offset roll turns in the reverse direction with respect to the web. In differential gravure the forward rotating gravure cylinder runs at a different speed than the backing or impression roll. However, the web does not have to be held against a backing or impression roll; it can also be unsupported, as in kiss coating. The coating liquid can be applied to the gravure roll by a number of methods, not just by the pan-fed system illustrated.

The three common cell patterns for the gravure cylinder are illustrated in Figure 14. The pyramidal and quadrangular cells are similar, except that the quadrangular has a flat, not a pointed, bottom in order to empty easier. The trihelical pattern consists of continuous grooves spiraling around the roll, usually at a 45° angle. The volume factor is the total cell volume per unit area, and has units of height, typically ranging from 4 to 300 μm. The fraction of the cell volume that transfers varies greatly, depending on the system. With high impression roll pressures, about 58% of the cell volume normally transfers. The cell pitch or count is the number of cells per centimeter measured perpendicular to the pattern and usually ranges from 4 to 160 cm^{-1}. The pattern is made by mechanical engraving, chemical etching, electromechanical engraving, or laser etching.

After the gravure cylinder is coated with coating liquid, the excess is doctored off, normally using a 01- to 0.4-mm spring steel blade. Usually the doctor blade makes a 55–70° angle with the incoming gravure roll surface and is oscillated 6–50 mm to give even wear and to dislodge dirt that could cause streaks. A reverse-angle doctor blade can also be used. It often makes an angle

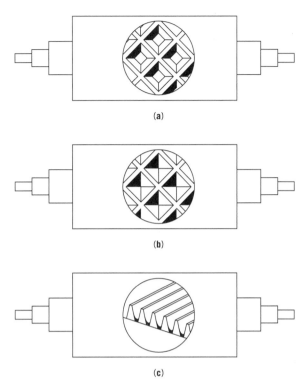

(a)

(b)

(c)

Fig. 14. Common cell patterns in gravure coating: (**a**) quadrangle; (**b**) pyramid; and (**c**) trihelical (2).

of 65–90° with the exiting surface. This blade does not have to be loaded against the cylinder face because fluid forces press the blade against the surface, and so the reverse blade can be made of softer materials, such as bronze or plastic. There is no need to oscillate this blade because in this position it cannot trap dirt; however, the standard blade is felt to do a better job of doctoring.

As with the flexible-blade coater, a softer doctor blade or one having a lower loading and an almost smooth cylinder with a shallow pattern allows excess liquid to pass through. A stiff, highly loaded blade against a cylinder having a large volume factor (cell volume per unit area) wipes the surface clean. The gravure roll has to be heavily loaded against the backup or impression roll in order to achieve good transfer to the web. The usual force is about 2000–20,000 N/m.

The most important factor in determining the transfer or web pickup is the gravure pattern design. The cell pitch controls the stability of web pickup. The leveling of the coating can be a problem. Large spacing between cells often results in printing of the cell pattern, rather than a uniform coating. Reverse and differential gravure tend to give better leveling. A smoothing bar can also be used.

A very useful new gravure coating technique is the Micro-gravure™ technique, which was introduced in the early 1990s. It is intended for low coating weight products on light gauge films for imaging, electronics, packaging, batteries, and other specialty applications. The unique features are the use of

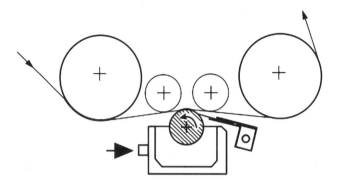

Fig. 15.   Side view sketch of Micro-Gravure™ coating head. Courtesy of Yasui-Seikei.

small diameter rolls, 20–50 mm versus 150–300 mm for conventional gravure. This results in a small stable bead, which when combined with reverse application gives very good quality and a low coating weight. A typical configuration is shown in Figure 15.

3.6. Dip Coating.   Dip coating is one of the oldest coating methods in use. Using continuous webs, the web passes under an applicator roll partially submerged in a pan of the coating fluid. The web is thus actually dipped into the coating solution. A doctor blade may be used to remove excess fluid if a reduction in the wet coating weight is desired. Otherwise the coverage is determined by coating speed and the characteristics of the liquid. The coverage increases with increasing viscosity and coating speed. Surface tension has a relatively small effect.

Dip coating is very commonly used for coating continuous objects that are not flat, such as fibers and for irregularly shaped discrete objects. Drops of coating at the bottom of dip-coated articles may be removed by applying electrostatic forces as the article is moved along a conveyor.

3.7. Extrusion.   Extrusion coating and slot coating are in principle very similar. In extrusion coating a high viscosity material, often a polymer melt, is forced out of the slot of the coating die onto a substrate where it is cooled to form a solid coating. As can be seen in Figure 16**a**, the highly viscous liquid does not wet the lips of the die. Similarly in slot coating, a relatively low viscosity liquid, usually under several thousand mPa · s (=cP), often a polymer solution, is forced out of the slot and onto the web. In slot coating the coating liquid does wet the lips of the die, as shown in Figure 16**b**. Some engineers use the terms slot coating and extrusion coating interchangeably.

Extrusion coating is often used in food packaging where vapor and oxygen barriers are required and heat sealability is desired. The expanding food packaging industry is the direct result of packaging improvements that can be attained from improving the surface and physical characteristics of a flexible web by extrusion coating.

Because of the high viscosities involved in extrusion coating, the coating die and the auxiliary equipment are massive. An extruder is needed to heat and melt the thermoplastic polymer, the die is heated by electric heaters, and the die also

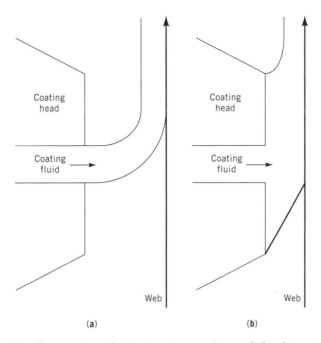

Fig. 16. Comparison of (**a**) extrusion coating and (**b**) slot coating.

contains adjusting bolts every 10 cm or so across the width that control the lip openings to try and obtain a uniform cross-web coverage. Internal choker bars controlled by bolts may also be used to adjust the uniformity. The bolts may be computer controlled. There may also be a laminating station to combine the plastic sheet with a substrate and to cool the laminate. The plastic may leave the die at about 175°C and may be about 0.5 mm thick. It is then elongated by the pulling effect of the faster-moving substrate which it joins in the pressure nip in the laminating section. The elongation reduces the width of the extruded film by perhaps 2–6 cm and reduces film thickness to approximately 12–25 μm before it makes contact with the substrate.

Good temperature control of the plastic and pressure control ahead of the coating die is important to the success of the coating. Variations in temperature lead to irregularities in the coating thickness both in the machine direction and across the web. Thickness variations in the cross-web direction can be reduced by adjusting the slot opening via the adjusting bolts and use of choker bars. The extruded film width is adjustable by external deckles to block off the exit of the die. In the laminator the nip helps to promote bonding, before chilling the molten plastic. The driven chill roll is chromium or nickel plated and can have a mirror, matte or an embossed surface. Once the extruded film passes through the laminating nip, it takes on the finish of the chill roll. The chill roll is 60–90 cm in diameter with perhaps a 120° wrap and utilizes refrigerated water to reduce the film temperature to about 65°C before the film is stripped. To improve adhesion of the extruded film to the substrate, adhesion-promoting "primers" are usually applied to the web before the laminator. Priming can be electrostatic (corona treatment), chemical, or in the form of ozone treatment. Coating weights are controlled by the line and extruder speeds, but in many cases the chill roll capacity limits the maximum thickness that can be obtained. Extrusion coating lines operate at speeds up to 1000 m/min and can apply 10–30 g/m^2 of coating.

3.8. Slot Coating.   Slot coating (Fig. 16**b**) involves a relatively low viscosity fluid, under perhaps 10,000 mPa · s (=cP) and uses much simpler equipment than extrusion coating. An ordinary pump or a pressurized vessel feeds the fluid through a flow meter and control valve to the coating die, which often operates at room temperature. If heating is required, water flowing through internal channels is usually adequate. Because of the simple rheologies of the fluids, the die can be designed to give uniform flow across the width with no adjustments. In fact, an adjustable die should be avoided. It is not difficult to design the die to give uniform flow, but it is very difficult to make the exactly correct adjustments. Because the viscosity is relatively low the pressures within the die are also relatively low, and the die can be much less massive than an extrusion die and still withstand the spreading forces.

Normally the web is supported by the backing roll in slot coating. However, for very thin coatings, under about 15-μm wet, the gap between the coating lips and the web becomes very tight, under about 100 μm, and the system becomes difficult to control and operate. The runout of the bearings can become a significant fraction of the gap. Dirt can hang up in the gap to cause streaks. If the web contacts the coating die the web can tear, causing a shutdown of the operation. Coating against unsupported or tensioned web should be used for very thin coatings. In this case web tension becomes an important variable.

In slot coating, bead vacuum is often used to increase the window of coatability, that is to allow thinner coatings and perhaps to allow coatings at higher speeds. A vacuum box is placed under the coating die and a vacuum of up to about 1000 Pa (4 in. of water) is pulled by a vacuum fan. Higher vacuums may be needed for higher viscosity liquids. There should be a tight vacuum seal against the sides or ends of the rotating backing roll, but no rubbing contact where the web enters the vacuum chamber in order to prevent scratches. The air in the vacuum chamber can resonate as in a musical instrument. These pressure fluctuations can cause chatter at wide gaps. The air leakage should be kept as small as possible to reduce the amplitude of the pressure fluctuations.

3.9. Curtain Coating.   Curtain coating is used to deliver coating liquid in a falling sheet or curtain to the substrate, which moves through the curtain at the coating speed. In one version a slot coating head is aimed downward and the coating emerges as a falling film or sheet as seen in Figure 17. The curtain thickness is controlled by the feed rate and by precise adjustments of the slot opening. The vertical distance of the coating die above the substrate can be adjusted. The falling curtain is protected from stray air movements by transparent enclosure sheets. Coating thicknesses as low as 12 μm are possible when coating with lacquers or with low viscosity wax melts, and are as low as 20 μm with hot-melt compositions of higher viscosity. There is no problem in obtaining heavier coating coverages.

Air-bubble entrapment may occur in the case of a gravity-applied continuous coating over an impermeable substrate. Bubbles may also be caused by moisture vaporization from the substrate. Remelting of the coating may minimize

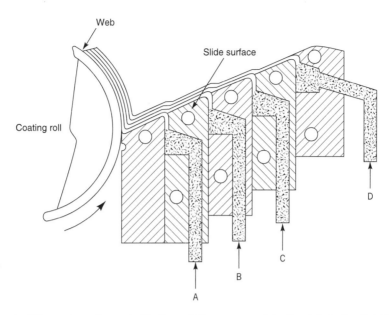

Fig. 17. A slide coater where A, B, C, and D correspond to the inlets for the liquids for layers 1, 2, 3, and the top layer, respectively (8,13).

the bubble defects. Curtain-coating equipment of this design is capable of operation at substrate speeds up to 500 m/min.

Curtain-coating equipment is also available in which the falling curtain is generated by overflow from an open weir. The coating is delivered to the open weir uniformly across its width by a pipe having diffuser jet openings. As the coating overflows the low side of the weir, it travels down a short flat skirt before dropping. The thickness of the falling curtain is adjusted by precise control of the rate of delivery of coating to the weir. Hot-melt coatings can also be applied by the open weir, as in the slot die. Because there is no close restriction to flow as in the slot die, the open weir does not tend to form scratches or coating streaks because of crusting or coating hang-up in the slot opening. When applying hot-melt coating formulations the coating supply is held in a reservoir at a temperature that does not thermally degrade the material during its residence. The coating is brought to this temperature using heat exchangers as it is pumped to the weir. Weir-type equipment is recommended for operation at substrate speeds up to 400 m/min. The coating fluid not carried away on the coated surface falls into a collection trough for recirculation.

Curtain coating is adaptable for coating irregularly sized sheets such as slotted cut-out corrugated carton blanks or sheets of plywood, as well as for continuous substrates. Coatings may also be applied to uneven geometric shapes such as blocks. The principal limitation of curtain coating is that a high flow rate of about 0.5–1.5 cm^3/(s·cm width) is needed to maintain an intact curtain. Usually about double this minimum is desirable. Thus, to obtain a thin coating, high coating speeds are required. Curtain coating is inherently a high speed process and the curtain will not form at low speeds or flow rates.

4. Multilayer Methods

4.1. Slide Coating. Slide coating is the primary method for simultaneously coating a multilayer structure. A slide coater, illustrated in Figure 17, can coat an unlimited number of top layers, 18 or more, simultaneously. Each layer flows out onto the slide yet does not mix with the other layers as they all flow together down the slide, across the gap, and onto the web, all in laminar motion. Slide coating is extensively used in coating photographic films and papers, both color and black and white. In color films, nine or more layers are coated simultaneously.

Instabilities can form on the slide in the form of interfacial waves, which may disturb the desired laminar flow and cause mixing. The closer the physical properties of all the layers are, the closer the system resembles a single layer in which internal waves do not form. The densities of the individual layers are always reasonably close to each other, and are usually not subject to control. The viscosities of adjacent layers should generally not vary by too much (14). It has been suggested that to avoid these waves the ratio of the viscosity of a layer to that of the adjacent lower layer should be more than 0.7 and less than 1.5 or 2 (15). However, if the ratio is above 10, waves again do not form. The bottom layer should have the lowest viscosity to reduce drag forces and allow higher coating speeds.

As with slot coating, a slight vacuum (up to 1 kPa for the usual low viscosity fluids) under the coating bead aids in coating by allowing thinner coatings and higher coating speeds. The bottom edge of the slide should be sharp and have a small radius of curvature, no more than about 50 μm or so, to pin the bottom meniscus and reduce the chance of cross-web barring or chatter.

4.2. Precision Multilayer Curtain. A variation of curtain coating can also be used to produce multilayer coatings. A slide is used to generate the multilayer structure which then flows over an added lip of the die to form a curtain. Edge guides are used to prevent the curtain from necking in because of surface tension. For precision coating the curtain has to be completely uniform across the width. Precision multilayer curtain coating is used to coat color photographic materials. This is illustrated in Figure 18. In most precision coatings the curtain is narrower than the web.

4.3. Slot Coating. Multiple layers can be coated simultaneously from a slot-coating die with multiple slots. The layers come together in the coating bead to form the final coating. Multiple layers in slot coating work well, but the die internals are complicated, especially when more than two layers are involved. This technique is particularly useful for coating of organic solvent systems, since the enclosed bead minimizes evaporation.

4.4. Extrusion Coating. In the most popular method of multilayer extrusion coating, the several layers come together before the coating die in a combining adapter or block, and then exit as one stream into the extrusion die (see Extrusion). The extrusion die is a standard single layer die, except that the feed port is rectangular to match the dimensions of the rectangular sandwich formed in the combining adapter. The separate layers remain distinct in the combining adapter and in the die. As long as the viscosity ratios are no greater than about 3:1 under processing conditions, the layer uniformity should be acceptable.

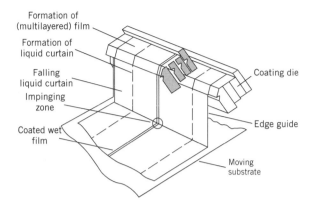

Fig. 18. Curtain-coating apparatus (13).

Multilayer extrusion provides the unique capability of producing layers of different resins to give superior functional properties. For example, an inexpensive resin can be used as the core of a three-ply extrudate, the outer plies being more expensive but also much thinner than if extruded alone.

5. Discrete Surface-Coating Methods

A variety of coating techniques are available to coat surfaces which are planar and have irregular surfaces.

5.1. Spray Coating. Coatings may be applied by spraying the coating material onto the object to be coated, which may be irregularly shaped with compound curves and with sharp edges. Many coating powders of suitable dielectric constant may be electrically polarized so that the powders are attracted to a grounded or oppositely charged surface. The object may then be heated to fuse the powders into a continuous film.

5.2. Dip Coating. The dip-coating technique described for webs can also be used to coat discrete surfaces such as toys and automotive parts. The item to be coated is suspended from a conveyor and dipped into the coating solution. The item is then removed; the coating drains and then levels to give the desired coverage. The object is then dried or cured in an oven.

5.3. Spin Coating. Spin coating is used to produce a thin uniform coating on discrete supports. In this process the coating fluid, usually a colloidal suspension, is placed on a horizontal substrate which rests on a rotating platform. The speed of the platform is increased to the desired level, which can be as high as 10,000 rpm. Centrifugal forces drive much of the coating off the support, leaving a thin, uniform film behind. In addition, the coating is drying during the process and as a result the viscosity increases, resistance to flow occurs, and a level thin coating is left. The coating chamber can provide hot air to the coating to dry or cure the remaining film. Additional coatings of different coating materials can be applied to develop a multilayer structure. This process is used to coat structures such as photomasks, magnetic disks, optical coatings, and a variety of layered products in the microelectronics industry.

5.4. Vacuum Deposition Techniques. Thin coatings are applied to a variety of substrates for use on semiconductors, ceramics, and electrooptical devices, using a wide variety of vacuum deposition techniques. Vacuum deposition is a rapidly advancing area of coating technology. In these processes the support to be coated is placed in a vacuum chamber which contains the coating material. Typically the coating material is a metal such as aluminum, gold, or tungsten. A high vacuum is then pulled; electrical energy or an electron beam is applied to heat the metal which evaporates off to deposit on the substrate. In sputtering, an ion beam is used to knock off atoms of the metal at lower temperatures. Individual supports, such as a target to be examined in a scanning electron microscope, or a continuous web, such as in making metallized poly-(ethylene terephthalate), can be used. The coatings can be continuous, or patterns or electrical circuits can be made if the support is masked.

There are several vacuum processes such as physical vapor deposition, chemical vapor deposition, sputtering, and anodic vacuum arc deposition. Materials other than metals, such as tetraethylorthosilicate, silane, and titanium aluminum nitride, can also be applied.

6. Patch Coating

It is sometimes necessary to coat patches of material on a web, such as coating the anode and cathode in batteries and in fuel cell membranes (Fig. 19). With these products an uncoated border is required around the coating to prevent a short circuit. Gravure coating is well suited for this purpose because the desired pattern can be etched into the gravure cylinder. Slot-coating techniques are also used (16). With slot coating there is the problem, however, of nonuniform coverage at the start and end of the patch, and the thicker edges along the sides. The start and end problems may be minimized by carefully controlling the flow with pumps and valves.

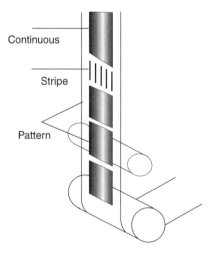

Fig. 19. Patch coating.

7. Coating Process Mechanisms

One of the principal advances in the coating process area in the 1980–1990s was the development of techniques to understand and define basic coatings mechanisms. This has led to improved quality and a wider range of utility for most coating techniques. This has involved the computer modeling of the coating process and the development of visualization techniques to actually see the flows in the coating process. The flow patterns predicted by the computer models can be verified by the visualization techniques.

Free surfaces and interfaces make the physics of coating flow systems extremely difficult to model by classical mathematical methods. As a result, coater designs and parameter ranges for defect-free coating have traditionally been determined through expensive and time-consuming statistical experimentation. Therefore coating developed largely as an art rather than a science. The ability to model the coating process by using modern methods of numerical and functional analysis, and to explain many of the complex mechanisms of coating instabilities and the resulting defects, is thus refreshing.

The most successful models are based on the finite element method. The flow is discretized into small subregions (elements) and mass and force balances are applied at each node. The result is a large system of equations, the solution of which usually gives the velocity and pressure of the coating liquid in each element and the location of the unknown free surfaces. The smaller the elements, the more the equations, which are often in the range from 10,000 to upwards of 100,000.

It is now possible to simulate steady transversely uniform flows of Newtonian or non-Newtonian liquids by using commercially available software packages such as FIDAP, NEKTON, FLUENT, PHOENICS, and POLYFLOW. Using these codes it is possible to locate regions of flow recirculation that may cause coating defects as a result of the increased residence time of solution. The free-surface handling capabilities of currently available commercial codes are limited to relatively simple steady flows and the transient response to specified transversely uniform disturbances. For a steady uniform flow to exist in nature, however, it should be able to recover from all small disturbances, such as building vibrations and the molecular fluctuations that are always present. Flow instabilities resulting in defects such as ribbing cannot be predicted by commercial software. Using more advanced methods developed first at the University of Minnesota and now in widespread use (17–20), it is now possible to predict most coating flow instabilities including bead break-up, flooding, cross-web barring (chatter), down-web ribbing, and diagonal chatter. It is also possible to follow the longtime development of the resulting defects and to explore parameter ranges of stable, defect-free operation (coating windows) at a fraction of the cost of the actual physical experiments. In a computer model the geometry can be quickly changed without having to construct expensive new parts. Considerable time and cost savings can be realized by optimizing coating systems computationally.

Experimental techniques to visualize flows have been extensively used to define fluid flow in pipes and air flow over lift and control surface of airplanes. More recently this technology has been applied to the coating process and it is now possible to visualize the flow streamlines (15,21). The dimensions of the flow field are small, and the flow patterns both along the flow and inside the flow are important. Specialized techniques involve generating small hydrogen bubbles using fine electrodes and injecting dyes in the regions of interest; optical sectioning is then required to observe and to photograph these flows.

A stereo zoom microscope having a very small depth of field and a clear window on the side of the applicator is employed. Fiber optic cables may be used for remote viewing. Accurate control of the light level and position is needed to reduce reflections that may mask the details of the flow field. The microscope is focused at varying points and flow nonuniformities are recorded. Using this technique, air entrainment, flow recirculation in the bead, curtain-coating formation, and the teapot effect have all been visualized for many types of slot, slide, and curtain coatings. This information leads to an improved understanding of the coating process. The same technology can also be applied to many other types of coating.

8. Drying and Solidification

The coating solution after application is in the liquid state and must be solidified. For coating with solvents, including water, this is brought about by removing the solvent, ie, drying. The drying process after the application of the coating is as important as the coating process itself. The properties of the coating are not complete until solidification has occurred. The coated film or web is transported through the dryer where the properties of the coating can either be enhanced or deteriorated by the drying process. Drying of coatings involves the removal of the inert inactive solvent used to suspend, dissolve, or disperse the active ingredients of the coating, which include polymeric binder, pigments, dyes, slip agents, hardener, coating aids, etc. Coating solvents range from the easy-to-handle water to flammable and toxic organic materials. Drying must occur without adversely affecting the coating formulation while maintaining the desired physical uniformity of the coating.

For polymer melts and for certain materials that gel on cooling, such as gelatin solutions, the temperature is lowered to solidify the coatings. With gelatin gels drying is still necessary. Similar hardware can be used for both heating and evaporation and for cooling, since they are both heat transfer devices.

While drying is a physical process involving only solvent removal, solidification can occur by cross-linking liquid monomer or liquid low molecular weight polymer. This can be accelerated by catalysts or can be accomplished by an electron-beam or uv radiation. This cross-linking process is called *curing*. Material coated from solution often also undergoes curing to improve the physical properties of the dried coating. Thus both curing and drying may occur in the dryer. This takes place with aqueous gelatin coatings which are cured using aldehyde cross-linkers. The cross-linking starts in the dryer.

The dryer provides heat to volatilize the solvent and a means to carry the solvent away from the coating. Efficient hardware is used to minimize energy costs. The dryers may be equipped with the appropriate pollution abatement devices to meet both OSHA and EPA standards. Dryers commonly use hot air both to provide heat and to carry away the solvent. The air may be heated by steam or by heat exchange with flue gases. Flue gases from the combustion of natural gas may be used directly in place of hot air. Infrared radiant energy from gas combustion or electric resistance heaters is sometimes used. Conduction heat transfer from heated drums is also used. The choice depends on availability of supply, the temperature range desired, and costs. Dryers can also use other sources of energy such as microwaves or radio-frequency waves. However, air is still needed to carry off the evaporating solvent. Radiant energy tends to be more expensive and is only used in special circumstances. If the coating can react with oxygen or if the solvents are flammable, inert gases such as nitrogen may be used in place of the air.

Because the evaporation of the solvent is an endothermic process, heat must be supplied to the system through conduction, convection, radiation, or a combination of these methods. The total energy flux into a unit area of coating, q_t, is the sum of the fluxes resulting from conduction, convection, and radiation.

Although heat transfer by conduction from heated drums is used extensively in the paper industry, convective heat transfer is very popular and used in most coating operations. Here the main focus is on convective heat transfer.

The rate of convective heating can be estimated as

$$q_{\text{convection}} = hA\Delta t$$

where A is the area (m^2), h is the heat-transfer coefficient [W/(m^2 · K)], q is the rate of heat transfer (W), and Δt is the temperature difference between the hot gas and the coating (K).

The heat-transfer coefficient is a key property of the dryer. It is controlled by the nozzle geometry and spacing, the distance from the web, and the velocity of the air. The evaporation rate is the rate at which heat is supplied for evaporation (heat also goes into heating the coating and the web) divided by the latent heat of vaporization of the solvent.

8.1. Air Impingement Dryers. Air impingement dryers, the most widely used for drying coated webs, basically consist of a heat source and heat exchangers (unless hot flue gas from combustion of natural gas is used), fans to move the air, ducts and nozzles or air delivery devices positioned close to the web, and solvent removal ducts. If all the air is recirculated, then equipment to remove solvent from the air is also provided. Figure 20 shows a typical dryer. In addition, there are controls for the air temperature and the air velocity from the nozzles and, in some cases, for the solvent level in the drying air. There may also be controls to keep the solvent concentration well below the lower explosive unit. Dryers often have separate sections or zones where the air temperature and velocity (and perhaps solvent level) can be controlled independently.

The dryer must also transport the web through the dryer using a combination of driven and idler rolls. The web path can be either horizontal or vertical, or, with the appropriate web-turning devices, fold back upon itself to conserve space. The idler rolls in single-sided dryers should be spaced so that there is enough wrap for the web to turn the rolls, and the coating should be kept within the effective area of the nozzles. Sometimes the web slips on the idler rolls and gets scratched. To prevent this the idler rolls can be driven by tendency drives

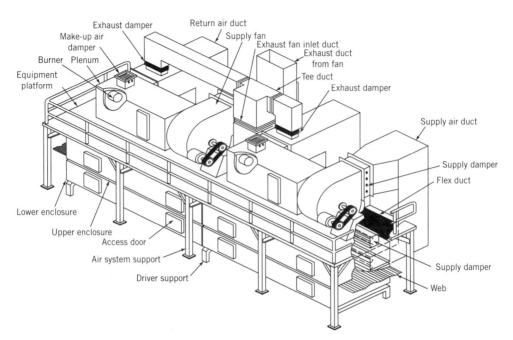

Fig. 20. Dryer components of a top mounted air system dryer. Courtesy of MEGTEC Systems.

which have two sets of bearings. The axle is driven at approximately the speed of the web, and the roll idles at this speed on a separate set of bearings. The driven web then easily brings the roll to the exact speed of the web. Tendency drives are needed for light webs. Typically the web should be within six to seven nozzle slot widths of the impingement nozzles.

In these single-sided dryers, the air impinges only on the coated side, heating and drying from that side only. The air can be delivered to the web from plenums with slots, with holes, or from specially designed nozzles. In a basic configuration the nozzles and idler rolls are contained in insulated boxes to minimize heat losses, solvent escape, and noise. The efficiency of the dryers depends on the heat transfer coefficient, the air usage, the temperatures used, and the solvent level. Wet coatings flow easily and the drying air should not disturb them.

The two-sided or floater dryer is now most often used. In this configuration the roll transport system in the dryer is replaced with air nozzles on the back side of the web so that the air transports and supports the web as well as heating and drying it from both sides. When impervious webs are used, while the heating is from both sides, the drying is only from the coated side. The two-sided heat transfer results in higher drying rates and thus shorter dryers, while eliminating problems of scratching from the idler rolls. However, more air is used.

Two types of floater nozzles are currently in use. One, based on the Bernoulli principle, is used in the airfoil flotation nozzles in which the air flows from the nozzle parallel to the web and the high velocities create a reduced pressure, which attracts the web while keeping it from touching the nozzles. The

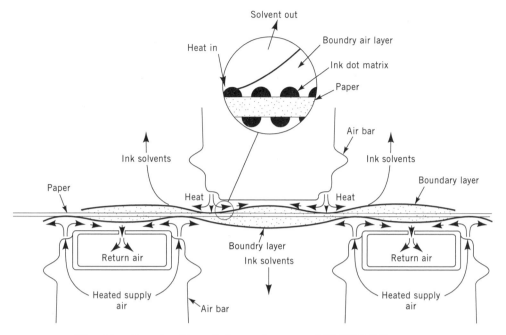

Fig. 21. Coanda effect air bars. Courtesy of MEGTEC Systems.

other uses the Coanda effect to create a flotation nozzle where the air is focused and follows the contours of surface of the air bar chamber, thus creating a pressure pad which supports the web, as shown in Figure 21.

Air flotation dryers have excellent heat-transfer coefficients, give very uniform drying across the web, and give excellent web stability. They can be used for a wide range of web types and tensions, and tend to be quieter, and thus pose less noise problems than the higher velocity single-sided dryers. Floater dryers are totally enclosed and compact so that they are clean and cause less dirt defects in the coating.

Single-sided dryer air velocities are 150–600 m/s, giving heat-transfer coefficients of 30–140 W/(m² · K). Floater dryers operate at slightly lower (150–500 m/s) slot velocities, and have higher [50–275 W/(m² · K)] heat-transfer coefficients based on the same single-sided web area.

8.2. Contact or Conduction Dryers. Coatings on webs, as well as sheets of newly formed paper, can be dried by direct contact with the surface of a hot drum. The drum is usually heated by steam. Here conduction is used to transfer the heat. Air still has to be supplied to carry away the solvent vapors. Drums can also be used to rapidly cool warm extruded films, to increase the viscosity, and to solidify the film.

8.3. Radiation Drying. Infrared or microwave radiation can supply concentrated energy to the web to evaporate the solvent, but air must still be used to carry away the solvent. These techniques provide a high heat input over short distances. Often Infrared is used in conjunction with convection dryers. It is often used at the start of drying to rapidly solidfy the coating, with the balance of drying done in convection oven. This can be cost efficient. Infrared heaters can be placed between the air nozzles. This increases the drying rate and thus the production rate without increasing the length of the dryer, and without requiring additional air handling systems. Use of Infrared heating is most effective in the first sections of the dryer, in the constant rate period, where the coating is coolest and most of the added heat, including that by Infrared radiation, goes to evaporate the solvent. In the falling rate period in most cases the drying rate is the rate of diffusion of the solvent to the surface, which is influenced only by the temperature. Infrared heating can only raise the temperature faster, which usually has just a minor effect.

8.4. Pollution Control. The solvent removed during drying is frequently a pollutant and the exhaust air must be treated to ensure that it meets government standards before being discharged to the atmosphere. The two basic approaches to treating the air are to recover the solvent for reuse and to convert it by burning to compounds which can safely be discharged. The basic solvent recovery systems involve condensation or adsorption in a charcoal bed. After recovery the solvent needs to be purified before reuse. For combustion of the exhaust sovent both thermal and catalytic systems are used. Pollution control systems are an essential part of the drying process and are available from dryer manufacturers. One should reduce the amount of volatile solvents in the coating process by coating as concentrated a coating fluid as possible. One should also investigate changing to a water-based system.

8.5. Modeling Convection Drying. Models of the drying process have been developed to estimate whether a particular coating can dry under the conditions of an available dryer. These models can be run on personal computers.

To model convection drying in the constant rate period both the heat transfer to the coated web and the mass transfer from the coating must be considered. The heat-transfer coefficient can be taken as proportional to the 0.78 power of the air velocity or to the 0.39 power of the pressure difference between the air in the plenum and the ambient pressure at the coating. The improvement in heat-transfer coefficients in dryers since the 1900s is shown in Figure 22. The mass-transfer coefficient for solvent to the air stream is related to the heat-transfer coefficient by the Chilton–Colburn analogy (23,24):

$$\frac{h}{\rho\, c_p k_m} = N_{Le}^{2/33}$$

where c_p is the heat capacity of the air $[J/(kg \cdot K)]$, h is the heat transfer-coefficient $[W/(m^2 \cdot K)]$, k_m is the mass-transfer coefficient $\{kg/[s \cdot m^2 \cdot (kg/m^3)]\}$, N_{Le} is the Lewis number, equal to the thermal diffusivity of the air divided by the mass diffusivity of solvent vapors in the air, and ρ is the density of the air (kg/m^3).

The total heat transfer to the coated web is then equated to the heat consumed by the evaporating solvent and the heat used in heating the web and the coating. This allows calculation of the temperature of the coated web (14,25). For single-sided drying, the equilibrium constant rate web temperature is the wet-bulb temperature of the air for that particular solvent. In the constant rate period of drying the coating behaves as if it were a pool of solvent. When dry patches appear on the surface, the rate of drying decreases and the falling rate period begins.

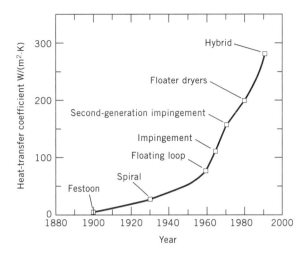

Fig. 22. Improvement in dryer heat-transfer coefficients over time (22). To convert $W/(m^2 \cdot K)$ to $Btu/(h \cdot ft^2 \cdot °F)$, multiply by 0.176.

Modeling the falling rate period is more difficult, because the drying rate then depends on the mechanisms occurring within the coating. In coating on impervious webs the rate-limiting process is diffusion; in porous coatings and coatings on porous paper it may be capillary action. In aqueous coatings most of the drying occurs in the constant rate period; for organic solvent systems most of the drying occurs in the falling rate period, to the extent that in some cases the constant rate period is over before the coated web enters the dryer. In the falling rate period all the solvent that reaches the surface evaporates; thus the rate of diffusion to the surface is the rate of evaporation (if diffusion is the transport mechanism).

The higher the air temperature, the more rapid the drying. However, there are temperature limitations both for the web and for the coating. Plastic films should not be heated above their glass-transition temperature (the softening point) to prevent distortion and stretching. The coatings themselves may have a maximum temperature above which the coating may degrade. Many photographic coatings, for example, should not be heated above about 50°C.

8.6. Rheology. Rheology is the science of deformation and flow of matter. Because the coating process creates shear and extensional stresses in the coating fluid, the rheological properties of coating liquids are important factors in the selection and successful running of a coating operation. The coating hardware exposes the coating solution to a wide range of shear rates (Table 2). As a first approximation, the extensional rates would be of the same order as the shear rates. Shear and extension affect the properties of the solution. Therefore, the rheological properties, the deformation and flow under stress, are important factors in the selection and successful running of a coating operation.

The shear or dynamic viscosity is the ratio of shear stress to shear rate, and measures the resistance of the fluid to flow while undergoing shear. The common unit of viscosity is centipoise (cP). One cP is the same as 1 mPa·s. A high viscosity

Table 2. **Coating Process Shear Rates**

	Shear rate, s^{-1}
Coating	
dip	10–100
roll, reverse	1,000–100,000
roll, forward	10–1,000
spray	1,000–10,000
slide	3,000–120,000
gravure, reverse	40,000–1,000,000
gravure, forward	10–1,000
slot die	3,000–100,000
curtain	10,000–1,000,000
blade	20–40,000
Ancillary operations	
simple mixing	0–100
high shear mixing	1,000–100.000
Measurement	
Brookfield	1–100
Haake	1–20,000
Cannon-Fenske glass	1–100

solution flows slowly. The shear rate is the rate of change of velocity with distance in the direction perpendicular to flow and can be crudely approximated by the coating speed divided by a coating gap. In coating flows it can reach values over 10^5 s^{-1}. In simple or Newtonian fluids the viscosity is a constant and does not change with shear rate. However, it is not constant in dilute polymer solutions, where the shape of the polymer molecules distort with shear. The spherical shape of a random coil becomes elongated in the direction of flow and so offers less resistance to flow. Thus the viscosity decreases with shear rate. At very low shear rates, such that the shape of the molecules has not yet changed, the viscosity is constant at its zero shear value. At high shear rates where the molecules are fully elongated, such that they cannot offer less resistance with increasing shear, the viscosity is again constant, now at its infinite shear value (assuming the molecule is not destroyed by the mechanical forces). Almost all coating fluids contain dissolved polymers and are shear thinning.

Extensional or stretching flows are also very important in coating. The extensional viscosity is the ratio to the tensile stress in the fluid to the extension rate. Fluids can support a tensile stress when they are in motion. If you put a finger in a jar of honey and withdraw it, a strand of honey will be carried along by the finger. If you stop moving the finger the honey will fall back in the jar. But while in motion the honey will be in tension. With Newtonian fluids the extensional viscosity is three times the shear viscosity. When polymer solutions are stretched slowly, the molecules can relax, disentangle, and slip past each other. At higher extension rates they do not have time to relax and disentangle, and the extensional viscosity increases. This aids the coating process, and polymer solutions are easier to coat than Newtonian fluids of the same low shear viscosity.

The extensional thickening of polymer solutions is one form of viscoelastic behavior. This ability to support a tensile stress can also be demonstrated in a tubeless syphon with dilute aqueous solutions of polymers such as polyacrylamide or polyethylene oxide. If you suck up solution with a medicine dropper attached to a water aspirator and then lift the dropper out of the solution, the solution will still be sucked up. In shear, viscoelastic fluids develop normal stresses, which causes rod climbing on a rotating shaft, as opposed to the vortex and depressed surfaces that form with Newtonian liquids. Polymer solutions and semiliquid polymers exhibit other viscoelastic behaviors, where, on short time scales, they behave as elastic solids. "Silly putty," a childrens toy, can be formed into a ball and will slowly turn into a puddle if left on a flat surface. But if dropped to the floor it bounces.

Concentrated dispersions may be shear thickening, as opposed to the shear thinning of dilute polymer solutions. Some materials, such as latex paints, tend to form a structure. As the structure breaks down with shearing action, the viscosity decreases. Such materials are thixotropic. Some fluids have a yield stress. A thorough characterization of the rheology may require a number of different measurements.

8.7. Surface Forces. Because fresh surface is created during coating, surface forces are involved. These are normally expressed as surface tension, in dyn/cm, which is identical to mN/m. Surface tension is identical to surface energy, expressed as erg/cm^2 or mJ/m^2. Whether we call it surface tension or surface energy is a matter of personal preference. Surface tension gives rise to a higher pressure on the concave side of a curved interface; thus the pressure on the inside of a drop is higher than on the outside. This higher pressure is called capillary pressure. It can be used to explain the shape of some coating beads.

In coating, the coating fluid should spread out on the support. For this to occur the surface tension of the fluid should be low and the surface energy of the support should be high. The contact angle in a drop of fluid on the surface, between the surface of the support and the surface of the fluid measured through the fluid, is a measure of the ability of the fluid to wet the surface. This contact angle should be low.

The surface energy of the support may be increased by oxidizing it, such as in a flame (flame treatment) or in an electrical discharge (corona treatment). Flame treatment is permanent, but for many polymers corona treatment is labile—after a matter of hours or days the surface energy decreases toward its original value. Because of this, corona treatment is done in-line, before the coating stand. High energy coatings may be applied to the support, such as the subbing often used on supports for photographic coatings.

The surface tension of a fluid containing surfactants varies with the age of the surface. As surfactant diffuses to the surface it lowers the surface tension; fresh surfaces have the higher surface tension of the solvent. The time for the surface to reach equilibrium varies from a number of milliseconds to several hours, depending on the system. The surface tension as a function of time is called the dynamic surface tension.

Similarly, the contact angle of a fluid on a support depends on whether the fluid is stationary or is moving, and at what speed. The contact angle for a stationary drop is lowest; as one coats faster in a coating operation the dynamic contact angle increases. When the coating speed increases to the point where the dynamic contact angle is 180°, air will be entrained and one can no longer coat.

8.8. Commercial Availability. All of the many types of coaters and dryers discussed herein are commercially available from many different vendors. These vendors usually have pilot facilities so that new coating and drying techniques can be easily tested. Contract coating companies, specializing only in coating, also exist.

BIBLIOGRAPHY

"Coating Process" in *ECT* 3rd ed., Vol. 6, pp. 386–426, by S. C. Zink; "Coating Process (Survey)" in *ECT*, 4th ed., Vol. 6, pp. 606–635, by Edward D. Cohen, E. I. du Pont de Nemours and Co., Inc., and Edgar B. Gutoff, Consultant; "Coating Process, Survey" *ECT* (online), posting date: December 4, 2000, by Edward D. Cohen, E. I. du Pont de Nemours and Co., Inc., and Edgar B. Gutoff, Consultant.

CITED PUBLICATIONS

1. E. D. Cohen and E. B. Gutoff, eds., *Modern Coating and Drying Technology*, Wiley-VCH, New York, 1992, pp. 64, 67, 83, 89, 92, 105, 120.
2. D. Satas, ed., *Web Processing and Converting Technology and Equipment*, Van Nostrand Reinhold Co., Inc., New York, 1984, pp. 32, 44.
3. D. Satas, ed., *Coatings Technology Handbook*, Marcel Dekker, Inc., New York, 1991.
4. H. L. Weiss, *Coating and Laminating Machines*, Converting Technology Co., Milwaukee, Wis., 1977.
5. E. D. Cohen and E. B. Gutoff, *Coating and Drying Defects: Troubleshooting Operating Problems*, John Wiley and Sons, Inc., New York, 1995.
6. S. F. Kistler and P. M. Schweizer, *Liquid Film Coating*, Chapman & Hall, London 1997.
7. R. J. Stokes and D. Fennell Evans, *Fundamentals of Interfacial Engineering*, Wiley-VCH, New York, 1997.
8. E. B. Gutoff and C. E. Kendrick, *AIChE J.* **33**, 141–145 (1987).
9. R. A. Buonopane, E. B. Gutoff, and M. M. I. Rimore, *AIChE J.* **32**, 682–683 (1986).
10. T. Hasegawa and K. Sorimachi, *AIChE J.* **39**, 935–945 (1993).
11. R. Hanumanthu, S. J. Gardner, and L. E. Scriven, *Paper 7a presented at the AIChE Spring National Meeting*, Atlanta, Ga., Apr. 17–21, 1994.
12. D. J. Coyle, in Ref. 1, pp. 63–109.
13. U.S. Pat. 2,761,410 (Sept. 4, 1956), J. A. Mercier and co-workers (to Eastman Kodak).
14. E. B. Gutoff, *Chem. Eng. Prog.* **87**(1), 73–79 (Feb. 1991).
15. L. E. Scriven and W. J. Suzynski, *Chem. Eng. Prog.* **86**(9), 24–29 (Sept. 1990).
16. U.S. Pat. 5,360,629 (Nov. 1, 1995), T. M. Milbourn and J. J. Barth, (to Minnesota Mining and Manufacturing Co.).
17. K. N. Christodoulou and L. E. Scriven, *J. Sci. Comput.* **3**, 355–406 (1988).
18. K. N. Christodoulou, S. F. Kistler, and P. R. Schunk, in Ref. 6, pp. 297–366.
19. L. E. Scriven, *Paper presented at AIChE Spring Meeting*, Atlanta, Ga., Mar. 1984.
20. L. E. Sartor, *Slot Coating: Fluid Mechanics and Die Design, Ph.D. dissertation*, University of Minnesota, 1990.
21. P. M. Schweizer, *J. Fluid Mech.* **193**, 285–302 (1988).
22. E. D. Cohen, in Ref. 1, pp. 267–296.
23. T. H. Chilton and A. P. Colburn, *Ind. Eng. Chem.* **26**, 1183 (1934).
24. R. H. Perry and C. H. Chilton, eds., *Chemical Engineers' Handbook*, 5th ed., McGraw-Hill, New York, 1973, p. 12–2.

25. S. F. Kistler and L. E. Scriven, *Paper presented at AIChE Spring Meeting*, Orlando, Fla., Mar. 1982.

EDWARD D. COHEN
Technical Consultant
EDGAR B. GUTOFF
Consulting Chemical Engineer

COATING PROCESSES, POWDER

1. Introduction

Powder coating is a process for applying coatings on a substrate using heat fusible powders. Materials used in the process are referred to as coating powders, finely divided particles of organic polymer, either thermoplastic or thermosetting, which usually contain pigments, fillers, and other additives. After application to the substrate, the individual powder particles are melted in an oven and coalesce to form a continuous film having decorative and protective properties associated with conventional organic coatings.

The origin of powder coating technology dates back to the late 1940s when powdered thermoplastic resins were applied as coatings to metal and other substrates by flame spraying. In this process, a powdered plastic was fed through a flame spraying apparatus where the plastic particles are melted and propelled by the hot gases to the substrate. A patent issued in Great Britain to Schori Metallising Process, Ltd. in 1950 described a process for forming a coating in which powdered thermoplastics were applied to a heated substrate by dipping or rolling the heated article in the plastic powder (1). This process was difficult to practice, however, and never achieved commercial success.

A major breakthrough in powder coating occurred in the mid-1950s, when Erwin Gemmer conceived the fluidized-bed coating process, in which a heated object is dipped into a fluidizing bed of powder. Gemmer was involved in developing flame spraying processes and materials in the laboratories of Knapsack-Griesheim (Hoechst) a manufacturer of specialty gases and was searching for a more efficient method than flame spraying for coating objects with powder. The first patent applications were filed in Germany in May 1953, and the basic patent was issued in September 1955 (2). The first U.S. patent was issued in 1958 (3), and the Polymer Corporation, Reading, Pa., acquired rights to the Knapsack-Griesheim patents. The Polymer Corporation mounted an aggressive effort to develop, license, and sell fluidized-bed coating technology in North America. However, acceptance of this coating process was rather slow. In 1960, the annual sales of coating powders in the United States were below 450 metric tons (mt) in part because of a lack of expertise in the methodology. In addition, the available powder coating materials were expensive, efficient production techniques had not been worked out, and volume of production was low.

Today, powder coating is widely accepted, with thousands of installations in the factories of original equipment manufacturers (OEMS) and custom coating job shops. It is the preferred method for coating many familiar items such as lawn and garden equipment, patio and other metal furniture, electrical cabinets, lighting, shelving and store fixtures, and many automotive components.

In the fluidized-bed coating process, the coating powder is placed in a container having a porous plate as its base. Air is passed through the plate causing the powder to expand in volume and fluidize. In this state, the powder possesses some of the characteristics of a fluid. The part to be coated, which is usually metallic, is heated in an oven to a temperature above the melting point of the powder and dipped into the fluidized bed where the particles melt on the surface of the hot metal to form a continuous film or coating. By using this process, it is possible to apply coatings ranging in thickness from $\sim$250–2500 μm (10–100 mils). It is difficult to obtain coatings thinner than $\sim$250 μm and, therefore, fluidized-bed applied coatings are generally referred to as thick-film coatings, differentiating them from most conventional paint-like thin-film coatings applied from solution or as a powder at thicknesses of 20–100 μm (0.8–4 mils) (see THIN FILMS, FILM FORMATION TECHNIQUES).

In the electrostatic spray process, the coating powder is dispersed in an air stream and passed through a corona discharge field where the particles acquire an electrostatic charge. The charged particles are attracted to and deposited on the grounded object to be coated. The object, usually metallic and at room temperature, is then placed in an oven where the powder melts and forms a coating. By using this process, it is possible to apply thin-film coatings comparable in thickness to conventional paint coatings, ie, 20–75 μm. A hybrid process based on a combination of high voltage electrostatic charging and fluidized-bed application techniques (electrostatic fluidized bed) has evolved as well as triboelectric spray application methods (see COATING PROCESSES, SPRAY COATINGS). Powder coating methods are considered to be fusion-coating processes; ie, at some time in the coating process the powder particles must be fused or melted. Although this is usually carried out in a convection oven, infrared (ir), resistance, and induction heating methods also have been used. Therefore, with minor exceptions, powder coatings are factory applied in fixed installations, essentially excluding their use in maintenance applications. Additionally, the substrate must be able to withstand the temperatures required for melting and curing the polymeric powder, limiting powder coating methods to metal, ceramic, and glass (qv) substrates for the most part, although recently some plastics and wood products have been powder coated successfully.

Compared to other coating methods, powder technology offers a number of significant advantages. These coatings are essentially 100% nonvolatile, ie, no solvents or other pollutants are given off during application or curing. They are ready to use, ie, no thinning or dilution is required. Additionally, they are easily applied by unskilled operators and automatic systems because they do not run, drip, or sag, as do liquid (paint) coatings. The reject rate is low, the finish is tougher and more abrasion resistant than that of most conventional paints (see PAINT). Thicker films provide electrical insulation, corrosion protection, and other functional properties. Powder coatings cover sharp edges for better corrosion protection. The coating material is well utilized: overspray can be collected and reapplied. No solvent storage, solvents dry off oven, or a mixing room are required. Air from spray booths is filtered and returned to the room rather than exhausted to the outside. Moreover, less air from the baking oven is exhausted to the outside, thus saving energy. Finally, there is no significant disposal problem because there is no sludge from the spray booth wash system. Any powder that cannot be reclaimed and must be discarded is not considered a hazardous waste under most environmental regulations. Whereas the terms coating powder and powder coating are sometimes used interchangeably, herein the term coating powder refers to the coating composition and powder coating to the process and the applied film.

Coating powders are frequently separated into decorative and functional grades. Decorative grades are generally finer in particle size and color and appearance are important. They are applied to a cold substrate using electrostatic techniques at a relatively low film thickness, eg, 20–75 μm. Functional grades are usually applied in heavier films, eg, 250–2500 μm using fluidized-bed, flocking, or electrostatic spray coating techniques to preheated parts. Corrosion resistance and electrical, mechanical, and other functional properties are more important in functional coatings (see CORROSION AND CORROSION CONTROL).

Coating powders are based on both thermoplastic and thermosetting resins. For use as a powder coating, a resin should possess: low melt viscosity, which affords a smooth continuous film; good adhesion to the substrate; good physical properties when properly cured, eg, high toughness and impact resistance; light color, which permits pigmentation in white and pastel shades; good heat and chemical resistance; and good weathering characteristics, ie, resistance to degradation by ultraviolet (uv) light, hydrolysis, and environmental pollutants. The coating powder should remain stable on storage at 25°C for at least six months and should possess a sufficiently high glass-transition temperature, T_g, so as to resist sintering on storage.

The volume of thermosetting powders sold exceeds that of thermoplastics by a wide margin. Thermoplastic resins are almost synonymous with fluidized-bed applied thick-film functional coatings and find use in coating wire, fencing, and corrosion resistant applications, whereas thermosetting powders are used almost exclusively in electrostatic spray processes and applied as thin-film decorative and corrosion resistant coatings.

Thermoplastic resins have a melt viscosity range that is several orders of magnitude higher than thermosetting resins at normal baking temperatures. (See Table 1). It is, therefore, difficult to pigment thermoplastic resins sufficiently to obtain complete hiding in thin films, yet have sufficient flow to give a smooth coating since incorporation of pigments reduces melt flow even further. In addition, thermoplastic resins are much more difficult to grind to a fine particle size than thermosetting resins, so grinding must usually be carried out under cryogenic conditions. Because powders designed for electrostatic spraying generally have a maximum particle size of ~75 μm (200 mesh), the thermoplastic powders are predominant in the fluidized-bed coating process where heavier coatings are applied and a larger particle size can be tolerated. Fluidized-bed powders typically contain only ~10–15% of particles <44 μm (325 mesh),

Table 1. **Physical and Coating Properties of Thermoplastic Powders**[a]

Property	Vinyls	Polyamides	Polyethylene	Polypropylene	PVDF[b]
melting point, °C	130–150	186	120–130	165–170	170
preheat/postheat temperatures, °C[c]	290–230	310–250	230–200	250–220	230–250
specific gravity	1.20–1.35	1.01–1.15	0.91–1.00	0.90–1.02	1.75–1.90
adhesion[d]	G–E	E	G	G–E	G
surface appearance[e]	smooth	smooth	smooth	smooth	sl OP
gloss, 60° meter	40–90	20–95	60–80	60–80	60–80
hardness, Shore D	30–55	70–80	30–50	40–60	70–80
resistance[e,f]					
impact	E	E	G–E	G	G
salt spray	G	E	F–G	G	G
weathering	G	G	P	P	E
humidity	E	E	G	E	G
acid[g]	E	F	E	E	E
alkali[g]	E	E	E	E	G
solvent[g]	F	E	G	E	G–E

[a] All powders require a primer and pass the flexibility test, which means no cracking under a 3-mm dia mandrel bend.
[b] Poly(vinylidene fluoride) = PVDF.
[c] Typical ranges.
[d] E = excellent; G = good.
[e] OP = orange–peel effect; sl OP = slight orange–peel effect.
[f] F = fair; P = poor.
[g] Inorganic; dilute.

whereas the high end of the particle-size distribution ranges up to ∼200 μm (70 mesh). Most thermoplastic coating powders require a primer to obtain good adhesion and priming is a separate operation that requires time, labor, equipment, and typically involves solvents. In automotive applications, some parts are primed by electrocoating. Primers are not usually required for thermosetting powder coatings.

2. Thermoplastic Coating Powders

Thermoplastic resins used in coating powders must melt and flow at the application temperatures without significant degradation. Attempts to improve the melt flow characteristics of a polymer by lowering the molecular weight and plasticizing or blending with a compatible resin of lower molecular weight can result in poor physical properties or a soft film in the applied coating. Attempts to improve the melt flow by increasing the application temperature are limited by the heat stability of the polymer. If the application temperature is too high, the coating shows a significant color change or evidence of heat degradation. Most thermoplastic powder coatings are applied between 200 and 300°C, well above the generally considered upper temperature limits for adequate heat stability. However, the application time is short, usually ≤5 min. The principal polymer types are based on plasticized poly(vinyl chloride) (PVC) [9002-86-27], polyamides (qv), and other specialty thermoplastics. Thermoplastic coating powders have one advantage over thermosetting coating powders: They do not require a cure and the only heating necessary is that required to complete melting or fusion of the powder particles. Thermoplastic resins have applications in coating wire, fencing, and other applications where the process involves continuous coating at high line speeds. Typical properties of thermoplastic coating powders are given in Table 1.

2.1. PVC Coatings. All PVC powder coatings are plasticized formulations [see VINYL POLYMERS, VINYL CHLORIDE and POLY(VINYL CHLORIDE)]. Without plasticizers (qv), PVC resin is too high in melt viscosity and does not flow sufficiently under the influence of heat to form a continuous film. Suspension and bulk polymerized PVC homopolymer resins are used almost exclusively because vinyl chloride–vinyl acetate and other copolymer resins have insufficient heat stability. A typical melt-mixed PVC coating powder formulation is given in Ref. 4. Dispersion grade PVC resin is added in a postblending operation to improve fluidizing characteristics (5). Additional information on the formulation and application of PVC coating powders can be found in Ref. 6. Whereas most PVC coating powders are made by the dry-blend process, melt-mixed formulations are used where superior performance, such as in outdoor weathering applications and electrical insulation, is required (see Fig. 1). Almost all PVC powder coatings are applied by the fluidized-bed coating process. Although some electrostatic spray-grade formulations are available, they are very erratic in their application characteristics. The resistivity of plasticized PVC powders is low compared to other powder coating materials and the applied powder quickly loses its electrostatic charge. Dishwasher baskets are coated with fluidized-bed PVC powder. Other applications are various types of wire mesh and chain-link fencing. PVC coatings have a very good cost/performance balance that is difficult to match with any of the other thermoplastic materials. Properly formulated PVC powders have good outdoor weathering resistance and are used in many applications where good corrosion resistance is required (see CORROSION AND CORROSION CONTROL). These coatings are also resistant to attack by most dilute chemicals except solvents. In addition, PVC coatings possess excellent edge coverage.

Powder coatings as a class are superior to liquid coatings in their ability to coat sharp edges and isolate the substrate from contact with corrosive environments. PVC coatings are softer and more flexible than any of the other powder coating materials. Primers used for PVC plastisols generally have been found suitable for powder coatings as well (7).

2.2. Polyamides. Coating powders based on polyamide resins have been used in fusion-coating processes from the earliest days. Nylon-11 [25587-80-9] has been used almost exclusively; however, coating powders also have been sold based on nylon-12 [24937-16-4]. The properties of these two resins are quite similar. Nylon-6 [25038-54-4] and nylon-6,6 [32131-17-2] are not used because the melt viscosities are too high.

Polyamide powders are prepared by both the melt-mixed and dry-blend process. In the latter, the resin is ground to a fine powder and the pigments are mixed in with a high intensity mixer (see Fig. 1). Melt-mixed powders have a higher gloss, eg, 70–90 on the Gardner 60° gloss meter, whereas dry-blended powders have a gloss in the range of 40–70. Because the pigment is not dispersed in the resin in the dry-blend process, it must be used at very low concentrations,

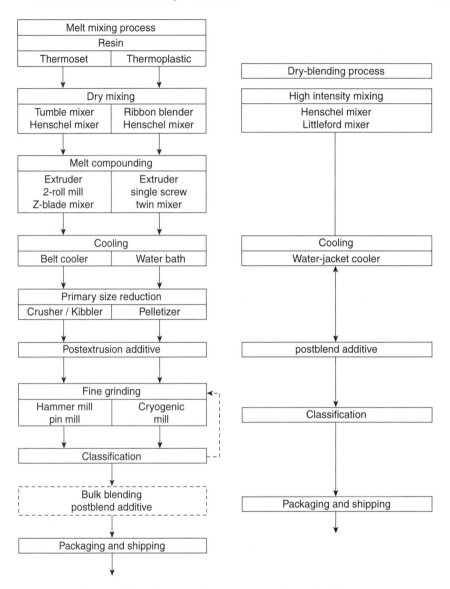

Fig. 1.   Flow diagram for coating powder manufacture.

usually <5%. Even in melt-mixed formulations, the concentration of pigment and fillers (qv) seldom exceeds ~20% of the composition.

Nylon coating powders are available for both electrostatic spray and fluidized-bed application. Nylon coatings are very tough, resistant to scratching and marring, have a pleasing appearance, and are suitable for food contact applications when properly formulated. These coatings are used for chair bases, hospital furniture, office equipment, knobs, handles, and other hardware. Because of expense, nylon is generally applied only to premium items. Nylon coatings have good solvent and chemical resistance and are used for dishwasher baskets, food trays, hot water heaters, plating and chemical-etching racks, and large diameter water pipes in power-generating stations. For maximum performance, a primer is used. Nylon coating powders are discussed in more detail in Ref. 8.

2.3. Other Thermoplastic Coating Powders. Coating powders based on polyethylene [9002-88-4] and polypropylene [9003-07-1] have been available for many years but have achieved limited commercial success (see OLEFIN POLYMERS). A primary problem in using polyolefin-based powders is poor adhesion to metal. However, ethylene copolymers functionalized with acrylic acid, sold under the tradename Envelon (9), and certain grades of ionomer resins (10) have been formulated into coating powders and are enjoying some measure of commercial success. Self-adhering clear coatings based on a combination of ionomer resins and high melt index ethylene/acrylic acid copolymers are described in a recent patent (11).

Thermoplastic polyester coating powders achieved some commercial success during the mid-1980s; however, these were eventually replaced by nylon coating powders in functional coatings and thermosetting polyester powders in decorative applications because of lack of any unique characteristics or price advantages (see POLYESTERS, THERMOPLASTIC).

Coating powders based on poly(vinylidine fluoride) (PVDF) [25101-45-5] are available and are used in architectural applications where long-term exterior performance is required. Most are modified with thermoplastic acrylic polymers (12) or other fluoropolymer resins containing comonomers to improve melt flow and application characteristics (13). A method for preparing pigmented PVDF powders that does not require melt compounding and cryogenic grinding is described in Ref. 14.

Several other thermoplastic powders are available based on specialty polymers such as ethylene-chlorotrifluoroethylene (E-CTFE) [25101-45-5], poly(phenylene sulfide) (PPS) [25212-74-2], and tetrafluoroethylene-ethylene copolymers [68258-85-5] (see FLUORINE COMPOUNDS, ORGANIC, POLYMERS CONTAINING SULFUR). Such powders are used in functional applications where resistance to corrosion and elevated temperatures are required. They are usually applied by fluidized-bed coating techniques but can also be applied by electrostatic techniques to a heated substrate (15). Extremely high application temperatures in the range of 250–350°C are required for these polymers because of high melting point and high melt viscosity.

3. Thermosetting Coating Powders

Thermosetting coating powders, with minor exceptions, are based on resins that cure by addition reactions. Thermosetting resins are more versatile than thermoplastic resins in the formulation of coating powders in that: many types are available varying in melt viscosity, functional groups, and degree of functionality; numerous cross-linking agents are available, thus the properties of the applied film can be readily modified; the resin/curing agent system possess a low melt viscosity allowing application of thinner, smoother films, and necessary level of pigments and fillers required to achieve opacity in the thin films can be incorporated without unduly affecting flow; gloss, textures, and special effects can be

Table 2. **Physical and Coating Properties of Thermosetting Powders**

Property	Epoxy	Polyurethane[a]	Polyester[b]	Hybrid	Acrylic[c]
fusion range, °C	90–200	160–220	160–220	140–210	100–180
cure time[d], min	1–30[e]	15–30	5–15	5–15	5–25
storage temp, °C[f]	30	30	30	30	20
adhesion[g]	E	G–E	G–E	G–E	G–E
gloss, 60° meter	5–95	5–95	40–95	20–95	80–100
pencil hardness[h]	H-4H	H-2H	H-4H	H-2H	H-2H
flexibility	E	E	E	E	F–P
resistance to:					
impact	E	G–E	G–E	G–E	F
overbake	F–P	G–E	E	G–E	G–E
weathering	P	G–E	G–E	P–F	G–E
acid[i]	G–E	F	G	G	F
alkali[i]	G–E	P	F	G	P
solvent	G–E	F	F–G	F	F

[a] Hydroxy function-blocked isocyanate cure.
[b] TGIC (triglycidyl isocyanurate) Hydroxy alkylamide cure.
[c] GMA(glycidyl methacrylate)type cured with DDA (dodecanedioic acid).
[d] Value is given at 160–200°C, unless otherwise indicated.
[e] At 240–135°C.
[f] Maximum value is given.
[g] E = excellent; G = good; F = fair; P = poor.
[h] Refers to highest degree of lead hardness at which coating can be marred.
[i] Inorganic; dilute.

produced by modifying the curing mechanism or through the use of additives; and manufacturing costs are lower because compounding is carried out at lower temperatures and the resins are friable and can be ground to a fine powder without using cryogenic techniques. The properties of thermosetting coating powders are given in Table 2.

Ideally, the appearance of a powder coating should equal that of a water borne or solution coating at the normal thickness range, eg, 20–60 µm (~1–2 + mils). While significant advances have been made in the formulation and application of powder coatings over the last 10 years, it is more difficult to apply powders uniformly in thin films that match the smoothness and appearance of conventional liquid finishes. However, the gap is closing rapidly. Automotive powder clear coats have been applied by BMW to several models on a production basis since 1998. As of May 2000, over 1000 car bodies per day are being powder coated on completely automated lines (but at a film thickness still higher than desired) (16).

In order to retain their particulate form and free-flow characteristics, coating powders must resist sintering or clumping during transportation, storage and handling. To maintain these properties, T_g of the formulated powder must be at a minimum of ~40°C and preferably >50°C. In the case of epoxy resins, because of their highly aromatic backbone, the necessary T_g is attained at a relatively low molecular mass. In contrast, polyester resins require linear comonomers to achieve the desired degree of flexibility leading to a lower T_g. Thus, to attain the desired T_g, higher molecular mass resins must be used resulting in

Table 3. **Melt Viscosity, Glass-Transition Temperature and Equivalent Weight of Various Thermosetting Resins**

Resin Type	Equiv. Weight	Tg, °C	Melt Viscosity, mPa·s (cP) at Indicated Temperatures, °C	
			175	200
epoxy resins				
"2" Type	600–750	50–60	500–1000	[a]
"3" Type	700–850	55–65	1000–2000	200–400
"4" Type	850–1000	60–70	2000–3000	1000
"7" Type	1500–2500	80–85	9000–10,000	10,000
polyester resins				
acid functional	750–2800	50–64		3000–6500
hydroxyl functional:				
general purpose	560–1870	50–55		3000–6000
high Tg.	1400–1800	60–70		6500–8500
acrylic resins				
acid functional	750–1600	55–68		3000–8000
hydroxyl functional	1250	60		5000
glycidyl methacrylate	510–560	39–56	390–470	[a]

[a] Too low to measure.

higher melt viscosities (17). At an equivalent range of T_g, polyester resins have a melt viscosity ~10 times higher than for epoxy resins (see Table 3).

The surface tension and melt viscosity are the main parameters affecting film formation and the flow of thermosetting coating powders (18). While a high surface tension promotes the coalescence and flow of molten powder particles, a low surface tension is necessary to wet the substrate. So-called, "flow control additives" are used in almost all coating powders to eliminate surface defects such as craters and pinholes. A more accurate term would be "surface tension modifiers", since this is their primary effect. They are believed to function by creating a uniform surface tension at the air–surface interface of the molten coating as a result of the partical compatibility (19). The most widely used flow control additives are acrylic oligomers (20). Most are primarily based on n-butyl acrylate [9003-49-0] and copolymers of ethylacrylate and 2 ethy-hexyl acrylate [26376-86-3] (see ACRYLIC ESTER POLYMERS).

While there is little difference between the surface tension lowering effects of various acrylic flow additives (21) monomer composition and molecular mass can have a significant effect on the flow and orange peel of the final coating (22). The relationship between flow and melt viscosity of the binder resin(s) is more obvious. Significant efforts on the part of resin manufacturers have been made to optimize the melt viscosity of the polymers while still maintaining the required T_g. Monomer composition (23), molecular mass, and functionality (24) are among the most important variables. The degree of reactivity between the resin and curative has a significant effect on flow and smoothness. For ideal flow and leveling of the coating, the time between melting of the powder and the start of cross-linking should be maximized. Most resins are intended for use with a particular curing agent, so the functionality of the resin is designed accordingly. Rapid heating of the applied powder layer, ie, high curing temperatures, results in the lowest level of melt viscosity, but also in a faster increase in viscosity once the cross-linking starts (25). Chemorheological measurements have proven useful in determining the most desirable resin/curing agent reactivity and functionality in specific thermosetting systems (26). Optimum curing conditions can be predicted based on the reaction kinetics of the coating powder (27). In addition to the chemical and rheological factors already mentioned, smoothness of the cured coating is also related to the structure of the electrostatically deposited powder layer. The initially applied powder layer is much more porous than would be expected from random close packing of particles and shows appreciable powder segregation and patterning. A comprehensive study of powder application, coalescence, and flow indicates orange peel, with its millimeter length scale and micron-scale amplitude, arises partly from incomplete leveling of the largest scales of unevenness originally present in the deposited powder layer (28). Previous studies have also noted the relationship between clusters or large agglomerates of particles and orange peel (29). Finer particle size powders eliminate some of the large agglomerates and result in smoother films (30), but finer particles are more cohesive, less free flowing, and more difficult to manipulate (31).

The higher degree of orange peel exhibited by powder coatings in comparison to conventional liquid coatings is a fundamental shortcoming. Determining the ultimate cause of this deficiency and correcting it will lead to even greater acceptance and use of powders in finishing processes.

3.1. Formulation. In many respects, the formulation of coating powders is similar to that for conventional paints. The resinous binder plays a major role in the basic properties of the final coating such as exterior durability, chemical resistance, flexibility and impact resistance, and, to some extent, appearance. Whereas in conventional paints solid ingredients such as pigments and fillers are dispersed in the liquid vehicle using a mill, in coating powders, the solid ingredients are dispersed in the molten binder in an extruder. Paint dispersions must be stabilized to prevent pigment agglomeration and settling. Flooding, floating, and pigment agglomeration are not a problem in a solid binder so rheological and dispersing additives are not necessary and are seldom used in coating powders. A significant formulating advantage for conventional paints is that there are few constraints on the T_g of the binder resins; they can even be liquids. In addition to the binder resin(s), curing agents, which can range from crystalline solids to polymers, flow agents, additives, pigments, and fillers are utilized in coating powders. An important formulating variable is the ratio of pigments and fillers to the binder, the pigment/binder or P/B ratio. Typical pigment/binder ratios used in formulating various types of coating powders are listed in Table 4. In general, the P:B ratios used in coating powders are much lower than for conventional paints since there are no other liquid ingredients present to wet out the pigments. Thus the pigment volume concentration in coating powders is correspondingly lower and seldom exceeds ~25%. Higher levels of pigments or fillers significantly reduce the flow out of coating powders resulting in a rough grainy coating.

In addition to the acrylate flow control additives, silicones, primarily the polyether modified types and fluoropolymer flow agents, are also used. Care must be exercised when using powders with differing flow additives or significantly different binder resins since cross-contamination can occur resulting in

Table 4. **Pigment: Binder Ratios for Various End Use Applications**

Pigment: binder ratio	% Pigment and fillers	Approximate PVC[a]	Application
0.01–0.10	1–9	0.3–2.9	transparent, flamboyant, clear metallic, special effects
0.1–0.2	9–17	2.9–5.7	high chroma, high DOI, automotive exterior durability, high flexibility
0.2–0.6	17–37.5	5.7–15	general purpose, appliance, furniture, fixtures, lawn and garden
0.6–1.0	37.5–50	15–23	low cost, anticorrosive, electrical insulation, pipe coatings
1.0–1.2	50–55	23–26.5	primers, textures, functional coatings, economy grades
>1.2	>55	>26.5	specialized applications, zinc rich primers

[a] Pigment volume concentration calculated using pigments and fillers having a specific gravity of 4.0.

loss of gloss, surface imperfections, and loss of smoothness. These incompatibilities arise from differences in the surface tension of the various powders (32). Low melting thermoplastic additives such as benzoin [119-53-9] are used in coating powders to promote bubble release and air entrapment (33). Many of the additives used in coating powders, such as uv absorbers, light and heat stabilizers, and mar and slip agents, have their basis in conventional coatings. However, others, such as electrostatic or tribocharging additives and postblend additives mixed with the finished powder to improve dry, free-flow characteristics (34) are unique to coating powders.

Many of the pigments used in conventional paints are also used in coating powders. These include inorganic pigments such as titanium dioxide, nickel/titanium rutile, iron oxides, and complex inorganic pigments. Lead- and cadmium-based pigments have not been used for many years in the United States and Europe and are being phased out in the rest of the world. Typical organic pigments used include phthalocyanine blues and greens, various azo types, quinacridones, carbazoles, diketo-pyrrolo pyrroles, among others. Generally, inorganic pigments are used to provide opacity, while organic pigments are used for their chroma or saturation. Because of their high surface area and high binder demand, organic pigments cannot be used at very high levels, ie, >5–6% before a noticeable reduction in flow and increase in orange peel of the coating starts to occur. Therefore, as a rule of thumb, the level of inorganic pigments should be maximized and the level of organic pigments should be minimized, while meeting the requirements for opacity and chroma. As noted, even inorganic pigments with their higher opacity and lower binder demand are not used at levels in coating powders as high as in paints. To obtain complete opacity with a white coating powder containing 30% TiO_2 (P/B ratio = 0.46) a film thickness of 89 μm is required. The addition of only 0.001% carbon black reduces the thickness required for complete hiding to 63 and 37 μm with the addition of 0.005% (35).

Fillers such as calcium carbonate, blanc fixe and barium sulfate [07727-43-7] and wollastonite [13983-17-0] are used in coating powders to modify gloss, hardness, permeability, and other coated film characteristics and to reduce costs (36). Clays and talcs are seldom used, except in textured coatings, because of their high binder demand and adverse affect on flow and smoothness. Silicas are usually avoided due to their abrasiveness during extrusion and grinding, with the exception of colloidal silica used as a postextrusion additive.

Matting or flattening agents are employed to control gloss, which is dependent on microscopic surface smoothness (37). Thus, nonmelting or incompatible thermoplastic resins of proper particle size such as Teflon and polypropylene are used to disrupt surface smoothness and reduce gloss (38). Similarly, incompatible waxes concentrate at the surface of the coating, also reducing gloss but resulting in a waxy feel, which is prone to showing blemishes, eg, fingerprints. Coarse grades of fillers such as barytes, calcium carbonate, or wollastonite are also used for gloss control, usually with other techniques. Curing agents having widely different reactivities cause a two-stage polymerization to occur, resulting in incompatible domains and impaired microscopic surface smoothness (39). In a similar fashion, low gloss is achieved by mixing two powders varying significantly in reactivity (40). Gloss can be controlled over a wide range using combinations of glycidyl methacrylate functional resins with acid functional polyester resins (41). In summary, there are many methods of gloss control available to the coating powder formulator covering the full range of gloss in both interior and exterior durable systems.

3.2. Special Finishes. Clear coatings are formulated using curing agents and flow additives, which have a high degree of compatibility with the resin. Conventional uv and hindered amine light stabilizers can be added to improve exterior durability. Metallic finishes can also be prepared but the metal flake must be added after the powder has already been ground to prevent break up of the metallic flakes and preserve the metallic appearance (42).

Hammertones, veins, and other special effects are prepared by the judicious addition of surface tension lowering ingredients, eg, silicones or flow control/ resin master batches, usually in conjunction with a dry blended metallic or mica flake pigment. Textured coatings are produced by controlling the flow and particle size of the powder particles as well as with nonmelting polymers of controlled particle size. Wrinkle finishes are obtained using selected curing agents and catalysts (43).

3.3. Epoxy Coating Powders. Thermosetting coating powders based on epoxy resins $C_{15}H_{16}O_2 \cdot (C_3H_5ClO)$, [25068-38-6] have been used longer than any other resin system. The reason that is solid epoxy resins were commercially available when thermosetting coating powders were being developed and had the necessary combination of low molecular mass, T_g, and melt viscosity (see Table 3). Further, a variety of latent curing agents were also known that allowed the development of stable, one component powders. Early efforts to develop powders based on dry-blend processing methods, such as by ball milling, were generally not commercially acceptable because the resultant coatings were low in gloss, lacked smoothness, and good appearance, especially with greater than minimal levels of pigments and fillers. These problems were overcome when powders were processed by melt mixing, eg, extrusion (see Fig. 1 and related text). The earliest powders were based on dicyandiamide [461-58-4] a latent curing agent. However, these powders were too slow curing, requiring 15–30 min at 200°C, to achieve full properties. A wide variety of catalysts for the dicyandiamide epoxy reaction were evaluated, but the clear choice after several years of trial and error were imidazoles, especially 2-methyl imidazole [693-90-1]. Dicyandiamide has a high melting point and limited solubility in epoxy resins. Dicyandiamide derivatives having aromatic substitution were developed, which are more compatible with epoxy resins, easier to disperse, and more reactive while still retaining their latency. Typical of these substituted dicyandiamide derivatives are o-tolyl biguanide [93-69-6] and 2,6-xylenyl biguanide (44). Highly reactive, compatible, and low melting curatives are also prepared by reacting an epoxy resin with excess imidazole. Another class of curatives, developed somewhat later, are the linear phenolics. These have the same structure as epoxy resins but are terminated with bisphenol A [80-05-07] rather than epoxide groups and contain significant levels of free bisphenol A. Since they have a functionality of only 2, the functionality of the epoxy resin in the binder must be increased, usually by blending with an epoxidized phenol novolak resin to give an average functionality of ~2.5–3.0. The epoxy/phenolic hydroxyl reaction is relatively slow for many applications so an imidazole catalyst is often included in the formulation (45). Most of these curatives are still used today in both decorative and functional epoxy coating powders and provide compositions that cure in the typical range of 10–15 min at 180–200°C or in the low temperature range of 15–20 min at 140–180°C.

Many other types of curing agents have been evaluated in formulating epoxy-based coating powders but have found use in only specialized applications. Conventional phenol or cresol novolak resin curatives impart a high degree of cross-link density but result in rather brittle coatings with undesirable color. Aromatic amines give very reactive, fast curing systems but are marginal in storage stability and their health and safety characteristics are questionable. Of course, primary aliphatic amines cannot be used because they react even at room temperature and cannot be compounded at elevated extrusion temperatures. Dihydrazides have also been evaluated but their cost/performance base can seldom justify their use. Many anhydride curing agents have been considered but only a few have found their way into commercial formulations. For a time, trimellitic anhydride [552-30-7] cured powders catalyzed with metalorganic salts were used in the formulation of very reactive, fast cure powders, such as used for coating concrete reinforcing bars (rebars) and pipe coatings (46). However, health and safety concerns led to the decline in use of this chemistry. Benzophenone tetracarboxylic dianhydride yields highly cross-linked coatings with very good heat and chemical resistance and is used in specialized functional applications (47).

Decorative epoxy powders are used in a wide variety of applications, eg, for lighting fixtures, garden equipment, motor control cabinets, and many automotive under the hood items including coil spring shock absorbers, mechanical parts, and even engine blocks. Low melt viscosity resins of the "3" type are most widely used in decorative applications (see Table 3). Type "4" resins and higher with a higher molecular mass and melt viscosity are more often used in functional applications such as for electrical insulation and corrosion resistance where thicker coatings are needed and a higher degree of edge coverage is necessary. Epoxy powders are used almost exclusively in rebar and pipe coatings (48). In outdoor applications, epoxy coatings chalk readily and lose gloss; however, they protect the substrate for many years. Figure 2 compares the gloss retention of epoxy coatings with other thermoset types on exterior exposure.

3.4. Epoxy-Polyester Hybrids.

A major class of interior grade coating powders is based on a combination of epoxy and acid functional polyester resins. As noted, epoxy resins cured with anhydrides have desirable properties but never gained a high degree of commercial acceptance. Most anhydrides are respiratory irritants and are difficult to work with. Also, they are hygroscopic and the reactivity of coating powders cured with anhydrides varies on storage depending on the ambient humidity and the degree of conversion of the anhydride to acid. On the positive side, anhydride cured powders possess heat resistance and good color stability in contrast to the tendency for most amine cured powders to discolor on exposure to heat or exterior exposure. Therefore, programs were initiated in the research facilities of polyester resin manufacturers to develop resins that retained the desirable characteristics of anhydrides while eliminating their undesirable characteristics. This work resulted in the development of acid functional, relatively linear, saturated polyester resins specifically designed for curing bisphenol A based epoxy resins. The original resins developed had an acid number of 70–80 (equiv. weight range 700–800) so they were used at 50:50 ratio with "3" type epoxy resins (see Table 3). The functionality is ~2.5–3, to provide good curing characteristics and cross-link density when used with the essentially bifunctional epoxy resins. Because polyester resins are less expensive than epoxy resins, higher equivalent weight polyester resins were subsequently developed, which are used at a stoichiometric ratio of 60:40 and even 70:30 ratios to epoxy resin. Properties of these polyester-epoxy hybrids are similar to those of a straight epoxy, but differ in several respects. The overbake resistance (resistance to color change after extending curing) and resistance to discoloration on exposure to sunlight is superior. Because the cross-link density for hybrid coating powders is generally less than for straight epoxies, cured hybrid coatings are inferior in solvent resistance and hardness. They are also somewhat inferior in salt spray and corrosion resistance. Polyester resins, having a higher melt viscosity than epoxy resins (see Table 3), result in the hybrids having more orange peel than epoxy based coatings, especially at the higher polyester/epoxy ratios.

The reaction rate between the carboxyl end groups of the polyester and the epoxide groups of the epoxy resin is generally quite slow, requiring a catalyst to obtain a practical baking time. Catalysts are frequently mixed with the polyester resin by the resin manufacturer. The ideal catalyst should exhibit good reactivity at the desired baking temperature, eg, 150–180°C, while providing good flow and shelf stability (50).Tertiary amines, amic acids, and quaternary phosphonium compounds are effective catalysts for the epoxy-carboxyl reaction (51,52).

Epoxy-polyester hybrid coatings are marginally better than straight epoxy based coatings in gloss retention on exterior exposure (Fig. 2) but generally are not recommended for exterior applications. For the most part, applications for the hybrid powders are the same as those for decorative epoxy coating powders. The latter are being replaced by the hybrid coating powders and are increasing in market share in the United States. In Europe, hybrid coating powders are the most widely used powder type (see Table 5).

3.5. Polyester–TGIC Cured.

A principal class of exterior durable powder coatings is based on acid functional, saturated polyester resins cured using TGIC [02451-62-9] (see Fig. 2). This system was first developed in Europe

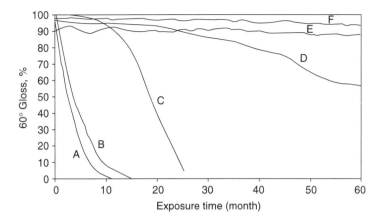

Fig. 2. Gloss retention in outdoor exposure in Florida for various powder coatings: A, epoxy; B, epoxy-polyester hybrid; C, polyester TGIC, HAA, and urethane (dark brown); D, super durable polyester TGIC, HAA, and urethane (dark brown); E, super durable polyester, clear, no light stabilizers; F, GMA acrylic, clear, light stabilized. (After Ref. 49,74).

in the early 1970s. The acid functional polyester resins used in TGIC cured coating powders are similar to those used in epoxy polyester hybrids. However, the resins for curing with TGIC have a higher equivalent weight, typically in the range of 1600–1900 and a lower degree of functionality. Thus most resins are used at a 93:7 ratio of resin to TGIC. Acid functional resins are normally prepared by a two-step process: the reaction of excess polyol and dibasic acids followed by esterification of the hydroxyl terminated resin using dibasic acids or anhydrides (53). This technique yields a resin where the functional groups are at the end of the molecule rather than occurring randomly along the polymer chain. The excellent exterior durability of polyester resins cured with TGIC is primarily a result of the nonaromatic structure of TGIC in contrast to that of the bisphenol A based epoxy resins used in polyester/epoxy hybrid coating powders. Monomer composition of the resin also plays a significant role. It is recognized that the exterior durability of polyester powder coatings is affected by factors other than resistance to uv radiation. Accumulation of moisture on the surface of the coatings, especially at elevated temperatures (darker colors) and exposure to oxygen leads to hydrolysis, oxidation, and degradation of the resin.

Table 5. **1998 Thermoset Powder Coatings Production by Resin Type**[a]

Resin type	North America (%)	Europe
epoxy	15.0	8.5
epoxy–polyester hybrid	35.7	54.0
polyester–carboxyl	22.6	29.0
polyester–hydroxyl	21.6	4.0
acrylic and other	5.1	4.5
total	*100.0*	*100.0*

[a] Ref. 119.

Environmental factors such as acid rain or alkaline bird droppings as well as cyclic heating and cooling also contribute to molecular breakdown and loss of properties. For these reasons, accelerated weather testing now frequently includes cyclic exposure to elevated temperatures and moisture as well as uv light (54).

The recent development of so-called "super durable polyesters" has validated the cyclic accelerated weathering approach. Super durable polyester resins are based almost exclusively on isophthalic acid while most standard resins contain terphthalic acid as well. Super durable polyesters are highly resistant to hydrolysis and degradation by uv light (55,56), which helps to account for their superior exterior durability. Compared with a standard TGIC cured polyester powder coating in a dark color, which loses ~50% of its original loss after ~2 years of Florida exposure, a super durable polyester cured with TGIC will last at least four times as long until the same loss of gloss occurs. Figure 2 is a plot of % gloss retained vs. original level, measured on a 60° gloss meter, comparing super durable polyesters with several other types of powders coatings. Super durable polyesters are generally inferior in flexibility and impact resistance but recent advances have shown significant improvements in this regard (57).

TGIC cured coating powders have gained a significant market share position in the exterior durable market in both Europe and North America. Many buildings coated with TGIC powders in various European locations from the early 1970s and later are still in good condition, exhibiting minimal corrosion and good retention of gloss and color. In the past 10 years, some concerns have developed over health and safety issues related to TGIC, especially with regard to mutagenic characteristics. In the United States, it is generally believed the hazards are adequately addressed by the Occupational Safety and Health Administration (OSHA) Hazard Communication Standard and the low exposure level of 0.05 mg/m^3 established (58). However, in Europe, the European Union (EU) ruled that the symbol T (Toxic, symbolized by a skull and crossbones) accompanied by the relevant Risk Phase, R46, relating to substances considered to cause heritable genetic damage, is to be used in labeling any product containing TGIC. This requirement went into effect in June 1998 and had the result of powder manufacturers replacing TGIC in their products. Since that time the market share of TGIC cured polyester coating powders has decreased significantly in Europe, being largely replaced by hydroxyalkylamide cured powders (59). Other glycidyl compounds such as a mixture of diglycidyl terphthalate [7195-44-0] and triglycidyl trimellitate [7237-83-4] (60) as well as tris (beta-methyl glycidyl) isocyanurate [26147-73-3] (61) are being evaluated as TGIC replacements as well.

3.6. Polyester—Hydroxyalkylamide Cured. HAA curatives were developed in the late 1970s and early 1980s (62). Evaluation as a curative in coating powders was described in early 1991 (63). The primary commercial product is bis (*N*,*N*-dihydroxyethyl) adipamide [6334-25-4] sold under the tradename Primid XL 552 by EMS Chemie. It reacts with acid functional polyester resins by esterification and the elimination of water. The toxicological profile of this curative is benign and it does not fall under the provisions of any current health and safety or environmental regulations. It has been shown that the ester linkage formed with Primid XL 552 and polyester resins has essentially the same characteristics as the TGIC/polyester bond (64). Practical experience has confirmed that polyester coating powders cured with Primid are essentially equivalent in weatherability and other properties to TGIC cured powders (65). Because the functionality of the HAA curative is ~4, the functionality of the polyester resins used with them has to be designed accordingly (66).

3.7. Urethane Polyesters. In the United States, the search for exterior durable coating powders led to technology based on hydroxyl functional polyester resins. The earliest curing agents evaluated were based on melamine-formalde-hyde resins, such as hexa(methoxymethyl) melamine (HMMM) [68002-20-0], which are widely utilized as curing agents in conventional paint systems (see AMINO RESINS). Coating powders based on this chemistry suffer limitations: the melamine resin depresses the T_g of the coating powder to the point where the powder sinters during storage, especially at elevated temperatures; and the methanol generated during the curing process becomes trapped in the film, especially at thicknesses above ~50 μm, resulting in a frosty or visually nonuniform surface. An amino resin, specifically developed for use in coating powders, tetramethoxymethyl glycouril [17464-88-9] (TMMGU), overcomes many of these disadvantages, but still requires the use of higher T_g resins and special acid catalysts (67). Coating powders based on this chemistry have not achieved a high level of commercial acceptance for general purpose use; however, with selected catalysts, attractive wrinkle finishes are produced (68,69) that find use in special applications. Curing agents based on polyisocyanates blocked with caprolactam [0105-60-2] (qv) give an excellent combination of properties in the final film. Because the unblocking reaction does not start to occur until ~160°C, the powder has a chance to flow out and give a smooth uniform film prior to any substantial cross-linking. Not all of the caprolactam evolves during the curing process and some remains in the film acting as a plasticizer. Thus, urethane polyesters yield a smoother, more orange peel free film than the TGIC polyesters and are more preferred in the United States and Japan than Europe (see Table 5).

The hydroxyl functional polyester resins used in this technology are similar in monomer composition to the acid functional polyesters and are based primarily on terephthalic acid [100-21-0], $C_8H_6O_4$, isophthalic acid [121-91-5], $C_8H_6O_4$, and neopentylglycol [126-30-7], $C_5H_{12}O_2$. These resins branched using trimellitic anhydride or trimethylol propane [77-99-6]. A variety of other polyols and dibasic acids are used to modify specific resin properties such as T_g, melt viscosity, curing characteristics, and others (70,71). Resins ranging in OH number from 30–300, equivalent weight of ~190–1870, are used in the formulation of urethane coating powders. Since the blocked isocyanate curatives are three to four times the cost of the polyester resin, the lower OH number resins are preferred for general purpose formulations, with a 50 OH number resin providing a good balance between raw material cost, physical properties and appearance. While the higher OH number resins require higher levels of curative, the final coatings have outstanding hardness and chemical resistance. Blends of high and low OH number resins and the stoichiometric level of curative form the basis of low gloss coatings. Hydroxyl functional super durable resins, similar in monomer composition to the acid functional resins cured with TGIC, are also available. When cured with isophorone diisocyanate (IPDI) based isocyanates, they possess exterior durability equal to that of the super durable/TGIC based coatings (see Fig. 2). The most commonly used curing agents are trimerized

IPDI, $C_{12}H_{18}N_2O_2$, [4098-71-9] blocked with caprolactam and the trimethylol propane adduct of IPDI blocked with caprolactam. Blocking agents, which unblock at lower temperatures, provide the basis for urethane powders that cure at lower temperatures (72). New curatives have been developed that do not rely on blocking agents to mask the reactivity of the isocyanate group. Most are based on the uretdione structure, a four-membered ring formed by the reaction of isocyanate groups with each other (73). While uretdione-based curatives have been available for many years, they have high melt viscosities and require higher temperatures for curing than the blocked isocyanates. Newer versions have overcome these problems to some extent (74).

Urethane polyesters have not received widespread commercial acceptance in Europe primarily because of the caprolactam (or other blocking agents) emitted during curing. Despite the development and commercial availability of the uretdione based curatives, which do not give off volatiles during cure, the European market for exterior grade polyester powders is still based primarily on acid functional resins (see Table 6).

3.8. Unsaturated Polyester Powders.   A special class of coating powders is based on unsaturated polyester resins. They are utilized in matched metal die molding operations such as those based on sheet molding compounds (SMC) and bulk molding compounds (BMC) where the mold is coated with the powder prior to placing the resin charge in the mold (see POLYESTERS, UNSATURATED). The powder melts and flows on the mold surface, and when the mold is closed, the powder reacts with the molding compound forming a coating on the molded part. This process is known as inmold coating. Unsaturated polyester resin powder coatings can provide a colored and finished exterior molded surface or a finish ready for painting. Normally, a primer/sealer must be applied to molded articles prior to painting. In addition to the unsaturated polyester resin, multifunctional resins prepared from unsaturated monomers such as triallyl cyanurate [101-37-1] or diallyl phthalate [131-17-9], suitable peroxide initiators (qv) or mixtures thereof, and mold release agents (qv), are used to formulate the coating powder (75).

3.9. Acrylic Powders.   Coating powders based on acrylic resins have been available in Europe, the United States, and Japan since the early 1970s but have not achieved significant commercial success until recently. However, since 1997 BMW has been applying an acrylic clear-coat powder to several models and currently (2000) is powder coating over 1000 cars/day (16). Acrylic-based

Table 6. **1999 World Thermoset Powder Coatings Production**[a]

Region	Metric tons	%
Europe—West and East	329,100	45.3
Far East	161,585	22.2
North America	175,600	24.2
South America	29,950	4.1
Rest of World	30,865	4.2
Total	*727,100*	*100.0*

[a] Ref. 118.

powders are also used in exterior trim (pigmented) and wheels (clear). The majority of clear coatings are based on a glycidyl functional methacrylic resin (GMA) cured with dodecanedioic acid [693-23-2] or a polyacid/anhydride polymer (76). Acrylic powders based on GMA resins have poor compatibility with epoxy- and polyester-based powders. If cross-contamination occurs, surface defects in the form of pinholes, craters, and excessive orange peel are common.

The GMA clear coatings have outstanding exterior durability, hence their acceptance in the automotive sector (see Fig. 2). Pigmented GMA acrylic powders have not found widespread use in market areas other than automotive, eg, architectural applications. Hydroxyl functional acrylic resins cured with blocked isocyanates or uretdione-based curatives have also found use in automotive applications and their exterior durability is only somewhat less than the GMA acrylic powders (77). Carboxyl functional acrylic resins are also commercially available. They can be cured with TGIC for exterior applications or with bisphenol A epoxy resins. The latter combinations are sometimes called acrylic/epoxy hybrid coating powders and are noted for their excellent hardness, stain, and chemical resistance. They find use primarily in appliance coatings (78).

3.10. Recent Developments.   As noted, powder coating technology has advanced to the point where powders are now routinely applied in critical automotive applications. Another market area where intense work is in progress is the development of powders having low temperature curing capabilities such that they can be used to coat wood, plastics, and other temperature sensitive substrates. A coating powder must have a high enough T_g and molecular mass so it does not sinter on the one hand while on the other hand a low melting point and melt viscosity is desirable for low temperature application and smoothness. It is a difficult task to balance these diverse requirements. This is especially true in the case of thermally cured powders compared with uv curable powders. In the latter case, the flow and leveling of the coating are separate from the curing reaction while with heat curable powders, cross-linking occurs even as melting and flow are in progress.

One technique for dealing with this low temperature cure/reactivity dilemma is to use a two component powder system. The resin and other binder components comprise one component and the curative and other nonreactive binder ingredients comprise, the other (79). In the case of uv curable powders, an unsaturated polyester resin is one of the major binder components but more reactive binder components such as oligomers with high allyl functionality (80) or vinylether/ester groups (81) must also be present. Crystalline resins with reactive methacrylyl groups have also been disclosed (82). A review of photoinitiators, additives, and other components of uv curable powders are given in (83,84).

4. Manufacture

The vast majority of thermosetting coating powders are prepared by melt mixing. Some thermoplastic powders are also produced by this method but most are manufactured by the dry blend process as shown in Figure 1. Production methods based on spray drying from solution (85) and precipitation from solution (86) have been evaluated but never achieved commercial success because of difficulties in solvent and/or water removal from the powders. Many types of coating powders are still manufactured in small batches, eg, 50–1500 kg due to differences in color or chemistry, where chemical processes are not economical.

4.1. Melt-Mixing. Dry ingredients, resins, curatives, additives, pigments, etc, are weighed into a batch mixer such as high speed impeller mixers, container mixers, horizontal plow mixers, or tumble mixers where they are thoroughly blended. High-speed impeller mixes such as the Henschel give the best distributive mixing and the cycle time is relatively short, eg, 1–2 min. However, these mixers have relatively small capacity, eg, 100–300 kg and many individual batches must be prepared for longer production runs. Horizontal plow mixers require a slightly longer mix cycle but have a larger capacity, 500–1000 kg. They also provide good distributive mixing but require longer cleaning times. Tumble mixers have a high capacity and are adequate for general purpose powders, but mixing cycles are long, 30–60 min, and distribution of ingredients is sometimes marginal. With container mixing systems, only one mixer is required and generally satisfactory distributive mixing is obtained. However, many mixing containers, which become the hopper for feeding the extruder, are required as well as a mixer for each size of mixing container (see MIXING AND BLENDING). The premix is then melt compound in a high shear extruder where the ingredients are compacted, the resin(s) melt, and individual components are thoroughly dispersed in the molten resin. These compounding machines generate sufficient heat through mechanical shear so that after start-up, little external heat needs to be supplied. Both single screw machines, with a reciprocating screw that intermeshes with fixed baffles in the barrel, such as the Buss Ko-Kneader and twin screw extruders, primarily corotating, intermeshing types as supplied by Krupp Werner Pfleider, Baker Perkins, and others are used (see PLASTICS PROCESSING). Residence time in the extruder is short, usually <1 min and melt temperatures is low, typically 90–120°C, slightly above the melting points of the resinous components. Because of these processing conditions, very little reaction between the thermosetting components occurs. In a study carried out based on epoxy resin compositions, it was determined that 6–11% of the epoxy groups initially present reacted during extrusion (87). Significant improvements have been made in both single and twin screw extruders and for a given size machine throughput rates have increased by a factor of 3–4 over the last 10 years or so. Product quality is more consistent as a result of improved process control (88). The molten compound is cooled rapidly by passing it through water-cooled nip rolls and subsequently onto a watercooled continuous stainless steel belt or drum. The cooled compound is broken into small chips, ~10–12 mm, suitable for fine grinding. Thermosetting resins are quite friable and are usually ground to final particle size in an air classifying mill. In this grinder, a blower generates an air stream through the mill in which the product is entrained and which also serves to remove the heat of grinding. A variable speed separator controls airflow in the grinding chamber so that only the particles with the desired particle size escape. The fine powder is separated from the air stream with a cyclone separator or bag house. Powders with a finer average particle size produce smoother coatings than powders with a larger particle size. While the addition of a colloidal silica or alumina to the powder, (either to the chips prior to grinding or to the ground powder), significantly improves the dry flow, handling, and transport characteristics (89), the presence of high levels of superfine particles adversely affects these same characteristics. Recent advances in grinding equipment include in-line air classifiers and baffles in the cyclone separators, which allow the production of powders with a narrower particle size distribution (90). This results in powders with a finer average particle size but without a significant increase in superfine particles, ie, those below $\sim$8–10 μm.

4.2. Dry Blending. Most plasticized PVC powders are prepared by a dry-blend process in which the plasticizers, stabilizers, pigments, and additives are absorbed on the porous PVC particles at elevated temperatures while they are being agitated in a high speed mixer (6). Other thermoplastic powders are pigmented in this fashion. Attempts to prepare thermosetting powders by a dry blend process have proven to be unsatisfactory because of the poor wetting and dispersion of pigments and the poor appearance of the subsequent coatings.

4.3. Recent Developments. A completely new process for the manufacture of coating powders has been recently developed by the Ferro Corp. It involves solvating resinous components in supercritical gas, typically carbon dioxide (CO_2), and dispersing the pigments and other solid ingredients with an impeller or dispersion blade. After dispersion is completed, the material is hydraulically atomized into a second vessel at a lower pressure. The CO_2 vaporizes and a combination of particles or easily grindable solid is obtained (91). A similar process for preparing coating powders is described in Ref. 92. The main advantage of these processes over conventional extrusion and grinding technology is that processing temperatures are lower, eg, 40–70°C, and it is possible to form powders directly. At present, there are no powders commercially available prepared by these methods.

Another recent development is the successful application of powders from aqueous dispersions. This process is being used to apply both primer surfacers and powder clear coats to the Mercedes A Class automobiles (93). An advantage of this process is that a very fine particle size powder can be used, 100% <10 μm, (94) and it can be applied using conventional wet spray equipment. The concept of applying powders by aqueous dispersion is not new (95) but this is the first time it has been commercially successful. Now that there are large volume applications for powder coating in the automotive sector as primer surfacers and clear coats, it is anticipated that novel methods for preparing either powders or powder dispersions utilizing efficient chemical processing methods will be developed.

5. Application Methods

5.1. Fluidized-Bed Coating. Fluidized-bed coating, the first significant commercial process for applying powdered polymeric materials to a substrate to form a uniform coating, is the method of choice for many applications where a heavy functional coating is required. The process is relatively simple. The main variables are the temperature of the part as it enters the fluidized bed, the mass of the part being coated, dip time, and postheat temperature. Other variables, such as motion of the part in the bed and the density and temperature of the powder in the bed, also affect the quality of the coating. The process is especially useful in coating objects having a high surface/mass ratio such as fabricated wire goods and expanded metal. Sharp edges and intersections are well covered because of the heavy film thickness, eg, 250–500 μm (10–20 mils) applied. The size of parts that can be coated is limited because the fluidized bed container must be large enough to readily accommodate them.

5.2. Electrostatic Fluidized-Bed Coating.   In an electrostatic fluidized bed, the fluidizing container and the porous plate must be constructed of a nonconductive material, usually plastic. Ionized air is used to fluidize and charge the powder. The parts to be coated are passed over the bed and charged powder is attracted to the grounded substrate. The rate of powder deposition varies significantly depending on the distance of the part from the fluidizing powder. Therefore, this process is usually utilized only when the object to be coated is essentially planar or symmetrical and can be rotated above the charged powder. Electrostatic fluidized-bed coating is an ideal method for continuously coating webs, wires, fencing, and other articles that are normally fabricated in continuous lengths and are essentially two dimensional. In a variation of this process, two electrostatic fluid beds are arranged back to back and the continuous web of material is passed between them, coating both sides simultaneously. Millions of lineal meters of window screen have been coated using this technique (96).

5.3. Electrostatic Spray Coating.   Electrostatic spray coating is the most widely utilized method for the application of powder coatings (see COATING PROCESSES, SPRAY COATINGS). In a typical high voltage system, powder is maintained in a fluidized-bed reservoir, injected into an air stream, and carried to the gun where it is charged by passing through a corona discharge field. The charged powder is transported to the grounded part to be coated through a combination of electrostatic and aerodynamic forces. Ideally, the powder should be projected toward the substrate by aerodynamic forces so as to bring the powder particles close to the substrate where electrostatic forces then predominate and cause the particles to be deposited. The powder is held by electrostatic forces to the surface of the substrate that is subsequently heated in an oven where the particles fuse and form a continuous film. The processes involved are powder charging, powder transport, adhesion mechanisms, back ionization, and self-limitation. As charged powder particles and free ions generated by the high voltage corona discharge approach the powder layer already deposited, the point is reached where the charge on the layer increases until electrostatic discharge occurs. At this point, any oncoming powder is rejected and loosely adhering powder on the surface falls off. It has been demonstrated that some imperfections in the final coating are a result of defects in the powder layer (97).

The characteristic of the electrostatic spray process to form self-limiting films enables operators to apply satisfactory coatings after only brief training and instruction. It is almost impossible to create runs, drips, or sags characteristic of spray-applied liquid finishes. Furthermore, the practical design of automatic spray installations is possible. Multiple electrostatic guns mounted on reciprocators are positioned in opposition to each other in an enclosed spray booth and parts to be coated are moved between the two banks of guns where a uniform coating of powder is applied. Oversprayed powder is captured in the reclaim system and reused. Powder coating booths have been designed with interchangeable filter units to facilitate change from one powder type of color to

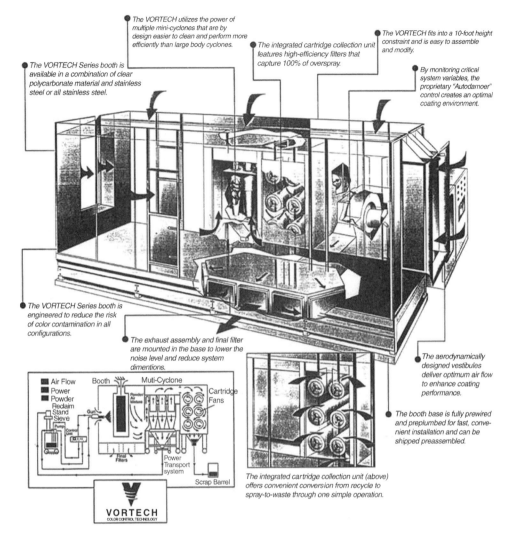

Fig. 3. Schematic diagram of an electrostatic powder spray system. Courtesy of ITW Gema.

another. A state-of-the-art automatic powder spray system using a combination of cyclone separators to reclaim the majority of over-spray powder and cartridge filters to capture the remaining powder is shown in Figure 3. The cyclone units are designed for quick cleaning and are much easier to clean than a bank of cartridge filters; quick color changes are facilitated. For very short runs, the cyclone separators can be removed and over-sprayed powder is not recovered.

One disadvantage of the electrostatic powder spray process using corona discharge guns is that a high voltage field is set up between the gun and the parts to be coated. Parts having deep angles or recesses are sometimes difficult to coat because of the Faraday cage effect. One method of overcoming the Faraday cage effect is by the use of triboelectric guns in which powder charging occurs by the frictional contact of the powder particles and the interior surface of the gun. Electrons are separated from the powder particles, which become positively charged and attracted to the substrate. Because there is no electrostatic field between the gun and the article being coated, a Faraday cage is not developed and particles are more readily able to penetrate into recessed areas of the substrate. There are a number of commercial powder coating systems using triboelectric charging technology, but this number is quite small compared to those based on the more conventional corona guns. The powder application rate of triboelectric guns is lower than for corona guns; additionally, whereas powders based on a wide variety of resins and formula types charge and apply readily with corona guns, only certain resin systems charge well in triboelectric guns. However, additives have been developed that improve the tribocharging characteristics of powders (98).

While the basic principles of the electrostatic powder spray process have not changed, many advances in the equipment and process design have been made to improve process efficiency and the quality of the finish (99). Significant improvements have been made in reducing the time to change types of powders or make color changes. Much effort has gone into improving the first pass transfer efficiency, ie, the ratio of powder deposited on parts to that sprayed (100). High transfer efficiencies mean less overspray to reclaim. Less powder to reclaim means the recovery system can be reduced in size and requires less air volume and energy to operate. Some of the means to improve transfer efficiency include: programmable logic controls to adjust the motion of the guns and spray patterns for the specific parts being coated and to recognize the parts passing through the booth, shutting off powder flow when a vacancy in the line occurs or the line stops (101); sensors that detect the amount of powder in the air and modulate the air flow according (102), and non conductive spray booths and filter belts to facilitate powder reclaim (103). In the ideal case of 100% transfer efficiency, a powder reclaim system would be unnecessary and no clean up required between color changes.

Improvements have also been made in materials. Stricter control of particle size distribution improves powder handling and transport characteristics, spray patterns, and transfer efficiency (104,105). Reduction of the superfine fraction, ie, particles below ~8–10 μm also results in less fines in the overspray powder and less load on the reclaim system (106). More spherical powders charge more efficiently and show improved spraying characteristics (107).

5.4. Recent Developments. A recent major development is the commercial application of powders to auto bodies as primer surfacers and clear coatings. Also, specialized coating lines have been developed for the application of powders to metal blanks and coil. In the case of blank coating, sheet stock is cut and punched and coated on a flat line, using either standard or tribocharging guns. Powders are applied at a lower thickness (~25–40 μm) and must cure rapidly, eg, 1–2 min, to accommodate the relatively high line speeds (108). Coated blanks are shipped to the manufacturer flat, and assembled there into the final shape, eg, a refrigerator box. This eliminates an in-house coating line.

The continuous powder coating of steel coils is being carried out at a number of locations in Europe and the United Stated (109). Lines are now operating on coils 60 in. (1.52 m) wide at line speeds of 250–300 ft/min (76–91 m/min) (110) and developments are in progress in which line speeds of up to 1000 ft/min (305 m/min) are expected (111). Many special finishes, such as textures, hammertones, and wrinkles, are possible with powder while smooth finishes are only possible with liquid coatings. Because of the high line speeds and volume of powder applied, a novel powder charging and application system was developed (112).

Specialized application equipment has also been designed to coat the interior of can bodies (113). Powders having a finer particle size are required as well (114). A significant new powder application method has been developed utilizing photocopying and laser printing technology. Designated electro-magnetic brush (EMB) technology, it is possible to apply very thin powder coatings continuously on flat stock, either metallic or nonmetallic. By using a clear uv curable powder on a 17-in. wide EMB machine, rolls of wallpaper were coated at a thickness of 5–7 μm (0.2–0.3 mils) at a line speed of 60 m/min (197 ft/min) (115).

5.5. Hot Flocking. Several nonfluidized-bed coating methods are based on contacting a preheated substrate with powder to form a coating. Although these techniques are not widely used, for certain parts they are the preferred method of application. For example, the coating of motor stators using a thermosetting powder provides primary insulation between the core and windings. The part is preheated to ~200°C and the powder is directed from a fluidized bed using an air venturi pump, similar to those used to supply powder from the reservoir to electrostatic guns, through flexible tubes and directed at the preheated part that is rotating on a mandrel. Multiple tubes, usually in pairs in opposition to each other, are normally used. In a similar fashion, the inside diameter (ID) of the pipe can be coated by entraining powder in an airstream and blowing it through a preheated section of pipe (116). Very uniform coatings in the range of 200–300 μm can be applied by this process to provide corrosion protection for drill pipe, and water injection and gathering pipe in diameters up to ~50 cm.

5.6. Metal Cleaning and Preparation. As in any finishing operation, the surface of the object to be coated must be clean, dry, and free from rust, mill scale, grease, oil, drawing compounds, rust inhibitors, or any soil that might prevent good wetting of the surface by the coating powder. Steel should be sandblasted or centrifugally blast-cleaned to give a near white finish (see METAL SURFACE TREATMENTS). Phosphate coatings are normally a pretreatment for most fabricated steel parts while nonchrome and organometallic conversion coatings are used on aluminum (117).

6. Economic Aspects

The worldwide market for coating powders increased at an annualized growth rate (AGR) of almost 10% during the 1990s in North America compared with a 12–13% AGR during the 1980s. Even though the rate of growth has decreased somewhat, it is estimated that powder coating represents only ~4–5% of the industrial paint market in North America. Globally, the penetration of powder coatings in the industrial paints sector is 6% with Europe leading at ~9%. By 2010, global penetration of powder coatings is expected to reach the 10% level (109).

While the primary driving force for growth during the 1980s was environmental regulations, growth during the 1990s has been driven additionally by superior performance and application economies. Powder coating is seen as an environmentally friendly technology because no solvents are present; volatile organic compounds (VOCs), are absent, for the most part, venting, filtering, and solvent recovery systems are not necessary; process air is recycled; and there is little waste since most overspray powder can be collected and reused. Further, any waste generated is not classified as a hazardous waste under current regulations (see HAZARDOUS WASTE TREATMENT). Permits required for new coating facilities or additions to existing lines are much easier to obtain when the installation utilizes powder coating processes. Future growth will largely depend on advances in both coating materials and application technology in comparison with advances in other environmentally friendly coating methods.

The worldwide production of coating powders in 1999 was estimated at 727,100 mt having a value of $803,861,000, a 6.9% increase over the quantity sold in 1998 (118) (see Table 5). This compares with a production of 236,000 mt in 1989, a growth rate of over 12%/year. Europe is still the largest market, accounting for almost half, followed by North America and the Far East, each with 20%+ market share.

An indication of the growth of the powder coating in North America is reflected in the membership statistics of the Powder Coating Institute, a trade organization representing the industry. From 1987 to 2000, the number of members who manufacture coating powders increased from 5 to 28; suppliers of powder application equipment from 3 to 8; custom powder coaters from 1 to 200; and raw material suppliers from 5 to 114 (118).

Thermoset decorative coatings are by far the largest segment of coating powder production accounting for >90% of the pounds produced. Other market segments are thermoplastic powders, essentially all of which are used in fluidized-bed coating and functional thermoset powders that find use in the pipe coating, rebar, and electrical insulation markets.

The distribution of coating powder production by resin type for several geographical areas is given in Table 6. The epoxy–polyester hybrid powder coatings account for the largest resin type in both Europe and North America. Increasing market share is mostly at the expense of the 100% epoxy-based powders, as a result of more favorable economics. In Europe, weatherable carboxyl polyesters cured with TGIC are being replaced with carboxyl functional polyesters cured with HAA, other glycidyl curatives and, to some extent, urethane polyesters. While the hydroxyl functional polyester resins cured with blocked isocyanates or uretdione curatives, ie, urethane polyesters, have a significant market share in the North American market, they are still a relatively small part of the market in Europe.

The vast majority of acrylic powders in both Europe and North America find use in the automotive industry, as clear coats for wheels and bodies and as pigmented coatings in exterior trim, "blackout" coatings, and primer surfacers.

The automotive market is the largest and fastest growing segment of the powder coating market in North America accounting for over 16% market share, followed by appliance coatings at 15%, architectural at 3%, lawn and garden at 7%, and general metal finishing at 58% (118).

7. Analytical Methods

Methods for evaluating the performance of powder coatings are the same as those used for conventional coatings. Test methods for coating powders include particle size distribution, powder free flow, sintering, fluidization characteristics, and others. They have been reviewed in detail and reported in the literature (120,121). In addition, the American Society for Testing and Materials (ASTM) has issued a comprehensive standard that covers the most important test methods for the evaluation and characterization of powder coatings (122).

8. Environmental and Energy Considerations

A significant factor contributing to the growth of powder coating processes has been the proliferation of federal, state, and local environmental regulations. Starting with the Clean Air Act of 1970, which defined Hazardous Air Pollutants (HAPs), VOCs, and set standards for nationwide air quality, many additional regulations have since been enacted. In nonattainment areas, localities where pollution levels persistently exceed National Ambient Air Quality Standards, the Best Available Control Technology (BACT) or Maximum Available Control Technology (MACT) may be necessary to reduce the level of pollutants to that required. Powder coating installations are generally accepted as meeting these levels of control technology. States are charged with the job of achieving compliance. Permits are required by states where new finishing operations are being added or in existing facilities. The presence or absence of VOCs and HAPs in coating materials has become a significant factor in the economic analysis in planning new finishing operations.

Since coating powders are 100% solid materials, they are essentially free of VOCs and HAPs, with the exception of trace quantities of monomers or, in some cases, solvents used in the manufacture of the raw materials (123). Caprolactam and other compounds used as blocking agents in blocked isocyanate curatives are emitted during curing. Typical levels in formulated urethane coating powders are in the range of 4–6%. While caprolactam is considered a VOC, it is not classified as an HAP. Powder coating remains the process of choice where VOCs must be reduced to the lowest possible levels, as in many OEM and automotive coating operations. Permits for additions to existing paint lines or installation of new systems are much easier to justify and obtain than when most liquid coatings are specified. In addition to the environmental advantages, the low volatile emissions of powder coatings during the baking operation has economic and energy saving advantages. Fewer air changes per hour in the baking oven are required for powder coatings than for solvent-based coatings, which saves fuel. Further, in the coating operation almost all powder is recovered and reused, resulting in higher material utilization, and waste minimization. The air used in the coating booths during application is filtered and returned to the workplace atmosphere, reducing heating and cooling demands. Additionally, because of the need for more sophisticated devices to control emission of VOCs in liquid systems, the capital investment to install a new powder coating line is becoming increasingly more economically favorable. The savings in material and energy costs of powder coating systems has been documented in a number of studies. An economic analysis worksheet for comparing the cost of operating a powder coating line compared with alternate systems can be found in Technical Brief No. 21 in Ref. 121.

The only components in a coating powder that might cause the waste to be classified as hazardous are certain heavy-metal pigments sometimes used as colorants. Lead (qv) and cadmium-based pigments (qv) are seldom used, however, and other potentially hazardous elements such as barium, nickel, and chromium are usually in the form of highly insoluble materials that seldom cause the spent powder to be characterized as a hazardous waste (124).

9. Health and Safety Factors

Any finely divided organic material can form ignitable mixtures when dispersed in air at certain concentrations. The most significant hazard in the manufacture and application of coating powders is the potential of a dust explosion (see POWDERS, HANDLING). The severity of a dust explosion is related to the material involved, its particle size, and concentration in air at time of ignition. The lower explosive limit (LEL) is the lowest concentration of a material dispersed in air that explodes in a confined space when ignited. The LEL for a number of epoxy and polyester powders was measured and found to be in the range of $0.039-0.085$ oz/ft^3 ($39-85$ g/m^3) (125). In powder coating installations, the design of the spray booth and duct work, if any, should be such that the powder concentration in air is always kept below the LEL employing a wide margin of safety. General safety considerations are detailed in Ref. 126. The use of flame detection systems in all automatic powder coating installations is required. These devices must respond within 0.5 s or less to arrest all powder flow in the system. If powder ignition should occur, flame detection sensors shut down the power to the system, halting powder spraying and air flow circulation. Another element of safety for a powder coating system is the design of the booth recovery equipment. Some recovery designs utilized today, such as illustrated in Figure 3, are configured so no external venting is required. However, when traditional cyclones or dust collectors, isolated by ducting, are utilized in the coating system, a pressure relief system is necessary. If these units are not located outside the building or properly vented, explosion suppression may also be required (see POLLUTION CONTROL METHODS). Furthermore, cyclones and dust collectors located inside the building should be near an outside wall and ductwork from the pressure relief vents should be directed through short runs, not exceeding 3 m when possible. Required explosion vent areas and other design considerations can be found in the literature (127). The spray guns, spray booth, duct work, dust collection, and powder reclaim system, as well as the work piece, must be properly grounded (128,129).

The health hazards and risk associated with the use of powder coatings must also be considered (see HAZARD ANALYSIS AND RISK ASSESSMENT). Practical methods to reduce employee exposure to powder such as the use of long sleeved shirts and gloves to prevent skin contact should be observed. Furthermore, exposure can be minimized by good maintenance procedures to monitor and confirm that the spray booth and dust collection systems are operating as designed. Ovens should be properly vented and operated under negative pressure so that any volatiles released during curing, eg, caprolactam, do not enter the workplace atmosphere.

In general, the raw materials used in the manufacture of powder coatings are relatively low in degree of hazard. None of the epoxy, polyester, or acrylic resins normally used in the manufacture of thermoset powder coatings are defined as hazardous materials by the OSHA Hazard Communication Standard. Most pigments and fillers used in powder coatings generally have no hazards other than those associated with particulates. Some epoxy curing agents are skin irritants; however, most of these characteristics are greatly diminished when these materials are compounded into the powder coating. In addition to being diluted, the materials are dispersed in a resinous matrix having a low degree of water solubility that appears to make them less biologically accessible. For example, anhydrides and anhydride adducts generally elicit a strong respiratory or eye irritant response. However, when powder coatings containing anhydride-based curing agents were tested in animal exposures, the coatings were found to be nonirritating to the skin, eye, and respiratory tract (130,131). Similarly, TGIC is a skin irritant, but formulated powders containing TGIC were not (132). Regulations in Europe require cautionary labeling of powders containing TGIC (see Sections under Thermosetting Coating Powders, and Polyester, TGIC Cured).

Although coating powders do not appear to pose significant hazards to personnel working with them, worker exposure should nevertheless be minimized. Coating powders should be treated as Particulates Not Otherwise Classified (PNOC) having a Threshold Limit Value–Time Weighted Average (TLV–TWA) of 10 mg/m^3 for total particulates (133). The TLV should be maintained primarily through environmental controls. Hoods and proper ventilation should be provided during handling and application of powders. When environmental control of dust cannot be maintained below the TLV, protective equipment such as dust and fume masks or externally supplied air respirators should be used (134).

BIBLIOGRAPHY

"Coating Processes, Powder Technology" in *ECT* 4th ed., Vol. 6, pp. 635–661, by D. S. Richart, Morton International Inc.; "Coating Processes, Powder Technology" in *ECT* (online), posting date: December 4, 2000, by Douglas S. Richart, Morton International Inc.

CITED PUBLICATIONS

1. U.K. Pat. 643,691 (Sept. 27, 1950), P. G. Clements (to Schori Metallising Process, Ltd.).
2. Ger. Pat. 933,019 (Sept. 15, 1955), E. Gemmer (to Knapsack-Griesheim, AG).
3. U.S. Pat. 2,844,489 (July 22, 1958), E. Gemmer (to Knapsack-Griesheim, AG).
4. U.S. Pat. 3,640,747 (Feb. 8, 1972), D. S. Richart (to the Polymer Corp.).
5. U.S. Pat. 3,264,271 (Aug. 2, 1966), H. M. Gruber and L. Haag (to the Polymer Corp.).

6. W. E. Wertz and D. S. Richart in E.J. Wickson ed., *Handbook of PVC Formulating*, John Wiley & Sons, Inc., 1993 pp. 771–781.
7. U.S. Pat. 3,008,848 (Nov. 14, 1961), R. W. Annonio (to Union Carbide Corp.).
8. D. S. Richart in M.I. Kohan, ed., *Nylon Plastics Handbook*, Hanser/Gardner Publications, Inc., 1995, pp. 253–269.
9. T. Glass and J. Depoy, Paper FC91-384 presented at *Finishing '91*, Sept. 23–25 Cincinnati Ohio, sponsored by SME Dearborn, Mich.
10. Du Pont Technical Bulletin, *Abcite Powder Coating Resins*, Sept. 1996.
11. WO 98/50475 (Nov. 12, 1998), J. M. McGrath, (to 3M Com., St. Paul. Minn.).
12. U.S. Pat. 4,770,939 (Sept. 13, 1988), W. Sietsess, T. M. Plantenga, and J. P. Dekerk (to Labofina, S.A.).
13. U.S. Pat. 5,599,874 (Feb. 4, 1997), E. Verwey, L. K. Rijkse, and M. Gillard (to Fina Research, S.A.).
14. U.S. Pat. 5,739,202 (April 14, 1998), R. L. Pecsok.
15. L. C. Stephans, *Materials Performance* **38**(6), 42–47 (1999).
16. R. Domitrz and H. Nowak, *Paint Coat Ind.* **XVI**(5), 86–94 (May 2000).
17. M. Y.H. Chang, *Paint Coat. Ind.* **XV**(10), 102–108 (Oct. 1999).
18. P. G. de Lange, *J. Coat. Technol.* **56**(717), 23–33 1984.
19. J. Hajas and H. Juckel, paper presented at the *Waterborne, Higher-Solids and Powder Coating Symposium*, Feb. 10–12, 1999, University of Southern Mississippi, pp. 273–283.
20. M. A. Grolitzer, *Am. Paint Coat. J.* **75**(46), 74–78 (April 1991).
21. M. Wulf, P. Uhlmann, S. Michel, and K. Grundke, *Prog. Org. Coat.* **38**(2000), 59–66 (2000).
22. S. A. Stachowiak, in G.D. Parfitt and A.V. Patsis, eds., *Organic Coatings Science and Technology*, vol. 5 Marcel Dekker, Inc., New York and Basel, 1982, pp. 67–89.
23. L. K. Johnson and W. T. Sade, *J. Coat. Technol.* **65**(826), 19–26 (1993).
24. T. Misev and E. Belder, *J. Oil Colour Chem. Assoc.* **72**(9), 363–368 (1989).
25. S. Gabriel, *J. Oil Colour Chem Assoc.* **58**, 52–61 (1975).
26. S. G. Yeates and co-workers, *J. Coat. Technol.* **68**(861), 107–114 (1996).
27. R. P. Franiau, *Paint India* **37**(9), 33–38 (1987).
28. Z. Huang, L. E. Scriven, H. T. Harris, and W. Eklund, paper presented at *Waterborne, Higher-Solids and Powder Coatings Sumposium*, University of Southern Mississippi, New Orleans, La., pp. 328–340, Feb. 5–7, 1997.
29. V. G. Nix and J. S. Dodge, *J. Paint. Technol.* **45**(586), 59–63 (Nov. 1973).
30. J. C. Kenny, T. Ueno, and K. Tsutsui, *J. Coat. Technol.* **68**(855), 35–43 (Apr. 1997).
31. P. R. Horinka, *Ind. Paint Powder* **71**(12), 26–30 (Dec. 1995).
32. A. J. Pekarik, paper presented at *Powder Coating 98, Formulation & Production Conference*, Indianapolis, Ind., The Powder Coating Institute, Alexandria, Va, 1998, pp. 103–115.
33. B. E. Maxwell, R. C. Wilson, H. A. Taylor, and D. E. Williams, "Understanding the Mode of Action of Benzoin in Powder Coatings", paper presented at the *26th International Conference in Organic Coatings*, Athens, Greece, 2000.
34. D. Fluck, J. Fultz, and M. Darsillo, *Paint Coat. Ind.* **XIV**(10), 214–220 (Oct. 1998).
35. K. Wolny, *The Hiding Powder of White Powder Coatings, Kronos Tech. Bulletin 6.16*, 1985.
36. D. S. Richart, *Powder Coating* **9**(1), 23–30 (Feb. 1998).
37. D. S. Richart, *Polymer Paint Colour J.* **188**(4408), 14–18 (Sept. 1998).
38. U.S. Pat. 4,242,253 (Dec. 30, 1980), M. D. Yallourakis (to E. I. du Pont de Nemours & Co., Inc.).
39. U.S. Pat. 3,947,384 (March 30, 1976), F. Schulde and co-workers (to Veba-Chemie AG).

40. U.S. Pat. 3,842,035 (Oct. 15, 1974), C. H.J. Klaren (to Shell Oil Co.).
41. E. Dumain, *Paint Coatings Ind.* **XV**(9), 52–58 (Sept. 1999).
42. D. P. Chapman, paper presented at the 18th annual *Waterborne, Higher-Solids and Powder Coating Conference*, Feb. 6–8, 1991, New Orleans, La., University of Southern Mississippi, pp. 339–346.
43. D. Foster, *Congress Papers Powder Coating Europe 2000* (PCE 2000) Amsterdam, The Netherlands, C.R. Vincentz Verlag, Hanover, Germany (Jan. 19–21, 2000), pp. 234–246.
44. U.S. Pat. 3,631,149 (Dec. 28, 1971), H. Gempeler and P. Zuppinger (to Ciba, Ltd.).
45. U.S. Pat. 4,122,060 (Oct. 24, 1978) M. Yallourakis (to E. I. du Pont de Nemours & Co., Inc.).
46. U.S. Pat. 3,477,971 (Nov. 11, 1969) R. Allen and W. L. Lantz (to Shell Oil Co.).
47. R. G. Doone, R. W. Tait, and A. P. Glaze, paper presented at *Powder Coating '96, Formulation and Production Conference*, The Powder Coating Institute, Alexandria, Va, 1996, pp. 39–44.
48. J. Didas, *Mater. Performance* **39**(6), 38–39 (June 2000).
49. Y. Merck, paper presented at *Powder Coating 2000 Formulators Technology Conference*, The Powder Coating Institute, Alexandria, Va., 2000, pp. 101–120; Y. Merck, Private communication, Nov. 2000.
50. R. van der Linde and E. G. Belder, in G.D. Parfitt and A.V. Patsis, eds., *Organic Coatings Science and Technology*, Vol. 5, Marcel Dekker Inc. New York, 1983, pp. 55–66.
51. S. P. Pappas, V. D. Kunz, and B. C. Pappas, *J. Coat. Technol* **63**(796), 39–46 (May 1991).
52. U.S. Pat. 3,792,011 (Feb. 2, 1974), J. D.B. Smith and R. N. Kauffman (to Westinghouse Electric Corp.).
53. U.S. Pat. 4,147,737 (April 3, 1979), A. J. Sein and co-workers (to DSM Resins).
54. Y. Merck, *Conference Papers "Powder Coatings What's Next"*, Birmingham, U.K., March 10, 1999, Paint Research Association, pp. 2–16.
55. R. R. Engelhardt, paper presented at the *Waterborne, Higher-Solids and Powder Coating Symposium*, Feb. 24–26, 1993, University of Southern Mississippi, New Orleans, La., pp. 549–561.
56. P. Loosen, J. M. Loutz, and co-workers, paper presented at *Powder Coating '96, Formulation and Production Conference*, Sept. 17, 1996, The Powder Coating Institute, Alexandria, Va., pp. 107–115.
57. Y. Merck, D. Maetens, L. Moens, and K. Buysens, *Euro. Coat. J.* 12/99, 18–24 (Dec. 1999).
58. *Threshold Limit Values for Chemical Substances and Physical Agents, 2000 TLV's and BEI's*, American Conference of Governmental Industrial Hygienists (ACGIH), Cincinnati, Ohio.
59. A. Pledger, paper presented at *Symposium on Powder Coatings*, Paper 11, Birmingham, U.K., April 4–5, 1995, Paint Research Association, 12 pp.
60. Eur. Pat. 0,536,085, A2 (Sept. 24, 1992), J. A. Cotting (to Ciba-Giegy AG).
61. U.S. Pat. 6,114,473 (Sept. 5, 2000), S. Miyake and co-workers (to Nissan Chem. Ind. Ltd.).
62. U.S. Pat. 4,076,917 (Feb. 28, 1978), G. Swift (to Rohm and Haas Co.).
63. K. Wood and D. Hammerton, paper presented at the *Waterborne, Higher-Solids and Powder Coating Symposium*, Feb. 6–8, 1991, University of Southern Mississippi, New Orleans, La., pp. 78–89.
64. A. Kaplan, "The Nature of the Primid-Bond", paper presented at the *1st International Primid Conference*, March 20–21, 1997, EMS Chimie, Flims Waldhaus, Switzerland.
65. M. Wenzler, "Primid in Use—7 Years Practical Experience", in Ref. 64.

66. D. Maetens and co-workers, *Euro. Coat. J.* 05/99, 26–33 (May 1999).
67. W. Jacobs and co-workers, paper presented at the *Waterborne, Higher-Solids and Powder Coating Symposium*, Feb. 26–28, 1992, University of Southern Mississippi, New Orleans, La., pp. 196–214.
68. W. Jacobs and co-workers, paper presented at the *Waterborne, Higher-Solids and Powder Coating Symposium*, Feb. 9–11, 1994, University of Southern Mississippi, New Orleans, La., pp. 629–652.
69. U.S. Pat. 5,695,852 (Dec. 9, 1997) D. S. Richart and C. P. Tarnoski (to Morton Int., Inc.).
70. M. Y. H. Chang, *Paint Coat. Ind.* **XV**(10), 98–108 (Oct. 1999).
71. H. B. Yokelson and co-workers, *Euro. Coat. J.* 5/98, 354–369 (May, 1998).
72. J. D. Pont, Proceedings of the Twenty-Sixth International *Waterborne, Higher-Solids and Powder Coating Symposium*, Feb. 10–12, 1999, University of Southern Mississippi, New Orleans, La., pp. 232–245.
73. M. Guida and J. V. Weiss, Proceedings of the Twenty-Second International *Waterborne, Higher-Solids and Powder Coating Symposium*, Feb. 22–24, 1995, University of Southern Mississippi, New Orleans, La., pp. 43–54.
74. H. U. Meier-Westhues, P. Thometzek, and J. J. Laas, *Farb Lack* **103**(4/97), 140–146 (April 1997).
75. U.S. Pat. 4,873,274 (Oct. 10, 1989), F. L. Cummings and G. D. Correll (to Morton Thiokol, Inc.).
76. Eur. Pat. 0,509,393 A1 (April 9, 1992) D. Fink and co-workers, (to Hoechst AG.).
77. C. Bowden, D. Ostrander, and S. Miller, *Metal Finishing* **97**(5), 14–18 (May 1999).
78. D. K. Moran and M. J. M. Verlaak, paper presented at the *Waterborne, Higher-Solids and Powder Coating Symposium*, Feb. 24–26, 1993, University of Southern Mississippi, New Orleans, La., pp. 497–507.
79. U.S. Pat. 5,907,020 (May 25, 1999), G. D. Correll and co-workers (to Morton International, Inc.).
80. U.S. Pat. 5,763,099 (June 9, 1998), T. A. Misev and co-workers (to DSM N.V.).
81. U.S. Pat. 5,703,198 (Dec. 20, 1997), F. Twigt and R. VanDerLinde (to DSM N.V.).
82. U.S. Pat. 5,639,560 (June 17, 1997) L. Moens and co-workers (to UCB S.A.).
83. U.S. Pat. 5,789,039 (Aug. 4, 1998) K. M. Biller and B. A. MacFadden (to Herberts Powder Coatings, Inc.).
84. R. Jahn and co-workers, *Congress Papers—Powder Coating Europe 2000* (PCE 2000), Amsterdam, The Netherlands, C. R. Vincentz Verlag, Harber, Germany, Jan. 19–21, 2000, pp. 309–317.
85. U.S. Pat. 3,561,003 (Feb. 2, 1971), B. J. Lanham and V. G. Hykel (to Magnavox Co.).
86. U.S. Pat. 3,737,401 (June 5, 1973), I. H. Tsou and J. W. Garner (to Grow Corp.).
87. B. Dreher, paper presented at XIVth FATIPEC Congress, 201–207, Budapest, June 4–9, 1978.
88. D. Mielcarek and K. Huber, paper presented at the *Waterborne, Higher-Solids and Powder Coating Symposium*, Feb. 24–26, 1993, University of Southern Mississippi, New Orleans, La., pp. 525–535.
89. D. Fluck, J. Fultz, and M. Darsillo, *Paint Coat. Ind.* **XIV**(10), 214–220 (Oct. 1998).
90. M. Giersemehl and G. Plihal, "Fine Grinding System with Impact Classifier Mill and Cyclone Classifier", *Powder Handling Processing* **11**(3) (July/Sept. 1999).
91. U.S. Pat. 5,399,597 (Mar. 21, 1995), F. Mandel, C. Green, and A. Scheibelhoffer (to Ferro Corp.).
92. U.S. Pat. 5,981,696 (Nov. 9, 1999), D. Satweeber and co-workers (to Herberts GmbH).
93. W. Kreis, "Meeting Requirements for Automotive Primer-Surfacer and Clearcoat", *Polmer Paint Colour J.* **188**, 4411 (Dec. 1998).

94. U.S. Pat. 5,379,947 (Jan. 10, 1995), C. F. Williams and M. A. Gessner (to BASF Corp.).

95. M. Gaschke and H. Lauterbach, paper presented at the *Waterborne, Higher-Solids and Powder Coating Symposium*, University of Southern Mississippi, New Orleans, La., March 10–12, 1980.

96. G. T. Robinson, *Prod. Finish.* 16 (Sept. 1976).

97. V. G. Nix and J. S. Dodge, *J. Paint Technol.* **45**(586), 59 (Nov. 1973).

98. U.S. Pat. 6,113,980 (Sept. 5, 2000), H. S. Laver, (to CIBA Specilty Chem. Co.).

99. S. Guscov, *Powder Coating* **11**(4), 22–31 (June 2000).

100. T. Rusk, N. Rajagopalan, and T. C. Lindsay, *Powder Coating* **11**(4), 33–42 (June 2000).

101. G. Stribling, L. Keen, and J. Trostle, *Conference Proceedings Powder Coating '98*, The Powder Coating Institute Indianapolis, In., Sept. 1998, pp. 229–248.

102. U.S. Pat. 6,071,348 (June 6, 2000), K. Seitz, M. Hasler, and H. Adams, (to Wagner Int. G).

103. U.S. Pat. 5,776,554 (July 7, 1998), C. R. Merritt and R. M. Thorn (to Illinois Tool Works, Inc.).

104. R. Deane, *Congress Papers, PCE 2000* 331–340.

105. P. R. Horinka, *Powder Coating* **6**(3), 69–75 (June 1995).

106. H. J. Lader, *Conference Proceedings Powder Coating 94*, Oct. 11–13, 1994, The Powder Coating Institute, Cincinnati, Ohio., pp. 254–267.

107. PCT WO 98/45356 (Oct. 15, 1998), G. Kodokian (to E. I. Du Pont de Nemours & Co., Inc.).

108. J. L. Quinn, *Metal Finishing* **96**(4), 10–13 (April 1998).

109. F. Busato, *Congress Papers PCE 2000* 409–417 (2000).

110. J. N. Pennington, *Modern Metals* **56**(8), 48–53 (Sept. 2000).

111. L. J. Black, *Metal Fin.* **97**(9), 67–69 (Sept. 1999).

112. U.S. Pat. 5,695,826 (Dec. 9, 1997), E. C. Escallon (to Terronics Development Corp.).

113. "Powder Coating in the Can," *Ind. Paint Powder* **75**(1), 20–24 (Jan. 1999).

114. U.S. Pat. 6,080,823 (June 27, 2000), L. Kiriazis (to PPG Industry Inc.).

115. P. H.G. Binda, *Congress Papers PCE 2000* 377–391 (2000).

116. U.S. Pat. 4,243,699 (Jan. 6, 1981), J. E. Gibson.

117. P. Droniau, T. Korner, and J. Kresse, *Congress Papers PCE 2000* 75–80 (2000).

118. "Summary Report Worldwide Powder Coating Markets", The Powder Coating Institute, Alexandria Va, 22314 (June 2000).

119. G. J. Bocchi, *Congress Papers PCE 2000* 23–36 (2000).

120. E. Bodner, *Euro. Coat. J.* 814–832 (Nov. 1993).

121. N. Liberto, ed. *Powder Coating—The Complete Finishers Handbook*, Appendix C, 2nd ed., The Powder Coating Institute, 1999, pp. 388–403.

122. ASTM D3451 "Practices for Testing Polymeric Powders and Powder Coatings", *1998 Annual Book of ASTM Standards*, Section 6.02, Paints, Related Coatings, and Aromatics, American Society for Testing and Materials, West Conshohocken, Pa., 19428.

123. T. Randoux and co-workers, *Euro. Coat. J.* 790–795 (Nov. 1975).

124. "Toxicity Characteristic Leaching Procedure", *Code of Federal Regulations*, 40 *CFR*, Part 261, Appendix II, Method 1311.

125. P. H. Dobson, "Safeguards in Powder Coating System Design", *Ind. Fin.* 77–82 (Sept. 1974).

126. D. R. Scarbrough, in *Fire Protection Handbook*, 17th ed. National Fire Protection Association (NEPA), Quincy, Mass., 1989, Chapt. 02–12.

127. *Guide for Explosion Venting*, Code No. 68 NFPA, 1988.

128. *Static Electricity*, Code No. 77, NFPA (1988).

129. *Spray Finishing Using Flammable and Combustible Materials*, Code No. 38, NFPA (1989).
130. J. F. Fabries and co-workers, *Toxicity of Powder Paints by Inhalation*, Report No. 1092/RI, Institut National de Recherche et de Sécurité (INRS) Department of Occupational Pathology, France, 1982.
131. *Power of Six Powder Paints to Cause Irritations and Allergies of the Skin*, INRS, May 1979.
132. *Toxicological Studies of Uralac Powder Coating Resin and Powders*, Scado B.V. (now DSM Resins), Zwolle, The Netherlands, 1979.
133. *Threshold Limit Values Biological Exposure Indices*, American Conference of Governmental Industrial Hygienists (ACGIH), Cincinnati, Ohio, 45240.
134. *OSHA Regulations*, 29 CFR, Section 1910.34.

GENERAL REFERENCES

D. A. Bate, *The Science of Powder Coatings*, Vol. 1, *Chemistry Formulation and Application*, 1990, Vol. 2 *Applications*, 1994. Sita Technology, London.
T. Misev, *Powder Coatings—Chemistry and Technology*, John Wiley & Sons, Inc., New York, 1991.
H. Jilek, *Powder Coatings*, Federation of Societies for Coatings Technology, Blue Bell, Pa., 1991.
N.P. Liberto, ed., *Powder Coating—The Complete Finishers Handbook*, 2nd ed. The Powder Coating Institute, Alexandria, Va. 1999.
D. Howell, *The Technology, Formulation and Application of Powder Coatings*, Sita Technology Ltd., London, Vol. 1, 2000.

Douglas S. Richart
D.S. Richart Associates

COATING PROCESSES, SPRAY

1. Introduction

A coating may be applied to articles, ie, workpieces, by spraying. This application method is especially attractive when the articles have been previously assembled and have irregularly shaped and curved surfaces. The material applied is frequently a paint (qv), ie, a combination of resin, solvent, diluent, additives, and pigment. The material can also be a hot thermoplastic, an oil, or a polymer dissolved in a solvent. Many types of spray equipment are available. Methods can be used in combinations, and most of the techniques can be used for simple one-applicator manual systems or in highly complex computer-controlled automatic systems having hundreds of applicators. In an automatic installation, the applicators can be mounted on fixed stands, reciprocating or rotating machines, or even robots (Fig. 1).

Fig. 1. An automatic spray-coating system, where air-automized electrostatic spray guns are mounted on reciprocators.

2. Atomization

2.1. Airless Atomization. In airless- or pressure-atomizing systems, the coating is atomized by forcing the coating (or the liquid) through a small-diameter nozzle under high pressure. The fluid pressure is typically between 5 and 35 MPa (700–5000 psi); fluid flow rates are between 150–1500 cm^3/min. In most commercial applications, a pump designed for the type of material sprayed is used to develop the high pressure. The pump can be mechanically, electrically, pneumatically, or hydraulically driven and the nozzle apertures have diameters ranging from 0.2 to 2.0 mm. The more viscous the fluid and the higher the desired liquid-flow rate, the larger the nozzle. As the fluid is forced through the nozzle, it accelerates to a high velocity and leaves the nozzle in a thin sheet or jet of liquid in the relatively motionless ambient air, producing a shear force between the fluid and air. The fluid is atomized by turbulent or aerodynamic disintegration, depending on the specific conditions (Fig. 2). The most common nozzles produce a long, narrow fan-shaped pattern of various sizes; others produce a solid or hollow cone. The pattern of a fan nozzle can split and form "fingers" if the pressure is not sufficient. Because the coating material is often abrasive, the nozzle is typically made from tungsten carbide (see CARBIDES).

Airless atomization generally produces a medium-to-coarse particle size. Using a given nozzle, the higher the fluid pressure, the finer the atomization. Airless atomization can be used to atomize a large amount of material, or can atomize at high flow rates that can be rapidly deposited on the workpiece with minimal overspray or misting and excellent penetration into recessed areas.

Fig. 2. Airless atomization process.

The flow rate is controlled by the nozzle size and the fluid pressure. The minimum nozzle size is determined by the size required to prevent plugging under operating conditions; minimum application pressure is determined by the required degree of atomization and the elimination of fingers.

A variation of airless atomization is called air-assisted airless. A small amount of compressed air at 35–170 kPa (5–25 psi) is introduced adjacent to the airless nozzle and impinges upon the thin sheet of fluid as it exits from the nozzle. This air aggravates the turbulence in the fluid and results in improved atomization at lower fluid pressures. Often, material that cannot be properly atomized using straight airless atomization can be using the air-assisted airless method. In some cases, the introduction of the air allows some control of the fan size.

2.2. Supercritical Atomization.   Atomization can be obtained by mixing a supercritical fluid (SCF) with the material to be atomized. This process reduces volatile organic compound (VOC) emissions as the SCF acts as a solvent and replaces some of the hydrocarbon solvents in the material (see SUPERCRITICAL FLUIDS).

The material sprayed is generally a very high solids or viscous material that is thinned with the SCF to a spray viscosity in a special mixing/metering system. This mixture is then sprayed in an airless-type spray gun specifically designed for the process. In addition to the pressure or airless-type atomization, there is secondary atomization that results when the SCF dissolved in the spray material changes to a gas and rapidly expands (Fig. 3). Thus atomization quality is excellent, and materials that cannot be atomized using other methods can be used.

An aerosol container can be considered a special application of airless atomization (see AEROSOLS). The pressure is usually supplied by a liquefied gas in the container at its equilibrium pressure. The material being sprayed has a very low viscosity to provide easy material flow through the feed tube and to permit fine atomization.

2.3. Air Atomization.   In an air atomizer, an external source of compressed air, usually supplied at pressures of 70–700 kPa (10–100 psi), is used

Fig. 3. Supercritical fluid atomization process.

to atomize the liquid. Air atomization is perhaps the most versatile of all the atomization methods. It is used with liquids of low to medium viscosity, and flow rates of 50–1000 cm^3/min are common. Medium-to-fine particle sizes are produced, and the resulting surface finish is very good. It is sometimes difficult to penetrate small recessed areas, however, because the atomization air forms a barrier in the recess that the coating particles must then penetrate. When higher air pressures are used, air atomization produces considerable misting and overspray, which can be a disadvantage under some conditions. Air-atomizing devices can be of internal- or external-mix design.

The most common type of air atomizer is an external-mix where the coating material and atomization air are mixed in the space in front of the nozzle and air cap. An annulus of air generally surrounds the fluid as it leaves the tip, and the shear stress between the fluid and the air causes the initial atomization. In addition to the annulus, numerous (in some cases, as many as 10) air holes are placed in the air cap to direct air jets that continue the atomization process, assist in keeping the cap clean, and help shape the spray. Air jets coming from holes located in two diametrically opposed "horns" or ears produce a fan-shaped pattern that is oriented 90° from the horns. Where a fan pattern is not required, special nozzles expel the coating from a circular annulus about 0.6 mm thick surrounded by an air annulus. A swirl imparted to the atomization air in the annulus results in a very efficient atomization process and a spray having low momentum and misting. This is especially useful when coating long, narrow objects or when the atomizers are reciprocated or rotated to blend the patterns from several atomizers. It is difficult to coat a flat surface manually using this type of spray pattern.

In an internal-mix nozzle, the atomization air is mixed with the coating material before being forced through a nozzle or tip. As the mixture of air and coating material passes through the nozzle, its pressure is significantly reduced, and the resulting expansion produces the atomization of the coating material. This is a very efficient method of atomization, but usually coating material builds up around the fluid tip where the atomization occurs. This material is then torn away without being properly atomized, producing slugs that may be deposited on the workpiece and blemish the finish. It is therefore necessary to clean the tip of an internal-mix air atomizer frequently.

A special case of air atomization is high volume low pressure (hvlp) spray. In this case the air pressure at the spray gun is less than 70 kPa (10 psig) and there are relatively large (up to 0.32 cm) holes in the air cap to easily pass the low pressure air. This type of atomizer produces a soft or slow moving spray and is generally considered to be rather efficient in depositing the material on the workpiece. However, the use of low pressure air for atomization usually limits the viscosity and/or flow rate of the material that can be atomized.

The fluid delivery in an air-spray system can be pressure or suction fed. In a pressure-fed system, the fluid is brought to the atomizer under positive pressure generated with an external pump, a gas pressure over the coating material in a tank, or an elevation head. In a suction system, the annular flow of air around the fluid tip generates sufficient vacuum to aspirate the coating material from a container through a fluid tube and into the air stream. In this case, the paint supply is normally located in a small cup attached to the spray device to keep the elevation differential and frictional pressure drop in the fluid-supply tube small.

Most industrial production systems use a pressure-feed system, whereas many touch-up and recoating operations use suction feed. In a pressure-feed system, the fluid-flow rate and the atomization air are controlled independently. This permits the fluid-flow rate to be set to the desired value within a very large range. The pressure of the automization air is then matched with the fluid-flow rate to give the desired fineness of atomization. In a suction-feed system, the coating flow rate is determined by the flow of the atomization air and the size of the fluid orifice. Generally, it is not possible to suction feed more material than can be atomized by the quantity of air being used. A suction-feed system works best with low viscosity fluids, as the pressure differential available to transport the fluid is small, and higher viscosity liquids generally do not have sufficient flow rate for practical applications.

2.4. Electrostatic Atomization.   The atomization of the coating material by electrostatic forces occurs when an electrical charge is placed on a filament or thin sheet of coating, and the mutual repulsion of the charges tears the coating material apart. For this process to produce acceptable atomization, the physical properties of the coating material must be within a relatively narrow range. The material is charged by an external source of high voltage, either prior to or as it is forced to flow over a knife edge, through a thin slot, or orifice, or it is discharged from the edge of a slowly rotating disk or bell (cup). The thin sheet or small diameter stringers or cusps of coating material are torn apart by the mutual repulsion of the charges on the material.

When a disk or bell is used, the coating material is fed near the center, and the rotation, generally 900–3600 rpm, provides a means of distributing the coating material to the edge of the device. At the edge, fluid surface tension or mechanical features of the rotating surface become the controlling factor, and the coating material comes off the surface in cusps. At higher rotation speeds, mechanical forces become significant and the stringers can break off at their base, resulting in larger particles being formed. Electrostatic atomization is limited to about 4 cm^3/min per centimeter of discharge length; voltages of 100–150 kV are used. This can be a very efficient coating method, but because of the required combination of low surface tension, low viscosity, and proper balance of electrical characteristics, this method of atomization has not been very successful for high solids or waterborne coatings (see COATINGS). Penetration into recessed areas is generally fair to poor, and excessive buildup of material on edges of the workpiece is possible.

2.5. Rotary Atomization. In rotary atomization, a bell (cup) or disk rotates at a speed of 10,000–40,000 rpm. In contrast to electrostatic atomization, mechanical forces dominate. The coating material is introduced near the center of the rotating device, and centrifugal force distributes it to the edge, where the material has an angular velocity close to that of the rotating member. As the coating material leaves the surface, its main velocity component is tangential, and it is spun off in the form of a thin sheet or small cusps as illustrated in Figure 4. The material is then atomized by turbulent or aerodynamic disintegration, depending on exact conditions. The diameter of a typical bell is 4–10 cm, whereas that of a disk is typically 10–25 cm. When a bell is used, the part to be coated is transported across the open face of the bell. A disk is generally used in an

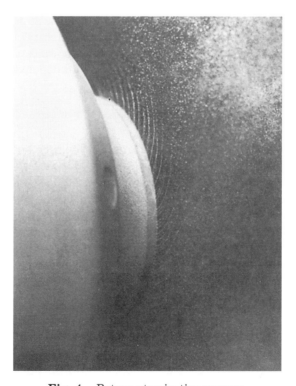

Fig. 4. Rotary atomization process.

omega-shaped loop with the centerline of the disk on the centerline of the loop. The disk has a 360° spray pattern and is reciprocated up and down to cover the length of the part.

Rotary atomization produces the most uniform atomization of any of the aforementioned techniques, and produces the smallest maximum particle size. It is almost always used with electrostatics and at lower rotational speeds the electrostatics assist the atomization. At higher rotational speeds the atomization is principally mechanical in nature and does not depend on the electrical properties of the coating material. If the viscosity of a coating material is sufficiently low that it can be delivered to a rotary atomizer, the material can generally be atomized. The prime mover is usually an air-driven turbine and, provided that the turbine has the required power to accelerate the material to the angular velocity, liquid-flow rates of up to 1000 cm^3/min can be atomized using an 8-cm diameter bell.

Rotary atomization produces an excellent surface finish. The spray has low velocity, which allows the electrostatic forces attracting the paint particles to the ground workpiece to dominate, and results in transfer efficiencies of 85–99%. The pattern is very large and partially controlled and directed by shaping air jets. The spray when using a metallic cup has relatively poor penetration into recessed areas. Excessive material deposited on the edges of the workpiece can also be a problem.

Recent developments in rotary atomization include the use of semiconductive composites (qv) for the rotary cup permitting the construction of a unit that does not produce an ignition spark when brought close to a grounded workpiece yet has the transfer efficiencies associated with a rotary atomizer. In addition, the use of the semiconductive material softens the electrostatic field and results in less edge buildup and better penetration into recess areas. Other systems use electronic means to effectively prevent arcing to grounded surfaces.

3. Electrostatic Spraying

Use of electrostatic spraying or electrostatic deposition increases the efficiency of material transfer to the workpiece (see COATING PROCESSES, POWDER TECHNOLOGY). The cost of solvents and coating materials and the emphasis on reducing emissions to the atmosphere have both increased dramatically since the late 1970s. These factors have effected an emphasis on increased transfer efficiency, ie, the fraction of the material removed from the coating bucket that is placed on the workpiece. The transfer efficiency is affected by the painting technique, workpiece geometry, the coating material, how the workpiece is presented to the atomizer, the ambient air movement, and other variables.

Electrostatic forces can be very effective in increasing the transfer efficiency. An electrical charge, usually negative, is placed on the coating material before atomization or as the coating particles are being formed. This is accomplished either by direct charging, where the coating material comes in contact with a conductor at high voltage, or by an indirect method, where the air in the vicinity of the coating particles is ionized and these ions then attach themselves to the coating particles. An external voltage source of 60–125 kV is normally used. A voltage gradient is established between the vicinity of the atomizer and the grounded workpiece by using the charged coating particles, charged metal atomizer, or an electrode near the atomizer as a local source of a high voltage field. An electrostatic force is exerted on each coating particle equal to the product of the charge it carries and the field gradient. The trajectory of the particle is determined by all the forces exerted on the particle. These forces include momentum, drag, gravity, and electrostatics. The field lines influencing the coating particles are very similar in arrangement to the alignment of iron particles when placed between two magnets. Using this method, coating particles that would normally pass alongside the workpiece are attracted to it, and it is possible to coat part or all of the back side of the workpiece.

Electrostatic spray atomizers are constructed from metal or nonconductive materials. A metal atomizer has sufficient electrical capacitance that when it is approached rapidly by a grounded object, eg, workpiece, an electrical arc may occur that can have sufficient energy to ignite certain solvent-air mixtures. A metal atomizer offers maximum ruggedness and efficiency but may present a fire hazard if not electronically protected. Thus this type of system often employs an electronic feedback system to reduce voltage and prevent arcing under these conditions. Most nonconductive atomizers are of a nonincendiary design; the rate of energy discharge has been specifically limited in such a way not to cause ignition. However, this type of atomizer is generally not as rugged as a metal atomizer, and in operation, the working voltages decrease, resulting in somewhat lower transfer efficiency.

All of the atomization techniques that produce a spray can be used with electrostatic spraying. Electrostatic atomization by definition uses electrostatic deposition. Furthermore, in rotary atomization the momentum toward the workpiece is relatively low and the transfer efficiency very poor if it is used without electrostatic deposition. Air and airless sprays also benefit from electrostatics in transfer efficiency. One problem for electrostatic spraying is that penetration into recessed areas is more difficult because the coating particles are attracted to the edges of the workpiece. The edges are closer to the high voltage source and therefore concentrate the field gradients. This problem can be overcome by reducing the voltage level or by using atomizers having high particle momentum, such as air or airless atomizers.

4. Economic Aspects

Spray equipment is marketed in a variety of ways in the United States. Several large manufacturers have broad product lines and sell equipment both directly to the user and through distributors. These companies can also provide custom engineered automatic systems for specific application. The largest U.S. manufacturers include Binks, Graco, ITW (DeVilbiss and Ransburg), and Nordson. There is also a multitude of smaller companies that provide a limited product line. Automatic spray systems can also be purchased through system houses that engineer an entire system, which might include the pretreatment system, ovens, and conveyor line.

Both individual components and small prepackaged systems are usually available from the many general distributors as are manual systems. Some manual systems are purchased in conjunction with automatic systems. Some of the larger distributors also custom design small automatic systems using standard components.

BIBLIOGRAPHY

"Coating Processes, Spray" in *ECT* 4th ed., Vol. 6, pp. 661–669, by K. J. Coeling, Nordson Corporation.

GENERAL REFERENCES

K. J. Coeling and T. Bublick, in J. I. Kroschwitz, ed., *Encyclopedia of Polymer Science and Engineering*, 2nd ed., Vol. 3, John Wiley & Sons, Inc., New York, 567–575.

R. P. Fraser, N. Dombrowski, and J. H. Routley, *Chem. Eng. Sci.* **18**, 339 (1969).

R. Ingebo, M. M. Elkotb, M. A. El-Sayed Mandy, and M. E. Montaser, *Proceedings of the 2nd International Conference on Liquid Atomization and Spray Systems*, Madison, Wis., 1982, 9–17, 107–115.

N. Dombrowski and W. R. Johns, *Chem. Eng. Sci.* **18**, 203 (1963).

D. Beadley, *J. Phys. D.* **6**, 1724 (1973).

J. Gretzinger and W. R. Marshall, Jr., *AIChE J.* **7**(2), 312 (June 1961).

R. Tholome and G. Sorcinelli, *Indust. Finish.* (Nov. 1977).

J. O. Hinte and H. Milborn, *J. Appl. Mech.*, 145 (June 1950).

N. Dombrowski and T. L. Lloyd, *Chem. Eng. J.*, 63 (1974).

K. J. Coeling, *Operating Characteristics of High Speed Centrifugal Atomizers*, The DeVilbiss Co., Toledo, Ohio, 1981.

F. A. Robinson, Jr., G. Pickering, and J. Scharfenberger, *Paint Con. '84*, Hitchcock Publishing Co., Carol Stream, Ill., 1984.

J. M. Lipscomb, *Surface Coating '83*, Chemical Coaters Association, *Finishing '83 Conference Proceedings*, Association for Finishing Processes of SME, 10–1, 10–10.

C. Dumouchel and M. Ledoux, *Proceeding of the 5th International Conference on Liquid Atomization and Spray Systems*, Garthersberg, Md., 1991, 157–164.

D. C. Busby and co-workers, *Supercritical Fluid Spray Application Technology: A Pollution Prevention Technology for the Future*, 218–239; *Proceedings of the 17th Water-Borne and High-Solid Coating Symposium*, New Orleans, La., 1990.

J. Schrantz and J. M. Bailey, *Indust. Finish.* (June 1989).

F. Robinson and D. Stephens, *Indust. Finish.* (Sept. 1990).

J. Wesson, *Prod. Finish.* (Aug. 1990).

K. J. COELING
Nordson Corporation

COATINGS

1. Introduction

Coatings are ubiquitous in an industrialized society. United States shipments of coatings in 1999 were $\sim 5.3 \times 10^9$ m^3 having a value of \$18 billion (1). Coatings are used for decorative, protective, and functional treatments of many kinds of surfaces. The low gloss paint on the ceiling of a room is used for decoration, but also diffuses light. Exterior automobile coatings fulfill both decorative and protective functions. Still others provide friction control on boat decks. Some coatings control the fouling of ship bottoms, others protect food and beverages in cans. Other coatings reduce growth of barnacles on ship bottoms, protect optical fibers against abrasion, etc.

Each year tens of thousands of coating types are manufactured. In general, these are composed of one or more resins, a mixture of solvents (except in powder coatings), commonly one or more pigments, and frequently one or more additives. Coatings can be classified into thermoplastic and thermosetting coatings. Thermoplastic coatings contain at least one polymer having a sufficiently high molecular weight to provide the required mechanical strength properties without further polymerization. Thermosetting coatings contain lower molecular weight polymers that are further polymerized after application in order to achieve desired properties. This article is organized to discuss succesively: Properties of coatings, components of coatings, application of coatings, end uses of coatings, and economic aspects.

2. Film Formation

Coatings are manufactured and applied as liquids and are converted to solid films after application to the substrate. In the case of powder coatings, the solid powder is converted after application to a liquid, which in turn forms a solid film. The polymer systems used in coatings are amorphous materials and therefore the term solid does not have an absolute meaning, especially in thermoplastic systems such as lacquers, most plastisols, and most latex-based coatings. A useful definition of a solid film is that it does not flow significantly under the pressures to which it is subjected during testing or use. Thus a film can be defined as a solid under a set of conditions by stating the minimum viscosity at which flow is not observable in a specified time interval. For example, it is reported that a film is dry-to-touch if the viscosity is greater than $\sim 10^6$ mPa $\cdot$ s$(= cP)$ (2). However, if the definition of dry is that the film resists blocking, ie, sticking together, when two coated surfaces are put against each other for 2 s under a mass per unit area of 1.4 kg/cm^3 (20 psi), the viscosity of the film has to be $> 10^{10}$ mPa $\cdot$ s$(=$cP$)$.

The viscosity of amorphous systems is a function of free-volume availability. The free volume of a material is the summation of the spaces or holes that exist between molecules of a material resulting from the impact of one molecule or molecular segment striking another. Such holes open and close as the molecules vibrate. Above the glass-transition temperature (T_g) the holes are large enough and last long enough for molecules or molecular segments to move into them. Free volume increases as temperature increases and the rate of volume increase is higher above T_g. An important factor affecting the free volume of a system is thus the difference between the temperature, T, and the T_g. The relationship between viscosity, η, and T_g is expressed in the Williams–Landel–Ferry (WLF) equation. Using so-called universal constants and assuming that viscosity at $T_g = 10^{15}$ mPa·s($=$ cP) the WLF equation, when η is in units of Pa·s, becomes equation 1:

$$\ln\eta = 27.6 + \frac{40.2(T - T_g)}{51.6 + (T - T_g)} \tag{1}$$

Using this equation, the approximate $(T-T_g)$ value required for a film of a thermoplastic copolymer to be dry-to-touch, ie, to have a viscosity of 10^6 mPa·s($=$ cP), can be estimated (3). The calculated $(T\text{-}T_g)$ for this viscosity is 54°C, which, for a film to be dry-to-touch at 25°C, corresponds to a T_g value of −29°C. The calculated T_g necessary for block resistance at 1.4 kgcm^{-3} for 2 s and 25°C, ie, $\eta = 10^{10}$ mPa, is 4°C. Because the universal constants in the WLF equation are only approximations, the T_g values are estimates of the T_g required. However, if parameters such as the mass per area applied for blocking were larger, the time longer, or the test temperature higher, the T_g of the coating would also have to be higher.

For practical coatings, it is not sufficient just to form a film; the film must also have a minimum level of strength depending on product use. Film strength depends on many variables, but one critical factor is molecular weight. For example, the acrylic polymers used in lacquers for refinishing automobiles must have a weight average molecular weight (M_w) >75,000. This required molecular weight varies according to the chemical composition of the polymer and the mechanical properties required for a particular application.

2.1. Solvent Evaporation from Solutions of Thermoplastic Polymers.

A solution of a copolymer of vinyl chloride (chloroethene) [75-01-4], vinyl acetate (acetic acid ethenyl ester) [108-05-4], and a hydroxy-functional vinyl monomer having a number average molecular weight (M_n) of 23,000 and a T_g of 79°C, gives coatings having good mechanical properties without cross-linking (4). A simple coating having only the resin and 2-butanone (methyl ethyl ketone, MEK) [78-93-3] as the sole solvent would give a polymer concentration of ~19 wt% solids or ~12 vol% in order to have a viscosity of ~100 mPa·s($=$ cP) for spray application. Because of the relatively high vapor pressure under application conditions, MEK evaporates rapidly and a substantial fraction of the solvent evaporates in the time interval between the coating leaving the orifice of the spray gun and arrival on the surface being coated. As the solvent evaporates, the viscosity increases and the coating reaches the dry-to-touch stage very rapidly after application and does not block under the conditions discussed. However, if the film is formed at 25°C, the dry film contains several percent retained solvent.

In the first stages of solvent evaporation from such a film, the rate of evaporation depends on the vapor pressure at the temperatures encountered during the evaporation, the ratio of surface area to volume of the film, and the rate of air flow over the surface. The rate of evaporation is essentially independent of the presence of polymer. However, as the solvent evaporates the T_g increases, free volume decreases, and the rate of loss of solvent from the film, at some point, becomes dependent not on how fast the solvent evaporates, but on how rapidly the solvent molecules can diffuse to the surface of the film. In this diffusion-control stage, solvent molecules must jump from free-volume hole to free-volume hole to reach the surface where evaporation can occur. As solvent loss continues, T_g increases, and free volume decreases. When the T_g of the remaining polymer solution approaches the temperature at which the film is being formed, the rate of solvent loss becomes very slow. If the film is being formed at 25°C from a solution of a resin having a higher T_g, eg, 79°C, loss of solvent becomes very slow when the T_g of the film exceeds 25°C and a significant amount of MEK remains in the hard, dry film indefinitely, acting as a plasticizer. In order to remove the last of the MEK in a short time, it is necessary to heat the film to a temperature significantly above the T_g of the solvent-free polymer.

The rate of solvent diffusion through the film depends not only on the temperature and the T_g of the film but also on the solvent structure and solvent–polymer interactions. The solvent molecules move through free-volume holes in the films and the rate of movement is more rapid for small molecules than for large ones. Additionally, linear molecules may diffuse more rapidly because their cross-sectional area is smaller than that of branched-chain isomers. For example, although isobutyl acetate (IBAc) [105-46-4] has a higher relative evaporation rate than *n*-butyl acetate (BAc) [123-86-4] IBAc diffuses more slowly out of a film of a nitrocellulose lacquer than BAc during the second stage of drying (5). Similarly, *n*-octane [111-65-9] diffuses more rapidly out of alkyd films than isooctane (2,2,4-trimethylpentane) [540-84-1] although isooctane has the higher relative evaporation rate (6).

Film thickness is a factor in solvent loss and film formation. In the first stage of solvent evaporation, the rate of solvent loss depends on the first power of film thickness. However, in the second stage when the solvent loss is diffusion rate controlled, it depends on the square of the film thickness. Although thin films lose solvent more rapidly than thick films, if the T_g of the drying film increases to ambient temperature during the evaporation of the solvent, then, even in thin films, solvent loss is extremely slow. Models have been developed to predict the rate of solvent loss from films as functions of the evaporation rate, thickness, temperature, and concentration of solvent in the film (7).

Thermoplastic polymer-based coatings have low solids contents because the relatively high molecular weight requires large amounts of solvent to reduce the viscosity to that required for application. Air pollution regulations limiting the emission of volatile organic compounds (VOC) and the increasing cost of solvents has led increasingly to replacement of such coatings with types that require less solvent for application.

2.2. Film Formation from Solutions of Thermosetting Resins. Substantially less solvent is required in formulating a coating from a low molecular weight resin that can be further polymerized to a higher molecular weight after application to the substrate and evaporation of the solvent. Theoretically, difunctional reactants could be used. However, this is not feasible for coatings where the close control of stoichiometric ratio and purity required to achieve a desired molecular weight reproducibly with difunctional reactants is impractical. Therefore, the average functionality must be >2 in order to ensure that the molecular weight of the final cured film is high enough for good properties. Not only should the average functionality be >2, it is usually preferable for the number of monofunctional molecules to be at a minimum because these terminate polymerization. If any of the resin molecules have no functional groups, they cannot react and remain in the film as a plasticizer. The reactions are commonly called cross-linking reactions. A cross-linked film not only has very high molecular weight it is also insoluble in solvents. For many applications, this solvent resistance is an advantage of thermosetting coatings over thermoplastic coatings.

The mechanical properties of the cross-linked film depend on many factors; two of the most important are the lengths of the segments between cross-links and the T_g of the cross-linked resin. Segment length depends on the average equivalent weight and the average functionality of the components and the fraction of cross-linking sites actually reacted. The size of the segments between cross-links is often expressed as cross-link density (XLD); the closer the cross-links, the higher the XLD. Everything else being equal, the higher the XLD, the higher the modulus, ie, the harder the film. The T_g of cross-linked polymers is controlled by four factors and corresponding interactions: T_g of the segments of polymer between cross-links, the cross-link density, the presence of dangling ends, and the possible presence of cyclic segments (8). The T_g of the polymer segments between cross-links is governed by the structure of the resin and cross-linking agent and by the ratio of these components. Because cross-links restrict segmental mobility, T_g increases as XLD increases. The parameter T_g also increases as the fraction of dangling ends decreases, ie, as cross-linking reactions proceed. Cyclization would be expected to restrict chain mobility, and hence to increase T_g.

In the initial steps of cross-linking low molecular weight resins, the molecular weight and XLD increase whereas the fraction of dangling ends decreases resulting in an increase in T_g. As the cross-linking reaction continues, a gel forms. Reaction does not stop at gelation but continues as long as there are functional groups to react and there is sufficient mobility in the matrix to permit the reactive groups to move into position for reaction. As the reaction continues, modulus above T_g increases and the film becomes insoluble in solvent. Solvent can still dissolve in a cross-linked film leading to swelling; the extent of swelling decreases as XLD increases.

A problem in thermosetting systems is the relationship between stability of the coating before application and the time and temperature required to cure the film after application. It is generally desirable to be able to store a coating for many months or even years without significant increase in viscosity that would result from cross-linking during storage. On the other hand, after application, the cross-linking reaction should proceed in a short time at as low a temperature as possible. Because reaction rates depend on the concentration of functional groups, storage life can be increased by using more dilute coatings, ie, adding more solvent increases storage life. When the solvent evaporates after application, the reaction rate increases initially. However, VOC regulations are forcing the use of less and less solvent, increasing the problem of storage stability.

The reaction rates are also dependent on the rate constants for the reactions at the temperatures of storage and curing, equation 2. The rate constant k changes most rapidly with temperature if the activation energy, E_a, of the reaction is high. However, the reaction rate is slow unless the preexponential term A is also large. Under the assumptions of required ratios of rate constants, rate equations and Arrhenius equations have been used to calculate what orders of magnitude of E_a and A are required to permit various combinations of storage times and curing temperatures (9). Such calculations show that to formulate a coating stable for 6 months at $30°C$, the calculated kinetic parameters become unreasonable if cure is desired in 30 min below $\sim120°C$. No known chemical reactions have a combination of E_a and A that would permit design of such a system.

$$\ln k = \ln A - E_a/RT \tag{2}$$

More reactive combinations can be used in two-package coatings where one package contains a resin with one of the reactive groups and the second package contains the component with the other reactive group. The packages are mixed shortly before use. Two-package coatings are used on a large scale commercially but are not generally desirable. They take extra time and are generally more expensive. Material is usually wasted and there is a chance of error in mixing. Two-package coatings are referred to as "2K coatings" and single-package coatings as "1K coatings".

Design of stable coatings that cure at lower temperatures or shorter times must be based on factors other than the kinetics of the cross-linking reaction. Several techniques are used: ultraviolet-curing systems; systems requiring an atmospheric component as a catalyst or reactant, eg, oxygen or moisture; use of a volatile inhibitor; use of a cross-linking reaction that is a reversible condensation reaction involving the loss of a volatile reaction product that includes some of this monofunctional volatile by-product as solvent in the coating; and use of catalysts or reactants that change phase over a narrow temperature range.

Before two functional groups can react, they must be in close proximity. As low molecular weight resins and cross-linking agents react, T_g increases and free volume decreases. If the cure temperature is at least somewhat higher than the T_g of the fully reacted system, the reaction can go to completion at rates governed by kinetic parameters. If the cure temperature is below the T_g of the fully reacted system, the reaction slows to a rate controlled by diffusion. If the cure temperature is $>50°C$ below the T_g of the fully reacted system, the reaction will stop before completion. A general review of the effect of variables on mobility control on reaction rates is available (10).

An example of the importance of free-volume availability on cross-linking has been reported in the evaluation of a trifunctional derivative of an aliphatic isocyanate that contains an aromatic ring, m-tetramethylxylidene diisocyanate (TMXDI) (1,3-bis(1-isocyanato-1-methylethyl)-benzene) [2778-42-9], as a cross-linking agent for hydroxy-functional resins (11). Because of steric effects, TMXDI is less reactive than the trifunctional derivatives of hexamethylene diisocyanate (HDI) (1,6-diisocyanatohexane) [822-06-0] but this can be overcome by catalyst composition and concentration. Although essentially complete reactions were obtained for films cured at elevated temperatures using a TMXDI/TMP (trimethylolpropane) (2-ethyl-2-(hydroxymethyl)-1,3-propanediol) [77-99-6] pre-polymer, when cured at 21°C the reaction was slow and essentially stopped at ~50% completion. The acrylic resin being cross-linked had been designed for use with a more flexible triisocyanate cross-linking agent derived from hexam-ethylene diisocyanate. In a series of acrylic resins having lower T_g, as the T_g decreased, the cross-linking reaction with TMXDI at 21°C was faster and went more nearly to completion. Using an appropriately designed acrylic resin and catalyst, TMXDI gave films that cured fully at room temperature. Resins and cross-linking agents must be selected or designed for use with each other, especially for curing systems used at ambient temperature.

2.3. Film Formation by Coalescence of Polymer Particles. Latex paints have low solvent emissions as well as many other advantages. A latex is a stabilized dispersion of high molecular weight polymer particles in water. Because the latex polymer is not in solution, the rate of water loss by evaporation is almost independent of concentration until near the end of the evaporation pro-cess. When a dry film is prepared from a latex, the forces that stabilize the dis-persion of latex particles must be overcome and the particles must coalesce into a continuous film. As the water evaporates, the particles come closer and closer together. As they approach each other, they can be thought of as forming the walls of capillary tubes in which surface tension leads to a force striving to col-lapse the tube. The smaller the diameter of the tube, the greater the force. When the particles get close enough together so that that force pushing them together exceeds the repulsive forces holding them apart, coalescence is possible. A sur-face tension driving force also promotes coalescence because of the decrease in surface area when the particles coalesce to form a film. Both factors have been shown to be important in film formation from latexes. Reference (12) gives a discussion of the various theories of the factors affecting film formation. Coales-cence, however, also requires that the polymer molecules in the particles be free to intermingle with those from adjoining particles. This movement can occur only if there are a sufficient number and size of free-volume holes in the polymer par-ticles into which the polymer molecules from other particles can move. In other words, the T_g of the latex particles must be lower than the temperature at which film formation is being attempted.

The rate of coalescence is controlled by the free-volume availability, which in turn depends mainly on $(T - T_g)$. The viscosity of the coalesced film is also dependent on free volume. If a film is to resist the mild blocking test described earlier, $(T - T_g)$ would have to be on the order of 21°C. If the film is to resist blocking at 40°C, the T_g would have to be >19°C. However, in many cases the paint must be formulated in such a way that it can be applied at a temperature as low as 5°C, and therefore, the T_g of the latex particles would have to be <5°C. This problem can be solved by adding a coalescing agent, such as 2,2,4-trimethyl-1,3-pentanediol monoisobutyrate (2-methylpropanoic acid 2,2,4-trimethyl-1,3-pentanediyl ester) [132503-14-1], tributyl phosphate [126-73-8], and 2-butoxyethanol [111-76-2], among others. The coalescing agent dissolves in the latex particles, acts as a plasticizer, increases free-volume availability, and hence permits film formation at a lower temperature. After the film has formed, the coalescing agent diffuses slowly to the surface and evaporates. Because the free volume in the film is relatively low, the last traces of coalescing solvent evaporate slowly.

Even though the films feel dry, they block and pick up dirt for a long time after application.

Coalescing solvents contribute to VOC emissions, there has been increasing emphasis on reducing or eliminating use of coalescing solvents. References (13) and (14) discuss some of the approaches. An approach is to use latexes in which the particles have outer shells of low T_g and interiors of higher T_g. The T_g of the outer shell is low, permitting coalescence at low temperatures, but in time, after application, intermixing occurs, leading to a higher average T_g and permitting block resistance at higher temperatures. Another approach is to use thermo-setting latexes that have a low T_g but attain adequate film properties by cross-linking. Hydroxy-functional latexes can be formulated with MF resins or a water-dispersible polyisocyanate for wood and maintenance coatings (15). Carboxylic acid functional latexes can be cross-linked with carbodiimides (16), or polyfunctional aziridines (17). m-Isopropenyl-α,α-dimethylbenzyl isocyanate (TMI) [1-(1-isocyanato-1-methylethyl)-3-(1-methylethenyl)benzene] [2094-99-7] reacts slowly with water and can be used to make thermosetting latexes (18).

Other thermosetting latexes cross-link at room temperature and are sto-rage stable. Carboxylic acid functional latexes can be cross-linked with β-(3,4-epoxycyclohexyl)ethyltriethoxysilane (19). A combination of amine-functional and epoxy-functional latexes gives stable one package coatings (20). A latex with allylic substitution cross-links on exposure to air (21). Hybrid alkyd/acrylic latexes are prepared by dissolving an oxidizing alkyd in the monomers used in emulsion polymerization (22). Stable thermosetting latexes can be prepared using triisobutoxysilylpropyl methacrylate as a comonomer (19).

Powder coatings form films by coalescence. Because the powder must not sinter during storage, the free volume at storage temperature must be suffi-ciently low to avoid coalescence at this stage. The T_g of powder coating particles is commonly of the order of 50–55°C. Higher T_g must be avoided because rapid coalescence after application requires that the $(T - T_g)$ under baking conditions be as large as possible without requiring excessively high baking temperatures. The problem is even more complex in thermosetting powder coatings where overly rapid cross-linking can impede coalescence and leveling. Other examples of coalescing systems include nonaqueous dispersion, plastisol, "water-reducible" resin, and electrodeposition coatings.

3. Flow

Rheological properties, ie, flow and deformation, of coatings have significant impacts on application and performance properties. The application and film for-mation of liquid coatings require control of the flow properties at all stages. The mechanical properties of the applied coating films are controlled by the viscoelas-tic responses of the films to stress and strain. A good overview of the field of rheology in coatings is available (23); a more extensive but somewhat dated discussion of the flow of coatings is also available (24).

3.1. Viscosity of Resin Solutions. The viscosity of coatings must be adjusted to the application method to be used. It is usually between 50 and 1000 mPa·s(= cP), at the shear rate involved in the application method used.

The viscosity of the coating is controlled by the viscosity of the resin solution, which is in turn controlled mainly by the free volume. The factors controlling free volume are temperature, resin structure, solvent structure, concentration, and solvent–resin interactions.

The temperature dependence of viscosity of resin solutions can be expressed by the WLF equation (eq. 3) where the reference temperature T_r is taken as the lowest temperature for which data are available and e_1, and e_2 are adjustable parameters (25).

$$\ln\eta = \ln \eta_{T_r} - \frac{c_1(T - T_g)}{c_2 + (T - T_g)} \tag{3}$$

This relationship has been shown to hold for a wide variety of coating resins and resin solutions over a wide range of concentrations. A simplification of equation 3, where T_g is the reference temperature is given in equation 4, which assumes that the viscosity at T_g is 10^{12} Pa·s and A and B are adjustable parameters.

$$\ln\eta = 27.6 - \frac{A(T - T_g)}{B + (T - T_g)} \tag{4}$$

Equation 4 does not model the relationship of viscosity with temperature quite as well as equation 3, but it is useful because it shows the relationship to T_g of the resin solution. The T_g of the resin solution is an important, but not singular, factor in controlling viscosity. The T_g of the resin solution depends on the T_g of the resin, the T_g of the solvent, the concentration of the resin solution, and the effects of resin–solvent interactions. The T_g of the resin depends on the molecular weight and the structure of the resin. The flow of solutions of coatings resins in good solvents in the viscosity range of 50–10,000 mPa·s(=cP) is Newtonian and log viscosity increases as the square root of the molecular weight. Much further work is needed to elucidate the solvent–resin interaction effects, but it appears that low viscosity hydrogen bond acceptor solvents give the greatest reduction in viscosity for polar substituted low molecular weight resins.

The viscosity of high solids coatings in the range of application varies more rapidly with temperature than is the case for conventional lower solids coatings (26). Hence, the viscosity is reduced more by using hot-spray systems that permit a further increase in solids. Within the range of 0.05–10 Pa·s, log viscosity varies approximately directly with concentration. The viscosity of higher viscosity solutions varies more steeply with concentration.

3.2. Viscosity of Systems with Dispersed Phases.   A large proportion of coatings are pigmented and, therefore, have dispersed phases. In latex paints, both the pigments and the principal polymer are in dispersed phases. The viscosity of a coating having dispersed phases is a function of the volume concentration of the dispersed phase and can be expressed mathematically by the Mooney equation (eq. 5):

$$\ln \eta = \ln \eta_e + \frac{K_E V_i}{1 - V_i/\phi} \tag{5}$$

where η_e is the viscosity of the external phase, K_E is a shape constant (2.5 for spheres), V_i is the volume fraction of internal phase, and ϕ is the packing factor, ie, the volume fraction of internal phase when the V_i is at the maximum close-packed state possible for the system. The Mooney equation assumes rigid particles having no particle–particle interaction. It fits pigment dispersions and latexes that exhibit Newtonian flow.

If there is particle–particle interaction, as is the case for flocculated systems, the viscosity is higher than in the absence of flocculation. Furthermore, a flocculated dispersion is shear thinning and possibly thixotropic because the floccules break down to the individual particles when shear stress is applied. Considered in terms of the Mooney equation, at low shear rates in a flocculated system some continuous phase is trapped between the particles in the floccules. This effectively increases the internal phase volume, and hence the viscosity of the system. Under sufficiently high stress, the floccules break up, reducing the effective internal phase volume and the viscosity. If, as is commonly the case, the extent of floccule separation increases with shearing time, the system is thixotropic as well as shear thinning.

Shear thinning systems that are generally also thixotropic, also result if the disperse phase particles are not rigid. In the shear field, the nonrigid particles are distorted, resulting in less crowding, and therefore lower viscosity. In terms of equation 5, K_E becomes smaller and the packing factor becomes larger. Hence, the viscosity of the system is lower as the shear rate increases and as the time at a given higher shear rate is extended. Emulsions, ie, dispersions of liquids in liquids, show thixotropic flow because the dispersed phase particles are fluid and can be distorted.

4. Mechanical Properties

Coating films should withstand use without damage. The coating on the outside of an automobile should not break when hit by a piece of flying gravel. The coating on the outside of a beer can must not abrade when cans rub against each other during shipment. The coating on wood furniture should not crack when the wood expands and contracts as a result of changing temperatures or swelling and shrinkage from changes in moisture content of the wood. The coating on aluminum siding must be flexible enough for fabrication of the siding and resist scratching during installation on a house. Reference (27) gives a methodology for considering the factors involved in service life prediction. A monograph discusses problems of predicting service lives and proposes reliability theory methodology for database collection and analysis (28).

Understanding relationships between composition and basic mechanical properties of films can provide a basis for more intelligent formulation. Reference (29) is a good review paper. In ideal elastic deformation a material elongates under a tensile stress in direct proportion to the stress applied. When the stress is released, the material returns to its original dimensions essentially instantaneously. An ideal viscous material elongates when a stress is applied in direct proportion to the stress, but does not return to its original dimensions when the stress is released. Almost all coating films are viscoelastic—they exhibit intermediate behavior.

Elastic deformation is almost independent of time and temperature. Viscous flow is time and temperature dependent; the flow continues as long as a stress is applied. Viscoelastic deformation is dependent on the temperature and the rate at which a stress is applied. If the rate of application of stress is rapid, the response can be primarily elastic; if the rate of application of stress is low, the viscous component of the response increases and the elastic response is lower. Similarly, if the temperature is low, the response can be primarily elastic, at a higher temperature, the viscous response is greater.

By dynamic mechanical analysis the elastic and viscous components of modulus can be separated. The higher the frequency of oscillation, the greater the elastic response—the smaller the phase angle; the lower the frequency, the greater the viscous response—the larger the phase angle.

Many coated products are subjected to mechanical forces either to make a product, as in forming bottle caps or metal siding, or in use, as when a piece of gravel strikes the surface of a car with sufficient force to deform the steel substrate. To avoid film cracking, the elongation-at-break must be greater than the extension of the film. Cross-linked coatings have low elongations-at-break when below T_g. Properties are affected by the extent to which cross-linking has been carried to completion. Incomplete reaction leads to lower XLD and, hence, lower storage modulus above T_g. The extent of reaction can be followed by determining storage modulus as a function of time (30). Thus, one can, at least in theory, design a cross-linked network to have a desired storage modulus above T_g by selecting an appropriate ratio of reactants of appropriate functionality.

An additional factor that can affect the mechanical properties of polymeric materials is the breadth of the T_g transition region (31). The same effect can be seen in tan δ plots, which exhibit various breadths. Broad tan δ peaks are frequently associated with heterogeneous polymeric materials. For thermosetting polymers, the T_g transition region is generally broader than for thermoplastics. Breadth of the distribution of chain lengths between cross-links is a factor, and blends of thermosetting resins such as acrylics and polyesters often display a single, broad T_g transition. Materials with broad and/or multiple T_g values have better impact resistance than comparable polymers with a sharp, single T_g.

When a cross-linked film on a metal substrate is deformed by fabrication, it is held in the deformed state by the metal substrate. As a result, there is a stress within the film acting to pull the film off the substrate. Stress within films can also arise during the last stages of solvent loss and/or cross-linking of films (32). It is common for coatings to become less flexible as time goes on. Particularly in air dry coatings, loss of the last of the solvent may be slow. If the cross-linking reaction was not complete, the reaction may continue decreasing flexibility. Another possible factor with baked films is densification. If a coating is heated above its T_g and then cooled rapidly, the density is commonly found to be lower than if the sample had been cooled slowly (33). During rapid cooling, more and/or larger free volume holes are frozen into the matrix. On storage, the molecules slowly move and free volume decreases, causing densification; it is also called physical aging.

Abrasion is the wearing away of a surface, marring is a disturbance of a surface that alters its appearance. A study of the mechanical properties of a series of floor coatings with known wear life concluded that work-to-break values best represented the relative wear lives (34). Studies of automobile clear coats have shown that wear resistance increases as energy-to-break of films increases (35). The coefficient of friction of a coating can affect abrasion resistance. Abrasion of the coating on the exterior of beer cans during shipment can be minimized by incorporation of a small amount of incompatible wax or fluorosurfactant in the coating. Another variable is surface contact area. Incorporation of a small amount of a small particle size SiO_2 pigment in a thin silicone coating applied to plastic eyeglasses reduces abrasion. The pigment particles reduce contact area, permitting the glasses to slide more easily over a surface.

Marring is a near-surface phenomenon; even scratches <0.5 µm deep can degrade appearance. Marring is a major problem with automobile clear topcoats. In going through automatic car washes, the surfaces of some clear coats are visibly marred and lose gloss (36). Plastic deformation and fracture lead to marring. The responses can be quantitatively measured by scanning probe microscopy (37). In general, MF cross-linked acrylic clear coats are more resistant to marring than isocyanate cross-linked coatings, but MF cross-linked coatings have poorer environmental etch resistance. Coatings can be made hard enough that the marring object does not penetrate into the surface, or they can be made elastic enough to bounce back after the marring stress is removed. If the hardness strategy is chosen, the coating must have a minimum hardness, however, such coatings may fail by fracture.

Field applications on a small scale and under especially stringent conditions accelerate possible failure. Traffic paints are tested by painting stripes across the lanes of traffic instead of parallel to traffic flow. Automobiles are driven on torture tracks with stretches of gravel, through water, under different climate conditions. Sample packs of canned goods are made; the linings are examined for failure and the contents evaluated for flavor after storage.

Many tests have been developed to simulate use conditions in the laboratory. An example is a gravelometer to evaluate resistance of coatings to chipping of automotive coatings when struck by flying gravel. Pieces of standard shot are propelled at the coated surface by compressed air under standard conditions. The tests have been standardized by comparison to a range of actual results and give reasonably good predictions of actual performance. A more sophisticated instrument, a precision paint collider, which permits variations in angle and velocity of impact and temperature has been described (38).

Many empirical tests are used to test coatings. In most cases, they are more appropriate for quality control than performance prediction. American Society for Testing and Materials (ASTM) tests are widely used (39). An excellent summary of the tests and discussion of their applications and significance is available (40).

5. Exterior Durability

For many coatings, an important performance requirement is exterior durability. There are many potential modes of failure when coatings are exposed outdoors. Commonly, the first indication of failure is reduction in gloss resulting from surface embrittlement and erosion leading to the development of roughness and cracks in the surface of the coating. In some cases, the next step is "chalking", ie, the erosion of resin from the surface of the coating leaving loose pigment particles on the surface. Chalking reduces gloss and, as chalking proceeds, can lead to complete film erosion and also to color change caused by the increased surface reflectance, which makes colors shift to light shades. Colors can also change if the pigment or pigment–resin combination undergoes photochemical degradation on exterior exposure. Photochemical oxidation and hydrolysis of the resins in coatings are common causes of failure. On exterior exposure, films may crack or check as a result of embrittlement, reducing the elongation-to-break. Such films crack as they undergo expansion and contraction. These changes may be caused by temperature changes, especially when the substrate and the coating have different coefficients of thermal expansion. Similar stresses on coatings can be caused by expansion and contraction of wood substrates with changes in moisture content of the wood.

Various kinds of chemical attack, such as those resulting from acid rain and bird droppings, can result in film degradation and discoloration. Retention of dirt from the atmosphere on the coating surface can lead to drastic color changes and blotchy appearance. Mildew can grow on the surface of many coating films, again leading to blotches of gray discoloration. In view of the large variety of exposure conditions and possible modes of failure, no laboratory test has been devised to predict field performance of coatings on exterior exposure. However, careful accumulation of actual field use results correlated with environmental, compositional, and application variables is the most useful way of understanding the causes of failures and, hence, of being able to forecast the possible performance of a new coating material.

Although many failure mechanisms are involved, the two most common modes are hydrolysis and photochemical oxidation by free-radical chain reactions. In general, resins that have backbone linkages that cannot hydrolyze provide better exterior durability than systems having, eg, ester groups in the backbone. Some esters are more resistant to hydrolysis than others. In general, esters of highly hindered alcohols such as neopentyl glycol (NPG) (2,2-dimethyl-1,3-propanediol) [126-30-7] are less easily hydrolyzed than those of less hindered alcohols such as ethylene glycol (1,2-ethanediol) [107-21-1]. Esters of isophthalic acid (1,3-benzenedicarboxylic acid) [121-91-5] are more resistant to hydrolysis in the range of pH 4–8 than esters of phthalic acid. Ester groups on acrylic polymers are highly resistant to hydrolysis.

Susceptibility to free-radical induced photoxidation varies with structure of the resins and pigments and, in some cases, with the interactions between pigment and resin. In general terms, resistance of resins to photochemical failure is related to the ease of abstraction of hydrogens from the resin molecules by free radicals. The greatest resistance is shown by fluorinated resins and silicone resins, especially methyl-substituted silicone resins. The greatest sensitivity to degradation is shown by resins having methylene groups between two double bonds; methylene groups adjacent to amine nitrogens, ether oxygens, or double bonds; and methine groups. Resins containing aromatic rings substituted with a heteroatom directly on the ring, such as bisphenol A epoxy resins and urethane resins based on aromatic isocyanates, are very susceptible to photochemical failure.

Pigment selection can also be critical in formulating for high exterior durability. Some pigments are more susceptible to color change on exposure than others. Some pigments act as photosensitizers to accelerate degradation of resins in the presence of uv and water. For example, anatase TiO_2 accelerates the chalking of coatings during exterior exposure. Rutile TiO_2 pigments with appropriate surface treatments, on the other hand, lead to little, if any, increase in chalking. It cannot be concluded that because one colored pigment gives greater exterior life in one resin system compared to another colored pigment the order of durability is the same in other resins systems. Reversals are fairly common.

The exterior durability of relatively stable coatings can be enhanced by use of additives. Ultraviolet absorbers reduce the absorption of uv by the resins, and hence decrease the rate of photodegradation. Further improvements can be gained by also adding free-radical trap antioxidants such as hindered phenols and especially hindered amine light stabilizers (HALS). A discussion of various types of additives is available (41).

It has also been found that there can be interactions between hydrolytic degradation and photochemical degradation. Especially in the case of melamine–formaldehyde cross-linked systems, photochemical effects on hydrolysis have been observed.

Although many variations in methods have been tried, no reliable laboratory test is available to predict exterior durability. Reference (42) recommends use of reliability theory using statistical distribution functions of material, process, and exposure parameters for predicting exterior durabilitty. For predicting performance of automotive base coat/clear coat systems four performance measurements: clear coat uva performance, clear coat and base coat HALS performance, ability of all coating layers to resist photoxidation, and clear coat fracture energy (43).

Exterior exposure of panels in various locations such as Florida is widely used to forecast performance with reasonable success. However, coatings that are sensitive to acid rain or to cracking on rapid temperature change might not perform as well in actual use as predicted by Florida exposure results. In the EMMAQUA tests (DSET Laboratories, Inc.), panels are exposed in Arizona on a machine that turns to keep mirrors that reflect the sunlight to the panel surface approximately normal to the sun. The machine increases the radiation intensity shining on the panels about sevenfold over direct exposure of panels in the same location. To simulate the effect of rain, the panels are sprayed with water each night. The test can provide useful guidance in a few weeks, especially if comparisons are based on exposure to equal intensities of uv radiation rather than for equal periods of time. Many other accelerated laboratory tests have been devised. However, reversals in performance between pairs of coatings in the field as compared with the laboratory results indicate that frequently these tests are not reliable predictors.

The most promising approach to laboratory techniques for predicting performance is to understand the mechanism of failure and then use instrumental methods to study the susceptibility of a coating to failure. The most powerful tool available now is the use of electron spin resonance (esr) spectrometry to monitor the rate of free-radical appearance and disappearance.

6. Adhesion

In most cases, it is desirable to have a coating that is difficult to remove from the substrate to which it has been applied. An important factor controlling this property is the adhesion between the substrate and the coating. The difficulty in removing a coating also can be affected by how difficult it is to penetrate through the coating and how much force is required to push the coating out of the way as the coating is being removed from the substrate as well as the actual force holding the coating onto the substrate. Furthermore, the difficulty of removing the coating can be strongly affected by the roughness of the substrate. If the substrate has undercut areas that are filled with cured coating, a mechanical component makes removal of the coating more difficult. This is analogous to holding two dovetailed pieces of wood together.

Surface roughness affects the interfacial area between the coating and the substrate. The force required to remove a coating is related to the geometric surface area, whereas the forces holding the coating onto the substrate are related to the actual interfacial contact area. Thus the difficulty of removing a coating can be increased by increasing the surface roughness. However, greater surface roughness is only of advantage if the coating penetrates completely into all irregularities, pores, and crevices of the surface. Failure to penetrate completely can lead to less coating-to-substrate interface contact than the corresponding geometric area and leave voids between the coating and the substrate, which can cause problems.

6.1. Adhesion to Metals. For interaction between coating and substrate to occur, it is necessary for the coating to wet the substrate. Somewhat oversimplified, the surface tension of the coating must be lower than the surface tension of the substrate. In the case of metal substrates, clean metal surfaces have very high surface tensions and any coating wets a clean metal substrate.

Penetration of the vehicle of the coating as completely as possible into all surface pores and crevices is critical to achieving good adhesion. This requires that the surface tension of the coating be low enough for wetting, but the extent of penetration is controlled by the viscosity of the continuous phase of the coating. Although broad rigorous scientific studies of the relationship between continuous-phase viscosity and adhesion have not been published, the importance of this relationship is evident from the formulating decisions made over many years in the manufacture of coatings having good adhesion performance. The critical viscosity is that of the continuous phase because many of the crevices in the surface of metal are small compared to the size of pigment particles. Because penetration takes time, the initial viscosity of the external phase should be low and the viscosity should be kept as low as possible for as long as possible. Slow evaporating solvents are best for coatings that are to be applied directly on metal. Systems that cross-link slowly minimize the increase of viscosity of the continuous phase. Because viscosity of the vehicle drops with increasing temperature, baking coatings can be expected to provide better adhesion than a similar composition coating applied and cured at room temperature.

Adhesion is strongly affected by the interaction between coating and substrate. On a clean steel substrate, hydrogen bond or weak acid−weak base interactions between the surface layer of hydrated iron oxide that is present on any clean steel surface and polar groups on the resin of the coating provide such interaction. It has been suggested that several polar groups spaced along a resin backbone, having some flexible units to permit facile orientation of the groups to the interface during coating application, and some rigid segments to promote only partial adsorption of the groups so that there can be interaction with the balance of the coating resin, can provide cooperative interactions that enhance adhesion. Conversely, adhesion can be adversely affected if additives having single-polar groups and long nonpolar tails are present in a coating. For example, dodecylbenzenesulfonic acid is a catalyst for MF resins. However, its use in coatings directly applied on steel can adversely affect adhesion. Presumably, the sulfonic acid groups interact strongly with the steel surface leading to a monolayer having a surface of long hydrocarbon chains. The effect is similar to trying to achieve adhesion to oily steel or polyethylene.

Fracture mechanics affect adhesion. Fractures can result from imperfections in a coating film that act to concentrate stresses. In some cases, stress concentration results in the propagation of a crack through the film, leading to cohesive failure with less total stress application. Propagating cracks can proceed to the coating–substrate interface, then the coating may peel off the interface, which may require much less force than a normal force pull would require.

Adhesion of coatings is also affected by the development of stresses as a result of shrinkage during drying of the film. For example, in the case of uv-cure coatings, where curing is achieved by photoinitiated free-radical polymerization of acrylic double bonds, a substantial volume reduction occurs in the fraction of a second required. This loss in volume leads to stresses in the film that partly supply the force needed to pull the film from the substrate. Hence, less external force must be applied to remove the film. Sometimes such stresses can be relieved by heating the coating to anneal it. Internal stresses can also result from solvent loss from a film and other polymerization reactions, as well as changes in temperature and relative humidity, particularly at temperatures below T_g. These stresses can affect coating durability, especially adhesion (44).

The formation of covalent bonds between resin molecules in a coating and the surface of the substrate can enhance adhesion. Thus, adhesion to glass is promoted by reactive silanes having a trimethoxysilyl group on one end that reacts with a hydroxyl group on the glass surface. The silanes have various functional groups that react with the cross-linking agent in the coating on the other end of the molecule (45).

6.2. Adhesion to Plastics and Coatings. Wetting can be a serious problem for adhesion of coatings to plastics. Some plastic substrates have such low surface tension that it may be difficult to formulate coatings having a sufficiently low surface tension to wet the substrate. Polyolefin plastics, in particular, are difficult to wet. Frequently, the surface of the polyolefin plastic must be oxidized to increase surface tension and provide groups to interact with polar groups on the coating resin. The surface can be treated using an oxidizing solution, by flame treatment or by exposure to a corona discharge. Difficulties in wetting plastics can also result from residual mold release agents on the surface of the plastic. See Ref. 46 for further discussion.

Adhesion to plastics can be enhanced if resin molecules from the coating can penetrate into the surface layers of the plastic. Maleic anhydride modified chlorinated polyolefin in xylene primers enhance the adhesion of coatings on thermoplastic polyolefins (TPO) automobile bumpers (47). Penetration would take place through free-volume holes; hence, raising the temperature above the T_g of the plastic substrate generally promotes adhesion. The T_g of the plastic can be lowered by penetration of solvent from the coating into the surface of the plastic. Solvent selection in formulating a coating can be critical. In selecting coatings solvents for application to articles fabricated from high T_g thermoplastics such as polystyrene and poly(methyl methacrylate), a solvent system having too high a rate of evaporation should not be used in order to avoid crazing, ie, cracking of the surface.

The same considerations apply to intercoat adhesion to other coatings as to plastics. A further design parameter is available in formulating primers. Because adhesion to rough surfaces is better than to gloss surfaces, primers are usually highly pigmented, and hence have a rough surface to promote adhesion of the topcoat to the primer surface. Adhesion of topcoats can be further enhanced if the primer is formulated using a Pigment Volume Concentration (PVC) higher than Critical PVC (CPVC) (see Section 11.2). A film from such a primer has voids into which resin solution from the topcoat can flow, providing a mechanical anchor between the topcoat and the primer. Because the voids in the primer coat are thus filled with vehicle from the topcoat, the effective PVC of the final primer layer of the overall coating is approximately equal to CPVC. The PVC of the primer should be only a little above CPVC. Loss of significant amounts of vehicle from the topcoat into the pores of the primer can increase the PVC of the topcoat, and hence reduce its gloss.

6.3. Testing for Adhesion. Because of the wide range of exposures to stresses in actual use, there is no really satisfactory laboratory test for the adhesion of a coating film to the substrate during use. A useful guide for an experienced coatings formulator is the use of a penknife to see how difficult it is to remove the coating and to observe its mode of failure. Many tests for adhesion have been devised (48,49). From the standpoint of obtaining a measurement related to the work required to separate the coating from the substrate, the direct pull test is probably the most widely used. The accuracy of the test is subject to considerable doubt and the precision is not very good. Even for experienced personnel, reproducibility variations of 15% or more must be expected. A compressive shear delamination test has proven to be useful in studying the adhesion and cohesion properties of clear coat–base coat–primer coatings on TPO (47). The most common specification test, cross-hatch adhesion, is of little value beyond separating systems having very poor adhesion from others.

7. Corrosion Control

An important function of many coatings is to protect metals, especially steel, against corrosion. Corrosion protection is required to protect steel against corrosion with intact coating films. Another objective is to protect the steel against corrosion even when the film has been ruptured.

7.1. Protection by Intact Films. In the case of intact films, the key factors responsible for corrosion protection are adhesion of the coating to the steel in the presence of water, oxygen and water permeability, and the resistance of the resins in the film to saponification (50). If the resins in the coating are adsorbed to cover the steel surface completely and if the adsorbed groups cannot be displaced by water, oxygen and water permeating through the film cannot contact the steel and corrosion does not occur. Only a monolayer is required to protect against corrosion, if the monolayer stays in place over the period of exposure.

The factors affecting adhesion are critical in controlling corrosion. Clean surfaces are critical. Sandblasting to white metal provides a good surface for application of corrosion-protective coatings if the coating is applied over the sand-blasted surface before it can become contaminated. Conversion coating having insoluble phosphate crystals improves adhesion, and hence corrosion protection. It has been found that in automobiles the quality of the conversion coating, which is affected by the quality of the steel, is the most important variable in corrosion control. Viscosity of the external phase of the coating applied to the steel surface can be a critical factor. Because viscosity drops as temperature increases in greater penetration of vehicles, superior corrosion protection observed with baking coatings. In the case of corrosion control, penetration into the micropores and crevices is especially critical because if those surfaces do not have resin adsorbed on them when the water and oxygen permeate through the film, corrosion begins. Corrosion leads to a solution of ionic materials in water under the coating film, establishing an osmotic cell. The osmotic driving force accelerates the permeation of water and the osmotic pressure provides a further force to displace the film from the surface to form blisters.

Because water permeates through any coating film, it is desirable to have groups adsorbed on the surface that cannot be displaced by water. Although there is some controversy as to the mode of action, it is a common observation that resins having multiple amine groups along the chain give better corrosion protection than those having multiple hydroxy groups, perhaps because amines are less readily displaced from the surface of steel by water. Even though a particular interaction between a polar group on a resin molecule and the surface of the steel or the phosphate groups of the conversion coating is displaced by water, if there are multiple groups from the same resin molecule adsorbed, there can be a cooperative enhancement of adhesion. If one group desorbs, other groups that can do so readsorb the molecule onto the substrate surface. Phosphoric acid partial esters, as in epoxy phosphates, give enhanced adhesion and resistance to displacement by water. Resins that have saponifiable groups in the backbones are particularly likely to show poor wet adhesion.

If a coating had perfect wet adhesion, no other factors would affect corrosion protection, but frequently in practice such a high degree of wet adhesion cannot be attained. Corrosion protection is enhanced if the permeability of the film to oxygen and water is low. A significant factor controlling these permeabilities is the free-volume availability in the film, which is in turn related to T_g. From the point of view of corrosion protection, it is desirable to have the T_g of films at least 50°C higher than the service temperatures of the coatings. This condition is difficult, if not impossible, to achieve using ambient cure coatings and for the highest corrosion protection, baking coatings are desirable whenever it is feasible to use them.

Permeability is controlled not only by the diffusion rate but also by the solubility of the diffusing molecules. There can be significant variations of the solubility of water in various films. Films containing highly polar groups have the highest tendency to dissolve water. Films made with chlorinated polymers such as chlorinated rubber, vinyl chloride copolymers, and vinylidene chloride copolymers dissolve very small amounts of water and are widely used as vehicles for topcoats for corrosion protection coating systems. Permeability of coatings to oxygen and water is also reduced by pigmentation. Lowest permeability results from having a PVC close to but not above CPVC. Platelet pigments such as leafing aluminum pigments and mica can orient parallel to the surface of a film during solvent evaporation, and therefore act to improve the barrier properties of coatings.

7.2. Protection by Nonintact Films.   It is also possible to achieve corrosion control by coatings after a film has been ruptured. Because the coatings used to achieve this control generally give poorer protection when their films are not ruptured, such systems should be used only when film rupture must be anticipated or when complete coverage of the steel interface cannot be achieved. There are two techniques used on a large scale: primers containing corrosion inhibiting pigments and zinc-rich primers.

The mechanism of action of corrosion inhibiting pigments is not completely understood, but it is generally agreed that they promote oxidation at the surface of anodic areas of the steel to form a barrier layer. This action is called passivation. For the pigments to be effective, they must have a minimum solubility in water. If the solubility is too high, however, they would be rapidly lost from the film by leaching with water and thus provide only short-term protection. Because the pigments are somewhat water soluble, their presence in an intact film can lead to blistering and loss of adhesion. Therefore, it is undesirable to use such pigments except when the need for protection by nonintact films is important.

The oldest corrosion inhibiting pigments are red lead Pb_3O_4 [1314-41-6], containing ~15% PbO, and zinc yellow $3ZnCrO_4 \cdot K_2CrO_4 \cdot Zn(OH)_2 \cdot 2H_2O$, [85497-55-8], commonly, but mistakenly, called zinc chromate. There is concern about toxic hazard with red lead. Even more serious, soluble chromates, such as are present in zinc yellow, are carcinogenic. Because there are no satisfactory laboratory tests to predict the effectiveness of coatings containing corrosion-inhibiting pigments, extended field experience is necessary to determine the utility of new pigments. Zinc–calcium molybdates, surface-treated barium metaborate, zinc phosphate, zinc salts of nitrophthalic acid, and calcium–barium phosphosilicates, and phosphoborates are pigments that have been recommended as corrosion inhibitors.

Zinc-rich primers can be very effective in protecting steel with nonintact films against corrosion. High contents of zinc metal powder are required in the primers; PVC of the zinc powder must exceed CPVC to permit the necessary electrical contact between the zinc particles and between zinc and steel. Also having the PVC above CPVC makes the coating porous so that water can enter the film permitting completion of the electrical circuit. The zinc becomes the anode and the steel the cathode of an electrolytic cell, the zinc acts as a sacrificial metal to protect the steel. Because a base, $Zn(OH)_2$, is generated, saponification-resistant binders are required. The most widely used systems are inorganic zinc-rich primers, also called zinc silicate primers. The vehicle is an alcoholic solution of partially polymermized tetraethyl orthosilicate (silicic acid tetraethyl ester) [78-10-4]. After application, atmospheric moisture continues the hydrolysis and completes the polymerization of the tetraethyl orthosilicate and zinc salts of the polysilicic acid form. In applying topcoats, penetration into the pores of the primer must be avoided. Increasingly, latex paints are being used as topcoats because they do not permit vehicle penetration into the pores.

It is desirable to clean a steel surface thoroughly before applying a coating, but it is not always possible. Sometimes it is necessary to apply coatings over oily, rusty steel. In this case, it is essential to have a vehicle that can displace and dissolve the oily contamination from the surface and have sufficiently low viscosity for a sufficiently long time after application for penetration through the rusty areas down to the surface of the steel. Drying oil primers pigmented with red lead are still widely used for this purpose. They have the very low surface tension necessary for wetting, and can dissolve the oil and penetrate through the rust. Although their wet adhesion to steel and saponification resistance are inferior to those of many other primers, it is better to have some vehicle on the surface of the steel rather than having no vehicle penetrate down through the rust to the steel surface. This inadequacy is partly offset by the use of red lead as a corrosion-inhibiting pigment.

Formulation of effective corrosion-resistant coatings is made difficult by the lack of a laboratory test that can provide reliable predictions of field performance. The most widely used test is exposure in a salt fog chamber. It has been shown repeatedly, however, that the results of such tests do not correlate with actual performance (51). Outdoor exposure of panels can provide useful data, especially in locations where salt spray occurs, but predictions of performance are not always satisfactory.

Useful guidance in evaluating wet adhesion can be obtained by checking adhesion after exposure in a humidity chamber. In some cases, cathodic disbonding tests may provide useful data (ASTM Standards G8-79, G19-83, and G42-80). Another approach to testing for delamination is the use of electrochemical impedance spectroscopy (eis) (52). Impedance is the apparent opposition to flow of an alternating electrical current, and is the inverse of apparent coating capacitance. When a film begins to delaminate there is an increase in apparent capacitance. The rate of increase of capacitance is proportional to the amount of surface area delaminated by loss of wet adhesion. High performance systems show slow rates of increase of capacitance so tests must be continued for long time periods. This method can be a powerful tool for study of the effect of variables on delamination.

An extensive survey of accelerated test methods for anticorrosive coating performance that emphasizes the need to develop more meaningful methods of testing has been published (53). The most powerful tool available is the accumulated material in data banks correlating substrate, composition, application conditions, and specifics of exposure environments with performance.

8. Resins for Coatings

8.1. Latexes. Latexes are aqueous dispersions of solid polymer particles, generally made by emulsion polymerization. Latex paints are sometimes called emulsion paints but that terminology should be avoided since there are paints made with emulsions of resin solutions. Reference (54) provides a broad discussion of latexes. Molecular weights of latex polymers is generally high, >1,000,000. However, the high molecular weight of the latex polymer does not affect the viscosity of the latex, making it possible to formulate relatively high solids coatings. Thus, film properties of thermoplastic coatings can be good. Latex coatings have low VOCs. Aqueous solvent free polyurethane dispersions are latexes but almost always called polyurethane dispersions, they are discussed in the section on waterborne urethanes.

Latexes are the principal vehicle of a large fraction of architectural coatings and a small but rapidly growing fraction of industrial and special purpose coatings. The largest volume of latexes are polymers of acrylic esters and almost all will form films at ambient temperatures. Minimum film forming temperatures (MFFT) are primarily determined by $(T - T_g)$ but are usually lower than the T_g because of plasticizing by water. By changing the monomer feed during emulsion polymerization it is possible to make latexes whose particles have higher T_g values in the center of the particles and lower T_g values in the outer shell. This permits making latexes that will form films at lower temperature related to the shell T_g polymer but after film formation exhibit properties intermediate to those from the low T_g shell and the higher T_g core (55).

Acrylic latexes are widely used in exterior latex coatings and in higher performance interior coatings. Vinyl acetate based latexes are widely used in interior coatings such as flat wall paint because of their lower cost. These latexes are copolymers to reduce MFFT, eg, vinyl acetate–butyl acrylate copolymers. Latex coatings form films by coalescence (see the section on Film Formation from Polymer Dispersions). It has been common to reduce film formation temperature by including a coalescing solvent in a formula. That is a solvent that dissolves in the latex particles reducing their T_g that slowly diffuses and evaporates out of the film after application.

There is increasing use of thermosetting latexes, ie, latexes having functional groups that can be cross-linked after application. Such latexes are prepared with lower T_g to permit lower temperature film formation and then cross-linked after application to increase their modulus. Some are used in two package coatings, eg, hydroxy-functional latexes with urea–formaldehyde or melamine–formaldehyde resins. Carboxylic-functional latexes can be cross-linked with carbodiimides (16) or polyimines (56). Use of β-(3,4-epoxycyclohexyl)-ethyltriethoxysilane as a cross-linker for carboxylic acid funtional latexes gives formulations that are stable for one year and films that cure at 115° (19).

Thermosetting latexes stable enough at ambient temperature for use in architectural coatings are more difficult to make. Several workers have prepared latexes with allylic substiutents that cure by autoxidation, for an example, see (22). Latexes made using triisobutoxysilylpropyl methacrylate as a coomonomer cross-link in 1 week with a tin catalyst and package stable for over 1 year (22).

8.2. Amino Resins. Melamine–formaldehyde (MF) resins are the most widely used cross-linking agents for baking enamels. They are made by reacting melamine (1,3,5-triazine-2,4,6-triamine) [108-78-1] and formaldehyde[50-00-0] followed by etherification of the methylol groups using an alcohol. Two classes of MF resins are used. Class I resins are made using excess formaldehyde and a high fraction of the methylol groups are etherified with alcohol. Commercial resins contain a range of compounds having a large fraction of the amine groups substituted with two alkoxymethyl groups; some oligomers have two or more triazine rings coupled with methylene ether groups. Various alcohols can be used. The largest volume Class I resins contain a high proportion of hexamethoxymethylmelamine (HMMM) (N,N',N''-hexamethoxymelamine).

Class II MF resins are made using a lower ratio of formaldehyde to melamine and a significant fraction of the nitrogens have one alkoxymethyl group and a hydrogen. An example of one compound present is N,N',N''-trimethoxymethylmelamine (TMMM). Class II resins have a larger fraction of oligomers.

The MF resins react with both hydroxyl and carboxylic acid groups and are used to cross-link any resins having such substituents. They also undergo self-condensation reactions, the Class II more rapidly than Class I. Class I resins tend to give cross-linked films having greater toughness and flexibility but Class II resins tend to cure at lower temperatures. Class I resins require strong acid catalysts such as sulfonic acids whereas Class II resin reactions are catalyzed by weaker acids such as carboxylic acids.

Work has been concentrated on increasing the solids content of coatings. Class I resins have lower molecular weights than Class II resins and provide higher solids. Lowest molecular weight resins are produced under conditions that minimize self-condensation and maximize HMMM content. Mixed methyl butyl ether Class I MF resins have lower T_g and viscosity and lead to coatings having somewhat lower VOC emissions. A further emphasis in research is to minimize emission of free formaldehyde and generally Class I resins are superior in this respect.

In high solids coatings, especially those made with low average functionality polyester resins, good film properties require closer control of curing time and temperature, catalyst concentration, and ratio of reactants than do the older, lower solids, higher average functionality systems. The potential problem is especially notable using Class II MF resins; there is greater latitude using the higher functionality Class I MF resins. Although it had been believed that steric hindrance limited the reaction of Class I resins so that only about one-half of the potential functional groups reacted, complete reaction has been demonstrated (57). Stoichiometric amounts of Class I MF resin and hydroxyl groups react, but it is frequently desirable to use a higher ratio of MF resin. The excess MF resin gives coatings having a higher modulus above T_g, apparently by increasing the extent of self-condensation. Another consideration is that in many cases the MF resin is less expensive than the coreactant.

In water-borne coatings, methyl ether MF resins are preferred because of greater water solubility compared to that of higher alcohol derivatives. Although either Class I or II MF resins can be used in water-borne coatings, Class I resins are preferred because of better storage stability. In anionic electrodeposition coatings, mixed methyl ethyl ether resins are commonly used for cross-linking the epoxy ester or other binder resin. The water solubility of these mixed ether MF resins is high enough for them to be incorporated into the system but sufficiently limited that essentially all of the MF resin dissolves in the epoxy ester aggregates rather than staying in solution in the continuous phase. This is essential to maintain balanced composition during application by electrodeposition.

Because of the high content of nitrogens having two alkoxymethyl groups, Class I MF resins require strong acid catalysts such as p-toluenesulfonic acid (pTSA) (4-methylbenzenesulfonic acid) [104-15-4]. The higher the concentration of catalyst the lower the curing temperature of the coating. However, the catalyst remains in the film and can catalyze hydrolysis of the cross-link bonds on exterior exposure, leading to film failure. Addition of sulfonic acid decreases storage stability. In order to avoid this problem, it is common to use blocked catalysts that are generally salts of a sulfonic acid and a volatile amine. It has been found that some grades of titanium dioxide pigment neutralize the catalyst and slow down curing as the coating is stored for longer times.

The choice of sulfonic acid can be critical in affecting film performance. Dinonylnapthalenedisulfonic acid [60223-95-2] gives superior film properties as compared to pTSA, perhaps because of greater solubility in the coating. Dodecylbenzenesulfonic acid [27176-87-0], which is also more soluble, gives superior properties when the coating is applied over primer, but poor adhesion results when applied directly to steel. Apparently the sulfonic acid is strongly adsorbed on the steel surface and the adhesion of the coating to this monolayer of catalyst is poor.

Other amino resins are used to a lesser degree in coatings. Urea–formaldehyde (UF) resins are used in some coatings for wood furniture because these resins cross-link at lower temperatures than MF resins and the higher water resistance and exterior durability that can be obtained using MF resins are not needed. Ethers of formaldehyde derivatives of 6-phenyl-1,3,5-triazine-2,4-diamine (benzoguanamine resins) [91-76-9] give coatings having superior resistance to alkalies and detergents and are used as cross-linkers for coatings used on home laundry machines and dishwashers. Ethers of formaldehyde derivatives of glycoluril (tetrahydroimidazo[4,5-d]imidazole-2,5-(1H,3H)-dione) [496-46-8] are useful for coatings requiring outstanding extensibility during forming. Amides react with formaldehyde to give methylol derivatives that in turn react with alcohols to give reactive ethers. Thermosetting acrylics having such reactive groups derived from acrylamide were among the first thermosetting acrylic resins.

8.3. Urethane Systems.   Isocyanates react with a wide variety of functional groups to give cross-links. The most widely used coreactants are hydroxy-functional polyester and acrylic resins. The isocyanate group reacts with the hydroxyl group to generate a urethane cross-link. The reaction proceeds relatively rapidly at ambient or modestly elevated temperatures. In addition to the low curing temperatures, significant advantages of urethane coatings are generally excellent abrasion and impact resistance combined with solvent resistance.

A variety of polyisocyanates are commercially available; those most widely used in coatings are toluene diisocyanate (TDI) (2,4-diisocyanato-1-methylbenzene) [584-84-9], 4, 4′-di(isocyanatophenyl)methane (MDI)(bis(4-isocyanatophenyl)-methane); HDI; isophorone diisocyanate (IPDI) (1-isocyanato-3-(isocyanato-methyl)-3,5,5-trimethylcyclohexane) [4098-71-9]; 4,4′-di(isocyanatocyclohexyl) methane (HMDI)[bis(4-isocyanatocyclohexyl)methane]; and TMXDI. Aromatic isocyanates such as TDI give films that discolor rapidly on exterior exposure. Aliphatic isocyanates give coatings having excellent exterior durability. TMXDI is classified as an aliphatic isocyanate because the NCO group is not directly on an aromatic ring.

Most diisocyanates are toxic, therefore many times it is safer to use polyisocyanates. The lower vapor pressure and skin permeability of trifunctional and polyfunctional isocyanates reduces the toxic hazard but, as for any other reactive coating, caution should still be exercised, especially when coatings are applied by spray techniques.

TDI and TMXDI are available as prepolymers derived by reaction of excess TDI or TMXDI with TMP and removing the excess diisocyanate by vacuum thin-film evaporation. The average NCO functionality is a little over 3. The polyisocyanates TDI, HDI, and IPDI are available as trimers having an isocyanurate ring, ie, substituted 1,3,5-triazine-2,4,6-($1H,3H,5H$)-triones. The average functionality is somewhat over 3. HDI is also available as a biuret derivative having an average functionality of $\sim$3. Polyisocyanates can also be made by the copolymerization of TMI [1-(1-isocyanto-1-methylethyl)-3-(1-methylethenyl)benzene] [2094-99-7] and acrylic esters (58).

The reaction of isocyanates and alcohols is too rapid to permit formulation of one-package stable coatings from polyols and polyisocyanates. Thus two-package coating systems are used on a large scale: one package contains the polyisocyanate and the other a polyhydroxy resin and the pigment(s). Reaction rates are controlled by catalysts. The most widely used catalysts are organotin compounds, most commonly dibutyltin dilaurate (DBTDL) [dibutyl bis[1-(oxododecyl)oxy]-stannate] [77-58-7]. Highly catalyzed systems cure rapidly at moderate temperature but have pot lives so short that they must be used in proportioning mixing spray guns so that the two packages are mixed just before spraying. Slower cure systems may have pot lives of several hours. Any polyhydroxy resin can be used; the most widely used are polyesters and acrylics.

A principal challenge is formulating to maximize pot life while still curing in a short time at low temperature, which is an increasing problem as coatings are formulated to higher and higher solids. The difficulty can be severe because in uncatalyzed systems, reaction rates are second order in respect to alcohol, so that the reaction rate slows down dramatically as the concentration of alcohol groups decreases. Use of tin catalysts like DBTDL is especially desirable because the catalyzed reaction rate with respect to alcohol has been shown to be one-half (59). By using DBTDL, pot life can be extended by including a volatile carboxylic acid, such as acetic acid [64-19-7], in the formula. After application the acetic acid evaporates and cure rate approaches that without the acetic acid. Similarly, 2,4-pentanedione increases pot life. The reaction rate of isocyanates and alcohols is affected by the media. Rates are slowest in strong hydrogen bond acceptor solvents and most rapid in hydrocarbon, especially aliphatic hydrocarbon, solvents. Therefore, hydrogen-bonding solvents and resins having as low hydrogen bond potential as possible are used (60).

In formulating air-dry coatings, one should select combinations of polyol and polyisocyanate such that the T_g permits complete reaction at curing temperature. Usually, some excess isocyanate is used because some isocyanate also reacts with water from the atmosphere. In aircraft coatings, it is common to use a 2:1 ratio of isocyanate to hydroxyl. The excess isocyanate reacts with water to yield an amine that reacts very rapidly with another molecule of isocyanate to give a urea cross-link.

Aliphatic amines in general react to rapidly with isocyanates to be used even in 2K coatings. However, sterically hindered amines have been developed that are useful. Michael addition products of diethyl maleate with a diamine such as bis(4-amino-3-methylcyclohexyl)methane, a diaspartate, is an example. They permit formulation of very high solid 2K coatings with excellent pot life, cure rate, and film properties (61). The reaction is catalyzed by carboxylic acids and water, organotin compounds retard the reaction. Dialdimines can also be used in 2K coatings. The aldimine groups are sufficiently stable to hydolysis that they react directly with isocyanates to give cyclic unsaturated substituted ureas (61).

Another class of urethane coatings is known as moisture cure coatings. These are stable 1K coatings based on isocyanate-terminated resins as the sole vehicle. The cross-links formed are urea groups. Because pigments have water adsorbed on the surface of the particles, these coatings are almost always clears, because the cost of drying pigment is usually prohibitive. Moisture cure clear coatings are used where high abrasion resistance is needed, such as on floors, bowling alleys, and pins.

Isocyanates are also used to make coating resins that do not cross-link through reactions of the isocyanate groups. These have the advantage that the toxic hazards associated with isocyanates are handled in the resin factory rather than by the coatings applicator. One class of such products is urethane oils, also called urethane alkyds and uralkyds. A diiosocyanate such as TDI reacts with partial glycerol esters of drying oils to yield polyurethanes analogous to alkyd resins. Like alkyds, urethane oils form dry films by autoxidation. The dry films exhibit superior resistance to hydrolysis and abrasion as compared to alkyd films. Many coating products for the do-it-yourself market labeled as varnish are urethane oils. Their properties are generally superior to varnishes as well as alkyds.

Carbamate-functional resins made by non-isocyanate processes are also used. Carbamate-functional acrylic resins cross-linked with MF resins are used in automotive clear coats that combine the environmental etch resistance of urethane coatings and the mar resistance of MF cross-linked hydroxy-functional acrylic resins coatings (62). In another approach, a carbamate-functional oligomer was synthesized by reacting a hyrdoxy acid, such as dimethylolpropionic acid (DMPA), with glycidyl neodecanate (63). Automotive clear coats formulated with Class I MF resin showed excellent performance. 85% NVV (non-volatile volume %) coatings could be applied with hot spray.

High molecular weight, linear, hydroxy-terminated polyurethanes are used in lacquers for topcoats for coated fabrics. Low molecular weight, hydroxy-terminated polyurethanes can be used with MF resins to replace polyesters to take advantage of the greater hydrolytic stability of the urethane groups as compared to ester groups and the greater abrasion resistance of the urethane coatings. These coatings are more expensive than polyesters, especially when made with aliphatic diisocyanates, and tend to give higher viscosity solutions than polyesters.

Blocked Isocyanates. Another important class of urethane coatings is based on blocked isocyanates. See (64) for an extensive review of the chemistry and applications for blocked isocyanates. Blocked isocyanates are made by reaction of a diisocyanate and a blocking agent giving a product that is quite stable in the presence of alcohols and water at ambient temperature, but reacts with hydroxyl or amine groups at elevated temperatures. Using blocked isocyanates and hydroxy-functional acrylic or polyester resins, stable one-package coatings can be formulated without need for moisture-free pigments and solvents and with substantially reduced toxic hazard. However, the coatings require relatively high temperature cures and, in some cases, the blocking agent that evolves during cure presents a pollution problem.

MEK oxime (MEKO) (2-butanone oxime) [96-29-7], phenols, alcohols, and caprolactam (hexahydro-2H-azepin-2-one) [105-60-2] are widely used as blocking agents. Phenol blocked isocyanates are used in wire coatings, curing temperatures are high. Alcohols are particularly useful in electrodeposition coatings since they provide excellent hydrolytic stability in the coating bath. The MEKO blocked isocyanates react with alcohols at 130–140°C in the presence of a catalyst, but still give reasonable storage stability.

Caprolactam blocked IPDI is widely used in powder coatings. Oligomeric uretdiones are increasingly used in powder coatings because there are no volatile by products. The reaction products of diisocyanates and diethyl malonate (propanedioic acid diethyl ester) [105-53-3] act as cross-linking agents for polyols at the lowest temperature of any commercial blocked isocyanates but they are not truly blocked isocyanates. Rather, they react with alcohols by transesterification to yield films cross-linked with ester and amide bonds instead of urethane cross-linked coatings (65).

Polyamines react with blocked isocyanates at lower temperatures than polyols. Oximes are used as blocking agents with polyamines in applications such as magnetic tape coatings where the curing temperature must be kept low to avoid heat distortion of the plastic tape substrate. The reactivity is too high to permit one-package coatings but the combination of amine and blocked isocyanate provides an adequate pot life to permit application by roll coating. Alcohol-blocked isocyanates are sufficiently stable at ambient temperatures in the presence of amines to be used in one-package coatings. They are used in cationic electrodeposition coatings where there is also need for long term stability in the presence of large amounts of water.

Waterborne Urethane Systems. There has been increasing effort to use waterborne urethane coatings. One approach is preparation of aqueous dispersions of polyurethanes. These materials are latexes but they are almost always called polurethane dispersions (PUDs). Most of them are acid substituted resins neutralized with an amine such as triethyl amine. Reference (66) reviews various approaches. Both thermoplastic and thermosetting PUDs are used. In such applications as fabric coatings and chip resistant primers for automobiles.

In recent years, increasing effort has been devoted to 2K waterborne urethane coatings. It had been assumed for many years that using free polyisocyanates in a waterborne coating was not feasible due to the high reactivity of isocyanates with water. Selection of components, formulation, and application methods are critical. They are now commercial on a large scale. Thermosetting

PUDs have hydroxyl groups as cross-link sites. For example, hydroxy-terminated PUDs using DMPA as one of the polyols in the reaction. The carboxylic
acid DMPA is so hindered that it reacts very slowly with the isocyanate. They
have excellent hydrolytic stability compared to water-reducible polyesters and
superior abrasion resistance. They can be cross-linked with polyisocyanates or
MF resins. Bayer Corporation was awarded a Presidential Green Achievement
Award in 2000 for their work with the systems. Reference (67) reviews the
literature.

8.4. Epoxy Resins. Epoxy resins are used to cross-link other resins
with amine, hydroxyl, carboxylic acid, and anhydride groups. The epoxy group,
properly called an oxirane, is a cyclic three-membered ether group. By far the
most widely used epoxy resins are bisphenol A (BPA) [4,4-(1-methylethylidene)-
bisphenol] [80-05-7], epoxy resins.

An important use for epoxy resins is as a component in two-package primers for steel. One package contains the epoxy resin and the other a polyfunctional amine. In coatings, generally low molecular weight polyamines are not
useful because the equivalent weight is so low that the ratio of the two components would be very high, increasing the probability of mixing ratio errors;
furthermore, low molecular weight amines tend to have greater toxic hazards.
Amine-terminated polyamides are widely used, they are sometimes called
amido-amines but frequently just polyamides. Amide groups do not react readily
with epoxy groups. Polyamides are made from long chain dibasic acids, such as
dimer acids, and a stoichiometric excess of a polyamine, such as diethylenetria-
mine (DETA) [N-(2-aminoethyl)-1,2-ethanediamine] [111-40-0]. The long aliphatic chain contributes flexibility to the cross-linked film. If more rigid films are
needed, so-called amine adducts are used. Amine adducts are the reaction products of a low molecular weight BPA epoxy resin and a large excess of polyamine
from which any unreacted polyamine is removed by vacuum thin-film evaporation. The product is a polyamine having substantially higher molecular and
equivalent weight and rigid aromatic rings. In designing combinations of
epoxy resin and amine for ambient temperature cure coatings, care must be
taken so that the T_g of the fully cured coating is not too much over the temperature at which the film is to be cured, otherwise, the cross-linking stops because of
restricted mobility in the film. Epoxy–amine coatings are particularly effective
as corrosion protective primers for steel because the amine groups resulting
from the cross-linking reaction promote adhesion to the steel in the presence
of water and because the cross-linked resins are completely resistant to
hydrolysis.

Epoxy resins also are used to cross-link phenolic resins. Such coatings are
used in interior can linings. The hydrolytic stability and adhesion of the coatings
are critical. Adhesion is further improved by the incorporation of a small amount
of epoxy phosphate in the coatings. Epoxy phosphates are made by reacting BPA
epoxy resins and small amounts of phosphoric acid and water. Complex reactions
occur including formation of partial phosphate esters of a primary alcohol from a
ring opening reaction with an epoxy group (68).

Epoxy resins are widely used in powder coatings. Probably the largest
volume usage is of BPA epoxy resins cross-linked with dicyanodiamide (cyano-
guanidine) [461-58-5]. Because BPA epoxy resins are easily photoxidized, they

are not useful in coatings requiring exterior exposure. Triglycidylisocyanurate (TGIC) [1,3,5-tris(oxiranylmethyl)-1,3,5-triazine-2,4,6($1H,3H,5H$)-trione] [2451-62-9] has been used carboxylic acid terminated polyesters in powder coatings that require exterior durability.

Epoxy resins are raw materials to make epoxy esters by reacting BPA epoxy resins and drying oil fatty acids; each epoxy ring can potentially react with two fatty acid molecules and hydroxyl groups on the backbone of the epoxy resin can also esterify. A low molecular weight BPA resin is initially reacted with BPA to make a higher molecular weight BPA resin and then the fatty acid is added directly to the reactor for the esterification step. Lower molecular weight resins give lower viscosity epoxy esters but the functionality is also low so that more rapid cross-linking is obtained for epoxy esters based on higher molecular-weight BPA resins. These synthetic drying oils provide coatings having better adhesion and substantially better saponification resistance than alkyd resins. Exterior durability of these materials is poor, and the principal use is in primers for steel.

Epoxy esters of fatty acids having some conjugated double bonds can be converted to water-reducible resins by reaction with maleic anhydride (2,5-furandione) [108-31-6] followed by amine neutralization. Baking primers made with such resins provide the same corrosion protection as conventional solventborne epoxy ester primers. For some years, this type of vehicle was used in anionic electrodeposition primers. However, for automobiles cationic electrodeposition primers are now preferred. The vehicles for cationic primers are proprietary but consist of the reaction product of epoxy resins and polyfunctional amines.

8.5. Acrylic Resins.   Acrylic resins are the largest volume class of coatings resins. Thermosetting acrylic resins are copolymers of acrylic or methacrylic esters and a hydroxy-functional acrylic ester. Other monomers such as styrene (ethenylbenzene) [100-42-5], vinyl acetate, and others, may be included in the copolymer. There is increasing emphasis on high solids and water-reducible types. The main advantages of acrylic coatings involve the high degree of resistance to thermal and photoxidation and to hydrolysis, giving coatings that have superior color retention, resistance to embrittlement, and exterior durability.

Hydroxy-functional thermosetting acrylics are widely used in baking enamels for automobile and appliance topcoats, exterior can coatings, and coil coating. Research efforts have been directed at increasing the solids content of such coatings while maintaining the excellent properties. In contrast to polyesters, where virtually all molecules have at least two hydroxyl groups, synthesis of very low molecular weight acrylic resins having an average functionality of two to three and containing few molecules that are nonfunctional or only monofunctional is difficult. Free-radical polymerization, the usual method for synthesizing thermosetting acrylics, results in a random distribution of the 2-hydroxyethyl methacrylate (HEMA) (2-methyl-2-propenoic acid 2-hydroxyethyl ester) [868-77-9] comonomer in the oligomer chains, and hence significant fractions of nonfunctional and monofunctional molecules unless the number average molecular weight is on the order of 3500 or higher and the average functional monomers per molecule is on the order of three or higher.

Various techniques have been studied to increase solids content. Hydroxy-functional chain-transfer agents, such as 2-mercaptoethanol [60-24-2], reduce the probability of nonfunctional or monofunctional molecules, permitting lower molecular-weight and functional monomer ratios. Making low viscosity acrylic resins by free-radical initiated polymerization requires the narrowest possible molecular weight distribution. This requires careful control of temperature, initiator concentration, and monomer concentrations during polymerization. The initiator structure may be critical; it should yield radicals that are least active in hydrogen-abstraction reactions. Azo initiators have been widely used. Most so-called high solids acrylic resin coatings contain only 45–50 vol% solids. However, there are proprietary acrylic resins available that can be used to make coatings having as high as 70 vol% solids.

Other functional groups than hydroxyls are also used. Carboxy acid functional acrylic resins are used with epoxy compounds as cross-linkers. Epoxy-functional acrylic resins are being used in powder clear coats for automobiles cross-linked with dicarboxylic acids such as dodecane dicarboxylic acid (69). Carbamate-functional acrylic resins are being used in automotive clear coats are cross-linked with MF resins, such coatings combine the mar resistance of urethane coatings and the environmental etch resistance of MF cross-linked resins (62).

Acrylic resins are, in general, more appropriate than polyesters for water-reducible baking coatings. Acrylic copolymers using acrylic acid(2-propenoic acid) [79-10-7] and HEMA as functional comonomers, prepared in an ether–alcohol solvent and partially neutralized with an amine, such as dimethylaminoethanol (DMAE) [108-01-0], are stable against both hydrolysis and alcoholysis. After pigmenting and adding MF resin, the coatings can be diluted with water and applied with relatively low VOC. Although such systems are commonly called water soluble, these acrylic resins are not truly soluble in water. On dilution with water, adequately stable dispersions of aggregates swollen with solvent and water are formed. As a result of the formation of these aggregates, the change in viscosity on dilution with water is abnormal. Initially the viscosity decreases; however, on further dilution, the viscosity increases. On still further dilution, the viscosity decreases rapidly. The pH after dilution is also abnormal, being on the order of 9 even with only 75% of the amount of amine required to neutralize the carboxylic acid groups on the resin. The viscosity at application solids is essentially independent of molecular weight. The morphology of such systems and the effects of variables on the abnormal viscosity and pH profiles obtained on dilution with water have been studied (70,71). Class I MF resins or polyisocyanates are used as cross-linking agents.

Another type of water-reducible acrylic resins is used in the interior linings of two-piece beverage cans. A graft copolymer of styrene, ethyl acrylate(2-propenoic acid ethyl ester) [140-88-5], and acrylic acid on a BPA epoxy resin is prepared in an ether–alcohol solvent (72). Benzoyl peroxide(dibenzoyl peroxide) [94-36-0] is used as the initiator that maximizes the opportunity for grafting. The product is a mixture of graft copolymer, unreacted epoxy resin, and ungrafted acrylic copolymer. The resin is partially neutralized with DMAE and diluted with water to spray application viscosity. Dilution with water gives a stable dispersion of resin aggregates swollen with solvent and water.

8.6. Polyester Resins. The term polyester is used in the coatings field almost entirely for low molecular weight hydroxy, or sometimes carboxylic acid, terminated oil-free polyesters. Polyesters have been a class of replacements for alkyd resins in MF cross-linked baking enamels. Hydroxy-terminated polyesters are also used with polyfunctional isocyanates in making air-dry and force-dry coatings as well as with blocked isocyanates in coatings for higher baking temperature and in powder coatings. Carboxylic acid-terminated polyesters are used predominantly with epoxy cross-linkers in powder coatings.

When adhesion directly to metal is required, polyesters are generally preferred over acrylic resins. When highest exterior durability is needed over primers, acrylic resins are generally preferred over polyesters because of the greater hydrolytic stability of acrylic resin coatings. In order to maximize the hydrolytic stability of polyesters for exterior durability, isophthalic acid (IPA) is commonly used instead of phthalic anhydride (PA) (1,3-isobenzofurandione) [85-44-0] in making these resins. An aliphatic acid, such as adipic acid (AA) (1,6-hexanedioic acid) [124-04-9] or dimer acids, is commonly used. The flexibility of the final film is partly controlled by the ratio of aromatic to aliphatic dibasic acids. The esters of highly substituted polyols such as neopentyl glycol (NPG), 1,4-dimethylolcyclohexane (cyclohexanedimethanol) (CHDM) [27193-25-5] and TMP are more hydrolytically stable than esters of simple polyols such as ethylene glycol. Mixtures of diol and triol are used to give an average functionality >2. The cross-link density of the final film is controlled by the equivalent weight, which is related to the average functionality and the number average molecular weight.

There has been significant progress in the development of polyesters for high solids coatings. In contrast to acrylic resins, the preparation of low molecular weight, and hence high solids polyesters, where substantially all of the molecules have a minimum of two hydroxyl groups is straightforward. The lower limit of the average molecular weight that is useful in baking systems is controlled by the volatility of the lowest molecular weight fractions. For conventionally prepared polyesters, the optimum number average molecular weight for baking enamels has been reported to be 800–1000 (73). Further improvements are achieved by synthetic techniques that give narrower molecular weight distributions (74). Because viscosity is increased by increasing the average number of functional groups per molecule, high solids polyesters are usually made with an average number of hydroxyl groups only a little over two. Low molecular weight, hydroxy terminated polyester diols and triols derived from the reaction of caprolactone (2-oxepanone) [502-44-3] and a diol or triol are commercially available, eg, Tone Polyols (75). Linear polyesters made by transesterification of 1,4-butanediol with a mixture of glutarate, adipate, and azeleate methyl esters give low viscosity, hydroxy-terminated polyesters with an average molecular weight of 680 (76). Such polyesters are used in mixtures having hydroxy-functional acrylic resins or somewhat higher molecular weight polyesters to increase solids.

Water-reducible polyester resins have terminal hydroxyl and carboxylic acid groups and an acid number of 40–60. To make such a resin reproducibly with a minimum risk of gelation, the reactivity of the different carboxylic acid groups must vary. The use of trimellitic anhydride (TMA) (1,3-dihydro-1,3-dioxo-5-isobenzofurancarboxylic acid) [552-30-7] near the end of the reaction at a lower temperature takes advantage of the higher reactivity of the anhydride group. Another method uses DMPA [3-hydroxy-2(hydroxymethyl)-2-methylpro-panoic acid] [4767-03-7] as part of the diol in making a polyester. The highly hindered carboxylic acid esterfies more slowly than the carboxylic acid groups from isophthalic and adipic acids.

The polyester is dissolved in an ether–alcohol such as 1-propoxy-2-propanol [1569-01-3]. A secondary alcohol should be used because primary alcohols lead to more rapid transesterification than secondary alcohols. An amine such as DMAE is used to neutralize the acid, pigment is dispersed in the resin solution, MF resin is added as are additives, and the coating is reduced to application viscosity with water. The resin forms aggregates swollen with solvent and water as a reason-ably stable dispersion. Use of cyclohexane dicarboxylic acid instead of isophthalic acid and 2-butyl-2-ethyl-1,3-propanediol instead of NPG gives a resin with some-what better hydrolytic stability (77). The storage life of such systems is adequate with careful invenory control. The problem of hydrolysis can be minimized by use of a powdered solid polyester. For example, a powdered solid ester prepared from IPA, AA, NPG, CHDM, hydrogenated BPA, and TMA can be stored and stirred into a hot aqueous solution of DMAE shortly before the coating is required (78).

Another potential problem with water-reducible polyesters is formation of some low molecular weight nonfunctional cyclic molecules. In baking ovens, small amounts of such polyesters gradually accumulate by condensation in cool spots in the oven and eventually sufficient resin can accumulate to drip on pro-ducts passing through the ovens. Water-reducible acrylic resins are used in much greater volumes because of the problems with water-reducible polyesters.

8.7. Alkyd Resins. Although no longer the principal class of resins in coatings, alkyds are still important and a wide range of types of alkyds are man-ufactured. Whereas some nonoxidizing alkyds are used as plasticizers in lac-quers and cross-linked with MF resins in baking enamels, the majority are oxidizing alkyds for use in coatings for air-dry and force-dry applications. The principal advantages of alkyds are low cost and relatively foolproof application characteristics, resulting from low surface tensions. The principal shortcomings of these resins are embrittlement and discoloration upon aging and relatively poor hydrolytic stability.

Emphasis is being placed on developing high solids alkyd systems, eg, by using longer oil alkyds. However, as oil length increases, air drying slows for two reasons: the average number of functional groups, ie, activated methylene groups between two double bonds, eg, $-CH=CHCH_2CH=CH-$, per molecule decreases because the molecular weight decreases, and the ratio of aromatic rings to long aliphatic chains is reduced, resulting in lower T_g. Solids can be increased at the same average molecular weight by reducing polydispersity, ie, the breadth of molecular weight distribution. However, the properties of films from such narrow molecular weight distribution alkyds are inferior to those of broader molecular weight distribution resins of similar composition (79). One approach to high solids alkyds is the partial replacement of solvent in a with a reactive diluent. For example, dicyclopentenyloxyethyl methacrylate [70191-60-5] (80), mixed amides from the condensation of acrylamide (2-propeneamide) [79-06-1], drying oil acid amides with hexamethoxymethylmelamine [3089-11-0], and trimethylolpropane triacrylate [37275-47-1] have been recommended.

Water-reducible alkyds give aqueous dispersions by neutralizing the free carboxylic acid groups with an amine or ammonia. Comparable drying performance to solvent-borne alkyds can be obtained. However, they are not widely used because of limited hydrolytic stability that reduces the storage life of coatings. Also, film properties tend to be poorer than those of solvent-borne alkyds, especially in air-dry systems. The slow loss of amine or ammonia leads to short term high sensitivity to water. Even in the fully dry films, the presence of unreacted carboxylic acid groups leads to films having comparatively poor water resistance limiting their usefulness.

Another approach to waterborne alkyds, more widely used in Europe than in the United States, is to emulsify alkyds (81). The emulsions are stabilized with surfactants and can be prepared with little, if any, solvent.

8.8. Other Coatings Resins. A wide variety of other resin types are used in coatings. Phenolic resins, ie, resins based on reaction of phenols and formaldehyde, have been used in coatings for many years. Use has been declining but there are still significant applications, particularly with epoxy resins in interior can coatings.

Silicone resins provide coatings having outstanding heat resistance and exterior durability. The cost is relatively high but for specialized applications these resins are important binders (82). Silicone-modified acrylic and polyester resins provide binders having intermediate durability properties and cost. Fluorinated polymers also exhibit outstanding durability properties. High cost, however, limits applicability.

Polyfunctional 2-hydroxyalkylamides serve as cross-linkers for carboxylic acid-terminated polyester or acrylic resins (83). The hydroxyl group is activated by the neighboring amide linkage. Solid grades of hydroxyamides are used as cross-linkers for powder coatings.

A range of acetoacetylated resins has been introduced (84). The acetoacetoxy functionality can be cross-linked with melamine–formaldehyde resins, isocyanates, polyacrylates, and polyamines. There is particular interest for possible corrosion protection on steel because the acetoacetoxy group can form coordination compounds with iron, perhaps enhancing the adhesion to steel surfaces.

9. Volatile Components

In most coatings, solvents are used to adjust the viscosity to the level required for the application process and to provide for proper flow after application (85). Most methods of application require coating viscosities of $50-1000$ mPa·s($=$cP). Many factors must be considered in the selection of the solvent or, more commonly, solvent mixtures. Except for water-borne systems, solvents are usually chosen that dissolve the resins in the coating formulation. Solubility parameters have been recommended as a tool for selecting solvents that can dissolve the resins. However, the concept of solubility parameters for the prediction of polymer solubility is an oversimplification and the old principle that like dissolves like is the most useful selection criterion. The problem of solvent selection is most difficult for high molecular weight polymers such as thermoplastic acrylics and nitrocellulose in lacquers. As molecular weight decreases, the range of solvents in which resins are soluble broadens. Even though solubility parameters are inadequate for predicting all solubilities, they can be very useful in performing computer calculations to determine possible solvent mixtures as replacements for a solvent mixture that is known to be satisfactory for a formulation. For resins that are soluble in esters and ketones, costs can generally be decreased by using mixtures that contain hydrocarbons and alcohols to replace part of the esters and ketones.

An important characteristic of solvents is rate of evaporation. Rates of solvent loss are controlled by vapor pressure and temperature, partial pressure of the solvent over the surface, and thus the air-flow rate over the surface, and the ratio of surface area to volume. Tables of relative evaporation rates, in which n-butyl acetate is the standard, are widely used in selecting solvents. These relative rates are determined experimentally by comparing the times required to evaporate 90% of a weighed amount of solvent from filter paper under standard conditions as compared to the time for n-butyl acetate. Most tables of relative evaporation rates are said to be at 25°C. This, however, means that the air temperature was 25°C, not that the temperature of the evaporating solvent was 25°C. As solvents evaporate, temperature drops; the drop in temperature is greatest for solvents that evaporate most rapidly.

When coatings are applied by spray gun the atomized particles have a very high ratio of surface area to volume, and hence solvent evaporation is much more rapid during the time when the atomized particles are traveling from the orifice of the spray gun to the surface being coated than from the film of the coating on the surface. Except in the case of some high solids coatings, this rapid evaporation of solvent during the atomized stage permits formulation and spray application of coatings that level well yet do not sag on vertical surfaces. By adjusting the combination of solvents to the particle size of the droplets obtained with a particular spray gun and the distance between the spray gun and the object being sprayed, the viscosity of the coating arriving at the surface can be adjusted to stay low enough to permit leveling, yet increase rapidly enough to avoid, or at least minimize, sagging. The control of sagging and leveling is also affected by film thickness. Because a thin film has a higher surface area to volume ratio than a thicker film, the concentration of solvent left in a thin film decreases more rapidly with time than the concentration in a thick film. The variations in solvent evaporation are so dependent on the particular application conditions that tables of relative evaporation rate are only useful as general guidelines. Final adjustment of solvent selection must be done under actual field conditions.

The loss of solvent during spray application of high solids coatings is usually less than that from conventional lower solids coatings. As a result, it is difficult to control sagging during spray application of many high solids coatings. The lower rate of solvent loss has not been fully explained. Because of the lower molecular weight and higher concentrations of resins in high solids coatings, the ratio of numbers of solvent molecules to resin molecules is lower than in lower solids coatings. Hence, vapor pressure depression is greater in the case of high solids coatings. However, this difference would not seem to be large enough to account for the very large differences in solvent losses that have been reported. Another factor may be that high solids coatings may reach a stage where solvent loss is controlled by diffusion rate much earlier than is the case in low solids coatings.

Toxic hazards, environmental considerations, flammability, odor, surface tension, and viscosity also affect solvent selection. In high solids coatings, the effect of solvent choice on viscosity can be critical. Because the resins used in high solids coatings tend to have polar functional groups, it is usually desirable to use hydrogen bond acceptor solvents so as to minimize hydrogen bonding between resin molecules, which tends to increase viscosity. Because VOC emission regulations are based on mass of solvent per unit volume of coating, it is important to take solvent density into consideration in comparing the effect of solvent selection on viscosity and choice of solvents.

10. Color and Appearance

Color and the related property of gloss are important to the decorative and, sometimes, functional aspects of coatings. Discusssion of the interrelationships between the three major variables, the object, the light, and the observer, that affect color is beyond the scope of this article. See (86) and the references cited therein for a broad discussion. The discussion of color here is limited to color matching.

10.1. Color Matching. Most pigmented coatings must be color matched to a standard. Poor color matching is a common source of customer complaints and problems and costs can be minimized if established specifications for the initial color match and for judging the acceptability of production batches are made. Acceptance of a color recommendation made by the coatings supplier effectively eliminates the time and cost involved in an initial color match and ensures selection of a pigment combination appropriate to the coating use. If, however, a coatings user provides a sample or standard for a color match, formulators need the following information.

1. *Possibility of a Spectral (Nonmetameric) Match*. Only if exactly the same pigments, including white and black if needed, can be used in establishing the new coating as were used in the customer's sample, can the fraction of light absorbed at each wavelength be identical to the sample. Although a color can usually be matched under one light source using different pigment combinations, it is only possible to match the color under all light sources if chemical compositions are identical. If this is not possible, the user must accept a metameric match; ie, the colors match under some light source but not under others. For example, if the coatings user has been using a coating that has lead-containing pigments and wishes to have a lead-free paint, only a metameric match is possible.

2. *Light Sources*. If the match is to be metameric, the coatings user and the supplier must agree on the light source(s) under which the color is to be evaluated. Furthermore, a decision should be made whether it is more desirable to have a close match under one light source without regard to how far off that match might be under other light sources, or to have only a fair match under several light sources.

3. *Gloss and Texture.* The color of a coating depends in part on gloss and surface texture. The light reaching the eye of an observer has been reflected both from the inside and the surface of the coating film. The light from within the film is "colored light", whereas the light reflected from the surface of the film is "white light". The color seen by the observer varies with the ratio of light from within and that from the surface of the film. At most angles of viewing, more light is reflected from a low gloss surface than from a high gloss surface. Therefore, at most angles of viewing, a low gloss coating having exactly the same colorant combination as a high gloss coating has a lighter color than the high gloss coating. It is impossible to match the colors of high gloss and low gloss coatings at all angles of viewing. Therefore, there must be agreement as to the gloss, and if the gloss of the color standard is different from the gloss desired for the new coating, the angles of illumination and viewing must be agreed upon. It is not possible to make even a metameric match of the color of some fabric samples with a coating under all angles of viewing because the colorants as well as the surface texture must be different. Claims to the contrary are misleading.

4. *Color Properties Required.* Colorants must be chosen that can meet performance requirements such as exterior durability and resistance to solvents, chemicals, and heat. Health and safety regulations may also affect colorant choice.

5. *Film Thickness and Substrate.* In most cases, a coating does not completely hide the substrate, and the color of the substrate affects the color of the applied coating. The extent of substrate effect depends on film thickness and is particularly important in applications such as can and coil coatings that are commonly <25 μm thick. A thin coating where the color was established over a gray primer does not match a standard applied over a red primer; a coating designed for a one-coat application on aluminum does not match the color standard when applied on steel.

6. *Baking Schedule.* Because resin color can be affected by heating, the color of the coating is affected by the time and temperature of baking, and the baking schedule must be specified. Color requirements for overbaking must also be established.

7. *Cost.* The color matcher should know any cost limitations. An important element affecting cost is the tolerance limits permitted in production.

8. *Tolerances.* The closeness of the color match required is important. In some cases, such as coatings for exterior siding or automotive topcoats, very close matches are required. In many other applications, users set tight tolerance limits even if they are not needed. Overly tight tolerance requirements raise costs without performance benefits. For coatings that are produced over time and have many repeat batches, the most appropriate way to set tolerances is by a series of limit panels. For example, for a deep yellow coating, the greenness and redness limits and limits of brightness and darkness are needed. Because colors may change on aging, spectrophotometric measurements need to be made of both the standard and limit panels, and Commission Internationale d'Eclairage (CIE) tristimulus values

should be calculated. Any coating giving a color that fell within this color space should acceptable. Attempts have been made to assign numerical values to color differences permitting specifications that would give a numerical definition of the allowable tolerance in a single number. This objective has not been attained, although progress has been made in developing equations that permit calculation of color difference numbers ΔE in which the scale is equal for all colors. The use of ΔE specifications is not desirable except for single colors, because a ΔE value of 1 is a tighter specification for some part of the color range than for other parts, even with 1976 CIE color difference equations. Furthermore, ΔE color difference specifications permit variation around the standard equally in any direction. In practice, coatings users are more concerned about color variation in one specific direction. For example, whites that are too blue are more acceptable, in general, than whites that are too yellow.

Once the laboratory establishes a color match, the factory should match the color in spite of the fact that batch-to-batch color variations are to be expected in the pigments as manufactured. Whenever possible, the color match should be made using at least four colorants, including black and/or white, if they are needed. Four colorants give the four degrees of freedom necessary to move in any direction in three-dimensional color space. Sometimes this is not possible. For example, because the type of color mixing involved in using mixtures of pigments is subtractive color mixing, if a bright color is desired that can be made only from a single pigment, eg, phthalocyanine green, there is no pigment available that could make a batch of phthalocyanine green that was grayer than the standard less gray. A large fraction of color matching, particularly when repeated production batches are made of the same color, is now performed using instrumental measurements and color matching computer programs.

Metallic colors for automotive topcoats are an example of critical formulation and color matching problems. These colors owe their popularity to the "color flop" as the viewing angle is changed. The change is in the opposite direction and larger than is the case when viewing ordinary high gloss coatings. The depth of the color is light when viewed from near normal, ie, perpendicular to the surface (face color), and darker when viewed from a wide angle from normal (flop color). A high degree of flop, ie, a large color change with a change in viewing angle, requires that three conditions be met: the gloss of the coatings must be high, the coating matrix in which the aluminum pigment is suspended must be essentially transparent, and the nonleafing aluminum pigment used must be oriented parallel or nearly parallel to the surface of the film. In order to have a transparent matrix, the pigment-free dry film must be completely transparent and the color pigment selection and dispersion must be such that there is no light scattering from the pigment. The aluminum pigment is oriented in the film as it is forming by film shrinkage because of evaporation of volatiles in the applied coating. The upper layers of the film increase in viscosity more rapidly than the lower layers. The extent of shrinkage and, hence, the degree of orientation decreases as the solids of the coating increase. The highest degree of orientation has been achieved for lacquers having volume solids on the order of 10%. Similar results were obtained using water-reducible acrylic coatings applied at ~18% volume solids. Conventional thermosetting acrylic enamels having a volume solids >30% give noticeably less flop. About 45% volume solids is the highest solids content that gives a significant degree of orientation of the aluminum pigment particles. The considerations in obtaining a pleasing appearance using the increasingly popular pearlescent pigments in automotive coatings are similar, because pearlescent pigments are flake pigments that must be oriented approximately parallel to the surface of the film.

10.2. Gloss.   Gloss is a complex phenomenon. See (87) for discussion of gloss and gloss measurement; (88) is an old but useful review of gloss of coatings. People commonly think that high gloss results from high reflectance of incident light from the surface but low gloss surfaces reflect more total light than high gloss surfaces. The high gloss surface, however, reflects most of the light at the specular angle and low gloss surfaces reflect at all angles. A rough surface has many small facets oriented randomly whereas a high gloss surface is more nearly planar. Most commonly, a rough surface is obtained by incorporation of a large fraction of pigment in the formula. Lower gloss will also result from a hazy pigment free film since light is also reflected back out of the film from the incompatible areas in the film with different refractive indices. Gloss is affected by the distance between the observer and the surface. For example, a coating with a fine wrinkle will have a low gloss when observed visually but under a microscope can be seen to be a wavy high gloss surface.

No fully satisfactory method of gloss measurement is available and no satisfactory rating scale for visual gloss has been developed. All people will agree as to which film is glossier if the differences are large but frequently disagree in ranking if the differences are small. Widely used gloss meters are limited because the reflection is only measured at the specular angle, the distance from the surface to the light meter is fixed. Furthermore, color affect the reflectance, dark colors absorb more light and therefore reflect less than light colors. To partially overcome this problem meters can be calibrated with white and black standards. Jet black gloss coatings reflect primarily at the specular angle because light is reflected only from the surface whereas in a white coating light is reflected from the surface but there is also diffuse reflectance from the white pigment in the coating. Colors of intermediated depths give intermediate reflectance results. The standards for the measurement are standardized with a sophisticated goniophotometer at NISH (89).

Distinctness-of-image gloss meters are more reliable for high gloss surfaces. The film is used as a mirror and distortion and blurring the image is compared to that of a mirror. New instruments are now available to measure reflectance at small angle increments without using an aperture in front of the detector (90). This permits separation of reflection from micro- and macroroughness giving a separate rating for gloss and the effect of surface variations such as orange peel or texture.

11. Pigments

Pigments in coatings provide opacity and color. Pigment content governs the gloss of the final films and can have important effects on mechanical properties.

Some pigments inhibit corrosion. Pigmentation affects the viscosity, and hence the application properties of coatings. Pigment manufacturing processes are designed to afford the particle size and particle size distribution that provide the best compromise of properties for that pigment. In the process of drying the pigment, the particles generally aggregate. The coatings manufacturer must disperse these dry pigment aggregates in such a way as to achieve a stable dispersion where most, if not all, of the pigment is present as individual particles. An excellent treatise on the chemistry, properties, and uses of pigments is available (91).

Pigments can be divided into four broad classes: white, color, inert, and functional pigments. The ideal white pigment, when dispersed in the coating, would absorb no visible light and would efficiently scatter light entering the film. Scattering leads to reflection of diffuse light back out of the film. The opacity of the film increases as light scattering increases. Light scattering increases rapidly as the difference in refractive index between the pigment particles and the binder increases. Light scattering is also affected by the particle size of the pigment and its concentration. Efficient scattering of light permits hiding the substrate under a film of coating by the thinnest film. Hiding is also increased by light absorption.

Because of its high refractive index and relatively low absorption of light, rutile TiO_2 [1317-80-2], a form of titanium dioxideTiO_2, [13463-67-7], is the most widely used white pigment. The optimum particle size of rutile TiO_2 for scattering visible light is 0.19 µm at 560 nm. Because light scattering drops off faster on the lower side of the optimum than on the higher side, commercial rutile TiO_2 is made with an average particle size of 0.22–0.25 µm. If aggregates of TiO_2 particles are not separated and stabilized as a dispersion of individual particles, light scattering is low. In the case of rutile TiO_2, hiding increases linearly with concentration in a dry coating film until the PVC exceeds ~10%. For a further increase in concentration, hiding increases less than proportionally. Finally, above ~22% PVC, hiding actually decreases because of less light scattering. Whereas the exact pigment content is system dependent, the most cost-efficient level of pigmentation using TiO_2 for white coatings is ~18%. Another crystal type of TiO_2, anatase [1317-70-0], is available; it absorbs even less visible light than rutile, and hence gives whiter films but has a lower refractive index so that it does not scatter light as efficiently. Titanium dioxide can promote the photodegradation of coatings on exterior exposure, leading to erosion of binder from the surface exposing loose pigment particles. This phenomenon is called chalking. Surface treatment using alumina and silica gives pigments that lead to minimal chalking.

Small air bubbles also scatter light because the refractive index of air is ~1.0, whereas the refractive index of most polymers is ~1.5. Air bubbles in films are sometimes useful in increasing opacity but the efficiency in scattering light is much less than for rutile TiO_2.

Color pigments selectively absorb some wavelengths of light more strongly than others. A wide variety of color pigments are used in coatings. The selection of pigments for a coating formulation is based among other factors on color, cost, transparency or opacity, durability, and resistance to heat, chemicals, and bleeding, ie, solubility in solvents. In contrast to white pigments, where there is an

optimum particle size for scattering, in the case of color pigments, smaller particle size leads to stronger light absorption. However, other properties are commonly adversely affected if the particle size is too small. For example, exterior
durability commonly decreases and solubility increases as particle size is
decreased. Pigments are produced having average particle sizes that provide
the best compromise of color strength and other properties.

Aluminum-flake pigments are used. There are two important classes: leafing and nonleafing aluminum pigments. Leafing aluminum pigments are surface
treated so that when a coating is applied, the platelets come to the surface of the
drying film, resulting in a barrier to permeation of water and oxygen and giving
high light reflectance so that the appearance approaches that of aluminum
metal. The nonleafing grade does not behave in this manner but tends to orient
within the film parallel to the surface.

Inert pigments, also called extenders and fillers, do not exhibit significant
absorption or scattering of light when incorporated into coatings. In most cases,
inert pigments are used to occupy volume in the coating composition. The viscosity and flow properties of coatings can be substantially affected by increasing
the volume of pigments dispersed in a coating. Many properties of the dried film
depend on the pigment volume concentration in the film. Although such effects
could be obtained using higher volumes of white and color pigments than needed
for opacity and color, the volume effects can generally be obtained by using less
expensive inert pigment and just sufficient white and color pigments required for
the opacity and color. A wide variety of clays, silica, and carbonates are used as
inert pigments.

The most widely used functional pigments are the corrosion-control
pigments. These pigments inhibit the corrosion of steel.

11.1. Pigment Dispersion.   The dispersion of pigments involves wetting, separation, and stabilization. Wetting, ie, displacement of air and water
from the surface of the pigment by the vehicle, requires that the surface tension
of the dispersion medium be lower than that of the pigment surface. Except when
the vehicle is water, wetting is rarely a problem. Wetting agents are required for
dispersion of low polarity surface pigments such as organics, in water, and occasionally for dispersion of pigments in organic media. The rate of wetting of
pigment is fastest when the viscosity of the vehicle is low.

Pigment aggregates are separated into individual particles by a variety of
dispersion equipment (24), which transmit shear stress of sufficient magnitude
to break up the aggregates. If little force is required, low viscosity dispersions
can be made using equipment that provides low shear stresses. The most widely
used equipment of this type is a high speed impeller in a tank. Pigments where
the aggregates are not easily separated require equipment that exerts a higher
shear stress on the aggregates, such as ball mills, sand mills, shot mills, attritor
mills, extruders, and others. For pigment aggregates that are even more difficult
to separate, equipment imparting even higher shear stress is needed, eg, dough
mixers and two-roll mills. Because this last type of equipment is expensive and
labor costs are high, use is limited to the preparation of dispersions from expensive pigments where ultimate color strength is important. This class of dispersion equipment is especially useful when the aggregates must be reduced to

the ultimate particle size in order to minimize light scattering, such as in transparent pigment dispersions for use in metallic coatings for automobiles.

Stabilization of the pigment dispersion is usually the most critical aspect of the process. If the dispersion is not fully stabilized, the separated pigment particles flocculate, ie, reagglomerate. Although flocculates are easily separated again by low shear stresses, flocculated dispersions are not desirable because the hiding and color strength are lower, reflecting the larger effective particle size of the flocculates in the film. Flocculation of the pigment in gloss coatings generally leads to reduction in gloss. Furthermore, at low shear rates the viscosity of flocculated systems is much higher than that of nonflocculated systems.

Pigment dispersions are stabilized by charge repulsion and entropic repulsion. Although both types of stabilization force may be present in most cases, for pigment dispersions in solvent-borne coatings entropic repulsion is usually the most important mechanism for stabilization. In solvent-borne coatings, entropic stabilization can generally be achieved by adsorption of resin on the pigment surfaces. The adsorbed layer of resin swells with solvent. It has been shown for a wide variety of pigments and binders, that a stable dispersion is obtained if the thickness of the adsorbed layer is at least 8–10 nm. If the thickness is less, flocculation occurs (92). The reduction in entropy involved in the compression of an adsorbed layer having an average layer thickness >10 nm requires a force greater than that imparted by the Brownian motion of the pigment particles. The principal factors controlling adsorbed layer thickness are molecular weight of the resin and the presence of multiple, but limited numbers of adsorbable groups scattered along the resin chain.

Solvent selection can sometimes affect the stability of a pigment dispersion. A change in solvent may swell the adsorbed layer further and promote stabilization. On the other hand, some solvents might be so strongly adsorbed on the pigment surface that the resin could not successfully compete with it despite having multiple interaction sites. Such a system would not lead to a stable dispersion because the adsorbed layer thickness of a solvent layer is <1 nm. Most conventional coatings are stabilized using the same resin system that has been selected for the coating binder. In some cases, a related resin having more functional groups or higher molecular weight has to be designed.

However, in high solids coatings, low molecular weight resins having a limited number of functional groups must be used, leading to thinner adsorbed layers.Therefore, flocculation is a more common problem than for conventional coatings. Dispersing agents have been designed having a low molecular weight, so there is a minimum effect on viscosity while stabilizing the dispersions. They have multiple functional groups are absorbed more strongly than conventional surfactants. Reference (93) discusses the design parameters for such dispersants.

Experimental determination of adsorbed layer thickness requires a laboratory effort appropriate for research studies but not for day-to-day formulation development. The Daniels flow point method (24) is a simple procedure that permits determination of whether a resin–solvent–pigment combination can give a stable dispersion and, if so, approximately what concentration of resin in the solvent is required to prevent flocculation. The procedure permits use of the lowest possible viscosity vehicle for pigment dispersion. This permits more rapid wetting by the vehicle and also permits higher pigment loading at the same viscosity. The most expensive equipment in a coatings factory is the dispersion equipment. Therefore, it is desirable to disperse as high a volume of pigment per unit time as possible.

Determination of the extent of dispersion is not a simple task. Color strength of colored pigment dispersions and scattering efficiency of white pigment dispersions are determined by comparing the tinting strength to a standard. Flocculation can be detected by color changes on application of low shear stress, rapid settling, or centrifugation to a bulky sediment, and by the presence of shear thinning flow. A steel bar having a tapered groove called a Hegman gauge is widely used for testing the degree of dispersion. A sample is placed in front of the deep end of the groove and the dispersion is drawn down to see at what gauge reading particles can be detected. The scale is 0–8 for a groove depth varying from 100 μm to 0 μm. The diameter of most white and color pigments is <1 μm. The procedure is of little practical value because it detects only the presence of relatively large-size aggregates and does not give any indication of whether most particles are present in their ultimate particle size. Furthermore, the Hegman gauge cannot detect the presence of flocculation. For research purposes, the most accurate technique available for quantifying pigment dispersion is infrared back scattering (94).

Dispersion of pigments for latex paints is the principal application of aqueous dispersions. Commonly, three surfactants are used in preparing the dispersion of the white and inert pigments: potassium tripolyphosphate, an anionic surfactant, and a nonionic surfactant. The final colored latex paint formulations, which are very complex, commonly contain seven pigments, some having high surface energies (inorganic pigments) and some having low surface energies (organic pigments). The latex polymer is present as a dispersion that must be stabilized against coalescence and flocculation, and latexes themselves commonly contain two or more surfactants and a water-soluble polymer. Generally, but not always, the latex particles have a low surface tension. Complexity is increased by the use of at least one and sometimes two water-soluble polymers in the latex paint formula that can adsorb on the surface of some of the pigment particles and the latex particles. Although wetting inorganic pigments with water usually presents no problem, many organic pigments require a wetting agent to displace the air from the surface of the pigment particles. All dispersions must be stabilized against flocculation. The appropriate surfactants, wetting agents, and water-soluble polymers are selected largely by trial and error. Accumulation of data banks of successful and unsuccessful combinations provides a valuable tool for more efficient formulation.

11.2. Pigment Volume Relationships. Pigmentation can have profound effects on the properties of coating films depending on the level of pigmentation of these films (95). Variations in these effects are best interpreted in terms of volume relationships rather than weight relationships. Pigment volume concentration is defined as the volume of pigment in a dry film divided by the total volume of the dry film, commonly expressed as a percentage.

As the volume of pigment in a series of formulations is increased, properties change, and at some PVC there is a fairly drastic change in a series of properties.

This PVC is defined as the critical pigment volume concentration (CPVC) for that system. The CPVC is the maximum PVC that can be present in a dry film of that system, having sufficient solvent-free resin to adsorb on all the pigment surfaces and fill all the interstices between the pigment particles. In other words, when the PVC is above the CPVC, there are voids in the film. Gloss decreases as PVC increases. Hiding generally increases as PVC increases but above CPVC there is a rapid increase in the rate of increase of hiding because of the presence of voids. In the unique case of rutile TiO_2, hiding passes through a maximum at $\sim$22%. If the PVC of rutile pigmented coatings were increased above CPVC, the hiding would again start to increase rapidly. Tinting strength of white paints increases rapidly above CPVC. Tensile strength of films increases with increasing PVC, passing through a maximum at CPVC. Stain resistance is poorer and the ease of removing stains becomes more difficult for coatings above CPVC. Blistering of films on wood is less likely to occur above CPVC. The intercoat adhesion to a primer is improved when the primer has a PVC > CPVC.

CPVC depends on the pigment and pigment combinations. Pigment density is an important variable in comparing weight data with volume data. The CPVC of dense pigments tends to be lower than might be expected looking at weight data. The CPVC is lower for smaller particle size pigments, because these have a higher volume proportion of adsorbed resin than larger particle size pigments. Broad particle size distribution systems pack more efficiently and have higher CPVC values. Flocculation results in lower CPVC in a dry film as compared to the same pigment combination having a well-stabilized dispersion. Each coating application has a PVC/CPVC ratio that provides the best compromise of properties for the use involved. By determining the appropriate PVC/CPVC ratio for some application, this ratio can be used as a starting point for formulation of other pigment combinations for that application.

Experimental determination of CPVC is time consuming. A reasonable approximation can be obtained by determining the pigment loading that gives a viscosity approaching infinity in a 100% solids system, such as linseed oil with a pigment or combination of pigments. When expressed as grams of linseed oil required per 100 g of pigment, this infinite viscosity value is called the oil absorption of the pigment. Oil absorption values can be converted to CPVC values using the densities of the pigment(s) and the oil. Because different pigments have different particle size distributions, mixtures give CPVC values that are higher than the average based on the CPVC values of the individual pigments. Equations have been developed to calculate CPVC of pigment mixtures from a combination of oil absorption values and particle size distributions of the individual pigments (96). Even if CPVC data are not available, a formulator can still make use of the concept by thinking in terms of volume rather than weight relationships. Volume relationships are particularly useful in maximizing the volume of inert pigment that can be used in a coating, which in turn minimizes cost while retaining the required physical properties.

Application of CPVC concepts to latex paints is controversial. Reference (97) provides a review of the literature. However, most workers agree that for a particular pigment combination the CPVC of a latex paint is always lower than that for the same pigment combination in a solution-based paint (98). This may result from the alignment of pigment and latex particles as the water evaporates in forming a film. Latexes of low T_g and small particle size give coatings having higher CPVCs than when higher T_g and/or larger particle size latexes are used. The addition of a coalescing agent to the formula tends to increase CPVC. To reduce cost of low gloss latex paints, the formula having the highest possible CPVC permits the incorporation of the largest volumes of low cost inert pigments.

12. Application Methods

Many methods are used to apply coatings. Many factors affect the choice of method: capital and operating costs, film thickness, appearance requirements, the structure of the object being coated, and VOC emissions. Reference (99) gives a review. Brushes and hand rollers are widely used for architectural coatings by do-it-yourselfers. Spray application is used for some architectural coatings but particularly for irregularly shaped industrial products. Several types of spray systems are used. Hand-held air spray guns are the least expensive and are most often used for objects not made on an assembly line. They are low cost but the transfer efficiency (fraction of the coating applied that goes on to the object) is low. Newer high volme, low pressure (HVLP) guns give much higher transfer efficiency and are increasingly required for refinishing automobiles to reduce VOC emissions. Airless spray systems cost more but apply coatings more rapidly and permit spraying into closed recesses. Electrostatic spray systems have higher transfer efficiencies and better wrap-around at a higher cost. There are two broad types of the spray systems: spray guns and spinning bells. Spray guns can be hand held or robot held. Bell systems are generally fixed or robot systems. The trend is to higher speed bells because they permit spraying higher viscosity coatings, hence reducing VOC emissions. Hot spray systems are also used to reduce VOC emissions. The coating is heated to 65°C, permitting a roughly 50% reduction in VOC emissions. The newest approach is supercritical fluid spraying (100). Supercritical liquid CO_2 is a solvent roughly comparable to xylene and is used to replace part of the solvent for spraying with airless guns. VOC emission reductions of 30–90% have been reported.

Dip coating gives very high transfer efficiency and complete coverage of all surfaces including those inaccessible to spraying and has a high capital cost but a very low operating cost. Film thicknesses at the top of the object are lower than at the bottom. Coatings must be very stable. The method is only applicable to long runs of the same coating. Waterborne coatings are desirable to reduce VOC and flammability. Flow coating is a variation in which the coating is squirted on the object and the excess is recirculated. Production line speeds can be higher and loss of solvent in minimized. This increases coating turnover reducing stability requirements and inventory cost. A related method is curtain coating. Coating is run through a slot while flat sheets are run under the slot and excess coating is recirculated. Film thickness is controlled by the rate of movement of the sheets under the slot. Capital cost is high but operating cost is low. Leveling is superior to that obtainable with roller coating.

Roll coating is also widely used for flat objects. There are two types: direct roll coating and reverse roll coating. Direct coating is applicable to sheets and reverse roll coating is applicable to continuous webs. Reverse roll coating is desirable where applicable since it gives superior leveling. The largest single use is for applying coil coatings.

13. Film Defects

Many kinds of defects can develop in a film during or after application. Reference (101) is a monograph about film defects.

13.1. Leveling. The most widely studied leveling problem has been leveling of brush marks. One proposal is that the driving force for leveling is surface tension (102). The formulator has little control over the variables except viscosity. This model provides satisfactory correlation between experimental data and predictions when the liquid film has Newtonian flow properties and sufficiently low volatility such that viscosity does not change. In most cases, viscosity changes due to solvent evaporation and the equation is not applicable. It has also been proposed that the surface tension differential is the major driving force for leveling in coatings with volatile solvents (103). Wet film thickness in valleys of brush marks is less than in the ridges; when the same amount of solvent evaporates per unit area of surface, the fraction of solvent that evaporates in the valleys is larger than in the ridges. As a result, the surface tension in the valleys is higher than on the ridges and surface tension differential flow drives coating from the ridges into the valleys. The extent of the flow driven by surface tension differential depends on the rate of evaporation of the solvent.

In spray application, surface roughness is called orange peel, which consists of bumps surrounded by valleys. Orange peel is encountered when spraying coatings that have solvents with high evaporation rates. Leveling of sprayed films can often be improved by addition of small amounts of silicone fluid that reduces surface tension. When one sprays a coating, initially the surface is fairly smooth then orange peel grows. The growth of orange peel results from a surface tension differential driven flow. The last atomized spray particles to arrive on the wet coating surface have traveled for a longer distance between the spray gun and the surface; hence, have lost more solvent, have a higher resin concentration and, therefore, a higher surface tension than the main bulk of the wet film. The lower surface tension wet coating flows up the sides of these last particles to minimize overall surface free energy. With the silicone fluid, the surface tension of the wet coating surface and the surface tension of the last atomized particles are uniformly low, there is no differential to promote growth of orange peel.

Electrostatically sprayed coatings are likely to show surface roughness. It has been suggested that the greater surface roughness results from arrival of the last charged particles on a coated surface that is quite well electrically insulated from the ground. These later arrivals may retain their charges sufficiently long to repel each other and thereby reduce the opportunity for leveling. It has also been suggested that when coatings are applied by high-speed bell electrostatic spray guns, differentials in the pigment concentration within the spray droplets may result from the centrifugal forces (104). These pigment concentration differences lead to rougher surfaces and reduction in gloss of the final films.

Leveling problems are particularly severe with latex paints. Latex paints, in general, exhibit shear thinning and rapid recovery of viscosity after exposure to high shear rates. Due to their higher dispersed phase content, the viscosity of latex paints changes more rapidly with loss of volatile materials than the viscosity of solventborne paints. The leveling is primarily surface tension driven, since surfactants give low surface tension to latex paints that is almost unchanged as water evaporates.

13.2. Sagging.   When a wet coating is applied to a vertical surface, gravity causes it to flow downward (sagging). Sagging increases with increasing film thickness and decreases with increasing viscosity. The commonly used test is a sag-index blade. A drawdown, which is a series of stripes of coating of various thickness, is made on a chart and placed in a vertical position. Sag resistance is rated by observing the thickest stripe that does not sag down to the next stripe. For research purposes, a sag balance gives more quantitative information (105).

In spray applied solvent solution coatings, sagging can generally be minimized while achieving adequate leveling by a combination of proper use of the spray gun and control of the rate of evaporation of solvents. Sagging of high solids solvent borne coatings is more difficult to control than with conventional solids coatings. While other factors may be involved, less solvent evaporates while atomized droplets are traveling between a spray gun and the object being coated (106). A factor is the colligative effect of the lower mole fractions of solvent(s) in a high solids coatings. While this effect slows solvent evaporation from a high solids coating, it is not large enough to account for the large differences in solvent loss that have been reported. High solids coatings may undergo transition from first to second-stage solvent loss with relatively little solvent loss as compared to conventional coatings (107). High solids polyesters are formulated at concentrations above the transition concentration where solvent loss rate becomes diffusion controlled (108). Also the transition points occur at higher solids with linear molecules such as n-octane [111-65-9] versus isooctane (2,2,4-trimethylpentane) [540-84-1], and n-butyl acetate [123-86-4] as compared to isobutyl acetate. It is necessary to make the systems thixotropic. For example, dispersions of fine particle size silicon dioxide, precipitated silicon dioxide, bentonite clay treated with a quaternary ammonium compound, or polyamide gels can be added to impart thixotropy. The problem of sagging in high solids automotive metallic coatings can be particularly severe. Even a small degree of sagging is very evident in a metallic coating, since it affects the orientation of the metal flakes. Use of SiO_2 to impart thixotropy is undesirable, since even the low scattering efficiency of SiO_2 is enough to reduce color flop in the coatings. Acrylic microgels have been developed that impart thixotropic flow using the swollen gel particles. Reference (109) discusses the rheological properties of the systems. In the final film, the index of refraction of the polymer from the microgel is nearly identical with that of the cross-linked acrylic binder polymer so that light scattering does not interfere with color flop.

Hot spraying helps control sagging. The coating cools on striking the object and the viscosity increase reduces sagging. Use of carbon dioxide under supercritical conditions is helpful in controlling sagging, since the CO_2 flashes off almost instantaneously when the coating leaves the orifice of the spray gun increasing viscosity. High-speed electrostatic bell application permits application of coatings at higher viscosity, which helps control sagging.

13.3. Crawling, Cratering, and Related Defects. If a coating is applied to a substrate that has a lower surface free energy, the coating will not wet the substrate. The mechanical forces involved in application spread the coating on the substrate surface, but since the surface is not wetted, surface tension forces tend to draw the liquid coating toward a spherical shape. Meanwhile, solvent is evaporating, and viscosity is increasing and flow stops resulting in uneven film thickness with areas having little, if any, coating adjoining areas of excessive film thickness. This behavior is called crawling. Crawling can result from applying a coating to steel with oil contamination on the surface. It is especially common in coating plastics. Crawling can also result from the presence in the coating of surfactant-type molecules that can orient rapidly on a highly polar substrate surface. Even though the surface tension of the coating is lower than the surface free energy of the substrate, it could be higher than the surface free energy of the substrate after a surfactant in the coating orients on the substrate surface. If one adds excess silicone fluid to a coating to correct a problem like orange peel, small droplets of insoluble fractions of the poly(dimethylsiloxane) can migrate to the substrate surface and spread on it, and the film crawls. Modified silicone fluids, such as polysiloxane–polyether block copolymers, have been developed that are compatible with a wider variety of coatings and are less likely to cause undesirable side effects. The effect of a series of additives on crawling and other film defects has been reported (110).

Cratering is the appearance of small round defects that look somewhat like volcanic craters on the surface of coatings. Cratering results from a small particle or droplet of low surface tension contaminant landing on the wet surface of a freshly applied film. Some of the low surface tension material dissolves in the adjacent film, creating a localized surface tension differential. This low surface tension part of the film flows away from the particle to cover the surrounding higher surface tension liquid coating. Loss of solvent increases viscosity, leading to formation of a characteristic crest around the pit of the crater. The user applying the coating should minimize the probability of low surface tension contaminants arriving on the wet coating surface. Spraying lubricating oils or silicone fluids on or near the conveyor causes craters. Presence of some contaminating particles cannot be avoided; so coatings must be designed to minimize the probability of cratering. Lower surface tension coatings are less likely to form craters. Alkyd coatings have low surface tensions and seldom give cratering problems. In general, polyester coatings are more likely to give cratering problems than acrylic coatings. Additives can be used to minimize cratering. Small amounts of silicone fluid generally eliminate cratering, excess silicone must be avoided. Octyl acrylate copolymer additives usually reduce cratering. A comparison of effects of additives on the control of defects such as cratering is available (111).

In roll coating tin plate sheets, the coated sheets are passed on to warm wickets that carry the sheets through an oven. In some cases, one can see a pattern of the wicket as a thin area on the final coated sheet. The heat transfer to the sheet is fastest where it is leaning against the metal wicket. The surface tension of the liquid coating on the opposite side drops locally because of the higher temperature. This lower surface tension material flows toward the higher surface tension surrounding coating. In spraying flat sheets as solvent evaporates, the coating is thickest at the edges and just in from the edge the coating is thinner than average. Solvent evaporates most rapidly from the coating near the edge, where the air flow is greatest. This leads to an increase in resin concentration at the edge and to a lower temperature. Both factors increase the surface tension there, causing the lower surface tension coating adjacent to the edge to flow out to the edge to cover the higher surface tension coating. Surface tension differential driven flow can also result when overspray from spraying a coating lands on the wet surface of a different coating. If the overspray has lower surface tension than the wet surface, cratering occurs. If the overspray has high surface tension compared to the wet film, local orange peeling results.

When coatings are applied on plastic sheets that are adhered to steel supports in an automobile, surface tension driven flow leads to defects at the boundary line above the steel supports called bondline readout (BLRO). A study has been published reporting an investigation of parameters affecting BLRO (112).

13.4. Floating and Flooding.   Floating is most evident in coatings pigmented with two pigments. A light blue gloss enamel panel can show a mottled pattern of darker blue lines on a lighter blue background. With a different light blue coating, the color pattern might be reversed. These effects result from pigment segregations that occur as a result of convection current flows driven by surface tension differentials while a film is drying. Rapid loss of solvent from a film during drying leads to considerable turbulence. Convection patterns are established whereby coating material flows up from lower layers of the film and circulates back down into the film. The flow patterns are roughly circular, but as they expand, they encounter other flow patterns and the convection currents are compressed. As solvent evaporation continues, viscosity increases and it becomes more difficult for the pigment particles to move. The smallest particle size, lowest density particles continue moving longest; the segregated pattern of floating results. Floating is particularly likely to occur if one pigment is flocculated and the other is a nonflocculated dispersion of fine particle size. If, in a light blue coating, the white pigment is flocculated and the blue is not, one will find darker blue lines on a lighter blue background. If the blue one is flocculated and not the white, there will be lighter blue lines on a darker blue background. Floating can occur without flocculation using of a combination of pigments with very different particle sizes and densities. When a fine particle size carbon black and TiO_2 are used to make a gray coating; the particle size of the TiO_2 is several times that of the carbon black and TiO_2 has about a fourfold higher density. A larger particle size black, such as lamp black, can be used to make a gray with a lower probability of floating. As with other flow phenomena driven by surface tension differentials, floating can be prevented by adding a small amount of a silicone fluid.

In flooding, the color of the surface is uniform but different than should have been obtained from the pigment combination used. One might have a uniform gray coating, but a darker gray than that expected from the ratio of black-to-white pigments. The extent of flooding can vary with the conditions encountered during application, leading to different colors on articles coated with the same coating. Flooding results from surface enrichment by one or more of the pigments in the coating. The stratification is thought to occur as a result of different rates of pigment settling within the film, which are caused by differences in pigment density and size or flocculation of one of the pigments. Flooding is accentuated by thick films, low vehicle viscosity, and low evaporation rate solvents. The remedies are to avoid flocculation and low density fine particle size pigments.

13.5. Wrinkling. A wrinkled coating shrivels or wrinkles into many small hills and valleys. Some wrinkle patterns are so fine that to the unaided eye, the film appears to have low gloss rather than to look wrinkled. However, under magnification, the surface can be seen to be glossy but wrinkled. In other cases, the wrinkle patterns are broad or bold and are readily visible. Wrinkling results when the surface of a film becomes high in viscosity while the bottom of the film is still relatively fluid. It can result from rapid solvent loss from the surface, followed by later solvent loss from the lower layers. It can also result from more rapid cross-linking at the surface of the film than in the lower layers of the film. Subsequent solvent loss or cure in the lower layers results in shrinkage, which pulls the surface layer into a wrinkled pattern. Wrinkling is more apt to occur with thick films than with thin films because the possibility of different reaction rates and differential solvent loss within the film increases with thickness.

Wrinkling can occur in uv curing of pigmented acrylate coatings with free-radical photoinitiators. High concentrations of photoinitiator are required to compete with absorption by the pigment. Penetration of uv through the film is reduced by absorption by the pigment as well as by the photoinitiator. There is rapid cross-linking at the surface and slower cross-linking in the lower layers of the film, resulting in wrinkling. The uv curing of pigmented cationic coatings, which are not air inhibited, is even more prone to surface wrinkling.

13.6. Popping. Popping is the formation of broken bubbles at the surface of a film that do not flow out. Popping results from rapid loss of solvent at the surface of a film during initial flash off. When the coated object is put into an oven, solvent volatilizes in the lower layers of the film, creating bubbles that do not readily pass through the high viscosity surface. As the temperature increases further, the bubbles expand, finally bursting through the top layer resulting in popping. The viscosity of the film meanwhile has increased enough so that the coating cannot flow together to heal the eruption. Popping can also result from entrapment of air bubbles in a coating. Popping can result from solvent that remains in primer coats when the topcoat is applied. In coating plastics, solvents can dissolve in the plastic, and then cause popping when a coating is applied over the plastic and then baked. Another potential cause of popping is evolution of volatile by-products of cross-linking.

Popping can be minimized by spraying more slowly in more passes, by longer flash-off times before the object is put into the oven, and by zoning the oven so that the first stages are relatively low in temperature. The probability of popping can also be reduced by having a slow evaporating, good solvent in the solvent mixture. This tends to keep the surface viscosity low enough for bubbles to pass through and heal before the viscosity at the surface becomes too high. Popping can be particularly severe with water-reducible baking enamels because slow loss of water during baking, especially with high T_g resins. In contrast to increased probability of popping with higher T_g water-reducible coatings, popping is more likely to occur with lower T_g latex polymers. Coalescence of the surface before the water has completely evaporated is more likely with a lower T_g latex.

13.7. Foaming. During manufacture and application, a coating is subjected to agitation and mixing with air, creating the opportunity for foam formation. In formulating a latex paint, an important criterion in selecting surfactants and water-soluble polymers as thickeners is their effect on foam stabilization (113). Acetylene glycol surfactants such as 2,4,7,9-tetramethyl-5-decyne-4,7-diol are reported to be effective surfactants that do not increase the viscosity of the surface of bubbles as much as surfactants such as alkylphenol ethoxylates (114).

A variety of additives can be used to break foam bubbles. Most depend on creating surface tension differential driven flow on the surface of bubbles. Silicone fluids, are effective in breaking a variety of foams, since their surface tension is low compared to almost any foam surface. Small particle size hydrophobic SiO_2 can also act as a defoamer and/or a carrier for active defoaming agents (114). Also, a small amount of immiscible hydrocarbon will often reduce foaming of an aqueous coating. Several companies sell lines of proprietary antifoam products and offer test kits with small samples of their products. The formulator evaluates the antifoam products in a coating with foaming problems to find one that overcomes the problem. While it is possible to predict which additive will break a foam in a relatively simple system, such predictions are difficult for latex paints because of the variety of components that could potentially be at the foam interface. The combination of surfactants, wetting agents, water-soluble polymers, and antifoam can be critical.

14. Product Coatings

About 30% of the total volume of coatings produced in the United States in 1999 (1.45×10^9 L) applied in factories to a very large variety of products ranging from automobiles to toys (1). They are often called original equipment market (OEM) coatings.

14.1. Coatings for Metal. A large fraction of the product coatings are applied to metal. The essential first step in metal coating is the preparation of the metal. Oil and related contaminants must be removed by detergent or solvent washing. Solvent degreasing is the most effective. In detergent washing, the last trace of detergent must be rinsed off before drying preparatory to painting. For best adhesion and corrosion resistance, the surface of the metal should be treated. For steel, phosphate conversion coating treatments are used. Aluminum, for applications when the product will not be exposed to salt, needs no treatment. Chromate conversion treatments have been used when salt exposure is possible but proprietary chromate-free treatments are now being used because of toxic hazards with chromates. The surface should be carefully rinsed before applying paint. Surface contaminants can result in crawling and/or blistering of the coating. Treated surfaces should not be touched. Water-soluble contaminants such as salts can lead to blistering after the coated product is put into service. Oils can lead to crawling of the wet coating film applied on top of the oil.

Primers. If reasonably high performance is required in the end product and unless cost is of paramount importance, a minimum of two coats, usually a primer and a topcoat, should be applied to metal. For highest performance, primer vehicles should provide good wet adhesion, be saponification resistant, and have low viscosity to permit penetration of the vehicle into microsurface irregularities in the substrate. Color, color retention, exterior durability, and other such properties are generally not important in primers. Resin systems such as those including BPA epoxy resins that provide superior wet adhesion can thus be used in spite of their poor exterior durability.

In order to provide a suitable surface for adhesion of the topcoat, the gloss of the primer should be low and the cross-link density should be as low as handling characteristics permit. In some cases, it is desirable to have the PVC > CPVC permitting good intercoat adhesion with the topcoat and relatively easy sanding. After penetration by the topcoat vehicle, the PVC becomes approximately equal to CPVC. High pigmentation is desirable because the least expensive component is inert pigment. Furthermore, high pigmentation reduces oxygen and water permeability of the final combined film as long as highly hydrophilic pigments are avoided. One drawback for high solids is that highly pigmented, high solids coatings have higher viscosities. Waterborne primers are replacing solventborne primers in many applications.

When feasible, baking primers are preferable since they provide better adhesion. A primer should be designed for baking at as high a temperature as possible because this promotes penetration into the conversion coating and micropores on the surface of the metal. The primer vehicle should be designed to provide maximum resistance to displacement by water, ie, to have good wet adhesion. Multiple polar groups spaced along the polymer backbone tend to promote wet adhesion and resins substituted with amine groups or phosphoric acid partial esters show enhanced wet adhesion. Cost is sometimes the dominant factor on composition and alkyds are still widely used as primer vehicles even though epoxy esters, epoxy/amine, or epoxy-phenolic-based primers generally provide better performance.

Electrodeposition Primers. Primers for automobiles and a significant part of primers for household appliances are applied by electrodeposition. Almost all electrodeposition primers are now cationic. The compositions are proprietary, but the vehicles in some primers are epoxy/amine resins neutralized with volatile organic acids such as lactic acid; an alcohol-blocked isocyanate is used as a cross-linking agent. The pigments are dispersed in the resin system and the coating is reduced with water. The amount of amine salt is such that a stable dispersion, not a solution, results in the aqueous phase. All of the resin, cross-linking agents, and pigments must be located in the dispersed phase aggregates, so that all components deposit at equal rates.

Recent emphasis has been directed to lowering baking temperatures. The two major approaches to the problem have been selection of blocking agents for the isocyanate cross-linking agents that unblock at lower temperatures and catalysts that reduce the temperature while retaining excellent hydrolytic stability. See (115) for a review.

In application, the automobile or other article to be coated is made the cathode in an electrodeposition system. A current differential on the order of 250–400 V is applied, which attracts the positively charged coating aggregates to the cathode. At the cathode, hydroxide ions from the electrolysis of water precipitate the aggregates on the surface of the metal. As the conveyor removes the coated product from the bath, residual liquid is rinsed off with water and the article is conveyed into a baking oven for a high temperature bake.

As the coating operation continues, acid accumulates in the electrodeposition bath. Thus pH must be controlled by the addition of acid-deficient make-up coating and by the removal of excess acid from the bath. Other soluble materials also tend to accumulate in the bath. These must be removed by continuous passage through ultrafiltration units. Additionally, the temperature must be closely controlled. Highly automated electrodeposition tanks are used. The high voltage application is required in order to have high throw power, ie, deposition on internal surfaces of the metal article in the short time that it is in the tank. High throw power without film rupture at the readily accessible coated surfaces requires a limited conductivity of the applied film and low conductivity of the continuous water phase. The low conductivity requirement is another reason that the efficient operation of the ultrafiltration process is essential. The voltage that can be used varies with the type of metal being coated. Whereas film rupture has been attributed to gas evolution under the film from the electrolysis of water, it appears that electric discharges through the film may be more important.

Cationic electrodeposition primers show substantially better corrosion protection than anionic ones. In anionic electrodeposition, the phosphate conversion coating on the steel partially dissolves with the hydrogen ions generated at the anode surface. However, in the case of cationic primers, if zinc–iron phosphate conversion coatings are used on steel and zinc–manganese–nickel conversion coatings are used on zinc-coated steels, the conversion coatings do not dissolve. Furthermore, the resin used in the cationic electrodeposition primer is generally saponification resistant and promotes good wet adhesion, which are critical requirements for corrosion protection.

Electrodeposition primers offer substantial advantages in some applications over conventional primers: The highly automated lines permit low operating costs; the VOC emissions are lower than from other primer systems; and areas not reached by spray application are coated, giving superior corrosion protection. There is also the advantage of essentially 100% utilization of the coating and no waste from overspray. Furthermore, electrodeposition gives uniform film thickness to the coating on all areas of the article, avoiding thin spots, fatty edges, sagging, etc.

The capital cost of application facilities is high for electrodeposition limiting this technique to high production articles. Because the film thickness of the electrodeposited primer is uniform, the surface of the coating in effect replicates any irregularities in the surface of the metal. If high gloss topcoats are to be applied over the primer, care must be taken in selecting steel and steel treatment procedures that permit adequate smoothness or a coat of primer-surfacer over the electrodeposited primer may be required.

It is more difficult to obtain adequate intercoat adhesion of a topcoat to an electrodeposited primer than to a spray-applied primer. This adhesion problem is partially because the PVC must be lower than CPVC. The lower pigment content gives a higher gloss surface, which reduces the opportunity for intercoat adhesion. Commonly, this problem is overcome by applying a coat of primer-surfacer to provide intercoat adhesion between the primer and topcoats, before application of the topcoat. A recent innovation has been the introduction of the use of two electrodeposition coatings for priming of automobiles. The first coat is an electroconductive coating. With this conductive primer another electrocoating can be applied as a second coat. The second is formulated to replace a spray coating of a primer-surfacer, which reduces application cost.

Topcoats. The selection of a topcoat depends on cost, method of application, product use and performance requirements, among other factors. As a result of increasingly stringent air quality standards and increased solvent costs, approaches to reduction of solvent emissions are being sought. Three approaches are being followed: high solids coatings, waterborne coatings, and powder coatings (for discussion of powder coatings see the section on Powder Coatings).

High Solids. There is no agreement on a definition of high solids coatings. For any particular use, it means higher solids than previously used in that application. In automotive acrylic metallic coatings, high solids usually is taken to be ~45 vol%. This limit results from the solvent level required to give sufficient shrinkage during film formation to orient the aluminum pigment to give the desired color effect. In the case of clear coats, ~70 vol% can be achieved with reasonable film properties. Urethane based coatings can be formulated with higher solids and generally give superior abrasion resistance than acrylic–MF coatings. For applications in which outstanding exterior durability is not required polyester coatings give satisfactory performance and can be formulated with up to ~70 vol% solids.

Whereas the main driving force behind the development of higher and higher solids coatings has been the reduction of VOC emissions, solvent cost is also a factor. A further advantage is that the same dry film thickness can be applied in less time. High solids coatings are made using lower molecular weight resins having fewer average functional groups per molecule as compared to conventional coatings. As a result, more complete reaction of the functional groups is necessary to achieve good film properties. Greater care is necessary for high solids coatings manufacturers to maintain close adherence to quality control standards. Additionally, application techniques, baking times, and temperatures for which the coating is designed, must be carefully adhered to. There is a relatively small window of cure. Shorter times or lower temperatures can result in undercure and the properties are more affected by overcure, in general, than conventional coatings (116). High solids coatings are likely to have high surface tensions, and hence are more likely to give film defects such as crawling and cratering than conventional coatings. Sagging of spray-applied high solids coatings is more difficult to control than for conventional coatings. See Section on Sagging.

Waterborne Coatings. Two classes of waterborne systems are used: water-reducible and latex systems. Water-reducible systems are used on a larger scale but the consumption of latex product coatings is increasing. When the coating is reduced with water, the polymer separates into relatively stable aggregates containing the pigment and cross-linking agent and these are swollen by solvent and water. In anionic systems, the resin is sufficiently substituted with

COOH groups to give the desired water dilutability after partially neutralizing with an amine. Acid numbers are in the range of 40–80. The coatings are cross-linked with MF resins or blocked isocyanates. Water-reducible urethane resins can be used with MF cross-linkers or as 2K clear coats with polyisocyanates to give coatings having superior abrasion resistance (117).

Because the molecular weight and average functionality of the resins used in water-reducible coatings are comparable to those of conventional solution thermosetting coatings, film properties obtained after curing are fully equivalent. Similarly, and in contrast to high solids coatings, they have tolerances to variations in cure conditions comparable to those of conventional solution coatings. Surface tensions are relatively low, and few problems with crawling and cratering are encountered. Because the solids are low, good orientation of aluminum pigment can be obtained and such water-reducible coatings are being used increasingly in base coats for automobiles along with a high solids clear coat (118).

There are disadvantages to the water-reducible coatings. Solids contents at application viscosities are relatively low, 18–25 vol%. Because most of the volatile material is water, VOC emissions are still lower than from most high solids coatings. The evaporation rate of water depends on the relative humidity, which can lead to inconsistent flow behavior during flash off. Above a critical relative humidity, water evaporates more slowly than the solvent; below this critical value the solvent evaporates more slowly than water. The viscosity of the coating depends strongly on the ratio of solvent-to-water as well as on the concentration, thus abnormal viscosity changes can occur when flash-off occurs above the critical relative humidity. Popping is more difficult to control in water-reducible baking enamels than in solvent-based counterparts, even when the relative humidity in the flash-off zone is well controlled. See the section on Popping.

Acrylic coatings of fairly similar composition can also be applied as topcoats directly on metal by anionic electrodeposition. Performance on aluminum is excellent, over steel some iron is dissolved at the anode, and anionic electrodeposition coatings tend to become discolored. Cationic electrodeposition topcoats can be made for use on steel, using, eg, 2-(dimethylamino)ethyl acrylate as a comonomer and a blocked aliphatic diisocyanate as the cross-linking agent for hydroxy-functional groups. Electrodeposition topcoats are particularly useful where it is difficult to achieve full, uniform coverage by other means such as on articles having sharp edges, eg, finned heat exchange units. The highly automated electrodeposition process can lead to significant cost reduction. It has been reported for cationic electrodeposition coating of air conditioners that replacing a former system of flow-coated primer and spray-applied topcoat required only one operator. In the former system, 50 people, including those doing the required touch up and repair were needed (119).

Until the 1980s, latex-based coatings were infrequently used in product coating applications. High gloss coatings cannot be made using latexes and transparent coatings are difficult if not impossible to make. Furthermore, the rheological properties of the coatings limit utility. Flow problems are least severe when coatings are applied by curtain coating or reverse roll coating because these two methods do not involve film splitting. The availability of associative thickeners that minimize flocculation of latex particles permits formulation of industrial coatings that are less thixotropic, and hence level better after application. Popping can cause difficult problems in applying thick coats of latex-based coatings. Low VOC emissions and excellent film properties resulting from the high molecular weight of the polymer are expected to lead to increases in the use of latexes in product coatings.

Coil Coatings. An important segment of the product coatings market is sold for application to coiled metal, both steel and aluminum (120). In this process, the metal is first coated and then fabricated into the final product rather than fabricating first into product. The method offers considerable economic advantages because coatings are applied by direct or reverse roll-coating at high (up to 400 m/min) speeds to wide (up to 3 m) coils in a continuous process. The metal is cleaned, conversion coated, and coated with primer (if desired) and topcoat in an in-line operation. Coatings can be applied to both sides of the metal during the same run through the coil coating line. The labor cost of coating application is much less than for application to a previously fabricated product. There is essentially 100% effective utilization of coating, and the loss by overspray involved in coating fabricated products is avoided.

Use of coil-coated stock reduces fire risk, and hence insurance costs for the metal fabricator. The problem of controlling VOC emissions is also avoided because no coating is done in the factory. The VOC emission control problems for the coil coater are minimal. The oven exhaust is used for the air needed for the gas heaters for the oven. The solvents are used as fuel rather than being allowed to escape into the atmosphere. Film thickness of the coatings is more uniform than can be applied by spray, dip, or brush coating of the final product. The coatings are all baked coatings, and when coil-coated metal is used, substantially greater exterior durability and corrosion protection can be achieved as compared with field-applied air-dry coatings.

There are limitations to the applicability of the process. The metal must be flat. Coil coating is economical only for long runs of the same color and quality of coated metal. Offsetting the economic advantages of coil coating is the higher cost of maintaining inventories of coated metal coils. Additionally, because the films must sometimes withstand extreme extensions in fabricating a final product, they must generally have long elongations-to-break, which may limit the range of other film properties that can be designed into the coating. For example, it may be difficult to achieve a high modulus at a low rate of application of stress while still having adequate elongation-to-break to permit fabrication without film cracking. Although some coating may be smeared over the cut edge of the metal when the coil is cut, the cut edge of the metal usually has little, if any, coating on it. These cut edges can be a weak spot for corrosion protection and can present an appearance problem if the edges are visible in the fabricated product. Although coatings can be made that can be welded through, they char from the intense heat and have, therefore, an unacceptable appearance if the weld marks are visible. Design of a final fabricated product must take the cut edge and weld marks into account, hiding and/or locating them in positions least likely to cause corrosion problems.

In order to have reasonable oven lengths for curing at high speeds, the coatings must cure in short times, <1 min. Therefore, the ovens are operated at high temperatures, $>290°C$. The short-time, high temperature cure schedules are difficult to duplicate in a coatings development laboratory. Substantial experience is required to make informed estimates of how to translate laboratory cure response to coating line cure response. This problem can be particularly acute for color matching. The ability to do satisfactory color matching for coil lines is a prerequisite to commercial success in supplying coil coatings. Because during start up, and sometimes in splicing a new coil onto the previous coil, the line must be slowed down, it is essential that there be a minimum change in color and film properties when the standard curing time is exceeded.

A wide range of resin compositions are used in coil coatings, depending on the product performance requirements and cost limitations. The lowest cost coatings are generally alkyd coatings. They are appropriate for indoor applications where color requirements are not stringent. Alkyd–MF coatings are sometimes used outdoors when long-term durability is not needed. Greater durability and better color retention on overbaking are obtained using polyester–MF. Polyesters usually provide better adhesion to unprimed metal than acrylic coatings. For two-coat systems, epoxy ester primers are widely used along with acrylic–MF topcoats. Blocked isocyanates can also be used for cross-linking the polyesters and acrylics but care is needed because urethanes decompose thermally and loss of film properties and discoloration on overbake may be severe. For this reason, it is necessary to use blocked diisocyanates that react at relatively low temperatures. Oximes are the most widely used blocking agents for coil-coating systems. Thermoplastic coatings based on vinyl plastisols and acrylic latexes resist severe deformation during fabrication and exhibit good exterior durability.

For greater exterior durability, silicone-modified polyesters or silicone-modified acrylic resins are used. Hydroxy-functional polyester or acrylic resins are partially reacted with a silicone intermediate having methoxy substituents on the siloxane resin chain. Cross-linking is completed during curing by reaction of the remaining methoxy substituents and hydroxyl groups on the polyester or acrylic resin. In order to avoid softening during prolonged exposure to high humidity, which can result in reversible hydrolysis of cross-links with groups having three oxygens on a silicone atom, it is common to use a small amount of an MF resin as a supplementary cross-linker. Silicone-modified coatings have longer exterior lifetimes than the corresponding unmodified polyester or acrylic coating. Silicone-modified coatings are more expensive. They are particularly useful in colored coatings where the color change can be substantial when chalking occurs. The white coated product in the same quality line may well be made without silicone modification to reduce cost, because color change resulting from the chalking of the white coating after exposure for a long period is less obvious. For still greater exterior durability, fluorinated polymer systems are used.

In the can industry, large-volume three-piece cans, used for packing many fruits and vegetables, are made from coil-coated stock. Oil-modified phenolic resins are used for coating tinplated steel coils on the side that becomes the interior of the can. Coil stock for making ends for two-piece beverage cans is coated with epoxy–MF, epoxy–phenolic, or a cationic uv cure epoxy coating.

14.2. Wood Products Coating. Furniture is an important class of wood products that is industrially coated. The appearance standards are set by the fine furniture industry where the flat areas are composed of plywood having high quality top veneer; the legs and rails are solid wood; and there are frequently carved wood decorative additions. The finishing process for fine furniture is long and requires significant artistic skill. If the final overall color of the furniture is lighter than the color of any part of the wood, the first step is bleaching using a solution of hydrogen peroxide in methanol [67-56-1]. The bleached wood, or unbleached for darker color finishes, is given a wash coat of size to stiffen the fibrils so they can be cleanly removed by sanding. The wood is then coated with stain, a solution of acid dyes in methanol, to give a desired overall color tone to the piece of furniture. A wash coat, generally a low solids vinyl chloride copolymer lacquer, is applied over the stain, partially to seal the stain in place but also to prepare the surface for the next operation, filling. The filler, a dispersion of pigments, usually in linseed oil with mineral spirits solvent, is sprayed over the whole piece of furniture. It is then wiped off, leaving filler only in the pores of the wood. The colors of fillers are commonly dark brown; the filler serves to emphasize the grain pattern in the hardwood veneer. Next, shading stains are selectively sprayed on the wood to give different colors to various sections. It is common to distress the surface to resemble antique furniture. For example, spots of black pigment stain can simulate India ink spots dropped from quill pens, and dark stains carefully applied into corners simulate dirt accumulated over many years. When this "art work" is completed, a sanding sealer is applied to immobilize the lower layers of the finish and to provide a surface that can be sanded smooth. Finally a topcoat is applied and polished smooth.

The primary binder that has been used for many years in the sanding sealer and the topcoat is nitrocellulose. Nitrocellulose lacquers supply a coating that brings out the natural beauty of the wood and gives a depth of finish that has not been matched with other systems. The film properties of the lacquers improve with increasing molecular weight of the nitrocellulose, whereas the solids of the lacquers decrease. Lacquers suitable for high quality furniture have volume solids <20%. Other components of the lacquer are plasticizers and hard resins to provide a combination of flexibility and hardness for sanding and polishing while retaining flexibility to withstand cracking as the wood in the furniture expands and contracts. Zinc stearate [557-05-1] is incorporated in the sanding sealer to reduce clogging of sand paper.

Topcoats generally include fine particle size silicon dioxide to reduce the gloss. In making a low gloss topcoat, it is essential to retain the transparency. Low gloss can be achieved using a minimum amount of fine particle size SiO_2 pigment. When the solvent evaporates from the low solids lacquer, convection currents carry the fine particle size SiO_2 to the surface. Hence, the PVC of the surface layer is relatively high, giving the low gloss, whereas the PVC of the entire film is low (2–4%) in order to maintain transparency. As higher and higher solids lacquers while maintaining clarity are attempted, low gloss becomes more and more difficult. The convection currents are not as strong using the lesser amounts of solvent.

Nitrocellulose lacquers offer other important advantages for furniture. There is latitude as to when a complete piece of furniture is rubbed after finishing, whereas thermosetting systems must be rubbed after some cross-linking but before cross-linking has advanced to the point that it cannot be rubbed without leaving permanent scratches. Using lacquers, the production line can be shut down at the end of a shift and pieces on the line can be rubbed the next day or after a weekend; using thermoset systems, the line must be stopped early or an extra shift must be brought in just for the rubbing. Because of warehousing space problems for furniture, factories frequently prefer to load furniture onto trucks almost directly off the finishing line. The finish must therefore withstand print tests equivalent to the pressures that are going to be encountered in shipping very soon after they are applied. This is a problem for water-based topcoats because the rate of loss of water depends on humidity. The loss of the last of the water from an almost dry film can be slow, resulting in delay before the furniture can be wrapped and shipped.

This finished furniture is a beautiful product attained at a high cost. Most good furniture is made at substantially lower cost with as little sacrifice in appearance as possible. The expensive hardwood veneer plywood is replaced by printed, coated particleboard. The artists' work of filling, shading, distressing, glazing, and so on needs only to be done once on a carefully selected beautiful wood laid up meticulously into fine patterns. The finished panels are then photographed to make a series of three gravure printing plates of the finest through the boldest parts of the pattern. The particleboard is coated with a uv cure filler and sanded smooth. A lacquer base coat is sprayed on, in a color corresponding to the first coat of stain in fine furniture finishing. Then three prints are applied by offset gravure printing using three carefully selected color inks. The panel is assembled into the furniture and finished with lacquers as in the standard process. Only experts can distinguish the prints from real wood by just looking at the surface.

Conventional nitrocellulose lacquer finishing leads to the emission of large quantities of solvents into the atmosphere. An approach to reducing VOC emissions is the use of supercritical carbon dioxide as a component of the solvent mixture (121). VOC emission reductions of 50% or more result. See the section Application Methods.

Many alternatives to nitrocellulose lacquers for topcoats are being used increasingly. Especially for lower cost furniture exposed to hard use such as motel and institutional furniture, alkyd–UF topcoats are used. The UF resins, in contrast to MF resins, are used because the former can be cured using butyl acid phosphate [1623-15-0] catalyst at room temperature or force dried at 60–70°C. These resins are applied at ~35 vol% solids and require about one-half of the spraying time of a nitrocellulose lacquer. They are also more difficult to repair and do not provide the depth of appearance of nitrocellulose lacquers.

Considerable efforts have been invested in developing waterborne coatings for wood furniture. The two approaches are water-reducible acrylic resins with UF cross-linkers and hydroxy-functional vinyl acetate copolymer latexes with UF cross-linking. The 2K waterborne urethane coatings with very low VOC are being increasingly adopted, especially for office furniture and kitchen cabinets (67). Waterborne systems would be used more widely if it were not for the effect of relative humidity on drying and the relatively long time for the coatings to develop print resistance. Because of grain raising caused by direct contact of water with wood, water systems are not suitable for the first layer of coating on the wood.

The uv-cure coatings are widely used in European furniture manufacture but have found more limited applications in the United States. Most U.S. furniture has a relatively low gloss finish, frequently has curved surfaces, and is finished after assembly; most European furniture has a relatively higher gloss finish with primarily flat surfaces, and the furniture is generally assembled after coating. The uv-cure coatings are more easily adapted to coating before assembly. Because of the pressure on the furniture industry to decrease VOC emissions, uv-cure coatings may become more widely adopted in the United States. The use of high pressure laminated plastics having wood grain reproductions can be expected to take a larger share of the furniture top market in response to VOC regulations.

The other principal component of industrial wood finishing is the panel industry. The highest cost segment of this industry, fine hardwood veneer paneling for executive office walls, is comparable to the fine furniture industry where similar nitrocellulose lacquer finishing systems are used. However, the bulk of the industry requires less expensive finishing operations. Some plywood is stained followed by roll coating with a low gloss nitrocellulose lacquer topcoat. It is common to print grain patterns from woods such as walnut onto inexpensive, relatively featureless veneers such as luan before applying a lacquer topcoat. Lacquer topcoats are being replaced by alkyd–UF coatings.

Large volumes of wood grain paneling are made from hardboard paneling. The coating systems are usually a base coat, three prints, and a clear topcoat. In higher style paneling, the hardboard is embossed in the pressing step involved in making the hardboard. For example, if a pecky cypress paneling is desired, the hardboard would be embossed with holes resembling those of pecky cypress. The first step in finishing such a board is to apply a so-called hole color with a brush roll rollercoater. Then the flat surfaces are precision coated with a base coat. The base coat color is the equivalent of the stain color if real wood were being used. Then three prints are applied from gravure rollers made by photographing carefully selected and finished real cypress. Finally, an overall low gloss topcoat is applied. Because hardboard can withstand contact with water and relatively high temperature baking, the coatings are most commonly water-reducible acrylic–MF coatings. Imitations can be made of any wood and also of brick paneling, marble paneling, and porcelain tile.

Factory coated hardboard is also used for exterior siding for homes. The largest volume of such hardboard is preprimed board. A primer is applied in the factory that is suitable for exterior exposures up to 6 months before painting with exterior house paint. While only one coat of latex paint need be applied, it has been reported that greater durability is obtained if a primer is applied after house construction using an acrylic latex topcoat (122). Fully prefinished hardboard is also sold commercially. The exterior durability of this paneling is superior to that of the field painted paneling, but the styling is more limited; it is used primarily in commercial buildings.

14.3. Radiation Curing. Radiation has several significant advantages over heat as the source of energy to carry out cross-linking reactions (123). Coatings can be designed that cure rapidly, in a second or less, at room temperature yet have relatively long storage lives. The energy requirements for curing are much lower than for thermal baking systems. Reactive monomers are used to provide the low viscosities for application. In some cases, the energy source is high energy electrons, but more commonly uv, radiation curing systems are used.

In uv-curing coatings, a photoinitiator is required in the formula, whereas in electron-cured coatings the energy is directly absorbed by the reacting molecules. Both free-radical and cationic initiated polymerization systems are employed for uv cures. Benzoin derivatives, benzil ketals, acetophenone derivatives, α-hydroxyalkylphenones, O-acyl-α-oximinoketones, and benzophenone (diphenylmethanone) [119-61-9] and 2-(dimethylamino)ethanol combinations are examples of photoinitiators (124).Vehicle systems are oligomers substituted with several acrylic ester groups mixed with low molecular weight difunctional or trifunctional acrylates and a monofunctional acrylate. Acrylic ester systems provide a much more reactive cure than methacrylate systems. Styrene is sometimes used as a monomer with unsaturated polyesters. However, cure is slower than for acrylic esters and significant amounts of styrene evaporate. Many pigments absorb uv radiation, thus most commercial uv-cure coatings are unpigmented, clear coatings. Printing inks, which are applied as very thin films, and particle board filler, where only inert pigments that absorb little uv are used, are examples of uv-cure pigmented systems.

Photoinitiators for cationic polymerization generate reactive electrophiles (acids). Three classes of cationic photoinitiators have been used commercially: triarysulfonium, diaryliodonium, and ferrocenium salts of very strong acids, such as hexafluorophosphoric acid (125). When the uv absorption of a potential photoinitiator is low, a strongly absorbing photosensitizer can also be used. The most widely used vehicles are epoxy resins diluted with monofunctional and difunctional epoxy compounds to adjust the viscosity. An important advantage of this type of system is that shrinkage is substantially less than in systems where acrylic double bonds are polymerized. Furthermore, the protons generated by the photoinitiator are more stable than the free radicals generated in radical systems. Hence, curing can continue after the article being coated has passed beyond the uv source. However, reaction rates are slower than for free-radical polymerized acrylate coatings. High speed curing can be carried out by combining uv exposure to release the acid catalyst and baking to give a rapid cross-linking reaction.

Cationic systems have fewer limitations. They are not air inhibited, the photoinitiators do not initiate photodegradation reactions, there is less shrinkage during curing, and the stability of the acids generated permits migration if the initiator through a pigmented film, allowing cure of somewhat thicker pigmented films. However, commercial adoption has been slow, possibly because of higher costs and slower cure rates at ambient temperatures.

There are many advantages to radiation curing, but there are also limitations. Only flat surfaces that can be passed under the focused uv source, or cylindrical surfaces that can be rotated under the source, can be practically cured by this method. Although it is desirable in many cases to make coatings that cure at low temperatures, in uv-cure systems, as in all other coatings, the T_g of the final cured film is limited to temperatures a little above the temperature of the film during curing by the limitation of available free volume on the cross-linking reactions. Radical initiated uv-cure systems are poor candidates for coatings requiring good exterior durability, because the residual photoinitiator initiate degradation reactions. Solvent-free pigmented uv-cure coatings are limited not only by the depth of penetration of the uv radiation but also by the effect of the pigmentation on flow properties. Even for gloss coatings, such as exterior white can coatings, the amount of pigmentation is sufficient to increase the viscosity enough to affect leveling adversely. As a result of the substantial shrinkage that occurs in a very short time in the uv curing of acrylate systems using radical photoinitiators, adhesion to smooth surfaces such as metals is generally reduced by the stresses in the film. Such stresses can frequently be relieved by heat treatment after curing. Also, costs tend to be relatively high.

Uses that have developed for uv curing reflect the special advantages of the system rather than replacement to reduce VOC emissions or energy consumption. Clear coatings on heat-sensitive flat plastic substrates, where rapid cure is needed, is an area where uv curing has found application, eg, uv-cure abrasion-resistant topcoats for vinyl plastic flooring. These coatings are made using acrylic ester-terminated polyurethanes as the oligomers. Other examples of application to heat-sensitive substrates are coatings for thin wood veneers used for door skins and coatings for paper.

The high solids of uv-cure systems can also be important in some cases. For example, uv curing fillers for particleboard for use in paneling and furniture has the advantage over conventional coatings that the high solids permit filling of the rough surface in one coat rather than using multiple coats of solvent-based systems. In printed circuit boards, uv-cure systems are widely used. Another application, which illustrates the advantages of high cure rates, is the use of uv cure clear coatings of glass optical fibers in waveguides.

14.4. Powder Coating.   Another coating technique that substantially reduces VOC emissions is powder coating. Worldwide production of powder coatings in 1999 was ~527,000 metric tons (126). A variety of coating types and application methods are used and powder coatings have been the most rapidly growing part of the coatings industry. There are three methods of application: passing heated objects to be coated into a fluidized bed of powder particles suspended in air, electrostatically spraying grounded articles at ambient temperature with powder, and flame spraying.

Thermoplastic powders are frequently applied by a fluidized-bed process, which results in relatively thick films. As the coating builds up, the temperature at the surface drops. The last particles picked up stick to the surface but do not completely fuse into the film. A conveyor transfers the article into an oven when fusion and leveling are completed. The polymers are usually vinyl chloride copolymers, polyolefins, or polyesters, especially scrap poly(ethylene) terephthalate.

Thermosetting powders are generally applied by electrostatic spray. The sprayed parts are carried into an oven for fusion, leveling, and cross-linking. The T_g of the powder coating before curing must be above the storage temperature in order to avoid sintering. As a result, the range of physical properties of the final films is limited. The high T_g of the powder before application requires high baking temperatures. In order to coalesce after application, the temperature must be sufficiently above the T_g for the free volume to be adequate for ready coalescence. The T_g of the resins used in powder coatings is controlled by the monomers and molecular weight. It has been reported that, at least in one case, it is advantageous to use higher molecular weight, more flexible resins because these can have adequate package stability but flow more easily at higher temperatures than a lower molecular weight resin of similar T_g with more rigid chains. The problem of coalescence and flow is further compounded by the decrease in free volume and increase in viscosity resulting from the cross-linking reaction that is simultaneously proceeding. Although leveling is adequate for many applications, it can be a limitation for powder coatings.

The range of suitable resin compositions is much narrower than in liquid coatings. The T_g of the powder must be sufficiently high to prevent sintering during storage. The components are blended and the pigments are dispersed in the vehicle by passing through a heated extruder that subjects the material to high shear stress. The cross-linking reaction must proceed slowly enough under the conditions encountered in the extruder that very little polymerization occurs during the process. Because some cross-linking does occur, the number of times that the same material passes through the extruder must be limited as well as the fraction of recycled material that is incorporated in any batch. These considerations place a premium on raw material control for resins as well as pigments. In solvent coatings, the final viscosity and application properties can be adjusted by the additions of solvents and varying their evaporation rates. In powder coatings, on the other hand, the application properties are governed by the resins alone.

The BPA epoxy resins, with dicyandiamide as the cross-linking agent, are widely used in coatings where exterior durability is not required. Phenolic resins are used as cross-linkers for epoxy resins when greater chemical resistance is required. So-called hybrid polyester powder coatings are based on carboxylic acid terminated polyesters and BPA epoxy resins. These have superior color retention at lower cost but still inferior exterior durability. For applications where exterior durability is required, triglycidyl isocyanurate (TGIC) is used along with carboxylic acid functional polyesters. However, concern about possible toxic hazards of TGIC has led to work on alternate systems. Tetra-(2-hydroxyalkyl)bisamides are used as cross-linkers for carboxylic acid functional resins (127). Tetramethoxymethylglycoluril is also used as a cross-linker for hydroxyfunctional polyesters (128).

Hydroxy-functional polyester or acrylic resins cross-linked with blocked isocyanates give powder coatings having superior exterior durability combined with abrasion resistance. Caprolactam blocked isophorone diisocyanate is the most widely used blocked isocyanate. Blocked isocyanate powder coatings tend to give better leveling than most other powder coatings perhaps because the caprolactam released by the unblocking reaction volatilizes only slowly. While the caprolactam is still present in the film, it can lower viscosity and may help promote coalescence and leveling. Polyfunctional uretdiones are increasingly used as blocking agents since there is no volatile blocking agents. See Ref. 129 for a review of blocked isocyanate powder coatings.

In flame spray powder coating, a thermoplastic powder is applied by propelling the powder through a flame in which it melts and is then deposited on the substrate to form a film. Poly(dimethyl terephthalate) powders are most commonly used. The largest advantage over other means of application is that it can be done in the field on objects like grain railcars. Also, since it is a nonelectrostatic application, nonconductors such as concrete, wood, and plastics can be coated (130).

Powder coatings have many advantages. The VOC emissions approach zero because no solvents are used. Compared to many solvent-borne baking coatings, fuel cost for heating ovens are low, even though the baking temperature may be higher. This results from the substantial increase in recirculation of the hot air in the ovens, because there is less need to exhaust air to keep the solvent concentration below the lower explosive limit. Lack of solvent permits even application of coatings to the inside of pipes without the solvent wash encountered when solvent coatings are applied to an enclosed area. As compared to electrostatic spray applied solvent-borne or waterborne coatings, utilization efficiency of powder coatings can be much higher because the oversprayed powder can be collected and reused. On the other hand, the overspray from liquid coatings must be caught in a water wash spray booth. This overspray cannot be directly recycled and generally the resulting sludge becomes a solid waste disposal problem.

There are disadvantages or limitations to powder coatings as well. Only substrates that can withstand the relatively high baking temperatures can be coated. Color changeover in the application line requires shutting down the line and cleaning the booth. Therefore, powder coating is most applicable to product lines where long runs of the same color are processed. Color matching is more difficult than with conventional coatings. Because the final coatings cannot be blended or shaded with tinting colors, the whole batch must be reworked if the color match is not satisfactory. By careful control of incoming raw materials and processing variables, satisfactory reproducibility of color matching for many applications is possible. Metallic colors showing the color change with angle of viewing such as are widely used in automotive topcoats cannot be made in powder coatings although other metallic colors can be obtained.

Examples of applications for powder coatings include pipe lining, coatings for rebars, clear coats and primer surfacers for automobiles, underbody automotive parts, wheels, garden tractors, home appliances, playground equipment, metal furniture, fire extinguishers, and many others. As VOC restrictions become more restrictive, powder coatings can be expected to increase in usage.

15. Architectural Coatings

About 45% of the volume of all coatings produced in the United States in 1999 were architectural coatings (2.16×10^9 L (1)). These coatings are designed to be applied to residences and offices, and for other light-duty building purposes. In contrast to most product coatings, they are designed to be applied in the field, in some cases by contractors but in large measure by do-it-yourself consumers. A wide range of products is involved.

15.1. Flat Wall Paint. The largest volume of architectural coatings is flat wall paint. In the United States, almost all flat wall paint is latex-based. Latex paints have the advantages of low odor, fast drying, easy clean-up when wet, durability of color and film properties, and lower VOC emissions as compared to oil- or alkyd-based paints. They are manufactured primarily as white base paints, which are tinted by the retailer to the color selected by the customer from large collections of color chips. Two or three base whites are made, one for pastel shades, one for deep shades, and sometimes one for medium color shades. This method of marketing has the advantage of being able to supply a wide choice of colors while carrying a relatively low inventory. It is critical to maintain the same level of hiding when formulating new white base paints. The tinting colors used by retail stores are designed to be used for a wide variety of formulas.

The performance of quality grades of latex flat wall paints is excellent in most respects, thus the emphasis in technical efforts is on cost reduction. In most cases, vinyl acetate copolymer latexes are used. When high scrub resistance is required a higher cost latex having a high content of acrylate ester monomers is used. The least expensive component in the paint, on a volume cost basis, is the inert pigment. Therefore, the inert pigment content is maximized. In order to have low gloss, flat paints have a PVC near the CPVC. In order to have the highest inert pigment content, formulas are developed having CPVC as high as possible by using a broad distribution of particle sizes.

In ceiling paint, paints having PVC higher than CPVC give excellent performance. The cost of such paints is lower because of the higher inert pigment content and, at the same time, hiding is better because the final film incorporates air voids. For ceiling paints, one coat hiding is particularly important. On a ceiling the reduction in stain and scrub resistances is not a significant disadvantage.

On a volume basis, the most expensive component of flat wall paints is the TiO_2. Therefore, significant efforts are applied to reach a standard hiding with the minimum possible level of TiO_2. The efficiency of hiding by TiO_2 decreases as the TiO_2 concentration is increased above $\sim$10 vol%; it becomes economically inefficient at roughly 18 vol% in the dry film. Above a PVC of $\sim$22, hiding actually decreases with increasing TiO_2 concentration. In the range of 15–18 vol% of TiO_2, the efficiency of hiding can be increased by using as part of the inert pigment some material having a particle size in the same range as that of the TiO_2. This permits substituting inexpensive inert pigment for some of the TiO_2 while maintaining hiding and tinting strength. A mathematical model has been developed to predict the optimum inert pigment size and concentration for a given formulation (131).

Another method of minimizing TiO_2 requirement is the use as pigments of high T_g latexes, the particles of which contain air voids, eg, Ropaque (Rohm and Haas Co.) (132). When used in paints having a PVC slightly lower than CPVC, the air voids inside the high T_g polymer particles in the film provide hiding and permit lower TiO_2 concentrations. Such paints can have stain and scrub resistances equal to other latex paints.

Most wall paints are applied by roller. During roller application, some latex paints give excess spatter, ie, small particles of paint are thrown into the air when the paint is applied. This is the result of the growth of the filaments that are produced by film splitting to the length where they break in two places rather than in one. Paints having high extensional viscosity exhibit this behavior (133). Extensional viscosity increases when high molecular weight water-soluble polymers with very flexible backbones are used as thickeners, leading to increased spattering. Minimum spattering is obtained with low molecular weight water-soluble thickeners having rigid segments in the polymer backbone, such as low molecular weight hydroxyethylcellulose.

A continuing challenge in latex paints of all kinds is to protect against the effects of bacteria in the can and mildew growth after application. Excellent housekeeping in a latex paint plant is essential to minimize the introduction of bacteria into the paint. A bacteriocide incorporated in the paint can kill the bacteria without destroying the enzyme, and viscosity can drop even using adequate bacteriocide in the paint. Bacteria can generate foul odors or, in extreme cases, sufficient gas pressure inside the cans to blow the can lids off. The more common problem is that bacteria feed on many water-soluble cellulosic thickeners used in latex paints. The enzymes split the cellulose molecules reducing the molecular weight, leading to a reduction in viscosity. It is also essential to avoid introducing enzymes that might have been generated through bacterial growth. Mildew can grow on almost any paint surface but latex paints are particularly susceptible, and therefore contain a fungicide. Many fungicides and bacteriocides are available. Testing is difficult because fungal and bacterial growth are so dependent on ambient conditions. See (134) for a review of additives.

15.2. Exterior House Paints. Latex paints dominate the exterior house paint market in the United States because of superior exterior durability of latex paints (resistance to chalking, checking, and cracking) compared to oil or alkyd paints. Another advantage of latex paints used on wood surfaces is that because of high moisture vapor permeability, they are much less likely to blister than oil or alkyd paints.

Another application where latex paints show outstanding performance is over masonry such as stucco or cinder block construction. This performance results from saponification resistance in the presence of the alkali from the cement. Because masonry surfaces are porous, having both small and large pores, the low viscosity external phase of a latex paint can penetrate rapidly into the small pores, causing a rapid increase in the viscosity of the remaining paint. The bulk paint, in turn, sinks into the larger holes more slowly than a solution-based paint. Thus, less latex paint is required to cover the same surface area as compared to alkyd paints. The penetration of various components of latex paint into such porous substrates has been studied (135).

There are limitations to the applicability of exterior latex house paints providing a continuing market for oil or alkyd exterior house paints. Because film formation from latex paints occurs by coalescence, there is a temperature limit, below which the paint should not be applied. This temperature can be varied by choice of the T_g of the latex polymer and the amount of coalescing agent in the formula. In the United States, most latex paints are formulated for application at temperatures >5–7°C. If painting must be done when the temperature is <5–7°C, oil or alkyd paint is preferable.

Another limitation is that latex paints do not give good adhesion when applied over a chalky surface such as weathered oil or alkyd paint. The latex particles are large compared to the pores between the "chalk" particles on the surface of the old paint. When a latex paint is applied, no vehicle penetrates between the chalk particles to the surface beneath and there is nothing to provide adhesion to the underlying substrate. This problem can be minimized by replacing part of the latex polymer solids with a modified drying oil emulsified into the latex paint. When the film dries, the emulsion breaks and the modified drying oil can penetrate between the chalk particles to the substrate. The exterior durability of the exterior latex paint is reduced by incorporating modified drying oil. It is common to use an oil- or alkyd-based primer over a chalky surface then apply a latex topcoat. The problem of adhesion to chalky surfaces is becoming less important with time because, as latex paints are more commonly used, there are fewer chalky surfaces to repaint.

Cost is an important factor in exterior house paints that are designed with low gloss because this permits higher pigment loading especially in high CPVC paints. Low gloss also reduces dirt pickup, which is a greater problem with latex paints than alkyd paints because they remain thermoplastic. Calcium carbonate pigments should be avoided. The latex paint film is permeable to water and carbon dioxide. Calcium carbonate can dissolve in the solution of carbonic acid and the soluble calcium bicarbonate leaches out to the surface of the film. The reaction is reversed when water evaporates, and the calcium bicarbonate decomposes depositing a frost of calcium carbonate on the surface of the film.

15.3. Gloss Enamels.   About one-half of the gloss paint or enamels sold are based on alkyd resins. Professional painters particularly favor the continued use of alkyd gloss paints. The need for reduction of VOC emission levels, especially in California, has led to efforts to increase the solids content of alkyd paints or overcome the disadvantages of latex gloss paints. It is not possible to make latex enamels that have as high a gloss as solution-based coatings. In solution-based coatings, gloss is enhanced during the solvent evaporation by the formation of a thin surface layer that has a lower pigment content than the average PVC of the coating as a whole. In many cases, there is a layer of roughly 1 μm on the surface of the gloss paint film that contains essentially no pigment particles. In latex paints, formation of such a clear surface layer is not possible, and the gloss is lower. The ratio of pigment to binder at the surface of a latex paint film can be decreased somewhat by using a finer particle size latex.

Also reducing the gloss of latex paints is the haze resulting from the incompatibility of surfactants with the latex polymer film and blooming of surfactant to the surface. Whereas this can be ameliorated by making latexes with as low a surfactant concentration as possible, it probably can never be completely eliminated. On the other hand, the far superior resistance of the latex polymer to photoxidation as compared to any alkyd leads to superior gloss retention by latex paints. The difference in gloss retention is particularly large in exterior applications. Furthermore, the latex paint films are more resistant to cracking, checking, and blistering.

An important limitation of gloss latex paints is not gloss, rather it is the greater difficulty of getting adequate hiding from one coat. In professionally applied paint, the cost of labor is higher than the cost of the paint. Alkyd paints, which provide hiding in one coat, are favored by painting contractors in many cases over latex paints that commonly require two coats. There are several factors involved in the hiding differences that exist between gloss latex and gloss alkyd paints. The volume solids of latex gloss paints are substantially lower than the volume solids of alkyd gloss paints. The solids of latex paints are commonly ∼33% compared to ∼67% for an alkyd paint. In order to apply the same dry film thickness, twice as much wet paint has to be applied. The main factor controlling the wet film thickness initially applied is the viscosity of the paint at the high shear rates experienced during brush application. In order to apply a thicker wet film, latex paint should have a higher viscosity at high shear rate than is appropriate for an alkyd paint. For many years the high shear viscosity of most latex gloss paints was lower than that of a corresponding alkyd paint. The film thickness applied is also affected by the judgment of the painter and the fact that the wet hiding of latex paints is substantially higher than the dry hiding. When the water (refractive index 1.33) evaporates from the film, the latex particles (refractive index ~1.5) coalesce. The number of interfaces for scattering light drops and hiding decreases. This effect is augmented by the larger difference in refractive index between rutile TiO_2 (refractive index 2.76) and water, which is the interface in the wet film, as compared to TiO_2 and the latex polymer that is the interface in the dry film. In the wet paint, the TiO_2 scatters light more efficiently because of the lower volume concentration as compared to the dry film. There is a similar difference in alkyd paints but the effect is smaller because the volume change during drying is less.

The largest factor affecting hiding by gloss latex paints is poor leveling. Assuming that a uniform dry film of 50 μm of a paint provides satisfactory hiding in some application, if the film has thinner areas of, eg, 35 μm, and thicker areas, eg, 65 μm, the hiding power of the uneven film is poor. Actually, the uneven film is even worse than a uniform 35-μm film, because the thick- and thin-film areas are right next to each other and the contrast in hiding is thus emphasized. The poorer leveling generally encountered for latex paints has several causes. First, in alkyd paints for brush application the solvent is very slow evaporating mineral spirits, whereas in latex paints the water evaporates more rapidly, unless the relative humidity is very high. As a result, the viscosity of the latex paint increases more rapidly after application. Second, the volume fraction of the internal phase in latex paints is much higher than in alkyd paints because for a latex paint both the binder and the pigment are in the dispersed phase. Therefore, the viscosity changes more rapidly with loss of volatiles than is the case for alkyd paints. In latex paint, leveling occurs only by flow driven by their low surface tension, whereas in the case of solvent-based brush applied paints the principal driving force is surface tension differential driven flow, which tends to lead to a more uniform film thickness (136).

The rheological properties of gloss latex paints influence leveling and hiding. Latex paints have exhibited a much higher degree of shear thinning than alkyd gloss paints, leading to paints having viscosity that is too low at high shear rate, and a subsequent applied film thickness that is too thin. At low shear rates, the viscosity is too high to permit adequate leveling. Latex paints recover viscosity rapidly after stopping a high shear rate. The reasons for the greater dependency of viscosity on shear rate in latex paints have not been fully elucidated. It appears, however, that it may at least partly result from flocculation of latex particles in the paint. Progress in minimizing this problem has been made by using associative thickeners in formulating latex paints. Many kinds of associative thickeners have been made that are all moderately low molecular weight, water-soluble polymers having occasional long-chain nonpolar hydrocarbon groups spaced along the backbone. Such thickeners reduce shear thinning of latex paints, perhaps by stabilizing the latex particles against flocculation. Latex paints formulated with associative thickeners have increased high shear viscosity allowing the application of thicker wet films. The thickeners afford reduced low shear rate viscosity and a slower rate of recovery of the low shear rate viscosity that improves leveling at the same time (137). The thicker wet film in itself also promotes leveling, because the leveling rate increases with the cube of wet film thickness.

Another shortcoming of latex paints, which is particularly evident in gloss paints, is the time required to develop full film properties. Latex paints dry to touch and even to handling much more rapidly than do alkyd paints, but the latex requires a much longer time to reach the full dry properties. For example, in blocking situations such as placing a heavy object on a newly painted shelf, or the problem of sticking windows and doors, latex paints require more time to develop block resistance than do alkyd paints. The initial coalescence of latex particles is rapid but the rate of coalescence is limited by free-volume availability. Because $(T - T_g)$ must be small, the free volume is small. This situation is helped by using coalescing solvents, but the loss of these solvents is diffusion rate controlled and the rate is affected strongly by free-volume availability.

Another important potential problem for use of gloss latex paints is adhesion to an old gloss paint surface when water is applied to the new dry paint film. Adequate adhesion to an old gloss paint surface is always a problem for any new coat of paint. It is essential to wash any grease off the surface to be painted and to roughen the surface by sanding. But after wetting with water, some latex paint films can be peeled off the old paint surface in sheets. The resistance to adhesion loss by wetting with water improves as the system ages, but for several weeks or even months wet adhesion can be a serious problem. Proprietary latexes have been developed that minimize this problem. It has been said that amine-substituted latex polymers exhibit greater resistance to loss of adhesion. Such polymers can be prepared using an amine-substituted acrylic monomer, such as 2-(dimethylamino)ethyl methacrylate, as a comonomer in preparing the latex or reacting the carboxylic acid groups on a latex polymer and hydroxyethylethyleneimine (1-aziridinoethanol) [1072-52-2].

16. Special Purpose Coatings

Special purpose coatings include those coatings that do not fit under the definition of architectural or product coatings. About 25% of the volume (1.22×10^9 L) of U.S. coatings production in 1999 falls into this category (1).

16.1. Maintenance Coatings. Heavy-duty maintenance coatings are applied to bridges, off-shore drilling rigs, chemical or petroleum refinery tanks, and similar structures. The applications are usually over steel at ambient temperatures and the objective is corrosion protection. In choosing a system, many factors must be taken into consideration: cleaning of the steel surface, film integrity, temperature of the painting operation, environment, materials, costs, etc.

For applications where the surface can be thoroughly cleaned and where maintenance of film integrity can be reasonably expected, it is most appropriate to use a primer without passivating pigments but having excellent wet adhesion.

For example, a two-package epoxy–amine coating would serve. This combination of epoxy resin and polyamine should be designed in such a way that the T_g of the fully reacted coating permits the reaction to go to completion in a reasonable time period (commonly a week) at the temperatures that prevail during the curing period. Epoxy–amine coatings have the further advantage that they can cure underwater, and hence are suitable for piers and off-shore installations. The topcoats should be designed to minimize oxygen and water permeability. Chlorinated polymers such as vinyl chloride copolymers, vinylidene chloride (1,1-dichloroethene) [75-35-4] copolymers, or chlorinated rubber are used in topcoats. Urethane coatings based on aliphatic isocyanates are used because of good abrasion resistance. Where applicable, use of leafing aluminum pigment in the final coat is desirable to provide a further barrier to water and oxygen permeation.

In situations where complete cleaning of the steel is not possible and where film ruptures are to be expected, primers having passivating pigments or zinc-rich primers are used. Cost is generally lower and mechanical integrity of the films is generally greater using passivating pigment primers. On the other hand, corrosion protection can generally be extended for longer time periods using zinc-rich primers. The substrate should be cleaned as well as possible before painting, generally by sandblasting, and primer should be applied as soon as possible after the surface has been sandblasted. Topcoats for passivating pigment primers are the same as for epoxy–amine primers. When using zinc-rich primers the effectiveness of the primer depends on maintaining its porous structure, therefore, the next coating should not penetrate down into the pores of the primer coat. For solvent-based topcoats, this is generally accomplished by applying a very thin film on the primer surface. The viscosity of a thin film increases rapidly as a result of fast solvent evaporation, minimizing penetration into the primer pores. The application requires highly skilled workmanship and careful inspection. Further topcoat can then be applied without concern for penetration into the primer.

Application of latex paints directly over a zinc-rich primer essentially eliminates the penetration problem. The latex polymer particles are large compared to the primer pores, and after coalescence the viscosity of the polymer is so high that it does not penetrate into the pores. The water and oxygen permeability of latex paints is generally higher than that of solvent-based paints. Thus, it is especially desirable to incorporate platelet pigments such as mica or leafing aluminum. Use of vinylidene chloride–acrylate ester copolymer latexes has been recommended because of the lower water permeability. Field performance using latex paints over zinc-rich primers in such applications as highway bridges has been reported to equal solvent-borne paint performance.

Not only are latex paints being used over zinc-rich primer, they are also increasingly being used directly on steel for corrosion protection. The steel should be thoroughly cleaned, eg, by sandblasting, and the first coat of paint applied quickly. A sandblasted steel surface is highly activated and can flash rust when water contacts the surface, ie, a layer of hydrated iron oxide forms almost instantly. This can be avoided when a latex paint is applied by incorporating a low volatility amine such as 2-amino-2-methyl-1-propanol [124-68-65] in the formula. Latexes have been designed that have superior wet adhesion to metal, eg, by using a small amount of 2-(dimethylamino)ethyl methacrylate as a comonomer. Because the size of the latex particles is large compared to the cross-section of surface pores in steel and the viscosity of the polymer after coalescence is high, the vehicle of latex paints cannot fully penetrate the micropores on the steel surface. Therefore, passivating pigments should be incorporated in the primer formula to provide corrosion protection in these areas. Passivating pigments for latex paints must be carefully selected. Solubility of the pigment must be high enough to provide the desired passivating action, but low enough that the ions do not destabilize the latex dispersion. Recommendations are that latex paints should not be applied on highway bridges when the temperature is <10°C and the relative humidity >75%.

16.2. Marine Coatings. Many marine coatings play a key role in extending the life of ships by corrosion protection (138). Products are similar to those used for other heavy-duty maintenance applications. Ship bottom coatings, designed to delay the growth of barnacles, algae, and other marine life on underwater hulls, are widely called antifouling paints. Cuprous oxide has been used as a toxicant in antifouling coatings, but it was replaced in large measure by toxicant pigments based on tributyltin derivatives because of the longer service life of the latter. The service life of antifouling paints was substantially extended by the development of polymers having tributyltin groups covalently bonded to the polymer. The polymers were designed so that organotin compounds were slowly released by hydrolysis (139). Because of the toxicity of tin, analogous copper based with degradable organic toxicants as supplementary antifouling agents are being used (140).

All toxicants used to control fouling are toxic to marine life in harbors, and although there is considerable controversy with regard to environmental risk, regulations are becoming more restrictive. Efforts are now concentrated on means other than toxicants to control marine growths on ship bottoms. Some progress has been reported using silicone coatings that have such low surface tensions that marine growth has difficulty adhering to the surface. For example, a silicone elastomer system has been commercialized for the fast ferry market; their speeds of >30 knots per hour displace fouling organisms that manage to attach to the surface (141).

16.3. Automobile Refinish Paints. Paint for application to automobiles after they have left the assembly plant (refinish paints) is a significant market. Although some of this paint is used for full repainting, especially of commercial trucks, most is used for repairs after accidents, commonly just one door or part of a fender, etc. In order to be able to serve this market, it is necessary to supply paints that match the colors of all cars and trucks, both domestic and imported, that have been manufactured over the previous 10 years or so. Repair paints for the larger volume car colors are manufactured and stocked, but for the smaller volume colors formulas are supplied by the coatings manufacturer to the paint distributor that permit a color match for any car by mixing standard bases.

The nitrocellulose lacquers used for primers are being replaced with waterborne primers such as 2K waterborne urethane coatings (67). Topcoats were formerly nitrocellulose lacquers but are now almost exclusively thermosetting enamels. In the United States, many shops cure the enamels at room temperature; in Europe, cure is commonly carried out at temperatures of 60–75°C. Most cars now have a base coat over the primer and a clear coat as the topcoat. Most base coats are metallic colors that have been applied with enamels wth ~40% volume solids. Increasingly waterborne base coats such as 2K waterborne urethane coatings or latex-based coatings. Solvent borne 2K clear coats such as polyisocyanate/aldimine or diaspartates are frequently used (60). The 2K waterborne urethane clear coats are being introduced.

When solvent borne coatings are used, HVLP are being increasingly used. These guns give better transfer efficiency, and hence lower paint usage and lower VOC emissions.

16.4. Other Special Purpose Coatings. Large volumes of traffic paint are used to mark the center lines and edges of highways. The white paints are pigmented with TiO_2 and the yellow paints have been pigmented with chrome yellow. Concern about the toxic hazard of using chrome yellow is leading to its replacement with organic yellow pigments. The paint must dry fast enough after application so that a car can drive over it within minutes. The most widely used vehicles are alkyds with chlorinated rubber and fast evaporating solvents. Immediately after application, glass reflector beads are applied. To an increasing degree, hot melt coatings are being applied and, in high traffic areas where the lifetime of paint is short, paints are being replaced with thermoplastic tapes. Other examples of special purpose coatings include aerosol-packaged coatings, swimming pool paints, and nonstick coatings for cooking utensils.

17. Economic Aspects

The value of the international coatings market in 1999 was $68.6 billion of which $23.4 billion were in Europe, $19.0 billion in North America, $15.5 billion in Asia, and $10.4 billion in the rest of the world (142).

The value of coatings has been growing but the volume of coatings has been growing slowly or declining. Because there are very large differences in the prices per unit volume, changes in product mix can give distorted comparisons of value/volume totals. A primary factor affecting both price and volume has been changes resulting from the effect of reducing VOC. For any particular use, there is usually a required dry film thickness for the coating after application. High solids coatings require lesser volumes of coating to achieve the same film thickness than conventional solvent content coatings. Therefore, for the same amount of applied coating, the volume of coating sold decreases. Because solvents are usually the lowest cost components of the coating, cost per unit volume increases.

Coating volume is also affected by other technologies that can substantially reduce the need for coatings. For example, in many cases coatings are not required on molded plastics products that have replaced coated metal products. High pressure laminates are increasingly used as furniture tops. In some cases, the replacements have been from one kind of coating to another. For example, recoated siding for residential housing, both metal and hardboard, have substantially reduced the potential market for exterior house paint for two reasons. The initial coating is sold industrially rather than as architectural coating and also the durability of the coatings can be much greater than that of field applied paint increasing the time interval before repainting. Such coated siding has to a degree been replaced in turn by uncoated vinyl siding.

The effect of environmental regulation and the increasing recognition of immediate or potential toxicity hazards of coating components has led to a technical revolution in the coatings field and will continue as strictness of environmental regulations increases. The number of companies has been declining rapidly due to mergers. A substantial fraction of the industry is concentrated in chemical companies and a few large independent companies. However, there are still a number of small- and medium-sized companies that generally serve specialized segments of the business or restricted geographical areas. It is estimated that 45% of the companies have <20 employees.

Whereas the larger companies are international in scope, imports and exports are relatively small. In general, requirements in different countries are quite different and there is generally a need for relatively close contact between consumer and supplier so that the U.S. industry faces little competition from imported coatings and, conversely, exports play a minor role in the field.

BIBLIOGRAPHY

"Coatings (Industrial)" in *ECT* 1st ed., Vol. 4, pp. 145–189, by H. C. Payne, American Cyanamid Co.; "Coatings, Industrial" in *ECT* 2nd ed., Vol. 5, pp. 690–716, by W. von Fischer, Consultant, and E. G. Bobaleck, University of Maine; in *ECT* 3rd ed., Vol. 6, pp. 427–445, by S. Hochberg, E. I. du Pont de Nemours & Co., Inc.; "Coatings, Resistant" in *ECT* 3rd ed., Vol. 6, pp. 455–481, by C. G. Munger, Consultant; "Coatings" in *ECT* 4th ed., Vol. 6, pp. 669–746, by Z. W. Wicks, Jr., Consultant; "Coatings" in *ECT* (online), posting date: December 4, 2000, by Z. W. Wicks, Jr., Consultant.

CITED PUBLICATIONS

1. A. H. Tullo, *Chem. Eng. News* **78**(41), 19 (2000).
2. H. Burrell, *Off. Dig Fed. Soc. Paint Technol.* **34**(445), 131 (1962).
3. Z. W. Wicks, Jr., *Film Formation*, Federation of Societies for Coatings Technology, Blue Bell, Pa., 1986.
4. W. P.Mayer and L. G. Kaufman, *FATIPEC Fed. Assoc. Tech. Ind. Paint Vernis Emaux Encres Impr. Eur. Coat. Cong. XVII* **I**, 110 (1984).
5. D. J. Newman and C. J. Nunn, *Prog. Org. Coat.* **3**, 221 (1973).
6. W. H. Ellis, *J. Coat. Technol.* **55**(696) 63 (1983).
7. H. P. Blandin, J. C. David, J. M. Vergnaud, J. P. Illien, and M. Malizewicz, *Prog. Org. Coat.* **15**, 163 (1987).
8. H. Stutz, K.-H. Illers, and J. Mertes, *J. Polym. Sci. B. Polym. Phys.* **28**, 1483 (1990).
9. S. P. Pappas and L. W. Hill, *J. Coat. Technol.* **53**(675), 43 (1981).
10. K. Dusek and I. Havlicek, *Prog. Org. Coat.* **22**, 145 (1993).
11. D. E. Fiori and R. W. Dexter, *Proc. Water-Borne Higher-Solids Coat. Symp.*, New Orleans, La. 1986, p. 186.
12. M. A. Winnik, in P. A. Lovell and M. S. El-Aasser, eds., *Emulsion Polymerization and Emulsion Polymers*, John Wiley & Sons, New York, 1997, pp. 467–518.

13. M. A. Winnik and J. Feng, *J. Coat. Technol.* **68**(852), 39 (1996).
14. S. A. Eckersley and B. J. Helmer, *J. Coat. Technol.* **69**(864), 97 (1997).
15. A. Trapani, K. Wood, T. Wood, and G. Munari, *Pitture Vernici Eur.* **71**(9), 14 (1995).
16. J. W. Taylor and D. W. Bassett, in J. E. Glass, ed., *Technology for Waterborne Coatings*, American Chemical Society, Washington, DC, 1997, p. 137.
17. G. Pollano, *Polym. Mater. Sci. Eng.* **77**, 73 (1996).
18. Y. Inaba, E. S. Daniels, and M. S. El-Aasser, *J. Coat. Technol.* **66**(833) 63 (1994).
19. M. J. Chen, and co-workers *J. Coat. Technol.* **69**(875) 49 (1997).
20. J. M. Geurts, J. J. G. S. van Es, and A. L. German, *Prog. Org. Coat.* **29**, 107 (1996).
21. M. J. Collins, J. W. Taylor, and R. A. Martin, *Polym. Mater. Sci. Eng.* **76**, 172 (1997).
22. T. Nabuurs, R. A. Baijards, and A. L. German, *Prog. Org. Coat.* **27**, 163 (1996).
23. P. A. Reynolds, in A. Marrion, ed., *The Chemistry and Physics of Coatings*, Royal Society of Chemistry, London, 1994.
24. T. C. Patton, *Paint Flow and Pigment Dispersion*, 2nd ed. Wiley-Interscience, New York, 1979.
25. Z. W. Wicks, Jr., and co-workers *J. Coat. Technol.* **57**(725) 51 (1985).
26. L. W. Hill and Z. W. Wicks, Jr., *Prog. Org. Coat.* **10**, 55 (1982).
27. R. A. Dickie, *J. Coat. Technol.* **64**(809) 61 (1992).
28. J. W. Martin, S. C. Saunders, F. L. Floyd, and J. P. Wineburg, *Methodologies for Predicting Service Lives of Coating Systems*, Federation of Societies for Coatings Technology, Blue Bell, Pa., 1996.
29. L. W. Hill, in J. V. Koleske, ed., *Paint and Coating Testing Manual*, 14th ed. ASTM, Philadelphia, Pa. 1995, p. 534.
30. D. J. Skrovanek, *Prog. Org. Coat.* **18**, 89 (1990).
31. M. B. Roller, *J. Coat. Technol.* **54**(691) 33 (1982).
32. D. Y. Perera and P. Schutyser, *FATIPEC Congress Book*, Vol. I, 1994, p. 25.
33. P. J. Greidanus, *FATIPEC Congress Book*, Vol. I, 1988, p. 485.
34. R. M. Evans, in R. R. Myers and J. S. Long, eds., *Treatise on Coatings*, Vol. 2, Part I, Marcel Dekker, New York, 1969, pp. 13–190.
35. K. L. Rutherford, R. I. Trezona, A. C. Ramamurthy, and I. M. Hutchings, *Wear* **203–204**, 325 (1997).
36. T. Hamada, H. Kanai, T. Koike, and M. Fuda, *Prog. Org. Coat.* **30**, 271 (1997).
37. F. N. Jones and co-workers, *Prog. Org Coat.* **34**, 119, (1998).
38. R. A. Ryntz, A. C. Ramamurthy, and J. W. Holubka, *J. Coat. Technol.* **67**(842) 23 (1995).
39. *Annual Book of Standards*, Vols. 06.01, 06.02, 06.03, ASTM, Philadelphia, PA., new editions available annually.
40. J. V. Koleske, *Paint and Coatings Testing Manual*, 14th ed., ASTM, Philadelphia, Pa., 1995.
41. A. Valet, *Light Stabilizers for Paints*, translated by M. S. Welling, Vincentz, Hamburg, Germany, 1997.
42. D. R. Bauer, *J. Coat. Technol.* **69**(864) 85 (1997).
43. J. L. Gerlock, A. V. Kucherov, and M. E. Nichols, *J. Coat. Technol.* **73**(918) 45 (2001).
44. D. Y. Perera, *Prog. Org. Coat.* **28**, 21 (1996).
45. E. F. Plueddemann, *Prog. Org. Coat.* **11**, 297 (1983).
46. R. A. Ryntz, *Painting of Plastics, Federation of Societies for Coatings Technology*, Blue Bell, Pa., 1994.
47. R. A. Ryntz, D. Britz, D. M. Mihora, and R. Pierce, *J. Coat. Technol.* **73**(921) 107 (2001).
48. T. R. Bullett and J. L. Prosser, *Prog. Org. Coat.* **1**, 45 (1972).
49. G. L. Nelson, Adhesion in (40) pp. 513–524.
50. W. Funke, *J. Coat. Technol.* **55**(705) 31 (1983).

51. R. Athey and co-workers *J. Coat. Technol.* **57**(726) 71 (1985).
52. E. P. M. van Westing, G. M. Ferrari, F. M. Geemen, and J. H. W. de Wit, *Prog. Org. Coat.* **23**, 89 (1993).
53. B. R. Appleman, *J. Coat. Technol.* **62**(787) 57 (1990).
54. M. S. El-Asser and P. A. Lovell, eds. *Emulsion Polymerization and Emulsion Polymers*, John Wiley & Sons, Inc., New York, 1997.
55. J. C. Padget, *J. Coat. Technol.* **66**(641) 89 (1994).
56. G. Pollano, *Polym. Mater. Sci. Eng.* **77**, 73 (1996).
57. L. W. Hill, *J. Coat. Technol.* **64**(808) 29 (1992).
58. U.S. Pat. 5,254,651 (Oct. 19, 1993) V. Alexanian, R. G. Lees, and D. E. Fiori, (to American Cyanamid Co.)
59. F. W. van der Weij, *J. Polym. Sci.: A. Polym. Chem.* **19**, 381 (1981).
60. N. Hazel, I. Biggin, J. Kersey, and C. Brooks, *Proc. Waterborne Higher-Solids Powder Coat. Symp.*, New Orleans, La., 1997, p. 237.
61. D. A. Wicks and P. E. Yeske, *Prog. Org. Coat.* **30**, 265 (1997).
62. U.S. Patent 5,605,965 (Feb. 25, 1997), J. W. Rehfuss and D. L. St. Aubin (to BASF Corp.).
63. M. L. Green, *J. Coat. Technol.* **73**(918) 55 (2001).
64. D. A. Wicks and Z. W. Wicks, Jr., *Prog. Org. Coat.* **36**, 148 (1999) and **41**, 1 (2001).
65. Z. W. Wicks, Jr., and B. W. Kostyk, *J. Coat. Technol.* **49**(634) 77 (1977).
66. J. C. Padget, *J. Coat. Technol.* **66**(839) 89 (1994).
67. Z. W. Wicks, Jr., D. A. Wicks, and J. W. Rosthauser, *Prog. Org. Coat.* in press.
68. J. L. Massingill, *J. Coat. Technol.* **63**(797) 47 (1991).
69. T. Agawa and E. D. Dumain, *Proc. Waterborne Higher-Solids Powder Coat. Symp.*, New Orleans, La., 1997, p. 342.
70. L. W. Hill and Z. W. Wicks, Jr., *Prog. Org., Coat.* **8**, 161 (1980).
71. Z. W. Wicks, Jr., E. A. Anderson, and W. J. Culhane, *J. Coat. Technol.* **54**(668) 57 (1982).
72. J. T. K. Woo and co-workers, *J. Coat. Technol.* **54**(689) 41 (1982).
73. S. N. Belote and W. W. Blount, *J. Coat. Technol.* **53**(681) 33 (1981).
74. J. D. Hood, W. W. Blount, and W. T. Sade, *J. Coat. Technol.* **58**(739) 49 (1986).
75. *TONE Polyols*, technical bulletin, Specialty Polymers and Composites Division, Union Carbide Corp., Stamford, Conn., 1986.
76. F. N. Jones, *J. Coat. Technol.* **68**(852) 25 (1996).
77. T. E. Jones and J. M. McCarthy, *J. Coat. Technol.* **68**(844) 57 (1995).
78. R. Engelhardt, *Proceedings of the Waterborne Higher-Solids Powder Coat. Symp.*, New Orleans, La., 1986, p. 14.
79. L. Kangas and F. N. Jones, *J. Coat. Technol.* **59**(744) 99 (1987).
80. D. B. Larson and D. B. Emmons, *J. Coat. Technol.* **55**(702) 49 (1983).
81. G. Ostberg and Bergenstahl, *J. Coat. Technol.* **68**(858) 39 (1996).
82. W. A. Finzel and H. L. Vincent, *Silicones in Coatings, Federation of Societies for Coatings Technology*, Blue Bell, Pa., 1996.
83. Z. W. Wicks, Jr., M. R. Appelt, and J. C. Soleim, *J. Coat. Technol.* **57**(726) 51 (1985).
84. T. Li and J. C. Graham, *J. Coat. Technol.* **65**(821) 64 (1993).
85. W. H. Ellis, *Solvents, Federation of Societies for Coatings Technology*, Blue Bell, Pa., 1986.
86. P. E. Pierce and R. T. Marcus, *Color and Appearance*, American Federation of Societies for Coatings Technology, Blue Bell, Pa., 1994.
87. H. K. Hammond, G. Kagle, and G. Kigle-Boeckler, Gloss, in (40), pp. 470–480.
88. U. Zorll, *Prog. Org. Coat.* **1**, 113 (1972).
89. M. E. Nadal and E. A. Thompson, *J. Coat. Technol.* **73**(917) 73 (2001).
90. K. B. Smith, *Surface Coat. Intl.* **80**, 573 (1997).

91. T. C. Patton, ed., *Pigment Handbook*, 3 Vol., Wiley-Interscience, New York, 1973; P. A. Lewis, ed., 2nd ed., Vol. 1, Wiley-Interscience, New York, 1989.

92. K. Rehacek, *Ind. Eng. Chem. Prod. Res. Dev.* **15**, 75 (1976).

93. H. J. W. van den Haak and L. L. M. Krutzer, *Prog. Org. Coat.* **43**, 56 (2001).

94. J. E. Hall, R. Bordeleau, and A. Brosson, *J. Coat. Technol.* **61**(770) 73 (1989),

95. R. S. Fishman, D. A. Kurtze, and G. P. Bierwagen, *Prog. Org. Coat.* **21**, 387 (1993).

96. G. P. Bierwagen, *J. Paint Technol.* **44**(574) 46 (1972).

97. G. del Rio and A. Rudin, *Prog. Org. Coat.* **28**, 259 (1996).

98. Ref. (24), p. 192.

99. S. B. Levinson, *Application of Paints and Coatings, Federation of Societies for Coatings Technolgy*, Blue Bell, Pa., 1988.

100. K. A. Nielsen and co-workers, *Proceeding of the Waterborne Higher-Solids Coat. Symp.*, New Orleans, La., (1995).

101. P. E. Pierce and C. K. Schoff, *Coating Film Defects, Federation of Societies for Coatings Technology*, Blue Bell, Pa., 1988.

102. S. E. Orchard, *Appl. Sci. Res.* **A11**, 451 (1962).

103. W. S. Overdiep, *Prog. Org. Coat.* **14**, 159 (1986).

104. K. Tachi, C. Okuda, and K. Yamada, *J. Coat. Technol.* **62**(791) 19 (1990).

105. W. S. Overdiep, *Prog. Org. Coat.* **14**, 1 (1986).

106. D. R. Bauer and L. M. Briggs, *J. Coat. Technol.* **56**(716) 87 (1984).

107. L. W. Hill and Z. W. Wicks, Jr., *Prog. Org. Coat.* **10**, 55 (1982).

108. W. H. Ellis, *J. Coat. Technol.* **53**(696) 63 (1983).

109. S. Ishikura, K. Ishii, and R. Midzuguchi, *Prog. Org. Coat.* **15**, 373 (1988).

110. R. Berndimaier and co-workers, *J. Coat. Technol.* **62**(790) 37 (1990).

111. L. R. Waelde, J. H. Willner, J. W. Du, and E. J. Vyskocil, *J. Coat. Technol.* **66**(836) 107 (1994).

112. R. H. J. Blunk and J. P. Wilkes, *J. Coat. Technol.* **73**(918) 63 (2001).

113. J. Schwartz and S. V. Bogar, *J. Coat. Technol.* **67**(840) 21 (1995).

114. W. Heilin, O. Klocker, and J. Adams, *J. Coat. Technol.* **66**(829) 47 (1994).

115. See (64), pp. 18–21.

116. D. R. Bauer and R. A. Dickie, *J. Coat. Technol.* **54**(685) 57 (1982).

117. Ref. (67) p.

118. C. B. Fox. *Proc. ESD/ASM Adv. Coat. Technol. Conf.* **1991**, 161 (1991).

119. T. J. Miranda, *J. Coat. Technol.* **60**(760) 47 (1988).

120. J. E. Gaske, *Coil Coatings*, Federation of Societies for Coatings Technology, Blue Bell, Pa., 1987.

121. K. A. Nielsen and co-workers, *Polym. Mater. Sci. Eng.* **63**, 996 (1990).

122. W. Bailey, et al., *J. Coat. Technol.* **62**(789) 133 (1990).

123. S. P. Pappas, ed., *Radiation Curing: Science and Technology*, Plenum Press, New York, 1992.

124. H. J. Hageman, *Prog. Org. Coat.* **13**, 123 (1985).

125. J. V. Crivello, *J. Coat. Technol.* **63**(793) 35 (1991).

126. R. Higgins, *Powder Coatings*, Campden Publishers Ltd., U.K. 1998, p. 77.

127. K. Kronberger, D. A. Hammerton, K. A. Wood, and M. Stodeman, *J. Oil Colour Chem. Assoc.* **74**, 405 (1991).

128. W. Jacobs, D. Foster, S. Sansur, and R. G. Lees, *Prog. Org. Coat.* **29**, 127 (1996).

129. Ref. (64) pp. 8–14.

130. T. A. Misev, *Powder Coatings Chemistry and Technology*, John Wiley & Sons, Inc., New York, 1991, p. 346.

131. J. Temperley, M. J. Westwood, M. R. Hornby, and L. A. Simpson, *J. Coat. Technol.* **64**(809) 33 (1992).

132. D. M. Fasona, *J. Coat. Technol.* **59**(752) 109 (1987).

133. J. E. Glass, *J. Coat. Technol.* **50**(640) 53, 61 (641) 56 (1978).
134. J. W. Gillatt, in D. R. Karsa and W. D. Davis eds., *Waterborne Coatings and Additives*, Royal Society of Chemistry, Cambridge, U.K. 1995, pp. 202–215.
135. D. Y. Perara, D. V. Eynde, and J.-M. Borsus, *J. Coat. Technol.* **73**(919) 89 (2001).
136. W. S. Overdiep, *Prog. Org. Coat.* **14**, 159 (1986).
137. P. A. Reynolds, *Prog. Org. Coat.* **20**, 393 (1992).
138. H. R. Bleile and S. Rodgers, *Marine Coatings*, Federation of Societies for Coatings Technology, Blue Bell, Pa., 1989.
139. C. M. Sghibartz, FATIPEC Fed. Assoc. Tech. Ind. Paint Vernis Emaux Encres Impr. Eur. Cont. Cong. XVII **IV**, 145 (1982).
140. J. E. Hunter, *Protective Coat. Eur.*, Nov., 16 (1997).
141. J. Millett and C. D. Anderson, *Proc. Fast Ferries '97 Conf.*, Sydney, Australia, 1997, p. 493.
142. P. G. Phillips in Ref. 1.

GENERAL REFERENCE

Z. W. Wicks, Jr., F. N. Jones, and S. P. Pappas, *Organic Coatings: Science and Technology*, 2nd ed., John Wiley & Sons, Inc., New York, 1999.

Zeno W. Wicks, Jr.
Consultant

COATINGS, ANTIFOULINGS

1. Introduction

When living organisms attach and grow on the underwater surfaces of ships there is either a loss of speed or an increase in propulsive energy required to counteract the speed loss: This presents an enormous economic problem. For example, if a large Container vessel succumbs to fouling it can add an extra $250,000/year to the fuel bill. The U.S. Navy (1) has estimated that they would have an increased annual fuel bill of $75–100 million if fouling was allowed to grow unchecked on all of the U.S. Fleets. The cost of removal of fouling in dry-dock, and reapplying a new antifouling coating system, can also be substantial (2).

Apart from the bottoms of commercial ships and boats, there are also other submerged surfaces on which fouling can create problems. Offshore oil platforms (which are designed to stay for long periods of time in the world's oceans), can become more susceptible to damage from the added weight that fouling contributes to the structure and from the increased resistance to tidal and water flow. Fouling growth in conduits for conveying cooling water to Power Stations can lead to serious and costly downtime for cleaning if fouling is allowed to build up. On yachts and pleasure craft, fouling is not only unsightly but also reduces the manoeuvrability and speed.

One of the most important fouling organisms is the barnacle. As barnacles grow they exert pressure on the surface to which they are attached and their basal edges, growing outward and downward, can penetrate and undermine protective coatings leading to premature corrosion and loss of structural integrity.

There are also marine organisms that can bore into underwater structures of wood, such as the pilings used for habor piers. The "shipworm" Teredo is a notorious example, which bores it way through the wood in which it lives thus weakening it considerably, and this has lead to the demise of many wooden ships (3).

2. Marine Biofouling

Many organisms can contribute to marine fouling communities; from microscopic bacteria and diatoms, through shelled invertebrates such as barnacles and tubeworms, to kelps >10 m long (2). Microfouling includes microbial organisms such as bacteria, fungi and microalgae (notably diatoms), and their secretions. Microfouling organisms are able to form tenacious films of exuded extracellular polymeric materials, which chelate inorganic ions (4). A wide range of factors affects the fouling rate and composition of microfouling communities, including water chemistry, water temperature, pressure, shear stress, and substratum composition and structure (4).

Although the number of species reported as fouling organisms is large and extremely diverse, with >4000 species recorded, this actually represents only a very small proportion of known species, even with dominant fouling groups such as barnacles, tubeworms, and algae (4). The Cirripedian barnacles are perhaps the best-adapted group of organisms, with >20% of known species recorded in fouling communities (5). However, the number of fouling species within a group may also belie the importance of those representatives. For example, although only a very small number of bivalve molluscs are known as foulers, mussels and oysters are among the most important fouling species worldwide.

On a worldwide scale, there is often similarity in a structure, and sometimes specific composition, of fouling communities. This is particularly the case in harbors and on vessel hulls, and is most likely due to biological adaptations of fouling species that facilitates their growth and survival in a range of environments and their ability to be transported around the world on vessel hulls. Examples can be seen in catalogues of marine fouling species (6) with many species identifiable in geographically disparate regions. In more pristine environments, such as encountered with offshore structures, the fouling community is likely to more closely reflect local biodiversity as seen on reefs and other natural hard substrates.

From studies on fouling composition and development at a number of sites scattered through the world's oceans, some general guidelines have been developed to enable the prediction of fouling severity in coastal and offshore environments (7). Coastal and open-sea fouling communities could be distinguished from each other, as could warm and cold water growth forms. Cold water fouling communities tend to be dominated by bulky growth forms, (mussels and kelps) near the surface and low profile, cementing calcareous growth forming near the bottom. In contrast, warm water fouling communities tend to have constantly low profile, hard shell growth forms from surface to bottom. In coastal areas, maximum fouling attachment is generally found close to shore, diminishing seaward and with increasing depth.

Within harbors and estuaries, although there are frequently structural similarities between fouling communities in different locations, the composition and severity of a fouling growth can vary immensely with latitude (largely due to variation in water temperatures), between seasons, from year to year, and site to site (8). Many factors contribute to this, including water temperature, depth, clarity, salinity, pollution levels and movement, and the proximity of brood sites. Even on a scale of several metres, variation in the distribution of larvae and spores in the water column can significantly affect the density or even presence of the adult form on a submerged surface (8).

3. The Biofouling Process

Three stages can be identified in the formation of the microbial film: conditioning colonization by "pioneer species", colonization by other microorganisms, and accumulation (2,4). Conditioning commences within seconds of the surface being immersed, with the formation of a film of both organic and inorganic matter adsorbed from the aquatic phase (9). This effectively generates a new substratum interface with altered physicochemical properties. Subsequent microbial colonization of this film is influenced by the composition of the conditioning film, the nature of the substratum, the nature of the aquatic phase, and the species composition of the microbial community in the aquatic phase.

The "pioneer" species are often very small; rod-shaped bacteria, that attach within several hours. Initial attachment is weak and reversible (adsorption), until the bacteria are able to secrete extracellular adhesive polysaccharide and secure nonreversible attachment (adhesion) (10). Once attached, these primary colonizers assimilate nutrients and synthesise new cellular and extracellular material that accumulates in the surface deposit (4). Attachment of secondary colonizers, including stalked or filamentous bacteria, diatoms, other microalgae, and protozoa, then proceed quite rapidly. Diatoms, that contribute much of the biomass in biofilms on illuminated surfaces in the sea (11), reach a surface by purely hydrodynamic means, then attach by a secretion of an adhesive polymer (12).

The biofilm surface is highly adsorptive and, although microorganisms and their remains make up the most conspicuous components in the deposit, varying amounts of organic secretions, trapped detritus, inorganic precipitates, and corrosion products, compose the bulk of the fouling layer. Formation of natural surface films and especially bacterial films can significantly change the adhesion force and interfacial energy of surfaces immersed in seawater (13). Measurements have shown that adhesion force is reduced by adsorption of natural conditioning films and enhanced by bacterial film formation.

Several days to weeks after a surface is first exposed, the last and longest phase of fouling colonization begins with the settlement, attachment, and growth of multicellular organisms (10). In the absence of any antifouling agent, the buildup of organisms will proceed until most of the bare surface is occupied (8). The major organizing factors influencing the department of fouling communities are recruitment of species onto a surface, competition between resident organisms, and disturbance by predation and/or environmental factors. Pioneering macrofoulers tend to be small, fast growing, and maturing species with extended periods of recruitment. In contrast, later spatial dominants within the community, such as solitary and colonial ascidians (sea squirts), are generally poor recruiters due to the short duration of their larval stage, but are competitively successful due to their large body size and extended longevity (8). Interactions between organisms that can influence the composition and structure of the biofouling community can include facilitation, in which resident species enhance the chances of subsequent colonizing species, inhibition, in which established species resist invasion, and tolerance, in which there is interaction between resident and colonizing species (8,10,14).

Algal spores generally rely on passive, random hydrological, and physical processes to deposit them on a surface and the surface texture of a substratum can be important in physically restraining the spores and allowing them to attach (8,14). Recruitment of fouling invertebrates onto a surface is a more complex process, in which several physical, chemical and biological factors interact (15). Recruitment requires temporal and spatial availability of larvae and appropriate chemical and physical cues to stimulate attachment and metamorphosis. Larvae seek a surface to which they can adhere, which is within appropriate environmental gradients, and that will allow the juveniles to grow to maturity and reproduce (8). Positioning in relation to food sources, light, water, velocity and temperature, turbulence, gravity, and hydrostatic pressure are major considerations. Some larvae show a strong response to settle near established individuals of the same species, particularly species such as barnacles, which rely on proximity of other adults for fertilization (8). Before permanently attaching to a surface, many invertebrate larvae explore a surface to determine its acceptability (16,17). During this phase, the larva need to maintain a hold on the surface, yet, if they find that the surface is unacceptable, they must be able to be released and carried to alternative settlement sites. Temporary attachment is by either a suction apparatus or a secreted sticky substance; permanent attachment by hardened or cured adhesive cement sometimes reinforced with calcareous deposits.

4. Historical Development of Antifoulings

Coatings to prevent fouling have been applied since antiquity (18). A very early record on the use of some form of paint on ship's hulls can be found in the translation from the Aramaic of a papyrus of ~ 412 BC concerning the repairs of a boat of that date (19, p. 48):

> "And that the arsenic and sulphur have been well mixed with Chian oil thou broughtest back on thy last voyage and the mixture evenly applied to the vessel's sides that she may speed through the blue waters freely and without impediment?"

In the third century BC, the ancient Greeks used tar and wax to coat ship bottoms. From the thirteenth to the fifteenth century, pitch, oil, resin, and tallow were used to protect ships. For his remarkable travels between 1405 and 1433, the Chinese Admiral Cheng Ho had the hulls of his junks coated with lime mixed with poisonous oil from the seed of *sryandra cordifolia* to protect the wood from worms (20,21). In his life of Columbus, Morison (22, p. 124) mentions that:

"All ships" bottoms were covered with a mixture of tallow and pitch in the hope of discouraging barnacles and teredos, and every few months a vessel had to be hove-down and graved on some convenient beach. This was done by careening her alternately on each side, cleaning off the marine growth, repitching the bottom and paying the seams."

During the following centuries, the main form of protection for wooden ships was copper sheathing or the use of a mixture containing sulfur and arsenic. It was not until the development of iron hulls that copper sheathing was abandoned because of serious galvanic problems (23).

In 1625, William Beale was the first to file a patent for a paint composition containing iron powder, copper, and cement. In 1670, Philip Howard and Francis Watson patented a paint consisting of tar, resin, and beeswax. In 1791, William Murdock patented a varnish mixed with iron sulfide and zinc powder, using arsenic as antifoulant (23).

The number of patents proliferated in the second half of the nineteenth century—more than 300 patents were registered by 1870—but all these paints had little to no effect over a very limited time. Mallet (24, p.129) stated:

"... it is probable that under no other head in the whole range of the Patent Office Records is such a mass of ignorance, absurdity and charlatanry exhibited, as in these antifouling patents. One or two of the best have proved palliatives (no more can be said for any of them), and are for want as yet of anything better, more or less in practical use. The writer refrains from particularizing those that to his observation seem best or next best; but the vast mass of these "compositions" are worthless or hurtful – several are more worthy of the name of "impositions"; and some even of the most recently patented are grotesque in their ignorant absurdity, – as for instance, one in which a farrago of the soluble drastic purgatives (such as colocynth) of the apothecary's shop is mixed up with incompatible resinous fluids, to scare away the unhappy zoöphytes."

The basic principle of these toxic antifouling paints, however, still holds today: A toxic substance is mixed into a resinous substance or binder to kill off fouling organisms by some kind of leaching mechanism (25,26). The latter is the key factor in the success of antifoulings and has always been the focus of the efforts of antifouling formulators. As Mallet (24, p. 120) put it:

"The necessary balance between adhesion and slow diffusion or washing away through the water of the poisonous soap is too delicate for practice. Either the soap adheres firmly and does not wash away enough to keep off fouling or it washes away so fast as soon to be all gone."

Salts of copper, arsenic, or mercury were popular biocides. Linseed oil, shellac, tar, and various kinds of resin were used as matrix and solvents included turpentine oil, naphtha, and benzene. In 1854, James McInnes patented the first practical composition to come into widespread general use. It used copper sulfate as toxin in a metallic soap composition, which was applied hot over a quick-drying priming paint of rosin varnish and iron oxide pigment. Soon after, a similar hot plastic paint known as "Italian Moravian" was developed that was a rosin and copper compound. It was one of the best paints of that time and was used well into the twentieth century. In 1863, James Tarr and Augustus Wonson were given a U.S. patent for antifouling paint using copper oxide and tar (23). In 1885, Zuisho Hotta was given the first Japanese patent for an antifouling paint made of lacquer, powdered iron, red lead, persimmon tannin, and other ingredients (27). These paints, although reasonably successful, were expensive and had a short life-span.

In 1906, the U.S. Navy decided to manufacture its own antifouling coatings and tested shellac and hot plastic paints at Norfolk Navy Yard (23). From 1911 to 1921, more experiments were performed both to find substitutes for scarce materials such as mercuric oxide and to improve the paints (23,28). In 1926, the U.S. Navy developed a hot plastic paint using coal tar or rosin as binder and copper or mercuric oxides as toxins. Hot plastic paint required some heating facility that made application difficult (23). Consequently, cold plastic paints were developed that dry by evaporation of the solvent and that were easier to apply (29). These paints effectively decreased fouling and the time between dry-docking for repainting was extended to 18 months.

It was only after the Second World War that major advances in antifouling coatings took place, leading to the currently used technologies.

5. Current Antifouling Technologies

There are currently only two principal ways that marine fouling is controlled on underwater hulls. The first of these is based on the historical method of dispersing a biocide in a binder system that is then released slowly from the coating surface once it is immersed in seawater. The second type does not use biocides, but relies on the surface being "nonstick". This is referred to as "foul release" technology.

5.1. Biocidal Antifoulings. There are two key factors in the development of a successful biocidal antifouling:

- The *toxicity* of the biocide, or biocide combinations.
- The *delivery mechanism* of the biocide(s) to the marine environment.

Biocide Toxicity. Following on from the historical use of copper as sheathing, copper compounds were the first biocides used in the large scale industrial production of antifouling paints, and they are still the most common biocides employed in antifoulings. The most commonly used copper compounds are cuprous oxide (Cu_2O), which is red, cuprous thiocyanate (CuSCN), a pale cream compound used for making brightly colored antifoulings, and metallic copper, either in the traditional sheet form or as a powder.

Copper by itself is, however, limited in its effectiveness. It tends to work well as a biocide against animal (shell) fouling, but algal (weed) fouling is more resistant to copper and therefore antifouling chemists have spent much time and effort searching for additional biocides that can be added to the copper to boost performance. These are referred to as boosting biocides. A key characteristic of these is that they should have very low seawater solubility (ideally <10 ppm) so that they are not released too quickly from the antifouling paint film. In the 1950s, mercury and arsenic compounds were commonly used as boosting biocides, but these were largely replaced in the 1960s by organotin compounds that came from the agricultural industry, where they were used as pesticides (30). The organotin compounds were found to be extremely effective against a very wide range of marine fouling species at very low concentrations. Since the discovery of organotin compounds there have been very few other boosting biocides developed. Not only has it proven to be very difficult to improve on the efficiency and effectiveness of the organotin boosters, but also the relatively small size of the overall antifouling market (∼50 million liters of paint worldwide, annually) makes it difficult to justify the very high cost of developing and registering new biocides. It can now cost in excess of $ 4 million to undertake all the human and environmental toxicity testing required for registration purposes. No return on this investment can occur until the product containing the biocide is itself registered, which can take many years to accomplish.

Some commonly used booster biocides currently in use are as follows (31):

- 2-methylthio-4-*tert*-butylamino-6-cycloproylamino-*s*-triazine (eg, Irgarol 1051 ex Ciba Speciality Chemicals).
- Dichlorophenyl dimethyl urea (eg, Diuron).
- Zinc hydroxypyridmethione (eg, Zinc Omadine from Arch Chemicals).
- Copper hydroxypyridmethione (eg, Copper Omadine from Arch Chemicals).
- 4,5-Dichloro-2*N*-octyl-4-isothiazol-3-one (eg, SeaNine 211 ex Rohm and Haas).
- *N*-Dimethyl-*N*-phenyl-*N*-fluorodicholoro-methylthiosulphamide (eg, Prevetol A4 ex Bayer).
- Tolylfluanid (eg, Preventol A5S ex Bayer).
- Zinc ethylene-1,2-bisdithiocarbamate (Zineb).

From the above, only those that show rapid degradation in sea water, and in sediments, will be likely to survive the close regulatory scrutiny to which they are being increasingly subjected.

Given the limited availability of new biocides, research and development in antifouling coatings is now focused on maximizing the efficiency of the few biocides that are available. This involves studies on both the synergies between the biocides, to get the maximum toxic effect from the minimum quantities, and studies on the best mechanism to control the release of the biocides, to maximize the lifetimes.

Biocide Release Mechanisms.   The standard method for measuring the biocide release from antifoulings is the leaching rate. This rate is defined as the amount of biocide released from a given surface area in a given time, and this is expressed as $\mu g/cm^2/day$.

The release mechanism itself depends on the technology of the coating system employed. There are three main technologies:

- Rosin based.
- Self-polishing copolymer (SPC).
- Hybrid SPC/rosin systems.

Rosin-Based Technologies.   Rosin, or rosin derivatives, are used to enable seawater to penetrate the antifouling coating and as it does so it allows release of the biocides by a diffusion process. This results in the pseudo-exponential leaching rate of biocides, with an excessive release in the early days of immersion that gradually falls over the following months to a level below which fouling will start to occur. This is shown in Figure 1.

Rosin comes from trees and is also commonly referred to as Gum Rosin or Wood Rosin or Tall Oil Rosin. It is widely used in the adhesives industry, and has a complex chemical makeup, which varies depending on the trees where it is grown and from which it is harvested. The major constituent of Rosin is abietic acid, which is slightly soluble in seawater (pH $\sim$8.2). It is this slight solubility that makes it suitable for use in antifouling coatings, since as it dissolves it enables the biocides to be released from the paint matrix. Modification of rosin can be carried out in a number of ways, such as hydrogenation or esterification.

Rosin by itself does not form durable films and it has to have other film-forming binder components (referred to as resins and plasticizers) added to give films of good mechanical strength. However, these added components are generally insoluble in seawater and their use has to be limited or the release of the biocides will be impaired. There is thus a careful balance needed between the amount of rosin necessary to get sufficient biocide release and the quantity of the other film-forming components needed to form tough and durable films.

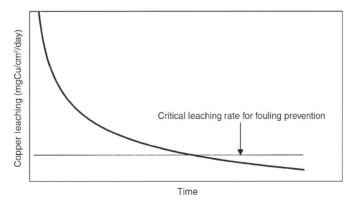

Fig. 1.   Exponential biocide release (rosin-based antifoulings).

Achieving the right balance has been a conundrum that has challenged antifouling chemists for >100 years (24, p. 120, 32).

A high rosin content means that the binder system is more soluble in seawater, whereas a low rosin content makes the binder system hard and very insoluble. The former types of antifouling are known as "soluble matrix" antifoulings, whereas the hard, insoluble types are known as "contact leaching" antifoulings.

(1) *Soluble matrix antifoulings.* Prior to the Second World War almost all antifouling paints were of the Soluble Matrix type. After this time, major improvements came about with the advent of a wide range of new industrial chemicals such as the synthetic petroleum-based polymers. As the chemical companies developed new synthetic polymers they were used to upgrade and improve Soluble Matrix antifoulings, a process that continues to this day. All Marine paint companies now market modern versions of the traditional Soluble Matrix antifoulings, with a wide variety of confusing nomenclatures to describe them, such as "Controlled Depletion Polymer" antifoulings, "Eroding" antifoulings, "Ablative" antifoulings, "Polishing" antifouling, "Self-polishing" antifoulings, and "Hydration" antifoulings (33). These all refer to the physical dissolution of the rosin-based antifouling systems, and should not be confused with the chemically controlled dissolution of SPC antifoulings, which are described in the section Self-Polishing Copolymer. Copper oxide is the biocide most commonly used in these Soluble Matrix paints, along with boosting biocides.

Although in theory these paints could dissolve fully over time releasing all the biocides contained in them, they do in fact become increasingly less soluble due to the build up of insoluble copper salts and other inert species (such as rosin impurities) at the surface. This results in the formation of a "spent" layer at the surface, free of biocide, which is commonly referred to as the "leached layer". Biocides from the depth of the film have to diffuse through this leached layer to reach the surface. The size of this leached layer increases with time, resulting in a decrease in the release rate of biocide according to Fick's laws of diffusion. This limits performance lifetime to a maximum 36 months between dry-dockings. In high fouling areas, Soluble Matrix antifoulings can succumb to fouling well before this time, especially if the copper leaching rate falls below the critical level, which is generally acknowledged to be $\sim 10 \ \mu g/cm^2/day$ (23).

(2) *Contact leaching antifoulings.* In the 1950s, attempts were made to increase the lifetime of soluble matrix antifouling coatings by increasing the biocide content. Such highly pigmented coatings required larger quantities of the inert resin and plasticizer film-forming components, with less rosin, thus making these coatings insoluble in seawater and hard. They became known as Contact Leaching antifoulings since the biocide particles were all in close contact with each other.

As with the Soluble Matrix antifoulings, it was recognized that the major barrier to extended lifetimes was the development of the leached layer at the surface, which acted as the rate-controlling and lifetime limiting step (30). As a way around this, it was found that the leached layer could be removed by in-water cleaning, without damaging the paint surface too extensively since Contact Leaching antifoulings form tough and hard films. These paints could thus be "reactivated" to extend their lifetime. Without the reactivation process in-service life was limited to 24 months.

Self-Polishing Copolymer.   The desirability of chemical mechanisms to control the release of biocides from antifouling coatings, rather than relying on a physical dissolution and diffusion process, long predates the means of their achievements. It was only in 1974 that the technology and opportunity finally came together with the introduction of tributyltin (TBT) and SPC antifoulings (30). Hydrolysis or ion exchange of an acrylic polymer at the surface of the antifouling makes the polymer soluble, resulting in biocide release without the use of Rosin. The control of surface solubility by this chemical reaction gives controlled, pseudo-zero-order release of the biocides, until all the paint is dissolved away. This results in a much more efficient use of the biocides, as shown in Figure 2.

The main improvements that TBT SPC antifoulings bought compared to the previously available rosin-based products were as follows:

- Improved antifouling performance.
- Extended in-service periods (up to 60 months).
- Reduced fuel consumption due to hull smoothness.
- Easier maintenance and repair.
- Bright and clean colors.

These benefits were rapidly recognized by yacht and ship owners and operators and SPC products soon came to dominate the market (34). Their biocidal effectiveness was the key to their success, but it also proved in the end to be their undoing since nontarget organisms (those not fouling the vessels) were found to be affected (35). In the early 1980s in France, a link between the use of TBT antifoulings on Yachts and the poor growth of oysters nearby was suggested (36). Subsequent research in other parts of the world demonstrated that the TBT from antifoulings was relatively persistent in seawater and therefore could have an impact on nontarget organisms. These findings resulted in government action to restrict their use. Initially, in many countries the use of TBT antifoulings

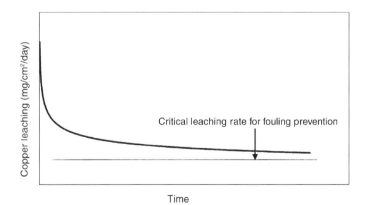

Fig. 2.   Controlled biocide release (self-polishing copolymer antifoulings).

was banned on vessels <25 m in length. This was targeted at the Yacht and Pleasure boat industry since TBT becomes the antifoulings used on these boats in marinas in enclosed waters was deemed to be the most damaging to nontarget organisms

The International Maritime Organisation (IMO), which was set up by the United Nations in 1958 to deal with maritime affairs on a global basis, established a Antifouling Working Group in 1990 to study the problem more widely. In 1997, the Marine Environment Protection Committee (MEPC) of IMO agreed a draft resolution calling for a worldwide ban on TBT antifoulings. This eventually resulted in the passing of an Antifouling Systems Convention at a Diplomatic Conference of the IMO in October 2001. This Convention calls for a worldwide ban on the application of organotins that are used as biocides in antifoulings on all vessels begining January 1, 2003, to be followed by a ban on the presence of organotins that are used as biocides in antifoulings on ships by January 1, 2008. Full details of this Convention can be found on the IMO website "http://www.imo.org" (37).

TBT SPC Antifoulings.   The chemistry of the organotins goes back as far as 1939 (38) and their antifouling possibilities to 1958 (39). The first TBT antifouling was introduced in 1968, in the Yacht market, with the TBT being added in pigment form as tributyltin fluoride (TBTF). Its use in this form, however, resulted in a concern for the health and safety of those handling the material, both at the manufacture stage and during application by the end-user. This concern led directly to the development of polymer-bound TBT, which greatly reduced this health risk.

It was discovered that when these TBT polymers were immersed in seawater, hydrolysis or ion exchange occurred at the surface, resulting in the polymer becoming soluble in seawater. As this process was repeated, the coating slowly dissolved away, without the need to use rosin, with the rate of dissolution being controlled by the quantity of TBT in the polymer.

This seawater reaction is confined to the top surface only, with the underlying bulk of the film remaining insoluble. This means that water is unable to penetrate into the depth of the film, and the biocide-depleted leached layer of an SPC coating is very thin and is generally <20 μ, even after several years of immersion. This is in marked contrast to the rosin-based soluble matrix and contact leaching antifoulings, that can have leached layer wells in excess of 50 μ.

It is the character of the Sn$-$O bond that is the key to the success of these TBT SPC antifoulings. In certain instances, this bond exhibits covalent behavior, as evidenced by a low dipole moment, but in other reactions, such as hydrolysis rate, it behaves as a characteristically ionic bond (40). In a low dielectric constant medium, it behaves covalently but in a high dielectric constant medium such as water it behaves ionically. This change in behavior with polarity is thought to be a consequence of the change in molecular shape from planar alkyl groups to trigonal pyramidal alkyl groups. The first ship trials with TBT SPC antifoulings were not very successful since the dissolution rate of the coatings, commonly referred to as the "Polishing Rate", was much too fast. It was found that the test patches dissolved away within 6 months, resulting in heavy fouling. However, it was soon realized (41) that the polishing rate could be controlled by varying either the polymer composition or some of the other key ingredients in the paint formulation, and therefore so long as enough paint was applied it would provide the necessary antifouling protection. For example, $100\,\mu$ at a low polishing rate was sufficient to provide fouling protection for up to 24 months, depending on in-service conditions (such as vessel speed, water temperature, vessel activity etc).

Initially, the main biocide used (in addition to the TBT polymers) in these SPC antifoulings was DDT, but this was soon replaced for environmental reasons by cuprous thiocyanate (CuSCN) and eventually by cuprous oxide (Cu_2O)—the latter only when in-can stability problems had been overcome. In the 1980s some manufaturers then replaced part of the cuprous oxide with a boosting biocide, Zineb (Zinc ethylene-1,2-bisdithiocarbamate), to improve slime control. This resulted in a formidable biocide "cocktail" with unparalleled fouling control, and so it was no surprise that these coatings came to dominate the antifouling market: >80% of all Marine antifouling sales were of the SPC type by the mid-1990s. This was helped by the fact that, as the technologies matured and competitive market forces were bought to bear, the costs were also significantly reduced.

As well as the superb ability to control fouling, TBT SPC antifoulings also kept surface roughness to a minimum by the polishing process. Asperities on the paint surface acquired during the painting process (such as runs and sags) were rapidly removed once the vessel was underway at sea, and this reduced the drag. Lowered fuel consumption resulted and this was just at the time when fuel prices rose dramatically following the 1973 Middle East war. Another major advance with TBT SPC antifoulings was the extension of in-service periods between dry-dockings from 36 to 60 months, which was possible since there was no decrease in biocide release rate over time since there was no leached layer build up. The in-service lifetime was determined solely by the original thickness applied. This gave ship owners the opportunity to keep their vessels at sea longer, thus increasing earning capacity.

The costs of maintaining and repairing SPC systems was also found to be lower than with traditional antifoulings since there was less detachment, and minimal deterioration due to weathering.

TBT-Free SPC Antifoulings. With the anticipated demise of TBT SPC antifoulings for environmental reasons, much R&D effort in the 1990s was directed at finding suitable alternatives. The first country to fully ban the use of antifoulings containing TBT was Japan, in 1991, and consequently most of the early TBT-free SPC systems came from the Japanese marine paint companies.

Three principal chemistry types have emerged, all based on acrylic copolymers, and with pendant groups that potentially undergo hydrolysis or ion exchange in seawater. These are all attempts to mimic the reaction of TBT acrylate copolymers in seawater:

TBT Acrylate: $Polymer-COO-Sn\,(Bu)_3 + NaCl \Leftrightarrow Polymer-COO-Na + Sn\,(Bu)_3-Cl$

Copper Acrylate: $Polymer-COO-Cu-R + NaCl \Leftrightarrow Polymer-COO-Na + R-Cu-Cl$

Zinc Acrylate: $Polymer-COO-Zn-R + NaCl \Leftrightarrow Polymer-COO-Na + R-Zn-Cl$

Silyl Acrylate: $Polymer-COO-Si\,(R)_3 + NaCl \Leftrightarrow Polymer-COO-Na + (R)_3\,Si-Cl$

These TBT-free SPC systems are at varying degrees of technical and commercial development, and there is some uncertainty regarding the claims made by their respective manufacturers that they have exactly the same benefits as can be obtained with TBT SPC systems (such as extended dry-docking intervals, self-polishing smoothness, and tough and durable films). However, the copper acrylate system has been proven to work on ships for the full 60 month in-service period required (42) and since the technology was first introduced in 1990 it has a large track record, of >6,000 vessel applications.

The main biocide used in all these TBT-free SPC systems is cuprous oxide (Cu_2O), which is added as a pigment during paint manufacture. In addition, boosting biocides are used to provide the necessary biocidal activity replacing that previously obtained from TBT. These boosting biocides are nonpersistent, and degrade rapidly once they leave the surface of the coating. Their rapid degradability means that they do not build up in the environment and therefore they should not affect nontarget organisms (43,44). Once the biocides have diffused away from the paint surface, they can degrade in the presence of light and/or bacteria into substances with very low toxicity and persistence, thus giving these systems a better environmental profile than the TBT SPC systems they replaced.

Hybrid SPC/Rosin Systems.   With some of the TBT-free SPC systems it was found that the hydrolysis or ion exchange reaction of the TBT-free acrylate copolymers is not as easily controlled as with the previous TBT SPC systems. The reaction is either too fast or too slow and so rosin has been added to provide a "backup" mechanism for the biocide release. This combination is a new technology type, and is referred to as a "Hybrid" technology. The use of rosin with these TBT-free SPC systems has the advantages of lowering cost, increasing the volume solids, giving improved overcoatability, and surface tolerance. However, the release of the biocides from these Hybrid SPC/rosin systems is not as well controlled as with the "pure" SPC systems and they do not have the same extended life time capability.

The price and performance of these Hybrid SPC/rosin systems is midway between that of the SPC and rosin-based "parent" systems. The higher the SPC polymer content the more SPC-like is the behavior of the product, and vice versa with the rosin content.

5.2. Foul Release Antifoulings.   From an environmental perspective, the most desirable approach to fouling control is one that does not rely on the release of biocides to achieve its effect. A plethora of ideas for how this can be achieved have been proposed, and numerous patents have appeared (30), but only the foul release or low adherence systems have been commercialised successfully. These operate by a "nonstick" principle, having surface characteristics that minimize the adhesion strength of fouling organisms, and any fouling that does settle is removed by hydrodynamic forces that are present on the hull when it moves through water.

The concept of low adherence to deter fouling was first considered in the nineteenth century (23, pp. 226–227), but it was not until the discovery of the fouling control properties of silicones (45) that commercial systems started to appear. These have been improved and refined since then and the majority of foul release systems currently available are silicone materials based on polydimethysiloxane (PDMS). The only other chemistry types to have been considered to any great extent have been fluorinated polymers (46,47), but they have not been commercialized to the same extent as the silicone systems.

The PDMS polymer has an extremely flexible backbone, with rotation around the [−O−Si−O] bond being very easy, resulting in a very low glass transition temperature (T_g). This allows the polymer chain to readily adapt the lowest surface energy configuration (48). Low surface energy alone, however, is not the only important criteria for success in deterrence of fouling settlement: coating thickness and elastic modulus (49,50) have also been shown to be important variables.

Smoothness is also a very important feature for an effective foul release coating. Surface energy, and the area available for adsorption, increases with roughness and it is well-known that fouling species prefer rough surfaces (51), an effect that is known as the *thigmotactic* nature of fouling settlement. It has been shown that the topography and texture of foul release silicone coatings is completely different from that of SPC antifoulings (18). Whereas typical SPC antifoulings have a "closed texture" with frequent peaks and troughs akin to a steep mountain range (Fig. 3), the foul release systems have a long wavelength "open texture" surface, similar to well-rounded hills and shallow valleys (Fig. 4).

In addition to surface topography, it has been demonstrated that enhancement of silicone foul release coatings can be accomplished by the incorporation of low molecular weight polymers or oils (45,52). It has been postulated that the surface structure of silicone foul release coatings is changed when the nonbonded oils migrate to the coating surface and increase the "slipperiness". Radio labeled studies on the fate of these oils has shown that they do not migrate from the coating surface to any great extent (53) and this is also confirmed by the successful in-service performance results of over 5 years (42).

Quantification of the foul release properties of coatings is done using the American Standard for Testing and Materials (ASTM) test D5618-94. This measures the force with which it is necessary to remove barnacles from the surface in

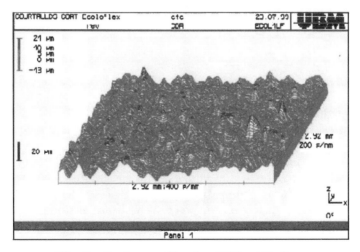

Fig. 3. Surface profile of an SPC antifouling.

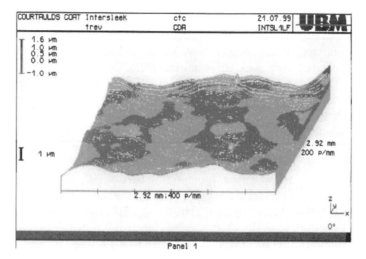

Fig. 4. Surface profile of foul release antifouling.

shear (54). From this measurement of the shear adhesion strength, it is possible
to predict the speed at which barnacles will release from surface of a vessel.
Towing experiments to verify these predictions have been carried out (55) and
these tests show that there is a good correlation between the predicted and
observed velocities at which fouling will release. It has been found that most
fouling species are removed at 15 knots or above (56), although slime fouling,
which stays close to the surface, can still remain even at speeds in excess of
30 knots.

Despite their beneficial environmental profile the market penetration of
foul release coatings has been limited. In part, this is due to the increased initial
cost of installation, but it also due to the fact that the majority of the world's fleet
(Crude Oil Tankers and Bulk Carriers) do not operate at high enough speeds or
activity for foul release coatings to perform at their best. As the technology
matures it is very likely that products will emerge that both work at lower
ship speeds, and are also less expensive to install.

6. The Future

As a result of the expected ban on the use of TBT antifoulings, there has been
increasing interest from innovators and academic institutions in solving the
challenge of fouling control (35). The use of "natural product" antifoulants has
received wide attention, along with attempts to mimic the surfaces of marine animals such as seals, dolphins, and whales. The successful commercial development of any such novel approach will require knowledge of biological,
chemical, and physical processes involved as well as an understanding of the
operational requirements of the end-users.

In the meantime, copper-based systems are certain to dominate the antifouling coatings market for the foreseeable future, but it is anticipated that

foul release systems will become increasingly important as the environmental pressure on the use of biocides increases.

BIBLIOGRAPHY

1. G. S. Bohlander, "Biofilm effects on drag: Measurements on ships", *RPaper no. 16, Polymers Marine Environ.*, 23–24 October, 1991.
2. J. A. Lewis, *Mater. Forum* **22**, 41 (1998).
3. O. D. Hunt, *Antifouling Research—Biological Approach*, International Paints Ltd. Technical Conference 1957, p. 343, 1957.
4. B. J. Little and J. R. DePalma, *Trans. Inst. Naval Arch.* **30**, 362 (1988).
5. H. S. A. Pratt, *A Manual of the Common Invertebrate Animals Exclusive of Insects*, P. Blakistons' Son & Co., Philadelphia, 1935.
6. R. L. Fletcher, *Catalogue of Marine Fouling Organisms*, Vol. 6, Algae, ODEMA, Brussels, Belgium, 1980.
7. J. R. DePalma, in R. F. Acker and co-workers, eds., *Proceedings of the 3rd International Congress on Marine Corrosion and Fouling*, Northwestern University Press, Chicago, p. 865, 1972.
8. M. D. Richmond and R. Seed, *Biofouling* **3**, 151 (1991).
9. R. P. Schneider, in S. Kjelleberg and P. Steinberg, eds., *Biofouling: Proceedings and Solutions*, Sydney, NSW, 14–15 April 1994, The University of New South Wales, 1994, p. 58.
10. M. Wahl, *Marine. Ecol. Prog. Ser.* **58**, 175 (1989).
11. D. A. Caron and J. M. Sieburth, *Appl. Enviro. Microbiol.* **41**, 268 (1981).
12. K. E. Cooksey, and B. Wigglesworth-Cooksey, *Biofouling* **5**, 227 (1992).
13. B. D. Johnson and K. Azetsu-Scott, *Am. Nat.* **111**, 1119 (1995).
14. T. A. Norton and R. Fretter, *J. Mar. Biol. Ass.* **61**, 929 (1981).
15. S. R. Rodriguez, F. P. Ojeda, and N. C. Inestrosa, *Mar. Ecol. Progr. Ser.* **97**, 193 (1993).
16. E. Lindner, in J. Costlow and R. C. Tipper, eds., *Marine Biodeterioration: An Interdisciplinary Study*, Naval Institute Press, Annapolis Me, p. 103, 1984.
17. C. Holmstrøm and S. Kjelleberg, *Biofouling* **8**, 147 (1994).
18. M. Candries, Drag, Boundary-Layer and Roughness Characteristics of Marine Surfaces Coated with Antifoulings, Ph.D. dissertation, Department of Marine Technology, University of Newcastle-upon-Tyne, Newcastle-upon-Tyne, U.K., 2001.
19. H. B. Culver and G. Grant, *The Book of Old Ships and Something of their Evolution and Romance.* Garden City, N.Y., Garden City Publishing Company Inc. 1924, 1935, pp. 306.
20. R. Dalley and D. J. Crisp, *Marine Biol. Lett.* **2**, 141 (1981).
21. R. Dalley, *Crustaceana* **46**, 39 (1984).
22. J. Needham, *Science and Civilisation in China*, Vol. 4, Part 3, Section 29, Cambridge University Press, 1971.
23. S. Seagrave, *Lords of the Rim*, Corgi Books, London, 1996.
24. S. E. Morison, *Admiral of the Ocean Sea*, Oxford University Press, 1942, p. 680.
25. Woods Hole Oceanographic Institution (WHOI, 1952). *Marine Fouling and its Prevention*. US Naval Institute, Annapolis, Ma. 1952.
26. R. Mallet, *Trans. Inst. Naval Arch.* **13**, 90 (1872).
27. V. B. Lewes, *Trea. Mater. Sci. Technol.* **28**, 89 (1889).
28. A. C. A. Holzapfel, *Trans. Inst. Naval Arch.* **46**, 252 (1904).
29. V. Bertram, Past, Present and Future Prospects of antifouling. *32nd WEGEMT School on Marine Coatings*, pp. 85–97. Plymouth, UK, July 2000, 10–14.

30. H. Williams, *J. Am. Soc. Naval Eng.* **35**, 357 (1923).
31. D. P. Graham, *Trans. Soc. Naval Arch. Marine Eng.* **55**, 202 (1947).
32. A. Milne, *Polymers Marine Environ.* 139 (Oct. 1991).
33. E. Bie Kjaer, *Prog. Org. Coatings* **20**, 339 (1999).
34. C. Anderson, Tin vs Tin-Free Antifoulings, *Proceedings of Protecting the Ship while Safeguarding the Environment*, April, 1995, pp. 5–6.
35. C. Anderson, *Protective Coatings Eur.*, 22 (April 1998).
36. C. Anderson, and A. Milne, *Paint Resin*, 13 (February 1984).
37. G. Swain, Redefining Antifouling Coatings. *Protective Coatings Europe*, July 1999, p. 18.
38. C. Alzieu, "TBT Detrimental Effects on Oyster Culture in France—Evolution since Paint Regulation", *Proceedings of Oceans 86, Organotin Symposium, Marine Technology Society*, Washington, D. C., Vol. 4, 1986, pp. 1130–1134.
39. IMO (2001), www. imo. org.
40. U. S. Pat. 2,253,128 (1939).
41. J. C. Montermoso, T. M. Andrews, and L. P. Marinelli, *J. Poly. Sci.* **32**, 523 (1958).
42. R. Subramanian and co-workers, ACS Divisional Organic Coating Plastics Chemistry Preprints, Vol. 40, p. 167, 1979.
43. GB Pat. 1,457,590 (April 3, 1974) A. Milne and G. Hails.
44. C. Anderson and J. E. Hunter, TBT-free Antifouling Coating Technologies and Performance: A Technical Review, *Proceedings from the International Symposium on Pollution Prevetion*, Miami, Fla., April 2001.
45. M. Callow and G. Willingham, *Biofouling* **10**, 239 (1996).
46. P. Turley and C. Waldron, "Smooth Sailing: The Role of Co-biocides in Marine Antifoulings", *Paint and Coatings Industry*, July 1999.
47. GB Pat. 1,470,465 (19 Jan 1976). A. Milne.
48. A. Milne, M. E. Callow, and R. Pitchers, in Evans and Hoagland, eds., *Algal Biofouling*, Elsevier, 1986.
49. A. Milne, "Roughness and Drag from the Marine Paint Chemist's Viewpoint", *Marine Roughness and Drag Workshop*, Paper 12. London.
50. J. R. Griffith, "The Fouling Release Concept: A Viable Alternative to Toxic Antifouling Coatings?" *Polymers in a Marine Environment, The Institute of Marine Engineers*, Paper 38, 1985.
51. R. F. Brady, J. R. Griffith, K. S. Love, and D. E. Field, "Non-toxic Alternatives to Antifouling Coatings", *Polymers in a Marine Environment, The Institute of Marine Engineers*, Paper 26, 1989.
52. R. F. Brady, *Chem. Ind.* **17**, 219, March 1997.
53. I. L. Singer, J. G. Kohl, and M. Patterson, *Biofouling* **6**, 301 (2000).
54. R. F. Brady and I. L. Singer, *Biofouling* **15**, 73 (2000).
55. D. J. Crisp, and G. Walker, "Marine Organisms and Adhesion", *Polymers in a Marine Environment, The Institute of Marine Engineers*, Paper 34, 1985.
56. T. B. Burnell, J. C. Carpenter, K. M. Carroll, J. Serth-Guzzo, J. Stein, K. E. Truby, M. Schultz, and G. W. Swain, "Advances in Nontoxic Silicone Biofouling Release Coatings", *Polymer Pre-Prints* **39**, 507 (1998).
57. K. Truby, C. Wood, J. Stein, J. Cella, J. Carpenter, C. Kavanagh, G. Swain, D. Wiebe, D. Lapota, A. Meyer, E. Holm, D. Wendt, C. Smith, and Montmarano, *Biofouling* **15**, 141 (2000).
58. M. P. Schultz, C. J. Kavanagh, and G. W. Swain, *Biofouling* **13**, 323 (1999).
59. B. S. Kovach, and G. F. Swain, "A Boat Mounted Foil to Measure the Drag Properties of Antifouling Coatings Applied to Static Immersion Panels", *Proceedings of the International Symposium on Sea Water Drag Reduction*, Newport, R. I., July 22–24, 1998, pp. 169–173.

60. J. W. Klinstra, K. C. Overbeke, H. Sonke, R. M. Head, and G. M. Ferrari, Critical Speeds for fouling removal from a silicone coating, *Program and Book of Abstracts from XIth International Congress on Marine Corrsion and Fouling*, San Diego, July 2002, p. 44.

COLIN ANDERSON
International Coatings, Ltd.

COATINGS FOR CORROSION CONTROL, ORGANIC

1. Introduction

Corrosion is a process by which materials, especially metals, are worn away by electrochemical and chemical actions. Metals have anodic and cathodic areas and in the presence of oxygen, water, and a conducting medium, corrosion results. The metal is oxidized to form metal ions at the anode and oxygen is reduced at the cathodes to form hydroxy ions. Since conductivity of the water is an important factor in the rate of corrosion, salts, such as sodium chloride, tend to increase the rate of corrosion.

The principal metals that corrode are steel and to a lesser extent aluminum. There are many kinds of steel; all are alloys of iron and carbon with other metals. Various kinds of steel corrode at different rates, depending on their composition and on the presence of mechanical stresses. The composition varies from location to location on the surface; as a result, some areas are anodic relative to other areas that are cathodic. Stresses and morphological structure of the metal surface can also be factors for setting up anode–cathode pairs. Cold-rolled steel has more internal stresses than hot-rolled steel and is generally more susceptible to corrosion; but, it is widely used because it is stronger. Internal stresses can also be created during fabrication or by the impact of a piece of gravel on an auto body.

Aluminum is higher in the electromotive series than iron and is more easily oxidized. Yet, aluminum generally corrodes more slowly than steel. A freshly exposed surface of aluminum oxidizes quickly to form a dense, coherent layer of aluminum oxide. On the other hand, aluminum corrodes more rapidly than iron under either highly acidic or highly basic conditions. Also, salt affects the corrosion of aluminum even more than it affects the corrosion of iron; aluminum corrodes rapidly in the presence of sea water.

Three strategies are employed to control electrochemical corrosion by coatings. First, one can cover the metal with a barrier coat to prevent water and oxygen from contacting the surface; second, one can suppress the anodic reaction, third, one can suppress the cathodic reaction.

In many cases, a metal object can be completely covered by intact coating films. However, there are situations where the metal surface cannot be completely covered or where an intact film is damaged during use. In such cases of nonintact films it is still essential to control corrosion.

2. Corrosion Protection by Intact Coatings

Coatings can be effective barriers to protect steel when it is anticipated that the coating can be applied to cover essentially all of the substrate surface and when the film remains intact in service. However, when it is anticipated that there will not be complete coverage of the substrate or that the film will be ruptured in service, alternative strategies using coatings that can suppress electrochemical reactions involved in corrosion may be preferable. It is seldom effective to try to use both strategies at the same time—one must choose one or the other.

2.1. Critical Factors. Until ~1950, coatings were generally believed to protect steel by acting as a barrier to keep water and oxygen away from the steel surface. Then it was found by Mayne (1) that the permeability of paint films was so high that the concentration of water and oxygen coming through the films would be higher than the rate of consumption of water and oxygen in the corrosion of uncoated steel. Mayne concluded that barrier action could not explain the effectiveness of coatings and proposed that the electrical conductivity of coating films is the variable that controls the degree of corrosion protection. Coatings with high conductivity would give poor protection, as compared to coatings of lower conductivity. It was confirmed experimentally that coatings having very high conductivity afforded poor corrosion protection. However, in comparisons of films with relatively low conductivity, little correlation between conductivity and protection has been found. It may be that high conductivity films fail because they also have high water permeability; some investigators believe that conductivity of all coatings is at least a factor in corrosion protection (2,3).

Current understanding of protection of steel against corrosion by intact films is based, to a significant degree, on the work of Funke (4–7). He found that an important factor not given sufficient emphasis in earlier work was adhesion of a coating to steel in the presence of water. Funke proposed that water permeating through an intact film could displace areas of film from steel. In such cases, the film shows poor *wet adhesion*. Water and oxygen dissolved in the water would then be in direct contact with the steel surface; hence, corrosion would start. As corrosion proceeds, ferrous and hydroxide ions are generated, leading to formation of an osmotic cell under the coating film. Osmotic pressure can provide a force to remove more coating from the substrate. Osmotic pressure can be expected to range between 2500 and 3000 kPa, whereas the resistance of organic coatings to deformational forces is lower, ranging from 6 to 40 kPa (4). Thus, blisters form and expand, exposing more unprotected steel surface. It has also been proposed that blisters can grow by a nonosmotic mechanism (8). The suggestion has been made that water absorbed by a coating induces in-plane compressive stress within the coating and elastically extends the interfacial bonds between the coating and the steel substrate (9). At a point of weak adhesion between the coating and the substrate, the stress can lead to disbondment. It has been demonstrated that the rate of growth of blisters is dependent on the modulus of the film, at a sufficiently high modulus blister growth is minimized (10).

In either osmotic or nonosmotic mechanisms, the key to maintaining corrosion protection by a coating is sufficient adhesion to resist displacement forces. Both mechanisms predict that if the coating covers the entire surface of the steel on a microscopic scale, as well as a macroscopic scale, and if perfect wet adhesion could be achieved at all areas of the interface, the coating would indefinitely protect steel against corrosion. It is difficult to achieve both of these requirements in applying coatings, so a high level of wet adhesion is important, but is not the only factor affecting corrosion protection by coatings. Funke found that in addition to wet adhesion, low water and oxygen permeability help increase corrosion protection (6). In any case, if wet adhesion is poor, corrosion protection is also poor. However, if the adhesion is fairly good, a low rate of water and oxygen permeation may delay loss of adhesion long enough so that there is adequate corrosion protection for many practical conditions.

Primers made with saponification resistant vehicles give better corrosion protection than primers made with vehicles that saponify readily (11,12). When water and oxygen permeate through a film and water displaces some of the adsorbed groups of the coating resin from the surface of the steel, corrosion starts. Hydroxide ions are generated at the cathodic areas. Hydroxide ions catalyze hydrolysis (saponification) of such groups as esters. If the backbone of the vehicle resin is connected by ester groups, hydrolysis results in polymer network degradation, leading to poorer wet adhesion and, ultimately, catastrophic failure.

2.2. Adhesion for Corrosion Protection. Adhesion, especially wet adhesion, is critical to corrosion protection. Good dry adhesion is required; if there is no coating left on the substrate, it cannot protect the steel. It has not been so obvious, however, that good wet adhesion is required. Good wet adhesion means that the adsorbed layer of the coating will not desorb when water permeates through the film.

The first step to obtain good wet adhesion is to clean the steel surface, especially to remove any oils and salts. A monograph on methods for cleaning and treating metals for coating is available (13). Zinc and iron phosphate conversion coatings have been the standard treatment for many years. As the last step (often called sealing) before drying, the treated metal is rinsed with a chromic acid solution. Waste disposal problems are severe, especially considering the chromic acid wash step. Soluble chromium(VI) compounds have been shown to be carcinogenic and are being replaced as rapidly as possible. There are also reported to be toxic hazards associated with soluble nickel compounds.

It is reported that addition of polyethyleneimine to a Ca−Zn phosphating treatment bath gives a satisfactory conversion treatment without a chromate rinse (14). It has been shown that the chromate rinse step over zinc phosphate can be replaced with a rinse of a dilute solution of trimethoxymethylsilane with enough H_2ZrF_6 to bring the pH down to ~4 gave even better performance as a treatment for steel to be coated (15). Another patent covers treatment with a zinc/manganese phosphate solution containing silicotungstates followed by a hexafluorozirconic acid rinse (16). A patent has been applied for covering use of highly purified zinc nitrite as the accelerator with a zinc/manganese phosphate conversion coating in treating steel for electrodeposition (17). A nickel and chromium(VI)-free treatment system for steel and galvanized steel is carried out by washing with an alkaline cleaner, rinsing, treating with hexafluorozirconic acid, rinsing, and then sealing with a solution of epoxy phosphate (18). A combination of 3-aminopropyltriethoxysilane, a water-dispersible SiO_2, and zirconylammonium carbonate has been patented for surface treatment of steel and galvanized steel (19).

For many applications, aluminum does not have to be treated for corrosion control due to the coherent aluminum oxide surface of the aluminum. But if exposure to salt is to be expected, the surface must be treated before applying a coating. Chromate surface treatments have been the standard in the industry but with the concern about carcinogenicity of chromium(VI), proprietary chromium free treatments have been developed. Patents (or patent applications) have been issued covering chromium(VI)-free conversion treatments: chromium(III) sulfate plus potassium hexafluorozirconate (20), an aqueous solution of potassium manganate, potassium fluoride, potassium hydroxide, sodium hydrosulfite, and ortho-phosphoric acid (21), vanadium tetrafluoroborate among other similar compounds (22), and chromium(III) sulfate plus potassium hexafluorozirconate (23).

A patent application for treatment of cadmium and zinc–nickel coated steel with basic chromium(III) sulfate and potassium hexafluorozirconate and potassium hexafluorosilicate has been filed (24).

Bis(trialkoxysilyl)alkanes (BTSE) are being investigated to treat the surface of steel. Clean steel is treated with an aqueous solution of bis(triethoxysilyl)-ethane. The BTSE reacts with water and hydroxyl groups on the steel to give a water-resistant anchor on the steel. After drying, the treated metal can be coated and baked. For some types of coatings it is desirable to react the BTSE treated steel with a reactive silane that reacts with other silanol groups of the BTSE and provides a functional group to react with a coating binder (25). A patent has been issued disclosing such a process (26). Another patent covers the use of a combination of BTSE and a ureidotriethoxysilane to treat steel, galvanized steel, zinc, and aluminum (27). It is reported that galvanized steel treated with zirconium nitrate followed by treating with aminoethylaminopropyltrimethoxysilane and coated with a polyester primer gives equal performance to commercial chromate primed metal (28). A review of potential uses of silane treatments for steel, galvanized steel, and aluminum is available (29).

An investigation of the effect of treatment of the surface of galvanized steel with rare earth nitrates has been published (30). Lanthanum nitrate was particularly effective. The results are based on electrochemical testing of samples without organic coatings. Tests with organic coatings will be required before the utility of the treatment can be evaluated.

After cleaning and treating, the surface should not be touched and should be coated as soon as possible. Fingerprints leave oil and salt on the surface. After exposure to high humidity, fine blisters can form, disclosing the identity of the miscreant by the fingerprints. A rusty handprint was once observed on a ship after only one ocean and lake passage (3). When coating surfaces near the ocean, it is critical to avoid having any salt on the metal surface when the coating is applied.

It is also critical to achieve as nearly complete penetration into the micropores and irregularities in the surface of the steel as possible. If any steel is left uncoated, when water and oxygen reach the surface, corrosion will start, generating an osmotic cell that can lead to blistering. An important factor for achieving penetration is that the viscosity of the external phase of the coating be as low as possible and remain low long enough to permit complete penetration. It is desirable to use slow evaporating solvents, slow cross-linking coatings, and, when possible, baking primers. Macromolecules may be large compared to the size of small crevices, so lower molecular weight components may give better protection.

Wet adhesion requires that the coating not only be adsorbed strongly on the surface of the steel, but also that it not be desorbed by water that permeates through the coating. Empirically, it is found that wet adhesion is enhanced by having several adsorbing groups scattered along the resin chain, with parts of the resin backbone being flexible enough to permit relatively easy orientation and other parts rigid enough to assure that there are loops and tails sticking up from the surface for interaction with the rest of the coating. Another reason baking primers commonly provide superior corrosion protection is that at the higher temperature, there may be greater opportunity for orientation of resin molecules at the steel interface. Amine groups are particularly effective polar substituents for promoting wet adhesion. Perhaps, water is less likely to displace amines than other groups from the surface. Phosphate groups also promote wet adhesion. For example, epoxy phosphates have been used to enhance the adhesion of epoxy coatings on steel (31). Carboxylic acid-substituted polymers, such as polyacrylic acid, also promote wet adhesion by forming salt groups (32). However, if the coating is ruptured, the ready availability of water and the base generated by corrosion make polyacrylic acid ineffective. There is need for further research to understand the relationships between resin structure and wet adhesion.

Saponification resistance is another important factor. Corrosion generates hydroxide ions at the cathode, raising pH levels as high as 14. Ester groups in the backbone of a binder can be saponified, degrading the polymer near the interface and reducing wet adhesion. Epoxy-phenolic primers are an example of high bake primers that are completely resistant to hydrolysis. In some epoxy-amine primers, there are no hydrolyzable groups. Amine-terminated polyamides, which are widely used in air dry primers to react with epoxy resins, have amide groups in the backbone that can hydrolyze. However, amides are more resistant to base-catalyzed hydrolysis than are esters. Alkyd resins are used when only moderate corrosion protection is required and low cost is important. Epoxy ester primers show greater resistance to saponification than do alkyd primers.

Water-soluble components that stay in primer films should be avoided because they can lead to blister formation. For example, zinc oxide is an undesirable pigment to use in primers. Its surface interacts with water and carbon dioxide to form zinc hydroxide and zinc carbonate, which are somewhat soluble in water and can lead to osmotic blistering. Passivating pigments cannot function unless they are somewhat soluble in water; their presence in coating films can, therefore, lead to blistering. Funke showed that hydrophilic solvents, which become immiscible in the drying film as other solvents evaporate, can be retained as a separate phase and lead to blister formation (4).

2.3. Factors Affecting Oxygen and Water Permeability.   Many factors affect permeability of coating films to water and oxygen (33). Water and oxygen can permeate, to at least some extent, through any amorphous polymer film, even though the film has no imperfections such as cracks or pores. Small molecules travel through the film by jumping from free volume hole to free volume hole. Free volume increases as temperature increases above glass-transition temperature T_g. Therefore, normally, one wants to design coatings with a T_g above the temperature at which corrosion protection is desired. Since cross-linking reactions become slow as the increasing T_g of the cross-linking polymer approaches the temperature at which the reaction occurs and become very slow at $T < T_g$, air dry films cannot have T_g values much above ambient temperatures. Water can act as a plasticizer for coatings such as epoxy-amines and polyurethanes; the swelling caused by the water increases internal stress that can lead to delamination (34). The internal stresses increase when a film is cycled through wet and dry stages. If the T_g is higher than the temperature of the water, water absorption is decreased, so internal stresses do not build up. The higher T_g values that can be reached with baked coatings may be another factor in their generally superior corrosion protection. In general, higher cross-link density leads to lower permeability. Both T_g and cross-link density affect other coating properties; so that some compromise between T_g and cross-link density and performance must be accepted.

Permeability is also affected by the solubility of oxygen and water in a film. The variation in oxygen solubility is probably small, but variation in water solubility can be large. Salt groups on a polymer lead to high solubility of water in films. This makes it difficult to formulate high-performance air dry, water-reducible coatings that are solubilized in water by amine salts of carboxylic acids. Although to a lesser degree than salts, resins made with polyethylene oxide backbones are likely to give high water permeabilities, as are silicone resins. On the other hand, water has low solubility in halogenated polymers; hence vinyl chloride and vinylidene chloride copolymers and chlorinated rubber are commonly used in formulating topcoats for corrosion resistance. Fluorinated polymers have low permeabilities and good wetting properties; hydroxy-functional poly(vinylidene fluoride) cross-linked with polyisocyanates is reported to give good corrosion protection even as a single coat (35).

Pigmentation can have significant effects on water and oxygen permeability. Oxygen and water molecules cannot pass through pigment particles; therefore, permeability decreases as pigment volume concentration (PVC) increases. However, if the PVC exceeds critical pigment volume concentration (CPVC), there are voids in the film, and passage of water and oxygen through the film is facilitated. Some pigments have high polarity surfaces that adsorb water, and in cases in which water can displace polymer adsorbed on such surfaces, water permeability can be expected to increase with increasing pigment content. Pigments should be used that are as free as possible of water-soluble impurities and use of hydrophilic pigment dispersants should be avoided, or at least minimized.

Pigments with platelet shaped particles can reduce permeability rates as much as fivefold when they are aligned parallel to the coating surface (5,36). A factor favoring alignment is shrinkage during solvent evaporation. Since oxygen and water vapor cannot pass through the pigment particles, the presence of aligned platelets can reduce the rate of vapor permeation through a film. The alignment is critical to the action of the platelets; if they are not aligned, permeability may be increased, especially if the film thickness is small relative to the size of the platelets. Mica, talc, micaceous iron oxide, glass flakes, and metal flakes are examples of such pigments. Aluminum flake is widely used; stainless steel and nickel platelets, while more expensive, have greater resistance to extremes of pH. When appearance permits, use of leafing aluminum pigment in the topcoat is particularly effective. Leafing aluminum is surface treated, so its surface free energy is very low. As a result, the platelets come to the surface during film formation, creating an almost continuous barrier. In formulating coatings with leafing aluminum, it is necessary to avoid resins and solvents that displace the surface treatment from the flakes.

A Monte Carlo simulation model of the effect of several variables on diffusion through pigmented coatings has been devised (37). The model indicates, as would be expected, that finely dispersed, lamellar pigment particles at a concentration near, but below CPVC, give the best barrier performance.

There are advantages to applying multiple layers of coatings. The primer can be designed so that it has excellent penetration into the substrate surface, has excellent wet adhesion, and saponification resistance without particular concern about other properties. The topcoat(s) can provide for minimum permeability and other required properties. The primer film does not need to be thick, as long as the topcoat is providing barrier properties; the lower limit is probably controlled by the need to assure coverage of the entire surface. Funke has reported good results with 0.2-µm primer thickness (7), or even a 10 nm layer of a wet adhesion promoting polymer (32). Another advantage of applying multiple coats is the decrease in probability that any area of the substrate will escape having any coating applied.

Film thickness affects the time necessary for permeation through films. Thicker films are expected to delay somewhat the arrival of water and oxygen at the interface, but are not expected to affect the equilibrium condition. The corrosion protection afforded by intact films would be expected to be essentially independent of film thickness. However, since film thickness affects the mechanical performance of films, there may be some optimum film thickness for the maintenance of an intact film. For example, erosion losses would take longer to expose bare metal as film thickness increases, but the probability of cracking on bending increases as film thickness increases. However, in air dry, heavy duty maintenance coatings, there is generally a film thickness, dependent on the coating that provides a more than proportional increase in corrosion protection relative to thinner films. Commonly, this film thickness is as much as 400 µm or more. Below certain coating thicknesses, there may be microscopic defects extending down through the film to the substrate. The film may look intact, but there may be microscopic defects that are large compared to the free volume holes through which permeation in fully intact films occurs. A potential source of such defects is cracks resulting from shrinkage of films as the last solvent is lost from a coating, with the T_g of the solvent-free system around ambient temperature. If the film is thick enough, such defects may not reach the substrate, hence substantially reducing passage of water and oxygen. This hypothesis is consistent with the general observation that greater protection is achieved by applying more coats to reach the same film thickness. In line with this proposal, the use of barrier platelet pigments permits a reduction in the required film thickness without loss of protection. The platelets may minimize the probability of defects propagating through the film to the substrate. Such defects are less likely to occur in baked films, and this may be another factor in the generally superior corrosion protection afforded by baked films, even though thinner film thicknesses are used.

3. Corrosion Protection by Nonintact Films

Even with coatings designed to minimize the probability of mechanical failure, in many end uses, there will be breaks in the films during their service life. There are situations in which it is not possible to have full coverage of all the steel surface. In such cases, it is generally desirable to design coatings to suppress electrochemical reactions, rather than primarily for their barrier properties.

3.1. Minimizing Growth of Imperfections. If there are gouges through the film down to bare metal, water and oxygen reach the metal and corrosion starts. If the wet adhesion of the primer to the metal is not adequate, water creeps under the coating, and the coating comes loose from the metal over a wider and wider area. Poor hydrolytic stability can be expected to exacerbate the situation. This mode of failure is called cathodic delamination. Control of cathodic delamination requires wet adhesion and saponification resistance. It has also been shown that blisters are likely to develop under a film near the location of a gouge (5,38).

When wet adhesion varies on a local scale, *filiform corrosion* can occur (39). It is characterized by development of thin threads of corrosion wandering randomly under the film, but never crossing another track. Formation of these threads often starts from the edge of a scratch. At the head of the thread, oxygen permeates through the film, and cathodic delamination occurs. The head grows following the directions of poorest wet adhesion. Behind the head, oxygen is consumed by oxidation of ferrous ions and ferric hydroxide precipitates, passivating the area, explaining why threads never cross. Since the ion concentration decreases, osmotic pressure drops, and the thread collapses, but it leaves a discernable rust track. Filiform corrosion can be difficult to see through pigmented films. Infrared (ir) thermography may be a useful way of detecting filiform corrosion of such films (40).

3.2. Passivating Primers. There are inhibitors that suppress corrosion. An important class acts by retarding the anodic reaction, they are called *passivators*. A passivator suppresses corrosion above some critical concentration, but may accelerate corrosion at lower concentrations by cathodic depolarization. This phenomenon is illustrated by the effect of oxygen concentration on corrosion rate. Increasing oxygen concentration up to ~12 mL/L increases corrosion rate because it acts to depolarize the cathode. That is, it oxidizes the hydrogen released at the cathode by electrolysis of water. At higher concentrations, more oxygen reaches the surface than is reduced by the cathodic reaction; beyond that concentration, oxygen is a passivating agent. The mechanism of passivation has not been fully elucidated. According to one theory, if the oxygen concentration near the anode is high enough, ferrous ions are oxidized to ferric ions soon after they are formed at the anodic surfaces. Since ferric hydroxide is less soluble in water than ferrous hydroxide, a barrier of hydrated ferric oxide forms over the anodic areas. The iron is said to be passivated.

Passivating pigments promote formation of a barrier layer over anodic areas, passivating the surface. To be effective, such pigments must have some minimum solubility. However, if the solubility is too high, the pigment would leach out of the coating film too rapidly, limiting the time that it is available to inhibit corrosion. For the pigment to be effective, the binder must permit diffusion of water to dissolve the pigment. Therefore, the use of passivating pigments may lead to blistering after exposure to humid conditions. Such pigments are most useful in applications in which the need to protect the steel substrate after film rupture has occurred outweighs the desirability of minimizing the probability of blistering. They are also useful when it is not possible to remove all surface contamination (blistering will probably occur anyway) or when it is not possible to achieve complete coverage of the steel by the coating.

The critical oxygen concentration for passivation depends on conditions. It increases with dissolved salt concentration and with temperature, and it decreases with increases in pH and velocity of water flow over the surface. At ~pH 10, the critical oxygen concentration reaches the value for air-saturated water (6 mL/L) and is still lower at higher pHs. As a result, iron is passivated against corrosion by the oxygen in air at sufficiently high pH values. It is impractical to control corrosion by oxygen passivation below ~pH 10, since the concentrations needed are in excess of those dissolved in water in equilibrium with air. However, a variety of oxidizing agents can act as passivators. Chromate, nitrite, molybdate, plumbate, and tungstate salts are examples. As with oxygen, a critical concentration of these oxidizing agents is needed to achieve passivation. The reactions with chromate salts have been most extensively studied. Partially hydrated mixed ferric and chromic oxides are deposited on the surface, where they presumably act as a barrier to halt the anodic reaction.

Certain nonoxidizing salts, such as alkali metal salts of boric, carbonic, phosphoric, and benzoic acids, also act as passivating agents. Their passivating action may result from their basicity. By increasing pH, they may reduce the critical oxygen concentration for passivation below the level reached in equilibrium with air. Alternatively, it has been suggested that the anions of these salts may combine with ferrous or ferric ions to precipitate complex salts to form a barrier coating at the anode. There is a possibility that both mechanisms may operate to some extent.

A fairly new approach to passivation is applying a film of electrically conductive polymer to a steel surface to protect it from corrosion. Polyphenyleneamine, commonly called polyaniline, is available commercially under the trade names Zypan (41) (DuPont), Versicon (Allied Signal Co.), and Panda (Monsanto Co.). It is said to be effective by leading to the formation of tight, very thin, metal oxide passivating layer on the surface of the metal. The polymer powders are insoluble in all solvents, nonfusible, and difficult to disperse because of high surface tension. Dispersions in a variety of vehicles are available. Reviews of the effects of conducting polymer coatings on metals are available (42,43). Another review of conducting polymer coatings emphasizes protection of aluminum with poly[2,5-bis(*N*-methyl-*N*-propylamino)phenylenevinylene] (44).

Many organic compounds are corrosion inhibitors for steel. Most are polar substances that tend to adsorb on high energy surfaces (38). Amines are particularly widely used. Clean steel wrapped in paper impregnated with a volatile amine or the amine salt of a weak acid is protected against corrosion. The reason for their effectiveness is not clear. They may act as inhibitors because they are bases and neutralize acids. It may be that amines are strongly adsorbed on the surface of the steel by hydrogen bonding or salt formation with acidic sites on the surface of the steel. This adsorbed layer then may act as a barrier to prevent oxygen and water from reaching the surface of the steel.

To protect uncoated steel and aluminum components during shipment and storage, strippable coatings can be used. For example, a coating comprised of two aqueous polyurethane dispersions with aminomethylpropanol is sprayed over the surfaces as a protective layer (45). The coating is easily removed and can be recycled.

Red lead pigment, Pb_3O_4 containing 2–15% PbO, has been used as a passivating pigment since the mid-nineteenth century. Red lead-in-oil primers are used for air dry application over rusty, oily steel. The mechanisms of action are not fully understood (38). They presumably include oxidation of ferrous ions to ferric ions followed by coprecipitation of mixed iron–lead salts or oxides. The somewhat soluble PbO raises the pH and neutralizes any fatty acids formed over time by hydrolysis of the drying oil. Toxic hazards of red lead restrict its use to certain industrial and special purpose applications, and regulations can be expected to prohibit its use.

The utility of chromate pigments for passivation is well established. Various mechanisms have been proposed to explain their effectiveness (38). All the proposed mechanisms require that the chromate ions be in aqueous solution. Like all passivators, chromate ions accelerate corrosion at low concentrations. The critical minimum concentration for passivation at 25°C is $\sim10^{-3}$ mol CrO_4^{2-}/L. The critical concentration increases with increasing temperature and increasing NaCl concentration. Sodium dichromate is an effective passivating agent, but would be a poor passivating pigment; its solubility in water (3.3 mol CrO_4^{2-}/L) is too high. It would be rapidly leached out of a film and would probably cause massive blistering. At the other extreme, lead chromate (chrome yellow) is so insoluble (5×10^{-7} mol CrO_4^{2-}/L) that it has no electrochemical action.

"Zinc chromates" have been widely used as passivating pigments. The terminology is poor, since zinc chromate itself is too insoluble and could promote corrosion, rather than passivate. Zinc yellow pigment is [$K_2CrO_4 \cdot 3Zn$-$3ZnCrO_4 \cdot Zn(OH)_2 \cdot 2H_2O$]. Zinc tetroxychromate [$ZnCrO_4 \cdot 4Zn(OH)_2$] has a solubility lower than desirable (2×10^{-4} mol CrO_4^{2-}/L), but is used in *wash primers*. Phosphoric acid is added to wash primers before application; it may be that this changes the solubility so that the chromate ion concentration is raised to an appropriate level. Strontium chromate ($SrCrO_4$) has an appropriate solubility in water (5×10^{-3} mol CrO_4^{2-}/L) and is sometimes used in primers, especially latex paint primers, in which the more soluble zinc yellow can cause problems of package stability.

It has been established that zinc chromates—and, presumably, other soluble chromates—are carcinogenic to humans. They must, therefore, be handled with appropriate caution. In some countries, their use has been prohibited; their use has been prohibited in Los Angeles and it is expected that this will be expanded statewide in California. Prohibition worldwide is probable in the future.

Substantial efforts have been undertaken to develop less hazardous passivating pigments (46). However, it is difficult to conclude from the available literature and supplier technical bulletins on how these pigments compare with each other and with zinc yellow. In some cases, a formulation that has been optimized for one pigment is compared to a formulation containing another pigment that may not be the optimum formulation for that pigment. (A common example is the substitution of one pigment for another on an equal weight basis, rather than formulating to the same ratio of PVC to CPVC; the results could be very misleading, since primer performance is quite sensitive to the PVC/CPVC ratio.) Much of the published data is based on comparing corrosion resistance in salt fog chamber tests (or other laboratory tests) rather than on actual field experience. A problem with accelerated tests is that an important factor in the performance of a pigment is the rate at which it is leached from a film. A pigment could be sufficiently insoluble that it is not fully dissolved in 1000–2000 h in an accelerated test but be fully dissolved in field applications where corrosion protection for several years is required. As discussed later, there is no laboratory test available that provides reliable predictions of field performance.

Basic zinc and zinc–calcium molybdates are said to act as passivating agents in the presence of oxygen, apparently leading to precipitation of a ferric molybdic oxide barrier layer on the anodic areas. Barium metaborate is the salt of a strong base and a weak acid. It may act by increasing the pH, thus lowering the critical concentration of oxygen required for passivation. To reduce its solubility in water, the pigment grade is coated with silicon dioxide. Even then, some workers feel that the solubility is too high for use in long-term exposure conditions. Zinc phosphate, $Zn_3(PO_4)_2 \cdot 2H_2O$, has been used in corrosion protective primers and may act by forming barrier precipitates on the anodic areas. There is a considerable difference of opinion as to its effectiveness. Calcium and barium phosphosilicates and borosilicates are being used increasingly; they may act by increasing pH. Calcium tripolyphosphate has also been recommended (47).

These pigments are all inorganic pigments; a wider range of potential oxidizing agents would seem to be available if organic pigments were used. An example of a commercially available organic pigment is the zinc salt of 5-nitroisophthalic acid. It is said to be as effective as zinc yellow at lower pigment levels. (Since over a long time, any effective passivating pigment will be lost by leaching, it seems doubtful that an equal lifetime could be achieved at a substantially lower pigment content.) The zinc salt of 2-benzothiazoylthiosuccinic acid has been recommended as a passivating agent.

Reference 48 reviews the subject of passivating pigments and reports that hybrid inorganic/organic pigments can equal the performance of strontium chromate. A patent discloses manufacture procedures for inorganic/organic anticorrosion pigments such as $Zn(NCN)_2/Zn(2\text{-mercaptobenzothiazole})$ (49).

3.3. Cathodic Protection by Zinc-Rich Primers. *Zinc-rich primers*
are another approach to protecting steel with nonintact coatings. They are
designed to provide the protection given by galvanized steel, but to be applied
to a steel structure after fabrication (50). The primers contain high levels of powdered zinc, >80% by weight is usual. On a volume basis, the zinc content exceeds
CPVC to assure good electrical contact between the zinc particles and with the
steel. Furthermore, when PVC is >CPVC, the film is porous, permitting water to
enter, completing the electrical circuit. The CPVCs of zinc powders vary, depending primarily on particle size and particle size distribution; values on the order of
67% have been reported (51). The zinc serves as a sacrificial anode, and zinc
hydroxide is generated in the pores. Since corrosion protection with zinc-rich primers, depends on maintaining the open pore structure of the primer so as to
maintain electrical conductivity, topcoats must not penetrate significantly into
the primer.

Vehicles for zinc-rich primers must be saponification resistant. Two classes
of binders are used, organic resins and inorganic resins. Alkyds are not appropriate resins for this application since they are readily saponified. Two package (2K)
epoxy/amine systems are commonly used. Zinc-rich primers with moisture-cure
urethanes are used for new construction when the metal surface is blast cleaned
to white metal conditions (52). Reference (53) provides a discussion of variables
in formulating such zinc-rich primers. Urethane coatings have adequate saponification resistance and have the advantage over epoxy/amine coatings of being
one package coatings. Water-borne urethane coatings are also available. For
example, a zinc-rich primer is formulated with an aqueous polyurethane dispersion (PUD) made from a polyether diol, N-(3-trimethoxysilylpropyl)aspartic acid
diethyl ester, and bis(4-isocyanatophenyl)methane diisocyanate (54).

However, the most widely used vehicles are tetraethyl orthosilicate and oligomers derived from it by controlled partial hydrolysis with a small amount of
water. Ethyl or isopropyl alcohol is used as the principal solvent, since an alcohol
helps maintain package stability. After application, the alcohol evaporates, and
water from the air completes the hydrolysis of the oligomer to yield a film of polysilicic acid partially converted to zinc salts. Cross-linking is affected by relative
humidity (RH); properties can be adversely affected if the RH is low at the time of
application. Such a primer is referred to as an *inorganic zinc-rich primer*. The
need to reduce VOC emissions has led to the development of water-borne zinc-
rich primers. These use sodium, potassium, and/or lithium silicate solutions in
water as the vehicle. In many applications, they have been found to be as useful
as solvent-borne zinc-rich primers. However, they have limitations: sensitivity to
high humidity, slow rate of evaporation, relatively poor wetting and flow. Properties are effected by the ratio of the silicate to alkali and particle size and distribution of the zinc dust (55). The chemistry of silicate binders is reviewed in (56).

Properly formulated and applied, zinc-rich primers are very effective in protecting steel against corrosion. Their useful lifetime is not completely limited by
the amount of zinc present, as one might first assume. Initially, the amount of
free zinc decreases from the electrochemical reaction; later, loss of zinc metal
becomes slow, but the primer continues to protect the steel. Possibly, the partially hydrated zinc oxide formed in the initial stages of corrosion of the zinc
fills the pores and, together with the remaining zinc, acts as a barrier coating

(57). It is also possible that the zinc hydroxide raises the pH to the level at which oxygen can passivate the steel.

Zinc is expensive, especially on a volume basis. Early attempts to replace even 10% of the zinc with low-cost inert pigment caused a serious decrease in performance, presumably due to decrease in metal to metal contact, even though the PVC was >CPVC. A relatively conductive inert pigment, iron phosphide (Fe_2P), has shown promise (58). It has been reported that in ethyl silicate-based coatings, up to 25% of the zinc can be replaced with Fe_2P; however, with epoxy-polyamide coatings replacement of part of the Zn with Fe_2P leads to a reduction in protection (59).

Zinc-rich powder coatings cannot be applied with the PVC above CPVC. Lower zinc content powder coatings have been used but corrosion protection is inferior to liquid applied coatings because of low penetration of water through the coating and inferior contact between zinc particles and between them and the steel substrate. It has been found that addition of sufficient carbon black increases electrical conductivity between zinc particles and with the substrate providing good performance (60).

Zinc-rich primers are frequently topcoated to minimize corrosion of the zinc, protect against physical damage, and improve appearance. Formulation and application of topcoats must be done with care. If the vehicle of the topcoat penetrates into the pores in the primer film, conductivity of the primer may be substantially reduced, rendering it ineffective.

4. Types of Coatings

4.1. Primers for Baked Coatings.
In most cases, baking primers are designed to be barrier coatings to minimize the risk of blistering. Primers formulated with epoxy and phenolic resins provide excellent corrosion over steel; they show good wet adhesion and are not subject to hydrolysis or saponification. Epoxy ester based primers cure at a somewhat lower temperature and approach epoxy-phenolic coatings in performance. Polyurethane primers can provide good properties. To minimize cost they are prepared with aromatic polyisocyanates but even then their cost tends to be higher than epoxy/amine primers. Alkyd resins are only used when performance requirements are limited. They have the advantage of low cost and ease of application but are subject to saponification.

Primers are generally formulated with PVC only slightly under CPVC. There are two resins for the high pigment loading. The pigment minimizes water and oxygen permeability especially if part of the pigment is a platy pigment such as mica or micaceous iron oxide. Also, the low gloss surface of a highly pigmented primer decreases the difficulty of having good intercoat adhesion with the topcoat. The high pigmentation of primers makes formulation of very high solids primers impossible. The upper limit of solids is ~70 vol%. As a result, the incentive to develop water-borne primers is high.

The most widely used water-borne primers are formulated with maleinized epoxy esters; most commonly dehydrated castor oil based epoxy esters. An epoxy ester is reacted with maleic anhydride (2,5-furandione) [108-31-6] that undergoes a Diels-Alder reaction with the conjugated double bonds of the dehydrated castor oil fatty acids. In this way, the acid groups are attached to the resin by carbon-to-carbon bonds and cannot be removed by hydrolysis. The maleated resin is reacted with an amine such as dimethylethanolamine and dispersed in water. Performance of primers based on this vehicle is fully comparable to that of solvent-borne epoxy ester primers.

Coil coating is sometimes an exception to the general statement that barrier coatings are used for corrosion protection with baked coatings. When coil coated steel is fabricated there are cut edges exposed. Corrosion can start at these cut edges. Epoxy ester primers with a chromate passivating pigment have been used. In a study of the effect of passivating pigment on cut edge corrosion of coil coated galvanized steel with a polyester/melamine–formaldehyde primer, it has been reported that $AlH_2P_3O_{10} \cdot 2H_2O$ (K White 84) and a proprietary pigment (Actinox-106) containing both PO_4 and MoO_3 anions provided protection close to that of strontium chromate (61).

4.2. Electrodeposition Coatings. Cationic electrodeposition coatings are used as primers on all automobiles and trucks as well as many other steel substrates. Reference (62) is an old but still useful review of cationic E-coats. A more recent review of electrodeposition coatings is given in (63).

The article to be coated is submerged on a conveyor in a tank and made the cathode of a electrolytic cell. The coating is an aqueous dispersion of a cationic resin system. The dispersed particles are electrophoretically attracted by the cathode at which OH^- ions are being generated by the electrolysis of water. The charges on the coating particles are neutralized leading to the deposition of a layer of coating on the cathode. It is critical that the binder be very resistant to hydrolysis or saponification.

The coatings must be designed so that all components are in the aggregate particles so that all components in composition are attracted to the electrode at the same rate. The electrolysis of water leads to the release of the neutralizing acid at the anode, which must be removed. The rate of deposition and deposition in restricted areas is affected by the conductivity of the aqueous phase. As the coated object on the conveyor moves out of the tank it must be rinsed so that the system is being continually being diluted. The dilution effect and the accumulation of acid can be partially made up for by adding new primer at higher solids and with a deficiency of acid. But also, an ultrafilter is required to remove excess water, soluble salts, and excess acid.

The rate of deposition increases with increasing voltage. As film thickness increases the rate of deposition decreases and finally stops as the electrical resistance increases. If the voltage is too high, hydrogen gas evolved below the surface of the coating erupts through the coating. If the conductivity of the aqueous phase is too high, the rate of deposition will decrease. The balance of conductivity and voltage is critical in controlling "throw power," that is the distance of deposition into a closed end narrow hollow in the object.

The coating composition consists of an amine modified 4,4'-(1-methylethylidene)bisphenol (BPA) [80-05-7] epoxy resin such as shown in Figure 1 with a blocked polyisocyanate as the cross-linker.

For many years, the blocking agent for the polyisocyanate was 2-ethylhexyl alcohol (2EH). Since 2EH blocked isocyanates have to be heated to temperatures in the range of 180°C for 30 min to cure, there has been a major amount of work

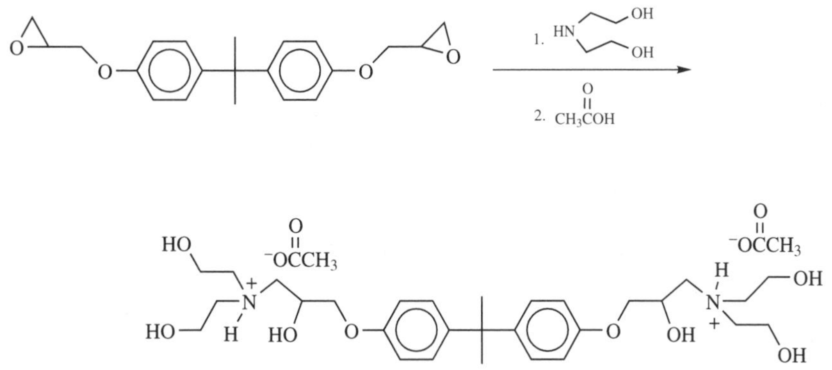

Fig. 1. Formation of water-dispersible BPA epoxy adduct.

devoted to studying other blocking agents. Butoxyethoxyethanol is an example of a blocking agent that permits cure at 150°C. See (64) for an extensive review of blocked isocyanates for E-coat.

The cross-linker for almost all E-coats is a blocked isocyanate usually derived from toluene diisocyanate (TDI), (2,4-diisocyanato-1-methylbenzene) [584-84-9], or bis(4-isocyanatophenyl)methane (MDI). Commonly, a half-blocked isocyanate such as shown in Figure 2 is reacted with the epoxy–amine compound before neutralization with acetic acid (or other low molecular weight carboxylic acid). Pigments are dispersed in the vehicle and the pigmented coating is dispersed in water.

There has also been substantial effort to find more effective catalysts for the cross-linking reaction. In early primers, it was common to use a combination of dibutyltin oxide and basic lead silicate or lead cyanamide as the catalyst. To remove the lead and lower cure temperatures many alternative catalysts have been patented (64). It is essential that a catalyst be very resistant to hydrolysis, eg, dibutyltin dilaurate dibutyl bis[1-(oxododecyl)oxy]-stannate (DBTDL) [77-58-7] is too easily hydrolyzed. Bis(trioctyltin) oxide is one example of a

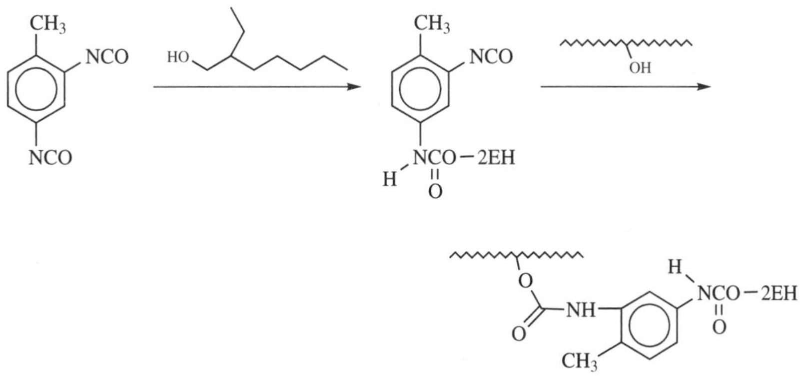

Fig. 2. Formation of half-blocked TDI-2EH adduct.

catalyst whose use in E-coats has been patented (65). See (64) for a review of catalysts for E-coating.

A patent has been applied for disclosing use of a carbamate-functional resin prepared by reacting a BPA epoxy resin with isophorone diisocyanate [1-isocyanato-3-(isocyanatomethyl)-3,5,5-trimethylcyclohexane] [4098-71-9] that has been partially reacted with hydroxypropylcarbamate. The resin is then reacted with dimethylethanolamine and formic acid to provide for water dispersibility. The cationic E-coat composition is formulated with this resin and butylated MF resin as the cross-linker. The coating has the advantage of curing in 30 min at a metal temperature of 100°C while retaining the property advantages resulting from the presence of carbamate (urethane) groups (66).

The corrosion resistance of automobiles coated by cationic E-coating is excellent. This results because of the strong interaction between the amine groups on the coating and the steel and its phosphate coating leads to excellent wet adhesion. Also, the film completely covers the surface and penetrates into all the micropores on the surface by the driving force of the electrocoating and the high temperature baking. The urethane cross-linked epoxy resin is almost impervious to saponification.

The film thickness applied is uniform, which means that any irregularities in the surface of the metal are copied in the surface of the primer coat. Also, since the pigmentation must be lower than usual in primers, adhesion to the surface of topcoats can be difficult to achieve. In general, a coat of a primer-surfacer is applied over the primer before topcoating. Another approach is to apply two E-coats, the first designed to provide the excellent wet adhesion to the steel, and the second one that is designed to provide a level surface to which topcoats will adhere. One example of a patent covering a conductive E-coat primer discloses a formulation using 2EH blocked TDI with an epoxy-amine resin and electroconductive black pigment (67).

4.3. Maintenance Paints.   Maintenance paints are coatings applied to field installations such as highway bridges, factories, and tank farms. Composition of the coating is a main variable but surface preparation and application procedures are also critical to performance. The most commonly used method of cleaning has been sandblasting, however, conventional sand blasting is being rapidly replaced by modifications that are less hazardous. Abrasive blasting with materials like steel grit or water-soluble abrasives like sodium bicarbonate or salt are replacing sand that can cause silicosis. Ultra-high pressure hydroblasting at pressures >175 mPa (25,000 psi) is very effective in removing oil and surface contaminants such as salt. Salt on the surface of steel to be painted is a major source of failure by osmotic blistering. Reference (68) reviews methods of testing the surfaces for salt content and effects on corrosion. Only a limited area should be cleaned at a time so that coating can be applied before contamination. A patent has been issued covering the replacement of grit blasting with a phosphoric acid wash, rinse, phosphate conversion treatment, rinse and a chrome-free sealer (69).

At least two coats, primer and topcoat, are applied; in some cases a primer, intermediate coat, and a topcoat are required. The primer provides the principal protection against corrosion but the other coats also help reduce water and oxygen permeability as well as protecting the primer and providing the desired surface properties. Three types of primers are used depending on the applications: barrier primers, zinc-rich primers, and primers containing passivating pigments.

Barrier Coating Systems. The primer must have excellent wet adhesion. It should have a low viscosity and slow evaporating solvents to permit penetration into microcracks and crevices in the steel surface. Amine and phosphate groups tend to promote wet adhesion. The T_g of the fully cured film should be only a little above the ambient temperature at which the film will be cured. If the T_g of the fully cross-linked film is significantly above ambient temperature, cure rate will slow and may well stop before full cure is attained.

Two package (2K) primers formulated with BPA or novolak epoxy resins and polyfunctional amines are widely used. They can provide excellent wet adhesion and do not saponify. To reduce VOC emissions, water-borne 2K epoxy/amine coatings have been developed whose properties approach those of solvent borne primers (70). Epoxy/amine primers have poor exterior durability and must be topcoated. Chlorinated rubber or vinyl chloride copolymers are desirable in topcoats because of their low permeability to water, however, VOC emissions are high.

Moisture-cure urethane primers can be used over old lead containing coatings reducing the hazards resulting from removal of the old coating. In contrast to many other types of finishes, good protection can be achieved by hand tool cleaning. Such cleaning removes loose old paint but the cleaning still leaves areas of adherent rust and grease, for good performance the viscosity of the external phase of a primer should be very low to permit penetration into the rust areas and crevices in the steel. Such primers are sometimes called sealers (71). A proprietary TDI/polyol prepolymer is reported to have much narrower molecular weight distribution than a conventional TDI/polyol prepolymer (72). Consequently, it has a lower viscosity, hence permitting higher solids primers to be formulated. The corrosion protection is also reported to be superior to that obtained using a conventional primer. A further advantage for moisture-cure urethanes for this application is that they react with water that usually is in and under the rust areas. The coatings are 1K in contrast to 2K epoxy primers that have been used. There are also the advantages that that they can be applied at low temperatures and relatively high humidity.

Micaceous iron oxide pigmented moisture-cure urethane primers have been recommended for application to steel surfaces requiring exposure to weather with air-borne pollution (73). They have outstanding chemical and abrasion resistance. They can be applied under cold, damp conditions. They are said to have good adhesion to damp surfaces. The coatings are not washed away by rain. Reference (74) provides a useful discussion of moisture-curing urethane coatings for maintenance coatings.

2K urethane primers are being successfully used, usually with 2K urethane topcoats. High solids coatings (>70% solids) are available. The 2K water-borne urethane coatings are now available and used on a wide scale. Reference (75) is a recent review of the field. A review of 2K water-borne urethane maintenance coatings has been published (76). Use of a proprietary water-reducible acrylic resin with conventional HDI based isocyanates is particularly recommended.

Since zinc rapidly develops strongly basic corrosion products, vehicles for primers galvanized steel must be very resistant to saponification. An extensive study of coatings on galvanized steel at Cape Kennedy, Florida for 4.5 years showed that the best performance with an epoxy–polyamide primer and a urethane topcoat (77).

Zinc-Rich Primers. Zinc-rich primers can provide excellent performance when a surface cannot be completely cleaned of rust or when, as is the case with latex coatings, complete penetration into surface pores cannot be achieved. There are three classes of zinc-rich primers: inorganic, organic, and water-borne, In the inorganic primers, the vehicle is partially hydrolyzed tetraethyl orthosilicate in an alcohol solvent. After application the solvent evaporates and reaction with water completes the polymerization. Cross-linking is affected by humidity. Organic primers are made with epoxy resins vehicles. They are in general not as effective as inorganic primers but have greater tolerance for incomplete removal of oils from the surface and better compatibility with topcoats. Water-borne primers are made with a vehicle of potassium, sodium, and/or lithium silicates with a dispersion of colloidal silica. Excellent performance on oil and gas production facilities in marine environments has been reported (78).

A challenge in the use of zinc-rich primers is the proper selection and application of topcoats. Activity of the primer depends on maintaining the porous structure that results from having the PVC > CPVC. If a topcoat penetrates into the pores, the required conductive contacts between zinc particles will be interrupted. Penetration can be minimized by spraying a very thin coat uniformly on the surface, in this way the solvent evaporates rapidly, hence the viscosity increases rapidly to minimize penetration. Applying such a coating uniformly requires a very skillful sprayer. After this first coat is applied, further topcoat can be applied without difficulty. Since latex paints have no resin in the continuous phase, there is less difficulty in applying them over zinc-rich primers. In California, extensive tests have shown that zinc-rich primers with a latex paint topcoat gave excellent protection of highway bridges. It was recommended that application should only be done when the temperature is >10°C and the relative humidity is <75%. Latex paints in general have high moisture permeabilities so it is particularly desirable to formulate with platy pigments. Vinylidene chloride/acrylic latexes have lower moisture permeability and have been used (79).

Passivating Primers. When film damage must be anticipated and when the substrate cannot be completely cleaned, primers with passivating pigments may be the primers of choice. Zinc yellow pigment was for many years the pigment of choice, but it is now known that it is a carcinogen. Since there is no way to evaluate the performance of new pigments in a laboratory, it has been difficult to establish satisfactory replacements. Basic zinc and zinc-calcium molybdates, calcium and barium phosphosilicates and borosilicates, and zinc 5-nitroisophthalate are examples of current pigments. The vehicles for these primers are epoxy-amines, epoxy esters, and urethanes.

In a study of chromium-free passivating pigments in water-borne epoxy/amine coatings, zinc iron phosphate, zinc aluminum phosphate, and zinc molybdenum phosphate were given the highest ratings (80).

Use of latex paints over sand blasted steel leads to "flash rusting." That is, there is virtually instantaneous rusting across the surface when the paint is applied. Commonly, 2-amino-2-methyl-propan-1-ol is added to a latex paint to eliminate the problem. Use of mercaptan substituted compounds as additives to prevent flash rusting has also been recommended (81).

A fairly new approach to achieving passivation is to apply a thin coat of an electrically conducting polymer to steel. A thin, dense layer of iron oxide forms on the steel surface that acts as a passivating layer. Perhaps the effect is enhanced by a strong attraction between amine groups on the polymer with the oxide layer. The corrosion resistance with various topcoats over polyaniline (Panda) films on cleaned steel has been reported (42). Excellent resistance to salt spray tests was obtained. A patent discloses the uses of a variety of dispersions of polyaniline with a variety of organic coatings (82). Another patent discloses 2K coatings formulated with one package containing a polyamide resin, polyaniline powder, and aluminum powder, the other with a BPA epoxy resin (83). In another patent, polyaniline coated nylon 12 polymer particles are dispersed in an epoxy ester resin (84).

4.4. Automotive Refinishing. When major damage has been experienced on a component, such as a door, it is common practice to use a replacement component that has been E-coated by the car manufacturer. Unfortunately, there are uncertified components that have only a black lacquer film on them, this should be removed. When bare metal has to be refinished wash primers are most often used. The binders of wash primers are poly(vinyl butyral) and a phenolic resin, they are generally pigmented with zinc tetroxychromate. Acid is added just before application and a wash coat (very thin coating) is applied to the bare metal. Wash primers give excellent adhesion for topcoats and provide excellent corrosion resistance. Such primers have very high VOC emissions but, except in California, their use has been allowed. Unpigmented wash primers have also been developed that offer properties very close to those with chromium.

4.5. Aircraft Coatings. Primers for exterior surfaces are 2K BPA epoxy/amino-amide-based coatings with a passivating pigment. Strontium chromate has been the pigment of choice. Interior components such as baggage areas are coated with a novolak epoxy amine coating pigmented with strontium chromate and generally not topcoated. Increasingly, water-borne epoxy primers are being used (85). 2K systems in which the amine is converted into a salt with a nitroalkane show performance equivalent to solvent-borne primers (86). The salt acts as an emulsifier and after application the nitroalkane evaporates, freeing the amine to react with the epoxy resin.

The U.S. Air Force has sponsored a major research program to develop chromate free corrosion resistant treatments for aluminum. One example of such work is the so-called sol–gel coatings. For example, aluminum is treated with a combination of tetraethyl orthosilicate and 3-aminopropyltriethoxysilane. Presumably the tetraethylorthosilicate forms Al–O–Si bonds at the surface and the aminosilane reacts with other OH groups from the silicate to provide amino groups substitution. The amine groups can react with the epoxy resins in the primer to provide adhesion between the primer and the treated metal surface. Reference (87) is a review of work on sol–gel treatments.

Another type of organically modified silicates prepared by hydrolysis of tetramethoxysilane and 3-glycidyloxypropyltrimethoxysilane using diamines such as ethylene diamine as a cross-linker has been studied as corrosion resistant coatings for aluminum (88). In another study, the effect of the ratio of water to tetramethylsilane and 3-glycidyloxypropyltrimethoxysilane in preparing a organically modified silicate was studied; it was found that relatively low ratios of water gave superior corrosion protection to aluminum (89).

Another approach has been the use of electrically conductive polymer coatings on aluminum. Aluminum coated with poly(3-octyl)pyrrole and topcoated with a urethane topcoat has shown promising results in preliminary evaluations (90). Hybrid organic/inorganic corrosion inhibiting pigments, such as $Zn(NCN)_2$/Zn(2-mercaptobenzothiazole)$_2$ have been reported to give equal performance to strontium chromate (48,49).

5. Evaluation and Testing

There is no laboratory test available that can be used to predict corrosion protection performance of a new coating system. This unfortunate situation is an enormous obstacle to research and development of new coatings, but it must be recognized and accommodated.

Use testing is the only reliable test of a coating system—ie, to apply it and then observe its condition over years of actual use. The major suppliers and end users of coatings for such applications as bridges, ships, chemical plants, and automobiles have collected data correlating performance of different systems over many years. These data provide a basis for selection of current coatings systems for particular applications. They also provide insight into how new coatings could be formulated to improve chances of success.

Simulated tests are the next most reliable for predicting performance. One common approach is to expose laboratory prepared panels on test fences in inland south Florida or on beaches in south Florida or North Carolina. The difficulties in developing tests to simulate corrosion in marine environments are discussed in (91). Test conditions must simulate actual use conditions as closely as possible. For example, exposure at higher temperatures may accelerate corrosion reactions; however, oxygen and water permeability can be affected by $(T-T_g)$. If actual use will be at temperatures below T_g, but the tests are run above T_g, no correlation should be expected.

Variables in preparation of test panels are frequently underestimated. The steel used is a critical variable (92). Also significant are how the steel is prepared for coating and how the coating is applied. Film thickness, evenness of application, flash off time, baking time and temperature, and many other variables affect performance. Results obtained with carefully prepared and standardized laboratory panels can be quite different than results with actual production products. In view of these problems, it is desirable, when possible, to paint test sections on ships, bridges, chemical storage tanks, etc, and to observe their condition over the years. The long times required for evaluation are undesirable, but the results can be expected to correlate reasonably with actual use.

Since wet adhesion is so critical to corrosion protection, techniques for studying wet adhesion can be very useful. *Electrochemical impedance spectroscopy* (EIS) is widely used to study coatings on steel. Many papers are available covering various applications of EIS; (93) and (94) are reviews that provide extensive discussion of the theory and interpretation of data. Impedance is the apparent opposition to flow of an alternating electrical current and is the inverse of apparent capacitance. When a coating film begins to delaminate, there is an increase in apparent capacitance. The rate of increase of capacitance is proportional to the amount of area delaminated by wet adhesion loss. High performance systems show slow rates of increase of capacitance, so tests must be continued for long time periods. Onset of delamination can be determined by EIS studies (95). As with many tests of coatings, results of EIS tests are subject to considerable variation; it has been recommended that a minimum of five replicate tests should be run (96). The method is very sensitive for detecting defects, but no information is obtained as to whether the defect is characteristic of the coating system or a consequence of poor application. A series of other problems involved in EIS are discussed in (97). The use of EIS, scanning acoustic microscopy (SAM), scanning vibrating electrode technique (SVET), and energy dispersive X-ray spectroscopy in studying blister formation of coatings on steel has been reported (98).

There have been many attempts to develop laboratory tests to predict corrosion protection by coatings, and these efforts continue. However, available tests have limited reliability in predicting performance; nonetheless, they are widely used. The most widely used test method for corrosion resistance is the salt spray (fog) test (ASTM Method B-117-95). Coated steel panels are scribed (cut) through the coating in a standardized fashion exposing bare steel and hung in a chamber where they are exposed to a mist of 5% salt solution at 100% relative humidity at 35°C. Periodically, the nonscribed areas are examined for blistering, and the scribe is examined to see how far from the scribe mark the coating has been undercut or has lost adhesion. It has been repeatedly shown that there is little, if any, correlation between results from salt spray tests and actual performance of coatings in use (99–103).

Many factors are probably involved in the unreliability of the salt spray test. Outdoor exposure can have a significant effect on film properties, and environmental factors such as acid rain vary substantially from location to location. The application of the scribe mark can be an important variable; narrow cuts generally affect corrosion less than broader ones. Also, if the scribe mark is cut rapidly, there may be chattering of cracks out from the main cut, whereas slow cutting may lead to a smooth cut. A passivating pigment with high solubility might be very effective in a laboratory test, but may provide protection for only a limited time under field conditions, owing to loss of passivating pigment by leaching.

Since with intact films, it is common for the first failure to be blister formation, humidity resistance tests are also widely used (ASTM Method D-2247-94). The face of a panel is exposed to 100% relative humidity at 38°C, while the back of the panel is exposed to room temperature. Thus, water continuously condenses on the coating surface. This humidity test is a more severe test for blistering than the salt fog test because pure water on the film generates higher osmotic pressures with osmotic cells under the film than the salt solution used in the salt fog test. It is common to run the test at 60°C "because it is a more severe test". The pitfalls of this approach are obvious in view of the previous discussion of the importance of $(T-T_g)$. Humidity tests do not predict the life of corrosion protection, but may provide useful comparisons of wet adhesion. Wet adhesion can be tested by scribing panels after various exposure times in a humidity chamber, immediately followed by applying pressure sensitive tape across the scribe mark, and then pulling the tape off the panel. A peel adhesion test for wet adhesion has been described (104). Wet adhesion can also be checked after storing panels in water (105).

It is often observed that alternating high and low humidity causes faster blistering than continuous exposure to high humidity. A possible explanation of this is that intermediate corrosion products form colloidal membranes, causing polarization and temporary inhibition of corrosion. The membranes are not stable enough to survive drying out and aging. Another factor may be the increase in internal stress that has been reported by cycling through wet and dry periods (34). A large number of humidity cycling tests has been described, commonly involving repeated immersion in warm water and removal for several hours. In some industries, such tests have become accepted methods of screening coatings, although their predictive value is questionable. Simply correlating them with salt fog tests proves little.

A testing regimen called *Prohesion* has been reported to correlate better with actual performance than the standard salt spray test (106). The procedure combines care in selection of substrates that will reflect real products, use of thin films, emphasis on adhesion checks, and a modified salt mist exposure procedure. Instead of 5% NaCl solution, a solution of 0.4% ammonium sulfate and 0.05% NaCl is used. Scribed panels are sprayed with the mixed-salt solution cycling over 24 h, six 3-h periods alternating with six 1-h drying periods using ambient air. During these cycles, water can penetrate through the film to a greater extent than in salt fog chamber testing in which the humidity is always at 100% and the 5% salt solution minimizes water penetration because of reverse osmotic pressure. The importance of the effect of weathering on corrosion protection has been emphasized. Including a QUV exposure cycle in the cycling regime is said to improve the reliability of accelerated testing (107). Another cycling test in which the automotive industry is gaining confidence is the Society of Automotive Engineers test SAE J-2334.

Neither salt fog nor humidity tests have good reproducibility. It is common for differences between duplicate panels to be larger than differences between panels with different coatings. Precision can be improved by testing several replicate panels of each system. (Commonly, decisions are based on the results from testing two or three panels.) The problem is further complicated by the difficulty of rating the degree of severity of failure. A rating system and an approach to statistical analysis of data have been published (108). The study was based on panels with an acrylic clear coating and a pigmented alkyd coating (neither coating would be expected to have good corrosion protection properties) exposed to 95% relative humidity at 60, 70, and 80°C. The times to failure were extrapolated down to ambient temperatures. In light of the effect of T_g on permeability, the validity of such extrapolations is doubtful.

A further problem of evaluating panels for corrosion protection is the difficulty of detecting small blisters and rust areas underneath a pigmented coating film without removing the film. Infrared thermography has been recommended as a nondestructive testing procedure (39).

A great deal of effort has been expended on electrical conductivity tests of paint films and electrochemical tests of coated panels. [See (109) for an extensive review.]

A variety of cathodic disbonding tests specifically for testing of pipeline coatings has been established by ASTM: G-8-96, G-42-96, and G-80-88 (reapproved 1992). In these tests, a hole is made through the coating and the pipe is made the cathode of a cell in water with dissolved salts at a basic pH. Disbonding (loss of adhesion) as a function of time is followed. While there is considerable variability inherent in such tests and their utility for predicting field performance is doubtful, useful guidance in following progress in modifying wet adhesion may be obtained. Such tests may be used more broadly than just for pipeline coatings. For a discussion of research on cathodic delamination, including investigation of the migration of cations through or under coating films, see (110).

The lack of laboratory test methods that reliably predict performance puts a premium on collection of databases permitting analysis of interactions between actual performance and application and formulation variables. It is especially critical to incorporate data on premature failures in the database. Availability of such a database can be a powerful tool for a formulator and may be especially useful in testing the validity of theories about factors controlling corrosion. In time, it may be possible to predict performance better from a knowledge of the underlying theories than from laboratory tests. Many workers feel this is already true in comparison with salt fog chamber tests.

BIBLIOGRAPHY

1. J. E. O. Mayne, *Official Digest* **24**(325), 127 (1952).
2. J. E. O. Mayne, in L. L. Shreir, ed., *Corrosion*, Vol. **2**, Butterworth, Boston, 1976, pp. 15:24–15:37.
3. H. Leidheiser, Jr., *Prog. Org. Coat.* **7**, 79 (1979).
4. W. Funke, *J. Oil Colour Chem. Assoc.* **62**, 63 (1979).
5. W. Funke, *J. Coat. Technol.* **55**(705) 31 (1983).
6. W. Funke, *J. Oil Colour Chem. Assoc.* **68**, 229 (1985).
7. W. Funke, *Farbe Lack* **93**, 721 (1987).
8. J. W. Martin, E. Embree, and W. Tsao, *J. Coat. Technol.* **62**(790) 25 (1990).
9. D. Y. Perera, *Prog. Org. Coat.* **28**, 21 (1996).
10. T.-J. Chuang, T. Nguyen, and S. Lee, *J. Coat. Technol.* **71**(895) 75 (1999).
11. J. W. Holubka, J. S. Hammond, J. E. DeVries, and R. A. Dickie, *J. Coat. Technol.* **52**(670) 63 (1980).
12. J. W. Holubka and R. A. Dickie, *J. Coat. Technol.* **56**(714) 43 (1984).
13. B. M. Perfetti, *Metal Surface Characteristics Affecting Organic Coatings*, Federation of Societies for Coatings Technology, Blue Bell, Pa., 1994.
14. U. B. Nair and M. Subbaiyan, *J. Coat. Technol.* **65**(819) 59 (1993).
15. N. Tang, W. J. Ooij, and G. Gorecki, *Prog. Org. Coat.* **30**, 255 (1997).
16. U.S. Pat. 6,395,105 (May 28, 2002) W. Wickelhaus, H. Endres, K.-H. Gottwald, H.-D. Speckmann, and J.-W. Brouwer (to Henkel K. A.)

17. U.S. Pat. Appl., 20,020,008,226 (Jan. 24, 2002) H. Chikara, K. Tsugi, Y. Kinase, T. Hata, E. Okuno, and T. Nagayama.

18. U.S. Pat. 6,312,812 (Nov. 6, 2001) B. T. Hauser, R. C. Gray, R. M. Nugent, Jr., and M. L. White (to PPG Industries Ohio, Inc.).

19. U.S. Pat. 6,475,300 (Nov. 5, 2002) T. Shimakura and co-workers (to Nippon Paint Co.).

20. U.S. Pat. 6,375,726 (Apr. 23, 2002) J. Green, M. Kane, and C. Matzdorf (to U.S. Navy).

21. U.S. Pat. Appl., 200,20,033,208 (Mar. 21, 2002) S. Krishnaswamy.

22. PCT Int. Appl., WO/0,228,550 (Apr. 4, 2002), S. E. Dolan, E. W. Sweet, W. N. Opdycke, and W. J. Wittke.

23. U.S. Pat. 6,375,726 (Apr. 23, 2002) C. Matzdorf, M. Kane, and J. Green, III (to U.S. Navy).

24. U.S. Pat. Appl., 20,020,053,301 (May 9, 2002) C. Matzdorf, M. Kane, and J. Green, III.

25. W. J. van Ooij and T. F. Child, *Chemtech*, Feb. 26 (1998).

26. U.S. Pat. 6,261,638 (July 11 2001) W. J. van Ooij, V. Subramanian, and C. Zhang (to University of Cincinnati).

27. U.S. Pat. 6,361,592 (Mar. 26, 2002) J. Song, N. Tang, K. Brown, and E. B. Bines (to Chemetall PLC).

28. M. E. Montemor and co-workers, *Prog. Org. Coat.* **38**, 17 (2000).

29. T. F. Child and W. J. van Ooij, *Trans. Inst. Metal Finishing* **77**(2) 64 (1999).

30. M. F. Montemor, A. M. Simoes, and M. G. S. Ferreira, *Prog. Org. Coat.* **44**, 111 (2002).

31. J. L. Massingill and co-workers, *J. Coat. Technol.* **62**(781) 31 (1990).

32. W. Funke, *Prog. Org. Coat.* **28**, 3 (1996).

33. N. L. Thomas, *Prog. Org. Coat.* **19**, 101 (1991).

34. O. Negele and W. Funke, *Prog. Org. Coat.* **28**, 285 (1995).

35. A. Barbucci, E. Pedroni, J. L. Perillon, G. Cerisola, *Prog. Org. Coat.* **29**, 7 (1996).

36. B. Bieganska, M. Zubielewicz, and E. Smieszek, *Prog. Org. Coat.* **16**, 219 (1988).

37. D. P. Bentz and T. Nguyen, *J. Coat. Technol.* **62**(783) 57 (1990).

38. H. Leidheiser, Jr., *J. Coat. Technol.* **53**(678) 29 (1981).

39. A. Bautista, *Prog. Org. Coat.* **28**, 49 (1996).

40. M. E. McKnight and J. W. Martin, *J. Coat. Technol.* **61**(775) 57 (1989).

41. E. I. DuPont de Nemours & Co., *Product Bulletin, Zypan*, Wilmington, Del., 1999.

42. G. M. Spinks, A. J. Dominis, G. G. Wallace, and D. E. Tallman, *J. Solid State Electrochem.* **6**(2) 85 (2002).

43. S. P. Sitram, J. O. Stoffer, and T. J. O'Keefe, *J. Coat. Technol.* **69**(866) 65 (1997).

44. P. Zares, J. D. Stenger-Smith, G. S. Ostrom, and M. H. Miles, *ACS Symposium Series, 735 (Semiconducting Polymers)* 1999, pp. 280–292.

45. U.S. Pat. 6482885 (Nov. 19, 2002) H.-P. Muller and co-workers (to Bayer AG).

46. A. Smith, *Inorganic Primer Pigments*, Federation of Societies for Coatings Technology, Blue Bell, Pa., 1988.

47. V. F. Vetere, M. C. Deya, R. Romangnoli, and B. del Arrio, *J. Coat. Technol.* **73**(917) 57 (2001).

48. J. Sinko, *Prog. Org. Coat.* **42**, 267 (2001).

49. U.S. Pat. 6139610 (Oct. 31, 2000) J. Sinko (to Wayne Pigment Corp.).

50. J. E. O. Mayne and U. R. Evans, *Chem. Ind.* **63**, 109 (1944).

51. M. Leclercq, *Mater. Technol.*, March, 57 (1990).

52. J. Schwindt, *J. Mater. Performance* **35**(12) 25 (1996).

53. M. Sonntag, *Polym. Colour J.* **172**(4081) 50 (1982).

54. R. R. Roesler and L. Schmalstieg, U.S. Pat. 6077902 (June 20, 2000) (to Bayer Corp.).

55. E. Montes, *J. Coat. Technol.* **65**(821) 79 (1993).
56. G. Parashar, D. Srivastava, and P. Kumar, *Prog. Org. Coat.* **42**, 1 (2001).
57. S. Feliu, R. Barajas, J. M. Bastidas, and M. Morcillo, *J. Coat. Technol.* **61**(775) 63, 71 (1989).
58. N. C. Fawcett, C. E. Stearns, and B. G. Bufkin, *J. Coat. Technol.* **56**(714) 49 (1984).
59. S. Feliu, Jr., M. Morcillo, J. M. Bastidas, and S. Feliu, *J. Coat. Technol.* **63**(793) 31 (1991).
60. H. Marchebois, S. Touzain, S. Joiret, J. Bernard, and C. Savall, *Prog. Org. Coat.* **45**, 415 (2002).
61. R. L. Howard, I. M. Zin, J. D. Scantlebury, and S. B. Lyon, *Prog. Org. Coat.* **37**, 83 (1999).
62. P. I. Kordomenos and J. D. Nordstrom, *J. Coat. Technol.* **54**(686) 33 (1982).
63. I. Krylova, *Prog. Org. Coat.* **41**, 119 (2001).
64. D. A. Wicks and Z. W. Wicks, Jr., *Prog. Org. Coat.* **41**, 1 (2001).
65. U.S. Pat. 6042893 (Mar. 28, 2000), E. C. Bossert, K. Cannon, W. D. Honnick, and W. Ranbom (to ELF Atochem North American Inc).
66. U.S. Pat. Appl. 20020082361 (June 27, 2002), T. S. December.
67. U.S. Pat. 4882090 (Nov. 21, 1989) W. Batzill and co-workers (to BASF Lacke & Farben AG).
68. B. A. Appleman, *J. Prot. Coat. Linings* **19**(5) 42 (2002).
69. U.S. Pat. Appl., 20,020,000,264 (Jan. 3, 2002) R. J. Zupancich.
70. A. Wegmann, *J. Coat. Technol.* **65**(827) 27 (1993).
71. B. P. Mallik, *Paintindia* **51**(9), 119 (2001).
72. J. Kramer, and S. Bassner, *Modern Paint Coat.*, June, 20 (1994).
73. C. Carter, *Anti-Corros. Methods Mater.* **33**(10) 12 (1986).
74. G. Gardner, *J. Prot. Coat. Linings* **13**(2) 34 (1996).
75. Z. W. Wicks, Jr., D. A. Wicks, and J. W. Rosthauser, *Prog. Org. Coat.* **44**, 161 (2002).
76. S. L. Bassner and C. R. Hegedus, *J. Prot. Coat. Linings* **13**(9) 52 (1996).
77. R. W. Drisco, *J. Prot. Coat. Linings* **12**(9) 27 (1995).
78. A. Szokolik, *J. Prot. Coat. Linings* **12**(5) 56 (1995).
79. H. R. Friedl and C. M. Keillor, *J. Coat. Technol.* **59**(748) 65 (1987).
80. B. del Amo, R. Romagnolli, C. Deya, and J. A. Gonzalez, *Prog. Org. Coat.* **45**, 389 (2002).
81. G. Reinhard, P. Simon, and U. Remmelt, *Prog. Org. Coat.* **20**, 383 (1992).
82. U.S. Pat. 5,721,056 (Feb. 24, 1998) B. Wessling (to Zipperling Kessler and Co.).
83. U.S. Pat. 6,231,789 (May 15, 2001) T. R. Hawkins and S. R. Geer (to Geotech Chemical Co.)
84. U.S. Pat. 6,060,116 (May 9, 2000) V. G. Kulkarni, J. Avlyanor, and T. Chen (to Americhem Inc.)
85. A. K. Chattopadhyay and M. R. Zenter, *Aerospace and Aircraft Coatings*, Federation of Societies for Coatings Technology, Blue Bell, Pa., 1990.
86. R. Albers, *Proc. Waterborne Higher-Solids Coat. Symp.*, New Orleans, La., 1983, pp. 130–143.
87. T. L. Metroke, R. L. Parkhill, and E. T. Knobbe, *Prog. Org. Coat.* **41**, 233 (2001).
88. T. L. Metroke, O. Kachurina, and E. T. Knobbe, *Prog. Org. Coat.* **44**, 185 (2002).
89. T. L. Metroke, O. Kachurina, and E. T. Knobbe, *Prog. Org. Coat.* **44**, 295 (2002).
90. V. J. Gelling and co-workers, *Prog. Org. Coat.* **43**, 149 (2001).
91. T. S. Lee and K. L. Money, *Mater. Perform.* **23**, 28 (1984).
92. R. G. Groseclose, C. M. Frey, and F. L. Floyd, *J. Coat. Technol.* **56**(714) 31 (1984).
93. J. R. Scully, Electrochemical Impedance Spectroscopy for Evaluation of Organic Coating Deterioration and Underfilm Corrosion—A State of the Art Technical

Review, Report No. DTNSRDC/SME-86/006, D. W. Taylor Naval Ship Research and Development Center, Bethesda, Md., 1986.
94. U. Rammelt and G. Reinhard, *Prog Org. Coat.* **21**, 205 (1991).
95. E. P. M. van Westing, G. M. Ferrari, F. M. Geenen, and J. H. W. de Wit, *Prog. Org. Coat.* **23**, 89 (1993).
96. W. S. Tait, *J. Coat. Technol.* **66**(834) 59 (1994).
97. H. J. Prause and W. Funke, *Farbe und Lack* **101**, 96 (1995).
98. I. Sekine, M. Yuasa, N. Hirose, and T. Tanaki, *Prog. Org. Coat.* **45**, 1 (2002).
99. R. Athey and co-workers, *J. Coat. Technol.* **57**(726) 71 (1985).
100. J. Mazia, *Met. Finish.* **75**(5) 77 (1977).
101. R. D. Wyvill, *Met. Finish.* **80**(1) 21 (1982).
102. W. Funke, in H. Leidheiser, Jr., ed., *Corrosion Control by Coatings*, Science Press, Princeton, N.J., 1979, pp. 35–45.
103. B. R. Appleman, *J. Prot. Coat., Linings* **9**(10) 134 (1992).
104. M. E. McKnight, J. F. Seiler, T. Nguyen, and W. F. Rossiter, *J. Prot. Coat. Linings* **12**, 82 (1995).
105. M. Hemmelrath and W. Funke, *FATIPEC Congress Book*, Vol. IV, 137 (1988).
106. F. D. Timmins, *J. Oil Colour Chem. Assoc.* **62**, 131 (1979).
107. B. S. Skerry and C. H. Simpson, *NACE Corrosion 91*, Cincinnati, Ohio (1991).
108. J. W. Martin and M. E. McKnight, *J. Coat. Technol.* **57**(724) 31, 39, 49 (1985).
109. J. N. Murray, *Prog. Org. Coat.* **30**, 225 (1997); **31**, 255 (1997); **31**, 375 (1997).
110. J. Parks and H. Leidheiser, Jr., *FATIPEC Congress Book*, Vol. II, 317, 1984.

GENERAL REFERENCES

111. R. A. Dickie and F. L. Floyd, eds., *Polymeric Materials for Corrosion Control*, ACS Symp. Ser. No. 322, American Chemical Society, Washington, D.C., 1986.
112. C. C. Munger, *Corrosion Prevention by Protective Coatings*, National Association of Corrosion Engineers, Houston, Tex., 1997.

ZENO W. WICKS, Jr.
Consultant

COATINGS, MARINE

1. Introduction

The marine environment is highly aggressive. Materials in marine service are constantly exposed to water, corrosive salts, strong sunlight, extremes in temperature, mechanical abuse, and chemical pollution in ports. This climate is very severe on ships, buoys, and navigational aids, offshore structures such as drilling platforms, and facilities near the shore such as piers, locks, and bridges.

Marine coatings are the most important, cost-effective means to preserve steel (qv) and other metals in the marine environment. These coatings impart physical and chemical properties, eg, antifouling, color, and slip resistance, to surfaces that cannot be obtained in any other way. Modern coatings are sophisticated mixtures of polymers and other chemicals, and they require careful control of surface preparation and application conditions. These materials are not as adaptable as earlier coatings to the variety of metals, design features, and surface conditions encountered in marine construction, but they provide long-lasting cost-effective protection when used as part of a corrosion control program (see CORROSION AND CORROSION CONTROL). In 1989 a total of 38×10^6 L of marine coatings were sold in the United States. Shipbuilding consumed 4×10^6 L, construction of pleasure boats required 3×10^6 L and 31×10^6 L were employed for maintenance and repair. The selection and application of marine coatings has become a highly specialized discipline in which governmental regulations are a dominant influence. Many paints (qv) and painting procedures used in the past are no longer permitted or are extremely costly to use.

1.1. Corrosion Control Plan. A corrosion control plan for each ship or structure, which is designed to control deterioration in the most economical and practical manner and to include all appropriate mechanisms for corrosion control, must be developed before construction begins. For steel in the marine environment, the chief methods available are protective coatings and cathodic protection. Cathodic protection is an electrical method of preventing metal corrosion in a conductive medium by placing a negative charge on the item to be protected. This protection mechanism is specifically designed as part of the total corrosion control system and protects only the submerged portions of steel ships and structures.

Materials most resistant to deterioration are chosen, the strength and available shapes of which have a critical influence upon design. Crevices, areas where water can collect, and sharp edges are to be avoided, and dissimilar metals must be electrically isolated from one another in order to prevent corrosion of the more anodic metal. Sharp edges cause paint to draw thin and should be removed by grinding or sanding. Welds have sharp projections that should be removed by grinding. Weld spatter should be scraped or ground from metal surfaces. Outside corners should be rounded, and inside corners should be filled because they provide a collection site for excess paint that may not fully cure. Crevices and pits should be filled with weld metal or caulking because they collect corrosive agents and accelerate deterioration.

1.2. Protective Coatings. Each coating in the protective coating system is designed to perform a specific function and to be compatible with the total system. Selection of a coating system is influenced by the chemical nature of the coating and by the conditions the coating is designed to resist. The identity of each material, the number of coats and the dry film thickness of each, the maximum times between blasting and coating and between coats, and the minimum time between application of the last coat and commencement of service need to be defined as do suitable surface preparation techniques, proper methods for the application of paint, and effective quality control procedures. The effectiveness of a coating is directly related to its ability to maintain adhesion to the substrate, its integrity, and its thickness. Areas that cannot be easily or safely repaired, especially those which require drydocking for repair, need to be given the best available coating.

Fouling organisms attach themselves to the underwater portions of ships and have a severe impact on operating costs. They can increase fuel consumption and decrease ship speed by more than 20%. Warships are particularly concerned about the loss of speed and maneuverability caused by fouling. Because fouling is controlled best by use of antifouling paints, it is important that these paints be compatible with the system used for corrosion control and become a part of the total corrosion control strategy.

2. Environmental Concerns

Local environmental regulations have significantly affected the production, transportation, use, and disposal of coatings.

2.1. Volatile Organic Compounds. As coatings dry, solvents are released into the atmosphere, where they undergo chemical reactions in sunlight and produce photochemical smog and other air pollutants (see AIR POLLUTION). As a general rule, the volatile organic compound (VOC) content of marine coatings is restricted to 340 g/L. In the locations where ozone (qv) levels do not conform to the levels established by the Environmental Protection Agency, regulations require an inventory of all coatings and thinners from the time they are purchased until they are used.

The VOC regulations have been the driving force behind the development of entirely new coatings technologies, the reformulation of coatings, and the creation of new surface preparation and paint application methods. High solids coatings, eg, those of epoxy and urethane, have displaced alkyds and are now the principal marine coatings, but some have short pot lives. Alkyd coatings have been extensively reformulated and are still important but dry more slowly than their forebears and are not used as extensively as in the 1970s. Coatings that contain high levels of solvents, such as vinyl and chlorinated rubber coatings, are disappearing from the marine industry.

VOC-conforming paints are more demanding than their predecessors, and attention to application conditions and techniques is imperative. The coatings can be more viscous and harder to apply, wet films have poorer leveling, and coatings of uniform thickness are more difficult to achieve.

2.2. Heavy-Metal Pigments. Lead (qv) and chromate pigments (qv), used for many years as corrosion inhibitors in metal primers and topcoats for marine coating systems, have been linked to adverse health and environmental effects (see CHROMIUM COMPOUNDS). Inhaling or ingesting droplets of lead- or chromate-containing coatings is a potential source of poisoning of paint applicators. Regulations concerning the removal of lead-based coatings require that existing coatings be analyzed before removal for toxic metals and that all debris be contained during removal. The air in the vicinity must also be monitored during removal to ensure safe conditions, and old paint and blasting abrasive must be disposed of as toxic waste.

Because of these concerns, lead- and chromate-containing pigments are not used in marine coatings. Chromate pigments, which contain the metal in the +6 valence state, are proscribed, but pigments containing chromium in the +3 valence state, such as chromium oxide, Cr_2O_3, are unregulated and continue to

Table 1. **Active Pigments in Anticorrosive Coatings**

Pigment	CAS Registry number	Molecular formula
	Prohibited	
red lead	[1314-41-6]	Pb_3O_4
white lead	[1344-36-1]	$PbCO_3$
lead chromate	[7758-97-6]	$PbCrO_4$
zinc chromate	[13530-65-9]	$ZnCrO_4$
strontium chromate	[7789-06-2]	$SrCrO_4$
basic lead silicochromate	[11113-70-5]	$PbCrO_4 \cdot nSiO_2$
	Allowed	
zinc oxide	[1314-13-2]	ZnO
zinc phosphate	[7779-90-0]	$Zn_3(PO_4)_2$
zinc phosphosilicate		$Zn_3(PO_4)_2 \cdot nSiO_2$
zinc molybdate	[13767-32-3]	$ZnMoO_4$
calcium borosilicate		$CaO \cdot B_2O_3 \cdot nSiO_2$
calcium phosphosilicate		$Ca_3(PO_4)_2 \cdot nSiO_2$

be used. The corrosion inhibitive pigments that were commonly used and those which have replaced them are listed in Table 1.

2.3. Organotins. In the mid-1970s compounds based on derivatives of triphenyl- or tributyltin (see TIN COMPOUNDS) known generically as organotins, were found to be much more effective than cuprous oxide paints in controlling fouling, and numerous products were introduced. These fell into two classes. Free-association coatings contained a tributyltin salt, eg, acetate, chloride, fluoride, or oxide, physically mixing into the coating. These were available in a variety of resins and characterized by a leach rate of organotin which is quite high when the coating is new, and which falls off rapidly until insufficient to prevent fouling. In contrast, copolymer coatings contain organotin which is covalently bound to the resin of the coating and is not released until a tin–oxygen bond hydrolyzes in seawater (1). This controlled hydrolysis produces a low and steady leach rate of organotin and creates hydrophilic sites on the binder resin. This layer of resin subsequently washes away and exposes a new layer of bound organotin. These coatings are also known as controlled release, self-polishing, or ablative coatings, and last for five years when applied at a dry-film thickness of 375 μm.

However, there is now considerable evidence that sufficiently high concentrations of organotins kill many species of marine life and affect the growth and reproduction of others. Thus many nations restrict organotin coatings to vessels greater than 25 meters in length. In the United States, laws prohibit the retail sale of copolymer paints containing greater than 7.5% (dry weight) of tin, and of free-association paints containing greater than 2.5% (dry weight) of tin, but do not restrict the size of the ship of which the paints may be applied (2). Organotins must be used on vessels with aluminum hulls, because copper is cathodic to aluminum and causes rapid pitting and perforation when used on an aluminum hull. Regulations to minimize the exposure of workers and protect the environment during the application and removal of organotin coatings also exist.

2.4. Abrasive Blast Cleaning. Removal of paint by abrasive blasting may lead to adverse health effects for workers who breathe dust formed during the operation. Regulations restrict blasting operations to such procedures as blasting within enclosures, using approved mineral abrasives, using a spray of water to reduce dust, and blasting with alternative materials such as ice, plastic beads, or solid carbon dioxide. A military specification (3) describes abrasives that are approved for use in U.S. naval shipyards. Limits are placed on carbonates, gypsum, and free silica, all of which are not abrasive but only contribute to dust, and on the amount of arsenic, beryllium, cadmium, lead, and 13 other toxic materials in blasting abrasives.

Debris from the removal of paint may contain lead, chromium, or other heavy metals. Collection of such debris is required to prevent release of these metals into the environment and to avoid exposure and contamination of workers. Blasting debris is contained in two ways: by use of containment systems, eg, screens, panels, tarpaulins, and shrouds, which enclose the removal area, and by paint removing machines equipped with vacuum collection devices. The latter include both powered mechanical tools, eg, grinders, brushes, and sanders, and self-contained abrasive blasting equipment. Blasting enclosures that draw in air help to contain particles of paint but do not ensure worker safety, and these are seldom more than 85% effective in preventing dust and debris from escaping from the system. In order to evaluate the efficiency of the containment method, periodic medical examinations of workers for respired air contaminants are required.

Hazardous waste generated by removal of toxic paints may be stored for only a limited time and must be disposed of in conformance with prevailing regulations. The amount of hazardous waste can be greatly reduced by cleaning and recycling the abrasive.

2.5. Reactive Coatings.   Coatings that cure by chemical reaction of two component parts are the most widely used in marine applications and protection of workers from the reactive ingredients is required. For example, urethane coatings may contain isocyanates that may cause respiratory difficulties, and epoxy coatings may contain glycidyl ethers which are skin sensitizers. This danger is diminished in modern coatings when the reactive groups are bound to oligomers having low vapor pressures and the likelihood of exposure to vapors is considerably reduced.

3. Surface Preparation for Marine Painting

Surface preparation, always important in obtaining optimal coatings performance, is critical for marine coatings (see METAL SURFACE TREATMENTS). Surface preparation usually comprises about half of the total coating costs, and if inadequate may be responsible for early coating failure. Proper surface preparation includes cleaning to remove contaminants and roughening the surface to facilitate adhesion.

3.1. Standards for Cleaned Steel Surfaces.   The most important standards (4) used to specify and evaluate cleaned steel surfaces are summarized in Table 2 in order of increasing cost. Photographic standards consistent with the written standards in Table 2 are also available (5).

Table 2. **Steel Surface Preparation Standards**[a]

Number[b]	Title	Intended use
SSPC-SP-1	Solvent Cleaning	removal of oil and grease prior to further cleaning by another method
SSPC-SP-2	Hand Tool Cleaning	removal of loose surface contaminants before spot repair
SSPC-SP-3	Power Tool Cleaning	removal of loose surface contaminants before spot repair
SSPC-SP-7	Brush-Off Blast	removal of loose surface contaminants before spot repair
SSPC-SP-6	Commercial Blast	for interior steel to be coated with alkyd paint
SSPC-SP-10	Near-White Metal Blast	for most exterior surfaces, decks, water, fuel tanks, etc
SSPC-SP-5	White Metal Blast	for the most demanding coating conditions and for those products, ie, inorganic zinc primers, thermal sprayed metals, and powder coatings, which require an uncontaminated surface

[a] Ref. 4.
[b] SSPC = Steel Structures Painting Council, Pittsburgh, Pa.

3.2. Abrasive Blasting. Blast cleaning using mineral abrasives (qv) is the preferred method for cleaning steel prior to applying marine coatings. Blasting not only provides the highest level of cleanliness but also roughens the surface to provide for good adhesion of the primer. As much blasting as possible is done in purpose-built enclosures to minimize the amount of particulates produced and to provide better and less costly cleaning. Shop blasting is accomplished by equipment having high speed rotating wheels that propel shot or grit abrasive onto steel. Portable closed-cycle vacuum blasting equipment is available for field use. Special machines have been made for steel decks and hulls which recycle the abrasive several times, saving costly abrasive and reducing the amount of blasting waste.

Abrasive blasting of steel ships using conventional equipment is usually accomplished at a nozzle pressure of about 700 kPa. Abrasives can completely remove rust, scale, dirt, and old coatings, but grease and oil are smeared and driven into the surface. Abrasive blasting must, therefore, be preceded by solvent cleaning if any grease or oil is present.

To reduce the amount of dust produced, water can be added to the abrasive from a circular water sprayer around the nozzle. Chemical corrosion inhibitors must be dissolved in the water to prevent flash rusting of the steel. Newer methods to reduce dust include the use of ice, solid carbon dioxide (dry ice), or plastic beads as abrasives. Blasting with dry ice is inexpensive and effective, but the accumulation of carbon dioxide must be avoided in enclosures. Plastic beads are inexpensive, but the cutting efficiency is low and paint removal is slow; the beads can be cleaned of paint particles and reused.

Softer metals such as aluminum and its alloys can be blast cleaned using abrasives that are not as hard as those used on steel. Garnet, walnut shells, corncobs, peach pits, glass or plastic beads, and solid carbon dioxide have been used successfully.

3.3. Other Cleaning Methods.   Solvent cleaning, ie, degreasing, is chiefly used to remove grease and oil. Solvent is applied to rags which are replaced when they become contaminated. The final rinse is always made using fresh solvent. Individual ship components can be solvent-cleaned by dipping in tanks of solvent.

Hand and power tool cleaning is used on ships mostly for spot repair of damaged areas. Hand tools include scrapers, wire brushes, and sanders. Electric and pneumatic power tools, which include grinders and needle guns, clean faster and more thoroughly than hand tools. Most power tools have vacuum lines connected to collect paint debris.

Blast cleaning with water, sometimes called hydroblasting, is used to remove marine fouling and sometimes to clean metal surfaces for coating. The water may be heated and detergent may be added to facilitate removal of oil, dirt, and marine slimes. Cleaning bare steel for coating may be achieved using pressures over 200 MPa (30,000 psi) and water volumes of only 8 to 56 L/min. Abrasive may be injected into the stream of water or used in a second operation to produce the rough surface needed for adhesion. Corrosion inhibitors must be used in the water, and extreme caution must be maintained using these high pressures.

Steam cleaning may also be used to remove grease and oil. On large surface areas such as the hulls of ships, steam cleaning is usually more economical and efficient than solvent cleaning. Detergents are sometimes brushed onto the hull before steam cleaning.

4. Types of Coatings

Coatings ingredients fall into four principal classes. Resins form a continuous solid film after curing, bind all ingredients within the film, and provide adhesion to the substrate (6). The properties of a coating are determined principally by the resins it contains. Pigments are metals or nearly-insoluble salts that impart opacity, color, and chemical activity to the coating. Solvents are used primarily to facilitate manufacture and application, but are lost by evaporation after application and are not a permanent part of the coating. Additives used in small (1–50 ppt) amounts give the coating such desirable additional properties as ease of manufacture, stability in shipment and storage, ease of application, or increased performance of the dried film.

These ingredients may be formulated to give coatings that protect against corrosion in different ways (7). Barrier coatings physically separate oxygen, water, ions, and other corrosive agents from the steel surface. Inhibitive coatings prevent corrosion by absorbing or neutralizing corrosive agents, or by slowly releasing protective ions. Sacrificial coatings contain a metal (usually zinc) that is oxidized more rapidly than steel, thereby providing protection for the substrate by electrochemical action. Conversion coatings chemically oxidize the surface of the substrate to a depth of 7–10 µm, producing a passive layer which resists corrosion better than the metal itself.

Modern marine coatings fall into eight generic categories, each named for the principal resin it contains. Significant variation in each category is achieved by varying pigments and other ingredients. The categories are discussed in rough order of importance.

4.1. Epoxies. Epoxy coatings are the workhorse materials for premium marine applications (8). High performance primers and anticorrosive coatings that conform to VOC regulations are widely available (9). The cured coatings are durable, tough, and smooth, and demonstrate excellent resistance to solvents, alkalies, and abrasion. Multiple-coat systems are applied before a preceding coat cures completely, in order to obtain chemical reaction between coats. Fully-cured epoxies have a hard finish that is difficult to topcoat. Epoxy resins (qv) photolyze in sunlight, leaving a dust of unbound pigment known as chalk. Thus they are always topcoated, usually using urethanes, alkyds, or vinyls, for exterior use.

The coatings are usually formulated using an epoxy resin in a first component and a polyamide, amine adduct, or polyamine curing agent in a second component. The coatings cure by a chemical reaction between the components, and curing time depends primarily on temperature. Epoxy–polyamide coatings can tolerate some surface dampness and contamination during application and a near white blast is satisfactory, although the best possible surface preparation is always desirable. Organic zinc primers containing about 30% epoxy and polyamide resins and about 69% zinc metal provide long-lasting corrosion protection but are not suitable for immersion service.

4.2. Urethanes. Urethane coatings are comparatively expensive and are used almost exclusively as topcoats over epoxy or inorganic zinc primers. Urethanes containing aliphatic polyisocyanates as curing agents are the best choice for excellent high gloss cosmetic topcoats where prolonged resistance to ultraviolet radiation and retention of appearance are important. Coatings containing aromatic polyisocyanates lose gloss and become yellow in sunlight, but give excellent service as tank linings and in other interior applications. Both types of curing agents produce tough, durable, smooth coatings with excellent resistance to chemicals and abrasion, and both can be formulated to give highly flexible elastomeric coatings if desired.

Urethane coatings are formulated in two components (10). The first contains a polyester polyol, pigments, additives, and solvents, and the second contains a polyisocyanate curing agent. Modern curing agents have low vapor pressures, which minimize worker exposure to isocyanate fumes. Yacht finishes that may be applied by brush are available for the individual user not wishing to apply the coating by spray. Use of urethanes requires careful attention to worker safety and application procedures but, when properly applied, they are the best finishes available for most exterior marine surfaces (see URETHANE POLYMERS).

4.3. Alkyds. Alkyd resins (qv) are polyesters formed by the reaction of polybasic acids, unsaturated fatty acids, and polyhydric alcohols (see ALCOHOLS, POLYHYDRIC). Modified alkyds are made when epoxy, silicone, urethane, or vinyl resins take part in this reaction. The resins cross-link by reaction with oxygen in the air, and carboxylate salts of cobalt, chromium, manganese, zinc, or zirconium are included in the formulation to catalyze drying.

Alkyd coatings were the standard products in the marine industry for atmospheric service until they were superseded by epoxies in the early 1970s. Alkyds dry reasonably fast, are easy to apply, and demonstrate good weathering in mild environments but are not suitable for immersion service. They have a high moisture vapor transmission rate and to be effective must contain inhibitive pigments (Table 1). The elimination of lead and chromate pigments has made the formulation of an effective alkyd coating challenging but not impossible.

4.4. Inorganic Silicate Coatings.   Inorganic silicate coatings are available for marine use in diverse formulations, which cure by different mechanisms. A silicate binder is formed when sodium, potassium, or lithium silicates in alkaline aqueous solution polymerize. Partially hydrolyzed ethyl silicate in an alcohol−water solution is also used in these coatings. Water is necessary for curing to take place but must evaporate before a film can form. High humidity, low temperature, or poor air circulation may retard evaporation and film formation, but very low humidity retards curing.

The coatings demonstrate excellent abrasion resistance, hardness, and toughness, but they are not flexible. Inorganic zinc-rich coatings containing more than 80 wt% metallic zinc in a silicate binder are used in automated blasting and priming operations as preconstruction primers for steel plate. They are also used near the seashore as primer coatings on bridges, electric power transmission towers, and structural steel but are not suitable for immersion service. A single coat of $75-125$ μm provides galvanic protection to steel. These coatings are almost always topcoated. Vinyls, epoxies, and urethanes are suitable, but alkyd coatings are not stable to the alkaline surface of zinc and should not be used as topcoats.

4.5. Vinyls.   Vinyl resins are thermoplastic polymers made principally from vinyl chloride; other monomers such as vinyl acetate or maleic anhydride are copolymerized to add solubility, adhesion, or other desirable properties (see MALEIC ANHYDRIDE, MALEIC ACID, AND FUMARIC ACID). Because of the high, from 4,000 to 35,000, molecular weights large proportions of strong solvents are needed to achieve application viscosities. Whereas vinyls are one of the finest high performance systems for steel, many vinyl coatings do not conform to VOC requirements (see VINYL POLYMERS).

Vinyl coatings are lacquers, that is, they form films solely by the evaporation of solvent. Thus throughout their lives they are soluble in the solvents used to apply them, allowing for good intercoat adhesion when solvents in a later coat soften the resins in an earlier coat. Vinyl coatings have been widely used on bridges, locks, ships, dams, and on- and off-shore steel structures. They are tough, flexible, adherent coatings. Excellent primer, anticorrosive, and antifouling formulations are available. The coatings have a low moisture vapor transmission rate, and inhibitive pigments are rarely used. These coatings are particularly useful where fast drying at low $(0-10°C)$ temperatures is required, and they require only a short curing time before being placed in service, but they do not tolerate surface moisture. Vinyls are excellent for immersion service only if applied over a near white or white blast. They have good weather resistance but are softened by heat and are not suitable for prolonged use above 65°C.

The wash primer is a special type of vinyl coating. This material contains a poly(vinyl butyral) resin, zinc chromate, and phosphoric acid in an alcohol-water solvent. The coating is so thin it is literally washed onto a freshly blasted steel surface, where it passivates the metal surface by converting it to a thin iron phosphate-chromate coating. The alcohol solvent makes it possible to apply the coating over damp surfaces. The coating forms the first coat of an all-vinyl system and can also be used to preserve a freshly-cleaned steel surface until an epoxy or other primer coat can be applied. These coatings contain chromates and are very high in VOC.

4.6. Chlorinated Rubber. Chlorinated rubber coatings are lacquers. These thermoplastic materials have low moisture vapor transmission and excellent acid, water, salt, and alkali resistance. They dry under very cold conditions but do not tolerate surface moisture. They are useful at extreme service temperatures (-35 to $120°C$) and are easily repaired. Toughness and high chemical resistance are similar to vinyls. Chlorinated rubber coatings are widely used in Europe and are common in the United Kingdom, where coatings pigmented with micaceous iron oxide are used by British Railways. They find fewer applications in the United States, primarily as topcoats on exterior steel exposed to high humidity.

4.7. Coal Tar. Coal-tar resins are made from processed coal-tar pitch (see TAR AND PITCH). They undergo rapid and severe cracking in sunlight and thus are suitable only for underground use on steel and saltwater immersion. Coal-tar resins are frequently combined with epoxy resins to add the chemical and abrasion resistance of the latter. Coal-tar epoxy coatings cured with polyamides are widely used on marine structures because of low water permeability. All of these are cost-effective coatings but are available only in black or dark shades and demand a white or near white blast for long life. The U.S. Navy does not use coal tar or coal-tar epoxy coatings because low levels of carcinogens may be present in processed coal tar.

4.8. Powder Coatings. Coating films can be formed from dry thermoplastic powder (see COATING PROCESSES, POWDER TECHNOLOGY). Small objects may be dipped in a fluidized bed of powder or grounded and coated by electrostatic spray; the object is then heated to fuse the particles and form a film. The powder may also be melted as it is sprayed; molten droplets coalesce and form a film on the object before cooling. Epoxy, polyester, and vinyl resins are widely used, and the particular properties of the coating depend on the type of resin. These coatings are used on electrical junction boxes, motor housings, hatches, and other small pieces of equipment, which are usually removed from the ship and coated in a shop.

5. Application Methods

The application of marine coatings is a critical factor in achieving maximum performance. Protective clothing and breathing equipment should be worn during application. Because of large surface areas, ships are usually spray painted (see COATING PROCESSES, SPRAY COATINGS). Three techniques are widely used: air, airless, and electrostatic spraying. In air spraying, paint is forced by 200–400 kPa (30–60 psi) of compressed air into a spray gun where a second stream of air atomizes the paint and carries it onto the surface. Paint losses can be as high as 40%

because some paint misses the surface (overspray) and some rebounds from the surface. Nearby objects must be protected from inadvertent painting.

Airless spray uses hydraulic pressure to deliver the paint. Paint is brought to the spray gun under 7–40 mPa (1000–6000 psi), where it is divided into small separate streams and forced through a very small orifice to produce the spray. Airless spray is faster, cleaner, and less wasteful than air atomization, but demands good technique because it delivers paint very quickly.

Electrostatic spraying is used in shops to coat conductive objects. It is very useful for odd-shaped objects such as wire fence, cables, and piping. An electrostatic potential of 60,000 volts on the object attracts oppositely-charged paint particles; the spray can wrap around and coat the side of the object opposite to the spray gun. This technique produces very uniform finishes and has the least paint loss of the three methods. However, it is slow, requires expensive equipment, produces only thin coats, and is sensitive to wind currents.

Manual painting occurs mostly during touch-up or repair, and is best suited for piping, railings, and other hard to spray places. The conventional tools for manual application are brushes, rollers, paint pads, and paint mitts. These methods are very slow but are suitable for unskilled applicators and allow the painter to work the coating deeply into the surface being painted.

6. Transfer Efficiency

Many components of ships and marine structures are now coated in the shop under controlled conditions to reduce the amount of solvents released into the atmosphere, improve the quality of work, and reduce cost. Regulations designed to limit the release of volatile organic compounds into the air confine methods of shop application to those having transfer efficiencies of 65%. Transfer efficiency is defined as the percent of the mass or volume of solid coating that is actually deposited on the item being coated, and is calculated as

$$\text{transfer efficiency } (\%) = \frac{\text{mass of solid coating on item} \times 100}{\text{mass of solid coating consumed}}$$

or

$$= \frac{\text{volume of solid coating on item} \times 100}{\text{volume of solid coating consumed}}$$

The principal factors affecting transfer efficiency are the size and shape of the object, the type of application equipment, the air pressure to the spray gun, and the distance of the spray gun from the object. The transfer efficiency becomes lower as the object becomes smaller or more complex. The transfer efficiency increases when the spray gun is brought closer to the object and when the atomizing pressure is reduced. The transfer efficiency of different types of application equipment in descending relative order is manual > electrostatic spray > airless spray > conventional atomized air spray.

7. Selection of Coatings

7.1. Underwater Hull. Hull coatings consist of two layers of an anticorrosive coating topped with one layer of an antifouling coating. The coating system must resist marine fouling, severe corrosion, the cavitation action of high speed propellers, and the high current densities near the anodes of the cathodic protection system. Epoxy and coal-tar epoxy systems are commonly used as anticorrosive coatings. The U.S. Navy uses two coats of epoxy polyamide paint (11), each 75 μm thick when dry. Epoxy and coal-tar epoxy systems are used extensively on commercial ships. Coal-tar epoxy systems (12) are usually applied in two coats to give a total dry film thickness of 200 μm.

Anticorrosive systems require an antifouling topcoat. Marine antifouling coatings (13) contain materials that are toxic to fouling organisms and are the only effective way to prevent the growth of marine organisms on the hull. The nature of the toxic substance is heavily regulated in Europe, Japan, and the United States. Arsenic, cadmium, and mercury are proscribed and organotins are severely restricted. Cuprous oxide has always been and remains the most widely used toxicant.

The system of hull coatings, including antifouling paint, must be compatible with the cathodic protection system. Thus the coating system must have good dielectric properties to minimize cathodic protection current requirements and must be resistant to the alkalinity produced by the electric current. The cathodic protection system should prevent corrosion undercutting of coatings that become damaged, and the current density should be able to be increased easily to meet the increased electrical current needed as the coating deteriorates.

Antifouling paints containing a vinyl-rosin base and cuprous oxide (14) were used beginning in the 1940s but are being discontinued because of their high VOC levels. They provided about two years of protection against fouling but needed to be cleaned about every three months, depending on operational schedules and the waters in which the vessel operated. Organotin antifouling paints effectively prevent fouling for much longer periods, and the copolymer paints containing covalently-bound tin furnish about five years of protection when applied at a dry film thickness of 375 μm (15 mils).

Hull coatings having low surface energies, known as fouling release coatings, provide fouling protection without the use of toxins (15). These coatings form only weak bonds with fouling organisms, and the fouling loses adhesion by its own weight or by the motion of the ship through the water. Heavily fluorinated urethane coatings were tested for some years, but toughened silicone coatings are now providing superior performance. Foulant release coatings are environmentally benign and promise extended service lives.

Epoxy and polyester systems filled with flake glass provide a finish that is tough and resistant to abrasion. One commercial system is filled with copper flakes to provide intrinsic antifouling action. These systems are applied at a total dry film thickness of about 625 μm and are used on pleasure boats.

7.2. Boottop and Freeboard Areas. The boottop is that part of the hull that is immersed when the ship is loaded and exposed when the ship is empty. The freeboard is the area from the upper limit of the boottop to the main deck. The boottop suffers mechanical damage from tugs, piers, and ice, and experiences intermittent wet and dry periods with nearly constant exposure to sunlight. Thus coatings for this area require resistance to sunlight and mechanical damage, good adhesion, and flexibility. Frequently the hull coating system is used in the boottop area, and one or two extra topcoats are applied for added strength.

In the freeboard areas, commercial ships use organic zinc-rich primers extensively and usually topcoat them with a two- or three-coat epoxy system. U.S. Navy ships use an organic zinc-rich primer, two to three coats of an epoxy-polyamide coatings, and a silicone-alkyd topcoat (16); the entire dry system is 150–225 μm thick.

7.3. Weather Decks.   Coatings for decks must be resistant to abrasion by pedestrians and small vehicles, and must be slip-resistant. Inorganic zinc primers overcoated with epoxy coatings for additional corrosion protection perform well on steel decks, or a multiple-coat all-epoxy system can be used. Nonskid coatings are used on aircraft carrier landing and hangar decks and in passageways of all ships to maintain traction during wet and slippery conditions. The coatings contain epoxy resins and a coarse grit and are applied using a roller over epoxy primers to produce a textured finish. Nonskid coatings are 6 to 10 mm thick when dry.

Aluminum and galvanized steel are widely used in equipment on weather decks. Historically, galvanized steel and aluminum surfaces have been treated using a thin coat of wash primer after cleaning and before coating. For environmental reasons, the wash primer is omitted and these metals are coated directly with an epoxy primer. An aliphatic polyurethane, alkyd or silicone-alkyd enamel provides improved weather resistance. Galvanizing can best be cleaned by water blasting if no rusting is present or a light brushoff blast can be used if rusting exists. The same blast technique can be used to clean aluminum surfaces. Alkyds must not be used on galvanized steel because zinc and moisture rapidly hydrolyze the resin, producing zinc soaps that destroy adhesion.

7.4. Superstructure.   Coatings for superstructures must have good resistance to sun, salt, and corrosion, and good gloss and appearance retention properties. In addition, coatings on antennas and superstructures must be resistant to acidic exhaust fumes and high temperatures. Deck hardware and machinery, masts, and booms are coated with an inorganic zinc primer, an intermediate coat of epoxy, and a finish coat of aliphatic polyurethane or silicone-alkyd enamel. Powder coatings are used effectively on antennas and other equipment on the superstructure. This equipment, as well as exhaust stacks, steam riser valves and piping, and other hot surfaces, can also be coated using thermal-sprayed aluminum. Powder coatings and sprayed metallic coatings (qv) are applied in shops under controlled conditions. Inorganic zinc primers are becoming widespread in new construction.

7.5. Tanks.   Coatings for liquid cargo tanks are selected according to the materials that the tanks (qv) are to contain. Tank coatings protect the cargo from contamination and must be compatible with the material carried. Epoxy systems are most frequently selected because they perform well with both aqueous and organic products. A carefully applied three-coat epoxy system having a dry-film thickness of 225–300 μm can be expected to last for 12 years.

Coatings for potable water tanks must not impart taste or odor to the water and must not allow corrosion products to enter the water. Epoxy coatings are usually used. In the United States these coatings must be approved by the National Sanitation Foundation, acting as agent for the U.S. Environmental Protection Administration. Complete cure is very important and up to two weeks at 20°C may be necessary.

Petroleum products far exceed all other substances carried aboard ship, as cargoes or as fuel. Fuel tanks have specialized requirements because they may be filled with seawater ballast after the fuel is consumed. The coating must resist attack by both fluids, and seawater is much more corrosive than hydrocarbons. A three-coat epoxy system totalling 250–300 μm dry thickness gives good service in U.S. Navy ships. Zinc primers are not permitted in fuel tanks because, in addition to being unsuitable for seawater immersion, zinc may dissolve in automotive or aviation fuels causing damage to the engines in which the fuel is subsequently used. The same three-coat epoxy system used in fuel tanks is used in a variety of other tanks, including ballast tanks, sanitary holding tanks, and hydraulic fluid reservoirs. Fluorinated polyurethane coatings pigmented with 24% of poly(tetrafluoroethylene) give exceptionally long service in fuel tanks (17).

7.6. Machinery Spaces, Bilges, and Holds. Machinery spaces and bilges are so inaccessible that surface preparation is a significant problem and damage to machinery that cannot be removed must be avoided. Chemical cleaning by aqueous citric acid solutions, followed by degreasing using a nonflammable solvent, is widely used. Surfaces are best protected using a two- or three-coat epoxy-polyamide system having a total thickness of 250–300 μm. Alkyd enamels perform well in dry machinery spaces. Holds for carrying cargo may be painted with either of these systems, but the epoxy system is preferred for chemical and abrasion resistance.

The interior of piping has been protected from corrosion and abrasion by aromatic amine-cured epoxy coatings (18). These coatings are forced through intact piping systems by compressed air and cure within 10 minutes, forming a hard impervious lining. They have been used to protect 70:30 and 90:10 copper:nickel pipes in sanitary systems from sulfide corrosion, and are also suitable for use in potable water piping systems.

7.7. Living Areas. Coatings for living areas must be easily cleaned and resistant to bacteria, soiling, and fire. Living areas are generally painted with nonflaming coatings, or with intumescent coatings which foam when heated and produce a thick char that lessens damage to the substrate. For ceilings and walls in living and sleeping areas the U.S. Navy uses coatings based on highly chlorinated alkyd resins (19) or on aqueous emulsions of vinylidene chloride (20). Epoxy systems are generally used in damp areas such as galleys, washrooms, and showers where moisture deteriorates enamels.

BIBLIOGRAPHY

"Coatings, Marine" in *ECT* 3rd ed., Vol. 6, pp. 445–454, by R. W. Drisko, U.S. Naval Civil Engineering Laboratory; in *Ect* 4th ed., Vol. 6, pp. 146–760, by Robert F. Brady, Jr., U.S.

Naval Research Laboratory and Richard W. Drisko, U.S. Naval Civil Engineering Laboratory.

CITED PUBLICATIONS

1. D. Atherton, J. Verborgt, and M. A. M. Winkeler, *J. Coatings Technol.* **51**(657), 88 (1979).
2. R. Abel, N. J. King, J. L. Vosser, and T. G. Wilkinson, in M. A. Champ, ed., *Proceedings of Oceans '86*, Marine Technology Society, Washington, D.C., 1986, p. 1314.
3. U.S. Military Specification MIL-A-22262A, *Abrasive Blasting Media, Ship Hull Blast Cleaning*, Feb. 6, 1987.
4. J. D. Keane and co-edits., *Systems and Specifications*, 6th ed., Steel Structures Painting Council, Pittsburgh, Pa., 1991, 9–48.
5. SSPC-VIS-1-89, *Visual Standard for Abrasive Blast Cleaned Steel*, Steel Structures Painting Council, Pittsburgh, Pa., 1989.
6. R. F. Brady, Jr., *J. Protective Coatings Linings* **4**(7), 42 (1987).
7. R. F. Brady, Jr., H. G. Lasser, and F. Pearlstein, in R. S. Shane and R. Young, eds., *Materials and Processes*, 3rd ed., Marcel Dekker, Inc., New York, 1985, 1267–1319.
8. R. F. Brady, Jr., *J. Protective Coatings Linings* **2**(11), 24 (1985).
9. R. F. Brady, Jr. and C. H. Hare, *J. Protective Coatings Linings* **6**(4), 49–60 (1989).
10. K. B. Tator, *J. Protective Coatings Linings* **2**(2), 22 (1985).
11. U.S. Military Specification MIL-P-24441A, *Paint, Epoxy Polyamide, General Specification For*, July 15, 1980.
12. Ref. 4, 233–239.
13. U.S. Military Specification DOD-P-24647, *Paint, Antifouling, Ship Hull*, Mar. 22, 1985.
14. U.S. Military Specification MIL-P-15931D, *Paint, Antifouling, Vinyl*, May 27, 1980.
15. R. F. Brady, Jr., J. R. Griffith, K. S. Love, and D. E. Field, *J. Coatings Tech.* **59**(755), 113 (1987).
16. U.S. Military Specification DOD-E-24635, *Enamel, Gray, Silicone Alkyd Copolymer, for Exterior Use*, Sept. 13, 1984.
17. J. R. Griffith and R. F. Brady, Jr., *Chemtech* **19**(6), 370 (1989).
18. R. F. Brady, Jr., *Surface Coatings Australia* **28**(5), 12 (1991).
19. U.S. Military Specification DOD-E-24607, *Enamel, Interior, Nonflaming, Chlorinated Alkyd Resin, Semigloss*, Oct. 13, 1981.
20. U.S. Military Specification DOD-C-24596, *Coating Compounds, Nonflaming, Fire-Protective*, Nov. 6, 1979.

GENERAL REFERENCES

H. R. Bleile and S. D. Rogers, *Marine Coatings*, Federation of Societies for Coatings Technology, Blue Bell, Pa., 1989.

R. F. Brady, Jr., "Marine Applications," in J. I. Kroschwitz, ed., *Encyclopedia of Polymer Science and Engineering*, Vol. 9, John Wiley & Sons, Inc., New York, 1988, 295–300.

J. D. Costlow and R. D. Tipper, eds., *Marine Biodeterioration: An Interdisciplinary Study*, Naval Institute Press, Annapolis, Md., 1984.

C. H. Hare, *The Painting of Steel Bridges*, Reichhold, New York, 1988.

J. D. Keane and co-eds., *Good Painting Practice*, 2nd ed., Steel Structures Painting Council, Pittsburgh, Pa., 1982.

J. D. Keane and co-eds., *Systems and Specifications*, 6th ed., Steel Structures Painting Council, Pittsburgh, Pa., 1991.

C. G. Munger, *Corrosion Protection by Protective Coatings*, National Association of Corrosion Engineers, Houston, Tex., 1984.

Z. W. Wicks, Jr., *Corrosion Protection by Coatings*, Federation of Societies for Coatings Technology, Blue Bell, Pa., 1987.

Journal of Coatings Technology, published monthly by the Federation of Societies for Coatings Technology, Blue Bell, Pa.

Journal of Protective Coatings and Linings, published monthly by the Steel Structures Painting Council, Pittsburgh, Pa.

ROBERT F. BRADY, JR.
U.S. Naval Research Laboratory

RICHARD W. DRISKO
U.S. Naval Civil Engineering Laboratory

COBALT AND COBALT ALLOYS

Cobalt [7440-48-4], a transition series metallic element having atomic number 27, is similar to silver in appearance. Cobalt was used as a coloring agent by Egyptian artisans as early as 2000 BC and cobalt-colored lapis or lapis lazuli was used as an item of trade between the Assyrians and Egyptians. In the Greco-Roman period cobalt compounds were used as ground coat frit and coloring agents for glasses. The common use of cobalt compounds in coloring glass and pottery led to their import to China during the Ming Dynasty under the name of Mohammedan blue.

The ancient techniques for mining cobalt and the use of cobalt compounds were lost during the Dark Ages. However, in the sixteenth century mining techniques became widely known through the works of Georgius Agricola, the German mineralogist. At that time cobalt was supplied as smalt or zaffre, the latter being a cobalt arsenide of sulfide ore that was roasted to yield a cobalt oxide. When fused with potassium carbonate to form a type of glass, zaffre became smalt. In the sixteenth century the ability of cobalt to color glass (qv) blue was rediscovered. Metallic cobalt, isolated in 1735 by a Swedish scientist, was established as an element in 1780.

Cobalt and cobalt compounds (qv) have expanded from use as colorants in glasses and ground coat frits for pottery (see COLORANTS FOR CERAMICS) to drying agents in paints and lacquers (see DRYING), animal and human nutrients (see MINERAL NUTRIENTS), electroplating (qv) materials, high temperature alloys (qv), hardfacing alloys, high speed tools (see TOOL MATERIALS), magnetic alloys (see MAGNETIC MATERIALS, BULK; MAGNETIC MATERIALS THIN-FILM), alloys used for prosthetics (see PROSTHETIC AND BIOMEDICAL DEVICES), and uses in radiology (see RADIOACTIVE

TRACERS). Cobalt is also used as a catalyst for hydrocarbon refining from crude oil for the synthesis of heating fuels (1,2) (see FUELS, SYNTHETIC; PETROLEUM).

1. Occurrence

Cobalt is the thirtieth most abundant element on earth and comprises approximately 0.0025% of the earth's crust (3). It occurs in mineral form as arsenides, sulfides, and oxides; trace amounts are also found in other minerals of nickel and iron as substitute ions (4). Cobalt minerals are commonly associated with ores of nickel, iron, silver, bismuth, copper, manganese, antimony, and zinc. Table 1 lists the principal cobalt minerals and some corresponding properties. A complete listing of cobalt minerals is given in Reference 4.

The cobalt resources of the United States are estimated to be about 1.3 million tons. Most of these resources are in Minnesota, but other important occurrences are in Alaska, California Missouri, Montana, and Idaho. Although large, most U.S. resources are in subeconomic concentrations. The identified world cobalt resources are about 11 milion tons. The vast majority of these resources are in nickel-bearing laterite deposits, with most of the rest occurring in nickel-copper deposits hosted in mafic and ultramafic rocks in Australia, Canada, and Russia, and in the sedimentary copper deposits of Congo (Kihshasa) and Zambia, In addition million of tons of hypothetical and speculative cobalt resources exist in manganese nodules on the ocean floor (5).

1.1. Future Sources. Lateritic ores (8) are becoming increasingly important as a source of nickel, and cobalt is a by-product. In the United States, laterites are found in Minnesota, California, Oregon, and Washington. Deposits also occur in Cuba, Indonesia, New Caledonia, the Philippines, Venezuela, Guatemala, Australia, Canada, and Russia (see NICKEL AND NICKEL ALLOYS).

The laterites can be divided into three general classifications: (1) iron nickeliferrous limonite which contains approximately 0.8–1.5 wt% nickel. The nickel to cobalt ratios for these ores are typically 10:1; (2) high silicon serpentinous ores that contain more than 1.5 wt% nickel; and (3) a transition ore between type 1 and type 2 containing about 0.7–0.2 wt% nickel and a nickel to cobalt ratio of approximately 50:1. Laterites found in the United States (9) contain 0.5–1.2 wt% nickel and the nickel occurs as the mineral goethite. Cobalt occurs in the lateritic ore with manganese oxide at an estimated wt% of 0.06 to 0.25 (10).

2. Properties

The electronic structure of cobalt is [Ar] $3d^7 4s^2$. At room temperature the crystalline structure of the α (or ε) form, is close-packed hexagonal (cph) and lattice parameters are $a = 0.2501$ nm and $c = 0.4066$ nm. Above approximately $417°C$, a face-centered cubic (fcc) allotrope, the γ (or β) form, having a lattice parameter $a = 0.3544$ nm, becomes the stable crystalline form. The mechanism of the allotropic transformation has been well described (6,11–13). Cobalt is magnetic up to $1123°C$ and at room temperature the magnetic moment is parallel to the c-direction. Physical properties are listed in Table 2.

Table 1. **Important Cobalt Minerals and Corresponding Properties**[a]

Mineral	CAS Registry number	Chemical formula	Crystalline form	Approximate hardness, Mohs'	Density, kg/m³	Cobalt, wt%	Location
			Arsenides				
smaltite	[12044-42-1]	$CoAs_2$	cubic	6.0	6.5	23.2	United States, Canada, Morocco
safflorite	[12044-43-8]	$CoAs_2$[bc]	orthogonal	5.0	7.2	28.2	Morocco, Canada
skutterudite	[12196-91-7]	$CoAs_3$	cubic	6.0	6.5	20.8	Ontario, Morocco
			Sulfides				
carrollite	[12285-42-6]	$CuCo_2S_4, CuS \cdot Co_2S_3, Co_3S_4$	cubic	5.5	4.85	38.7	Congo, Zambia
linnaeite	[1308-08-3]	Co_3S_4	cubic	5	4.5	48.7	Congo
siegenite	[12174-56-0]	$(Co,Ni)_3S_4$				26.0	United States
cattierite	[12017-06-0]	CoS_2[b]					Congo
			Arsenide-sulfide				
cobaltite	[1303-15-7]	$CoAsS$	cubic	6	6.5	35.5	United States, Canada, Australia
			Oxide				
asbolite	[12413-71-7]	$CoO \cdot 2MnO_2 \cdot 4H_2O$	ore	1–2	1.1		Congo, Zambia
erythrite	[149-32-6]	$3CoO \cdot As_2O_5 \cdot 8H_2O$	ore	2	3	29.5	
heterogenite	[12332-83-0]	$CuO \cdot 2Co_2O_3 \cdot 6H_2O, Co_2O_3 \cdot H_2O$[d]	ore	4	3.5	57	Congo
sphaerocobaltite	[14476-13-2]	$CoCO_3$	ore			49.6	Congo, Zambia

[a] Refs. 6 and 7.
[b] May also include some nickel.
[c] May also include some nickel and some iron.
[d] $CuO \cdot 2Co_2O_3 \cdot 6H_2O$ and $CoO \cdot 3Co_2O_3 \cdot CuO \cdot 7H_2O$ are also present.

209

Table 2. **Properties of Cobalt**

Property	Value		
at wt	58.93		
transformation temperature, °C	417		
heat of transformation, J/g[a]	251		
mp, °C	1493		
latent heat of fusion, ΔH_{fus}, J/g[a]	259.4		
bp, °C	3100		
latent heat of vaporization, ΔH_{vap}, J/g[a]	6276		
specific heat, J/(g · °C)[a]			
15–100°C	0.442		
molten metal	0.560		
coefficient of thermal expansion, °C^{-1}			
cph at RT	12.5		
fcc at 417°C	14.2		
thermal conductivity at RT, W/(m · K)	69.16		
thermal neutron absorption, Bohr atom	34.8		
resistivity, at 20°C[b], 10^{-8} Ω · m	6.24		
Curie temperature, °C	1121		
saturation induction, $4\pi I_S$, T[c]	1.870		
permeability, μ			
initial	68		
max	245		
residual induction, T[c]	0.490		
coercive force, A/m	708		
Young's modulus, GPa[d]	211		
Poisson's ratio	0.32		
hardness,[f] diamond pyramid, of % Co	99.9	99.98[e]	
at 20°C	225	253	
at 300°C	141	145	
at 600°C	62	43	
at 900°C	22	17	
strength of 99.9% cobalt, MPa[g]	as cast	annealed	sintered
tensile	237	588	679
tensile yield	138	193	302
compressive	841	808	
compressive yield	291	387	

[a] To convert J to cal, divide by 4.184.
[b] Conductivity = 27.6% of International Annealed Copper Standard.
[c] To convert T to gauss, multiply by 10^4.
[d] To convert GPa to psi, multiply by 145,000.
[e] Zone refined.
[f] Vickers.
[g] To convert MPa to psi, multiply by 145.

Many different values for room temperature mechanical properties can be found in the literature. The lack of agreement depends, no doubt, on the different mixtures of α and γ phases of cobalt present in the material. This, on the other hand, depends on the impurities present, the method of production of the cobalt, and the treatment.

The hardness on the basal plane of the cobalt depends on the orientation and extends between 70 and 250 HK. Cobalt is used in high temperature alloys of the superalloy type because of its resistance to loss of properties when heated to fairly high temperatures. Cobalt also has good work-hardening characteristics, which contribute to the interest in its use in wear alloys.

Whereas finely divided cobalt is pyrophoric, the metal in massive form is not readily attacked by air or water or temperatures below approximately 300°C. Above 300°C, cobalt is oxidized by air. Cobalt combines readily with the halogens to form halides and with most of the other nonmetals when heated or in the molten state. Although it does not combine directly with nitrogen, cobalt decomposes ammonia at elevated temperatures to form a nitride, and reacts with carbon monoxide above 225°C to form the carbide Co_2C. Cobalt forms intermetallic compounds with many metals, such as Al, Cr, Mo, Sn, V, W, and Zn.

Metallic cobalt dissolves readily in dilute H_2SO_4, HCl, or HNO_3 to form cobaltous salts (see also COBALT COMPOUNDS). Like iron, cobalt is passivated by strong oxidizing agents, such as dichromates and HNO_3, and cobalt is slowly attacked by NH_4OH and NaOH.

Cobalt cannot be classified as an oxidation-resistant metal. Scaling and oxidation rates of unalloyed cobalt in air are 25 times those of nickel. The oxidation resistance of Co has been compared with that of Zr, Ti, Fe, and Be. Cobalt in the hexagonal form (cold-worked specimens) oxidizes more rapidly than in the cubic form (annealed specimens) (3).

The scale formed on unalloyed cobalt during exposure to air or oxygen at high temperature is double-layered. In the range of 300 to 900°C, the scale consists of a thin layer of the mixed cobalt oxide [1308-06-1], Co_3O_4, on the outside and a cobalt(II) oxide [1307-96-6], CoO, layer next to the metal. Cobalt(III) oxide [1308-04-9], Co_2O_3, may be formed at temperatures below 300°C. Above 900°C, Co_3O_4 decomposes and both layers, although of different appearance, are composed of CoO only. Scales formed below 600°C and above 750°C appear to be stable to cracking on cooling, whereas those produced at 600–750°C crack and flake off the surface.

3. Source and Supplies

World production and reserves are listed in Table 3 (5).

The United States is the world's largest consumer of cobalt. With the exception of negligible amounts of by-product cobalt produced as intermediate products from mining operations, the United States did not mine or refine cobalt in 2000 (5).

Projects in Australia and Uganda will increase production. Cobalt supply is expected to increase 3–6%. Demand for cobalt depends on world economic conditions. Supply is expected to increase faster than demand.

The United States government maintained significant quantities of cobalt metal in the National Defense Stockpile (NDS) for military, industrial, and essential civilian use. Since 1993, sales of excess cobalt for NDS has contributed to U.S. and world supplies. U.S. government has set a disposal limit for cobalt of 2720 tons during fiscal 2001 (5).

In 1999, two Internet web sites were established for selling cobalt. A cobalt producer established one site and the other was established by a brokerage firm (14). In 2000, OMG also opened a cobalt web site.

Table 3. **World Mine Production, Reserves, and Reserve Base**[a]

| | Mine production[b] | | | |
	1999	2000[c]	Reserves[b]	Reserve base
United States				860,000
Australia	4,100	5,700	880,000	1,300,000
Canada	5,300	5,000	45,000	260,000
Congo (Kinshasa)[d]	7,000	7,000	2,000,000	2,500,000
Cuba	2,200	2,300	1,000,000	1,800,000
New Caledonia[e]	1,000	230,000	860,000	
Philippines	na	na	na	400,000
Russia	3,300	4,000	140,000	230,000
Zambia	4,700	4,000	360,000	540,000
other countries	2,300	3,200	90,000	1,200,000
World total (rounded)	29,900	32,300	4,700,000	9,900,000

[a] Data in metric tons.
[b] na = not available.
[c] Estimated.
[d] Formerly Zaire.
[e] Overseas territory of France.

4. Processing

4.1. Sulfide Ores.   In the ores from the Congo, cobalt sulfide as carrollite is mixed with chalcopyrite and chalcocite [21112-20-9]. For processing, the ore is finely ground and the sulfides are separated by flotation (qv) using frothers. The resulting products are leached with dilute sulfuric acid to give a copper–cobalt concentrate that is then used as a charge in an electrolytic cell to remove the copper. Because the electrolyte becomes enriched with cobalt, solution from the copper circuit is added to maintain a desirable copper concentration level. After several more steps to remove copper, iron, and aluminum, the solution is treated with milk of lime to precipitate the cobalt as the hydroxide.

Zambian copper sulfide ores are leaner in cobalt and are, therefore, concentrated twice. In the first stage, the bulk of the copper ore is floated off in a high lime circuit. The carrollite is not carried by the lime, but is recovered in a second pass using different flotation methods. The second concentration, which contains 25% Cu, 17% Fe, and 3.5–4% Co, undergoes a sulfatizing roasting to convert the cobalt to water-soluble cobalt sulfate. The iron and copper in the concentrate form insoluble oxides and sulfates. After roasting, the matte is leached with water and filtered. After removing the last of the copper, milk of lime is added to precipitate the cobalt hydroxide which is filtered and dissolved in sulfuric acid. The resulting solution is then used as an electrolyte from which cobalt is electrodeposited. The cobalt thus produced is marketed either in a granulated form or still attached to the cathode (see METALLURGY, EXTRACTIVE).

4.2. Arsenic-Free Cobalt–Copper Ores.   The arsenic-free cobalt ores of the Congo are treated by smelting. Ores such as heterogenite which have high cobalt content can be sent directly to an electric furnace in lump form.

The fines must be sintered before being charged. Smelting the cobalt–copper feed along with lime and coke produces a slag and two alloys.

4.3. Arsenic Sulfide Ores. The high grade ores of Morocco are magnetically separated to give arsenides and oxides (see SEPARATION, MAGNETIC). Ores that are most concentrated in the arsenides are subjected to an oxidizing roast to remove the arsenic. The concentrates are used as feed in a blast furnace, resulting in speiss, matte, and perhaps boullion. The cobalt-containing speiss is then crushed and roasted and the roasted speiss is treated with sulfuric acid and the solids removed and roasted. After the iron compounds are precipitated with sodium chlorate and lime, the cobalt–nickel solution is treated with sodium hypochlorite to precipitate a cobalt hydrate.

Pressure-acid leaching was used to extract cobalt from Blackbird mine ores before its closing in 1974. The result was a very fine cobalt powder which was subjected to a seeding process to produce cobalt granules. Leaching methods are also used in the refinement of lateritic ores.

4.4. Lateritic Ores. The process used at the Nicaro plant in Cuba requires that the dried ore be roasted in a reducing atmosphere of carbon monoxide at 760°C for 90 minutes. The reduced ore is cooled and discharged into an ammoniacal leaching solution. Nickel and cobalt are held in solution until the solids are precipitated. The solution is then thickened, filtered, and steam heated to eliminate the ammonia. Nickel and cobalt are precipitated from solution as carbonates and sulfates. This method (9) has several disadvantages: (1) a relatively high reduction temperature and a long reaction time; (2) formation of nickel oxides; (3) a low recovery of nickel and the contamination of nickel with cobalt; and (4) low cobalt recovery. Modifications to this process have been proposed but all include the undesirable high 760°C reduction temperature (10).

A similar process has been devised by the U.S. Bureau of Mines (9) for extraction of nickel and cobalt from United States laterites. The reduction temperature is lowered to 525°C and the holding time for the reaction is 15 minutes. An ammoniacal leach is also employed, but oxidation is controlled, resulting in high extraction of nickel and cobalt into solution. Mixers and settlers are added to separate and concentrate the metals in solution. Organic strippers are used to selectively remove the metals from the solution. The metals are then removed from the strippers. In the case of cobalt, spent cobalt electrolyte is used to separate the metal-containing solution and the stripper. Metallic cobalt is then recovered by electrolysis from the solution. Using this method, 92.7 wt% nickel and 91.4 wt% cobalt have been economically extracted from domestic laterites containing 0.73 wt% nickel and 0.2 wt% cobalt (9).

4.5. Deep Sea Nodules. Metal prices influence the type of extraction process used for sea nodules. Whereas there are those typically rich in manganese, most of the mining and refining expenses must be met by the price of commodity metals such as copper and cobalt rather than by an abundant metal like manganese. It has been suggested that a method be used that would selectively remove cobalt, nickel, and copper, leaving manganese stored in the tailings for future use (14,15). This method involves smelting of the reduced new nodules to produce a manganiferrous slag and an alloy containing the metals. The alloy is converted into a matte by oxidation and sulfidation. An oxidative pressure leach is used to produce a purified solution from which the metal is obtained

(14). Research is continuing in the areas of pyrometallurgy, hydrometallurgy, and electrometallurgy to find more efficient extraction techniques (16–19) (see OCEAN RAW MATERIALS).

5. Economic Aspects

In 2000, approximately 45% of U.S. cobalt use was in superalloys, which are used primarily in aircraft gas turbine engines, 9% was in cemented carbides for cutting and wear-resistant applications, 9% was in magnetic alloys, and the remaining 37% in various chemical uses. The total estimated value of cobalt consumed in 2000 was 300×10^6 (5).

Table 4 gives U.S. consumption of cobalt between 1998 and 2000 by form. Table 5 gives consumption according to products.

Table 4. **U.S. Consumption of Cobalt, t**[a]

Form	1998	1999	2000[b]
metal	4,240	3,780	3,170
scrap	3,080	2,720	2,460
chemical compounds[c]	1,860	1,910	1,700
Total	*9,180*	*8,420*	*7,320*

[a] Ref. 5.
[b] Jan.–Nov. 2000.
[c] Includes oxide.

Table 5. **U.S. Reported Consumption of Cobalt by End Use**[a,b]

Use	1998	1999	2000[c]
steel	134	154	129
superalloys	410	3,830	3,280
magnetic alloys	771	794	690
other alloys[d]	421	291	W
cemented carbides[e]	844	755	705
chemical and ceramic uses[f]	—	2,530	2,250
Miscellaneous and unspecified	2,900	64	266
Total	*9,810*	*8,420*	*7,320*

[a] Refs. 5,14.
[b] W Withheld to avoid disclosing company proprietary data; included with "Miscellaneous and unspecified," Data are rounded to no more than three significant digits; may not add to totals shown.
[c] May include revisions to prior months, Jan.–Nov. 2000.
[d] Includes diamond bit matrices, cemented and sintered carbides, and cast carbide dies or parts.
[e] Includes diamond bit matrices, cemented and sintered carbides, and cast carbide dies or parts.
[f] Includes catalysts, driers in paints or related usage, feed or nutritive additive, glass decolorizer, ground coat frit, pigments, and other uses.

During much of its history, the price of cobalt metal was set primarily by the producers. In the 1990s, the African producers lost much of their influence on cobalt prices (20). This was the result of reduced production in Zambia and Zaire (now the Democratic Republic of the Congo). Free market prices can change rapidly. Free market can originate from producers, government stockpile releases, or consumers with excess metal. Prices reflect overall supply and demand, but can change if there is a perception of short term availability. Sudden changes are not always evident.

A crisis in the 1970s led to an increase in prices (ie, cessation of sales from stockpiles, invasion of mines in Zaire). In the 1980s to mid-1990, Zaire and Zambia were successful in returning stability to cobalt prices. Free market prices ended during the second half of 1990. Prices from 1990 on reflect changes in supply/demand, political unrest in leading producer countries, selling from stockpiles, delayed purchases, etc. (21).

During the first ten months of 2000, average spot price of electrolytic cobalt varied between $6.12–7.94/kg ($13.50 to 17.50/lb) (5).

6. Analytical Methods

There have been numerous analytical procedures developed for the identification and determination of cobalt metal in a variety of matrices. Some of the most popular instrumental and wet chemical techniques are briefly summarized in this section. The complete methods are documented in Analytical Chemistry reference texts or have been published as ASTM procedures.

6.1. Wet Chemical Methods. For qualitative spot test the 1-nitroso-2-naphthol method and the ammonium thiocyanate method are used (22). The compleximetric titration-EDTA (27) method was developed for the determination of cobalt in cobalt paint driers (23). A gravimetric method is based on a cobalt complex with α-nitroso naphthol (24). Electrogravimetry can also be used (25).

6.2. Instrumental Methods. Atomic absorption spectrophotometry (30) (aas) method was developed to determine low concentration of cobalt in metal mixtures. Both aqueous and organic solutions can be analyzed, however analyses in aqueous solutions are less problematic and more common (26). The inductively coupled plasma spectrophotometry technique is somewhat comparable to aas. Its advantage is a broad linear range, long term stability and a greater sensitivity. A factor of 10 improvement is achieved for the detection limit of cobalt (27). The technique is susceptible to emission interferences. A photometric method was developed for the determination of cobalt in stainless steel in the range of 0.01 to 0.3% (28). The optical emission spectrophotometry – point to plane technique was developed for cast metal which can be machined to be compatible with the instrument (29). In the ion-exchange–Potentiometric titration method, cobalt is separated from interfering elements by selective elution from an anion-exchange column using hydrochloric acid. The potentiometric titration is performed using a platinum and saturated calomel electrode (30). The x-ray emission spectrophotometry method was developed for the analysis of stainless steel. The concentrations of the elements are determined by relating the measured radiation of unknown samples to analytical curves prepared from standard reference materials (31).

7. Environmental Concerns

Cobalt is ubiquitous in the environment and occurs naturally in many different chemical forms. The most important natural sources of cobalt to the environment are soil and dust, seawater, volcanic eruptions, and forest fires (32). The most important synthetic sources of cobalt are the byprocesses of burning coal and oil; exhaust from cars and other vehicles; industrial processes that use cobalt and compounds; and sewage sludge from cities (32).

Relatively small amounts of cobalt are released to the environment from manufacturing operations. Cobalt is not currently mined in the United States, and users of cobalt and cobalt compounds obtain the material through import and by recycling scrap metal. Because of cobalt's intrinsic value, the overwhelming majority of cobalt wastes are recycled. Based on recent Toxics Release Inventory (TRI) information collected by the U.S. Environmental Protection Agency (EPA), more than 93% of all cobalt wastes generated by reporting industries during 1996, the most recent reporting year, were recycled. Much of the remainder was treated. Less than 1% of cobalt waste was released to the environment (33).

General population exposure to cobalt is very low. Cobalt has been detected in the ambient air at levels ranging from 0.4 to 1.0 ng/m^3, although higher transient concentrations have been detected in some industrial areas. Elemental cobalt is insoluble. Cobalt compounds vary in their solubility, however, and cobalt has been detected in surface waters. The median concentration of cobalt in U.S. surface waters has been found to average 2 µg/L. Wastewaters from refining and other industrial activities may contain higher concentrations (34). Bioaccumulation factors for cobalt range from 100 to 4000 in marine fish and 4 to 1000 in freshwater fish (34).

Cobalt and compounds are subject to the toxic chemical release reporting requirements of the Emergency Planning and Community Right-to-Know Act (EPCRA) (35). Under EPCRA, owners and operators of certain facilities that manufacture, process, or otherwise use cobalt or cobalt compounds must report their releases to all environmental media. Cobalt compounds also are hazardous substances subejct to the comprehensive Environmental Response, Compensation, and Liability Act (CERCLA or Superfund) (36). Cobalt and its compounds additionally are regulated by the Clean Water Effluent Guidelines for a number of industrial point sources, including the following: nonferrous metals manufacturing (37) inorganic chemicals manufacturing (38) and battery manufacturing (39).

8. Health and Study Factors

Cobalt is essential for human health. Low levels of cobalt, as part of the vitamin B_{12} complex, are necessary to maintain good health, and the Food and Drug Administration has recognized a number of cobalt compounds as safe for use in materials that come in contact with food (40). Cobalt also stimulates the production of red blood cells and, accordingly, has been widely used as a treatment for anemia, particularly in pregnant women. For these reasons, cobalt is unlikely to produce adverse health effects in the general population at levels typically found in the environment.

General population exposure to cobalt is very low. Workers in certain occupations are exposed to higher concentrations. The main route of absorption of cobalt during occupational exposure is via the respiratory tract, due to inhalation of dusts, fumes, or mists containing cobalt or cobalt compounds. Occupational exposures occur principally through the production of cobalt powder, or in hard metal production, processing, and use (41). Workers in the hard metal industry, in particular, have been found to be exposed to concentrations of cobalt in the workplace ranging from 1 to 300 $\mu g/m^3$ (42).

The principal occupational health problem associated with exposure to cobalt is pulmonary effects, particularly in the hard metal industry where cobalt-containing dust is generated. Because the dust in such industries always contains agents in combination with cobalt (tungsten carbide and other substances such as tungsten, titanium carbide, tantalum carbide, and vanadium carbide), it is unclear whether cobalt is solely responsible for the observed health effects. EPA has concluded that long-term exposure to cobalt by inhalation causes respiratory effects, such as irritation, wheezing, and fibrosis (43). Nevertheless, the respiratory effects of exposure to cobalt and its compounds depend to a large extent on the form of cobalt involved, and bronchial obstruction and interstitial lung fibrosis have been principally found in workers exposed to very fine dust.

To address the potential inhalation toxicity of cobalt and cobalt compounds, the Occupational Health and Safety Administration (OSHA) has established a permissible exposure limit (PEL) in the workplace of 0.1 mg/m^3 for cobalt metal, dust, and fume as an 8-h time weighted average (44). This OSHA standard is based on concerns for cobalt's potential pulmonary toxicity. Similarly, the American Conference of Governmental Industrial Hygienists (ACGIH) recently established a threshold limit value (TLV) of 0.02 mg/m^3 for elemental cobalt and inorganic cobalt compounds to protect against potential pulmonary toxicity and sensitivity reactions (45). The National Institute of Occupational Safety and Health (NIOSH) has established a risk level of 0.5 mg/m^3 for exposure to cobalt and cobalt compounds. In the United Kingdom, the Advisory Committee on Toxic Substances of the Health and Safety Commission also has established a maximum exposure limit of 0.05 mg/m^3 following an examination of cobalt-related health effects information.

The carcinogenic toxicity potential of cobalt and cobalt compounds is well characterized. IARC, for example, published in 1991 a monograph on the carcinogenic risks to humans from cobalt and cobalt compounds (46). Following a comprehensive review of animal and human studies, IARC concluded that there is inadequate evidence for the carcinogenicity of cobalt and cobalt compounds in humans, but that there is sufficient evidence for the carcinogenicity of cobalt metal powder and cobalt oxides in experimental animals. Based on this evidence, IARC identified cobalt and its compounds as Class 2B potential human carcinogens (47).

ACGIH also has concluded that cobalt and cobalt compounds are carcinogenic in animals. ACGIH assigned cobalt and its compounds an A3 classification (48). Likewise, the German Commission has identified cobalt in the form of respirable dust/aerosols to be a potential human carcinogen (49). Similarly, the Registry of Toxic Effects of Chemical Substances (RTECs) published by NIOSH identifies cobalt as a Group 2B potential carcinogen based on the IARC Monograph (50).

The National Toxicology Program (NTP) released a draft report on the carcinogenicity of one highly soluble cobalt compound, cobalt sulfate heptahydrate (51). The NTP draft report indicates an increased incidence of certain neoplasms in the respiratory tracts of rats and mice exposed to high concentrations of the compound by inhalation. It is unclear, however, whether the findings represent cobalt's potential carcinogenicity, or nonspecific responses to the physical acidic sulfate salt solution. It is also unclear whether the result observed following exposure to a highly soluble cobalt compound would be observed in animals exposed to elemental cobalt, or other far less soluble cobalt compounds.

9. Uses

9.1. As Metal. The largest consumption of cobalt is in metallic form in magnetic alloys, cutting and wear-resistant alloys, and superalloys. Alloys in this last group are used for components requiring high strength as well as corrosion and oxidation resistance, usually at high temperatures.

During World War II German scientists developed a method of hydrogenating solid fuels to remove the sulfur by using a cobalt catalyst (see COAL CONVERSION PROCESSES). Subsequently, various American oil refining companies used the process in the hydrocracking of crude fuels (see CATALYSIS; SULFUR REMOVAL AND RECOVERY). Cobalt catalysts are also used in the Fisher-Tropsch method of synthesizing liquid fuels (52–54) (see FUELS, SYNTHETIC).

Cobalt–molybdenum alloys are used for the desulfurization of high sulfur bituminous coal, and cobalt–iron alloys in the hydrocracking of crude oil shale (qv) and in coal liquefaction (7).

9.2. As Salts. The second largest use of cobalt is in the form of salts (see COBALT COMPOUNDS), which have the largest application as raw material for electroplating (qv) baths and as highly effective driers for lacquers, enamels, and varnishes. Addition of cobalt salts to paint greatly increases the rate at which paint (qv) hardens. Cobalt oxide colors glass pink or blue depending on the environment of the CoO_x molecule within the glass. The pink colors are formed in boric oxide or alkali-borate glasses. The cobalt concentration determines the intensity of the color obtained, eg, glass used in making foundryman's goggles requires 4.5 kg of cobalt for every ton of glass. By contrast, the glass used in making decorative bottles requires only about 280 g/t. Cobalt is also used to decolorize soda–lime–silica glass. Pottery enamels react very similarly because the fusible enamels are forms of fusible glasses (see ENAMELS, PORCELAIN OR VITREOUS). Colors varying from blue to black can be obtained, depending on the oxides added to the frit to improve the adherence of porcelain enamel to steel sheet metal.

Combinations of cobalt compounds are also used as ceramic pigments (qv) ranging in colors of violet, blue, green, and pink (54–56).

Radioactive cobalt, ^{60}Co, produced by bombarding stable ^{59}Co with low energy neutrons, has application in radiochemistry, radiography, and food sterilization (57–59) (see FOOD PROCESSING; RADIOISOTOPES; STERILIZATION TECHNIQUES).

^{60}Co has many applications as a strong γ source. Cancer therapy machines, using ^{60}Co as their source, treat more than one million people around the world each year (60). ^{60}Co is also used in γ-imaging medical diagnostic cameras. Its industrial applications include thickness gauges, liquid flow measurement, and level control.

One of its most widespread uses is in sterilization facilities. Most medical products and supplies, including sutures, masks, gloves, dressings, scalpel blades, catheters, and syringes, are prepackaged and then sterilized by ^{60}Co irradition. It is also used for blood products and pharmaceuticals that cannot tolerate sterilization by heat or chemical processes.

Food irradiation is an application of ^{60}Co with great potential, but growth in this area is slow because of the public perception of radiation. Nevertheless, food irradiation has been approved for use in 40 countries, and is endorsed by the UN World Health Organization, the American Medical Association, the Institute of Food technologists, the Science Council of Canada, Health and Welfare Canada, and the U.S. Food and Drug Administration (FDA). It has several advantages over conventional food preservation techniques (1) it does not leave additives in the product, as some chemical preservatives do; (2) fruits and vegetables retain their taste, texture, and nutritional value, unlike the results of conventional heating, freezing, additives, drying, or powdering; and (3) it is proven effective in neutralizing common foodborne pathogens such as *Campylobacter* (for which it is the only known control in poultry meat), *Cryptosporidium Escherichia coli* (for which it is the most effective control in ground beef), *Listeria*, *Salmonella*, and *Toxoplasma*. Approval of food irradiation proceeds on an individual-case basis, so it is slow. In January 1998, the FDA approved the irradiation of red meat (61).

Many of the uses of ^{60}Co can also be provided by ^{137}Cs (half-life 30 years, γ-ray energy 662 keV), a reactor fission product, but ^{60}Co is by far the most widely used of the two.

Cobalt is an essential ingredient in animal nutrition. It has been shown that animals deprived of cobalt show signs of retarded growth, anemia, loss of appetite, and decreased lactation. Dressing the top soil of pastures with cobalt increases the cobalt content of the vegetation. Cobalt is known to be necessary to the synthesis of vitamin B_{12} [68-19-9] (see VITAMINS), a lack of which has been linked to pernicious anemia in humans. Cobalt is also used as the target material in electrical x-ray generators (see X-RAY TECHNOLOGY).

10. Cobalt Alloys

Pure metallic cobalt has a solid-state transition from cph (lower temperatures) to fcc (higher temperatures) at approximately 417°C. However, when certain elements such as Ni, Mn, or Ti are added, the fcc phase is stabilized. On the other hand, adding Cr, Mo, Si, or W stabilizes the cph phase. Upon fcc-phase stabilization, the energy of crystallographic stacking faults, ie, single-unit cph inclusions that impede mechanical slip within the fcc matrix, is high. Stabilizing the cph phase, however, produces a low stacking fault energy. In the case of high stacking fault energy, fcc stabilization, only a few stacking faults occur, and the ductility of the alloy is high. When foreign elements are dissolved throughout the matrix lattice, the mechanical slip is generally impeded. This results in increased hardness and strength, a metallurgical phenomenon known as solid-solution hardening.

Mechanical properties depend on the alloying elements. Addition of carbon to the cobalt base metal is the most effective. The carbon forms various carbide phases with the cobalt and the other alloying elements (see CARBIDES). The presence of carbide particles is controlled in part by such alloying elements such as chromium, nickel, titanium, manganese, tungsten, and molybdenum that are added during melting. The distribution of the carbide particles is controlled by heat treatment of the solidified alloy.

Cobalt alloys are strengthened by solid-solution hardening and by the solid-state precipitation of various carbides and other intermetallic compounds. Minor phase compounds, when precipitated at grain boundaries, tend to prevent slippage at those boundaries thereby increasing creep strength at high temperatures. Aging and service under stress at elevated temperature induce some of the carbides to precipitate at slip planes and at stacking faults thereby providing barriers to slip. If carbides are allowed to precipitate to the point of becoming continuous along the grain boundaries, they often initiate fracture (see FRACTURE MECHANICS). A thorough discussion of the mechanical properties of cobalt alloys is given in References 62 and 63 (see also REFRACTORIES).

10.1. Cobalt-Base Alloys.   As a group, the cobalt-base alloys may generally be described as wear-resistant, corrosion-resistant, and heat-resistant, i.e., strong even at high temperatures. Table 6 lists typical compositions of cobalt-base alloys in these application areas. Many of the alloy properties arise from the crystallographic nature of cobalt, in particular its response to stress; the solid-solution-strengthening effects of chromium, tungsten, and molybdenum; the formation of metal carbides; and the corrosion resistance imparted by chromium. Generally, the softer and tougher compositions are used for high temperature applications such as gas-turbine vanes and buckets. The harder grades are used for resistance to wear.

Historically, many of the commercial cobalt-base alloys are derived from the cobalt–chromium–tungsten and cobalt–chromium–molybdenum ternaries first investigated at the turn of the twentieth century. The high strength and stainless nature of the binary cobalt–chromium alloy, and powerful strengthening agents, tungsten and molybdenum, were identified early on. These alloys were named Stellite alloys after the Latin *stella* for star because of their starlike luster. Stellite alloys were first used as cutting tools and wear-resistant materials.

Following the success of cobalt-base tool materials during World War I, these alloys were used from about 1922 in weld overlay form to protect surfaces from wear. These early cobalt-base hardfacing alloys were used on plowshares, oil well drilling bits, dredging cutters, hot trimming dies, and internal combustion engine valves and valve seats. In 1982, approximately 1500 metric tons of

Table 6. **Compositions of Cobalt-Base Wear-Resistant Alloys, wt %[a]**

Alloy trade name	Cr	W	Mo	C	Fe[b]	Ni	Si	Mn
Cobalt-base wear-resistant alloys[c]								
Stellite 1	31	12.5	1[b]	2.4	3	3[b]	2[b]	1[b]
Stellite 6	28	4.5	1[b]	1.2	3	3[b]	2[b]	1[b]
Stellite 12	30	8.3	1[b]	1.4	3	3[b]	2[b]	1[b]
Stellite 21	28		5.5	0.25	2	2.5	2[b]	1[b]
Haynes alloy 6B	30	4	1	1.1	3	2.5	0.7	1.5
Tribaloy T-800	17.5		29	0.08[b]				
Stellite F	25	12.3	1[b]	1.75	3	22	2[b]	1[b]
Stellite 4	30	14.0	1[b]	0.57	3	3[b]	2[b]	1[b]
Stellite 190	26	14.5	1[b]	3.3	3	3[b]	2[b]	1[b]
Stellite 306[e]	25	2.0		0.4		5		
Stellite 6K	31	4.5	1.5[b]	1.6	3	3[b]	2[b]	2[b]
Cobalt-base high temperature alloys[d,f]								
Haynes alloy 25 (L605)	20	15	0.10		3	10	1[b]	1.5
Haynes alloy 188	22	14		0.10	3	22	0.35	1.25
MAR-M alloy 509[g]	22.5	7		0.60	1.5	10	0.4[b]	0.1[b]
Cobalt-base corrosion-resistant alloys[h]								
MP35N, multiphase alloy	20		10			35		
Ultimet[i]	25.5	2	5	0.08[b]	3[j]	9		

[a] Where the balance of the alloy consists of cobalt.
[b] Value given is maximum value.
[c] Stellite and Tribaloy are registered trademarks of Stoody Deloro Stellite.
[d] Haynes is a registered trademark of Haynes International, Inc.
[e] Also contains 6 wt% niobium.
[f] MAR-M-Alloy is a registered trademark of Martin-Marietta.
[g] Also contains 3.5 wt% tantalum, 0.2 wt% titanium, and 0.5 wt% zirconium.
[h] MP35N is a registered trademark of Standard Pressed Steel Co.
[i] Also contains 0.1 wt% nitrogen.
[j] Value given is not necessarily maximum.

cobalt-base alloys were sold for the purpose of hardfacing, one-third of this quantity being used to protect valve seating surfaces for both fluid control and engine valves.

In the 1930s and early 1940s, cobalt-base alloys for corrosion and high temperature applications were developed (64). Of the corrosion-resistant alloys, a cobalt–chromium–molybdenum alloy having a moderately low carbon content was developed to satisfy the need for a suitable investment cast dental material (see DENTAL MATERIALS). This biocompatible material, which has the tradename Vitallium, is used for surgical implants. In the 1940s this same alloy underwent investment casting trials for World War II aircraft turbocharger blades, and, with modifications to enhance structural stability, was used successfully for many years in this and other elevated-temperature applications. This early high temperature material, Stellite alloy 21, is in use in the 1990s predominantly as an alloy for wear resistance.

10.2. Cobalt-Base Wear-Resistant Alloys. The main differences in the Stellite alloy grades of the 1990s versus those of the 1930s are carbon and tungsten contents, and hence the amount and type of carbide formation in the microstructure during solidification. Carbon content influences hardness, ductility, and resistance to abrasive wear. Tungsten also plays an important role in these properties.

10.3. Types of Wear. There are several distinct types of wear that can be divided into three main categories: abrasive wear, sliding wear, and erosive wear. The type of wear encountered in a particular application is an important factor influencing the selection of a wear-resistant material.

Abrasive wear is encountered when hard particles, or hard projections on a counter-face, are forced against and moved relative to a surface. In alloys such as the cobalt-base wear alloys which contain a hard phase, the abrasion resistance generally increases as the volume fraction of the hard phase increases. Abrasion resistance is, however, strongly influenced by the size and shape of the hard-phase precipitates within the microstructure, and the size and shape of the abrading species (see ABRASIVES).

Sliding wear is perhaps the most complex in the way different materials respond to sliding conditions. The metallic materials that perform best under sliding conditions are cobalt-based either by virtue of oxidation behavior or ability to resist deformation and fracture. Little is known of the influence of metal-to-metal bond strength during cold welding. For materials such as the cobalt-base wear alloys having a hard phase dispersed throughout a softer matrix, the sliding-wear properties are controlled predominantly by the matrix. Indeed, within the cobalt alloy family, resistance to galling is generally independent of hard particle volume fraction and overall hardness.

Four distinct forms of erosive wear have been identified: solid-particle erosion, liquid-droplet erosion, cavitation erosion, and slurry erosion.

The abrasion resistance of cobalt-base alloys generally depends on the hardness of the carbide phases and/or the metal matrix. For the complex mechanisms of solid-particle and slurry erosion, however, generalizations cannot be made, although for the solid-particle erosion, ductility may be a factor. For liquid-droplet or cavitation erosion the performance of a material is largely dependent on ability to absorb the shock (stress) waves without microscopic fracture occurring. In cobalt-base wear alloys, it has been found that carbide volume fraction, hence, bulk hardness, has little effect on resistance to liquid-droplet and cavitation erosion (65). Much more important are the properties of the matrix.

10.4. Alloy Compositions and Product Forms. The nominal compositions of various cobalt-base wear-resistant alloys are listed in Table 6. The six most popular cobalt-base wear alloys are listed first. Stellite alloys 1, 6, and 12, derivatives of the original cobalt–chromium–tungsten alloys, are characterized by their carbon and tungsten contents. Stellite alloy 1 is the hardest, most abrasion resistant, and least ductile.

Stellite alloy 21 differs from the first three alloys in that molybdenum rather than tungsten is used to strengthen the solid solution. Stellite alloy 21 also contains considerable less carbon. Each of the first four alloys is generally used in the form of castings and weld overlays. Haynes alloy 6B differs in that it is a wrought product available in plate, sheet, and bar form. Subtle compositional differences between alloy 6B and Stellite alloy 6, eg, such as silicon control, facilitate processing. The advantages of wrought processing include greatly enhanced ductility, chemical homogeneity, and resistance to abrasion.

The Tribaloy alloy T-800, is from an alloy family developed by DuPont in the early 1970s, in the search for resistance to abrasion and corrosion. Excessive amounts of molybdenum and silicon were alloyed to induce the formation during solidification of hard and corrosion-resistant intermetallic compounds, known as Laves phase. The Laves precipitates confer outstanding resistance to abrasion, but limit ductility. As a result of this limited ductility the alloy is not generally used in the form of plasma-sprayed coatings.

The physical and mechanical properties of the six commonly used cobalt wear alloys are presented in Table 7. In the case of the Stellite and Tribaloy alloys this information pertains to sand castings. Notable are the moderately high yield strengths and hardnesses of the alloys, the inverse relationship between carbon content and ductility in the case of the Stellite alloys, and the enhanced ductility imparted to alloy 6B by wrought processing. Typical applications of the cobalt wear-resistant alloys are given in Table 8. Generally, the alloys are used in moderately corrosive and/or elevated-temperature environments.

10.5. Cobalt-Base High Temperature Alloys. For many years, the predominant user of high temperature alloys was the gas-turbine industry. In the case of aircraft gas-turbine power plants, the chief material requirements were elevated-temperature strength, resistance to thermal fatigue, and oxidation resistance. For land-base gas turbines, which typically burn lower-grade fuels and operate at lower temperatures, sulfidation resistance was the primary concern. The use of high temperature alloys (qv) in the 1990s is more diversified, as more efficiency is sought from the burning of fossil fuels and waste, and as new chemical processing techniques are developed.

Table 7. **Mechanical and Physical Properties of Cobalt-Base Wear-Resistant Alloys**

Property	Alloy[a]					
	1	6	12	21	6B	T-800
hardness, Rockwell	55	40	48	32	37[b]	58
yield strength, MPa[c]		541	649	494	619[b]	
ultimate tensile strength, MPa[c]	618	896	834	694	998[b]	
elongation, %	<1	1	<1	9	11	
thermal expansion coeff, μm/(m·°C)						
20–100°C	10.5	11.4	11.5	11.0	13.9[d]	
20–500°C	12.5	14.2	13.3	13.1	15.0[d]	12.6
20–1000°C	14.8		15.6		17.4[d]	15.1
thermal conductivity, W/(m·K)					14.8	14.3
specific gravity	8.69	8.46	8.56	8.34	8.39	8.64
electrical resistivity, $\mu\Omega$·m	0.94	0.84	0.88		0.91	
melting range, °C						
solidus	1255	1285	1280	1186	1265	1288
liquidus	1290	1395	1315	1383	1354	1352

[a] See Table 6.
[b] 3.2 mm (1/8 in.) thick sheet.
[c] To convert MPa to psi, multiply by 145.
[d] Starting temperature of 0°C.

Table 8. Applications of Cobalt-Base Wear-Resistant Alloys

Applications	Stellite alloys[a]	Forms	Mode of degradation
	Automotive industry		
diesel engine valve seating surface	6, F	weld overlay	solid particle erosion, hot corrosion
	Power industry		
control valve seating surfaces	6, 21	weld overlay	sliding wear, cavitation erosion
steam turbine erosion shields	6B	wrought sheet	liquid droplet erosion
	Marine industry		
rudder bearings	306	weld overlay	sliding wear
	Steel industry		
hot shear edges	6	weld overlay	sliding wear, impact, abrasion
bar mill guide rolls	12	weld overlay	sliding wear, impact, abrasion
	Chemical processing industry		
control valve seating surfaces	6	weld overlay	sliding wear, cavitation erosion
plastic extrusion screw flights	1, 6, 12	weld overlay	sliding wear, abrasion
pump seal rings	6, 12	weld overlay	sliding wear
dry battery molds	4	casting	abrasion
	Pulp and paper industry		
chainsaw noses	6B, 6	wrought sheet weld overlay	sliding wear, abrasion
	Textile industry		
carpet knives wool spinning	6K, 12	wrought sheet weld overlay	abrasion
	Oil and gas industry		
rotary drill bearings	190	weld overlay	abrasion, sliding wear
	Aircraft industry		
helicopter rotor blade erosion shields	6B	sheet	abrasion

[a] See Table 6.

Cobalt-base alloys are not as widely used as nickel and nickel–iron alloys in high temperature applications. Nevertheless, cobalt-base high temperature alloys play an important role because of excellent resistance to sulfidation and strength at temperatures exceeding those at which the γ'- and γ''-precipitates in the nickel and nickel–iron alloys dissolve. Cobalt is also used as an alloying element in many nickel-base high temperature alloys. The various types of iron-base, nickel-base, and cobalt-base alloys for high temperature application are discussed in Reference 66. Nickel-base and cobalt-base castings for high temperature service are also covered.

10.6. Alloy Compositions and Product Forms.   Stellite 21, an early type of cobalt-base high temperature alloy, is used primarily for wear resistance.

The use of tungsten rather than molybdenum, moderate nickel contents, lower carbon contents, and rare-earth additions typify cobalt-base high temperature alloys of the 1990s as can be seen from Table 5.

Haynes alloys 25, also known as L605 and 188, are wrought alloys available in the form of sheets, plates, bars, pipes, and tubes together with a range of matching welding products for joining purposes. MAR-M alloy 509 is an alloy designed for vacuum investment casting. Selected mechanical properties of these three alloys are given in Table 9.

10.7. Cobalt-Base Corrosion-Resistant Alloys. Although the cobalt-base wear-resistant alloys possess some resistance to aqueous corrosion, they

Table 9. **Properties of Cobalt-Base High Temperature Alloys**

	Alloy[a]		
Property	25	188	MAR-M509
yield strength, MPa[b]			
at 21°C	445[c]	464[d]	585[e]
at 540°C		305[f]	400[d]
tensile strength, MPa[b]			
at 21°C	970[c]	945[d]	780[e]
at 540°C	800[g]	740[f]	570[e]
1000-h rupture strength, MPa[b]			
at 870°C	75	70	140
at 980°C	30	30	90
elongation, %	62	53[c]	3.5[e]
thermal expansion coeff, μm/(m·K)			
from 21−93°C	12.3	11.9	
from 21−540°C	14.4	14.8	
from 21−1090°C	17.7	18.5	
thermal conductivity, W/(m·K)			
at 20°C	9.8[h]	10.8	
at 500°C	18.5[i]	19.9	
at 900°C	26.5[j]	25.1	
specific gravity	9.13	8.98	8.86
electrical resistivity, μΩ·m	0.89	1.01	
melting range, °C			
solidus	1329	1302	1290
liquidus	1410	1330	1400

[a] See Table 6.
[b] To convert MPa to psi, multiply by 145.
[c] 3.2 mm (1/8 in.) thick.
[d] Sheet 0.75−1.3 mm (0.03−0.05 in.) thick.
[e] As cast.
[f] Sheet, heat treated at 1175°C for 1 h with rapid air cool.
[g] Sheet, heat treated at 1230°C for 1 h with rapid air cool.
[h] At 38°C.
[i] At 540°C.
[j] At 815°C.

are limited by grain boundary carbide precipitation, the lack of vital alloying elements in the matrix after formation of the carbides or Laves precipitates, and, for the case and weld overlay materials, by chemical segregation in the microstructure. By virtue of homogeneous microstructures and lower carbon contents, the wrought cobalt-base high temperature alloys, which typically contain tungsten rather than molybdenum, are even more resistant to aqueous corrosion, but still fall well short of the nickel–chromium–molybdenum alloys in corrosion performance. To satisfy the industrial need for alloys which exhibit resistance to aqueous corrosion yet share the attributes of cobalt as an alloy base, ie, resistance to various forms of wear and high strength over a wide range of temperatures, several low carbon, wrought cobalt–nickel–chromium–molybdenum alloys are produced. The compositions of two of these are presented in Table 5. In addition, the cobalt–chromium–molybdenum alloy Vitallium is widely used for prosthetic devices and implants owing to excellent compatibility with body fluids and tissues.

The two corrosion-resistant alloys presented in Table 5 rely on chromium and molybdenum for their corrosion resistance. The corrosion properties of Ultimet are also enhanced by tungsten. Both alloys are available in a variety of wrought product forms: plates, sheets, bars, tubes, etc. They are also available in the form of welding (qv) consumables for joining purposes.

10.8. Mechanical Properties.   An advantage of the two corrosion-resistant alloys is that they may be strengthened considerably by cold working. MP35N alloy is intended for use in the work-hardened or work-hardened and aged condition, and the manufacturers have supplied considerable data concerning the mechanical properties of the alloy at different levels of cold work. Some of these data are given in Table 8.

10.9. Uses.   Applications of both these alloys include pump and valve components and spray nozzles. MP35N alloy is also popular for fasteners, cables, and marine hardware.

10.10. Economic Aspects.   With cobalt historically being approximately twice the cost of nickel, cobalt-base alloys for both high temperature and corrosion service tend to be much more expensive than competitive alloys. In some cases of severe service their performance increase is, however, commensurate with the cost increase and they are a cost-effective choice. For hardfacing or wear applications, cobalt alloys typically compete with iron-base alloys and are at a significant cost disadvantage.

BIBLIOGRAPHY

"Cobalt and Cobalt Alloys" in *ECT* 1st ed., Vol. 4, pp. 189–199, by G. A. Roush, Mineral Industry; in *ECT* 2nd ed., Vol. 5, pp. 716–736, by F. R. Morral, Cobalt Information Center, Battelle Memorial Institute; in *ECT* 3rd ed., Vol. 6, pp. 481–494, by F. Planinsak and J. B. Newkirk, University of Denver. "Cobalt and Cobalt Alloys" in *ECT* 4th ed., Vol. 6, pp. 760–777, by F. Galen Hodge, Haynes International, Inc.; "Cobalt and Cobalt Alloys" in *ECT* (online), posting date: December 4, 2000, by F. Galen Hodge, Haynes International Inc.

CITED PUBLICATIONS

1. R. S. Young, *Cobalt*, American Chemical Monograph Series, No. 149, American Chemical Society, Rhinehold, New York, 1960.
2. *Cobalt Monograph*, Cobalt Information Center, Brussels, 1960.
3. A. H. Hurlich, *Met. Progr.* **112**(5), 67 (Oct. 1977).
4. C. S. Hurlbut Jr., *Dana's Manual of Mineralogy*, 17th ed., John Wiley & Sons, Inc., New York, 1966.
5. K. B. Shedd, "Cobalt," *Mineral Commodity Summaries*, U.S. Geological Survey, Reston, VA, Jan. 2001.
6. J. W. Christian, *Proc. R. Soc.* **206A**, 51 (1951).
7. *U.S. Bureau of Mines Minerals Yearbook*, Vol. **III**, U.S. Bureau of Mines, Washington, D.C., 1971, 1974.
8. C. Chandra, *Characterization of Lateritic Nickel Ores by Electron-Optical and X-Ray Techniques*, Ph.D. dissertation, University of Denver, Denver, Colo., 1976.
9. R. E. Siemens, *Process for Recovery of Nickel from Domestic Laterites*, presented at the 1976 Mining Convention, U.S. Bureau of Mines, 1976.
10. L. F. Power and G. H. Geiger, *Miner. Sci. Eng.* **9**(1), 32 (1977).
11. J. B. Hess and C. S. Barrett, *Trans. Am. Inst. Min. Met. Eng.* **194**, 645 (1952).
12. H. Bibring and F. Sebilleal, *Rev. Met.* **52**, 569 (1955).
13. A. Seeger, *Z. Metallkunde* **47**, 653 (1956).
14. R. Sridhar, W. E. Jones, and J. S. Warner, *J. Met.* **28**(4), 32 (1976).
15. Ref. 5, Feb. 2000.
16. J. C. Agarwal and co-workers, in Ref. 19, p. 24.
17. M. Wadsworth, *J. Met.* **28**(3), 4 (1976).
18. P. Duby, in Ref. 16, p. 8.
19. P. Tarassoff, in Ref. 16, p. 11.
20. M. G. Manzone, in Ref. 16, p. 16.
21. *Cobalt Prices in 1998*, U.S. Geological Survey, Reston, VA, 1999.
22. *Eng. Mining J.* **180**(3), 138 (1979); 112 (1980); *Metal Bulletin Handbook*, 73 (1981); 51 (1982).
23. F. J. Welcher, ed., *Standard Methods of Chemical Analysis*, 6th Vol. **2**, Part A, p. 56.
24. *Annual Book of ASTM Standards*, Section 6, Vol. 06.03 Designation D 2373-85 p. 470.
25. Ref. 22, p. 883.
26. Ref. 22, p. 717.
27. *Annual Book of ASTM Standards*, Section 11, Vol. 11.01, Designation D 3558-90 p. 472.
28. M. Walsh and M. Thompson, *A Handbook of Inductively Coupled Plasma Spectroscopy*, Methuen, N.Y., 1983.
29. Ref. 22, p. 716.
30. *Annual Book of ASTM Standards*, Section 3, Vol. 3.06 Designation E 607-80 p. 245.
31. *Annual Book of ASTM Standards*, Section 3, Vol. 3.05 Designation E 351-88a p. 437.
32. *Annual Book of ASTM Standards*, Section 3, Vol. 3.06 Designation E 572-88 p. 241.
33. Agency for Toxic Substances and Disease Registry, *Toxicological Profile for Cobalt* **2**, 79 (1992).
34. EPA, *1996 Toxics Release Inventory: Public Data Release—Ten Years of Right-to-Know* 118–119 (May 1998).
35. Ref. 32, p. 79.
36. 40 CFR 1/2 372.65.
37. 40 CFR 1/2 302.4.

38. CFR Part 421, Subpart U (primary nickel and cobalt subcategory) and Subpart AC (secondary tungsten and coblat subcategory).

39. CFR Part 415, Subpart BM (cobalt salts production subcategory).

40. CFR Parts 461, Subpart A (cadmium subcategory).

41. 21 CFR 1/21/2175.105, 177.2420, and 181.25.

42. International Agency for Research on Cancer (IARC), *Monograph on the Evaluation of Carcinogenic Risks to Humans: Chlorinated Drinking-Water; Chlorination By-Products; Some Other Halogenated Compounds; Cobalt and Cobalt Compounds*, Vol. **52** 1991, p. 402.

43. Ref. 32, p. 93.

44. EPA Office of Air Quality Planning and Standards, Cobalt and Compounds at 1, http://www.epa.gov/ttnuatw1/hlthef/cobalt.html.

45. CFR 1/2 1910.1000 Table Z-1. **54** *Fed. Reg.* 2332, 2664 (Jan. 19, 1989). *AFL-CIO v. OSHA*, 965 F.2d 962 (11th Cir. 1992).

46. ACGIH, *1999 TLVs and BEIs*, pp. 27, 99.

47. Ref. 45, note 1.

48. Ref. 41, pp. 449, 450.

49. ACGIH, *1999 TLVs and BEIs*, 27, 79.

50. Cobalt und Seine Verbindungen (in Form atembarer Staube/Aerosole), *in MAKWerte*, Toxikologisch-arbeitsmedizinische Begrundungen.

51. National Library of Medicine, RTECs Database, last updated July 31, 1998.

52. NTP, Draft Technical Report on the Toxicology and Carcinogenesis Studies of Cobalt Sulfate Heptahydrate in F344/N Rats and B6C3F1 Mice (Inhalation Studies), Dec. 1996.

53. Ger. Pat. 1,012,124 (July 11, 1957), B. Lopmann.

54. U.S. Pat. 3,576,734 (Apr. 27, 1971), H. L. Bennett (to Bennett Engineering Co.).

55. M. A. Aglan and H. Moore, *J. Soc. Glass Tech.* **39**, 351 (1955).

56. J. Berk and J. De Jong, *J. Am. Ceram. Soc.* **41**, 287 (1958).

57. J. C. Richmond and co-workers, *J. Am. Ceram. Soc.* **36**, 410 (1953).

58. A. Charlesby, *Nucleonics* **14**, 82 (Sept. 1956).

59. E. B. Darden, E. Maeyens, and R. C. Bushland, *Nucleonics* **12**(10), 60 (1954).

60. A. E. Berkowitz, F. E. Joumot, and F. C. Nix, *Phys. Rev.* **98**, 1185 (1954).

61. *What is the Health Benefit of Radiation?*, Canadian Nuclear Association fact sheet, Jan. 1998.

62. *Nuclear News* **40**(1), 55 (1998).

63. C. T. Sims, N. S. Stoloff, and W. G. Hagel, eds., *Superalloys II*, John Wiley & Sons, Inc., New York, 1987, p. 135.

64. N. J. Grant and J. R. Lane, *Trans. ASM* **41**, 95 (1949).

65. R. D. Gray, *A History of the Haynes Stellite Company*, Cabot Corp., Kokomo, Ind., 1974.

66. K. C. Antony and W. L. Silence, *ELSI-5 Proceedings*, University of Cambridge, UK, 1979, p. 67.

67. *ASM Metals Handbook*, Vol. **1**, 10th ed., ASM International, Materials Park, Ohio, 1990.

F. Galen Hodge
Haynes International, Inc.

Larry Dominey
OM Group, Inc., (retired)

COBALT COMPOUNDS

1. Introduction

Cobalt [7440-48-4] forms numerous compounds and complexes of industrial importance. Nonmetallic cobalt usage represents 30% of total cobalt produced (1). Cobalt, at wt 58.933, is one of the three members of the first transition series of Group 9 (VIIIB). The electronic configuration is $[Ar]3d^7 4s^2$. There are thirteen known isotopes, but only three are significant: ^{59}Co is the only stable and naturally occurring isotope; ^{60}Co has a half-life of 5.3 years and is a common source of γ-radioactivity; and ^{57}Co has a 270-d half-life and provides the γ-source for Mössbauer spectroscopy (see COBALT AND COBALT ALLOYS).

Cobalt exists in the +2 or +3 valence states for the majority of its compounds and complexes. A multitude of complexes of the cobalt(III) ion [22541-63-5] exist, but few stable simple salts are known (2). Werner's discovery and detailed studies of the cobalt(III) ammine complexes contributed greatly to modern coordination chemistry and understanding of ligand exchange (3). Octahedral stereochemistries are the most common for the cobalt(II) ion [22541-53-3] as well as for cobalt(III). Cobalt(II) forms numerous simple compounds and complexes, most of which are octahedral or tetrahedral in nature; cobalt(II) forms more tetrahedral complexes than other transition-metal ions. Because of the small stability difference between octahedral and tetrahedral complexes of cobalt(II), both can be found in equilibrium for a number of complexes. Typically, octahedral cobalt(II) salts and complexes are pink to brownish red; most of the tetrahedral Co(II) species are blue (see COORDINATION COMPOUNDS).

Cobalt metal is significantly less reactive than iron and exhibits limited reactivity with molecular oxygen in air at room temperature. Upon heating, the black, mixed valence cobalt oxide [1308-06-1], Co_3O_4, forms; at temperatures above 900°C the olive green simple cobalt(II) oxide [1307-96-6], CoO, is obtained. Cobalt metal reacts with carbon dioxide at temperatures greater than 700°C to give cobalt(II) oxide and carbon monoxide.

In the absence of complexing agents and in acidic solution the cobalt(II) hexaaquo ion [15276-47-8] oxidizes with difficulty.

$$Co(H_2O)_6^{3+} + e^- \longrightarrow Co(H_2O)_6^{2+} \qquad E^0 = 1.86 \text{ V}$$

Indeed the cobalt(III) ion is sufficiently unstable in water to result in release of oxygen and formation of cobalt(II) ion. Under alkaline conditions the oxidation is much more facile and in the presence of complexing agents, eg, ammonia or cyanide, the oxidation may occur with ease or even spontaneously.

$$CoO(OH)(s) + H_2O + e^- \longrightarrow Co(OH)_2(s) + OH^- \qquad E^0 = 0.17 \text{ V}$$

$$Co(NH_3)_6^{3+} + e^- \longrightarrow Co(NH_3)_6^{2+} \qquad E^0 = 0.1 \text{ V}$$

$$Co(CN)_6^{3-} + H_2O + e^- \longrightarrow [Co(CN)_5(H_2O)]^{3-} + CN^- \qquad E^0 = -0.80 \text{ V}$$

2. Preparation and Properties

2.1. Cobalt(II) Salts.

Cobalt(II) acetate tetrahydrate [71-48-7], $Co(C_2H_3O_2)_2 \cdot 4H_2O$, occurs as pink, deliquescent, monoclinic crystals. It can be prepared by reaction of cobalt carbonate or hydroxide and solutions of acetic acid, by reflux of acetic acid solutions in the presence of cobalt(II) oxide, or by oxygenation of hot acetic acid solutions over cobalt metal. The tetrahydrate is soluble in water, alcohol, and acidic solutions. Dehydration of the crystals occurs at about 140°C. It is used as a bleaching agent (see BLEACHING AGENTS) and drier in inks (qv) and varnishes, and in pigments (qv), catalysis (qv), agriculture, and the anodizing industries (see DRYING).

The mauve colored cobalt(II) carbonate [7542-09-8] of commerce is a basic material of indeterminate stoichiometry, $(CoCO_3)_x \cdot (CO(OH)_2)_y \cdot zH_2O$, that contains 45–47% cobalt. It is prepared by adding a hot solution of cobalt salts to a hot sodium carbonate or sodium bicarbonate solution. Precipitation from cold solutions gives a light blue unstable product. Dissolution of cobalt metal in ammonium carbonate solution followed by thermal decomposition of the solution gives a relatively dense carbonate. Basic cobalt carbonate is virtually insoluble in water, but dissolves in acids and ammonia solutions. It is used in the preparation of pigments and as a starting material in the preparation of cobalt compounds.

Cobalt(II) acetylacetonate [14024-48-7], cobalt(II) ethylhexanoate [136-52-7], cobalt(II) oleate [14666-94-5], cobalt(II) linoleate [14666-96-7], cobalt(II) formate [6424-20-0], and cobalt(II) resinate can be produced by metathesis reaction of cobalt salt solutions and the sodium salt of the organic acid, by oxidation of cobalt metal in the presence of the acid, and by neutralization of the acid using cobalt carbonate or cobalt hydroxide.

Cobalt(II) chloride hexahydrate [7791-13-1], $CoCl_2 \cdot 6H_2O$ is a deep red monoclinic crystalline material that deliquesces. It is prepared by reaction of hydrochloric acid with the metal, simple oxide, mixed valence oxides, carbonate, or hydroxide. A high purity cobalt chloride has also been prepared electrolytically (4). The chloride is very soluble in water and alcohols. The dehydration of the hexahydrate occurs stepwise:

$$CoCl_2 \cdot 6H_2O \xrightarrow[-4H_2C]{50°C} CoCl_2 \cdot 2H_2O \xrightarrow[-H_2O]{90°C} CoCl_2 \cdot H_2O \xrightarrow[-H_2O]{140°C} CoCl_2$$

The anhydrous chloride is blue, ie, tetrahedral cobalt, and commonly used as a humidity indicator in desiccants (qv).

Cobalt(II) hydroxide [1307-86-4], $Co(OH)_2$, is a pink, rhombic crystalline material containing about 61% cobalt. It is insoluble in water, but dissolves in acids and ammonium salt solutions. The material is prepared by mixing a cobalt salt solution and a sodium hydroxide solution. Because of the tendency of the cobalt(II) to oxidize, antioxidants (qv) are generally added. Dehydration occurs above 150°C. The hydroxide is a common starting material for the preparation of cobalt compounds. It is also used in paints and lithographic printing inks and as a catalyst (see PAINT).

Cobalt(II) nitrate hexahydrate [10026-22-9], $Co(NO_3)_2 \cdot 6H_2O$, is a dark reddish to reddish brown, monoclinic crystalline material containing about 20%

cobalt. It has a high solubility in water and solutions containing 14 or 15% cobalt are commonly used in commerce. Cobalt nitrate can be prepared by dissolution of the simple oxide or carbonate in nitric acid, but more often it is produced by direct oxidation of the metal with nitric acid. Dissolution of cobalt(III) and mixed valence oxides in nitric acid occurs in the presence of formic acid (5). The trihydrate forms at 55°C from a melt of the hexahydrate. The nitrate is used in electronics as an additive in nickel–cadmium batteries (qv), in ceramics (qv), and in the production of vitamin B$_{12}$ [68-19-9] (see VITAMIN B$_{12}$).

Cobalt(II) oxalate [814-89-1], CoC$_2$O$_4$, is a pink to white crystalline material that absorbs moisture to form the dihydrate. It precipitates as the tetrahydrate on reaction of cobalt salt solutions and oxalic acid or alkaline oxalates. The material is insoluble in water, but dissolves in acid, ammonium salt solutions, and ammonia solution. It is used in the production of cobalt powders for metallurgy and catalysis, and is a stabilizer for hydrogen cyanide.

Cobalt(II) phosphate octahydrate [10294-50-5], Co$_3$(PO$_4$)$_2 \cdot$ 8H$_2$O, is a red to purple amorphous powder. The product is obtained by reaction of an alkaline phosphate and solutions of cobalt salts. The material is insoluble in water or alkali, but dissolves in mineral acids. The phosphate is used in glazes, enamels, pigments (qv) and plastic resins, and in certain steel (qv) phosphating operations (see ENAMELS, PORCELAIN OR VITREOUS).

Cobalt(II) sulfamate [1407-41-5], Co(NH$_2$SO$_3$)$_2$, is generally produced and sold as a solution containing about 10% cobalt. The product is formed by reaction of sulfamic acid and cobalt(II) carbonate or cobalt(II) hydroxide, or by the aeration of sulfamic acid slurries over cobalt metal. Cobalt(II) sulfamate is used in the electroplating (qv) industry and in the manufacture of precision molds for record and compact discs (see INFORMATION STORAGE MATERIALS).

Cobalt(II) sulfate heptahydrate [10026-24-1], CoSO$_4 \cdot$ 7H$_2$O, is a reddish pink monoclinic crystalline material that effloresces in dry air to form the hexahydrate. The dehydration–decomposition occurs according to the following:

$$CoSO_4 \cdot 7H_2O \xrightarrow[-H_2O]{41°C} CoSO_4 \cdot 6H_2O \xrightarrow[-5H_2O]{71°C} CoSO_2 \cdot H_2O \xrightarrow[-H_2O]{250°C} CoSO_4$$

$$3\,CoSO_4 \xrightarrow{710°C} Co_3O_4 + 3\,SO_2 + O_2$$

Cobalt(II) sulfate can be prepared by solution of cobalt(II) carbonate, cobalt(II) hydroxide, or cobalt(II) oxide in sulfuric acid. The digestion of the metal in sulfuric acid solution is assisted by air sparging. Also, cobalt(III) and mixed valence oxides in the presence of formic acid can be dissolved in sulfuric acid solution (5). High concentration (6) and high purity (7) cobalt sulfate solutions have been prepared electrolytically. Cobalt sulfate heptahydrate and cobalt(II) sulfate monohydrate [10124-43-3] are the most economical sources of cobalt ion and are used in feed supplements (see FEEDS AND FEED ADDITIVES) as well as in the electroplating industry, in storage batteries, in porcelain pigments, glazes, and as a drier for inks.

2.2. Cobalt Fluorides. *Cobalt Difluoride.* Cobalt difluoride [10026-17-2], CoF$_2$, is a pink solid having a magnetic moment of $4,266 \times 10^{-23}$ J/T (4.6 Bohr magneton) (8) and closely resembling the ferrous (FeF$_2$) compounds. Physical

Table 1. **Physical Properties of the Cobalt Fluorides**

Parameter	Cobalt difluoride[a]	Cobalt trifluoride
molecular weight	96.93	115.93
melting point, °C	1127	926
solubility, g/100 g[b]		
water	1.36	dec
anhydrous HF	0.036	
density, g/cm^3	4.43	3.88
ΔH_f, kJ/mol[c]	−672	−790
ΔG_f, kJ/mol[c]	−627	−719
S, J/(mol · K)[c]	82.4	95
C_p, J/(mol · K)[c]	68.9	92

[a]The bp of CoF_2 is 1739°C.
[b]CoF_2 is also soluble in mineral acids.
[c]To convert J to cal, divide by 4.184.

properties are listed in Table 1. Cobalt(II) fluoride is highly stable. No decomposition or hydrolysis has been observed in samples stored in plastic containers for over three years.

CoF_2 is manufactured commercially by the action of aqueous or anhydrous hydrogen fluoride on cobalt carbonate (see COBALT COMPOUNDS) in a plastic, ie, polyethylene/polypropylene, Teflon, Kynar, rubber, or graphite-lined container to avoid metallic impurities. The partially hydrated mass is lavender pink in color. It is dried at 150–200°C and then pulverized to obtain the anhydrous salt. A very high (99.9%) purity CoF_2 having less than 0.05% moisture content has also been prepared by reaction of $CoCO_3$ and liquid hydrogen fluoride. This is a convenient synthetic route giving quantitative yields of the pure product. The reaction of $CoCl_2$ and anhydrous HF is no longer commercially practical because of environmental considerations. The various hydrates, eg, the cobalt(II) fluoride dihydrate [13455-27-1], $CoF_2 \cdot 2H_2O$, cobalt(II) fluoride trihydrate [13762-15-7], $CoF_2 \cdot 3H_2O$, and cobalt(II) fluoride tetrahydrate [13817-37-3], $CoF_2 \cdot 4H_2O$, have been obtained by the reaction of freshly prepared oxide, hydroxide, or carbonate of cobalt(II) and aqueous hydrogen fluoride (9).

Cobalt difluoride, used primarily for the manufacture of cobalt trifluoride, CoF_3.

Cobalt Trifluoride.   Cobalt(III) fluoride [10026-18-3] or cobalt trifluoride, CoF_3, is one of the most important fluorinating reagents. Physical properties may be found in Table 1. It is classified as a hard fluorinating reagent (10) and has been employed in a wide variety of organic and inorganic fluorination reactions. CoF_3, a light brown, very hygroscopic compound, is a powerful oxidizing agent and reacts violently with water evolving oxygen. It should be handled in a dry box or in a chemical hood and stored away from combustibles, moisture, and heat. The material should not be stored in plastic containers for more than two years. The crystals possess a hexagonal structure.

Cobalt trifluoride is readily prepared by reaction of fluorine (qv) and $CoCl_2$ at 250°C or CoF_2 at 150–180°C. Direct fluorination of CoF_2 leads to quantitative yields of 99.9% pure CoF_3 (11).

CoF_3 is used for the replacement of hydrogen with fluorine in halocarbons (12); for fluorination of xylylalkanes, used in vapor-phase soldering fluxes (13); formation of dibutyl decalins (14); fluorination of alkynes (15); synthesis of unsaturated or partially fluorinated compounds (16–18); and conversion of aromatic compounds to perfluorocyclic compounds (see FLUORINE COMPOUNDS, ORGANIC). CoF_3 rarely causes polymerization of hydrocarbons. CoF_3 is also used for the conversion of metal oxides to higher valency metal fluorides, eg, in the assay of uranium ore (19). It is also used in the manufacture of nitrogen fluoride, NF_3, from ammonia (20).

2.3. Cobalt Oxides. Cobalt(II) oxide [1307-96-6], CoO, is an olive green, cubic crystalline material. The product of commerce is usually dark gray and contains 75–78 wt% cobalt. The simple oxide is most often produced by oxidation of the metal at temperatures above 900°C. The product must be cooled in the absence of oxygen to prevent formation of Co_3O_4. Cobalt(II) oxide is insoluble in water, ammonia solutions, and organic solvents, but dissolves in strong mineral acids. It is used in glass (qv) decorating and coloring and is a precursor for the production of cobalt chemicals.

Cobalt(II) dicobalt(III) tetroxide [1308-06-1], Co_3O_4, is a black cubic crystalline material containing about 72% cobalt. It is prepared by oxidation of cobalt metal at temperatures below 900°C or by pyrolysis in air of cobalt salts, usually the nitrate or chloride. The mixed valence oxide is insoluble in water and organic solvents and only partially soluble in mineral acids. Complete solubility can be effected by dissolution in acids under reducing conditions. It is used in enamels, semiconductors, and grinding wheels. Both oxides adsorb molecular oxygen at room temperatures.

2.4. Cobalt Carbonyls. Dicobalt octacarbonyl [15226-74-1], $Co_2(CO)_8$, is an orange-red solid that decomposes in air. It is prepared by heating cobalt metal to 300°C under 20–30,000 kPa (3–4000 psi) of carbon monoxide, by reduction of cobalt(II) carbonate with hydrogen under pressure at high temperatures, or by heating a mixture of cobalt(II) acetate and cyclohexane to 160°C in the presence of carbon monoxide and hydrogen at 30,000 kPa (4000 psi) pressure. The octa-carbonyl is reduced with sodium amalgam and acidified to yield tetracarbonylhy-dridocobalt [16842-03-8], $HCo(CO)_4$, a yellow liquid that is an active oxo catalyst (see CARBONYLS; OXO PROCESS).

3. Economic Aspects

Approximately 51% of U.S. cobalt use was in superalloys, 8% was in cemented carbides, 19% was in various other metallic uses, and the remaining 22% was in a variety of chemical uses (1) (see Table 2). Tables 3 and 4 give U.S imports for consumption of cobalt and U.S. exports of cobalt materials (21).

4. Analytical Methods

Typical analyses of selected cobalt compounds are given in Table 5.

4.1. Separation. 1-Nitroso-2-naphthol [131-91-9], $C_{10}H_7NO_2$, can be used for the preliminary separation of cobalt from other metals by extraction

Table 2. **U.S. Reported Consumption of Cobalt,**
by End Use, t Contained Cobalt[a,b]

End use	2001	2002
steel	624	543
superalloys	4,850	3,970
magnetic alloys	472	374
other alloys[c]	661	W
cemented carbides[d]	720	747
chemical and ceramic uses[e]	2,100	1,860
miscellaneous and unspecified	63	300
Total	*9,490*	*7,790*

[a]Ref. 21; W = Withheld to avoid disclosing company proprietary data; included with "Miscellaneous and unspecified."
[b]Data are rounded to no more than three significant digits; may not add to totals shown.
[c]Includes nonferrous alloys, welding materials, and wear-resistant alloys.
[d]Includes diamond tool matrices, cemented and sintered carbides, and cast carbide dies or parts.
[e]Includes catalysts, driers in paints or related usage, feed or nutritive additive, glass decolorizer, ground coat frit, pigments, and other uses.

into chloroform (22,23). Cobalt can be separated from cadmium, lead, and zinc by extraction using dithizone [60-10-6], $C_{13}H_{12}N_4S$, at pH 6–10 followed by hydrolysis in dilute HCl. The more stable cobalt complex remains in the organic extract. The formation of stable anionic halide complexes allow for the separation from nickel. Extraction of the complexes is effected by amines such as triisooctylamine, trioctyl-methylammonium salts, or strongly basic anionic resins. Nickel and chromium do not form anionic chloride complexes and are not extracted under conditions of the experiment. Copper, zinc, and iron can be eluted from basic anionic resins using 0.01 M HCl whereas cobalt is eluted with 4 M HCl. Cobalt can also be separated from nickel as the sulfate salt by extraction with bis(2,4,4-trimethylpentyl)phosphinic acid. The blue thiocyanate complex of cobalt can be extracted into a mixture of ether and amyl alcohol to bring about separation from nickel and iron(III) in the presence of citrate ion (24). Manganese can be separated from cobalt by chlorination at a pH of 2.0. Cobalt sulfate is separated from a variety of metals using a chelating ion-exchange (qv) resin (25).

Cobalt(II) can be separated from cobalt(III) as the acetylacetonate (acac) compounds by extraction of the benzene soluble cobalt(III) salt (26). Magnesium hydroxide has been used to selectively adsorb cobalt(II) from an ammonia solution containing cobalt(II) and cobalt(III) (27).

4.2. Determination. Pure cobalt compounds can be assayed by EDTA titration at 40°C using hexamine [100-97-0], $C_6H_{12}N_4$, buffer (pH = 6) to a xylenol orange endpoint. Cobalt and nickel can be determined by cyanometry (28) or potentiometric titration using ferricyanide (29) can be used in the presence of large amounts of nickel, zinc, or copper. Colorimetry can be used to determine large amounts of cobalt by formation of the blue thiocyanate complex,

Table 3. **U.S. Imports for Consumption of Cobalt, by Country**[a,b]

Period and country of origin	Metals[c] Quantity, kg	Value, $[e]	Oxides and hydroxides Quantity, kg	Value, $[e]	Salts and compounds[d] Quantity, kg	Value, $[e]	Cobalt content, year to date[f]
Australia	2,970	42,200					162,000
Belgium	43,300	686,000	31,000	396,000			804,000
Brazil	49,000	860,000					269,000
Canada	56,000	926,000					818,000
China			596	17,900			40,500
Congo (Kinshasa)							271,000
Finland	62,200	1,070,000	33,500	370,000	182,000	554,000	1,840,000
France	2,910	150,000	4,060	84,600			55,200
Germany							45,800
Japan					4,000	37,400	22,800
Korea, Republic of	54	3,440					669
Morocco							56,000
Netherlands							54,400
Norway	51,000	782,000					1,590,000
Philippines					9,070	65,300	33,700
Russia	17,900	252,000					952,000
South Africa	20,000	274,000	1,360	21,300			230,000
Sweden	1,330	46,200					26,900
Switzerland							3
Uganda							44,000
United Kingdom			6,480	148,000			148,000
Zambia							278,000
Total	*307,000*	*5,090,000*	*77,000*	*1,040,000*	*195,000*	*657,000*	*7,740,000*

[a]Ref. 21.
[b]Data are rounded to no more than three significant digits; may not add to totals shown.
[c]Unwrought cobalt, excluding alloys; includes cobalt cathode and cobalt metal powder.
[d]Includes cobalt acetates, cobalt carbonates, cobalt chlorides, and cobalt sulfates.
[e]Customs value.
[f]Oct. 2002.
Source: U.S. Census Bureau, with adjustments by the U.S. Geological Survey.

$Co(SCN)^{2-}$;[4]. An excellent discussion of interferences and their elimination is available (30). Colorimetric methods using nitroso-naphthols (31,32) or nitroso-R salt [525-05-3], $C_{10}H_5NNa_2O_8S_2$ (33,34) are quite specific for cobalt, but not particularly sensitive. 2-(5-Bromo-2-pyridylazo)-5-diethylaminophenol (5-Br-PADAP) is a very sensitive reagent for certain metals and methods for cobalt have been developed (35). Nitroso-naphthol is an effective precipitant for cobalt(III) and is used in its gravimetric determination (36,37). Atomic absorption spectroscopy (38,39), x-ray fluorescence, polarography, and atomic emission spectroscopy are specific and sensitive methods for trace level cobalt analysis (see SPECTROSCOPY, OPTICAL; TRACE AND RESIDUE ANALYSIS).

Table 4. **U.S. Exports of Cobalt Materials**[a]

Period	Unwrought cobalt, powders, matte, waste and scrap[b]		Oxides and hydroxides		Salts and compounds[c]		Total cobalt content for the month[e,f]	Cobalt content, year to date[c,f]
	Quantity, kg	Value, $[d]	Quantity, kg	Value, $[d]	Quantity, kg	Value, $[d]		
2001:								
November	306,000	4,520,000	41,600	581,000	7,770	34,300	338,000	3,000,000
December	182,000	3,220,000	32,500	701,000	9,110	77,100	207,000	3,210,000
January–December	2,240,000	7,300,000	1,260,000	10,200,000	254,000	1,980,000	3,210,000	XX[g]
2002:								
January	127,000	2,870,000	25,600	379,000	14,100	98,500	149,000	149,000
April	161,000	3,690,000	35,500	483,000	19,000	97,200	191,000	662,000
July	69,700	1,680,000	62,800	1,150,000	12,200	52,700	118,000	1,200,000
November	173,000	3,520,000	47,000	741,000	61,700	233,000	221,000	1,92,000

[a]Ref. 21; data are rounded to no more than three significant digits; may not add to totals shown.
[b]May include other intermediate products of cobalt metallurgy and unwrought cobalt alloys.
[c]Cobalt acetates and cobalt chlorides.
[d]Free alongside ship (f.a.s.) value.
[e]Estimated from gross weights.
[f]Year to date may include revisions to prior months.
[g]XX, not applicable.

Source: U.S. Census Bureau, with adjustments by the U.S. Geological Survey.

Table 5. **Analysis of Cobalt Compounds**[a]

Assay, wt%	Acetate tetrahydrate	Carbonate	Chloride hexahydrate	Hydroxide	Nitrate	Mixed oxide[b]	Sulfate hexahydrate	Sulfate feed grade
Co	23.5	46.0	24.8	61.0	20.3	71.0	21.0	21.0
Ni	0.04–0.10	0.1–0.3	0.07–0.1	0.1–0.2	0.03–0.10	0.15–0.35	0.05	0.10
Fe	0.005–0.015	0.1–0.6	0.005	0.01–0.10	0.001–0.002	0.05–0.20	0.001–0.002	0.01
Cu	0.0005		0.002	0.01	0.004	0.008–0.02	0.001–0.004	0.005
Mn	0.001–0.01	0.006		0.015	0.005	0.03–0.3	0.002	0.005
Pb		0.01	0.002	0.005	0.001	0.02	0.001	0.001
HCl insols[c]				0.02		0.01		
H$_2$O insols[c]	0.007–0.015				0.01		0.05	0.10
ABD[d], kg/m^3	940	660	1000	350	950	1500	1000	1000

[a]All materials are technical-grade cobalt(II) compounds unless indicated.
[b]Material is Co$_3$O$_4$.
[c]Insols = insoluable material.
[d]ABD = apparent bulk density.

237

5. Health and Safety Factors

Cobalt is one of twenty-seven known elements essential to humans (40) (see MINERAL NUTRIENTS). It is an integral part of the cyanocobalamin [68-19-9] molecule, ie, vitamin B_{12}, the only documented biochemically active cobalt component in humans (41,42) (see VITAMINS, VITAMIN B_{12}). Vitamin B_{12} is not synthesized by animals or higher plants, rather the primary source is bacterial flora in the digestive system of sheep and cattle (43). Except for humans, nonruminants do not appear to require cobalt. Humans have between 2 and 5 mg of vitamin B_{12}, and deficiency results in the development of pernicious anemia. The wasting disease in sheep and cattle is known as bush sickness in New Zealand, salt sickness in Florida, pine sickness in Scotland, and coast disease in Australia. These are essentially the same symptomatically, and are caused by cobalt deficiency. Symptoms include initial lack of appetite followed by scaliness of skin, lack of coordination, loss of flesh, pale mucous membranes, and retarded growth. The total laboratory synthesis of vitamin B_{12} was completed in 65–70 steps over a period of eleven years (44). The complex structure was reported by Dorothy Crowfoot-Hodgkin in 1961 (45) for which she was awarded a Nobel prize in 1964.

Cobalt compounds can be classified as relatively nontoxic (46). There have been few health problems associated with workplace exposure to cobalt. The primary workplace problems from cobalt exposure are fibrosis, also known as hard metal disease (47,48), asthma, and dermatitis (49). Finely powdered cobalt can cause silicosis. There is little evidence to suggest that cobalt is a carcinogen in animals and no epidemiological evidence of carcinogenesis in humans. The LD_{50} (rat) for cobalt powder is 1500 mg/kg. The oral LD_{50} (rat) for cobalt(II) acetate, chloride, nitrate, oxide, and sulfate are 194, 133, 198, 1700, 5000, and 279 mg/kg, respectively; the intraperitoneal LD_{50} (rat) for cobalt(III) oxide is 5000 mg/kg (50).

Several nonoccupational health problems have been traced to cobalt compounds. Cobalt compounds were used as foam stabilizers in many breweries throughout the world in the mid to late 1960s, and over 100 cases of cardiomyopathy, several followed by death, occurred in heavy beer drinkers (51,52). Those affected consumed as much as 6 L/d of beer (qv) and chronic alcoholism and poor diet may well have contributed to this disease. Some patients treated with cobalt(II) chloride for anemia have developed goiters and polycythemia (53). The impact of cobalt on the thyroid gland and blood has been observed (54).

For cobalt fluorides , ACGIH TLV, NIOSH TWA are both 2.5 mg (F)/m^3 (55).

The ACGIH TLV for cobalt metal, dust, and fumes is 0.02 mg (Co)/mg^3, the OSHA PEL is 0.05 mg/m^3 (55).

6. Uses

The uses of cobalt compounds are summarized in Table 6.

6.1. Cobalt in Catalysis. About 80% of cobalt catalysts are employed in three areas: (1) hydrotreating/desulfurization in combination with molybdenum for the oil and gas industry (see SULFUR REMOVAL AND RECOVERY); (2) homogeneous catalysts used in the production of terphthalic acid or dimethylterphthalate (see

Table 6. **Uses of Cobalt Compounds**[a]

Compound	CAS Registry number	Molecular formula	Molecular weight	Uses
cobalt(III) acetate	[917-69-1]	$Co(C_2H_3O_2)_3$	235.9	catalyst, powder
cobalt(II) acetate tetrahydrate	[71-48-7]	$Co(C_2H_3O_2)_2 \cdot 4H_2O$	248.9	drier for lacquers and varnishes, sympathetic inks, catalyst, pigment for oil cloth, mineral supplement, anodizing agent
cobalt(II) acetylacetonate	[14024-48-7]	$Co(C_5H_7O_2)_3$	355.9	vapor plating of cobalt, catalyst synthesis enamels
cobalt(II) aminobenzoate		$Co(C_7H_6NO_2)_2$	194.9	tire cord adhesion
cobalt(II) ammonium sulfate	[13586-38-4]	$CoSO_4 \cdot (NH_4)_2SO_4 \cdot 6H_2O$	394.9	catalyst, plating solutions
cobalt(II) bromide	[7789-43-7]	$CoBr_2$	218.7	catalyst, hydrometers
cobalt(II) carbonate	[513-79-1]	$CoCO_3 \cdot Co(OH)_2$	118.9	pigment, ceramics, feed supplement, catalyst, fodder fat stabilizer
cobalt(II) carbonate (basic)	[7542-09-8]	$2CoCO_3 \cdot Co(OH)_2 \cdot H_2O$	348.7	chemicals
dicobalt octacarbonyl	[15226-74-1]	$Co_2(CO)_8$	341.8	catalyst, powder
cobalt(II) chloride	[7791-13-1]	$CoCl_2 \cdot 6H_2O$	237.9	chemicals, sympathetic inks, hydrometers, plating baths, metal refining, pigment, catalyst, dyestuffs, magnetic recording materials, moisture indicators
cobalt(II) citrate	[18727-04-3]	$Co_3(C_6H_5O_7)_2 \cdot 2H_2O$	590.7	therapeutic agents, vitamin preparations, plating baths
cobalt(II) fluoride	[10026-17-2]	CoF_2	96.7	fluorinating agent
cobalt(II) fluoride tetrahydrate	[13817-37-3]	$CoF_2 \cdot 4H_2O$	168.7	catalyst
cobalt(III) fluoride	[10026-18-3]	CoF_3	115.6	fluorinating agent

Table 6 (*Continued*)

Compound	CAS Registry number	Molecular formula	Molecular weight	Uses
cobalt(II) fluorosilicate hexahydrate	[15415-49-3]	$CoSiF_6 \cdot 6H_2O$	308.3	ceramics
cobalt(II) formate	[6424-20-0]	$Co(CHO_2)_2 \cdot 2H_2O$	184.9	catalyst
cobalt(II) hydroxide	[1307-86-4]	$Co(OH)_2$	92.9	paints, chemicals, catalysts, printing inks, battery materials, powder
cobalt(II) iodide	[15238-00-3]	CoI_2	312.7	moisture indicator
cobalt(II) linoleate	[14666-96-7]	$Co(C_{18}H_{31}O_2)_2$	616.9	paint and varnish drier
cobalt(II) naphthenate		$Co(C_{11}H_{10}O_2)_2$	406.9	catalyst, paint and varnish drier, tire cord adhesive, antistatic adhesive
cobalt(II) nitrate hexahydrate	[10026-22-9]	$Co(NO_3)_2 \cdot 6H_2O$	290.9	pigments, chemicals, ceramics, feed supplements, catalysts, battery materials
cobalt(II) 2-ethyl-hexanoate	[136-52-7]	$Co(C_{18}H_{15}O_2)_2$	344.9	paint and varnish drier, adhesion additive
cobalt(II) oleate	[14666-94-5]	$Co(C_{18}H_{33}O_2)_2$	620.9	paint and varnish drier, antistatic adhesive
cobalt(II) oxalate	[814-89-1]	CoC_2O_4	146.9	catalysts, cobalt powders
cobalt phthalocyanine		$Co(pc)$	572.9	coating, conducting polymer
cobalt(II) potassium nitrite	[17120-39-7]	$K_3Co(NO_2)_6 \cdot 1.5H_2O$	478.9	pigment
cobalt(II) resinate		$Co(C_{44}H_{62}O_4)_2$	1366.9	paint and varnish drier, catalyst
cobalt Schiff-base complexes				oxygen-sensing and oxygen-indicating agents
cobalt(II) stearate	[13586-84-0]	$Co(C_{18}H_{35}O_2)_2$	624.9	paint and varnish drier, tire cord adhesive
cobalt(II) succinate trihydrate	[3267-76-3]	$Co(C_4H_4O_4) \cdot 3H_2O$	228.9	therapeutic agents, vitamin preparations

Table 6 (*Continued*)

Compound	CAS Registry number	Molecular formula	Molecular weight	Uses
cobalt(II) sulfamate		$Co(NH_2SO_3) \cdot 3H_2O$	208.9	plating baths
cobalt(II) sulfate	[10026-24-1]	$CoSO_4 \cdot xH_2O$	154.9	chemicals, ceramics, pigments, plating baths, dyestuffs, magnetic recording materials, anodizing agents, corrosion protection agent
cobalt(II) sulfide	[1317-42-6]	CoS	90.0	catalysts
cobalt(II) thiocyanate	[3017-60-5]	$Co(CNS)_2 \cdot 3H_2O$	174.9	humidity indicator, drug testing
cobalt(II) aluminate	[13820-62-7]	$CoAl_2O_4$	176.7	pigment, catalysts, grain refining
cobalt(II) arsenate	[24719-19-5]	$Co_3(AsO_4)_2 \cdot 8H_2O$	598.5	pigment for paint, glass, and porcelain, vapor plating
cobalt(II) chromate	[24613-38-5]	$CoCrO_4$	174.8	pigment
cobalt(II) ferrate	[12052-28-7]	$CoFe_2O_4$	234.5	catalyst, pigment
cobalt(II) manganate	[12139-69-4]	$CoMn_2O_4$	232.7	catalyst, electrocatalyst
cobalt(II) oxide	[1307-96-6]	CoO	74.9	chemicals, catalysts, pigments, ceramic gas sensors, thermistors
cobalt oxide	[1308-06-1]	Co_3O_4	240.7	enamels, semiconductors, solar collectors pigments, magnetic recording materials
cobalt dilanthanum tetroxide	[39449-41-7]	La_2CoO_4	400.7	anode, catalyst
tricobalt tetralanthanum decaoxide	[60241-06-7]	$La_4Co_3O_{10}$	892.3	catalyst
lithium cobalt dioxide	[12190-79-3]	$LiCoO_2$	97.8	battery electrode
sodium cobalt dioxide	[37216-69-6]	$NaCoO_2$	113.8	battery electrode
dicobalt manganese tetroxide	[12139-92-3]	$MnCo_2O_4$	236.7	catalyst

Table 6 (*Continued*)

Compound	CAS Registry number	Molecular formula	Molecular weight	Uses
dicobalt nickel tetroxide	[12017-35-5]	$NiCo_2O_4$	240.5	anode, catalyst
lanthanum cobalt trioxide	[12016-86-3]	$LaCoO_3$	245.8	oxygen electrode
cobalt(II) phosphate	[10294-50-5]	$Co_3(PO_4)_2 \cdot 8H_2O$	510.9	glazes, enamels, pigments, steel pretreatment
cobalt(II) tungstate	[12640-47-0]	$CoWO_4$	305.7	paint and varnish drier

*a*Reprinted with permission (56).

PHTHALIC ACID AND OTHER BENZENE POLYCARBOXYLIC ACIDS); and (*3*) the high pressure oxo process for the production of aldehydes (qv) and alcohols (see ALCOHOLS, HIGHER ALIPHATIC; ALCOHOLS, POLYHYDRIC). There are also several smaller scale uses of cobalt as oxidation and polymerization catalysts (57–59).

Cobalt's ability to actively catalyze reactions comes from one or more of the following properties (56): (*1*) the facile redox properties of the element, the ability to form stable species in multiple oxidation states, and the ease of electron transfer between the oxidation states via unstable intermediates or free radicals; (*2*) the ability to form complexes with suitable donor groups and the relative stabilities of those complexes; (*3*) the ability of cobalt to form complexes of varying coordination number and to exist in equilibrium in more than one stereochemical form; (*4*) the ability to undergo equilibrium decomposition reactions such as $Co(RCOO)_2 \rightleftarrows Co^{2+} + 2\ RCOO^-$ in polymerization systems. The decomposition products may take part in the catalysis; (*5*) cobalt complexes are able to undergo ligand exchange reactions which can be important in assisting catalysis. This is an example of vitamin B_{12}'s activity as a catalyst; and (*6*) cobalt oxides and sulfides, because of lattice vacancies, bond energies, or electronic effects, are active heterogenous catalysts.

The breadth of reactions catalyzed by cobalt compounds is large. Some types of reactions are hydrotreating petroleum (qv), hydrogenation, dehydrogenation, hydrodenitrification, hydrodesulfurization, selective oxidations, ammonoxidations, complete oxidations, hydroformylations, polymerizations, selective decompositions, ammonia (qv) synthesis, and fluorocarbon synthesis.

Hydrotreating Catalysts. A preliminary step in refining of crude oil feedstocks is the removal of metals, significant reduction of the organic sulfur and nitrogen constituents, and reduction of the molecular weight of the high molecular-weight fraction. This step is called hydrotreating and it typically employs a catalyst of cobalt and molybdenum on a high surface area alumina or silica support. Hydrotreating of a nominal 4000 cubic meters (25,000 barrel) per day operation produces 200 tons per day of sulfur and deposits ca 450 kg of metal, primarily vanadium and nickel, on the catalyst (60) which must be sulfur tolerant.

The primary reactions occurring during hydrotreating are (56) desulfurization of sulfides, polysulfides, mercaptans, and thiophene as exemplified by

$$R{-}S{-}R' + 2\,H_2 \longrightarrow RH + R'H + H_2S$$

denitrification of pyridine and pyrrole resulting in the corresponding alkane and ammonia; deoxygenation of phenol; hydrogenation of alkenes and aromatics, eg, of butadiene and benzene; and hydrocracking of alkanes such as n-decane

$$C_{10}H_{22} + H_2 \longrightarrow 2\,C_5H_{12}$$

Hydrodesulfurization, by far the most common hydrotreating reaction, usually occurs in the presence of a cobalt–molybdenum catalyst support on alumina. Hydrodesulfurization constitutes the largest single use of cobalt in catalysis (qv). These catalysts are not easily poisoned, are regenerable numerous times, and usable for several years. The catalysts are prepared by precipitation of cobalt and molybdate solutions in the presence of an alumina support, by mixing the dried components in powder form, by precipitation of the three components, or by impregnation of the support with solutions of cobalt salts and molybdic acid or ammonium molybdate. The material is calcined at about 600°C to produce oxides of the cobalt–molybdenum–alumina followed by activation by reduction-sulfidation to yield the cobalt–molybdenum sulfide catalyst. The activity of the catalyst depends on several factors: type of support, ratio of cobalt to molybdenum, content of cobalt and molybdenum, impregnation procedure, calcination conditions, and reduction–sulfidation (61,62). The cobalt content is commonly between 2.5 and 3.7 wt%.

Homogeneous Oxidation Catalysts. Cobalt(II) carboxylates, such as the oleate, acetate, and naphthenate, are used in the liquid-phase oxidations of p-xylene to terephthalic acid, cyclohexane to adipic acid, acetaldehyde (qv) to acetic acid, and cumene (qv) to cumene hydroperoxide. These reactions each involve a free-radical mechanism that for the cyclohexane oxidation can be written as

initiation $\qquad\qquad C_6H_{12} + R\cdot \longrightarrow C_6H_{11} + RH$

chain $\qquad\qquad C_6H_{11} + O_2 \longrightarrow C_6H_{11}OO\cdot$

$$C_6H_{11}OO\cdot + C_6H_{12} \longrightarrow C_6H_{11}OOH + C_6H_{11}$$

decomposition $\qquad C_6H_{11}OOH \longrightarrow C_6H_{10}O + H_2O$

$$Co^{2+} + C_6H_{11}OOH \longrightarrow Co^{3+} + C_6H_{11}O\cdot + OH^- \longrightarrow C_6H_{12}$$

$$Co^{3+} + C_6H_{11}OOH \longrightarrow Co^{2+} + C_6H_{11}OO\cdot + H^+ \longrightarrow chain$$

Cobalt-catalyzed oxidations form the largest group of homogeneous liquid-phase oxidations in the chemical industry.

Oxo or Hydroformylation and Hydroesterification.   Reactions of alkenes with hydrogen and formyl groups are catalyzed by $HCo(CO)_4$

$$HCo(CO)_4 + RCH{=}CH_2 \longrightarrow RCH_2CH_2Co(CO)_4$$

$$RCH_2CH_2Co(CO)_4 + CO \longrightarrow RCH_2CH_2COCo(CO)_4$$

$$RCH_2CH_2COCo(CO)_4 + H_2 \longrightarrow RCH_2CH_2COH + HCo(CO)_4$$

and hydroesterification occurs according to

$$R{-}CH{=}CH_2 + R'OH + CO \longrightarrow RCH_2CH_2COOR'$$

The oxo reactions occur under 10,000–30,000 kPa (1450–4000 psi) of synthesis gas at temperatures of 75–200°C. Oxo syntheses account for the third largest use of cobalt in catalysis.

Cobalt compounds are used as catalysts in several processes of lesser industrial importance. Fischer-Tropsch catalysts produced from cobalt salts produce less methane than the corresponding iron catalysts. The chemistry of carboxylation reactions catalyzed in the presence of cobalt is also rich (63). Cobalt cyanide can be used to hydrogenate a broad range of organic and inorganic materials. Exhaust gas purification including oxidation of carbon monoxide and nitrogen monoxide is effected in the presence of Co_3O_4 (see EXHAUST CONTROL, INDUSTRIAL). The oxide is also effective in the production of nitric acid (qv) from ammonia. Cobalt carboxylates are used in the adhesive industry as cross-linking and polymerization catalysts (see ADHESIVES).

Cobalt in Driers for Paints, Inks, and Varnishes.   The cobalt soaps, eg, the oleate, naphthenate, resinate, linoleate, ethylhexanoate, synthetic tertiary neodecanoate, and tall oils, are used to accelerate the natural drying process of unsaturated oils such as linseed oil and soybean oil. These oils are esters of unsaturated fatty acids and contain acids such as oleic, linoleic, and eleostearic. On exposure to air for several days a film of the acids convert from liquid to solid form by oxidative polymerization. The incorporation of oil-soluble cobalt salts effects this drying process in hours instead of days. Soaps of manganese, lead, cerium, and vanadium are also used as driers, but none are as effective as cobalt (see DRYING).

A film of drying oils undergoes a series of reactions, but the primary reaction is autooxidation polymerization where oxygen attacks a carbon adjacent to a double bond forming a hydroperoxide. Once peroxide formation begins, dissociation into free radicals occurs giving polymers having carbon–carbon, carbon–oxygen–carbon, alcohol, ketone, aldehyde, and carboxylic acid linkages (56). The reactions depend on drying conditions and properties of the dried film may change.

The concentration of drier also effects the rate and physical characteristics of the dried film. In dryer combinations cobalt is considered a surface oxidation catalyst, whereas driers such as lead are considered a through drier or polymerization catalyst. Lead, cerium, manganese, and iron soaps are often used in combination with cobalt to slow down the rate of drying. Zinc and calcium soaps are not active as driers, but can also be used as cobalt diluents to slow down the drying rate. The mixture of metal soaps determines the rate of drying as well as the properties of the final dried film. Coordination driers such as aluminum and zirconium soaps aid the cross-linking polymerization process and are always used along with an oxidative polymerization catalyst.

For water-based alkyd paints, greater (0.2% cobalt on a resin basis) concentrations of drier are required than for other systems because the reaction of the drier with water decreases the activity of the catalyst. The cobalt content of oil-based paint formulations is usually 0.01–0.05% cobalt. Although the concentration of cobalt in the formulations is small, the large volume of paints, inks, and varnishes constitute a significant use for cobalt chemicals.

6.2. Cobalt as a Colorant in Ceramics, Glasses, and Paints. Cobalt(II) ion displays a variety of colors in solid form or solution ranging from pinks and reds to blues or greens. It has been used for hundreds of years to impart color to glasses and ceramics (qv) or as a pigment in paints and inks (see COLORANTS FOR CERAMICS). The pink or red colors are generally associated with cobalt(II) ion in an octahedral environment and the chromophore is typically $Co–O_6$. The tetrahedral cobalt ion, $Co–O_4$ chromophore, is sometimes green, but usually blue in color.

Cobalt pigments are usually produced by mixing salts or oxides and calcining at temperatures of 1100–1300°C. The calcined product is then milled to a fine powder. In ceramics, the final color of the pigment may be quite different after the clay is fired. The materials used for the production of ceramic pigments are

Color	Ingredients
purple	$Co_3O_4–Al_2O_3–MgO$
blues	$Co_3O_4–SiO_2–CaO$
Thenard blue	$Co_3O_4–Al_2O_3$
blue-greens	$Co_3O_4–Cr_2O_3–Al_2O_3$
blue-black	$Co_3O_4–Fe_2O_3–Cr_2O_3$
black	$Co_3O_4–Cr_2O_3–Fe_2O_3–MnO_2–Ni_2O_3$

Paint pigments do not change colors on application. Other common colors are violet from cobalt(II) phosphate [18475-47-3], pink from cobalt and magnesium oxides, aureolin yellow from potassiuim cobalt(III) nitrite [13782-01-9], $KCo(NO_2)_4$, and cerulean blue from cobalt stannate [6546-12-5]. Large quantities of cobalt are used at levels of a few ppm to decolorize or whiten glass and ceramics. Iron oxide or titanium dioxide often impart a yellow tint to various domestic ware. The cobalt blue tends to neutralize the effect of the yellow.

Cobalt is used in ceramic pigments and designated as underglaze stains, glaze stains, body stains, overglaze colors, and ceramic colors. The underglaze is applied to the surface of the article prior to glazing. The glaze stain uses cobalt colorants in the glaze. A body stain is mixed throughout the body of the ceramic. Overglaze colors are applied to the surface and fired at low temperatures. Ceramic colors are pigments used in a fusible glass or enamel and are one of the more common sources of the blue coloration in ceramics, china, and enamel ware.

Cobalt oxides are used in the glass (qv) industry to color or decolorize. Usually the cobalt is tetrahedrally coordinated and the color that it imparts to the glass is blue. Some special low temperature glasses contain pink, octahedrally, coordinated cobalt. These pink glasses turn blue on heating. The blue color from cobalt is very stable and its intensity easily controlled by doping rates. A discernable tint can be noticed for glass containing only a few parts per million of cobalt ion. The more typical blue cobalt glass usually contains closer to 200 ppm cobalt.

Cobalt is used as a blue phosphor in cathode ray tubes for television, in the coloration of polymers and leather goods, and as a pigment for oil and watercolor paints. Organic cobalt compounds that are used as colorants usually contain the azo (64) or formazon (65) chromophores.

6.3. Miscellaneous Uses. *Adhesives in the Tire Industry.* Cobalt salts are used to improve the adhesion of rubber to steel. The steel cord must be coated with a layer of brass. During the vulcanization of the rubber, sulfur species react with the copper and zinc in the brass and the process of copper sulfide formation helps to bond the steel to the rubber. This adhesion may be further improved by the incorporation of cobalt soaps into the rubber prior to vulcanization (66,67) (see Tire cords).

Adhesion of Enamel to Steel. Cobalt compounds are used both to color and to enhance adhesion of enamels to steel (68). Cobalt oxide is often incorporated into the ground frit at rates of 0.5–0.6 wt%, although levels from 0.2 to 3 wt% have been used. The frit is fired for ten minutes at 850°C to give a blue enamel that is later coated with a white cover coat.

Agriculture and Nutrition. Cobalt salts, soluble in water or stomach acid, are added to soils and animal feeds to correct cobalt deficiencies. In soil application the cobalt is readily assimilated into the plants and subsequently made available to the animals (69). Plants do not seem to be affected by the cobalt uptake from the soil. Cobalt salts are also added to salt blocks or pellets (see Feeds and feed additives).

Electroplating. Cobalt is plated from chloride, sulfate, fluoborate, sulfamate, and mixed anionic baths (70). Cobalt alloyed with nickel, tungsten, iron, molybdenum, chromium, zinc, and precious metals are plated from mixed metal baths (71,72). A cobalt phosphorus alloy is commonly plated from electroless baths. Cobalt tungsten and cobalt molybdenum alloys are produced for their excellent high temperature hardness. Magnetic recording materials are produced by electroplating cobalt from sulfamate baths (73) and phosphorus-containing baths or by electroless plating of cobalt from baths containing sodium hypophosphite as the reducing agent. Cobalt is added to nickel electroplating baths to enhance hardness and brightness or for the production of record and compact discs (74).

Electronic Devices. Small quantities of cobalt compounds are used in the production of electronic devices such as thermistors, varistors, piezoelectrics (qv), and solar collectors. Cobalt salts are useful indicators for humidity. The blue anhydrous form becomes pink (hydrated) on exposure to high humidity. Cobalt pyridine thiocyanate is a useful temperature indicating salt. A conductive paste for painting on ceramics and glass is composed of cobalt oxide (75).

Batteries and Fuel Cells. Cobalt salts are used as activators for catalysts, fuel cells (qv), and batteries. Thermal decomposition of cobalt oxalate is used in the production of cobalt powder. Cobalt compounds have been used as selective absorbers for oxygen, in electrostatographic toners, as fluoridating agents, and in molecular sieves.

Cure Accelerator. Cobalt ethylhexanoate and cobalt naphthenate are used as accelerators with methyl ethyl ketone peroxide for the room temperature cure of polyester resins.

BIBLIOGRAPHY

"Cobalt Compounds" in *ECT* 1st ed., Vol. 4, pp. 199–214, by S. B. Elliott, Ferro Chemical Corp. and C. Mueller, General Aniline & Film Corp.; in *ECT* 2nd ed., Vol. 5, pp. 737–748, by F. R. Morral, Battelle Memorial Institute; in *ECT* 3rd ed., Vol. 6, pp. 495–510, by F. R. Morral; in *ECT* 4th ed., pp 778–793, by Wayne Richardson, CP Chemicals, Inc. "Cobalt Compounds" under "Fluorine Compounds, Inorganic" in *ECT* 1st ed., Vol. 6, p. 693, by F. D. Loomis; "Cobalt" under "Fluorine Compounds, Inorganic" in *ECT* 2nd ed., Vol. 9, pp. 582–583, by W. E. White; in *ECT* 3rd ed., Vol. 10, pp. 717–718, by D. T. Meshri, Advance Research Chemicals Inc.; in *ECT* 4th ed., Vol. 11, pp. 336–338, by Dayal T. Meshri, Advance Research Chemicals, Inc.; "Cobalt Compounds" in *ECT* (online), posting date: December 4, 2000, by H. Wayne Richardson, CP Chemicals, Inc.

CITED PUBLICATIONS

1. K. B. Shedd, "Cobalt", *Mineral Commodity Summaries*, U.S. Geological Survey, Reston, Va., Jan. 2003.
2. W. Levason and C. A. McAuliffe, *Coord. Chem. Rev.* **12**, 151 (1974).
3. A. Werner, *Ber.* **46**, 3674 (1913); **47**, 1964, 1978(1914); *Z. Anorg. Chem.* **3**, 267 (1893); *Alfred Werner and Cobalt Complexes, Werner Centennial ACS Monograph Series, Advances in Chemistry Series*, Vol. 62, Washington, D.C., 1966.
4. Jpn. Kokai Tokkyo Koho, JP 55 158280(1980),(Sumitomo Metal Mining).
5. Czech. CS 140,580(1971), V. Stastny and J. Sedlacek.
6. Jpn. Kokai Tokkyo Koho, JP 59 83785(1984), (Nippon Soda).
7. Jpn. Kokai Tokkyo Koho, JP 57 57876(1982), (Sumitomo Metal Mining).
8. A. G. Sharp, *Quart. Rev. Chem. Soc.* **11**, 49 (1957).
9. I. G. Ryss, *The Chemistry of Fluorine and its Inorganic Compounds*, State Publishing House for Scientific and Chemical Literature, Moscow, 1956, Eng. Trans. ACE-Tr-3927, Vol. II, Office of Technical Services, U.S. Department of Commerce, Washington, D.C., 1960, 659–665.
10. D. T. Meshri and W. E. White, "Fluorinating Reagents in Inorganic and Organic Chemistry," in the *Proceedings of the George H. Cady Symposium, Milwaukee, Wis.*, June 1970; M. Stacy and J. C. Tatlow, *Adv. Fluorine Chem.* **1**, 166 (1960).
11. E. A. Belmore, W. M. Ewalt, and B. H. Wojcik, *Ind. Eng. Chem.* **39**, 341 (1947).
12. R. D. Fowler and co-workers, *Ind. Eng. Chem.* **39**, 292 (1947).
13. Eur. Pat. Appl EP 281,784 (Sept. 14, 1988), W. Bailey and J. T. Lilack (to Air Products and Chemicals, Inc.).
14. U.S. Pat. 4,849,553(July 18, 1989), W. T. Bailey, F. K. Schweighardt, and V. Ayala (to Air Products and Chemicals, Inc.).

15. Jpn. Kokai Tokkyo Koho JP 03 167,141 (July 19, 1991), H. Okajima and co-workers (to Kanto Denka Kaggo Co. Ltd.).
16. U.S. Pat. 2,670,387 (Feb. 23, 1954), H. B. Gottlich and J. D. Park (to E. I. du Pont de Nemours & Co., Inc.).
17. D. A. Rausch, R. A. Davis, and D. W. Osborn, *J. Org. Chem.* **28**, 494 (1963).
18. Ger. Pat. DD 287,478(Feb. 28, 1991), W. Radeck and co-workers (to Akademie der Wissenschaften der DDR).
19. R. Hellman, Westinghouse Corp., Cincinnati, Ohio, private communication, Jan. 1989.
20. Jpn. Kokai Tokkyo Koho, JP 03,170,306 (July 23, 1991), S. Lizuka and co-workers (to Kanto Denka Kogyo Co. Ltd.).
21. K. B. Shedd, "Cobalt", *Mineral Industry Survey*, U.S. Geological Survey, Reston Va., March 2003.
22. Z. Marczenko, *Separation and Spectrophotometric Determination of Elements*, Ellis Horwood Ltd., West Sussex, UK,1986.
23. E. B. Sandell,*Colorimetric Determination of Traces of Metals*,Interscience Publishers, New York, 1959.
24. N. S. Bayliss and R. W. Pickering. *Ind. Eng. Chem., Anal. Ed.* **18**, 446 (1946).
25. T. Jeffers and M. Harvey, Report No. 8927, U.S. Bureau of Mines, Washington, D.C., 1985.
26. J. E. Hicks, *Anal. Chem. Acta* **45**, 101 (1969).
27. I. V. Melikhov, M. Belousova, and V. Peshkova, *Zh. Analit. Khim.* **25**, 1144 (1970).
28. B. S. Evans, *Analyst* **62**, 363 (1937).
29. B. Bagshawe and J. Hobson, *Analyst* **73**, 152 (1948).
30. G. Charlot and D. Bezier, *Quantitative Inorganic Analysis*, Methuen and Co., Ltd., London, 1957, p. 410.
31. E. Boyland, *Analyst* **71**, 230 (1946).
32. J. Yoe and C. Barton, *Ind. Eng. Chem., Anal. Ed.* **12**, 405 (1940).
33. H. Willard and S. Kaufmann, *Anal. Chem.* **19**, 505 (1947).
34. T. Ovenston and C. Parker, *Anal. Chem. Acta* **4**, 142 (1950).
35. J. Zbiral and L. Sommer, *Z. Anal. Chem.* **306**, 129 (1981).
36. C. Mayr and W. Prodinger, *Z. Anal. Chem.* **117**, 334 (1939).
37. H. Brintzinger and R. Hesse, *Z. Anal. Chem.* **122**, 241 (1941).
38. A. Varma, *CRC Handbook of Atomic Absorption Analysis*, Vol. 1, CRC Press, Boca Raton, Fla., 1984, 435–449.
39. W. Slavin, *Graphite Furance AAS, A Source Book*, Perkin-Elmer, Ridgefield, Conn., 1984.
40. W. Mertz, *Nutrition Today* **18**, 26 (1983).
41. D. Dolphin, ed., B_{12}, Vols. 1 and 2, John Wiley & Sons, Inc., New York, 1982.
42. R. S. Young, *Cobalt in Biology and Biochemistry*, Academic Press, Inc., London, 1979.
43. J. D. Donaldson, S. J. Clark, and S. M. Grimes, *Cobalt in Medicine, Agriculture and the Environment*, monograph series, Cobalt Development Institute, London, 1986.
44. J. N. Krieger, *Chem. Eng. News* (Mar. 12, 1973).
45. P. Senhard and D. Hodgkin, *Nature* **192**, 937 (1961); D. Crowfoot-Hodgkin, *Proc. Royal Soc. (London)* **A288**, 294 (1965).
46. E. Matromatteo, *Am. Ind. Hyg. Assoc. J.*, 29 (1986).
47. H. Jobs and C. Ballhausen, *Vertrauensartz Krankenkasse* **8**, 142 (1940).
48. A. P. Wehner, R. H. Busch, R. J. Olson, and D. K. Craig, *Am. Ind. Hyg. Assoc. J.* **38**, 338 (1977).
49. D. Munro-Ashman and A. J. Miller, *Contact Dermatitis* **2**, 65(1976).

50. G. Speijers, E. Krajnc, J. Berkvens, and M. Van Logten, *Food Chem. Toxicol.* **20**, 311 (1982); J. Llobet and J. Domingo, *Rev. Esp. Fisiol.* **39**, (1983); W. Frederick and W. Bradley, *Ind. Med.* **14**, 482 (1946).

51. Y. L. Morin, A. R. Foley, G. Martineau, and J. Roussel, *Can. Med. Assoc.* **97**, 881 (1967).

52. Y. L. Morin and P. Daniel, *Can. Med. Assoc. J.* **97**, 926 (1967).

53. R. T. Gross, J. P. Kriss, and T. H. Spaet, *Pediatrics* **15**, 284 (1955).

54. A. G. Cecutti, in G. Tyroler and C. Landolt, eds., *Extractive Metallurgy of Nickel and Cobalt*, The Metallurgical Society, Warrendale, Pa., 1988.

55. R. J. Lewis, Sr., *Sax's Dangerous Properties of Industrial Materials*, 10th ed., Vol. 2, John Wiley & Sons, Inc., New York, 2000.

56. J. D. Donaldson, S. J. Clark, and S. M. Grimes, *Cobalt in Chemicals, The Monograph Series*, Cobalt Development Institute, London, 1986; Guilford, Surrey, 2003.

57. B. Delmon, *Proceedings of the International Conference on Cobalt Metallurgy and Uses*, ATB Metallurgy, Brussels, Belgium, 1981.

58. E. de Bie and P. Doyen, *Cobalt* (1962).

59. P. Granger, *Proceedings of the 2nd International Conference, Cobalt Metallurgy and Uses, Venice, 1985*, Cobalt Development Institute, London, 1986.

60. B. G. Silbernagel, R. R. Mohan, and G. H. Singhal, in T. Whyte, R. Dalla Betta, E. Derouane, and R. Baker, eds., *Catalytic Materials: Relationship Between Structure and Reactivity*, ACS Symposium Series No. 248, ACS, Washington, D.C., 1984, 91–99.

61. C. Wivel, B. Clausen, R. Candia, S. Moerup, and H. Topsoe, *J. Catal.* **87**, 497 (1984).

62. L. Petrov.C. Vladdy, D. Shopov, V. Friebova, and L. Beranek, *Coll. Czech. Chem. Commun.* **48**, 691 (1983).

63. H. Alper, *Adv. Organomet. Chem.* **19**, 190 (1981).

64. R. Price, in K. Ventkataraman, ed., *Chemistry of Synthetic Dyes*, Academic Press, Inc., New York, 1970, p. 303.

65. A. Nineham, *Chem. Rev.* **55**, 355 (1955).

66. W. Van Ooij, *Rubber Chem. Technol.* **52**, 605 (1979).

67. G. Haemers, *Rubber World* **182**, 26 (1980).

68. K. Bates, *Enameling, Principles & Practice*, Funk & Wagnalls, Ramsey, N.J., 1974.

69. E. Underwood, *Trace Elements in Humans and Animal Nutrition*, Academic Press, Inc., New York, 1971.

70. W. H. Safranek, *The Properties of Electrodeposited Metals and Alloys*, 2nd ed., Elsevier, New York, 1986.

71. W. Betteridge, *Cobalt and Its Alloys*, Ellis Horwood, Ltd., Chichester, UK, 1982.

72. R. Brugger, *Nickel Plating: A Comprehensive Review of Theory, Practice, Properties, and Applications Including Cobalt Plating*, Draper, Teddington, UK, 1970.

73. T. Chen and P. Caralloti, *Appl. Phys. Lett.* **41**, 206 (1982).

74. Jpn. Pat. 62083486 (1987), S. Takei (to Seiko Epson Corp.).

75. U.S. Pat. 4, 317, 749 (1982), J. Provance and K. Allison(to Ferro Corp.).

H. WAYNE RICHARDSON
CP Chemicals, Inc.

DAYAL T. MESHRI
Advance Research Chemicals, Inc.

COFFEE

1. Introduction

Coffee was originally consumed as a food in ancient Abyssinia and was presumably first cultivated by the Arabians in ~575 AD (1). By the sixteenth century it had become a popular drink in Egypt, Syria, and Turkey. The name coffee is derived from the Turkish pronunciation, kahveh, of the Arabian word *gahweh*, signifying an infusion of the bean. Coffee was introduced as a beverage in Europe early in the seventeenth century and its use spread quickly. In 1725, the first coffee plant in the Western hemisphere was planted on Martinique, West Indies. Its cultivation expanded rapidly and its consumption soon gained wide acceptance.

Commercial coffees are grown in tropical and subtropical climates at altitudes up to ~1800 m; the best grades are grown at high elevations. Most individual coffees from different producing areas possess characteristic flavors. Commercial roasters obtain preferred flavors by blending or mixing the varieties before or after roasting. Colombian, and washed Central American or East African coffees are generally characterized as mild, winey-acid, and aromatic; Brazilian coffees as heavy body, moderately acid, and aromatic; and African and Asian robusta coffees as heavy body, neutral to slightly earthy, slightly acid, and slightly aromatic. Premium coffee blends contain higher percentages of Colombian, East African, and Central American coffees.

2. Green Coffee Processing

The coffee plant is a relatively small tree or shrub belonging to the family Rubiaceae. It is often controlled to heights between 3 and 5 m. Two main species are in cultivation and of importance worldwide; *Coffea arabica* (also referred to as milds, which accounts for 70–80% of world production) and *Coffea Canephora* (also referred to as robustas and accounting for most of the remaining word's production) are the main varieties. *Coffea liberica* and >20 others comprise the remaining species. Each of the commercial species include several varieties.

Following spring rains, the plant produces white flowers. After fertilization, the flower drops and is followed by a "coffee button", which matures into a fruit approximately the size of a small cherry. The coffee bean is the endosperm part of the seed. Due to the nature of climate and species, subsequent rains may produce additional flowering and subsequent fruit set. Because of this, the fruit on the tree may include under ripe, ripe (red, yellow, and purple color) and over ripe cherries. Coffee may be selectively picked (ripe only), or strip picked (predominately ripe plus some underripe and overripe). Selective picking during harvest to maximize the yield of the more desirable ripe fruit is considered optimum.

Green coffee processing is effected by one of three methods: the wet method (washed or semiwashed, which appears to be growing in popularity) and the natural or dry process method.

In the wet (fully washed) method (practiced in Colombia and many Central American origins), the harvested ripe coffee cherry is passed through a tank for washing that removes stones and other foreign matter. The coffee is then passed through a depulper, removing the outer covering or skin and most of the pulp. Some pulp mucilage remains on the parchment layer covering the bean. This remaining pulp is removed by utilizing a fermentation process. Depending on the amount of mucilage remaining on the bean, fermentation may last from 12 h to several days. Because excessive fermentation may cause development of undesirable characters in the flavor and odor of the beans, enzymes may be added to speed the process. The fermentation breaks down the remaining pulp so it can then be easily washed away.

The wet (semiwashed) method is similar. Here, a partial or reduced fermentation after hulling may be used. The coffee is then washed to remove the remaining pulp (other variations include aggressively mechanically washing the beans to remove the pulp without fermentation).

From both wet methods, the resultant beans (covered with parchment) will be dried to uniform moisture using natural or mechanical methods before going through the milling process. The mechanical methods have gained in popularity despite higher costs. They are faster and not dependent on weather conditions. A potential negative of this method is the use of wood fuel for the dryers that may impart smoke/ash flavors to the coffee.

After drying, the coffee is further mechanically processed to remove the parchment and silverskin before grading. The wet process generally produces coffee of more uniform quality and flavor.

The dry method is favored by Brazil and other countries where available water is limited and a relatively uniform flowering allows strip picking with a predominance of ripe coffee. This method involves setting the harvested coffee cherries on a patio to dry naturally (mechanical methods can be used but are much more costly). The coffee is periodically turned to ensure uniform drying and to help prevent the occurrence of mold. Once proper drying has occurred (7–21 days based upon conditions) the resultant coffee may then be stored or hulled to remove the husk, parchment, and silverskin.

After being processed by any of the above methods, the coffee may need to be further dried to bring it to the desired 11–13% moisture range.

Coffee prepared by either method is then machine graded using oscillating screens, density separating tables, and airveyors to separate the coffee into varying sizes. The beans are then further processed to remove foreign matter, and any undesirable and damaged beans. This is accomplished by hand picking, machine separating, or color sorting (Sortex or Deltron electronic color sorters may sometimes be used) or any combination of the aforementioned to achieve the desired result.

Unroasted green coffee, once processed and properly stored, can be expected to maintain its quality for 6–12 months.

Coffee is valued and offered for sale based on its grade as determined by the number of imperfections contained in a 300-g sample (imperfections are considered to be beans that are black, broken, insect damaged or otherwise unsound, and foreign material). Coffee processors and roasters will also grade the coffee based on final moisture and bean color as well as the cup quality of the prepared beverage.

Consumers more conscious of health and also those interested in conservation have driven niche markets for coffee that has been organically grown, farms that are sustainable, and fair traded coffee.

Organic coffee is coffee that has been certified by an acceptable agency as being grown under conditions considered to be organic in nature and have not used synthetic fertilizers and pesticides.

Sustainable coffee farms employ techniques beneficial to the environment. Effective land management, the use of fertilizers and pesticides that are safe for animal and bird populations, and following clean water practices ensures farms remain sustainable for future generations.

Fair traded coffee is coffee that has a guaranteed minimum price, provides credit to producers, and establishes long-term relationships directly with cooperatives. Fair traded coffee encourages organic and sustainable farming by providing financial incentives to the farmers and growers through the elimination of the intermediaries (middlemen).

3. Coffee Chemistry

3.1. Chemical Composition of Green Coffee.

The composition of the green coffee bean is complex; carbohydrate polymers and protein make up ~60% of the bean, the rest consists of low molecular weight compounds of various types, including lipids, acids and sugars. The composition can vary according to species, variety, growing environment, postharvest handling (including wet and dry processing), and storage time, temperature, and humidity. Table 1 summarizes the analyses of robusta and arabica green beans [data from various sources including (1,2)].

3.2. Chemistry of Coffee Constituents.

Chlorogenic acids. The chlorogenic acids result from the esterification of the hydroxyl groups at the 3, 4, and 5 positions of quinic acid with various phenolic acids, the principal two of which are caffeic and ferulic acid. In green coffee, the 3 [327-97-9], 4 [905-99-7], and 5-caffeoylquinic acid [906-33-2] (CQA) isomers are present; the 5-isomer has the highest content. Also present are 3 [1899-29-2], 4, and 5-feruloylquinic acid [40242-06-6] (FQA), the latter dominates. Iso-chlorogenic acids, which are diesters, ie, 3,4 [14534-61-3]; 3,5 [2450-53-5], or 4,5 [57378-72-0] linked phenolic acids, occur at lower levels. Other phenolic acids, eg, *p*-coumaric acid [501-98-4], have also been found as quinic acid esters in coffee beans. Robusta and arabica beans differ in their chlorogenic acid content and composition; the former contains somewhat more, mainly due to higher levels of FQA and iso-CQAs (3). FQA is particularly important with respect to bean roasting behavior as it leads to higher yields of guaiacol derivatives, which contribute to the spicy, medicinal, phenolic flavor characteristic of robustas.

Acids. Apart from the chlorogenic acids, the principal acids in coffee beans are, in order of decreasing content: citric, malic, quinic, phytic, and phosphoric (4). As the pH of green coffee aqueous solutions is in the region of 6, the

Table 1. **Typical Analyses of Green Coffee** (% Dry Weight Basis)

Constituent	Type[a]	
	Robusta	Arabica
lipids	11.5 (9–13)	16 (15–18)
ash	5.0 (4.5–5.3)	4.7 (4.2–5.2)
caffeine	2.3 (1.8–2.8)	1.3 (1.2–1.5)
chlorogenic acids	10.5 (9–11)	9 (8–10)
other acids	3.0 (2.7–3.3)	3.5 (3.2–3.8)
trigonelline	1.0 (0.7–1.2)	1.3 (1.1–1.5)
protein	12 (11–14)	11 (10–13)
free amino acids	0.2	0.2
sucrose	4 (2–5)	7 (5–8.5)
other sugars	0.5 (0.2–1.0)	0.5 (0.1–0.8)
polymeric carbohydrate		
mannan	22	22
arabinogalactan	17	14
cellulose others	8	8
other compounds	2.0	2.0

[a] Robustas generally have lower levels of lipid, trigonelline [535-83-1], sucrose [57-50-1], and phytic acid [83-86-3] but higher levels of caffeine [58-08-2], chlorogenic acids (mainly 5-caffeoylquinic acid [906-33-2]) and arabinogalactan [9036-66-2] than arabicas.

acids occur almost completely in the salt form with potassium as the dominant cation.

Alkaloids. Caffeine, due to its physiological properties, is the best known component of the coffee bean. In the pure form it has a bitter taste, but it is not considered to be the main source of the bitterness perceived in coffee beverages. It forms complexes with chlorogenic acids. Robusta caffeine levels, at ~2%, are about twice as high as those in arabica beans. Trigonelline tends to be present at a higher level in arabica beans.

Lipids. The coffee oil fraction consists of ~75% triglycerides, 20% free and esterified diterpene alcohols, 5% free and esterified sterols, and a small quantity of other lipid types. The principal bound fatty acids are palmitic (C16, 35–41%), linoleic (C18:2, 37–46%), oleic (C18:1, 8–12%) and stearic (C18, 7–11%) [57-11-4]. The diterpene alcohols are principally cafestol [469-83-0] and kahweol [6894-43-5], the only structural difference is an unsaturated instead of a saturated C–C linkage in the latter. Arabica and robustas show differences in their lipid profiles: The arabicas have higher total lipid contents (~15 vs 10% in robusta), and contain more kahweol, which is present in only low levels in robusta beans (5). Only robustas contain the diterpene, 16-*O*-methylcafesteol [108214-28-4], which because of its stability to the roasting process has been proposed as a means of detecting robusta beans in arabica blends.

The surface of green coffee beans contains a cuticular wax layer (0.2–0.3% wt. basis). The wax contains insoluble hydroxytryptamides derived from 5-hydroxytryptamine [61-47-2] and saturated C18, C20, and C22 fatty acids.

Carbohydrates. Sucrose [57-50-1] is the principal low molecular weight carbohydrate with the content in arabica beans about twice that of robustas. Glucose [50-99-1] and fructose [57-48-7] are the main sugars in the immature seeds but levels decrease during maturation and are very low in the mature beans (6). Other sugars are present at only minimal levels.

The polysaccharides constitute the major fraction in the coffee bean and together with protein they form the basis of the cell wall material in the bean. The major polysaccharides are arabinogalactan [9036-66-2], mannan [9052-06-6], and cellulose [9004-34-6] (7). The arabinogalactan is a high molecular weight polymer with a β(1–3)-linked galactan backbone with frequent side chains containing arabinose [147-81-9], galactose [59-23-4], and glucuronic acid [6556-12-3] units (8). The mannan is a linear, lower molecular weight polymer. Its linear structure allows it to form hydrogen-bonded tight structures, which are responsible for the hardness of the beans. Due to the one unit galactose side chains substituted at ~5–10% of the main chain mannose units, this polymer is sometimes termed "galactomannan" in coffee. Cellulose is found in most plant material. There is evidence of small quantities of other polymers, including one of a pectin-type (8). Polysaccharide contents of green beans are similar. Robustas tend to have 2–3% more arabinogalactan than arabicas and this is mainly responsible for their slightly higher total polysaccharide content.

Proteins. Protein content of robustas tends to be slightly higher than that of arabicas. The total amino acid composition, ie, that released by acid hydrolysis, is similar for both types. Free amino acid levels in green beans are of the order of 0.2% and vary significantly according to samples (9). This is to be expected due to the significant protease enzyme activity in green coffee.

3.3. Roast Coffee Composition.

The pleasant taste and aroma characteristics as well as the brown color of roast coffee are developed during the roasting process. The chemical and physical changes associated with this process are very complex. In the first stage of roasting, loss of free water (~12% in the green bean) occurs. In the second stage, chemical dehydration, fragmentation, recombination, and depolymerization reactions occur. Many of these reactions are of the Maillard type and lead to the formation of lower molecular weight compounds such as carbon dioxide, free water, and those associated with flavor and aroma as well as higher molecular weight compounds termed melanoidins that are responsible for the brown color. The chemical reactions of roasting are exothermic and cause a rapid rise in temperature, usually accompanied by a sudden expansion or puffing of the beans with a volume increase of 50–100% depending on variety and roasting conditions. The loss of carbon dioxide and other volatile substances, as well as the water loss produced by chemical dehydration during the second stage, accounts for most of the 2–6% dry weight roasting loss. Most of the lipid, caffeine, inorganic salts, and polymeric carbohydrate survive the roasting process.

Table 2 indicates some of the chemical changes that occur in arabican green coffee as a result of processing. Data are presented as a range and typical values (when appropriate) compiled from the literature and other sources and reflect the fact that the chemistry is highly dependent on degree of roast and starting material. The principal water-soluble constituents, ~25% of the green coffee, are involved. These include some of the protein, the free amino acids, organic acids,

Table 2. **Approximate Analyses of Roasted, Brewed, and Instant Coffee, % Dry Wt. Basis**

Constituent	Roasted	Brewed	Instant
lipids	17 (16–20)	0.2^a	0.1
ash	4.5 (4–5)	15	8
caffeine	1.2 (1.0–1.6)	5	2.5 (2–3)
chlorogenic acids	3.5 (1.5–5)	15	5 (3–9)
other acids	3 (2.5–3.5)	10	5.5 (4–8)
trigonelline	0.8 (0.5–1.0)	2 (1–2.5)	1.5 (1–2)
protein	8–10	5	3 (2–5)
sucrose	0.1 (0–0.3)	0.2 (0–0.5)	0.1 (0–0.3)
reducing sugars	0.1 (0–0.3)	0.2 (0–0.5)	2 (1–5)
polymeric carbohydrate[b]			
mannan	22	1	15 (10–18)
arabinogalactan	13	2	17 (15–20)
cellulose	8	0	0
other compounds[c]	18	45	40

[a] Maximum level in brew (filtered), espresso levels higher, even higher levels in "Scandanavian" brews.

[b] Much of polymer is converted in Instant Coffee processing to lower molecular weight (oligosaccharides) material.

[c] Most of "other" material is in the form of "browning products', such as the melanoidins.

and inorganic salts. Most sucrose disappears early in the roast. Its decomposition products include low molecular weight acids, oxygen-containing heterocylic compounds such as 5-hydroxymethylfurfural [67-47-0] and the volatile and non-volatile flavor contributing compounds, which are formed as a result of Maillard reactions with free and bound amino acids.

Roasting denatures and insolubilizes much of the protein. Some of the constituent amino acids in the protein are destroyed during the latter stages of roasting and thus contribute to aroma and flavor development. Analysis of the protein amino acids in green and corresponding roasted coffee show marked decreases in arginine, cysteine, lysine, serine, and threonine in both Arabica and Robusta types after roasting. Alanine, glycine, leucine, glutamic acid, and phenylalanine are relatively stable to roasting (10). Cysteine [52-90-4] and methionine [63-68-3] are the probable sources of the many potent sulfur compounds found in coffee aroma, eg, mercaptans, organic sulfides, and thiazoles. Other amino acids are capable of generating aromatic compounds such as pyrazines, pyrroles, etc.

About 50–80% of the trigonelline is decomposed during roasting. Trigonelline is a probable source for niacin [59-67-6] but also a source of some of the aromatic nitrogen compounds such as pyridines, pyrroles, and bicyclic compounds found in coffee aroma (11). Certain acids, such as acetic, formic, propionic, quinic, and glycolic, are formed or increase upon roasting, while other acids present initially in the green coffee, such as the chlorogenic acids, citric, and malic, decrease in level with increasing degree of roast (12). The composite of acids in brew of a lightly roasted coffee contribute to the taste quality termed "fine acidity". Brews of darkly roasted coffees are less acidic. Slight cleavage of the triglycerides and the diterpene and sterol esters occurs during roasting. It is likely that some oxidation of lipid components is initiated during the roasting stage.

However, some of the Maillard products generated upon roasting may act as antioxidants, slowing down the deterioration of the lipids upon storage. The aromatic, oil-soluble aroma compounds slowly partition into the oil phase after the roasted bean is allowed to cool and equilibrate.

Aroma. The chemistry of aroma compound formation is complex. In the early and middle stages of roasting, a significant portion of these compounds is derived from carbohydrates (mainly sucrose and arabinose from arabinogalactan), proteinaceous material (including amino acids and peptides), and other bean components (chlorogenic acids, trigonelline, organic acids, lipids, etc). Reactions involve degradation to form reactive products, which include saturated and unsaturated aldehydes, ketones, dicarbonyls, amines, and hydrogen sulfide, or interactions via nonenzymatic browning reactions. As roasting progresses, the precursor profile changes (less reactive small molecules) and structurally more complicated sulfur-, oxygen-, and nitrogen-containing heterocyclic compounds are formed with the brown colored polymeric melanoidins dominating. Table 3 lists the number of aroma compounds found in coffee (presently ~850) according to substance class. The identification of so many compounds can be attributed to major advances in instrumental analysis, in particular, high-resolution gas chromatography (HRGC) and mass spectrometry (MS).

In order to identify which of this large number of compounds are the major contributors to the coffee aroma (the "potent odorants"), a number of specialized analytical approaches have been utilized (13). Coffee aroma is extracted and separated by HRGC; aroma potency is then determined by means of aroma extraction dilution analysis (ADEA) or combined hedonic and response measurements (CHARM). Highly volatile aroma compounds are detected by gas chromatography–olfactometry of headspace samples (GCOH). Following enrichment and identification, aroma activity values of the aroma compounds (OAVs) are determined. Finally, the relative importance of the aroma compounds is

Table 3. **Aroma Compounds of Roasted Coffee**[a]

Class of Compound	Number
hydrocarbons	80
alcohols	24
aldehydes	37
ketones	85
carboxylic acids	28
esters	33
pyrazines	86
pyrroles	66
pyridines	20
other bases (eg, indoles)	32
sulfur compounds	100
furans	126
phenols	49
oxazoles	35
others	20
total	*841*

[a] See Ref. 13.

determined by sensory evaluation of the impact of compound omission from a synthetic blend of all the compounds. As a result of such studies, only 20–30 compounds are deemed important and it has even been suggested that only ~15 compounds have major impact on the final aroma (13). These compounds include four alkyl pyrazines, four furanones, 2-furfurylthiol, 4-vinylguaiacol, acetaldehyde, propanol, methylpropanol, and 2- and 3-methylbutanol.

3.4. Chemistry of Brewed Coffee. The chemistry of brewed coffee is dependent on the extraction of water-soluble and hydrophobic aromatic components from the coffee cells and lipid phase, respectively. Factors that affect extraction and flavor quality of brewed coffee are degree of roast; blend composition; grinding technique; particle size and density; water quality; water to coffee ratio; and brewing technique or device, such as drip filter, percolator, or espresso, which defines the water temperature, steam pressure, brewing time, water recycle, etc. Extraction yields of home brewing range from ~9 to 28% and typically ~23% dry basis roasted and ground (R&G) (14). The trend in the United States, with some notable exceptions, has been toward weaker brew strengths, with typical recipes of ~5 g of R&G coffee per 6 oz cup (brew solids concentration ~0.7%), compared to ~10 g a generation ago (brew solids ~1.2%). In comparison, espresso typically uses ~8–12 g of coffee per 2 oz of beverage (brew solids ranging from ~3 to 5%). As espresso is brewed rapidly under steam pressure (brew time ranging from ~15 to 35 s), it contains a relatively large amount of oil droplets (~0.1–0.2% basis brew) and suspended colloidal solids (~0.3% basis brew) that contribute to the greater turbidity and mouthfeel of this beverage compared to filter brewed coffee (15).

3.5. Instant Coffee. The chemistry of instant or soluble coffee is dependent on the R&G blend and processing conditions. This is indicated in Table 2 by the wide range of constituents. In addition to the atmospherically extractable solids found in brewed coffee, commercial percolation generates water-soluble carbohydrate by hydrolysis that contributes to the yield. This additional carbohydrate includes the sugars, arabinose, mannose, and galactose; oligosaccharides derived from mannan and arabinogalactan; and the partially hydrolyzed polysaccharides, mannan, and arabinogalactan. It improves the drying properties and retention of volatiles by the extract, and reduces hygroscopicity. These water-soluble carbohydrates formed the basis for the first 100% pure instant coffee developed by General Foods Corp. in the late 1940s.

4. Roasted and Ground Coffee Processing and Packaging

The main processing steps in the manufacture of roast and ground coffee products are blending, roasting, grinding, and packaging. Green coffee is shipped in burlap bags (60–70 kg) or in bulk containers (16,500–18,000 kg). Prior to processing, the green coffee is mechanically cleaned to remove string, lint, dust, husk, and other foreign matter. Coffee of different varieties or from different sources may be blended before or after roasting at the option of the manufacturer.

4.1. Roasting Technology. Commercial roasting is generally by hot combustion gases in either rotating cylinders or fluidized-bed systems. Though steam pressure, infrared (ir) and microwave roasting systems are found in the patent literature none are believed to be used to a significant extent. Roasting times in batch cylinders were traditionally 8–15 min, whereas much of the coffee produced today is roasted in batch or continuous fluidized beds in only 0.5–4 min (16–18). In either case, the initial step of roasting is a moisture elimination and uniform heating step. The onset of roasting reactions has been reported to be as low as 151°C and when the bean temperature exceeds ~165°C, the reactions have switched from endothermic to exothermic and the roast has begun to fully develop. This stage is generally accompanied by a noticeable crackling sound like that of corn popping and the beans swell to as much as twice their unroasted volume. The final bean temperature of 185–250°C is determined by the flavor development desired for the finished goods, whether a blend or individual varieties are roasted. A water or air quench terminates the roasting reaction. Most, but not all, of any water added is evaporated from the heat of the beans. Theoretically, ~700 kJ is needed to roast 1 kg of coffee beans. A roaster efficiency of 75% or more (933 kJ/kg) is possible with recirculation of the roaster gas, whereas older, nonrecirculating units operate with an efficiency as low as 25% (2800 kJ/kg). Though cupping for flavor is the ultimate test, the acceptability of the roast is often judged off-line by a photometric reflectance measurement on a ground sample of the bean, and adjustments of the temperature controls that initiate the quenching end point. Typically, the bean temperature measurement is used as the process control parameter, but at least one fluid bed unit is controlled via roasting gas exhaust temperature. Attempts to utilize *in situ* photometric reflectance measurements for process control have met with limited success due to the varying nature of the agricultural commodity and the fact that these systems quickly get coated with an oily brown residue. State-of-the art roaster controls can also be set to provide a specific heat input profile or to store and replicate an historical pattern. The desired patterns are chosen to deliver cup quality and strength for a given consumer recipe. The faster fluidized-bed roasting processes (batch or continuous) are the basis of high yield coffee products. These units are generally operated at lower air temperature, ie, 185–400°C vs 425–490°C, resulting in a more uniform roast throughout the bean, an increase in extractable soluble solids of 20% or more, and higher aroma retention. The higher circulation rate of roaster gas required for fluidization increases heat transfer to allow a faster roast at lower temperature. Exhausted roaster gas often must be incinerated in an afterburner for environmental pollution purposes. The use of IR heat (gas-fired ceramics) or microwave energy to speed up the roasting process while providing a more even roast has also been patented (19,20).

The roasted and quenched beans are air cooled and conveyed to storage bins for moisture and temperature equilibration before grinding, this storage step, called curing, also allows the carbohydrate matrix to harden properly before grinding. Residual foreign matter (mostly stones), which may have passed through the initial green cleaning step, is removed in transit to the storage bins by means of a high velocity air lift that leaves the heavier debris behind. The roasted beans may flow by gravity or be airveyed to the grinders.

4.2. Grinding. Grinding of the roasted coffee beans is tailored to the intended method of beverage preparation. Average particle size distributions range from very fine (500 µm or less) to very coarse (1100 µm). The classic way to measure and control particle size distribution has been via use of stacked standard sieves. Many commercial roasters now use laser light scattering devices to provide a quick and reproducible measure of this critical parameter. A finer grind will allow greater solids extraction, but may slow the brewing process because of increased flow resistance and reduced wettability. Most coffee is ground in multistep steel roll mills in order to produce the most desirable particle size distribution. After passing through cracking rolls, the broken beans are fed between two more rolls, one of which is cut or scored longitudinally; the other, circumferentially. The paired rolls operate at speeds designed to cut rather than crush the cracked particles. For finer grinds, a second pair of more finely scored rolls running at higher speed is positioned below the first set. Some coffee is flaked, to increase extractability without slowing the brewing speed, by passing it through closely spaced smooth rolls after grinding (21). Unlike grinding, flaking is a crushing step that disrupts the cellular structure to reduce the diffusion path for extraction. Flaking can increase extraction yield by as much as 15% regardless of the roasting profile. A normalizer/homogenizer mixing section, generally an integral part of the grinder, assures a uniform particle distribution and may be used to increase density before packaging. Like flaking, attempts to stretch the bean even further by more severe crushing have met with limited consumer acceptance due to the need for a consumer behavior change with the denser ground product.

4.3. Packaging. Most roasted and ground coffee sold directly to consumers in the United States is vacuum-packed with 0.33–0.37 kg in 10.3-cm (4 1/16 in.) diameter metal cans or with three times as much in 15.7-cm (6 3/16 in.) diameter metal cans. Whether packing 0.33 or 0.37 kg of coffee, the 10.3-cm diameter cans generally have a fill volume of ~1000 cm^3. Ground coffee pack density is managed via modern roasting and grinding technology to deliver a full can with the desired flavor profile. After roasting and grinding, the coffee is conveyed, usually by gravity, to weighing-and-filling machines that achieve the proper fill by tapping or vibrating. A loosely set cover is partially crimped. The can then passes into the vacuum chamber maintained at 3.3 kPa (25 mm Hg) absolute pressure or less. The cover is clinched to the can cylinder wall and the can moves through an exit valve or chamber. This process removes 95% or more of the oxygen from the can. Polyethylene snap caps for reclosure are placed on the cans before they are stacked in cardboard cartons for shipping. A case or tray usually contains 12 of the smaller diameter cans or 6 of the larger diameter, and a production packing line usually operates at a rate of 250–350 smaller cans per minute or ~100–150 larger cans per minute. Though other can sizes are present in the marketplace the growth of grocery club stores and supercenters has helped to drive the bulk of the volume to the larger packs both in the United States and in Canada.

Vacuum-packed coffee retains a high-quality rating for at least 1 year. The slight loss in fresh roasted character that occurs is because of chemical reactions with the residual oxygen in the can and previous exposure to oxygen prior to packing (22).

Coffee vacuum-packed in flexible, bag-in-box packages has gained wide acceptance in Europe and is the prevalent format for prepackaged ground coffee. The inner liner, usually a plastic-laminated aluminum foil, is formed into a hard brick shape during the vacuum process (23). In the United States, a similar printed multilaminated flexible structure is used to form the brick pack that is sold as is at retail. The aluminum foil layer of these types of packages provides a barrier to moisture and oxygen similar to that of a metal can.

Inert gas flush packing in plastic-laminated or metalized nylon pouches, although less effective than vacuum packing, can remove or displace 80–90% of the oxygen in the package. These packages offer satisfactory shelf-life and are sold primarily to food service operators (restaurants, office coffee service, caterers, and institutions).

An appreciable amount in Europe and more often than before coffee in the United States, is distributed as whole beans, which are ground in stores or by consumers in their homes. Whole-bean roasted coffee remains fresh longer than ground coffee. The specialty gourmet shop trade based on this system has continued to grow significantly in the United States since the early 1980s. Nearly 20% of the coffee now consumed in the United States is now classified as premium whole bean or ground gourmet coffee, whereas 10 years ago this category was not considered a significant factor in the market and was not tracked separately (24). Fifty eight percent of gourmet coffee consumption is away from home in specialty gourmet coffee shops, at work, while traveling, and in other foodservice outlets. Unprotected whole bean coffee, in bins or bags, can be considered as little different from fresh if it is used within 10–12 days and significant flavor differences can take as long as 40 days to be noticeable (25).

Packaging of roasted whole beans in foil laminate barrier bags generally is carried out with the use of a one-way valve that allows carbon dioxide gas released from the beans to escape and prevent air and moisture from entering the package. This permits packing the coffee soon after roasting and a nitrogen purge or vacuum evacuation of the pack before sealing assures that long-term oxygen exposure is minimized. Whole bean coffee packed in this manner would be expected to be of excellent quality at least for the 1 year claimed for roast and ground coffee.

The one-way valve is now often used with ground coffee in flexible foil laminate pouches. A common packing scheme calls for filling the pouch, evacuating the pack to minimize retained oxygen, and then backflushing with nitrogen or another inert gas before sealing. The ground coffee continues to respire significant amounts of carbon dioxide and carbon monoxide over the first weeks after packing such that these pouches are soft rather than rigid like the brick pack described above. This system allows one to pack the coffee sooner after grinding, with the potential to retain more of the desirable coffee aromatics.

5. Instant Coffee Processing and Packaging

Instant coffee is the dried water-extract of ground, roasted coffee. Although used in Army rations as early as the U.S. Civil War, the popularity of instant coffee as a grocery product grew only after World War II, coincident with improvements in manufacturing methods and consumer trends toward convenience. Extensive patent literature dates back to 1865. Instant coffee products represented only

7% of the coffee consumed in the United States in 2001, less than one-half that consumed in 1991 in both absolute and percentage terms (24).

Green beans for instant coffee are blended, roasted, and ground similarly to those for roasted and ground products. A concentrated coffee extract is normally produced by pumping hot water through the coffee in a series of cylindrical percolator columns. The extracts are further concentrated prior to a spray- or freeze-drying step, and the final powder is packaged in glass or other suitable material. Some soluble coffees, both spray- and freeze-dried, are manufactured in producing countries for export.

5.1. Blend/Roast/Grind. Blends of Brazilian, Central American, and Colombian milds as well as African, Asian, and Brazilian robustas are prepared to achieve desired flavor characteristics. The batch- or continuous-type roasters used for roasted and ground coffee also are used for instant coffee. Grinding of roasted beans for an instant coffee process is adjusted to suit the type of commercial percolation system to be used. The average particle size is generally larger than that used for domestic brewing to avoid excessive pressure drops across the percolator columns. Similarly, very fine particles are avoided.

5.2. Extraction. Commercial extraction equipment and conditions have been designed to obtain the maximum yield of soluble solids with the desired flavor character. Conceptually, most commercial systems can be represented by a series of countercurrent batch extractors. The freshwater feed, at pressures well above 1 atm and temperatures high enough to hydrolyze the coffees' polysaccharides to oligosaccharides, contacts the most spent coffee grounds. During the final extraction stage, fresh ground coffee is contacted with an extract of these oligosaccharides at temperatures near the atmospheric boiling point.

Significant factors influencing extraction efficiency and product quality are grind size, feed water temperature and temperature profile through the system, percolation time, coffee/water ratio, premoistening or wetting of the ground coffee, design of extraction equipment, and flow rate of extract through the percolation columns.

Percolation trains consisting of 5–10 columns are the norm. Height/diameter ratios usually range from 4:1 to 7:1. To improve extraction, the ground coffee may be steamed or wetted. Feed water temperatures ranging from 154 to 182°C are common and the final extract exits at 60–82°C. To minimize flavor and aroma loss prior to drying, the effluent extract may be cooled in a plate heat exchanger. The yield, a function of the properties of the particular blend and roast, the operating temperatures, and the percolation time, is generally controlled through adjustment of the soluble solids drawn off from the final stage. Extraction yield is calculated from both the weight of extract collected and the soluble solids concentration as measured by specific gravity or refractive index. Soluble yields of 24–48% or higher on a roasted coffee basis are possible. Robusta coffees give yields ~10% higher than arabica because of a higher level of available polysaccharides and caffeine. The latest technology in thermal extraction of spent grounds provides roasted yields in excess of 60% (26).

Extract is stored in insulated tanks prior to drying. Because high soluble solids concentration is desirable to reduce aroma loss and evaporative load in the driers, most processors concentrate the 15–30% extract to 35–55% prior to drying (27). This may be accomplished by vacuum evaporation or freeze concentration. Clarification of the extract, normally by centrifugation, may be used to assure the absence of insoluble fine particles.

The flavor of instant coffee can be enhanced by recovering and returning to the extract or finished dry product some of the natural aroma lost in processing. The aroma constituents from the grinders, percolation vents, and evaporators may be added directly or in concentrated or fractionated form to achieve the desirable product attributes.

5.3. Drying.   The criteria for good instant coffee drying processes include minimization of loss or degradation of flavor and aroma, uniformity of size and shape in a free-flowing form, acceptability of the bulk density for packaging, product color acceptability, and moisture content below the level required to maintain shelf-stability (<5%). Operating costs, product losses, and capital investment are also considerations in the selection of a drying process.

Spray Drying and Agglomeration.   Most instant coffee products are spray-dried. Stainless steel towers with a concurrent flow of hot air and atomized extract droplets are utilized for this purpose. Atomization, through pressure nozzles, is controlled based on selection of the nozzles, properties of the extract, pressures used, bulk density, and capacity requirements. Low inlet air temperatures (200–280°C) are preferred for best flavor quality. The spray towers must be provided with adequate dust collection systems such as cyclones or bag filters. The dried particles are collected from the conical bottom of the spray drier through a rotary valve and conveyed to bulk storage bins or packaging lines. Processors may screen the dry product to assure a uniform particle size distribution.

Most spray-dried instant coffees have been marketed in a granular form, rather than the small spherical spray-dried form, since the mid-1960s. The granular appearance is achieved by steam fusing the spray-dried material in towers similar to the spray drier. Belt agglomerators are also common.

Freeze Drying.   Commercial freeze drying of instant coffee has been a common practice in the United States since the mid-1960s. The freeze-drying process provides the opportunity to minimize flavor degradation due to heat (28).

Sublimation of ice crystals to water vapor under a very high vacuum, ~67 Pa (0.5 mm Hg) or lower, removes the majority of the moisture from the granulated frozen extract particles. Heat input is controlled to assure a maximum product end point temperature below 49°C. Freeze drying takes significantly longer than spray drying and requires a greater capital investment.

5.4. Packaging.   In the United States, instant coffee for the consumer market is usually packaged in glass jars containing from 56 to 340 g of coffee. Larger units for institutional, hotel, restaurant, and vending machine use are packaged in bags and pouches of plastic or laminated foil. In Europe, instant coffee is packaged in glass jars or foil-laminated packages.

Protective packaging is primarily required to prevent moisture pickup. The flavor quality of regular instant coffee changes very little during storage. However, the powder is hygroscopic and moisture pickup can cause caking and flavor impairment. Moisture content should be kept <5%.

Many instant coffee producers in the United States incorporate natural coffee aroma in coffee oil into the powder. These highly volatile and chemically unstable flavor components necessitate inert-gas packing to prevent aroma deterioration and staling from exposure to oxygen.

6. Decaffeinated Coffee Processing

Decaffeinated coffee products represented 9% of the coffee consumed in 2001 in the United States; less than one-half that consumed in 1991 (24). Decaffeinated coffee was first developed commercially in Europe ∼1900. The process as described in a 1908 patent (29) consists of first, moisturizing green coffee to at least 20% to facilitate transport of caffeine through the cell wall, and then contacting the moistened beans with solvents.

Until the 1980s, synthetic organic solvents commonly were used in the United States to extract the caffeine, either by direct contact as above or by an indirect secondary water-based system (30). In each case, steaming or stripping was used to remove residual solvent from the beans and the beans were dried to their original moisture content (10–12%) prior to roasting.

In the 1980s, manufacturers' commercialized processes that utilized either naturally occurring solvents or solvents derived from natural substances to position their products as naturally decaffeinated. The three most common systems use carbon dioxide under supercritical conditions (31), oil extracted from roasted coffee (32), or ethyl acetate, an edible ester naturally present in coffee (33). Specificity for caffeine and caffeine solubility is key to selection and system design. Because caffeine can be selectively removed from water extracts of green beans by activated charcoal, several processes that utilize water or recycled green coffee extract have been described and are also considered natural decaffeination. If water is used, the green coffee extract produced is externally decaffeinated and the noncaffeine solids containing important flavor precursors are reabsorbed before drying and roasting. In what is commonly advertised as the "Swiss Water Process", the use of recycled green extract obviates the need for a separate reabsorption step as the caffeine deficient green extract selectively leaches caffeine. Preabsorbing sugar on activated charcoal to improve its specificity also has been commercialized (34). The degree of decaffeination, based on comparison to the starting material, is controlled using known time–temperature relationships for the particular process for each bean type.

In all the above mentioned processes of coffee decaffeination, changes occur that affect the roast flavor development. These changes are caused by the pre-wetting step, the effects of extended (4 h plus) exposure at elevated temperature as required to economically extract the caffeine from whole green beans, and the postdecaffeination drying step.

To make an instant decaffeinated coffee product, the decaffeinated roast and ground coffee is extracted in a manner similar to nondecaffeinated coffee. Alternatively, the caffeine from the extract of untreated roasted coffee is removed by using the solvents described previously.

7. Economic Importance

Coffee ranks second only to petroleum as the worlds largest traded commodity. The total world production of green coffee in the 1999–2000 growing season was 113.5 million bags a (35); exportable production was 88.2 million bags (Table 4) with an export value of $7.5 billion (36). Of particular interest is the rapid climb

Table 4. **World Production of Green Coffee, 1999–2000**[a]

Country	Exportable production[b]	%
Brazil	18,000	20.4
Vietnam	10,660	12.1
Colombia	7,982	9.0
Ivory Coast	5,640	6.4
Indonesia	5,225	5.9
Mexico	5,138	5.8
India	4,070	4.6
Guatemala	3,964	4.5
Uganda	3,017	3.4
Honduras	2,803	3.2
El Salvador	2,445	2.8
Costa Rica	2,347	2.7
Peru	2,341	2.7
Ethiopia	2,200	2.5
Kenya	1,662	1.9
Papua New Guinea	1,385	1.6
Cameroon	1,270	1.4
Ecuador	985	1.1
Tanzania	823	0.9
Zaire	550	0.6
Madagascar	371	0.4
others	5,377	6.1
Totals	*88,255*	*100.0*

[a] Ref. 37.
[b] Thousands of 60-kg bags.

of Vietnam to become the no. 2 producer of coffee surpassing all producers except Brazil.

From 1999–2000, the United States import from producing countries totaled 21.1 million bags of green coffee equivalent. This includes 19.1 million bags of green coffee, 0.7 million bags of roasted coffee, and 1.3 million bags of soluble coffee with a total value of ~$1.8 billion (38) (Table 5).

Table 5. **World Imports of Green Coffee (Selected Countries), 2000**

Country	Imports[a]	%
United States[b]	21,095	23.9
Germany[c]	14,442	16.4
France[c]	6,506	7.4
Japan[c]	7,360	8.3
Italy[c]	6,335	7.2
Spain[c]	3,819	4.3
United Kingdom[c]	3,000	3.4
Canada[c]	3,140	3.6
Belgium/Luxembourg[c]	3,273	3.7
All Others[c]	19,285	21.8
Totals	*88,255*	*100.0*

[a] Thousands of 60-kg bags.
[b] Ref. 39.
[c] Ref. 40.

8. Coffee Biotechnology

Biotechnology has made rapid progress in recent years. There has been considerable interest in applying biotechnology to coffee through both conventional breeding using tissue culture techniques and molecular markers, and through the application of genetic modification (GM) technology.

8.1. Advances in Coffee Breeding and Culture Methods. Application of modern plant breeding methods to coffee improvement, including production of F_1 hybrids, has led to improvements in crop performance (41). Breeding targets remain a combination of agronomic factors such as disease resistance and yield, and quality, eg, cup-quality and caffeine content. Disease resistance targets include coffee leaf rust, coffee berry disease, and nematodes. Clonal propagation methods have been used to improve multiplication efficiency of robusta coffee with the added advantage of maintaining uniformity. Somoclonal variation, a natural phenomenon that occurs during the tissue culture technique somatic embryogenesis, has been used to generate new sources of genetic variation in coffee. For example, the arabica variety Bourbon LC was developed through somoclonal variation of the parent *Laurina* line, a natural mutation with a caffeine content 50% lower than typical arabica varieties (42). The result is a new variety with superior agronomic performance than the parent line but retaining the reduced caffeine content. The variety has now been well characterised at both the agronomic and molecular level. The variety is genetically distinct from its parent as well as other common arabica varieties (43).

8.2. Application of Molecular Markers to Coffee Breeding. Like other tree-based crops, coffee's long generation time make it a good candidate for application of marker-assisted selection in breeding. Programs seeking to identify genes controlling traits such as host resistances to diseases and pests, caffeine content and cup quality, and to develop molecular markers for these traits are underway. Mapping populations based on interspecific crosses [eg, *C. liberica* X *C. pseudozanhuebariae*, (44)] have helped identify a number of genes controlling caffeine content. A similar approach is being used to identify candidate genes controlling biochemical pathways that make a major contribution to cup quality such as chlorogenic acid accumulation (45).

8.3. Advances and Applications of Coffee Molecular Biology. *Coffee Transformation and Genetic Modification.* Coffee transformation using Agrobacterium sp. vectors has been reviewed recently (46). Arabica coffee has proved to be more amenable to transformation than robusta coffee. There are also strong variety effects on transformation and regeneration efficiency (47). However, this has not prevented the development of the genetically modified coffees described in the following section.

Isolation of Genes from Coffee. Rapid progress has been made in isolating and identifying genes from coffee. The NCBI gene databank (GenBank; www.ncbi.nlm.nih.gov) currently holds ~200 isolated sequences from coffee. These can be grouped as coding genes (eg, enzymes) and genetic markers [eg., internally transcribed spacers (ITS) and microsatellite sequences]. In addition, at least two groups have prepared expressed sequence tag (EST) libraries from coffee tissue at different stages of development (48,49). These libraries are a valuable resource for understanding the link between biology and quality.

Table 6. **Summary of Major Functional Genes Cloned from Coffee**

Gene	Application	Reference
xanthosine-N^7- methyltransferase	caffeine-free coffee	53
ACC synthase and ACC oxidase	controlled-ripening coffee	54
11S storage protein	manipulation of bean biochemistry	57
seed-specific promoters	manipulation of bean biochemistry	57
coffee mannanase	soluble coffee processing, manipulation of bean biochemistry	58

Genetic marker sequences can be used to develop markers for breeding purposes or for use in studying the genetic diversity of coffee species (50). However, it is the application of coding genes that is of immediate interest for improving coffee agronomy and quality. Table 6 summarizes the main developments in this area. Of particular interest is the application of GM technology to produce caffeine-free coffee and controlled-ripening coffee (51,52). Genes coding for xanthosine-N^7-methyltransferase, ACC synthase, and ACC oxidase, respectively, were isolated and used to transform arabica coffee to block synthesis of these key enzymes of caffeine and ethylene biosynthesis. This is claimed to block caffeine accumulation (53) and fruit ripening (54). The latter is of interest as it allows mechanical harvesting of uniformly developed fruit followed by synchronous ripening of the coffee berries by application of exogenous ethylene. This has already been applied to other climacteric fruits such as tomatoes (55). A third application has been introduction of the *Cry1Ac* gene for *Bacillus thuringiensis toxin* (Bt) into robusta coffee to confer resistance to coffee berry borer (56). Trees are now in field trials in French Guyana. Looking to the future, seed-specific promoters (the molecular 'switches' that turn genes 'on'), seed storage proteins and germination-specific mannanases have now been isolated from coffee (57,58). These hold potential to manipulate coffee bean biochemistry and quality by genetic transformation.

Despite the technical progress that has been made, there are currently no GM coffee varieties in the marketplace. There has been much adverse publicity surrounding GM foods generally, and GM coffee in particular (59). Until there are clear consumer benefits, this is unlikely to change.

Process Biotechnology. Biotechnology has been applied to roast and ground and soluble coffee processing. The flavor profile of roast and ground coffee has been manipulated by treating green or partially roasted coffee with a cocktail of hydrolytic enzymes (60). The process is claimed to generate a novel and pleasant flavor profile. Fermentation of a green coffee extract has been used to produce flavour compounds, principally diacetyl [431-03-08, 21]. Treating steam-expanded coffee with a mixture of hydrolytic enzymes is claimed to increase yield in soluble coffee processing (61). Since last reviewed, enzymes technology has been applied to improving the solubility of polysaccharides extracted during soluble coffee processing. By using fungal mannanase [from *Aspergillus* sp, (62)] or a coffee mannanase sourced from either coffee beans or produced using a recombinant bacterium (58), the claimed benefit is reduced sedimentation during high-temperature extraction/cooling cycles. It is likely that there will be increased activity in this area as a wider range of coffee and food-grade enzymes become available.

9. Coffee Regulations and Standards

Soluble and roast and ground coffee is covered by a range of international and national legislation (eg, EEC, United States, Germany) and voluntary Codes of Practice (eg, National Coffee Association). The standards and regulations provide definitions for green and processed coffee and potential health risks (eg, ochratoxin in coffee, decaffeination of coffee). Additionally, the International Standards Organization (ISO) maintains standards for coffee and coffee-based products (63). The following list of regulations is by no means meant to be complete, but it covers the major points of general interest.

9.1. Roast and Ground Coffee. Roast and ground coffee in regulated at the national level, which leads to a range of regulations from very strict definition of roast coffee (eg, Austria, Germany) to an absence of regulation (eg, United Kingdom). The exception to this is regulation governing decaffeinated coffee that states maximum caffeine content of 0.1% db and states maximum limits for solvent residues. (In the United States, decaffeinated signifies 97% caffeine removal.)

Soluble Coffee. Soluble coffee is regulated within the EU (64). The regulations cover extraction solvent (water only) and hydrolysis methods (the use of acids and bases is forbidden). Composition is also regulated. For example, soluble coffee powder must contain not less than 95% coffee-based matter and may not contain >12% edible sugars. In addition, decaffeinated soluble coffee must not contain >0.3% caffeine.

Coffee Consumption and Health. Coffee contains between 1.2 and 3.4% db caffeine, a pharmacologically active alkaloid. The stimulant and diuretic effects of caffeine are well documented. A small number of people are sensitive to caffeine so that consumption in these individuals can cause anxiety, restlessness, sleeping difficulties, headache, and palpitations of the heart. This findings has lead to many studies over the years seeking to determine a link between coffee consumption and physiological conditions. However, extensive reviews of the available literature show no negative effects from moderate coffee consumption of between 3 and 5 cups a day (65).

Ochratoxin in Coffee. Ochratoxin A (OTA) is a by-product of mould growth on a range of foodstuffs including cereals, fruits, and coffee. OTA has been shown to be a kidney toxin in several species. In rats and mice it has been shown to be a carcinogen and teratogen. Europe is the main region where OTA occurs in food, with cereals contributing 60% of dietary intake. Of the remainder, wines contribute 25% of intake and coffee and grape juice contributes 5–7% each. Thus, coffee is not the major contributor of dietary OTA. Despite this, there is legislation in force within Europe (eg, Italy, Romania) at the national level that restricts OTA in green coffee and/or finished products. There are moves to introduce EU-wide legislation. As a point of reference, the Joint/World Health Organization (FAO/WHO) Expert Committee on Food

Additives (JEFCA) set a tolerable weekly intake for OTA of 100 g/kg body weight per week in 1995 (66).

10. Coffee Substitutes

Coffee substitutes, which include roasted chicory, chick peas, cereal, fruit, and vegetable products, have been used in all coffee consuming countries. Although consumers in some locations prefer the noncoffee beverages, they are generally used as lower cost beverage sources. Additionally, it is not unusual for consumers in some of the coffee producing countries to blend coffee with noncoffee materials.

Chicory is harvested as fleshy roots that are dried, cut to uniform size, and roasted. Chicory contains no caffeine, and on roasting develops an aroma compatible with that of coffee. It gives a high yield, ~70%, of water-soluble solids with boiling water and can also be extracted and dried in an instant form. Chicory extract has a darker color than does normal coffee brew (67).

BIBLIOGRAPHY

"Coffee" in *ECT* 1st ed., Vol. 4, pp. 215–223, by L. W. Elder, General Foods Corp.; "Coffee, Instant" in *ECT* 1st ed., Suppl. 2, pp. 230–234, by H. S. Levenson, Maxwell House Division of General Foods Corp.; "Coffee" in *ECT* 2nd ed., Vol. 5, pp. 748–763, by R. G. Moores and A. Stefanucci, General Foods Corp.; in *ECT* 3rd ed., Vol. 6, pp. 511–522, by A. Stefanucci, W. P. Clinton, and M. Hamell, General Foods Corp.; in *ECT* 4th ed., Vol. 6, pp. 793–811 by G. Wasserman, H. D. Stahl, W. Rehman, P. W. Heilmann, Kraft General Foods Corporation; "Coffee" in *ECT* (online), posting date: December 4, 2000, by Gerald Wasserman, Howard D. Stahl, Warren Rahman, Peter Whitman, Kraft General Foods Corporation.

CITED PUBLICATIONS

1. R. J. Clarke and O. G. Vitzthum, eds., *Coffee, Recent Developments*, Blackwell Science, London, U.K., 2001, Chapt. 2–4.
2. R. J. Clarke and R. Macrae, eds., *Coffee*, Vol. 1, Chemistry, Elsevier Applied Science Publishers, Ltd., Barking, U.K., 1985, Chapt. X.
3. M. N. Clifford in Ref. 2, pp. 153–202.
4. A. Scholze and H. G. Maier, *Z. Lebensmittel. Unters. Forsch.* **178**, 5 (1984).
5. K. Speer and I. Kolling-Speer, in Ref. 1, pp. 33–49.
6. J. R. Rogers, S. Michaux, M. Bastin, and P. Buchell, *Plant Science* **149**, 115 (1999).
7. A. G. W. Bradbury, in Ref. 1, pp. 1–17.
8. R. J. Redgwell, D. Curti, M. Fischer, P. Nicolas, and L. B. Fay, *Carbohydrate Res.* **337**, 239 (2001).
9. U. Arnold, E. Ludwig, R. Kuhn, and U. Moeschwitzer, *Z. Lebensmittel. Unters. Forsch.* **199**, 22 (1994).
10. R. MacRae, in Ref. 2. pp. 141–142.
11. R. MacRae, in Ref. 2. pp. 125–135.
12. H. H. Balzer, in Ref. 1, pp. 18–30.

13. W. Grosch, in Ref. 1, pp. 68–89.
14. G. Pictet, in Ref. 2, Vol. 2. *Technology*, 1987, pp. 221–225.
15. M. Petracco, in Ref. 1, pp. 140–164.
16. U.S. Pat. 4,322,447 (Mar. 20, 1982), M. Hubbard (to Hills Bros. Coffee Co.).
17. U.S. Pat. 4,737,376 (Apr. 12, 1988), L. Brandlein and co-workers (to General Foods Corp.).
18. U.S. Pat. 4,988,590 (Jan. 29, 1990), S. E. Price and co-workers (to Procter & Gamble Co.).
19. U.S. Pat. 4,860,461 (Aug. 29, 1989), Y. Tamakiand co-workers (to Pokka Corp. and NGK Insulators Ltd.).
20. U.S. Pat. 4,780,586 (Oct. 25, 1988), T. Le Viet and T. Bernard (to Nestec S.A.).
21. U.S. Pat. 3,615,667 (Oct. 26, 1971), F. M. Joffe (to Procter & Gamble Co.).
22. W. Clinton, *"Evaluation of Stored Coffee Products,"* Proceedings of the 9th Colloquium of ASIC, London, 1980, p. 273.
23. A. L. Brody, *Food and Flavor Section*, Arthur D. Little, Inc., Cambridge, Mass.; *Flexible Packaging of Foods*, CRC Press, Division of the Chemical Rubber Co., Cleveland, Ohio, 1970, pp. 41–42.
24. *National Coffee Drinking Trends*, National Coffee Association, New York, 2001.
25. R. Clarke in Ref. 2. p. 206.
26. Eur. Pat. Appl. 0363529A3 (June 10, 1988), H. D. Stahl and co-workers (to General Foods Corp.).
27. U.S. Pat. 4,107,339 (Aug. 15, 1978), B. Shrimpton (to General Foods Corp.).
28. U.S. Pat. 3,438,784 (Apr. 15, 1969), W. P. Clinton and co-workers (to General Foods Corp.).
29. U.S. Pat. 897,763 (Sept. 1, 1908), J. F. Meyer (to Kaffee-Hag).
30. U.S. Pat. 2,309,092 (Jan. 26, 1943), N. E. Berry and R. H. Walters (to General Foods Corp.).
31. U.S. Pat. 4,820,537 (Apr. 8, 1989), S. Katz (to General Foods Corp.).
32. U.S. Pat. 4,465,699 (Aug. 14, 1964), F. A. Pagliaro and co-workers (to Nestlé SA).
33. U.S. Pat. 4,409,25 (Oct. 11, 1983), L. R. Morrison, Jr. (to Procter & Gamble Co.).
34. W. Heilmann, in *Proceedings of the 14th Colloquium of ASIC*, San Francisco, Calif., 1991, pp. 349–356.
35. Green Coffee: Total Production in Selected Countries Horticultural and Tropical Products Division, FAS/USDA, December 2001.
36. Coffee: ICO Monthly and Composite Indicator Prices on the New York Market Horticultural and Tropical Products Division, FAS/USDA December 2001.
37. Green Coffee: Exportable Production in Specified Countries: Horticultural and Tropical Products Division, FAS/USDA December 2001.
38. Coffee: ICO Monthly and Composite Indicator Prices on the New York Market Horticultural and Tropical Products Division, FAS/USDA, December 2001.
39. U.S. *Coffee Imports by Type and Origin Source*: U.S. Department of Commerce Horticultural and Tropical Products Division, FDA/USDA, December 2001.
40. Coffee: *Specified Country Imports Source*: U.S. Department of Commerce Horticultural and Tropical Products Division, FDA/USDA, December 2001.
41. H. A. M. Van der Vossen in Ref. 1, pp. 184–201.
42. M. R. Sondahl "Coffee breeding assisted by somoclonal variation: case of 'Bourbon LC' variety", *Proceedings of the 19th Colloquium of ASIC*, Trieste, Italy, 2001.
43. S. Zezlina, M. Soranzio, P. Rovelli, M. A. Kriger, M. R. Sondahl, and G. Graziosi, "Molecular characterisation of the cultivar Bourbon LC", *Proceedings of the 18th Colloquium of ASIC*, Helsinki, Finland, 1999, pp. 314–321.
44. P. Barre, S. Akaffou, J. Charrier, A. Hamon, and M. Noirot, *Theor. and Appl. Gen.* **96**(2), 306 (1998).

45. C. Campa, H. Chrestin, A. de Kockko, C. Bertrand, T. Leroy, and M. Noirot, "Towards Identification and Characterisation of Candidate Genes Involved in Coffee Cup Quality", *Proceedings of the 19th Colloquium of ASIC*, Trieste, Italy, 2001.

46. J. Spiral, T. Leroy, M. Paillard, and V. Petiard, *Biotechnol. Agr. Forestry* **44**, 55, (1999).

47. J. I. Stiles Ref. 1, pp. 224–234.

48. A. Pallavicini, L. Del Terra, B. De Nardi, P. Rovelli, and G. Graziosi, "A catalogue of genes expressed in *Coffea arabica* L.", in Ref. 5 *Proceedings of the 19th Colloquium of ASIC*, Trieste, Italy, 2001.

49. P. Marraccini, C. Allard, M. L. André, C. Courjault, C. Garborit, N. Nicolas, A. Meunier, S. Michaux, V. Petit, P. Priyond, J. W. Rogers, and A. Deshayes, "Update on Coffee Biochemical Compounds, Proteins and Gene Expression during Bean Maturation and in Other Tissues.", *Proceedings of the 19th Colloquium of ASIC*, Trieste, Italy, 2001.

50. F. Anthony, B. Bertrand, O. Quiros, A. Wilches, J. Berthaund, and A. Charrier, *Euphytica* **118**, 53, (2001).

51. S. Moisyadi, K. R. Neupane, and J. I. Stiles, *Acta Horticult* **461**, 367, (1997).

52. K. R. Neupane S. Moisyadi, and J. I. Stiles, "Cloning and Characterisation of Fruit-Expressed ACC Synthase and ACC Oxidase from Coffee", in Ref. 42, pp. 322–326.

53. International Pat. WO 98/42848 (Oct. 1, 1998), J. I. Stiles and co-workers (to University of Hawaii).

54. International Pat. WO 98/06852 (Feb. 19, 1998), J. I. Stiles and co-workers (to University of Hawaii).

55. M. Nagata, H. Mori, Y. Tabei, T. Sato, M. Hirai, and H. Imaseki, *Acta Horticult.* **394**, 213, (1995).

56. B. Perthuis, R. Philippe, J. L. Pradon, M. Dufour, and T. Leroy, "Premiéres observations sur la résistance au champ de plants de *Coffea canephora* génétiquement modifiées contre la mineuse des feuilles *Perileucoptra coffeella* Guérin-Méneville", *Proceedings of the 19th Colloquium of ASIC*, Trieste, Italy, 2001.

57. International Pat. No. WO 99/02688 (to Societe de Produits Nestle SA).

58. International Pat. WO 99/02688 (Jan. 21, 1999), P. Marraccini and co-workers (to Societe de Produits Nestle SA).

59. 'Robbing coffee's cradle', ActionAid briefing, 2001 (www.actionaid.org).

60. U.S. Pat. 4904484 (Feb. 27, 1990), L. E. Smith and T. N. Asquith (to Proctor & Gamble Co.).

61. U.S. Pat. 4867992 (Sept. 19, 1989), B. Boniello and co-workers (to General Foods Corp.).

62. U.S. Pat. 4983408 (Jan. 8, 1991), R. Colton.

63. U.S. Pat. 5714183 (Feb. 3, 1998), P. Nicols and co-workers (to Nestec SA).

64. R. J. Clarke in Ref. 1, pp. 235–237.

65. Directive 1999/4/EC of the European Parliament, 1999.

66. B. Schilter, 'Health Benefits of Coffee', *Proceedings of the 19th Colloquium of ASIC*, Trieste, Italy, 2001.

67. J. I. Pitt 'The Importance of Ochratoxin A in Foods: Report of the 56th Meeting of JEFCA', *Proceedings of the 19th Colloquium of ASIC*, Trieste, Italy, 2001.

68. Ref. 2, Vol 5, *Related Beverages*.

GENERAL REFERENCES

R. J. Clarke and O. G. Vitzthum, eds., *Coffee, Recent Developments*, Blackwell Science, London, U.K., 2001.

R. J. Clarke and R. Macrae, eds., *Coffee*, Vol. 1, *Chemistry*, 1985; Vol. 2, *Technology*, 1987; Vol. 3, *Physiology*, 1988; Vol. 4, *Agronomy*, 1988; Vol. 5, *Related Beverages*, 1987; Vol. 6, *Commercial and Technico-Legal Aspects*, 1988; Elsevier Applied Science Publishers, Ltd., Barking, U. K. An excellent reference series.

M. Sivetz and N. Desrosier, *Coffee Technology*, AVI Publishing Co., Inc., Westport, Conn., 1979. Somewhat dated but good reference for the basic coffee processing technology.

GERALD S. WASSERMAN
ALLAN BRADBURY
THEODORE CRUZ
SIMON PENSON
Kraft Foods Corporation

COLLOIDS

1. Introduction

Matter of colloidal size, just above atomic dimensions, exhibits physicochemical properties that differ from those of the constituent atoms or molecules yet are also different from macroscopic material. The atoms and molecules of classical chemistry are extremely small, usually having molar masses <1000 g/mol and measurable by freezing point depression. Macroscopic particles fall into the realm of classical physics and can be understood in terms of physical mechanics. Residing between these extremes is the colloidal size range of particles whose small sizes and high surface area-to-volume ratios make the properties of their surfaces very important and lead to some unique physical properties. Their solutions may have undetectable freezing point depressions, and their dispersions, even if very dilute, may sediment out very slowly, and not be well described by Stokes' law. Whereas the particles of classical chemistry may have one or a few electrical charges, colloidal particles may carry thousands of charges. With such strong electrical forces, complete dissociation is the rule rather than the exception. In addition, the electric fields can strongly influence the actions of neighbouring particles. In industrial practice it is very common to encounter problems associated with colloidal sized particles, droplets, or bubbles.

The field began to acquire its own identity when Graham coined the term colloid in 1861 (1–3). Since that time the language of colloid science has evolved considerably (4–6) and makes two principal distinctions: lyophobic (thermodynamically unstable) and lyophilic (thermodynamically stable) colloidal dispersions. Examples of lyophobic and lyophilic colloidal dispersions are suspensions of gold particles and surfactant micelles in solution, respectively. Colloidal particles (or droplets or bubbles) are defined as those having at least one dimension between ~1 nm and 1 μm. In dealing with practical applications, the upper size limit is frequently extended to tens or even hundreds of micrometres. For example, the principles of colloid science can be usefully applied to emulsions whose droplets

Table 1. **Types of Colloidal Dispersion**

Dispersed phase	Dispersion medium	Name	Examples
liquid	gas	liquid aerosol	fog, mist
solid	gas	solid aerosol	smoke, dust
gas	liquid	foam	soap suds
liquid	liquid	emulsion	milk, mayonnaise
solid	liquid	sol, suspension	ink, paint, gel
gas	solid	solid foam	polystyrene foam, pumice stone
liquid	solid	solid emulsion	opal, pearl
solid	solid	solid suspension	alloy, ruby-stained glass

exceed the 1 µm size limit by several orders of magnitude (ie, in cases for which the surface properties of the dispersed phase dominate). Simple colloidal dispersions are two-phase systems, comprising a dispersed phase of small particles, droplets, or bubbles, and a dispersion medium (or dispersing phase) surrounding them. In modern practice, the terms lyophilic and lyophobic (especially hydrophilic and hydrophobic) are often used to characterize surfaces in addition to colloidal dispersions. This sometimes leads to confusing usage. For example, a clay dispersion in water could be classified as a lyophobic colloid with hydrophilic surfaces.

A variety of types of colloidal dispersions occur, as illustrated in Table 1. In practice, many colloidal dispersions are more complex and are characterized by the nature of the continuous phase and a primary dispersed phase, according to the designations in Table 1.

One reason for the importance of colloidal systems is that they appear in a wide variety of practical disciplines, products, and processes. The colloidal involvement in a process may be desirable, as in the stabilizing of emulsions in mayonnaise preparation, or undesirable, as in the tendency of very finely divided and highly charged particles to resist settling and filtration in water treatment plants. Examples of the variety of practical problems in colloid chemistry include control of filtration operations, breaking of emulsions, regulating foams, preparing catalysts, managing fluid flow, and cleaning surfaces (see Table 2).

The variety of systems represented or suggested by Tables 1 and 2 underscores the fact that the problems associated with colloids are usually interdisciplinary in nature and that a broad scientific base is required to understand them completely. A wealth of literature exists on the topic of colloidal dispersions, including a range of basic reference texts (7–11), dictionaries (4–6,12), and treatises on the myriad of applied aspects, of which only a few are cited here (13–24).

2. Preparation and Stability of Dispersions

2.1. Preparation. Colloidal dispersions can be formed either by nucleation with subsequent growth or by subdivision processes (7,8,11,25,26). The nucleation process requires a phase change, such as condensation of vapor to

Table 2. **Some Occurrences of Colloids**

Field	Liquid aerosol	Solid aerosol	Foam	Emulsion	Suspension	Solid foam	Solid emulsion	Solid suspension
environment and meteorology	fog, mist, cloud, smog	volcanic smoke, dust, smog	polluted river foams	water/sewage treatment emulsions, oil spill emulsions	river water, glacial runoff			
foods			champagne, soda and beer heads, whipped cream, meringue	milk, butter, mayonnaise, cheese, cream liqueurs	jellies	leavened breads		
geology, agriculture, and soil science	crop sprays		foam fumigant, insecticide and herbicide blankets	insecticides and herbicides	quicksand, clay soil suspensions	pumice stone, zeolites	opal, pearl	pearl
manufacturing and materials science	paint sprays	sand blasting	foam fractionation, pulping black liquor foam	polishes	ink, gel, paints, fiber suspensions	polystyrene foam, polyurethane foam	high impact plastics, alkaline battery fill	stained glass, ceramics, cement, plastics, catalysts, alloys, composites
biology and medicine	nasal sprays	airborne pollen, inhalant drugs	vacuoles, insect excretions, contraceptive foam	soluble vitamin and hormone products, biological membranes, blood	liniment suspensions, proteins, viruses	loofah plant		wood, bone

273

Table 2 (*Continued*)

Field	Liquid aerosol	Solid aerosol	Foam	Emulsion	Suspension	Solid foam	Solid emulsion	Solid suspension
petroleum production and mineral processing			refinery foams, flotation froths, fire extinguishing foams, explosion suppressant foam	oilfield emulsions, asphalt emulsion	drilling fluids, drill cuttings, mineral slurries, process tailings		oil reservoir	
home and personal care products	hair spray		shampoo suds, shaving cream, contraceptive foams, bubble bath foam	hair and skin creams and lotions		sponges, carpet underlay, cellular foam insulation		bakelite products

274

Table 3. **Industrially Produced Colloidal Materials and Related Processes**

Mechanism	Examples[a]
vapor→liquid→solid ↓→→→↑	oxides, carbides via high intensity arc; metallic powders via vaccum or catalytic reactions
vapor + vapor →solid	chemical vapor deposition, radio frequency-induced plasma, laser-induced precipitation
liquid→solid	ferrites, titanates, aluminates, zirconates, molybdates via precipitation
solid→solid	oxides, carbides via thermal decomposition

[a] Ref. 27,28.

yield liquid or solid, or precipitation from solution. Some mechanisms of such colloid formation are listed in Table 3. The subdivision process refers to the comminution of particles, droplets, or bubbles into smaller sizes by applying high shearing forces, using devices such as a propeller-style mixer, colloid mill or ultrasound generator. A complex technology has developed to conduct and to control comminution and size-fractionation processes. Mathematical models are available to describe changes in particle size distribution during comminution, but these are generally restricted to specific processes. Comprehensive reviews of the developments in preparing colloidal solids by subdivision should be consulted for further details (27,28) (see POWDERS, HANDLING, DISPERSION OF POWDERS IN LIQUIDS). A wide range of techniques is now available including, eg, atomizers and nebulizers of various designs used to produce colloidal liquid or solid aerosols, and emulsions having relatively narrow size distributions. Monosized powders and monodispersed colloidal sols are frequently used in many products, eg, pigments, coatings, and pharmaceuticals.

Colloidal suspensions of uniform chemical and phase composition, particle size, and shape are now available for many elements (including sulfur, gold, selenium, and silver, carbon, cobalt, and nickel), many inorganic compounds (including halide salts, sulfates, oxides, hydroxides, and sulfides), and many organic compounds [including, poly(vinyl acetate), polystyrene, poly(vinyl chloride), styrene-butadiene rubber, poly(acrylic acid), polyurea, poly(styrene)- poly(acrylate), and poly(methacrylate)-poly(acrylate)].

2.2. Stability. A complete characterization of colloid stability requires consideration of the different processes through which dispersed species can encounter each other: sedimentation (creaming), aggregation and coalescence. Sedimentation results from a density difference between the dispersed and continuous phases and produces two separate layers of dispersion that have different dispersed phase concentrations. One of the layers will contain an enhanced concentration of dispersed phase, which may promote aggregation. Aggregation is when two or more dispersed species clump together under the influence of Brownian motion, sedimentation, or stirring, possibly touching at some points, and with virtually no change in total surface area. Aggregation is sometimes referred to as flocculation or coagulation (although in specific situations these latter terms can have slightly different meanings). In aggregation, the species retain their identity but lose their kinetic independence since the aggregate moves as a single unit. Aggregation of droplets may lead to coalescence and the formation of larger droplets until the phases become separated. In coalescence thin film drainage occurs, leading to rupture of the separating film, and two or more particles, droplets or bubbles fuse together to form a single larger unit, reducing the total surface area. In this case the original species lose their identity and become part of a new species. In emulsions, inversion can take place, in which the emulsion suddenly changes form, from oil-in-water (O/W) to water-in-oil (W/O), or vice versa. For example, butter results from the creaming, breaking, and inversion of emulsified fat droplets in milk. Kinetic stability can thus have different meanings. A colloidal dispersion can be kinetically stable with respect to coalescence but unstable with respect to aggregation. Or, a system could be kinetically stable with respect to aggregation but unstable with respect to sedimentation. In summary, lyophobic colloids are thermodynamically unstable, but may be relatively stable in a kinetic sense, and it is crucial that stability be understood in terms of a clearly defined process.

3. Dispersed Species Characterization and Sedimentation

The characterization of colloids depends on the purposes for which the information is sought because the total description would be an enormous task. Among the properties to be considered are the nature and/or distributions of purity, crystallinity, defects, size, shape, surface area, pores, adsorbed surface films, internal and surface stresses, stability, and state of agglomeration (27,28).

3.1. Surface Area, Porosity, and Permeability.
Some very interesting and important phenomena involve small particles and their surfaces. For example, SO_2 produced from mining and smelting operations that extract metals such as Cu and Pb from heavy metal sulfide ores can be oxidized to SO_3 in the atmosphere, thus contributing to acid rain problems. The reaction rate depends not only on the concentration of the SO_2 but also on the surface area of any catalyst available, such as airborne dust particles. The efficiency of a catalyst depends on its specific surface area, defined as the ratio of surface area to mass (17). The specific surface area depends on both the size and shape and is distinctively high for colloidal sized species. This is important in the catalytic processes used in many industries for which the rates of reactions occurring at the catalyst surface depend not only on the concentrations of the feed stream reactants but also on the surface area of catalyst available. Since practical catalysts are frequently supported catalysts, some of the surface area is more important than the rest. Since the supporting phase is usually porous the size and shapes of the pores may influence the reaction rates as well. The final rate expressions for a catalytic process may contain all of these factors: surface area, porosity and permeability.

The total surface area can be estimated by measuring the amount of gas needed to form an adsorbed monolayer by physical adsorption (29,30), in which case the number of molecules adsorbed divided by the area per molecule yields the surface area. The classic models for physical adsorption are those of Langmuir (monomolecular adsorption and constant ΔH_{ads}, independent of the extent of surface coverage) and Brunauer, Emmett, and Teller (BET, multilayer adsorption and several ΔH_{ads} components); while many other models are available as well (8,10,17). These models require a knowledge of the area each molecule occupies on the surface. Chemisorption can be studied in the same fashion as described for nitrogen adsorption but using a gas that is chemisorbed, ie, that bonds chemically. This yields a specific chemisorption surface area. In evaluating supported catalyst samples, both kinds of surface area may need to be measured since particle size changes would be reflected in the specific surface area while site deactivation might change only the specific chemisorption area.

The Langmuir and BET equations work well with nonporous solids, but not as well for porous solids because the pores influence the local numbers of adsoprtion layers formed. By using adsorption gases of different molecular size or by varying the temperature, pores of different size will be accessible to the adsorbing molecule. Coupling this with the appropriate mathematical interpretation allows for the determination of solid porosity using BET analysis. The porosity of a solid (pore volume divided by bulk volume) is most easily determined by the imbibition method, frequently used in the petroleum industry. A sample of solid is dried and weighed, then saturated with a wetting liquid (often water or heptane) under vacuum. The pore volume accessible is calculated by a material balance. In prepared catalysts, the pore sizes may be quite uniform. However, in most naturally occurring materials there is a wide range of pore sizes. The actual pore size distribution can be obtained from methods such as porosimetry, in which a nonwetting liquid (usually mercury) is pumped into a solid sample (7,8,10,17,29,31). According to the Laplace equation each increment of applied pressure will cause only pores down to a certain size to be filled, and employing a series of pressure increments allows the pore size distribution to be obtained.

The ease with which a fluid can flow through a porous medium, permeability, can be determined through the measurement of pressure drop (Δp) across the porous medium under steady flow. The intrinsic permeability (k) is defined by Darcy's law and is given by $k = (Q/A)(\eta L/\Delta p)$, where Q is the discharge flow rate, A is the cross-sectional area normal to the main flow direction, η is the flowing fluid viscosity, L is the length of the flow path (sample), and Δp is the pressure gradient along the medium (15). Mercury porosimetry can also be used to assess permeability (17,29).

3.2. Size and Size Distribution. If the sizes of particles, droplets, bubbles, or their aggregates in a colloidal dispersion are large enough, then optical microscopy can be used to determine the shape, size, and size distribution, but if they are smaller then ~0.5 µm they will not be resolved in a typical optical microscope. Confocal scanning laser microscopes can extend this resolution down to ~0.1 µm. For dispersions of such smaller-size species, the most direct methods include scanning and transmission electron microscopy. Adaptations, such as cryogenic-stage scanning electron microscopy can be used for emulsions and foams (32,33).

Small, dispersed particles or droplets cause the "cloudy" or "milky" appearance of such diverse colloids as dust clouds, rain clouds, suspended sediment in a river, and milk. This appearance is due to light scattering. When a beam of light enters a colloidal dispersion some light is absorbed, some is scattered, and some is transmitted. The intensity of the scattered light depends largely on the size and shape of the colloidal species, and on the difference in refractive index between the phases. Light scattering analysis is very powerful because it can yield the complete dispersed-phase size distribution. This is important in emulsions that are commonly, but generally incorrectly, characterized in terms of a specified droplet size whereas there is inevitably a size distribution. The theory underlying the determination of size distribution for a colloidal dispersion is quite involved (8,34). Rayleigh theory predicts that larger particles scatter more light than do smaller ones. Since the scattering intensity is proportional to $1/\lambda^4$ then blue light ($\lambda = 450$ nm) is scattered much more than red light ($\lambda = 650$ nm). With incident white light a scattering material will, therefore, tend to appear blue when viewed at right-angles to the incident light beam, and red when viewed end-on. Thus the sky can appear blue overhead while the sun appears yellowish-red when viewed across the horizon as it is rising or setting. When a test tube containing a dilute suspension of particles so small that they would be invisible under the light microscope is held up to the light it may appear to have a blue colour due to Rayleigh scattering. This phenomenon provides a way to indirectly observe particles that would otherwise be invisible. In the darkfield microscope, or ultramicroscope, the light scattered by small particles is viewed against a dark background. This method is applied in observing the electrophoretic motions of colloidal particles (35).

Other indirect techniques for determining colloidal species' size or size distribution include sedimentation–centrifugation, conductometric techniques, X-ray diffraction, gas and solute adsorption, ultrafiltration, diffusiometric, and ultrasonic methods (7,8,31,36). Care must be taken in selecting an indirect method since these require assumptions about either the real size distribution, the shape, or the process on which the analysis is based. For example, conductometric "sensing zone" equipment relies on the assumption of sphericity, which is reasonable for emulsion droplets but may not hold for particles in a suspension. Similarly, light-scattering techniques are reliable only if the particle shape and refractive index is known or assumed, and adsorption analyses rely on model adsorption isotherms, the uniformity of particle size and porosity, and the orientation of adsorbed species. In all cases, one must be careful that sample preparation techniques do not change the size distribution (32). Typically, more than one method is needed to characterize size and/or size distribution properly (see SIZE MEASUREMENT OF PARTICLES).

Nearly all colloidal systems undergo some aggregation leading to a distribution of aggregate sizes. Ultramicroscopy is the preferred method for measuring the rate and/or extent of aggregation because it is direct. The indirect methods listed above can also be used when tailored to suit the specific colloidal system in hand. More than one technique is required to assess the state of aggregation when a wide range of colloidal dimensions exists. If aggregates larger than approximately 5 μm are present, the aggregate-size distribution can be evaluated using classical techniques such as sieving (27,28,37), as long as such methods do not themselves induce or break aggregates.

3.3. Sedimentation. Whereas the use of light scattering to determine a complete size range distribution is quite involved, if the particles or droplets are not too small, a simpler approach can be used that yields an approximate average size. This is done by measuring settling velocities. Consider the small particles in a dust cloud, or suspended sediment in a river. It is principally the small size that keeps these particles from rapidly settling out. If a particle or droplet is placed in a fluid it will fall, or sediment out, if its density is greater than that of the fluid. The driving force is that of gravity. In the Stokes model, the terminal settling velocity is proportional to gravity and the square of particle size and inversely proportional to the fluid viscosity (7,8,31). This assumes that the species is uncharged and spherical, the situation being more complicated for charged and/or asymmetric particles. Further, if the particle concentration is high then the particles don't sediment independently but are influenced by the motions of surrounding particles producing slower, hindered, settling. Sedimentation under gravity is only practical down to ~ 1 μm diameter, but the centrifuge or ultracentrifuge can be used to study sedimentation of colloidal systems since the added centrifugal forces can be employed to overcome the mixing tendencies of diffusion and convection. Centrifugal force, like gravitational force, is proportional to the mass but the coefficient is not the acceleration due to gravity (g) but rather the square of the angular velocity (ω) times the distance of the particle from the axis of rotation (x). Since $\omega^2 x$ is substituted for g in the governing equation, one speaks of multiples of g in a centrifuge. For example, using a conventional laboratory bench-top centrifuge capable of applying thousands of "gs" one can reduce the time needed to sediment out 0.2 μm particles, from an aqueous suspension, to ~ 20 min compared with the 16 days that would be needed to achieve the same sedimentation from a standing column. In an ultracentrifuge, even greater centrifugal forces ($\sim 40,000\ g$) can be employed.

4. Rheology

For calculations involving pumping and mass transfer, industrial process streams can sometimes be treated as simple, "single phase" fluids that obey Newton's law of viscosity, $\tau = \eta\dot{\gamma}$, in which the shear stress, τ, is given as a linear function of the shear rate, $\dot{\gamma}$, with the proportionality constant being the viscosity, η. In fact many industrial process streams occur as colloidal dispersions, introducing the complication that in many cases viscosity is not expressed by a single number at constant temperature and pressure, but also depends on whether the material is flowing, and even its recent history (see RHEOLOGICAL MEASUREMENTS). Due to polydispersity, high dispersed phase content, mutual orienting and/or structure formation of the dispersed species under flow a non-Newtonian dispersion exhibits a viscosity that is not constant, but is itself a function of the shear rate, [ie, $\tau = \eta\ (\dot{\gamma})\ \dot{\gamma}$]. The function itself can take many forms (38–43). Many different terms are used to express specific kinds of viscosities, including absolute, apparent, differential, intrinsic, reduced, relative, and inherent viscosity. These are defined elsewhere, as is an entire lexicon of terms used to describe the different rheological classifications of colloidal dispersions (4–6,39,42). Typical rheological classifications are listed in Table 4. Some descriptions appropriate to different yield stresses and some approximate values of shear rate appropriate to various industrial processes are given in reference (44).

　　A colloidal system can exhibit several of these characteristics at once. For example, paint must be plastic and thixotropic so that it will flow when brushed on and (only) immediately after brushing (for a smooth finish); a further benefit

Table 4. **Typical Rheological Classifications**

Rheological classification	Description	Examples
pseudoplastic (shear-thinning)	as shear rate increases viscosity decreases.	in paint, a suspension of pigment particles in a liquid, irregular particles can align to match the induced flow, lowering the viscosity.
dilatant (shear-thickening)	as shear rate increases viscosity increases	in the "drying" of wet beach sand when walked on, a dense packing of particles occurs. Under low shear the particles can move past each other, whereas under high shear the particles wedge together such that the fluid can't fill the increased void volume.
pseudoplastic with yield stress (plastic)	pseudoplastic or Newtonian flow begins only after a threshold shear stress, the yield stress, is exceeded.	in an oil well drilling mud the interparticle network offers resistance to any positional changes. Flow only occurs when these forces are overcome.
thixotropic	time-dependent pseudoplastic flow. At constant applied shear rate viscosity decreases. In a flow curve hysteresis occurs.	in bentonite clay "gels" which "liquefy" on shaking and "solidify" on standing there is a time-dependent aligning to match the induced flow. After the shear rate is reduced it takes some time for the original alignments to be restored.
rheopectic	time-dependent dilatant flow. At constant applied shear rate viscosity increases. In a flow curve hysteresis occurs.	a suspension that sets slowly on standing but quickly when gently agitated due to time-dependent particle interference under flow.
rheomalaxic	time-dependent behavior in which shear rate changes cause irreversible changes in viscosity.	an emulsion that when sheared inverts to a higher (or lower) viscosity emulsion, and does not reinvert when the shear is removed.

is that vigorous mixing readily disperses the pigments which then stay dispersed for some time when standing (high yield stress). Finally, shortly after brushing-on the paint should cease to flow so that it doesn't "run".

A range of methods are available for making rheological measurements (39–42) (see RHEOLOGICAL MEASUREMENTS). A frequently encountered problem involves knowing the particle–droplet–bubble size and concentration in a dispersion and the need to predict the suspension, emulsion or foam viscosity. Many equations have been advanced for this purpose. In the simplest case, a colloidal system can be considered Einsteinian. Here, the viscosity of the colloidal system depends on that of the continuous phase, η_0, and the volume fraction of colloid, ϕ, according to the Einstein equation, which was derived for a dilute suspension of noninteracting spheres:

$$\eta = \eta_0 \left(1 + 2.5 \; \phi\right)$$

This relationship forms the basis for the use of volume fraction as the theoretically favored concentration unit in rheology. In practice once ϕ reaches between 0.1 and 0.5 dispersion viscosity increases and can also become non-Newtonian (due to particle/droplet/bubble "crowding," or structural viscosity). The maximum volume fraction possible for an internal phase made up of uniform, incompressible spheres is 0.74, although emulsions and foams with an internal volume fraction of >0.99 can exist as a consequence of droplet–bubble distortion.

Many empirical and theoretical modifications have been made to Einstein's equations. A useful extension to dilute suspensions of anisotropic particles, such as clays, is given by the Simha Equation, which is approximately

$$\eta = \eta_0 \, (1 + a\phi/1.47b)$$

where a is the major particle dimension and b the minor particle dimension. Many of the other viscosity equations are empirical extensions of Einstein's equation for a dilute suspension of spheres, including virial expansions such as

$$\eta = \eta_0 \, (1 + \alpha_0\phi + \alpha_1\phi^2 + \alpha_2\phi^3 + \cdots)$$

These equations usually apply if the particles or droplets are not too large, and if there are no strong electrostatic interactions. Additional equations are tabulated elsewhere (6,45).

Size distribution also has an important influence on viscosity. For electrostatically or sterically interacting drops, emulsion viscosity will be higher when droplets are smaller. The viscosity will also be higher when the droplet sizes are relatively homogeneous, ie, when the drop size distribution is narrow rather than wide. The rheological properties also depend on any specific interactions among the colloidal species, the dispersing medium, and the solute additives, ie, salts, surfactants, and polymers.

There are many important influences of rheology in industrial practice. From Stokes' law the terminal settling velocity is inversely proportional to the viscosity of a colloidal dispersion (see SEDIMENTATION), which has a direct impact on sedimentation in, eg, treatment of waste water (see WATER, MUNICIPAL WATER TREATMENT) and on mineral fractionation and/or flotation (47,48). Another major application area is transport behavior, involving the pumping of fluid systems containing colloids, such as in extrusion in the polymer industry, the processing of gelatinous foods and cosmetic items, the fabrication of high performance materials in the ceramic and metallurgical industries, transportation in the petroleum industry, and the preparation and handling of pigment slurries in the paint industry. The prediction and control of suspension, emulsion, and foam rheology, especially the thixotropic and dilatant tendencies, is primarily important for these and other uses.

5. Interfacial Energetics

In colloidal dispersions, a thin intermediate region or boundary, known as the interface, lies between the dispersed and dispersing phases. Each interface has a certain free energy per unit area that has a great influence on the stability and structure of the colloidal dispersion, and that has a great influence in practical areas such as mineral flotation, detergency and waterproofing.

5.1. Surface and Interfacial Tensions. For a liquid exposed to a gas the attractive van der Waals forces between molecules are felt equally by all molecules except those in the interfacial region. This makes the latter molecules tend to move to the interior and causes the interface to contract spontaneously. This is the reason droplets of liquid and bubbles of gas tend to adopt a spherical shape. For two immiscible liquids a similar situation applies, except it may not be so immediately obvious how the interface will tend to curve but there will still be an imbalance of intermolecular forces and a configuration that minimizes the interfacial free energy, or interfacial tension. Surface tension can be thought of as either the contracting force around the perimeter of a surface, or as the surface free energy associated with area change. Emulsions and foams represent colloidal systems in which interfacial properties are very important because emulsified droplets and dispersed gas bubbles have large interfacial areas, so that even a modest interfacial energy per unit area can become a considerable total interfacial energy. There are many methods available for the measurement of surface and interfacial tensions (46,49–51).

5.2. Pressure and Curved Surfaces. Interfacial tension causes a pressure difference to exist across a curved surface such that the pressure inside a bubble or drop exceeds that outside. The pressure difference is given, in terms of the principal radii of curvature and surface or interfacial tension, by the Young-Laplace equation (8,10). For spherical droplets of liquid in a gas, $\Delta p = 2\gamma/R$, ie, the pressure difference, Δp, varies inversely with the radius, R. Thus the vapor pressure of a drop should become higher as the drop becomes smaller. This is shown by the Kelvin equation (8,31), which gives the pressure, at equilibrium, above a spherical surface of given radius, r, and surface tension, γ, as $RT \ln (p/p_0) = 2V_L\gamma/r$, where p_0 is the normal vapor pressure and V_L is the molar volume of the liquid. By replacing pressure with activity of dissolved solute and relating activity, in turn, to molar solubility the Kelvin equation can be used to describe a number of supersaturation phenomena including supercooled vapors and supersaturated solutions.

5.3. Contact Angle and Wettability. When a drop of liquid is placed on a solid surface the liquid may form a bead on the surface, or it may spread to form a film. A liquid having a strong affinity for the solid will seek to maximize its contact (interfacial area) and form a film. A liquid with much weaker affinity may form into a bead. This affinity is termed the wettability (10,25,26). To account for the degree of spreading, the contact angle, θ, is defined as the angle, measured through the liquid, that is formed at the junction of three phases, eg, at the S-L-G junction. Whereas interfacial tension is defined for the boundary between two phases, contact angle is defined for a three-phase junction. If the interfacial forces acting along the perimeter of the drop are represented by the interfacial tensions, then an equilibrium force balance is given by Young's equation as, $\gamma_{L/G} \cos \theta = \gamma_{S/G} - \gamma_{S/L}$. The solid is completely wetted if $\theta = 0$ and only partially wetted otherwise. Although in theory complete nonwetting would be $\theta = 180°$ this is not seen in practice and values of $\theta > 90°$ are considered to represent "nonwetting" whereas values of $\theta < 90°$ are often considered to represent "wetting". This rather arbitrary assignment is based on correlation with the visual appearance of drops on surfaces.

An example is provided in enhanced oil recovery. In an oil-bearing reservoir the relative oil and water saturations depend on the distribution of pore sizes in the rock, the pressure in a pore, the interfacial tension and contact angle according to the Young-Laplace and Young equations. The same relationships determine how water or other fluids can be injected to change pressure, interfacial tension, and/or contact angle and thereby change the liquid saturations and increase oil recovery (13,52).

Some compounds, like short-chain fatty acids, can be soluble in both water and oil because one part of the molecule has an affinity for oil (the nonpolar hydrocarbon chain) and one part has an affinity for water (the polar group). The energetically most favorable orientation for these molecules is at an interface so that each part of the molecule can reside in the solvent medium for which it has the greatest affinity. These molecules that form oriented monolayers at interfaces show surface activity and are termed surfactants (see SURFACTANTS). Some consequences of surfactant adsorption at a surface are that it causes a reduction in surface tension and an alteration in the wettability of the surface. Surfactant molar masses range from a few hundreds up to several thousands of g/mol.

Surfactants can be used to selectively alter wettability. For example, in mineral flotation surfactant can be added to adsorb on metal ore particles increasing the contact angle, so they attach to gas bubbles. The surfactant is chosen so that it will not adsorb much on silicates, so the latter do not attach to gas bubbles. The surfactant may also stabilize a foam containing the desired particles, thereby facilitating their recovery as a particle-rich froth that can be skimmed. Flotation processes thus involve careful modification of surface tension and wettability.

Another problem in colloids is detergency, which involves the action of surfactants (originally soaps were used) to alter interfacial properties so as to promote dirt or oil removal from solid surfaces. The detergent's role is to alter interfacial tensions in order to reduce the amount of mechanical energy required to dislodge the dirt. If the dirt is solid then it is a simple matter of wettability alteration.

Surfactants play an important role in the formation and stability of emulsions and foams. Surfactant adsorption at fluid interfaces can, eg, lower interfacial tension, increase surface elasticity, increase electric double-layer repulsion (ionic surfactants), lower the effective Hamaker constant, and sometimes increase surface viscosity. For emulsions, the nature of the surfactant can also determine the arrangement of the phases (ie, which phase will form the dispersed versus continuous phase). Several empirical rules and scales have been developed for categorizing emulsifying agents, including the oriented wedge and Bancroft theories, the hydrophile–lipophile balance, HLB, and the phase inversion temperature, PIT (13,26,45). Although there are exceptions to each of these rules, they remain useful for making initial predictions.

5.4. Lyophilic Colloids. Another key feature of surfactants is that above a certain solution concentration they form organized aggregates called micelles (19,26) in which the lipophilic parts of the surfactants associate in the interior of the aggregate, leaving hydrophilic parts to face the aqueous medium. A solution of micelles is a good example of a thermodynamically stable lyophilic colloidal dispersion. The concentration at which micelle formation becomes significant is called the critical micelle concentration (CMC). The CMC is a property of the surfactant and several other factors, since micellization is opposed by thermal and electrostatic forces. A low CMC is favoured by increasing the lipophilic part of the molecule, lowering the temperature, and adding electrolyte. Compilations of CMC values are given in references (19,53).

The ability of biological amphiphilic molecules to aggregate into spherical and nonspherical clusters, ie, vesicles, may have been important for the development of early living cells (54). Cellular biological membranes in plants and animals share features with these colloidal systems, although the molecular and hierarchical membrane structures, their hydration, and their dynamic properties are complex (54–56). The macroscopic nature of concentrated gels, such as lubricating greases formed by dispersing short-chain surfactants, eg, lithium 12-hydroxystearate, in mineral oil (57), is akin to the behavior of biological amphiphiles, being also dependent on self-assembly mechanisms. The associations between fibrous clusters, the length of threadlike surfactant strands, and the density of their contact points (cross-links) govern the grease's shear-resistance (57).

Microemulsions, like micelles, are considered to be lyophilic, stable, colloidal dispersions. In some systems, the addition of a fourth component, a cosurfactant, to an oil–water–surfactant system can cause the interfacial tension to drop to near-zero values, easily on the order of $10^{-3} - 10^{-4}$ mN/m, allowing spontaneous or nearly spontaneous emulsification to very small drop sizes, $\sim$10 nm or smaller. The droplets can be so small that they scatter little light, the emulsions appear to be transparent and do not break on standing or centrifuging. Unlike coarse emulsions, microemulsions are thought to be thermodynamically stable. The thermodynamic stability is frequently attributed to transient negative interfacial tensions, but this, and the question of whether microemulsions are really lyophilic or lyophobic dispersions are areas of some discussion in the literature. As a practical matter, microemulsions can be formed, have some special qualities, and can have important applications in areas such as enhanced oil recovery, soil and aquifer remediation, foods, pharmaceuticals, cosmetics, herbicides and pesticides (13,16,45,58–60).

6. Electrokinetics

6.1. Charged Interfaces. Most substances acquire a surface electric charge when brought into contact with a polar medium such as water. The origin of the charge can be ionization, as when carboxyl and/or amino functionalities ionize when proteins are put into water; ion adsorption, as when surfactant ions adsorb onto a solid surface; ion dissolution, as when Ag^+ and I^- dissolve unequally when AgI is placed in water; or ion diffusion, as when a clay particle is placed in water and the counterions diffuse out to form an electric double layer. In the AgI example, the ions Ag^+ and I^- will be potential determining because either may adsorb at the interface and change the surface potential. Conversely, indifferent ions, such as Na^+ and NO_3^-, will not change the surface potential.

Surface charge influences the distribution of nearby ions in a polar medium: ions of opposite charge (counterions) are attracted to the surface while those of like charge (coions) are repelled. Together with mixing caused by thermal motion a diffuse electric double layer is formed. The electric double layer (EDL) can be viewed as being composed of two layers: an inner layer that may include adsorbed ions and a diffuse layer where ions are distributed according to the influence of electrical forces and thermal motion. Gouy and Chapman proposed a simple quantitative model for the diffuse double layer assuming an infinite, flat, uniformly charged surface, and ions that can be regarded as point charges (also that solvent effects arise only through a uniform dielectric constant, and that the electrolyte is symmetrical [z-z]). The surface potential is designated as ψ° and the potential at a distance x as ψ. The Poisson-Boltzmann equation comes from a combination of the Boltzmann distribution of concentrations of ions (in terms of potential), the charge density at each potential (in terms of the concentration of ions), and the Poisson equation (describing the variation in potential with distance). Given the physical boundary conditions, assuming low surface potentials, and using the Debye-Hückel approximation yields,

$$d^2\Psi/dx^2 = [e^2\Psi/(\in kT)]\Sigma_i c_i z_i^2$$

If we now define the cluster of constants as $\kappa^2 = (e^2/\epsilon kT)\Sigma_i c_i z_i^2$ (having units of distance^{-2}) then this can be simplified to

$$d^2\Psi/dx^2 = \kappa^2\Psi \qquad \text{or} \qquad \Psi = \Psi^\circ \exp(-\kappa x)$$

This is the Debye-Hückel expression for the potential at a distance from a charged surface. The parameter $1/\kappa$ is called the double layer thickness and is given for water at 25°C by $\kappa = 3.288 \sqrt{I}$ (nm^{-1}), where $I = (1/2)\Sigma_i c_i z_i^2$. For 1-1 electrolyte $1/\kappa$ is 1 nm for $I = 10^{-1}\ M$, and 10 nm for $I = 10^{-3}\ M$.

There remain some problems. In order to handle higher (more practical) potentials, the Gouy-Chapman theory can be applied, but with a more complicated result. One such result is

$$\Upsilon = \Upsilon^\circ \exp(-\kappa x)$$

where Υ is a complex ratio involving Ψ given by

$$\Upsilon = \frac{\exp[ze\Psi/2kT] - 1}{\exp[ze\Psi/2kT] + 1}$$

At low surface potentials this equation reduces to the Debye-Hückel expression above. Second, an inner layer exists because ions are not really point charges and an ion can only approach a surface to the extent allowed by its hydration sphere. The Stern model incorporates a layer of specifically adsorbed ions bounded by a plane—the Stern plane. In this case, the potential changes from ψ° at the surface, to $\psi(\delta)$ at the Stern plane, to $\psi = 0$ in bulk solution.

6.2. Electrokinetic Phenomena. Electrokinetic motion occurs when the mobile part of the electric double layer is sheared away from the inner layer (charged surface). There are several types of electrokinetic measurements, electrophoresis, electroosmosis, streaming potential, sedimentation potential, and two electroacoustical methods. The first four methods are described in references (35,61). Of these the first finds the most use in industrial practice. The electroacoustical methods involve detection of the sound waves generated when dispersed species are made to move by an imposed alternating electric field, or vice versa (62). In all of the electrokinetic measurements either liquid is made to move across a solid surface or vice versa. Thus the results can only be interpreted in terms of charge density (σ) or potential (zeta potential, ζ) at the plane of shear. The location of the shear plane is generally not exactly known and is usually taken to be approximately equal to the potential at the Stern plane, $\zeta \approx \psi(\delta)$. Several methods can be used to calculate zeta potentials (11,35,61).

7. Colloid Stability

A consequence of the small size and large surface area in colloids is that quite stable dispersions of these species can be made. That is, suspended particles may not settle out rapidly and droplets in an emulsion or bubbles in a foam may not coalesce quickly. Charged species, when sedimenting, present a challenge to Stokes' law because the smaller counterions sediment at a slower rate than do the larger colloidal particles. This creates an electrical potential that tends to speed up the counterions and slow down the particles. At high enough electrolyte concentrations the electric potentials are quickly dissipated and this effect vanishes.

Although some lyophobic colloidal dispersions can be stable enough to persist for days, months, or even years, they are not thermodynamically stable. Rather, they possess some degree of kinetic stability, and one must consider the degree of change and the timescale in the definition of stability. Having distinguished coalescence and aggregation as processes in which particles, droplets, or bubbles are brought together with and without large changes in surface area respectively, it is clear that there can be different kinds of kinetic stability. Finally, stability depends upon how the particles interact when this happens, since encounters between particles in a dispersion can occur frequently due to diffusion (as in Brownian motion), sedimentation, or stirring.

7.1. Electrostatic and Dispersion Forces. Several repulsive and attractive forces operate between colloidal species and determine their stability (7,8,10,25,31,54). In the simplest example of colloid stability, particles would be stabilized entirely by the repulsive forces created when two charged surfaces approach each other and their electric double layers overlap. The overlap causes a Coulombic repulsive force acting against each surface, and that will act in opposition to any attempt to decrease the separation distance. One can thus express the Coulombic repulsive force between plates as a potential energy of repulsion. There is another important repulsive force causing a strong repulsion at very small separation distances where the atomic electron clouds overlap, called the Born repulsion.

There also exist dispersion, or London-van der Waals forces that molecules exert towards each other. These forces are usually attractive in nature and result from the orientation of dipoles, whether dipole–dipole (Keesom dispersion forces), dipole–induced dipole (Debye dispersion forces), or induced dipole–induced dipole (London dispersion forces). Except for quite polar materials the London dispersion forces are the more significant of the three. For molecules, the force varies inversely with the sixth power of the intermolecular distance, whereas for particles, etc, the force varies approximately inversely with interparticle distance.

7.2. DLVO Theory. Derjaguin and Landau (63), and Verwey and Overbeek (64) developed a quantitative theory for the stability of lyophobic colloids referred to as DLVO theory. It was known from experiment that classical colloids (AgI, Au) coagulated quickly at high electrolyte concentrations and slowly at low concentrations with a very narrow electrolyte concentration range over which the transition from kinetically stable to kinetically unstable occurred. Thus a critical coagulation concentration (CCC) could be defined. Using DLVO theory one can calculate the energy changes that take place when two particles approach each other by estimating the potential energies of attraction (London-van der Waals dispersion, V_A) and repulsion (electrostatic including Born, V_R) versus interparticle distance. These are then added together to yield the total interaction energy, V_T. The theory has been developed for two special cases, the interaction between parallel plates of infinite area and thickness, and the interaction between two spheres. The original calculations of dispersion forces employed a model due to Hamaker although more precise treatments now exist (54).

The parameter V_R decreases exponentially with separation distance having a range about equal to κ^{-1} while V_A decreases inversely with separation distance. Figure 1 shows how van der Waals forces can predominate at small and large interparticle distances. Repulsive forces can predominate at extremely small (Born) and intermediate (electric double layer) separation distances. If the colloidal species are charged and have an interfacial potential of $\sim$25–50 mV, the DLVO model predicts for binary particle interactions that a substantial repulsive potential energy barrier will inhibit the close approach of the particles, thereby stabilizing them against aggregation (V_{max} in Fig. 1). The primary maximum usually ensures stability, if its magnitude (V_{max}) exceeds the range 10–15 kT; smaller barriers lead to irreversible aggregation in the primary minimum (7,8,11,31,35,63,64). The secondary minimum (V_{min}) can promote a loose, easily reversible aggregation of particles, if its magnitude is on the order of 10 kT or more (54). In clay colloids, this is part of the explanation for flocculation as distinguished from coagulation. The overall energy barrier to coagulation in the primary minimum is given by $V_{barrier} = V_{max} - V_{min}$, where the primary minimum represents the potential energy at contact. Stability ensues if the magnitude of $V_{barrier}$ exceeds 10–15 kT and is therefore large compared to the thermal energy of the particles. Note that the primary minimum is a finite number because of the contributions that come into play for very small particle separations, ie, the Born repulsion (54).

The classic DLVO models are for flat planes and spheres, but more complex shapes arise in practice. For example, there will be some distortion of originally spherical emulsion droplets as they approach each other and begin to seriously

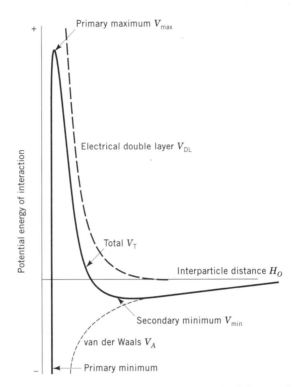

Fig. 1. Potential energies of interaction between two colloidal particles as a function of their distance of separation, for electrical double layers due to surface charge (V_{DL}), London-van der Waals dispersion forces (V_A), and the total interaction (V_T).

interact, causing a flattening. The model has been extended to systems with particles that differ in size, shape, and chemical composition (63), and to those with particles that have an adsorbed layer of ions (7,8,10,11,31,35,63,64), as depicted in Figure 2.

The role of electrostatic repulsion in the stability of suspensions of particles in nonaqueous media is not entirely clear. In attempting to apply DLVO theory it can be difficult to judge the electrical potential at the surface, the appropriate Hamaker constant, and the ionic strength to be used for the nonaqueous medium. The ionic strength will be low so the electric double layer will be thick, the electric potential will vary slowly with separation distance, and so will the net electric potential as the double layers overlap. As a result the repulsion between particles can be expected to be weak (70).

Returning to aqueous systems, it will be apparent that the DLVO calculations can become quite involved, requiring considerable foreknowledge about the systems of interest. There are empirical "rules of thumb" that can be used to give a first estimate of the degree of colloidal stability a system is likely to have if the Zeta potential is known. For example, in the case of colloids dispersed in aqueous solutions, one such rule stems from observations that the colloidal particles are quite stable when the Zeta potential is ~30 mV (positive or negative) or more, and quite unstable due to aggregation when the Zeta potential is

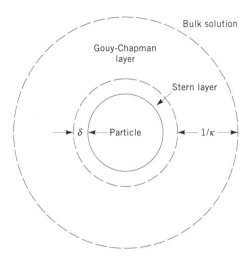

Fig. 2. Schematic diagram of a suspended colloidal particle, showing relative locations of the Stern layer (thickness, δ) that consists of adsorbed ions and the Gouy-Chapman layer ($1/\kappa$), which dissipates the excess charge, not screened by the Stern layer, to zero in the bulk solution (69). In the absence of a Stern layer, the Gouy-Chapman layer dissipates the surface charge.

between $+5$ and -5 mV. The transition from stable dispersion to aggregation usually occurs over a fairly small range of electrolyte concentration. This makes it possible to determine aggregation concentrations, often referred to as critical coagulation concentrations (CCC). The Schulze-Hardy rule summarizes the general tendency of the CCC to vary inversely with the sixth power of the counterion charge number (for indifferent electrolyte). The ability to predict CCCs was the first success of DLVO theory.

7.3. Steric and Hydrodynamic Effects. Additional influences on dispersion stability beyond those accounted for by the DLVO theory, like steric, surface hydration, and hydrodynamic effects have received considerable attention over the past several decades (54). More generally, the stability of a dispersion can be enhanced (protection) or reduced (sensitisation) by the addition of material that adsorbs onto particle surfaces. Protective agents can act in several ways. They can increase double layer repulsion if they have ionisable groups. The adsorbed layers can lower the effective Hamaker constant. An adsorbed film may necessitate desorption before particles can approach closely enough for van der Waals forces to cause attraction.

Long-chain surfactants and natural and synthetic high molecular weight polymers can be adsorbed at the surfaces of dispersed species such that a significant amount of adsorbate extends out from the surfaces. In this situation, an entropy decrease can accompany particle approach providing a short-range stabilization mechanism referred to as steric stabilization. Molecular structure and solvation, adsorption layer thickness and hydrodynamic volume, and temperature determine the effectiveness of steric stabilization (71–75). Finally, the adsorbed material may form such a rigid film that it poses a mechanical barrier to droplet coalescence. Sensitizing agents are the opposite of protective agents.

Again there are several possible mechanisms of action. If the additive is oppositely charged to the dispersed particles then decreased double layer repulsion will result. In some kinds of protecting adsorption a bilayer is formed with the outer layer having lyophilic groups exposed outward; the addition of enough additive to form only the single layer will have lyophobic groups oriented outward with a sensitising effect. If the additive is of long chain length, sometimes a bridging between particles occurs. Colloidal destabilization by electrolytes and bridging flocculation by polymers have been addressed both experimentally and theoretically. Comprehensive reviews on the relevant phenomena that derive from soluble and adsorbed polymers are available (71–75).

Oilfield W/O emulsions may be stabilized by the presence of a protective film around the water droplets. Such a film can be formed from the asphaltene and resin fractions of the crude oil. When drops approach each other during the process of aggregation, the rate of oil film drainage will be determined initially by the bulk oil viscosity, but within a certain distance of approach the interfacial viscosity becomes important. A high interfacial viscosity will significantly retard the final stage of film drainage and promote kinetic emulsion stability. If the films are viscoelastic, then a mechanical barrier to coalescence will be provided, yielding a high degree of emulsion stability (76,77).

Another example can be found in the field of water and wastewater treatment. Water treatment, whether for drinking water or for disposal of industrial wastes involves the removal of suspended solids usually silt, clay, and organic matter. The electric charge on the solids is often sufficiently negative to yield a stable dispersion that settles slowly and is difficult to filter. The solution to this problem is to reduce the Zeta potential to values that permit rapid coagulation increasing both sedimentation and filterability. A first step toward coagulating the suspension might be to add aluminum sulfate (alum), from which the trivalent aluminum ions will have a powerful effect on the Zeta potential (according to the Schulze-Hardy rule). Figure 3**a** shows an example of this effect. In practice,

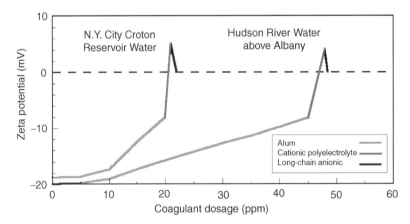

Fig. 3. Illustration of Zeta potentials and coagulation of solids in New York City water treatment through sequential additions of aluminum sulfate (alum), cationic polyelectrolyte, and anionic polymer. [Adapted from reference (78). Used with permission, L.A. Ravina, Zeta-Meter, Inc., Staunton, Va.]

however, the alum required to reduce the Zeta potential to below about -10 mV or so reduces the solution pH too much (unreacted alum becomes carried to other parts of the plant and forms undesirable precipitates). A second step can be introduced, in which a cationic polyelectrolyte is added to reduce the Zeta potential to about zero without changing the pH. As a final step a high molecular weight anionic polymer may be added (MW 500,000–1,000,000, or more) whose molecules can bridge between agglomerates yielding very large, rapid settling flocs. Figure 3**b** shows how two New York water samples were treated in this way.

8. Kinetic Properties

The best experimental technique for monitoring colloid stability is usually dictated by the nature of the specific colloidal material and the dispersing medium. In principle, any distinctive physical property of the colloidal system in question can be used, at least empirically, to monitor changes in the dispersed state. The more complex a system is (chemically or with respect to its particulate heterogeneity), the less likely it is that a single property uniquely and completely describes changes in the colloidal state. Aggregation and/or coalescence of colloidal material can be monitored by a wide variety of techniques including light scattering, neutron scattering, microscopy, rheology, conductivity, filtration, sedimentation, and electrokinetics.

Encounters between colloidal species can occur frequently due to any of Brownian motion, sedimentation, or stirring. If velocity or shear gradients are present and are sufficiently large, the frequency of collisions depends on the volume fraction of solids and the mean velocity gradient. Assuming that sedimentation is slow compared to other collision mechanisms, the overall aggregation rate, $-dN/dt$, is

$$-dN/dt = k_d N^2 + k_s N$$

where N is the number concentration of dispersed species, k_d and k_s are the respective rate constants corresponding to diffusion-controlled and shear-induced collision processes, and the minus sign denotes that the number concentration decreases with time, t. The constants, k_d and k_s, depend on particle–droplet–bubble properties such as: chemical composition of the bulk and surface phases, dielectric constant, dipole moment, size, size distribution, shape, surface charge, solid-phase distribution within particles, and particle anisotropy. Properties of the liquid-dispersing medium that contribute significantly to the values of these rate constants are dielectric constant, dipole moment, and the ability to dissolve electrolytes and polymers, in addition to those properties cited earlier. The k_d term usually dominates in quiescent systems containing submicrometre particles. The full expression for $(-dN/dt)$ and its use are treated in more detail elsewhere (7,8,31,79).

Chemical reactions can also affect the k_d and k_s terms and thereby influence or control colloidal stability (80). Pertinent examples are dissolution, precipitation, hydrolysis, precipitation, and chemical complexing. The last reaction may involve either simple species, eg,

$$Al^{3+} + SO_4^{2-} \rightleftharpoons AlSO_4^+$$

Table 5. **Representative Solution and Surface Equilibria Influencing Colloidal Stability**

Solution	Surface analogue
Hydrolysis	
$CH_3(CO)OCH_3 + H_2O \rightleftharpoons CH_3(CO)OH + CH_3OH$	$M_2O + H_2O \rightleftharpoons 2\,MOH$
$PO_4^{3-} + H_2O \rightleftharpoons HPO_4^{2-} + OH^-$	
$MOH + H_2O \rightleftharpoons MOH_2^+ + OH^-$	
Dissociation	
$Al(OH)_3 \rightleftharpoons Al^{3+} + 3OH^-$	$MOH_2^+ \rightleftharpoons MO^- + 2\,H^+$
$C_6H_5(CO)OH \rightleftharpoons C_6H_5(CO)O^- + H^+$	$MOH \rightleftharpoons MO^- + H^+$
Dissolution	
$ZnC_2O_4\,(S) \rightleftharpoons Zn^{2+} + C_2O_4^{2-}$	
$Al(OH)_3(S) + OH^- \rightleftharpoons AlO_2^- + 2H_2O$	
Complexation	
$Cu^{2+} + 4OH^- \rightleftharpoons Cu(OH)_4^{2-}$	$MO^- + Na^+ \rightleftharpoons MO^-Na^+$
$n\text{-}C_{12}H_{25}N(CH_3)_2 + HCl \rightleftharpoons n\text{-}C_{12}H_{25}NH^+(CH_3)_2Cl^-$	$MOH + HCl \rightleftharpoons MOH_2^+Cl^-$

or complicated solutes such as $Al_8(OH)_{20}^{4+}$, chelated metals (77), synthetic and natural polymers (25,71–75), or a variety of surfactants and dispersants (19,81). Many of the possible bulk solution chemical reactions that influence colloidal stability, along with specific sample reactions and their general interfacial analogues, are listed in Table 5.

9. Applications

9.1. General Uses. Diffusion, Brownian motion, sedimentation, electrophoresis, osmosis, rheology, mechanics, interfacial energetics, and optical and electrical properties are among the general physical properties and phenomena that are primarily important in colloidal systems (7,8,27,28,31). Chemical reactivity and adsorption often play important, if not dominant, roles. Any physical and chemical feature may ultimately govern a specific industrial process and determine final product characteristics and colloids are deemed either desirable or undesirable based on their unique physiochemical properties.

Although colloids may be undesirable components in industrial systems, particularly as waste or by-products and, in nature, in the forms of fog and mist, they are desirable in many technologically important processes such as mineral beneficiation and the preparation of ceramics, polymers, composite materials, paper, foods, textiles, photographic materials, drugs, cosmetics, and detergents. The remainder of this section specifies some applications for colloidal solids, liquids, and gases and illustrates how colloids affect many technologically important systems in a positive manner.

9.2. Colloidal Solids. Some uses of solid colloids include reinforcement aids in metals, ceramics, and plastics; as adhesion promoters in paints and thermoplastics; as nucleating agents in cloud seeding; as activated powder catalysts; as thickening agents in gels and slurries; and as abrasives in toothpastes (7,8,27,28,31,37,82–85). When used as reinforcement agents, the colloidal particles may be spherical, angular, fibrillar, or flake-shaped. Alumina and thoria are used to reinforce aluminum and nickel, respectively, by providing obstacles to the movement of dislocations in the metals, and zirconia and silicon carbide to reinforce a variety of ceramics (eg, alumina, silicon nitride, and glass, by inhibiting the propagation and opening of cracks in the matrix). (See COMPOSITE MATERIALS.) Stable but ordered suspensions can be regarded as precursory systems for ordered, prefired, ceramic components; outlets for such systems include various processing techniques such as slip, tape, freeze, pressure, centrifugal, and ultrasonic casting, as well as isostatic, and hot pressing (86,87). Asbestos, crystalline silicas, and organic solids are added to concrete to improve its strength by providing an interlocking particulate structure within the concrete matrix (83); asbestos, various oxides, and carbon black are added to reinforce polymers by inducing a stiffened or high yield matrix (27,28,80–83).

The magnitude of the strengthening often depends on particle shape. Fibrillar fillers are used as discontinuous fibers in metals and plastics, eg, in epoxy resin (80–82) and as whiskers in ceramics (88). Glass and aluminum oxide are common fillers that are occasionally pretreated with a polymeric or metallic coating; silica and various clays are also used. The mechanism by which ceramics and metals are reinforced often involves precipitation of colloidal material during thermal treatment of the matrix composition, eg, TiO_2 in borosilicate glasses designed for enamels (89). Strengthening mechanisms (27,28) include precipitation and dispersion when the reinforcing phase is metallic and the toughened materials are metals or ceramics. When inorganic, nonmetallic ceramics strengthen metals or polymers (45), the mechanism may be dispersion or reinforcement, for example, by cross-linking. Reinforcement implies a higher volume fraction than in dispersion hardening. Fillers can be added not only to improve mechanical properties such as impact strength, fracture toughness, and tensile strength of structural ceramics, but also to enhance optical properties, as is done for colored glasses containing colloidal gold or crystalline, chromium-based oxides.

Particle suspensions have long been of great practical interest because of their widespread occurrence in everyday life. Some important kinds of familiar suspensions include those occurring in foods (batters, puddings, sauces), pharmaceuticals (cough syrups, laxative suspensions), household products (inks, paints, "liquid" waxes) and the environment (suspended lake and river sediments, sewage). In the petroleum industry alone, suspensions may be encountered throughout each of the stages of petroleum recovery and processing, including migrating fine solid suspensions during secondary and enhanced oil recovery, dispersions of asphaltenes in crude oils, produced (well-head) solids in oil recovery, drilling muds, well stimulation and fracturing suspensions, well cementing slurries, and oilfield surface treatment facility sludges.

Other applications of colloidal solids include the preparation of rigid, elastic and thixotropic gels (80–82), aerogel-based thermal insulators (81), and surface coatings (25,27,28,81). Commercial uses of silica gel and sol–gel processing (90) often focus on rigid gels having 20–30 vol% SiO_2. The principal interparticulate forces in a rigid gel are chemical and irreversible, and the colloid improves the gel's mechanical strength. Elastic gels are commonly associated with cellophane, rubber, cellulosic fibers, leather, and certain soaps. Many thixotropic gels and surface coatings contain colloidal solids, eg, clays, alumina, ferric oxide, titania, silica, and zinc oxide. Consumer and industrial pastes belong to this category; putty, dough, lubricating grease, toothpaste, and paint are some examples.

9.3. Colloidal Liquids.   These fluids are commonly used in the form of emulsions by many industries. Permanent and transient antifoams consisting of an organic material, eg, polyglycol, oils, fatty materials, or silicone oil dispersed in water, is one application (10,91,92) that is important to a variety of products and processes: foods, cosmetics, pharmaceuticals, pulp and paper, water treatment, and minerals beneficiation. Other emulsion products (see Table 2) include foods, insecticides and herbicides, polishes, drugs, biological systems, asphalt paving emulsions, personal care creams and lotions, paints, lacquers, varnishes, and electrically and thermally insulating materials.

Emulsions may be encountered throughout all stages of the process industries. For example, in the petroleum industry both desirable and undesirable emulsions permeate the entire production cycle, including emulsion drilling fluid, injected or *in situ* emulsions used in enhanced oil recovery processes, wellhead production emulsions, pipeline transportation emulsions, and refinery process emulsions (13). Such emulsions may contain not just oil and water, but also solid particles and even gas, as occur in the large Canadian oil sands mining and processing operations (13–15).

Some emulsions are made to reduce viscosity so that an oil can be made to flow. Emulsions of asphalt, a semisolid variety of bitumen dispersed in water, are formulated to be both less viscous than the original asphalt and stable so that they can be transported and handled. In application, the emulsion should shear thin and break to form a suitable water-repelling roadway coating material. Another example of emulsions that are formulated for lower viscosity with good stability are those made from heavy oils and intended for economic pipeline transportation over large distances. Here again the emulsions should be stable for transport but will need to be broken at the end of the pipeline.

Some special problems arise at sea, when crude oil is spilled on the ocean a slick is formed that spreads out from the source with a rate that depends on the oil viscosity. With sufficient energy an O/W emulsion may be formed, which helps disperse oil into the water column and away from sensitive shorelines. Otherwise, the oil may pick up water to form a W/O emulsion, or mousse ("chocolate mousse"). These mousse emulsions can have high water contents and have very high viscosities, and with weathering, they can become semisolid and considerably more difficult to handle. The presence of mechanically strong films makes it hard to get demulsifiers into these emulsions, so they are hard to break.

Deemulsification, ie, the breaking of emulsions, is an important process, with the oil industry being a common one in which the process is often critical. The stability of emulsions is often a problem. Demulsification involves two steps. First, there must be agglomeration or coagulation of droplets. Then, the agglomerated droplets must coalesce. Chemical and particulate agents that displace the surfactant and permit an unstabilized interface to form are used for this purpose.

Only after these two steps can complete phase separation occur. It should be realised that either step can be rate determining for the demulsification process. This is a large subject in its own right; see references (13,16,91–93).

Other common applications of colloidal liquids include liquid aerosols, such as those occurring in the areas of environment (fog, mist, cloud, smog), agriculture (crop sprays), manufacturing (paint sprays), medicine (nasal sprays) and personal care (hair spray).

9.4. Colloidal Gases. Fluid foams are commonplace in foods, shaving cream, fire-fighting foam, mineral flotation, and detergents (10,65,94–96). Thus, in view of the fact that the concentration of bubbles greatly affects the properties of foams, the production, dispersion, and maintenance of colloidal gas bubbles are basic to foams and related materials. Often, natural and synthetic soaps and surfactants are used to make fluid foams containing colloidal gas bubbles. These agents reduce the interfacial tension and, perhaps, increase the viscosity at the gas–liquid interface, making the foam stable. Also, some soluble proteins that denature upon adsorption or with agitation of the liquid phase can stabilize foams by forming insoluble, rigid layers at the gas–liquid interface (91).

A class of enhanced oil recovery processes involves injecting a gas in the form of a foam. Suitable foams can be formulated for injection with air–nitrogen, natural gas, carbon dioxide, or steam (14,16). In a thermal process, when a steam foam contacts residual crude oil, there is a tendency to condense and create W/O emulsions. On the other hand, in a nonthermal process, the foam may emulsify the oil itself (now as an O/W emulsion), which is then drawn up into the foam structure; the oil droplets eventually penetrate the lamella surfaces, destroying the foam (14).

Microfoams (also termed colloidal gas aphrons) comprise a dispersion of aggregates of very small foam bubbles in aqueous solution. They can be created by dispersing gas into surfactant solution under conditions of very high shear. The concept is that, under the right conditions of turbulent wave break-up, one can create a dispersion of very small gas bubbles, each surrounded by a bimolecular film of stabilizing surfactant molecules. Under ambient conditions the bubble diameters are typically in the range 50–300 µm. There is some evidence that such microfoams tend to be more stable than comparable foams that do not contain the bimolecular film structure (97–99). Some interesting potential applications have been reported: soil remediation (98,100–103) and reservoir oil recovery (95,104,105), but the literature for these applications is sometimes inconsistent; more work needs to be done in this area.

Some agents will act to reduce the foam stability of a system (termed foam breakers or defoamers) while others can prevent foam formation in the first place (foam preventatives, foam inhibitors). There are many such agents, Kerner (65) describes several hundred different formulations for foam inhibitors and foam breakers. In all cases, the cause of the reduced foam stability can be traced to changes in the nature of the interface, but the changes can be of various kinds. The addition to a foaming system of any soluble substance that can become incorporated into the interface may decrease dynamic foam stability if the substance acts in any combination of: increase surface tension, decrease surface elasticity, decrease surface viscosity or decrease surface potential. Such effects may be caused by a cosolubilization effect in the interface or by a partial or even complete replacement of the original surfactants in the interface. Some branched, reasonably high molecular mass alcohols can be used for this purpose. Not being very soluble in water they tend to be adsorbed at the gas–liquid interface, displacing foam promoting surfactant and breaking or inhibiting foam. Alternatively, a foam can be destroyed by adding a chemical that actually reacts with the foam-promoting agent(s). Foams may also be destroyed or inhibited by the addition of certain insoluble substances such as a second liquid phase or a solid phase. Antifoaming and defoaming represent a large subject on their own [see (106,107)].

Other common applications of colloidal gases include solid foams, such as those occurring in the areas of food (leavened breads), geology (pumice stone, zeolites), manufacturing (polystyrene foam, polyurethane foam), and personal care (synthetic sponges). Solid foams such as polyurethane foam contain dispersed gas bubbles that are often produced via viscoelastic polymer melts within which gas, eg, carbon dioxide, bubbles are nucleated (66).

10. Chemical and Surface Analysis

Any classical wet-chemical analyses or instrumental techniques that are routinely used to analyze the bulk composition of solids and liquids are, in principle, also suitable for colloids. The available instrumental methods range, eg, from spectrographic analysis for chemical composition, which is limited to crude estimates for impurities, to Raman spectroscopic identification of chemical functionalities, which can be very accurate (see SPECTROSCOPY, OPTICAL). These techniques can also be used to analyze adsorbed layers if they can be quantitatively desorbed and collected for study. Surface-chemical analyses that cannot be conducted using conventional methods designed for bulk materials are usually accomplished by optical, diffraction, and spectroscopic techniques; these are often applied under conditions of an ultrahigh vacuum. The spectroscopies measure the responses of solid surfaces to beams of electrons, ions, neutral species, and photons (see SURFACE AND INTERFACE ANALYSIS). Each spectroscopy has unique attributes (10), making it suitable for certain colloids but unsuitable for others. Although too numerous to list here, descriptions and tabulations of surface techniques and their synonyms and acronyms are given in references (6,10,67,68). A wide range of information can be obtained including surface and adsorbed layer compositions (including heterogeneity); surface morphology, atom packing, and structure; and surface reactions and their kinetics. Advances in these fields continue at a significant rate (see ANALYTICAL METHODS, TRENDS).

11. Hazards of Colloidal Systems

The occurrence of some materials in the form of a colloidal dispersion can introduce or enhance safety hazards. Considering that the dispersion of a material down to colloidal size results in a high specific surface area and colloidal chemical reactivity may differ considerably from that of the identical macroscopic material with less surface area. This is particularly important if the colloidal surface is easily and rapidly oxidized. Dust explosions and spontaneous combustion are potential dangers whenever certain materials exist as finely divided dry matter exposed to oxidizing environments (69,116) (see POWDERS, HANDLING AND COATING PROCESSES, POWDER TECHNOLOGY). Dispersions of charged colloidal particles in non-aqueous media occur throughout the petroleum industry. The flow of petroleum fluids in tanks or pipelines, combined with the low conductivity of the petroleum fluids themselves, can allow the build-up of large potential gradients and a separation of charges. Sufficient charging for there to be an electrostatic discharge can cause an explosion (70).

Health problems can be caused by solids and liquids suspended in air or water. Specific potential hazards have been associated with a diverse spectrum of colloidal materials, including chemicals, coal, minerals, metals, pharmaceuticals, plastics, and wood pulp. Limits for human exposure for many particulate, hazardous materials are published (67,68). The effects of the colloidal solids and liquids that comprise smog are widespread and well known. Liquid droplets may also constitute a hazard; eg, smog can contain sulfuric acid aerosols. Elements such as lead, zinc, and vanadium, that are released into the atmosphere as vapors, can subsequently condense or be removed as solid particulates by rain (108,109). Similarly, exposure to airborne pollutants found indoors and in confined spaces, many of which are particulates or microbes of colloidal size, can lead to complex physiological responses. This hazard ranges from short-term allergic reactions (eg, pollen and household dust causing asthma and hay fever) to long term, possibly fatal effects (eg, silicosis, asbestosis, and black lung disease).

There is a need to understand the environmental properties and risks associated with any large volume chemicals. The mass of surfactants that could ultimately be released into the environment, for example, is significant. A 1995 estimate of the global use of linear alkylbenzenesulfonates, alcohol ethoxylates, alkylphenol ethoxylates, alcohol sulfates, and alcohol ether sulfates totaled 3 million tonnes (110). Surfactant usage in industry will probably increase as new applications are found. The toxicity and persistence of surfactants is now fairly predictable for a variety of environmental situations (108,111,112) and although surfactants are not generally viewed as a serious threat to the environment (108) they can exhibit considerable toxicity to aquatic organisms. In addition to products, many industries produce waste containing significant amounts of suspended matter for which treatment incurs significant technological challenges and costs (113–115). Trace metals that are commonly found as suspended matter in the chemical form of hydrous oxides and other insoluble matter are tabulated elsewhere (69,116). Large fractions of readily hydrolyzable metals exist as adsorbed species on suspended (colloidal) solids in fresh and marine water systems and can also be anticipated to exist in industrial wastewater. Continued research is needed to understand the health hazards linked to colloidal pollutants.

12. Emerging Areas in Colloid Science

Advances on the theoretical front continue to be aided by advances in bulk and surface analytical technology (composition) and in physical surface characterization technology (topology, structure, and forces). Applications of atomic force microscopy in particular continue to expand into diverse areas, including magnetic force microscopy (117).

Much is known about colloids, their formation, properties and applications, but considerably more surely remains unknown. In particular, the full potential to control colloids is not presently realized. There are several types of mixed colloids that are only poorly understood. For example, the properties of colloids in which more than one type of colloidal particle is dispersed may be dominated by the behavior of the minor dispersed phase component. The nature and properties of colloids within colloids, such as suspended solids in the dispersed phase of an emulsion, or emulsified oil within the aqueous lamellae of a foam, are only beginning to be understood (13–15).

Nanotechnology refers to the production of materials and structures in the 0.1–100 nm range by any of a variety of nanoscale physical and chemical methods (see NANOTECHNOLOGY). This is a rapidly growing area of materials science. As the size range indicates, there is an overlap between nanotechnology and colloid science since they share some similarity of scale, both dealing with matter having dimensions of tens and hundreds of nm. Although many areas of nanotechnology do not directly deal with colloidal dispersions, such as nanotubes, nanoelectronic devices, other areas do, eg, the use of colloidal ink dispersions in robocasting to build near-nanometer scale three-dimension structures. Examples of work at the interface between these fields is provided by current research aimed at controlling the solubility of nanotubes (118), and in the fabrication of nano-engineered films on colloidal particles (119).

Smart colloids are colloidal dispersions for which certain properties, such as size and structure, can be altered by changing an external influence, such as temperature. Example: cross-linked polymer gels of poly(N-isoproplyacrylamide). Such polymer systems can swell or shrink in response to temperature changes, and are also termed "thermo-shrinking polymers." (See 120–122).

13. Acknowledgment

The author acknowledges the valuable contribution made by Alan Bleier (Oak Ridge National Laboratory) in his article of the same title published in the 1993 edition of this encyclopedia, from which some materials, including several figures and tables, have been retained in the present version.

BIBLIOGRAPHY

"Colloids" in *ECT* 3rd ed., Suppl. Vol., pp. 241–259, by Alan Bleier, Massachusetts Institute of Technology; in *ECT* 4th ed., Vol. 6, pp. 812–840, by Alan Bleier, Oak Ridge National Laboratory; "Colloids" in *ECT* (online), posting date: December 4, 2000, by Alan Bleier, Oak Ridge National Laboratory.

CITED PUBLICATIONS

1. M. Kerker, *J. Colloid Interface Sci* **116**(1), 296 (1987).
2. I. Asimov, *Words of Science and the History Behind Them*, Riverside Press: Cambridge, Mass., 1959.

3. H. Freundlich, *Colloid and Capillary Chemistry*, English translation of 3rd. ed., Methuen, London, 1926.

4. P. Becher, *Dictionary of Colloid and Surface Science*; Dekker, New York, 1990.

5. L. L. Schramm, *The Language of Colloid and Interface Science*, American Chemical Society, Washington, and Oxford University Press, New York, 1993.

6. L. L. Schramm, *Dictionary of Colloid and Interface Science*, John Wiley & Sons, Inc., New York, 2001.

7. D. J. Shaw, *Introduction to Colloid and Surface Chemistry*, 3rd ed., Butterworths, London, 1980.

8. P. Hiemenz and R. Rajagopalan, *Principles of Colloid and Surface Chemistry*, 3rd. ed., Dekker, New York, 1997.

9. D. Myers, *Surfaces, Interfaces, and Colloids*; 2nd ed., Wiley-VCH, New York, 1999.

10. A. W. Adamson, *Physical Chemistry of Surfaces*; 5th ed., John Wiley & Sons, Inc., New York, 1990.

11. H. R. Kruyt, *Colloid Science, Volume I, Irreversible Systems*, Elsevier, Amsterdam, The Netherlands, 1952.

12. S. P. Parker, ed., *McGraw-Hill Dictionary of Scientific and Technical Terms*, 3rd ed., McGraw-Hill, New York, 1984.

13. L. L. Schramm, ed., *Emulsions: Fundamentals and Applications in the Petroleum Industry*, American Chemical Society, Washington and Oxford University Press, New York, 1992.

14. L. L. Schramm, ed., *Foams: Fundamentals and Applications in the Petroleum Industry*, American Chemical Society, Washington and Oxford University Press, New York, 1994.

15. L. L. Schramm, ed., *Suspensions: Fundamentals and Applications in the Petroleum Industry*, American Chemical Society, Washington, and Oxford University Press, New York, 1996.

16. L. L. Schramm, ed., *Surfactants: Fundamentals and Applications in the Petroleum Industry*, Cambridge University Press, Cambridge, U.K., 2000.

17. S. Lowell and J. E. Shields, *Powder Surface Area and Porosity*, Chapman and Hall, London, 1991.

18. D. Myers, *Surfactant Science and Technology*, VCH, New York, 1988.

19. M. J. Rosen, *Surfactants and Interfacial Phenomena*, 2nd ed., John Wiley & Sons, Inc., New York, 1989.

20. M. El-Nokaly and D. Cornell, eds., *Microemulsions and Emulsions in Foods*, American Chemical Society, Washington, D.C., 1991.

21. R. Beckett, ed., *Surface and Colloid Chemistry in Natural Waters and Water Treatment,* Plenum Press, New York, 1990.

22. H. van Olphen, *An Introduction to Clay Colloid Chemistry*, 2nd ed., Wiley-Interscience, New York, 1977.

23. M. M. Rieger and L. D. Rhein, eds., *Surfactants in Cosmetics*, 2nd ed., Dekker, New York, 1997.

24. E. Dickinson, *An Introduction to Food Colloids*, Oxford University Press, New York, 1992.

25. G. D. Parfitt, ed., *Dispersion of Powders in Liquids*, 3rd ed., Applied Science, London, 1981.

26. S. Ross and I. D. Morrison, *Colloidal Systems and Interfaces*, John Wiley & Sons, Inc., New York, 1988.

27. C. R. Veale, *Fine Powders*, John Wiley & Sons, Inc., New York, 1972.

28. J. K. Beddow, *Particulate Science and Technology*, Chemical Publishing, New York, 1980.

29. M. J. Jaycock and G. D. Parfitt, *Chemistry of Interfaces*, Ellis Horwood, Chichester, U.K., 1981.

30. G. D. Parfitt and K. S. W. Sing, eds., *Characterization of Powder Surfaces*, Academic Press, New York, 1976.

31. R. D. Vold and M. J. Vold, *Colloid and Interface Chemistry*, Addison-Wesley, Reading, Mass., 1983.

32. R. J. Mikula and V. A. Munoz, in L. L. Schramm, ed., *Surfactants: Fundamentals and Applications in the Petroleum Industry*, Cambridge University Press, Cambridge, U.K., 2000, pp. 51–77.

33. L. L. Schramm, S. M. Kutay, R. J. Mikula, and V. A. Munoz, *J. Petrol. Sci. Eng.* **23**, 117 (1999).

34. M. Kerker, *The Scattering of Light and Other Electromagnetic Radiation*, Academic Press, New York, 1969.

35. R. J. Hunter, *Zeta Potential in Colloid Science*, Academic Press, New York, 1981.

36. A. J. Babchin, R. S. Chow, and R. P. Sawatzky, *Adv. Colloid Interface Sci.* **30**, 111 (1989).

37. C. Orr, *Particulate Technology*, Macmillan, New York, 1966.

38. Th. G. M. van de Ven, *Colloidal Hydrodynamics*, Academic Press, Inc., London, 1989.

39. R. W. Whorlow, *Rheological Techniques*, Ellis Horwood, Chichester, U.K., 1980.

40. J. R. Van Wazer, J. W. Lyons, K. Y. Kim, and R. E. Colwell, *Viscosity and Flow Measurement*, John Wiley & Sons, Inc., New York, 1963.

41. C. C. Mill, ed., *Rheology of Disperse Systems*, Pergamon Press, Inc., Elmsford, New York, 1959.

42. A. G. Fredrickson, *Principles and Applications of Rheology*, Prentice-Hall, Englewood Cliffs, N.J., 1964.

43. K. Walters, *Rheometry: Industrial Applications*, Research Studies Press, New York, 1980.

44. H. A. Barnes and S. A. Holbrook, In P. Ayazi Shamlou, ed., *Processing of Solid–Liquid Suspensions*, Butterworth Heinemann, Oxford, U.K., 1993, pp. 222–245.

45. L. E. Murr, *Interfacial Phenomena in Metals and Alloys*, Addison Wesley, London, 1975.

46. W. D. Harkins and A. E. Alexander, in A. Weissberger, ed., *Physical Methods of Organic Chemistry*, Interscience, New York, 1959, pp. 757–814.

47. J. Leja, *Surface Chemistry of Froth Flotation*, Plenum Press, New York, 1982.

48. L. L. Schramm, *J. Can. Petrol. Technol.* **28**, 73 (1989).

49. J. F. Padday, in E. Matijevic, ed., *Surface and Colloid Science*, Vol. 1, Wiley-Interscience, New York, 1969, pp. 101–149.

50. C. A. Miller and P. Neogi, *Interfacial Phenomena Equilibrium and Dynamic Effects*, Dekker, New York, 1985.

51. A. I. Rusanov and V. A. Prokhorov, *Interfacial Tensiometry*, Elsevier, Amsterdam, The Netherlands, 1996.

52. L. W. Lake, *Enhanced Oil Recovery*, Prentice Hall, Englewood Cliffs, N.J., 1989.

53. P. Mukerjee and K. J. Mysels, *Critical Micelle Concentrations of Aqueous Surfactant Systems*, NSRDS-NBS 36, National Bureau of Standards, Washington, D.C., 1971.

54. J. N. Israelachvili, *Intermolecular and Surface Forces*, 2nd ed., Academic Press, New York, 1992.

55. G. Cevc and D. Marsh, *Phospholipid Bilayers*, John Wiley & Sons, Inc., New York, 1987; D. Marsh, *Handbook of Lipid Bilayers*, CRC Press, Boca Raton, Fla., 1990.

56. R. Miller and G. Kretzschmar, *Adv. Colloid Interface Sci.* **37**, 97 (1991).

57. J. Prost and F. Rondelez, *Nature (London)* **350**, 11 (1991).

58. D. O. Shah, ed., *Macro-and Microemulsions: Theory and Applications*, American Chemical Society, Washington, D.C., 1985.

59. D. A. Sabatini, R. C. Knoz, and J. H. Harwell, eds., *Surfactant-Enhanced Subsurface Remediation Emerging Technologies*, American Chemical Society, Washington, D.C., 1995.

60. M. El-Nokaly and D. Cornell, eds., *Microemulsions and Emulsions in Foods*, American Chemical Society, Washington, D.C., 1991.

61. A. M. James, in R. J. Good and R. R. Stromberg, eds., *Surface and Colloid Science*, Vol. 11, Plenum, 1979, pp. 121–186.

62. A. J. Babchin, R. S. Chow, and R. P. Sawatzky, *Adv. Colloid Interface Sci.* **30**, 111 (1989).

63. B. V. Derjaguin and L. D. Landau, *Acta Physiochem. URSS* **14**, 633 (1941); B. V. Derjaguin, N. V. Churaev, and V. M. Miller, *Surface Forces*, Consultants Bureau, New York, 1987.

64. E. J. Verwey and J. Th. G. Overbeek, *Theory of the Stability of Lyophobic Colloids*, Elsevier, Amsterdam, The Netherlands, 1948.

65. H. T. Kerner, *Foam Control Agents*, Noyes Data Corp., Park Ridge, New Jersey, 1976.

66. F. W. Billmeyer, Jr., *Textbook of Polymer Science*, 2nd ed., Wiley-Interscience, New York, 1971.

67. D. P. Woodruff and T. A. Delchar, *Modern Techniques of Surface Science*, Cambridge University Press, New York, 1986.

68. G. A. Somorjai, *Introduction to Surface Chemistry and Catalysis*, Wiley-Interscience, New York, 1994.

69. S. Budavari, M. J. O'Neil, A. Smith, P. E. Heckelman, and J. f. Kinneary, *The Merck Index*, 11th ed., Merck, Rahway, New Jersey, 1989, p. MISC-53.

70. I. D. Morrison, *Colloids Surfaces* **71**, 1 (1993).

71. T. Sato and R. Ruch, *Stabilization of Colloidal Dispersions by Polymer Adsorption*, Marcel Dekker, New York, 1980.

72. Th. F. Tadros, ed., *The Effect of Polymers on Dispersion Properties*, Academic Press, London, 1982.

73. Y. S. Lipatov and L. M. Sergeeva, *Adsorption of Polymers*, Halsted, New York, 1974.

74. C. A. Finch, ed., *Chemistry and Technology of Water-Soluble Polymers*, Plenum Press, New York, 1983.

75. B. Vincent, in Th. F. Tadros, ed., *Solid/Liquid Dispersions*, Academic Press, Inc., 1987, pp. 147–162.

76. R. J. R. Cairns, D. M. Grist, and E. L. Neustadter, in A. L. Smith, ed., *Theory and Practice of Emulsion Technology*, Academic Press, New York, 1976, pp. 135–151.

77. T. J. Jones, E. L. Neustadter, and K. P. Whittingham, *J. Can. Petrol. Technol.* **17**, 100–108 (1978).

78. Zeta-Meter applications brochures, Zeta-Meter Inc., New York, 1982.

79. W. Stumm and J. J. Morgan, *Aquatic Chemistry*, 2nd ed., John Wiley & Sons, Inc., New York, 1981.

80. E. Matijević, in K. J. Mysels, C. M. Samour, and J. H. Hollister, eds., *Twenty Years of Colloid and Surface Chemistry*, American Chemical Society, Washington, D.C., 1973, p. 283.

81. *McCutcheon's Detergents and Emulsifiers*, MC Publ. Co., Glen Rock, N.J., 1980.

82. Z. D. Jastrzebski, *The Nature and Properties of Engineering Materials*, 2nd ed., John Wiley & Sons, Inc., New York, 1977.

83. J. E. Gordon, *The New Science of Strong Materials*, 2nd ed., Princeton University Press, Princeton, N.J., 1976.

84. A. G. Guy, *Essentials of Materials Science*, McGraw-Hill Book Co., Inc., New York, 1976.

85. S. J. Lefond, ed., *Industrial Minerals and Rocks*, 4th ed., American Institute of Mineralogy and Metallurgy, Petroleum Engineering, Inc., New York, 1975.

86. F. Y. Wang, *Ceramic Fabrication Processes*, Academic Press, Inc., New York, 1976.

87. J. S. Reed, *Introduction to the Principles of Ceramic Processing*, Wiley-Interscience, New York, 1988.

88. A. Kelly and R. B. Nicholson, eds., *Strengthening Methods in Crystals*, Applied Science Publishing, London, 1971.

89. W. D. Kingery, H. K. Bowen, and D. R. Uhlmann, *Introduction to Ceramics*, 2nd ed., Wiley-Interscience, New York, 1976.

90. C. J. Brinker, and G. W. Sherer, *Sol–Gel Science: The Physics and Chemistry of Sol–Gel Processing*, Academic Press, San Diego, Calif., 1990.

91. C. G. Sumner, *Clayton's The Theory of Emulsions and Their Technical Treatment*, 5th ed., Blakiston, New York, 1954.

92. K. J. Lissant, *Demulsification. Industrial Applications*, Dekker, New York, 1983.

93. P. Becher, *Emulsions: Theory and Practice*, 3rd ed., American Chemical Society and Oxford University Press, New York, 2001.

94. D. H. Everett, *Basic Principles of Colloid Science*, Royal Society of Chemistry, London, 1988.

95. A. K. Mirzadjanzade, I. M. Ametov, A. A. Bokserman, and V. P. Filippov, in *Proceedings, 7th European IOR Symp.*, Moscow, Russia, 1993, pp. 27–29.

96. E. Dickinson, ed., *Food Emulsions and Foams*, Royal Society of Chemistry, London, 1987.

97. F. Sebba, *J. Colloid Interface Sci.* **35**, 643 (1971).

98. F. Sebba, *Foams and Biliquid Foams-Aphrons*, John Wiley & Sons, Inc., New York, 1987.

99. F. Sebba, *Chem. Ind.* **4**, 91 (1985).

100. D. Roy, K. T. Valsaraj, and A. Tamayo, *Sep. Sci. Technol.* **27**, 1555 (1992).

101. D. Roy, K. T. Valsaraj, W. D. Constant, and M. Darji, *J. Hazardous Mat.* **38**, 127 (1994).

102. D. Roy, R. R. Kommalapati, K. T. Valsaraj, and W. D. Constant, *Water Res.* **29**, 589 (1995).

103. D. Roy, S. Kongara, and K. T. Valsaraj, *J. Hazardous Mat.* **42**, 247 (1995).

104. M. V. Enzien, D. L. Michelsen, R. W. Peters, J. X. Bouillard, and J. R. Frank, in *Proceedings, 3rd Int. Symp. In-Situ and On-Site Reclamation*, San Diego, Calif., 1995, pp. 503–509.

105. A. K. Mirzadjanzade, I. M. Ametov, A. O. Bogopolsky, R. Kristensen, U. N. Anziryaev, S. V. Klyshnikov, A. M. Mamedzade, and T. S. Salatov, in *Proceedings, 7th European IOR Symp.*, Moscow, Russia, 1993, pp. 469–473.

106. J. J. Bikerman, *Foams*, Springer-Verlag, New York, 1973.

107. J. J. Bikerman, *Foams, Theory and Industrial Applications*, Reinhold, New York, 1953.

108. A. E. Martell, *Pure Appl. Chem.* **44**, 81 (1975).

109. F. T. Mackenzie and R. Wollast, in E. D. Goldberg, ed., *The Sea*, Vol. 6, Wiley-Interscience, New York, 1977, p. 739.

110. L. N. Britton, *J. Surf. Det.* **1**, 109–117 (1998).

111. L. N. Britton, in L. L. Schramm, ed., *Surfactants, Fundamentals and Applications in the Petroleum Industry*, Cambridge University Press, Cambridge, U.K., 2000, pp. 541–565.

112. C. M. Maddin, Preprint, *International Conference Health, Safety and Environment*, Society of Petroleum Engineers, Richardson, Tex., SPE paper 23354, 1991.

113. J. Wei, T. W. F. Russel, and M. W. Swartzlander, *The Structure of the Chemical Processing Industries*, McGraw-Hill, New York, 1979.
114. M. P. Freeman and J. A. Fitzpatrick, eds., *Physical Separations*, Engineering Foundation, New York, 1980.
115. A. J. Rubin, ed., *Chemistry of Wastewater Technology*, Ann Arbor Sci., Ann Arbor, Mich., 1978.
116. R. C. Weast and M. J. Astle, eds., *Handbook of Chemistry and Physics*, 62nd ed., Chemical Rubber Company, Boca Raton, Fla., 1981, p. D-101.
117. M. Rasa, B. W. M. Kuipers, and A. P. Philipse, *J. Colloid Interface Sci.* **250**, 303–315 (2002).
118. R. Dagani, *Chem. Eng. News* **80**(28), 38 (2002).
119. G. B. Sukhorukov, in D. Möbius and R. Miller, eds., *Novel Methods to Study Interfacial Layers*, Elsevier, New York, 2001, pp. 383–414.
120. E. S. Matsuo and T. Tanaka, *Nature (London)* **358**, 482 (1992).
121. M. Snowden, *Science Spectra.* **6**, 32 (1996).
122. M. Snowden, M. J. Murray, and B. Z. Chowdry, *Chem. Ind.*, 15 July, 531 (1996).

LAURIER L. SCHRAMM
Saskatchewan Research Council

COLOR

1. Introduction

Being a perception, it is difficult to give a simple definition of color. We could say that perceived color is the part of perception that is carried to the eye from our surroundings by various wavelengths (or frequencies) of light. First, this involves the nature and spectral power distribution in the light from illuminating light sources. Next, there are several often interrelated processes derived from the interaction of the illumination with matter, including absorption, reflection, refraction, diffraction, scattering, and fluorescence. Finally, there is the perception system, involving the eye and the transmission system from eye to brain, leading to the final interpretation reached in the brain.

Perceived color depends on the spectral distribution of the color stimulus, on the size, shape, strucure, and surround of the stimulus area, on the state of adaptation of the observer's visual system, and on the observer's experience of the prevailing and similar conditions of observation., These last are complex processes, involving psychological as well as physiological factors. As one example, a specific shade of green may have a quite different meaning for a jungle resident than it has for a desert dweller. The color perceived is affected by the shape, size, texture, and gloss of the surface being viewed as well as by adjacent colors, recently seen colors, etc. There are also differences in perception for the different viewing modes discussed below.

As with any other perception, such as pain, one cannot know precisely what color another individual perceives, but the development of an agreed terminology of the stimuli that lead to color perception based on common experience has led to a satisfactory science of color description and measurement.The precise measurement of color is of significance in many branches of science and technology. It serves as a record for archival description, for standardization purposes, and for matching and controlling the many colorful products of commerce. In a field that has changed significantly even in the last decade, six books can be particularly recommended for further details (1–6). Additional books for a well-rounded basic library on color might include References 7–14. The latest Commission Internationale de l'Éclairage (CIE) publications (13) should be consulted for the current definitive word.

2. Color Fundamentals

An immediate complexity is illustrated in the two early and apparently incompatible theories of color vision. Trichromatic theory, first proposed in 1801 by Thomas Young and later refined by Hermann von Helmholtz, postulated three types of color receptors in the eye. This explained many phenomena, such as various forms of color blindness, and was confirmed in 1964, when three types of blue-, green-, and red-sensitive cones were reported to be present in the retina. Yet Ewald Hering's 1878 opponent theory, which used three pairs of opposites, light–dark, red–green, and blue–yellow, also offered much insight, including the explanation of contrast and afterimage effects and the absence of some color combinations such as reddish greens and bluish yellows. In the modern zone theories, it is now recognized that the data from three trichromatic detectors in the eye are processed on their way to the brain into opponent signals, thus removing the apparent inconsistencies.

In color technology and measurement, both types of approaches are used. Color printing, eg, generally employs three colors (usually plus black), and the ever useful CIE system was based on experiments in which colors were matched by mixtures of three primary color light beams, blue, green, and red. Yet transmitted television signals are based on the opponent system, with one intensity and two color-balance signals, as are the modern representations of color, such as the CIELAB and related color spaces based on red–green and yellow–blue opponent axes.

2.1. Light and Color. Visible light is that part of the electromagnetic spectrum, shown in Figure 1, with wavelengths between the violet limit of 400 nm and the red limit at about 700 nm. Depending on the observer, light intensity, etc, typical values for the spectral colors are blue, 450 nm; green, 500–550; yellow, 580; orange, 600; and red, 650. There are also the nonspectral colors purple, magenta, brown, etc, as well as mixtures of all of these with black and white. It has been estimated that an individual with normal color vision can distinguish a total of 7 million different colors.

In 1666, Sir Isaac Newton first split white light with a prism into its component colors, the spectrum, and he assigned the colors red, orange, yellow, green, blue, indigo, and violet, using just seven colors, possibly by analogy with the seven notes of the musical scale. The eye, however, functions quite differently from the ear, which is able to perceive the sound from individual

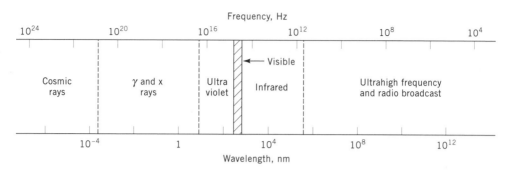

Fig. 1. The electromagnetic spectrum (2).

instruments sounding together, while the eye always perceives only a single color at a given point, whether this be spectrally pure orange, an equivalent mixture of yellow and red, an equivalent mixture of green and red, etc. Newton did recognize that his "rays ... are not coloured. In them there is nothing else than a certain power ... to stir up a sensation of this or that colour." Nevertheless, terms such as "orange light" are commonly used and need not produce any confusion if this qualification is kept in mind.

The spectral color sequence, joined by some nonspectral colors such as purple and magenta, is one of several attributes used in descriptions of color, variously designated hue, chromatic color, dominant wavelength (or simply, but imprecisely and quite unsuitably for the present purposes, color). A second attribute is saturation, chroma, tone, or purity, which gives a measure of how little or much gray (or white or black) is present. Thus a mixture of pure spectral blue with gray gives an unsaturated blue, transforming into gray (or white or black) as the amounts of additive are increased. For any given hue and saturation, there can be different levels of brightness, lightness, luminance, or value, completing the three parameters normally required to specify color.

Color in the full technical sense encompasses a multidimensional space defined at a minimum by the foregoing three parameters with the addition of factors such as the color environment, gloss, reflectancy, and translucency. An example that clarifies this point is that a strong orange becomes, when the brightness is reduced compared to adjacent areas, but with constant hue and saturation, not a weak orange but a brown. Thus to obtain brown one either mixes orange and black paints or one views an orange surface through a hole in a more brightly illuminated screen (so that the orange in appearing to be of low brightness, is then perceived as brown). In the same way, a greenish yellow can be perceived as having an olive color.

The appearance of color clearly depends significantly on the exact viewing circumstances. Normally one thinks of viewing a colored object under some type of illumination, the object mode; then there is the surface mode, eg, when light is perceived to be reflected from an object's surface; viewing a light source there is the illuminant or luminous mode; viewing through a hole in a screen there is the aperture mode, and so on. Perception differs significantly in these modes. In the object mode, the eye–brain has the ability to compensate for a wide range of illuminants (white sun, blue sunless sky, reddish incandescent lamp, or candle) and

Table 1. **Object Mode Perceptions**[a]

Object	Dominant perception	Secondary attributes[b]
opaque metal, polished	specular reflection	reflectivity, gloss, hue
opaque metal, matte	diffuse reflection	hue, saturation, brightness, gloss
opaque nonmetal, glossy	diffuse and specular reflections	hue, saturation, gloss, brightness
opaque nonmetal, matte	diffuse reflection	hue, saturation, brightness
translucent nonmetal	diffuse transmission	translucency, hue, saturation
transparent nonmetal	transmission	hue, saturation, clarity

[a]Ref. 1.
[b]In approximate sequence of importance.

infer something very close to the true color. At the same time, various surface effects enter, as shown in Table 1. Even a nonmetallic object with a deeply colored surface will reflect almost pure illuminant in the glare of a glancing angle if it is glossy. The perceived color is influenced by the presence of adjacent colors in the object mode, and several additional color-appearance phenomena also can influence color perception. One, two, or more additional parameters (1) may be required in addition to the customary three for a full specification of the perceived color, but fortunately this is unusually not necessary.

2.2. Interactions of Matter with Light.   In the most generalized interaction of light with matter the many phenomena of Figure 2 are possible. In absorption, electrons are excited by the absorbed photons and their energy may subsequently appear as heat or as fluorescence, an additional emitted light at a lower energy, ie, longer wavelength. This effect is utilized in fluorescent whitening agents used in detergents, paper, textiles, etc. Scattering may derive from irregularities of the surface, diffuse, or matte as distinguished from glossy

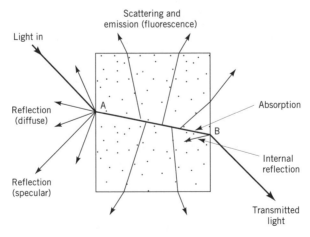

Fig. 2.   The adventures of a beam of light passing through a block of partly transparent substance (2).

or specular reflection. Interference, eg, the diffraction grating effect, derives from regularly repeated patterns. Irregularities in the interior, depending on the size and geometry, may involve Rayleigh or Mie-type scattering and then lead to translucency or opacity if there is pronounced multiple scattering, as well as color at times.

3. Color Vision

3.1. The Eye.
In vision, light passes through the cornea, the transparent outer layer of the eye, through the lens and the aqueous and vitreous humors, and is focused onto the retina. The iris, forming the pupil, acts as a variable aperture to control the amount of light that enters the eye, varying from $\sim f/2.5$ to $f/13$ with a 30:1 light intensity ratio. The two humors serve merely as neutral transmission media and to keep the eyeball distended. The retina is a layer ~ 0.1 mm thick that contains the light-sensitive rods and cones. Only the rods function in low levels of illumination of about <1 lux, providing an achromatic, noncolor image. The rod spectral response is shown in Figure 3a. The cones are of three types, designated B, G, and R in this figure. Although these are often spoken of as blue-, green-, and red-detecting cones, such designations are incorrect. As can be seen, each set of cones is sensitive to a wide range of wavelengths with extensive overlap. Appropriate designations are short-, medium-, and long-wavelength sensitive cones, but the B, G, and R labels provide a convenient mental picture useful as long as the actual functioning is kept in mind.

The distribution of rods and cones is shown in Figure 3b centered about the fovea, the area of the retina that has the highest concentration of cones with essentially no rods and also has the best resolving capability, with a resolution about one minute of arc. The fovea is nominally taken as a 5° zone, with its central 1° zone designated the foveola. There are ~ 40 R and 20 G cones for each B

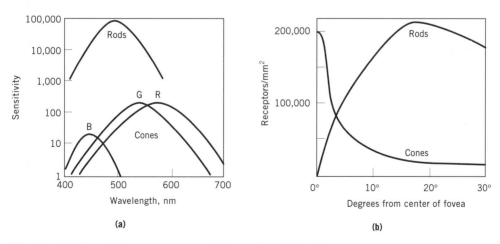

(a)   (b)

Fig. 3. (a) The relative sensitivities of the rods and the three sets of cones in the human eye (2). (b) The distribution of rods and cones in the central part of the human retina. The x axis is marked in degrees from the center of the fovea.

cone in the eye as a whole, whereas in the fovea there are almost no B cones. A result of this is that color perception depends on the angle of the cone of light received by the eye. Extremely complex chemistry is involved in the stimulation of opsin molecules, such as the rhodopsin of the rods (2).

The trichromatic theory, subsequently confirmed by the existence of the three sets of cones, must be combined with the opponent theory, which is involved in the signal sent along the retinal pathway from the eye to the brain. A third approach, the appearance theory (4) or the retinex theory, must be added to explain color constancy and other effects. As one example of this last, consider an area in a multicolored object such as a Mondrian painting perceived as red when illuminated with white light. If the illumination is changed so that energy reflected by this same area is greater at shorter wavelengths than the energy reflected at longer wavelengths (as would be the case for green), this area is still perceived as red within the overall visual context. If all other colors are now covered so that only our "red" area is visible, corresponding to the aperture mode, then this area is perceived as green. Clearly, all three approaches must be melded to give a full description of color perception, a process that is not yet fully understood.

In addition to this color constancy phenomenon, there are several other well-known effects that can influence the perception of color. In simultaneous contrast phenomena, there are effects from both luminosity differences and color differences across a boundary. The Bezold-Brücke effect involves a change in hue with luminance, a shift toward the blue end of the spectrum as the luminance is increased. Colors also appear more desaturated at both very high and very low luminances than at intermediate values. In the Helmholtz-Kohlrausch effect there is an increase in the apparent luminance as reflected light increases in spectral purity at constant luminosity. Finally, there is the well-known aftereffect, when prolonged viewing of a color distorts the next-viewed color in the direction of the complementary color of the first.

3.2. Color Vision Defects.   Anomalous color vision is present, eg, if one of the three sets of cones is inoperative (dichromacy) or defective (anomalous trichromacy). This affects 2–3% of the population with males more prone because these defects reside on the X chromosome, with only one present in males but two in females. Eye specialists have standard tests for detecting these and other defects. Summaries of this whole field are available (eg, see Ref. 6).

4. Color Order Systems

Many one-, two-, and three-dimensional systems have been developed over the years to order colors in a systematic way and provide specimen colors for visual comparison. Coordination has now been achieved with computer programs between essentially all of these systems, including the CIE systems described below, and conversions can easily be made between them.

4.1. The Munsell System.   The best known and most widely used color order system is the Munsell system (10,14,15), developed by the artist A. H. Munsell in 1905 and modified over the years. This is a three-dimensional space shown in Figures 4 and 5 on the color plate. Munsell value V (or lightness)

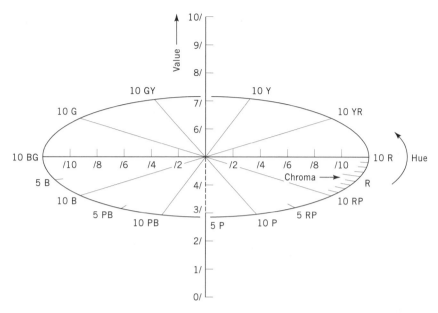

Fig. 4. The coordinate system of the Munsell color order system: R, red; Y, yellow; G, green; B, blue; and P, purple (1).

is used as the vertical axis with 0/ for black at the bottom and 10/ for white at the top. Radially there is the Munsell chroma C with /0 at the center and maximum / 10 or higher at the periphery. The Munsell hue H at the periphery uses the five principal hues: red, yellow, green, blue, and purple and the five adjacent binary combinations such as BG, with 10 steps within each, for a total of 100 hue steps as in Figures 4 and 5. These 100 steps can be further subdivided. A full designation for an orange school bus in the usual HVC sequence and using interpolation might then be 9.5YR 7/9.25. The Munsell system was renotated in 1943 to make it more uniform and consistent with the CIE system. The *Munsell Book of Color* (15) consists of ~1500 painted paper chips available in both glossy and matte versions; there are also textile color collections (15), now discontinued. With interpolations some 100,000 colors can be distinguished within the Munsell system by visual comparison under carefully standardized viewing conditions.

4.2. Other Color Order Systems. The Natural Color System (16), abbreviated NCS, developed in Sweden is an outgrowth of the Hesselgren Color Atlas, and uses the opponent color approach. Here colors are described on the basis of their resemblances to the basic color pairs red-green and blue–yellow, and the amounts of black and white present, all evaluated as percentages. Consider a color that has 10% whiteness, 50% blackness, 20% yellowness, and 20% redness; note that the sum is 100%. The overall NCS designation of this color is 50, 40, Y50R indicating in sequence the blackness, the chromaticness (20 + 20), and the hue (50% on the way from yellow to red; the sequence used is Y, R, G, B, Y).

The Ostwald Color System (17) is a nonequally spaced system and is no longer published, as is also true of the *Maerz and Paul Dictionary of Color*

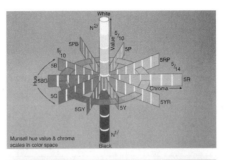

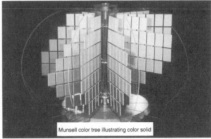

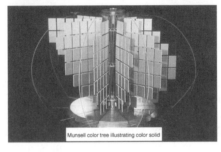

Fig. 5.   Munsell hue, value, and chroma scale arrangement. Courtesy of Gretag Macbeth.

(18). There are also the Colorcurve, the Coloroid, and the German DIN systems, among others. The Optical Society of America has published the OSA Uniform Color Scale System (19) with 558 equally spaced color chips. This uses a designation such as 2:5:3, with the first number (ranging from −7 to +5) specifying the lightness, the second (−6 to +15) giving the blue-yellow content, and the third (−10 to +6) for red–green. For many purposes a much simpler set of 267 color regions as provided by the ISCC-NBS Centroid System, a joint project of the Inter-Society Color Council and the USA National Bureau of Standards, now NIST, (20) is convenient. This uses color names with adjectival modifications of the form vivid blue, brilliant yellow-green, light yellowish brown, and so on.

Many systems of limited range or of lower dimensionality are used in industry for the designation and control of color circumstances where a limited scale is suitable. Examples are the systems for describing plant tissues; soils; and skin, hair, and eye colors. There are no <19 color scales for yellow in oil, fat, varnish, etc, under names such as Gardner, Saybolt, Lovibond, and Hellige (5). Then there is the D, E, F... scale used for near-colorless diamonds devised by the Gemological Institute of America. These systems are well known within highly specialized industries and many have been standardized by industry organizations such as ASTM, TAPPI, and Federal Tests. A summary is given in Reference 5.

5. Basic Colorimetry

The International Commission on Illumination (abbreviated CIE from the French name) over the years has recommended a series of methods and standards in the field of color; for a history of this process see Reference 21.

Consider light of a certain spectral energy distribution falling on an object with a given spectral reflectance and perceived by an eye with its own spectral response. To obtain the perceived color stimulus, it is necessary to multiply these factors together as in Figure 6. Standards are clearly required for both the observer and the illuminant.

5.1. The CIE Standard Observer. The CIE standard observer is a set of curves giving the tristimulus responses of an imaginary observer representing an average population for three primary colors arbitrarily chosen for convenience. The 1931 CIE standard observer was determined for $2°$ foveal vision, while the later 1964 CIE supplementary standard observer applies to a $10°$ vision; a subscript 10 is usually used for the latter. The curves for both are given in Figure 7 and the differences between the two observers can be seen in Table 2. The standard observers were defined in such a way that of the three primary responses $\bar{x}(\lambda)$, $\bar{y}(\lambda)$, and $\bar{z}(\lambda)$, the value of $\bar{y}(\lambda)$ corresponds to the spectral photopic luminous efficiency, ie, to the perceived overall lightness of an object.

CIE used the 1931 CIE standard observer to establish a color representation system in which the hue and saturation could be represented on a two-dimensional diagram. Three tristimulus values X, Y, and Z are first obtained, based on the standard observer, so that the hue and saturation of two objects

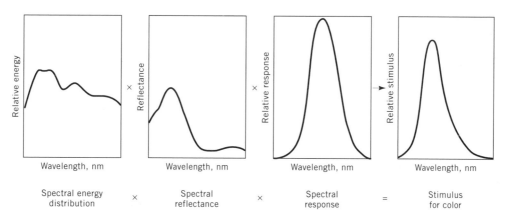

Fig. 6. The stimulus perceived as color is made up of the spectral power (or, as here, energy) curve of a source times the spectral reflectance (or transmittance) curve of an object times the appropriate spectral response curves (one shown here) of the eye (1).

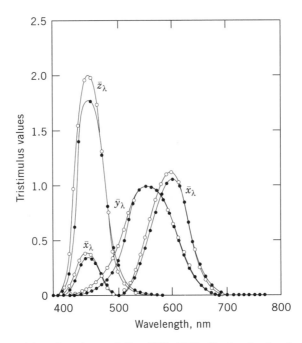

Fig. 7. Color matching functions of the CIE 1931 2° standard colorimetric observer (−●−), and the CIE 1964 10° supplementary standard colorimetric observer (−○−) (1).

match if they have equal values of these three parameters. Each of these is defined, following the concept of Figure 6, in the form:

$$X = k \int S(\lambda)R(\lambda)\bar{x}(\lambda)d\lambda$$

where $S(\lambda)$ is the spectral power distribution of the illuminant, $R(\lambda)$ is the spectral reflectance factor of the object, and $\bar{x}(\lambda)$, $\bar{y}(\lambda)$, and $\bar{z}(\lambda)$ are the color matching functions of one of the standard observers of Figure 7. The constant k is defined

Table 2. **Spectral Chromaticity Coordinates**[a]

Wavelength, nm	x	y	x_{10}	y_{10}	u'	v'	u'_{10}	v'_{10}
400	0.1733	0.0048	0.1784	0.0187	0.2558	0.0159	0.2488	0.0587
450	0.1566	0.0177	0.1510	0.0364	0.2161	0.0550	0.1926	0.1046
500	0.0082	0.5384	0.0056	0.6745	0.0035	0.5131	0.0020	0.5477
550	0.3016	0.6923	0.3473	0.6501	0.1127	0.5821	0.1375	0.5789
600	0.6270	0.3725	0.6306	0.3694	0.4035	0.5393	0.4088	0.5387
650	0.7260	0.2740	0.7137	0.2863	0.6005	0.5099	0.5700	0.5145
700	0.7347	0.2653	0.7204	0.2796	0.6234	0.5065	0.5863	0.5121

[a]Values for the 2° CIE chromaticity coordinates x, y and the 10° coordinates x_{10}, y_{10}; the 2° and 10° red-green metric chromaticity coordinates u' and u'_{10} and the 2° and 10° yellow-blue metric coordinates v' and v'_{10}.

in terms of $S(\lambda)$, the spectral power distributions of the illuminant as

$$k = 100 \Big/ \int S(\lambda)\bar{y}(\lambda)d\lambda$$

so that Y for a perfectly reflecting diffuser is 100.

5.2. Chromaticity Diagrams. The CIE 1931 chromaticity diagram uses the chromaticity coordinates:

$$x = X/(X+Y+Z); \qquad y = Y/(X+Y+Z); \qquad z = Z/(X+Y+Z)$$

It is not actually necessary to specify z, since $x + y + z = 1$. The two-dimensional presentation of x and y is shown in Figure 8 and in Figure 9 on the color plate. Here Newton's pure spectral colors in fully saturated form follow the horseshoe-shaped curve with wavelengths from 400-nm violet to 700-nm red. Coordinates for these spectral colors at 50-nm intervals are given in Table 2 for both 2° and 10° observers. Closing the curves in these figures is the dashed line from red to blue, including the saturated nonspectral purples and magentas.

Any straight line passing through the central achromatic point marked W for white (standard daylight D_{65} in this instance) connects complementary colors,

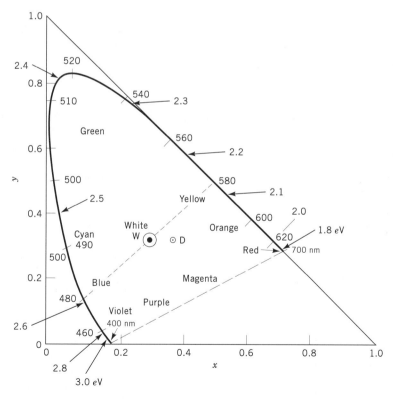

Fig. 8. The 1931 CIE x,y chromaticity diagram showing wavelengths in nm and energies in eV. The central point W (for white) corresponds to standard daylight D_{65} (2).

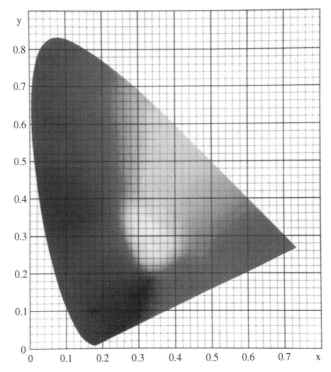

Fig. 9. An artist's representation of where the colors lie on the 1931 CIE *x, y* chromaticity diagram. The CIE does not asociate specific colors with regions on this diagram. Courtesy of Minolta.

such as the line connecting 480-nm blue and 580-nm yellow shown. These two colors will together give white in amounts given by the law of the lever lengths (length WY of the blue and length BW of the yellow). Saturated orange at 620 nm becomes unsaturated when mixed with white as at D. The color of D can be specified by its *X, Y,* and *Z* tristimulus values, by the chromaticity coordinates $x = 0.4$, $y = 0.3$ together with the lightness *Y* in the frequently used *x, y, Y* designation, or it could be described by *Y* together with the dominant wavelength of 620 nm, with 25% purity, since D corresponds to three parts of W and one part of 620-nm orange. If a point is in the lower right part of Figure 8, where the dominant hue is on the nonspectral blue to red join, then there is no dominant wavelength; instead the line is extended upward through the W point to reach the spectral curve at the complementary dominant wavelength, such as 510c nm.

One of several defects of the chromaticity diagram of Figures 8 and 9 is that the minimum distinguishable colors are not equally spaced, ie, that equal changes in *x, y,* and *Y* do not correspond to equally perceived color differences. Very obviously, the greens occupy a disproportionately large area in these figures. Many transformations have been studied to adjust this, but none is perfect. Probably, the most useful employs L^* (the perceived lightness) and the CIE 1976

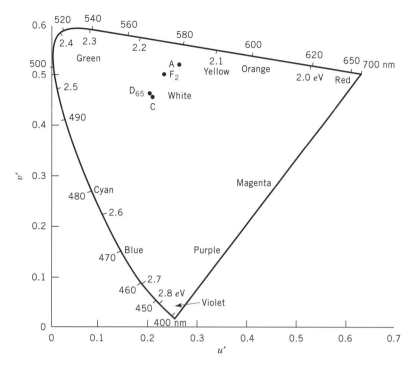

Fig. 10. The currently recommended 1976 CIE uniform u', v' chromaticity diagram showing wavelength in nm and energies in eV; A, C, and D_{65} are standard illuminants and F_2 is a typical fluorescent lamp.

Chromaticity Coordinates u' and v' as shown in Figure 10 obtained from

$$u' = 4X/(X + 15Y + 3Z) = 4x(-2x + 12y + 3)$$

$$v' = 9Y/(X + 15Y + 7Z) = 9y(-2x + 12y + 3)$$

5.3. Standard Illuminants. Three of many sources with quite different energy distributions, which the eye nevertheless accepts as white, are the daylight, incandescent light, and fluorescent lamp light curves shown in Figure 11. The chromaticity diagram is convenient for representing light sources as in Figure 10. One particular use is to describe ideal black body colors, the sequence black, red, orange, yellow, white, and bluish white exhibited as any object is heated up under idealized conditions as described in Section 8.1 below. For real incandescent objects, ie, nonideal black bodies, that ideal black body temperature is used which gives the closest visual color match.

Standardized light sources are clearly desirable for color matching, particularly in view of the phenomenon of illuminant metamerism described below. Over the years, CIE has defined several standard illuminants, some of which can be closely approximated by practical sources. In 1931, there was Source A, defined as a tungsten filament incandescent lamp at a color temperature of

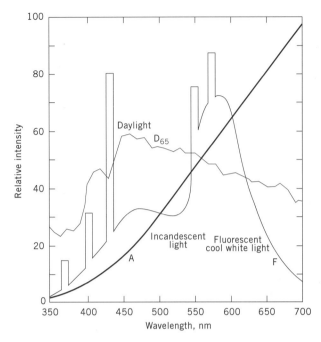

Fig. 11. Energy distributions of CIE standard illuminant A, a tungsten incandescent lamp; a cool white fluorescent lamp F; and CIE standard illuminant D_{65} that approximates average daylight (2).

2854 K. Sources B and C used filtering of source A to simulate noon sunlight and north sky daylight, respectively, but are no longer used. Subsequently a series of D illuminants was established to better represent natural daylight. Of these the most important is Illuminant D_{65}. A series of F illuminants pertain to fluorescent tube lamp sources.

There are ASTM (American Society for Testing and Materials, West Conshohocken, Pa.) standards promulgated for various aspects of color measurement; these are cited in many textbooks (eg, see Ref. 1). Both CIE and ASTM continue to promulgate improved standards and techniques from time to time; this is a field that has not yet reached full maturity.

5.4. Gloss and Opacity. Attributes such as gloss, transparency, translucency, opacity, haze, and luster may apply to some materials (Table 1), and these are relevant in that they may influence the judgment of color differences. As one example, gloss can produce veiling reflections that change the apparent contrast. When present, these attributes can be measured using specialized approaches. Six types of gloss are distinguished (5): specular gloss, sheen, contrast gloss or luster, absence-of-bloom gloss, distinctness-of-image gloss, and surface uniformity gloss. Opacity can be measured by the contrast ratio method, using the reflectance with both a white and a black backing. Standardized procedures for some of these methods have been established by organizations such as ASTM and TAPPI (5).

5.5. Light Mixing. Light or additive mixing applies to light beams, such as stage lighting or on a television screen. A beam of light that appears to match orange at 600 nm when projected onto a white surface may consist of pure 600-nm light or of one of many types of mixtures, eg, of red, orange, yellow, and green, the energy distribution of which has its center of gravity at 600 nm. White results from mixtures of two complementary colors, as discussed in the Section Chromaticity Diagrams, or when any suitable set of three-color beams of the appropriate intensity are mixed, in essence reconstructing the full spectrum in some instances. On the chromaticity diagram of Figures 8 and 9, the condition for equal intensity beams to produce white is that W lies at the center of gravity of the triangle formed by the three sources. A suitable set of primary additive light beams is red, blue, and green, each being near a corner of Figures 8 and 9. Any color within the triangle formed by the three primary colors, designated the gamut, can be reproduced. Red and green by themselves add to give yellow, red and blue give purple and magenta, and blue and green give blue-green and cyan, as can be established by tie lines on Figure 8. It is important to distinguish magenta from red and cyan from blue to avoid confusing the additive from the subtractive system described next.

5.6. Colorant Mixing. A colorant, whether a dye dissolved in or absorbed by a medium or pigment particles dispersed in a paint, produces color by absorbing and/or scattering part of the transmitted or reflected light. Consider a pigment that when illuminated with white light matches 580-nm yellow. It does so because it absorbs wavelengths complementary to yellow, ie blue; more specifically when the center of gravity of the energy distribution of the absorbed wavelengths falls at the 480-nm blue that is complementary to the 580-nm yellow. If only absorption is present, the Beer-Lambert law applies:

$$A(\lambda) = \log 1/T(\lambda) = \sum_i a_i(\lambda)bc_i$$

where A is the absorbance, T is the transmittance, $a_i(\lambda)$ is the absorptivity or specific absorbance of absorber i at wavelength λ, b is the length of the absorbing path, and c_i the concentration of the absorber. When colorants are mixed, they function by each independently absorbing light and the subtractive mixing rules merely specify this additivity. Here the result of mixing three primary colorants is to absorb all light and produce black if sufficiently concentrated. In actual practice a fourth pigment, usually a white or black opacifier, needs to be added to ensure opacity. The preferred primary colorants are the complementary colors of the corners of Figures 8 and 9, namely, cyan, yellow, and magenta (unfortunately usually designated blue, yellow, and red when taught in elementary school). Combining yellow and cyan colorants then produces green, yellow and magenta give red, and cyan plus magenta give blue. Mixtures of light beams in additive mixing are fully predictable because they give straight lines on the chromaticity diagram. Mixtures of pigments usually give curves on the chromaticity diagram and are therefore not simply predictable. It should also be noted that artists do not in fact mix primary colors on their palette, since this would limit the saturation: no mixture of blue and yellow paints can match the intensity available in the pigment chrome green.

When both absorption and scattering are present, the Beer-Lambert law must be replaced by the Kubelka-Munk equation employing the absorption

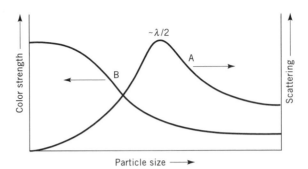

Fig. 12. Curve A, the schematic variation of scattering, and B, color strength for pigment particles of various sizes (2).

and scattering coefficients K and S, respectively. This gives the reflectivity R_∞

$$(1 - R_\infty)^2/2R_\infty = K/S = \sum_i c_i K_i(\lambda) \Big/ \sum_i c_i S_i(\lambda)$$

where c_i is the concentration and $K_i(\lambda)$ and $S_i(\lambda)$ are the specific absorbance and scattering parameters, respectively, of absorber and scatterer i at wavelength λ.

The color of an opaque paint depends both on the size of the pigment particles and on refractive index considerations. The scattering is maximum, as shown in Curve A of Figure 12, when the particle size is about one-half of the wavelength of light, ie, in the 200–350-nm range; this is desired for opacity in typical inorganic pigments with refractive index much larger than that of the medium. With organic pigments, where the refractive index is not very different, scattering is replaced by absorption as the principal interaction; here the color strength or absorptive power increases as the particle size is decreased to a much smaller size, as shown in curve B of Figure 12. Unfortunately, too small a particle size may lead to increased chemical reactivity and a reduction of the lightfastness, requiring compromises.

The ready availability of computers has led to the detailed analysis of the colorant formulation problems faced every day by the textile, coatings, ceramics, polymer, and related industries. The resulting computer match prediction has produced improved color matching and reductions in the amounts of colorants required to achieve a specific result with accompanying reductions of cost. Detailed treatments have been given for dyes and for pigments (eg, see 9).

5.7. Metamerism.   There are several types of metamerism, the phenomenon in which two objects perceived as having a perfect color match under one set of conditions are found to differ in color under other conditions. Most common is illuminant metamerism that occurs when a change in illuminant is the cause. This originates from the above described situation that a given visually identified hue can be caused by many different stimuli, eg, by an object reflecting, say, only pure spectral orange, or red plus yellow, etc. A change in illuminant to one having more energy in the red, for example, would leave the perceived color of the pure spectral orange unchanged, but would make the red plus yellow combination appear to be a more reddish orange. Another related cause would be the presence of an ultraviolet component in one of the sources (usually actual daylight) causing a fluorescent object to emit light in addition to that reflected compared to a visible-light-only source such as an incandescent lamp. Note that the standard illuminant D_{65} specifies significant intensity at wavelengths less than 400 nm in the uv.

Observer metamerism derives from the significant differences in spectral response found among persons even with normal color vision. The variation in cone concentration and cone-type distribution discussed above results in field size metamerism. Here an example is the difference between the 2° and 10° response curves of the standard observers of Figure 7. Finally, there can also be a change in perceived color with a change in viewing angle, as with some metallic paints and with interference-based color-changing inks.

Quantification of metamerism is difficult and its avoidance is a principal aim in color technology. The car where body color, upholstery, and plastic parts match in the lit showroom should not show clashing colors in daylight. The obvious but rarely applicable solution is to use the same pigments in all parts. Some illuminant metamerism is almost unavoidable; the aim of the color expert is to keep it within acceptable limits, ie, to achieve adequate color consistency.

6. The Measurement of Color Differences

The 1931 CIE system with its x, y, Y chromaticity diagram system is useful for general colorimetry and color matching, but it suffers several drawbacks, including that it is not suitable for color-difference measurements. These problems were partly corrected in the 1976 CIE u', v' chromaticity diagram to achieve a more uniform color space. Both the x, y, and u', v' systems are still basically two-dimensional, with the lightness Y or L^* added. A series of transformations led to modified systems that are fully three-dimensional with the length of any lines, however, oriented within this space, more closely representing perceived color differences. Some of these are opponent related. Two systems were agreed upon, designated the CIELUV and CIELAB color spaces, and have come into widespread use.

6.1. The 1976 CIELUV Color Space. The CIEUV space, properly designated CIE $L^* u^* v^*$, uses a white object or light source designated by the subscript n as the reference standard and employs the transformations:

$$L^* = (116Y/Y_n)^{1/3} - 16; \ Y/Y_n > 0.008856$$

$$L^* = (903.3Y/Y_n); \ Y/Y_n \leq 0.008856$$

$$u^* = 13L^*\left(u' - u'_n\right)$$

$$v^* = 13L^*\left(v' - v'_n\right)$$

where L^* is the perceived lightness.

The CIELUV space preserves a property of the CIE 1931 chromaticity space which is important in the field of color reproduction, eg, in the television industry. This is the characteristic that the chromaticities of additive mixtures of color stimuli lie on the straight line connecting the chromaticities of the component stimuli; this is true of the 1976 metric chromaticity diagram but not of the CIELAB space that follows.

6.2. The 1976 Cielab Color Space.   Defined at the same time as the CIELUV space, the CIELAB space, properly designated CIE $L^*a^*b^*$, is a non-linear transformation of the 1931 CIE x, y space. It also uses the metric lightness coordinate L^*, together with

$$a^* = 50\left[(X/X_n)^{1/3} - (Y/Y_n)^{1/3}\right]$$

$$b^* = 200\left[(Y/Y_n)^{1/3} - (Z/Z_n)^{1/3}\right]$$

These equations apply for X/X_n, Y/Y_n, and Z/Z_n all >0.008856. For $X/X_n \leq 0.008856$, the term $(X/X_n)^{1/3}$ is replaced by $[7.787(X/X_n) + 16/116]$ and similarly for Y and for Z in these two equations.

This transformation results in a three-dimensional space that follows the opponent color system with $+a^*$ as red, $-a^*$ as green, $+b^*$ as yellow, and $-b^*$ as blue. CIELAB is closely related to the Munsell system of Figure 4 as well as to the older Adams-Nickerson and other spaces of the L,a,b type, which it replaced (1,5).

The CIELAB coordinates $L^*a^*b^*$, either in that form or in the $L^* C_{ab}^*$, h_{ab} form discussed below, are the most commonly used color descriptors in the field of paints, pigments, textiles, paper, ceramics, polymers, and most other opaque to transparent substances.

6.3. The 1976 CIE Metric Color Spaces.   Both the CIELUV and CIELAB spaces can have their Cartesian coordinates converted to cylindrical coordinates, called metric or hue-angle coordinates, with L^* unchanged. These coordinates are designated CIE $L^* C_{uv}^* h_{uv}$ and CIE $L^* C_{ab}^* h_{ab}$, respectively. The metric hue-angles are given as

$$h_{uv} = \tan^{-1}(v^*/u^*) \qquad h_{ab} = \tan^{-1}(b^*/a^*)$$

and the metric chromas

$$C_{uv}^* = \left[(u^*)^2 + (v^*)^2\right]^{1/2} \qquad C_{ab}^* = \left[(a^*)^2 + (b^*)^2\right]^{1/2}.$$

The close analogy of h_{ab} to the Munsell hue and of C_{ab}^* to the Munsell chroma of Figure 4 is evident; L^* is also closely related to the Munsell value. The set of four opponent (or psychological) primary colors based on CIELAB is red, blue, green, and yellow; these are not the identical colors to those used in the additive and sbtractive primaries of Sections Light Mixing and Colorant Mixing. Closely related are the many types of color wheels, used for pedagogical purposes but never actually used by artist.

In the system derived from CIELUV it is also possible to specify a metric saturation with reference to a standard white designated by subscript n:

$$S_{uv} = 13\left[(u^* - u_n^*)^2 + (v^* - v_n^*)^2\right]^{1/2}$$

A saturation correlate cannot be given for the CIE $L^* C_{ab}^* h_{ab}$ space.

6.4. Hunter *L,a,b* and Other Color Spaces.

The CIELAB and CIE-LUV color spaces were the outgrowth of a large and complex group of interrelated early systems and have replaced essentially all of these. An important early system was the 1942 Hunter *L,a,b* group of color spaces (1). This was the earliest practical opponent-based system that is still occasionallyt used. In this system, for illuminant C and the 2° standard observer:

$$L = 10Y^{1/2} \qquad \text{(lightness coordinate)}$$

$$a = 17.5(1.02X - Y)/Y^{1/2} \qquad \text{(red–green coordinate)}$$

$$b = 7.0(Y - 0.847Z)/Y^{1/2} \qquad \text{(yellow–blue coordinate)}$$

There are other equations for other illuminants and other observers (1) and also various modifications for special conditions.

6.5. Color Difference Assessment.

Early color difference scales include those of Judd-Hunter, Macadam, Adams-Nickerson, ANLAB, and ANLAB40. All of these have limitations in some way or another; they are described in most texts (1,3–5). Each applies only to the precise conditions used in their determination and interconversion is not possible; they are falling out of use.

In the CIELAB and CIELUV color spaces, the difference between a batch sample and a reference standard designated with a subscript *s*, can be designated by its components, eg, $\Delta L^* = L^* - L_s^*$.

The three-dimensional total color differences are given by Euclidian geometry as the 1976 CIE $L^* a^* b^*$ and 1976 CIE $L^* u^* v^*$ color difference formulas:

$$\Delta E_{ab}^* = \left[(\Delta L^*)^2 + (\Delta a^*)^2 + (\Delta b^*)^2 \right]^{1/2}$$

$$\Delta E_{uv}^* = \left[(\Delta L^*)^2 + (\Delta u^*)^2 + (\Delta v^*)^2 \right]^{1/2}$$

In CIE metric coordinates, either for CIELAB or CIELUV, ΔL^* and ΔE^* are the same, $\Delta C^* = C^* - C_s^*$, $\Delta h = h - h_s$, and $\Delta H^* = [(\Delta E^*)^2 - (\Delta L^*)^2 - (\Delta C^*)]^{1/2}$. The last of these, the metric hue difference ΔH^*, is preferred to Δh, since the latter is in degrees rather than in units compatible with ΔL^* and ΔC^*, as in ΔH^*.

An example may clarify this system. Consider a red apple with CIELAB coordinates measured as $L^* = 41.75$, $a^* = 45.49$, $b^* = 9.61$. This converts to metric as $L^* = 41.75$, $C^* = 46.49$, $h = 13.25$ and, incidentally, to Munsell 10RP 4/10 (recall that the sequence is hue, value, chroma). A second apple has $L^* = 49.23$, $a^* = 40.13$, $b^* = 12.20$, $C^* = 41.94$, $h = 18.79$, and Munsell 2.5R 5/9. Taking the differences $\Delta L^* = +7.48$, ie, the second apple is brighter (note Munsell values are 4 and 5, respectively); $\Delta a^* = -5.36$, ie, more green (or less red); $\Delta b^* = +2.59$, ie, more yellow (or less blue); $\Delta C^* = -4.55$, ie, less saturation (note Munsell chromas are 10 and 9, respectively); $\Delta h = +5.54$, ie, larger hue angle that means more yellow in this instance (note Munsell hue 2.5R is $2\frac{1}{2}$ steps from 10RP as in Figure 4 or $2.5 \times 360/100 = 9°$ away). The total three-dimensional color difference $\Delta E^* = 9.56$. The metric hue difference can also now be calculated as $\Delta H^* = 3.84$.

The lightness difference $\Delta L^* = +7.48$ is the dominant component in the total color difference $\Delta E^* = 9.56$ and the hue difference $\Delta H^* = 3.84$ is less important than the chroma difference $\Delta C^* = -4.55$.

The color difference magnitudes derived from the CIE as well as from other color-space and color-ordering systems do not agree as well as could be desired with one another or with the visually perceived differences; they cannot be inter-converted by a constant factor in general. This is probably the least satisfactory part of colorimetry. Many factors contribute to this. In an ideal three-dimensional color space, the region that is not distinguishably different from a given point would be a sphere. In actual practice this is not achieved fully in any color space; these regions are the so-called MacAdam ellipses (1,3–5) on two-dimensional chromaticity diagrams; they are ellipsoids in the three-dimensional color spaces.

The most recently promulgated improved color-difference equation is CMC(l:c). This quite complex set of equations uses correction factors to reduce the previous deficiencies (1). Work continues on such improvements (eg, see Ref. 1) and the CIEDE2000 color-difference equation set is currently under review but has not yet been promulgated.

$$\Delta E_{\text{CMC}(l:c)} = \left[\left(\frac{\Delta L^*}{lS_L}\right)^2 + \left(\frac{\Delta C_{ab}^*}{cS_C}\right)^2 + \left(\frac{\Delta H_{ab}^*}{S_H}\right)^2\right]^{1/2}$$

where

$$S_L = 0.04097\, L^*/(1 + 0.01765L^*)$$

unless

$$L^* < 16 \quad \text{then}$$

$$S_L = 0.511$$

$$S_C = \{0.0638\, C_{ab}^*/(1 + 0.0131\, C_{ab}^*)\} + 0.638$$

$$S_H = S_C(TF + 1 - F)$$

$$F = \{(C_{ab}^*)^4/[(C_{ab}^*)^4 + 1900]\}^{1/2}$$

$$T = 0.38 + [0.4 \cos(h_{ab} + 35)]$$

unless h_{ab} is between $164°$ and $345°$, then

$$T = 0.56 + [0.2 \cos(h_{ab} + 168)]$$

L^*, C_{ab}^*, and h_{ab} are calculated from the CIELAB L^*, a^*, and b^* values of the standard. Constants l and c are defined by the user and weight the importance of lightness and chroma relative to hue.

A perceived color difference varies with the mode (object, illuminant, aperture); the texture (glossy, rough, metallic, etc); size, flatness, and transparency characteristics of objects; the level, color, and geometry (point-source versus diffuseness) of the illumination; the presence of uv light, fluorescence, and polarized light; and the nature and color of the surroundings and background. Finally, the various metamerisms discussed above are at work as well as observer experience and adaptation to the observing situation. Nevertheless the system does work and color difference measurements can be used successfully in actual practice if all these parameters are controlled.

Finally, it cannot be overemphasized that despite instrumental measurements and data manipulations, it is the perception of the eye that still is the final arbiter as to whether or to what degree two colors match. Instrumental methods do serve well for the typical industrial task of maintaining consistency under sufficiently well-standardized conditions; however, a specific technique may not serve in extreme or unusual conditions for which it was not designed.

7. Color Measuring Instruments

There has been a tremendous change in the last two decades as computers have taken over the tedious calculations involved in color measurement. Indeed, microprocessors either are built into or are connected to all modern instruments, so that the operator may merely need to specify, for example, x, y, Y or L^*, a^*, b^* or L^*, C^*, h, either for the $2°$ or the $10°$ observer, and for a specific standard illuminant, to obtain the desired color coordinates or color differences, all of which can be stored for later reference or computation. The use of high intensity filtered xenon flash lamps and array detectors combined with computers has resulted in almost instantaneous measurement in most instances.

7.1. Measurement Conditions. In 1968, the CIE recommended the four geometries of Figure 13 for reflectance measurements. In the first $0°/45°$ or normal/$45°$ geometry, the illumination is normal to the sample and the detector is at $45°$. In the second $45°/0°$ geometry the two conditions are reversed. These techniques usually give the same results except, eg, when polarized light interactswith oriented metallic flakes. In the third $0°$/diffuse or normal/diffuse geometry the incident light beam impinges normally onto the sample and the reflected and scattered light is collected by an integrating sphere. In the last diffuse/$0°$ geometry the incident light is first diffused by the integrating sphere before interacting with the sample, with the detector normal to the sample. Integrating sphere instruments in the normal/diffuse geometry are often operated in a "*near-normal*" configuration. Here an offset of $6°-8°$ is used and a specular port is provided at a symmetrical position By filling this port with a light-absorbing trap instead of the usual sphere material, the effect of specular reflection can be eliminated instead of being included in the measurement.

The illumination used is usually filtered white light to approximate daylight. The detector may respond to all light, may use a monochromator, a diode array, a rotating wedge interference filter, or band pass filters. Often, several photodetectors are covered with filters approximating the CIE color matching functions $\bar{x}$, $\bar{y}$, $\bar{z}$ or a suitable linear combination of these. Problems arise when fluorescence is present. Ideally, one would use a spectrofluorimeter, which contains two monochromators, one before and one after the sample.

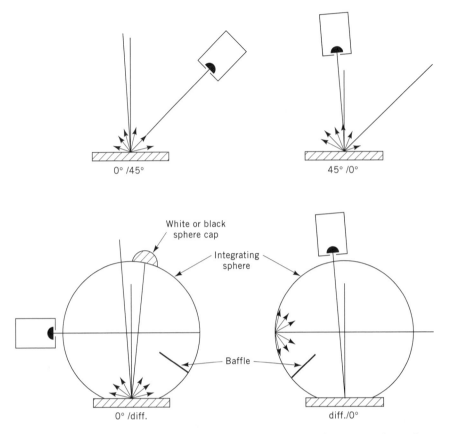

Fig. 13. CIE recommended geometries for illumination and viewing for reflectance-factor measurements (8).

This permits the separation of the fluorescent and nonfluorescent components. More usually D_{65} illumination is used, in which case the detected intensity at the fluorescence wavelength may well be higher than the intensity of that wavelength in the illuminant. Since the fluorescence usually varies strongly with the amount of uv in the illuminant, it may be difficult to obtain meaningful results. Here, as elsewhere there are ASTM standards (eg, see Ref. 1).

For transmittance measurements there are two principal geometries, one with a full collection of all transmitted light by an integrating sphere, the other with both illuminating and detector beams collimated. Translucency can be measured by two reflectance measurements, one with a white background, the other with a black one. Goniophotometric instruments are used to measure specular reflections, important in glossy and metallic objects and coatings, as well as in color-changing paints, containing oriented pigment flakes that show interference.

7.2. Instruments. Spectrophotometers are the most sophisticated color measuring instruments and provide the most detailed and accurate information. They may provide continuous spectral data of reflectance and transmittance against wavelength or use up to 20-nm wavelength steps, with high precision in reflectance (down to 0.01%) and tristimulus values (down to 0.01). There may be dual beams with one for the sample and a second as reference for the most stable and precise operation. Many spectrophotometers use an integrating sphere, although some use other geometries or permit alternative ones. Some of these have additional capabilities such as limited gloss measurements; some are compact, battery-powered, portable units with built-in microprocessors.

High end spectrophotometers may use xenon flash lamps and large diode detector arrays while lower end units may use light emitting diodes and small arrays. Slightly less sophisticated are spectrocolorimeters that determine spectral response curves for the computation of colorimetric values, but from which the spectral curve itself is not available.

Colorimeters, also known as tristimulus colorimeters, are instruments that do not measure spectral data but typically use four broadband filters to approximate the $\bar{y}, \bar{z}$ and the two peaks of the $\bar{x}$ color-matching functions of the standard observer curves of Figure 7. They may have lower accuracy and be less expensive, but they can serve adequately for some industrial color control functions.

A goniospectrophotometer measures the angular dependence of the spectral variation of light reflection and scattering. More widely used are abridged goniospectrophotometer or multiangle goniophotometers that may have three-to-five set geometries. Such measurements are important in characterizing metallic and pearlescent finishes, eg. A variety of goniophotometers, reflectometers, gloss meters, haze meters, etc, are available, as are computers and software for acquiring and manipulating data from various instruments and for converting among the various color-order and color-measuring systems. Most modern instruments either have such data manipulating capabilities built in or may connect directly to a computer.

Companies that manufacture color-measuring instruments and other color-related products include (in alphabetical order): BYK-Gardner Instruments (www.byk-gardner.com), Lawrenceville, NJ (subsidiary of BYK-Chemie, Geretsried, Germany); Color-Tec (www.color-tec.com), Clinton, NJ; Datacolor (www.datacolor.com), Lawrenceville, NJ (subsidiary of Eichhof Holding AG, Lucerne, Switzerland); Gretag Macbeth LLC (www.gretagmacbeth.com), New Windsor, NY; Hunter Associates Laboratory (www.hunterlab.com), Reston, Va; Minolta (www.minoltausa.com), Ramsey, NJ (subsidiary of Konica Minolta, Osaka, Japan); Pantone (www.pantone.com), Carlstadt, NJ; X-rite (www.xrite.com) Grandville, MD.

Sample preparation is always critically important. Calibration with standard reflectance and transmittance samples should be routinely used for optimum results in spectrophotometry and colorimetry. Calibration of the wavelength and photometric scales is also advisable. The calibration of a white reflectance standard in terms of the perfect reflecting diffuser is important, as is the use of diagnostic tiles for tristimulus colorimetry. A collaborative reference program is available on instrument performance (22).

Color-order systems, such as the many Munsell collections available from Gretag Macbeth, have been described previously. Essential for visual color matching is a color-matching booth. A typical one, such as the Gretag Macbeth Spectralite, may have available a filtered 7500 K incandescent source equivalent to north-sky daylight, 2300 K incandescent illumination as horizon sunlight, a cool-white fluorescent lamp at 4150 K, and an uv lamp. By using the various illuminants, singly or in combination, the effects of metamerism and fluorescence can readily be demonstrated and measured. Every user should be checked for color vision deficiencies.

8. The Fifteen Causes of Color

No less than 15 distinct chemical and physical mechanisms explain the various causes of color, ordered into five groups as in Table 3. (A detailed book-length treatment may be found in Ref. 2.) In the first group, covered by quantum theory, there are incandescence, simple electronic excitations, and vibrational and rotational excitations. Most chemical compounds contain only paired electrons that

Table 3. Fifteen Causes of Color

Cause	Examples
Vibrations and simple excitations	
incandescence	flames, lamps, carbon arc, limelight
gas excitations	vapor lamps, flame tests, lightning, auroras, some lasers
vibrations and rotations	water, ice, iodine, bromine, chlorine, blue gas flame
Transitions involving ligand field effects	
transition-metal compounds	turquoise, chrome green, rhodonite, azurite, copper patina
transition-metal impurities	ruby, emerald, aquamarine, red iron ore, some fluorescence and lasers
Transitions between molecular orbitals	
organic compounds	Most dyes, most biological colorations, some fluorescence and lasers
charge transfer	blue sapphire, magnetite, lapis lazuli, ultramarine, chrome yellow, Prussian blue
Transitions involving energy bands	
metals	copper, silver, gold, iron, brass, pyrite, ruby glass, polychromatic glass, photochromic glass
pure semiconductors	silicon, galena, cinnabar, vermillion, cadmium yellow and orange, diamond
doped semiconductors	blue and yellow diamond, light-emitting diodes, some lasers and phosphors
color centers	amethyst, smoky quartz, desert amethyst glass, some fluorescence and lasers
Geometrical and physical optics	
dispersive refraction	prism spectrum, rainbow, halos, sun dogs, green flash, fire in gemstones
scattering	blue sky, moon, eyes, skin, butterflies, bird feathers, red sunset, Raman scattering
interference	oil slick on water, soap bubbles, coating on camera lenses, some biological colors
diffraction	diffraction gratings, opal, aureole, glory, some biological colors, most liquid crystals

require very high energies to become unpaired and form excited energy levels; this requires uv, hence there is no visible absorption and no color. Absorption color can, however, be derived from the easier excitation of lower energy unpaired electrons in transition-metal compounds and impurities, covered by ligand-field theory in the second group. Absorptions from paired electrons can be shifted into the visible by increasing the size of the region over which the electrons are localized, as in organic compounds, covered by molecular orbital theory in the third group; this also explains various forms of charge transfer. In the fourth group, there is color in metals and alloys as well as in semiconductors such as yellow cadmium sulfide, both pure and doped, covered by band theory; this also covers color centers. In the final group, there are four color-causing mechanisms explained by geometrical and physical optics.

8.1. Color from Incandescence. Any object emits light when heated, with the sequence of blackbody colors, black, red, orange, yellow, white, and bluish-white as the temperature increases. The locus of this sequence is shown on a chromaticity diagram in Figure 14. A blish-white hot is the hottest possible, corresponding to an infinite temperature; terms such as blue-hot and green-hot sometimes seen are spurious.

The distribution of energy under idealized conditions is given by Planck's equation in which the energy E_R radiated into a hemisphere in W/cm^2 in wavelength interval $d\lambda$ at wavelength λ at T in K is

$$E_R = 37415 \, d\lambda/\lambda^5 \left[e^{(14338/\lambda T)} - 1 \right]$$

The total energy emitted E_T in W/cm^2 is given by Stefan's law as

$$E_T = 5.670 \times 10^{-12} \, T^4$$

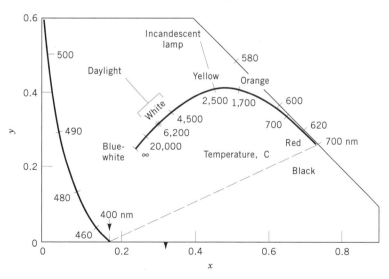

Fig. 14. Blackbody colors shown on the 1931 CIE x,y chromaticity diagram of Figures 8 and 9; the asterisk is standard illuminant D$_{65}$ at 650 K (2).

and Wien's law gives the peak wavelength λ_m in nm as

$$\lambda_m = 2,897,000/T$$

Real or gray bodies deviate from these ideal blackbody values by the λ-dependent emissivity, but the color sequence remains essentially the same. This mechanism explains the color of incandescent light sources such as the flame in a candle, tungsten filament light bulb, flash bulb, carbon arc, limelight, lightning in part, and the incandescent part of pyrotechnics (qv).

8.2. Color from Gas Excitation.   When the atoms in a gas or vapor are excited, eg, by an electric discharge, electrons on the atoms are elevated to higher energy states. On their way back to the ground state they can emit light, the energies being limited by the quantum states available and by the selection rules that control the probabilities of the various possible transitions. Examples are efficient gas discharge light sources such as sodium and mercury vapor lamps, the latter emission also being present in fluorescent lamps in part; lightning in part; auroras, excited by energetic particles from the sun, in part; the Bunsen-burner flame tests for Na, Li, Sr, etc, where the excitation is thermal; and the electrically excited gas lasers such as the helium–neon laser.

8.3. Color from Vibrations and Rotations.   Vibrational excitation states occur in H_2O molecules in water. The three fundamental frequencies occur in the ir at >2500 nm, but combinations and overtones of these extend with very weak intensities just into the red end of the visible and cause the complementary blue color of water and of ice when viewed in bulk [any green component present derives from algae, etc. The attribution of the blue color to reflection from the sky on water or to the presence of gas bubbles in ice is spurious (2)]. This phenomenon is normally seen only in H_2O, where the lightest atom H and very strong hydrogen bonding combine to move the fundamental vibrations closer to the visible than in any other material.

Electronic energy levels in molecules are often modified by vibrational and rotational excitations. In iodine vapor this mechanism is involved in the intense purple color, as well as in the much weaker colors of bromine and chlorine, both in the condensed and gaseous states. Blue and green emissions from energy levels in the unstable CH and C_2 molecules occur in the premixed region of candle and gas flames. Such emissions also occur in some auroras, particularly in the broad bands from the excited N_2^* molecule. Finally, the slow phosphorescence from triplet states in organic molecules usually involves vibrational states.

8.4. Color from Transition-Metal Compounds and Impurities.   The energy levels of the excited states of the unpaired electrons of transition-metal ions in crystals are controlled by the field of the surrounding cations or cationic groups. From a purely ionic point of view, this is explained by the electrostatic interactions of crystal-field theory; ligand-field theory is a more advanced approach also incorporating molecular orbital concepts.

Consider a crystal of corundum [1302-74-5], pure Al_2O_3. Each Al is surrounded by six O ligands in the form of a slightly distorted octahedron. All electrons are paired and there are no absorptions in the visible region, and hence no color. If a few percent of the aluminum atoms are replaced with chromium, the result is the red mineral, gem, laser, and maser crystal ruby [12174-49-1]. The term diagram, giving the effect of different strength ligand fields on the ^2E, $^4T_1$, and $^4T_2$ excited energy levels with respect to the ground-state $^4A_2$ level of the three unpaired electrons in the $3d$ orbitals of Cr^{3+}, is shown at (**a**) in Figure 15. The 2.23 eV ligand field in ruby results in the energy level scheme at (**b**) in this figure, with absorption of light in the violet and green as at (**c**), leading to the intense red color with a weak blue component in transmission. There is also the well-known red fluorescence of ruby as the system returns to the ground state after visible light or uv absorption and heat emission, as shown at (**b**) and (**c**). The change in chemistry from Al_2O_3 to the $Be_3Al_2Si_6O_{18}$ of beryl [1302-52-9] does not change the nature or the symmetry of the Al-surrounding ligands but reduces the ligand field some 8% to 2.05 eV. In Cr^{3+}-containing beryl the result is the scheme at (**d**) in Figure 15, with both ^4T levels shifted to slightly lower energy levels. The result is the intense green color of emerald [12415-33-7], with essentially the same red fluorescence as in ruby.

The unexpected situation intermediate between red ruby and green emerald is chrysoberyl [1304-50-3], $BeAl_2O_4$, called alexandrite [12252-02-7] when containing Cr^{3+}. Here an intermediate crystal field of 2.17 eV produces evenly balanced red and green transmissions, resulting in the alexandrite effect with a red color perceived when this material is viewed in red-rich incandescent light and a green color when viewed in blue- and green-rich daylight and fluorescent light. This same change occurs as the chromium content of ruby is increased in the solid-solution series Al_2O_3:Cr_2O_3, the Cr_2O_3 end member being the pigment chrome green [1308-38-9] with a ligand field near 2.07 eV. At ~25% Cr_2O_3 the color changes from red via gray to green. In the intermediate region, pressure shifts the color to red by shortening the bonds and increasing the ligand field in piezochromism, while temperature has the reverse effect, shifting the color to green in thermochromism (see CHROMOGENIC MATERIALS).

The ligand field decreases in the spectrochemical sequence $CN^- > NH_3 > O^{2-} > H_2O > F^- > Cl^- > Br^- > I^-$ and increases in isoelectronic sequences such as $V(II) < Cr(III) < Mn(IV)$. Whereas in the $3d$ transition elements the color varies strongly with the symmetry and magnitude of the ligand field, in the lanthanides and actinides the unpaired electrons in the $4f$ and $5f$ shells, respectively, are shielded by the outer electrons so that there are only minor energy level changes and, therefore, color changes with the ligand-field. Most ligand-field transitions are formally forbidden by the selection rules, hence transitions have low oscillator strengths and colors tend to be weak. That is why well-colored red ruby and green emerald need to contain several percent of Cr.

Idiochromatic (self-colored) transition-metal compounds occur where the transition-metal ions are an essential part of the structure and contribute to the nature of the ligand field. Examples in addition to the pigment chrome green Cr_2O_3 are purple chrome alum $KCr(SO_4)_2 \cdot 12H_2O$; pink rhodonite [14567-57-8], $MnSiO_3$; green vitriol, $FeSO_4 \cdot 7H_2O$, and yellow goethite [1310-14-1], $FeO(OH)$; blue cobalt oxide CoO [1307-96-6] and pink sphaerocobaltite [14476-13-2], $CoCO_3$; green bunsenite [1313-99-1], NiO; yellow tenorite [1317-92-6], CuO, blue azurite [1319-45-5], $Cu_3(CO_3)_2(OH)_2$, turquoise [1319-32-0], $CuAl_6(PO_4)_4(OH)_8 \cdot 5H_2O$, and the various green salts present in copper patina. Allochromatic impurities may modify these colors.

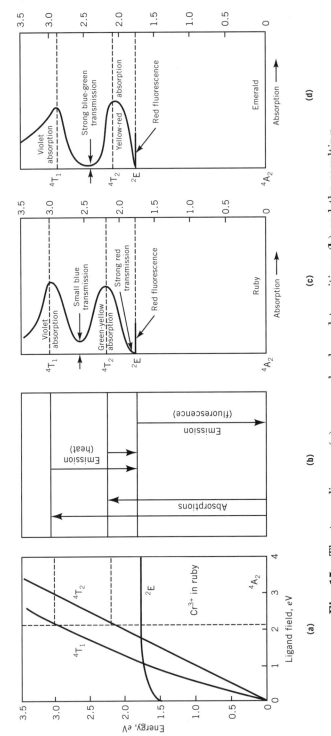

Fig. 15. The term diagram (**a**), energy levels and transitions (**b**) and the resulting absorption and emission of ruby (**c**) and emerald (**d**) (2).

330

Allochromatic (other-colored) transition-metal compounds, involve small amounts of these same transition elements but in the ligand field of the host lattice. Examples in addition to the above discussed chromium-containing ruby, emerald, and alexandrite are red beryl containing manganese; the iron-containing green or blue beryl aquamarine [1327-51-1] and many brown and red iron-containing minerals such as sandstone and red iron ore; the intense blue cobalt glass; a green vanadium -containing form of emerald; and purple neodymium - containing yttrium aluminum garnet YAG [12005-21-9], $Y_3Al_5O_{12}$. Some of these, such as ruby and Nd:YAG, serve as the active media of optically pumped crystal lasers. The absorptions and fluorescence emissions from ligand field energy levels tend to be relatively narrow in crystals. In glasses, where disorder leads to a range of ligand fields, these absorptions and emissions are much broader, as in the Nd:glass used in lasers such as the NOVA thermonuclear fusion lasers. In the decolorizing of glass, the greenish color caused by iron impurities is removed by adding $Mn^{IV}O_2$ [1313-13-9]. This acts in two ways: it reduces some of the green-producing Fe(II) to the yellow-producing but weaker colorant Fe(III) while forming some Mn(III). This latter also produces a purple color; since this is complementary to the green of Fe(II), it results in an inconspicuous very pale gray.

8.5. Color in Organic Compounds. In organic molecules, particularly those containing conjugated chains of alternating single and double bonds, eg, as in the polyenes, $H_3C(-CH=CH-)_nCH_3$, the double bonding p_z orbitals have a variety of excited states. For $n = 2$ in 2,4-hexadiene [592-46-1], the transition between the HOMO (highest occupied molecular orbital) and the LUMO (lowest unoccupied MO) is in the uv, hence no color. With $n = 8$ in 2,4,6,8-decatetraene [2423-96-3] this transition has shifted down into the violet end of the visible part of the spectrum, thus producing a complemenmtary yellow color. The electronic energy levels are also modified by vibrational effects, resulting in transitions to triplet states and leading to phosphorescence from the very slow forbidden transition back to the singlet ground state.

Examples of polyene-type colorants are the orange β-carotene pigment [7235-40-7] of carrots, the pink carotenoids of flamingos, and the yellow crocin [42553-65-1] present in saffron, the pollen of the crocus *sativa*. Cyclic but non-benzenoid conjugated colored systems include the green chlorophyll [1406-65-1] of the vegetable kingdom, the red-to-brown hemoglobins and porphyrins of the animal kingdom, and the related blue dye copper phthalocyanin [147-14-8].

The conjugated chromophore (color-causing) system can be extended by electron-donor groups such as $-NH_2$ and $-OH$ and by electron-acceptor groups such as $-NO_2$ and $-COOH$, often used at opposite ends of the molecule. An example is the aromatic compound alizarin [72-48-0], also known since antiquity as the red dye madder.

alizarin   indigo

Another ancient dye is the deep blue indigo [482-89-3]; the presence of two bromine atoms at positions marked by an * gives the dye Tyrian purple [19201-53-7] once laboriously extracted from certain sea shells and worn by Roman emperors.

Organic colors caused by this mechanism are present in most biological colorations and in the triumphs of the dye industry (see AZINE DYES; AZO DYES; FLUORESCENT WHITENING AGENTS; CYANINE DYES; DYE CARRIERS; DYES AND DYE INTERMEDIATES; DYES, ANTHRAQUINONE; DYES, APPLICATION AND EVALUATION; DYES, NATURAL; DYES, REACTIVE; POLYMETHINE DYES; STILBENE DYES; and XANTHENE DYES). Both fluorescence and phosphorescence occur widely and many organic compounds are used in tunable dye lasers such as rhodamine B [81-88-9], which operates from 580 to 655 nm.

rhodamine B

8.6. Color from Charge Transfer.   This mechanism is best approached from MO theory, although ligand-field theory can also be used. There are several types of color-producing charge-transfer (CT) processes.

Consider corundum, Al_2O_3 containing both Fe^{2+} and Ti^{4+} substituting on adjacent Al sites. The transfer of one electron can occur with the absorption at about 2.2 eV, resulting in the production of Fe^{3+} and Ti^{3+}. This light absorption results in the color of blue sapphire [1317-82-4]. The reverse transition restores the initial state with the production of heat. Since all CT transitions are fully allowed, only a few hundredth of one percent of the impurities involved are required for an intense color. The process can also be termed electron hopping or photochemical redox. Another example of this heteronuclear intervalence CT is the gray to black color of most moon rock, again from the Fe—Ti combination.

In homonuclear intervalence CT, atoms of the same element but on different sites A and B interact as in

$$Fe_A^{2+} + Fe_B^{3+} \longrightarrow Fe_A^{3+} + Fe_B^{2+}$$

Idiochromatic examples of this are black magnetite [1309-38-2], Fe_3O_4, otherwise written as $Fe(II)O \cdot Fe_2(III)O_3$ and the analogous red Mn_3O_4; and the pigments Prussian blue [14038-43-8] and Turnbull's blue [25869-98-1], both $Fe_4(III)$-$[Fe(II)(CN)_6]_3$. Allochromatic examples are widespread in the mineral field, with $Fe^{2+} + Fe^{3+}$ being involved in blue and green tourmaline [1317-93-7], blue iolite (cordierite) [12182-53-5], etc.

Metal←ligand or cation←anion CT is present in the yellow to orange chromates and dichromates such as $K_2Cr(VI)O_4$ [7789-00-6] and $K_2Cr_2(VI)O_7$ [7778-50-9]; these contain no unpaired electrons but the color derives from transfer of electrons from O^{2-} to the otherwise very highly charged Cr^{6+}. Once again the transitions are fully allowed and intense colors occur as in permanganates such as $KMnO_4$ and the pigments chrome yellow [1344-37-2], $PbCrO_4$, and red ochre (mineral hematite) [1317-60-8], Fe_2O_3. The reverse ligand—metal or anion—cation CT is not common but occurs in the yellow liquid $Fe(CO)_5$ [13463-40-6], where low-energy π-orbitals of CO can accept electrons from the Fe.

Anion—anion CT occurs in the pigment ultramarine [57455-37-5] (mineral lapis lazuli [1302-85-8]), which contains $(S_3)^-$ groups having a total of 19 electrons in molecular orbitals. Among these orbitals there is a strong transition giving an absorption band at 2.1 eV and leading to the deep blue color.

Acceptor—donor CT occurs, eg, in the solution of iodine in benzene, where an electron can transfer from the π-electron system in benzene to the I_2 molecule. Organic dyes containing both donor and acceptor groups can also be approached from this viewpoint.

8.7. Color in Metals and Alloys. This is the first of four mechanisms best approached from an energy-band point of view. The two equivalent equal-energy molecular orbitals of two hydrogen atoms combining to form H_2 split to form a lower energy bonding orbital containing the two electrons and a higher energy empty antibonding orbital, also capable of holding two electrons. In an n-valent metal the $n \times 10^{23}$ or so outermost electrons, n from each atom, are again equivalent and of equal energy, forming an essentially continuous band of energy states. The $n \times 10^{23}$ or so delocalized electrons available fill the band from the bottom up to the Fermi level E_f, (Fig. 16). Electrons in the band can absorb photons at any energy as shown, but the light is so intensely absorbed that it can penetrate to a depth of only a few atoms, typically less than a wavelength. Being an electromagnetic wave, the light induces electrical current on the metal surface, which immediately reemits the light, resulting in metallic

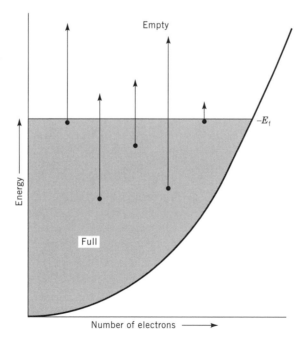

Fig. 16. A metal can absorb some light at any energy, but immediately reemits most of it (2).

luster and metallic reflection. The same principles apply to alloys and to some metal-like compounds such as the "fool's gold" pyrite [1309-36-0], FeS_2, which contains both localized and delocalized electrons.

Using the complex refractive index $N = n + iK$, where $i = \sqrt{-1}$ and K is the absorption coefficient, the reflectivity R of metals and alloys is given by

$$R = 100[(n-1)^2 + K^2]/[(n+1)^2 + K^2]$$

where K varies with the wavelength. This variation originates from the nature of the various orbitals that originally combined to form the density of states diagram. Most metals have more complex shapes than the simple parabolic type of Figure 16, which applies to alkali metals; the variation of the efficiency of the reflection process with light energy then controls the color. If all energies are equally efficiently reflected, the almost colorless reflections of clean iron [7439-89-6], mercury [7439-97-6], and silver [7440-22-4] result. However, if the efficiency decreases with increasing energy, there is reduced reflection at the blue end of the spectrum, resulting in the yellow of gold [7440-57-5], brass [12597-71-6], and pyrite, and the reddish color of copper [7440-50-8].

The direct light absorption of a metal in the absence of reflection is observed only rarely. When gold is beaten into gold leaf <100 nm thick, a blue-green color is seen by transmitted light. When in the form of colloidal particles, eg, the 10-nm diameter gold particles in deep red ruby glass, complex Mie scattering theory may be used or the situation can be treated as a bounded plasma resonance. The color produced by such particles also depends on their shape, a characteristic used to produce colored images in polychromatic glass. Here a

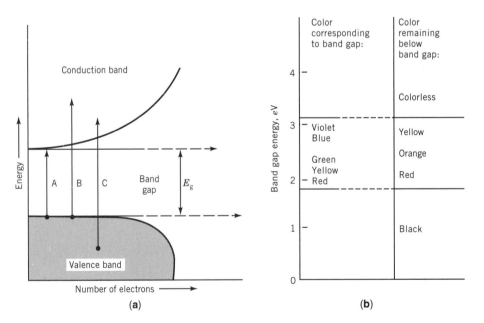

Fig. 17. The absorption of light in a band-gap material (**a**) and the variation of color with the size of the band gap (**b**) (2).

Table 4. **Color of Some Band-Gap Semiconductors**

Substance	CAS Registry number	Mineral name	Pigment name	Band gap, eV	Color
C	[7782-40-3]	diamond		5.4	colorless
ZnO	[1314-13-2]	zincite	zinc white	3.0	colorless
CdS	[1306-23-6]	greenockite	cadmium yellow	2.6	yellow
$CdS_{1-x}Se_x$	[12656-57-4]		cadmium orange	2.3	orange
HgS	[19122-79-3]	cinnabarite	vermilion	2.0	red
HgS	[23333-45-1]	metacinnabar		1.6	black
Si	[7440-21-3]			1.1	black
PbS	[12179-39-4]	galena		0.4	black

glass containing both silver and cerium [7440-45-1] yields on exposure to light the transformation:

$$Ag^+ + Ce^{3+} \longrightarrow Ag + Ce^{4+}$$

with the formation of silver particles; a complex technology causes the aspect ratio of these particles to vary with wavelength and thus produce colored images. Closely related are photochromic sunglasses, where exposure of silver halide particles to intense sunlight produces metallic silver and darkening, with reversal when the intensity of the light is reduced (see CHROMOGENIC MATERIALS, PHOTOCHROMIC).

8.8. Color in Pure Semiconductors. In some materials a gap is present in the band, so that a lower energy valence band is separated from a higher energy conduction band by a band gap of energy E_g as at the left in Figure 17. When there are exactly four valence electrons per atom, the conduction band is exactly full and the energy band is exactly empty. Since most of these materials do not conduct electricity well, the metallic reflection is absent and excitation of electrons by light, as in Figure 17, results in absorption. If the band gap is very small, all light is absorbed and the color is black. If the band gap is very large, then light in the visible region cannot be absorbed and the material is colorless. At a band gap corresponding to 2.2 eV (560 nm) violet, blue, and green are absorbed, leaving a complementary orange transmission color. The full sequence is black, red, orange, yellow, colorless as can be readily deduced from Figure 17. Examples of some band-gap materials are given in Table 4.

8.9. Color in Doped Semiconductors. Consider a diamond crystal [7782-40-3], otherwise colorless because of its large energy band gap, with 10 ppm substitutional nitrogen atoms added. Each N has one more electron than the C it replaces and will donate this electron to a donor level within the band gap that is located 4 eV below the conduction band; this level is broadened by various causes, including thermal vibrations. Absorption of light can now occur, with the excitation of an electron from the donor level into the conduction band from 2.2 eV up, leading to the yellow color of natural and synthetic N-containing diamonds. The analogous presence of boron atoms produces a blue color, as in the famous Hope diamond. In this case, B has one less electron than the C it replaces and each atom forms one hole in an acceptor level within the band gap, located 0.4 eV above the valence band. Light is absorbed by electrons being excited from the valence band into the acceptor level holes, thus producing the blue color as well as electrical conductivity since the excitation energy involved is so small.

Donor and acceptor levels are the active centers in most phosphors, as in zinc sulfide [1314-98-3], ZnS, containing an activator such as Cu and various coactivators. Phosphors are coated onto the inside of fluorescent lamps to convert the intense uv and blue from the mercury emissions into lower energy light to provide a color balance closer to daylight as in Figure 11. Phosphors can also be stimulated directly by electricity as in the Destriau effect in electroluminescent panels and by an electron beam as in the cathodoluminescence used in television and cathode ray display tubes and in (usually blue) vacuum-fluorescence alphanumeric displays.

Some impurities can form trapping levels within the band gap, as shown in Figure 18. Absorption of light can move electrons into the traps, as shown by the arrows. Energy E_b is required to release the electron from the trap into the conduction band and permit it to decay by one of several possible light-emitting paths not shown. If E_b is small so that room temperature thermal excitations can release the electrons slowly, then phosphorescence results. If E_b is a little larger, then ir may permit the escape, as in an ir-detecting phosphor that first has been activated to load up the traps.

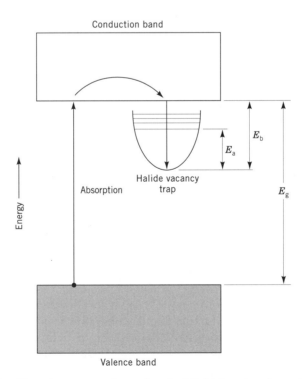

Fig. 18.   Trapping energy from absorbed light in a band-gap trap (2).

Finally, an electric current can produce injection luminescence from the recombination of electrons and holes in the contact zone between differently doped semiconductor regions. This is used in light-emitting diodes (LED, usually red), in electronic displays, and in semiconductor lasers.

8.10. Color from Color Centers. Although ligand-field theory can also be used, this mechanism is best approached from band theory. Consider a vacancy, eg, a missing Cl^- ion in a KCl crystal produced by irradiation, designated an F-center (from the German *Farbe* for color). An electron can become trapped at the vacancy and this forms a trapped energy level system inside the band gap just as in Figure 18. The electron can produce color by being excited into an absorption band such as the E_a transition, which is 2.2 eV in KCl and leads to a violet color. In the alkali halides $E_a = 0.257/d^{1.83}$, where E_a is in eV and d is the anion–cation distance in nanometer (nm). In addition to irradiation, F centers in halides can also be produced by solid-state electrolysis, by in-diffusion of metal, or by growth in the presence of excess metal. Many other centers exist in the alkali halides, such as the M center, consisting of two adjacent F centers; the F′, an F center that has trapped two electrons; and the V_K, two adjacent F centers with only one trapped electron between them. All of these are electron color centers, where an electron is present in a location where it is not normally found.

An example of a hole color center is smoky quartz [14808-60-7]. Here irradiation (either produced by nature or in the laboratory) of SiO_2 containing trace amounts of Al ejects an electron from an oxygen adjacent to the Al or, in customary nomenclature: $[AlO_4]^{5-} \rightarrow [AlO_4]^{4-} + e^-$; the ejected electrons are trapped elsewhere in the crystal, eg, at K^+ or H^+ impurities. The electron-deficient $[AlO_4]^{4-}$ hole color center has excited energy levels and produces light absorption leading to the smoky color. When iron is present in quartz, the result is yellow citrine [14832-92-9], containing $[Fe(III)O_4]^{5-}$; irradiation now analogously produces under certain circumstances the deep purple of amethyst [14832-91-8], which is colored by the $[Fe(III)O_4]^{4-}$ hole color center.

In general, for irradiation-produced color centers, either the hole center or the electron center can be the light-absorbing species, or even both. If the electron is released from the electron center by heat, then bleaching occurs by its recombination with the hole center to restore the preirradiation state. If this bleaching energy, E_b in Figure 18, is small, then the color center will fade either from room-temperature thermal excitation or from light absorption itself; if larger, it then requires higher temperatures as in the 300–500°C required to fade smoky quartz or amethyst. Color center transitions are fully allowed, leading to large oscillator strengths and intense colors even at very low concentration.

Additional examples of color centers are old Mn-containing glass turned purple by irradiation with the mechanism: $Mn^{2+} \rightarrow Mn^{3+} + e^-$; even uv can produce this change in old sun-exposed bottles then termed desert amethyst glass. In many minerals, the mechanism is unknown, such as the following where F indicates fading at room temperature or in light, and S stands for a light-stable color: purple fluorite, calcium fluoride [7789-75-5] (blue john) (S); two types of yellow sapphire (F and S); blue (S) and two types of brown (F and S) topaz [1302-59-6]; deep blue Maxixe beryl (F); and irradiated yellow, green, blue, and red diamond (all S; these are different from the impurity-caused colors covered in the previous section).

Unusual is some hackmanite [1302-90-5], $Na_4Al_3Si_3O_{12}(Cl,S)$, which can have a deep magenta color as mined but fades in the light; the color can be restored in some material with uv exposure or simply by storing in the dark, where the reaction is $S_2^{2-} \rightarrow S_2^- + e^-$. The hole center S_2^- absorbs at 3.1 eV (400 nm) and the electron combines with a vacancy to form an F center absorbing at 2.35 eV (530 nm); it is the combination of both absorption bands that produces the color. Light absorption while showing the color also reverses this process, as does heat, both causing the color to fade.

8.11. Color from Dispersive Refraction. This mechanism involves the variation of the refractive index n with wavelength λ, given by the Sellmeier dispersion formula:

$$n^2 - 1 = a\lambda^2/(\lambda^2 - A^2) + b\lambda^2(\lambda^2 - B^2) + \cdots$$

where $A, B, \ldots$ are the wavelengths of individual ir, vis, and uv absorptions and $a, b, \ldots$ are constants representing the strength of these absorptions. Three terms are normally sufficient for an excellent fit in the visible region.

Even a transparent colorless material has uv absorptions (from electronic excitations) and ir absorptions (from atomic and molecular vibrations), which result in the decrease of n with wavelength in the visible region as seen in Figure 19. Only a vacuum has no absorption and no dispersion; neither of these is fundamental since each can be viewed as producing the other, with the Kramers-Kronig relationships as the connection. As a result Newton's prism produces the spectrum, crystal glassware and faceted gemstones show fire, raindrops produce the primary and secondary rainbows (higher orders can be seen only in the laboratory), and ice crystals produce various colored halos around the sun and moon, as well as sundogs. The green flash, rarely seen

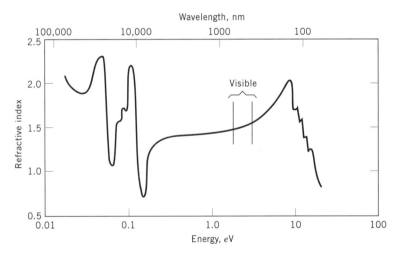

Fig. 19. The dispersion curve of a colorless soda-lime–silicate crown glass where the wt% of $Na_2O = 21.3$; $CaO = 5.2$; and $SiO_2 = 73.5$ (2).

when the sun sets, results when the density gradient of the atmosphere acts as a prism, separating the spectral colors; since violet and blue are strongly scattered from the following mechanism, green is the last spectral color seen. Anomalous dispersion, where n varies anomalously in the vicinity of an absorption band in the visible, can be classified in this section.

8.12. Color from Scattering. Particles that are small compared to the wavelength of light produce scattering, the intensity of which is proportional to λ^{-4}, as shown by Lord Rayleigh. This happens not only from particles but even from refractive index variations from density fluctuation in gases, thermal vibrations in crystals, and chemical composition variations in glasses and polymers. As a result, violet at the 400-nm limit of visibility is scattered 9.38 times as much as red at the 700-nm limit. Scattered light therefore appears violet to blue, as in the blue of the clear sky, while the remaining light is usually orange to red, as in the setting sun where the effects of dust add to the atmosphere's density fluctuations. The exact color also depends on the shape of the scattering particles. One speaks of Rayleigh scattering and of Tyndall blues after an early investigator. The scattered light is polarized with an intensity variation of $(1 + \cos^2 \theta)$ with angle θ from the light beam. When the particle size becomes equal to or a little larger than the wavelength, the complex Mie scattering theory must be used; this type of scattering can produce almost any color. At even larger sizes scattering becomes noncolor-selective and produces only white, as in fogs and clouds.

Rayleigh scattering produces, in addition to the blue sky and red sunset, most blue colors in bird feathers and butterfly wings, the blue iris in eyes, cold blue skin, blue cigarette smoke, blue moonstone, and the rare blue moon from forest fire oil-droplet haze. (The supposed blue moon explanation based on calendric occurences is spurious.) A dark background is necessary for intense blues, as in the black of outer space for the blue sky and the melanin backing for the blue iris of the eye (additional yellow-to-brown melanin at the front of the iris leads to green and brown colors). Also included in this group are colors produced by effects such as second harmonic generation and parametric oscillation in nonlinear materials and various inelastic processes such as the Raman, Brillouin, polariton, magnon, Thompson, and Compton scatterings. Mie scattering from metallic particles in glass, such as in ruby glass, can be considered here or under the metals of Section Color of Metals.

8.13. Color from Interference. Under this heading is considered only that color produced by interference that does not involve any diffraction; the combination of these two processes is considered in the following section.

When two light waves of the same frequency interfere, anything from constructive reinforcement to destructive cancellation can occur, depending on the phase relationship. This is present in a wedge-shaped film, eg, where the light reflected from the front surface interferes with that reflected from the back surface, the phase difference depending on the refractive index and the thickness of the film. With monochromatic light this usually produces light and dark fringes as in interferometers, while with white light the color sequence of Newton's colors is produced. Starting with the thinnest film, this sequence is black, gray, white, yellow, red (end of the first order), violet, blue, green, yellow, orange-red, violet (end of the second order), blue, green, yellow, red, etc.

Newton's colors are seen in the tapered air gap between touching nonflat sheets of glass; cracks in a transparent medium such as glass or a crystal; soap bubbles; oil slicks on water; thin tarnish coatings on substances such as the mineral bornite and on ancient buried glass; supernumerary rainbow fringes; antireflection coatings on camera lenses, etc; the wings of house and dragon flies; and in some beetle wing cases as in Japanese beetles. When intense colors occur, they appear metalliclike in nature and are usually termed iridescent; these effects are intensified by a dark backing. They also can be intense in some multiple layer structures that are present in interference filters; the naturally occurring multilayered mineral labradorite and layer-structured pearls; fish scales and imitation pearls based on fish scale essence, both involving guanidine flakes; peacock, hummingbird, and some other bird feather colors; the luster of hair and nails; and in the metalliclike reflections from the eyes of cats and many other nocturnal animals.

Other interference-produced colors falling into this section include doubly refracting materials such as anisotropic crystals and strained isotropic media between polarizers, as in photoelastic stress analysis and in the petrological microscope.

8.14. Color from Diffraction.   Diffraction refers to the nonrectilinear propagation of light, eg, when the edge of an object produces an interrupted wavefront and light bends around the edge; interference can then occur between the bent and the undisturbed parts of the light. Light and dark fringes are formed from monochromatic light and colored fringes from white light in Fresnel diffraction from a collimated beam of light passing through a small slit, as well as in Fraunhofer diffraction when the image of such a slit is focused with a lens. This produces the corona, a set of colored rings around bright lights; the corona aureole, a disk of bluish light seen around the sun behind a thin cloud (the corona of the sun is a different phenomenon); and fog- and cloud-related effects such as the glory and Bishop's ring.

Diffraction from two-dimensional diffraction gratings follows a variant of Bragg's law (2) and produces spectral displays as in diffraction grating spectroscopes; some beetles and snakes; phonograph records and compact disks; a distant street lamp seen through a cloth umbrella; and in the play of color seen in the natural three-dimensional diffraction grating opal. Liquid crystals (qv) of the cholesteric or chiral nematic type also function to diffract light as in liquid crystal thermometers, thermography, and in mood jewelry, where a change in temperature alters the grating spacing and hence the color of a cholesteric mesophase supported between pieces of glass or plastic.

BIBLIOGRAPHY

See "Color Measurement" in the Kirk-Othmer *Encyclopedia of Chemical Technology* (*"ECT"*) 1st ed., Vol. 4, pp. 242–251, by G. W. Ingle, Monsanto Chemical Co.; "Color" in *ECT* 2nd ed., Vol. 5, pp. 801–812, by G. W. Ingle, Monsanto Chemical Co.; in *ECT* 3rd ed., Vol. 6, pp. 523–548, by F. W. Billmeyer, Jr., Rensselaer Polytechnic Institute; in *ECT* 4th ed., Vol. 6, pp. 841–876, by K. Nassau, AT&T Bell Laboratories; "Color" in *ECT* (online), posting date: December 4, 2000, by K. Nassau, AT&T Bell Laboratories.

CITED PUBLICATIONS

1. R. S. Berns, *Billmeyer and Saltzman's Principles of Color Technology*, 3rd ed., John Wiley & Sons, Inc., New York, 2000.
2. K. Nassau, *The Physics and Chemistry of Color*, 2nd ed., John Wiley & Sons, Inc., New York, 2001.
3. R. G. Kuehni, *Color*, John Wiley & Sons, Inc., New York, 1997.
4. R. W. G. Hunt, *Measuring Color*, 3rd ed., Newpro, U.K., Ltd., 1996.
5. R. S. Hunter and R. W. Harold, *The Measurement of Appearance*, 2nd ed., John Wiley & Sons, Inc., New York, 1987; Table A7.
6. P. K. Kaiser and R. M. Boynton, *Human Color Vision*, 2nd ed., Optical Society of America, Washington, D.C., 1996.
7. M. S. Cayless and A. M. Marsden, *Lamps and Lighting*, Edward Arnold, London, 1983.
8. K. L. Kelly and D. B. Judd, *Color: Universal Language and Dictionary of Names*, NBS Special Publication 440, U.S. Government Printing Office, Washington D. C., 1976.
9. R. G. Kuehni, *Computer Colorant Formulation*, D. C. Heath & Co., Lexington, Mass., 1975.
10. A. H. Munsell, *A Color Notation*, 14th ed., Macbeth Division of Kollmorgen Instruments Corp., Newbaugh, N.Y., 1990.
11. W. N. Sproson, *Colour Science in Television and Display Systems*, Hilger, Bristol, U.K., 1983.
12. G. Wyszecki and W. S. Stiles, *Color Science*, 2nd ed., John Wiley & Sons, Inc., New York, 2000.
13. *CIE Publication 15.2 Colorimetry*, 2nd ed., Central Bureau of CIE, Vienna, Austria, 1986; Available from U.S. National Committee CIE, c/o National Institute of Standards and Technology, Washington, D.C.
14. J. Long and J. T. Luke, *The New Munsell Student Color Set*, 2nd ed., Fairchild Books, 2001.
15. *Munsell Book of Color*, glossy or matte finishes, and other collections, Macbeth Division of Kollmorgen Instruments Corp., Baltimore, Md., 1929 on, long discontnued.
16. A. Hård and L. Sivik, *Color Res. Appl.* **6**, 129 (1981).
17. E. Jacobson, *Basic Color: An Interpretation of the Ostwald Color System*, P. Theobald, Chicago, Ill., 1948.
18. A. Maerz and M. R. Paul, *A Dictionary of Color*, McGraw-Hill Book Co., New York, 1930.
19. D. L. MacAdam, *J. Opt. Soc. Am.* **64**, 1691 (1974); *Uniform Color Scales Committee Samples*, Optical Society of America, Washington, D.C., 1977.
20. *ISCC-NBS Centroid Color Charts, NBS Standard Reference Material No. 2106*, National Institute of Standards and Technology, Washington, D.C., 1965.
21. W. D. Wright, *The Measurement of Color*, 4th ed., Adam Hilger, London, 1969.
22. *Color and Appearance Collaborative Reference Program*, Collaborative Testing Services, Inc., Herndon, Va., previously *MCCA-NBS Collaborative Reference Program on Color and Color Differences*, National Institute of Standards and Technology, Washington, D.C., 1991.

GENERAL REFERENCE

K. Nassau, "Colour" in, *Encyclopedia Britannica*, 15th ed., Macropedia Vol. 4, pp. 595–604, 1988 on.

KURT NASSAU
Consultant

COLORANTS FOR CERAMICS

1. Introduction

Any product that depends on aesthetics for consideration for purchase and use will be improved by the use of color. Hence, many ceramic products, such as tile, sanitary ware, porcelain enameled appliances, tableware, and some structural clay products and glasses, contain colorants.

For both economic and technical reasons, the most effective way to impart color to a ceramic product is to apply a ceramic coating that contains the colorant. The most common coatings, glazes, and porcelain enamels are vitreous in nature. Hence, most applications for ceramic colorants involve the coloring of a vitreous material.

There are a number of ways to obtain color in a ceramic material (1). First, certain transition-metal ions can be melted into a glass or dispersed in a ceramic body when it is made. Although suitable for bulk ceramics, this method is rarely used in coatings because adequate tinting strength and purity of color cannot be obtained this way.

A second method to obtain color is to induce the precipitation of a colored crystal in a transparent matrix. Certain materials dissolve to some extent in a vitreous material at high temperatures, but when the temperature is reduced, the solubility is also reduced and precipitation occurs. This method is used to disperse nonoxide precipitates of gold, copper, or cadmium sulfoselenide in bulk glass. In coatings it is used for opacification, the production of an opaque white color. Normally, some or all of the opacifier added to the coating slip dissolves during the firing process and recrystallizes upon cooling. For oxide colors other than white, however, this method lacks the necessary control for reproducible results and is seldom used.

The third method to obtain color in a vitreous matrix is to disperse in that matrix an insoluble crystal or crystals that are colored. The color of the crystal is then imparted to the transparent matrix. This method is the one most commonly used to introduce color to vitreous coatings.

2. Colored Bodies and Glasses

The addition of oxides to ceramic bodies and to glasses to produce color has been known since antiquity (2). The use of iron and copper oxides predates recorded history. Cobalt was introduced into Chinese porcelain ~700 AD. Chromium compounds have been used since 1800 AD.

The colors obtained depend primarily on the oxidation state and coordination number of the coloring ion (3). Table 1 lists the solution colors of several ions in glass. All of these ions are transition metals; some rare-earth ions show similar effects. The electronic transitions within the partially filled d and f shells of these ions are of such frequency that they fall in that narrow band of frequencies from 400 to 700 nm, which constitutes the visible spectrum (4). Hence, they are suitable for producing color (qv).

Table 1. **Solution Colors in Glass**

Ion	Number of d Electrons	Color
Ti^{3+}	1	blue-green
V^{3+}	2	gray
V^{4+}	1	blue
Cr^{3+}	3	dark green
Mn^{3+}	4	purple
Fe^{3+}	5	yellow-brown
Fe^{2+}	6	light blue or green
Co^{3+}	6	pink
Co^{2+}	7	dark blue
Ni^{2+}	8	gray
Cu^{2+}	9	light blue

Decolorizing is sometimes desirable (5). In the manufacture of glass, the presence of iron as an impurity cannot be completely avoided. Iron imparts an unacceptable dirty brown color to the glass. If the iron can be oxidized through the use of additives such as arsenic oxide [1327-53-3] (As_2O_3), cerium oxide [1306-38-3] (CeO_2), or manganese oxide [1313-13-9] (MnO_2), the discoloration can be diminished through chemical means. Alternatively, the effect of the iron can be negated through the addition of ingredients that themselves produce complementary colors in the glass. Materials such as selenium [22541-48-6], cobalt oxide [1308-06-1] (Co_3O_4), neodymium oxide [1313-97-9] (Nd_2O_3), and manganese oxide are used for this purpose. The addition of one of these materials produces a slightly darker, but more neutrally colored glass, which is more acceptable visually.

3. Precipitation Colors

Several colors in bulk glass can be produced by precipitation processes. One such technique involves developing a colloidal suspension in the glass matrix (5). Metals, such as gold, silver, and copper (or possibly Cu_2O), and nonoxide pigments, such as the cadmium sulfoselenides, produce strong colors when precipitated from a colloidal suspension in glass (Table 2).

To produce this color, the glass is melted, formed, and annealed as is usually done in making glass, which is followed by reheating to a temperature

Table 2. **Colloidal Precipitation Colors**

Crystal	CAS Registry number	Color
Au	[7440-57-5]	gold
Ag	[7440-22-4]	gray
Cu	[7440-50-8]	red
CdS	[1306-23-6]	yellow
Cd(S,Se)	[a]	red, orange

[a] Cadmium sulfoselenides (CdSSe) [11112-63-3], (Cd_2SSe) [12214-12-9].

where nuclei of the desired material are formed. The temperature is then adjusted to permit these nuclei to grow to an optimum size for the development of the particular color. This process is known as striking the color.

Most striking colors are obtained in glasses containing 10–20 wt% K_2O, 10–22 wt% ZnO, and 50–60 wt% silica. Calcium oxide (CaO) and B_2O_3 may also be present (6). To this batch is added 1–3 wt% CdS, CdSe, and/or CdTe. Melting must be under neutral or mildly reducing conditions. Otherwise, S, Se, and Te will be oxidized to SO_2, SeO_2, or TeO_2, which are colorless.

After casting, the glasses solidify to colorless glass. When reheated at 550–700°C, the glasses precipitate minute crystals of a cadmium sulfoselenide, which causes the absorption edge to move to longer wavelengths, producing colors from yellow to orange, to red, to maroon.

The color obtained is a function of both the composition and the particle size of the precipitated crystals. A redder color results from both increased selenium to sulfur ratio and from larger crystals, caused by a more severe heat treatment. Hence, it is possible to make, a series of color filter types from the same glass by controlled reheating.

A related but different type of colored glass is the ruby glass (6). Ruby glasses can be made in several base glass systems by adding to the batch 0.003–0.1 wt% of a noble metal salt, ie, copper(II) chloride [7447-39-4], silver nitrate [19582-44-6], or gold chloride [13453-07-1], together with a reducing agent such as stannous chloride or antimony oxide. When cooled from the melt, they are usually colorless or weakly colored. Reheating results in a more or less intense red for copper, yellow-brown for silver, and red for gold. The color comes from a colloidal dispersion of metal particles that have precipitated from the glass.

4. Opacification

Whiteness or opacity is introduced into ceramic coatings by the addition to the coating formulation of a substance that will disperse in the coating as discrete particles that scatter and reflect some of the incident light(7). To be effective as a scatterer, the discrete substance must have a refractive index that differs appreciably from that of the clear ceramic coating, because the greater the difference in index of refraction between the matrix and the scattering phase, the greater the degree of opacity. The refractive index of most glasses is 1.5–1.6, and therefore the refractive indexes of opacifiers must be either greater or less than this value. As a practical matter, opacifiers of high refractive index are used. Some possibilities include tin oxide [18282-10-5] (SnO_2), with $n_D = 2.04$, zirconium oxide [1314-23-4] (ZrO_2), with $n_D = 2.40$, zirconium silicate [10101-52-7] ($ZrSiO_4$), with $n_D = 1.85$, titanium oxide [13463-67-7] (TiO_2), with $n_D = 2.5$ for anatase [1317-70-0], and $n_D = 2.7$ for rutile [1317-80-2].

In glazes fired at temperatures >1000°C, zircon [14940-68-2] is the opacifier of choice (8). It has a solubility of ~5% in many glazes at high temperature, and 2–3% at room temperature. A customary mill addition would be 8–10% zircon. Thus most opacified glazes contain both zircon that was placed in the mill and went through the firing process unchanged, and zircon that dissolved in the molten glaze during firing but recrystallized on cooling.

The solubility of zircon in glazes is a function of the glaze composition (9). Zinc oxide (ZnO) lowers the solubility, promoting opacification where SrO increases solubility.

In porcelain enamels and in glazes firing under $1000°C$, TiO_2 in the anatase crystal phase is the opacifying agent of choice (10,11). Because it has the highest refractive index, TiO_2 is the most effective opacifying agent. However, at temperatures of $\sim850°C$, anatase inverts to rutile in silicate glasses. Once inverted to rutile, TiO_2 crystals are able to grow rapidly to sizes that are no longer effective for opacification. Moreover, because the absorption edge of rutile is very close to the visible, as the rutile particles grow the absorption edge extends into the visible range, leading to a pronounced cream color. Thus while TiO_2 is a very effective opacifier at lower temperatures, it cannot be used $>1000°C$.

The solubility of TiO_2 in molten silicates is $\sim8-10\%$. At room temperature this solubility is reduced to $\sim5\%$. Thus when using TiO_2 as an opacifier, substantial amounts, $\sim15\%$, must be used.

5. Ceramic Pigments

The principal method of coloration of ceramic coatings is dispersal of a ceramic pigment in a vitreous matrix. To be suitable as a ceramic pigment, a material must possess a number of properties (12) that fall in two categories: strength of pigmentation and stability. The first requirement is high tinting strength or intensity of color. It is also desirable that the color be pure and free of grayness or muddiness. Related to the intensity is the diversity of colors obtainable with similar and compatible materials, since many users blend two or more colors to obtain the final shade desired.

The other important property of a ceramic pigment is stability under the high temperatures and corrosive environments encountered in the firing of glazes. The rates of solution in the molten glass should be very low in spite of very fine particle sizes (~10 μm). Neither should gases be given off as a result of the contact of the ceramic pigment with the molten vitreous material.

Another desirable property for a ceramic color is a high refractive index. For example, valuable pigments are based on spinels [1302-67-6] ($n_D = 1.8$) and on zircon ($n_D = 1.9$), but no valuable pigments are based on apatite ($n_D = 1.6$), even though the lattice of apatite is as versatile for making ionic substitutions as that of spinel.

5.1. Manufacturing. Although a number of different pigment systems exist, most are prepared by similar manufacturing methods. The first step in pigment manufacture is close control over the selection of raw materials. Most of these raw materials are metallic oxides or salts of the desired metals. Considerable differences in the required chemical purity are encountered, ranging from impure natural minerals to chemicals of industrial-grade purity. In pigment manufacture, purity does not always equal quality. Often less pure raw materials may prove superior in the production of a given pigment.

The raw materials are weighed and then thoroughly blended. The reaction forming the pigment crystal occurs in a high temperature calcining operation. The temperature may range from 500 to 1400°C depending on the particular system. Normally, air is the atmosphere of choice because the pigments must ultimately be stable in a molten oxide coating, so materials sensitive to oxygen have limited use. During the calcination process, any volatiles are driven off and the pigment crystal is developed in a sintering reaction. Some of these reactions occur in the solid state, but many involve a fluid-phase mineralizing pathway. Although a few large volume pigments may be made in rotary kilns, most pigments are placed in saggers of 22–45 kg (50–100 lb) capacity for the calcining operation.

Following calcination, the product may require milling to reduce particle size to that necessary for use. This size reduction may be carried out either wet or dry. If there are soluble by-products, a washing operation may also be required. It is almost always necessary to break up agglomerates by a process such as micronizing.

It has been found that there are several advantages to modification of pigments by addition to the calcined product of a small quantity of a dispersing agent (13).

5.2. Pigment Systems.   Most of the crystals used for ceramic pigments are complex oxides, owing to the great stability of oxides in molten silicate glasses. Table 3 lists these materials. Blues, yellows, browns, pinks, blacks, and grays are available in high purity (14). Greens and purples are available in fair purity. However, the purity of red and orange is very poor. Therefore, the one significant exception to the use of oxides is the family of cadmium sulfoselenide red pigments. This family is used because the red and orange colors obtained cannot be obtained in oxide systems; thus it is necessary to sustain the difficulties of a nonoxide system.

Table 3 is arranged by crystal class (16). The crystal class of a given pigment is determined almost solely by the ratio of the ionic sizes of the cation and the anion and their respective valences. Hence, for any given stoichiometry and ionic size ratio, only one or two structures will be possible. In some classes (spinel, zircon), a wide range of colors is possible within the confines of that class. Pigments within a given class usually have excellent chemical and physical compatibility with each other. This compatability is important, as most applications involve the mixing of two or more pigments to achieve the desired shade.

Table 3. **Classification of Mixed-Metal Oxide Inorganic Pigments According to Crystal Class**[a]

Pigment name	CAS Registry number	Formula	DCMA number[b]
	Baddeleyite		
zirconium vanadium yellow baddeleyite	[68187-01-9]	$(Zr,V)O_2$	1-01-4
	Borate		
cobalt magnesium red-blue borate	[68608-93-5]	$(Co,Mg)B_2O_5$	2-02-1

Table 3 (*Continued*)

Pigment name	CAS Registry number	Formula	DCMA number[b]
Corundum–hematite			
chromium alumina pink corundum	[68187-27-9]	$(Al,Cr)_2O_3$	3-03-5
manganese alumina pink corundum	[68186-99-2]	$(Al,Mn)_2O_3$	3-04-5
chromium green-black hematite	[68909-79-5]	$(Cr,Fe)_2O_3$	3-05-3
iron brown hematite	[68187-35-9]	Fe_2O_3	3-06-7
Garnet			
victoria green garnet	[68553-01-5]	$3CaO \cdot Cr_2O_3 \cdot 3SiO_2$	4-07-3
Olivine			
cobalt silicate blue olivine	[68187-40-6]	Co_2SiO_4	5-08-2
nickel silicate green olivine	[68515-84-4]	Ni_2SiO_4	5-45-3
Periclase			
cobalt nickel gray periclase	[68186-89-0]	$(Co,Ni)O$	6-09-8
Phenacite			
cobalt zinc silicate blue phenacite	[68412-74-8]	$(Co,Zn)_2SiO_4$	7-10-2
Phosphate			
cobalt violet phosphate	[13455-36-2]	$Co_3(PO_4)_2$	8-11-1
cobalt lithium violet phosphate	[68610-13-9]	$CoLiPO_4$	8-12-1
Priderite			
nickel barium titanium primrose priderite	[68610-24-2]	$2NiO \cdot 3BaO \cdot 17TiO_2$	9-13-4
Pyrochlore			
lead antimonate yellow pyrochlore	[68187-20-2]	$Pb_2Sb_2O_7$	10-14-4
Rutile–cassiterite			
nickel antimony titanium yellow rutile	[71077-18-4]	$(Ti,Ni,Sb)O_2$	11-15-4
nickel niobium titanium yellow rutile	[68611-43-8]	$(Ti,Ni,Nb)O_2$	11-16-4
chromium antimony titanium buff rutile	[68186-90-3]	$(Ti,Cr,Sb)O_2$	11-17-6
chromium niobium titanium buff rutile	[68611-42-7]	$(Ti,Cr,Nb)O_2$	11-18-6
chromium tungsten titanium buff rutile	[68186-92-5]	$(Ti,Cr,W)O_2$	11-19-6
manganese antimony titanium buff rutile	[68412-38-4]	$(Ti,(Ti,Mn,Sb)O_2$	11-20-6
titanium vanadium antimony gray rutile	[68187-00-8]	$(Ti,V,Sb)O_2$	11-21-8
tin vanadium yellow cassiterite	[68186-93-6]	$(Sn,V)O_2$	11-22-4
chromium tin orchid cassiterite	[68187-53-1]	$(Sn,Cr)O_2$	11-23-5
tin antimony gray cassiterite	[68187-54-2]	$(Sn,Sb)O_2$	11-24-8

Table 3 (*Continued*)

Pigment name	CAS Registry number	Formula	DCMA number[b]
manganese chromium antimony titanium brown rutile	[69991-68-0]	$(Ti,Mn,Cr,Sb)O_2$	11-46-7
manganese niobium titanium brownrutile	[70248-09-8]	$(Ti,Mn,Nb)O_2$	11-47-7
	Sphene		
chromium tin pink sphene	[68187-12-2]	$CaO \cdot SnO_2 \cdot SiO_2:Cr$	12-25-5
	Spinel		
cobalt aluminate blue spinel	[68186-86-7]	$CoAl_2O_4$	13-26-2
cobalt tin blue-gray spinel	[68187-05-3]	Co_2SnO_4	13-27-2
cobalt zinc aluminate blue spinel	[68186-87-8]	$(Co,Zn)Al_2O_4$	13-28-2
cobalt chromite blue-green spinel	[68187-11-1]	$Co(Al,Cr)_2O_4$	13-29-2
cobalt chromite green spinel	[68187-49-5]	$CoCr_2O_4$	13-30-3
cobalt titanate green spinel	[68186-85-6]	Co_2TiO_4	13-31-3
chrome alumina pink spinel	[68201-65-0]	$Zn(Al,Cr)_2O_4$	13-32-5
iron chromite brown spinel	[68187-09-7]	$Fe(Fe,Cr)_2O_4$	13-33-7
iron titanium brown spinel	[68187-02-0]	Fe_2TiO_4	13-34-7
nickel ferrite brown spinel	[68187-10-0]	$NiFe_2O_4$	13-35-7
zinc ferrite brown spinel	[68187-51-9]	$(Zn,Fe)Fe_2O_4$	13-36-7
zinc iron chromite brown spinel	[68186-88-9]	$(Zn,Fe)(Fe,Cr)_2O_4$	13-37-7
copper chromite black spinel	[68186-91-4]	$CuCr_2O_4$	13-38-9
iron cobalt black spinel	[68187-50-8]	$(Fe,Co)Fe_2O_4$	13-39-9
iron cobalt chromite black spinel	[68186-97-0]	$(Co,Fe)(Fe,Cr)_2O_4$	13-40-9
manganese ferrite black spinel	[68186-94-7]	$(Fe,Mn)(Fe,Mn)_2O_4$	13-41-9
chromium iron manganese brownspinel	[68555-06-6]	$(Fe,Mn)(Fe,Cr,Mn)_2O_4$	13-48-7
cobalt tin alumina blue spinel	[68608-09-3]	$CoAl_2O_4$	13-49-2
chromium iron nickel black spinel	[71631-15-7]	$(Ni,Fe)(Cr,Fe)_2O_4$	13-50-9
chromium manganese zinc brown spinel	[71750-83-9]	$(Zn,Mn)Cr_2O_4$	13-51-7
	Zircon		
zirconium vanadium blue zircon	[68186-95-8]	$(Zr,V)SiO_4$	14-42-2
zirconium praseodymium yellow zircon	[68187-15-5]	$(Zr,Pr)SiO_4$	14-43-4
zirconium iron coral zircon	[68187-13-3]	$(Zr,Fe)SiO_4$	14-44-5

[a] Refs. 1,13.
[b] Dry Color Manufacturers Association, Arlington, Va.

Red Pigments.   There are no oxides that can be used to give a true red pigment that is stable to the firing of ceramic coatings. Hence, orange, red, and dark red colors are obtained by the use of cadmium sulfoselenide pigments (7,17–19). The cadmium sulfoselenides are a group of pigments based on solid solutions of cadmium selenide, cadmium sulfide, and/or zinc sulfide.

The synthetic pigment is produced by one of several related procedures. The best quality product is made by reaction of an aqueous solution of $CdSO_4$ or $CdCl_2$ with a solution of an alkaline metal sulfide or H_2S. Zn, Se, or Hg may be added to the CdS to produce shade variations. After precipitation, the color is filtered, washed, and calcined in an inert atmosphere at 500–600°C.

The colors come in a range of shades from primrose yellow through yellow, orange to red, and maroon. The primrose or light yellow color is produced by precipitating a small amount of ZnS with the CdS. The orange, red, and maroon shades are made by incorporating increasing amounts of selenium compounds with the CdS. An orange pigment is obtained at a CdS to CdSe ratio of ~4:1. A red pigment is obtained at a ratio of CdS/CdSe of 1.7:1. A deeper red is formed at a ratio of 1.3:1. A very deep red-maroon is manufactured at a ratio of 1:1.

These pigments require a glaze specially designed for their use. The glaze will contain only small amounts of PbO, B_2O_3, or other aggressive fluxes, because strong fluxes react with the selenium to form black lead selenide. The glaze should have a low alkali content. It should contain a few percent of CdO so that its chemical potential for Cd, relative to Cd in the pigment, is reduced. The glaze must also be free of strong oxidizing agents, such as nitrates, which hasten the breakdown of the cadmium sulfoselenide pigment.

The resistance of these materials to firing temperature is definitely limited. They can be fired to ~1000°C. Hence, they are limited to use in porcelain enamels and in low firing artware glazes.

These sulfoselenide pigments have good resistance to alkali solutions, but poor resistance to even dilute acids. Owing to the latter, together with the high toxicity of cadmium, cadmium sulfoselenide pigments should not be used in applications that will come in contact with food and drink. They must also be handled with great care to avoid the possibility of ingestion.

In an attempt to extend the firing range of these colors, the inclusion pigments (12,20) have been developed. In these pigments cadmium sulfoselenides are incorporated within a clear zircon lattice. The superior stability of zircon is thus imparted to the pigment. Colors from yellow to orange-red are available. Deep red is not available, and the purity of these colors is limited.

Pink and Purple Pigments.   Although red is not available in oxide systems, pink and purple shades are obtained several ways. One such system is the chrome alumina pinks (19,21). Chrome alumina pinks are combinations of ZnO, Al_2O_3, and Cr_2O_3. Depending on the concentration of zinc, the crystal structure may be either spinel or corundum. The latter is analogous to the ruby.

In general, a ceramic coating formulated for use of a chrome alumina pink should be free of CaO, low in PbO and B_2O_3, and with a surplus of ZnO and Al_2O_3. Using an improper coating may lead to a brown instead of a pink. Sufficient ZnO must be in the coating to prevent the glaze from attacking the pigment and removing zinc from it. An excess of alumina prevents the molten coating from dissolving the pigment.

A related but somewhat stronger pink pigment is the manganese alumina pink (7,19). This composition is formulated from additions of manganese oxide and phosphate to alumina and produces a very pure clean pigment. It requires a zinc-free coating high in alumina. Unfortunately, this pigment cannot be made without creating some serious pollution problems and there is question as to its continued availability.

The most stable pink pigment is the iron-doped zircon system (12,19,22,23). This pigment is made by calcining a mixture of ZrO_2, SiO_2, and iron oxide at a stoichiometry to produce zircon. This pigment is sensitive to details of the manufacturing process, so one manufacturer's product may not duplicate another's (24,25). Color variations extend from pink to coral. The pigments are stable in all coating systems, but those without zinc are bluer in shade.

The chrome–tin system is the only family to produce purple and maroon shades, as well as pinks. The system can be defined as pigments that are produced by the calcination of mixtures of small amounts of chromium oxide with substantial amounts of tin oxide. In addition, most formulations contain substantial amounts of silica and calcium oxide.

The chemistry of these materials is complex (19,26). If one mixes $\sim$90% of tin oxide with small amounts of Cr_2O_3 and either CaO or CeO_2, together with B_2O_3 as a mineralizer, one obtains a purple or orchid shade. The crystal structure is cassiterite. This pigment is a solid solution of chromium oxide in tin oxide. Although this is not the crystal structure of most chrome–tin pinks, residual amounts are present in almost all cases. This residual amount of chromium-doped tin oxide gives most chrome tin pinks a gray or purple overtone.

Most chrome–tin pinks also contain substantial amounts of CaO and SiO_2. Only in the presence of these materials can pink, red, or maroon shades be obtained. Here, the crystal structure is tin sphene ($CaO–SnO_2–SiO_2$) with Cr_2O_3 dissolved as an impurity.

The color of this pigment depends on the ratio of Cr_2O_3 to tin oxide. With high chrome contents, the pigments are green. As the chrome is reduced, the color becomes purple at a Cr_2O_3/SnO_2 ratio of 1:15, red at a ratio of 1:17, maroon at a ratio of 1:20, and pink at a ratio of 1:25.

These pigments must be used in a coating low in ZnO and high in CaO. Some SnO_2 should be used as an opacifier, in order to increase the strength and stability of the pigment.

Gold purple, often called Purple of Cassius, is a tin oxide gel colored by finely divided gold (7). It has good coverage and brilliance in low temperature coatings such as porcelain enamels. It is a very expensive pigment, because of its difficult preparation as well as the price of gold.

Brown Pigments. The most important brown pigments used in ceramic coatings are the zinc iron chromite spinels (19,27,28). This pigment system produces a wide palette of tan and brown shades. It can be controlled with reasonable care to produce uniform, reproducible pigments. In this pigment, the Cr_2O_3 is found on the octahedral sites of the spinel structure; the ZnO is found on the tetrahedral sites and the iron oxide is distributed in such a way as to fulfill the stoichiometry of the structure. Consequently, adjustment of the formula does result in alteration of the shade. For example, minor additions of NiO to this system produce a dark chocolate brown. The presence or absence of iron on the tetrahedral site affects the yellowness of the shade. Because they are comparatively low in price, these pigments are the browns selected for most applications.

Two systems closely related to the zinc iron chromites have been developed to improve the stability and firing range of brown pigments (19). The first of these is the zinc iron chrome aluminate pigment. It is a hybrid of the zinc iron chromite brown and the chrome alumina pink. It produces warm and orange brown shades of improved firing stability. The pigment must be used in coatings high in ZnO and Al_2O_3, low in CaO.

The other related pigment is the chrome iron tin brown, often called a tin tan. It must always be used in a ZnO containing coating, since the pigment requires ZnO from the coating to react during the firing process to produce a zinc iron chromite pigment. This pigment is characterized by excellent stability at low concentrations. Thus it makes an excellent toner for tan and beige shades in blends with pink pigments.

The last brown pigment to be considered is the iron manganese brown, which is the deep brown associated with electrical porcelain insulators and with artware and bean pots. In many glazes, the presence of manganese will cause poor surface and unstable color. Hence, the use of this pigment is limited to dark colors on products where glaze surface quality requirements are modest.

Yellow Pigments.   There are several systems for preparing a yellow ceramic pigment. Moreover, there are valid technical and economic reasons for the use of a particular yellow pigment in a given application (19). The pigments of greatest tinting strength, the lead antimonate yellows, cadmium sulfide, and the chrome titania maples, do not have adequate resistance to molten ceramic coatings. Thus other systems must be used if the firing temperature is >1000°C.

Zirconia vanadia yellows are prepared by calcining ZrO_2 with small amounts of V_2O_5 (19,29,30). Small amounts of Fe_2O_3, with or without TiO_2, may be used to alter the shade from lemon yellow to orange yellow. In ceramic coatings, zirconia vanadia yellows are usually weaker than tin vanadium yellows and dirtier than praseodymium zircon yellows. However, they are economical pigments for use with a broad range of coatings firing at temperatures >1000°C. They are brighter and stronger in glazes low in PbO and B_2O_3.

Tin vanadium yellows are prepared by introducing small amounts of vanadium oxide into the cassiterite structure of SnO_2(19,31). Tin vanadium yellows develop a strong color in all ceramic coating compositions. They are very opaque pigments, requiring little further opacification. The pigments are not stable under reducing conditions. They are incompatible with pigments containing Cr_2O_3. However, the primary deterrent to their use is the very high cost of the SnO_2, their principal component.

The praseodymium zircon pigments are formed by calcination of ~5% of praseodymium oxide with a stoichiometric zircon mixture of ZrO_2 and SiO_2 to yield a bright yellow pigment (12,19,31,32). The crystal structure is zircon. These pigments have excellent tinting strength in coatings fired to as high as 1280°C (cone 10). They can be used in almost any ceramic coating, although preferably with zircon opacifiers. They are compatible with most other pigments, particularly other zircon and zirconia pigments.

For applications fired <1000°C, the tinting strength of the lead antimonate pigment is unsurpassed, except by the cadmium sulfoselenides (19). These

Pb–Sb pigments are very clean and bright and have good covering power, requiring little or no opacifier. Their primary limitation is their instability above 1000°C, followed by volatilization of the Sb_2O_3. Substitutions of CeO_2, Al_2O_3, or SnO_2 are sometimes made for a portion of the Sb_2O_3 to improve stability, but these are just palliatives. Hence, their use is limited to coatings such as porcelain enamels, which fire at temperatures <1000°C. Moreover, such pigments are formulated with toxic materials, and great care is required in their use.

An orange-yellow pigment is formed when Cr_2O_3 is added together with Sb_2O_3 and TiO_2 to form a doped rutile (19,33). This material gives an orange yellow or maple shade useful at lower firing temperatures. Like the lead antimony yellows, this material also decomposes ~1000°C. It has substantial use in enamels, where it is the basis for some of the more important appliance colors.

Cadmium sulfide yellow can be considered for the brightest low temperature applications (16,19). It is a very bright, clean orange yellow. Primrose yellow and light yellow shades are made by precipitating small amounts of ZnS with the CdS. All the limitations of the cadmium sulfoselenide reds discussed above apply to the cadmium sulfide yellow.

Green Pigments. Just as there are several alternative yellow pigments, there are several alternative green pigments (19,34). Formerly, the chromium ion was the basis for green pigments. Green Cr_2O_3 itself may be used in a few applications. This procedure, however, has several limitations. There is some tendency for pure Cr_2O_3 to fume and volatilize during firing of the coating, leading to absorption of chromium into the refractory lining of the furnace. Chromium oxide is incompatible with tin oxide, reacting to form a pink coloration. The coating must not contain ZnO, because ZnO in the coating reacts with Cr_2O_3 to produce an undesirable dirty brown color.

Higher quality results are obtained if chromium oxide is used as a constituent in a calcined pigment. One such system is the cobalt zinc alumina chromite used to produce blue-green pigments. These pigments are spinels in which varying amounts of cobalt and zinc appear in the tetrahedral sites and varying amounts of alumina and chromium oxide appear in the octahedral sites. By using higher concentrations of Cr_2O_3 and lower amounts of CoO greener pigments are obtained. Conversely, by lowering the amounts of Cr_2O_3 and raising the amount of CoO, shades from blue-green to blue are obtained. These pigments should only be used in strong masstones. In low concentrations they give an undesirable dirty gray color.

Victoria green is prepared by calcining silica and a dichromate with calcium carbonate to form the garnet $3CaO–Cr_2O_3–3SiO_2$. This pigment gives a transparent bright green color. It tends to blacken if applied too thinly. It is not satisfactory for opaque glazes or pastel shades because then it has a gray cast. It can be used only in zinc-free coatings with high CaO content, which is a difficult, costly pigment to manufacture correctly.

Because of all the difficulties with the use of chromium-containing pigments and because there is a definite limitation on the brilliance of green pigments made with chromium, many green ceramic glazes are now made with zircon pigments (12,19). The cleanest, most stable greens are obtained today by the use of blends of a zircon vanadium blue and a zircon praseodymium yellow. The bright green shades are obtained from a mixture of about two parts of the yellow pigment to one part of the blue pigment.

The final green to be discussed involves copper compounds for low firing-temperature applications. The use of copper is of little interest to most industrial manufacturers, but the colors obtained from them are of great interest to artists because of the many subtle shades that can be obtained. These subtle shade variations arise because the pH of the glaze used has a particular effect on the colors obtained. If the coating is acidic, a beautiful green color is developed; but if the coating is alkaline, a turquoise blue color results. The copper oxide dissolves in the coating producing a very transparent color. Copper oxide glazes are limited to 1000°C because of copper volatilization. Also, copper oxide renders all lead-containing glazes unsafe for use with food and drink (35).

Blue Pigments. The traditional way to obtain blue in a ceramic coating is with cobalt, which has been used as a solution color since antiquity (19,36). Today, cobalt may react with Al_2O_3 to produce the spinel $CoAl_2O_4$ or with silica to produce the olivine Co_2SiO_4. The silicate involves higher concentrations of cobalt, with only modestly stronger color. In porcelain enamels and glass colors, pigments based on cobalt continue to be fully satisfactory both for stability and for tinting strength. At the higher temperatures used for ceramic glazes, difficulties arise from partial solution of the pigment, and diffusion of cobalt oxide in the glaze, causing a defect called cobalt bleeding.

In glazes, the cobalt pigments have been largely replaced by pigments based on vanadium-doped zircon (12,19,37,38). These pigments are turquoise in shade and are less intense than the cobalt pigments. Therefore, they are not applicable when the greatest tinting strength is required or when a purple shade is called for. Where they are applicable they give vastly improved color stability.

The zircon vanadium blue pigment is made by calcining a mixture of zirconia, silica, and vanadia in the stoichiometry of zircon and in the presence of a mineralizer. Mineralizers, selected from various halides and silicohalides, facilitate the transport of the silica during the reaction forming the pigment (24). The strongest pigments are formed at the stoichiometry of zircon, using such mineralizers as will facilitate the various transport processes, incorporating the maximum amount of vanadium into the zircon structure.

5.3. Black Pigments. Black ceramic pigments are formed by calcination of several oxides to form the spinel structure (19,39,40). The formulation of black illustrates the flexibility of the spinel structure in incorporating various chemical entities. The divalent ion may be cobalt, manganese, nickel, iron, or copper. The trivalent ions may be iron, chromium, manganese, or aluminum. The selection of a particular black pigment depends somewhat on the specific coating material with which the pigment is to be used. Care must be taken to see that the pigment does not show a green, blue, or brown tint after firing. Of particular importance is the tendency of some glazes to attack the pigment and release cobalt. Thus in some cases it is desirable to use a cobalt-free pigment.

The prototype black is a cobalt iron chromite. In some systems, however, it will have a slight greenish tint. In zinc-containing glazes, a black with some nickel oxide would be recommended. For a black with a slightly bluish tint, a formula containing manganese and higher cobalt would be recommended. For a black with a brownish tint, a complex formula containing cobalt, iron, nickel, manganese, alumina, and chrome would be recommended.

When a cobalt-free system is needed, there are three possibilities. The copper chromite black pigment is a spinel, which is suitable for use in coatings firing below 1000°C, such as porcelain enamels. The chromium black hematite is an inexpensive system that is suitable for use in zinc-free coatings (41,42). If ZnO is present, it will react to form a brown color. For glazes firing >1000°C, and containing ZnO, the chromium iron nickel black spinel can be used (39,43). This pigment can be used in most glazes and on firing schedules as high as cone 10.

5.4. Gray Pigments. It might be expected that the easiest way to obtain a gray pigment would be to dilute a black pigment with a white opacifier. However, it is very difficult to provide an even color, free of specking, with this technique. Usually, it is preferable to select a compound that has been formulated to give a gray color.

It is easiest to obtain a uniform gray color when a calcined pigment is used that is based on zirconia or zircon as a carrier for various ingredients of blacks such as Co, Ni, Fe, and Cr oxides (19). This pigment is called cobalt nickel gray periclase. For underglaze decorations it is possible to prepare a very beautiful deep gray color by dispersing antimony oxide in tin oxide. The limitation on the use of this material is the high cost of the SnO_2, which limits its use to special effects.

6. Economic Aspects

Owing to the limited market and the variety and complexity of the products, ceramic pigments are manufactured by specialist firms, not by the users. The principal producers are Ceramic Color and Chemical Corp., Englehard Corp., Ferro Corp., General Color and Chemical Corp., Mason Color and Chemical Corp., and Pemco Corp. Estimated annual production is ~2500–3000 metric tons. This figure does not include some of the same and similar products manufactured for use in non-ceramic applications. The costs of ceramic pigments range from $10 to $60/kg or higher, depending on the elemental composition and the required processing. The most expensive pigments are those containing gold and the cadmium sulfoselenides.

7. Health and Safety Factors

Properly handled, ceramic colorants should not cause unacceptable problems of health and safety. Preventive measures to avoid inhalation of fine particulate matter should invariably be used. Care should be taken to avoid ingestion of pigments by thorough washing before eating or smoking. Particular care should be taken in handling cadmium sulfoselenide pigments and lead antimonate pigments, which are highly toxic if ingested or inhaled (44).

When these pigments are used with lead-containing glazes, care should be exercised to use lead-safe glaze materials (see LEAD COMPOUNDS, INDUSTRIAL TOXICOLOGY).

8. Use of Pigments in Coatings

There are several additional factors that must be considered in selecting pigments for a specific coating application (44). These factors include processing stability requirements, pigment uniformity and reproducibility, particle size distribution, dispersibility, and compatibility of all materials to be used.

8.1. Processing Stability.   A significant limitation on the selection of ceramic pigments is the set of processing conditions imposed during coating application and firing (44). An engobe or body stain must be stable to the bisque fire, usually between cone 7 (1225°C) and cone 11 (1300°C). An underglaze color, or a colored glaze, must be stable to the glost fire, usually between cone 06 (1000°C) and cone 4 (1200°C), and to corrosion by the molten glaze ingredients. An overglaze or glass color needs only to be stable to the decorating fire by which it is applied, usually between cone 020 (625°C) and cone 016 (775°). More important here is corrosion by the molten flux used in the application.

The stability of the various pigments is discussed above for those pigments with limited stability. Detailed information is available (44).

8.2. Uniformity and Reproducibility.   For most ceramic pigments rapid, uniform, and reproducible conversion to the desired product requires great care in production (44). Adjustment of each lot to standard, using toners, is usually required. The Victoria green garnet, the manganese alumina pink corundum, and the chrome–tin pink sphene are noteworthy for their difficulty in making reproducible product.

If a small amount (<5% of the blend) of a strong pigment is used as a component in a blend, it will be difficult to obtain sufficiently uniform mixing to avoid specking. It is preferable to use larger concentrations of a less intense pigment.

Some pigments are sensitive to details of the glaze application and firing procedures. With these pigments it may be difficult to maintain uniformity, even within a given lot of material. The Victoria green garnet, the copper greens, and the cadmium sulfoselenides are particularly sensitive to these problems.

8.3. Particle Size.   Most calcined ceramic pigments are in the 1–10-μm range in mean particle size, with no residue on a 325 mesh (44 μm) screen. The selection of an optimum particle size distribution is a compromise between considerations of dissolution rate, agglomeration of the pigment, loss of strength on milling, uneven surface smoothness, and pigment strength (44–47). The optimum particle size is the largest size that gives adequate dispersion and adequate strength in letdowns.

8.4. Dispersibility.   Pigments modified with a dispersion additive take less time and energy to disperse in a coating (13,48). The equipment for blunging is simpler and less expensive than ball mills. Color correction is simplified and settling is minimized. Color strength in letdowns is often improved (46).

8.5. Compatibility.   A ceramic pigment must function as a component in a glaze or porcelain enamel system. Hence, it must be compatible with the other components, ie, the glaze itself, the opacifier(s), and other additives (44). There is a large variability in glaze–pigment interaction during firing. Some pigments, such as the zircon compounds, are relatively inert in conventional glazes. Other pigments are much more reactive. In particular, it is important to prevent reaction of a pigment with a glaze component to produce a more stable pigment

(49). Hematite pigments react with divalent ions in the glaze to form spinels. Victoria Green garnet reacts with divalent ions to form spinels plus silicates.

Probably the most important glaze consideration is the presence or absence of ZnO in the glaze (19). The manganese alumina pink corundum, chromium, green–black hematite, Victoria green garnet, chrome–tin orchid cassiterite, and chrome–tin pink sphene are not stable in the presence of zinc oxide [1314-13-2], (ZnO). The iron brown hematite, chrome alumina pink spinel, iron chromite brown spinel, zinc ferrite brown spinel, and zinc iron chromite brown spinel require high ZnO concentration. High calcium oxide concentration is required for adequate stability of Victoria green garnet and chrome–tin pink sphene. Calcium oxide should be avoided when using chrome alumina pink spinel, zinc ferrite brown spinel, and zinc iron chromite brown spinel. Pigments containing chromium(III) oxide are incompatible with pigments containing tin oxide.

The presence or absence of PbO in a glaze affects some pigments (50). Victoria green and cobalt black pigments are stronger in a high PbO glaze. Zircon vanadium blue, zirconia vanadium yellow, and chrome–tin pink pigments are only suitable for low PbO or lead-free glazes. Mixed zircon greens, zircon iron pink, zinc iron chromite brown, and zirconia gray pigments are stronger in low PbO and lead-free glazes.

BIBLIOGRAPHY

"Colors for Ceramics and Glass" in ECT 1st ed., Vol. 4, pp. 276–287, by W. A. Weyl and R. R. Shively, Jr., The Pennsylvania State College; "Colors for Ceramics" in ECT 2nd ed., Vol. 5, pp. 845–856, by W. A. Weyl, The Pennsylvania State University; "Colorants for Ceramics" in *ECT* 3rd ed., Vol. 6, pp. 549–561, by E. E. Mueller, Alfred University.; "Colorants for Ceramics" in *ECT* 4th ed., Vol.6, pp. 877–891, by Richard A. Eppler, Eppler Associates; "Colorants for Ceramics" in *ECT* (online), posting date: December 4, 2000, by Richard A. Eppler, Eppler Associates.

CITED PUBLICATIONS

1. A. Burgyan and R. A. Eppler, *Am. Ceram. Soc. Bull.* **62**(9), 1001 (1983).
2. P. M. Rice, *Pottery Analysis*, University of Chicago Press, Chicago, Ill., 1987, pp. 331–346.
3. W. D. Kingery, H. K. Bowen, and D. R. Uhlmann, *Introduction to Ceramics*, John Wiley & Sons, Inc., New York, 1976, pp. 677–689.
4. A. Paul, *Chemistry of Glasses*, Chapman & Hall, Ltd., London, 1982, pp. 233–251.
5. S. R. Scholes and C. H. Greene, *Modern Glass Practice*, 7th ed., Cahners Publishing Co., Boston, Mass., 1975, pp. 302–329.
6. W. Vogel (trans. N. Kriedl), *Chemistry of Glass*, American Ceramic Society, Columbus, Ohio, 1985, pp. 163–177.
7. R. A. Eppler, *Ullmann's Encyclopedia of Industrial Chemistry*, Vol. A5, VCH Publishers, Inc., Weinheim, West Germany, 1986, pp. 163–177.
8. F. T. Booth and G. N. Peel, *Trans. J. Br. Ceram. Soc.* **58**(9), 532 (1959).
9. D. A. Earl and D. E. Clark, *Ceram. Eng. Sci. Proc.* **21**(2), 109 (2000).
10. R. D. Shannon and A. L. Friedberg, *Univ. Ill. Eng. Exp. Sta. Bull.* **456**, 49 (1960).

11. R. A. Eppler, *J. Am. Ceram. Soc.* **52**(2), 89 (1969).
12. R. A. Eppler, *Am. Ceram. Soc. Bull.* **56**(2), 213, 218, 224 (1977).
13. T. D. Wise, S. H. Murdock, and R. A. Eppler, *Ceram. Eng. Sci. Proc.* **12**(1–2), 275, 194 (1991).
14. R. A. Eppler and D. R. Eppler, *Ceram. Eng. Sci. Proc.* **15**(1), 281 (1994).
15. *DCMA Classification and Chemical Description of the Mixed Metal Oxide Inorganic Colored Pigments*, 2nd ed., Metal Oxides and Ceramic Colors Subcommittee, Dry Color Manufacturer's Association, Arlington, Va., 1982.
16. R. A. Eppler, *J. Am. Ceram. Soc.* **66**(11), 794 (1983).
17. R. A. Eppler and D. S. Carr, in *Proceedings of the 3rd International Cadmium Conference, International*, Lead Zinc Research Organization, New York, 1982, pp. 31–33.
18. U.S. Pat. 2,643,196 (June 23, 1953), B. W. Allan and F. O. Rummery (to Glidden Co.); U.S. Pat. 2,777,777 (Jan. 15, 1957), (to Glidden Co.).
19. R. A. Eppler, and D. R. Eppler, *Glazes and Glass Coatings*, American Ceramic Society, Westerville, Ohio, 2000.
20. H. D. deAhna, *Ceram. Eng. Sci. Proc.* **1**(9–10) 860 (1980); U.S. Pat. 4,482,390 (Nov. 13, 1984), A. C. Airey and A. Spiller (to British Ceramic Research Association Ltd.).
21. R. L. Hawks, *Am. Ceram. Soc. Bull.* **40**(1), 7 (1961).
22. U.S. Pat. 3,189,475 (June 15, 1965), J. E. Marquis and R. E. Carpenter (to Glidden Co.).
23. U.S. Pat. 3,166,430 (Jan. 19, 1965), C. A. Seabright (to Harshaw Chemical Co.).
24. R. A. Eppler, *J. Am. Ceram. Soc.* **53**(8), 457 (1970).
25. C.-H. Li and R. A. Eppler, *Ceram. Eng. Sci. Proc.* **13**(1–2), 109 (1992).
26. R. A. Eppler, *J. Am. Ceram. Soc.* **59**(9–10), 455 (1976).
27. J. E. Marquis and R. E. Carpenter, *Am. Ceram. Soc. Bull.* **40**(1) 19 (1961).
28. S. H. Murdock and R. A. Eppler, *J. Am. Ceram. Soc.* **71**(4), C212 (1988).
29. C. A. Seabright and H. C. Draker, *Am. Ceram. Soc. Bull.* **40**(1), 1 (1961).
30. F. T. Booth and G. N. Peel, *Trans. J. Br. Ceram. Soc.* **61**(7), 359 (1962).
31. E. H. Ray, T. D. Carnahan, and R. M. Sullivan, *Am. Ceram. Soc. Bull.* **40**(1), 13 (1961).
32. R. A. Eppler, *Ind. Eng. Chem. Prod. Res. Dev.* **10**(3), 352 (1971).
33. *The Colour Index*, 3rd ed., Society of Dyers & Colourists, Bradford-London, UK, 1971.
34. P. Henry, *Am. Ceram. Soc. Bull.* **40**(1), 9 (1961).
35. *Lead Glazes for Dinnerware*, International Lead Zinc Research Organization, New York, 1970.
36. R. K. Mason, *Am. Ceram. Soc. Bull.* **40**(1), 5 (1961).
37. U.S. Pat. 2,441,447 (May 11, 1948), C. A. Seabright (to Harshaw Chemical Co.); U.S. Pat. 3,025,178 (Mar. 13, 1962), (to Harshaw Chemical Co.).
38. T. Demiray, D. K. Nath, and F. A. Hummel, *J. Am. Ceram. Soc.* **53**(1), 1 (1970).
39. R. A. Eppler, *Am. Ceram. Soc. Bull.* **60**(5), 562 (1981).
40. W. F. Votava, *Am. Ceram. Soc. Bull.* **40**(1), 17 (1961).
41. S. H. Murdock and R. A. Eppler, *Am. Ceram. Soc. Bull.* **68**(1), 77 (1989).
42. A. Escardino, S. Mestre, and A. Barba, *J. Am. Ceram. Soc.* **83**(1), 29 (2000).
43. U.S. Pat. 4,205,996 (June 3, 1980), R. A. Eppler, (to SCM Corp.).
44. R. A. Eppler, *Am. Ceram. Soc. Bull.* **66**(11), 1600 (1987).
45. S. H. Murdock, T. D. Wise, and R. A. Eppler, *Am. Ceram. Soc. Bull.* **69**(2), 228 (1990).
46. S. H. Murdock, T. D. Wise, and R. A. Eppler, *Ceram. Eng. Sci. Proc.* **10**(1–2), 55 (1989).
47. C. Decker, *Effects of Grinding on Pigment Strength in Ceramic Glazes*, presented at the 93rd annual meeting, American Ceramic Society, Cincinnati, Ohio, May 1, 1991 *Ceram. Eng. Sci. Proc.* (1992).
48. A. Sefcik, *Ceram. Eng. Sci. Proc.* **12**(1–2), 173 (1991).

49. D. R. Eppler, and R. A. Eppler, *Ceram. Eng. Sci. Proc.* **18**(2), 139 (1997).
50. R. A. Eppler and D. R. Eppler, *Ceram. Eng. Sci. Proc.* **13**(1–2), 338 (1992), and, *Ceram. Eng. Sci. Proc.* **14**(1–2), 137, (1993).

RICHARD A. EPPLER
Eppler Associates

COLORANTS FOR PLASTICS

1. Introduction

Color has been important to society since the beginning of time. From cave walls that told stories of past life to today's automobiles color has influenced the lives of every person on earth. In fact, even those who are color deficient or totally color blind are influenced and/or make special accommodations for color or their lack of the ability to perceive color. We control our movements on the roadways when encountering a traffic light that shows green, yellow, or red. The color blind person even though being unable to distinguish green, yellow, or red will make judgments on the basis of the lightness or other cues he/she sees and then makes decisions to go or stop. Therefore, color impacts every one of us regardless of our individual capabilities and/or disabilities. In the final analysis, we chose the color of our homes, in and out, the clothes we wear, the cosmetics we use, the automobiles we drive, the art we collect, the people we chose to associate with, and the places we chose to live all on the basis of how we decide on colors. We chose products or objects of various colors because we like them!

Once over the hurdle of color for its pleasure and appearance, the issue becomes one of performance and economics. Any color system in any product for industry and/or commerce must first be able to withstand the fabrication of the product. Next, the product must be able to endure the requirements of its end use environment for the designed life of the product. Finally, the economics must be such that the desirable products in the marketplace are within the economic reach of the consumer.

This brings us to the essence of coloring and colorants. Incorporating colorants into plastics may be illustrated by saying that the colorants must meet "zero or no reaction chemistry". If any chemical reaction takes place any time in a products fabrication and/or lifetime a color change is most likely to take place and the product will have failed to meet its technical and commercial design objectives.

Historically, colorants are divided into three distinct groups: organic pigments, inorganic pigment and soluble dyes. One can make the general distinction that pigments are particulate materials that should remain physically unchanged in a plastic system. Soluble dyes, on the other hand, go into solution in the plastic system, are at the molecular level, and have no particulate identity. Unfortunately, this distinction is not as clear-cut as one would like. Inorganic pigments are thought to be inherently insoluble. Indeed, most organic pigments are essentially insoluble. However, there are situations where under very specific circumstances an organic pigment may become soluble or partially soluble causing product failure. Soluble dyes to perform properly must be totally soluble in the plastic system to be effective and meet their design criteria.

The following are definitions refined from numerous sources and modified to reduce the definitions to plain words, which are understandable. These definitions are not posed to be scientifically and in totality exact since some liberty was used to make the definitions understandable to the novice reader. They are posed only to give the reader an initial understanding of what these materials are.

1.1. Organic Pigments. An organic pigment is made up of carbon, hydrogen, oxygen, and sulfer and nitrogen in combinations of atoms bonded together. Sulfur, chlorine, bromine, fluorine, calcium, strontium, barium, and manganese may also be present. The molecules are crystalline in nature and selectively absorb light to provide color. The organic pigment must remain in particulate form to perform its color functionality.

1.2. Inorganic Pigment. An inorganic pigment is made up of a combination of metalloid or metallic elements combined with oxygen, sulfur, and/or selenium. The molecules will selectively absorb light to provide color. The inorganic pigment must remain in its original particulate form to perform its color functionality.

1.3. Soluble Dye (Dyestuff). Soluble dyes dissolve in the plastic system losing any crystalline characteristics and operate at the molecular level. Since soluble dyes properly dissolved in a system have no particulate characteristics they impart color only through selective absorption. The soluble dye must be 100% in solution to perform as designed.

These truncated definitions should bring one to the conclusion that pigments can vary from almost transparent (slightly hazy to translucent) to totally opaque, while soluble dyes, used properly, will impart complete transparency.

1.4. Colorant Forms. Pigments and soluble dyes predominantly leave their manufacturing sites as powders. There may be a number of intermediate steps, however, to improve value before the colorant reaches a company for fabrication into a useful consumer product.

1.5. Dry Color Pigment (Raw Color Pigment). Dry colorants are usually added to powder polymers since the particle sizes of the powder polymers and the dry colorants are reasonably close together. The dry colorants would be added as single colorants and/or blends of dry colorants formulated to match a target color. This is a significant consideration as adding a powder colorant to pellets, cubes, or other much larger polymer granules may lead to separation during processing resulting in streaks and nonuniform color. Maximizing the efficiency of the dry colorant is also an important issue. The ability to disperse dry colorants uniformly and totally during the final fabrication step is difficult to impossible. A preliminary cost calculation may tend to indicate economies with this approach, however, additional costs are most likely to occur through rejected parts, difficult color control, and other processing issues that will consume any projected economies.

2. Dispersions

2.1. Dry Color Concentrates. Colorant dispersions are designed to provide highly efficient colorant use and uniformity of color in fabricated plastic products. The colorant dispersions bring value added effects to the plastics processes avoiding many of the problems associated with dry colors. The unit cost of colorant materials will be higher, but the finished product or system cost usually is lower and results in a superior product. Dispersions for plastics fall into two definite categories. They are solid pellets of various sizes and shapes, and liquids or pastes. The solid dispersions are further divided into cube pellets (~1/8 to 1/4 in.), spherical pellets (~1/8 in. diameter), microbeads (~1/64 in. or smaller in diameter), and cryogenically ground material (random shapes of various very small sizes). The most popular method to process these materials is twin-screw corotating compounding extruders. Other methods are also utilized such as single screw compounding extruders, Banbury high-intensity mixers, and continuous high-intensity mixers. Liquid dispersions, on the other hand, are colorants dispersed in a liquid vehicle that is part of the plastic system or is compatible with the system. Organic colorants can range from ~20 to 60% in dry concentrate while inorganic colorants can range from 50 to 85%. Soluble dyes in dry concentrates will usually range between 15 and 30% colorant loadings. Particular situations may suggest ranges outside these generalized loadings for uncommon special situations.

No matter the equipment used, the dispersion process must accomplish a number of tasks. First is distributive mixing followed by dispersion of the colorants. The dispersion process is designed to break up agglomerates of loosely held colorant particles, reduce aggregates of tightly held particles as much as possible (1), and remove adsorbed air from particle surfaces (2) (Fig. 1).

Care must be taken since the assumption that more dispersion energy is better may not produce the best product. Two issues should be considered. If additional color intensity or strength results from more dispersion energy input, the colorants have not been dispersed and developed in the original dispersion step. If a hue, chroma, and value change results from additional dispersion energy, pigment degradation and/or an unwanted chemical reactions occur both of which indicate the processing conditions are to severe. In almost every instance, all three; hue, chroma, and value will change at the same time. It is possible as the dispersion process goes forward the color properties will increase. However, it is possible to overdevelop or disperse colorants to the point that the color value will decrease.

2.2. Liquid Colorant Concentrates. Liquid color concentrates play the same role as the dry concentrate using a different method to introduce colorants into plastics. The two significant differences are that the carrier is a liquid rather than a solid polymer. The liquid vehicle opens the door to an entirely different dispersion technique. Dispersion can be very intense and superior to the dry color concentrates in terms of reducing aggregates and agglomerates.

Loadings in liquid colorant concentrates can range from as low as 20% to as high as 95%. The methods for obtaining these high loadings of colorants are high intensity rotating disk dispersers; sand mills, and three roll ink mills.

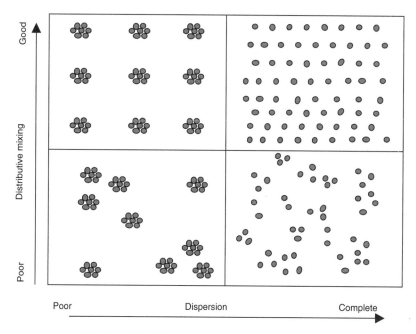

Fig. 1. Distributive mixing versus dispersion.

The potential for very high loadings in liquid colorant concentrates may lead to cost savings per unit product produced (Fig. 2).

3. Colorant Properties

The properties of colorants can have a profound impact on the plastic being colored. There are a number of general property issues that should be considered each and every time an individual colorant is added to a system. General summaries of these properties are covered here.

1. *Dispersibility*. Colorants must be able to be uniformly dispersed in any plastic system. If the system is unable to tolerate the shear energy required to disperse a given colorant, no matter how desirable the colorant may be, it is not usable in that system. Dispersed colorants as solid or liquid concentrates may provide the avenue to use colorants not meeting the initial dispersion requirements.

2. *Lightfastness*. Lightfastness is a function of exposure to ultraviolet (uv) and possibly visible (vis) radiation and humidity, but no contact with liquid water. A rule of thumb description might be to think of lightfastness as indoor exposure.

3. *Migration*. Migration is defined as the movement of particulate colorants or additives in a finished product. Even though many assume a particulate material is "locked in" in a solid plastic, movement can take place when

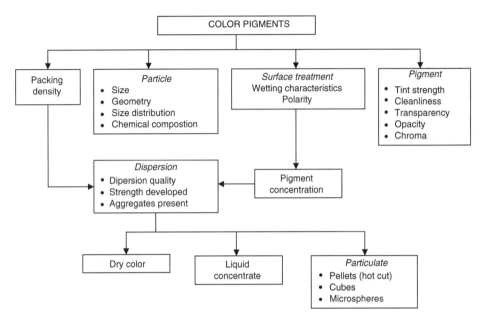

Fig. 2. Physical and dispersion properties of pigments.

conditions permit. This mobility is thought by some to be blooming and/or bronzing. Usually, it can be detected over time by the presence of a metallic appearing dry film on the surface of a part, which can be mechanically removed.

4. *Chemical Resistance*. Chemical resistance means the colorant(s) do not get involved chemically or react in any way as a result of mechanical mixing, heat, shear energy, attack by chemicals externally applied, uv, vis, infrared (ir) radiation, gases, aerosols, or any miscellaneous items. Any time a colorant and/or additives in a plastic system are active chemically, a color change can be expected. Therefore, the goal is to have the colorants function without being chemically active.

5. *Blooming*. Blooming is sometimes called sublimation. It usually involves low molecular weight colorants that literally evaporate out of a plastic system then condenses on the surface of the plastic product. This condensate can usually be removed mechanically. The process is one that takes place over time.

6. *Toxicity*. This issue is easy to declare, but difficult to achieve. Unanimity of which colorants are toxic to society is difficult to obtain since opinions vary widely. It can be said that no colorants or combination of colorants should present a danger to health during manufacture, processing, and fabrication, or during the life cycle of the product.

7. *Solvent Resistance*. Colorants must resist attack by solvents, which might cause them to partially or totally go into solution. If this happens there is usually a color change accompanying the solubility. Many organic pigments and, of course, soluble dyes are subject to this phenomenon.

It should be noted that unexpected solubility of colorants may come from the polymer as a solvent at normal and/or at some elevated threshold temperature. Also forgotten many times is that water is or can be an active solvent!

8. *Environmental*. This issue is one of doing nothing that results in a hazard to the world around us. It is different from toxicity, which deals with hazards to human health. There is concern that plastic products that are discarded may be harmful to the environment and the creatures and growing things that inhabit our environment. This describes just one illustration of a complex issue. For our specific purpose, the consideration is that colorants, however, they may get into the environment cause no harm to plants and animals.

9. *Compatibility*. This issue seems self-explanatory, however many misunderstand it. Some colorants, mainly organics and soluble dyes, have a negative effect on some polymers where they may be introduced. Some colorants may become soluble or partially soluble in some polymers and in other cases; eg, colorants that promote polymer degradation are issues that must be understood by the technician working with them.

10. *Batch Uniformity*. Any user of colorants may expect batch No. 10 or higher to be identical to batch No. 1. Typical manufacturing of colorants suggests making subsequent batches absolutely identical to the first batch is unrealistic. However, a tolerance can be obtained that satisfies the needs of a project. The issue here is not to demand or expect consistent uniformity beyond the capabilities of the manufacturing systems involved.

11. *Particle Integrity*. Integrity of a colorant particle refers to a colorants ability to withstand fracture during intensive dry mixing, high shear rates during compounding, and other physical abuse during fabrication. Some colorant particles are thought to be "brittle". These colorants are susceptible to fracture anywhere during processing. If a colorant particle is, in fact, fractured, a color change will result. It may be assumed chemical degradation has taken place while in fact a mechanical problem exists. No solution is possible unless the problem is correctly identified as either a chemical or a mechanical one.

12. *Particle Size*. Colorant particle size and size distribution are paramount for color and performance. Particle size and distribution of colorants are directly related to the ability of the colorant to selectively absorb and scatter light. The ability of a colorant to scatter light results in opacity or transmission of light. The ability to selectively absorb light results in the observed color. Many applications for colorants can fail if the particle size is not well matched to design properties of the system. An example of a miss-match might be a large size particle colorant used in a fiber application. In this case, fiber breakage can be expected due to the large particle weakening a fiber since it occupies a significant cross-section of the fiber thereby making it very weak and subject to breaking. This can occur during fiber manufacture and/or during later operations.

13. *Heat Stability*. Heat stability should be understood as a time–temperature relationship. Most pigments and soluble dyes have a limit to their

endurance to heat over time. It is possible that a low temperature history over a long time may degrade a colorant resulting in a color change. On the other hand, the same colorant may tolerate a very high temperature of short duration successfully, which means no color change. Typically, technologists will test the heat stability of a colorant in a 5-min exposure at a given temperature just below where degradation takes place. It should be noted holding a colorant in a plastic machine barrel for a 5-min residence time usually does this. Note this type of testing is in the absence of air, whereas a test that exposes the colorant polymer combination to air (oxygen) may result in a totally different result. This issue is further complicated in that the results of a colorant tested in multiple plastic materials may also give widely varying results for each polymer.

14. *Weatherability*. Weatherability measures the ability of a colorant in its polymer system to withstand the effects of uv, vis, and ir radiation in the presence of heat cycling, direct contact with water, humidity, gases, aerosols, particulates, and the other pollutants occurring in an outdoor environment. These can vary significantly depending on location such as a clean mountain exposure compared to an industrial exposure with many pollutants, where some maybe unknown.

15. *Bleeding or Migration*. Bleeding, sometimes called contact bleeding, is defined as a colorant in a polymer, which is in intimate contact with a similar or different polymer due to colorant solubility moving from the initial polymer to the second polymer resulting in discoloration. This movement may be associated with noncolorant additives in either or both polymers.

16. *Electrical*. Many plastic products are designed to function in an electrical environment where the ability to conduct a charge or provide insulation properties are paramount. Colorants can have a significant effect on these properties. This can be by design or unwanted consequences. Carbon based or metallic pigments can provide significant opportunities to conduct a charge. On the other side of the issue where insulating qualities are desired, colorants can reduce the insulating properties. Many colorants contain metal salts intentionally or due to inadequate manufacturing procedures. If this is the case, the colorant can contribute to the conduction of an electrical charge. This issue is further complicated due to the recent emergence of conducting polymers. Polymeric or plastic materials that actually conduct an electrical charge at some level are commercially available.

17. *Fire Resistance*. Fire resistance addresses the issue of will colorants in a plastic increase or decrease the inherent resistance for the plastic to burn. Inorganic colorants do not contribute to the burning of plastic materials. Organic colorants are combustible in themselves. However, the contribution is usually quite small or nonexistent. One positive factor at work is that, in most cases, the amount of colorant present may be so small that its contribution is not significant. Flame and/or fire resistance is not an issue that can be ignored. In any system where fire resistance must be considered, the contribution by colorants good and bad must be accounted for.

18. *Availability*. This should be self-evident to most. Color technologists are drawn to what's new! This has its merits since the first out in the market-place with a new and different product stands to reap benefits. However, the down side of this position is that in some cases the continued reliability of the source and the uniformity (No. 10) of the product over time needs to be closely determined. Products are removed from the market on occasion, not because the product is faulty, but because the confidence of a continuing supply of uniform colorant and/or colorant raw materials cannot be assured.

19. *Economics–Cost*. The best product in the world is useless if the market cannot afford it. Particularly in the field of coloring plastics, the colorist focuses on making the best match to meet all the requirements. Then and only then are economics and costs addressed. One should always remember that the cost of producing a colored plastic product always has the economics in the overall equation. To ignore this issue is to invite disaster.

20. *Color Values*. Color values of individual colorants play an important role. The colorist needs to understand issues such as, but not limited to, hue, undertone color, and strength. An example might be trying to match a yellow shade (undertone color) red using a blue shade (undertone color) red. Chances are under these stated conditions a close color match is unattainable. Strength of a colorant deals with just how much color is visually delivered per unit weight per unit cost. The colorist must understand and be aware of these issues if success is going to be achievable.

21. *Ease of Cleanup*. This in many cases is a minor issue. It can be important particularly in a production scenario where many color changes are made on a given production setup. Ease of cleanup has many variables, however, a combination of a low specific gravity (density) coupled with a low apparent bulk density indicates a dusty colorant. This is just one example of this issue. There are many more depending on the processes involved.

4. Organic Pigments

There are hundreds of organic pigment chemical families in commerce. Additionally, manufacturers compete with one another directly or with variations of the most popular chemical types. This makes organic pigment selection a complex process. In many cases, there is no right choice, just a choice. Therefore, basic information presented here will allow the technologist to direct colorant selection with a basic understanding of the issues involved (Fig. 3).

Table 1 lists selected organic pigments for plastics. This list cannot include each and every organic pigment that has utility in plastics. To do so would mean listing the hundreds of chemical types and the thousands of variations of these organic colorants. The list is an overview of what is possible. Individual coloring projects may have to search for similar pigments to satisfy specific requirements.

Table 2 offers and additional guide to organic pigment selection.

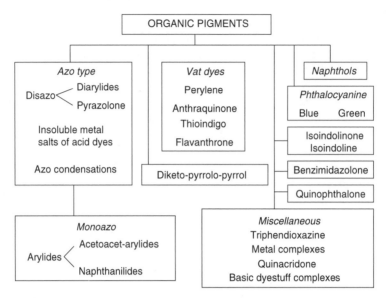

Fig. 3. Organic pigments.

Table 1. **Selected Organic Pigments for Plastics**

Colour Index name	Pigment chemical types	Colour Index number	CAS Registry number
Vat Yellow 1	Flavanthrone Yellow	70600	[475-71-8]
Vat Yellow 20	Anthrapyrimidine Yellow	68420	[4216-01-7]
Pigment Yellow 13	Diarylide Yellow AAMX	21100	[5102-83-0]
Pigment Yellow 14	Diarylide Yellow AAOT	21095	[5468-75-7]
Pigment Yellow 17	Diarylide Yellow AAOA	21105	[4531-49-1]
Pigment Yellow 62:1	Monoazo Yellow, Calcium Salt	13940:1	[12286-66-7]
Pigment Yellow 83	Diarylide Yellow 83	21180	[5567-15-7]
Pigment Yellow 93	Disazo Condensation Yellow	20710	[5580-57-4]
Pigment Yellow 95	Disazo Condensation Yellow	20034	[5280-80-8]
Pigment Yellow 97	Permanent Yellow FGL	11767	[12225-18-2]
Pigment Yellow 138	Quinophthalone Yellow	56300	[56731-19-2]
Pigment Yellow 139	Isoindoline Yellow	56298	[36888-99-0]
Pigment Yellow 150	Metal Complex Yellow	12764	[68511-62-6]
Pigment Yellow 155	Bisacetoacetarylide	N/A	[68516-73-4]
Pigment Yellow168	Monoazo Yellow, Calcium Salt	13960	[71832-85-4]
FD&C Yellow #5	Tartrazine Yellow	19140	[1934-21-0]
Pigment Yellow 191	MonoazopyrazoloneYellow, Calcium Salt	18795	[129423-54-7]
Pigment Red 48:1	Permanent Red 2B, Barium Salt	15865:1	[7585-41-3]
Pigment Red 48:2	Permanent Red 2B, Calcium Salt	15865:2	[7023-61-2]
Pigment Red 48:3	Pemanent Red 2B, Strontium Salt	15865:3	[15782-05-5]
Pigment Red 53	Red Lake C	15585	[2092-56-0]
Pigment Red 122	Quinacridone Magenta	73900	[980-26-7]
Pigment Red 123	Perylene Red	71145	[24108-89-2]
Pigment Red 144	Disazo Condensation Red	71145	[12225-02-4]
Pigment Red 166	Disazo Condensation Red	20730	[1225-04-6]

Table 1 (*Continued*)

Colour Index name	Pigment chemical types	Colour Index number	CAS Registry number
Pigment Red 168	Brominated Anthanthrone Orange	59300	[4378-61-4]
Pigment Red 177	Anthraquinone Red	65300	[4051-63-2]
Pigment Red 179	Perylene Red	71130	[5521-31-3]
Pigment Red 190	Perylene Red	71140	[6424-77-7]
Pigment Red 202	Quinacridone Magenta	73907	[68859-50-7]
Pigment Red 220	Disazo Condensation Red	20055	[57971-99-0]
Pigment Red 221	Disazo Condensation Red	20065	[61815-09-6]
Pigment Red 254	Diketo-pyrrolo-pyrrol	56110	[84632-65-5]
Pigment Red 255	Diketo-pyrrolo-pyrrol	N/A	N/A[a]
Pigment Red 264	Diketo-pyrrolo-pyrrol	N/A	[120500-90-5]
Pigment Red 272	Diketo-pyrrolo-pyrrol	561150	N/A[a]
Vat Orange 7	Perinone Orange	71105	[4424-06-0]
Vat Orange 15	Anthramide Orange	69025	[2379-78-4]
Pigment Orange 13	Pyrazolone Orange	21110	[3520-72-7]
Pigment Orange 34	Disazo Condensation Orange	21115	[15793-73-4]
Pigment Orange 43	Perinone Orange	71105	[42612-21-5]
Pigment Orange 64	Azoheterocycle Orange	12760	[75102-84-2]
Pigment Orange 71	Diketo-pyrrolo-pyrrol	561200	N/A[a]
Pigment Orange 73	Diketo-pyrrolo-pyrrol	561170	N/A[a]
Vat Blue 4	Indanthrone Blue	69800	[81-77-6]
Pigment Blue 15:1	Phthalocyanine Blue	74160:1	[147-14-8]
Pigment Blue 15:3	Phthalocyanine Blue	74160:3	[147-14-8]
Pigment Blue 15:4	Phthalocyanine Blue	74160:4	[147-14-8]
Pigment Blue 25	Napthol Blue AS	21180	[10127-03-4]
Pigment Violet 23	Carbazole Dioxazine Violet	51319	[6358-30-1]
Pigment Green 7	Phthalocyanine Green	74260	[1328-53-6]
Pigment Green 36	Phthalocyanine Green	74265	[14302-13-7]
Pigment Violet 19	Quinacridone Violet	46500	[1047-16-1]
Pigment Violet 19	Quinacridone Violet	73900	[1047-16-1]
Pigment Violet 29	Perylene Violet	71129	[12236-71-4]
Pigment Blue 60	Indanthrone Blue	69800	[81-77-6]
Pigment Violet 32	Benzimidazolone Violet	12517	[12225-08-0]

[a] NA = Not available.

5. Inorganic Pigments

These pigments are unique in that they all contain a metal in their composition. Inorganic pigments are essentially insoluble; therefore, migration and bleeding are nonexistent. Another characteristic worthy of considering is that lightfastness, weatheribility, chemical resistance, heat stability, and opacity usually come as a package. What this means is if a particular inorganic pigment exhibits excellent heat stability, eg, the chances that its other properties will probably be good to excellent as well. Note that naturally mined-oxide pigments have little or no utility in plastics since the colorant properties are inferior to the synthetic varieties. These natural colorants do find utility in some artist colors. The basic information that follows should serve as an initial guide in the usage of inorganic pigments (Fig. 4, Table 3).

Table 2. Typical Organic Pigment Applications

	Thermoplastics[a]														Thermoset[a]				
	GPPS	LDPE	HDPE	PP	ABS	Acrylic	Acetal	Flex. PVC	Rigid PVC	Poly-amide	Cell-ulose	Poly-ester	Polycar-bonate	Fluoro	Epoxy	Poly-ester	Phenolic	Sili-cone	Poly-urethane
Violets																			
quinacridones	L	W	W	W	L	W	L	W	L	L	L	L	N	N	L	L	L	L	L
dioxazines	W	W	W	W	W	W	W	W	W	L	L	L	N	N	L	W	L	L	W
Blues																			
phthalocyanines	W	W	W	W	W	W	W	W	W	L	W	W	W	N	W	W	L	L	W
indanthrones	W	W	W	W	L	W	L	W	W	N	W	L	L	N	N	L	L	L	L
Greens																			
phthalocyanines	W	W	W	W	W	W	W	W	W	L	W	W	W	N	W	W	W	W	W
Yellows																			
disazo condensations	W	W	W	W	W	W	L	W	L	L	W	W	N	N	W	W	L	L	L
diarylides[b]	W	W	W	L	L	L	N	W	L	L	W	W	N	N	W	L	L	N	L
flavanthrones	L	L	L	L	N	L	N	W	W	N	W	W	N	N	W	W	N	N	W
isoindolinones	W	W	W	W	W	L	N	W	W	L	W	W	L	N	W	W	W	L	W
hansas	L	N	N	N	N	N	N	N	L	N	L	L	N	N	L	L	W	L	N
Oranges																			
disazo condensations	W	W	W	W	L	W	W	W	L	N	W	W	N	N	W	W	W	W	W
diarylides[b]	L	L	L	L	L	L	N	L	N	N	L	L	N	N	L	L	L	L	W
pyrazolones	W	W	W	W	L	L	N	L	N	N	L	L	N	N	L	L	L	L	W
isoindolnones	W	W	W	W	L	N	W	L	L	N	W	W	N	N	W	W	L	W	W
anthanthrones	W	W	W	W	L	N	L	W	W	N	W	W	N	N	L	W	W	L	W
diketo-pyrrolo-pyrrolos	L	W	W	W	L	L	L	W	W	N	N	L	N	N	L	L	L	L	L
Reds																			
permanent red 2B's	W	W	W	L	N	L	N	W	L	N	W	N	N	N	L	L	N	L	L
pigment scarlets	W	W	W	L	N	L	N	W	L	N	W	L	N	N	L	L	N	L	L
red lake C's	W	W	W	W	L	W	N	L	L	L	L	L	L	N	W	L	L	W	L
perylenes	L	W	W	W	W	W	W	W	W	L	W	L	L	L	L	W	L	L	W
quinacridones	L	W	W	L	L	W	W	L	L	L	W	W	N	N	W	L	W	W	W
disazo condensations	W	W	W	W	W	W	W	W	L	L	W	W	N	N	W	W	W	W	W
diketo-pyrrolo-pyrrols	L	W	W	W	L	L	L	W	W	L	L	L	L	L	L	L	L	L	L

[a] W = widely used, L = limited use (testing suggested), N = not used.
[b] Note: Diarylides may decompose when processed > 200°C (392°F), which may release toxic substances.

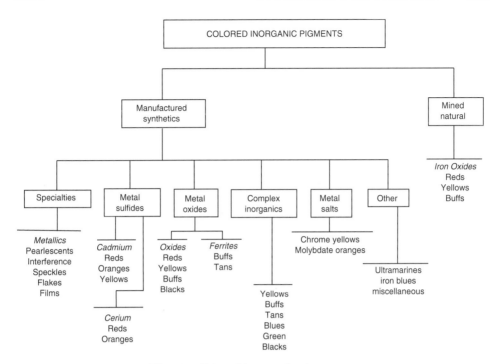

Fig. 4. Colored inorganic pigments.

Table 3. **Selected Inorganic Pigments for Plastics**[a]

Colour Index name	Pigment name	Colour Index number	CAS Registry number
Pigment Black 7	Carbon Black	77266	[1333-86-4]
Pigment Black 11	Iron Oxide Black	77499	[12227-89-3]
Pigment Black 12	Iron Titanate Brown	77543	[68187-02-0]
Pigment Black 26	Manganese Ferrite Black	77494	[68186-94-7]
Pigment Black 27	Iron Cobalt Chromite Black	77502	[68186-97-0]
Pigment Black 28	Copper Chromate Black	77428	[68186-91-4]
Pigment Black 30	Chrome Iron Nickel Black	77504	[71631-15-7]
Pigment Brown (Black) 35	Iron Chromate Black	77501	[68187-09-7]
Pigment White 6	Titanium Dioxide White (Anatase)	77891	[13463-67-7]
Pigment White 6	Titanium Dioxide White (Rutile)	77891	[13463-67-7]
Pigment Brown 6	Iron Oxide Brown	77491 and 77492 and 77499	[52357-70-7]
Pigment Brown 6	Iron Oxide Buff	77491	[52357-70-7]
Pigment Brown 11	Iron/Zinc/Magnesium Oxide Tan	77495	[64294-89-9]
Pigment Brown 24	Chrome Antimony Titanate Buff	77310	[68186-90-3]
Pigment Brown 31	Zinc Ferrite Buff	77496	[68187-51-9]
Pigment Brown 33	Zinc Iron Chromite Brown	77503	[68186-88-9]
Pigment Brown 35	Iron Chromite Brown	77501	[68187-09-7]
Pigment Brown 39	Chrome Manganese Zinc Brown	77312	[71750-83-9]

Table 3 (*Continued*)

Colour Index name	Pigment name	Colour Index number	CAS Registry number
Pigment Brown 40	Manganese Chrome Antimony Titanate Brown	77897	[69991-68-0]
Pigment Brown 45	Manganese Tungsten Titanate Brown	778965	[144437-66-1]
Pigment Yellow 34	Lead Chromate Yellow	77600 and 77603	[1344-37-2] and [7758-97-6]
Pigment Yellow 35	Cadmium Sulfide Yellow	77205	[8048-07-5] and [12442-27-2]
Pigment Yellow 37	Cadmium Sulfide Yellow	77199	[1306-23-6] and [68859-25-6]
Pigment Yellow 42	Iron Oxide Yellow	77492	[51274-00-1]
Pigment Yellow 53	Nickel Antimony Titanate Yellow	77788	[8007-18-9]
Pigment Yellow 119	Zinc Ferrite Brown	77496	[68187-51-9]
Pigment Yellow 157	Nickel Barium Titanate Yellow	77900	[68610-24-2]
Pigment Yellow 161	Nickel Niobium Titanate Yellow	77895	[68611-43-8]
Pigment Yellow 162	Chrome Niobium Titanate Yellow	77896	[68611-42-7]
Pigment Yellow 163	Chromium Tungsten Titanate Yellow	77897	[68186-92-5]
Pigment Yellow 164	Manganese Antimony Titanate Buff	77899	[68412-38-4]
Pigment Yellow 184	Bismuth Vanadate Yellow	N/A	[14059-33-7]
Pigment Yellow 189	Nickel Tungsten Titanate Yellow	77902	[69011-05-8]
Pigment Orange 20	Cadmium Sulfoselenide Orange	77202	[12656-57-4]
Pigment Orange 21	Chrome Orange	77601	[1344-38-3]
Pigment Red 101	Iron Oxide Red	77491	[1309-37-1]
Pigment Red 104	Lead Molybdate Orange	77605	[12656-85-8]
Pigment Red 104	Lead Molybdate Red	77605	[12656-85-8]
Pigment Red 108	Cadmium Sulfoselenide Red	77202 and 77196	[58339-34-7]
Pigment Red 108	Cadmium Sulfoselenide Orange	77202 and 77196	[58339-34-7]
Pigment Red 259	Ultramarine Pink	77007	[12769-96-9]
Pigment Green 17	Chromium Oxide Green	77288	[1308-38-9] and [68909-79-5]
Pigment Green 18	Hydrated Chromium Oxide Green	77289	[1201-99-9]
Pigment Green 26	Cobalt Chromite Green	77344	[68187-49-5]
Pigment Green 50	Cobalt Titanate Green	77377	[68186-85-6]
Pigment Blue 27	Ferriferrocyanide Blue (Iron Blue)	77510	[14038-43-8]
Pigment Blue 28	Cobalt Aluminate Blue	77346	[1345-16-0]
Pigment Blue 29	Ultramarine Blue	77007	[57455-37-5]
Pigment Blue 36	Cobalt Chromite Blue	77343	[68187-11-1]
Pigment Blue 36:1	Zinc Chrome Cobalt Aluminate Blue	77343:1	[74665-01-3]
Pigment Blue 72	Cobalt Zinc Aluminate Blue	77347	[68186-87-8]
Pigment Violet 14	Cobalt Phosphate Violet	77360	[13455-36-2]
Pigment Violet 15	Ultramarine Violet	77007	[12769-96-9]
Pigment Violet 16	Manganese Violet	77742	[10101-66-3]
Pigment Violet 47	Cobalt Lithum Phosphate Violet	77363	[68610-13-9]

[a] Note: Some colorants have multiple entries for Colour Index and CAS Registry Numbers. Variations within the colorant chemistry account for these multiple entries.

Table 4 offers usage information to act as an initial guide toward inorganic pigment selection.

6. Soluble Dyes

Soluble dyes present a unique situation to the colorist. There are many chemical types of soluble dyes to choose from similar to organic and inorganic pigments. The different types add confusion as most chemical types of dyes deliver a full range of colors. This full range is available in each chemical type that figuratively covers the entire visual color spectrum. Therefore, soluble dye selection is critically and fundamentally based upon what chemical type performs the best in a system and the target color desired. This is not to say these issues are not important for organic and inorganic pigments. They are a distinct issue that comes with soluble dyes is that they, by being soluble in a polymer system, are totally transparent. This difference brings to bear a visible difference not enjoyed by pigments, which have some translucency or opacity. Simply stated; pigments scatter some light, soluble dyes do not (Fig. 5).

Figure 5 displays the wide range of soluble dye chemical type available. Only a selected number from this list make up the major usage, which simplifies the selection issues.

Widely used solvent dye chemical types include: anthrapyridone; anthraquinone; azo; nigrosine; perinone; pyrazolone; quinoline; quinophthalone; and xanthene.

Table 5 offers selection data to assist the colorist in making the initial selection when considering soluble dyes for a coloring of plastics project.

Table 6 offers usage information to act as an initial guide toward soluble dye selection.

7. Effect Colorants

Effect colorants cover a wide gamut of optical effects. They include, but are not limited to, metallic, pearlescent, fluorescent, phosphorescent, speckles, marble, granite, and others that will appear in the market place from time to time. There is such a wide variety in this group it is impossible to characterize them as done with the organic, inorganic pigment, and soluble dye groups.

Metallic pigments are normally thought of as aluminum flakes of various particle sizes. In fact, this is mainly true. The aluminum particles that are very small will exhibit a gray metallic sheen with no particular specular high lights. The larger sizes, sometimes called flake or dollars, can give attractive sparkling like effects. The addition of zinc and/or copper alloyed to the aluminum will exhibit brass, bronze, and copper appearing effects. A negative effect of these pigments is that they, as platelet shaped particles, will tend to align with polymer flow in injection molding and extrusion. This alignment with flow may show as undesirable visible flow lines in the product. The larger size particles tend to minimize this undesirable effect. Proper mold and die design can minimize and/ or eliminate this appearance defect.

Table 4. **Typical Inorganic Pigment Applications**

	Thermoplastics[a]														Thermoset[a]				
	GPPS	LDPE	HDPE	PP	ABS	Acrylic	Acetal	Flex. PVC	Rigid PVC	Poly-amide	Cellu-lose	Poly-ester	Polycar-bonate	Flouro	Epoxy	Polyester	Phenolic	Silicone	Polyur-ethane
Violets																			
ultramarines	W	W	W	W	W	L	L	W	L	L	W	W	L	N	W	W	W	W	W
manganeses	W	W	W	W	L	L	N	W	W	N	L	W	N	N	W	W	L	L	W
complex inorganics	W	W	W	W	W	W	L	W	W	L	W	L	L	W	W	W	W	W	W
Blues																			
irons	L	W	L	L	N	N	N	L	L	N	L	L	N	N	L	L	L	L	L
ultramarines	W	W	W	W	W	W	W	W	W	W	W	W	W	N	N	W	W	W	W
manganeses	W	W	W	W	W	W	W	W	N	W	W	W	W	N	W	W	W	W	W
complex inorganics	W	W	W	W	W	W	W	W	W	W	W	W	W	W	W	W	W	W	W
Greens																			
chromes	W	W	W	L	L	L	N	L	L	N	L	W	N	N	L	W	L	L	L
hydrated chomium oxides	W	W	W	W	L	L	N	W	L	N	L	L	N	N	W	W	L	W	W
chromium oxides	W	W	W	W	W	W	W	W	W	W	W	W	W	W	W	W	L	L	W
complex inorganics	W	W	W	W	W	W	L	W	W	W	W	W	W	W	W	W	W	W	W
Yellows																			
lead chromates	W	W	W	W	L	N	N	W	W	N	W	W	L	N	W	W	W	N	W
iron oxides	W	L	L	L	N	N	N	L	N	N	W	W	N	N	W	W	L	W	L
cadmiums	W	W	W	W	W	N	N	W	W	W	W	W	W	W	W	W	L	L	W
complex inorganics	W	W	W	W	W	W	W	W	W	W	W	W	W	W	W	W	W	W	W
Oranges																			
lead molybdates	W	W	W	W	L	L	N	W	W	N	W	W	N	N	W	W	W	N	W
cadmiums	W	W	W	W	W	W	W	W	W	W	W	W	W	W	W	W	W	W	W
iron oxides	W	L	L	L	N	N	N	W	L	N	W	W	N	N	W	W	L	L	L
Reds																			
ultramarines	W	W	W	W	W	L	L	W	W	L	W	L	L	N	W	W	W	N	W
cadmiums	W	W	W	W	W	W	W	W	W	W	W	W	W	W	W	W	W	W	W
iron oxides	W	W	W	W	W	L	N	W	L	N	W	W	N	N	W	W	W	L	L

[a] W = widely used, L = limited use (testing suggested), N = not used.

372

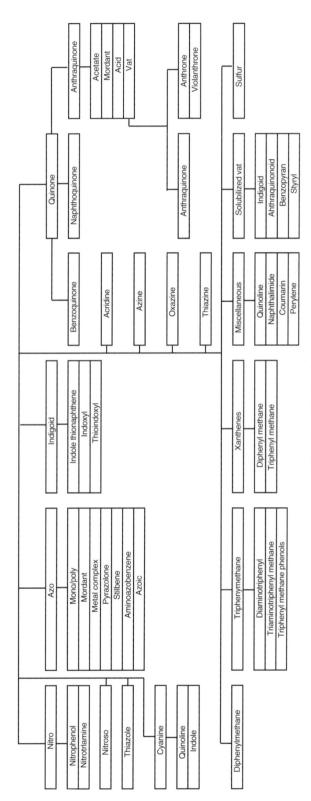

Fig. 5. Soluble dyes.

Table 5. **Selected Solvent Dyes for Plastics**

Colour Index name	Solvent dye chemical types	Colour Index number	CAS Registry number
Solvent Black 3	Disazo Black	26150	[4197-25-5]
Solvent Blue 35	Anthraquinone Blue	61554	[17354-14-2]
Solvent Blue 36	Anthraquinone Blue	61551	[14233-37-5]
Solvent Blue 58	Anthraquinone Blue	617043	[61814-09-3]
Solvent Blue 59	Anthraquinone Blue	61552	[6994-46-3]
Solvent Blue 70	Phthalocyanine Blue	N/A	[12237-24-0]
Solvent Blue 101	Anthraquinone Blue	615670	[6737-68-4]
Solvent Blue 102	Anthraquinone Blue	61501	[15403-56-2]
Solvent Blue 104	Anthraquinone Blue	61568	[116-75-6]
Solvent Blue 128	Anthraquinone Blue	N/A	[18038-99-8]
Solvent Green 3	Anthraquinone Green	61565	[128-80-3]
Solvent Green 5	Perylene Green (Yellow)	59075	[2744-50-5]
Solvent Green 28	Anthraquinone Green	625580	[71839-01-5]
Solvent Orange 7	Monoazo Orange	12140	[3118-97-6]
Solvent Orange 60	Perinone Orange	564100	[6925-69-5]
Solvent Orange 63	Thioxanthene Orange	68550	[16294-75-0]
Solvent Orange 107	Polymethine Orange	N/A	[185766-20-5]
Vat Red 1	Thioindigoid Red	73360	[2379-74-0]
Vat Red 41	Thioindigoid Red	73300	[522-75-8]
Solvent Red 23	Disazo Red	26100	[85-86-9]
Solvent Red 24	Disazo Red	26105	[85-83-6]
Solvent Red 26	Disazo Red	26120	[4477-79-6]
Solvent Red 1	Monoazo Red	12150	[1229-55-6]
Solvent Red 52	Anthrapyridone Red	68210	[81-39-0]
Solvent Red 111	Anthraquinone Red	60505	[82-38-2]
Solvent Red 135	Perinone Red	564120	[20749-68-2]
Solvent Red 149	Anthrapyridone Red	674700	[21295-57-8]
Solvent Red 155	Anthraquinone Red	N/A	[110616-99-4]
Solvent Red 168	Anthraquinone Red	60510	[1096-48-6]
Solvent Red 169	Anthraquinone Red	605060	[27354-18-3]
Solvent Red 172	Anthraquinone Red	607280	[68239-61-2]
Solvent Red 179	Perinone Red	564150	[6829-22-7]
Solvent Red 195	Monoazo Red	N/A	[164251-88-1]
Solvent Red 196	Benzopyran Red	505700	[52372-36-8]
Solvent Red 197	Benzopyran Red	505720	[52372-39-1]
Solvent Red 207	Anthraquinone Red	617001	[15958-68-6]
Solvent Red 242	Thioindigoid Red	73300	[522-75-8]
Vat Violet 1	Violanthrone Violet	60010	[1324-55-6]
Solvent Violet 11	Anthraquinone Violet	61100	[128-95-0]
Solvent Violet 13	Anthraquinone Violet	60725	[81-48-1]
Solvent Violet 14	Anthraquinone Violet	61705	[8005-40-1]
Solvent Violet 36	Anthraquinone Violet	N/A	[61951-89-1]
Solvent Violet 38	Anthraquinone Violet	615655	[63512-14-1]
Solvent Violet 59	Anthraquinone Violet	62025	[6408-72-6]
Solvent Yellow 14	Monoazo Yellow	12055	[842-07-9]
Solvent Yellow 16	Monoazo Yellow	12700	[4314-14-1]
Solvent Yellow 33	Quinoline Orange	47000	[8003-22-3]
Solvent Yellow 72	Azo Yellow	127450	[61813-98-7]
Solvent Yellow 93	Pyrazolone Yellow	48160	[4702-90-3]
Solvent Yellow 98	Thioxanthene Yellow	56238	[12671-74-8]
Solvent Yellow 114	Quinoline Yellow	47020	[7576-65-0]
Solvent Yellow 160	Coumarin Yellow	N/A	[35773-43-4]
Solvent Yellow 163	Anthraquinone Yellow	58840	[13676-91-0]

Table 5 (*Continued*)

Colour Index name	Solvent dye chemical types	Colour Index number	CAS Registry number
Solvent Yellow 176	Quinoline Red	47023	[10319-14-9]
Solvent Yellow 179	Styryl Yellow	N/A	[80748-21-6]
Solvent Yellow 185	Coumarin Yellow	551200	[27425-55-4]
Disperse Yellow 241	Pyridone Yellow	N/A	N/A

Aluminum-based pigments are not the only metallic pigments used in polymers. Appearance is the most visible, but not the only attribute of some specific metallics. An example might be stainless steel flakes or short wires. These materials, as an illustration, may bring corrosion resistance and/or electrical conductivity to a plastic product. Other metal-based materials are available for aesthetic as well as physical property enhancement.

Pearlescent (nacreous) pigments are sophisticated synthetic inorganic pigments. The original pearlescent pigments were made from materials extracted from fish scales. The physical and performance properties of these natural materials did not meet the demands of plastic processing and product life requirements. Most synthetic pearlescent pigments are mica-based flakes, however, new flake materials based on silicas and other inorganic materials are appearing in the market place. Depositing a very thin layer of titanium dioxide on very carefully selected and sized flakes produces the pearl effect. The observed pearl effect is a result of the interference, scatter, and transmission of white light. A similar effect pigment uses a very thin layer of iron oxide, titanium dioxide, and/or combinations of both producing varied color effects. Adjusting the thickness of the deposited layer provides selective color interference, scatter, and transmission resulting in colors ranging from metallic, red, blue, green, and copper colors.

Fluorescent and phosphorescent colorants fall into the luminescent colorant family. These colorants are soluble dyes or are soluble dyes converted into pigmentary forms. They are unique. They accept uv light reemitting that uv light as visible light is usually similar in hue to the spectral or diffuse pigment color observed. This results in colorants radiating visible light enhancing the base color. If viewed where no uv light is present, even the fluorescent or phosphorescent colorants appear dull and unattractive. The performance difference between fluorescent and phosphorescent colorants is time related. Fluorescent colorants radiate their visible light instantaneously. Phosphorescent colorants release their visible light energy over time resulting in a significant time delay before fading completely. These colorants are known more generally as "glow-in the-dark colorants".

Speckles and granite effects are created by a number of techniques. The speckle appearance is usually obtained by adding large size granular material of different colors during processing of the plastic into consumer products. Theses granular materials are, but not limited to, granulated colored thermoset plastics. Other particulate nonpolymeric materials may also be used. The main issues with these materials are that they must be compatible, remain

Table 6. Typical Soluble Dye Applications

	Thermoplastics[a]													Thermoset[a]					
	GPPS	LDPE	HDPE	PP	ABS	Acrylic	Acetal	Flex. PVC	Rigid PVC	Poly-amide	Cellu-lose	Poly-ester	Polycar-bonate	Fluoro	Epoxy	Polyester	Phenolic	Silicone	Polyur-ethane
Violets																			
anthraquinones	W	N	N	L	W	W	N	N	L	W	L	W	W	L	L	L	N	N	L
Blues																			
anthraquinones	W	N	N	L	W	W	N	N	L	W	L	W	W	L	L	L	N	N	L
Greens																			
anthraquinones	W	N	N	L	W	W	N	N	L	W	L	W	W	L	L	L	N	N	L
Yellows																			
anthraquinones	W	N	L	L	W	W	N	N	L	W	L	W	W	L	L	L	N	N	L
azos	W	N	L	L	N	W	N	N	N	N	L	W	W	N	N	N	N	N	N
methines	W	N	N	N	W	W	N	N	N	L	N	W	W	L	N	N	N	N	N
pyrazolones	W	N	N	N	W	W	N	N	N	N	N	W	W	N	N	N	N	N	N
pyridones	W	N	N	N	L	W	N	N	N	N	N	W	W	N	N	N	N	N	N
quinolines	W	N	N	N	W	W	N	N	N	N	L	W	W	L	L	L	N	N	N
quinophthalones	W	N	N	N	W	W	N	N	N	L	L	W	W	L	L	L	N	N	N
xanthenes	W	N	N	N	L	W	N	N	N	L	L	W	W	N	N	N	N	N	N
Oranges																			
azos	W	N	N	L	L	W	N	N	N	N	L	W	W	N	N	N	N	N	N
perinones	W	N	N	W	W	W	N	N	N	W	L	W	W	N	L	L	N	N	L
polymethines	W	N	W	W	W	W	N	N	N	N	L	W	W	N	N	N	N	N	N
thioxanthenes	W	N	N	L	L	W	N	N	N	N	N	W	W	N	N	N	N	N	N
Reds																			
anthraquinones	W	N	N	W	W	W	N	N	L	W	L	W	W	L	L	L	N	N	L
azos	W	N	L	L	L	W	N	N	N	N	L	W	W	N	L	N	N	N	N
benzopyrans	W	N	W	W	W	W	N	N	N	L	N	W	W	N	N	N	N	N	N
indigoids	W	N	L	L	L	W	N	N	N	N	N	W	W	N	N	N	N	N	N
perinones	W	N	W	W	W	W	N	N	N	W	L	W	W	N	L	L	N	N	L

[a] W = widely used, L = limited use (testing suggested), N = not used.

particulate, and not be soluble in the polymer system. Different effects are obtained through particle coloring, particle size, and particle size distribution.

Marble effects are produced differently in thermo and thermoset systems. A typical method for thermoplastics is to add pellets of the contrast or highlight color at the level needed for the desired effect. These pellets usually are a slightly incompatible polymer and/or have a slightly higher melting point. These characteristics retard or prevent homogenous mixing or dispersion of the highlight polymer during processing resulting in highlight streaks and swirls or a marbleized effect. In thermoset plastics, which are for the most part liquids, the highlight color liquid, which is identical or compatible with the base liquid, is carefully mixed into the system by various procedures that avoid complete mixing or dispersion. This leaves highlight color streaks and swirls that are ultimately cured with the base material. The result is the desired color effect contained in a completely cured homogeneous integral product.

Powdered thermoset systems such as melamines requiring a marble or other effects are usually made by imbedding a printed inlay into or under the surface of the product. The product then undergoes the cure process. The inlay highlight pattern is visible giving the effect and is protected by a layer of the cured thermoset polymer covering the inlay.

8. Special or Unusual Affect Colorants

These groups of colorants are difficult to classify. Some are pigments or distinctly different materials and others are soluble dyes. Every colorant brings an extraordinary visual effect not found in typical colorants.

Die cut or small strips of cellophane and/or mylar films exhibit striking visual effects. The thin films are vacuum metalized, color coated, or both. The die cut films can be any shape such as stars, rounds, squares, crescents, or any shape imaginable. The only criterion for use is that the die cut films meets all the specifications required for pigments and dyes.

Special light interference colorants take advantage of their ability to use light interference techniques to vary the perceived color depending on the viewing angle or conditions. Thus, as an example, at one viewing angle a product may appear green and as the viewing angle changes the color shifts to red. This dramatic change is known as "color travel". These colorants have significant visual impact drawing attention to products. These special colorants, however, are very costly. Some of these special colorants are priced at hundreds of dollars per pound.

Chemiluminescent colorants are a group of chemicals that produce luminescence as a result of a chemical interaction with an oxidizer, usually a peroxide. An example would be the emergency chemical luminescent light sticks that produce a green fluorescence when the stick is bent releasing the oxidizer or the luminescent novelties found at carnivals. Other common colors produced by this system are yellow and red.

A natural luminescent light not normally thought of as a commercial product is bioluminescence. This is typically produced by luciferin in the bodies of

A. Industry is mature
B. Diketo-pyrrollo-pyrrole (DPP) new
C. Entirely new chromophores unlikely
D. Priority efforts to lower costs
E. Physical property focus?
   1. Dispersion improvements
   2. Physical properties
      Heat stability (incremental)
      Light stability (incremental)
      Weatherability (incremental)
      Strength (incremental)
   3. Focus on hue improvements
      Red
      Oranges
      Yellows
F. Market focus on (heavy) metal replacement
G. Tailoring colorants to meet application needs

Fig. 6. The future of organic pigments.

glowworms, fireflies, and some species of fungi. Plant genetics using gene transplants are a recent example.

Thermochromic colorants are prevalent in society today. These colorants, usually soluble dyes, are found in digital thermometer tapes and printings where the devices are used to detect whether frozen products have defrosted and spoiled or the opposite, where products if frozen become defective.

Photochromic colorants are soluble dyes that change hue when subjected to particular kinds of light. An example is the eyeglass lens that darkens when exposed to uv light in an outdoor situation.

Electroluminescent colorants are materials that emit visible light when an electrical charge or voltage is applied. The typical LED is an excellent example of this phenomenon.

There are others such as piezoluminescent and hydrochromatic materials that can produce strange color effects that do find small uses in commerce.

A. Industry is mature
B. Entirely new chromophores unlikely
C. Priority efforts to lower costs
D. Physical property focus?
   1. Properties
      Heat stability (incremental)
      Light stability (incremental)
      Weatherability (incremental)
      Strength (incremental, + 5−15%)
      Improved dispersion
   2. Focus on hue improvements
      Higher chroma
      Cleaner hues
      Larger color pallete
E. Market focus on nonheavy metal colorants
F. Tailoring colorants to meet application needs

Fig. 7. The future of inorganic pigments.

A.　Industry is mature
B.　Entirely new chromophores unlikely
C.　Significant number of offshore manufacturers
　　1. India
　　2. Mainland China
　　3. Minor domestic manufacture
D.　Improvement focus?
　　1. Properties
　　　　Heat stability (little or none)
　　　　Light stability (little or none)
　　　　Weatherability (little or none)
　　　　Improve chemical purity (substantial)
　　2. Market focus
　　　　Replace heavy metal colorants
　　　　Supplement organic colorants to reduce cost
　　3. Market and exploit color values
　　　　Brightness
　　　　Transparency
　　　　Economics
　　4. Focus on market growth

Fig. 8.　The future of soluble dyes.

9. Future Considerations

The future considerations for organic and inorganic pigments and soluble dyes are outlined in Figures 6–8.

BIBLIOGRAPHY

"Colorants for Plastics" in *ECT* 3rd ed., Vol. 6, pp. 597–617, by T. G. Webber, Consultant, in *ECT* 4th ed., pp. 944–965, Gary Beebe, Rohm and Haas Company. "Colorants for Plastics" in *ECT* (online): posting date: December 4, 2000, by Gary Beebe, Rohm and Haas Company.

CITED PUBLICATIONS

1. T. B. Reeve and W. L. Dills, *Principles of Pigment Dispersion in Plastics*, 28th ANTEC Preprint, Society of Plastics Engineers, Inc. Brookfield, Conn. 1970, p. 574.
2. R. J. Kennedy and J. F. Murray, *Internal Pigmentation of Low Shrink Polyester Molding Compositions with Flushed Pigments*, 33rd ANTEC Preprint, Society of Plastics Engineers, Inc., Brookfield, Conn. 1975, p. 148.

GENERAL REFERENCES

Colour Index, and its Additions and Amendments, 3rd ed., Society of Dyers and Colourists, London, and American Association of Textile Chemists and Colorists, Durham, N.C.

Raw Materials Index, Pigments, National Paint & Coatings Association (NPCA), 1500 Rhode Island Avenue, N.W., Washington, D.C. 20005.

W. Herbst and K. Hunger, *Industrial Organic Pigments*, VCH Verlagsgesellschaft mbH, D-6940 Weinheim, Federal Republic of Germany, 1993.

G. Buxbaum, *Industrial Inorganic Pigments*, VCH Verlagsgesellschaft mbH, D-69451 Weinheim, Federal Republic of Germany, 1993.

T. C. Patton, *Pigment Handbook, Volume I*, John Wiley & Sons, Inc., New York, 1973.

Engelhard Corporation, private communication, 2003.

Polysolve, Inc., private communication, 2003.

Robert A. Charvat
Charvat and Associates, Inc.

COMBINATORIAL CHEMISTRY

1. Introduction

Thirty-two years ago, J.J. Hanak (1) reported a novel methodology for rapidly screening new electronic materials using thin films composed of two different metal oxides. Hanak demonstrated that, within the constraints of a physical measuring probe, a large number of different compositions could be measured *on a single, small sample*, ie, within the infinitely varying concentration gradient of the two-component mixture. This methodology represented a vastly increased speed for screening new electronic materials, however, a variety of technical and cultural issues precluded broad use until B. van Dover and L. Schneemeyr at Lucent Technologies reported (2) the discovery of new dielectric materials using similar methods as Hanak.

In 1995, a team of researchers at Lawrence Berkeley National Laboratory (LBNL) reported (3) the rapid fabrication of tens, hundreds, and, eventually, tens of thousands in two-dimensional arrays of discrete microscaled samples (pixels) using lithography methods developed for the electronics industry. Importantly, the team developed technologies for the rapid characterization of entire arrays (library) by using a matrix of sensors that corresponded to the samples deposited on the silicon wafer. The researchers commercialized these methodologies, hardware, and software through the start-up firm Symyx Technologies (4) starting in 1997. Symyx developed and marketed "combinatorial methods", or high throughput experimentation (HTE), for advanced materials, and created a new paradigm in materials research for the chemical process industry (CPI) and the advanced materials producers. This discontinuity, or step-change, reflected an earlier response by the pharmaceutical industry toward significant market demands for new products—reduced product innovation cycle time, increased return on research and development (R&D) investment, and industry consolidation. Symyx Technologies facilitated implementation efforts throughout the chemicals and advanced materials industries, with the result that the methodology known as "combinatorial chemistry" is now relatively ubiquitous in companies conducting research in advanced materials, catalysts, and polymers.

Many of the methodologies adopted by advanced materials researchers were developed for drug discovery in the pharmaceuticals industry. Drug discovery had entered the new paradigm of combinatorial chemistry and high throughput screening (HTS) in the 1980s, led independently by Furka (5) and Geysen & co-workers (6). The major industry driver was to develop new therapeutics with very tight time- and cost constraints. Traditional techniques of synthesizing and characterizing synthetic targets one-by-one were too slow. Combinatorial chemists indicated that, in theory, the number of potential drug targets—small organic molecules containing C, H, N, and O atoms —approaches 10^{50}, although the number of compounds considered useful is probably closer to 10^{10}–10^{15} (7), and that the only way to screen this diversity was by using massively parallel synthesis and characterization techniques. A number of review articles have covered these pioneers and subsequent developments (8). The commercial importance, and acceptance, of combinatorial methods became obvious when, in 1994 and 1995, Eli Lilly acquired Sphinx for $80 million, Glaxo plc (now GlaxoSmithKline, GSK) acquired technology leader Affymax for $533 million, and Marion Merrell Dow bought Selectide for $58 million (9). Experimental throughputs using 96-well titre plates were on the order hundreds to thousands of reactions per day (a "hit" in the discovery phase). By 1999, the ability to robotically synthesize and characterize 1 million distinct organic compounds per year was realized by some pharmaceutical companies, driving down R&D costs per sample by two orders of magnitude, to $1 or less per "hit".

Today, many chemical and advanced materials companies have implemented some form of HTE in their discovery research phases, through internal investment, mergers, and acquisitions. The market drivers—global pressures for higher performance specialty materials and higher profits on commodity materials—have pressed these industries to increase productivity in their new product discovery and process development phases. HTE for materials often has little similarity with methods developed in the drug discovery arena, however, researchers have quite effectively leveraged many methods and tools from the pharmaceutical applications (10). This is evident in the area of industrial catalysis, where HTE utilized in the discovery and process development phases have cut concept-to-launch cycle times in half; this represents significant cost savings as well as commercial advantages such as market penetration and intellectual property position.

2. The New Paradigm for Materials Research

2.1. Application Areas. Many of the market factors that influenced drug discovery are presently driving a need for reducing the cycle time for the discovery and process development of new advanced materials and lower cost chemical products such as pharmaceutical intermediates, fine and specialty chemicals, and commodity chemicals and materials. The transfer of technology from drug discovery has resulted in the development of HTE for inorganic materials, fine and specialty chemicals, and advanced materials. Because HTE techniques are especially suited to complex mixtures containing many different components and/or processing conditions, this methodology lends itself to the discovery of new, higher performance materials that contain multicomponent formulations, for eg, the dopants in electronic materials, or polymer blends in engineering plastics. In addition, HTE permits the screening of compositions that would not otherwise be attempted, ie, it maximizes serendipity. For example, Symyx Technologies Inc. (11) reported the identification of a ternary fuel cell catalyst composition (M-M'-M'') of high selectivity and conversion containing transition metals that show little or no catalytic activity in the three possible binary mixtures (M-M', M-M'', M'-M''). Broad economic benefits are envisioned from the downstream impact of these methodologies, as indicated in Table 1.

High throughput experimentation (also known as combinatorial methodologies) utilizes synthesis and analysis procedures wherein "libraries" of many tens, hundreds, or thousands of discrete samples are fabricated, processed, and characterized in parallel in hours and days, rather than the months and years, at a fraction of the cost of traditional serial approaches (12). An alternative methodology developed by Hanak (13) that is now referred to as "compositional spread", fabricates continuous multidimensional gradients of different materials, for example by codeposition of two or more components at different concentrations onto a substrate surface. More recently, van Dover and Schneemeyer (14) reported the discovery of new dielectric materials by vacuum sputtering three

Table 1. **Application Areas Open to High Throughput Experimentation**

Application areas	Technical challenges	Impacted products	Economic benefits
• industrial chemicals and monomers	• faster catalyst screening • capability of screening • extremely diverse combinations of catalyst ingredients	• industrial chemicals • engineering thermoplastics • other plastics	• lower cost • lower energy usage • new products based on newly affordable raw materials

• polymers
  – catalysts
  – polymer blends
  – surface modifiers
	• faster screening
	• process optimization
		• engineering plastics
		• thermoplastics
			• new products
			• new markets such as automotive glazing
			• reduced domestic energy consumption

• ceramics	• thermal barrier coating • optimization • electronic properties • higher strength	• aircraft engines • advanced power machines • conductors • semiconductors • dielectrics • machine tools	• higher engine temperatures • increased service life, and reduction of downtime • higher component densities and speeds • machining of new alloys, increased productivity

different organometallic precursors into a "continuous compositional phase spread" thin film that varies by composition along the $x-y$ plane. New materials with unique chemical, optical, and electronic properties have been discovered, using a variety of parallel screening approaches, for new products (15). Industrial, academic, and Federal laboratories have adopted HTE in polymers (16–20), biomaterials (11), hybrid organic–inorganic materials (22), phosphors (23), and optoelectronic devices such as ZnO-based compound semiconductors for light-emitting diodes (LEDs) (24,25).

One market factor that might correlate with the movement of HTE methods is the current level of spending for R&D in these sectors. Implementation of HTE technologies is expensive; sufficient R&D funds must be available, and acceptable returns on the investment may be precluded in some industries by their profit structure. Aerospace, automotive, biomedical, telecommunications, and computers are the principal end-users of the chemicals and materials sectors. Semiconductor and electronic hardware sectors are materials and device producers (OEMs), and analytical instruments and software sectors provide technology infrastructure. The annual R&D budgets for these firms are an indicator of the leverage that HTE methods can have: assuming a 5% penetration into the infrastructure ($9.7 billion in 1997) and supplier sector R&D ($13.4 billion), ~$1.2 billion in R&D budgets might be assigned to high throughput methods development with a potential downstream increase of $905 billion in sales, excluding pharmaceuticals (26).

2.2. Methodology. The conversion of the traditional discovery process to high throughput experimentation was enabled by the convergence of several technologies: computer software (data warehousing and query engines, molecular modeling, machine control software, and statistics) running on more cost-effective computers; robotics; MEMS (microelectromechanical systems) technologies; and sensors. In effect, research was transformed from the traditional "serial" processes into parallel "factory" processes. Industrial sectors outside of the pharmaceutical industry recognized the attributes of the new drug discovery paradigm and implemented high throughput screening for discovering new materials and developing new processes. Traditional, serial discovery processes rely on the preparation and characterization of individual samples from bench scale (milligrams or grams) to pilot scale (grams to kilograms). The HTE methods can more rapidly sample the same preparation and characterization space *in parallel* using automated laboratory instrumentation at the microgram-to-milligram scale. An emerging trend is to use HTE in process development, as several firms are utilizing multiple microreactors in parallel to scale-up from the discovery phases to manufacturing scales. Eventually, HTE will be applied to product development and customer service when it becomes more automated, simpler, and faster, resulting in a greater probability of commercialization and market penetration. For example, the results from mining the data of an extensive R&D database for polymer blend formulations could be fed into a rapid-screening system that identifies process chemistries in response to customer specifications.

The materials and chemicals industries can leverage, to some extent, the tools developed for drug discovery, ie, from being focused on solution-state synthesis to solid-state materials fabrication. The challenges to the materials scientist arise from the increased degrees of freedom associated with a larger number of parameters that define the design, fabrication, and analysis of a materials library compared to a drug discovery library. The processing of materials typically involves energetic reaction environments, with temperature requirements of hundreds or thousands of degrees Celsius (°C), and pressures of hundreds or thousands of kilopascals (kP). Microscale solid-state samples may also be subject to significant influence from the substrate onto which they are deposited—interfacial effects such as diffusion can produce phenomena that are not reproducible in bulk samples of manufacturing scale. Thus, advanced materials suffer from "scalability", or differences in properties observed in microscale vs. lab-, pilot-, or commercial scale. Finally, solid-state compositions may develop into different (kinetically controlled) metastable structures depending on processing and testing conditions. Most practitioners of HTE validate their sample libraries at every possible process stage ("hit" or "lead") by using control samples, calibrated reference standards, or comparison with bulk samples.

A general process flow for materials discovery is shown in Figure 1.

- Target Definition utilizes expert opinion, hypothesis, market need, and knowledge based on computational chemistry to develop an experimental target. This aspect of research has not changed, however a knowledge of when to use HTE methodologies and its issues might provide a greater scope of experiments to be completed.

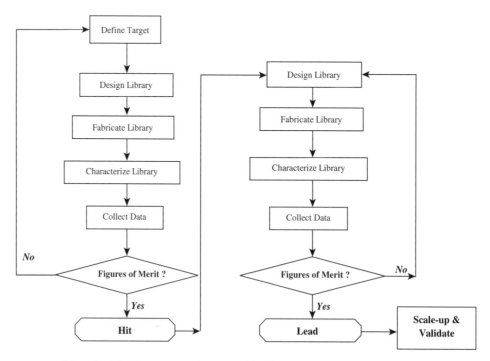

Fig. 1. The key steps of a typical high throughput RD&E process.

- Library Design with computational inputs such as quantitative structure–property relationships (QSPR) and molecular modeling, relies heavily on statistical methods such as Design of Experiments (DOE) (27) or evolutionary approaches (28) to reduce the number of samples in defined sample spaces within the experimental universe or to direct screening to other spaces within the universe. Due to the potentially high number of candidates available in HTE methodology, the design of the sample library requires rational chemical synthesis or process information to reduce the number of samples and experiments without increasing the probability for endless searches, false positives, or false negatives. New tools will have to be developed to enable the integration of this information into molecular- and property-modeling engines. The increased amount of data that can be input into computational engines will require a significant increase in speed, bandwidth and storage. Advanced, high speed quantum calculation programs such as Wavefunction, Inc.'s program SPARTAN (29) and Accelrys' program Cerius (30) will require interfacing with databases and experimental design programs; advances in experimental strategy for dealing with large, diverse chemical spaces, using space-filling experimental designs, predictive algorithms, and optimization techniques.

- Library Fabrication involves the automated deposition and/or processing of an n-dimensional matrix of physical samples. Sample fabrication is highly dependent on specific application areas and will be detailed in later sections.

- Library Characterization in a parallel or massively parallel mode involves the use of robotics and sensors to rapidly and automatically analyze the library of targets for desired properties. It is important to distinguish the characterization of the materials from the measure of their performance. Characterization is highly dependent on specific application areas and will be detailed in later sections.

- Data Collection and Analysis uses data base and artificial intelligence tools—"informatics"—expanded into the more complex realm of materials properties. Informatics can be defined as the computer software that collects and stores raw data and converts that data into information in such a way that it is easily interpreted by researchers; "intelligence" converts information into knowledge, and although there are strong efforts in artificial intelligence, its practice has not yet become widespread (31,32). An informatics engine sits at the front end of the HTE installation as an input into experimental design. It provides hardware control and collects analytical data from instrumentation; it stores and manages large data bases, either external or internal to the organization; and it provides a suitable human-computer interface for visualizing the data to yield knowledge. The underlying software technology must be able to define profitable experimental spaces, permit visualization of complex relationships from large volumes of multidimensional data, and correlate target materials with properties to permit data base queries from a broad spectrum of data mining engines and the development of structure–property relationships. This requires interfacing with data visualization tools at the back end, and database search and experimental design engines on the front end, while remaining interoperable with enterprise-wide systems for knowledge management and maintaining control of experimental hardware. The technology challenges have been identified by several workshops and roadmapping exercises (26):

- Integrated packages linking modeling, development and management of databases and search engines, hardware control, data visualization, and logistics. Database search engines need to be interoperable with the diverse flavors of databases currently in use;
- Development of Quantitative Structure Property Relationships (QSPR) for materials. This would permit the prediction of advanced materials with known or proposed composition or structure;
- Developing relationships between chemistry, processing, microstructure, metastable states, etc, would enable the design of new materials from atomic level chemistry;
- Development of a query language that links the data with many different query methods;
- Assembly of a high performance data mining toolbox that extends a database management system with additional operators;
- Connection to the diverse metrics in materials design, where important properties are sensitive to broad ranges of length or time scales, eg, from 10^{-9} to 10^2 m and nanoseconds to years;
- Development of tools to present complex, multidimensional data relationships to the human interface. HTE establishes the new paradigm for the researcher who now must interpret data surfaces and not just data points.

A list of software challanges is given in Table 2. A brief list of companies selling software into this arena is included in Table 3.

Table 2. **Technology Challenges—Software**

Software technology	Challenge
Library design	
statistics	• development of higher order designs
modeling	• (comprehensive review by NSF on computation chemistry is pending)
literature/patent data bases	• query languages; visualization; integration
Informatics	
QSPR (structure–property predictions)	• property prediction, large-scale correlations, integration into experimental design tools
data base query engines	• new languages, genetic programs
	• inter-operability, enterprise-wide integration

Table 3. **Software Companies**[a]

Company	Web site
Accelrys (in June, 2001 combined Oxford Molecular, Molecular Simulations Inc., Synomics Ltd., Genetics Computer Group, Synopsys Scientific Systems)	*http://www.accelrys.com*
Advanced Chemistry Development	*http://www.acdlabs.com*
CambridgeSoft	*http://www.camsoft.com*
Cambridge Crystallographic Data Centre	*http://www.camsoft.com*
Chemical Abstracts Service	*http://www.cas.org*
Daylight Chemical Information Systems	*http://www.daylight.com*
SciVision	*http://www.scivision.com*
Spotfire	*http://www.spotfire.com*
Tripos	*http://www.tripos.com*

[a]Ref. 24.

2.3. Commercial Environment. Materials manufacturers are utilizing several business scenarios to obtain HTE capabilities: (*1*) by developing internal capabilities; (*2*) by contracting with service providers having a core competency in high throughput discovery methods; (*3*) by developing an independent consortium or alliance partnership with individual tools providers; and (*4*) by various combinations of these scenarios. These events signal a clear trend that the materials industry is beginning to outsource (subcontract) their front-end discovery efforts to smaller external entities. A number of chemical and materials-producing companies have implemented HTE by internal growth (Table 4), most commonly realizing adequate return on assets by placing it in a corporate or central R&D facility to benefit more than one business unit. There are currently seven companies known to be either performing front-end R&D on a contract basis, or developing integrated systems for large materials manufacturers utilizing HTE R&D (Table 5).

There have been a variety of models for entering the area of combinatorial catalysis through strategic partnerships. Two notable consortia have developed:

1. Combicat, the European Consortium on Combinatorial Catalysis, was formed in August 2001 in Budapest, Hungary (23). Current industrial members include Eni Technologie, DSM Research B.V., Engelhard de Meern, AMTEC GmbH, and Millennium Pharmaceuticals Ltd. Academic partners include the Institut Francais du Petrole, the Dutch Energy Research Foundation and Environment Department of Fuels Conversion, Institut de Recherches sur la Catalyse (Villeurbanne), the Berlin Institute for Applied Chemistry, and the Universidad Politecnica de Valencia. The objectives of the consortium are to develop both tools and methodologies, with the goal of achieving a yearly throughput of >100,000 samples (discovery phase) leading to 1–5 catalysts entering commercial development. The effort will encompass sample fabrication through data handling and informatics issues. Catalyst discovery will focus on relatively generic processes: the oxidative dehydrogenation of ethane, water–gas shift reaction, the

Table 4. **Companies Implementing Internal HTE Efforts**

Company	Application area	Partner/supplier
3M	polymers	self
Akzo-Nobel	catalysts	Avantium Technologies
Albemarle	catalysts cocatalysts polymer additives	self
Avery Dennison	polymer coatings	self
BASF	catalysts advanced materials polymers	hte GmbH Symyx Technologies
Bayer AG	advanced materials	Symyx Technologies
BP-Amoco	catalysts	self
Celanese	catalysts	Symyx Technologies
Ciba Specialties	pigments polymer additives	Symyx Technologies
Dow Chemical	polyolefin catalysts specialty chemical catalysts	Symyx Technologies
DSM	catalysts	Cambridge Combinatorial
E. I. DuPont de Nemours	catalysts	self
Eastman Chemical Company	catalysts	self
Energiser	battery materials	Symyx Technologies
ExxonMobil	polyolefin catalysts	Symyx Technologies
General Electric CR&D	catalysts, polymers	self
ICI	catalysts	Symyx Technologies
Lonza	specialty materials	Symyx Technologies
Rohm & Haas	polymers	self
TexacoChevron	catalysts	hte GmbH
Unilever	personal care polymers	Symyx Technologies
UOP LLC	catalysts	self, with SINTEF
W. R. Grace	catalysts	Avantium Technologies

isomerization of ethylbenzene into *p*-xylene, the alkylation of toluene with methanol into styrene, and the selective hydrogenation of crotonaldehyde.

2. The Measurements and Standards Laboratories (MSL) of the National Institute of Standards and Technology (NIST) (Gaithersburg, Md.) have initiated an industry consortium in the area of polymer science (34). The NIST MSL are uniquely qualified to participate in this area: NISTs broad capabilities for scientific research and technology development address major challenges of high throughput methods implementation.

3. The NIST Advanced Technology Program (ATP) is supporting research within the NIST MSL in the areas of polymer array scaffolds for tissue engineering, two-dimensional infrared (ir) array detection of inorganic and organic substrates, X-ray and microwave measurement of dielectric ceramic thin films, analysis of dopants in compound semiconductor thin films, X-ray studies of supported catalysts, and genetic programming of data query engines. The largest contribution ATP has made in this field is through cost-shared funding to the private sector for a variety of research projects in catalysis, polymer coatings, and electronic materials.

Table 5. **High Throughput Methods Providers for Advanced Materials**

Company	Parent	Financing/partnerships, major clients
Symyx Technologies	(A. Zafaroni, etc)	IPO-11/20/99
	(Santa Clara)	Hoechst AG, Celanese, Bayer AG, BASF, B.F. Goodrich, Dow Chemical Co., Unilever, Argonaut (equipment manufacture and distribution), Agfa, Applied Biosystems, Merck, Prolinx, ExxonMobil, Instituto Mexicano del Petroleo (IMP)
hte GmbH (Heidelberg) hte North America (San Diego)	MPI-Kohlenforschung, BASF	MPI-Kohlenforschung, BASF (client), private financing, MSI/Pharmacopoeia
Physical Sciences Division, SRI International (Menlo Park)	SRI International	Internal, clients
Thales Technologies (Zurich)	ETH-Swiss Federal Institute of Technology (Zurich)	N/A[a]
CombiPhos Catalysts, Inc. (Princeton)	Dupont de Nemours (Wilmington)	N/A[a]
NovoDynamics (Ann Arbor, MI)	-Catalytica Advanced Technologies/NovoTech -Nonlinear Dynamics	N/A[a](Announced June 2001)
Torial LLC	UOP LLC/SINTEF	N/A
Avantium NV (Amsterdam) Avantium Inc. (Columbia, MD)	Shell Chemical Co. (Amsterdam)	Universities of Delft, Twente, Eindhoven; the government of The Netherlands; GSE, Inc.; W.R. Grace, others

[a]Not available = N/A.

3. High Throughput Experimentation for Industrial Catalysis

3.1. Economic Benefits of HTE Methods in the Chemicals Industry.
The U.S. chemical process industry has annual revenues of over $400 billion, and its products enter many industry sectors, with economic impact of >$1.2 trillion in downstream product sales and service. Catalysts play a key role in the manufacture of well >$3 trillion in goods and services of global gross domestic product annually. Royalties and fees from technology process licensing in chemicals, polymers, and refining exceed $3.5 billion annually and catalysts sold in the merchant market exceeded $9 billion last year (35). Production of chemicals from industrial catalysts provide significant value to the United States in comparison to the global economy, as shown in Tables 6 and 7. Catalyst market is listed in Table 8.

Table 6. **Catalysts for Chemicals Production**
(1997 Demand, 000s mt)

Process	North America	Rest of world
aromatics:	37	123
organic synthesis	276	174
oxidation	64	324
$NH_3/H_2/CH_4$	135	555
hydrogenation	88	230
dehydrogenation	6	135
Total	*606*	*1,541*

Table 7. **Chemicals Produced by Catalysts**
(1997 Demand, 000s mt)

Product	North America	Rest of world
aromatics	13,563	31,404
organic synthesis	4,226	7,326
oxidation	85,721	200,381
steam reforming	27,420	135,570
hydrogenation	39,044	96,760
dehydrogenation	5,852	14,492
Total	*175,826*	*485,933*

Global market pressures force the chemicals industry to continuously improve the productivity of their manufacturing processes and their research and development (R&D) facilities. In the last two decades, the market drivers—shorter innovation cycle times, higher product quality at lower cost, with reduced environmental impact from manufacturing—have constrained industrial R&D to near-term, applied research and process development at the expense of discovery and invention (36), resulting in a relatively flat R&D funding trend.

The penetration of HTE methods in industry can be measured by the amount of research money expended on this technology. Modest growth might also be expected because of the lower revenues from materials-based industries, where profits and revenues are significantly lower, and development times are shorter, compared to the pharmaceutical industry (37,38).

Table 8. **Catalyst Market, North America ($ millions)**[a]

Market	1995–1996	1998	2001
refining	791	813	909
chemicals	681	720	766
polymerization	380	436	520
emissions control	1070	1201	1390
Total	*2922*	*3170*	*3585*

[a]See Ref. 4.

Another measure of the growth of HTE methods is the commercial impact of the materials discovered using this technology. For the end-user, there will be nothing to distinguish a material discovered using HTE from a material discovered using "traditional" technology. Certainly, over time, it is likely that materials discovery using HTE itself will become the "traditional" research methodology (39).

With rising economic demands for higher efficiency and productivity in research and development, HTE methods are increasingly being implemented to bring more catalysts per unit time to the marketplace. High throughput automated synthesis and advanced screening technologies are now being applied to the discovery of more efficient homogeneous as well as heterogeneous catalysts and materials. The HTE process allows the exploration of large and diverse compositional and parameter spaces by establishing an integrated workflow of rapid parallel or combinatorial synthesis of large numbers of catalytic materials, subsequent high throughput assaying of these compounds and large-scale data analysis. The number of experiments that can be screened has risen by orders of magnitude, and has resulted in a much higher probability of discovering new catalysts or materials.

Demands for more effective catalysts in the polymer, specialty chemical, and environmental markets are driving growth by 5%/year, and catalyst suppliers are reacting to lowered sales in the commodity markets through consolidations and strategic alliances with process developers, suppliers, and customers. High throughput experimentation is speeding up the discovery process with the expectation that new catalysts and corresponding intellectual property will be produced faster and more efficiently (40). This has been facilitated since many of the HTE methodologies and tools for the CPI have leveraged the developments made in HTS and combinatorial chemistry in the pharmaceutical industry.

In an effort to define what potential value HTE methods could bring to a chemical company, Busch (41) surveyed 30 experts from the plastics industry to define the effort, costs, and potential profits in discovering a mid-level "engineering polymer" such as polycarbonate, with global annual consumption of around 500 million kilograms and a market opportunity of $2 billion. Busch observed that, using traditional laboratory methods, the first year for revenue generation is 12 years following the discovery phase, and financial break-even occurred on average 17 years after ideation; a net present value (NPV) of $10 million was assumed at project start. The HTE methods, on the other hand, reduced the time interval for both the discovery phase and scale-up by 1 year each, and this earlier market entry granted the plastic a 15% price premium; even with $6 million in depreciated capital outlays for equipment, the present value was calculated to be $37 million. Busch concluded that, at least for a high valued polymer, HTE methods are financially viable to implement based on return on investment.

Similar studies have been completed using discounted cash flow (DCF) in a pro forma estimate of NPV for traditional vs. HTE methods applied to the discovery phase of a pharmaceutical intermediate. Using a published case study (42) as a basis for time and personnel reductions, and a generic pro forma spreadsheet (43), the difference between a 10- and a 6-year commercialization period resulted in increased NPV from ~$200 to >$250 million. This difference was attributable solely to the decreased time in the discovery phase; the general feeling of this industrial community is that additional value would be generated using HTE methods to reduce the time spent in the process development and pre-commercialization stages.

Neither of these studies take into account intangible productivity gains (44) introduced by HTE methods. For example, the ability to scan literally 100–1000 times more samples allows a trained research team to explore more and different areas of chemistry that would not otherwise be possible under typical industrial time constraints. As indicated earlier, Symyx reported examples of ternary catalysts with substantially improved performance over binary catalyst mixtures as reported in the open literature and patents. Pioneering researchers are showing that HTE increases serendipity, as it is a powerful method to generate and evaluate a large number of catalysts in very short time (45). Recently, both UOP LLC (46) and DuPont (47) announced the commercial utilization of new catalysts developed with HTE methods in significantly less time than using classical methods. UOP confirmed that their commercialization cycle time has been reduced from a period of eight to ten years using traditional methods to a period of three-to-five years using HTE.

The entry costs for obtaining HTE capabilities can be very high—in the neighborhood of $10–20 million for very high throughput systems. Many companies, such as UOP and DuPont, have centralized HTE facilities that were developed internally to service their business units. Alternatively, firms of all sizes are exploiting alliances with external contract research firms, such as Symyx Technologies, hte (Heidelberg, San Diego), Thales Technologies (Zurich), CombiPhos Catalysts (Princeton) and Avantium Technologies (Amsterdam), as a route to obtaining competitive advantages. The subcontactors exercise business models based on sharing intellectual property and/or equity with partners to develop methodologies, technology, and applications, or are obtaining royalties from licensing agreements.

As noted in the previous section, Combicat, the European Consortium on Combinatorial Catalysis, was formed in August 2001 in Budapest, Hungary (48). Eleven European partners from seven countries shall assess the potential of combinatorial chemistry (HTE) applied to five generic heterogeneous catalyst systems (the selective hydrogenation of crotonaldehyde; water–gas shift reaction; dealkylation–isomerization; alkylation of toluene with methanol; and oxidative dehydrogenation of ethane) and facilitate the development of precompetitive tools and methodologies across the members. Members include academic and industrial research institutions, experienced small- and medium-sized enterprises (SMEs), catalyst manufacturers, end users, and engineering companies.

The period 2000–2002 has shown significant advances in the development of HTE approaches to the discovery and optimization of catalysts. The highlights include the development of modular approaches to the synthesis of libraries of organometallic complexes and catalysts, novel indirect screening methods, and automated syntheses and screening of high density solid-state libraries (49).

This article provides an overview of selected advances in this rapidly growing field over the past several years, advances that have taken place in academia and industry. Review articles provide more detail in different areas of HTE applied to catalyst discovery (50–62) and a web site (http://www.highthroughputexperimentation.com/links/CatMat.htm) is focused on different application areas of catalysis.

3.2. Application Areas. *Chemical Processing and Refining Catalysts.* Preliminary work in HTE methods applied to homogeneous organometallic catalysis began in 1998, utilizing instrumentation from the pharmaceutical sector. The pharmaceutical tools had previously been applied almost exclusively to aqueous solution-based systems, and modification was needed by early adopters to synthesize libraries under anaerobic conditions using organic solvents (64–67). Instrument manufacturers currently offer robotic sample handling to accommodate anaerobic conditions and higher pressures and temperatures (68), which has opened the methodology up to additional commercial research facilities to both rapidly synthesize organic ligands and organometallic complexes and to screen them for catalytic activity in chemical reactions.

Supported homogeneous catalysts were also studied using traditional methods analogous to pharmaceutical combinatorial chemistry, eg, novel phosphine and phosphine oxide ligands were prepared on polymeric supports for metal-complexed catalysts (69–71). Albemarle (Baton Rouge, La.) has developed new catalysts for the Suzuki coupling reaction of aryl bromides using palladium–phosphine and palladium–ligand complexes (ligand = heterocyclic carbenes and diazabutadiene) (72). Crabtree, Janda, Finney, and others reported the combinatorial synthesis of biomimetic catalysts using metal complexes attached to supports using linking ligands, where binding to an active site is controlled via substituents, that is by molecular recognition (73).

Heterogeneous catalysts account for ~80% of the global catalyst market. As shown in Figures 2–4, there is a substantial market for industrial catalysts in the areas of refining, chemical intermediates, polymer manufacture, and emissions control catalysts for environmental applications. The primary chemical processes are production of aromatics, oxidation catalysis, hydrogenation, miscellaneous organic chemical intermediates, ammonia production, and dehydrogenation of olefins. And would have been the leading application for HTE were it not for the significant technical barriers (see below) that slowed the use of HTE methods for these applications. Symyx Technologies (74), hte GmbH (75) and

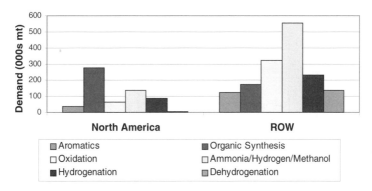

Fig. 2. Catalyst demand for chemicals production (1997), North America vs. rest of world (Row).

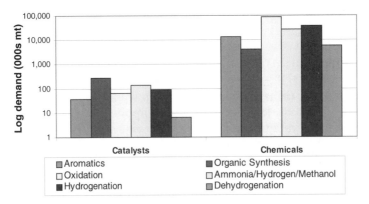

Fig. 3. Catalyst demand vs. chemical production, North America (1997).

others (76,77) reported findings for a variety of reactions using supported catalysts for chemicals processing. Symyx has entered into a discovery and licensing agreement to develop novel catalysts for the manufacture of chemical intermediates with Lonza, where Symyx has validated chemistry platforms to discover a broad range of catalysts, including chiral chemistry, hydrogenation, aromatic cross-coupling, Friedel Crafts catalysis and selective oxidation.

Refining Catalysts. The United States demand for petroleum refining catalysts was ~$1,075 million in 2000, and is expected to rise 3.5% annually to $1290 million by 2005 (2,561 million kg rising to a volume of 3067 million kg) (78). Market activity is summarized in Table 9.

Increased demand is mainly attributed to the increasing need for environment-friendly refinery and chemical products (79), especially for reformulated and other less polluting gasolines—a result of proposed gasoline sulfur limits, declining crude quality, and a move to new, higher value products. Broader trends in the crude oil industry also play a role, especially pricing, industry consolidation, and the quality of crude stocks.

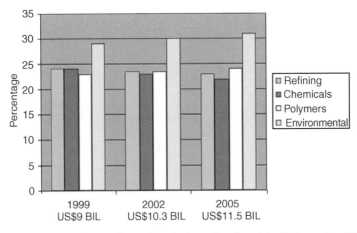

Fig. 4. Catalyst demand vs. chemical production, North America (1997).

Table 9. **Catalytic Refining and Chemicals Production (million lb), North America (1997)**[a]

Application	2000	2005 (est.)
catalytic cracking	520	545
alkylation	5030	5490
hydrotreating	50	55
hydrocracking	10	15
reforming	5	5
others	20	25

[a]See Ref. 45.

High throughput synthesis of catalytic zeolites has been reported by Bein and others (80). The research effort in high throughput synthesis and screening of heterogeneous catalysts at UOP and the Norwegian research institute SINTEF has focused on the hydrothermal synthesis of microporous materials and subsequent deposition of supported noble metal catalysts for fluid bed reactors. An automated reactor developed by SINTEF allows 48 zeolite syntheses to be carried out simultaneously (81). The supports are analyzed using X-ray diffraction (XRD) and scanning electron microscope (SEM) in a high throughput process. The analysis gives them information about zeolite structure as well as particle size and shape. Catalysts are tested for performance in reactors that simulate production-scale equipment. For example, UOP developed a small array of fluid bed reactors that were used to discover and scale up catalysts that will be commercialized in 2002 (48). UOP/SINTEF have utilized HTE to screen >2500 catalysts in 6 months, compared to previous, traditional catalyst discovery programs at UOP that could screen 271 catalysts in 3 years (83).

Fuel Cell Catalysts. A significant growth area for catalysts includes fuel cells for stationary and vehicular electrical power. In a fuel cell, hydrogen fuel is fed into the anode, oxygen from air is fed into the cathode, and the reaction is completed via an internal proton-conducting membrane and external electron-carrying circuit to generate water, heat, and electricity.

Because there is already an extensive infrastructure for natural gas for stationary source applications, and gasoline, alcohol, propane, liquefied natural gas (LNG) and diesel fuel for mobile applications, conversion of hydrocarbons to hydrogen is preferred over building a new infrastructure to support hydrogen distribution. Catalysts are therefore a crucial part of fuel cell systems because they convert hydrocarbons into hydrogen inside a secondary unit called a reformer, enabling the fuel cell to generate electrochemical energy from hydrocarbons. The reformer could be attached directly to a fuel cell or could be in a centralized location for distribution to multiple fuel cells. Relatively clean hydrogen is required for fuel cells, and its production from hydrocarbons is well known, yet there remain significant technical challenges, namely being able to provide enough heat; removal of sulfur from the hydrocarbons before it reaches the reformer; use of more robust catalysts; and removal of carbon monoxide impurities.

The catalyst industry has been pursuing fuel cell reformer catalyst technology, however, the open literature for using HTE methodologies in this area is not extensive. Johnson Matthey has been conducting research in fuel cell catalysts using HTE for some time, and has an agreement with energy company TXU Europe (Wherstead, U.K.) and fuel cell developer Energy Systems (West Palm Beach, Fla.) to construct and evaluate a fuel cell system for residential applications (84). Fagan, of DuPont Experimental Station, reported the rapid automated screening of modified electrode surfaces for electrochemical activity (85). Mallouk, at the Pennsylvania State University, reported (86) using arrays of electrodes that contained hundreds of unique combinations of five elements (Pt, Ru, Os, Ir, and Rh) screened for activity as oxygen reduction and hydrogen oxidation catalysts; he found that the ternary catalyst $Rh_{4.5}Ru_4Ir_{0.5}$ is significantly more active than the previously described Pt_1Ir_1 bifunctional catalyst. Symyx Technologies has reported (87) some activity in this area, and a general review article of this field was written by Service (88). Among the objects of the invention are the preparation of catalysts based on platinum, ruthenium, and palladium, which have a high resistance to poisoning by carbon monoxide thereby improving the efficiency of a fuel cell, decreasing the size of a fuel cell and reducing the cost of operating a fuel cell. The inventions (89) are directed to ternary and quaternary metal alloy compositions, eg, PtRuPd and PtRuPdOs.

Polymerization Catalysts. Higher performing plastics are making polymerization catalysis the fastest growing segment in the catalyst industry. Freedonia reported that demand for metallocene and other next-generation polymerization catalysts is growing fast, although conventional catalysts, including Ziegler-Natta catalysts, and initiators, accounted for >90% of U.S. polymerization catalyst demand in 1998 (89). Metallocene (single-site) catalysts are expected to displace Ziegler-Natta catalysts by the end of the year 2010, with prices lowered from the increased economies of scale. Metallocene catalyst demand is being driven principally by demand for polyethylene that has improved performance characteristics that can be obtained from these new catalysts. A 9%/year increase in demand for polyolefins in the United States, to >$1 billion/year in 2003, is estimated, with global demand reaching 5.9 million tons in 2010.

Polymerization catalysts received early attention from researchers utilizing HTE methods for both discovery and for process development in the area of olefin polymerization. Symyx Technologies developed alliances with leading chemical and polymer manufacturers such as BASF, Bayer, Celanese, Lonza, Sumitomo, and Unilever, to develop new catalysts and processes (90). Symyx is working with Dow Chemical in the area of non-metallocene single-site catalysts for ethylene–styrene copolymers and polypropylene (91); with ExxonMobil to develop petrochemical catalysts; and with ICI for the discovery of specialty polymers and other materials for high performance coatings, binding agents for textiles, and specialty adhesives (92).

Extensive studies have been conducted independently by Symyx (93). A solid-phase protocol has been developed that allows for the parallel synthesis, screening, and chemical encoding of nickel(II) and palladium(II) olefin-polymerization catalysts. These catalysts display activity profiles comparable to the analogous homogeneous catalyst systems prepared by traditional methods. A chemical encoding strategy has also been developed that enables the chemical history of pooled solid-phase catalysts to be evaluated.

Table 10. **Environmental Catalysts**

Market ($ million)	1999	2005 (est.)
motor vehicle	$1588.0	$1965.0
stationary	$855.0	$1063.0
other	$170.0	$526.0

Albemarle intends to develop new families of activators for metallocenes and other single-site catalysts using HTE (60). Although there have been many years of research in this area, the precise structure of methylaluminoxane (MAO), used as an activator for metallocene single site catalysts, is unknown and the manner in which it interacts with metallocenes is poorly understood. It is widely believed that actual utilization of the MAO under catalytic conditions is under 10%, and large excesses are usually used in processes to ensure activity. Albemarles' goals are to make the activators highly efficient, stable, lower-cost, and a "drop-in" replacement into existing production processes.

Environmental Catalysts.   Environmental catalysts are multifunctional, supported catalysts containing noble metals that reduce nitrogen oxides (NOx) and oxidize carbon monoxide and unburned hydrocarbons in mobile sources (automobiles and trucks). In stationary sources (power plants and refineries), environmental catalysts reduce sulfur oxides (SOx). The worldwide emission-control catalyst market was $2.6 billion in 1999, and it should grow 4%/year, to reach $3.6 billion by 2005 (95). Table 10 summarizes the markets for environmental catalysts.

Within the United States, demand for emission control catalysts is being driven by U.S. Environmental Protection Agency (EPA) regulations limiting emissions of pollutants from stationary sources that will go into effect in May 2003. Forty-eight catalytic hydrodesulfurization plants in 34 states will have to be expanded or upgraded, with costs estimated to be $3.5 billion (95).

New EPA regulations aimed at reducing sulfur content of gasoline and diesel fuel for automotive and truck applications drives a substantial R&D effort in the United States for new refinery catalysts as well as sulfur-tolerant automotive and truck catalysts (64,95). The regulations have translated into a global effort to develop lowered sulfur content in fuels since the new emissions catalysts are easily poisoned by sulfur. The international effort is also focused on catalysts with reduced sulfur sensitivity that will significantly reduce NO_x emissions from gasoline and diesel engines. Lean burn (high oxygen/fuel ratio) catalysts for gasoline-powered vehicles is something of a Holy Grail for catalyst researchers, and this area has yet to see a commercial product despite intensive efforts over many years. For this reason, research using HTE methodologies applied to lean burn catalysts has obtained more attention than other areas of environmental catalysis.

The German firm Degussa Metals Catalysts Cerdec (dmc^2), a major automotive catalyst producer, is spending ~5% of its $4.1 billion in sales on traditional and HTE catalyst research and development (96). Another firm with German roots, hte GmbH (Heidelberg and San Diego), is investigating the catalytic reduction of NOx under lean burn conditions, where current catalysts lack sufficient durability, activity and/or selectivity for commercial automotive applications (62). In October, 2001, the U.S. DOE awarded Engelhard (New Jersey) a total of $4.5 million in funds for the development of fuel processors for transportation applications and the use of combinatorial catalysis for the reduction of nitrogen oxide emissions in collaboration with General Motors, ExxonMobil, and Los Alamos National Laboratory (97).

Senkan and co-workers at UCLA reported (98) the selective catalytic reduction of nitrogen oxide (NO) by propane by screening a library of 56 quaternary Pt-Pd-In-Na catalyst mixtures prepared by robotically impregnating γ-Al$_2$O$_3$ pellets with various solutions. The group evaluated a microreactor array using mass spectrometry, however, the performance of the entire library under lean burn conditions was poor.

Symyx reported the testing of a 56-member catalyst library consisting of silver-, cobalt-, copper-, and indium-impregnated alumina support in the reduction of NO$_x$ by propane under lean conditions in the temperature range 400–500°C. The catalyst data was acquired by a system of 64 microreactors in parallel (99). In another study of de-NO$_x$ catalysts (100), Symyx applied an HTE approach to the design and screening of the exchange cations in ion-exchanged ZSM-5 by investigating the adsorption energies of NO and water. There are no reports that the catalysts will be commercialized.

3.3. Methodologies.   The application of high throughput discovery and process development for chemicals and materials will drive the enabling integration of a hardware- and software-based infrastructure toward specific product applications. The long-term vision shared by many researchers is to have high throughput research become part of expanded enterprise-wide systems that include tools for hardware interfaces, technology assessment/decision, and logistics (Fig. 5). Because HTE is currently highly capital intensive, with start-up

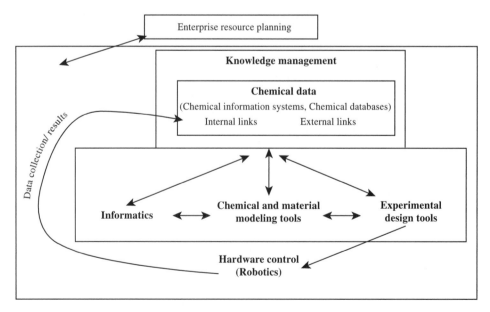

Fig. 5.   The integration of software systems.

costs in the range of \$8–20 million, discontinuous innovation in generic and/or modular hardware and software technologies will be necessary to drive down costs and facilitate its implementation in the industrial sectors that have lower returns on R&D investment, such as exist in CPI and materials sectors.

The methodologies inherent in high throughput experimentation for catalysts can be categorized according to design of the experimental space (Library Design), data capture and information retrieval (Informatics), library synthesis or fabrication, and library characterization.

Library Design.   The number of possible combinations of parameters relevant for catalysts is too large to try them all, even using very high throughput methods. A methodology for defining boundary sets for reaction conditions and constituents is therefore necessary before libraries are fabricated. The design phase is based on the expert knowledge of the researchers along with outside information, eg, patent and literature databases, previous experimentation, or molecular modeling. Rational chemical synthesis or process information needs to be input to reduce the number of samples and experiments without increasing the probability for endless searches, false positives, or false negatives. Technological challenges include new tools that enable the integration of this expert knowledge and knowledge assets into molecular- and property-modeling engines. The increased amount of data that can be input into computational engines will require a significant increase in speed, bandwidth and storage.

Currently, there are advanced, high speed quantum calculation programs (101) such as Wavefunction, Inc.'s program SPARTAN and Accelrys' Cerius that attempt to interface with databases and experimental design programs. For example, Accelrys' simulation software provides unique tools for studying diffusion and adsorption in microporous materials. In the Insight II product line, scientists can construct models of zeolite structure using the structure determination tools. An ensemble of trial guest molecules can be generated and used to probe the zeolite lattice. A detailed picture of the factors affecting diffusion and sorption can thus be built up—assisting zeolite selection and design prior to synthesis and testing, and helping to understand the results of such tests. Cerius2 predicts adsorption sites and binding energies of molecules through the C2·Sorption module. Diffusion and the dynamic behavior of the framework can be studied using the C2·Minimizer and C2·Dynamics modules, together with the zeolite force fields in the C2·Open Force Field.

The goal of materials researchers is to develop experimental strategies for dealing with large, diverse chemical spaces, using designs that explore all of the experimental space, predictive algorithms, and optimization techniques. Baerns utilized an evolutionary method for selection and optimization of heterogeneous catalytic materials that was developed and validated for the oxidative dehydrogenation of propane using various oxides (V_2O_5, MoO_3, MnO_2, Fe_2O_3, GaO, MgO, B_2O_3, La_2O_3) as primary components for the generation of catalytic materials (102). Briefly, the method relies on random selection of generations of catalysts using Monte Carlo techniques. Future generations of catalysts are created through random mutation of previous generations to give variations in the catalyst composition. In practice, one needs between 30 and 100 generations to find the best catalyst (103).

Fagan demonstrated an analogy for homogeneous catalysis based on the models of diversity used in drug discovery. A 96-member "pyridine" library consisting of both rationally chosen and "random" members was used to screen Ullmann ether forming reactions (104). He constructed a large library of ligands to screen for lead homogenous catalysts, then screened highly diverse compounds in an experimental space based on performance closest to a lead catalyst. He found that using a diversity model produced a larger fraction of ligands of increased performance than did the "rational" methods used traditionally by catalyst chemists.

"Hill Climbing" is a classic method for developing diversity where the direction of change of catalyst composition, eg, is calculated by finding the optimum in a multidimensional plot. This can be the fastest road to success, but one needs a good starting point, otherwise, there is a risk of obtaining a local maximum and never finding the best catalyst (105).

Informatics.   Informatics software contains integrated packages linking modeling, development and management of data bases and search engines, hardware control, data visualization, and logistics. Data base search engines need to be interoperable with the diverse flavors of databases currently in use.

Other challenges have been noted in the community (106–108):

- Development of QSPR for materials will permit predictions of compositions based on needed properties.

- Acceleration of the design process from atomic level chemistry to engineering design by developing relationships between chemistry, processing, microstructure, etc, and processing involving metastable states, etc.

- Development of a query language for linking many different methods for querying the data with appropriate query optimization methods.

- Assembly of a high performance data mining toolbox that extends a database management system with additional operators.

- Connection to the diverse metrics in materials design, where important properties are sensitive to numerous ranges of length or time scales.

- Development of tools to present complex, multi-dimensional data relationships to the human interface. HTE establishes the new paradigm for the researcher who now must interpret data surfaces and not just data points.

Symyx's Renaissance Software Components includes Library Studio (chemical equations, mapping, recipe file databasing and export); Impressionist (hardware automation); Epoch (high throughput instrument control, data acquisition, viewing, workflow and data management software); Oracle database; and PolyView (searching and viewing data). Symyx has obtained a European patent that claims on computer- implemented methods, programs and apparatus for designing a combinatorial library of materials (109).

Synthesis and Library Fabrication.   In the pharmaceutical industry, new drug candidate molecules undergo chemical analyses such as gas chromatography, nuclear magnetic resonance (nmr) spectrometry, mass spectroscopy, fluorescence spectroscopy, while "performance" is measured by the response of a biological receptor (*in vivo* or *in silico*). Catalysts, on the other hand, require performance-based characterizations to correlate to manufacturing realities, in addition to analyses for catalyst structure and composition. Analysis for catalyst performance (selectivity and conversion) requires measurement of products produced by the catalyst, correlated (indexed) back to known or presumed catalyst composition and structure. This may be especially difficult in heterogeneous or supported catalysts for at least two reasons:

1. Since microscale solid-state samples are influenced by the substrate onto which they are deposited, interfacial effects such as metal diffusion into a substrate or activity at certain crystalline defects may not be reproducible with bulk catalyst samples obtained from large-scale production. Thus, catalyst development using microscale screening can suffer from "scalability" issues, nonlinear correlations between microscale results and observed labor pilot-scale (bulk) properties at fundamental levels.
2. Solid-state compositions may develop into different (kinetically controlled) metastable structures depending on processing and testing conditions. Therefore, sample libraries must undergo validation at every process step.

Automated hardware for processing catalysts differs from small molecule drug discovery tools. Materials processing typically involves more energetic reaction environments than pharmaceutical processing. The leverage of tools developed for drug discovery, eg, activity-focused/solution-state systems, to solid-state materials, is challenging due to the diversity of potential design, fabrication, and analysis parameters. The ability to utilize data obtained from libraries of microscale samples will require better scientific understanding in the areas of interfacial solid-state interactions and composition–structure–property relationships. The transfer of traditional *serial* research methodologies to multidimensional *parallel* methodologies will require the integration of previously diverse computational and characterization tools for sample library design, sample library fabrication, characterization, and informatics. Other technical needs have been identified by industry (Table 11) (106,110).

Hardware technologies for HTE catalyst library fabrication and analysis are emerging from primary applications in other arenas. Micromachines and microreactor technologies (MRT), based on microelectromechanical systems (MEMS), will address the need for higher library densities, facilitating reduced raw materials costs for library fabrication, and economies of scale and modularity in laboratory instrumentation. The Institut für Mikrotechnik Mainz in Germany (111) and the Zentrum für Werkstoffe in der Mikrotechnik, Karlsruhe (112) are two of several European groups working on microreactor technologies. Research in the United States is centered primarily at the Pacific Northwest National Laboratory (113), the Massachusetts Institute of Technology, Microsystems Technology Laboratory and at Oak Ridge National Labs. Collaboration between DuPont Experimental Station and Massachusetts Institute of Technology has produced a circuit board-level discovery plant containing modular, socket-borne catalytic reactors with integrated fluidics control, heat transfer, separation, and mixing for high throughput discovery and process development (114).

Table 11. **Technical Challenges for Catalyst Testing**

Technology	Challenge
library screening	
thermal properties	• thermal conductivity
optical characterization	• fluorescence, luminescent properties
mechanical properties	• ablation
chemical properties	• molecular weight distribution
	• polymer architecture/morphology
	• turn-over, selectivity, conversion
library processing	
control of physical environment	• control of temperature and pressure over library array with control over individual sample sites or wells
control of chemical properties	• sample size-control of diffusion, mass transport properties, etc
	• interactions with catalyst support or solvent
library fabrication	
micro/ink jet	• reproducible drop size, consistent composition
robotic pipeting of nanoaliquots	
vacuum deposition	• reduction of cross-talk between samples, across substrate surface
thin-films sputtering, etc	
microreactor technologies (MRT)	
systems integration	• development of standards for modular component interconnections
	• on-chip sensors and electronics
reactor design	• process control devices tuned to MRT
	• modeling for fluid flow, heat and mass transfer
manufacturing processes	• substrates: glass, polymer, metal
surface micromachining using wet or dry chemical etching, laser ablation, mechanical micromilling, LIGA processes	

Fabrication of solid-state materials libraries may require deposition of metal oxides onto substrates, using, eg, microjet, laser ablation, or vacuum deposition (PVD, CVD) of oxide precursor followed by annealing, oxidation, reduction, etc. Heterogeneous and homogeneous catalysts are also made with robotic solution transfer using equipment modified from drug discovery applications. While fully automated library fabrication is desired, many laboratories manually prepare libraries under inert atmosphere conditions using standard weighing and solution transfer techniques if their throughput does not justify the expense of robotics.

In an effort to address the specific needs of the catalyst R&D community, Symyx Technologies developed two microreactor systems. One product consists of 12 or more concentric microreactors in plug-flow configuration machined into a cylindrical metal block. A second offering was made with Argonaut Technologies to commercialize Symyx's Endeavor eight-cell, continuous-stirred parallel pressure reactor that was introduced by Symyx in 1999 (115). The Endeavor was the first multireactor platform for running catalyst reactions either in solution or suspension with input of reagents and output of samples during runs. The Endeavor holds eight 15-mL reaction vessels offering individual control on temperature (up to 200°C) and pressure (500 psi) conditions with various mixing options.

Other manufacturers provide multireactor systems with automation. Altimira Instruments (formerly Zeton-Altimira) introduced the "Celero" automated reactor manifold for the chemical, environmental, and other industries for high throughput screening of catalysts. Chemspeed Ltd. (Switzerland) offers a line of Manual Synthesis Workstations and Automated Synthesis Workstations (ASW) based on a Gilson sample processor platform that can hold up to 6 syringes for liquid handling. The ASW can deliver reagents while shaking, heating, or cooling (−70°–150°C). These workstations are adapted in the factory for use at elevated temperature and pressure for screening of homogeneous catalysts (116).

Beginning in the late 1970s, parallel plug-flow reactors were used for heterogeneous catalyst testing by several researchers (83), however, substantial growth emerged in the late 1990s as high throughput screening of catalysts became of interest to manufacturers of pharmaceutical intermediates (118).

A 49-channel parallel reactor for the high throughput screening of heterogeneous catalysts has been developed by the German firm hte GmbH (Heidelberg and San Diego) (119). The reactor consists of a stainless steel body connected to a multiport valve via capillary tubes. Almost any technique can be used for product analysis, eg, gas chromatography–mass spectroscopy (GC–MS). The reactor has been shown to give data of a quality nearly that obtained from a conventional single-tube plug-flow reactor. This reactor can achieve a test throughput of 49 samples per day.

Claus and co-workers (120) reported studies on the reproducibility of miniaturized screening systems for heterogeneous catalysts. The group studied two different reactor configurations with different degrees of miniaturization for gas-phase reactions: a monolithic reactor system and a system based on MRT. In both cases, a scanning mass spectrometer analyzed for conversion in three reactions, the oxidation of methane, the oxidation of CO, and the oxidative dehydrogenation of i-butane.

Most catalyst screening devices are based on small plug-flow reactors. UOP LLC patented a novel miniature fluid bed reactor useful for screening multiple heterogeneous catalysts in parallel (121) as well as process development of catalysts for hydrotreating and hydrocracking using countercurrent flow reactors in parallel (122).

In addition to the fabrication of catalyst libraries based on the types of metals, research has also focused on the fabrication of libraries of support materials for heterogeneous catalysts. Zeolite synthesis requires demanding conditions: temperatures above the normal boiling point of the solvent, high pressures, and high pH. Akporiaye and co-workers reported in 1998 the parallel synthesis and crystallization of 100 zeolite samples that were subsequently analyzed by conventional X-ray diffraction techniques (123). Since that time, SINTEF and UOP LLC formed an alliance and have reported the automated synthesis of zeolites using HTE methodologies (124).

Bein and co-workers developed a parallel reactor/recrystallizing system having nine or 18 chambers with volumes of either 150 or 300 L in which six blocks can be processed simultaneously. Automatic X-ray diffraction (XRD) and/or SEM are used to characterize the samples (125).

Library Characterization.   Optical techniques have proliferated in the parallel characterization of HTE libraries since the human eye, or CCD electronics, can quickly detect colors and patterns. Basically, reactants were placed into multiwell plates containing catalysts supported on polymer beads. Using a variety of optical techniques such as fluorescence or thermography, those catalysts having highest activity could be identified. This method was a "parallel analysis", and in some cases faster than other serial techniques, and it served as a proof-of-concept that catalyst libraries could be screened in parallel.

One of the earliest examples of optical screening was in the use of thermography at ir-frequencies. Detection of catalyst activity can take advantage of optical emissions from exothermic reactions. In a typical experiment, a catalyst is placed into a solution containing the reactants. A charged coupled device (CCD) or other ir sensor records the evolution of heat coming from the beads (after subtracting background or latent emissions). One can subsequently determine which of the beads was most active by correlating the position of the beads to the thermal signature (126–128). The methodology has been applied to heterogeneous catalysts and to homogeneous catalysts that have been attached to a support.

Fourier transform infrared imaging (FTIR) has also been used for characterization of heterogeneous catalyst libraries. FTIR imaging combines the chemical specificity and high sensitivity of ir spectroscopy with the ability to rapidly analyze multiple samples simultaneously. FTIR spectrometers require much shorter scan times with equal or better data quality than scanning instruments (129).

Fluorescence has also been used to screen for catalyst activity. For example, Hoveya reported the simultaneous labeling of beads with a fluorescent sensor. The active catalysts are observed with a spectrometer (130). Orschel and co-workers reported the use of laser-induced fluorescence imaging for naphthalene oxidation in 15-member libraries of heterogeneous catalysts. Laser ablation inductively coupled plasma mass spectrometry (ICPMS) was used to confirm the composition of individual samples from the libraries (131).

Senkan and co-workers pioneered the use of resonance enhanced multiphoton ionization (REMPI) for the high throughput testing of catalysts libraries as well as array microreactors (132). Impregnated support pellets were evaluated in a parallel array of microreactors in which the effluent was evaluated using mass spectrometry during laser irradiation. The technique involves the selective photoionization of reaction products over spatially addressable catalyst cluster rows using tunable ultraviolet (uv) lasers under REMPI conditions, followed by the detection of photoelectrons by an array of microelectrodes.

Electrochemical catalysts have been screened using HTE methods. Mallouk and co-workers used fluorescent acid-base indicators to optically detect activity in a 645-member array testing the electrooxidation of methanol. Subsequent screening indicated that a quaternary catalyst, 44% platinum/41% ruthenium/ 10% osmium/5% iridium, was significantly more active than a traditional binary

50:50 platinum/ruthenium in a direct methanol fuel cell operating at 60°C, even though the latter catalyst had about twice the surface area of the former (133).

High-throughput synthesis and screening of mixed-metal oxide libraries for ethane oxidative dehydrogenation to ethylene have been developed by researchers at Symyx Technologies. A catalyst library was prepared on a 15-cm quartz wafer using vacuum deposition, and the product gases from each sample were sequentially exposed to ethane and scanned using a probe that led to a photo thermal deflection spectrometer and a mass spectrometer. The researchers observed that a ternary Cr-Al-Nb oxide was most active if it contains ~4% Nb (134).

Reetz reported the use of two linked gas chromatography instruments for the high throughput screening of catalysts based on enantioselectivity. Reported in the literature are several examples, including the combination of high-performance liquid chromatography (HPLC), uv/vis spectroscopy and circular dichroism (CD). Analyses of both enzyme- and transition metal-catalyzed reactions have been reported (135).

4. Polymeric Materials

As applied to polymeric materials research, combinatorial, or HTE, methodologies allow efficient characterization of polymer properties and optimization of polymer processing parameters in addition to synthesis of new polymers and accelerated development of new materials. This approach has led to scientific discoveries related to polymer material properties. Consequently, a new feature for polymer materials is that combinatorial and high throughput methods are rapidly extending knowledge discovery into fundamental polymer science as well as industrially important application areas like organic light emitting diodes and coatings.

4.1. Application Areas and Needs. Fundamental characterization of polymers is driven by their applications in structural materials, packaging, microelectronics, coatings, bioengineering, and nanotechnology. Current trends in advanced materials for such applications demand finer control of chemistry, polymer interfaces, morphology, surface properties, multicomponent mixtures, composites, and thin films. Polymeric systems are complex due to phase transitions, reactions, transport behavior, and interfacial phenomena that occur during synthesis and processing. In addition, a large number of variables control these phenomena, including structure, composition, solvent, temperature, annealing history, pressure, and thickness. Conventional microscopy, spectroscopy, and analytical tools for polymer characterization were designed for detailed characterization over a limited set of variable combinations. This approach can be used to test hypotheses efficiently if the most relevant variable combinations are known a priori or can be predicted from theory. However, the complex phenomena and large variable spaces present in multicomponent, multiphase, and interfacial regions in polymers strain the capabilities of conventional one-sample for one-measurement polymer characterization. Hence, the need to develop high throughput techniques for efficient synthesis and characterization of complex polymeric systems. For polymers, the advantages of high throughput methods include efficient characterization of novel regimes of thermodynamic and kinetic behavior, rapid testing and identification of structure–property relationships, testing hypotheses for accelerated development of functional materials, and reduced experimental variance (many measurements at same environmental conditions). High throughput screening has been successfully used for measurements of phase behavior in polymer blend, block copolymer ordering, polymer dewetting, polymer crystallization and organic light emitting devices (OLED).

4.2. Methodologies and Techniques.   There has been an increasing effort to apply HTE to materials science, indicated by recent reports of high throughput approaches to developing and characterizing a wide range of inorganic (135–145) and organic/polymeric materials (144,146–156).Unfortunately, the widespread adaptation of HTE to polymers research has been hindered by a lack of techniques for preparing polymer libraries with systematically varied composition (Φ) thickness (h), and processing conditions, eg, temperature (T). In addition, much of polymer characterization instrumentation is not suitable for high throughput screening. For this reason, a number of research groups have begun developing library preparation and screening approaches suited for polymer characterization. This article reviews these recent developments in polymer library preparation and high throughput screening methods. Several novel methods are presented below for the preparation of polymer film libraries with continuous gradients in T, h, and surface energy (E). By focusing on characterization, we have omitted discussion of the many successful examples of high throughput synthesis, experimental design, and informatics previously developed for polymers. These encompass a large variety of materials including biodegradable polymers (146,147,151) support materials for organic synthesis, (150) chemical sensors (148,152) and dendrimers (149). The reader is referred to other sources for a review of these topics (144,157).

Library Fabrication.   Techniques have been developed at the National Institute of Standards and Technology (NIST) for preparing continuous gradient polymer libraries with controlled variations in temperature, composition, thickness, and substrate surface energy (155,156,158). These libraries are then used to characterize fundamental properties including polymer blend phase behavior, thin-film dewetting, block copolymer order–disorder transitions, and polymer crystallization. In contrast to the discrete libraries used in pharmaceutical, catalysis, and many inorganic HTE approaches, the deposition of films with continuous gradients in Φ, h, T, and E is a convenient and elegant method for preparing polymer high throughput libraries.

Thickness Gradient Libraries.   A velocity-gradient knife coater (155,156, 158) depicted in Figure 6, was developed to prepare coatings and thin films containing continuous thickness gradients. A polymer solution is spread under a knife-edge onto a substrate at constant acceleration. The velocity-gradient results in dried films with controllable thickness gradients. By using several h-gradient films with overlapping gradient ranges, thickness-dependent phenomena can be investigated from nanometers to micrometers. Figure 6 shows a photograph of a polymer gradient, in which the h-gradient produces a continuous change in color and brightness of light reflected from the film–substrate interface.

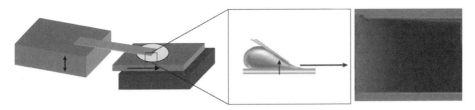

Fig. 6. Schematic of the knife-edge polymer film coating apparatus (left) and a polystyrene coated film with a gradient of film thickness (right).

Composition Gradient Libraries. Three steps are involved in preparing composition gradient films: gradient mixing (Fig. 7**a**), gradient deposition (Fig. 7**b**), and film spreading (Fig. 7**c**), discussed in detail elsewhere (155). Two pumps introduce and withdraw polymer solutions A and B to and from a small mixing vial at rates I and W. Initially loaded with pure B solution, the infusion of solution A causes a time-dependent gradient in composition in the vial. A small amount of this solution is continuously extracted with an automated sample syringe. At the end of the sampling process, the sample syringe contains a solution of polymers A and B with a gradient in composition along the length of the syringe needle. The rates I and W control the slope of the composition gradient, which is linear only if $I = (W + S)/2$ and sample time determines the endpoint composition. The gradient solution (Fig. 7**b**) is deposited as a thin stripe on the substrate, and then spread as a film orthogonal to the composition gradient using a knife-edge coater. The solvent evaporates, resulting in a film with a continuous linear gradient in composition from polymer A to polymer B. Any remaining solvent is removed under vacuum during annealing.

Temperature Gradient Libraries. To explore a large T range, the h- or Φ-gradient films are annealed on a T-gradient heating stage (Fig. 8), with the T-gradient *orthogonal* to the h- or Φ-gradient. This custom aluminum T-gradient stage, described in detail in previous publications (155,157) uses a heat source

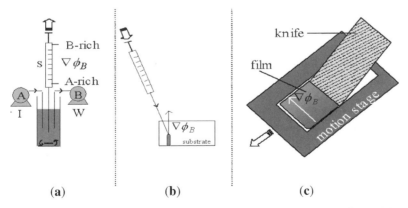

Fig. 7. Schematic of the composition gradient deposition process involving (**a**) gradient mixing, (**b**) deposition of stripe, and (**c**) film spreading. Adapted with permission from Ref. (156).

Fig. 8. Picture of the temperature gradient stage for processing of gradient polymer films.

and a heat sink to produce a linear gradient ranging between adjustable end-point temperatures.

End-point temperatures typically range from 160 ± 0.5 to $70.0 \pm 0.2°C >$ 40 mm, but are adjustable within the limits of the heater, cooler, and maximum heat flow through the aluminum plate. To minimize oxidation and convective heat transfer from the substrate, the stage is sealed with an O-ring, glass plate, and vacuum pump. Each two-dimensional $T–h$ or T-Φ parallel library contained $\sim$1800 or 3900 state points, respectively, where a "state point" is defined by the T, h, and Φ variation Δ over the area of a $200 \times$ optical microscope image: $\Delta T = 0.5°C$, $\Delta h = 3$ nm, and $\Delta \Phi = 0.02$. These libraries allow T-, h-, and Φ-dependent phenomena, eg, dewetting, order–disorder, and phase transitions, to be observed *in situ* or postannealing with relevant microscopic and spectroscopic tools.

Surface Energy Gradients.   In many polymer coating and thin-film systems, there is considerable interest in studying the film stability, dewetting, and phase behavior as a function of surface energy. A gradient-etching procedure has been developed to produce substrate libraries with surface energy continuously varied from hydrophilic-to-hydrophobic values (159). The gradient-etching procedure involves immersion of a passivated Si–H/Si substrate (Polishing Corporation of America) into a 80°C Piranha solution at a *constant immersion rate*. The Piranha bath etches the Si–H surface and grows an oxide layer, $SiO_x/SiOH$, at a rate dependent on T and the volume fraction H_2SO_4. A gradient in the conversion to hydrophilic $SiO_x/SiOH$ results because one end of the wafer is exposed longer to the Piranha solution (159). Another procedure for varying substrate energy uses composition-gradient self-assembled monolayers (SAMs) (160,161). In this procedure alkane thiolates with different terminal groups, eg, $-CH_3$ and $-COOH$, diffuse from *opposite ends* of a polysaccharide matrix deposited on top of a gold substrate. Diffusion provides for the formation of a SAM with a concentration gradient between the two thiolate species from one end of the substrate to another, resulting in controllable substrate energy gradients.

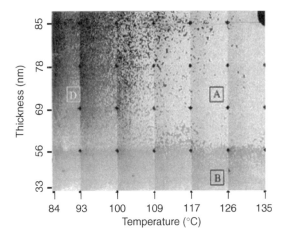

Fig. 9. Picture of water droplets on a uv radiation induced gradient energy SAM surface showing change in droplet spreading due to increasing hydrophobicity from left to right.

More recently, the use of uv radiation to create gradients in surface energy of SAM coated layers (Fig. 9) has been shown to be effective and a convenient methodology for obtaining gradients in surface energy in a controllable manner as compared to the more complicated diffusion approach.

4.3. Characterization of Polymer Library Properties. *Thin-Film Dewetting.* The wetting and dewetting of thin polymer films is of profound importance to advanced materials including microelectronics, optical communications, and nanotechnology. Figure 10 shows a composite of optical microscope images of a T–h library of polystyrene (Goodyear, $M_w = 1900$ g/mol, $M_w/M_n = 1.19$) on a $SiO_x/SiOH$ substrate (162,163).

The thickness ranges from 33 to 90 nm and temperature T ranges from 85 to 135°C. The images (see Fig. 10), taken 2 h after initiation of dewetting, show

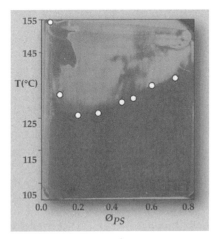

Fig. 10. Composite of optical images of a T–h film dewetting combinatorial library of PS ($M_w = 1800$ g/mol) on silicon, $t = 2$ h.

wetted and dewetted regimes that are visible as dark and bright regions, respectively, to the unaided eye.

Repeated examination of high throughput T–h libraries at thicknesses ranging from (16 to 90 nm) indicates three distinct thickness regimes with different hole nucleation mechanisms. This analysis also allowed the first observation of a T, h superposition for heterogeneous-nucleated dewetting rates, reflecting variations in the film viscosity with T and h.

Phase Behavior.   The phase behavior and related microstructure of polymer blends is of critical importance in many engineered plastics, but determination of phase behavior for new blends remains a tedious task. Figure 10 presents a photograph of a temperature-composition library of a polystyrene/polyvinylmethylether (PS/PVME) blend after 16 h of annealing. The lower critical solution temperature (LCST) cloud point curve can be seen with the unaided eye as a diffuse boundary separating one-phase and two-phase regions. Cloud points measured on bulk samples with conventional light scattering (white points) agree well with the cloud point curve observed on the library (20) (Fig. 11). The diffuse nature of the cloud point curve reflects the natural dependence of the microstructure evolution rate on temperature and composition. The high throughput technique employing T–Φ polymer blend libraries allows for rapid and efficient characterization of polymer blend phase behavior (cloud points) in orders of magnitude less time than with conventional light scattering techniques.

Block Copolymer Segregation and Surface Morphology.   The morphology of symmetric diblock copolymer thin films has been studied extensively with traditional techniques (164–166). These materials hold potential as templates for nanostructural patterning of surfaces. Because of the thickness dependence in these systems, fundamental investigation of block copolymer segregation is ideal for h-gradient using HTE GMBH methods (157,167). Films of symmetric polystyrene-b-poly(methyl methacrylate) (PS-b-PMMA) with h-gradients were produced using the knife-edge flow coating technique and annealed at 170°C for 30 h to allow lamellar organization. Figure 12 presents optical micrographs showing morphological changes associated with the addition of two lamellae to the surface of the film as h increases.

Fig. 11.   Digital optical photographs of a PS/PVME T–ϕ library after 91 min of annealing, showing the LCST cloud point curve visible to the unaided eye. White circles are conventional light-scattering cloud points measured on separate uniform samples. Picture adapted from cover of *Macromolecules*, 2001.

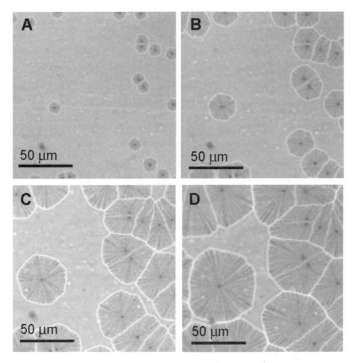

Fig. 12. OM images of block-copolymer lamellae of PS–PMMA cast as a continuous gradient in thickness on a silicon oxide substrate and annealed above its glass temperature (T_g).

This finding is in agreement with previous (non-HTE) work showing that lamellae with thickness equal to the equilibrium bulk lamellar thickness L_o form parallel to the substrate (165–167). The morphology evolves from a smooth film to circular islands to a bicontinuous hole/island region to circular holes back to a smooth film from left to right in the micrograph and repeats twice. This result represents the first observation of stable bicontinuous morphologies in segregating block copolymer thin films (157,168). Another novel observation enabled by HTE methods is the wide thickness range over which the smooth film regions persist.

Polymer Crystallization. Crystal growth rates of isotactic polystyrene were measured as a function of temperature and film thickness on continuous gradient films (Fig. 13).

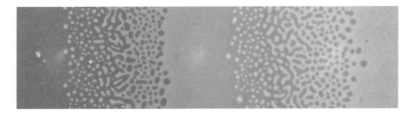

Fig. 13. Optical images of ipS crystallizing at $T = 170°C$.

The measured rates as a function of temperature agree with those obtained previously. A decrease in free energy (G) was also observed in progressively thinner regions of the film below $h = 50$ nm. These results validate the high throughput/high throughput approach to investigating polymer crystallization in thin films, accelerating the pace of experimentation by increasing both the material and processing parameter space available for sampling (168). Furthermore, the dependence of G on h and T was explored in two dimensions to a degree previously considered unfeasible.

Organic Light-Emitting Diodes. Here, we mention two characterization studies of the optimization of organic light emitting diodes (OLEDs). Schmitz and co-workers (153,169–171) used a masked deposition technique to produce thickness gradients in both the organic hole transport layers and the inorganic electron transport and emitting layer. The OLEDs with single-gradient and orthogonal two-dimensional gradient structures were produced in order to evaluate the effects of the various layer thicknesses on the device efficiency. An optimal thickness for both the hole and electron transporting layers was reported. Likewise Gross and co-workers reported the use of high throughput methods to investigate the performance of doped (oxidized) π-conjugated polymers in OLEDs (154). In these devices, the polymers serve as hole transport layers, but an energy barrier for hole injection exists between the polymeric material and the inorganic anode. By varying the oxidation level of the polymer, this energy barrier can be reduced to lower the device working voltage. The effect of oxidation was studied by electrochemically treating the polymer to create a continuous gradient in the oxidation level of the polymer. A gradient in thickness was created orthogonal to the gradient in oxidation to explore variations of both properties simultaneously. For this reason, this study represents a cross between both high throughput synthesis (oxidation steps) and process characterization (thickness gradient deposition). The gradient libraries were characterized by monitoring the efficiency and onset voltage of OLEDs fabricated on the gradients.

Industrial Polymer Coatings and Technology. A number of reports have been published on application of high-throughput approaches for polymer coatings from the industrial arena. Vratsanos and co-workers (172) describe a high throughput screening approach for latex development in architectural coatings. In this, a highly effecient approach for testing abrasion resistance of formulated paints was designed and built. Optical technology is used to detect wear-through of architectural paints on standard plastic Laneta panels. This automated instrument dispenses scrub media and water as specified and records tge endpoints and the development of abraded areas of multiple samples. Images of the panels can be saved digitally for future use. Good correlation is seen between the endpoints found from standard ASTM testing and those using the approach and technique. Schrot and co-workers (173) also describe HTE materials research to accelerate coatings development. They describe how materials as diverse as catalysts and polymers, to formulations and coatings can be developed using combinatorial methods. However, the transition from high throughput drug screening to materials research requires major changes, eg, fast characterization of bulk properties; mechanical hardness instead of molecular assays; modified parallel reactors that can operate at high temperature and pressure, handling viscous materials, aggressive fluids and fast miniaturized materials synthesis. In a comprehensive and well-illustrated article, Wicks and Bach (174) describe the approaching revolution for coatings science brought in by high throughput screening for formulations. The article summarizes the current situation and practices in coatings laboratories, lessons from the pharma world, current situation in materials science, applications of these (combi, HTE) techniques to coatings science and limitations and applicability of the technology. Chisolm and co-workers (175) describe the development of a "combinatorial factory" capable of preparing and testing over 100 coatings per day and the components of such a factory. These include (1) an automated system to prepare liquid coating formulations; (2) a novel coating application process capable of making high dimension arrays of coatings of controlled thicknesses; (3) curing of the coating arrays either thermally or with uv light; (4) testing of the coatings using newly developed high throughput screening methods; and (5) a data handling process to quickly identify the most promising coatings produced. Ramsey and co-workers (176) at ORNL have developed lab on a chip that can analyze a volume of liquid 10,000 times or more smaller than that used in a conventional analytical instrument. In the lab chips, molecules rather than electrons flow through labyrinths of tiny channels and chambers outfitted with valves, filters, and pumps. Smaller volumes permit faster mixing because molecules do not have so far to travel. Chemicals in a liquid droplet can be mixed and separated very rapidly. In fact, Ramsey and his co-worker Jacobson separated two species in <1 ms— 100,000 times faster than conventional methods.

5. High Throughput Screening: Inorganic Materials

5.1. Introduction. The use of complex inorganic materials is increasing in advanced electronic, optoelectronic, magnetic, and structural applications. These materials are typically composed of three or more elements and may have a crystalline structure, where atoms are in a periodic arrangement, or an amorphous structure, where atoms have no long-range order. Inorganic materials are used in a variety of forms, including thin films and coatings, bulk masses, and powders. The fabrication processes typically involve a number of variables, such as temperature, pressure and atmosphere. It is necessary to search large composition and process spaces to optimize the properties and structure of an inorganic material for a specific application or to discover a new material. High throughput experimentation (HTE) methods are ideally suited to such searches of multiparameter space.

5.2. Applications. The use of HTE methods to increase the efficiency of materials discovery was first reported by Hanak (177) in 1970. Despite his predictions that these methods could increase research productivity by 750-fold for ternary and higher order systems, the application of combinatorial methods to materials research was not reported again for 25 years. In 1995, Xiang and co-workers (178) demonstrated that known high temperature superconducting compounds could be readily identified using HTE methods. Subsequently, combinatorial studies have been performed on materials for electronic, magnetic, photonic, and structural applications, examples of which are given in Table 12.

Table 12. **Application Areas for HTE Studies of Inorganic Materials**

Application area	Material class	References
electronic devices	superconductor	178
magnetic data storage	magnetoresistive	179
displays	phosphors	180,181
microwave devices	dielectrics	182
memory devices	dielectrics	183,184
photonic devices	semiconductors	185
optical data storage	phase change	186
jet engines	superalloys	187
molecular sieves	microporous	188

5.3. Tools and Methodologies. A number of novel library fabrication and characterization techniques have been developed specifically for HTE studies of inorganic materials. Some tools used in conventional studies have been adapted or modified to increase fabrication or measurement throughput.

Library Fabrication. The primary combinatorial variable in most inorganic libraries is chemical composition; additional processing parameters such as temperature, pressure, and atmosphere may also be varied during library fabrication. The techniques that have been used to synthesize thin film, bulk and powder libraries are summarized in Table 13.

The physical vapor deposition techniques for thin-film library fabrication include sputtering, electron-beam evaporation, pulsed-laser deposition, and laser molecular beam epitaxy. In each of these techniques, material is vaporized from a target or source by a high energy process and is deposited on a substrate. The energy source may be an ion gun, an electron gun, or a laser. Two different types of libraries can be formed by physical vapor deposition methods, depending on whether a codeposition or a sequential deposition approach is employed. During codeposition, material is simultaneously vaporized from two or more sources, resulting in a film with a continuously graded composition, also referred to as a continuous composition spread. A schematic diagram of a codeposition system with three targets is shown in Figure 14.

Codeposited libraries have been fabricated by magnetron sputtering, radio-frequency (rf) sputtering, and electron beam evaporation. Composition-spread libraries have also been synthesized by sequential deposition of layers of differing composition, as illustrated in Figure 15 for a pulsed-laser deposition process.

Here, a linear thickness gradient of material is deposited by moving a shutter across the sample; another target of differing composition is then rotated into the laser beam and the second material is deposited as the shutter traverses the sample; and so on. Two-dimensional shutters have been employed to fabricate more chemically complex libraries (183). The resulting multilayer films are postannealed to homogenize the composition through the film thickness and allow reaction of the constituents. Uniform composition films can be formed without postannealing by depositing layers of submonolayer thickness sequentially in a rotating target pulsed-laser deposition system (200). Libraries composed of arrays of cells, each with a discrete composition, have been fabricated by a

Table 13. **Library Fabrication Techniques for Inorganic Materials**

Material form	Fabrication method	Material system	References
thin film	sputtering	Mo–Nb	177
	(*sequential*	Y–Ba–Cu–O,	183
	deposition with masks)	Bi–Sr–Ca–Cu–O	
		Li–metal–Co–O	179
		doped oxides	181
		doped Ba–Sr–Ti–O	182
	sputtering		
	(*codeposition*)	Zn–Sn–Ti–O	184
		Hf–Sn–Ti–O	189
		Ge–Sb–Te	186
		Tb–Ni–Fe–Co	190
	electron-beam evaporation		
	(*sequential*	doped oxides	180
	deposition with masks)		
	pulsed laser deposition		
	(*sequential*	doped Ba–Sr–Ti–O	183
	deposition with masks)		
		Ba–Sr–Ca–Ti–O	191
	pulsed-laser deposition		
	(*sequential*	Ba–Sr–Ti–O, Sn–In–Zn–O	184
	submonolayer deposition)		
	laser molecular beam epitaxy		
	(*sequential deposition*	Ba–Ti–O/Sr–Ti–O	192,193
	with masks)	superlattices	
		doped ZnO	185
	chemical solution deposition	Pb–Zr–Ti–O	194
bulk	melting or floating zone technique	Al–Co	195
		doped Y_2O_3	195
		SiO_2-based glasses	195
	hydrothermal synthesis	Al–P–O	188
	Diffusion multiple approach	Fe–Ni–Mo	187
		Fe–Cr–Mo–Ni, Ni–Al, Ni–Pt	196
	powder inkjet delivery	doped oxides	197
		CuCeO, LaSrCoO	198
	micropipette solution deposition	doped oxides	199

sequential deposition process. In this method, one material is deposited at a time through a mask that permits the material to be put down only onto selected regions of the substrate. A second material is then deposited through a different mask, etc for all of the component materials in the library. Sequential deposition with masks has been used in rf sputtering, electron-beam evaporation, pulsed-laser deposition, and laser molecular beam epitaxy. These library films require postannealing for homogenization and reaction.

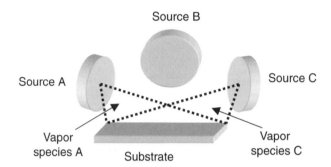

Fig. 14. Schematic diagram of a codeposition system with three sources A, B, and C. The distribution of vapor species from two of the three sources is shown; the concentration of the species from a source is highest at the edge closest to the source and decreases with distance away from the source. The deposited films have a graded composition.

Film libraries composed of discrete cells of differing composition and thickness have also been fabricated by a liquid solution approach. As illustrated in Figure 16, liquid droplets of metallorganic solutions A and B of differing concentrations are dispensed onto a substrate, spun at very high speeds (up to 10,000 rpm) to form thin liquid films, and pyrolyzed to solidify the films.

This sequence can be repeated to build up different thicknesses of films. The array of films must then be annealed to react the constituents.

Two approaches have been used for the synthesis of bulk libraries of continuously graded composition. Both methods involve placing blocks of differing composition in intimate interfacial contact. In the diffusion multiple block method, the blocks are heated to a high temperature to allow thermal inter-

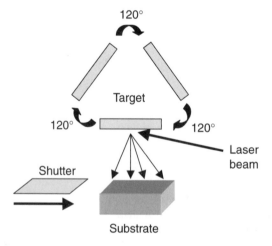

Fig. 15. Schematic diagram of a pulsed-laser deposition system with three rotatable targets and a movable shutter. A linear thickness gradient of material from one target is deposited by moving the shutter across the substrate. Repeating this procedure with the other two targets results in a composition spread film library.

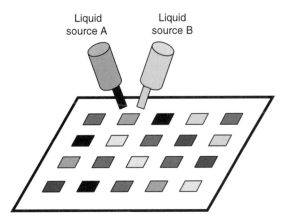

Fig. 16. Schematic diagram of metalorganic liquids A and B being dispensed in differing ratios onto a substrate to fabricate a film library with cells of different compositions.

diffusion; a multiple block composed of four different metals is illustrated in Figure 17.

In the melting or floating zone method shown in Figure 18, the end of one block is heated locally above the melting point; this molten zone is subsequently moved along the length of the block one or more times.

The resulting composition gradient in the solidified block after each pass is determined by the segregation coefficient of each constituent, defined as the ratio of the concentrations of the constituent in the freezing solid and in the molten zone.

Libraries of powders or microporous solids have been fabricated by dispensing liquid reagents into an array of small volume (<0.5 mL) wells via ink-jetting

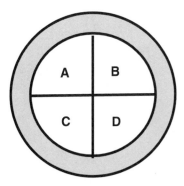

Fig. 17. Top view of a diffusion multiple composition of four materials A, B, C and D. The quarter cylinder pieces of each material are packed into a cylindrical shell and heated to high temperature to cause interdiffusion across the interfaces, thereby producing a composition spread bulk sample. In the diffusion multiple method, the blocks are heated to a high temperature to allow thermal interdiffusion; a multiple composed of four different metals is illustrated.

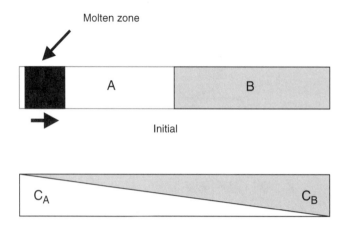

Fig. 18. Illustration of the melting or floating zone method. Two blocks of materials A and B are joined; a small region melted by localized heating is then traversed across the blocks by either moving the block through a stationary heater or by moving the heater over a fixed block. After multiple passes, the two materials mix, resulting in gradients of concentration of A (C_A) and B (C_B) across the block.

or micropipetting processes. An automated scanning multihead inkjet delivery system is shown in Figure 19.

Predetermined amounts of different reagents are dispensed into each well to form an array of samples with systematically varying composition. The library is heated to a low temperature to vaporize the solvent, and then heated at elevated temperatures to promote reaction of the constituents. In an automated hydrothermal method used to synthesize microporous materials, reagents are

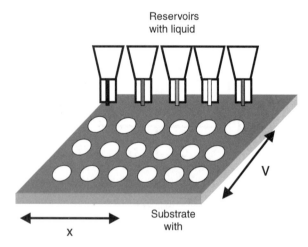

Fig. 19. Schematic diagram of an inkjet delivery system with five reservoirs, each containing a different liquid reagent. Small volumes of each liquid are dispensed into a well with a pump, and the substrate is scanned above the reservoirs in both the x and y directions. The system can be fully automated.

ink-jetted into wells in an autoclave block and processed at elevated temperatures and pressures. The product is washed by filtration and centifuged to remove the liquid.

High Throughput Characterization. Ideally, characterization methods in HTE involve the measurement of a small area or volume of material at a very high measurement rate, or throughput. The measurements may be made sequentially (in serial) or simultaneously (in parallel). In serial mode, one point is measured at a time as the sample is translated on an x–y stage, resulting in an array of data points. In parallel mode, a large area, often the entire sample, is measured simultaneously. Imaging techniques are inherently high throughput since data are collected in parallel from a large area, resulting in a map of information. For any measurement technique, rapid data analysis methods must also be developed to extract the required information from the results. In order to compare the performance of the elements in a given library, it is desirable to formulate a figure of merit that includes the critical properties. For example, the figure of merit for a capacitor application requiring high dielectric constant and high breakdown voltage can be defined as CV_{br}/A, where C/A is the capacitance per unit area and V_{br} is the breakdown voltage (184). For some applications, the figure of merit may simply be one measured parameter, such as resistivity for superconducting materials. Composition and structure of inorganic materials, which often have a strong effect on the properties, must also be determined in library characterization.

Several methods have been used for high throughput electrical measurements. The four-point probe contact technique illustrated in Figure 20 is a common method for measuring electrical resistance (201), particularly for low resistance samples.

In this method, current is applied to a sample through two of the contacts, and the voltage drop is measured across the other two contacts. This four-point configuration eliminates contact resistance, which can dominate electrical

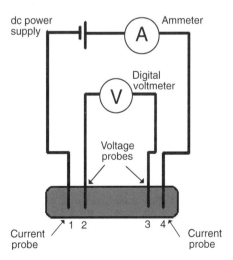

Fig. 20. Conventional four-point contact probe method for electrical resistance measurements. Current is applied across two electrodes (1,4) and voltage is measured across the other two electrodes (2,3).

measurements. A parallel version of this method containing an array of miniature probes was developed to measure the temperature-dependent resistivity in
a superconducting film library (178). A similar parallel approach was used to
measure resistance as a function of temperature and applied magnetic field in
a magnetoresistive film library (179).

Dielectric properties of insulating materials are evaluated by capacitance–
voltage measurements (8). One common method for contacting the sample uses a
liquid mercury probe as the electrode. A scanning mercury probe technique was
developed to measure serially the dielectric constant and breakdown voltage in
high permittivity dielectric libraries (184).

Dielectric constant and loss at microwave frequencies can be mapped by a
scanning evanescent microwave microscopy technique. This near-field microscopy method provides dielectric information with spatial resolution extending
from the microscopic (202) to the atomic scale (203), and has been applied to
dielectric thin-film libraries (182). In this technique, microwave radiation is
coupled evanescently to the sample surface using a sharp proximal probe that
is part of a resonant cavity or a transmission line structure. Analysis of the
reflected and transmitted signals yields quantitative dielectric constant and
loss maps for a library sample.

High throughput photoluminescence and birefringence measurements have
been used to evaluate the optical properties of combinatorial library samples.
Photoluminescence is an optical phenomenon resulting from the excitation of
an electron to a higher electronic state by the absorption of a photon from a
high energy light source, typically a laser (204). Light is emitted when the electron drops back down to a lower electronic state. The emitted light or luminescence has a specific wavelength determined by the energy difference between the
two electronic states; thus, the light has high spectral clarity of a specific color,
such as red, green or blue. Combinatorial phosphor libraries have been screened
for luminescence efficiency and color by a photoluminescence parallel imaging
technique (180). In this method, the entire library is illuminated with an ultraviolet lamp, yielding a color intensity map of visible emission that can be quantitatively analyzed using materials with known efficiency as calibration
standards. Improved quantitative analysis of the excitation and emission spectra
was obtained using a scanning spectrophotometer (205). This technique measures the absorption, transmission, or reflectance of light from a material as a
function of wavelength.

Electrooptic measurement methods probe the effect of an electric field on
the optical properties of a material (206). A high throughput birefringence technique was developed to measure the electrooptic coefficient, defined as the
change in refractive index with applied electric field (207). In this method, a
row of parallel electrodes is deposited on a library, and a combination of direct
current and alternating current electric fields is applied across each pair of electrodes. The laser light illuminating the sample interacts with the applied electric
field, and the two electric components of the light, one parallel to the applied field
and one perpendicular to the applied field, are analyzed. The resulting signal is
used to calculate the electrooptic coefficients.

X-ray diffraction is commonly used to study the structure of inorganic materials (208). In this technique, the angle of the sample (theta) and the angle of the

detector (two-theta) are scanned, and the intensity at each value of two-theta is measured. The resulting data are then used to identify the crystalline phases present in the sample. X-ray fluorescence (208) measurements are used to determine the chemical composition of multicomponent materials. Standards of known composition are measured first for calibration purposes and are used to quantify the composition of an unknown sample. Measuring X-ray diffraction or X-ray fluorescence on a submillimeter area in a combinatorial library requires the use of a microbeam (209) or a microfocusing method, and a high energy X-ray source such as a synchrotron radiation beam to generate sufficiently intense signals for the analysis of small volumes of material.

5.4. Summary. There is a wealth of opportunities for applying HTE methods to inorganic materials. A wide variety of methods have been used to fabricate thin-film, bulk, and powder libraries of varying composition. High throughput characterization tools to measure electrical, optical and structural properties have been developed for inorganic libraries. These fabrication and analysis tools will facilitate the discovery and optimization of multi-component inorganic materials for advanced applications.

6. The Path Forward

6.1. Methodologies. The future of high throughput experimentation (HTE) will require that the basic underlying software technology must be capable of defining profitable experimental spaces; visualizing complex data relationships; and of correlating high throughput experimentation target materials with properties to permit database queries from a broad spectrum of data mining engines and the development of structure–property relationships. This requires interfacing with data visualization tools at the back end and statistical experimental design engines on the front end while remaining compliant with enterprise-wide systems for knowledge management and maintaining control of experimental hardware. The integration of Informatics, Modeling, and Design are high risk opportunities.

Informatics. Integrated packages will need to link modeling, development and management of databases and search engines, hardware control, data visualization, and logistics. More specifically (210):

- Database search engines will need to be interoperable with the diverse flavors of databases currently in use.
- Development of QSPR for materials. The term QSAR is used in organic combinatorial synthesis;
- Acceleration of the design process from atomic level chemistry to engineering design by developing relationships between chemistry, processing, microstructure, etc, and processing involving metastable states, etc.
- Development of a query language for linking many different methods for querying the data with appropriate query optimization methods.
- Assembly of high performance data mining toolboxes that extend database management systems with additional operators.

- Connections to the diverse metrics in materials design, where important properties are sensitive to numerous ranges of length or time scales, eg, from 10^{-9} to 10^2 m and nanoseconds to years;
- Development of tools to present complex, multidimensional data relationships to the human interface; HTE establishes a new paradigm for the researcher who now must interpret data surfaces and not just data points.

Design of Experiments.   Due to the potentially high number of candidates available from combinatorial methodologies, the design of the sample library requires rational chemical synthesis or process information to reduce the number of samples and experiments without increasing the probability for endless searches, false positives, or false negatives.

New tools will have to be developed to enable the integration of information into molecular- and property-modeling engines. The increased amount of data that can be input into computational engines will require a significant increase in speed, bandwidth and storage. Advanced, high speed quantum calculation programs such as Wavefunction, Inc.'s program SPARTAN (211) and Acellrys' (formerly MSI) program Cerius (212) will require interfacing with databases and experimental design programs. Advances in experimental strategy for dealing with large, diverse chemical spaces, using space-filling experimental designs, predictive algorithms, and optimization techniques will be critical.

6.2. Hardware.   Micromachines and microreactor technologies (MRT) based on microelectromechanical systems (MEMS) and nanoelectromechanical systems (NEMS) will address the need for higher library densities to facilitate reduced raw materials costs for library fabrication and economies of scale and modularity in laboratory instrumentation. Construction of solid-state libraries currently requires automated deposition onto substrates, using, eg, microjet, laser ablation, vacuum deposition using pulsed vapor deposition (PVD) or chemical vapor deposition (CVD), or microfluidics in a lab-on-chip application. Microscale methods have already been commercialized for health care diagnostics with liquid samples, and this technology will penetrate the CPI and materials arena with systems utilizing solution state reaction substrates with homogeneous catalysts. There is a growth of activity in this area, with links to the HTE community, especially for drug discovery applications.

There is a growing international effort to understand MRT for distributed manufacturing of chemicals and for high throughput screening. Principal components of MRT are analogous to large-scale machines—reagent mixers, reactors, distribution and separation. The key technical challenges for MRT are in component integration, development of modular systems, and development of engineering tools to design and build microscopic systems. Since fluid flow in these devices is laminar, a significant challenge is in understanding fluid dynamics in these systems. Clearly, the convergence of the microelectronics technologies with microreactor technologies will accelerate in the near future, and are being tracked with interest by many parties for their economic potential. The current technology leaders for chemical process development and high throughput screening are the German Institut für Mikrotechnik Mainz (213). Research in the United States is centered primarily at the Pacific Northwest National Laboratory (214) , The Oak Ridge National Laboratory (215) and at the

Massachusetts Institute of Technology Microsystems Technology Laboratory (216). A collaboration between DuPont Experimental Station and Massachusetts Institute of Technology (MIT) has produced a circuit board-level discovery plant containing multiple socket-borne, interchangeable catalytic reactors with integrated fluidics control, heat transfer, separation, and mixing for combinatorial discovery and process development (217). The U.K. Lab on a Chip Consortium

Table 14. Technology Challenges

Software

technology	challenge
library design	
statistics modeling	• development of higher order designs
literature/patent	• (review by NSF is pending)
databases	• query languages; visualization; integration
informatics	
QSPR (structure-property predictions)	• property prediction, large-scale correlations, integration into experimental design tools
database query engines	• new languages, genetic programs
	• interoperability, enterprise-wide integration
Hardware	
technology	challenge
screening	
thermal properties	•electrical and thermal conductivity
optical characterization	• fluorescence, luminescent properties, X-ray diffraction
mechanical properties	• modulus, tensile strength, impact resistance
electrical properties	• capacitance, conductance
chemical properties	• molecular weight
	• polymer architecture/morphology
	• catalyst turnover, selectivity, conversion
processing	
control of physical environment	• control of temperature and pressure over library array with control over individual sample sites or wells
	• sample size—control of interfacial diffusion, mass transport properties, etc
deposition	
thermally driven (e-beam, laser, etc)	• delivery of finite samples or composition spreads of consistent, known composition at microscopic sizes: reproducibility
laser ablation	
microjet	
vacuum deposition	
robotic pipetting of nanoaliquots	• reduction of cross-talk of sample properties through diffusion, etc. across substrate surface
micro-reactor technologies	
systems integration	• development of standards for modular component interconnections
	• process control devices tuned to MRT
	• modeling for fluid flow, heat and mass transfer
reactor design	• substrates: glass, polymer, metal
manufacturing processes	• surface micromachining using wet or dry chemical etching, laser ablation, mechanical micromilling, LIGA processes

(218) is a miniaturization research project, which started in early 1999 with government funding of £1.33 million matched by £1.8 million from U.K. industry. Seven universities and twelve companies are participants and, over the next 30 months, will develop and commercialize microdevices for chemical analysis and synthesis (219). Academic and industrial partners will study reactions in an on-chip environment—how do conditions affect reaction efficiency compared with bulk reactions—as well as development of an infrastructure required for the commercial exploitation. Some of the technology challenges are shown in Table 14.

Microscale sensors.   High throughput methods will drive the development of advanced sensors and sensor arrays. The current technology relies on contact and noncontact methods of characterization and external control of process conditions. A significant impetus for developing microscale sensors has been the U.S. Department of Defense (220), primarily in response to battlefield detection of chemical and biological warfare agents and portable power generation, and the National Aeronautics and Space Administration (NASA) for microrobotic space exploration. The CPI sector is moving toward smaller, more integrated sensing devices in process control of manufacturing sites. Robotics, next-generation "titer plates", lab-on-a-chip designs, and rapid scanning devices for HTE innovation, likely with application-driven tools targeting specific physical properties that can be analyzed at microscopic levels. Automated library processing is especially challenging for new materials development since samples within a library may require different or nonequilibrium processing parameters across the library. Therefore, with the impetus toward MRT lab-on-chip devices, integration will drive the development of foundry methods to produce on-chip optical sources (eg, semiconductor lasers) as well as detectors. Finally, the capability for interpretating bulk characteristics from microscale will place increasing pressure on computational tools such as QSPR.

Success using HTE in the CPI and materials sectors has already resulted in the reduction of idea-to-commercialization cycle times to 3 to 5 years from 7 to 10 years. HTE has already reached commercial validation.

7. Acknowledgments

Certain commercial equipment, instruments, or materials are identified in this paper in order to specify the experimental procedure adequately. Such identification is not intended to imply recommendation or endorsement by the National Institute of Standards and Technology, nor is it intended to imply that the materials or equipment identified are necessarily the best available for the purpose.

BIBLIOGRAPHY

1. J. J. Hanak, *J. Mater. Sci.* **5**, 964 (1970).
2. R. B. van Dorn, L. F. Schneemeyer, and R. M. Fleming, *Nature (London)* **392**, 162 (1998).
3. G. Briceno, H. Chang, X. Sun, P. G. Schultz, and X.-D. Xiang, *Science* **270**, 273 (1995); E. Danielson, J. H. Golden, E. W. McFarland, C. M. Reeves, W. H. Weinberg, and X. D. Wu, *Nature (London)* **389** 944 (1997); E. Danielson, M. Devenney, D. M.

Giaquinta, J. H. Golden, R. C. Haushalter, E. W. McFarland, D. M. Poojary, C. M. Reeves, W. H. Weinberg, and X. D. Wu, *Science* **279**, 837 (1998).

4. U.S. Pat. 6,048,469 (Jan. 30, 1998) X.-D. Xiang, X. Sun and P. G. Schultz (The Regents of the University of California); U.S. Pat. 6, 004,617 (Dec. 21, 1999), U.S. Pat. 5,985,356 (Nov. 16, 1999), P. G. Schultz, X.-D. Xiang, and I. Goldwasser.

5. A. Furka, http://szerves.chem.elte.hu/Furka accessed on 3/15/02 (1982); R. Frank, W. Heikens, G. Heisterberg-Moutsis, and H. Blecker, *Nucleic. Acid. Res.* **11**, 4365 (1983).

6. H. M. Geysen, H. M. Rodda, and T. J. Mason, in R. Porter and J. Wheelan, eds., *Synthetic Peptides as Antigens*, Ciba Foundations Symposium, John Wiley & Sons, Inc., New York, 1986, pp. 131–149; H. M. Geysen, S. J. Rodda, and T. J. Mason, *Mol. Immunol.* **23**, 709 (1986); H. M. Geysen, S. J. Rodda, T. J. Mason, G. Tribbick, and P.G. Schoofs, *J. Immunol. Method* **102**, 259 (1987).

7. M. Geysen, personal communication.

8. A perspective on the historical development of high throughput screening for drug discovery: M. Lebl, *J. Combinatorial Chem.* **1**, 3 (1999); W. Warr "Combinatorial Chemistry and High Throughput Screening" (http://www.warr.com/ombichem.html accessed on March 20, 2002).

9. A. M. Thayer, *Chem. Eng. News* Feb 12, 1996.

10. P. Fairley, *Chem. Week* 18 (Aug. 12, 1998); P. Fairley and A. Scott, *Chem. Week* 27 (Aug. 11, 1999); A. Wood and A. Scott, *Chem. Week* 39 (Aug. 9, 2000); J. R. Engstrom and W. H. Weinburg, *AIChE J.* **46**, 2 (2000); R. Dagani, *Chem. Eng. News* 59 (Aug. 27, 2001); S. Borman, *Chem. Eng. News* 49 (Aug. 27, 2001); M. Lonergan, *Chem. Eng. News* 230 (March 26, 2001); K. Watkins, *Chem. Eng. News* 30 (Oct. 22, 2001).

11. B. Jandeleit, D. J. Schaefer, T. S. Powers, H. W. Turner, W. H. Weinberg, *Ang. Chem. Int. Ed. Engl.* **38**(17), 2495 (1999); J. R. Engstrom and W. H. Weinberg, *AIChE J.* **46**, 2-5 (2000).

12. A. Wood and A. Scott, *Chem. Week*, 39 (Aug. 9, 2000).

13. J. J. Hanak, *J. Mater. Sci.* **5**, 964 (1970); J. J. Hanak, *Vide-Science Technique Appl.* **30**(175), 11 (1975).

14. R. B. van Dorn, L. F. Schneemeyer, and R. M. Fleming, *Nature (London)* **392**, 162 (1999).

15. Symyx Technologies, Inc., Santa Clara, Calif. (accessed from http://www.symyx.com); Symyx Technologies Inc.10Q Filings, 2002 (accessed from http://www.sec.gov).

16. G. Briceno, H. Chang, X. Sun, P. G. Schultz, and X. D. Xiang, *Science* **270**, 273 (1995); U.S. Pat. 6,395,850 (May 28, 2002), D. Charmot and H. T. Chang (to Symyx Technologies, Inc.).

17. D. Akporiaye and co-workers, *Microporous Mesoporous Mater.* **48**(1–3), 367 (2001).

18. C. H. Reynolds, *Abs. Pap. Am. Chem. Soc.* **221**, 53-BTEC, Part 2 (2001).

19. R. A. Potyrailo, W. G. Morris, B. J. Chisholm, R. J. Wroczynski, and W. P. Flanagan, of Ref. 9.

20. A. P. Smith and co-workers, *Phys. Rev. Lett.* **8701**, 5503 (2001); T. J. Prosa, B. J. Bauer, and E. J. Amis, *Macromolecules* **34**, 4897 (2001); J. C. Meredith and co-workers, *Macromolecules* **33**, 9747 (2000); J. C. Meredith, A. Karim, and E. J. Amis, *Macromolecules* **33**, 5760 (2000).

21. J. Kohn, *Abstr. Pap. Am. Chem. Soc.* **219**, U547–U547 Part 2 (2000); see also http://www.njbiomaterials.org/.

22. D. Loy, "Combinatorial Materials Discovery: Organic-Inorganic Materials" http://www.sandia.gov/inorganic-organic-materials-group/comb.htm, accessed March 25, 2002.

23. X.-D. Xiang, X. Sun, G. Briceno, Y. Lou, K. A. Wang, H. Chang, W. G. Wallace-Freedman, S.-W. Chen, and P. G. Schultz, *Science* **268**, 1738 (1995); U.S. Pat. 6,203,726 (March 20, 2001), E. Danielson, M. Devenney, D. M. Giaquinta (to Symyx

Technologies, Inc.); U.S. Pat. 6,013,199 (Jan. 11, 2000), E. McFarland, E. Danielson, M. Devenney, C. Reaves, D. M. Giaquinta, D. M. Poojary, X. D. Wu, and J. H. Golden (to Symyx Technologies, Inc.).

24. G. E. Jabbour and Y. Yoshioka, of Ref. 9.

25. H. Koinuma and co-workers, *Appl. Phys. A-Mater.* **69**, S29 (1999); Y. Matsumoto and co-workers, *Jpn. J. Appl. Phys.* **2**(38), L603 (1999).

26. J. Hewes, *High Throughput Methodologies for Chemicals and Materials Research, Development, & Engineering*, White Paper, National Institute of Standards and Technology (April, 2000) accessible from http://www.atp.nist.gov/atp/ccmr/ ccmr_off.html; J. Hewes "Economic Impact of Combinatorial Materials Science on Industry and Society," in R. A. Potyrailo and E. J. Amis, eds., *High Throughput Analysis: A Tool for Combinatorial Materials Science*, Kluwer Academic/Plenum Publishers, Inc., New York, in press.

27. For example, C. Daniel, *Applications of Statistics to Industrial Experimentation*, John Wiley & Sons, Inc., New York, 1976.

28. D. Wolf, O. V. Buyevskaya, and M. Baerns, *Appl. Catal. A—Gen.* **8**, 63 (2000).

29. Wavefunction, Inc.'s program SPARTAN, ref. http://www.Wavefunction.com.

30. Accelrys' Cerius, ref. http://www.acellrys.com.

31. J. Devaney, J. Hagedorn, S. Satterfield, B. am Ende, and H. Hung, *[Conf.]* Workshop on Combinatorial Methods for Materials: Systems Integration in High Throughput Experimentation, Nov 15, 2000, Los Angeles California; J. Devaney, J. Hagedorn, O. Nicolas, G. Garg, A. Samson, and M. Michel, *[Conf.]* 15th Annual International Parallel & Distributed Processing Symposium, IPDPS 2001, Workshop on Biologically Inspired Solutions to Parallel Processing Problems, April 23, 2001, San Francisco (accessed from http://math.nist.gov/mcsd/savg/papers/allpapers.html on 4/08/02).

32. See also W. Warr and co-workers http://www.warr.com/.

33. C. Mirodatos and L. Savary, *[Conf.]* COMBI 2000, Jan 20–23, 2000, San Diego, CA (accessed from http://www.ec-combicat.org/).

34. E. Amis and co-workers, Polymers Division, Materials Science and Engineering Laboratory, NIST; E. Heilweil and co-workers. Optical Technology Division, Physics Laboratory, NIST; S. Stranick and co-workers, Surface and Microanalysis Division, Chemical Sciences and Technology Laboratory, NIST; D. Kaiser and co-workers, Ceramics Division, Materials Science and Engineering Laboratory, NIST; C. Hand-werker and co-workers, Metallurgy Division, Materials Science and Engineering Laboratory, NIST; D. Fischer, Ceramics Division, Materials Science and Engineering Laboratory, NIST, Brookhaven National Laboratory; J. Devaney and co-workers, High Performance Computing Division, Information Technology Laboratory, NIST. See also http://polymers.msel.nist.gov/combi and http://www.atp.nist.gov.

35. G. Graff , *Chem. Week* **163**(40), 31 (2001).

36. Anonymous, *Chem. Business Newsbase (Cambridge)*, (Feb 26, 2002).

37. Anonymous, *Chem. Business Newsbase (Cambridge)*, 1, (Feb 26, 2002); G. Graff, *Chem. Week* **163**(40), 31 (2001); Anon., *Chem. Business Newsbase (Cambridge)*, 1 (Aug 20, 2001); D. Richards, *Chem. Market Rep.* **260**(8), 4 (2001); C. Caruana, *Chem. Week* **162**(41), 40–42; W. Byfleet, *Hart's European Fuels News*, 1 (June 13, 2001); Anonymous, *Sulphur, London*, 28 (May/June 2001); H. W. Wilson, *Oil Gas J.* **98**(41), 64; I. Lerner, *Chem. Market Rep.* **257**(25), 28.

38. Anonymous, *Catalysts and Specialty Chemicals*, The Catalyst Group, (retrieved from http://www.catalystgrp.com/catalystsandchemicals.html).

39. Business Communications Company, Inc., Norwalk, CT 06855, (retrieved from http://www.bccresearch.com/editors/RC-229.html on May 3, 2001).

40. P. J. Cong and co-workers, *Angew. Chem Int. Ed. Engl.* **40**, 484 (1999); W. H. Weinberg, B. Jandeleit, K. Self, and H. Turner, *Current Opinion Solid State and Mater. Sci.* **36**, 104 (1998).

41. J. Busch, *Research-Technology Management*, (April–May, 2001) pp. 38–45.

42. S. Thomke and co-workers, *Res. Policy* **27**, 315 (1998).

43. F. P. Boer, *The Valuation of Technology: Business and Financial Issues in R&D*, John Wiley & Sons, Inc., New York, 1999.

44. B. M. Werner and W. E. Souder, *Research-Technology Management*, (March-April, 1997) pp. 34–42, and references cited therein; B. M. Werner and W. E. Souder, *Research-Technology Management*, May–June, 1997, pp. 28–32, and references cited therein.

45. Anonymous, *Chem. Business Newsbase* (Aug. 20, 2001).

46. (a) J. Holmgren *[Conf.]* Combi 2002 (Knowledge Foundation), (Jan. 21, 2002), San Diego, Calif; (b) U.S. Pat. 6,368,865 (April 9, 2002), I. Dahl, A. Karlsson, D. E. Akporiaye, R. Wendelbo, K. M. Vanden Bussche, and G. P. Towler (to UOP LLC).

47. G. Y. Li, in Ref. 12a.

48. C. Mirodatos, *[Conf.]* Combi2000 (Knowledge Foundation) (Jan. 20, 2000), San Diego, Calif; C. Mirodatos, *Actual Chim.* Sept. 9, 35 (2000).

49. A. Hagemeyer and co-workers, *Appl. Catalysis A-General* **221**, 23 (2001).

50. I. E. Maxwell, *Nature (London)* **394**(6691), 325 (1998), S. M. Senkan, *Nature (London)* **394**(6691), 350 (1998); M. B. Francis, T. F. Jamison, and E. N. Jacobsen, *Curr. Opin. Chem. Biol.* **2**(3), 422 (1998).

51. A. H. Tullo, *Chem.Eng. News* **79**, 38 (2001).

52. L. Resconi, L. Cavallo, A. Fait, and F. Piemontesi, *Chem.Rev.* **100**, 1253 (2000).

53. S. D. Ittel, L. K. Johnson, and M. Brookhart, *Chem. Rev.* **100**, 1169 (2000).

54. G. W. Coates, *Chem. Rev.* **100**, 1223 (2000).

55. S. M. Senkan and S. Ozturk, *Angew. Chem. Int. Ed. Engl.* **28**, 791 (1999).

56. W. F. Maier, *Angew. Chem. Int. Ed. Engl.* **21**, 1216 (1998).

57. R. Schlogl, *Angew. Chem. Int. Ed. Engl.* **21**, 2333 (1998).

58. V. V. Guliants, *Current Developments in Combinatorial Heterogeneous Catalysis*, Elsevier Science, New York, 2001.

59. P. P. Pescarmona, J. C. van der Waal, I. E. Maxwell, and T. Maschmeyer, *Catal. Lett.* **63**, 1 (1999).

60. J. M. Newsam and F. Schuth, *Biotechnol. Bioeng.* **61**, 203 (1999).

61. E. G. Derouane, *Combinatorial Catalysis and High Throughput Catalyst Design and Testing*, Kluwer Academic Publishers, Inc., Boston, 2000.

62. H. E. Tuinstra and C. L. Cummins, *Adv. Mat.* **12**, 1819 (2000).

63. The Catalyst Group (http://www.catalystgrp.com).

64. D. Richards, *Chem. Market Rep.* **260**, 32 (2001); W. Byfleet, *Hart's Eur. Fuels News* **5**(2001).

65. J. A. Loch and R. H. Crabtree, *Pure Appl. Chem.* **73**(1), 119 (2001).

66. T. Berg, A. M. Vandersteen, and K. D. Janda, *Bioorg. Med. Chem. Lett.* **8**, 1221 (1998); U.S. Pat. 6,316,616 (Nov. 13, 2001), E. N. Jacobsen and S. M. Sigman, (to the President and Fellows of Harvard College).

67. P. J. Fagan and E. Hauptman *[Conf.]* University of Delaware at Newark, Oct. 18, 1999; G. Li and P. J. Fagan, Ref. 14; P. J. Fagan, E. Hauptman, R. Shapiro, and A. Casalnuovo, *J. Am. Chem. Soc.* **122**(21), 5043 (2000); G. Li, Ref. 12a; G. Y. Li, P. J. Fagan, and P. L. Watson, *Angew. Chem.Int. Ed. Engl.* **40**(6), 1106 (2001).

68. Current equipment manufacturers for homogeneous catalyst development include: Argonaut Technologies (U.S.); Mettler-Toledo Bohdan (U.S.); Advanced ChemTech (UK); Chemspeed, Ltd. (Switzerland); Zeton Altimira (U.S.); and Symyx Technologies, Inc. (U.S.).

69. U.S. Pat. 6,350,916 (Feb. 26, 2002), A. Guram and X. Bei (Symyx Technologies); U.S. Pat. 6,339,157 (Jan. 15, 2002), X. Bei and A. Guram (Symyx Technologies); U.S. Pat. 6,316,663 (Jan. 13, 2001), A. Guram, C. Lund, H. W. Turner, and T. Uno (to Symyx Technologies); U.S. Pat. 6,268,513 (July 13, 2001), X. Bei and A. Guram

(Symyx Technologies); U.S. Pat. 6,265,601 (July 24, 2001) X. Bei and A. Guram (to Symyx Technologies); U.S. Pat. 6,242,623 (June 5, 2001), W. H. Weinberg, E. McFarland, I. Goldwasser, T. Boussie, H. Turner, J. A. M. VanBeek, V. Murphy, and T. Powers (to Symyx Technologies); U.S. Pat. 6,248,540 (June 19, 2001) T. Boussie, V. Murphy, and J. A. M. van Beek (to Symyx Technologies); U.S. Pat. 6,225,487 (May 15, 2001) A. Guram (to Symyx Technologies); U.S. Pat. 6,177,528 (Jan. 23, 2001), A. M. LaPointe, A. Guram, T. Powers, B. Jandeleit, T. Boussie, and C. Lund, (to Symyx Technologies); U.S. Pat. 6,124,476 (Sept. 26, 2000) A. Guram, X. Bei, T. Powers, B. Jandeleit, and T. Crevier (to Symyx Technologies); U.S. Pat. 6034240 (March 7, 2000), T. Pointe (to Symyx Technologies); U.S. Pat. 6,030,917 (Feb. 29, 2000), W. H. Weinberg, E. McFarland, I. Goldwasser, T. Boussie, H. Turner, J. Van-Beek, V. Murphy, and T. Powers, (to Symyx Technologies); R. Drake, R. Dunn, D. C. Sherrington, and S. J. Thomson, *Combinatorial Chem. High Throughput Screening* **5**(3), 201 (2002).

70. W. A. Herrmann, V. P. W. Bohm, F. A. Rampf, and T. Weskamp, *Abstracts of Papers of the American Chemical Society* **219**, 43-inor (2000); V. P. W. Bohm, T. Weskamp, C. W. K. Gstottmayr, and W. A. Herrmann, *Angew. Chem. Int. Ed. Engl.* **39**(9), 1602 (2000).

71. J. A. Loch and R. H. Crabtree, *Pure Appl. Chem.* **73**(1), 119 (2001).

72. H. Shea, Ref. 12.

73. M. B. Francis, N. S. Finney, and E. N. Jacobsen, *J. Am. Chem. Soc.* **118**(37), 8983 (1996); A. M. Vandersteen, H. Han, and K. D. Janda, *Mol. Diversity* **2**(1/2), 89 (1996); R. H. Crabtree, J. A. Loch, K. Gruet, D. H. Lee, and C. Borgmann, *J. Organomet. Chem.* **600**(1–2), 7 (2000); K. Severin, *Chem.-Eur. J.* **8**(7), 1515 (2002); T. J. Colacot, *[Conf.]* CombiCat2001 (The Catalyst Group), November 15–16, 2001, Philadelphia.

74. U.S. Pat. 6,395,552 (05/28/02), R. B. Borade, D. Poojary, and X. P. Zhou (Symyx Technologies); U.S. Pat. 6,362,309 (03/26/02), C. Lund, K. A. Hall, T. Boussie, V. Murphy, and G. Hillhouse (Symyx Technologies); U.S. Patent 6,355,854, (03/12/02) Y. Liu; U.S. Pat. 6,149882 (11/21/00) S. Guan, L. VanErden, R. C. Haushalter, X. P. Zhou, X. J. Wang, and R. Srinivasan (Symyx Technologies); U.S. Pat. 5,959,297 (09/28/99) W. H. Weinberg, E. W. McFarland, P. Cong, and S. Guan (Symyx Technologies).

75. R. Schlögl, *Catalysis Today* **62**, 91 (2000).

76. U. Rodemerck, P. Ignaszewski, M. Lucas, P. Claus, M. Baerns, *Topics Catalysis* **13**, 249 (2000).

77. K. E. Simons, *Topics Catalysis* **13**, 201 (2000).

78. Anonymous, *Asia Intelligence Wire* from FT Information, Aug 25, 1999.

79. Anonymous, *Oil Gas J. (Tulsa, Ok.)* **98**(41), 64 (2000).

80. J. M. Newsam, T. Bein, J. Klein, W. F. Maier and W. Stichert, *Microporous Mesoporous Mater.* **48**(1–3), 355 (2001); K. Choi and co-workers, *Angew. Chem. Int. Ed. Engl.* **38**(19), 2891 (1999); T. Bein, *Angew. Chem. Int. Ed. Engl.* **38**(3), 323 (1999); R. Lai, B. S. Kang, and G. R. Gavalas, *Angew. Chem. Int. Ed. Engl.* **40**(2), 408 (2001).

81. D. E. Akporiaye, I. M. Dahl, A. Karlsson, and R. Wendelbo, *Angew. Chem. Int. Ed. Engl.* **37**(5) (1998).

82. U.S. Pat. 6,368,865 (April 9, 2002) I. M. Dahl, A. Karlsson, D. E. Akporiaye, R. Wendelbo, K. M. Vanden Bussche, and G. P. Towler (UOP LLC); U.S. Pat. 6,342,185 (Jan. 29, 2002) I. M. Dahl, A. Karlsson, D. E. Akporiaye, K. M. Vanden Bussche, and G. P. Towler (to UOP LLC); U.S. Pat. 6,327,334 (Dec. 4, 2001) R. C. Murray, Jr., C. M. Bratu, G. J. Lewis (UOP LLC).

83. J. Holmgren, CombiCat2001 *[Conf.]*, Philadelphia, November 15–16, 2001.

84. T. R. Ralph *[Conf.]*, North American Catalyst Society Annual Meeting, June 3–8, 2001, Toronto.

85. M. G. Sullivan, H. Utomo, P. J. Fagan, and M. D. Ward, *Anal. Chem.* **71**(19), 4369 (1999).

86. G. Y. Chen and co-workers, *Catal. Today* **67**(4), 341 (2001); Y. P. Sun, H. Buck and T. E. Mallouk, *Anal. Chem.* **73**(7), 1599 (2001); U.S. Pat. 6,284,402 (09/04/01) T. E. Mallouk, B. C. Chan, E. Reddington, A. Sapienza, G. Chen, E. Smotkin, B. Gurau, R. Viswanathan, and R. Liu (to Pennsylvania State University); Y. P. Sun and co-workers, *Anal. Chem.* **73**(7), 1599 (2001); B. Gurau and co-workers, *J. Phys. Chem. B* **102**(49), 9997 (1998); E. Reddington and co-workers, *Science* **280**(5370), 1735 (1998); B. C. Chan, E. Reddington, A. Sapienza, and J. S. Yu, "Combinatorial Discovery and Optimization of Anode Electrocatalysts for Direct Methanol Fuel Cells," in *Fuel Cell: Clean Energy For Today's World*, Courtesy Associates, Palm Springs, Calif., 1998.

87. WO Pat. 0,069,009 (Nov. 16, 2000), A. Gorer (Symyx Technologies, Inc); WO Pat. 0,054,346 (Sept. 14, 2000), A. Gorer (Symyx Technologies, Inc.); WO Pat. 0054346 (Sept. 14, 2000), A. Gorer (Symyx Technologies, Inc.).

88. R. F. Service, *Science* **280**(5370), 1690 (1998).

89. Freedonia Group, *Catalysis*, Freedonia Group, Cleveland, Ohio (2001).

90. P. Fairley, *Chem. Week* **160**(26), 43 (1998); Anonymous, *Business Wire* (2002); H. E. Tuinstra and C. L. Cummins, *Adv. Materials* **12**(23), 1819 (2000); Symyx Technologies 10Q Report of May 7, 2002 to the U.S. Securities Exchange Commission (obtained from http://www.sec.gov).

91. Anonymous, *Chem. Business Newsbase (Cambridge)*, October 30, 2000.

92. Anonymous, *Business Wire*, New York, March 29, 2001.

93. T. R. Boussie, V. Murphy, K. A. Hall, C. Coutard, C. Dales, M. Petro, E. Carlson, H. W. Turner, and T. S. T. I. Powers, *Tetrahedron* **55**(39), 11699 (1999); T. R. Boussie, C. Coutard, H. Turner, V. Murphy, and T. S. T. I. Powers, *Angew. Chem. Int. Ed. Engl.* **37**(23), 3272 (1998).

94. A. H. Tullo, *Chem. Eng. News* **79**(43), 38 (2001); H. A. Shea, *Chem. Eng. News* **79**(13), 235 (2001); Anonymous, *Chem. Business Newsbase (Cambridge)*, March 2, 2001; Anonymous, *Chem. Business Newsbase (Cambridge)*, Feb. 23, 2001; Anonymous, "Research award for Albemarle: Catalysts," *Chem. Business Newsbase (Cambridge)*, Nov. 9, 2000.

95. Anonymous, *Chem. Business Newsbase (Cambridge)*, Aug. 25, 1999.

96. J. Bechtel, D. Demuth, K. E. Finger, S. Schunk, W. Stichert, W. Strehlau1, A. Sundermann, and J. M. Newsam, *[Conf.]* CombiCat 2000 (The Catalyst Group), Philadelphia, Pa., November 18, 2000.

97. R. Farrauto, *[Conf.]*, International Symposia on Chemical Reaction Engineering, January 7, 2001, Houston, Tex.; G. Koermer, *[Conf.]* NACS2001, June 3–8, 2001, Toronto; http://www.engelhard.com/AnnualReport.

98. K. Krantz, S. Ozturk, and S. Senkan, *Catal. Today* **62**, 281 (2000).

99. A. Richter and co-workers, *Appl. Catalysis B-Environ.* **36**(4), 261 (2002).

100. K. Yajima and co-workers, *Appl. Catalysis A-Gen.* **194**, 183 (2000).

101. http://www.wavefunction.com; http://www.accelrys.com

102. M. Baerns, D. Wolf, G. Grubert, M. Langpape, N. Dropka, and M. Holena *[Conf.]* 2001 National Meeting of the AIChE, Oct. 2001.

103. O. V. Buyevskaya and D. Wolf, M. Baerns, *Appl. Catal.* **200**, 63 (2000); M. T. Reetz, *Angew. Chem. Int. Ed. Engl.* **40** 284 (2001).

104. P. J. Fagan, E. Hauptman, R. Shapiro, and A. Casalnuovo, *J. Am. Chem. Soc.* **122**, 5043 (2000).

105. J. N. Cawse *[Conf.]* AIChE National Meeting, Los Angeles, CA, November 15, 2000; J. N. Cawse, *Acct. Chem. Res.* **34**(3), 213 (2001); J. N. Cawse, ed., *Strategies of Catalyzed Optimization in High-Throughput Experimentation in Combinatorial Experimental Design*, John Wiley & Sons, New York, in press.

106. J. Hewes, *High Throughput Methodologies for Chemicals and Materials Research, Development, and Engineering*, White Paper, National Institute of Standards and Technology, March 21, 2000 (http://www.atp.nist.gov/www/ccmr/ccmr_off.htm).

107. J. M. Newsam, S. M. Levine, D. King-Smith, D. Demuth, and W. Strehlau, *Abs. Pap. Am. Chem. S.* **219**, 13-mtls (2000).

108. L. A. Harmon, A. J. Vayda and S. G. Schlosser, *Abs. Pap. Am. Chem. S.* **219**, 12-mtls (2000); L. A. Harmon, A. J. Vayda, S. G. Schlosser, *Abs. Pap. Am. Chem. S.* **221**, 67-BTEC Part 2 (2001).

109. Eu. Pat. 1,164,514 (12/19/2001) P. Cohan, S. D. Lacy, A. L. Safir, L. Van Erden, P. Wang, E. W. McFarland, and S. J. Turner (Symyx Technologies, Inc.).

110. *Technology Vision 2020: A Strategic Plan For The U.S. Chemical Industry*, [*Conf.*] June 2, 2000 at National Institute of Standards and Technology, Gaithersburg, Md., available from http://www.cstl.nist.gov/div837/Division/combi.pdf.

111. W. Ehrfeld and co-workers, *Microsystem Technol.* **7**(4), 145 (2001); K. Benz and co-workers, *Chem. Eng. Tech.* **24**(1), 11 (2001); H. D. Bauer and co-workers, *Synth. Metals* **115**(1–3), 13 (2000).

112. E. Smela, *J. Micromech. Microeng.* **9**(1), 1 (1999); Y. S. S. Wan, J. L. H. Chau, A. Gavriilidis, and K. L. Yeung, *Microporous Mesoporous Mater.* **42**(2–3), 157 (2001); A. R. Noble-Luginbuhl, R. M. Blanchard, and R. G. Nuzzo, *J. Am. Chem. S.* **122**(16), 3917 (2000); K. Kusakabe, S. Morooka and H. Maeda, *Kor. J. Chem. Eng.* **18**(3), 271 (2001).

113. T. A. Ameel and co-workers, *J. Propul. Power* **16**(4), 577 (2000); A. Y. Tonkovich, and co-workers, *Chem. Eng. Sci.* **54**(13–14), 2947 (1999).

114. I. M. Hsing, R. Srinivasan, M. P. Harold, K. F. Jensen, and M. A. Schmidt, *Chem. Eng. Sci.* **55**(1), 3 (2000); S. K. Ajmera, M. W. Losey, K. F. Jensen, and M. A. Schmidt *AICHE J.* **47**(7), 1639 (2001); M. W. Losey, M. A. Schmidt, and K. F. Jensen, *Ind. Eng. Chem. Res.* **40**(12), 2555 (2001); S. L. Firebaugh, K. F. Jensen, and M. A. Schmidt, *J. Microelectromechan. Systems* **10**(2), 232 (2001); K. F. Jensen, *Chem. Eng. Sci.* **56**(2), 293 (2001); K. F. Jensen, *AICHE J.* **45**(10), 2051 (1999).

115. Anon., "Argonaut and Symyx Announce Collaboration," *Business Wire (New York)*, Aug. 17, 1999; http://www.argotech.com/.

116. http://www.chemspeed.com/applications.html.

117. U.S. Pat. 4,014,657 (March 29, 1977), V. M. Gryaznov, V. S. Smirnov, and S. I. Aladyshev (Union of Soviet Socialist Republics); U.S. Pat. 4,099,923 (July 7, 1978), E. Milberger (The Standard Oil Company); U.S. Pat. 5,489,726 (Feb. 6, 1996), A. Huss and I. I. Rahmim, P. Wood (Mobil Oil Corporation).

118. U.S. Pat. 5,304,354 (April 19, 1994), C. M. Finley and C. L. Kissel (Baker Hughes Incorporated); U.S. Pat. 5,612,002 (March 18, 1997), D. R. Cody and co-workers (Warner-Lambert Company).

119. C. Hoffmann, H. W. Schmidt, and F. Schuth, *J. Catalysis* **198**, 348 (2001); S. Thomson, C. Hoffmann, S. Ruthe, H. W. Schmidt, and F. Schuth, *Appl. Catalysis A-Gen.* **220**, 253 (2001).

120. P. Claus, D. Honicke, and T. Zech, *Catal. Today* **67**, 319 (2001).

121. U.S. Pat. 4,099,923 (July 11, 1978), E. C. Milberger and O. H. Solon (The Standard Oil Company); U.S. Pat. 6,368,865 (April 9, 2002), I. M. Dahl and co-workers (UOP LLC).

122. U.S. Pat. 6,312,586 (Nov. 6, 1999), T. N. Kalnes, S. R. Dunne, and V. P. Thakkar (to UOP LLC).

123. D. E. Akporiaye and co-workers, *Angew. Chem. Int. Ed. Engl.* **37**, 609 (1998); D. E. Akporiaye and co-workers, *Angew. Chem. Int. Ed. Engl.* **37**, 3369 (1998).

124. S. Zillman, "UOP Accelerates Combinatorial Chemistry Research with Sintef Alliance," (accessed from http://www.sintef.no/units/chem/catalysis_oslo/sinte-fuopalliance.html, 6/28/1999).

125. T. Bein, J. M. Newsam, J. Klein, W. F. Maier, and W. Stichert, *Microporous Meso-porous Mater.* **48**(1–3), 355 (2001); K. Choi, D. Gardner, N. Hilbrandt, and T. Bein, *Angew. Chem. Int. Ed. Engl.* **38**(19): 2891 (1999).

126. F. C. Moates, M. Somani, J. Annamalai, J. T. Richardson, D. Luss, and R. C. Willson, *Ind. Eng. Chem. Res.* **35**, 4801 (1996); A. Holzwarth, H.-W. Schmidt, and W. F. Meier, *Angew. Chem. Int. Ed. Engl.* **37**, 2644 (1998); M. T. Reetz, M. H. Becker, K. M. Kuhling, and A. Holzwarth, *Angew. Chem. Int. Ed. Engl.* **37**, 2647 (1998); S. J. Taylor and J. P. Morkin, *Science* **280**, 267 (1998); M. T. Reetz, M. H. Becker, M. Liebl, and A. Furstner, *Angew. Chem. Int. Ed. Engl.* **39**, 1236 (2000).

127. O. Lavastre and J. P. Morken, *Angew. Chem. Int. Ed. Engl.* **38**, 3163 (1999).

128. J. Le Bars, T. Haussner, J. Lang, A. Pfaltz, and D. G. Blackmond, *Adv. Synth. Catal.* **343**, 207 (2001).

129. C. M. Snively, G. Oskarsdottir, and J. T. I. Lauterbach, *Catal. Today* **67**, 357 (2001).

130. M. Orschel, J. Klein, H. W. Schmidt, W. F. Maier, *Angew. Chem. Int. Ed. Engl.* **38**, 2791 (1999).

131. H. Su, Y. J. Hou, R. S. Houk, G. L. Schrader, and E. S. Yeung, *Anal. Chem.* **73**, 4434 (2001).

132. S. M. Senkan, *Nature (London)* **394**, 350 (1998); S. M. Senkan and S. Ozturk, *Angew. Chem. Int. Ed. Engl.* **38**, 791 (1999). S. Senkan, K. Krantz, S. Ozturk, V. Zengin, and I. Onal, *Angew. Chem. Int. Ed. Engl.* **38**, 2794 (1999); K. Krantz, S. Ozturk, and S. Senkan, *Catal. Today* **62**, 281 (2000); U. Rodemerck, D. Wolf, O. V. Buyevskaya, P. Claus, S. Senkan, and M. Baerns, *Chem. Eng. J.* **82**, 3 (2001); http://www.seas.ucla.edu/~senkan.

133. E. Reddington, A. Sapienza, B. Gurau, R. Viswanathan, S. Sarangapani, E. S. Smotkin, and T. E. Mallouk, *Science* **280**(5370), 1735 (1998).

134. Y. M. Liu, P. J. Cong, R. D. Doolen, H. W. Turner, and W. H. Weinberg, *Catal. Today* **61**(1–4), 87 (2000); U.S. Pat. 5,959,297 (Sept. 28, 1999) Weinberg , and co-workers (Symyx Technologies, Inc.).

135. M. T. Reetz, *Angew. Chem. Int. Ed. Engl.* **41**(8), 1335 (2002); M. T. Reetz, K. M. Kuhling, S. Wilensek, H. Husmann, U. W. Hausig, and M. Hermes, *Catal. Today* **67**(4), 389 (2001); M. T. Reetz, K. M. Kuhling, H. Hinrichs, and A. Deege, *Chirality* **12**, 479 (2000).

136. K. Kennedy, T. Stefansky, G. Davy, V. F. Zackay, and E. R. Parker, *J. Appl. Phys.* **36**, 3808 (1965).

137. J. J. Hanak, *J. Mat. Sci.* **5**, 964 (1970).

138. N. K. Terrett, *Combinatorial Chemistry*, Oxford, 1998.

139. X.-D. Xiang, X. Sun, G. Briceno, Y. Lou, K.-A. Wang, H. Chang, W. G. Wallace-Freedman, S.-W. Chen, and P. G. Schultz, *Science* **268**, 1738 (1995).

140. E. Reddington, A. Sapienza, B. Gurau, R. Viswanathan, S. Sarangapani, E. Smotkin, and T. Mallouk, *Science* **280**, 1735 (1998).

141. J. Wang, Y. Yoo, C. Gao, I. Takeuchi, X. Sun, H. Chang, X.-D. Xiang, and P. G. Schultz, *Science* **279**, 1712 (1998).

142. X.-D. Sun and X. X.-D., *Appl. Phys. Lett.* **72**, 525 (1998).

143. E. Danielson, M. Devenney, D. M. Giaquinta, J. H. Golden, R. C. Haushalter, E. W. McFarland, D. M. Poojary, C. M. Reaves, W. H. Wenberg, and X. D. Wu, *Science* **279**, 837 (1998).

144. B. Jandeleit, D. J. Schaefer, T. S. Powers, H. W. Turner, and W. H. Weinberg, *Angew. Chem. Int. Ed. Engl.* **38**, 2494 (1999).

145. B. Jandeleit, D. J. Schaefer, T. S. Powers, H. W. Turner, and W. H. Weinberg, *Angew. Chem. Int. Ed. Engl.* **38**, 2494 (1999).

146. J. Klein, C. W. Lehmann, H.-W. Schmidt, and W. F. Maier, *Angew. Chem. Int. Ed. Engl.* **37**, 3369 (1998).

147. S. Brocchini, K. James, V. Tangpasuthadol, and J. Kohn, *J. Biomed. Mater. Res.* **42**, 66 (1998).
148. T. A. Dickinson, D. R. Walt, J. White, and J. S. Kauer, *Anal. Chem.* **69**, 3413 (1997).
149. G. R. Newkome, C. D. Weis, C. N. Moorefield, G. R. Baker, B. J. Childs, and J. Epperson, *Angew. Chem. Int. Ed. Engl.* **37**, 307 (1998).
150. D. J. Gravert, A. Datta, P. Wentworth, and K. D. Janda, *J. Am. Chem. Soc.* **120**, 9481 (1998).
151. C. H. Reynolds, *J. Comb. Chem.* **1**, 297 (1999).
152. T. Takeuchi, D. Fukuma, and J. Matsui, *Anal. Chem.* **71**, 285 (1999).
153. C. Schmitz, P. Posch, M. Thelakkat, and H. W. Schmidt, *Macromol. Symp.* **154**, 209 (2000).
154. M. Gross, D. C. Muller, H. G. Nothofer, U. Sherf, D. Neher, C. Brauchle, and K. Meerholz, *Nature (London)* **405**, 661 (2000).
155. J. C. Meredith, A. Karim, and E. J. Amis, *Macromolecules* **33**, 5760 (2000).
156. J. C. Meredith, A. P. Smith, A. Karim, and E. J. Amis, *Macromolecules* **33**, 9747 (2000).
157. Certain equipment and instruments or materials are identified in the paper in order to adequately specify the experimental details. Such identification does not imply recommendation by the National Institute of Standards and Technology, nor does it imply the materials are necessarily the best available for the purpose.
158. A. P. Smith, J. Douglas, J. C. Meredith, A. Karim, and E. J. Amis, *Phys. Rev. Lett.* **87**, 15503 (2001).
159. K. Ashley, J. C. Meredith, A. Karim, and D. Raghavan, *Polym. Int.*, in press.
160. B. Liedberg and P. Tengvall, *Langmuir* **11**, 3821 (1995).
161. J. Genzer and E. J. Kramer, *Europhys. Lett.* **44**, 180 (1998).
162. According to ISO 31-8, the term "molecular weight" has been replaced by "relative molecular mass," M_r. The number average molecular mass is given by M_n. We use the conventionally accepted symbol M_w for weight average molecular mass.
163. J. C. Meredith, A. Karim, and E. J. Amis, in R. Malhotra ed., *ACS Symposium Series: Combinatorial Approaches to Materials Development*, American Chemical Society, Washington, D.C., 2001.
164. H. Hasegawa and T. Hashimoto, *Macromolecules* **8**, 589 (1985).
165. T. P. Russell, G. Coulon, V. R. Deline, and D. C. Miller, *Macromolecules* **22**, 4600 (1989).
166. A. M. Mayes, T. P. Russell, P. Bassereau, S. M. Baker, and G. S. Smith, *Macromolecules* **27**, 749 (1994).
167. A. P. Smith, J. Douglas, J. C. Meredith, A. Karim, and E. J. Amis, *J. Polym. Sci. B: Polym. Phys.* **39**, 2141 (2001).
168. K. L. Beers, J. F. Douglas, E. J. Amis, and Alamgir Karim, *ACS Preprint*, Fall 2001.
169. C. Schmitz, M. Thelakkat, and H. W. Schmidt, *Adv. Mat.* **11**, 821-26 (1999).
170. C. Schmitz, P. Posch, M. Thelakkat, and H. W. Schmidt, *Phys. Chem. Chem. Phys.* **1**, 1777 (1999).
171. C. Schmitz, M. Thelakkat, and H. W. Schmidt, *Adv. Mater.* **11**, 821 (1999).
172. L. A. Vratsanos, M. Rusak, K. Rosar, T. Everett, and M. Listemann (Air Products Polymers, LP and Air Products and Chemicals, Inc., Allentown, Pa.) Athens Conference on Coatings: Science and Technology, Proceedings, 27th, Athens, Greece, July 2–6 (2001), pp. 435–442. Publisher: Institute of Materials Science, New Paltz, N.Y. CODEN: 69CGM9.
173. W. Schrot, S. Lehmann, J. Hadeler, G. Oetter, G. Dralle-Voss, E. Beck, W. Paulus, S. Bentz (BASF Aktiengesellschaft, Ludwigshafen, Germany) Athens Conference on Coatings: Science and Technology, Proceedings, 27th, Athens, Greece, July 2–6 (2001), pp. 283–296. Publisher: Institute of Materials Science, New Paltz, N.Y. CODEN: 69CGM9.

174. D. A. Wicks, H. Bach (Coatings and Colorants Division, Bayer Corporation, Pittsburgh, Pa.) *Proceedings of the International Waterborne, High-Solids, and Powder Coatings Symposium*, 2002, 29th 1-24. CODEN: PIWCF4.

175. B. Chisolm, P. Radislav, J. Cawse, R. Shaffer, M. Brennan, C. Molaison, D. Whisenhunt, W. Flanagan, D. Olson, J. Akhave, D. Saunders, A. Mehrabi, and M. Licon (GE Corporate Research and Development, Niskayuna, NY; and Avery Dennison, USA) *Progress in Organic Coatings* (2002), 45 (2-3), 313-321 CODEN: POGCAT; See also *Analytical Chemistry* (2002), 74, 5676-5680; and *Progress in Organic Coatings* 45 (2002), pp. 313–321.

176. See eg, C. Jacobson, R. Hergenrsder, L. B. Koutny, and J. M. Ramsey, *Anal. Chem.* **66**, 1114 (1994); and J. M. Ramsey, S. C. Jacobson, and M. R. Knapp, *Nature Med.* **1**, 1096 (1995).

177. J. J. Hanak, *J. Mat. Sci.* **5**, 964 (1970).

178. X.-D. Xiang, X. Sun, G. Briceno, Y. Lou, K.-A. Wang, H. Chang, W. G. Wallace-Freedman, S.-W. Chen, and P. G. Schultz, *Science* **268**, 1738 (1995).

179. G. Briceno, H. Chang, X. Sun, P. G. Schultz, and X.-D. Xiang, *Science* **270**, 273 (1995).

180. E. Danielson, J. H. Golden, E. W. McFarland, C. M. Reaves, W. H. Weinberg, and X. D. Wu, *Nature (London)* **389**, 944 (1997).

181. X.-D. Sun, C. Gao, J. Wang, and X.-D. Xiang, *Appl. Phys. Lett.* **70**, 3353 (1997).

182. H. Chang, C. Gao, I. Takauchi, Y. Yoo, J. Wang, P. G. Schultz, X.-D. Xiang, R. P. Sharma, M. Downes, and T. Venkatesan, *Appl. Phys. Lett.* **72**, 2185 (1998).

183. Takeuchi, H. Chang, C. Gao, P. G. Schultz, X.-D. Xiang, R. P. Sharma, M. J. Downes, and T. Vanketesan, *Appl. Phys. Lett.* **73**, 894 (1998).

184. R. B. van Dover, L. F. Schneemeyer, and R. M. Fleming, *Nature (London)* **392**, 162 (1998).

185. Y. Matsumoto, M. Murakami, Z. Jin, A. Ohtomo, M. Lippmaa, M. Kawasaki, and H. Koinuma, *J. Appl. Phys.* **38**, L603 (1999).

186. R. Cremer, S. Kyrsta, D. Neuschutz, M. Laurenzis, P. H. Bolivar, and H. Kurz, in G. E. Jabbour and H. Koinuma, eds., *Combinatorial and Composition Spread Techniques in Materials and Device Development II,* SPIE, San Diego, Vol. 51, 2001.

187. J.-C. Zhao, *Adv. Engr. Mater.* **3**, 143 (2001).

188. K. Choi, D. Gardner, N. Hilbrandt, and T. Bein, *Angew. Chem. Int. Ed.* **38**, 2891 (1999).

189. L. F. Schneemeyer, R. B. van Dover, and R. M. Fleming, *Appl. Phys. Lett.* **75**, 1967 (1999).

190. S. E. Russek, W. E. Bailey, G. Alers, and D. L. Abraham, *IEEE Trans. Magnetics* **37**, 2156 (2001).

191. H. Chang, I. Takeuchi, and X.-D. Xiang, *Appl. Phys. Lett.* **74**, 1165 (1999).

192. H. Koinuma, T. Koida, T. Ohnishi, D. Komiyama, M. Lippmaa, and M. Kawasaki, *Appl. Phys. A* **69**, 929 (1999).

193. T. Ohnishi, D. Komiyama, T. Koida, S. Ohashi, C. Stauter, H. Koinuma, A. Ohtomo, M. Lippmaa, N. Nakagawa, M. Kawasaki, T. Kikuchi, and K. Omote, *Appl. Phys. Lett.* **79**, 536 (2001).

194. G. He, T. Iijima, H. Funakubo, and Z. Wang, in G. E. Jabbour and H. Koinuma. Eds., *Combinatorial and Composition Spread Techniques in Materials and Device Development II,* SPIE, San Diego, Vol. 27, 2001.

195. M. Th. Cohen-Adad, M. Gharbi, C. Goutaudier, and R. Cohen-Adad, *J. Alloys Compound* **289**, 185 (1999).

196. J.-C. Zhao, *J. Mater. Res.* **16**, 1565 (2001).

197. X.-D. Sun, K.-A. Wang, Y. Yoo, W. G. Wallace-Freedman, C. Gao, X.-D. Xiang, and P. G. Schultz, *Adv. Mater.* **9**, 1046 (1997).

198. H. M. Reichenbach and P. J. McGinn, *J. Mater. Res.* **16**, 967 (2001).

199. K.-S. Sohn, E. S. Park, C. H. Kim, and H. D. Park, *J. Electrochem. Soc.* **147**, 4368 (2000).
200. H. M. Christen, S. D. Silman, and K. S. Harshavardhan, *Rev. Sci. Instr.* **72**, 2673 (2001).
201. A. Moscovici, in J. I. Kroschwitz and M. Howe-Grant, eds., *Kirk-Othmer Encyclopedia of Chemical Technology*, 4th ed., John Wiley & Sons, Inc., New York, 14, 1995, p. 632.
202. Y. Cho, A. Kirihara, and T. Saeki, *Rev. Sci. Instrum.* **67**, 2297 (1996).
203. S. J. Stranick and P. S. Weiss, *J. Phys. Chem.* **98**, 1762 (1994).
204. C. B. Guillaume, in G. L. Trigg, ed., *Encyclopedia of Applied Physics*, VCH Publishers, Inc., New York **13**, 1995, p. 497.
205. J. Wang, Y. Yoo, C. Gao, I. Takeuchi, X. Sun, H. Chang, X.-D. Xiang, and P. G. Schultz, *Science* **279**, 1712 (1998).
206. D. A. B. Miller, in S. P. Parker, ed., *McGraw-Hill Encyclopedia of Science & Technology*, 8th ed., McGraw-Hill, New York, 1997, p. 291.
207. J. Li, F. Duewer, C. Gao, H. Chang, X.-D. Xiang, and Y. Lu, *Appl. Phys. Lett.* **76**, 769 (2000).
208. R. A. Sparks, in J. I. Kroschwitz and M. Howe-Grant, eds., *Kirk-Othmer Encyclopedia of Chemical Technology*, John Wiley & Sons, Inc., New York, 25, 1998, p. 735.
209. E. D. Issacs, M. Marcus, G. Aeppli, X.-D. Xiang, X.-D. Sun, P. Schultz, H.-K. Kao, G. S. Cargill III, and R. Haushalter, *Appl. Phys. Lett.* **73**, 1820 (1998).
210. J. Cawse, Industry White Paper to the NIST Advanced Technology Program (1998).
211. http://www.wavefunction.com.
212. http://www.accelrys.com.
213. W. Ehrhard and co-workers (http://www.fzk.de/pmt/); W. Ehrhard and co-workers, *Proceedings of the 2000 AIChE Spring Meeting, 4th International Conference on Microreaction Technology*, March 6, 2000.
214. R. S. Wegeng and co-workers, *Proceedings of the 2000 AIChE Spring Meeting, 4th International Conference on Microreaction Technology*, March 6, 2000.
215. M. Ramsey and co-workers, Oak Ridge National Laboratory, http://www.ornl.gov/ORNLReview/meas_tech/shrink.htm.
216. K. F. Jensen and co-workers, Massachusetts Institute of Technology; http://www-mtl.mit.edu/mtlhome/; Gleason *et al.*, (http://web.mit.edu/cheme/www/People/Faculty).
217. J. F. Ryley and co-workers, *Proceedings of the 2000 AIChE Spring Meeting, 4th International Conference on Microreaction Technology*, March 6, 2000.
218. P. Fairley, *Chem. Week* (27) August 11, 1999.
219. Ref. http://www.chemsoc.org/networks/locn/whoswho.htm.
220. DARPA MicroFLUMES Program (http://www.darpa.mil/MTO/mFlumes/index.html).

JOHN D. HEWES
Honeywell International, Inc.

DEBRA KAISER
ALAMGIR KARIM
ERIC AMIS
National Institute of Standards and Technology

COMBUSTION SCIENCE AND TECHNOLOGY

1. Introduction

Fuel combustion is a complex process, the understanding of which involves knowledge of chemistry (structural features of the fuel), thermodynamics (feasibility and energetics of the reactions), mass transfer (diffusion of fuel and oxidant molecules), reaction kinetics (rate of reaction), and fluid dynamics of the process. Therefore, the design of combustion systems involves utilizing information and data generated in a range of disciplines. Often, the design of practical combustion systems is based on experience rather than on fundamental mechanistic understanding. However, for certain fuels such as methane, the combustion mechanism is better understood than for other more complex fuels such as coal. To accommodate the variety of approaches used to solve practical problems, this article is divided into two subsections: combustion science and combustion technology.

2. Combustion Science

2.1. Definitions and Terminology. *Higher Heating Value.* The heating value of a fuel is the amount of heat released during its combustion and is expressed in two forms: higher heating value (HHV) or gross calorific value and lower heating value (LHV) or net calorific value. The higher heating value of a fuel is the heat of combustion at constant pressure and temperature (usually ambient) determined by a calorimetric measurement in which the water formed by combustion is completely condensed (1–3). The lower heating value is the similarly measured or defined heat of combustion in the absence of water condensation. The higher heating value is most often used in combustion and flame calculations. The difference between HHV and LHV is numerically equal to the corresponding enthalpy difference, due to latent heat of vaporization. For example, in the complete combustion of methane with O_2 at 298 K, $CH_4(g) + 2\,O_2(g) \longrightarrow CO_2(g) + 2\,H_2O\ (g)$, for one mole of fuel the enthalpy change is $\Delta H = -802.1$ kJ/mol. This value is equal to the lower heating value of methane, and the higher heating value for the above reaction is obtained by adding the heat of vaporization of water at 298 K (44.01 kJ/mol). Thus the higher heating value of methane is $802.1 + 2(44.01) = 890.1$ kJ/mol of CH_4 (212.7 kcal/mol). The evaluation of ΔH for use in thermochemical calculations is more generally performed with the use of tables such as the Joint Army Navy Air Force (JANAF) Tables.

FAR or AFR. The composition of a mixture of fuel and air or oxidant is often specified according to the Fuel to Air Ratio (FAR), and can be expressed on a mass, molar, or volume basis. The FAR is normalized to the stoichiometric composition by defining the equivalence ratio ϕ as in equation 1, where

m_f = mass of fuel, kg; and m_o = mass of oxidizer, kg.

$$\phi = \frac{\left(\dfrac{m_f}{m_o}\right)}{\left(\dfrac{m_f}{m_o}\right)_{stoich}} \tag{1}$$

If $\phi < 1$, the mixture is said to be lean (in fuel) and the products of combustion contain unreacted or excess O_2. If $\phi > 1$ the mixture is said to be rich (in fuel) and the products of combustion contain CO and possibly H_2 because of incomplete combustion caused by the oxygen deficiency (1–3).

Flammability Limits.   Any given mixture of fuel and oxidant is flammable (explosive) within two limits referred to as the upper (rich) and lower (lean) limits of flammability (Fig. 1). Most mixtures are flammable when the fuel to air volume ratio lies between 50 and 300% of the stoichiometric ratio (2). The stoichiometric ratio is the exact theoretical ratio of fuel to air required for complete combustion. The flammability region widens with increasing temperature, and usually with increasing pressure, although the effect of pressure is less predictable. The region within the flammability limits of a fuel–air mixture can be divided into two subregions, the slow oxidation region and the explosion region, separated by the spontaneous ignition temperature. To determine this temperature for liquid fuels, standard tests are used in which the liquid fuel is dropped into an open-air container heated to a known temperature. The lowest temperature at which visible or audible evidence of combustion is observed is defined as the spontaneous ignition temperature (1).

Flash Point.   As fuel oil is heated, vapors are produced which at a certain temperature "flash" when ignited by an external ignition source. The flash point is the lowest temperature at which vapor, given off from a liquid, is in sufficient quantity to enable ignition to take place. The flash point is in effect a measure of the volatility of the fuel. The measurement of flash point for pure liquids is relatively straightforward. However, the measured value may depend slightly on the

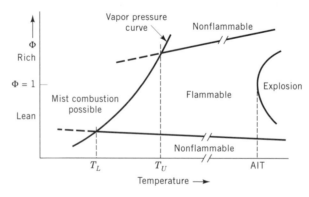

Fig. 1.   Effect of temperature on limits of flammability of a pure liquid fuel in air, where T_L = lean (or lower) flash point; T_U = rich (or upper) flash point; and AIT = auto ignition temperature (4,5).

method used, especially for liquid mixtures, since the composition of the vapor evolved can vary with the heating rate. Special problems arise when the liquid contains a mixture of fuels and/or fuels and inhibitors. In these cases the vapor composition will be different from the liquid composition. Consequently, successive tests with the same sample can lead to erroneous results because the composition of the liquid in the sample holder changes with time as a result of fractional distillation (4–7).

Ignition. To understand the phenomenon of ignition it is necessary to consider the following concepts: ignition source, gas temperature, flame volume, and presence of quench wall surfaces. In general, there are two main methods of igniting a flammable mixture. In the self-ignition method, the mixture is heated slowly so that the vapor released as the temperature is raised ignites spontaneously at a particular temperature. In the forced ignition method, a small quantity of combustible mixture is heated by an external source and the heat released during the combustion of this portion results in propagation of a flame. The external ignition source can be an electric spark, pilot flame, shock wave, etc. For ignition to take place the following conditions should be satisfied: (1) the amount of energy supplied by the ignition source should be large enough to overcome the activation energy barrier; (2) the energy released in the gas volume should exceed the minimum critical energy for ignition; and (3) the duration of the spark or other ignition source should be long enough to initiate flame propagation, but not too long to affect the rate of propagation. Ignition models fall into two categories: the thermal model explains the ignition as resulting from supplying the mixture with the amount of heat sufficient to initiate reaction. In the chemical diffusion model the main role of the ignition source is attributed to the formation of a large number of free radicals in the preheat zone, where their diffusion to the surrounding region initiates the combustion process. The thermal model is applied more widely in the literature and shows better agreement with experimental data.

At ambient temperatures, reaction rates for gaseous mixtures of fuel and oxidant are extremely slow. As temperature is increased gradually, slow oxidation begins and as a result of the exothermic reactions that occur, the temperature keeps increasing. With further temperature increase, the reaction rate suddenly increases causing rapid combustion reactions to occur. This is providing, of course, that the rate of heat release is greater than the rate of heat loss through the container walls. The temperature at which the heat released by the reaction exceeds the heat loss is commonly referred to as the ignition temperature. The spontaneous ignition temperature (SIT), on the other hand, is the lowest temperature at which ignition occurs. Increasing the pressure results in a decrease in the spontaneous ignition temperature (1) (Fig. 2 and Table 1). This temperature is highly dependent on the material of construction, apparatus configuration, and test procedure, therefore reported test values vary widely (2,6,7).

From the chemical diffusion model standpoint, the usual values of ignition temperatures of mixtures often do not correlate well with other flame properties. Nevertheless, the following highly simplified, even speculative, qualitative description may be useful in thinking about flames. Early in the ignition or induction period, the attack of fuel by O_2 is initially slow, but generates free radicals (OH, H, O, HO_2, hydrocarbons) and other intermediate species (CO, H_2, and

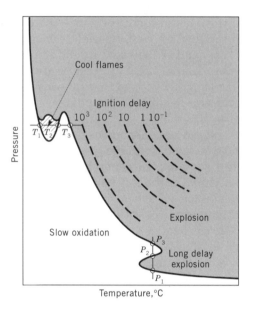

Fig. 2.   Slow oxidation, spontaneous ignition, and explosion as a function of pressure and temperature variations in hydrocarbon mixtures (1).

partial oxidation and decomposition products of hydrocarbons, if present). For some time, there may be little or no temperature rise, the energy essentially being stored in the free radicals. In this stage the reactions may be similar to those that occur in very slow and nearly isothermal oxidation and cool flames observed in very rich hydrocarbon–O_2 mixtures, usually at lower temperatures and pressures. These reactions often stop at the production of stable oxidation and decomposition products, such as aldehydes, peroxides, lower hydrocarbons, etc, but under some conditions may lead to explosions. In a variety of chain

Table 1. **Spontaneous Ignition Temperatures**[a]

Fuel	SIT, K
propane	767
butane	678
pentane	558
hexane	534
heptane	496
octane	491
nonane	479
decane	481
hexadecane	478
isooctane	691
kerosine (JP-8 or Jet A)[b]	501
JP-3	511
JP-4	515
JP-5	506

[a]Ref. 1.
[b]See AVIATION AND OTHER GAS TURBINE FUELS.

reactions, the fuel and any intermediates are rapidly attacked by radicals, some of which are also undergoing very fast chain-branching reactions. Evidence for the behavior of such free radicals is provided by studies of the mechanism of H_2/O_2 and similar explosions. These free radicals can multiply rapidly and attain high transient concentrations, initiating chain reactions that can eventually lead to the explosion of the flammable mixture. They are eventually consumed to form stable species by three-body recombination reactions in which most of the heat release occurs. An example of such a reaction, where M is any atom or molecule in the gas, is represented by

$$H + OH + M \longrightarrow H_2O + M \qquad \Delta H = -498 \text{ kJ/mol (119 kcal/mol)}$$

Numerous other reactions of this type, including a sequence involving HO_2, exist and the relative importance of each reaction depends on the mixture composition. The rates of these termolecular processes increase with increasing pressure but have little or no temperature dependence.

When the partial pressures of the radicals become high, their homogeneous recombination reactions become fast, the heat evolution exceeds heat losses, and the temperature rise accelerates the consumption of any remaining fuel to produce more radicals. Around the maximum temperature, recombination reactions exhaust the radical supply and the heat evolution rate may not compensate for radiation losses. Thus the final approach to thermodynamic equilibrium by recombination of OH, H, and O, at concentrations still many times the equilibrium value, is often observed to occur over many milliseconds after the maximum temperature is attained, especially in the products of combustion at relatively low (<2000 K) temperatures.

The radicals and other reaction components are related by various equilibria, and hence their decay by recombination reactions occurs in essence as one process on which the complete conversion of CO to CO_2 depends. Therefore, the hot products of combustion of any lean hydrocarbon flame typically have a higher CO content than the equilibrium value, slowly decreasing toward the equilibrium concentration (CO afterburning) along with the radicals, so that the oxidation of CO is actually a radical recombination process.

In most hydrocarbon flames there are many radical consumption reactions that interfere with the chain-branching reactions, and the peak radical concentrations are therefore always lower than in analogous flames of H_2, H_2/CO, or moist CO. Consequently, the overall reaction rates for hydrocarbon flames at a given temperature are lower than the overall reaction rates for H_2, H_2/CO, or moist CO flames. In general, variations in the concentrations of such radicals are believed to interfere with the flame chemistry and can account for inhibition of flames by various additives, notably halogen-containing substances. Such additives can narrow flammability limits and reduce burning velocities, even when they cause little or no reduction in the ignition temperature. Those additives containing bromine are particularly effective, and brominated organic compounds are extensively used in extinguishing devices to suppress unwanted diffusion flames (fires). Although most of the chemistry summarized here has evolved from the study of flames and explosions in premixtures, it also applies semiquantitatively to many nonpremixed systems.

The fundamental parameters in the two main methods of achieving ignition are basically the same. Recent advances in the field of combustion have been in the development of mathematical definitions for some of these parameters. For instance, consider the case of ignition achieved by means of an electric spark, where electrical energy released between electrodes results in the formation of a plasma in which the ionized gas acts as a conductor of electricity. The electrical energy liberated by the spark is given by equation 2 (1), where V = the potential, V; I = the current, A; Θ = the spark duration, s; and t = time, s.

$$E = \int_0^Q VI \, dt \qquad (2)$$

The electrical energy is rapidly transformed into thermal energy, and because the temperature of the ionized gas is generally above 300 K, the ignition delay time is short compared with the spark duration, Θ. Ignition only takes place if the electrical energy exceeds the critical value, E_c, and if this energy is liberated within a critical volume, v_c (1,8,10,11).

Ignition can also be produced by a heated surface. During the process of heat transfer from a hot surface to a flammable mixture, reactions are initiated as the temperature rises and the combination of additional heat transfer from the surface and heat release by chemical reactions can lead to ignition of the mixture.

In experiments in which a heated cylinder was introduced into a propane–air mixture, it was found that approximately 11 μs were required for ignition. During this time, heat released by chemical reactions is negligible. This time interval before ignition is referred to as the ignition delay or induction time and is a function of the physical properties of the fuel, the rate of heat conduction, and the pressure of the combustion chamber (1,3,8,10,11). Ignition is characterized by a rapid change in temperature (eq. 3), where t_{ignition} = ignition delay time, s; E_a = activation energy, J/mol; R = universal gas constant, J/(K·mol); and T_i = ignition temperature, K.

$$t_{\text{ignition}} \, \alpha \, \exp\left(\frac{E_a}{RT_i}\right) \qquad (3)$$

For liquid fuels, ignition delay times are of the order 50 μs at 700 K and 10 μs at 800 K. At low temperatures most of the ignition delay is the result of slow, free-radical reactions, and a distinction between the initiation and explosion periods within the ignition delay time can be made. With increasing ignition temperature for a given mixture, these times become comparable and at temperatures as high as 1500 K, both times may be of the order of 10^{-4} s. Consequently, the reaction zone in the flame of a mixture is observed to be one continuous event (12–14).

Another important concept is that of the critical ignition volume. During the propagation of the combustion wave, the flame volume cannot continually grow beyond a critical value without an additional supply of energy. The condition that controls the critical volume for ignition is reached when the rate of increase of flame volume is less than the rate of increase of volume of the combustion products. In this condition a positive exchange of heat between the flame and the fresh mixture is achieved.

For a point spark source, the flame volume is initially spherical and the critical ignition volume is determined by calculating the rate of change of flame volume with respect to radius compared to the rate of change of volume of the combustion products (eq. 4)

$$\frac{d}{dr}\left(\frac{4}{3}\pi\left((r+e)^3 - r^3\right)\right) \le \frac{d}{dr}\left(\frac{4}{3}\pi r^3\right) \tag{4}$$

where r = radius of flame, m; e = thickness of flame front, m; and r_c = radius of the critical spherical volume for ignition, m. This gives equations 5 or 6.

$$2\,er + e^2 \le r^2 \tag{5}$$

$$r_c = e + \sqrt{2} \tag{6}$$

For a line spark source, the flame volume is initially cylindrical with the cylinder length equal to the separation distance between the electrodes. Thus, for a cylindrical flame, $r_c = e$, and the critical ignition volumes are equation 7 for a spherical flame and equation 8 for a cylindrical flame where v_c = critical ignition volume, m³/kg; e = thickness of flame front, m; and d = flame height, m.

$$v_c = \frac{4\pi e^3 \left(1 + \sqrt{2}\right)^3}{3} \tag{7}$$

$$v_c = \pi e^2 d \tag{8}$$

In order to control and monitor the amount of energy required to achieve ignition, the concept of the minimum ignition energy as the smallest quantity of energy that will ignite a mixture is defined (15–17). The minimum ignition energy is calculated using the assumption that sufficient energy must be supplied from the exterior to the critical volume, v_c, to raise the mixture from its initial temperature, T_i, to the flame temperature, $T_f(1)$. The critical energy is given by equation 9: where E = critical energy, J; v_c = critical ignition volume, m³/kg; ρ = density, kg/m³; c_p = specific heat capacity at constant pressure, J/(kg · K); T_f = flame temperature, K; and T_i = initial temperature, K.

$$E = v_c\,\rho\,c_p\left(T_f - T_i\right) \tag{9}$$

Because of heat being transferred to the exterior of the critical volume, only a fraction of the total spark energy, E, is utilized as ignition energy.

Minimum ignition energies are determined experimentally by means of a combustion bomb with two electrodes. Both the ignition energy and the quenching distance can be varied in this device. By using high performance (usually air-gap) condensers, the majority of the stored energy will appear in the spark gap.

The stored energy can be calculated using equation 10 (1,2,9), where E = total spark energy released/stored, J; C = capacity of the condenser; and V = voltage just before the spark is passed through the gas, V.

$$E = \frac{CV^2}{2} \tag{10}$$

The spark must always be produced by a spontaneous breakdown of the gas because an electronic firing circuit or a trigger electrode would either obviate the measurement of spark energy or grossly change the geometry of the ignition source (1,2,9).

Once ignition has been achieved, flame propagation can be controlled. This is achieved by the use of the concept of the quenching distance, defined as the minimum orifice diameter, wall separation, or mesh spacing just sufficient to prevent flame propagation (Fig. 3). Experiments carried out on spark ignition of mixtures between parallel plates show that, with sufficiently powerful sparks, a flame develops in the immediate neighborhood of the spark but does not propagate through the mixture unless the plates are separated by more than the quenching distance (1,2).

Cool Flames.   Under particular conditions of pressure and temperature, incomplete combustion can result in the formation of intermediate products such as CO. As a result of this incomplete combustion, flames can be less exothermic than normal and are referred to as cool flames. An increase in the pressure or

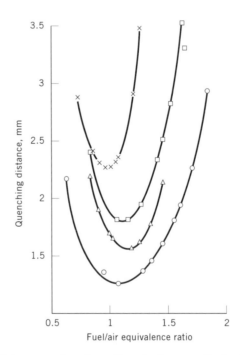

Fig. 3. Quenching distance as function of equivalence ratio for hydrocarbon mixtures with air (1), where x = methane, □ = propane, △ = propylene, and ○ = ethylene.

temperature of the mixture outside the cool flame can produce normal sponta-neous ignition (1).

Flame Temperature. The adiabatic flame temperature, or theoretical flame temperature, is the maximum temperature attained by the products when the reaction goes to completion and the heat liberated during the reaction is used to raise the temperature of the products. Flame temperatures, as a func-tion of the equivalence ratio, are usually calculated from thermodynamic data when a fuel is burned adiabatically with air. To calculate the adiabatic flame temperature (AFT) without dissociation, for lean to stoichiometric mixtures, complete combustion is assumed. This implies that the products of combustion contain only carbon dioxide, water, nitrogen, oxygen, and sulfur dioxide.

Actual temperatures in practical flames are lower than calculated values as a result of the heat losses by radiation, thermal conduction, and diffusion. At high temperatures, dissociation of products of combustion into species such as OH, O, and H reduces the theoretical flame temperature (7). Increasing the pres-sure tends to suppress dissociation of the products and thus generally raises the adiabatic flame temperature (4).

Adiabatic flame temperatures agree with values measured by optical tech-niques, when the combustion is essentially complete and when losses are known to be relatively small. Calculated temperatures and gas compositions are thus extremely useful and essential for assessing the combustion process and predict-ing the effects of variations in process parameters (4). Advances in computational techniques have made flame temperature and equilibrium gas composition calcu-lations, and the prediction of thermodynamic properties, routine for any fuel-oxidizer system for which the enthalpies and heats of formation are available or can be estimated.

Flame Types and Their Characteristics. There are two main types of flames: diffusion and premixed. In diffusion flames, the fuel and oxidant are separately introduced and the rate of the overall process is determined by the mixing rate. Examples of diffusion flames include the flames associated with can-dles, matches, gaseous fuel jets, oil sprays, and large fires, whether accidental or otherwise. In premixed flames, fuel and oxidant are mixed thoroughly prior to combustion. A fundamental understanding of both flame types and their struc-ture involves the determination of the dimensions of the various zones in the flame and the temperature, velocity, and species concentrations throughout the system.

The development of combustion theory has led to the appearance of several specialized asymptotic concepts and mathematical methods. An extremely strong temperature dependence for the reaction rate is typical of the theory. This makes direct numerical solution of the equations difficult but at the same time accurate. The basic concept of combustion theory, the idea of a flame moving at a constant velocity independent of the ignition conditions and determined solely by the properties and state of the fuel mixture, is the product of the asymptotic approach (18,19). Theoretical understanding of turbulent combustion involves combining the theory of turbulence and the kinetics of chemical reactions (19–23).

Laminar Premixed Flames. The structure of a one-dimensional premixed flame is well understood (1). By coupling the rate of heat release from chemical reaction and the rate of heat transfer by conduction with the flow of the unburned mixture, an observer moving with the wave would see a steady laminar flow of unburned gas at a uniform velocity, S_u, into the stationary wave or flame. Hence S_u is defined as the burning velocity of the mixture based on the conditions of the unburned gas. The thickness of the preheating zone and the equivalent reaction zone is found to be inversely proportional to the burning velocity. By considering the heat release from the chemical reaction, it is possible to calculate the thickness of the effective reaction zone. For example, at atmospheric pressure, for hydrocarbon flames, the thicknesses of the preheat and reaction zones are found to be 0.7 and 0.2 mm, respectively (1–3). A propagating, premixed flame is in essence a thin wave in which the temperature rise is continuous and rapid. The thicknesses of the wave depend on ϕ and on pressure; for a typical hydrocarbon–air mixture at $\phi = 1$ at 101.3 kPa (1 atm), the entire rise, from the initial temperature to the adiabatic temperature, occurs in 0.01 mm, and the concentration and temperature gradients are accordingly very steep.

S_u has been approximated for flames stabilized by a steady uniform flow of unburned gas from porous metal diaphragms or other flow straighteners. However, in practice, S_u is usually determined less directly from the speed and area of transient flames in tubes, closed vessels, soap bubbles blown with the mixture, and, most commonly, from the shape of steady Bunsen burner flames. The observed speed of a transient flame usually differs markedly from S_u. For example, it can be calculated that a flame spreads from a central ignition point in an unconfined explosive mixture such as a soap bubble at a speed of $(\rho_o/\rho_b)S_u$, in which the density ratio across the flame is typically 5–10. Usually, the expansion of the burning gas imparts a considerable velocity to the unburned mixture, and the observed speed will be the sum of this velocity and S_u.

By applying the conservation equations of mass and energy and by neglecting the small pressure changes across the flame, the thickness of the preheating and reaction zones can be calculated for a one-dimensional flame (1).

There are a number of sources of instability in premixed combustion systems (23,24).

1. System instabilities involve interactions between flows in different parts of a reacting system. Although these instabilities can cause turbulence, they cannot be analyzed on their own and the whole system should be considered as a unit (11,25).

2. Acoustic instabilities involve the interactions of acoustic waves with the combustion processes. These instabilities are of high frequency and can be significant in certain situations but can successfully be avoided by suitable design (11,24).

3. Taylor instabilities involve effects of buoyancy or acceleration in fluids with variable density; a light fluid beneath a heavy fluid is unstable by the Taylor mechanism. The upward propagation of premixed flames in tubes is subject to Taylor instability (11).

4. Landau instabilities are the hydrodynamic instabilities of flame sheets that are associated neither with acoustics nor with buoyancy but instead involve only the density decrease produced by combustion in incompressible flow. The mechanism of Landau instability is purely hydrodynamic. In principle,

Landau instabilities should always be present in premixed flames, but in practice they are seldom observed (26,27).

5. Diffuse-thermal instabilities involve the relative diffusion reactants and heat within a laminar flame. These are the smallest-scale instabilities (11).

Laminar flame instabilities are dominated by diffusional effects that can only be of importance in flows with a low turbulence intensity, where molecular transport is of the same order of magnitude as turbulent transport (28). Flame instabilities do not appear to be capable of generating turbulence. They result in the growth of certain disturbances, leading to orderly three-dimensional structures which, though complex, are steady (1,2,8,9).

Turbulent Premixed Flames. Combustion processes and flow phenomena are closely connected and the fluid mechanics of a burning mixture play an important role in forming the structure of the flame. Laminar combusting flows can occur only at low Reynolds numbers, defined as

$$RE = \frac{\rho u d}{\mu} \tag{11}$$

where, ρ = density, kg/m^3; u = velocity, m/s; d = diameter, m; and μ = kinematic viscosity, kg/ms. When $Re > Re_{cr}$ the laminar structure of a flow becomes unstable and when the Reynolds number exceeds the critical value by an order of magnitude, the structure of the flow changes. Along with this change of structure from an orderly state to a more chaotic state, the following parameters begin to fluctuate randomly: velocity, pressure, temperature, density, and species concentrations. Overall, with increasing Reynolds number, laminar flow becomes unstable as a result of these fluctuations and breaks down into turbulent flow. Laminar and turbulent flames differ greatly in appearance. For example, while the combustion zone of a laminar Bunsen flame is a smooth, delineated and thin surface, the analogous turbulent combustion region is blurred and thick (1).

Turbulent flame speed, unlike laminar flame speed, is dependent on the flow field and on both the mean and turbulence characteristics of the flow, which can in turn depend on the experimental configuration. Nonstationary spherical turbulent flames, generated through a grid, have flame speeds of the order of or less than the laminar flame speed. This turbulent flame speed tends to increase proportionally to the intensity of the turbulence.

In high speed dusted, premixed flows, where flames are stabilized in the recirculation zones, the turbulent flame speed grows without apparent limit, in approximate proportion to the speed of the unburned gas flow. In the recirculation zones the intensity of the turbulence does not affect the turbulent flame speed (1).

In the reaction zone, an increase in the intensity of the turbulence is related to the turbulent flame speed. It has been proposed that flame-generated turbulence results from shear forces within the burning gas (1,28). The existence of flame-generated turbulence is not, however, universally accepted, and in unconfined flames direct measurements of velocity indicate that there is no flame-generated turbulence (1,2).

The balanced equation for turbulent kinetic energy in a reacting turbulent flow contains the terms that represent production as a result of mean flow shear, which can be influenced by combustion, and the terms that represent mean flow dilations, which can remove turbulent energy as a result of combustion. Some of the discrepancies between turbulent flame propagation speeds might be explained in terms of the balance between these competing effects.

To analyze premixed turbulent flames theoretically, two processes should be considered: (*1*) the effects of combustion on the turbulence, and (*2*) the effects of turbulence on the average chemical reaction rates. In a turbulent flame, the peak time-averaged reaction rate can be orders of magnitude smaller than the corresponding rates in a laminar flame. The reason for this is the existence of turbulence-induced fluctuations in composition, temperature, density, and heat release rate within the flame, which are caused by large eddy structures and wrinkled laminar flame fronts.

A unified statistical model for premixed turbulent combustion and its subsequent application to predict the speed of propagation and the structure of plane turbulent combustion waves is available (29–32).

Laminar Diffusion Flames.   Generally, a diffusion flame is defined as one in which no mixing of the fuel and oxidant takes place prior to emission from the burner. However, it can also be defined as a flame in which the mixing rate is sufficiently slow compared to the reaction rate, that the mixing time controls the burning rate. Since a continuous spectrum of flames exists between the perfectly premixed flame and the diffusion flame, the term diffusion flame is reserved for those flames in which there is a total separation between fuel and oxidant (1,5). Some of the practical advantages of diffusion flames include high flame stability, safety (as there is no need for the storage of the combustible mixture), intense radiation and hence high heat exchange with the surroundings. It is difficult to give a general treatment of diffusion flames largely because no simple measurable parameter, analogous to the burning velocity for premixed flames, can be used to characterize the burning process (3).

Many different configurations of diffusion flames exist in practice (Fig. 4). Laminar jets of fuel and oxidant are the simplest and most well understood diffusion flames. They have been studied exclusively in the laboratory, although a complete description of both the transport and chemical processes does not yet exist (2).

The discussion of laminar diffusion flame theory addresses both the gaseous diffusion flames and the single-drop evaporation and combustion, as there are some similarities between gaseous and liquid diffusion flame theories (2). A frequently used model of diffusion flames has been developed (34), and despite some of the restrictive assumptions of the model, it gives a good description of diffusion flame behavior.

The Displacement Distance theory suggests that since the structure of the flame is only quantitatively correct, the flame height can be obtained through the use of the displacement length or "displacement distance" (35,36) (eq. 12), where h = flame height, m; V = volumetric flow rate, m^3/s; and D = diffusion coefficient.

$$h = \frac{V}{2\pi D} \tag{12}$$

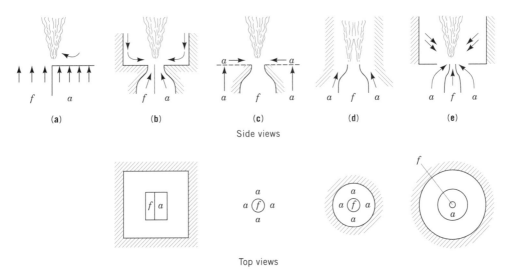

Fig. 4. Various configurations (**a–e**) used to obtain gaseous diffusion flames where a = air and f = fuel (33).

In addition to the Burke and Schumann model (34) and the Displacement Distance theory, a comprehensive laminar diffusion flame theory can be written using the equations of conservation of species, energy, and momentum, including diffusion, heat transfer, and chemical reaction.

Evaporation and burning of liquid droplets are of particular interest in furnace and propulsion applications and by applying a part of the Burke and Schumann approach it is possible to obtain a simple model for diffusion flames.

Combustion chemistry in diffusion flames is not as simple as is assumed in most theoretical models. Evidence obtained by adsorption and emission spectroscopy (37) and by sampling (38) shows that hydrocarbon fuels undergo appreciable pyrolysis in the fuel jet before oxidation occurs. Further evidence for the existence of pyrolysis is provided by sampling of diffusion flames (39). In general, the preflame pyrolysis reactions may not be very important in terms of the gross features of the flame, particularly flame height, but they may account for the formation of carbon while the presence of OH radicals may provide a path for NO_x formation, particularly on the oxidant side of the flame (39).

Combustion chemistry in diffusion flames also accounts for the smoke formation process. Characteristic behavior of smoking diffusion flames includes: (1) the first appearance of smoke is generally at the flame tip, the width of the smoke trail increasing as the amount of smoke increases; (2) the appearance and quantity of smoke increases with increasing flame height; (3) the smoking tendency changes with oxidant flow rate and the oxygen content of the oxidant stream; (4) the smoking tendency is a function of fuel type; and (5) the smoking tendency generally increases with increasing pressure (39).

Diffusion Flames in the Transition Region. As the velocity of the fuel jet increases in the laminar to turbulent transition region, an instability develops at the top of the flame and spreads down to its base. This is caused by the shear forces at the boundaries of the fuel jet. The flame length in the transition region is usually calculated by means of empirical formulas of the form (eq. 13): where l = length of the flame, m; r = radius of the fuel jet, m; v = fuel flow velocity, m/s; and C_1 and C_2 are empirical constants.

$$l = \frac{r}{C - \dfrac{C_2}{v}} \tag{13}$$

Turbulent Diffusion Flames. Laminar diffusion flames become turbulent with increasing Reynolds number (1,2). Some of the parameters that are affected by turbulence include flame speed, minimum ignition energy, flame stabilization, and rates of pollutant formation. Changes in flame structure are believed to be controlled entirely by fluid mechanics and physical transport processes (1,2,9).

Consider the case of the simple Bunsen burner. As the tube diameter decreases, at a critical flow velocity and at a Reynolds number of about 2000, flame height no longer depends on the jet diameter and the relationship between flame height and volumetric flow ceases to exist (2). Some of the characteristics of diffusion flames are illustrated in Figure 5.

The preferred models for predicting the behavior of turbulent-free shear layers involve the solution of the turbulent kinetic energy equation in order to obtain the local turbulent shear stress distribution (1,2,9). These models are ranked according to the number of simultaneous differential equations that need to be solved. The one-equation model considers the turbulent kinetic energy equation alone, whereas the two-equation model considers the turbulent kinetic energy equation plus a differential equation for the turbulence length scale, or equivalently, the dissipation rate for turbulent kinetic energy. These equations are solved along with the conservation equations (momentum, energy, and species) to model turbulent flows.

The physics and modeling of turbulent flows are affected by combustion through the production of density variations, buoyancy effects, dilation due to heat release, molecular transport, and instability (1–3,5,8). Consequently, the conservation equations need to be modified to take these effects into account. This modification is achieved by the use of statistical quantities in the conservation equations. For example, because of the variations and fluctuations in the density that occur in turbulent combustion flows, density weighted mean values, or Favre mean values, are used for velocity components, mass fractions, enthalpy, and temperature. The turbulent diffusion flame can also be treated in terms of a probability distribution function (pdf), the shape of which is assumed to be known a priori (1).

In general, comprehensive, multidimensional modeling of turbulent combustion is recognized as being difficult because of the problems associated with solving the differential equations and the complexities involved in describing the interactions between chemical reactions and turbulence. A number of computational models are available commercially that can do such work. These include FLUENT, FLOW-3D, and PCGC-2.

The various studies attempting to increase our understanding of turbulent flows comprise five classes: moment methods disregarding probability density functions, approximation of probability density functions using moments,

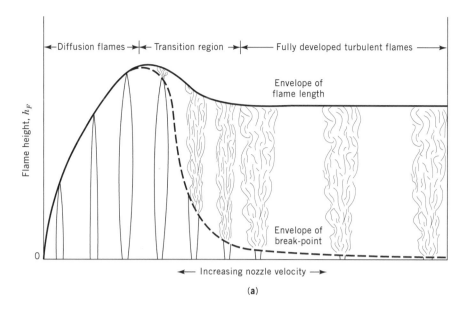

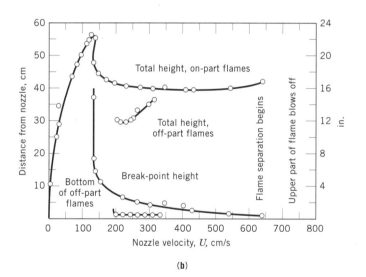

Fig. 5. Effects of nozzle velocity on flame appearance in laminar and turbulent flow: (**a**), flame appearance; (**b**), flame height and break-point height (40).

calculation of evolution of probability density functions, perturbation methods beginning with known structures, and methods identifying coherent structures. For a thorough review of turbulent diffusion flames see References (41–48).

2.2. Fundamentals of Heterogeneous Combustion.   The discussion of combustion fundamentals so far has focused on homogeneous systems. Heterogeneous combustion is the terminology often used to refer to the combustion of liquids and solids. From a technological viewpoint, combustion of liquid hydrocarbons, mainly in sprays, and coal combustion are of greatest interest.

Most theories of droplet combustion assume a spherical, symmetrical droplet surrounded by a spherical flame, for which the radii of the droplet and the flame are denoted by r_d and r_p, respectively. The flame is supported by the fuel diffusing from the droplet surface and the oxidant from the outside. The heat produced in the combustion zone ensures evaporation of the droplet and consequently the fuel supply. Other assumptions that further restrict the model include: (1) the rate of chemical reaction is much higher than the rate of diffusion and hence the reaction is completed in a flame front of infinitesimal thickness; (2) the droplet is made up of pure liquid fuel; (3) the composition of the ambient atmosphere far away from the droplet is constant and does not depend on the combustion process; (4) combustion occurs under steady-state conditions; (5) the surface temperature of the droplet is close or equal to the boiling point of the liquid; and (6) the effects of radiation, thermodiffusion, and radial pressure changes are negligible.

In order to obtain an expression for the burning rate of the droplet, the following parameters are needed: physical constants such as the specific heat, and the thermal conductivity of the droplet, the radius of the flame, and the temperature of the flame. To determine these quantities, heat conduction, diffusion, and the kinetics of the chemical processes associated with droplet combustion need to be analyzed. This is achieved mathematically by solving the equations of mass continuity, mass continuity for components, and the energy equation. The solving of these equations can be facilitated if the following simplifying assumptions are made: the flame surrounding the droplet is a diffusion flame and, by definition, is formed where the fuel and oxidant meet in stoichiometric proportions; the temperature of this flame is very close to the adiabatic flame temperature; and the heat required for evaporation of the droplet and the heat loss to the surroundings through the burned gas are small and can therefore be neglected. These equations are usually solved in spherical coordinates for a one-dimensional case. However, since the flame is relatively thick, and the droplet is relatively small, the one-dimensional model of the process may not be a particularly accurate representation. Nevertheless, the values obtained for burning rates provide useful information (9).

The burning rate of the droplet (kg/s), and its rate of change of radius are related by:

$$m_e = -4\pi r_d^2 \rho_f \frac{dr_d}{dt} \tag{14}$$

where m_e = burning rate of the droplet, kg/s; ρ_f = density of fuel, kg/m^3; and r_d = radius of droplet, m. This equation can be simplified, assuming $r_t/r_d \simeq$ constant and $m_e \simeq r_d$.

$$\frac{d(d_d^2)}{dt} = K = \text{constant} \tag{15}$$

Hence, the constant K is termed the "burning-constant of the droplet." Integration of the equation 15 produces the droplet burning law:

$$d_d^2 = d_0^2 - Kt \tag{16}$$

where d_0 = the initial diameter of the droplet, m, and t = burnout time, s.

The amount of data available on droplet combustion is extensive. However, the results can be easily summarized, because the burning rate constants for the majority of fuels of practical interest fall within the narrow range of 7 to 11×10^{-3} cm^2/s. An increase in oxygen concentration results in an increase in the burning rate constant. If the burning takes place in pure oxygen, the values for burning rate are increased by a factor of about 2.0, compared to when the burning takes place in air (9).

The convective gas flow around a burning particle affects its burning rate. It has been postulated that in the absence of convection, the burning rate is independent of pressure. Forced convection, on the other hand, is believed to increase the burning rate.

During the final stages of the combustion of a droplet, coke remains, and although it represents a relatively small percentage of the mass of the original oil droplet, the time taken for the heterogeneous reaction between the oxygen-depleted combustion air and the coke particle is generally the slowest of all the combustion steps (9).

The reaction between a porous solid, such as a coke sphere, and a gas, such as oxygen, occurs in the following stages: (1) the main reactant species diffuse thoroughly through the boundary layer toward the solid surface and the products of reaction diffuse away from the surface; (2) diffusion and simultaneous chemical reaction take place within the pores of the solid proceeding from the external surface toward the interior, and gaseous products diffuse in the opposite direction; and (3) at the participating surfaces, the reacting gas chemisorbs, some intermediate species are formed, then the final products of the reaction desorb from the surface. Thus the observed reaction rate is a function of the individual resistances–boundary layer diffusion, pore diffusion, and chemical kinetics, and the rate controlling process is the slowest of them or a combination of these processes. Even though a number of gas reactions may take place at the surface of a burning carbonaceous solid, the reaction forming CO is most often assumed, $C + 1/2\,O_2 \longrightarrow CO$. Coke combustion is treated mathematically like char combustion.

In a practical combustion chamber, the droplets tend to burn in the form of sprays, hence it is important to understand the fundamentals of sprays. In the most simple case, a fuel spray suspended in air will support a stable propagating laminar flame in a manner similar to a homogeneous gaseous mixture. In this case, however, two different flame fronts are observed. If the spray is made up of very small droplets they vaporize before the flame reaches them and a continuous flame front is formed. If the droplets are larger, the flame reaches them before the evaporation is complete, and if the amount of fuel vapor is insufficient for the formation of a continuous flame front, the droplets burn in the form of isolated spherical regions. Flames of this type are referred to as heterogeneous laminar flames. Experimental determination of burning rates and flammability limits for heterogeneous laminar flames is difficult because of the motion of droplets caused by gravity and their evaporation before the arrival of the flame front.

In modern liquid-fuel combustion equipment the fuel is usually injected into a high velocity turbulent gas flow. Consequently, the complex turbulent flow and spray structure make the analysis of heterogeneous flows difficult and a detailed analysis requires the use of numerical methods (9).

The combustion of a coal particle occurs in two stages: (1) devolatilization during the initial stages of heating with accompanying physical and chemical changes, and (2) the subsequent combustion of the residual char (49). The burning rate or reactivity of the residual char in the second stage is strongly dependent on the process conditions of the first stage. During pulverized coal combustion the devolatilization step usually takes about 0.1 s and the residual char combustion takes on the order of 1 s. Since char combustion occurs over ~90% of the total burning time, its rate can affect the volume of the combustion chamber required to attain a given heat release and combustion efficiency. The rate of devolatilization, and the amount and nature of volatiles can significantly affect the ignition process and hence the onset of char combustion. During devolatilization or thermal decomposition, rupture of various functional groups bonded to the macromolecular structure leads to the evolution of gases and the formation and opening of pores. Also, depending on its thermoplastic properties, a coal particle may undergo softening and swelling resulting in a change in size and physical characteristics. Diffusion of oxygen to, and within, a char particle depends on its physical structure and accessible surface area (char morphology). Attempts have been made to explain the combustion rates of chars in terms of their morphologies (50–52).

The ignition mechanism is rather complex and is not well understood in terms of actually defining the ignition temperature and reaction mechanisms. The ignition temperature is known, however, not to be a unique property of the coal and depends on a balance between heat generated and heat dissipated to the surroundings around the coal particle. Measuring the ignition characteristics is complicated by the fact that they are strongly dependent on the physical arrangement of the particles, eg, single particle, clouds of coal dust, or coal piles. Reported ignition temperatures range from 303 to 373 K in the case of spontaneous ignition of coal piles at ambient temperature to 1073–1173 K in the case of single coal particle ignition (53,54). Characteristics such as coal type, particle size and distribution of mineral matter, and experimental conditions such as gas temperature, heating rate, oxygen, and coal dust concentration are some of the important factors that influence values obtained for ignition temperatures. Ignition of coal particles can occur either homogeneously, with ignition of the volatiles released and the subsequent ignition of the char surface, or heterogeneously, with both volatiles and the char surface igniting simultaneously. Heating rate and particle size affect the mode of ignition. The early theories of ignition were based on an energy balance per unit volume of reactive mixture (55,56). Later, a generalized theory of flame propagation in laminar coal dust flames was proposed (57). Good reviews in this area include References (58–60).

A variety of techniques has been used to determine ignition temperatures: fixed beds, the crossing point method, the critical air blast method (61), photographic techniques (62,63), entrained flow reactors (64), electric spark ignition (65), luminous glow observations (66), plug flow reactors (67), shock tubes (68), and thermogravimetric analysis (69). The techniques mostly used are constant temperature methods (66,70,71) in which coal particles are introduced into a preheated furnace maintained at a fixed temperature. If ignition of the coal particles is not observed by the appearance of a glow, flame, or sharp temperature rise, the test is then conducted at a higher temperature and the procedure is repeated until the ignition of the sample is observed. It has been a matter of controversy as to whether the value of the critical temperature determined using this method for these low temperature tests represents the ignition tendency or the combustibility of the coal during the high heating rates encountered in a pulverized coal flame.

The structure of residual char particles after devolatilization depends on the nature of the coal and the pyrolysis conditions such as heating rate, peak temperature, soak time at the peak temperature, gaseous environment, and the pressure of the system (72). The oxidation rate of the char is primarily influenced by the physical and chemical nature of the char, the rate of diffusion and the nature of the reactant and product gases, and the temperature and pressure of the operating system. The physical and chemical characteristics that influence the rate of oxidation are chemical structural variations, such as the concentration of oxygen and hydrogen atoms (73–75), the nature and amount of mineral matter (76–79), and physical characteristics such as porosity, particle size, and accessible surface area. The rate of diffusion of the reactant gas is governed by the temperature and pressure of the operating system.

The rate limiting step in the combustion of char is either the chemical kinetics (adsorption of oxygen, reaction, and desorption of products) or diffusion of oxygen (bulk and pore diffusion). Variations in the reaction rate with temperature for gas–carbon reactions have been grouped into three main regions or zones depending on the rate limiting resistance (80) (Figs. 6 and 7).

The overall reaction rate is determined by equation 17,

$$q = \frac{1}{\dfrac{1}{k_{\text{diff}}} + \dfrac{1}{k_s}}\, p_g \tag{17}$$

where q = rate of removal of carbon atoms per unit external surface area, kg/m^{2}s; p_g = partial pressure of oxygen in the free stream, Pa; k_{diff} = diffusional

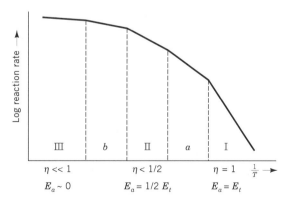

Fig. 6. The three ideal zones (I–III) representing the rate of change of reaction for a porous carbon with increasing temperature where a and b are intermediate zones, E_a is activation energy, and E_t is true activation energy. The effectiveness factor, η, is a ratio of experimental reaction rate to reaction rate which would be found if the gas concentration were equal to the atmospheric gas concentration (80).

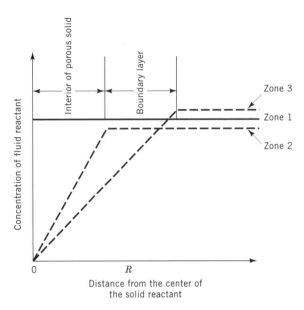

Fig. 7. The concentration of the reacting fluid as a function of the distance from the center of the solid reactant in various reaction zones.

reaction rate coefficient, kg/(m^2 · Pa); and k_s = surface reaction rate coefficient, kg/(m^2 · Pa).

In Zone 1, the low temperature zone, the reaction rate is slow and the concentration of the reactant gas is uniform throughout the interior of the solid. The overall rate is controlled by the chemical reactivity of the solid, the kinetics are not influenced by mass transfer, and the kinetics are the intrinsic kinetics. The measured activation energy is equal to the true activation energy. In a higher temperature zone, Zone 2, the concentration of the gaseous reactant becomes zero at a point somewhere between the surface and the center of the particle, and the reaction rate is controlled by both the chemical kinetics and the diffusion rate of the gaseous reactant. The measured activation energy in this zone is one-half of the true activation energy. At still higher temperatures, in Zone 3, the chemical reactivity is so high that oxygen is consumed as soon as it reaches the surface of the solid and the concentration of the gaseous reactant approaches zero at the surface, indicating that the reaction rate is controlled by bulk diffusion of the reacting gas. The measured activation energy approaches a value of zero in this zone.

Burning times for coal particles are obtained from integrated reaction rates. For larger particles (>100 μm) and at practical combustion temperatures, there is a good correlation between theory and experiment for char burnout. Experimental data are found to obey the Nusselt "square law" which states that the burning time varies with the square of the initial particle diameter ($t_c \sim D_o^{2.0}$). However, for particle sizes smaller than 100 μm, the Nusselt relationship seems to predict higher burning times than those observed and it has been noted that burning times are proportional to the initial diameter of the particles raised to the power of 1.4 ($t_c \sim D_o^{1.4}$) (54).

3. Combustion Technology

Technology addresses the more applied, practical aspects of combustion, with an emphasis on the combustion of gaseous, liquid, and solid fuels for the purpose of power production. In an ideal fuel burning system (1) there should be no excess oxygen or products of incomplete combustion, (2) the combustion reaction should be initiated by the input of auxiliary ignition energy at a low rate, (3) the reaction rate between oxygen and fuel should be fast enough to allow rapid rates of heat release and it should also be compatible with acceptable nitrogen and sulfur oxide formation rates, (4) the solid impurities introduced with the fuel should be handled and disposed of effectively, (5) the temperature and the weight of the products of combustion should be distributed uniformly in relation to the parallel circuits of the heat absorbing surfaces, (6) a wide and stable firing range should be available, (7) fast response to changes in firing rate should be easily accommodated, and (8) equipment availability should be high and maintenance costs low (49–51).

3.1. Combustion of Gaseous Fuels. In any gas burner some mechanism or device (flame holder or pilot) must be provided to stabilize the flame against the flow of the unburned mixture. This device should fix the position of the flame at the burner port. Although gas burners vary greatly in form and complexity, the distribution mechanisms in most cases are fundamentally the same. By keeping the linear velocity of a small fraction of the mixture flow equal to or less than the burning velocity, a steady flame is formed. From this pilot flame, the main flame spreads to consume the main gas flow at a much higher velocity. The area of the steady flame is related to the volumetric flow rate of the mixture by equation 18 (81,82)

$$\dot{V}_{\mathrm{mix}} = A_f S_u \qquad (18)$$

where $\dot{V}_{\mathrm{mix}}$ = volumetric flow rate, m^3/s; A_f = area of the steady flame, m^2; and S_u = burning velocity, m/s.

The volumetric flow rate of the mixture is, in turn, proportional to the rate of heat input (eq. 19):

$$\dot{V}_{\mathrm{mix}} \cdot (HHV) = \dot{Q} \qquad (19)$$

where $\dot{V}_{\mathrm{mix}}$ = volumetric flow rate, m^3/s; HHV = higher heating value of the fuel, J/kg; and $\dot{Q}$ = rate of heat input, J/s.

In the simple Bunsen flame on a tube of circular cross-section, the stabilization depends on the velocity variation in the flow emerging from the tube. For laminar flow (parabolic velocity profile) in a tube, the velocity at a radius r is given by equation 20:

$$v = \mathrm{const}\left(R^2 - r^2\right) \qquad (20)$$

where v = laminar flow velocity, m/s; R = tube radius, m; r = flame radius, m; and const = experimental constant.

The maximum velocity at the axis is twice the average, whereas the velocity at the wall is zero. The effect of the burner wall is to cool the flame locally and decrease the burning velocity of the mixture. This results in flame stabilization. However, if the heat-transfer processes (conduction, convection, and radiation) involved in cooling the flame are somehow impeded, the rate of heat loss is decreased and the local reduction in burning velocity may no longer take place. This could result in upstream propagation of the flame.

To make the flame stable against the flow in a thin annular region near the rim, the flow velocity v_r should be made equal to the burning velocity at some radius r. This annulus serves as a pilot and ignites the main flow of the mixture, ie, the flame gradually spreads toward the center. In most of the mixture flow, $v_r > S_u$ which results in a stable flame. With increasing mixture flow, the height and area of the flame increase. Measurement of the area of a stable Bunsen flame is the basis for the method most commonly used to determine S_u (81–83).

By feeding the mixture through a converging nozzle, the velocity profile may be made nearly flat or uniform. A Bunsen flame in such a flow has a smaller range of stability but the mechanism is essentially the same and the flame very closely approximates a cone. If the apex angle of the flame is Θ, then S_u can be obtained from equation 21

$$S_u = v_r \sin\left(\frac{\Theta}{2}\right) \tag{21}$$

where S_u = burning velocity, m/s; v_r = mixture velocity at the nozzle exit, m/s; and Θ = the apex angle of the flame in degrees.

If the tube diameter is appreciably larger than the quenching distance, S_u will exceed v_r in some parts of the flowing mixture due to a lack of quenching, and the flame will then propagate down the tube as far as there is mixture to consume. This undesirable condition is referred to as flashback. If, on the other hand, v_r exceeds S_u in the mixture flow, the flame lifts from the port and blows off. This condition is referred to as blowoff and like flashback should be avoided (Fig. 8). The velocity gradient at the wall, g_w, is defined as

$$g_w = \left(\frac{dv}{dr}\right)_{r=R}$$

For example, in a laminar or Poiseuille flow in a round tube of radius R, $g_w = 4g_{av}/R$, and for a given initial mixture composition, flashback (or blowoff) will occur at the same value of g_w in tubes of various sizes, whereas the corresponding average velocity at flashback (or blowoff) is proportional to R. Typical velocity gradient values for stoichiometric methane–air flames are at flashback about 400 s^{-1} and at blowoff 2000 s^{-1}. Thus, if the mixture is burned on a 1-cm diameter tube, the average velocity of flashback is $400 \times 0.5/4 = 50$ cm/s and at blowoff the average velocity is 250 cm/s, or the range of stability would be roughly one and a half to seven times the burning velocity. At flashback, g_w is at a maximum around $\phi = 1$. If the burner is operated such that the surrounding inert atmosphere is inert, g_w is at a maximum at blowoff.

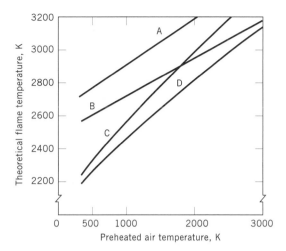

Fig. 8. Flashback and blowoff critical velocity gradients for a natural gas flame (10), where A is $N_2/O_2 = 2$, $p = 1.01$ MPa (10 atm); B, $N_2/O_2 = 2$, $p = 101.3$ kPa (1 atm); C, air, $p = 1.01$ MPa; and D, air, $p = 101.3$ kPa.

The behavior of rich mixtures is complicated by the entrainment of air at the burner port that sustains combustion of hot combustion products of the primary flame near the port. The blowoff velocity is found to increase continuously with ϕ, or richer mixtures are more stable with respect to blowoff. They also have a lesser tendency toward flashback. Hence, a Bunsen flame has more latitude for stable operation if the primary mixture is rich. For this reason many appliance burners that involve assemblies of such flames are routinely adjusted by first making the primary mixture so rich that soot just forms in the burned gas (yellow-tipping), and then increasing the air until the yellow luminosity disappears. The primary equivalence ratio is then perhaps 1.5 or more; the rich products of that primary flame are burned in the secondary diffusion flame in the surrounding air, or the faintly luminous outer mantle of a Bunsen flame.

Most of the commercial gas–air premixed burners are basically laminar-flow Bunsen burners and operate at atmospheric pressure. This means that the primary air is induced from the atmosphere by the fuel flow with which it mixes in the burner passage leading to the burner ports, where the mixture is ignited and the flame stabilized. The induced air flow is determined by the fuel flow through momentum exchange and by the position of a shutter or throttle at the air inlet. Hence, the air flow is a function of the fuel velocity as it issues from the orifice or nozzle, or of the fuel supply pressure at the orifice. With a fixed fuel flow rate, the equivalence ratio is adjusted by the shutter, and the resulting induced air flow also determines the total mixture flow rate. The desired air–fuel volume ratio is usually seven or more, depending on the stoichiometry. Burners of this general type with many multiple ports are common for domestic furnaces, heaters, stoves, and for industrial use. The flame stabilizing ports in such burners are often round but may be slots of various shapes to conform to the heating task.

Atmospheric pressure industrial burners are made for a heat release capacity of up to 50 kJ/s (12 kcal/s), and despite the varied designs, their principle of stabilization is basically the same as that of the Bunsen burner. In some cases the mixture is fed through a fairly thick-walled pipe or casting of appropriate shape for the application and the desired distribution of the flame. The mixture issues from many small and closely spaced drilled holes, typically 1–2 mm diameter, and burns as rows of small Bunsen flames. It may be ignited with a small pilot flame, spark, or heated wire, usually located near the first holes to avoid accumulation of unburned mixture before ignition. The rated heat release for a given fuel–air mixture can be scaled with the size and number of holes. For example, for 2-mm diameter holes it would be 10–100 J/(hole) or, in general, 0.3–3 kJ/cm^{2}s (72–720 cal/cm^{2}s) of port area, depending on the fuel. The ports may also be narrow slots, sometimes packed with corrugated metal strips to improve the flow distribution and reduce the tendency to flashback.

Gas burners that operate at high pressures are usually designed for high mixture velocities and heating intensities and therefore stabilization against blowoff must be enhanced. This can be achieved by a number of methods such as surrounding the main port with a number of pilot ports or using a porous diaphragm screen.

In order to achieve high local heat flux the port velocity of the mixture should be increased considerably. In burners that achieve stabilization using pilot ports, most of the mixture can be burned at a port velocity as high as 100 S_u to produce a long pencil-like flame, suitable for operations requiring a high heat flux.

High local heat flux can also be obtained with Bunsen flames using mixtures with high burning velocities as in H_2/O_2 and C_2H_2/O_2 torches. Their stabilization mechanism is essentially the same as it is for slower burning mixtures but the port or nozzle of the torch is usually much smaller, in part to avoid turbulence in the mixture flow. The consequences of flashback are also much more severe since most such mixtures are detonatable, and the premixing chamber or tube must accordingly be more rugged. For this reason, many large hydrogen–oxygen and hydrocarbon–oxygen flames are not premixed. They actually consist of assemblies of closely spaced diffusion flames, produced from separately fed but contiguous fuel and oxidizer flows. In such surface-mixing burners, the surface is an array of very small and closely packed alternating fuel and oxidizer ports. The arrangement and number of the ports and the complexity of the required manifolding of the reactant passages vary with the application and the desired geometry of the burned gas flow. With a very fast burning fuel-oxidizer combination, the individual diffusion flames may be so short that the assembly approximates a large, flat, premixed flame, as is the case with some rocket engine injectors (84,85). The most frequently used modern burners are the circular type, capable of burning oil and gas.

It is often desired to substitute directly a more readily available fuel for the gas for which a premixed burner or torch and its associated feed system were designed. Satisfactory behavior with respect to flashback, blowoff, and heating capability, or the local enthalpy flux to the work, generally requires reproduction as nearly as possible of the maximum temperature and velocity of the burned gas, and of the shape or height of the flame cone. Often this must be done precisely, and with no changes in orifices or adjustments in the feed system.

If the substitute fuel is of the same general type, eg, propane for methane, the problem reduces to control of the primary equivalence ratio. For nonaspiring burners, ie, those in which the air and fuel supplies are essentially independent, it is further reduced to control of the fuel flow, since the air flow usually constitutes most of the mass flow and this is fixed. For a given fuel supply pressure and fixed flow resistance of the feed system, the volume flow rate of the fuel is inversely proportional to $\sqrt{\rho_f}$. The same total heat input rate or enthalpy flow to the flame simply requires satisfactory reproduction of the product of the lower heating value of the fuel and its flow rate, so that $WI \equiv Q_p/\sqrt{\rho_f}$ remains the same. WI is the Wobbe Index of the fuel gas, and is a commonly used criterion for interchangeability in adjusting the composition of a substitute fuel. The units of WI are variously given, but, if used consistently, are unimportant since only ratios of the Wobbe Indices are ordinarily of interest. Sometimes ρ_f is taken as the specific gravity relative to some reference gas, eg, air, or average molecular weights may be used.

The Wobbe Index criterion also applies to substitution with aspirating or atmospheric pressure burners in which the volume flow of primary air induced by momentum exchange with the fuel increases with $\sqrt{\rho_f}$. Because the volumetric air requirement for a given ϕ is nearly proportional to the heating value of the fuel, an adjustment of Q_p and ϕ_f to the same WI results in about the same stoichiometry of the primary flame. For example, if propane is an available substitute for methane or natural gas, it is common practice to prepare a mixture of approximately 60% propane–40% air (which of course is well above the upper flammability limit) to use as the fuel supply; though the heating value of the mixture is $1.53/\sqrt{2.36} = 1.0$ so that its Wobbe Index is the same. Though there would be slight differences in the stoichiometry of the flame, arising from the air mixed with the propane, and in S_u of the final mixture, substitution with the same supply conditions would be quite satisfactory. On the other hand, if a mixture of the same heating value as that of methane were used (39.2% propane in air) at the same supply pressure, the flame would be much leaner and generally unsatisfactory (84,85).

There are direct substitutions of possible interest that would not be feasible without drastic changes in the feed system or pressure. Thus if the available substitute for natural gas is, eg, a manufactured gas containing much CO, there would almost always be a mismatch of the WIs unless the fuel could be further modified by mixing with some other gaseous fuel of high volumetric heating value (propane, butane, vaporized fuel oil, etc). Moreover, if there are substantial differences in S_u, eg, as a result of the presence of considerable H_2 as well as CO in the substitute gas, the variation in flame height and flashback tendency can also make the substitution unsatisfactory for some purposes, even if the WI is reproduced. Refinements and additional criteria are occasionally applied to measure these and other effects in more complex substitution problems (10,85).

Turbulence in the flow of a premixture flattens the velocity profile and increases the effective burning velocity of the mixture; eg, at a pipe-Reynolds number of 40,000 the turbulent burning velocity is several times the laminar burning velocity and it can be perhaps fifty times larger at very high Reynolds numbers. A turbulent flame is always somewhat noisy, the apparent flame surface becomes diffuse owing to the fluctuations in the actual or flame surface about its average position, and its stability tends to be less predictable. The instantaneous flame surface may be thought of as wrinkled by velocity variations in turbulent flow, or by the average distribution over a greater thickness (or time). Although the resulting enhancement of the mixture consumption rate may be considerable, turbulence is often considered undesirable in Bunsen-type flames. For this and other reasons, a large number of burner ports of small characteristic dimension, rather than a single large port, are frequently used to assure laminar flow to the individual flames. However, turbulence has an essential role in facilitating the mixing of fuel, oxidizer, and flame products, and serves an important function in the various types of flame-stabilizers of practical importance.

3.2. Combustion of Liquid Fuels.

There are several important liquid fuels, ranging from volatile fuels for internal combustion engines to heavy hydrocarbon fractions, sold commercially as fuel oils. The technology for the combustion of liquid fuels for spark-ignition and compression-ignition internal combustion engines is not described here. The emphasis here is primarily on the combustion of fuel oils for domestic and industrial applications.

In general, the combustion of a liquid fuel takes place in a series of stages: atomization, vaporization, mixing of the vapor with air, and ignition and maintenance of combustion (flame stabilization). Recent advances have shown the atomization step to be one of the most important stages of liquid fuels combustion. The main purpose of atomization is to increase the surface area to volume ratio of the liquid. This is achieved by producing a fine spray. The finer the atomization spray the greater the subsequent benefits are in terms of mixing, evaporation, and ignition. The function of an atomizer is twofold: atomizing the liquid and matching the momentum of the issuing jet with the aerodynamic flows in the furnace (86–88). Some of the more common designs of atomizers are shown in Figure 9.

Atomizers for large boiler burners are usually of the swirl pressure jet or internally mixed twin-fluid types, producing hollow conical sprays. Less common are the externally mixed twin-fluid types (89,90).

The principal considerations in selecting an atomizer for a given application are turn-down performance and auxiliary costs. An ideal atomizer would possess all of the following characteristics: ability to provide good atomization over a wide range of liquid flow rates; rapid response to change in liquid flow rate; freedom from flow instabilities; low power requirements; capability for scaling (up or down) to provide desired flexibility; low cost, light weight, easily maintained, and easily removed for servicing; and low susceptibility to damage during manufacture and installation (89,90). There are differences in the structures of the sprays between atomizer types which may affect the rate of mixing of the fuel droplets with the combustion air and, hence, the initial development of a flame.

For distillate fuels of moderate viscosity 30 mm^2/s (= cSt), at ordinary temperatures, simple pressure atomization with some type of spray nozzle is most commonly used. Operating typically with fuel pressure of 700–1000 kPa (7–10 atm) such a nozzle produces a distribution of droplet diameters from 10–150 μm. They range in design capacity of 0.5–10 cm^3/s or more and the

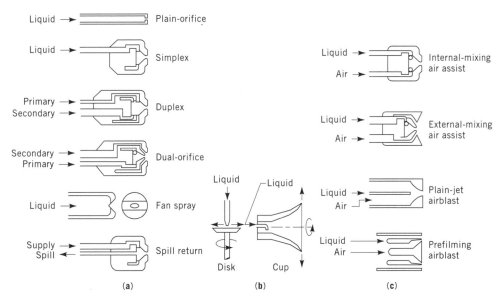

Fig. 9. (**a**) Pressure atomizers; (**b**) rotary atomizer; and (**c**) twin-fluid atomizers (89,90).

pumping power dissipated is generally less than 1% of the corresponding heat release rate. A typical domestic oil burner nozzle uses about 0.8 cm^3/s of No. 2 fuel oil at the design pressure. Although pressure-atomizing nozzles are usually equipped with filters, the very small internal passages and orifices of the smallest units tend to be easily plugged, even with clean fuels. With decreasing fuel pressure the atomization becomes progressively less satisfactory. Much higher pressures are often used, especially in engine applications, to produce a higher velocity of the liquid relative to the surrounding air and accordingly smaller droplets and evaporation times. Other mechanical atomization techniques for the production of more nearly monodisperse sprays or smaller average droplet size (spinning disk, ultrasonic atomizers, etc) are sometimes useful in burners for special purposes and may eventually have more general application, especially for small flows.

Conventional spray nozzles are relatively ineffective for atomizing fuels of high viscosity, such as No. 6 or residual fuel oil (Bunker C) and other viscous, dirty fuels. In order to transfer and pump No. 6 fuel oil, it is usually heated to about 373 K, at which temperature its viscosity is about 40 mm^2/s. Relatively large nozzle passages and orifices are necessary to accommodate the possibility of suspended solids. Atomization of such fuel is often accomplished, or at least assisted, by the use of atomizing air, pumped at high velocity through adjacent passages in or around the liquid injection ports. Much of the relative velocity required to shear the liquid and form droplets is thus provided by the atomizing air; its mass flow is usually comparable with the fuel flow and thus a small fraction of the stoichiometric combustion air, although it is sometimes called primary air. In a typical high pressure, air-atomizing nozzle designed for injecting residual oil in a gas turbine combustor, the atomizing air is supplied by an auxiliary compressor with a power usage of about 1% of the combustion heat release rate.

Dry steam, if available, may also be used in a similar way, as is common practice in the furnaces of power plant boilers using residual oil (89,90).

Air atomization with low pressure and relatively low velocity air is also used in some burners for low viscosity distillate oils. In most aircraft gas turbines some, or even a large part, of the atomization is done in this way by a small fraction of the warm, compressed combustion air supplied in swirling flow around the fuel nozzles. Imparting swirl to at least some of the air flow around fuel injectors of all types is a common feature of many burners and combustors; in some, swirl is introduced on a larger scale in all of the primary combustion air. The velocity gradients or shear in the resulting vortexlike flow promote mass transfer or mixing, including the recirculation of hot products of combustion to the rich mixture or suspension in the low pressure core that contributes to stabilization of the primary combustion zone (25). The angular or swirl velocity imparted to the air and the strength of such flows are of course limited by the available pressure drop; eg, in gas turbine combustors the allowable pressure loss is usually <4% of the absolute pressure.

Combustion of fuel oil takes place through a series of steps, namely vaporization, devolatilization, ignition, and dissociation, which finally lead to attaining the flame temperature. Vaporization and devolatilization of the fine spray of fuel droplets take place as physiochemical processes in the combustion chamber. The vaporization temperature for fuel oil is in the range of 311–533 K, depending on the grade of the fuel. Devolatilization takes place at about 700 K. The final flame temperature attained is between 1366–1918 K. Complete combustion of an oil droplet can occur in 2–20 ms depending on the size of the droplet. A typical characteristic of an oil flame is its bright luminous nature which is the result of incandescent carbon particles in the fuel rich zone.

The study of the combustion of sprays of liquid fuels can be divided into two primary areas for research purposes: single-droplet combustion mechanisms and the interaction between different droplets in the spray during combustion with regard to droplet size and distribution in space (91–94). The wide variety of atomization methods used and the interaction of various physical parameters have made it difficult to give general expressions for the prediction of droplet size and distribution in sprays. The main fuel parameters affecting the quality of a spray are surface tension, viscosity, and density, with fuel viscosity being by far the most influential parameter (95).

The following general expression (eq. 22) is commonly used to describe the droplet size distribution in a spray:

$$dn = ar\,\alpha \exp(-br\,\beta)\,dr \qquad (22)$$

where dn = number of droplets with radii between r and $r + dr$. The constants a, b, α, and β are independent of r and are usually determined empirically. The best-known special case of this general equation is the Rosin-Rammler distribution (95–97).

Theoretical modeling of single-droplet combustion has provided expressions for evaporation and burning times of the droplets and the subsequent coke particles. A more thorough treatment of this topic is available (88,91–93,98).

Experimental techniques used for studying the combustion of single droplets can be divided into three groups: suspended droplets, free droplets, and porous droplets, with ongoing research in all three areas (98).

3.3. Combustion of Solid Fuels.   Solid fuels are burned in a variety of systems, some of which are similar to those fired by liquid fuels. In this article the most commonly burned solid fuel, coal, is discussed. The main coal combustion technologies are fixed-bed, eg, stokers, for the largest particles; pulverized-coal for the smallest particles; and fluidized-bed for medium size particles (99,100) (see COAL).

Fixed-Bed Technology.   Fixed-bed firing of coal by means of stokers consists of a solid bed of large (2–3 cm) coal particles on moving grates with combustion air passing through the grates and ash removal from the end of the grate. The use of a grate limits the application of this technique to small units, as grates are restricted to a maximum size of about 100 m^2 for structural reasons. Mechanical stokers can be classified into the following groups, based on the method of introducing fuel to the furnace: spreader stokers, underfeed stokers, water-cooled vibrating-grate stokers, and chain-grate and traveling-grate stokers. For a thorough review of stokers see Reference 81.

Pulverized-Coal Firing.   This is the most common technology used for coal combustion in utility applications because of the flexibility to use a range of coal types in a range of furnace sizes. Nevertheless, the selection of crushing, combustion, and gas-cleanup equipment remains coal dependent (54,100,101).

Prior to being fed to a pulverized fuel burner, coal is ground to a size generally specified such that at least 70% passes a 200 mesh screen (75 μm) and less than 2% is retained on a 52 mesh screen (300 μm). The top size is determined by the classifying component of the crushing mill, oversize material being retained for further grinding (54, 100, 101).

Suspensions of pulverized coal or coal dust in air can be explosive, hence, it is essential to have adequate guidelines and procedures to ensure safe and stable operation during pulverized-coal (PC) firing. Some of these guidelines include: (*1*) coal dust should never be allowed to accumulate except in specified storage facilities; (*2*) coal and dust suspensions in air should not exist except in drying, pulverizing, or burner equipment and the necessary transportation ducts; (*3*) the furnace and its setting should be purged before introducing any light or spark; (*4*) before introducing the fuel into the furnace, a lighted torch or spark-producing device should be in operation; (*5*) a positive flow of secondary air should be maintained through the burners into the furnace and up the stack; and (*6*) a positive flow of primary air-coal to the burner should be maintained (81,82).

As for oil and gas, the burner is the principal device required to successfully fire pulverized coal. The two primary types of pulverized-coal burners are circular concentric and vertical jet-nozzle array burners. Circular concentric burners are the most modern and employ swirl flow to promote mixing and to improve flame stability. Circular burners can be single or dual register. The latter type was designed and developed for NO$_x$ reduction. Either one of these burner types can be equipped to fire any combination of the three principal fuels, ie, coal, oil and gas. However, firing pulverized coal with oil in the same burner should be restricted to short emergency periods because of possible coke formation on the pulverized-coal element (71,72).

The self-igniting characteristics of pulverized coal vary from one coal to another, but for most coals it is possible to maintain ignition without auxiliary fuel when firing above the capacity of the boiler. The igniters may have to be activated in the following cases: (*1*) when firing pulverized coal with volatile matter less than about 25%, (*2*) when firing excessively wet coal, and (*3*) when feeding coal sporadically into the pulverizer (81,102–104).

Compared to natural gas and oil, complete combustion of coal requires higher levels of excess air, about 15% as measured at the furnace outlet at high loads, and this also serves to avoid slagging and fouling of the heat absorption equipment.

The process of coal ignition in the flame involves a number of steps. Initially, the pulverized coal is heated by convection as the flame jet entrains and mixes with the furnace gases and also by radiation from the hotter furnace gases. On heating to temperatures above about 773 K, the coal starts to decompose, and evolves a mixture of combustible gases such as CO, H_2, and hydrocarbons (C_nH_m) as well as noncombustible gases such as CO_2 and H_2O. At temperatures of about 1173 K most of the volatile matter has been evolved and, given adequate mixing of air in the jet, its combustion will sustain the ignition of the flame. The char residue remaining after the devolatilization is then burned relatively slowly in the flame and furnace. Char combustion has been the subject of intensive investigation since the early 1930s and is one of the least understood areas in coal combustion. Good reviews on char combustion are available (49,105,106). For more information on the industrial applications of coal, References (107–110) are recommended and for a thorough review of coal devolatilization see References (111–114) (see COAL CONVERSION PROCESSES).

As pulverized-coal combustion potentially has a significant impact on the environment, the 1980s saw the employment of techniques such as coal washing and beneficiation to reduce the emissions of fly-ash, SO_x, and water-soluble metallic oxides. Fly-ash emissions can be reduced by means of electrostatic precipitators and fabric fillers, with efficiencies higher than 99.8%. SO_x emissions are reduced considerably by means of gas scrubbing which employs water slurries of lime and limestone. Staged combustion is an effective method of reducing NO_x emissions from coal combustion. In this method, combustion takes place in two zones: a low temperature fuel-rich zone and a high temperature fuel-lean zone, where hydrocarbons and CO are afterburned. In the fuel-rich zone the fuel bound nitrogen is converted to N_2 rather than to NO_x as in the case at high values of ϕ.

The environmental impact associated with pulverized-coal firing has given rise to efforts to develop other combustion technologies such as fluidized beds or the use of coal-water slurry fuels (CWSF), which can be burned as substitutes for certain liquid fuels (115–117). CWSFs were developed as alternatives to more expensive and increasingly scarce conventional hydrocarbon fuels. In their most common form, they are composed of 70–75% by weight of coal (usually high volatile A bituminous), 24–29% water, and 1% chemical additives. The principal potential market for CWSF is as a replacement for residual fuel oil, ie, heavy fuel oil and No. 6 fuel oil, in utility and large industrial boilers. CWSFs offer all the advantages of liquid fuels in addition to the cost advantages associated with the use of coal. The main challenge in the utilization of CWSFs is obtaining stable mixtures that can be successfully atomized and burned. Much research has been carried out in the area of CWSF atomization but this phenomenon is far from being well understood. On the combustion front, novel techniques such as the coupling of a high intensity acoustic field have been employed to enhance the convective processes occurring during the combustion of CWSF (116,117).

Fluidized-Bed Technology. In fluidized-bed combustion of coal, air is fed into the bed at a sufficiently high velocity to levitate the particles. This velocity is referred to as the minimum fluidizing velocity, u_{mf}. At this velocity, the volume occupied by the bed increases abruptly and the bed exhibits some of the characteristics of a fluid. The two predominant designs of fluidized beds are bubbling and recirculating, with most theories of fluidization being based on the simpler bubbling bed concept.

Fluidized combustion of coal entails the burning of coal particles in a hot fluidized bed of noncombustible particles, usually a mixture of ash and limestone. Once the coal is fed into the bed it is rapidly dispersed throughout the bed as it burns. The bed temperature is controlled by means of heat exchanger tubes. Elutriation is responsible for the removal of the smallest solid particles and the larger solid particles are removed through bed drain pipes. To increase combustion efficiency the particles elutriated from the bed are collected in a cyclone and are either re-injected into the main bed or burned in a separate bed operated at lower fluidizing velocity and higher temperature.

Fluidized beds are ideal for the combustion of high sulfur coals since the sulfur dioxide produced by combustion reacts with the introduced calcined limestone to produce calcium sulfate. The chemistry involved can be simplified and reduced to two steps, calcination and sulfation.

$$Calcination$$
$$CaCO_3 \longrightarrow CO_2 + CaO$$

$$Sulfation$$
$$SO_2 + CaO + \tfrac{1}{2} O_2 \longrightarrow CaSO_4$$

The main steps associated with coal combustion (heating, devolatilization, volatiles combustion, and char burnout), occur sequentially to some extent; however, there is always some overlap between the stages. Char burnout is the slowest step so there is practical interest in determining the factors that influence its rate. In order to determine the char combustion rate and time, it is necessary to understand the interaction between the rate of oxygen diffusion to the reacting surface and the inherent chemical kinetics of char oxidation. In the case of fluidized beds the use of a simplified rate coefficient overestimates the burnout time substantially. The enhancement of mass transfer through the boundary layer as the result of an applied velocity must be considered in order to predict char combustion times under conditions relevant to fluidized-bed combustion. Char combustion in fluidized beds is believed to be controlled by both diffusional and chemical kinetic parameters, ie, mixed control. This indicates that models attempting to predict char burnout times in fluidized beds must consider both oxygen diffusion rates and inherent chemical kinetics (117–120).

The main stages of coal combustion have different characteristic times in fluidized beds than in pulverized coal combustion. Approximate times are a few seconds for coal devolatilization, a few minutes for char burnout, several minutes for the calcination of limestone, and a few hours for the reaction of the calcined limestone with SO_2. Hence, the carbon content of the bed is very low (up to 1% by weight) and the bed is 90% CaO in various stages of reaction to $CaSO_4$. About 10% of the bed's weight is made up of coal ash (91). This distribution of 90/10 limestone/coal ash is not a fixed ratio and is dependent on the ash content of the coal and its sulfur content.

Devolatilization and combustion occur close to the coal inlet tubes. However, because of rapid mixing in the bed the composition of the solids in the bed may be assumed to be uniform.

Atmospheric Pressure Fluidized-Bed Boilers. A typical bubbling fluidized-bed is usually 1.2 m deep in its expanded or fluidized condition. Normally, the heat-transfer surface is placed in the bed in the form of a tube handle to achieve the desired heat balance and bed operating temperature. For fuels with low heating values the amount of surface can be minimal or absent. Coal-fired bubbling-bed boilers normally incorporate a recycle system that separates the solids leaving the economizer from the gas and recycles them to the bed. This maximizes combustion efficiency and sulfur capture. Normally, the amount of solids reacted is limited to about 25% of the combustion gas weight. For highly reactive fuels this recycle system can be omitted. Bubbling-beds that burn coal usually operate in the range of 2.4 to 3 m/s superficial flue gas velocity at maximum load. The bed material size is 590 μm and coarser, with a mean size of about 1000–1200 μm.

Circulating fluidized-beds do not contain any in-bed tube bundle heating surface. The furnace enclosure and internal division wall-type surfaces provide the required heat removal. This is possible because of the large quantity of solids that are recycled internally and externally around the furnace. The bed temperature remains uniform, because the mass flow rate of the recycled solids is many times the mass flow rate of the combustion gas. Operating temperatures for circulating beds are in the range of 816 to 871°C. Superficial gas velocities in some commercially available beds are about 6 m/s at full loads. The size of the solids in the bed is usually smaller than 590 μm, with the mean particle size in the 150–200 μm range (81).

Some of the advantages of fluidized beds include flexibility in fuel use, easy removal of SO_2, reduced NO_x production due to relatively low combustion temperatures, simplified operation due to reduced slagging, and finally lower costs in meeting environmental regulations compared to the conventional coal burning technologies. Consequently, fluidized-bed combustors are currently under intensive development and industrial size units (up to 150 MW) are commercially available (Fig. 10).

The modeling of fluidized beds remains a difficult problem since the usual assumptions made for the heat and mass transfer processes in coal combustion in stagnant air are no longer valid. Furthermore, the prediction of bubble behavior, generation, growth, coalescence, stability, and interaction with heat exchange tubes, as well as attrition and elutriation of particles, are not well understood

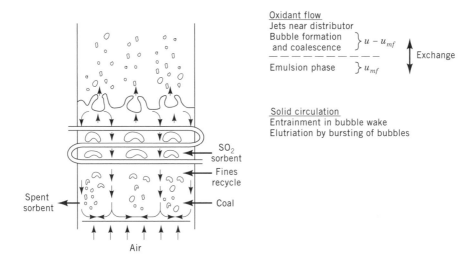

Fig. 10. The main processes taking place in a fluidized bed (92). Heat transfer to immersed tubes is 30% by radiation.

and much more research needs to be done. Good reviews on various aspects of fluidized-bed combustion appear in References 121 and 122 (Table 2).

3.4. Design Considerations in Fossil Fuel Combustion Systems.

One of the most important considerations in the design of a combustion chamber for a boiler is the fuel that is to be burned in the chamber (see FURNACES, FUEL-FIRED). Although all fuels burn and release heat during combustion, the rate at which a fuel burns and releases heat, and the impurities associated with the fuel have to be considered.

Furnaces for Oil and Natural Gas Firing. Natural gas furnaces are relatively small in size because of the ease of mixing the fuel and the air, hence the relatively rapid combustion of gas. Oil also burns rapidly with a luminous flame. To prevent excessive metal wall temperatures resulting from high radiation rates, oil-fired furnaces are designed slightly larger in size than gas-fired units in order to reduce the heat absorption rates.

Furnaces for Pulverized Coal Firing. The main differences between boilers fired with coal and those fired with oil or natural gas result from the presence of mineral matter in coals. The volume of the coal-fired furnace is higher because of the longer residence time required for the complete combustion of coal particles, the requirement of a controlled combustion rate to reduce NO_x formation, the provision for a larger heat-transfer surface area resulting from decreased heat-transfer rates because of ash deposits on the surfaces, and increased spacing of heat-transfer tubes to reduce flue gas velocities and thereby erosion of heat-transfer surfaces. Even when firing coal, depending on the reactivity of the coal (rank), the size of the combustion chamber required can vary (Fig. 11).

The combustion chamber of a modern steam generator is a large water-cooled chamber in which fuel is burned. Firing densities are important to ensure

Table 2. **Operating Conditions for an Atmospheric Pressure Fluidized-Bed Combustor**[a]

Process	Representative value	Comments
excess air, %	≈30	values selected to maintain CO emissions at acceptable levels
bed height, m	≈1.5	trade-off between pressure drop and gas residence time
bed temperatures, K	≈1100	higher values favor higher combustion efficiencies, higher rates of NO reduction by char; high values reduce SO_2 capture
calcium/sulfur (stoichiometric ratio)	≈2.5	value may be reduced by use of more active stone, lower gas velocities, deeper beds
gas velocity, m/s	≈2	high values are favored by high energy release rates per unit plan area; lower values are favored by higher efficiencies of combustion and SO_2 capture
sorbent particle top size, mm	≈3	low values are favored by shorter reaction times; size determined by consideration of elutriation at operating velocities
coal particle top size, mm	≈3	trade-off between elutriation rate and reaction time
sorbent residence time, s	≈5 × 10⁴	determined by stone reactivity and Ca/S ratio
gas residence time in bed, s	≈¼	an additional residence time of 1–2 s is available in the free board
coal particle burning time, s	≈400	dependent on temperature, size, and excess air
solid circulation time, s	≈2	short relative to solid reaction time

[a]Ref. 122.

that the chamber wall metal temperatures do not exceed the limits of failure of the tubes. Firing densities are expressed in two ways: volumetric combustion intensities and area firing intensities. The volumetric combustion intensity is defined by equation 23,

$$I_v = \frac{J_f h_f}{V_c P} \tag{23}$$

where I_v = volumetric combustion intensity, kJ/(m³ · h); J_h = fuel feed rate, kg/h; h_f = heating value of the fuel, kJ/kg; and V_c = volume of the combustion chamber, m³.

The area firing density is defined by equation 24,

$$I_a = \frac{J_f h_f}{A_c} \tag{24}$$

where, I_a = area firing density, kJ/(m² · h); J_h = fuel feed rate, kg/h; h_f = heating value of the fuel, kJ/kg; and A_c = cross−section of a fluid-bed distributor plate or grate area of a mechanical stoker or plan area in pulverized-coal combustors. Table 3 provides some design parameters for fossil fuel burners.

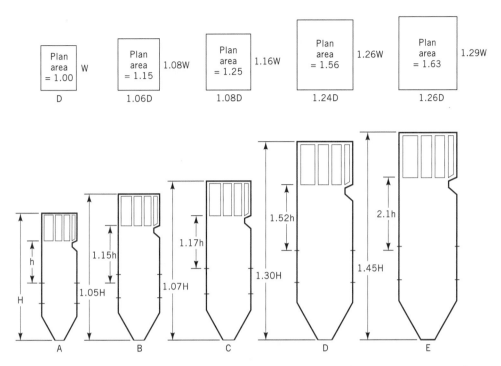

Fig. 11. Effect of coal rank on furnace sizing (constant heat output) (82), where W = width, D = depth, and h and H are the heights indicated. A represents medium volatile bituminous; B, high volatile bituminous or subbituminous; C, low sodium lignite; D, medium sodium lignite; and E, high sodium lignite.

Table 3. **Comparison of Design Parameters for Fossil Fuel Boilers**

			Coal		
Parameter	Gas	Oil	Grate	Fluid bed	Pulverized coal
heat rate, mW (t)	0.03–3000	0.03–3000	0.3–30	up to 30	30–3000
volumetric combustion intensity, kW/m^3	250–450	250–450	250–750^a	up to 2000^a (based on bed volume)	150–250
area combustion intensity, kW/m^2	280–500	280–500	2000	3000	7500
fuel firing density, kg/m^{3}h			30–100	≈250	15–30
kg/m^{2}h	6–11	6–11	40–250	up to 500	up to 1000
practical combustion temperature, °C	1000–1600	1100–1700	1200–1300	850–950	1600–1700
combustion time, s	10×10^{-1}	$20-25 \times 10^{-1}$	up to 5000	100–500	≈1–2
particle heating rate, °C/s			<1	10^3–10^4	10^4–10^5

^{a}Based on the total combustion volume which includes space between the bed and the convective tubes.

3.5. Environmental Considerations.   Atmospheric pollutants released by combustion of fossil fuels fall into two main categories: those emitted directly into the atmosphere as a result of combustion and the secondary pollutants that arise from the chemical and photochemical reactions of the primary pollutants (see AIR POLLUTION).

The main combustion pollutants are nitrogen oxides, sulfur oxides, carbon monoxide, unburned hydrocarbons, and soot. Combustion pollutants can be reduced by three main methods depending on the location of their application: before, after, or during the combustion. Techniques employed before and after combustion deal with the fuel or the burned gases. A third alternative is to modify the combustion process in order to minimize the emissions.

Nitrogen Oxides.   From the combustion of fuels containing only C, H, and O, the usual air pollutants or emissions of interest are carbon monoxide, unburned hydrocarbons, and oxides of nitrogen (NO_x). The interaction of the last two in the atmosphere produces photochemical smog. NO_x, the sum of NO and NO_2, is formed almost entirely as NO in the products of flames; typically 5 or 10% of it is subsequently converted to NO_2 at low temperatures. Occasionally, conditions in a combustion system may lead to a much larger fraction of NO_2 and the undesirable visibility thereof, ie, a very large exhaust plume.

NO is formed to some extent from N_2 and O_2 in flame products when N atoms are produced at a significant rate. Above 1700 K, the important step in the much studied Zeldovitch (thermal) mechanism is the production of N atoms by:

$$O + N_2 \longrightarrow NO + N$$

This is followed by a very fast reaction:

$$N + O_2 \longrightarrow NO + O$$

When $[NO] \ll [NO]_{eq}$, as is usually true in practice, its formation is essentially irreversible, and its rate is proportional to $[NO][N_2]$ with a large temperature dependence, an activation energy of 316 kJ/mol (75.5 kcal/mol). Unfortunately, the rate becomes appreciable just in the range of typical hydrocarbon–air flame conditions. If it is also assumed that $[O] = [O]_{eq}$, the observed rate in most lean-flame products in which N_2 is roughly 75 mol % of the gas can be approximated by

$$\frac{d[NO]}{dt} = \left(\frac{3.3 \times 10^{18}}{T}\right)\exp\left(\frac{-68700}{T}\right)(x_{O_2})^{1/2}$$

Here, x_{O_2} is the mole fraction of O_2 in the products at temperature T, and the rate is given in ppm/ms. The exponential implies a large effective activation energy of 570 kJ/mol, the sum of that for the O–N_2 reaction and half the dissociation energy of O_2. In typical hydrocarbon–air flames, the rate of NO formation by the thermal mechanism can be shown to be about 8 ppm/ms, or in a 10 ms residence time the thermal NO would be about 80 ppm. If preheating the mixture were to raise the gas temperature by 100 K, the rate of NO production would be nearly tripled, making the NO concentration unacceptable. Conversely, the rate can be reduced by the same amount by a 100 K reduction in temperature by precooling or heat abstraction from the flame itself, or by dilution of the mixture with excess air, steam, or other inert gas such as recirculated, relatively cool exhaust gas. Control of thermal NO_x thus involves reduction of the maximum attainable temperature, or the residence time at high temperature, or both. Such measures, however, always entail some compromise in stability and control, and possibly also in the efficiency of the combustion process. The afterburning of CO tends to be quenched by rapid temperature reduction, and the resulting increase in the emission of CO must be balanced against the desired NO_x reduction. Heat abstraction or cooling of the flame always occurs to some extent by radiation from the highly luminous flames produced by pulverized-coal or oil combustion, as is typical in boilers and similar furnaces. When heat is rejected, eg, by a boiler fluid, and not returned or recuperated to the unburned mixture, the maximum temperature and thermal NO_x formation will be reduced. An extension of this effect has been applied to achieve low NO_x emissions in some furnaces and boilers in which combustion occurs in a very rich, relatively low temperature primary stage, followed by heat abstraction by convection as well as radiation to reduce the gas enthalpy (two-stage combustion). Secondary products and any excess air are then introduced to complete the combustion and, owing to the previous heat transfer, the maximum temperature attainable in that stage will never approach the adiabatic flame temperature. Much soot, which is responsible for the radiative heat loss, may be present in the rich primary flame products. To avoid smoke from such two-stage processes, care must be taken to assure its oxidation in the second stage. In practice $[O]/[O]_{eq}$ is seldom unity as assumed. Though $[O]$ is decreased in the burned gas, its average value may be several times $[O]_{eq}$ and NO formation may be correspondingly higher than predicted from the $[O]_{eq}$. Similarly, the very high radical concentration, eg, $[O]$, in the reaction zone of a flame often leads to almost instantaneous NO production, even though the temperature is still relatively low and the residence time is relatively short. Other fast reactions involving transient flame species producing N atoms, for example, $CH + N_2 \longrightarrow HCN + N$, can also contribute to the production of some NO. In any case, the NO inevitably formed by these species is called prompt NO. The total concentration of prompt NO is usually not large, 10–50 ppm depending on the composition of the flame, but is significant if very low NO_x emission levels are sought.

A different and often more serious source of NO_x is chemically bound nitrogen in the fuel, eg, NH_3, amines, nitrites, pyridine, a fraction of which is always converted to NO. Most coals contain at least 1% N, of which 50% or more is retained in gaseous or liquid fuels derived from coal. The N content of distillate and gaseous petroleum fuels from most sources is usually very low, but it can reach 0.5% and if such a fuel were burned with air under stoichiometric or lean conditions, the conversion of its fuel N to NO in the flame would yield up to 400 ppm of NO in the burned gas. NO_x production levels from fuel N and NO_x control measures are now well established from correlations of data from flames and combustors. Formation occurs in the flame reaction zone by OH radical oxidation of intermediate species formed by the decomposition of the fuel N, eg, NH_2, NH, N, which if unoxidized would in a short time simply form stable N_2.

Whether NO or N_2 formation prevails depends on the flame conditions as well as on the concentrations of the intermediates. At high levels of fuel N, NO can also be converted directly to N_2. In general, the yield (mol of NO/mol of fuel N) is much higher in lean and stoichiometric flames than in rich flames, and asymptotically approaches unity at low fuel N concentrations. It decreases with increasing fuel N, and at some level that depends on ϕ, the NO concentration becomes constant, albeit rather high, and any additional fuel N is converted rapidly to N_2, from which NO can then be formed only by the relatively slow thermal process with oxygen.

Thus if combustion can be effected in two stages, with or without the intermediate heat rejection for thermal NO_x control discussed above, the conversion of fuel N to NO can be largely circumvented by first, a primary stage at $\phi = 1.5 - 2$ with a modest residence time to allow formation of N_2 in the hot primary products, followed by rapid addition of secondary air to complete the combustion at an effective $\phi \approx 0.8$. There will of course be a maximum in the temperature near $\phi = 1$ in the course of the secondary air addition, but if the residence time at that condition is minimized, the production of thermal NO will also be minimized.

Combustion system developments for reducing NO_x formation include: low NO_x burners, staged burning techniques, and flue gas recirculation (FGR). Some postcombustion techniques for reducing NO_x include: selective noncatalytic reduction (SNCR) and selective catalytic reduction (SCR). In either technology, NO_x is reduced to nitrogen (N_2) and water (H_2O) through a series of reactions with a chemical agent injected into the flue gas. The most common chemical agents used commercially are ammonia and urea for SNCR and ammonia for SCR systems. Most ammonia-based systems have used anhydrous ammonia (NH_3) as the reducing agent. However, because of the hazards of storing and handling NH_3, many systems use aqueous ammonia at 25–28% concentration. Urea can be stored as a solid or mixed with water and stored in solution (81).

These ideas form the basis of most approaches to NO_x control with N-containing fuels. In principal, they are readily applicable to the modification of certain combustors in which the desired divisions in the combustion process exist for other reasons. Although such improvements have been demonstrated, it is difficult in practice to make the required revisions in the air and fuel distribution without adverse effects on other emissions or on performance. It has also been shown that when steam is used to reduce thermal NO_x production, the formation of NO_x from fuel N is enhanced, or the reduction is less than otherwise expected.

Sulfur Oxides.   Oxides of sulfur are also pollutants of concern. When present in the fuel as inorganic sulfides or organic compounds, sulfur is converted almost completely to SO_2 in the products of complete combustion. There are no known techniques for the elimination of this conversion process in flames, and emission control measures necessarily involve either desulfurization of the fuel or removal of the SO_2. Up to 10% of the SO_2 is oxidized to SO_3 at low temperatures in most combustion processes, and the total sulfur oxides emission is often given as the sum of the concentrations of SO_2 and SO_3, or SO_x. The presence of H_2O in the combustion products is inevitable and at low temperatures SO_3 combines with H_2O to form H_2SO_4, which is both a highly corrosive agent to heat exchange surfaces and a highly undesirable stack emission.

Soot. Emitted smoke from clean (ash-free) fuels consists of unoxidized and aggregated particles of soot, sometimes referred to as carbon though it is actually a hydrocarbon. Typically, the particles are of submicrometer size and are initially formed by pyrolysis or partial oxidation of hydrocarbons in very rich but hot regions of hydrocarbon flames; conditions that cause smoke will usually also tend to produce unburned hydrocarbons with their potential contribution to smog formation. Both may be objectionable, though for different reasons, at concentrations equivalent to only 0.01–0.1% of the initial fuel. Although their effect on combustion efficiency would be negligible at these levels, it is nevertheless important to reduce such emissions.

Neither soot nor unburned hydrocarbons are found in the products of a lean or stoichiometric premixed flame with air or O_2, although hydrocarbons may be formed or survive unburned in lean flames partially quenched, eg, by the cold wall of a combustion chamber. A moderately rich flame also should yield only the water-gas equilibrium products (and N_2); but with increasing equivalence ratio, at some ϕ still well below the upper flammability limit, the burned gas becomes faintly luminous with precipitated soot particles that increase in number density and luminous intensity with further increase in ϕ. The appearance of the condensed phase (soot) is connected with appreciable nonequilibrium concentrations of fairly stable low molecular-weight hydrocarbons, notably acetylene, from the decomposition of unoxidized fuel. These polymerize (condense with elimination of H_2) to form high molecular-weight products, mostly with ring structures. If these intermediates are not consumed, eg, by OH, in simultaneous and competing oxidation reactions, they grow until nucleation, and eventually precipitation occurs as visibly radiating soot, typically over a period of several milliseconds after the main flame reaction zone. The particle growth and competing oxidation then continue in the burned gas.

The composition, properties, and size of soot particles collected from flame products vary considerably with flame conditions and growth time. Typically the C–H atomic ratio ranges from two to five and the particles consist of irregular chains or clusters of tiny spheres 10–40 nm in diameter with overall dimensions of perhaps 200 nm, although some may agglomerate further to much larger sizes.

Whether soot particles form at all and grow depends on ϕ, the fuel type, and other variables, eg, the growth is easier and faster at a given ϕ with fuels of higher C–H ratio and at elevated pressure. Given sufficient residence time to attain equilibrium at the burned gas condition, the soot and hydrocarbons would eventually be consumed. In practice, their rapid oxidation occurs in a secondary flame in which the hot primary products are burned with the required excess air, which is added by diffusion or by more intensive mixing. However, if a large excess of air is added too rapidly, the cooling can, in effect, quench the oxidation of both unburned hydrocarbons and the accompanying soot, which would then persist as visible smoke. The blackness of the smoke depends on the size and number density of the particles when quenched, their further aggregation, etc. Some may also survive as much smaller invisible particles, or condensation nuclei. For thorough reviews on the mechanisms of pollutant formation see References (123–125).

Diffusion Flame Chemistry. Since most combustion systems employ mixing-controlled diffusion flames, which are characterized by very high pollutant emissions, it is imperative to look into the chemistry occurring in diffusion flames. In a typical diffusion flame the mixture composition in the reaction zone is close to the stoichiometric proportion and the temperature is at a maximum resulting from the large volume of this zone, thus NO_x production is favored. If, however, the surrounding gas cools the combustion products rapidly, further reactions of CO and NO are eliminated. This fixes the concentrations of these pollutants at unfavorable levels. Furthermore, the fuel diffuses into the combustion zone through the burned gases and thus is heated in the absence of oxygen. This creates ideal conditions for the formation of soot and the reduction of the CO_2 produced in the combustion zone to CO. Additionally, diffusion flames have low combustion intensity and efficiency and hence release large amounts of unburned hydrocarbon emissions. In general, despite the fact that the structure of the diffusion flame is more complex and difficult to analyze, the same basic description of soot formation and oxidation should apply to diffusion flames as for premixed flames.

Emissions Control. From the combustion chemistry standpoint, lean mixtures produce the least amount of emissions. Hence, one pollution prevention alternative would be to use lean premixed flames. However, lean mixtures are difficult to ignite and form unstable flames. Furthermore, their combustion rates are very low and can seldom be applied directly without additional measures being taken. Consequently the use of lean mixtures is not practical.

Another potential solution is the use of catalytic combustors, which produce extremely low levels of emissions by the use of combustion catalysts such as platinum. The main disadvantage of catalytic combustors, however, is their high cost.

More advanced techniques for emissions control include electrical or plasma jet augmentation of flames based on radical production. Since in two-phase, heterogeneous combustion the flames are always diffusion flames on the microscale, ie, the individual droplets or particles burn as diffusion flames, and since at the characteristic times for evaporation, decomposition and burning of individual particles can be comparable with the characteristic times for mixing and pollutant formation, prevaporization or gasification of the fuel can reduce pollutant emissions. For this reason catalytic systems for liquid-fuel decomposition and coal gasification are being considered seriously as alternatives to conventional combustion technology (126–128).

BIBLIOGRAPHY

"Fuels (Combustion Calculation)," in *ECT* 1st ed., Vol. 6, pp. 913–935, by H. R. Linden, Institute of Gas Technology; in *ECT* 2nd ed., Vol. 10, pp. 191–220, by D. M. Himmelblau, The University of Texas; "Burner Technology," in *ECT* 3rd ed., Vol. 4, pp. 278–312, by G. E. Moore, Consultant; in *ECT* 4th ed., Vol. 6, pp. 1049–1092, by R. Sharifi, S. A. Pisupati, and A. W. Scaroni, Pennsylvania State University.

CITED PUBLICATIONS

1. N. Chigier, "Energy," *Combustion and Environment*, McGraw-Hill, Inc., New York, 1981.

2. W. Bartok and A. F. Sarofim, *Fossil Fuel Combustion: A Source Book*, John Wiley & Sons, Inc., New York, 1991.

3. J. A. Barnard and J. N. Bradley, *Flame and Combustion*, 2nd ed., Chapman and Hall, London and New York, 1985.

4. W. E. Baker, P. A. Cox, P. E. Westine, J. Kulesz, and R. A. Strehlow, *Explosion Hazards and Evaluation*, Elsevier, Amsterdam, The Netherlands, 1983.

5. R. A. Strehlow, *Combustion Fundamentals*, McGraw-Hill, Inc., New York, 1984.

6. D. J. McCracken, *Hydrocarbon Combustion and Physical Properties*, Rep. No. 1496, Ballistic Research Laboratories, Sept. 1970.

7. M. G. Zabetakis, *Flammability Characteristics of Combustible Gases and Vapors*, Technical Bulletin, U.S. Bureau of Mines, Washington, D.C., 1965, p. 627.

8. I. Glassman, *Combustion*, Academic Press, Inc., New York, 1987.

9. J. Chomiak, *Combustion; A Study in Theory, Fact and Application*, Gordon and Breach Science Publishers, Montreux, Switzerland, 1990.

10. B. Lewis and G. von Elbe, *Combustion, Flames and Explosion of Gases*, 2nd ed., Academic Press, Inc., New York, 1951.

11. F. A. Williams, *Combustion Theory*, The Benjamin/Cummings, Publishing Co. Inc., Menlo Park, Calif., 1985.

12. K. K. Kuo, *Principles of Combustion*, John Wiley & Sons, Inc., New York, 1986.

13. R. C. Flagen and J. H. Seinfeld, *Fundamentals of Air Pollution Engineering*, Prentice-Hall, Inc., Englewood Cliffs, N.J., 1988.

14. H. B. Palmer and J. M. Beer, eds., *Combustion Technology: Some Modern Developments*, Academic Press, New York and London, 1974.

15. M. V. Blank, P. G. Guest, G. von Elbe, and B. Lewis, *Third Symposium, Combustion, Flame and Explosion Phenomena*, Williams & Wilkens, Baltimore, Md., 1949.

16. H. C. Barnett and R. R. Hibbard, *Basic Considerations in the Combustion of Hydrocarbon Fuels with Air*, NASA Technical Report, 1959, p. 1300.

17. H. F. Colcote, C. A. Gregory, C. M. Barnett, and R. B. Gilmer, *Ind. Eng. Chem.* **44**, 2656 (1952).

18. Ya. B. Zeldovich, G. I. Brenblatt, V. B. Librovich, and G. M. Mackhviladze, *The Mathematical Theory of Combustion and Explosions*, Consultants Bureau, 1985.

19. V. R. Kuznetsov and V. A. Sabel'nikov, *Turbulence and Combustion*, Hemisphere Publishing Corp., Washington, D.C., 1990.

20. M. Y. Hussaini, A. Kumar, and R. G. Voigt, *Major Research Topics in Combustion*, Springer-Verlag, Inc., New York, 1992.

21. C.-M. Brauner and C. Schmidt-Laine, *Mathematical Modeling in Combustion and Related Topics*, Martinus Hijhoff Publishers, Dordrecht, The Netherlands, 1988.

22. R. M. C. So, H. C. Mongia, and J. H. Whitelaw, *Turbulent Reactive Flow Calculations*, special issue of *Combustion Science and Technology*, Gordon and Breach Science Publishers, Inc., Montreux, Switzerland, 1988.

23. F. A. Williams, in W. Bartok and A. F. Sarofim, eds., *Turbulent Reacting Flows In Fossil Fuel Combustion: A Source Book*, John Wiley & Sons, Inc., New York, 1991.

24. F. A. Williams, in J. H. S. Lee and C. M. Cuirao, eds., *Laminar Flame Instability and Turbulent Flame Propagation, In Fuel-Air Explosions*, University of Waterloo Press, Waterloo, Ontario, Canada, 1982.

25. J. M. Beer and N. A. Chigier, *Combustion Aerodynamics*, Applied Science, London; John Wiley & Sons, Inc., New York, 1972.

26. L. D. Landau, *Acta. Phys. Chim. (USSR)* **19**, 77 (1944).

27. L. D. Landau and E. M. Lifschitz, *Mechanics of Continuous Media*, Moscow, Russia, 1953; Eng. trans. Addison-Wesley Publishing Co., Inc., Reading, Mass., 1959.

28. A. C. Scurlock and J. H. Grover, *4th International Symposium on Combustion*, The Combustion Institute, Pittsburgh, Pa., 1953, p. 645.

29. K. N. C. Bray, *Equations of Turbulent Combustion*, Report No. 330, University of Southampton, UK, 1973.
30. K. N. C. Bray, *17th International Symposium on Combustion*, The Combustion Institute, Pittsburgh, Pa., 1978, 57–59.
31. K. N. C. Bray and P. A. Libby, *Phys. Fluids* **19**, 1687–1701 (1976).
32. K. N. C. Bray and J. B. Moss, *Comb. Flame* **30**, 125 (1977).
33. H. C. Hottel, in Ref. 28, p. 97–113.
34. S. P. Burke and T. E. W. Schumann, *Ind. Eng. Chem* **20**, 998–1004 (1928).
35. W. Jost, *Explosion and Combustion Processes in Gases*, McGraw-Hill, Inc., New York, 1946.
36. W. Jost, *Diffusion*, Steinkopff, Darmstadt, Germany, 1957.
37. A. G. Gaydon and H. G. Walfhard, *Flames*, 2nd ed., Macmillan, New York, 1960.
38. T. Takeno and Y. Kotani, in L. A. Kennedy, ed., *Turbulent Combustion, Prog. Astronaut. Aeronaut.*, Vol. 58, AIAA, 1978, 19–35.
39. M. Gerstein, *Diffusion Flames*, in Ref. 2.
40. H. C. Hottel, *3rd International Symposium on Combustion*, The Combustion Institute, Pittsburgh, Pa., 1948, 254–266.
41. P. A. Libby and F. A. Williams, eds., *Turbulent Reacting Flows*, Springer-Verlag, Berlin and New York, 1980, 1–43, 219–236.
42. A. M. Mellor, in Ref. 31, 377–387.
43. F. A. Williams, in W. E. Stewart, W. H. Ray, and C. C. Conley, eds., *Current Problems in Combustion Research, Dynamics and Modeling of Reactive Systems*, Academic Press, Inc., New York, 1980.
44. P. A. Libby and F. A. Williams, *AIAA J.* **19**, 261–274 (1981).
45. P. A. Libby and F. A. Williams, *Annu. Rev. Fluid Mech.* **8**, 351–379 (1976).
46. R. W. Bilger, *Energy and Combustion Science*, Student ed. 1, Pergamon Press, Oxford, 1979, 109–131.
47. D. B. Spalding, *Some Fundamentals of Combustion*, Butterworth & Co., Ltd., London, 1955.
48. R. Borghi, *Turbulent Combustion Modeling, Progress in Energy and Combustion Science*, Vol. 14, No. 4, Pergamon Press, Elmsford, N.Y., 1988, 245–292.
49. I. W. Smith, *19th International Symposium on Combustion*, The Combustion Institute, Pittsburgh, Pa., 1982, p. 1045.
50. R. B. Jones, B. B. McCourt, C. Morley, and K. King, *Fuel* **64**, 1460 (1985).
51. N. Oka, T. Murayama, H. Matsuoka, and S. Yamada, *Fuel Process. Tech.* **15**, 213 (1987).
52. M. Shibaoka, *Fuel* **48**, 285 (1969).
53. J. M. Kuchta, V. R. Rowe, and D. S. Burgess, U.S. Bureau of Mines Report 8474, Washington, D.C., 1980, p. 1.
54. R. H. Essenhigh, in *M. A. Elliot, eds.*, Fundamentals of Coal Combustion, In Chemistry of Coal Utilization, 2nd Suppl. Vol., John Wiley & Sons, Inc., New York, 1981.
55. N. N. Semenov, *Chemical Kinetics and Chain Reactions*, Oxford University Press, London, 1935.
56. F. Kamenetski, *Diffusion and Heat Exchange in Chemical Kinetics*, Princeton University Press, Princeton, N.J., 1955; trans. from Russian by Thanel.
57. K. Annamalai and P. Durbetaki, *Comb. Flame* **29**, 193 (1977).
58. K. Annamalai, *Trans. ASME* **101**, 576 (1979).
59. L. D. Smoot, M. D. Horton, and G. A. Williams, *16th International Symposium on Combustion*, The Combustion Institute, Pittsburgh, Pa., 1976, p. 375.
60. R. H. Essenhigh, M. Misra, and D. Shaw, *Comb. Flame* **77**, 3 (1989).
61. H. E. Blayden, W. Noble, and H. L. Riley, *Gas J.* **2**, 81 (1934).
62. H. K. Griffin, J. P. Adams, and D. F. Smith, *Ind. Eng. Chem.* **21**, 808 (1929).

63. G. P. Ivanova and V. Babii, *Thermal Eng.* **13**, 70 (1966).
64. A. A. Orning, *Proceedings of Conference on Pulverized Fuel*, Vol. 1, Institute of Fuel, London, 1947, p. 45.
65. D. L. Carpenter, *Comb. Flame* **1**, 63 (1957).
66. H. M. Cassel and I. Liebman, *Comb. Flame* **3**, 467 (1959).
67. J. B. Howard and R. H. Essenhigh, *Comb. Flame* **9**, 337 (1965).
68. M. A. Nettleton and R. Stirling, *Comb. Flame* **22**, 407 (1974).
69. L. Tognotti, A. Malotti, L. Petarca, and S. Zanelli, *Comb. Flame* **44**, 15 (1985).
70. M. R. Chen, L. S. Fan, and R. H. Essenhigh, *20th International Symposium on Combustion*, The Combustion Institute, Pittsburgh, Pa., 1984, p. 1513.
71. C. P. Gomez, *Ignition of and Combustion of Coal and Char Particles; a Differential Approach*, M.S. dissertation, Pennsylvania State University, University Park, Pa., 1982.
72. A. W. Scaroni, M. R. Khan, S. Eser, and L. R. Radovic, *Ullmann's Encyclopedia of Industrial Chemistry*, Vol. A7, VCH Publishers, New York, 1986, p. 245.
73. J. D. Blackwood and F. K. McTaggart, *Aust. J. Chem.* **12**, 533 (1959).
74. R. G. Jenkins, S. P. Nandi, and P. L. Walker, Jr., *Fuel* **52**, 288 (1973).
75. D. W. Van Krevelen, *Coal Typology–Chemistry–Physics–Constitution*, Elsevier Publications, New York, 1961, p. 219.
76. E. J. Badin, *Coal Combustion Chemistry—Correlation Aspects*, Elsevier, New York, 1984, Chapt. 6, p. 68.
77. D. W. McKee, in P. L. Walker, Jr. and P. A. Thrower, eds., *Chemistry and Physics of Carbon*, Vol. 16, Marcel Dekker, Inc., New York, 1981, p. 1.
78. B. A. Morgan and A. W. Scaroni, *International Conference on Coal Science*, International Energy Agency, Sydney, Australia, 1985, p. 347.
79. P. L. Walker, Jr., M. Shelef, and R. A. Anderson, in Ref. 79, Vol. 4.
80. P. L. Walker, Jr., F. Rusinko, Jr., and L. G. Austin, in D. D. Eley, P. W. Selwood, and P. B. Weisz, eds., *Advances in Catalysis*, Vol. 2, Academic Press, Inc., New York, 1959, p. 133.
81. S. C. Stultz and J. B. Kitto, eds., *Steam, Its Generation and Use*, 40th ed., Babcock and Wilcox Co., Barberton, Ohio, 1992.
82. J. E. Singer, *Combustion-Fossil Power Systems*, Combustion Engineering, Windsor, Conn., 1981.
83. American Gas Association, *Gas Engineers Handbook*, Industrial Press, New York, 1978.
84. S. S. Penner and B. P. Mullins, *Explosions, Detonations, Flammability and Ignition*, AGARD Monograph, Pergamon Press, Inc., New York, 1959.
85. M. W. Thring, *The Science of Flames and Furnaces*, John Wiley & Sons, Inc., New York, 1962.
86. A. Williams, *Comb. Flame* **21**, 1 (1973).
87. A. Williams, *Combustion of Sprays and Liquid Fuels*, Elek Science, London, 1976.
88. A. Williams, *Combustion of Liquid Fuels Sprays*, Butterworths & Co., Ltd., London, 1990.
89. A. H. Lefebvre, *Gas Turbine Combustion*, Hemisphere Publishing Corp., Washington, D.C., 1983.
90. A. H. Lefebvre, *Atomization and Sprays*, Hemisphere Publishing Corp., Washington, D.C., 1989.
91. A. C. Fernandez-Pello and C. K. Law, in Ref. 51, p. 1037.
92. A. L. Randolf, A. Markino, and C. K. Law, *21st International Symposium on Combustion*, The Combustion Institute, Pittsburgh, Pa., 1988, p. 601.
93. J. C. Lasheras, L. T. Yap, and F. L. Dryer, in Ref. 72, p. 1761.

94. T. Niioka and J. Sato, in Ref. 94, p. 625.
95. A. H. Levebvre, *Airblast Atomization, Progress in Energy and Combustion Science*, Vol. 6, Pergamon Press, Oxford, UK, 1980, 233–261.
96. N. K. Rizk and A. H. Levebvre, *J. Fluids Eng.* **97**(3), 316–320 (1975).
97. A. Rizkalla and A. H. Levebvre, *AIAA J.* **21**(8), 1139–1142 (1983).
98. C. J. Lawn and co-workers, *The Combustion of Heavy Fuels Oils*, in C. J. Lawn, ed., *The Principles of Combustion Engineering for Boilers*, Academic Press, Inc., New York, 1987.
99. T. F. Wall, *The Combustion of Coal as Pulverized Fuel Through Swirl Burners*, in Ref. 100.
100. M. A. Field, D. W. Gill, B. B. Morgan, and P. J. W. Hawksley, *Combustion of Pulverized Coal*, BCURA, UK, 1967.
101. H. H. Lowry, ed., *Chemistry of Coal Utilization*, Suppl. Vol., John Wiley & Sons, Inc., New York, 1963.
102. L. D. Smoot and D. T. Pratt, eds., *Pulverized Coal Combustion and Gassification*, Plenum Press, New York, 1979.
103. L. D. Smoot, *Coal and Char Combustion*, in Ref. 2.
104. B. R. Cooper and W. A. Ellingson, *The Science and Technology of Coal Utilization*, Plenum Press, New York, 1984.
105. M. F. R. Mulcahy and I. W. Smith, *Rev. Pure. Appl. Chem.* **19**, 81 (1969).
106. N. M. Laurendeau, *Prog. Energy Comb. Sci.* **4**, 221 (1978).
107. *Modern Power Station Practice*, 2nd ed., Pergamon Press, Oxford, UK, 1971.
108. L. D. Smoot and P. J. Smith, *Coal Combustion and Gasification*, Plenum Press, New York, 1985.
109. A. Stambuleanu, *Flame Combustion Process in Industry*, Abacus Press, Tunbridge Wells, UK, 1979.
110. *Prog. Energy Comb. Sci., Special Issue* **10**, 81–293 (1984).
111. J. B. Howard, *Fundamentals of Coal Pyrolysis and Hydropyrolysis*, in Ref. 56.
112. C. Y. Wen and E. Stanley Lee, eds., *Coal Conversion Technology*, Addison-Wesley Publishing Co., Reading, Mass. 1979.
113. R. H. Essenhigh, in Ref. 61, p. 372.
114. A. W. Scaroni, P. L. Walker, and R. G. Jenkins, *Fuel* **60**, 70–76 (1981).
115. P. Ramachandran, A. W. Scaroni, and R. G. Jenkins, *I. Chem., E. Symp. Ser.* **107**, 128–219 (1987).
116. P. Ramachandran, A. W. Scaroni, G. Reethof, and S. Yavuzkurt, *13th International Conference on Coal and Slurry Technology*, Slurry Technology Association, 1988, 241–247.
117. G. Huang and A. W. Scaroni, *Fuel* **71**, 159–164 (1992).
118. J. R. Howard, ed., *Fluidized Beds: Combustion and Applications*, Elsevier, New York, 1983.
119. R. Schweiger, *Fluidized Bed Combustion and Application Technology, The 1st International Symposium*, Hemisphere Publishing Corp., New York, 1987.
120. H. R. Hoy and D. W. Jill, *The Combustion of Coal in Fluidized Beds*, in Ref. 100.
121. H. B. Palmer and C. F. Cullis, *The Formation of Carbon from Gases*, in P. L. Walker, ed., *Chemistry and Physics of Carbon*, Marcel Dekker, New York, 1976.
122. S. Kaliaguine and A. Mahay, eds., *Catalysts on the Energy Scene*, Elsevier, New York, 1984.
123. J. M. Beer, in Ref. 59, 439–460.
124. A. F. Sarofim and J. M. Beer, in Ref. 30, 189–204.
125. K. C. Taylor, *Automobile Catalytic Converters*, Springer-Verlag, New York, 1984.
126. H. G. Wagner, in Ref. 30, p. 3.
127. K. H. Homann, in Ref. 70, p. 857.

128. P. Kesselring, *Catalytic Combustion*, in F. Weinberg, ed., *Advanced Combustion Methods*, Academic Press, Inc., New York, 1986.

Reza Sharifi Sarma
V. Pisupati
Alan W. Scaroni
Pennsylvania State University

COMPUTER-AIDED CHEMICAL ENGINEERING

1. Introduction

In the 1950s, the chemical industry was the first civil industry to make extensive use of computers. At that time, computers were relatively more expensive (a mainframe might cost US$1 million). However, they were inexpensive compared to the cost of a chemical plant that would typically exceed US$100 million. The cost of a mainframe could be repaid in its first application to process design. Distillation columns were being designed by computer in the 1950s and process simulation programs were available by the early 1960s. The first commercial simulation package was probably PACER, which was well established by the mid-1960s. The first on-line computer controlled processes were commissioned in the late 1950s. Thus, Computer-Aided Chemical Engineering has nearly 50 years of history.

Computers are now cheap and there is one on the desk of every engineer. Computer aids are used at every stage from conception through design to operation. Virtually all chemical engineering is now "computer-aided chemical engineering".

At the conceptual stage, software is used to plan and analyze laboratory experiments. Computer programs are used to estimate chemical and physical properties for chemical species for which experimental data are lacking. Software tools, such as computer-aided molecule design (CAMD), are employed to devise chemicals with desired properties. CAMD is increasingly relevant as the chemical industry moves to the position that its primary goal is to sell effects rather than chemicals. Thus, it sells detergents, solvents, fuels and fibers and only incidentally sells specific chemicals that have these properties. Process synthesis is used for the conceptual design of processes that will manufacture the desired chemical species and mixtures.

At the design stage, processes (developed manually or by computer synthesis) are simulated in detail to ensure safe and economic operation. The designs may also be optimized to minimize costs, maximize profits or meet environmental or safety criteria. Computer programs are used for both the process and mechanical design of individual unit operations. The safety and environmental impact of the proposed processes is assessed using appropriate software.

A wide range of computer tools may be applied to operating plant. Computers may be applied on-line or off-line. On-line applications include regulation control and optimizing control of continuous processes. On-line computer-control is also applied for scheduling batch and semibatch processes. From the earliest applications, on-line computers have been used for safety monitoring and rapid automatic shutdown. Most current on-line software includes such emergency response facilities.

Off-line optimization of operating variables can give substantial benefits. Processes rarely operate exactly as designed, even when the design is optimized. Uncertainties allowed for in design are largely resolved in operation. Reduced uncertainty enables more accurate simulation than was possible at the design stage. Greater accuracy enables optimization of operating policy to exploit favorable outcomes to the uncertainties and to mitigate the consequences of unfavorable outcomes. Typically, 20–50 variables are available for optimization. Beyond 2 or 3 optimization variables, it is impossible to determine optimal conditions manually, and the optimum may even be counterintuitive. There are examples where off-line optimization has doubled or tripled the operating margins for processes. The impact on plant profitability can thus be high. Off-line studies are also undertaken to assess potential process improvement, eg, through modification of operating schedules or through retrofit design.

In practice, an engineer may have up to 200 different computer programs that can be applied to aid the efficient design and operation of chemical processes. These programs must give correct answers and the answers must be correctly interpreted. A typical range of computer software employed by a large manufacturing company is given in Table 1. The AIChE on-line directory (1) lists a wide range of available commercial software.

The objective of this chapter is to assist both chemical engineers developing software and chemical engineers using software written by others.

The section Developing Engineering Software covers program development. It is aimed primarily at engineers writing relatively small programs, or contributing specialist modules for larger programs. This section provides guidance on commonly occurring problems, and includes material important to engineering software that is not usually covered in software engineering or numerical analysis texts. At the same time, introductory material and references are provided for specialist chemical engineering software engineers.

The section Using Engineering Software is designed for engineers using third-party software. It describes steps necessary to ensure that software is used effectively and it refers to more comprehensive works on the topic.

The section Current Advances briefly describes some areas of active research in computer-aided process engineering (CAPE).

This chapter does not describe particular chemical engineering software in detail. The breadth of material (as indicated in Table 1) makes such a description beyond the scope of a single chapter. Neither does the chapter give a comprehensive exposition of software validation or numerical analysis. There is extensive literature available for specialists in this area and some of this material is referred to. Nevertheless, the material presented here is intended to be sufficient for chemical engineers involved in computer-aided chemical engineering as an incidental part of their work. In-depth discussion of on-line computer control is

Table 1. **Examples of Computer-Aided Engineering Software Summary Table**

Data Correlation and Prediction of Physical and Chemical Properties
Fitting experimental data for physical properties. Predicting physical properties
Fitting experimental reaction equilibrium and rate data
Prediction of flammability, toxicity, and other data important for safety computation
Prediction of ozone depletion, greenhouse effect, and other data for environmental studies
Prediction of equipment failure and repair rates

Unit Operations
Heat exchanger process and mechanical design (includes multi pass, multi fluid, tube, plate). Fired heaters
Evaporator design
Design of pressure vessels according to various national and international standards
Simulation and design of distillation columns, batch, and continuous distillation
Absorption and stripping column simulation and design
Design of column internals, packed columns, valve tray, sieve tray, and bubble-cap
Reboiler and condenser design
Compressor and expander design and simulation
Liquid–liquid extractor simulation and design
Modeling Rankine cycle systems
Gas and steam turbine modeling
Analysis and performance of agitated vessels
Pressure drop calculations for liquids and for gases in isothermal and adiabatic flow
Two-phase pressure drop calculations in pipes and conduits
Two-phase choked flow
Instability in two-phase flow
Surge analysis
Estimation of tube vibration
Tubular and fluidised bed reactor simulation and design
Fluidized bed drier modeling
Driers, indirectly heated, directly heated. Spray driers
Restrictor orifice design
Combustion calculations
Non-Newtonian flow and heat transfer calculations
Furnace design and radiant heat transfer
3-D fluid flow, heat, and mass transfer through Computational Fluid Dynamics

Environmental Calculations
Dispersion of gases, aerosols and smokes from stacks, ruptures, and fires
Dry and rain-enhanced deposition of aerosols and smokes from plumes
Leaching from landfill sites and dispersion of leakages through groundwater
Modeling river networks for accumulation of pollutants
Concentration of pollutants in land and marine life (vegetable, animal and microbial)
Integrated effect of releases to air, water, and land

Safety Studies
Hazard analysis, fire and explosion, toxic chemical release
Bursting disk and pressure relief valve computations
Adiabatic and isothermal relief in piping networks
Design and simulation of flare release systems

Process Availability and Reliability
Plant availability computed from equipment failure and maintenance statistics
Availability with stand-by equipment

Process Simulation
Simulation and design of steady-state processes
Data reconciliation (estimation of statistically most likely performance from measurements)
Simulation and design of unsteady processes: Control, start-up, shut-down, upset conditions
Design of batch and semi batch processes; dedicated, multi purpose and multi-product plant

Table 1 (*Continued*)

Discrete dynamic simulation of batch and semi-batch processes
Simulation of linked distillation columns
Simulation of heat-exchanger networks
Simulation of evaporator trains
Site simulation: energy use, utilities requirements, major intermediates (e.g., HCl)
Economic evaluation of plants and sites
Optimization of design and operating conditions

On-line Computation

On-line optimization to compensate for performance, market, and raw material changes
On-line data reconciliation
On-line regulation control to ensure stable performance in the face of disturbances
On-line fault diagnosis
Condition monitoring (prediction of equipment deterioration from noise etc. measurements)
Optimization of start-up, shut-down and load change trajectories
On-line scheduling, e.g., of batch-process operation

Aids to the Design Process

Intelligent piping and instrumentation diagram systems
Design rationale systems (knowledge-based design)
Interchange software based on Process industries STEP standards (ISO 10303-221)

Process Synthesis

Minimal energy or utility requirements for a process
Minimal energy or utility requirements for a site
Optimal heat-exchanger network design
Optimal separator network design
Optimal process design by process synthesis

deliberately avoided (although an important part of the subject) because it is covered in a separate chapter in this encyclopedia on Process Control . Many of the principles employed to produce quality software and quality-assured results from software are common to other areas of Quality Assurance, for which the reader is referred to the chapter in this encyclopedia on Quality Control.

2. Developing Engineering Software

This section presents general principles in developing engineering software, which should:

1. Meet defined engineering goals.
2. Be based on sound principles of chemistry and physics.
3. Be testable and maintainable.
4. Be tested to ensure that it correctly codes the models on which it is based, for example that it is dimensionally consistent.
5. Take account of the finite precision of computer arithmetic to give numerically accurate results.

This section covers program design, preparation and testing, and covers aspects of numerical analysis relevant in developing engineering software. The general principles are illustrated with examples from well-known chemical engineering topics.

2.1. Program Design. We present guidelines broadly based on the European Space Agency (ESA) software engineering standards (2). These guidelines are easily adaptable to specific engineering applications. Professional software companies may apply more detailed and formal tools. For example, the Unified Modeling Language (3) presents a formal object-oriented approach to software design, backed by commercial software tools. A number of established texts [eg, NIST (4) and W. Perry (5)] give guidance for professional software engineers. Tanzio (6) puts these validation methods in a chemical engineering context. The general approach given here is consistent with such more detailed tools. Figure 1 presents an outline of the software design process.

The initial stage is to specify exactly what the user wants, the User Requirements. Where the software automates a procedure well known to the intended end-users, these requirements should be drawn up in consultation with the end-users. Where the software can give functionality beyond the experience of current users, the input of innovators in the field is required.

The rationale (reasoning) behind each requirement should be recorded to ensure that the program is written efficiently. For example, there may be a requirement that a simulation should be capable of dealing with two-phase gas/liquid flow. If the intended use is to simulate the boiling and evaporation of water to feed steam turbines, the whole range from 100% liquid to 100% vapor must be covered. Vapor/liquid equilibrium must also be covered and an approach using steam thermodynamics might be appropriate. On the other hand, the intended use may be the simulation of flow in gas pipelines in which small levels of liquid contamination might occur. This requirement is easier to meet with simpler, faster, and more easily tested software. The provision of the rationale for each requirement enables the programmer to provide the most appropriate tools and gives scope to meet the requirement in alternative ways.

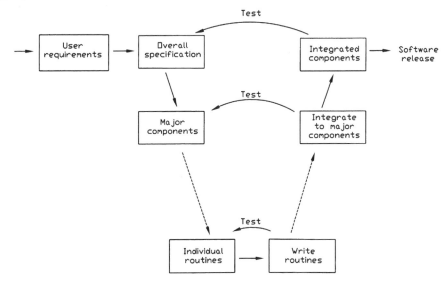

Fig. 1. Software design process.

The User Requirements cover:

The functionality, that is the chemical engineering problems to be solved.

The user interface, ie, how the user will interact with the program (the appearance of any windows, the use of buttons and keys etc).

The environment in which the program will be used. Will it be stand-alone, accessed over the web or used in some other way, eg, as an additional function to be attached to a spreadsheet?

The programs to which it must be interfaced. Does it import data from another program and/or export results to a further program?

The services that it must use. For example, must it employ a defined physical properties package?

The User Requirements prioritize the required functionality. *Essential* requirements must be met in the first release; without this functionality, the software does not meet its basic goals. *Desirable* requirements would add value to the program, but could be deferred or omitted. Desirable features may be further prioretized, ranging from those that a majority of users would wish to see in an early release, to those that would add only marginal value. If they are easily met, desirable features may appear in the first release. A User Requirements document should include features that were considered and rejected for good reason.

The User Requirements should translate directly into tests that will verify that (at least for the specific values tested) the software meets the requirements. These tests form part of a validation plan for the software.

The overall software package is divided into its major components. These software components may be subroutines, classes (objects), components dynamically linked by middleware, or separate programs that will be invoked as needed. The interfaces between these components must be defined so that they can work together. At this stage, separate acceptance tests for each major software component are defined.

The major software components are further subdivided and acceptance tests for each subcomponent defined. The extent of subdivision depends on the size and complexity of the program.

When the program is written, each component is verified against the pre-defined tests before it is integrated into the final program. Similarly, subcomponents are tested both before and after integration. The test procedure is an integral part of the program design.

Before programming starts, a "build" sequence is defined. The build methodology provides early delivery of a simple program offering a subset of the requirements. The program is enhanced in each subsequent build as more functionality is added. For example, a simulation may initially work only for single-phase mixtures with simple ideal thermodynamics and a small subset of chemical components. Each build is tested and evaluated as it is developed. There will be a number of builds before the first version meeting the essential user requirements is released.

The build approach has benefits both in meeting delivery dates and in improved program quality. Delivery dates are improved because any delay in the first build signals problems that can be identified and corrected early. A schedule without early delivery may enable such software problems to remain unrecognized for a long period. Quality is improved because experience with early builds enables the User Requirements to be refined. Iterative refinement gives a better final product. Actual use of a program (even a version with limited functionality) is more effective in highlighting opportunities and difficulties than any paper study. These early releases also give an opportunity to involve the intended final users where, previously, the specification was largely drawn up by specialist innovators.

The build discipline extends beyond the initial release into the support and development phase. User experience provides suggestions for enhancements and uncovers bugs that escaped pre-release testing. In response to this experience, the User Requirements document is updated, the changes prioritized, and new releases planned each with its own test and build schedule. The prioritization of changes depends on the intended end-use of the program. For example, a bug that causes a program crash requires urgent correction in an on-line control application. However, an engineering design program bug that gives answers leading to unsafe designs requires more urgent correction than a bug that causes occasional program crashes.

Each build and release should be archived so that they remain accessible over extended periods. (For example, in a plant upgrade, users might wish to compare results with results they obtained 5 years earlier. Additionally, enhancements may introduce new bugs, and it may be desirable to go back to the last bug-free version.) Versions of major components should similarly be separately archived. Dynamically linked components may follow separate, unsynchronized, build and release patterns.

An effective software design and validation procedure thus lasts throughout the life of the software.

2.2. Programming Languages and Modeling Systems.

Chemical engineering software may be written in a conventional high-level language, an artificial intelligence (AI) language, a general-purpose or specific equation-based modeling system, or using spreadsheet tools. Many engineers construct simulations using modeling tools rather than write special-purpose programs. Simulations developed using these tools must be designed and tested as for conventional programs. There is the same scope for logical and numerical errors.

A separate list of references is given for programming languages and modeling systems. The general-purpose languages described below are not referenced because they are so well known that a wide choice of texts is available from most libraries and other book suppliers.

General-Purpose Languages. Popular high-level languages include C/C++, Java, Fortran90, Delphi, Smalltalk, Eiffel, Ada and BASIC. Most of these languages are now object-oriented. Such object-oriented languages group methods and data in a way that is natural to the engineer. Indeed, the original object-oriented language, Simula, see Birtwhistle and co-workers (7), was developed specifically for simulation. It is no coincidence that Birtwhistle came from a background of simulation in the chemical industry.

The most used languages for large-scale engineering software are Fortran and C++, with most new software written in C++. The evolution of C++ from C results in alternatives for many common features. For example, there are four subtly different array types, a "vector", a "valarray", a fixed-bound directly declared array, and a pointer to dynamically allocated memory. The programmer can also create array-like container classes. Confusion between alternatives gives scope for subtle bugs. Later facilities frequently provide safer alternatives to earlier relatively risky methods. In recognition of the risks, Stroustrup (8) repeatedly emphasizes safe programming methods. A simple subset of C++ is recommended emphasizing safety rather than speed. For example, use only "vector" for arrays.

Languages Designed for Artificial Intelligence. AI languages include LISP and Prolog, and can be used for rule-based programming. They may be convenient for implementing standards that are presented as rules. For more general programs, the rules may be heuristic (based on experience) or based on fundamental science. Where they are based on experience, applications should preferably be restricted to the systems for which the experience was gained. Interrogation facilities should be provided to display the rules employed and the conclusions may need to be tested by conventional modeling tools.

Equation-Based Modeling Systems. Equation-based systems range from equation manipulation systems such as Mathematica and MathCad, through general tools for solving and optimizing problems defined as equations (eg, GAMS) to tools specifically for the chemical engineer, such SpeedUp, and gProms.

Equation manipulation systems are rarely used for large engineering calculations. The difficulty is that there is no control over the numerical characteristics of the resulting equations. These systems are more generally used in developing algorithms. The manipulated equations are examined before the results are incorporated into an assignment-type program or an equation-oriented modeling system.

The remaining equation-based systems do not rearrange individual equations. For example, they cannot convert $a = \ln(b)$ to $b = \exp(a)$. Large sparse sets of algebraic equations are solved by forming a local linearization and solving the local linearization. The equations are re-linearized at the resulting solution point and the iteration continued until a solution is obtained. Differential equations are solved numerically using robust integration methods.

Spreadsheet Tools. Spreadsheets, such as Excel and Lotus 1-2-3, offer many attractions to the engineer. A calculation can be put together quickly and a variety of graphical output is immediately available without special programming. Most spreadsheets also have standard database interfaces for input and output of extensive data.

Spreadsheets have the disadvantage that the formulas tend to be hidden and conditional expressions, required to maintain numerical accuracy, are difficult to program. The selection of cells as matrix sizes are changed also depends on the way that the programmer handles cell references and errors can result. Most importantly, a user may inadvertently overwrite a formula cell with a value. In all these cases, there is the danger that plausible but wrong answers can be obtained. Consequently, it is difficult to apply quality assurance measures to spreadsheet calculations.

It is recommended that spreadsheet computations are restricted to simple accounting-type operations that are easily checked by hand. Relatively complex computations should be undertaken with a conventional programming language.

If required, a conventional program can be linked to a spreadsheet to get the benefits of spreadsheet input–output and of the postprocessing tools that become immediately available.

2.3. Programming.   Programs should be written to minimize programming errors. This section concentrates specifically on techniques for minimizing incorrect results and run-time failures. The section on numerical analysis deals with numerical errors in otherwise correct programs. The topics covered in this section are

Dimensional consistency of programs.

Limiting side-effects.

Limiting run-time.

Limiting use of computer memory.

Arithmetic failures.

Division into component parts.

Checking data.

Handling uncertainty.

Minimizing error messages.

Dimensional Consistency.   Every assignment and comparison in a chemical engineering program should be dimensionally consistent. Thus, any program that assigns a velocity to a mass is certainly wrong. Errors such as writing

$$E = 0.5 * m * v$$

when the intended assignment is

$$E = 0.5 * m * v * v$$

cannot be detected by any of the standard high level compilers. Such mistakes do not cause the program to crash and rarely give obvious run-time errors. In the above example, the error is not obvious unless v differs considerably from unity. Such errors can be detected by checking that every term on the right-hand side of an assignment or comparison has the same dimensions and that the dimensions are the same as on the left-hand side. Tools are being developed for automating such tests before compilation, but generally the tests must be done manually.

Alternatively, dimensional consistency can be checked dynamically (at run time). Any computer language permitting overloading (including all object-oriented languages) permits the definition of new data types. A "dimensioned" data type, which consists of the floating point value plus the dimensions, is needed. For example, with dimensions in the sequence mass (M), length (L), time (T), temperature (Θ), a force of 27.9 may be recorded as $\{27.9, 1, 1, -2, 0\}$. The significance is 27.9 units, with dimension MLT^{-2}.

In this system, the standard mathematical operations are modified so that multiplication multiplies the values and adds the dimensions. Addition, comparison, and assignment check that the dimensions are consistent and generate a run-time error for any inconsistency. Other operations are similarly modified.

The technique is slow, but gives a thorough check. The final release version may have the checking facility removed.

As a further assurance, engineering programs should be written in dimensionally consistent units. Thus, do not mix viscosity in poise (the cgs unit) with pascal-seconds ($\text{kg} \cdot \text{m}^{-1} \cdot \text{s}^{-1}$, the SI unit). A dimensionally consistent program will work equally well in the cgs, SI(MKS), or fps systems. It avoids dimensional constants embedded in the program. Unit conversion should be applied at input and output, not within the program itself.

Side Effects.   Side effects occur as follows. A statement is written such as

$$y = \text{somefunction}(x)$$

This operation alters "y". There is a suspicion that "x" may also be altered. No other variables should be altered. If, on calling "somefunction", an unrelated variable z is altered, the change is known as a side effect. Side effects give error-prone programs and make debugging and upgrading difficult. Where side effects are excluded, a problem in the above statement can be localized to "somefunction". That is then the only function that needs checking. If side effects are allowed, every function that might contain "z" has to be checked as does every line of code in which z appears. If a function containing z also contains side effects, the number of lines of code that need checking are further multiplied. It is thus strongly recommended that side effects be avoided.

Some languages strictly forbid side effects, to the extent of not allowing changes in x. However, all popular languages allow side effects. Fortran COMMON allows every program module to access variables listed after a COMMON label. It is not necessary to put these variables in argument lists to access them. Assigning values to COMMON variables thus creates a side effect. COMMON was recommended in the early days of Fortran because it gave better run-time efficiency than passing parameters as subroutine or function arguments. Fortran90 discourages the use of COMMON and provides safer alternatives. C++ provides both pointers and global variables. Pointers enable several variables to refer to the same memory location. Thus allocating a value to one of the variables alters all the others. Reference variables can provide similar confusion. If not declared "constant", global variables can be accessed from anywhere. Thus, any routine that is called can alter one or more global variables (C++ global variables are similar to Fortran COMMON in this respect). These altered global variables can have unsuspected effects elsewhere in the program. Pointers were recommended in the early days of C to avoid copying whole structures. Run-time efficiency was thereby improved. In engineering programs, the time saving is likely to be minimal; there are many more operations performed on the elements of a structure than merely copying them. The major current use for C++ pointers is accessing components dynamically linked by middleware.

Polymorphism provides a concise way of performing related tasks. (Polymorphism provides a common interface capable of invoking a variety of related behaviors that do not need to be defined in advance.) However, its use employs side effects, and it is deliberately excluded from some languages (Fortran90). It should be used only when the alternative would be a more complex program.

All side effects should be used sparingly and commented whenever they are used.

Limiting Run Time. There should be a known upper limit on run time for all engineering software. On-line programs should be strictly timed in advance so that there is always slack time and interrupts cannot accumulate without limit. Possible long run times for off-line programs should be noted in the documentation and a warning message displayed before the time-consuming part of the calculation commences.

Run-time should be estimated from the data values read. Where theory is deficient, the engineering programmer should produce an experimental correlation of run time versus problem size. It is then possible to set a fixed upper limit on number of iterations that will give an acceptable run time on most computers. Where, because of the limit, the program runs out of iterations, the programmer can provide a meaningful message related to the engineering problem being solved. In most cases, an estimate of the result can also be given. Unless the program has its own stop button, a program stopped by the user can provide no meaningful message.

Limiting Use of Computer Memory. As for run time, it is should be possible to make a conservative estimate of how much memory will be required as a function of data values. An upper limit for space required should be set in advance so that memory overflow is avoided. Where additional space is taken at each iteration (eg, in branch and bound optimization), an iteration count can limit memory used. A run-time error message from the computer operating system when computer memory is exhausted is of no value to the user. However, the programmer can provide a meaningful message related to the engineering problem. Where possible, it is more efficient to allocate the maximum memory needed at the beginning of the computation rather than resize arrays as the computation proceeds.

Arithmetic Failures. Arithmetic errors include divide by zero, square root, or logarithm of a negative number, and exponential overflow. Each potential failure should be tested before the expression is computed. If part-way through an iteration, the error should be suppressed (see section on Minimizing Error Messages) otherwise a meaningful error message should be displayed (for example, crossover in the computation of log−mean temperature). Most arithmetic failures can be avoided by appropriate numerical analysis (see section on Numerical Analysis) or program organization (see section on Arranging Expressions for Computation). An engineering computer program should never fail with a message generated by the computer operating system; such failures do not help the end-user.

Division into Component Parts. Programs divided into logical self-contained component parts are quicker to write and easier to test. However, component size must be chosen carefully. The benefits of componentization are lost with excessively large components. On the other hand, a program divided into excessively small components is dominated by interface programming. The interface programming introduces more lines of code and more potential bugs. It also increases run time and obscures the program logic. Considerations of program size and run-time should not dominate in deciding component size. It takes many man-years to write a megabyte of object code. The bulk of memory is taken by numerical data, text, and particularly graphics data. Any marginal space saved by sharing object code between programs would be completely swamped by a saving that could be made by displaying a simpler graphic. Similarly, time savings by in-line coding (repeating source code in each routine to avoid the time taken to call common code through an interface) are negligible. The next hardware release will make much greater time savings. Time reduction can best be achieved by attention to the algorithm. For example, employ a polynomial algorithm rather than an NP algorithm, and amongst polynomial algorithms, choose one second order, rather than third order, in problem size. Component size should thus be made on the basis of clarity, simplicity, and maintainability rather than run-time and size.

Components can be provided as classes or functions that are linked as integral parts of the compiled program, or can be dynamically linked at run time. Dynamically linked components are favored where they are obtained from third-party sources. Dynamic linking is also favored for components with a distinct use that may be shared by several independent programs. Dynamic linking allows components to be developed and released independently of the main program. However, the resulting lack of release synchronization makes quality assurance more difficult, particularly when there are large numbers of dynamically linked components. There is a risk that users will employ a different set of components than has been tested with the delivered program. Consequently, dynamically linked components should be employed only when there is a strong case for using them. Object-oriented programming languages are designed to be modular, and small components can be written as classes and bound into several different programs. The majority of components are best provided as classes and functions within such object-oriented programming languages.

Checking Data. Erroneous output from a well-written program is most likely to be the result of erroneous data entered by the user. Programs should be made resistant to erroneous data both by checking the magnitudes of the values input and by checking the consistency of the data. An error message should be issued if the sign is wrong. For example, the sign convention for the "B" coefficient in the Antoine equation differs in different databanks. The coefficient should be checked to ensure that the sign is consistent with the convention in the program.

The most common data input error is confusion over units and dimensions. These errors can be trapped by checking that the numbers are within reasonable bounds. For example, a program designed to deal with liquid hydrocarbons should warn against density >3000 kg/m^3 or <300 kg/m^3. The density of water is 1000 kg/m^3, 1.0 g/cm^3 and 62.5 lb/ft^3. Consequently, this check will detect incorrect units for any common liquid. Similarly, incorrect conversion between metric units can be detected because most conversions introduce a power-of-10 error. Issue a warning rather than an error in case a user wants to model less common species. It is the end user's responsibility to check all warning messages (see section on Using Engineering Software). Additional tests are available for specific properties. For example, gas specific heats can be checked using the ratio C_p/C_v, with $1.0 < C_p/C_v < 1.7$. For gases, the Prandtl number is well predicted from the ratio of specific heats, which gives a further check on the consistency of heat capacity, viscosity, and thermal conductivity data. There are also a number of rigorous tests for thermodynamic consistency of data.

For data generated by the program, messages giving likely error bounds should be produced. Such messages should be given both for estimated data (eg, physical properties computed by the group contribution method) and for default values. Such warning messages serve the additional purpose of reminding the user that a default has been used. For example, a program designed for hydrocarbon liquids might employ a mean default viscosity, but generate a message indicating the possible range from lightest to heaviest.

Handling Uncertainty.   Chemical engineers employ many semiempirical correlations, the potential errors in which should be notified to users. These correlations give uncertain predictions, even with accurate data. For example, pressure drop calculations are based on the work of Stanton and Pannell (9) updated by Moody (10). Turbulent heat transfer correlations are derived from the work of Dittus and Boulter (11), updated by McAdams (12). There is considerable experimental scatter about the empirical charts and equations put forward by these authors. It is recommended that programs should not add safety margins to account for the correlation uncertainties. The cautious "safe" bound depends on the user's application and it is the user's responsibility to apply safety margins. It is the programmers' responsibility to make the uncertainties clear to the end user. Provision should also be made for the user to explore the effect of uncertainties on the results of the computation. In many cases, the effect can be simulated without direct access to the model or its built-in parameters. For example, the user can investigate the effect of an uncertain heat transfer coefficient by altering the corresponding area. In other cases, it may be necessary to give the user direct access to uncertainties in the model. For example, the user can be provided with a parameter that is set to 0.0 for the most likely result. It is set 1.0 for the high result at the 90% probability level and -1.0 for the low result at the 90% probability level.

Minimizing Error Messages.   Most chemical engineering computations are iterative. The end-user is only concerned with the correctness of the final result. Error and warning messages should be suppressed until the final iteration. Such messages should then be output where they can be archived as part of a decision audit trail created by the end user. Where potential failure conditions arise during the iteration (eg, underflow or overflow), a suitable value should be generated to allow the iteration to continue in anticipation that the error will disappear as the solution is approached. The values generated should avoid introducing function discontinuities, which can have an adverse effect on convergence.

2.4. Numerical Analysis.   A program that gives wrong answers or fails is unacceptable. This section introduces general principles and useful tools for avoiding or minimizing these problems. The principles are illustrated through specific examples. We concentrate on very simple cases that will be faced by the majority of engineer programmers. eg, we consider evaluation of expressions that might come to 0/0. This situation potentially arises wherever there is a division. There are many numerically difficult cases that are not treated in this chapter, eg, solution of stiff differential equations (those with a very wide range of time constants) or large sets of ill-conditioned algebraic equations. Programmers are referred to specialist texts on numerical analysis in such cases, eg, Epperson (13). Chemical engineers are most likely to include third party routines rather than write such specialist methods themselves. There are a number of sources for such routines, eg, the HSL Library (14), the NAG Library (15), and Numerical Recipes (16). The basic advice provided here for the simple cases still applies for the more difficult problems; eg, it is still necessary to avoid numerical errors in evaluating expressions and to check on convergence.

Four areas that can give rise to numerical problems are covered.

1. Expressions that evaluate to 0/0.
2. Convergence of simple iterations. Both single variable and multivariable iteration is considered.
3. Solution of equations and matrix inversion.
4. Numerical solution of differential equations.

Expressions That Evaluate to 0/0.   Consider a program for designing a countercurrent heat exchanger. The program employs a log–mean temperature difference. The simple method of programming the mean is to put:

```
deltaTmean = (deltaT1 − deltaT2)/ln(deltaT1/deltaT2)
```

Where

```
deltaT1, deltaT2 and deltaTmean
```

are the temperature differences at the two ends and the log-mean of these two differences.

The expression is impossible to compute if one of the following conditions apply:

1. The temperature differences have opposite signs at the two ends. Such a crossover is physically impossible. The program will fail with a "logarithm of negative number" error.
2. The temperature difference is zero at one or other end. The program will fail with a "divide by zero" error

   ```
   (deltaT2 zero) or ''logarithm of zero'' error (deltaT1 zero).
   ```

3. The temperature differences are the same at the two ends. The ratio

   ```
   deltaT1/deltaT2
   ```

   is unity, the logarithm of which is zero. The program then fails with a "divide by zero" error. This condition is thermodynamically most favorable and is trivial to compute by hand.

   In addition to the above obvious failures, significant errors arise when

   ```
   deltaT1 and deltaT2
   ```

   are similar. Consider the following case:

4. ```deltaT1 differs from deltaT2```

   by a small fractional difference δ and the machine precision is ε. For floating point arithmetic, ε is a fractional error almost independent of the absolute size of

`deltaT1 or deltaT2`.

The proportional error in both the numerator and the denominator is then ε/δ. These errors rarely cancel. If δ is sufficiently small, the error can exceed 100%. This situation is not as rare as it might seem. For example, an optimization can gradually bring the two temperature differences together until the error condition arises. This error is worse than the errors 1–3 because it gives a wrong answer with no warning.

Cases 3 and 4 are both numerical problems resulting from the finite precision of computer arithmetic. Cases 1 and 2 are more fundamental, and are treated below under "arranging expressions for computation". Solving case 4 automatically solves case 3 and it can be treated as follows. Writing the temperature differences as u and v, and the mean as y, gives:

$$y = (u - v)/\ln(u/v) \tag{1}$$

The general approach is to expand the terms that approach zero as a series. The denominator then becomes

$$\ln(u/v) = 2[(u - v)/(u + v) + \{(u - v)/(u + v)\}^3/3 + \cdots] \tag{2}$$

Substituting equation 2 into equation 1 gives:

$$y = 0.5(u + v)/[1 + \{(u - v)/(u + v)\}^2/3 + \cdots] \tag{3}$$

Equations 1 and 3 are alternative ways of computing the mean. For small values of the difference, the term in curly brackets is given by

$$(u - v)/(u + v) = \delta/2$$

When δ is small, it is apparent that equation (3) gives the correct result. For larger values, the truncation error steadily increases. Equation 1 has the opposite behavior; for small values of δ, round-off gives erroneous results, but it is accurate when δ is large. Figure 2 shows the dependence of round-off and truncation error on δ for single precision on an IBM compatible PC. To maintain maximum precision throughout, equation 3 should be used when it is more accurate than equation 1. The proportional errors are, respectively.

Equation 1, round-off error resulting from finite precision in computer arithmetic:

$$r = \mathrm{abs}(\varepsilon/\delta)$$

Equation 2, truncation error in taking only a finite number of terms in an infinite series:

$$t = \delta^2/12 \tag{4}$$

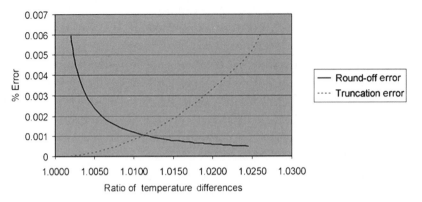

Fig. 2.   Errors in computing log–mean temperature differences.

Equation 4 is derived from equation 3 by taking only 1 term in the series so that the first term ignored is $\{(u-v)/(u+v)\}^2/3$. Equation 3 is preferred to equation 1 when $t < r$. Namely when

$$\mathrm{abs}(\delta^3) < 12\epsilon \qquad (5)$$

Taking only the first term of equation 3 gives the following simple expression:

$$y = 0.5(u+v)$$

Thus, for small temperature-difference ranges, it is more accurate to use the arithmetic mean as an estimate of the logarithmic mean than to evaluate the logarithmic mean directly.

The value of ε is obtainable either from documentation on the compiler used, or directly from facilities available as part of the computer language employed. For example, in C++, the value is given by

$$\mathrm{eps} = \mathrm{numeric\_limits::epsilon}();$$

and

$$\mathrm{eps} = \mathrm{numeric\_limits::epsilon}();$$

depending on whether single or double precision computation is employed. On a PC, the value of

eps

for single precision is 1.19207e-7. The worst error using equation 5 is then

$$r = t = \epsilon^{2/3}/12^{1/3} = 1.058e - 5$$

Thus, even always choosing the best method of computing the mean still gives a maximum error 100 times worse than machine precision. Figure 2 shows the truncation and round-off errors plotted against ratio of temperature differences for a PC in single precision. This figure shows that it is more accurate to use the arithmetic mean up to differences in excess of 1%.

In double precision, the value of eps for a PC is 2.22045e-16, and the worst error is 1.6e-11. In this case, the computer precision is reduced by a factor of 100,000.

If the resulting precision is not adequate, the error can be further reduced by taking the first two terms of equation 3. Equation 3 can now be used over a wider range up to

$$\text{abs}(\delta^5) < 40\epsilon$$

On a PC, the maximum error is then given by

$$r = t = \epsilon^{4/5}/40^{1/5} = 1.382e - 6 \text{ (single precision) or } 1.1435e$$
$$- 13 \text{ (double precision)}$$

The loss in precision is then reduced to a factor of 10 in single precision and a factor of 100 in double precision.

These considerations of numerical precision apply equally to equation-based systems. Such systems should employ a log–mean function coded as above, or an approximate log–mean that cannot accumulate numerical error, eg, the Underwood mean

$$y = [(u^{1/3} + v^{1/3})/2]^3 \tag{6}$$

Care must be taken in implementing equation 6 to ensure that the cube root function employed is capable of computing the roots of negative numbers.

This example illustrates a general point in engineering software. Wherever there could be a divide by zero error, check to see whether there is a definite value at the zero point. If there is, large errors are likely when the numerator and denominator are near zero. It is then necessary to make a series expansion; in many cases of both the denominator and the numerator. The resulting expansion can be used to find the range in which it is better to use the first few terms of the expansion rather than use the expression that evaluates to 0/0. It is also possible to determine how many terms of the expansion to employ to achieve a desired precision.

Convergence of Simple Iterative Schemes.   Iterative methods are used to solve many engineering problems. For a single variable problem, successive estimates can be written

$$x + e_1, x + e_2, x + e_3, x + e_4, \cdots$$

Where x is the correct result and e is the error. The term $(x + e_1)$ is the first guess of the solution. For n th order convergence, successive errors are given by

$$e_{i+l} = ke_i^n$$

where k is a constant coefficient. If, for any i, $ke_i^n < e_i$, all subsequent errors decrease and the scheme converges.

First order convergence $(n = 1)$. is commonly found in chemical engineering computations. (It only takes one first-order step in an otherwise second-order scheme to reduce the order of convergence.) For first order convergence,

$$e_j = k^j e_o$$

Many chemical engineering programs accept convergence if

$$f = \text{abs}\,(x_i - x_{i-1}) = \text{abs}\,(k^i e_0 - k^{i-1} e_0) < c$$

where c is some convergence criterion.

The corresponding error may, however, be much larger than c. Thus

$$e_i = kf/(1 - k) \tag{7}$$

The rate of convergence (measured by k) is likely to change from data set to data set. On slowly converging data sets, eg, with $k = 0.9$, the error is likely to be large compared to the convergence criterion c. The recommended strategy is to compute k (from f_i/f_{i-1}) and average it over several iterations. The error given by equation 7 can then be estimated and a consistent precision accepted on convergence.

Where the magnitude of k consistently decreases from iteration to iteration, convergence is better than first order. For higher-order convergence, equation 7 gives a conservative estimate of residual error (thus, the actual error is less than the value computed by the equation).

For a multi-variable problem with m variables, f in equation 7 is replaced by

$$f_i = \sqrt{\sum_j (x_{j,i} - x_{j,i-1})^2 / m}$$

e_i is then a measure of the mean error of the variables. For each iteration, k is computed from

$$k = f_{i+1}/f_i$$

Where higher order convergence is achieved, this procedure gives a conservative estimate of residual errors.

If e_i cannot be reduced to zero, it indicates either that the set of equations has no solution or that there is an accumulation of numerical error. In either case, the numerical methods discussed in the following sections may be applied.

Solution of Equations and Matrix Inversion. There is a frequent requirement in chemical engineering to solve sets of equations. For example, applications arise in modeling complex multiple reaction processes, in statistical analysis fitting experimental data, and in balancing flowsheets including recycles. Frequently, the equations are nonlinear; such sets of nonlinear equations are solved by successive linearization. Thus, solution of sets of linear equations is central to much chemical engineering computation.

Where a single nonlinear equation is solved, the programmer will normally write the solution routine. The solution point should first be bounded by two points with residuals of opposite sign. These bounds should be reduced at each iteration as new points are computed. If a next iterate indicates a solution outside the bounds, the basic iteration should be replaced by a simple division algorithm (eg, halving) that generates a potential solution within the bounds. This fall-back algorithm will be invoked when the local linearization has a low, or zero, slope and thus predicts a solution far from the current point. Bounding in this way thus automatically avoids divide by zero. It is equally applicable to Newton's method (the linearization is the tangent line at the most recently computed point) or the secant method where the linearization is a line joining two points on the nonlinear residual curve.

Nonspecialist engineers will solve large sets of nonlinear equations using third party tools that have been optimized to minimize accumulation of numerical errors. However, when used as part of a larger iterative scheme, such third party tools can fail either because a set of equations having no solution has been generated, or because of accumulation of numerical errors. Steps can be taken to minimize such failures and we outline here some of the steps that can be taken.

It is first necessary to have some understanding of the problem characteristics that lead to solution failures. There are two cases in which solution methods frequently fail. The first is when the equations are redundant; the second is when they are inconsistent. The cases can be illustrated by reference to the following sets of equations.

Consider the equations

$$2x_1 + 3x_2 = 2$$
$$2x_1 + 3x_2 = 2$$

and

$$2x_1 + 3x_2 = 3$$
$$2x_1 + 3x_2 = 1$$

The first set shows redundancy (the same equation is repeated twice) and there are an infinite number of solutions. The second shows inconsistency and there are no solutions. In an iterative scheme to solve a nonlinear set of equations, it would be hoped that this situation would not occur at the final solution point. (Such inconsistent problems can easily arise, eg, in modeling systems including distillation. During iteration, it can occur that 100% of a steadily produced or fed component exits from the top and is recycled. There can be no material balanced solution until other values have been modified so that <100% is recycled.) We require a strategy that will allow the iteration to continue. A suitable strategy is to modify the equations as follows. Write the set of equations

$$2x_1 + 3x_2 = 3 + \epsilon_1$$
$$2x_1 + 3x_2 = 1 + \epsilon_2$$

Now solve the problem that $\Sigma \epsilon_i^2$. is minimized. This step converts an inconsistent set of equations to a set that can be solved or (as in this case) a set that contains redundancy. Redundancy can be removed by adding the constraint that Σx_i^2 is minimized. This constraint must have a much lower weight than the constraint to minimize the errors in the right-hand sides, or an incorrect solution will be obtained when a correct solution is possible. Depending on the physical situation being modeled, this strategy can be modified to produce answers with any desired properties without compromising a correct solution where that is possible. Modifying the problem can thus eliminate difficulties for third-party solvers.

The most frequently used approach to solving sets of linear equations is LU factorization [see *Perry's Chemical Engineers' Handbook* (17)]. Whether or not the inverse of the matrix of x coefficients is also produced, the method is third order in problem size. Thus, run time is proportional to the cube of number of equations. The discussion above relates primarily to this solution strategy. However, a number of chemical engineering programs employ an alternative method when solving sets of nonlinear equations which, in principle, is faster. It is noted that the matrix to be inverted only changes slightly at each successive linearization. Instead of completely reinverting the matrix, the inverse can be directly updated (eg, by quasi-Newton or by methods that exactly replace one row or column). These updates are second order, so that the relevant linear equations can be solved by a second-order rather than a third-order method. With >10 equations, the method can be significantly faster. A disadvantage of the updating method is that the quality of the inverse can progressively deteriorate as errors accumulate. This section presents a simple procedure for checking and refining inverse matrices.

The most direct test that matrix B is the inverse of matrix A is to form the difference

$$D = AB - 1$$

$$D_{ij} = \sum_k A_{ik}B_{kj} \qquad i \neq j$$

$$D_{ii} = \sum_k A_{ik}B_{ki} - 1$$

The size of d can be measured in various ways. The simplest is to find d, the root-mean-square size of the elements. Thus, for an $m \times m$ matrix

$$d = \sqrt{\sum_i \sum_j D_{ij}^2 / m}$$

If d is comparable with machine precision, B is a good estimate of the inverse of A. If d is too large, an improved estimate of B is given by

$$B' = B(1 - D)$$

If $d < 1/m$, it is guaranteed that $d' < d$, where d' is found from the matrix AB'. This convergence criterion is conservative and B' may be better than B, even when the condition is not met. The refinement can be applied iteratively to achieve sufficient accuracy from any suitable inverse estimate. The method is simple and safe to program because it requires no divisions. It is, however, third order and slower than recomputing the inverse by LU factorization.

Similar methods can be employed for handling matrix inversions that arise in nonlinear optimization and statistical analysis.

To avoid numerical errors during a larger iteration, it may be desirable to convert non-invertable matrices to invertable matrices. Similar adjustments can be made to those described above for solving redundant or inconsistent equations.

Solution of Differential Equations.   Numerical analysis textbooks describe a number of robust methods for solving differential and partial differential equations and many of these methods are available in computer program libraries, such as those referenced above. It is beyond the scope of this chapter to classify and describe these methods. In general, it is recommended to consult such specialist sources in developing solutions to engineering problems. However, chemists and engineers frequently employ simple discretization when solving problems such as tracing composition changes through a plug-flow packed-bed reactor, or the progress of a reaction in a well-stirred batch. This simple discretization is equivalent to a forward-difference (forward Euler) solution procedure. The poor numerical performance of the method is well known. We present here, a modification that can improve the performance of simple discretization methods without employing more powerful library methods. The Central difference (or Tapezoidal rule) method is described.

In order to solve a single differential equation to give f as a function of t, the forward difference method works as follows: The parameter f and its derivative df/dt are computed at time t. An estimate of f at the next time increment is then

$$f_{t+\delta t} = f_t + (df/dt)_t \delta t \tag{8}$$

The value of f at time $(t + \delta t)$ enables the derivative at this point to be computed and the integration continued. The procedure is simple and requires no iteration. However, it has several disadvantages. Even without accumulating numerical error, it can predict unstable oscillatory behavior for systems that are stable in practice.

Improved performance can be obtained by replacing the derivative computed at t by the mean of the derivatives computed at t and $(t + \delta t)$. The new formula is

$$f_{t+\delta t} = f_t + 0.5[(df/dt)_t + (df/dt)_{t+\delta t}]\delta t \tag{9}$$

The computation proceeds as follows. The value at time $(t + \delta t)$ is computed as for the forward difference case. The derivative at the new point is computed and substituted into equation 9 to give an improved estimate. The iteration is continued to convergence. The value of f that solves the equation may be obtained by established methods such as the Secant or Newton method, or, if the derivative expression is sufficiently simple, analytically.

Where there are a large number of elements in f, the central difference method thus requires the solution of a set of (in general) nonlinear algebraic equations. Process simulators include powerful built-in methods for solving such equations. These built-in solvers have been successfully employed to model three-dimensional (3D) reaction and heat transfer using central differences within a conventional dynamic process simulator.

The benefits of central difference are that the same precision can be achieved with fewer (longer) steps and the method does not show spurious unstable behavior.

Where error checking is not a part of the integration routine employed, the precision of integration should be checked. For an nth order integration, the error is given by

$$\epsilon = k\Delta x^n$$

where ϵ is the error and Δx the step-length for integration. For a first-order integration, the error per step varies as the square of step length. The total error is the error-per-step multiplied by the number of steps. The number of steps is inversely proportional to step length. Hence, the net effect is that total error is directly proportional to step length.

A simple check that can be put into any engineering program is to repeat the integration with twice the step length. Two error estimates result:

$$y_1 = y + k\Delta x$$

and

$$y_2 = y + 2^n k\Delta x^n$$

Combining the two equations, gives

$$y_1 = y + (y_2 - y_1)/(2^n - 1) \tag{10}$$

The error in the most accurate integration is thus less than or equal to the difference between the two estimates. Adding this double step-length integration increases computational time by 50% for simple integration and by 25% for two-dimensional (2D) integration. The cost in computational time is thus relatively low in order to provide a measure of quality control for the integration. Where there is confidence in the order of integration, equation 10 can be used to obtain a better estimate of y. (For $n = 2$, this is known as Richardson's h-squared method). Where there is doubt about the order, it is cautious to assume a higher order for predicting a solution but a lower order for estimating the error. Theoretically, forward difference integration is first order and central difference integration is second order.

2.5. Arranging Expressions for Computation. The general principles to be applied in arranging equations for computation are as follows:

1. The equations should be as nearly linear as possible over as wide a range as possible.

2. The equations should be computable for all values of the right-hand side variables.

3. Explicitly computable expressions should be preferred to expressions that need to be solved iteratively.

Linearity is required because expressions frequently form part of larger iterative schemes. Such iterations are nearly always solved by local linearization (eg, Newton–Raphson iteration). The size of the region within which rapid convergence is achieved is determined by the size of the region within which the relevant equations are (nearly) linear. These considerations apply equally to conventional (assignment-type) programs and to equation-based modeling systems.

If the right-hand side (or equation) is not computable, no iterative scheme can make sensible progress to a converged solution.

Explicit expressions are preferred both because they are faster and because the risk of an iteration failing to converge is eliminated.

In order to put the equations into the best form for computation, it may be necessary to derive them from first principles rather than employ a conventional textbook formula. The general principles will be illustrated by reference to two specific examples, heat exchanger simulation and isothermal flash simulation.

For heat exchanger simulation, the equation most often seen in chemical engineering literature is

$$Q = UA\Delta T$$

where ΔT is the logarithmic-mean temperature difference. This equation has been used extensively in simulation, optimization and even process synthesis studies. The heat exchanger is simulated by estimating one of the exchanger outlet temperatures, computing the other by heat balance, and hence determining the log–mean temperature difference. The log–mean temperature difference is used to compute the heat transferred. The outlet temperatures are then recomputed and the iteration continued. This iteration is seen explicitly in modular simulators but, although present, may not be immediately obvious in equation-based simulators.

An improved formulation can be obtained by noting that the log–mean temperature is derived from an analytical solution of an ideal countercurrent (or cocurrent) exchanger with constant physical properties. This analytical solution can be employed directly to avoid the temperature iteration, thus

$$T_{\text{out}} = aT_{\text{in}} + (1 - a)t_{in} \tag{11}$$

where

$$a = p(1 - S)/(1 - pS)$$
$$p = \exp\{UA(S - 1)/(M_h C_h)$$
$$S = M_h C_h / M_c C_c$$

where T is hot stream temperature, t cold stream temperature, M is mass flow rate, C is specific heat, and subscripts and c and h correspond to cold and hot streams.

There is a similar equation for outlet cold temperature. When $S \approx 1$, equation 11 shows the same numerical problems as for equation 1 when u $\approx$ v. A similar expansion is required to compute a for this case.

We have achieved the following goals:

1. T and t are linear functions of the inlet temperatures
2. The temperatures are computable for all right-hand side variables. Specifically, the temperature cross-over problems associated with computing log–mean temperatures are eliminated.
3. The expressions for outlet temperature are explicit and not iterative.

Equation 11 is particularly useful in computing the temperature distribution in a network with fixed fluid flows. The temperature distribution is determined by linear equations and the complete temperature distribution can be obtained quickly and noniteratively.

For nonconstant physical properties, the equation gives the same errors as are incurred by employing the log–mean temperature. (It is based on exactly the same treatment, so this behavior is to be expected.)

The log–mean temperature is more frequently seen in the literature because it is suitable for hand "design" calculations. The computer-aided approach is to design through simulation. Thus, a design is hypothesized that is then simulated. Sizes and operating conditions are adjusted to improve performance.

For the heat exchange problem, it is possible to generate equations that are exactly linear in the required (temperature) variables. Where such an exact linearization is not possible, it is still desirable to make the equations as nearly linear as possible. The following example illustrates application of the principle to such a problem, namely, the simulation of a simple isothermal flash.

In the simplest isothermal flash simulation, an ideal mixture of known composition and feed rate is fed to a vessel held at constant temperature. The feed splits into liquid and vapor phases, and the objective is to determine the proportion and composition of each phase. In the simplest case, the vapor mole fraction, y_i, is computed from the liquid mole fraction, x_i, by the k-value relationship assuming that k-values depend only on temperature, thus

$$y_i = k_i x_i$$

An iterative solution is employed starting with an initial estimate of the liquid fraction, u. Material balance gives

$$z_i = ux_i + (1 - u)y_i = ux_i + (1 - u)k_i x_i \qquad (12)$$

where z is the feed mole fraction.

Rearranging equation 12 gives

$$x_i = z_i / [u(1 - k_i) + k_i] \qquad (13)$$

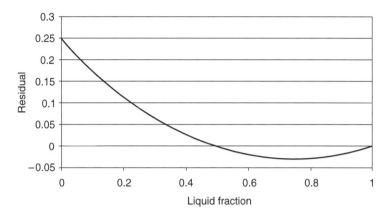

Fig. 3.   Isothermal flash residuals: Liquid mole fractions sum to 1.

The liquid mole fractions sum to 1. Hence, equation 13 gives

$$1 = \Sigma z_i / [u(1 - k_i) + k_i] \qquad (14)$$

Equation 14 is a single variable equation for u. The solution can be substituted into equation 13 to obtain the liquid mole fractions, and hence also the vapor mole fractions. The flash can thus be fully computed.

Equation 14 does not meet our criterion of a near-linear relationship. Figure 3 shows a typical plot with a solution at $u = 0.5$. It is seen that there is a spurious solution at $u = 1.0$. Any iterative scheme shows poor convergence properties if the initial estimate is >0.5, and it can converge to the wrong solution. These difficulties are equally experienced in conventional programs and in equation-based modeling systems.

A more nearly straight line is obtained by taking the difference between the vapor and liquid mole fractions, which results in equation 15:

$$0 = \Sigma (k_i - 1) z_i / [u(1 - k_i) + k_i] \qquad (15)$$

The corresponding curve is shown in Figure 4. Differentiation of equation 15 shows that it is monotonically positive, and nearly linear, for all compositions and k-values.

Equation 15 gives the additional benefit that the end points correspond to the relevant dew and bubble-point conditions. Thus, if the function is zero at $u = 0.5$, the mixture is all-liquid, just at its bubble point. A positive value corresponds to a single-phase liquid below its bubble-point. Similarly, if the function is zero at $u = 1.0$, the mixture is all-vapor at its dew point, and negative values correspond to a gas-phase above its dew point.

The linearity of equations can often be improved by judicious choice of variables. For example, in computing bubble and dew points, the relationship between the logarithm of the total pressure and $1/T$ is more nearly linear than the relationship between P and T. In other cases, it is often better to chose partial pressures than total pressure and mole fraction as variables.

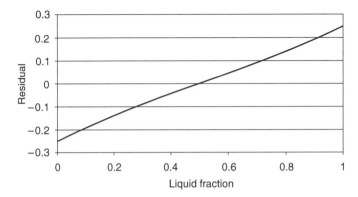

Fig. 4.   Isothermal flash residuals: Sums of liquid and vapor mole fractions equal.

3. Using Engineering Software

The moral and professional responsibility for engineering decisions rests with the engineer making the decisions rather than the authors of any software employed. Usually, the legal responsibility also rests with the engineer making the decisions. Many of the topics introduced in this section are described more fully by Best et al. (18).

Engineers using computer programs written by others must thoroughly understand the application to which the program is applied. They should provide a decision audit trail so that all recommendations can be checked. The audit trail should include

1. A clear statement of the problem being tackled.
2. A statement of the assumptions made and their justification.
3. A review of the software applicable to the problem.
4. Identification of the chemical species that might arise.
5. Review of the data sources and the range of temperature, pressure and composition over which they are valid.
6. Review of the models employed by the software, their validity and applicability.
7. Estimation of the errors that might be introduced through the data or the models.
8. Sensitivity tests to assess the consequences of possible data or model errors.
9. The alternative solutions that have been generated.
10. A critical assessment of the performance and risks of the alternative solutions.
11. The recommended solution.

The audit trail should also include all error messages generated by the software used and a critical assessment of the implications of the messages.

The statement of the problem is the vital starting point for the study. Economic, safety, and environmental goals should be clearly stated. For chemical or petroleum production, requirements for product purity and production rate should be stated and the reasons for these requirements recorded. For example, the end uses of the product may be given and the consequences of impurities noted.

The software available for the study should be listed. One study might employ up to 30 separate programs. These may range from simplified modeling and synthesis software for generating a range of possible process variants, through detailed process simulation, to software for mechanical design, safety assessment and environmental impact assessment. Each potential program should be assessed against criteria of suitability for purpose; these criteria are discussed in more detail below. The extent to which each program has been validated should also be noted. General-purpose programs obtained from third-party suppliers are generally well validated. However, they will not have been validated for the specific problem to which they are to be applied. The end-user thus has to assess the relevance of the prior validation and may need to formulate further tests.

All software needs to be validated by the user as well as the writer (6,18). The documentation should fully describe the models employed and give clear guidance on use of the program. If this information is deficient, it is indicative that the program may also be deficient. Where the company employing the engineer has in-house standards, company-validated software should be employed. The engineer should clearly understand the phenomena being modeled by the software. It is not possible to take responsibility for decisions made in ignorance. Companies should retain a consultant, or in-house expert, on any software that they use. In this way, all users can consult experts who fully understand the software employed. The program should be verified for simpler systems for which results are known in advance. The validation is then gradually elaborated until the required results are obtained.

An important part of any study is to identify the chemical species that might occur; no computer program can model components omitted from the data. Minor components can build up if there is no way of discharging them. There are cases (particularly with batch process) where minor components have caused explosions and toxic releases. Such minor components can also effect the properties of mixtures, eg, causing or breaking azeotropes.

For each chemical component, the reliability of the data available should be reviewed and data sources identified. Where the properties are represented by parameters in correlating equations, the valid range of the correlations should be established to ensure that they apply to the specific problem to be solved. Error bands for the values predicted by the correlations should be established. Where values outside the fitted range are required, extrapolations should be based on fundamental thermodynamic principles and application of theory with a sound scientific basis, such as the kinetic theory of gases. Many correlating equations are polynomials with little sound scientific basis. An extrapolation based on scientific principles will be more reliable than extrapolating a polynomial formula from a database. Where values are extrapolated, error bounds should also be set by reference to scientific principles. Particular care should be taken when conditions near to the critical point arise. Similar considerations apply to data that is estimated when experimental data is lacking. It should not be assumed that data from well-known reference books are reliable. These books expressly deny responsibility for errors. Data from such sources should be checked for thermodynamic consistency and for consistency with data for similar chemical species. Similar considerations apply to data and correlations supplied with commercial software. The reliability of such data should be determined either by entering a confidentiality agreement with the software supplier or by comparing predictions with experiment and data from other sources.

It is recommended that physical property predictions are displayed graphically to provide clearer comparison than is available from a table of figures. Mixture properties need to be assessed in the same way. In studying binary mixtures, validation should include very low concentrations of each component. Simplified theory is available for low concentrations and discontinuity errors (where the mixture equations differ from the pure component equations) are more likely to be apparent.

Correlations and data should be applied only under the conditions for which they have been established. For example, if the correlations have been derived for non polar mixtures, they are unlikely to be valid for polar mixtures. The use of computers does not obviate the necessity for experimental investigation. Where a mixture has not been studied previously, experimental confirmation of predicted properties may be needed.

The models employed should be similarly validated. Software suppliers should give full details of the models that they employ. The applicability of the models needs to be checked. For example, if the pressure of a flowing gas changes by >10 or 20%, compressible flow models are needed. As for physical property data, models should be checked for continuity at points where the model might change. For example, at low liquid concentrations, the pressure drop for a two-phase gas–liquid mixture should approach the pressure drop for a pure gas for low liquid concentrations.

Start-up as well as steady performance may have to be assessed. For example, a significant number of environmental discharge contraventions occur during start-up when catalysts have not reached operating temperature and separation systems are not working to full efficiency. Care must be taken that non-steady simulation models are adequate for modeling start-up. Dynamic models developed for regulation control studies are rarely adequate for start-up and shut-down studies.

Every engineering system is built in the face of uncertainty. It is the engineer's responsibility to ensure that the uncertainties are understood and scoped. Uncertainties arise in the data and in the models within the programs employed. Uncertainties extend to the possible hazards, reliability, and environmental impact of the process studied. Once the uncertainties have been identified, sensitivity tests can be undertaken to assess their impact. The consequences can be classified into safety hazards, environmental impact, operability, and economic impact. The uncertainties should include allowance for operational flexibility. Sensitivity tests can be integrated to indicate the combined effect of multiple uncertainties (see the section Integrating Multiple Uncertainties).

Alternative solutions should be considered. For example, in generating a process design, alternative processes should be generated, with alternative flow-sheet structures and alternative unit operations. In many countries, it is a requirement that genuine alternatives must be evaluated and the alternative with the best environmental performance selected. The proper evaluation of these alternatives is thus a vital part of the decision audit trail leading to the final design. The impact of uncertainties on the alternatives should also be evaluated; some alternatives may be more sensitive to uncertain data than others.

Brief notes on the special considerations in using spreadsheets and neural networks are appended to this section, as is a procedure for integrating the effects of multiple uncertainties.

3.1. Use of Spreadsheets.
As discussed in the section Programming Languages, quality assurance of results obtained from spreadsheets requires special attention. In particular, there is the potential for the user to inadvertently change formulas in cells. Strict guidelines on spreadsheet use should be issued. Caution is required for safety-critical applications. The decision audit trail should include a full copy of the spreadsheet program, not just its results.

3.2. Neural Networks.
Neural Networks contain a large number of fitted parameters. Statistically, the larger the number of parameters, the less significance any such fit has. In many cases, the statistical significance of Neural Net predictions is minimal. Consequently, the Net may be an effective method of interpolating the conditions in which it was trained but it may not be a good basis for predicting behavior outside the training region. Applications in control may be valid because there is a constant stream of data enabling retraining to be undertaken and the data is likely to span the conditions being predicted. In other applications, however, engineers should use neutral networks with caution.

3.3. Integrating Multiple Uncertainties.
Extensive sensitivity tests are recommended above. Sensitivity tests are normally undertaken by simulation with the "best estimate" values of the uncertain parameters and with perturbed estimates. In this way, the sensitivity of performance to the assumptions can be assessed. The uncertainties should include both data uncertainties and model uncertainties (see the section Handling Uncertainty). Where specific interactions are important, users may explore perturbing several uncertain parameters in the same simulation. However, the combinatorial problem of exploring all possible combinations of uncertainties makes it impracticable to explore more than a few multiple uncertainty cases. There is much research on handling multiple uncertainties, particularly optimal design under uncertainty. However, there is no widely agreed tool. In this section, we present a tool that has been employed successfully in a number of studies and is simple to apply. In the future, better tools may be available.

The expressions given below are applicable to models that, in the region of the expected solution, can be approximated to cubic expressions (including interaction terms). The uncertain parameters are distributed independently and symmetrically. The expected value of a performance measure, y_e, can be then obtained from

$$y_e = y_0 + \Sigma[y\{\Delta x_i\} + y\{-\Delta x_i\})/2 - y_0](\sigma_i/\Delta x_i)^2 \qquad (16)$$

The variance, σ_y^2, of the performance measure is given by

$$\sigma_y^2 = y_0^2 - y_e^2 + \Sigma[(y^2\{\Delta x_i\} + y^2\{-\Delta x_i\})/2 - y_0^2](\sigma_i/\Delta x_i)^2 \qquad (17)$$

Where, y may be any uncertain performance measure (component concentration, stream temperature etc). The parameter y_0 is the value of y computed with all uncertain parameters at their expected values. $y\{\Delta x_i\}$ is the value of y computed with all uncertain parameters at their expected values except for variable number i, which has its value at $(x_{0i} + \Delta x_i)$. The summations are over all the uncertain parameters. The parameter σ_i is the standard deviation of parameter i.

Where there are n uncertain parameters, $(2n + 1)$ simulations are required to compute all the uncertain outcomes. Thus, the total number of computations is independent of the number of performance measures to be assessed.

In engineering design, most parameters have independent uncertainties. However, interdependencies can arise. For example, an experimentally measured rate constant (K) may typically be computed using

$$K = Ae^{-E/RT} \qquad (18)$$

In this expression, both the pre-exponential A and the activation energy E will be uncertain. Note that high reaction rates can be obtained either by increasing A or by decreasing E. Consequently, there is uncertainty as to whether there is a high pre-exponential or a low activation energy, and this uncertainty will be higher than the uncertainty in the predicted rate constant. An error ellipse can be drawn around the best estimate of the two parameters. It will show a large probability that the pre-exponential is higher than the expected value and the activation energy is also higher, but there is a low probability that the pre-exponential is high and the activation energy is low. There are two ways of treating such parameters, the values of which cannot be estimated independently. The first is to replace A and E in the study by a variable that goes along the major axis of the error ellipse and a variable that goes along the minor axis. These two variables are statistically independent. The second approach is to put

$$K = K_0(1 + s)$$

where K_0 is computed from equation 18 and s measures the scatter of the experimental observations about the predicted line. In this case, the two correlated uncertain parameters are replaced by one uncertain parameter, s, with a mean value of zero.

The uncertain distribution functions for most parameters found in engineering studies are, to sufficient accuracy, symmetrical. However, there is obvious asymmetry in parameters such as a mass transfer coefficients. These may have large uncertainties, but they clearly cannot be negative. It is found that the logarithms of such parameters are statistically sufficiently symmetrically distributed. Such logarithmic transformations can be applied equally to parameters x or y in equations 16 and 17. In most engineering problems, estimates of standard deviation are little more than guesses. In comparison, the form of the error distribution is a relatively second-order consideration and does not need to be treated in more detail.

This approach to integrating uncertainties can be built into software. However, more often it will be used as part of the quality assurance tests employed by the end user. For example, if the engineer records sensitivity results in a spreadsheet, equations 16 and 17 can be used to generate additional columns giving the integrated effect of uncertainty.

4. Current Advances

There are number of areas of computer-aided chemical engineering that are gaining in importance as the pressure for a more environmentally friendly chemical industry grows. This survey does not claim to be comprehensive. It gives a few of the important developing areas, namely, computer-aided molecule design, process synthesis, and flexible design.

4.1. Computer-Aided Molecule Design. CAMD has been developing rapidly during the last 10–15 years with an increasing number of successful applications. Techniques such as the group contribution method enable the properties of molecules to be predicted before they have been synthesised. CAMD exploits these abilities to design molecules that have desired properties. Recent developments and applications are given by Harper and Gani (19). Initial applications have been in the development of selective solvents; particularly non halogenated solvents with reduced toxicity and reduced ozone depletion potential. Potential applications include the development of economic, effective, safer, and less polluting chemical products.

4.2. Computer-Aided Process Synthesis. Process syntheses has been studied for over 30 years but, until recently, applications have been limited to energy reduction studies. Process synthesis differs from process optimization in the variables selected for optimization. Process optimization adjusts only continuously variable parameters such as lengths, temperatures, and pressures. Process synthesis optimizes also discrete variables. Recent advances in computer hardware and software promise a much wider range of application.

In a sense, all process design is process synthesis. Thus, the designer selects

The sequence of operations.

The choice of unit operations (eg, extractive distillation or liquid–liquid extraction).

The selection of hot- and cold-stream heat exchange matches.

The choice of separating agents.

In industry, design is still usually undertaken by hypothesizing a flowsheet that is incrementally improved as a result of simulation and other studies. However, Johns (20) summarizes a range of current computer-aided synthesis techniques that show promise of getting better results by automating more of the conceptual design process. Many give a feasible design as a direct output of optimization. Others, particularly Pinch Technology (21), deliver performance targets as goals for subsequent detailed design. Rigorous integer optimization is difficult, and is not practicable for many industrially relevant problems. Artificial intelligence methods aim to solve problems not treatable by rigorous optimization. Such heuristic (AI) methods develop good, although not necessarily optimal, designs. Automatic optimization, whether mathematical or heuristic, can only consider a limited number of performance criteria (eg, just cost). In practice, a design will be judged also by other criteria such as operability, safety, and environmental impact, many of which cannot be adequately included as constraints or objectives in the optimization. It is, therefore, desirable that the optimization produces a number of alternative flowsheet structures that can be evaluated against these additional criteria.

Robust simulation models are required for both optimization and synthesis. The optimizer is likely to generate sizes and/or operating conditions that are outside the normal simulation range and it is important to avoid run-time errors. Optimizations should not set constraints that may be difficult to meet. For example, instead of setting a minimal product purity constraint, it may be better to set a realistic cost penalty for purity shortfall. The optimization will be easier (because the function will be smooth up to the desired purity). Furthermore, if the purity constraint cannot be met, the user has a meaningful result, instead of a failure message.

The flowsheet structures obtained by process synthesis are frequently insensitive to uncertain commercial and technical data. Failure to achieve an optimal flowsheet structure is more likely to result from shortcomings in the optimizer than uncertainties in the data. Where the choice between two structures is sensitive to an uncertain parameter, it is desirable to consider both alternatives against other criteria not included in the optimization. This insensitivity enables simplified models to be employed for determining the range of likely process structures. Each structure is then subsequently simulated and optimized using relatively rigorous models. Simplified models can be made more resistant to run-time error and can be made more than a thousand times faster than rigorous models. Tools are available for tuning simplified models against rigorous models.

The benefit of computer-aided process synthesis is that very large numbers (eg, millions) of process alternatives can be implicitly evaluated. The evaluation identifies processes that are economic and have reduced environmental emissions. It is impracticable to evaluate such a large range of alternatives by hand. Regulatory authorities increasingly demand that processes are evaluated to ensure that there are not competitive alternatives with lower environmental impact. Process synthesis enables these demands to be met in a rigorous manner.

4.3. Flexible Design.   Parameters such as physical size are expensive to change after a process plant is built. However, flow rates, temperatures, and pressures can be changed. Thus, if the desired purity cannot be achieved at the nominal throughput, it may be possible to achieve it at a reduced throughput. Flexible design recognizes that operation can be optimized after production starts and that retrofit modification is possible where initial production targets cannot be met. It recognizes that it may be uneconomic to design to ensure that, even under the worst combination of uncertain outcomes, target production rate is met. Market size is often one of the most uncertain parameters on which a chemical engineering design is based. As the market builds up, there may be the opportunity to debottleneck a plant that initially underperforms. Flexible design explores the trade-off between applying excessive design margins and running a risk that a target production rate cannot be met.

Flexible design requires optimization under uncertainty. In such optimization, it is recommended that hard constraints on production rates are avoided. Instead, it should be recognized that there is flexibility in operation policy and realistic penalties should be applied for shortfalls that might occur under unfavorable outcomes to the uncertain parameters. This approach also avoids sharp discontinuities in the cost function that makes optimization more difficult.

There is active research in automated methods for optimal design under uncertainty. The technology is, however, not yet available for use outside the relevant research schools.

5. Conclusions

The use of computer aids does not reduce the responsibility of engineers. Indeed, with fewer simplifying assumptions, there is a greater requirement that the engineer has a fundamental understanding of the technology. Furthermore, more detailed models demand more extensive data that also needs to be obtained and critically assessed. The speed and accuracy of the computations makes it practicable to design better processes. Processes can be more thoroughly evaluated and their risks and opportunities more thoroughly assessed. Recent developments promise better, cleaner, safer processes.

BIBLIOGRAPHY

"Computer-Aided Engineering" in *ECT* 4th ed., Vol. 7, pp. 128–163, by M. T. Tayyabkhan, Tayyabkhan Consultants, Inc. and H. Britt, Aspen Technology, Inc., "Computer-Aid Engineering" in *ECT* (online), posting date: December 4, 2000, by M. T. Tayyabkhan, Tayyabkhan Consultants, Inc. and H. Britt, Aspen Technology, Inc.

CITED PUBLICATIONS

1. American Institute of Chemical Engineers, On-line Software Directory, http://www.cepmagazine.org/features/software/.
2. C. Mazza, J. Fairclough, B. Melton, D. DePablo, A. Scheffer, R. Stevens, M. Jones, and G. Alvin, *Software Engineering Guides*, Pearson Education, Harlow, UK, 1995.
3. G. Booch, I. Jacobson, and J. Rumbaugh, *Unified Modeling Language User Guide*, Addison Wesley Longman Publishing Co., Reading, Mass, 1998.
4. *NIST, Reference Information for the Software Verification and Validation Process*, NIST Special Publication 500–234, U. S. Dept of Commerce, Washington, D.C. Mar 1996.
5. W. E. Perry, *Effective Methods for Software Testing*, Wiley, New York, 1995.
6. M. Tanzio, Validate your Engineering Software, *Chem Eng Prog* **97** (7), 64, 2001.
7. G. Birtwhistle, O. J. Dahl, B. Myhrhaug, and K. Nygaard, *Simula – Begin*, Chatwell-Bratt, Bromley, UK, 1979.

8. B. Stroustrup, *The C++ Programming Language*, 3rd edn, Addison-Wesley, Reading, Mass., 1997.
9. T. Stanton and J. Pannell, *Philos. Trans. R. Soc.* **214**, 199, (1914).
10. L. F. Moody, *Trans Am Soc Mech Eng.* **66**, 671, (1944).
11. F. W. Dittus and L. M. K. Boelter, *University of Berkeley, Pubs Eng.* **2**, 443, (1930). Reprinted in *Int. Comm. Heat Mass Transfer* **12**, 3, (1985).
12. W. H. McAdams, *Heat Transmission*, 2nd ed., McGraw Hill Book Co., Inc., New York, 1942.
13. J. F. Epperson, *An Introduction to Numerical Methods and Analysis*, Wiley, New York, 2001.
14. HSL (formerly the Harwell Subroutine Library), *AEA Technology Engineering Software*, Harwell, UK.
15. NAG Library, The Numerical Algorithms Group Ltd, Oxford, UK.
16. W. H. Press, S. A. Teukolsky, and B. P. Flannery, *Numerical Recipes in C++*, Cambridge University Press, Cambridge, UK, 2002. (Also available in Fortran 77 and Fortran 90).
17. R. H. Perry and D. W. Green (eds), *Perry's Chemical Engineers' Handbook*, 7th Edition, McGraw-Hill, New York, 1997.
18. R. Best, G. Goltz, J. Hulbert, A. Lodge, F. A. Perris, and M. Woodman, *The Use of Computers by Chemical Engineers*. IChemE (CAPE Subject Group), Rugby, UK, 1999.
19. P. M. Harper and R. Gani, *Comp Chem Eng.* **24**, 677, (2000).
20. W. R. Johns, *Chem. Eng. Prog* **97** (4), 59, April 2001.
21. *User Guide on Process Integration for the Efficient Use of Energy*, IChemE, Rugby, UK (1992).

References to Programming Languages and Modeling Systems

LISP: P. Graham, *The ANSI Common Lisp Book*, Prentice Hall, Inc., Englewood Cliffs. N.J., 1995.
Prolog: I. Bratko, *Prolog Programming for Artificial Intelligence*, Longman, Reading, Mass, 2000.
Mathematica: S. Wolfram, *The Mathematica Book*, Cambridge University Press, Cambridge, UK, 1999.
MathCAD: R. W. Larsen, *Introduction to MathCAD 2000*, Prentice-Hall, Inc., Englewood Cliffs, N. J., 2001.
GAMS: *General Algebraic Modeling System*, GAMS Development Corporation, Washington, D.C.
SpeedUp: Aspen Technology Inc, Cambridge, Mass.
gProms: Process Systems Enterprises Ltd, London, UK.
Excel: Microsoft Corporation, Redmond, Washington.
Lotus 1-2-3: Lotus Corporation, Cambridge Mass.

W. R. Johns
Chemcept Limited

CONDUCTING POLYMERS

1. Introduction

The discovery that polyacetylene could be prepared with a high electronic conductivity initiated a major research effort on organic conducting polymers, and was recognized by the award of the Nobel Prize for Chemistry in 2000 to Alan Heeger, Alan MacDiarmid, and Hideki Shirakawa. A variety of organic conducting polymer materials has now been developed for applications ranging from electromagnetic shielding and "plastic electronics" to light-emitting devices and corrosion-inhibiting paints.

This article gives an overview of the synthesis, properties, and applications of redox dopable electronically conducting polymers, updating the previous article in the fourth edition of *ECT* by Reynolds and co-workers (1). Inorganic conducting polymers also exist, eg poly(sulfur nitride), which was discovered before the organic conducting polymers, and found to have a high conductivity (2). However, these materials will not be discussed here because of their lack of development for applications; a brief summary of their electrical conduction properties is given in an earlier review (3). Carbon nanotubes can be regarded as conducting polymers composed entirely of carbon, but are usually discussed in conjunction with fullerenes (see NANOTECHNOLOGY). We consider only intrinsically conductive polymers—excluding polymer composite materials containing metal particles or carbon black, where the electrical conductivity is the result of percolation of conducting filler particles in an insulating matrix (4) or tunneling between the particles. Ionically conducting polymers, where charge transport is the result of the motion of ions and is thus a problem of mass transport (5), are not discussed here.

The key feature of the electrically conductive organic polymers is the presence of conjugated bonds with π-electrons delocalized along the polymer chains. In the undoped form, the polymers are either insulating or semiconducting with a large band gap. The polymers are converted to the electrically conductive, or doped, forms via oxidation or reduction reactions that form delocalized charge carriers. Charge balance is accomplished by the incorporation of an oppositely charged counterion into the polymer matrix. The conductivity is electronic in nature and no concurrent ion motion occurs in the solid state. The redox doping processes are reversible and can be accomplished electrochemically. Note that during electrochemical switching, ions do move into and out of the polymers as charge-balancing species for the charge carriers on the polymer backbone. Most applications of conducting polymers utilize their electronic (electrically conducting and optical) properties, but some (eg, battery or sensor electrodes) involve their ionic properties.

2. Synthesis of Electrically Conductive Polymers

A number of synthetic routes have been developed for the preparation of conjugated polymers. The diversity has been driven by the desire to examine many different types of conjugated polymers and attempts to improve material properties. Material property enhancement has centered on the synthesis of polymers that are processible in various forms. In some cases, high molecular weight precursor polymers are prepared that can be converted into the conjugated form after processing. The conjugated polymer itself can also be soluble or fusible and is processed prior to doping. Five primary classes of conjugated polymers have been shown to exhibit high levels of electrical conductivity in the doped state. These include polyacetylenes, polyarylenes and polyheterocycles, poly(arylene vinylenes), and polyanilines. In addition, a number of multicomponent materials, usually polymer blends and composites, have been prepared in which at least one of the components is a conducting polymer.

2.1. Polyacetylenes. The first report of the synthesis of a strong, flexible, free-standing film of the simplest conjugated polymer, polyacetylene [26571-64-2], $(CH)_x$, was made in 1974 (6). The process, known as the Shirakawa technique, involves polymerization of acetylene on a thin-film coating of a heterogeneous Ziegler-Natta initiator system in a glass reactor, as shown in equation 1.

$$n\ HC \equiv CH \quad \xrightarrow[\text{Ti (OC}_4\text{Hg)}_4]{\text{Al (C}_2\text{H}_5)_3} \quad \text{\Large$\{\!\!\sim\!\!\}$}_n \tag{1}$$

The resulting porous, fibrillar polyacetylene film is highly crystalline, so is therefore insoluble, infusible, and otherwise nonprocessible. It is also unstable in air in both the conducting and insulating forms.

Much effort has been expended toward the improvement of the properties of polyacetylenes made by the direct polymerization of acetylene (7). Variation of the type of initiator systems (8,9), annealing or aging of the catalyst (10,11), and stretch orientation of the films (12,13) has resulted in increases in conductivity and improvement in the oxidative stability of the material. The improvement in properties is due to producing a polymer with fewer defects.

Even with improvement in properties of polyacetylenes prepared from acetylene, the materials remained intractable. To avoid this problem, soluble precursor polymer methods for the production of polyacetylene have been developed. The most highly studied system utilizing this method, the Durham technique, is shown in equation 2.

This method involves the metathesis polymerization of 7,8-bis(trifluoromethyl)tricyclo[4.2.2.02.5]deca-3,7,9-triene (**1**), to form an acetone-soluble precursor polymer (14,15). After purification, thin films are cast from solution and the polymer is converted to polyacetylene by thermal treatment (16). This technique is useful only for the production of thin films because of the high exothermicity of the thermal conversion, which is a potential explosion hazard.

However, the films produced in this manner are much less fibrillar and of higher density than the films made by the Shirakawa technique. Variable morphologies are also available using this method; cast films have a much higher amorphous content than Shirakawa films, but stretching of the precursor polymer before or during thermal treatment yields highly oriented and crystalline materials (17–19).

A drawback to the Durham method for the synthesis of polyacetylene is the necessity of elimination of a relatively large molecule during conversion, which can be overcome by the inclusion of strained rings into the precursor polymer structure. This technique was developed in the investigation of the ring-opening metathesis polymerization (ROMP) of benzvalene (20).

Copolymerizations of benzvalene with norbornene have been used to prepare block copolymers that are more stable and more soluble than the polybenzvalene (21). Other copolymerizations of acetylene with a variety of monomers and carrier polymers have been employed in the preparation of soluble polyacetylenes (22–26). In most cases, the resulting copolymers exhibit poorer electrical properties as solubility increases.

2.2. Polyphenylenes. Poly(p-phenylene) [25190-62-9] (PPP), synthesized using direct polymerization methods, yields oligomers of a largely intractable material (27–30). Although it is generally difficult to use intractable polymers in practical applications, sintering techniques are available that may make this polymer technologically useful. Electrochemical polymerization affords one route to produce intractable conducting polymers as useful films (31). Soluble precursor methods have been applied to PPP syntheses to circumvent this problem. Bacterial oxidation of benzene using the microorganism pseudomonas putida to prepare 5,6-cis-dihydroxycyclohexa-1,3-diene (**2**), followed by esterification and free-radical polymerization, has made it possible to form a soluble PPP precursor as shown in equation 3 (32). A variety of esters (diacetate, dibenzoate, etc) have been investigated to yield high molecular weight, processible precursor polymers. Thermal conversion of thin films and fibers produce fully aromatic PPP. The cis diesters were also prepared using conventional chemical routes (33). The polymers were found to contain 10–15% 1,2 linkages rather than the desired all 1,4-linked structure.

$$(3)$$

In both studies described, the resulting PPP was completely intractable. Preparation of a soluble and processible form of PPP has been accomplished using dialkyl-substituted benzenes in the polymerization (34,35). Originally, Grignard coupling of alkyl-substituted 1,4-dibromobenzenes was employed, but low molecular weights were obtained because of loss of chain end functionality. Subsequently, an A–B step-growth polymerization technique was developed, and the resulting PPP had a significantly higher degree of polymerization. Other organometallic routes to produce PPP and its derivatives have been investigated (36,37).

2.3. Polypyrroles. Heterocyclic monomers, such as pyrrole and thiophene, form fully conjugated polymers with the potential for doped conductivity when polymerization occurs in the 2,5 positions as shown in equation 4. The heterocycle monomers can be polymerized by an oxidative coupling mechanism, which can be initiated by either chemical or electrochemical means.

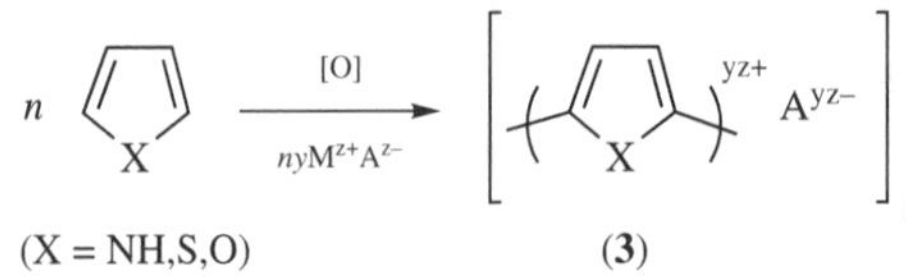

$$n \underset{X}{\bigcirc} \xrightarrow[nyM^{z+}A^{z-}]{[O]} \left[\underset{X}{\bigcirc} \right]^{yz+} A^{yz-} \bigg]_n \qquad (4)$$

$$(X = NH,S,O) \qquad\qquad (3)$$

The electrochemical polymerization of pyrrole is generally believed to follow a radical step-growth mechanism (38). The process is illustrated in Figure 1. The monomer is oxidized at the anode to form radical cations, which quickly couple and eliminate two protons to rearomatize. The pyrrole dimer thus formed is more easily oxidized than the pyrrole monomer, and is reoxidized to allow further coupling reactions to proceed. As the chain length of the growing oligomer increases, it becomes insoluble and deposits on the surface of the cell anode as a black film, where solid-state polymerization continues to occur. The polymer is further oxidized to a positively charged state (**3**) (redox doped) and, to compensate for this charge, negatively charged ions from the electrolyte are incorporated into the film. To avoid corrosion problems, the anode material for oxidative polymerization is normally a noble metal such as platinum, but surprisingly successful electropolymerization has also been obtained on a few commodity metals such as iron, steel, or even aluminum (39,40). Studies have indicated that coupling is primarily at the 2 and 5 positions, with small amounts at the 3 position, which leads to structural disorder in the final polymer (41–44). Pyrrole can also be polymerized using chemical oxidants such as $FeCl_3$ in the presence of an electrolyte to form the conducting polymer in the form of a powder or coating.

Significant variations in the properties of polypyrrole [30604-81-0] are controlled by the electrolyte used in the polymerization. Monoanionic, multianionic,

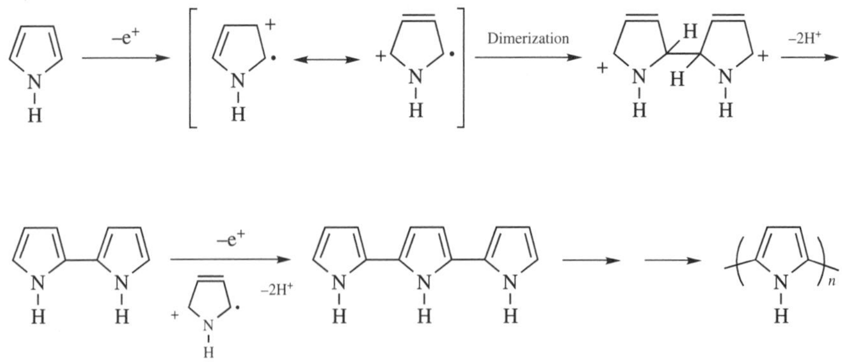

Fig. 1. Mechanism of electrochemical polymerization of pyrrole.

and polyelectrolyte dopants have been studied extensively (38,45–49). Properties can also be controlled by polymerization of substituted pyrrole monomers, with substitution being at either the 3 position (**4**) (50–52) or on the nitrogen (**5**) (53–55). An interesting approach has been to substitute the monomer with a group terminated by an ion, which can then act as the dopant in the oxidized form of the polymer forming a so-called self-doped system such as the one shown in (**6**) (56–58).

Both electrochemically and chemically polymerized unsubstituted poly-pyrroles are intractable in both the conducting and insulating forms. Some of the above substituted polypyrroles are appreciably soluble in common solvents, but generally an increase in solvent processibility is accompanied by a loss of conductivity.

2.4. Polythiophenes. In contrast to the intractability of many polypyr-roles, a substantial number of substituted polythiophenes have been found to be processible both from solution and in the melt. The most studied of these systems are the poly(3-alkylthiophenes) (P3ATs).

Electrochemical synthesis of P3ATs was accomplished in 1986 (59). In the same year, a technique was reported for the chemical synthesis of P3ATs as shown in equation 5.

$$THF = tetrahydrofuran \tag{5}$$

The synthesis involves the nickel-catalyzed coupling of the mono-Grignard reagent derived from 3-alkyl-2,5-diiodothiophene (60). Also in that year, transition-metal halides, ie, $FeCl_3$, $MoCl_5$, and $RuCl_3$, were used for the chemical oxidative polymerization of three-substituted thiophenes (61). Poly(3-hexylthiophene) prepared with an $FeCl_3$ oxidant is highly thermoplastic and soluble, with a conductivity of 40 S/cm when doped by I_2 vapor (62); however, it is difficult to remove the last trace of Fe residues from this product, and this was found to have an adverse effect on its electronic device properties (63). Much more substantial decreases in conductivity were noted when branched side chains were present in the polymer structure, due to twists in the backbone causing a substantial loss of conjugative delocalization (64).

By contrast, significant increases in conductivity (>1000 S/cm) have been reported for regioselectively coupled P3ATs. McCullough and co-workers

(65,66) used an organometallic coupling route (equation 6) to produce highly regioregular polymers. In place of the traditional Grignard route, they prepared 2-bromo-3-alkylthiophene, converted it regioselectively at low temperature into 2-bromo-5-(bromomagnesio)-3-alkylthiophene and polymerized this with Ni(dppe)Cl$_2$ [dppe = 1,3-bis(diphenylphosphino)ethane] into 98–100% regioregular P3AT products with number-average molecular masses of (20–40) × 10^3. An alternative method described by Chen and co-workers used 2,5-dibromo-3-alkylthiophene as a starting point; it was reacted with highly active "Rieke zinc" at −78°C to form 2-(bromozincio)-3-alkyl-5-bromothiophene and 2-bromo-3 alkyl-5-(bromozincio)thiophene (equation 7). This unpromising mixture of isomers was regioselectively coupled by Ni(dppp)Cl$_2$ [dppp = 1,3-bis(diphenylphosphino)propane] to produce near-perfect polymer chains with high molecular mass (67,68). Subsequently, Budd and co-workers (69) showed that up to 90% regioregularity could be obtained in a single step by the addition of divalent nickel or cobalt compounds to the reaction mixture for ordinary oxidative polymerisation, at ∼0°C; this method has the advantage of producing P3ATs of high conductivity (up to 200 S/cm for cast, undrawn films doped with iodine), and of being readily scaled-up for commercial production.

$$(6)$$

LDA = lithium diisopropylamide

$$(7)$$

Polythiophenes with substituents other than alkyl groups at the 3 position have been prepared by the polymerization of substituted monomers. Many of these polymers have been substituted alkylthiophenes (**7**), where example side chains are (R =) $-C_6H_5$ (70–73), $-OCH_3$ (50), $-NHC(O)(CH_2)_{10}CH_3$ (50), and $-OSO_2(CH_2)_3CH_3$ (74). Poly(3-alkoxythiophenes) (**8**) (75–79) and poly(3-alkylthiothiophenes) (**9**) (80–82) have been prepared by both chemical and electrochemical methods.

(**7**)   (**8**)   (**9**)

There are now many other poly(3-substituted thiophenes) and poly(3,4-disubstituted thiophenes), synthesized for a wide variety of scientific and technological purposes (83,84). Among the former group may be mentioned the poly(3-perfluoroalkylthiophenes), which are hydrophobic, "electron deficient", and hard to dope p-type (see the section on Doping Processes); they have doping potentials $\sim$ 0.3–0.4 V higher than those of P3ATs. In contrast, poly(3-fluoroalkyloxythiophenes) have similar oxidation potentials to P3ATs and show fairly high conductivities (1–10 S/cm) (85,86). The "self-doped" poly(3-thiophene-β-ethanesulfonate) (**10**) (56,87), its propane analogue (88), and poly(3-thiopheneacetic acid) (**11**) (89) are hydrophilic and in some cases water soluble. They appear to exist as micellar dispersions in aqueous media, from which they can be cast as conductive thin films.

(**10**)   (**11**)

A substantial amount of research has been done on poly(3,4-disubstituted thiophenes) (84). Those having two separate substituents at the 3 and 4 positions generally suffer from steric crowding, which leads to twisting of the backbone, and hence poor π-overlap, large energy gap, and low conductivity. In contrast, those with both substituents in a single ring have high conductivities, and frequently have even smaller energy gaps than P3ATs (see the section on Optical Properties). This finding is partly due to the impeded ability of the ring atoms to cause steric hindrance between adjacent monomer units, and partly to the annelation forcing an increased double-bond character upon the links between the monomers. The principle was first demonstrated in 1984 (90) with the classic example of poly(isothianaphthene) (**12**), which has an energy gap (E_g) of $\sim$1 eV; the polymer may be prepared in a single step by reacting phthalic anhydride with P_4S_{10} (91). Since 1984 there has been a great deal of research to produce other small-gap conducting polymers (92), including polythiophenes such as poly(2,3-dihexylthieno[3,4-b]pyrazine) (**13**) (93), the "tethered" polybithiophene (**14**) (94), and the alternating copolymer (**15**) (95); the last-named have E_g values of 0.9, 0.8, and 0.3 eV, respectively.

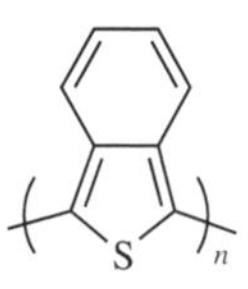

(12) 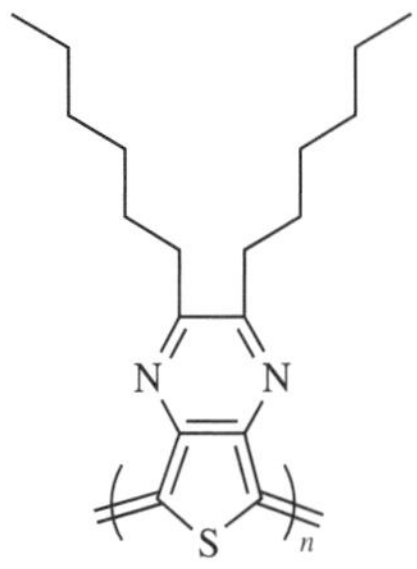   (13) 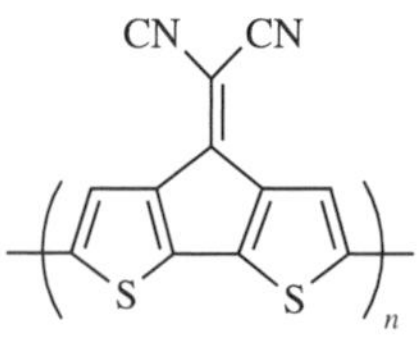   (14) 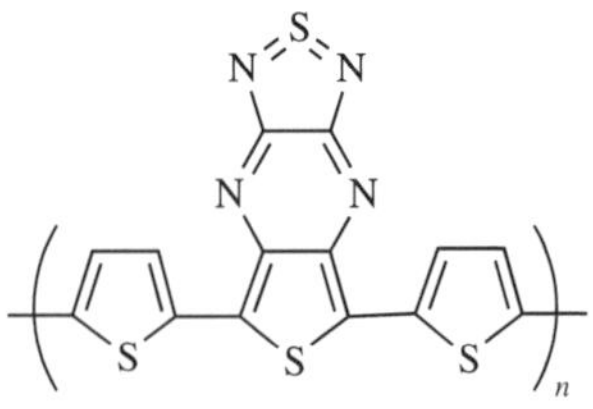   (15)

Poly(3,4-ethylenedioxythiophene) or PEDOT (**16**) (96) shows a similar effect of substitution on band gap ($E_g \sim 1.5$ eV), and the ether groups in the dioxin ring enable the polymer to complex with metal cations. The idea of polydentate complexation by a poly(3,4-substituted thiophene) has been extended to include polymers with crown ether moieties (97–99), which exhibit cation-selective optical properties.

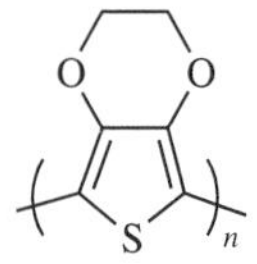

(16)

2.5. Poly(arylene vinylenes).   The use of the soluble precursor route has been successful in the case of poly(arylene vinylenes), both those containing benzenoid and heteroaromatic species as the aryl groups. The simplest member of this family is poly(p-phenylene vinylene) [26009-24-5] (PPV). High molecular weight PPV is prepared via a soluble precursor route (100,101). The method involves the synthesis of the bis(sulfonium) salt from 1,4-bis(chloromethyl)benzene, followed by a sodium hydroxide induced elimination polymerization reaction at 0°C to produce an aqueous solution of a polyelectrolyte precursor polymer (**17**). This polyelectrolyte is then processed into films, foams, and fibers, and converted to PPV thermally (eq. 8).

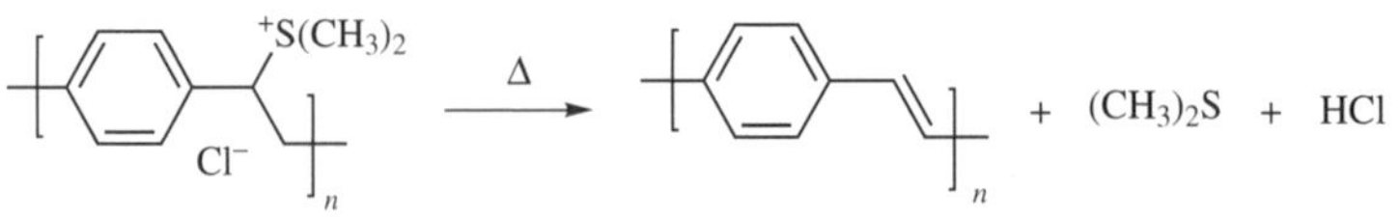

$$+ (CH_3)_2S + HCl \qquad (8)$$

(17)

The nature of the sulfonium structure affects the yield and quality of the resulting PPV, and it has been found that use of cyclic sulfonium structures (**18**) is preferable (102). With cyclic sulfonium polyelectrolytes, more efficient elimination of sulfur and the counterion occurs during thermal conversion, so fewer sp^3 defects are present in the final PPV.

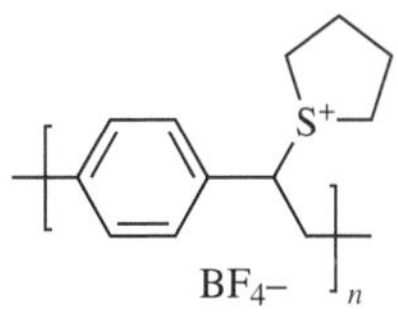

(**18**)

Substituted PPVs have been prepared using similar techniques, especially those with alkoxy substituents on the aromatic ring (103–106). The advantage of long-chain alkoxy (butoxy or hexyloxy) substituents is that not only is the precursor polyelectrolyte soluble, but after conversion the substituted PPV is also soluble (107,108). Heteroaromatic ring structures can also be incorporated into poly(arylene vinylene) structures using the same precursor polymer method shown for PPV. Poly(thienylene vinylene) (109–113) and poly(furylene vinylene) (114,115) have been prepared in this manner. In addition, alkoxy-substituted poly(thienylene vinylenes) (114,116) and poly(pyrrylene vinylenes) (117) have been synthesized. Various copolymers containing phenylene, thienylene, and furylene moieties have been produced (115,118,119). The benzannelated analogue of PPV, poly(1,4-naphthalenevinylene), has also been prepared by a cyclic sulfonium precursor route (120).

Alternative precursor routes to PPV and its derivatives have been investigated, eg, a bromo precursor that thermally eliminates HBr to produce trifluoromethyl-substituted PPV (121). Nonprecursor routes have also proven to be valuable, eg, ring-opening metathesis polymerization (ROMP), especially for the study of polymers with very high molecular mass (122–124).

2.6. Polyanilines. Polyaniline [25233-30-1] (PAni) is commonly prepared by polymerization of aniline using $(NH_4)_2S_2O_8$ in HCl (125,126). As prepared, it has structure (**19**) known as emeraldine hydrochloride. In this form, PAni is highly conductive but completely insoluble. When emeraldine hydrochloride is deprotonated with NH_4OH, the highly soluble emeraldine base (**20**) is produced. It is processible from organic solvents such as aqueous acetic acid (127). It must then be treated with HCl to regenerate the insoluble conducting form of the polymer.

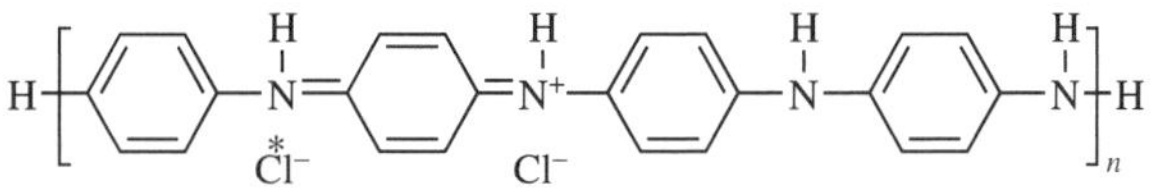

(**19**)

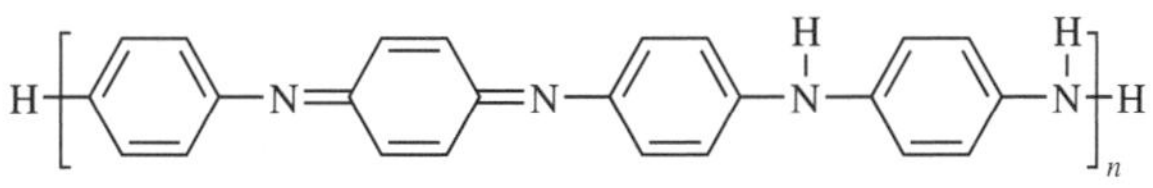

(**20**)

As was the case for polythiophenes, substitution along the PAni backbone has been utilized as a means of improving processibility. Many derivatives are known for PAni because of the possibility of substitution of the monomers at either the main-chain nitrogen atoms, or on the aromatic ring (128–136). Substituents studied have included alkyl, aryl, sulfonyl, and amino groups. The presence of a substituent dramatically decreases the yield of the polymerization and polymer molecular weight, the extent of the effect being proportional to substituent size. Most of the simply substituted (eg, alkyl) PAnis are insoluble as prepared, but treatment with base yields polymers that are highly soluble in common organic solvents. However, the sulfonated, self-doped PAnis have some solubility in water and polar organic solvents (eg, N-methylpyrrolidinone) in the conductive state (137,138). Poly(metanilic acid) (**21**) was found to be soluble in water and organic solvents, and is reportedly n-dopable (139).

(**21**)

A much-improved combination of high conductivity and excellent processibility has come about through the use of counterion induced solubilization, especially through the use of sulfonate counterions such as 4-dodecylbenzenesulfonate (140) and camphorsulfonate (141).

2.7. Polypyridine and Ladder-Structured Polyquinoxalines.   Unlike most other heterocyclic monomers for conducting polymers, pyridine cannot be polymerized directly by oxidation. However, polypyridine, poly(2,5-pyridinediyl), (**22**) can be prepared by chemical or electrochemical reductive debromination of 2,5-dibromopyridine (142). Polypyridine is somewhat soluble in polar solvents such as formic acid and it is found to be relatively electron accepting; hence it can be doped n-type much more easily than p-type, and it has been found to have a doped conductivity in the n-type form of ~0.1 S/cm. A 95% regioregular polymer has been synthesized (143), and it displays significantly higher conductivities. The isomer poly(2,2'-bipyridine-5,5'-diyl) (**23**) (144) has a similar conductivity to irregular polypyridine; it also complexes strongly with metal ions such as Ru(II), Os(II), and Rh(III), and the luminescent and conductive properties of these complexes have been the subject of keen interest (145).

(**22**)   (**23**)

Ladder polymers are those having double strands rather than single linkages between the monomer units; conjugated ladder polymers are of considerable interest through being expected to have small energy gaps (146); some of them, like polypyridines, also have potential commercial interest as possible electron-transporting layers in light-emitting devices. The ladder polyquinoxalines (**24**) and their oligomers (147–150) can be synthesized by Stille-type condensation reactions (eq. 9) between derivatives of 1,2,4,5-benzenetetramine and 2,5-dihydroxy-1,4-benzoquinone, taking care to keep the reactants in solution. They may be regarded as ladder-structured analogues of polyanilines, in common with which they possess great chemical stability. Other conducting ladder polymers have been synthesized by a variety of methods (150).

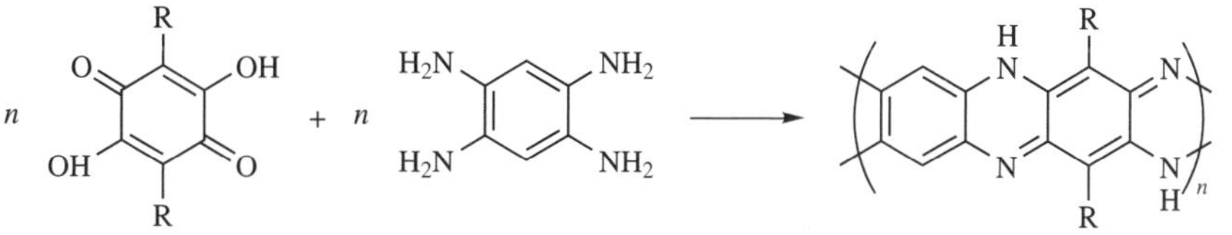

R = H, alkyl, alkylamino

$$(9)$$

(24)

2.8. Liquid Crystalline Conducting Polymers (LCCPs). Since the electronic properties of conjugated polymers are sensitive to the extent of π-orbital delocalization, and hence to chain alignment and planarity, there has been considerable interest in the effects of solvation (in solution) and of drawing (in the solid state). An alternative method of modifying the backbone conformation is to use the self-organizing properties of liquid crystals (LCs) to control it, either by blending or by chemical functionalization of the conducting polymer with mesogenic groups. This offers potential benefits from switchable electronic properties or from the use of electric or magnetic fields as an aid to orienting conjugated polymers during processing.

Park and co-workers (151) used the alignment capability of LCs to produce partially aligned conducting polymers in the 1980s; they polymerized acetylene with a catalyst trapped in a magnetically aligned LC matrix. Subsequently, polythiophenes with surfactant-like substituents were shown to have lyotropic LC properties and to be magnetically alignable in the presence of solvent (152). Later, polythiophene (153,154), polypyrrole (155,156), and poly(p-phenylene) (157) were functionalized with LC substituents in attempts to produce thermotropic LCCPs. Most of these were β-substituted polyaromatics with side groups consisting of short alkylene or oxyalkylene spacer moieties (typically 6–8 atoms long), terminated by common mesogens such as cyanobiphenyl. Polymerization was most commonly carried out by chemical oxidation using FeCl$_3$, although reductive dehalogenation using Ni(0) as a catalyst was used for the synthesis of LC polyphenylenes.

The first processible thermotropic LC polyanilines (**25**) were produced by Gabaston (158) and N-substituted polypyrroles by Ibison (159) and Hasegawa and co-workers (160). These all show typical LC self-organization and an ability to be switched or aligned by the application of electric or magnetic fields.

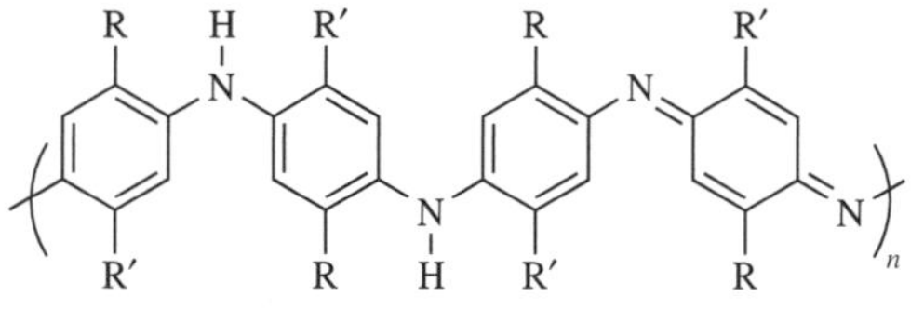

R = H, allkyl   (25)
R′ = -O(CH$_2$)$_x$ O-Ph-Ph-CN,
     -O(CH$_2$)$_x$O-Ph-Ph-CO-OAr

LCCPs based on poly(phenylene vinylenes) are now being studied as emissive materials in organic light-emitting devices (O-LEDs; see the section Polymer OLEDs and Diode Lasers) to produce polarized light emission. Apart from having useful bulk alignment properties, some LCCPs have been found to be microprocessible by scanning the surface of cast, amorphous films with a finely focused laser beam (161). In this way, local conductivity enhancements of five orders of magnitude have been obtained.

The orienting effects of LCs on conducting polymers should be even more potent for ferroelectric LCCPs, and the synthesis of such a polymer (based on polyacetylene) has been reported (162).

In the syntheses of all the LCCPs above, the mesogenic substituents sterically impede the coupling of monomers in solution and, consequently, the time scale required for polymerization is considerably longer for LCCPs than for most conducting polymers; hence it is not easy to produce high molar mass polymer by conventional step-growth reactions. Nevertheless, by optimizing the conditions and concentrations, average molar masses of $(10-20) \times 10^4$ may be obtained.

2.9. Conducting Polymer Blends, Composites, and Colloids. Incorporation of conducting polymers into multicomponent systems allows the preparation of materials that are electroactive and also possess specific properties contributed by the other components. Dispersion of a conducting polymer into an insulating matrix can be accomplished as either a miscible or phase-separated blend, a heterogeneous composite, or a colloidally dispersed latex. When the conductor is present in sufficiently high composition, electron transport is possible.

There are several approaches to the preparation of multicomponent materials, and the method utilized depends largely on the nature of the conductor used. In the case of polyacetylene blends, *in situ* polymerization of acetylene gas into a polymeric matrix has been a successful technique. Low density polyethylene (163) and polybutadiene (164) have both been used in this manner.

Because of the aqueous solubility of polyelectrolyte precursor polymers, another method of polymer blend formation is possible. The precursor polymer is codissolved with a water-soluble matrix polymer, and films of the blend are cast. With heating, the fully conjugated conducting polymer is generated to form the composite film. This technique has been used for poly(arylene vinylenes) with a variety of water-soluble matrix polymers, including polyacrylamide, poly(ethylene oxide), polyvinylpyrrolidinone, methylcellulose, and hydroxypropylcellulose (165). These blends generally exhibit phase-separated morphologies.

The true thermoplastic nature of poly(3-alkylthiophenes) and of some polyanilines, ie, solubility and fusibility, allows the use of compounding methods commonly used in the plastics industry for the preparation of composites of these polymers. The polymers can be codissolved with a matrix polymer, then processed from organic solution. Again the resulting blends are phase separated (70,166–172), but if the composition of conducting polymer is high enough, the conducting component forms the matrix with the insulating polymer dispersed within it, and high conductivity is possible. Melt processing has also been applied to poly(3-alkylthiophenes) (173–175). Films and sheets of blends of poly(3-octylthiophene) with polystyrene, polyethylene, poly(ethylene-*co*-butyl acrylate), and poly(ethylene-*co*-vinyl acetate) have been prepared using compression molding and film-blowing extrusion techniques.

Electrochemical polymerization of heterocycles is useful in the preparation of conducting composite materials. One technique employed involves the electropolymerization of pyrrole into a swollen polymer previously deposited on the electrode surface (176–178). This method allows variation of the physical properties of the material by control of the amount of conducting polymer incorporated into the matrix film. If the matrix polymer is an ionomer such as Nafion (179–181) it contributes the dopant ion for the oxidized conducting polymer and acts as an effective medium for ion transport during electrochemical switching of the material. Other matrix polymers used for polypyrrole have included cross-linked poly(vinyl alcohol) (PVAL) (182), poly(arylether ketone) (183), polyoxyphenylene (184), polyimide (185–187), poly(vinyl chloride-*co*-vinyl acetate) (PVC–PVA) (188), and hydrogels (189). Polythiophenes have also been polymerized electrochemically in analogous fashion, the hosts including polyamide (190) and natural and synthetic rubbers (191).

A number of heterocyclic polymers have been formed chemically within host polymer matrices; the most widely used monomer was pyrrole, which has been polymerized into highly conductive composites after sorption into polyethylene (192,193), poly(methyl methacrylate) (193), polypropylene (193,194), polyurethane (195), PVC–PVA (188), polycarbonate (196), epoxy resins (193,197), and polyvinylpyrrolidinone (198).

Conducting polymer composites have also been formed by coelectrodeposition of matrix polymer during electrochemical polymerization. Because both components of the composite are deposited simultaneously, a homogeneous film is obtained. This technique has been utilized for both neutral thermoplastics such as PVC, as well as for a large variety of polyelectrolytes (46–49,199–201). When the matrix polymer is a polyelectrolyte, it serves as the dopant species for the conducting polymer, so there is an intimate mixing of the polymer chains and the system can be appropriately termed a molecular composite.

The preparation of molecular composites by electropolymerization of heterocycles in solution with polyelectrolytes is an extremely versatile technique, and many polyelectrolyte systems have been studied. The advantages of this method include the use of aqueous systems for the polymerization. Also, the physical and mechanical properties of the overall composite depend on the properties of the polyelectrolyte, so material tailorability is feasible by selection of a polyelectrolyte with desirable properties.

Polypyrrole has been used in the preparation of colloidally dispersed latexes for processing into conducting polymer blends. Electrochemical polymerization of pyrrole in the presence of latex particles having a high concentration of bound sulfonate or sulfate groups on their surface results in the formation of thick heterogeneous films. These films are then ground, solvent swollen, and sheared to yield dispersions from which conducting films can be cast or spin coated (202). Chemical polymerization of pyrrole or aniline in aqueous solutions of sterics stabilizers such as methylcellulose or PVA also results in colloidal systems suitable for the casting of films of conducting polymer blends (203–206).

By using related chemistry, conducting-polymer-coated particles of colloidal dimensions (micro- or nanometer scale) have been prepared via chemical polymerization of pyrrole or aniline adsorbed onto finely divided inorganic powders. The resulting coated particles are eminently suitable for incorporation into a wide variety of blends or composites with conventional thermoplastic polymers. Maeda and Armes (207) produced the first conductive polypyrrole-silica nanocomposites, and Perruchot and co-workers (208) improved the surface concentration of conductive component by the use of a common silane coupling agent, aminopropyltriethoxysilane (APS) prior to the introduction of monomer.

3. Doping and Optical Properties

In order to induce high electrical conductivity in organic conjugated polymers, charge carriers must be introduced. These charge carriers are created by removing electrons from, or adding electrons to, the delocalized π-electron network of the polymer, creating a conducting unit that is now a polymeric ion rather than a neutral species. The charges introduced are compensated by ions from the reaction medium. This process is called doping by analogy to the changes that occur in inorganic semiconductors upon addition of small quantities of electronic defects. However, it proceeds through a different mechanism and is more precisely termed a redox reaction (12,173). The doping level, or ratio of charge carriers per polymer repeat unit, is generally between 0.2 and 0.4 in most polyarylenes (174) and can be determined by measuring the content of charge-balancing counterions. The ability to control the electrical properties of conducting polymers over wide ranges, by adjusting the redox doping level, has created interest in these materials for a number of emerging applications (15).

3.1. Charge Carriers in Conducting Polymers. Metals have unpaired electrons and their highest occupied electronic levels are half-occupied. Electrical conductivity results from the fact that the electrons can move readily under an electric field since there is no forbidden gap between the highest occupied and lowest unoccupied electronic levels. Conductivity is limited, therefore, by defects in the lattice and vibrational distortions (also called phonons). Since this phonon activity increases with elevated temperatures, the conductivity of metals increases as the temperature is decreased. However, the electrons in conjugated organic polymers, as in inorganic semiconductors, are paired, creating a gap between the highest occupied levels (the valence band) and the lowest unoccupied levels (the conduction band). The energy difference between these bands gives rise to the intrinsic insulating or semiconducting properties of conjugated organic polymers. The moderate conductivity of these materials is a result of thermal excitation of valence electrons into the conduction band. Therefore, the conductivity of semiconductors and conjugated organic polymers increases with increasing temperature in the neutral or undoped state.

The mechanism for the conductivity increase resulting from doping in inorganic semiconductors involves the formation of unfilled electronic bands. Electrons are removed from the top of the valence band during oxidation, called p-type doping, or added to the bottom of the conduction band during reduction, termed n-type doping. Extension of this argument to the case of conjugated organic polymers was found to be inaccurate as the conductivity in many conducting polymers was found to be associated with spinless charge carriers.

In situ electron spin (epr)/electrochemistry techniques have shown that the conducting entity in polyacetylene (175), polypyrrole (176), polythiophene (177,178), and poly(*p*-phenylene) (179) can be spinless, although evidence exists for mixed-valence charge carriers as well (180).

The conductivity increase following doping in conjugated polymers is explained in terms of local lattice distortions and localized electronic states. In this case, the valence band remains full and the conduction band remains empty so that there is no appearance of metallic character. When the polymer chain is redox doped, a lattice distortion results and the equilibrium geometry for the doped state is different than the ground-state geometry. This is evident in many small organic molecules, eg, the ground-state geometry of biphenyl is benzenoid but the geometry of its radical cation is quinoidal (181).

Many conjugated polymers have nondegenerate ground states that behave similarly upon redox doping. When one electron is removed (or added) to the polymer chain a radical cation (or anion) is formed. This results in a lattice distortion that leads to an upward shift of the highest occupied molecular orbital (HOMO) and a downward shift in the lowest unoccupied molecular orbital (LUMO). Although the radical ion is expected to be delocalized over the entire polymer chain, the species is localized, with a localized lattice distortion creating a localized electronic state. Since the geometry of the chain between the radical and the ion must be distorted, and the energy of the distorted geometry is normally higher than that of the ground state, separation, and delocalization of the radical ion results in energetically unfavorable further lattice distortions. This radical ion associated with a lattice distortion is called a polaron (182) as shown in Figure 2 for PPP. A similar argument can be used when a second electron is removed from (or added to) this site. The resulting species is a dication (or dianion), with a lattice distortion separating the two charges. This species is termed a bipolaron and is also illustrated in Figure 2.

Although this model can be applied to most conducting polymers, polyacetylene is a special case because of the degeneracy of its ground state. The energy of the distorted geometry, where the single and double bonds are simply reversed, is equivalent to the energy of the initial structure (Fig. 3). This fact allows the charges formed during the doping of polyacetylene to readily separate

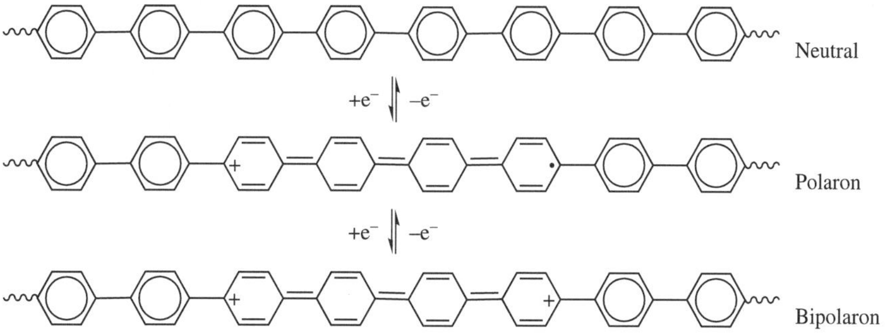

Fig. 2. Lattice distortions associated with the neutral, polaron, and bipolaron states in poly(*p*-phenylene).

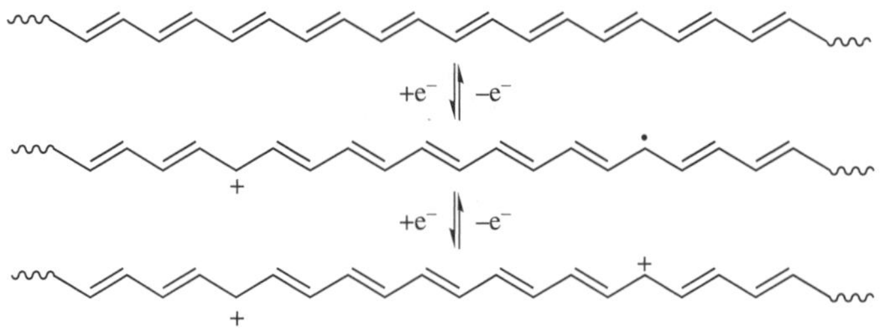

Fig. 3. Lattice distortions associated with the neutral and soliton states in poly-acetylene.

because there is no increase in distortion energy. This type of charge carrier is called a soliton and is unique to conjugated polymers, that have degenerate ground states such as polyacetylene (16,183).

3.2. Doping Processes. Redox doping can be accomplished through both chemical and electrochemical means. Vapor-phase doping of neutral polyacetylene has been carried out using AsF_5 and I_2 as oxidants. During this process, the dopant is reduced to form a charge-balancing ion such as AsF_6^- and I_3^-. Polypyrrole can be doped with oxygen because of the low oxidation potential of the neutral polymer (16). Solution doping of many conjugated polymers in suspension or, in the case of processible derivatives, in solution, has been accomplished using a variety of oxidizing and reducing agents. Typical chemical dopants include $FeCl_3$, $NOPF_6$ (184), and sodium naphthalide. Conducting polymers that are synthesized using oxidative coupling techniques are obtained in the oxidized form.

These conjugated polymers can be chemically and electrochemically reduced and reoxidized in a reversible manner. In all cases, the charges on the polymer backbone must be compensated by ions from the reaction medium, which are then incorporated into the polymer lattice. The rate of the doping process is dependent on the mobility of these charge compensating ions into and out of the polymer matrix.

Electrogenerated conducting polymer films incorporate ions from the electrolyte medium for charge compensation (185). Electrochemical cycling in an electrolyte solution results in sequential doping and undoping of the polymer film. In the case of a p-doped polymer, oxidation of the film results in the formation of cations on the polymer backbone and the introduction of anions into the film. Since the ions that are transported into the polymer matrix generally make up 20–40% of the doped polymer, the choice of electrolyte has a marked influence on the mechanical and electrochemical properties of the film. Tetra-alkylammonium salts are often utilized because of their high solubility in electrochemical solvents and their large electrochemical window. Their perchlorate, fluoroborate, and hexafluorophosphate salts are commonly used in the electrochemical doping of polythiophenes and polypyrroles. However, superior film-forming properties have been observed in the electrosynthesis and doping of polypyrrole using organic sulfonates, such as toluenesulfonate and other arylsulfonates (186).

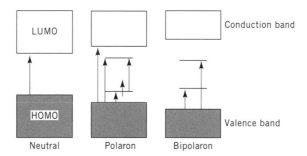

Fig. 4. Band diagram of a nondegenerate ground-state conducting polymer.

In the special case when polyelectrolytes (187) or multivalent anions (188) are used as the electrolyte during electrochemical polymerization, the dopant anion is entrapped because of its large size and multiple charges. As such, it cannot be transported into and out of the films. Cations from the electrolyte medium move into the film during electrochemical reduction to compensate the negative charges on the polyelectrolyte when the electroactive polymer is in its neutral form. Conversely, when the electroactive polymer is oxidized, the cations are expelled into the electrolyte medium and the negative charges on the polyelectrolyte are free to compensate the newly formed positive charges on the conducting polymer backbone.

3.3. Optical Properties. The energy difference between the valence band and the conduction band in conjugated polymers is referred to as the band gap and can be determined using optical spectroscopy. In the neutral (or insulating) form, conducting polymers exhibit a single electronic absorption in either the visible or ultraviolet (uv) region attributed to the electronic transition between the HOMO and LUMO levels as shown in Figure 4. The band gap is generally determined from the onset of this transition. This figure shows a band diagram of a conjugated polymer with a nondegenerate ground state, and the possible transitions in the neutral, polaron, and bipolaron states. Table 1 displays optically determined band gaps for several conducting polymers.

In the doped form, transitions between the band edges and newly formed intragap electronic states are observed in the optical spectra of conducting polymers. When polarons are present as charge carriers, an additional transition is apparent that corresponds to the electronic transitions between the two gap

Table 1. **Optical Band Gaps for Conjugated Polymers**

Polymer	Band gap, eV	Reference
polyacetylene	1.4	209
polypyrrole	2.5	210
polythiophene	2.0	211
poly(3-methylthiophene)	2.2	212
polyfuran	2.7	213
poly(3-hexylfuran)	2.5	213,214
poly(p-phenylene vinylene)	2.4	215
poly(p-phenylene)	3.0	216
polly(isothianaphene)	1.0	90

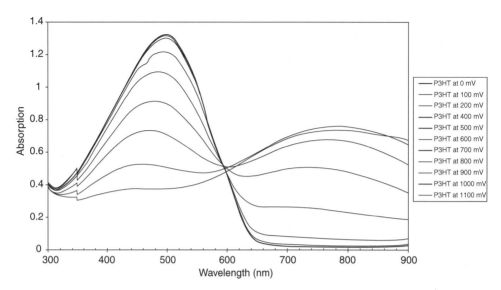

Fig. 5. Evolution of optical spectra of poly(3-hexylthiophene) during electrochemical doping.

states. Since the intragap electronic states are taken from the band edges, the band gap increases with increasing doping levels. Also, since a bipolaron creates a larger lattice distortion than a polaron, the gap states are further away from the band edges in the bipolaron model.

The changes in the optical absorption spectra of conducting polymers can be monitored using optoelectrochemical techniques. The optical spectrum of a thin-polymer film, mounted on a transparent electrode, such as indium tin oxide (ITO) coated glass, is recorded. The cell is fitted with a counter and reference electrode so that the potential at the polymer-coated electrode can be controlled electrochemically. The absorption spectrum is recorded as a function of electrode potential, and the evolution of the polymer's band structure can be observed as it changes from insulating to conducting (16).

An optoelectrochemical spectrum of poly(3-hexylthiophene), showing the changes in the band structure at various applied voltages, is shown in Figure 5 (217). The spectrum at low voltages contains a single absorption at ~500 nm, corresponding to a band gap transition. At higher voltages an absorption at 740–780 nm appears, which corresponds to a transition between gap states, indicating the presence of polaronic charge carriers. At higher voltages the band gap absorption diminishes significantly and decreases in wavelength.

Since the formation of these types of charge carriers is essential for electrical conductivity in conjugated organic polymers, an important factor in the structural design of conducting polymers is the ease with which they can be oxidized or reduced. The ionization potential of a class of conducting polymers can be altered by modification of the chemical structure. Substitution of the polymer with electron-donating groups has been shown to lower the ionization potential for *p*-type doping. The increased electron density along the conjugated system allows for easier removal of electrons during oxidation.

Another important factor affecting the electronic properties is the steric barrier to planarity along the polymer chain. Since polyheterocycles and polyarylenes must adopt a planar geometry in the ionized state to form quinoid-like segments, steric factors that limit the ability of the polymer to adopt geometries that are planar with respect to adjacent rings have a detrimental effect on the electronic properties (184).

In the case of polythiophenes, eg, the band gaps of alkoxy-substituted derivatives, substituted in either the 3 position (197) or both the 3 and 4 positions (198), are lower than that of the parent polymer, however the band gaps of 3-alkyl derivatives are slightly higher than polythiophene (190). The electron-donating ability of the alkoxy groups are able to outweigh the steric interactions they create. Conversely, the steric factors involved with the alkyl groups at the 3 position of the thiophene ring create geometric strains that outweigh the contribution of these substituents to the electron density of the polymer chain.

The structural effects on the band gap of conjugated organic polymers have been used to create materials with unique optical properties. Design of low band gap polymers has centered on decreasing the energy difference between the ground state and distorted geometries. One example of such a polymer is poly-(isothianaphthene) (**26**) (195). The six-membered ring fused to the thiophene backbone becomes aromatic when the thiophene moiety assumes a quinoid-like geometry (**27**). This significantly reduces the energy difference between these two geometries and yields a polymer with a band gap of 1.0 eV.

(**26**)   (**27**)

When doped, low band gap polymers have optical transitions in the infrared (ir) region of the spectrum, and therefore transmit more visible light in the conducting form than in the insulating form. This feature enables this class of conducting polymers to be investigated for a number of optical applications where both electrical conductivity and optical transparency are desired.

4. Electrical Conduction Properties

The electrical conductivity σ of a material, which is the inverse of its specific resistivity ρ, is a measure of a material's ability to transport electrical charge. It is determined by measuring the electrical resistance R of a sample of length L and cross-sectional area A through which the current is passing:

$$\sigma = 1/\rho = L/RA$$

The resistivity and conductivity are intrinsic properties of the material, ie, independent of the sample dimensions, unlike resistance that depends on sample size. Electrical conductivity is commonly measured in units of S/cm (ie, $\Omega^{-\varnothing}$cm^{-1}). The most common method of measuring conductivity is the 4-probe method, in which the outer pair of contacts are used to feed current into the sample, while the inner pair is used to measure the voltage developed (218). The Van der Pauw method and later modifications (219–221) are also used.

Quite generally, the magnitude of the conductivity of organic conducting polymers depends not only on the doping level, but also on the degree of disorder and structural aspects of the polymer. These include main-chain structure and π-overlap, molecular weight and polydispersity, interchain interactions controlled by both main- and side-chain structures, and chain orientation that can be affected by sample processing conditions. Sample morphology plays a key role in limiting or assisting in charge transport. Important factors are the ratio of crystalline-to-amorphous material content, and whether the polymer has a bulk dense morphology or a highly open fibrillar morphology.

No reproducible observation of superconductivity has so far been made in organic polymers (an initial report of superconductivity in a polythiophene derivative is now discredited). Superconductivity does, however, occur in the inorganic polymer poly(sulfur nitride) at temperatures below about 0.3 K (222).

4.1. Metallic Polymers. A compilation of measurements of the room-temperature conductivity of several of the common organic conducting polymers (223) is shown in Figure 6.

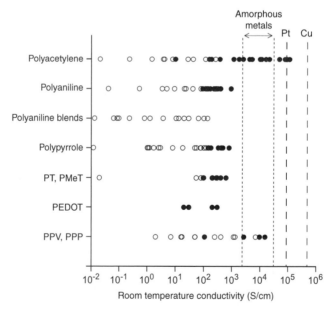

Fig. 6. Measured magnitudes of the room-temperature conductivity of different conducting polymers: polyacetylene (PA) (224–233) polyaniline (PAni) (234–239) and its blends (240–242), polypyrrole (PPy) (243–247), polythiophene (PT) (248,249) and its derivatives poly(3-methyl thiophene) (PMeT) (250,251) and poly(3,4-ethylenedioxythiophene) (PEDOT) (96), poly(p-phenylene) (PPP) (252) and poly(p-phenylene vinylene) (PPV) (253,254). The solid circles indicate metallic polymer samples and the open circles semiconducting samples. The conductivities of copper, platinum and typical conventional amorphous metals are indicated for comparison.

It is clear that the conductivity reaches remarkably high values considering that the density of carriers is much less than that in conventional metals. Conductivities of 10,000–30,000 S/cm for stretched polyacetylene are widely reproduced, while the highest reported conductivities are 80,000–150,000 S/cm (225,230,231). These conductivities exceed that of copper at room temperature on a weight basis, given that the density of copper is 8.9 gm/cm^3, while that of polyacetylene is only 1.1 gm/cm^3. The highest conductivities are for polyacetylene stretched by a factor of ~6 to improve alignment of the polymer chains— the conductivity perpendicular to the stretch direction is much lower. The doping levels of $FeCl_4^-$ or I_3^- ions for maximum conductivity are ~8% of the number of carbon atoms (225,255).

Apart from the high conductivity values, the metallic character of these highly conducting polymers is demonstrated most decisively by the observation (Fig. 7) that their conductivity remains nonzero as the temperature tends to absolute zero. In contrast, semiconductors show zero conductivity at the absolute zero of temperature owing to the gap between valence and conduction bands or localization of states. (We have used this as the criterion to separate metallic and semiconducting polymers in Fig. 6.) Further evidence is provided by the observation of a metallic Pauli susceptibility (256), a linear electronic specific heat term (257), and a linear diffusion thermoelectric power (223,258), all of which are signatures of delocalized metallic states at the Fermi level.

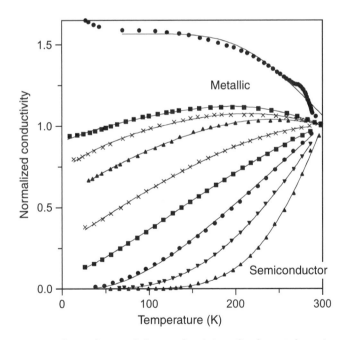

Fig. 7. Temperature dependence of the conductivity of polyacetylene (normalized to its value at room temperature), showing the transition from disordered semiconductor to metallic behavior as doping level increases. The data sets are for polyacetylene doped with ClO_4^- (224) (top two data sets [●, ■]), $FeCl_4^-$ (225) (third from top [×]) or $MoCl_6^-$ (226) (the other data sets). The lines are fits to the heterogeneous disorder model (223) as mentioned in the text.

In spite of this metallic character, the temperature dependence of the conductivity shows a predominantly nonmetallic sign, ie, the conductivity increases as temperature increases, as illustrated in Figure 7. Only near room temperature in the most highly conducting samples does the temperature dependence change to the expected metallic sign [a rare exception is a perchlorate-doped polyacetylene sample, which shows a metallic sign for the conductivity temperature dependence down to very low temperatures (224)]. A similar pattern of temperature dependence is seen for polyaniline, while for polypyrrole, polythiophene, and PPV the nonmetallic sign for temperature persists even up to room temperature (note, however, that the best of these polymers are also clearly metallic in that their conductivity is still nonzero in the zero-temperature limit).

The origin of this mixed metallic–nonmetallic behavior lies in the incomplete crystallinity of the conducting polymers, which is a feature of almost all polymers (259). Detailed studies show that highly conducting polyacetylene is ~90% crystalline with crystallite size of the order of 10 nm (260). The anomalous nonmetallic temperature dependence of conductivity arises from the contribution to resistance of the disordered nonmetallic "barriers" between the metallic crystallites [as shown by the fits in Fig. 7, a heterogeneous disorder model (223) gives a good account of the experimental data for the polymers with nonzero conductivity in the zero-temperature limit]. An interesting consequence of this is that the intrinsic conductivity of the metallic polyacetylene crystallites must exceed that of copper. The remarkably large magnitude of this conductivity is ascribed to the quasi-one-dimensional character of conduction along the polymer chains, which reduces the usual scattering of the metallic charge carriers by lattice vibrations (3,261,262).

This picture of disordered regions around metallic crystalline regions appears to be a general feature of highly conducting polymers (223,232,234, 258,263), with polypyrrole showing a greater overall degree of disorder than polyaniline and polyacetylene.

4.2. Semiconducting Polymers.   As the doping level of conducting polymers is reduced, the conductivity changes from mixed metallic–nonmetallic behavior to disordered semiconductor behavior, as clearly shown by the lower data sets in Figure 7 that have zero conductivity in the zero-temperature limit. For a wide variety of semiconducting polymers, the shape of the temperature dependence agrees with that for variable-range hopping conduction (ie, tunneling between localized electronic states assisted by thermally excited lattice vibrations) (223). This finding illustrated by the fits of the hopping expression (264) to the lower curves for semiconducting polymers in Figure 7. It appears that conduction is dominated by hopping between the states created in the semiconductor gap by the doping process (233).

Dispersions of metallic conducting polymer particles in an insulating polymer matrix show a similar temperature-dependent conductivity (242), which is consistent with charging-limited tunneling between mesoscopic metallic islands (265). Owing to the similarity of the predicted conductivity for this mechanism and variable-range hopping, it is difficult to identify the precise nature of the tunneling conduction mechanism in semiconducting polymers.

In high electric fields, the conductivity of conducting polymers shows significant nonohmic behavior, ie, a nonlinear current–voltage (I–V) characteristic.

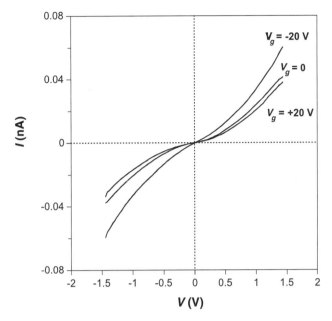

Fig. 8. The I–V characteristic of a polyacetylene nanofibre (lightly doped with iodine) of diameter 4.4 nm, with different gate voltages V_g applied to the Si back gate at a temperature of 233 K (266).

Of particular interest is the I–V characteristic of polyacetylene nanofibres (such nanofibers have potential applications in nanoscale electronics). Electric fields $>10^5$ V/cm can easily be applied to the sample when placed across electrodes only 100–200-nm apart. An example is shown in Figure 8, where the electrodes are deposited on a SiO_2 layer on a Si substrate that acts as a back gate (266). It can be seen that a negative gate voltage significantly changes the current, demonstrating operation of the polyacetylene nanofibre as a field-effect transistor (FET).

5. Stability

The rapid development of many processible conducting polymers that can now be obtained as films, fibers, and molded shapes, suggests that they will be quite useful in a number of applications. One of the concerns presently being addressed is the stability of the conductive polymers during these specific uses.

Although polyacetylene has served as an excellent prototype for understanding the chemistry and physics of electrical conductivity in organic polymers, its instability in both the neutral and doped forms precludes any useful application. In contrast to polyacetylene, both polyaniline and polypyrrole are significantly more stable as electrical conductors. When addressing polymer stability it is necessary to know the environmental conditions to which it will be exposed; these conditions can vary quite widely. For example, many of the electrode applications require long-term chemical and electrochemical stability at room temperature while the polymer is immersed in electrolyte. Aerospace applications, on the other hand, can have quite severe stability restrictions with testing carried out at elevated temperatures and humidities.

The intrinsic stability of undoped conjugated polymers (even polyacetylene) is quite high, but in most practical applications the polymers are doped and exposed to moist air. Herein lie the main stability problems, all of which can be related to the electronic properties of the polymers. Oxygen has a high electron affinity, and forms a charge-transfer complex with polymers such as polyacetylene. This "doping" process leads to the production of the superoxide ion $(O_2)^-$, which is an aggressive oxidant and rapidly leads to irreversible oxidation, chain scission, and loss of conductivity (267). Polymers with a large energy gap such as polyphenylene have high ionisation potentials and cannot be doped by oxygen; they are extremely resistant to oxidation. Polythiophenes, polyanilines, and polypyrroles have an intermediate behavior: All are somewhat susceptible to oxygen doping in their neutral states, but when doped (*p*-type) they have high ionisation potentials and are completely resistant to oxygen. However, a combination of oxygen and a nucleophilic agent, such as water or ammonia, can have a synergic effect due to the loss of dopant, low level incorporation of oxygen and then irreversible degradation and loss of conjugation; the environmental stability can then depend greatly on the ease of removal of the original dopant that is the first stage of the degradation process.

It is well known that the doped polypyrroles exhibit *some* of the best stabilities of electrical conductivity. Typically, poly(pyrrole tosylate) undergoes about a 10% conductivity decrease in a year at room temperature in air. The stability of the conductivity of polypyrrole and a series of poly(3-alkylthiophenes) has been compared in detail (201,202) under systematically controlled environmental and thermal conditions. When poly(pyrrole tosylate) is exposed to a dry N_2 atmosphere at 120°C, there is no evident conductivity change for at least 24 h. Poly(3-alkyl-thiophene tetrachloroferrates), on the other hand, show distinct conductivity decreases because of a thermal undoping process (203). The conductivity change is a function of the length of the pendent alkyl chain, with higher stability exhibited by polymers having the shortest alkyl chains. For example, poly(butylthiophene) exhibits a one-half order of magnitude drop in conductivity in the same amount of time that the conductivity of poly(3-octylthiophene) drops two orders of magnitude.

Poly(pyrrole tosylate) exhibits a slightly lower stability in air at 120°C with conductivity changes in the order of 10% in 24 h, as a result of reactions with environmental O_2 and H_2O. To circumvent this, a study of the mechanical integrity and thermal stability of polypyrroles containing a broad range of dopant ions (204) showed that encapsulation of the conducting polymer in an epoxy matrix led to a great improvement in stability. This finding is illustrated in Figure 9, where an electrochemically prepared poly(pyrrole tosylate) film and a polypyrrole-coated textile were encapsulated and their electrical resistivities monitored during exposure to 80°C and 95% relative humidity (rh) for 1 month. After an initial conductivity change, the film maintains a high conductivity throughout the entire time. The polypyrrole-coated textile, although exhibiting no change in conductivity over the entire experiment, has a higher overall resistivity than the film.

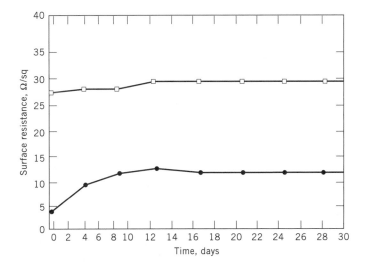

Fig. 9. Thermal stability of epoxy-encapsulated poly(pyrrole tosylate) film, dark squares, 0.5 Ω/sq, and polypyrrole-coated textiles, open squares, 20 Ω/sq, with exposure to 80°C and 95% rh.

PAni is rendered conductive by protonation rather than oxidation of a polyene; consequently its backbone has N^+ rather than C^+ sites in the doped state, and it has far more resistance to nucleophilic attack than a carbon. This confers a very high environmental stability at ambient temperatures. The thermal stability of doped PAni is high, and it is the loss of volatile dopant that leads to conductivity loss and subsequent oxidation at high temperatures (268,269). This PAni–HCl quite readily loses HCl on heating to 130°C, but the use of tosylate or dodecylbenzenesulfonate dopants confers a high thermal stability because the corresponding sulfonic acids are quite nonvolatile.

6. Applications

The novel and varied electrical, electrochemical and chemical properties of conducting polymers are leading to their development for a large variety of applications. In many cases the polymers have unique advantages over other materials, eg, low density, mechanical flexibility, tunable optical properties, and their ability to be functionalized for diferent purposes (270). There are of course disadvantages: eg, the poor stability of polyacetylene mentioned in the previous section severely limits its usefulness for applications.

The range and versatility of conducting polymers in technology is demonstrated in the following sections covering some of the main applications. Where possible, references to books or journal articles are given for each application, and relevant company websites are mentioned for detailed information on specific products and brand names.

6.1. Electromagnetic Shielding and Charge Dissipation. One of the first applications was the general area of electromagnetic shielding (271).

The surface-confined chemical oxidative polymerization of pyrrole and aniline has been used by Milliken Corp. (Spartanburg, South Carolina) to produce textile materials with electrically conductive polymer coatings (272,273). Exposure of various fabrics, including nylons, polyester, and quartz, to an appropriate oxidant leads to a preferential surface adsorption. Subsequent addition of dopant anion and monomer leads to polymerization on the fiber surfaces and formation of coherent, well adhered, conductive polymer coatings ~1 μm in thickness. Polypyrrole-coated fabric can be used as radar camouflage (218,272).

Polyaniline has been used as a charge disspation layer for electron-beam lithography (274,275), as a removable SEM discharge layer (276), as a conducting electrode for electrolytic metallization of copper on through-holes in circuit boards (277), and to provide a solderable finish in printed circuit board (PCB) technology (278).

The processibility of some of the conducting polymers has enabled them to be prepared in forms applicable to typical industrial processes for applications. For example, polyaniline has been dispersed in PVC to produce a family of conducting blends (278,279) marketed by Ormecon Chemie (www.ormecon.com.de), a subsidiary of Zipperling Kessler and Co. (www.zipperling.com.de). Thin films of polyaniline form lightly colored transparent films for antistatic coatings.

6.2. Corrosion-Inhibiting Paints. An application of PAni is its use to inhibit corrosion (280). Studies have shown reduction of corrosion of steel plates exposed to brine and acid environments when coated with a thin layer of PAni (281). The mechanism of this effect could be that the PAni leads to the formation of a thin, uniform layer of unhydrated iron oxide that acts as a barrier to the rusting process (278,282). Polyaniline-based anticorrosion paints have been developed by Ormecon and are in commercial use (283).

6.3. Charge Storage Batteries. The reversible redox chemistry of conductive polymers has allowed the materials to be used as electrodes in rechargeable storage batteries (284). Initially, it was hoped that the low density of polymer electrodes would yield lightweight high-energy-density batteries. This has not been realized because the polymers have low charge densities relative to metal electrodes. The charging and discharging reactions (redox doping and undoping) create charged carriers along the conjugated backbones with concurrent insertion or emission of charge-balancing counterions from the electrolyte. As such, there is no deposition or dissolution of the electrode material and it can be dimensionally stable. This is one limitation of metal electrodes in battery cells where repeated deposition and dissolution cause morphological features at the electrode surface that can separate and break from the electrode. In some instances, high-surface-area dendrites can form on metal surfaces, ultimately leading to an electrical contact between the two electrodes.

Conductive polymer batteries were developed by VARTA/BASF who investigated a series of polypyrrole-based cells (285). Bridgestone/Sienko market a polyaniline/lithium coin-type, or button, cell (286). These cells exhibit near 100% Coulombic efficiencies up to cell voltages of 3.5 V, with recommended use voltages of 3.0 V. At this cell voltage the battery exhibits discharge characteristics of 3.0 mAh or 60 Ah/kg. These cells have cycle lives that are a function of discharge depth. At low discharge depths (<10%) the battery could be recharged up to 10,000 times. These cells proved to be quite stable with storage.

6.4. Actuators. Conducting polymers can be used in electromechanical applications (270), since they show a volume change of up to 10% on oxidation or reduction as ions and solvent move into the polymeric structure (286,287). Bilayer constructions can act as artificial muscles (289,290), and the compatibility of conducting polymers with the human body makes implants feasible (291). Polypyrrole–gold bilayer microactuators that bend out of the plane of the wafer can be used to position other microcomponents (292).

6.5. Chemical and Biochemical Sensors. The sensitivity of the electrical properties of conductive polymers to chemical stimuli means that they have proved useful in a number of sensing applications (293). One early program carried out at Allied Signal proposed the use of conductive polymers in remotely readable indicators (294). Conductivity changes induced in the conductive polymer could be read externally and the history of the sample known. Systems designed to detect time–temperature, temperature limit, radiation dosage, mechanical abuse, and chemical exposure were developed.

Conductive polymers can change their electrical conductivities by many orders of magnitude when exposed to specific gaseous reagents (295,296). The reactivity of polypyrrole to NH_3, NO_2, and H_2S was tested to examine this phenomena. These sensors proved to be quite sensitive with large resistance changes occurring with exposure to an atmosphere containing 0.01–0.1% of the reactive gas. In all three cases, the reactions were shown to be reversible as removal of the gas from the atmosphere led to recovery of the original conductivity. "Electronic noses" to detect odors at low concentrations have been marketed, eg, by Osmetech (www.osmetech.plc.uk).

Entrapment of biochemically reactive molecules into conductive polymer substrates can be used to develop electrochemical biosensors (297). This has proven especially useful for the incorporation of enzymes that retain their specific chemical reactivity. Electropolymerization of pyrrole in an aqueous solution containing glucose oxidase (GO) leads to a polypyrrole in which the GO enzyme is codeposited with the polymer. These polymer-entrapped GO electrodes have been used as glucose sensors. A direct relationship is seen between the electrode response and the glucose concentration in the solution that was analyzed with a typical measurement taking between 20 and 40 s.

6.6. Gas Separation Membranes. Studies suggest that, because of the fixed ionic sites along the polymer backbones, conducting polymers are useful as gas separation membranes (298–301). The ability to switch the polymers between charge states (doped and undoped) suggests that external control of gas permeability and selectivity may be possible. Studies of polyaniline (299) showed it served as an excellent membrane material for the separation of the hydrogen–nitrogen, oxygen–nitrogen, and carbon dioxide–nitrogen gas pairs with extremely high selectivities. By utilizing dopants of varied sizes the separation properties of the membranes to different gases can be controlled.

6.7. Electrochromic Applications. During electrochemical switching, distinct changes occur in the optical properties (color and extinction coefficient) of conducting polymer films. This has led many investigators to examine the feasibility for use in electrically switchable, electrochromic, or "smart windows" with polymer layers that darken on application of a voltage (302). As exemplified by polyisothianaphthene (90), thin polymer films are relatively transparent when held in the conducting state and opaque in the insulating state. By using the concepts originally put forth for polyisothianaphthene, a number of low band gap polymers have been synthesized with similar properties.

6.8. Polymer OLEDs and Diode Lasers.   The application of a high electric field across a thin conjugated polymer film showed the materials to be electroluminescent (303–305). Until recently the development of electroluminescent displays has been confined to the use of inorganic semiconductors and a limited number of small molecule dyes as the emitter materials (see CHROMOGENIC MATERIALS, ELECTROCHROMIC).

One of the most exciting developments in applications of conducting polymers (306–313) is the use of their electroluminescent properties in organic light-emitting diodes (OLEDs). (Note that OLEDs also include devices based on organic molecules rather than polymers, so OLEDs based on conducting polymers are identified as "polymer OLEDs" or "PLEDs".)

In the basic construction of an OLED based on an electroluminescent polymer, electrons are injected at a cathode and holes at an anode, and their recombination in the bulk polymer leads to the emission of visible photons with an energy roughly equal to that of the polymer's energy gap. A material such as ITO glass is used as the anode, since its electron energies are comparable to those in the valence band of a luminescent polymer. It serves the dual purpose of injecting holes into the device and (being transparent) of allowing the light generated to escape and be seen. Other materials such as thin layers of silicon have been used for this purpose with greater injection efficiency but generally poorer transmission of light. The cathode material is a low work function metal, whose Fermi energy roughly matches the energy of the polymer's conduction band; this serves to inject electrons into the device. The emissive polymer is usually PPV or a soluble alkyl-substituted derivative, although regioregular poly(alkylthiophenenes) such as poly(3-hexylthiophene) have shown good electroluminescent properties for red/ir light.

The initial efficiency of polymer LEDs based on the above principle was minute (0.001%), but a level as high as 10% now seems feasible. These improvements have been brought about by ensuring the efficient and symmetrical injection and transport of carriers into the emissive region. Reactive metals such as calcium have been used as cathodes with improved electron-injection efficiency compared to aluminium, although Heeger and co-workers (314) have demonstrated high efficiency with an aluminium alloy cathode combined with a surfactant layer. A hole transporting layer such as polyaniline is normally interposed between the anode and the emissive polymer, and an efficient electron transporter based on poly(methyl methacrylate) or a polydiazole is sandwiched between the emissive layer and the cathode (315).

Polymer OLEDs have the advantage of flexibility, coverage of a wide area, low power consumption, high intensity at low voltages, fast response times, thinness, and wide viewing angles. The thin film of light-emitting polymer is typically sandwiched between a transparent electrode and a metal electrode. Companies involved in development include Cambridge Display Technologies (CDT), Uniax, Covion, Philips, Toshiba, and others.

In 2002, Philips claimed the first high volume production polymer LED product (the battery charging indicator for an electric razor), based on technology developed by CDT, and further products are planned. Toshiba has demonstrated a 17-in. diagonal full color OLED, also using CDT technology, and commercial display products from other companies are expected (316–319).

6.9. Photovoltaic Cells. The inverse effect to electroluminescence, ie, the generation of electricity from incident light (the photovoltaic effect), also operates in conducting polymers, which can therefore be used to make photovoltaic cells (320–322). Quantum efficiencies of up to 29% with overall power conversion of approximately 2% (for a simulated solar spectrum) have been obtained (320) using polythiophene as the hole acceptor and a PPV derivative as the electron acceptor. Hybrid photovoltaic cells incorporating fullerenes and conducting polymers are under development (323,324), and solar cells using inorganic CdSe nanorods in semiconducting organic polymers could make solar power more affordable (325).

6.10. Plastic Electronics. Electrolytic capacitors using polypyrrole as a solid-state electrolyte can achieve very high values of capacitance (218,326), producing a very thin dielectric layer without short circuits.

The first conducting polymer FET transistor (327) has been followed by more promising developments (328). Cheap disposable all-polymer integrated circuits (ICs) being developed by Philips (329) are likely to find application in smart labels to replace bar codes in supermarkets, eg, that allow automatic pricing without the need to unload goods from the trolley. "Line patterning" of conducting polymer circuits could lead to very inexpensive production (330–332). In this process, a circuit pattern printed on a substrate is exposed to fluid containing a conducting polymer that reacts differently with substrate and printed line. After evaporation of the solvent, the lines can be removed by sonication, leaving a conducting polymer circuit.

7. Acknowledgments

We acknowledge the authors (John R. Reynolds, Andrew D. Child, and Melinda B. Gieselman) of the previous article on this subject in ECT, from which part of the material has been retained. We thank many collaborators and colleagues for discussions, and Dr. Ben Chapman for his assistance in the final stages. ABK thanks the Marsden Fund administered by the Royal Society of New Zealand for support.

BIBLIOGRAPHY

"Polymers, Conductive" in *ECT* 3rd ed., Vol. 18, pp. 755–793, by C. B. Duke and H. W. Gibson, Xerox Corp.; "Electrically Conductive Polymers," in *ECT* 4th ed., Vol. 9, pp. 61–88, by J. R. Reynolds and A. D. Child, University of Florida and M. B. Gieselman, 3M Specialty Adhesives and Chemicals; "Electrically Conductive Polymers" in *ECT* (online), posting date: December 4, 2000, by John R. Reynolds, Andrew D. Child, University of Florida, Melinda B. Gieselman, 3M Specialty Adhesives and Chemicals.

CITED REFERENCES

1. J. R. Reynolds, A. D. Child, and M. B. Gieselman, *Electrically Conductive Polymers,* ECT 4th ed., Vol. 10, 1994, pp. 61–85.
2. V. V. Walatka, M. M. Labes, and J. H. Perlstein, *Phys. Rev. Lett.* **31**, 1139 (1973).
3. A. B. Kaiser, *Rep. Prog. Phys.* **64**, 1 (2001).
4. W. M. Wright and G. W. Woodham, in J. M. Margolis, ed., *Conductive Polymers and Plastics*, Chapman and Hall, New York, 1989.
5. A. J. Polak, in Ref. 4.
6. T. Ito, H. Shirakawa, and S. Ikeda, *J. Polym. Sci., Polym. Chem. Educ.* **12**, 11 (1974).
7. H. Shirakawa, *Handbook of Conducting Polymers*, 2nd ed., Marcel Dekker, New York, 1998, p. 197.
8. A. Munardi, R. Aznar, N. Theophilou, J. Sledz, F. Schue, and H. Naarmann, *Eur. Polym. J.* **23**, 11 (1987).
9. I. Kminek and I. Trekoval, *Makromol. Chem. Rapid Commun.* **7**, 53 (1986).
10. H. Tanaka and T. Danno, *Synth. Met.* **17**, 545 (1987).
11. K. Akagi, K. Sakamaki, H. Shirakawa, and H. Kyotani, *Synth. Met.* **69**, 29 (1995).
12. H. Naarman and N. Theophilou, *Synth. Met.* **22**, 1 (1987).
13. G. Lugli, U. Pedretti, and G. Perego, *J. Polym. Sci. Polym. Lett. Ed.* **23**, 129 (1985).
14. J. H. Edwards and W. J. Feast, *Polymer* **21**, 595 (1980).
15. J. H. Edwards, W. J. Feast, and D. C. Bott, *Polymer* **25**, 395 (1984).
16. P. J. S. Foot, P. D. Calvert, N. C. Billingham, C. S. Brown, N. S. Walker, and D. I. James, *Polymer* **27**, 448 (1986).
17. G. Lieser, G. Wegner, R. Weizenhofer, and L. Brombacher, *Polym. Prepr., Am. Chem. Soc. Div. Poly. Chem.* **25**, 221 (1984).
18. C. S. Brown, M. E. Vickers, P. J. S. Foot, N. C. Billingham, and P. D. Calvert, *Polymer* **27**, 1719 (1986).
19. D. D. C. Bradley, R. H. Friend, T. Hartmann, E. A. Marseglia, M. M. Sokolowski, and P. D. Townsend, *Synth. Met.* **17**, 473 (1987).
20. T. M. Swager, D. A. Dougherty, and R. H. Grubbs, *J. Am. Chem. Soc.* **110**, 2973 (1988).
21. T. M. Swager and R. H. Grubbs, *J. Am. Chem. Soc.* **111**, 4413 (1989).
22. J. C. W. Chien, G. E. Wnek, F. E. Karasz, and J. A. Hirsch, *Macromolecules* **14**, 479 (1981).
23. M. E. Galvin and G. E. Wnek, *Polym. Bull.* **13**, 109 (1985).
24. M. Aldissi and A. R. Bishop, *Polymer* **26**, 622 (1985).
25. J. A. Stowell, A. J. Amass, M. S. Beevers, and T. R. Farren, *Makromol. Chem.* **188**, 2535 (1987).
26. G. L. Baker and F. S. Bates, *Macromolecules* **17**, 2619 (1984).
27. P. Kovacic and M. B. Jones, *Chem. Rev.* **87**, 357 (1987); and references cited therein.
28. C. E. Brown and co-workers, *J. Polym. Sci., Polym. Chem. Educ.* **24**, 255 (1986).
29. T. Yamamoto, Y. Hayashi, and Y. Yamamoto, *Bull. Chem. Soc. Jpn.* **51**, 2091 (1978).
30. S. A. Arnautov and V. M. Kobryanskii, *Makromol. Chem. Phys.* **201**, 809 (2000).
31. V. M. Kobryanskii and S. A. Arnautov, *Synth. Met.* **55**, 1371 (1993).
32. D. G. H. Ballard, A. Curtis, I. M. Shirley, and S. C. Taylor, *Macromolecules* **21**, 294 (1988).
33. P. R. McKean and J. K. Stille, *Macromolecules* **20**, 1787 (1987).
34. M. Rehahn, A.-D. Schlüter, G. Wegner, and W. J. Feast, *Polymer* **30**, 1054 (1989).
35. Ibid., p. 1060.
36. W. Heitz, *Angew. Makromol. Chem.* **223**, 135 (1994).
37. A. D. Schluter, *Handbook of Conducting Polymers*, 2nd ed., Marcel Dekker, New York, 1998, p. 209.

38. A. F. Diaz and B. Hall, *IBM J. Res. Dev.* **27**, 342 (1983).
39. M. Schirmeisen and F. Beck, *J. Appl. Electrochem.* **19**, 401 (1989).
40. J. Rodriguez, T. F. Otero, H. Grande, J. P. Moliton, A. Moliton, and T. Trigaud, *Synth. Met.* **76**, 301 (1996).
41. A. F. Diaz, A. Martinez, K. K. Kanazawa, and M. Salmon, *J. Electroanal. Chem.* **130**, 181 (1981).
42. M. Salmon, A. F. Diaz, J. A. Logan, M. Krounbi, and J. Bargon, *Mol. Cryst. Liq. Cryst.* **83**, 265 (1982).
43. T. C. Clarke, J. C. Scott, and G. B. Street, *IBM J. Res. Dev.* **27**, 313 (1983).
44. J. C. Scott, P. Pfluger, M. T. Krounbi, and G. B. Street, *Phys. Rev. B* **28**, 2140 (1983).
45. K. J. Wynne and G. B. Street, *Macromolecules* **18**, 2361 (1985).
46. L. F. Warren and D. P. Anderson, *J. Electrochem. Soc.* **134**, 101 (1987).
47. D. T. Glatzhofer, J. Ulanski, and G. Wegner, *Polymer* **28**, 229 (1987).
48. T. Shimidzu, A. Ohtani, T. Iyoda, and K. Honda, *J. Electroanal. Chem.* **224**, 123 (1987).
49. M. B. Gieselman and J. R. Reynolds, *Macromolecules* **23**, 3118 (1990).
50. M. R. Bryce, A. Chissel, P. Kathirgamanathan, D. Parker, and N. R. M. Smith, *J. Chem. Soc., Chem. Commun.*, 466 (1987).
51. J. Ruhe, T. Ezquerra, and G. Wegner, *Makromol. Chem., Rapid Commun.* **10**, 103 (1989).
52. H. Masuda, S. Tanaka, and K. Kaeriyama, *J. Chem. Soc., Chem. Commun.*, 725 (1989).
53. A. Deronzier and J.-C. Moutet, *Acc. Chem. Res.* **22**, 249 (1989).
54. G. Bidan and M. Guglielmi, *Synth. Met.* **15**, 49 (1986).
55. J. R. Reynolds, P. A. Porapatic, and R. L. Toyooka, *Macromolecules* **20**, 958 (1987).
56. A. O. Patil, Y. Ikenoue, F. Wudl, and A. J. Heeger, *J. Am. Chem. Soc.* **109**, 1858 (1987).
57. P. G. Pickup, *J. Electroanal. Chem.* **225**, 273 (1987).
58. J. R. Reynolds, N. S. Sundaresan, M. Pomerantz, S. Basak, and C. K. Baker, *J. Electroanal. Chem.* **250**, 355 (1988).
59. M. Sato, S. Tanaka, and K. Kaeriyama, *J. Chem. Soc., Chem. Commun.*, 873 (1986).
60. R. L. Elsenbaumer, K.-Y. Jen, and R. Oboodi, *Synth. Met.* **15**, 169 (1986).
61. R. Sugimoto, S. Takeda, H. B. Gu, and K. Yoshino, *Chem. Express* **1**, 635 (1986).
62. S. Marchant and P. J. S. Foot, *Chemtronics* **5**, 131 (1991).
63. M. S. A. Abdou, X. Lu, Z. W. Xie, F. Orfino, M. J. Deen, and S. Holdcroft, *Chem. Mater.* **7**, 631 (1995).
64. M. Lemaire, R. Garreau, F. Garnier, and J. Roncali, *New J. Chem.* **11**, 703 (1987).
65. R. D. McCullough, R. D. Lowe, M. Jayaraman, and D. L. Anderson, *J. Org. Chem.* **58**, 904 (1993).
66. R. D. McCullough and co-workers, in L. R. Dalton and C. Lee, eds., *Electrical, Optical and Magnetic Properties of Organic Solid State Materials*, Vol. 328 and references therein.
67. T.-A. Chen and R. D. Riecke, *J. Am. Chem. Soc.* **114**, 10087 (1992).
68. X. Wu, T.-A. Chen, and R. D. Riecke, *Macromolecules* **28**, 2101 (1995); and references therein.
69. U.K. Pat. GB 2, 304, 346B (1998) D. Budd, R. Davis, and P. J. S. Foot, PCT/GB97/ 00742 (Baxenden Ltd., UK).
70. S. Hotta, S. D. Rughooputh, and A. J. Heeger, *Synth. Met.* **22**, 79 (1987).
71. K. Kaeriyama, M. Sato, and S. Tanaka, *Synth. Met.* **18**, 233 (1987).
72. M. Sato, S. Tanaka, K. Kaeriyama, and F. Tomanaga, *Polymer* **28**, 107 (1987).
73. J. Roncali, R. Garreau, D. Delabouglise, F. Garnier, and M. Lemaire, *Synth. Met.* **28**, C341 (1989).

74. A. O. Patil, *Synth. Met.* **28**, C495 (1989).

75. S. Tanaka, M. Sato, and K. Kaeriyama, *Polym. Commun.* **26**, 303 (1985).

76. K. Kaeriyama, S. Tanaka, M. Sato, and K. Hamada, *Synth. Met.* **28**, C611 (1989).

77. M. Feldhues, G. Kaempf, H. Litterer, T. Mecklenburg, and P. Wegener, *Synth. Met.* **28**, C487 (1989).

78. G. Daoust and M. Leclerc, *Macromolecules* **24**, 455 (1991).

79. S.-A. Chen and C.-C. Tsai, *Macromolecules* **26**, 2234 (1993).

80. R. L. Elsenbaumer, K.-Y. Jen, G. G. Miller, and L. W. Shacklette, *Synth. Met.* **18**, 277 (1987).

81. J. P. Ruiz, K. Nayak, D. S. Marynick, and J. R. Reynolds, *Macromolecules* **22**, 1231 (1989).

82. J. R. Reynolds, Jose P. Ruiz, Fei Wang, Cynthia A. Jolly, Kasinath Nayak, and Dennis S. Marynick, *Synth. Met.* **28**, 621 (1989).

83. J. Roncali, *Chem. Rev.* **92**, 711 (1992).

84. J. Roncali, *Handbook of Conducting Polymers*, 2nd ed., Marcel Dekker, New York, 1998, p. 311.

85. A. El Kassmi, W. Buchner, F. Fache, M. Lemaire, *J. Electroanal. Chem.* **326**, 357 (1992).

86. L. Robitaille and M. Leclerc, *Chem. Mater.* **59**, 301 (1992).

87. Y. Ikenoue, N. Outani, A. O. Patil, F. Wudl, and A. J. Heeger, *Synth. Met.* **30**, 305 (1989).

88. Y. Ikenoue, Y. Saida, M. Kira, H. Tomozawa, H. Yashima, and M. Kobayashi, *J. Chem. Soc., Chem. Commun.*, 1694 (1990).

89. P. Bauerle, *Adv. Mater.* **2**, 490 (1990).

90. F. Wudl, M. Kobayashi, and A. J. Heeger, *J. Org. Chem.* **49**, 3382 (1984).

91. D. Vanderzande, R. Van Asselt, I. Hoogmartens, J. Gelan, P. E. Froehling, M. Aussems, O. Aagaard, and R. Schellekens, *Synth. Met.* **74**, 65 (1995).

92. M. Pomerantz, *Handbook of Conducting Polymers*, 2nd ed., Marcel Dekker, New York, 1998, p. 277.

93. M. Pomerantz, B. Chaloner-Gill, L. O. Harding, J. J. Tseng, and W. J. Pomerantz, *Synth. Met.* **55**, 960 (1993).

94. J. P. Ferraris and T. L. Lambert, *J. Chem. Soc., Chem. Commun.*, 752 (1991).

95. S. Tanaka and Y. Yamashita, *Synth. Met.* **69**, 599 (1995).

96. A. N. Aleshin, R. Kiebooms, and A. J. Heeger, *Synth. Met.* **101**, 369 (1999).

97. J. Roncali, H. S. Li, and J. Roncali, *J. Electroanal. Chem.* **95**, 8983 (1990).

98. P. Bauerle and S. Scheib, *Adv. Mater.* **5**, 848 (1993).

99. M. J. Marsella and T. M. Swager, *J. Am. Chem. Soc.* **115**, 12214 (1993).

100. D. R. Gagnon, J. D. Capistran, F. E. Karasz, and R. W. Lenz, *Polym. Bull.* **12**, 293 (1984).

101. I. Murase, T. Ohnishi, T. Noguchi, M. Hirooka, and S. Murakami, *Mol. Cryst. Liq. Cryst.* **118**, 333 (1985).

102. R. W. Lenz, C.-C. Han, J. Stenger-Smith, and F. E. Karasz, *J. Polym. Sci., Polym. Chem. Ed.* **26**, 3241 (1988).

103. I. Murase, T. Ohnishi, T. Noguchi, and M. Hirooka, *Synth. Met.* **17**, 639 (1987).

104. S. Antoun, D. R. Gagnon, F. E. Karasz, and R. W. Lenz, *Polym. Bull.* **15**, 181 (1986).

105. K.-Y. Jen, L. W. Shacklette, and R. L. Elsenbaumer, *Synth. Met.* **22**, 179 (1987).

106. P. L. Burn, D. D. C. Bradley, A. R. Brown, R. H. Friend, and A. B. Holmes, *Synth. Met.* **41–43**, 261 (1991).

107. C.-C. Han and R. L. Elsenbaumer, *Synth. Met.* **30**, 123 (1989).

108. S. H. Askari, S. D. Rughooputh, and F. Wudl, *Synth. Met.* **29**, E129 (1989).

109. K.-Y. Jen, M. Maxfield, L. W. Shacklette, and R. L. Elsenbaumer, *J. Chem. Soc., Chem. Commun.*, 309 (1987).

110. PCT Int. Pat. WO 88 00,954 (1988), K.-Y. Jen, R. L. Elsenbaumer, and L. W. Shacklette.
111. I. Murase, T. Ohnishi, T. Noguchi, and M. Hirooka, *Polym. Commun.* **28**, 229 (1987).
112. Ger. Pat. DE 3,704,411 (1987), I. Murase, T. Ohnishi, and T. Noguchi.
113. S. Yamada, S. Tokito, T. Tsutsui, and S. Saito, *J. Chem. Soc., Chem. Commun.*, 1448 (1987).
114. K.-Y. Jen, T. R. Jow, and R. L. Elsenbaumer, *J. Chem. Soc., Chem. Commun.*, 1113 (1987).
115. K.-Y. Jen, R. Jow, L. W. Shacklette, M. Maxfield, H. Eckhardt, and R. L. Elsenbaumer, *Mol. Cryst. Liq. Cryst.* **160**, 69–77 (1988).
116. K.-Y. Jen, H. Eckhardt, T. R. Jow, L. W. Shacklette, and R. L. Elsenbaumer, *J. Chem. Soc., Chem. Commun.*, 215 (1988).
117. M. A. Takassi, W. Chen, R. E. Niziurski, M. P. Cava, and R. M. Metzger, *Synth. Met.* **36**, 55 (1990).
118. J.-I. Jin, H.-K. Shim, and R. W. Lenz, *Synth. Met.* **29**, E53 (1989).
119. H.-K. Shim, R. W. Lenz, and J.-I. Jin, *Makromol. Chem.* **190**, 389 (1989).
120. J. D. Stenger-Smith, T. Sauer, G. Wegner, and R. W. Lenz, *Polymer* **31**, 1632 (1990).
121. A. B. Holmes, A. C. Grimsdale, X. -C. Li, S. C. Moratti, F. Cacialli, J. Grüner, and R. H. Friend, *Synth. Met.* **76**, 165 (1996).
122. E. Thorn-Csanyi and H. D. Hohnk, *J. Mol. Catal.* **76**, 101 (1992).
123. A. Kumar and B. E. Eichinger, *Makromol. Chem., Rapid Commun.* **13**, 311 (1992).
124. A. Greiner, *Adv. Mater.* **5**, 477 (1993).
125. A. G. MacDiarmid, J.-C. Chiang, A. E. Richter, N. L. D. Somasiri, and A. J. Epstein, in L. Alcacer, ed., *Conducting Polymers*, Reidel, Dordrecht, Holland, 1987, p. 105.
126. K. L. Tan, B. T. G. Tan, S. H. Khor, K. G. Neoh, and E. T. Kang, *J. Phys. Chem. Solids* **52**, 673 (1991).
127. M. Angelopoulos, A. Ray, A. G. MacDiarmid, and A. J. Epstein, *Synth. Met.* **21**, 21 (1987).
128. E. M. Genies, M. Lapkowski, and J. F. Penneau, *J. Electroanal. Chem.* **249**, 97 (1988).
129. G. D'Aprano, M. Leclerc, and G. Zotti, *J. Electroanal. Chem.* **351**, 145 (1993).
130. G. D'Aprano, M. Leclerc, G. Zotti, and G. Schiavon, *Chem. Mater.* **7**, 33 (1995).
131. Y. Wei, W. W. Focke, G. E. Wnek, A. Ray, and A. G. MacDiarmid, *J. Phys. Chem.* **93**, 495 (1989).
132. J. Guay, M. Leclerc, and L. H. Dao, *J. Electroanal. Chem.* **251**, 31 (1988).
133. M. Leclerc, J. Guay, and L. H. Dao, *Macromolecules* **22**, 641 (1989).
134. L. H. Dao, J. Guay, M. Leclerc, and J.-W. Chevalier, *Synth. Met.* **29**, E377 (1989).
135. L. H. Dao, J. Guay, and M. Leclerc, *Synth. Met.* **29**, E383 (1989).
136. J.-W. Chevalier, J.-Y. Bergeron, and L. H. Dao, *Polym. Commun.* **30**, 308 (1989).
137. J. Yue and A. J. Epstein, *J. Am. Chem. Soc.* **112**, 2800 (1990).
138. X.-L Wei, Y. Z. Wang, S. M. Long, C. Bobeczko, and A. J. Epstein, *J. Am. Chem. Soc.* **118**, 2545 (1996).
139. K. Krishnamoorthy, A. Q. Contractor, and A. Kumar, *J. Chem. Soc., Chem. Commun.*, 240 (2002).
140. Y. Cao, P. Smith, and A. J. Heeger, *Synth. Met.* **48**, 91 (1992).
141. R. Menon, Y. Cao, D. Moses, and A. J. Heeger, *Phys. Rev. B* **47** (1993).
142. T. Yamamoto, T. Ito, and K. Kubota, *Chem. Ind.*, 337 (16 May 1988).
143. T. Yamamoto, T. Nakamura, H. Fukumoto, and K. Kubota, *Chem. Lett.*, 502 (2001).
144. T. Yamamoto and co-workers, *Chem. Lett.* 223 (1990).
145. F. Barigelletti and L. Flamigni, *Chem. Soc. Rev.* **29**, 1 (2000).
146. M. Kertesz, *Macromolecules* **28**, 1475 (1995).
147. L. R. Dalton and co-workers, *Polymer* **28**, 543 (1987).

148. L. P. Yu and L. R. Dalton, *Macromolecules* **23**, 3439 (1990).
149. P. J. S. Foot, V. Montgomery, and P. Spearman, *Mater. Sci. Forum* **191**, 251 (1995).
150. U. Scherf, *J. Mater. Chem.* **9**, 1853 (1999).
151. Y. W. Park and co-workers, *Synth. Met.* **17**, 539 (1987).
152. M. Aldissi, *Mol. Cryst. Liq. Cryst.* **160**, 121 (1988).
153. M. R. Bryce and co-workers, *Synth. Met.* **39**, 397 (1991).
154. N. Koide, *Mol. Cryst. Liq. Cryst. A* **261**, 427 (1995).
155. F. Vicentini and co-workers, *Liq. Crystals* **19**, 235 (1995).
156. P. J. Langley, F. J. Davis, and G. R. Mitchell, *J. Chem. Soc., Perkin Trans. II*, 2229 (1997).
157. V. Percec and co-workers, *Macromolecules* **32**, 2597 (1999).
158. L. I. Gabaston, P. J. S. Foot, and J. W. Brown, *J. Chem. Soc., Chem. Commun.*, 429 (1996).
159. P. Ibison, P. J. S. Foot, and J. W. Brown, *Synth. Met.* **76**, 297 (1996).
160. H. Hasegawa, M. Kijima, and H. Shirakawa, *Synth. Met.* **84**, 177 (1997).
161. U.K. Patent, GB 2 318 119B, (2000) P. Ibison and co-workers.
162. K. Akagi, H. Goto, and H. Shirakawa, *Synth. Met.*, **84**, 313 (1997).
163. M. E. Galvin and G. E. Wnek, *J. Polym. Sci., Polym. Chem. Ed.* **21**, 2727 (1983).
164. M. F. Rubner, S. K. Tripathy, J. Georger, and P. Cholewa, *Macromolecules* **16**, 870 (1983).
165. J. M. Machado, J. B. Schlenoff, and F. E. Karasz, *Macromolecules* **22**, 1964 (1989).
166. L. W. Shacklette, C. C. Han, and M. H. Luly, *Synth. Met.* **57**, 3532 (1993).
167. L. M. Gan and co-workers, *Polymer Bull.* **31**, 347 (1993).
168. T. Vikki and co-workers, *Macromolecules* **29**, 2945 (1996).
169. J. Stejskal and co-workers, *Polymer International* **44**, 283 (1997).
170. M. Zilberman, G. I. Titelman, A. Siegmann and co-workers, *J. Appl. Polymer Sci.* **23**, 243 (1997).
171. W. A. Gazotti, Jr., R. Faez, and M-A. De Paoli, *Eur. Polymer. J.* **35**, 35 (1999).
172. H. Morgan, P. J. S. Foot, and N. W. Brooks, *J. Mater. Sci.* **36**, 5369 (2001).
173. J.-E. Österholm and co-workers, *Synth. Met.* **28**, C435 (1989); ibid., C467 (1989).
174. H. Isotalo and co-workers, *Synth. Met.* **28**, C461 (1989).
175. J.-O Nilsson and co-workers, *Synth. Met.* **28**, C445 (1989).
176. S. E. Lindsey and G. B. Street, *Synth. Met.* **10**, 67 (1984).
177. M.-A. DePaoli, R. J. Waltman, A. J. Diaz, and J. Bargon, *J. Polym. Sci., Polym. Chem. Ed.* **23**, 1687 (1985).
178. O. Niwa, M. Kakuchi, and T. Tamamura, *Polym. J.* **19**, 1293 (1987).
179. F.-R. F. Fan and A. J. Bard, *J. Electrochem. Soc.* **133**, 301 (1986).
180. R. M. Penner and C. R. Martin, *J. Electrochem. Soc.* **133**, 310 (1986).
181. G. Nagasubramanian, S. Di Stefano, and J. Mascanin, *J. Phys. Chem.* **90**, 447 (1986).
182. R. Gangopadhyay and A. De, *Sensors Actuators B* **77**, 326 (2001).
183. F. Selampinar, U. Akbulut, and L. Toppare, *Polym.—Plast. Technol. Eng.* **37**, 295 (1998).
184. J. Aguilar-Hernandez, J. Skarda, and K. Potje-Kamloth, *Synth. Met.* **95**, 197 (1998).
185. F. Selampinar and co-workers, *Polym. Sci. Polym. Chem.* **35**, 3009 (1997).
186. F. Selampinar, U. Akbulut, and L. Toppare, *Synth. Met.* **84**, 185 (1997).
187. idem., *Macromol. Sci.—Pure Appl. Chem.* **A33**, 309 (1996).
188. N. Balci and co-workers, *J. Appl. Polym. Sci.* **64**, 667 (1997).
189. K. Gilmore and co-workers, *Polymer Gels Netw.* **2**, 135 (1994).
190. F. Vatansever and co-workers, *Polym. Int.* **41**, 237 (1996).
191. S. Yigit and co-workers, *Synth. Met.* **79**, 11 (1996).
192. B. Aydinli, L. Toppare, and T. Tincer, *J. Appl. Polym. Sci.* **72**, 1843 (1999).
193. M. Omastova and co-workers, *Synth. Met.* **81**, 49 (1996).

194. J. P. Yang and co-workers, *Polymer* **37**, 793 (1996).
195. Y. Lee and J. Kim, *Polym.—Korea* **22**, 953 (1998).
196. U. Gessler, M. L. Hallensleben, and L. Toppare, *Adv. Mater.* **3**, 104 (1991).
197. J. Fournier, G. Boiteux, G. Seytre, and G. Marichy, *Synth. Met.* **84**, 839 (1997).
198. A. Mohammadi and co-workers, *J. Polymer Sci. Pol. Chem.* **32**, 495 (1994).
199. R. E. Noftle and D. Pletcher, *J. Electroanal. Chem.* **227**, 229 (1987).
200. L. L. Miller and Q.-X Zhou, *Macromolecules* **20**, 1594 (1987).
201. T. Shimidzu, A. Ohtani, and K. Honda, *J. Electroanal. Chem.* **251**, 323 (1988).
202. S. J. Jasne and C. K. Chiklis, *Synth. Met.* **15**, 175 (1986).
203. S. P. Armes, J. F. Miller, and B. Vincent, *J. Colloid Interfac. Sci.* **118**, 410 (1987).
204. S. P. Armes, M. Aldissi, M. Hawley et al., *Langmuir* **7**, 1447 (1992).
205. H. Nakao, T. Nagaoka, and K. Ogura, *Anal. Sci.* **13**, 327 (1997).
206. R. Gangopadhyay, A. De, and G. Ghosh, *Synth. Met.* **123**, 21 (2001).
207. S. Maeda and S. P. Armes, *J. Mater. Chem.* **4**, 935 (1994).
208. C. Perruchot and co-workers, *Synth. Met.* **113**, 53 (2000).
209. A. O. Patil, A. J. Heeger, and F. Wudl, *Chem. Rev.* **88**, 183 (1988).
210. J. R. Reynolds, *J. Molec. Elec.* **2**, 1 (1986).
211. T. C. Chung, J. H. Kaufman, A. J. Heeger, and F. Wudl, *Phys. Rev. B* **30**, 702 (1984).
212. S. N. Hoier, D. D. Ginley, and S.-M Park, *J. Electrochem. Soc.* **135**, 91 (1988).
213. S. Wang, T. Kawai, K. Yoshino, K. Tanaka, and T. Yamabe, *Jpn. J. Appl. Phys.* **29**, L2264 (1990).
214. K. Nishioka, S. Wang, and K. Yoshino, *Synth. Met.* **41–43**, 815 (1991).
215. K. F. Voss, C. M. Foster, L. Smilowitz, D. Mihailovic, S. Askari, G. Srdanov, Z. Ni, S. Shi, A. J. Heeger, and F. Wudl, *Phys. Rev. B* **43**(6), 5109 (1991).
216. H. Eckhart, L. W. Shacklette, K.-Y Jen, and R. L. Elsenbaumer, *J. Chem. Phys.* **91**, 1303 (1989).
217. C. M. Pratt, Ph.D. Thesis, Kingston University (2003).
218. S. Roth, One-Dimensional Metals, VCH, Weinheim (1995).
219. Van der Pauw, *Philips Res. Repts.* **13**, 1 (1958).
220. Van der Pauw, *Philips Res. Repts.* **16**, 187 (1961).
221. H. C. Montgomery, *J. Appl. Phys.* **42**, 2971 (1971).
222. R. L. Greene, G. B. Street, and L. J. Suter, *Phys. Rev. Lett.* **34**, 577 (1975).
223. A. B. Kaiser, *Adv. Mater.* **13**, 927 (2001).
224. Y. W. Park, E. S. Choi, and D. S. Suh, *Synth. Met.* **96**, 81 (1998).
225. J. Tsukamoto, *Adv. Phys.* **41**, 509 (1992).
226. Y. W. Park, *Synth. Met.* **45**, 173 (1991).
227. F. Körner, G. Thummes, and J. Kötzler, *Synth. Met.* **51**, 31 (1992).
228. Y. Nogami, H. Kaneko, H. Ito, T. Ishiguro, T. Sasaki, N. Toyota, A. Takahashi, and J. Tsukamoto, *Phys. Rev. B* **43**, 11829 (1991).
229. R. Zuzok, A. B. Kaiser, W. Pukacki, and S. Roth, *J. Chem. Phys.* **95**, 1270 (1991).
230. T. Schimmel, W. Riess, J. Gmeiner, G. Denninger, M. Schwoerer, H. Naarmann, and N. Theophilou, *Solid State Commun.* **65**, 1311 (1988).
231. N. Basescu, Z.-X Liu, D. Moses, A. J. Heeger, H. Naarmann, and N. Theophilou, *Nature (London)* **327**, 403 (1987).
232. K. Ehinger and S. Roth, *Philos. Mag. B* **53**, 301 (1986).
233. A. J. Epstein, H. Rommelman, R. Bigelow, H. W. Gibson, D. M. Hoffmann, and D. B. Tanner, *Phys. Rev. Lett.* **50**, 1866 (1983).
234. J. P. Travers, B. Sixou, D. Berner, A. Wolter, P. Rannou, B. Beau, B. Pépin-Donat, C. Barthet, M. Guglielmi, N. Mermilliod, B. Gilles, D. Djurado, A. J. Attias, and M. Vautrin, *Synth. Met.* **101**, 359 (1999).
235. S. J. Pomfret, P. N. Adams, N. P. Comfort, and A. P. Monkman, *Polymer* **41**, 2265 (2000).

236. E. R. Holland, S. J. Pomfret, P. N. Adams, and A. P. Monkman, *J. Phys.: Condens. Matter* **8**, 2991 (1996).
237. C. O. Yoon, J. H. Kim, H. K. Sung, J. H. Kim, K. Lee, and H. Lee, *Synth. Met.* **81**, 75 (1996).
238. A. Raghunathan, G. Rangarajan, and D. C. Trivedi, *Synth. Met.* **81**, 39 (1996).
239. Z. H. Wang, J. Joo, C.-H Hsu, and A. J. Epstein, *Synth. Met.* **68**, 207 (1995).
240. G. Du, J. Avlyanov, C. Y. Wu, K. G. Reimer, A. Benatar, A. G. MacDiarmid, and A. J. Epstein, *Synth. Met.* **85**, 1339 (1997).
241. B. Sixou, C. Barthet, M. Guglielmi, and J. P. Travers, *Phys. Rev. B* **56**, 4604 (1997).
242. C. K. Subramaniam, A. B. Kaiser, P. W. Gilberd, C.-J Liu, and B. Wessling, *Solid State Commun.* **97**, 235 (1996).
243. N. T. Kemp, A. B. Kaiser, C.-J Liu, B. Chapman, A. M. Carr, H. J. Trodahl, R. G. Buckley, A. C. Partridge, J. Y. Lee, C. Y. Kim, A. Bartl, L. Dunsch, W. T. Smith, and J. S. Shapiro, *J. Polym. Sci. B: Polym. Phys.* **37**, 953 (1999).
244. T. H. Gilani and T. Ishiguro, *J. Phys. Soc. Jpn.* **66**, 727 (1997).
245. W. P. Lee, Y. W. Park, and Y. S. Choi, *Synth. Met.* **84**, 841 (1997).
246. K. Sato, M. Yamaura, T. Hagiwara, K. Murata, and M. Tokumoto, *Synth. Met.* **40**, 35 (1991).
247. D. S. Maddison, R. B. Roberts, and J. Unsworth, *Synth. Met.* **30**, 47 (1989).
248. J. A. Reedijk, H. C. F. Martens, S. M. C. van Bohemen, O. Hilt, H. B. Brom, and M. A. J. Michels, *Synth. Met.* **101**, 475 (1999).
249. S. Masubuchi and S. Kazama, *Synth. Met.* **74**, 151 (1995).
250. D. Berner, J.-P Travers, and D. Delabouglise, *Synth. Met.* **101**, 377 (1999).
251. S. Masubuchi, T. Fukuhara, and S. Kazama, *Synth. Met.* **84**, 601 (1997).
252. L. W. Shacklette, H. Eckhardt, R. R. Chance, G. G. Miller, D. M. Ivory, and R. H. Baughman, *J. Chem. Phys.* **73**, 4098 (1980).
253. M. Ahlskog, M. Reghu, A. J. Heeger, T. Noguchi, and T. Ohnishi, *Phys. Rev. B* **53**, 15529 (1996).
254. J. M. Madsen, B. R. Johnson, X. L. Hua, R. B. Hallock, M. A. Masse, and F. E. Karasz, *Phys. Rev. B* **40**, 11751 (1989).
255. R. H. Baughman, N. S. Murthy, G. G. Miller, and L. W. Shacklette, *J. Chem. Phys.* **79**, 1065 (1983).
256. A. J. Epstein, H. Rommelman, M. A. Druy, A. J. Heeger, and A. G. MacDiarmid, *Solid State Commun.* **38**, 683 (1981).
257. D. Moses, A. Denenstein, A. Pron, A. J. Heeger, and A. G. MacDiarmid, *Solid State Commun.* **36**, 219 (1980).
258. Y. W. Park, A. J. Heeger, M. A. Druy, and A. G. MacDiarmid, *J. Chem. Phys.* **73**, 946 (1980).
259. M. P. Stevens, *Polymer Chemistry: An Introduction*, 2nd ed., Oxford University Press, Oxford, 1990, p. 91.
260. J. P. Pouget, Z. Oblakowski, Y. Nogami, P. A. Albouy, M. Laridjani, E. J. Oh, Y. Min, A. G. MacDiarmid, J. Tsukamoto, T. Ishiguro, and A. J. Epstein, *Synth. Met.* **65**, 131 (1994).
261. L. Pietronero, *Synth. Met.* **8**, 225 (1983).
262. S. Kivelson and A. J. Heeger, *Synth. Met.* **22**, 371 (1988).
263. A. J. Epstein, in P. Bernier, S. Lefrant, and G. Bidan, eds., *Advances in Synthetic Metals*, Elsevier, Amsterdam, The Netherlands, 1999, p. 349.
264. N. F. Mott and E. A. Davis, *Electronic Processes in Non-Crystalline Materials*, 2nd ed., Clarendon Press, Oxford, 1979.
265. P. Sheng, *Philos. Mag.* **65**, 357 (1992).
266. J. G. Park, G. T. Kim, V. Krstic, S. H. Lee, B. Kim, S. Roth, M. Burghard, and Y. W. Park, *Synth. Met.* **119**, 469 (2001).

267. N. C. Billingham, P. D. Calvert, P. J. S. Foot, and F. Mohammad, *Polymer Degradation Stability* **19**, 323 (1987).

268. D. Poussin, H. Morgan, and P. J. S. Foot, *Polymer Inter.* **52**, 433 (2003).

269. A. A. Pud, *Synth. Metals* **66**, 1 (1994).

270. G. G. Wallace, G. M. Spinks, P. R. Teasdale, *Conductive Electroactive Polymers*, Technomic, Lancaster, Pa., 1997.

271. N. F. Colaneri and L. W. Shacklette, *IEEE Trans. Intr. Meas.* **41**, 291 (1992).

272. R. V. Gregory, W. C. Kimbrell, and H. H. Kuhn, *Synth. Met.* **28**, C823 (1989).

273. R. V. Gregory, W. C. Kimbell, and H. H. Kuhn, *J. Coated Fabrics* **20**, 1 (1991).

274. M. Angelopoulos, J. M. Shaw, R. D. Kaplan, and S. Perreault, *J. Vac. Sci. Tech. B* **7**, 1519 (1989).

275. M. Angelopoulos, *IBM J. Res. Develop.* **45**, 1 (2001).

276. M. Angelopoulos and J. M. Shaw, *Microelect. Eng.* **13**, 515 (1991).

277. W. S. Huang, M. Angelopoulos, J. R. White, and J. M. Park, *Mol. Cryst., Liq. Cryst.* **189**, 227 (1990).

278. B. Wessling, *Chem. Innov.* **31**, 34 (2001).

279. B. Wessling, *Synth. Met.* **85**, 1313 (1997).

280. D. W. Berry, *J. Electrochem. Soc.* **132**, 1022 (1985).

281. W.-K Lu, R. L. Elsenbaumer, and B. Wessling, *Synth. Met.* **71**, 2163 (1995).

282. B. Wessling, *Adv. Mater.* **6**, 226 (1994).

283. J. Posdorfer and B. Wessling, *Fresenius J. Anal. Chem.* **367**, 343 (2000).

284. M. D. Levi, Y. Gofer, and D. Aurbach, *Poly. Adv. Tech.*, **13**, 697 (2003).

285. D. Naegele and R. Bittihn, *Solid State Ionics* **28–30**, 983 (1988).

286. T. Matsunaga, H. Daifuku, T. Nakajima, and T. Kawagoe, *Polymers Adv. Technol.* **1**, 33 (1990).

287. L. W. Shacklette, M. Maxfield, S. Gould, J. F. Wolf, T. R. Jow, and R. H. Baughman, *Synth. Met.* **18**, 611 (1987).

288. R. H. Baughman, *Makromol. Chem. Macromol. Symp.* **51**, 193 (1991).

289. T. F. Otero, J. Rodriguez, E. Angulo, and C. Santamaria, *Synth. Met.* **55**, 3715 (1993).

290. Q. Pei and O. Inganas, *Synth. Met.* **55**, 3718 (1993).

291. E. Smela, *Adv. Mat.* **15**, 481 (2003).

292. E. W. H. Jager, E. Smela, and O. Inganas, *Science* **290**, 1540 (2000).

293. L. Dai, P. Soundarrajan, and T. Kim, *Pure Appl. Chem.* **74**, 1753 (2002).

294. R. H. Baughman, R. L. Elsenbaumer, Z. Iqbal, G. G. Miller, and H. Eckhardt, *Proceedings of the International Winter School on Conducting Polymers*, Kirchberg, Austria, March 1987.

295. G. Gustafsson and I. Lundstrom, *Proceedings of the Second International Meeting on Chemical Sensors*, Bordeaux, France, July 1986.

296. J. Janata and M. Josowicz, *Nature Mat.* **2**, 19 (2003).

297. N. C. Foulds and C. R. Lowe, *J. Cem. Soc., Faraday Trans. 1* **82**, 1259 (1986).

298. W. Liang and C. R. Martin, *Chem. Mater.* **3**, 390 (1991).

299. M. R. Anderson, B. R. Mattes, H. Reiss, and R. B. Kaner, *Science* **252**, 1412 (1991).

300. J. Sarrazin, M. Persin, and M. Cretin, *Macromol. Sym.* **188**, 1 (2002).

301. J. Pellegrino, *Adv. Memb. Tech. Ann. N. Y. Acad. Sci.* **984**, 289 (2003).

302. D. R. Rosseinsky and R. J. Mortimer, *Adv. Mat.* **13**, 783 (2001).

303. J. H. Burroughes and co-workers, *Nature (London)* **347**, 539 (1990).

304. D. Braun, A. J. Heeger, and H. Kroemer, *J. Electron. Mater.* **20**, 945 (1991).

305. Y. Ohmori, M. Uchida, K. Muro, and K. Yoshino, *Jpn. J. Appl. Phys.* **30**, L1938 (1991).

306. M. Granstrom, K. Petritsch, A. C. Arias, A. Lux, M. R. Andersson, and R. H. Friend, *Nature (London)* **395**, 257 (1998).

307. A. Kraft, A. Grimsdale, and A. B. Holmes, *Angew. Chem. Int. Ed.* **37**, 402 (1998).

308. R. H. Friend, R. W. Gymer, A. B. Holmes, J. H. Burroughes, R. N. Marks, C. Taliani, D. D. C. Bradley, D. A. Dos Santos, J. L. Bredas, M. Logdlund, and W. R. Salaneck, *Nature (London)* **397**, 121 (1999).
309. M. T. Bernius, M. Inbasekaran, J. O'Brien, and W. Wu, *Adv. Mater.* **12**, 1737 (2000).
310. P. K. H. Ho, S. Thomas, R. H. Friend, and N. Tessler, *Science* **285**, 233 (1999).
311. I. D. Rees, K. L. Robinson, A. B. Holmes, C. R. Towns, and R. O'Dell, *MRS Bull.* **27**, 451 (2002).
312. C. D. Muller, A. Falcou, N. Reckefuss, M. Rojahn, V. Wiederhirn, P. Rudati, H. Frohne, O. Nuyken, H. Becker, and K. Meerholz, *Nature (London)* **421**, 829 (2003).
313. A. Holmes, *Nature (London)* **421**, 800 (2003).
314. Yong Cao, Gang Yu, and Alan J. Heeger, *Synth. Met.* **102**, 881 (1999).
315. M. S. Weaver and co-workers, *Thin Solid Films* **273**, 39 (1996).
316. M. D. McGehee and A. J. Heeger, *Adv. Mater.* **12**, 1655 (2000).
317. C. Kallinger, M. Hilmer, A. Haugeneder, M. Perner, W. Spirkl, U. Lemmer, J. Feldmann, U. Scherf, K. Müllen, A. Gombert, and V. Wittwer, *Adv. Mat.* **10**, 920 (1998).
318. N. Tessler, *Adv. Mat.* **11**, 363 (1999).
319. G. Kranzelbinder and G. Leising, *Rep. Prog. Phys.* **63**, 729 (2000).
320. M. Grandstrom, K. Petritsch, A. C. Arias, A. Lux, M. R. Anderson, and R. H. Friend, *Nature (London)* **395**, 257 (1998).
321. N. S. Saraciftci, *Curr. Opin. Solid-State Mater. Sci.* **4**, 373 (1999).
322. G. G. Wallace, P. C. Dastoor, D. L. Officer, and C. O. Too, *Chem. Innov.* **30**, 14 (2001).
323. G. Yu, J. Gao, J. C. Hummelen, F. Wudl, and A. J. Heeger, *Science* **270**, 1789 (1995).
324. C. J. Brabec, F. Padinger, J. C. Hummelen, R. A. J. Janssen, N. S. Saraciftci, *Synth. Met.* **102**, 861 (1999).
325. W. Huynh, J. J. Dittmer, and A. P. Alivisatos, *Science* **295**, 2425 (2002).
326. J. S. Miller, *Adv. Mater.* **5**, 671 (1993).
327. J. H. Burroughes, C. A. Jones, and R. H. Friend, *Nature (London)* **335**, 137 (1988).
328. H. Sirringhaus, N. Tessler, and R. H. Friend, *Science* **280**, 1741 (1998).
329. G. H. Gelinck, T. C. T. Geuns, and D. M. de Leeuw, *Appl. Phys. Lett.* **77**, 1487 (2000).
330. I. Chun, D. H. Reneker, H. Fong, X. Fang, J. Dietzel, N. B. Tan, and K. Kearns, *J. Adv. Mat.* **31**, 36 (1999).
331. A. G. MacDiarmid, *Angew. Chem. Int. Ed. Engl.* **40**, 2591 (2001).
332. D. Hohnholz and A. G. MacDiarmid, *Synth. Met.* **121**, 1327 (2001).

PETER J. S. FOOT
Materials Research Group, Kingston University

ALAN B. KAISER
MacDiarmid Institute for Advanced Materials and
Nanotechnology and Victoria University of Wellington

CONTROLLED RELEASE TECHNOLOGY, AGRICULTURAL

1. Pesticide Controlled Release Formulations

Controlled release formulations (CRF) aim to make available pesticides at rates appropriate for efficient control of pests under field conditions. These formulations are combinations of the pesticidal active agent with inert materials that protect, and release, the active agent according to the pest control needs. A depot or reservoir of active agent, within the releasing device or particle, is released at a defined rate, or variable rate, into the environment over a specified period. The releasing systems are usually solid and can vary in size from microparticles to large devices several centimetres across. However, the aspect that differentiates CRF from conventional formulations such as emulsifiable concentrates, wettable powders, soluble liquids, water-dispersible granules, etc, is time and the kinetics of release are central to CRF. In contrast, in the case of conventional formulations, complete availability of the active agent is usually considered to be immediate or rapid following deployment.

Controlled release formulations can be used with a wide range of pesticides, including inorganic substances, conventional low molecular weight organic substances, high molecular weight substances such as peptides or proteins, microbials (such as mycopesticides), and semiochemicals (which, eg, pheromones, modify pest behavior). Applications may be found in agriculture, veterinary, and public health sectors, and may be aimed at controlling a variety of pest organisms such as insects, mites, rodents, nematodes, weeds, and microorganims as well improving crop production with plant growth regulators. Within the term of "controlled release" there exist a variety of release types such as extended, slow, fast, delayed, programmed, sustained, pulsed, etc.

As for all pesticide formulations, CRF need to be applied, or placed, in the field appropriately for targeting the pests. In crop protection, this usually means application to the crop, or crop area, by means that achieve good distribution. Such distribution depends on how the pesticide moves to the target organism following application and often needs small particle size to provide this. Thus, the standard application methods of spraying and granules are important in agricultural pesticide delivery, which in turn limits the device size of the controlled release systems deployed. Greatest advances in CRF in agriculture have thus been found with sprayable, and to a lesser extent, granular methods. More specialized methods, eg, based on pheromones or baits, have been commercially possible using larger devices (1).

Most CRF are based upon macromolecules (usually polymers) as the inert components (sometimes combined with clays, salts, etc). The reason is because large molecules tend not to move in the environment (being often water insoluble and nonvolatile) and they can entrap small and large molecules such as pesticides. Thus, formulation with polymers provides the construct needed to entrap the pesticide and to build into the resulting depot device the mechanism for reliable release rates. Such polymers are best degradable so as to remove them from the environment following their use. In selecting polymers for formulation, cost is of great importance in agriculture, when compared to medical drug delivery systems where benefits are considered more commercially valuable.

Thus, controlled release technology aims to manipulate the bioavailability of the pesticide in the local environment following application (2). This approach has many benefits compared to conventional formulations that include increased safety to the environment, workers, and consumers. Lower concentrations of released pesticide in the environment leads to reduced losses, such as leaching, evaporation, degradation, and binding. Reduced losses may mean better pest control, less nontarget impacts, reduced crop phytotoxicity, and safer formulations. Numerous benefits have been given on behalf of controlled release formulations, including

Protection of active ingredients from environmental degradation.

Manipulation of bioavailability and persistence.

Reduction of toxicity and operator hazards.

Reduction of phytotoxicity to seeds and crops.

Improved selectivity between target and nontarget organisms and usage in integrated pest management (IPM).

Reduction in repellency (also reduction in odors).

Allows coformulation, especially of incompatible pesticides (eg, of chemical and microbial pesticides).

Permits elimination of solvents.

Improves formulation of actives with phase changes near ambient temperatures.

Improves handling qualities of formulations and ease of cleaning sprayers.

Possible reduced application rates.

However, these advantages have been known for some decades [in fact, an early publication on a site specific release formulation of an insecticide dates to 1948 (3)] but their exploitation has been slow to develop in commercial practice. The first microcapsule formulation came on the market in 1974 (4); since then uptake represents only a small portion of total pesticide formulations. This is in contrast to the drug sector where controlled release and delivery has been rapidly expanding.

This slow uptake in the pesticide market may be related to the increased costs of the new technology on a product basis (but not on an "effect" basis) and there is a need for a change in attitude to pest control. However, there are also technical problems involved in active agent delivery in the open environment that may restrict extensive uptake. Renewed interest in good pesticide delivery could be prompted by the inexorable increase in limitations on pesticide numbers and their use as the U.S. Environmental Protection Agency (EPA), and other organizations, phase out more pesticides. New active ingredients coming to the market are currently fewer and registration is slow and expensive. In this situation, the commercial benefits of new safer controlled release formulations may be starting to outweigh the perceived disadvantages and the usual route of new molecule introduction.

2. Principles of CRF for Use in The Environment

2.1. Pesticide Delivery. As with all treatments, based on biologically active molecules, targeting is fundamental to its success. If the substances do not reach the pest, then logically no control will be achieved. Delivery to the target pest is considered both in time and place; sometimes the pesticide moves toward the target [eg, with foliar herbicides (5)] and sometimes the target moves toward the pesticide [residual insecticides (6)]. Thus, the efficiency of this delivery process can be defined as the ratio of the amount of pesticide reaching the pest divided by the amount applied, per unit cropping area. For many pest organisms the amount of pesticide needed for control can be ascertained, thus giving some idea of the efficiency of the delivery process. This delivery process is the sum of the placement methods, ie, spraying, granules, bait, etc, and the subsequent movement of the pesticide combined with the movement and growth of the pest itself. Sometimes, the pesticide is activated following application, in which it is chemically transformed into a more pesticidal substance. Transfer of the pesticide occurs through contact and also in mobile phases such as air and water.

2.2. Losses and Half-Life. During delivery, the pesticide is lost by a multitude of processes. The most rapid loss mechanisms cause the greatest amount of loss in which the pesticide is removed from the cropping location. These include spray drift, evaporation, leaching, run-off, sorption, dispersal, and dilution below active concentrations. Slower loss mechanisms include degradation of the pesticide caused by light (photodegradation), by biological processes (especially by microorganisms in soil) and chemical processes (such as hydrolysis and oxidation) (7). Degradation produces breakdown products (metabolites) that may be more or less toxic than the parent molecule and may be more or less prone to movement away from the application site. The environmental hazard of the metabolites may be greater than the pesticide but usually degradation represents a reduction of the pesticidal activity and overall toxicity. Alternatively, the pesticide, and its metabolites, may be bound into plants or into the organic matter (or clay) of soil.

The tendency for any pesticide to be degraded is characteristic of its molecular properties and can be expressed in the DT_{50} (time for 50% disappearance) or $t_{1/2}$ (half-life), typical values for agricultural soils. This value is based on a pseudo-first-order kinetic for loss. Pesticides with long half-lives, which persist for long periods are more effective in pest control, and are thus more efficient in contacting the pest, than those with short half-lives. However, such pesticides are less desirable environmentally, as long persistence can allow the substance to migrate, or otherwise cause detrimental impacts, such as by entering the food chain if having a high partition coefficient. In contrast, other bioactives may have short half-lives, thus requiring large application rates to provide an adequate period of effective control. This can be demonstrated in Figure 1, where the durations of control provided by two pesticides, P-1 with a half-life of 15 days and P-2 with a half-life of 50 days, are shown. Both require a minimum active level of 1 mg to provide control and the two plots "A" compare the periods of control, on a logarithmic basis. It can be seen that the faster degrading pesticide P-1 requires an initial application of 1000 mg, whereas the slower degrading P-2

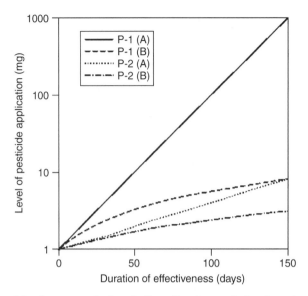

Fig. 1.   Relationships between the level of application and the duration of action of two pesticides, P-1 and P-2, with differing half-lives, for conventional (**A**) and controlled release (**B**) formulations.

only needs 8 mg to be effective up to 150 days. In this case, the high initial concentration in the environment forms a reservoir or depot albeit exposed to degradation and losses that is available above the level required for control (ie, 1 mg) for the duration of the control period. Thus, the proportion of pesticide that is lost (ie, wasted) increases as the half-life decreases.

2.3. Controlled Delivery.   As pesticide loss is concentration dependent, reducing environmental concentrations will reduce losses. If the concentration at the target pest could be kept at the minimum (or just above) for effective pest control by continuous supplementing for that portion lost or dissipated then the overall losses could be minimized (8). Keeping this minimum for the duration of control needed would represent the ideal approach with highest possible level of efficiency of delivery.

To maintain the concentration at the target, pesticide needs to be supplied at the same rate at which it is dissipated. Thus, the supply can be set equal to the loss, as in the following equation:

$$\text{Rate of supply } \frac{dS}{dt} = \frac{dM}{dt} \quad \text{rate of loss} \tag{1}$$

where $S =$ pesticide supplied and $M =$ the amount at time, t. By approximating the loss processes occurring in the environment, the rate of loss at any time is directly proportional to the amount of the pesticide

$$\text{The rate loss } \frac{dM_t}{dt} = -kM_t \tag{2}$$

where M_t is the amount of the pesticide at time t and k is the loss coefficient. After integration this gives the relationship:

$$\ln \frac{M_t}{M_0} = -kt \tag{3}$$

and M_0 is the amount of pesticide applied.

The time taken for the initial pesticide application M_0 to dissipate and fall to the minimum level required at the pest for control t_m is

$$t_m = \frac{1}{k} \ln \frac{M_0}{M_m} \tag{4}$$

which is used to plot the logarithmic lines in Figure 1 (lines "A").

If the pesticide is delivered from a formulation at a continuous rate to replace that which is lost in the environment

$$\frac{dS}{dt} = kM_t \tag{5}$$

$$dS = kM_t \, dt \tag{6}$$

The incremental change is thus given by

$$\frac{M_0 - M_m}{M_m} = kt_m \tag{7}$$

By using this relationship, the amount released from the formulation to replace that lost and to prevent the environmental amount to falling <1 mg can be calculated (8). This provides the curves "B" in Figure 1 for each of the two pesticides. It can be seen that the amount needed to give 150-days control has now fallen to 7.9 mg for P-1 (from 1000 mg) and to 3.1 mg for P-2 (from 8 mg). The areas between the two sets of curves logarithmically represents the pesticide that is lost and that only serves the purpose of a degradable reservoir. The amounts saved at shorter periods are less than for 150 days. By comparing the two sets of curves, the potential for saving is substantially greater for the pesticide with the short half-life. In fact, using an ideal controlled delivery system for this pesticide produces an efficiency over the 150 days equivalent of using the second pesticide with the longer half-life.

Controlled release formulations, combined with other aspects of pesticide application, thus offer the feasibility of improving pesticide delivery, reducing losses and benefitting the environment. The above model is based on many assumptions, including a constant and uniform environment. The agricultural situation is characterized by continual fluctuation and thus the theoretical objective can only be partially realized. However, it does demonstrate that compounds of short persistence (such as insect pheromones) may be used effectively in place of long persistent compounds when appropriately formulated. Loss kinetics for any individual pesticide vary depending on the environmental location considered; eg, loss by evaporation or photodegradation at the surface of plants may be much more rapid than published data giving half-life values for soils (7).

3. Types of Formulation: Physical, Chemical, Biological

3.1. Physical Systems, Matrix, and Reservoir.

The role of delivery of pesticides is becoming more recognized as it has been a much neglected part of pest management. In a recent book arising from the IUPAC meeting on pesticides, the position of delivery was placed second only to the discovery of new biologically active molecules. Indeed, with the widespread advent of proteinaceous pesticides, the ability to deliver these to the crop plant using genetically modified varieties has reached the level of 100% efficiency, but not with 100% bioavailability or delivery to the pest organism. However, this ability does not apply to all pest problems and situations and it is desirable to maintain a multiplicity of pest management methods (including conventional pesticides) and controlled release formulations as described here have an important role in good delivery. In terms of IPM, controlled delivery has a major contribution to the combination of pesticides with biocontrol methods when compared to conventional formulations.

For environmental application of pesticide delivery, controlled release formulations have been traditionally divided into chemical and physical types (9). More recently, a third approach has appeared, biological, partly in response to delivery requirements for genetically engineered pesticides. The types of controlled release formulations described to date can be categorized as follows:

Chemical
Backbone linking.
Side-chain bonding.
Matrix degradation.
Carrier molecules such as cyclodextrins.

Physical
Reservoir
with membrane (micro- and macroencapsulation, coated solids, laminates and large devices).
without membrane (hollow fibers, porous solids, and foams, gels, osmotic pumps).
Monolith or matrix (films, paint, sheets, slabs, pellets, strips, granules, microparticles, powders, and microspheres).

Biological
Living, or dead, cells (microorganisms) as delivery mechanisms.

All formulation types have been prepared and tested but not all have reached commercial practice. The more important types are described in the following, but this description is not comprehensive and the opportunities are only

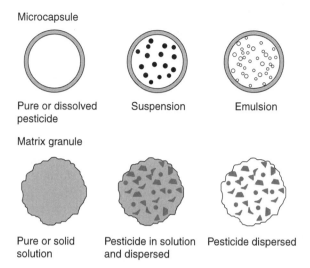

Fig. 2. Various configurations of capsule and matrix formulations.

limited by innovation and development of new approaches (10). The basic configurations of CRF are given in Figure 2.

3.2. Kinetics and Characteristics. There is a great deal of variation in the release kinetics of pesticides from the various formulation types described above. However, based on mathematical treatment the main types of release kinetics are represented in Figure 3. In order to discern the kinetics, the rate-controlling step needs to be identified (11). This should be done under controlled conditions, ie, in the laboratory, but it does not necessarily follow that this will be

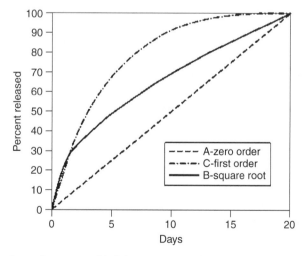

Fig. 3. Cumulative release provided by various release kinetics. (A) Constant release, independent of time (zero order) such as that possible from a membrane reservoir device free of lag time or initial burst effects, (B) matrix or monolithic sphere with square root time release, (C) first-order release.

true in the environment where the formulation is to be used. To study the basic kinetics, ie, those independent of the environment, the formulation has to be placed under "sink" conditions. These prevent the released pesticide from accumulating in the immediate vicinity of the surface of the formulation and slowing the release rate. The type of test environment will depend on the interfacial transport mechanism, especially movement into aqueous media or into the vapor phase or uptake by a biological system (12). Ideally, release kinetics ought to be determined under field conditions, but this often presents insurmountable problems (especially those resulting from the variability of the open environment) and instead the biological response to the released pesticide is observed over time to validate the performance of any formulation.

3.3. Mechanisms of Release. *Constant Release (Zero Order).* In the Figure 3, Line A represents release from a reservoir system with a large core relative to the wall mass. This could be a microcapsule releasing by steady-state diffusion through a uniform nonerodible wall. Transport through the polymer membrane (or matrix) occurs by a dissolution–diffusion process, where the active ingredient first dissolves in the polymer and then diffuses across the polymer to the external surface where the concentration is lower. The diffusion is in accordance with Fick's first law:

$$J = -Ddc_m/dx \tag{8}$$

where J is the flux of pesticide, D is the diffusivity, and dc_m/dx is the concentration gradient of the active ingredient. The rate remains constant as long as the internal and external concentrations of the pesticide and the concentration gradient are constant. A lag phase may occur while the system reaches this steady state (7).

Release by Erosion. The rate is independent of the concentration of pesticide remaining in the device. This zero order is also typical of certain surface erodible devices but their geometry is important and only laminar shapes produce a true constant rate as the device is eroded from one or both faces (2). Cylindrical, spherical, and irregular granular shapes provide a decreasing rate (as these particles lose surface area as erosion proceeds) and the overall release rate can be sustained only if hollow (concave) surfaces are available. These erodible systems have not been substantially exploited in pesticide delivery, mainly due to cost. Erosion can occur by dissolution of surface polymer or by degradation of the matrix, the best examples of which are the many polyesters such as *d,l*-polylactic acid or polyhydroxybutyrate. In this case, erosion may be through bulk hydrolysis of the polymer (13).

Release From Reservoir Systems. Most controlled release systems, including microcapsules, are positively rate dependent on temperature that makes effective delivery during cool night periods when interspersed with hot days (such as that required for pheromone release for control of nocturnally mating insect pests) problematic. For microcapsules, if the activity of the pesticide within the reservoir decreases then the release rate will also decrease; this effect will vary according to the capsule dimensions. This leads to a consideration of the polydispersity, or the range of sizes in a given number of microparticles. Although the release rate from individual particles may be constant, the duration of this release will varying according to the size of each particle. Thus small particles become depleted before large particles and the overall release from a population of particles will decrease with time. This has been shown in laboratory and field tests where overall first-order rates from microcapsules (Penncap-M; 20–40 μm) have been observed (14).

Release from Polymeric Matrices. In nonsurface erodible matrix systems, diffusion of the active ingredient occurs from the interior of the particle to the surface. This gives rise to a declining rate of release according to the square root of time ($t^{-1/2}$) as shown by curve B in Figure 3. In practice, the approximate Higuchi model (15) applies and is true up to 60–70% of release, typically from a sphere or microsphere. The pesticide may be dissolved or dispersed in the polymer; for dissolved pesticide the second phase of release is by first-order kinetics. For dispersed pesticide, the $t^{-1/2}$ kinetics last for almost all the release. These kinetics are a special case of the generalized description (16) of proportional release (at time t) from matrix or monolithic devices, as follows:

$$\frac{M_t}{M_\infty} = kt^n \qquad (9)$$

where k is a constant incorporating characteristics of the polymer and the pesticide, and n is the diffusional exponent and indicative of the transport mechanism. In these cases, $n = 0.5$ and indicates Fickian diffusion as the rate-controlling step in release.

3.4. Swellable Matrices. In matrix systems, where water uptake or swelling can occur, such as may be possible in moist soil or water, the rate-controlling step may be solid-state diffusion or relaxation of the polymer by incoming water or a combination (17). Thus, the time exponent of the equation characterizing the release rate (18) may vary from 0.5 (square root of time) for Fickian diffusion to 1.0 (zero order) for swelling according to the nature of the matrix and the pesticide. Generally, the higher the water solubility of the pesticide, the faster will be its rate of release. Less polar molecules with high partition coefficients tend to transport slower. In the case of irregular particles, such as granules, and where polydispersity exists, the overall time exponent will typically be less than the corresponding value for microspheres. For diffusion controlled systems, a typical low value is $n = 0.43$ (18).

First-Order Release. Finally, in situations where a chemical reaction liberates the active species, or where boundary conditions are rate limiting, the rate of release depends on the concentration in the solid phase (19), and first-order kinetics are seen as in Figure 3 as curve C. Where more than one mechanism (including diffusion) operates, then complex release patterns occur.

4. Design and Preparation of Controlled Release Formulations

4.1. Chemical Methods. Chemical methods involve the formation of a chemical bond with the pesticide and another molecule; this bond is then broken in the field to allow the release of the pesticide. The bond energy relates to the ease of breaking and thus the rate of release of the pesticide (20). Where the structure of the pesticide permits, it can be homopolymerized through a condensation reaction and the pesticide forms the backbone of a resulting high molecular weight polymer, which is in effect a polymeric propesticide. In the environment, this polymer depolymerizes to release the original pesticide, usually from each end of the chain (unzipping). Often breakdown of such homopolymers is slow and there is need for copolymerization with another appropriately functional monomer. Pesticides capable of homo- or copolymerization are few and include those containing functional groups such as amino, hydroxyl, and carboxyl groups.

A second approach to chemical-based formulations is where the pesticide is attached to a side chain of a high molecular weight polymer or macromolecule. This polymer may be either preformed, and the pesticide is then bound to appropriate side-chain functional groups, or the pesticide is first attached to a polymerizable monomer that is subsequently polymerized to yield the pesticide bound polymer (21). Again the release rate will depend on the energy of the bond holding the pesticide to the polymer that then undergoes scission to release the pesticide moiety (20).

The third approach to chemical-based release of pesticides is where the active is trapped in a network of a cross-linked polymer. Chemical breakdown of this polymer then allows the release of the pesticide. This mechanism incorporates physical processes of diffusion within the release mechanism.

The first two of the chemical release mechanisms usually involve covalent bonding of the pesticide and the formation of a molecular species different to the original structure. As a result of the registration requirements of new pesticide molecules, this approach implies considerable additional costs that outweigh the putative formulation benefits. Thus, true chemical approaches are often proscribed in favor of physical methods.

4.2. Physical Methods. Physical methods are divided into two general approaches. The pesticide is entrapped within a physical structure either at a molecular or microdomain level or the pesticide in the form of a reservoir is enclosed within a polymeric envelope (2). In the first, the pesticide is mixed with the polymer (or other material with high energy density) to form a monolithic structure or matrix. Release is normally by means of diffusion through the matrix or dissolution and erosion of the matrix. In the second approach, structures are based upon a reservoir of the pesticide enclosed by the polymer, from nanoscale up to centimeter-sized devices. The shapes of these devices are varied and include spherical such as microcapsules, and laminar or layered structures with the reservoir bounded by permeable membranes. These membranes provide a permeable barrier that controls the release rate. Other mechanisms of release include capsule rupture and erosion of the membrane.

As these "physical" methods provide the most important technologies for CRF of pesticides they will be presented in more detail in the this section.

4.3. Reservoir Based Formulations with Membrane. In this method, a reservoir or depot of the pesticide is bounded by a polymeric membrane, which protects (and separates) it from the environment and also provides a mechanism for its release. Thus, the specifications for the chemical nature and structure of this membrane are critical in the performance of such formulations. This makes for exacting requirements in the manufacturing processes if the desired release rates are to be consistently obtained in the field.

The method most suitable for use for pesticides is microencapsulation, where particle sizes are of the order of 10–100 µm and can be delivered by standard agricultural spraying. Microencapsulation has been defined (22) as the placing of a layer on the surface of a single *liquid* droplet. Conversely, coating refers to the covering of a single *solid* particle, whereas a matrix particle contains the solid or liquid active agent dispersed throughout a binding material. Even though these matrix particles are usually granular, they can be similar in size to microcapsules and are also intended for spraying as suspensions, they should be considered as matrices, and are covered below.

4.4. Microencapsulation. Microencapsulation has now been commercially practiced for >30 years, following the first application of the technology to carbon-less copying paper. Pesticide formulations based on microcapsules appeared in 1974 with the product Penncap-M containing the insecticide methyl parathion (4). Since then, many microcapsule suspension formulations have been introduced and form the major group of CRF.

Production of microcapsules is based on three main methods (23). The oldest, that of phase separation or coacervation, uses emulsification to produce core droplets containing the pesticide dispersed in a immiscible phase in which the wall material is dissolved, but then precipitates around the core droplets. Interfacial encapsulation is done by emulsifying or dispersing the pesticide solution in a continuous phase and a polymerization reaction takes place at the interface. Finally, in the physical methods the wall material is spread around the pesticide containing core to make the microcapsule.

Microcapsule Preparation by Interfacial Polymerization. The polymer forming the wall of the microcapsule can be made by addition or condensation poymerization or by *in situ* condensation polymerization. Addition polymerization using unsaturated monomers and free-radical generating catalysts may start with the monomer in the pesticide-containing dispersed oil phase and the water-soluble catalyst in the aqueous phase. Other combinations of monomer and catalyst distributed between the two phases are possible but less practical. Pesticide impurities can interfere with the polymerization producing unsatisfactory capsules (24).

The condensation route to wall polymers is the best method for pesticide encapsulation. In this process the two reactive monomers, one dissolved in each of the two phases (of the emulsified oil/pesticide in water), polymerize at the interface and generate the wall material. Typically the oil-phase monomers are polyfunctional isocyanates (A) or acid chlorides (B) and the water phase reactants are polyalcohols or amines. Compounds sufficiently reactive are chosen such that when they meet at the interface the condensation polymer forms the capsule wall (Figure 4). Alternatively, the two reactants (a diol and a diisocyanate) and a low boiling solvent make up the oil phase of the emulsion along with the pesticide. When heated, the solvent evaporates bringing the monomers together at the droplet surface to form the capsule wall (25).

The resulting polyamide wall tends to be weak and soft, but the polyurea and polyester produce tough strong materials (26). Other combinations of

$$
\begin{array}{ccccc}
-\text{NCO} & + & -\text{NH} & \longrightarrow & -\text{NH-CO-NH}- \\
\text{(A) polyisocyanate} & & \text{polyamine} & & \text{polyurea wall}
\end{array}
$$

$$
\begin{array}{ccccc}
-\text{COCl} & + & -\text{NH} & \longrightarrow & -\text{CONH}- \\
\text{(B) diacid chloride} & & \text{diamine} & & \text{polyamide wall}
\end{array}
$$

$$
\begin{array}{ccccc}
-\text{COCl} & + & -\text{OH} & \longrightarrow & -\text{COO}- \\
\text{(B) diacid chloride} & & \text{diamine} & & \text{polyester wall}
\end{array}
$$

Fig. 4. Various routes to capsule wall formation through condensation polymerization.

reactants give tough and strong polyurethane walls [polyamine and bis-(haloformate) or polyol and polyisocyanate] or epoxy walls (amine and epoxide).

For *in situ* condensation polymerization, only the oil-phase isocyanate reactant is used (27). When the emulsion is heated the isocyanate reacts with water at the interface to form an amine that then reacts in turn with remaining isocyanate. The resulting polyurea wall material formed is thin and strong providing good release properties for environmental applications. The permeability (the product of the diffusion coefficient and the solubility coefficient) of the wall can be varied by incorporating cross-linking monomers into the oil phase. A typical monomeric system is toluenediisocyanate (TDI) and polymethylene-polyphenylisocyanate (PAPI), a multifunctional monomer that causes cross-linking of the wall polymer (see Fig. 5). Both isocyanate monomers react with water at their own rates.

Forming the wall is only part of the successful microcapsule formulation. Recombination during microcapsule formation or subsequent storage to give large irregular shapes is a problem. Protective colloids offset this and reduce loss of the active ingredient into the continuous phase. Commercial colloids

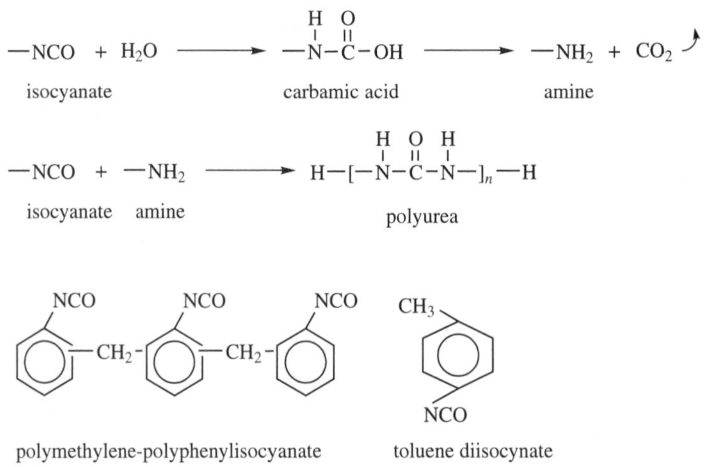

Fig. 5. Monomers and polymer forming reaction for *in situ* microcapsule wall formation.

used include poly(methyl vinyl ether/maleic anhydride) cross-linked with poly-(vinyl alcohol), styrene/maleic anhydride coplymers, vinylpyrrolidone/vinyl acetate copolymers, vinylpyrrolidone/styrene copolymer, and lignin sulfonate. A typical capsule suspension formulation made in this way may have up to 60% of the active ingredient, solvent (up to 20%), polymer (5–10%), protective colloids (1–20%), emulsifiers (1–5%), ultraviolet (uv)-protectant (0–5%), buffer (5%), viscosity/structure modifiers (2–10%), and water to make up the bulk (24).

Microcapsule Preparation by Phase-Separation Methods. The earlier methods used for pesticide microencapsulation were based on phase separation. There are two main approaches to coacervation, based on phase separation. Simple coacervation is when an aqueous solution of a hydrophilic polymer separates into two phases (solid and liquid) on addition of salt, alcohol, or other water-miscible solvents (27). An example would be the phase separation from an aqueous solution of poly(vinyl alcohol) by a nonsolvent such as propyl alcohol or by a salt solution, such as sodium sulfate. Complex coacervation occurs when polymers in solution with opposite electric charge come together and separate from the water. The most common example of complex coacervation is based on gelatin and gum arabic, two natural hydrophilic colloids. The process involves gelatin solution mixed with the core material in oil that is emulsified and the gum arabic added. The emulsion is heated and the phase separation is induced by dilution, reducing pH, or cooling. If the pH is reduced <4.5, the gelatin is then below its isoelectric point (pH 4.5) and its charge becomes positive; it then reacts with the gum arabic that has a residual negative charge and coats the oil droplets. Alternatively, phase separation can be caused by dilution. The emulsion would then be cooled and treated with formaldehyde to cross-link and strengthen the capsule wall.

Phase separation can also be produced from solutions of polymers in organic solvents. By addition of a nonsolvent for the polymer to the solution containing the core material the polymer will precipitate around the emulsified core to form microcapsules (28). This can allow for the encapsulation of aqueous solutions or suspensions of pesticides. For example, such an aqueous solution can be emulsified in oil containing the dissolved polymer. Addition of the nonsolvent to the oil phase separates out the polymer that can then form the wall around the water droplets.

Microcapsule Preparation by Physical Methods. These work by passing the two phases, core and wall material, through a small opening such that the wall material coats the core (29). This coating can be achieved using biliquid extrusion nozzles or with centrifugation in which the two liquids pass through many orifices. These allow the wall material to be cooled or dried after leaving the nozzle thus forming a rigid wall structure. These processes generally give high wall to core ratios and cannot be used to prepare very small microcapsules (<100 m).

4.5. Coating of Solid Particles. Covering a solid particle with a polymeric wall is usually referred to as coating although the product may be termed a microcapsule. Various methods may be used with degrees of uniformity of the wall structure (30). Pan coating is well established in which the core particles (>1–2 mm) are tumbled in a rotating drum while the coating solution is sprayed slowly; warm air circulates to remove the solvent. For smaller particles, a

Table 1. **Acute Mammalian Toxicities for Encapsulated and EC Formulations of Methyl Parathion and Diazinon**[a]

Formulation	Rat oral LD_{50} mg/kg	Rabbit dermal LD_{50} mg/kg
diazinon (EC)	350	600
Knox Out 2FM	>21,000	>10,000
methyl parathion (EC)	25	400
Penncap M	600	>5,450

[a] After Ref. 39.

fluidized bed is needed. The core particles (down to 100–150 μm) are fluidized in a rising air current and the coating solution slowly sprayed into the bed. Spray drying in which the core material and the coating solution is atomized and the droplets dried rapidly in hot air gives poorer quality of encapsulation. There are numerous other methods for encapsulation, many specialized for specific applications.

4.6. Examples of Microcapsule Formulations. Even though the first microcapsule formulation (using phase-separation technology) was introduced into commerce in 1960 for the purpose of releasing ink in carbonless copying, it was not until 1974 that the first pesticide microcapsule appeared (14). This was Penncap M, methyl parathion encapsulated within a polyamide/polyurea wall material, prepared from the reaction of sebacoyl chloride and polymethylene polyphenylisocyanate with ethylenediamine and diethylenetetramine, suspended in water (240 g/L). This product showed reduced toxicity and extended insect control, and was followed by a similar formulation based on diazinon for indoor control of cockroaches. Superior and extended pest control was achieved, compared to conventional formulations. Mammalian toxicity was reduced as can be seen for these microcapsule formulations in Table 1. These pioneering formulations were followed by similar types using other insecticides such as ethyl parathion, permethrin, cypermethrin, and chlorpyriphos.

Microcapsule formulations have been made, based on pesticides, for control on crops and soils, on timber and other surfaces for structural and indoor pests, on seeds and livestock. A few examples of microcapsule formulations from the 60 or more currently available worldwide are in Table 2. An electron micrograph of a pesticide microcapsule formulation is shown in Figure 6. Each particular

Table 2. **Selection of Some of the Microcapsule Formulations**

Trade name	Active ingredient	Wall material	Company
Penncap M	methyl parathion	polyamide/polyurea	Atochem
Knox Out 2FM	diazinon	polyamide/polyurea	Atochem
Micro-Sect	pyrethrin/synergist	polyurea	3M
Kareit MC	fenitrothion	polyurethane	Sumitomo
Sumithion MC	fenitrothion	polyurethane	Sumitomo
Lumbert	fenitrothion	polyurethane	Sumitomo
Icon	lambda-cyhalothrin	polyurea	Syngenta
Karate Zeon	lambda-cyhalothrin	polyurea (thin wall, low cross-linking)	Syngenta

Fig. 6. Electron micrograph of a pesticide microcapsule formulation.

formulation is designed for its specific application and method of use. Variables that can be exploited for this purpose include capsule size and size range, wall thickness, wall permeability and strength (provided by degree of cross-linking), nature of wall material, adjuvants and other formulation constituents. Release is usually through diffusion of the active agent through the capsule wall but other mechanisms can be used, such as rupture triggered by mechanical, erosion or degradation, thermal processes, or by osmotic swelling. An example of how encapsulation variables can influence release has been provided by Tsuji for a fenvalerate formulation (31). This polyurethane microcapsule was prepared by interfacial polymerization of polyisocyanate and ethylene glycol and to have various wall thicknesses and mass median diameters.

Efficacy against the important brassica pest, diamond-back moth (*Plutella xylostella*), was assessed and it was found that the LC_{50} value decreased for the larger capsules (diameter D) when wall thickness (T) was constant and for the thinner wall when the diameter was kept constant. For any batch of microcapsules, the ratio of diameter (expressed as the mass median diameter in μm) to the wall thickness (in μm) can be determined. This value D/T can be interpreted as relating to the strength of the capsules. As the ratio D/T increased, the 48 h LD_{50} value decreased. This means that the rupture of the capsules, by insects or other factors, is the important factor in biological efficacy. The availability of the insecticide depends on the strength of the capsule and thus the persistence of action of such microcapsule formulations will depend on having an optimum diameter/wall ratio with the wall thickness, neither too thin or too thick. Other similar microcapsules (MC) formulations for use in agriculture have been made based on insecticides fenitrothion and fenpropathrin, that also showed better safety to sensitive crops and to nontarget organisms such as fish. In extending this technology, formulations based on permethrin were developed for use in transplanted rice, and on fenitrothion and fenobcarb for aerial application in rice for bug and planthopper control.

The mechanism of release for these capsules, ie, by rupture or breakage, is suitable for controlling insect pests on surfaces. Formulations were developed for the control of cockroaches (based on fenitrothion or cyphenothrin), for termites (based on fenitrothion either applied to surfaces or incorporated into the glue of plywood), and for mosquitoes and flies on aggressive surfaces such as cement (based on fenitrothion and lambda-cyhalothrin). In the case of cockroach control, the pick up of microcapsules and the efficient delivery (often by grooming and ingestion of dust plus capsules) can give control of individuals resistant to diazinon or fenitrothion.

Formation of microcapsules by *in situ* interfacial polymerization (where the monomers are entirely in the oil phase of the capsule core) yields microcapsules with a high core/wall ratio and a bilayer wall with an outer layer (~0.05 μm) and an inner reinforcing spongy layer (0.5 μm). This method has been used to encapsulate a range of insecticides, pheromones, and herbicides, many of which have been available commercially (32). The capsule size may be varied from submicron to 100-μm diameter and the permeability selected for rapid or slow release of the pesticide. Release is by diffusion through the wall rather than rupture. For an effective formulation, the capsule suspension formed after polymerization needs protective stabilizers, dispersants, flow etc, to provide a high active ingredient content with good shelf-life and acceptable handling at dilution.

Applications for MC formulations include seed treatment (especially where an insecticide may be phytotoxic at dosages required), soil treatment of insecticides and herbicides, and treatment of surfaces for cockroach and mosquito control.

4.7. Laminate Formulations.   The laminate system comprises a reservoir layer of pesticide-containing polymer sealed between two other plastic layers (33). The two outer layers of this multilaminate structure protect and release the active ingredient by diffusion driven by the concentration gradient. Often one of the layers is impermeable and functions as a support for adhesion to suitable surfaces. At the surface, the pesticide is continually removed by evaporation, degradation, leaching, or by mechanical contact by humans, insects, moisture, wind, dust, or other agents.

The form and structure of the laminate varies according to the active agent and the intended application. The laminate may be used as a sheet for covering surfaces or may be cut into strips, ribbons, wafers, flakes, confetti, or even into granules or sprayable powders. Laminate strips (2.5 × 10 cm) consisting of a reservoir of an insecticide in poly(vinyl chloride) (PVC) on a base of impermeable Mylar sheet and covered with a 0.127-mm layer of PVC have been developed for indoor cockroach control (Hercon Insectape). The insecticides include chlorpyrifos, diazinon, and propoxur. The tape is intended to be affixed to cockroach frequenting surfaces and provide control for up to 3–5 months, especially valuable in areas where spraying is not desirable.

The laminate tape may also be used in its strip form as part of a collar for control of ticks and fleas on pet animals. Durations of control of these pests have been demonstrated to be up to 8 months for tick control.

An important application is for release of insect pheromones and attractants for insect control. A combination of the insecticide propoxur with the cockroach attractant periplanone-B provides 1 month of control in a laminate bait strip. Delivery of volatile compounds to the atmosphere surrounding crops is a crucial part of the mating disruption technique for many insect pests (34). The number and disposition of devices releasing the volatile pheromones depends on the pest, crop, and other environmental factors. Laminates can thus be dispersed in the crop as individual large devices (adhesive strips) or as small flakes or confetti applied by aircraft (applied with adhesive to ensure retention towards the top of the crop canopy) according to the control requirements. Among a number of crop pests, an important use has been for pink bollworm in cotton. The use of a plastic film for controlling release of pheromones can take the form of the laminate, or film enclosing a reservoir of the active agent on a porous substrate, or even in the form of polyethylene bags, vials, tubes, and caps.

4.8. Reservoir Based Formulations Without Membrane. Reservoir systems that lack a bounding membrane to protect and regulate the release are usually designed for liquid actives that are volatile. The liquid is held in place through capillary forces and is released in the vapor phase. The rate of evaporation is regulated by diffusion of the vapor through the static air phase above the liquid surface. The simplest example is the hollow fiber, which is a fine polymeric capillary closed at one end and filled (or partially filled) with the liquid active (35). This diffuses through the air column to the opening from where it disperses. This method has been mostly developed to deliver many of the volatile sex pheromones for insect pest control to maintain a minimum concentration in the air surrounding the crop to be protected.

Operating by a similar process are the porous and foam polymers. The active is held in the pores of the structure and released by diffusion through the pores to the surface where it moves away from the particle. The diffusion is driven by evaporation or by dissolution in environmental water that penetrates the porous particle. Highly absorbent polymers such as Culigel have been developed for delivery in water bodies for mosquito control (36).

4.9. Matrix Formulations. These formulations, also known as monolithic, consist of a uniform continuous phase with the pesticide dissolved or dispersed throughout. Their preparation is generally easier, requiring less process control, but can exhibit a rich variety of release types according to the material and structure of the matrix. An almost endless selection of materials are available for the matrix. Elastomers (rubbers) as well as thermoplastics and thermosetts can be used and many applications for pesticides have been developed using the technologies of the rubber and plastics industries (9). Generally, a range of additives such as plasticizers, light protectants, pigments, antioxidants, processing aids, etc, are usually included (37). The products can be produced in a number of forms or shapes, especially sheets, ropes, extruded cylinders, slabs, and granules. As release is inversely related to device size, the production of simple powders involves difficulty in uniformity as well as minimal reduction in release kinetics compared to conventional formulations. Many of the large formulation types are, or have been, popular for aquatic applications, such as for insect and mollusc vectors of human disease causing organisms, with few applications in agriculture. Examples have been larvicidal sheets containing temephos, malathion, or chlorpyrifos using polyamide, PVC, polyethylene, and polyurethane. Current survivors of this approach are thermoplastic formulations of chlorpyrifos (Dursban 10CR), of dichlorvos (No Pest PVC strips), and of tributyltin fluoride (Ecopro 1330) for controlling freshwater snails (*Biomphalaria glabrata*) the vector of *Schistosoma mansoni*, the causal agent of bilharzia (38). Monolithic systems such as these require appropriate plasticizers to promote migration of the pesticide to the surface, even so a substantial proportion will still remain entrapped when the rate of release declines below effective levels. This is less of a problem with elastomers where diffusion is faster (due to lower intermolecular forces) at similar temperatures.

4.10. Pesticide-Containing Films. In the agricultural field, the use of plastic mulch and plastic films for plant growing has become widespread, as it advances and enhances cropping when temperatures are low. It also encourages pests, particularly weeds and disease causing agents. The use of pesticides in these conditions can cause problems (eg, crop phytotoxicity), not least a result of the need for reapplication after the film has been laid. Incorporation of the pesticide into the film obviates both of these problems. Release of the pesticide can occur predominantly on the underside of the film and transported away (to the soil/crop) by condensation (39). Pesticides can be incorporated into agricultural films prior to their being formed by blown extrusion or into other forms such as sheets, tapes, cords and ropes, and chopped pieces such as confetti. Herbicides incorporated into ethylene–vinyl acetate copolymer EVA films are said to reduce by 2–4 times the amounts needed in covering early season vegetables such as cabbage, sweet corn, and celery. Using coating processes, films may be applied to seeds (40) that provides a very effective means of controlled delivery of pesticides to the seed region and to the emerging seedling.

4.11. Matrix Particles. Small particles based on a matrix can range in size from powders (microparticles) to granules (fine to macrogranules) to pellets (41). Microspheres can be considered as the matrix equivalent to microcapsules. Most controlled release granules are matrix based, although some have a solid core or reservoir of pesticide (with a coating).

4.12. Microparticles. Size matters; release rates depend on surface area, ie, a function of the square of the radius of a spherical particle and thus larger particles release for longer and are able to manipulate the external availability of the pesticide. Small microparticles are therefore limited in their scope for controlling release but can be used in traditional spraying of dispersions onto soils and crops as well as for seed dressing. Suspension concentrate formulations of matrix microparticles have been developed based on various rosins, phenolic resins, waxes, and bitumens. These have focused on lipophilic pesticides such as trifluralin and chlorpyrifos, and reductions in volatility have been demonstrated (41).

4.13. Granules. Whereas microparticles refer to sizes up to 100-μm, granules are typically 0.5–2.0 mm (fine granules 0.3–1.0 mm, microgranules 0.1–0.6 mm, macrogranules 2–6 mm). Granular controlled release formulations can be achieved by coating as well as from matrices. Although controlled release granules have not been as popular as microcapsules, there have been significant developments for applications to soil especially where extended control is required.

Cane grubs are a widespread pest problem in sugar cane growing and attack the roots of the plant over long periods. Conventional control has used persistent organochlorine insecticides applied at or soon after planting to optimize placement and residual protection. The introduction of long release granule formulations of short-lived insecticides such as chlorpyrifos (suSCon Blue) allowed the phase out of the organochlorines while giving good protection to the sugar

Table 3. **Acute Oral Toxicities of Controlled Release (suSCon) Granules Compared to Technical Grade Pesticides**[a]

Product	Acute oral rat LD_{50}
suSCon Blue (140 g/kg chlorpyrifos)	>1000 mg/kg
Technical chlorpyrifos	135–165 mg/kg
Marshal suSCon (100 g/kg carbosulfan)	>1000 mg/kg
Technical carbosulfan	185–250 mg/kg
suSCon Fu Ming (100 g/kg phorate)	319 mg/kg
Technical carbosulfan	1.6–3.7 mg/kg
G22001 (140 g/kg parathion)	578 mg/kg (male)
Technical parathion	3.6 (female) 13 (male) mg/kg

[a] After Ref. 43.

cane (42). The formulation is a 2-mm diameter extruded cylindrical granule of polyethylene containing 140 g of active ingredient per kilogram with an additive that sustains release by pore formation, through leaching by soil moisture. Single applications at planting can provide up to 3-years protection through the first harvesting and to several follow-on ratoon crops. This approach to pest control has since been extended to other problems, especially of concealed insects where systemic insecticides can be used effectively. Shoot borers in sugar cane and forestry, termites and weevils in forestry and ornamentals, borers and nematodes in a number of crops are examples of potential targets.

In addition to improved delivery and replacing persistent organochlorine pesticides with nonpersistent compounds controlled release granules also provide reduced toxicity of the product. For example, the acute oral toxicity of these granules are much reduced compared to unformulated or conventional sprayable formulations, as shown in Table 3.

Other granule formulations are based on biodegradable polymers that also offers the possibility of using wastes and byproducts from biological industries such as farming and forestry. New uses for cornstarch developed by the USDA included processes for the formulation of pesticides based on cross-linking of starch (43). The method involves the use of a corotating twin-screw extruder for mixing and gelatinizing starch with water (90–95°C), introducing the pesticide and providing the extrudate, which is then cut and dried to give the pelleted product. Hydrogen bonding occurs between the starch molecules (a process called retrogradation) to give a water-insoluble matrix entrapping the pesticide. Release occurs following soil placement by swelling. Evaluation of this granule formulation, especially of herbicides, has shown good efficacy at the same time reducing environmental losses and detrimental effects on surface and groundwater quality (44). In field experiments, significant reductions in herbicide volatility, surface runoff losses, and leaching have been observed compared to commercial formulations. These effects are particularly noticeable very shortly after application, when commercial formulations make the herbicide rapidly available at high concentrations.

Another natural polymer type abundantly available is the polyphenolic lignin. This aromatic macromolecule, which occurs in terrestrial plants, is obtained as a water-insoluble waste or byproduct from the pulping of wood. Its natural protective attributes, which it contributes to the success of plants, can be exploited to protect, and deliver, pesticides (45). It can be melt processed with compatible pesticides (similar polarity) by blending or screw extrusion to provide granules or powders. Release of pesticides from the granules declines with time and depends on diffusion kinetics according to a swelling-diffusion model (46). Formulations based on a wide range of soil-applied pesticides have been evaluated and field trialed extensively. For example, carbofuran-containing lignin-based granules in tropical flooded rice gave good control of virus disease (through controlling the insect vector of the virus) but using one-third of the amount of pesticide compared to conventional formulations (47). This approach afforded safe handling to applicators and reduced risks (to bare feet) during the transplanting of the rice seedlings. A wide range of pesticides have been controlled release formulated by this method.

Other matrix methods for the preparation of granules can use gelating polymers such as alginic acid and other polyelectrolytes. This approach effectively entraps pesticides through cross-linking with polyvalent ions such as Ca^{2+}; combined with adsorbents, useful release profiles can be obtained (48). This method of preparation, using mild conditions in aqueous media at ambient temperatures, is used for formulating microbial pesticides to protect and extend the active lives and release of propagules of such living pesticidal agents (49). A wide range of bacteria, fungi, viruses, protozoa, and nematodes have been formulated by this method, and related methods, to provide effective sustained formulations.

5. Biological Methods

Finally, the use of living cells (eg, yeast) as encapsulating materials has been under investigation for many years. The problems associated with the encapsulation of pesticides within preformed cells have been overcome by using proteinaceous pesticides such as the toxin from *Bacillus thuringiensis* (Bt). The genes for the production of the toxin have been introduced into the soil bacterium *Pseudomonas fluorescens*, the toxin is expressed and is seen as a crystalline inclusion. Following production by fermentation, the cells are killed and fixed to provide the capsule formulation that is registered for use on brassicas (50).

BIBLIOGRAPHY

"Controlled Release Technology, Agricultural," in *ECT* 4th ed., Vol. 7, pp. 251–274, by Harvey M. Goertz, The O. M. Scott and Sons Company; "Controlled Release Technology, Agricultural" in *ECT* (online), posting date: December 4, 2000 by Harvey M. Goertz, The O. M. Scott and Sons Company.

CITED PUBLICATIONS

1. B. A. Leonhardt, in R. M. Wilkins, ed., *Controlled Delivery of Crop-Protection Agents*, Taylor and Francis, London, 1990, pp. 169–190.

2. D. H. Lewis and D. R. Cowsar, in H. B. Scher, ed., *Controlled Release Pesticides*, American Chemical Society, Washington, D.C., 1980, pp. 1–16.
3. W. E. Ripper, R. M. Greenslade, J. Heath, and K. Barker, *Nature* **161**, 484 (1948).
4. DeSavigny and E. E. Ivy, in J. E. Vandegaer, ed., *Microencapsulation Processes and Applications*, Plenum Press, New York, 1974.
5. F. R. Hall, in R. M. Wilkins, ed., *Controlled Delivery of Crop-Protection Agents*, Taylor and Francis, London, 1990, pp. 3–22.
6. D. I. Gustafson, in R. M. Wilkins, ed., *Controlled Delivery of Crop-Protection Agents*, Taylor and Francis, London, 1990, pp. 23–42.
7. C. S. Hartley and B. J. Graham-Bryce, *Physical Principles of Pesticide Behaviour*, Vols. 1 and 2, Academic Press, New York, 1980.
8. G. G. Allan, C. S. Chopra, J. F. Friedhoff, R. I. Gara, M. W. Maggi, A. N. Neogi, S. C. Roberts, and R. M. Wilkins, *Chem. Tech.* **3**, 171 (1973).
9. A. F. Kydonieus, ed., *Controlled Release Technologies: Methods, Theory and Applications*, Vols. 1 and 2, CRC Press, Boca Raton, Flor., 1980.
10. H. B. Scher, ed., *Controlled-Release Delivery Systems for Pesticides*, Marcel Dekker, Inc., New York, 1999.
11. R. W. Baker and H. K. Lonsdale, in A. C. Tanquary and R. E. Lacy, eds., *Controlled Release of Biologically Active Agents*, Plenum Press, New York, 1974, pp. 15–71.
12. G. Pfister and M. Bahadir, in R. M. Wilkins, ed., *Controlled Delivery of Crop-Protection Agents*, Taylor and Francis, London, 1990, pp. 279–309.
13. A. Gopferich, *Macromolecules* **30**, 2598 (1997).
14. K. L. Smith, in H. B. Scher, ed., *Controlled-Release Delivery Systems for Pesticides*, Marcel Dekker, Inc., New York, 1999, pp. 137–149.
15. T. Higuchi, *J. Pharm. Sci.* **50**, 874 (1961).
16. P. L. Ritger and N. A. Peppas, *J. Controlled Release* **5**, 23 (1987).
17. L. R. Sherman, *J. Appl. Polym. Sci.* **27**, 997 (1983).
18. P. L. Ritger and N. A. Peppas, *J. Controlled Release* **5**, 37 (1987).
19. A. N. Neogi and G. G. Allan, in A. C. Tanquary and R. E. Lacy, eds., *Controlled Release of Biologically Active Agents*, Plenum Press, New York, 1974, pp. 195–223.
20. G. G. Allan, C. S. Chopra, A. N. Neogi, and R. M. Wilkins, *Nature (London)* **234**, 349 (1971).
21. A. Akelah, *Mater. Sci. Eng. C* **4**, 83 (1996).
22. R. E. Sparks and I. C. Jacobs, in H. B. Scher, ed., *Controlled-Release Delivery Systems for Pesticides*, Marcel Dekker, Inc., New York, 1999, pp. 3–29.
23. (a) J. E. Vandegaer, ed., *Microencapsulation Processes and Applications*, Plenum Press, New York, 1974. (b) A. Kondo, *Microcapsule Processing and Technology*, Marcel Dekker, Inc., New York, 1979.
24. M. Gimeno, *J. Environ. Sci. Health* **B31**, 407 (1996).
25. H. B. Scher, in H. B. Scher, ed., *Controlled Release Pesticides*, American Chemical Society, Washington, D.C., 1980, pp. 126–144.
26. P. Chamberlain and K. C. Symes, in D. R. Karsa and R. A. Stephenson, eds., *Encapsulation and Controlled Release*, Royal Society of Chemistry, Cambridge, U.K., 1993, pp. 131–140.
27. G. J. Marrs and H. B. Scher, in R. M. Wilkins, ed., *Controlled Delivery of Crop-Protection Agents*, Taylor and Francis, London, 1990, pp. 65–90.
28. U.S. Pat. 3, 173, 878 (1965), Z. Reyes.
29. J. T. Goodin and G. R. Somerville, in J. E. Vandegaer, ed., *Microencapsulation Processes and Applications*, Plenum, New York, 1974, pp. 155–163.

30. G. B. Beestman, in H. B. Scher, ed., *Controlled Release Pesticides*, American Chemical Society, Washington, D.C., 1980, pp. 31–54.
31. K. Tsuji, in H. B. Scher, ed., *Controlled-Release Delivery Systems for Pesticides*, Marcel Dekker, Inc., New York, 1999, pp. 55–85.
32. H. B. Scher, M. Rodson, and K.-S. Lee, *Pestic. Sci.* **54**, 394 (1998).
33. A. G. Kydonieus, in H. B. Scher, ed., *Controlled Release Pesticides*, American Chemical Society, Washington, D.C., 1980, pp. 152–167.
34. C. C. Doane, in H. B. Scher, ed., *Controlled-Release Delivery Systems for Pesticides*, Marcel Dekker, Inc., New York, 1999, pp. 295–317.
35. T. W. Brooks, in A. E. Kydonieus, ed., *Controlled Release Technologies: Methods and Applications*, Vol. 2, CRC Press, Boca Raton, Flor., 1980, pp. 165–193.
36. R. Levy, N. A. Nichols and T. W. Miller, *Pro. Intern. Symp. Control. Rel. Bioact. Mater.* **20**, 212 (1993).
37. N. F. Cardarelli, in A. E. Kydonieus, ed., *Controlled Release Technologies: Methods and Applications*, Vol. 1, CRC Press, Boca Raton, Flor., 1980, pp. 73–128.
38. A. R. Quisumbing and A. F. Kydonieus, in R. M. Wilkins, ed., *Controlled Delivery of Crop-Protection Agents*, Taylor and Francis, London, 1990, pp. 43–61.
39. N. F. Cardarelli and S. V. Kanakkanatt, in H. B. Scher, ed., *Controlled Release Pesticides*, American Chemical Society, Washington, D.C., 1980, pp. 60–73.
40. M. Bahadir, and G. Pfister, *Controlled Release, Biochemical Effects of Pesticides, Inhibition of Plant Pathogenic Fungi*, Springer-Verlag, Berlin, 1990, pp. 1–64.
41. D. J. Park, W. R. Jackson, I. R. Mckinnon and M. Marshall, in H. B. Scher, ed., *Controlled-Release Delivery Systems for Pesticides*, Marcel Dekker, Inc., New York, 1999, pp. 89–136.
42. N. Boehm and T. P. Anderson, in R. M. Wilkins, ed., *Controlled Delivery of Crop-Protection Agents*, Taylor and Francis, London, 1990, pp. 125–147.
43. M. E. Carr, R. E. Wing, and W. M. Doane, *Cereal Chem.* **68**, 262 (1991).
44. M. V. Hickman, G. D. Vail, and M. M. Schreiber, *Weed Tech.* **13**, (1999).
45. R. M. Wilkins, in H. B. Scher, ed., *Controlled-Release Delivery Systems for Pesticides*, Marcel Dekker, Inc., New York, 1999, pp. 195–222.
46. A. Ferraz, J. A. Souza, F. T. Silva, A. R. Cotrim, A. R. Gonçalves, R. E. Bruns, R. M. Wilkins, *J. Agric. Food Chem.* **45**(3), 1001 (1997).
47. R. M. Wilkins, E. A. Batterby, G. B. Aquino, and J. Valencia, *Econom. Entomol.* **77**, 495 (1984).
48. A. B. Pepperman and J-C. W. Kuan, *J. Controlled Release* **26**, 21 (1993).
49. W. J. Connick, Jr., in B. Cross and H. B. Scher, eds., *Pesticide Formulations: Innovations and Developments*, ACS Symposium Series 371, American Chemical Society, Washington, D.C., 1988, pp. 241–250.
50. F. H. Gaertner and L. Kim, *Trends Biotechnol.* **3**, 4 (1988).

GENERAL REFERENCES

A. F. Kydonieus, ed., *Controlled release technologies: methods, theory and applications*, CRS Press, Boca Raton, Flor., 1980.

H. B. Scher, ed., *Controlled-release delivery systems for pesticides*, Marcel Dekker, New York, 1999.

R. M. Wilkins, ed., *Controlled Delivery of Crop Protection Agents*, Taylor and Francis, London, 1990.

The Proceedings of the International Symposium on the Controlled Release of Bioactive Materials, published annually since 1973 (Controlled Release Society, Deerplain, Ill.)

and the Volumes on *Pesticide Formulations and Application Systems* (eg, Vol. 13, ASTM STP 1183, 1993; American Society for Testing and Materials, Philadelphia, Pa.) are useful sources, as are some issues of the following journals;

Journal of Controlled Release
Journal of Microencapsulation
Journal of Agricultural and Food Chemistry
Chemosphere
Pest Management Science (formerly Pesticide Science)

RICHARD M. WILKINS
Newcastle University

COORDINATION COMPOUNDS

1. Introduction

(Normal) Coordination Compounds: A coordination compound typically consists of a metal atom or ion (Lewis acid) surrounded by a number of electron-pair donors (Lewis bases) called ligands.

Inverse Coordination Compounds: More rare are the coordination compounds with inversed polarity, in which anions (Lewis bases) are surrounded by ligands with positively polarized atoms (Lewis acids).

Coordination compounds of the first type, also known as metal complexes, are pervasive throughout chemistry, biochemistry, and chemical technology. The metallic elements, which constitute 80% of the Periodic Table, exhibit predominantly coordination chemistry. Whereas the ligands may have charges equal in magnitude and opposite in sign to the charge of the metal ion, in which case a neutral coordination compound or inner complex results, often the metal plus ligands result in a charged entity or complex ion. In these situations, the coordination compound may include neutral ligands or counterions, either simple or complex. Generally, coordination compounds have properties, which are unique relative to both the ligands and the metal ion itself.

Coordination compounds are used as catalysts in nature, ie, metal enzymes, and in industry in situations where the metal ion or ligand would not work alone. The ligands often modify the properties of the metals or metal ions and vice versa. For example, the metal deactivators in gasoline modify the chemistry of dissolved copper to the extent that the copper does not promote gum formation. Complexation of iron in heme allows us to render iron ions soluble in human blood, and thus they are bioavailable for oxygen transport in living organisms. Some ligands react with different metal ions such that selective analysis of metallic elements is possible through coordination followed by solvent extraction, spectroscopy, gravimetry, electrochemistry, etc. Furthermore, complexes of copper and zinc in water allow brass to be electroplated. Conversely, metals are sometimes used to modify the properties of ligands. For example, azo dyes are metalated to give more permanence and/or alter color tones; the coordination of zinc to bactericides modifies the properties of the bactericides. The modification includes template and neighboring group effects, promotion of nucleophilic substitution, enhanced ligand acidity, and strain modification (1). Common geometries for coordination numbers from two through nine are shown in Figure 1, with International Union of Pure and Applied Chemistry (IUPAC) (2) and American Chemical Society (ACS) (3) polyhedral symbols for the geometries provided in Table 1.

Supramolecular compounds are a recent development of coordination chemistry. Their formation is based on the principle of molecular recognition, and they are characterized by the energy and information contained in the formed bonds as well as in their selectivity. Metal and ligand must be complementary in their stereochemical preferences, steric constraints, and coordination energies (thermodynamically). This systematic was called the key-and-lock principle by Emil

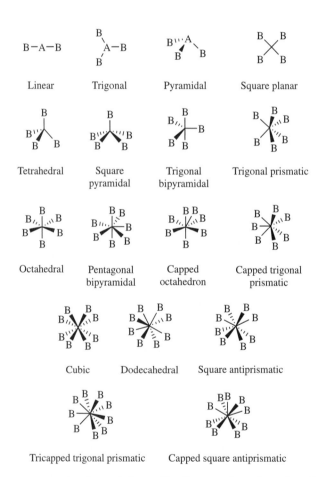

Fig. 1. Common geometries for metal coordination numbers from two through nine (see Table1), A represents the metal and B, a ligand donor atom. For higher coordination numbers, A is in the center of the structure. The principal axis orientation is vertical.

Table 1. **Polyhedral Symbols for Common Coordination Geometries**[a]

Coordination geometry	Coordination number	Polyhedral symbols IUPAC[b]	ACS[c]
linear	2	L-2	L-2
triangular plane	3	TP-3	TP-3
tetrahedron	4	T-4	T-4
square plane	4	SP-4	SP-4
trigonal bipyramid	5	TBPY-5	TB-5
square or tetragonal pyramid	5	SPY-5	SP-5
octahedron	6	OC-6	OC-6
trigonal prism	6	TPR-6	TP-6
pentagonal bipyramid	7	PBPY-7	PB-7
octahedron, face capped	7	OCF-7	OCF-7
trigonal prism, square face monocapped	7	TPRS-7	TPS-7
cubic	8	CU-8	CU-8
dodecahedral	8	DD-8	DD-8
square antiprism	8	SAPR-8	SA-8
trigonal prism, square face tricapped	9	TPRS-9	TPS-9

[a]See Figure 1.
[b]Ref. 2.
[c]Ref. 3.

Fischer in 1894. Supramolecular chemistry and the corresponding compounds will be treated in more detail in another article of this series, and only a few examples will be given in this article (4,5).

2. Definitions

2.1. Classification of Ligands.
The ligands can be classified as follows:

Unidentate ligands coordinate to the metal ion via one donor atom featuring at least one electron pair. Typical unidentate ligands are H_2O, NH_3 or phosphanes. They can act as terminal or as bridging ligands. In the latter case, they usually possess more than one lone electron pair, so that each pair can coordinate to a different metal ion.

Chelating [(greek) = pincer, claw] ligands possess more than one donor atom and can coordinate a metal ion by two or more donor atoms. Usually, the more stable five- and six-membered rings are formed on coordination. Coordination compounds with chelating ligands are usually more stable than unidentate ligands if the same donor atoms are concerned. This chelate effect is due to a gain in enthalpy ($\Delta H < 0$) as well as in entropy ($\Delta S > 0$), the latter being predominant. Typical chelating ligands are bidentate ethylene diamine, polydentate polyethyleneglycol dimethyl ethers, as well as the hexadentate ethylenediaminetetraacetate, known as EDTA^{4-} (**1**).

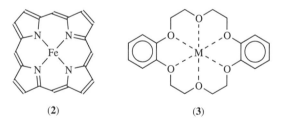

(**1**)

The latter plays a predominant role for the efficient complexation of metal ions with an octahedral coordination sphere, and is used to prevent precipitation of metal salts, ie., in water treatment (bathroom cleaners, eg.) for complexing calcium and magnesium ions, to prevent blood clots, to remove heavy metals from the body on poisoning, to solubilize iron in plant fertilizers, or to remove the iron taste from mayonnaise.

Macrocyclic ligands are chelating ligands, that are closed to a ring. Compounds in which the metal ion is coordinated by a macrocycle are more stable than compounds in which the metal ion is coordinated by the analogue open chelating ligand. The gain in energy is mainly entropic and due to the preorganization of the ligand. A typical example for a macrocyclic ligand is the porphyrin [101-60-6] (**2**), whose derivative forms the heme with iron in human blood, and cyclic polyethers such as dibenzo-[18]-crown-6 [14187-32-7] (**3**), which are used as extraction agents for metal ions and in biomimetics as compound mimicking ion transport through a cell membrane. The size of the macrocyclic ligand can be used as a selective tool in extraction and separation processes in industry.

(**2**)   (**3**)

Cryptates are macropolycyclic ligands, and their size plays an important role as to which metal ions they are capable of complexing. The coordination compounds formed are usually more stable than the corresponding macrocyclic ones. This phenomenon is called the cryptate effect. A typical example is the [2,2,2]-cryptate [23978-09-8] (**4**), the numbers in brackets indicating the number of oxygen atoms present in the chain connecting two bridgehead nitrogen atoms.

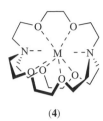

(4)

Inversing the polarity of the donor atoms of a ligand will lead to coordination compounds of anions. One of the most common examples is the protonation or quaternization of macrocyclic amines, leading to the formation of Lewis acid nitrogen atoms. The interaction of such a ligand with anions is, however, generally weaker than the one found in metal coordination compounds. Chelate, macrocyclic, and cryptate effects are observed as well in these inverse anion coordination compounds, but are less important than in metal complexes.

2.2. Nomenclature. Coordination compounds are named systematically beginning with the ligands. Ligands are listed alphabetically using prefixes that are Latin based (di, tri, tetra, etc, for 2, 3, 4, etc) for identical ligands, or Greek based (bis, tris, tetrakis, etc, for 2, 3, 4, etc) for identical multicomponent or complicated ligands that are placed in parentheses. The name of the metal follows, also using di, tri, etc, prefixes if polynuclear, or ending in -ate if the overall species is anionic. Then, in parentheses, comes either the overall charge of the species, if any, using an Arabic number followed by a plus or minus sign (Ewens-Bassett nomenclature), or the oxidation number of the metal represented by a Roman numeral or Arabic zero (Stock nomenclature). Examples of simple coordination species, including names, are given in Table 2.

In chemical formulas, the coordination compound or metal complex is written in square brackets with the charge of the complex given as a superscript outside at the end of the compound if written alone. Bridging ligands are designated as μ or μ^n, where n is the number of metal atoms bonded by the bridging ligand if n is >2. Such bridging ligands donate electron pairs to two or more metal ions simultaneously; eg, the $W_2Cl_9^{3-}$ ion of $K_3W_2Cl_9$ has three bridging chloro ligands, and the potassium salt can be named as potassium tri-μ-chloro-hexachloroditungstate(3−) [23403-17-0], $K_3W_2Cl_9$. Alternatively, the Ewens-Bassett (3−) can be replaced by (III) if the Stock nomenclature is used. The direct metal–metal bonded $[Re_2Cl_8]^{2-}$ ion is named octachlorodirhenate(2−)(Re−Re) [19584-24-8] [or (III) may be used instead of (2−)]. The 1990 IUPAC *Nomenclature of Inorganic Chemistry* volume recommends that chelating ligands having 2, 3, 4, 5, 6, etc, donors bonded to the same metal atom be termed didentate, tridentate, tetradentate, pentadentate, hexadentate, etc, analogous to the prefixes used to designate identical ligands bonded to a metal center (2,6,7). More complicated nomenclature referring to heteronuclear polyanions, ring and chain compounds, intercalation, and polymer compounds can be found in the 2000 IUPAC *Nomenclature of Inorganic Chemistry II* (7).

Isomerism. Two or more different compounds having the same formula are called isomers. Two principal types of isomerism are known among

Table 2. **Coordination Compounds Possessing Unidentate Ligands**

Name[a]	CAS Registry number	Molecular formula	Coordination number	Typical use
Common coordination numbers				
dicyanoargentate(I)	[15391-88-5]	$[Ag(CN)_2]^-$	2	recovery of silver
tetraamminecopper (II)	[16828-95-8]	$[Cu(NH_3)_4]^{2-}$	4	dissolution of copper (II) in basic soln
tetracarbonylnickel (0)	[13463-39-3]	$[Ni(CO)_4]$	4	separation of nickel from metals
hexafluorosilicate (IV)	[17084-08-1]	$[SiF_6]^{2-}$	6	dissolution of silicon (IV) in hydrolytic media
Less common coordination numbers				
tricyanocuprate(I)	[16593-63-8]	$[Cu(CN)_3]^{2-}$	3	brass plating bath component
pentacarbonyl-manganate(-I)	[14971-26-7]	$[Mn(CO)_5]^-$	5	composite polymer–metal particle synthesis
heptafluorozirconate (IV)	[27679-73-8]	$[ZrF_2]^{3-}$	7	fluoride extractive metallurgy
octacyanotungstate (IV)	[18177-17-8]	$[W(CN)_8]^{4-}$	8	light sensitzer for TiO_2
nonaaquaneodymium (III)	[54375-24-5]	$[Nd(H_2O)_9]^{3-}$	9	aqueous ion in Nd laser synthesis

[a]Stock nomenclature.

coordination compounds: stereoisomerism and structural isomerism. Stereoisomers have the same atoms, same sets of bonds, but differ in the relative orientation of these bonds. They can be divided into geometrical isomers (possible for square planar and octahedral complexes, but not tetrahedral ones), and optical isomers (possible for tetrahedral and octahedral compounds, but not square planar ones). Structural isomers can involve coordination, ionization, hydrate, or linkage isomerism.

Octahedral arrangements are most common for coordination compounds, so a short overview of the most important possible stereoisomers will be given (2). Octahedral coordination compounds with at least two unidentate ligands can have two isomers, one called cis if the two unidentate ligands are placed in vicinal positions, the other called trans for the two ligands in opposite sites of the octahedron. A tridentate ligand can coordinate to a metal ion by occupying three positions forming a triangular face of the octahedron, and is then called *facial* (fac). If the ligand coordinates in a linear way, it is called *meridional* (mer). Three bidentate ligands lead to a helicoidal coordination of the metal ion. The left- and right-turning helices are called by the Greek letters, lambda ($\lambda-$) and delta ($\Delta-$) forms, respectively. These prefixes are placed before the chemical names and formula of the compounds. See Figure 2.

Examples for structural isomerism concerning coordination, ionization, hydrate, or linkage are in the order: $[Co(NH_3)_6][Cr(C_2O_4)_3]$ and $[Co(C_2O_4)_3]$

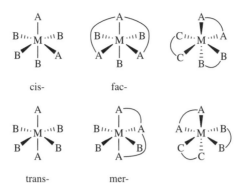

cis-   fac-

trans-   mer-

Fig. 2. Most frequent isomeries for octahedral complexes with mono-, tri-, and bidentate ligands.

$[Cr(NH_3)_6]$, $[Pt(NH_3)_3Br]NO_2$ and $[Pt(NH_3)_3NO_2]Br$, $[Cr(H_2O)_4Cl_2]$ Cl·2H$_2$O (bright-green), $[Cr(H_2O)_5Cl]Cl_2$·H$_2$O (gray-green) and $[Cr(H_2O)_6]Cl_3$ (violet), and $[Co(NH_3)_5(ONO)]$ and $[Co(NH_3)_5NO_2]$ with the NO_2 group once linked via the oxygen, once via the nitrogen atom.

Depending on the isomers, the compounds can have very different chemical and physical properties.

3. Examples of Coordination Compounds

3.1. Hydrides. Whereas dihydrogen was formerly thought to be bonded to metals only as the dissociated hydrido (H$^-$) species, a number of η^2-H$_2$ complexes are known (8). The η^2-H$_2$ designation implies the H–H bond is still intact and both hydrogen atoms are coordinated to the metal. Discovery of these dihydrogen species altered the thinking of chemists not only with regard to hydrogenation processes but even regarding the very nature of bonding between metals and ligands (8). Transition metal hydrides can be used as ligands for other metal ions, ie, in $\{[(PPh_3)_3IrH(\mu\text{-}H)_2]Ag\}(CF_3SO_3)$ (9).

3.2. Groups 1 (I A) and 2 (II A) Coordination Compounds. Ethers, polyethers, and especially the cyclic polyethers are ideal ligands for coordinating the alkali and alkaline earth metal ions. The macrocycle 18-crown-6 and its derivatives show bonding constants that increase in the order Li$^+$ < Na$^+$, Cs$^+$ < Rb$^+$ < K$^+$.Depending on the size of the cryptant, certain metal ions can be extracted via size exclusion. In biology, coordination of group, 1 (IA) (Na$^+$, K$^+$) and Group 2 (II A) metal ions (Mg^{2+}, Ca^{2+}) with cyclic polypeptides play an important role in the metal-ion transport through cell membranes. Such biological systems, ie, nonactine, can be modeled by macrocyclic polyethers or calixarenes (3) (10). See Figure 3.

Coordination compounds of Group 1 (IA) with main group element ligands can also afford polynuclear species, which are best studied for the smaller ions Li$^+$, Na$^+$, Mg^{2+}, and Ca^{2+}, due to their use in other fields of chemistry such as organic synthesis. They possess mainly polymeric ladder or cluster-type cage structures (11–13).

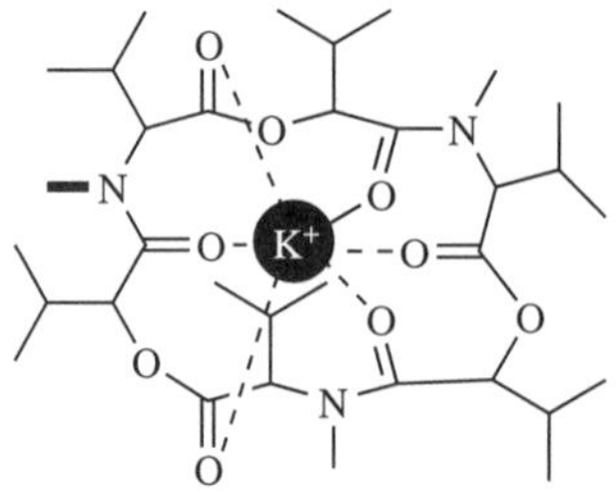

Fig. 3. Structure of the potassium complex with [D-hydroxyisovalericacid-*N*-methyl-L-valine]$_3$, the natural antibiotic enniatin B [917-13-5].

Coordination compounds of beryllium, a highly toxic element to living organisms, have been studied recently due to the increasing use of beryllium in materials like windows for X-rays or alloys with improved properties, and exhibit mainly a coordination number of 4 for the metal ion (14).

3.3. Main Group Metal Coordination Compounds. Main group elements show a neat tendency to form element–element bonds, and this is expressed in their coordination compounds. The more electronegative the elements, the better donor ligands they are in their anionic form, and the less tendency they show for being coordinated themselves.

Group 13 (III A). Organometallic compounds are as well known as coordination compounds of the electron poor Group 3 (III B) ions with classical Lewis base donor ligands, such as in the boron–nitrogen compounds (aminoboranes) or the $[Al_{77}\{N(SiMe_3)_2\}_{20}]^{2-}$ anion with a metalloid center. Transition-metal complexes with boron are well described in a recent review (15). RAl or RGa [R = Cp*, C(SiMe$_3$)$_3$ or other bulky fragments] can be used as ligands on transition-metal ions such as chromium, iron, nickel, and cobalt (16). The Tl$^+$ ion has a similar ionic radius to K$^+$. It may replace the latter in certain enzymes and is very toxic.

Group 14 (IV A). Carbon has a chemistry on its own, termed "organic chemistry" and is mainly based on covalent bonds. However, it may also act as a ligand. Metals bonded to hydrocarbons through carbon atoms are normally termed organometallics. Some examples will be given in the transition-metal examples. The metal fullerenes are newer (17). A large number of metal encapsulated fullerenes, ie, three-dimensional (3D) C$_x$ polytopal cage derivatives, have been observed, such as the number of M@C$_{60}$ (**5**) species. The @ indicates the metal is inside the C$_{60}$ polytope, known as a fullerene. Smaller M@C$_{28}$ species for M = Ti, Zr, Hf, and U, are thought to be stabilized by these tetravalent metals. External mono- and multimetallic fullerene complexes, eg, [(C$_6$H$_5$)$_3$P]$_2$Pt(η^2-C$_{60}$) (**6**) or [C$_{60}\{M(PEt_3)_2\}_6$], M = Ni, Pd, Pt (17) are also known where the η^2 indicates that two of the carbons in the C$_{60}$ are bonded to the platinum. Additionally, alkali metal doped fullerenes having several metal atoms per fullerene show interesting properties. For example, Rb$_3$C$_{60}$ becomes superconducting at 28 K. Another class of 3 D structures are the icosohedral metallocarbohedranes such as M$_8$C$_{12}$, where M = Ti, Zr, or Hf.

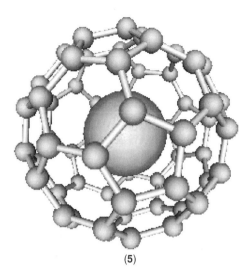

(5)

and

$(C_6H_5)_3P$
$(C_6H_5)_3P$ Pt

(6)

SnCl$_4$ [7646-78-8] is a good Friedel-Crafts catalyst, and [Pb$_2$(OOCCH$_3$)$_4$] is used in chemistry as a strong but selective oxidizing agent. Polymeric fluorosilanes Si$_{16}$F$_{34}$, tin clusters [Sn$_3$(OH)$_4$]$^{2+}$, and lead clusters [Pb$_6$O(OH)$_6$]$^{4+}$ show the general tendency to form element-element bonds.

Group 15 (V A). Nitrogen-containing compounds are usually good ligands in coordination compounds, whereas nitrogen itself or its derivatives are rarely found complexed by ligands. So are the phosphines. For the heavier Group 5 (VB) elements, complexes do exist, such as [Sb(C$_2$O$_4$)$_3$]$^{3-}$ with an incomplete pentagonal-bipyramidal structure with a lone pair at one axial position. The tartrate complexes of antimony(III) have been used medicinally as "tartar emeric" for >300 years. Only bismuth features a real cationic chemistry in this group, and hexameric species predominate in solution, ie, [Bi$_6$O$_6$(OH)$_3$]$^{3+}$. The coordination chemistry of Bi(III) is surveyed by Briand and co-workers (18).

Group 16 (VI A). Whereas compounds of oxygen with more electronegative elements are rare, some inverse coordination compounds are known in which an oxide (O^{2-}) or peroxide (O$_2^{2-}$) anion is coordinated octahedrally by six cations such as alkali (Li$^+$) or alkaline earth metal (Ba^{2+}) cations in cluster compounds. On the other hand are the oxygen atoms containing ligands well suited to act as terminal and/or bridging ligands. Dioxygen itself can be found as a ligand, and can coordinate to a metal ion in four ways: terminally (= end-on, like in hemoglobin), side-on, linearly, or side-on bridging.

Sulfur, selenium, and tellurium form homo- and heteropolycations as well as polyanions with themselves. The former can react with Lewis bases to form monomeric species, the latter are ideal Lewis-basic ligands for transition-metal ions of the second and third row, due to their soft character.

Group 17 (VII A).   Whereas fluoride, chloride, and bromide are strong ligands for many metal ions, iodide in its elemental form can act as a ligand for polar molecules as it is very soft and thus polarizable. Its charge-transfer complexes can be written as $I_2^- X^+$ (for X = element less electronegative than iodide, ie, S). It also has a tendency to form polyanions with itself, or the other halides. The heavier halogen atoms can be coordinated by oxygen to form hypohalite, halites, halates, and perhalates. Inverse complexes, in which anions, especially halides, are coordinated by ligands, are of interest in water treatment. Recent developments are the synthesis of amide groups containing macrocyclic and cryptate-like species that can be attached to the surface of a substrate in order to extract anions (19).

Group 18 (VIII A).   The noble gases are usually inert, however, some rare coordination compounds are known that involve fluoride and oxide anions. Thus, complexes like $[Xe_2F_3]^+$ can be isolated and are used as fluoride transfer agents. The compound Rb_2XeF_8 is stable and decomposes only above 400°C. Recently, more classical coordination compounds like $[C_5F_5N-XeF]^+$ [114481-53-7] (**7**) were found to be stable at −30°C and could be isolated as $[AsF_6]^-$ salts from BrF_5 solution. Even organometallic coordination compounds like $[C_6F_5Xe]$ exist at low temperatures. For the heavier noble gas elements, $[HRnO_3]^+$, $[HRnO_4]^-$, and $[RnO_3F]$ have been reported (20).

(**7**)

3.4. Transition-Metal Coordination Compounds.   *Group 3 (III B) + Lanthanum + Actinium.*   Group 3 (III B) elements can have very high coordination numbers, up to 12 for lanthanum in $[La_2(SO_4)_3 (H_2O)_9]$, 10 in $[Y(NO_3)_5]^{2-}$, and 9 for $[Sc(NO_3)_5]^{2-}$, the nitrate ligands binding as bidentate ligands except for one in the latter compound. This is due to their hard character, their ionic radius being rather small, and the ion being highly charged.

Group 4 (IV B).   Still highly charged in their higher oxidation state, these metal ions prefer a high coordination number, ie, eight in $Ti(NO_3)_4$. Clusters via bridging ligands are known as well, such as in $[Ti(OEt)_4]_4$. Octahedral coordination is preferred for the oxidation states + III and + II. Nitrogen-donor ligands such as bipyridine can stabilize unusual oxidation states of Group 4 (IV B) metals, such as 0, -I, and -II, as in $Li_2[M(bpy)_3]$ formally (bpy = 2,2′-bipyridine.

These metals are then so strongly reducing that they can attack the ligands. The Ti(IV) coordination compounds are used in the Ziegler-Natta catalysis for the industrial synthesis of polypropylene.

Group 5 (V B).   Having a wide range of oxidation states at their disposition, the coordination geometry varies from trigonal planar to square antiprismatic. These elements show a tendency to form polyoxometallates and clusters with bridging halide ligands. Metal organic compounds are known as well. The cluster chemistry of these and Group 6 (VI B) elements is vast, and was recently resumed by Prokopuk and Shriver (21).

Group 6 (VI B).   A wide redox chemistry allows for coordination compounds with coordination numbers from 3 to 12 and 13. Similar to Group 5 (V B) elements, organometallic compounds are found as well as polyoxometallates with as much as 248 molybdenum atoms. The synthesis of unnatural amino acids is a very special example for the use of (chiral) organochromium carbenes (22).

Heteropolymetallates are known for their colorful species and applications in the synthesis of tungsten bronzes. Low-valent compounds compensate their electron poor configuration by sharing electrons in a metal–metal bond. Metal clusters containing molybdenum are found in numerous enzymes. In particular, the Mo–S bonds play an important role in biological systems, and therefore, the chemistry of Mo–S containing compounds has been widely investigated (23).

Group 7 (VII B).   With a wide range of oxidation states, the possible coordination numbers in coordination compounds vary from 4 to 11. Mono- and polynuclear clusters are formed for halides as ligands, and mainly octahedral coordination compounds exist for these metal ions. Metal–metal bonds can be found in polynuclear clusters with a mainly octahedral geometry at each metal ion.

Group 8 (VIII).   Coordination compounds with coordination numbers of 3–10 are known for these metal ions. For M(II)/M(III), the redox potentials change dramatically with the ligands used for their coordination. This implies changes in the ligand field and color transitions. Thus, in the late eighteenth century, cobalt coordination compounds were discovered to possess a variety of different colors depending on the ligands and isomers. Rhodium and iridium complexes find their application as catalysts in organic and pharmaceutical synthesis. Metal complexes of the cobalt group were found to activate dioxygen, which makes them potential candidates for biological applications (24). For metal ions with a d^4, d^5, d^6, or d^7 electronic configuration, the possibility of spin crossover is given for coordination compounds, making them interesting materials for information storage and switches. The Ni(II) coordination compounds are very frequent, and are interesting due to their magnetic properties. Macrocyclic N-donor ligands are known to stabilize Ni(III) and Ni(I), allowing a wide redox chemistry. Such compounds are used in catalytic reactions such as olefin epoxidation and electrochemical reduction of alkyl halide and carbon dioxide. Partially oxidized cyano coordination compounds of Pt(II), the so-called Krogmann's and Wolfram's salts, were shown to possess a one-dimensional (1 D) structure with conducting properties along the polymer axis. A similar structure is proposed for the so-called platinum blue "Platinblau" (25). Pt(II) and Pd(II) usually possess a coordination number of 4 in the form of quadratic planar complexes.

The trans-effect, a phenomenon in which the ligand in the trans position to a studied ligand is destabilized, has been investigated on behalf of the platinum compounds. In biochemistry, iron-containing proteins, namely, hemoglobine, myoglobine, cytochrome, and iron–sulfur proteins, play the role of oxygen transporters and storage as well as electron- transfer systems. Cobalt as a biologically essential element is found coordinated in vitamin B12. The organometallic chemistry of these elements is also a very wide subject, mainly covering carbonyl- and cyclopentadienyl-containing coordination compounds. Alkene and alkyne complexes of many of these metal ions find applications as catalysts.

Group 9 (VIII). Coordination numbers for Group 9 (VIII) elements vary from 2 to 12. Coordination compounds of M(I, II, III) are known. Copper plays an important role in biochemistry, where it is found coordinated by hemocyanine, the blood of many mollusks. Blue proteins are known in plants in which redox chemistry of Cu(II)/Cu(I) is involved (26). Silver finds its application in black and white photography. Most interesting is the coordination chemistry of gold clusters with metal–metal bonds that can be used to make gold colloid particles whose color is dependent of the particle size.

Group 10 (VIII). Coordination numbers between 2 and 8 are found for these elements. Their coordination chemistry is characterized by the M(II) cations, however, Hg(I) species are known in halide complexes. Zinc is one of the most important metal ions in biological systems. For example, in carboxypeptidase A, it catalyzes the hydrolysis of terminal peptide bonds during digestion. Compounds of cadmium and mercury, however, are among the most toxic ones for living organisms.

3.5. Rare Earth Metal Coordination Compounds. *Lanthanides.* Lanthanide coordination chemistry has been rediscovered lately as their coordination compounds can exhibit interesting properties such as luminescence and paramagnetism. Lanthanide ions have been mostly studied in their Ln^{3+} form for their luminescence properties. Coordination numbers vary from 6 to 10, ligands usually exchanging rapidly in solution (27–29). The chemistry of Ln^{2+} was manily studied at the solid state for the properties of doped-solid state materials as phosphors, but their solution chemistry is evolving (30).

Actinides. The state-of-the-art coordination chemistry of these radioactive elements has been recently resumed by Kharisov and Mendez-Rojas (31). Extraction of actinide ions from waste waters has been studied with calixarenes and porphyrins (28,32). Plutonium is one of the best studied elements of the 5*f* element's row, especially concerning its biouptake via trihydroxamate desferrioxamine siderophore complexes of plutonium, which are recognized by microbial metal-siderophore binding sites (33).

4. Coordination Theories and Bonding

The quest for a comprehensible theory of coordination chemistry has given rise to the use of valence-bond, crystal-field, ligand-field, and molecular orbital theories. Ligand-field theory incorporates covalency with the electrostatic crystal field. The symmetry-induced separation of energy levels that is perceived from the chemical and physical properties of coordination compounds is inherent in all

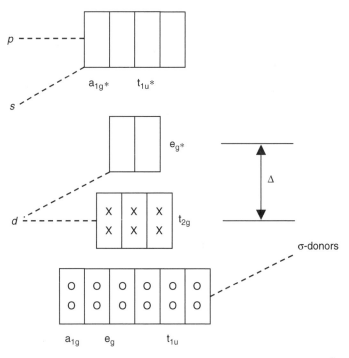

Fig. 4. Simplified molecular orbital diagram for a low spin octahedral d^6 complex, such as $[Co(NH_3)_6]^{3+}$, where $\Delta =$ energy difference; a, e, and t may be antisymmetric (subscript μ, ungerade) or centrosymmetric (subscript g, gerade) symmetry orbitals. (see text).

four theories, but symmetry effects are more apparent in the latter three. Excellent references on symmetry and group theory are available (34). Molecular orbital treatments of complexes range from the very simple semiempirical, such as Hückel, angular overlap, etc, through so-called *ab initio* methods (35,36).

A simplified energy-level diagram for octahedral complexes, the most common ones, is shown in Figure 4. The 12 electrons designated by circles are the electron-pair donors from the ligand-donor atoms; the electrons designated by x represent electrons from the d orbitals of a transition-metal ion. For octahedral metal ions having >3, but <8, d electrons, choices of spin states exist that are important in the kinetic, thermodynamic, spectroscopic, and magnetic properties of the octahedral species (see applications). In general, low spin octahedral complexes have a maximum number of electrons paired, and thus fewer antibonding electrons. Therefore, unless electron–electron repulsions predominate, the low spin state is more thermodynamically stable. Furthermore, low spin complexes are more kinetically inert.

Ligands having more than one electron pair can interact in a pi (π) sense as well as in the simple Lewis-base electron-pair sigma (σ) sense. Examples of π-donors include halide ions, oxygen donors, some nitrogen donors, such as $NH_2{}^-$, $N_3{}^-$, or β-ketoimines, and some sulfur donors, eg, S^{2-} and RS^-. Ligands such as the cyanide ion (cyano ligand) and carbon monoxide are considered to be π acceptors (π-acids) as a result of empty π^* levels that have appropriate symmetry for π-type overlap with filled or partially filled metal orbitals, eg, the t_{2g} level of octahedral complexes (see CARBONYLS). This type of π interaction, also known as back-donation, helps relieve the buildup of electron density on the central metal atom that results from the ligand σ donation. Therefore, π-acceptor ligands stabilize low oxidation states.

Other types of bonding include donation by ligand π orbitals, as in the classical Zeiss's salt ion $[Pt(\eta^2\text{-}CH_2{=}CH_2)Cl_3]$ [12275-00-2] and sandwich compounds such as ferrocene. Another type is the delta (δ) bond, as in the $[Re_2Cl_8]^{2-}$ [19584-24-8] ion (**8**), which consists of two $ReCl_4$ squares with the Re–Re bonding and eclipsed chlorides. The Re–Re d bond makes the system quadruply bonded and holds the chlorides in sterically crowded conditions.

(**8**)

Numerous other coordination compounds contain two or more metal atoms having metal–metal bonds (37) and correspond therefore to cluster compounds in their strictest sense.

5. Properties

5.1. Stability.   The thermodynamic stability of coordination compounds in solution has been extensively studied (38). The equilibrium constants may be reported as stability or formation constants.

$$M + n\,L \rightleftharpoons ML_n$$

This compound, ML_n, has a cumulative stability constant β_n related to the activities a of the species by

$$\beta_n = a_{ML_n}/\left(a_M a_L^n\right)$$

and the stepwise constant K_n

$$K_n = a_{ML_n}/(a_{ML_{n-1}}a_L)$$

Alternatively, instability or dissociation constants are sometimes used to describe compounds, and caution is necessary when comparing values from different sources.

Attempts have been made to categorize the interactions between metal ions and ligands. Whereas all metal ions interact more strongly with fluoride than with chloride in the gas phase, in aqueous solution a number of exceptions occur. Metal ions that have the normal (class **a**) aqueous solution stability order of $F \gg Cl > (Br > I$ also have $N \gg P > As > Sb$ and $O \gg S > Se > Te$ donor stability order (39). The inverse (class **b**) aqueous solution stability

Table 3. Classification of Donors, Acceptors, and Solvents Based on Polarizability

Class a, Nonpolarizable or Hard	Borderline	Class b, Polarizable or Soft[a]
	Metallic ions	
$H^+, Li^+, Na^+, K^+, Rb^+,$	Cs^+	
$Be^{2+}, Mg^{2+}, Ca^{2+}, Sr^{2+}, Ba^{2+}$		
$Sc^{3+}, Y^{3+}, La^{3+}, (lanthanides)^{3+}$		
$Ti^{4+}, Zr^{4+}, Hf^{4+}$		
$VO^{2+}, Cr^{3+}, UO_2{}^{2+}, Pu^{4+},$		
$Mn^{2+}, Fe^{3+}, Co^{3+}$	Fe^{2+}, Co^2	$Co(CN)_5^{2-}, Co(DMG))_2CH_3^+$
$Rh(NH_3)_5^{3+}, Ir(NH_3)_5^{3+}$		$Rh(CN)_5^{2-}, Ir(CN)_5^{2-}, Rh^+, Ir^+$
	Ni^{2+}, Cu^2	$Pd^{2+}, Pt^{2+}, Cu^+, Ag^+, Au^{+,3+}$
$Al^{3+}, Ga^{3+}, In^{3+}$	Zn^{2+}	$Cd^{2+}, Hg_2^{2+}, Hg^{2+}, Tl^{+,3+}$
$Si^{4+}, Sn^{2+}, 4^+, As^{3+}$	Pb^{2+}	
metals in oxidation state $4+$ or higher[b]		metal in oxidation state 0 or lower
	Ligands	
F^-, O donors		H^-, CO, CO^-, R^-
		alkide ions, olefins
		aromatics
	N donors	P, As, Sb, Bi donors
	Cl^-, Br^-	S, Se, Te donors, I^-
	Solvents	
HF, H_2O, ROH	NH_3	$(CH_3)_2CO, (CH_3)_2SO, (CH_2)_4SO_2,$
		nitroparaffins, DMF

[a]DMG = dimethylglyoxime; DMF = dimethylformamide.
[b]Exceptions exist if a large number of soft bases are coordinated to the metal.

donor orders are $F \ll Cl < Br < I$, $N \ll P > As > Sb$, and $O \ll S \cong Se \cong Te$. This inverse order has long been considered related to the polarizability of the ligand.

Bases of low polarizability such as fluoride and the oxygen donors are termed hard bases. The corresponding class **a** cations are called hard acids; the class **b** acids and the polarizable bases are termed soft acids and soft bases, respectively. The general rule that hard prefers hard and soft prefers soft prevails. A classification is given in Table 3. Whereas the divisions are arbitrary, the trends are important. Attempts to provide quantitative gradations of "hardness and softness" have appeared (40). Another generality is the usual increase in stability constants for divalent $3d$ ions that occurs across the row of the Periodic Table through copper and then decreases for zinc (41).

Chelating or multidentate ligands are usually more stable than the analogous unidentate ligands, and this gives rise to the chelate effect. This effect is particularly evident for the formation of five-membered rings, anionic ligands, and aromatic donors. Larger rings, unless rigid, lose appreciable entropy on coordination; anionic ligands displace large quantities of oriented solvent molecules; and aromatic ring donors are normally rigid. Detailed quantitative considerations between similar systems is often difficult, however, because the ΔH and ΔS changes are normally a very small fraction of the enthalpies involved in the coordination bonds.

5.2. Steric Selectivity. In addition to the normal regularities that can be rationalized by electronic considerations, steric factors are important in coordination chemistry. To illustrate, at $100°C$ 8-hydroxyquinoline, or 8-quinolinol

(Hq) [148-24-3], precipitates both Mg^{2+} and Al^{3+} from aqueous solution as hydrated $Mg(q)_2$ (formulated as $Mg(q)_2(H_2O)_2$ [56531-18-1]) and as $Al(q)_3$ [2085-33-8], respectively. 2-Methyl-8-hydroxyquinoline [826-81-3] (**9**),

(**9**)

however, precipitates only Mg^{2+} from aqueous solution. The 2-methyl group prevents three of the methyl-substituted ligands from coordinating to Al^{3+} in water. By varying ring sizes and donor atoms of the macrocyclic crown ethers and the fused ring cryptates, different metal ions can be selectively accommodated. The crown ethers, such as 18-crown-6 or eicosahydrodibenzo[*b,k*][1,4,7,10,13,16]hexaoxacyclooctadecin [14187-32-7] (2) are highly selective for K^+ relative to Na^+. However, the cryptate 4,7,13,16,21,24-hexaoxa-1,10-diazabicyclo[8.8.8]hexacoxane [23978-09-8] (4) is selective for the sodium ion (see CATALYSIS, PHASE-TRANSFER).

Another example of steric selectivity involves the homopoly and heteropoly ions of molybdenum, tungsten, etc. Each molybdenum(VI) and tungsten(VI) ion is octahedrally coordinated to six oxygen (oxo) ligands. Chromium(VI) is too small and forms only the well-known chromate-type species having four oxo ligands. The ability of other cations to participate in stable heteropoly ion formation is also size related.

Coordination stereochemistry (including various forms of isomerization) is an area of significant research interest. This aspect of coordination is important for stereospecific catalytical applications.

5.3. Reactivity.   Coordination species are often categorized in terms of the rate at which they undergo substitution reactions. Complexes that react with other ligands to give equilibrium conditions almost as fast as the reagents can be mixed by conventional techniques and are termed labile. Included are most of the complexes of the alkali metals, the alkaline earths, the aluminum family, the lanthanides, the actinides, and some of the transition-metal complexes. On the other hand, numerous transition-metal complexes that kinetically resist substitution reactions are termed inert. These terms refer to substitution reactivity and not to thermodynamic properties. As an illustration, the reaction rates, k, and activation parameters, ΔH^*, (S^*, and ΔV^*, of solvent-exchange for some of the aquated cations of the first-row transition elements are listed in Table 4. Although vanadium(II) and nickel(II) react more slowly than the other ions of oxidation state II listed, they are labile by the operational definition. On the other hand, chromium(III) and cobalt(III) complexes are inert. Iron(III), however, is labile even though iron lies between chromium and cobalt in the Periodic Table.

In general, octahedral complexes of transition-metal ions possessing 0, 1, or 2 electrons beyond the electronic configuration of the preceding noble gas, ie, d^0, d^1, d^2 configurations, are labile. The d^3 systems are usually inert; the relative lability of vanadium(II) may be charge and/or redox related. However, high spin

Table 4. **Water Exchange Rates and Activation Parameters of Hexaaqua Complexes at 25°C**^{a,b}

Ion	k_1, s^{-1}	ΔH^*, kJ/mol^c	ΔS^*, J/(K · mol)c	ΔV^*, cm^3/mol	Electron configuration^d
V^{2+}	8.9×10^1	62	-0.4	-4.1	d^3
Cr^{2+}	7×10^9				d^4
Mn^{2+}	2.1×10^7	33	6	-5.4	d^5
Fe^{2+}	4.4×10^6	41	21	3.8	d^6
Co^{2+}	3.2×10^6	47	37	6.1	d^7
Ni^{2+}	3.2×10^4	57	32	7.2	d^8
Cu^{2+}	8×10^9	20			d^9
Ti^{3+}	1.8×10^5	43	1	-12.1	d^1
V^{3+}	5.0×10^2	49	-28	-8.9	d^2
Cr^{3+}	2.4×10^{-6}	109	12	-9.6	d^3
Fe^{3+}	1.6×10^2	64	12	-5.4	d^5
$CrOH^{2+}$	1.8×10^{-4}	110	55	2.7	d^3
$FeOH^{2+}$	1.2×10^5	42	5	7.0	d^5

^{a}Ref. 1, unless noted otherwise.
^{b}For the reaction $M(H_2O)_6^{n+} + H_2^{18}O \longrightarrow M(H_2O)_5(H_2^{18}O)^{n+} + H_2O$.
^{c}To convert from J to cal, divide by 4.184.
^{d}The de^4 (Cr^{2+}) and d^9(Cu^{2+}) complexes have distorted octahedra.

d^4, d^5, and d^6 species, which possess 4, 5, and 4 unpaired electrons, respectively, are labile, as are d^7 through d^{10} octahedral complexes. In addition to the inert d^3 systems, low spin d^4, d^5, and d^6 complexes are inert to rapid substitution. The d^8 species are the least labile of the configurations classed as labile.

Spin-paired octahedral d^6 ions and spin-free octahedral d^8 ions appear to react by largely dissociative (D or I_d) reactions, eg,

$$ML_n \longrightarrow ML_{n-1} + L \quad \text{followed by} \quad ML_{n-1} + L' \longrightarrow ML_{n-1}L'$$

although d^6 complexes THAT are just barely spin paired, such as tri-1,10-phenanthroline-iron(II) [14708-99-7] and ethylenediaminetetraacetatocobalt(III) [15136-66-0], appear to undergo associative (A or I_a) reactions, with good nucleophiles.

$$ML_n + L' \longrightarrow ML_nL' \longrightarrow ML_{n-1}L' + L$$

Discernible associative character is operative for divalent 3d ions through manganese and the trivalent ions through iron, as is evident from the volumes of activation in Table 4. However, deprotonation of a water molecule enhances the reaction rates by utilizing a conjugate base π- donation dissociative pathway. As can be seen from Table 4, there is a change in sign of the volume of activation ΔV^*. Four-coordinate square-planar molecules also show associative behavior in their reactions.

For many species, the effective atomic number (EAN) or 18- electron rule is helpful. Low spin transition-metal complexes having the EAN of the next noble gas (Table 5), which have 18 valence electrons, are usually inert, and normally

Table 5. Complexes Having Effective Atomic Numbers (EAN) of a Noble Gas^a

Name	CAS Registry number	Molecular formula	Coordination number	EAN
nickel tetracarbonyl	[13463-39-3]	$Ni(CO)_4$	4	36
iron pentacarbonyl	[13463-40-6]	$Fe(CO)_5$	5	36
chromium hexacarbonyl	[13007-92-6]	$Cr(CO)_6$	6	36
hexaammineplatinum(IV)	[16893-12-2]	$Pt(NH_3)^{4+}{}_6$	6	86
tricarbonyldichlorobis-(triphenylphosphine)-molybdenum(II)	[17250-39-4]	$Mo(CO)_3Cl_2$ $[P(C_6H_5)_3]_2$	7	54
tetrakis(8-quinolino-lato)·tungsten(IV)	[17499-74-0]	$W(C_9H_6NO)_4$	8	86
nonahydridorhenate(VII)	[44863-47-0]	$[ReH_9]^{2-}$	9	86

^{a}18-Electron species.

react by dissociation. Each normal donor is considered to contribute two electrons; the remainder are metal valence electrons. Sixteen-electron complexes are often inert, if these are low spin and square planar, but can undergo associative substitution and oxidative–addition reactions.

Oxidation–Reduction.   Redox or oxidation(reduction reactions are often governed by the hard–soft base rule. For example, a metal in a low oxidation state (relatively soft) can be oxidized more easily if surrounded by hard ligands or a hard solvent. Metals tend toward hard-acid behavior on oxidation. Redox rates are often limited by substitution rates of the reactant so that direct electron transfer can occur (42). If substitution is very slow, an outer sphere or tunneling reaction may occur. One-electron transfers are normally favored over multielectron processes, especially when three or more species must aggregate prior to reaction. However, oxidative addition

$$[Ir^I(P(C_6H_5)_3)_2(CO)Cl] + CH_3I \longrightarrow [Ir^{III}(P(C_6H_5)_3)_2(CO)(CH_3)ClI]$$

oxygen atom transfer

$$[M^{n+}L_x] + ClO^- \longrightarrow [M^{n+2}OL_x] + Cl^-$$

where M is a metal ion of charge n^+ and L is a ligand, are considered complementary two-electron transfers that are second only to one-electron processes in general. The distinction between atom and electron transfer is academic when a bridging atom moves from one species to another during a redox reaction:

$$[Co(NH_3)_5Cl]^{2+} + [Cr(H_2O)_6]^{2+} \xrightarrow{H_2O} [Cr(H_2O)_5Cl]^{2+} + [Co(H_2O)_6]^{2+} + 5\,NH_3$$

In effect, the atom is transferred by internal sphere.

Coordination species that coordinate an organometallic ligand in catalytic processes often cycle between 16- and 18-electron systems, although ligand substitution also participates. For example, the first step in the well-studied olefin hydrogenation by chlorotris–(triphenylphosphine)rhodium(I) [14694-95-2] is now thought to be solvation followed by oxidative addition of hydrogen because the phosphine is somewhat labile and the solvated species adds hydrogen $\sim 10^7$ times as fast as the original complex and goes from a 16-electron rhodium(I) species to an 18- electron rhodium(III) species (43):

$$[RhCl\{P(C_6H_5)_3\}_3] + \text{solvent} \longrightarrow [RhCl\{P(C_6H_5)_3\}_2(\text{solv})] + P(C_6H_5)_3$$

$$[RhCl\{P(C_6H_5)_3\}_2(\text{solv})] + H_2 \longrightarrow [Rh(H)_2Cl\{P(C_6H_5)_3\}_2(\text{solv})]$$

where solv is solvent. An olefin, R, such as cyclohexene, substitutes for the solvent on the dihydride.

$$[Rh(H)_2Cl\{P(C_6H_5)_3\}_2(\text{solv})] + \text{olefin} \longrightarrow [Rh(H)_2Cl\{P(C_6H_5)_3\}_2(\eta^2 - R)] + \text{solvent}$$

This equilibrium is followed by a rate-determining tautomeric shift ($18 \rightarrow 16$ electrons) to give a coordinated alkyl group:

$$[Rh(H)_2Cl\{P(C_6H_5)_3\}_2(\eta^2 - R)] \longrightarrow [Rh(RH)(H)Cl\{P(C_6H_5)_3\}_2]$$

where RH = alkyl. These last two steps are often called an olefin insertion. Addition of solvent gives the 18-electron complex, $[Rh(R')(H)Cl(P(C_6H_5)_3)_2(\text{solv})]$, and reductive elimination produces the original 16-electron solvated catalyst:

$$[Rh(R')(H)Cl(P(C_6H_5)_3)_2(\text{solv})] \longrightarrow [RhCl(P(C_6H_5)_3)_2(\text{solv})] + R'H$$

This catalytic cycle is related to some stereoselective industrial catalysis.

Photochemistry. Substitution rates of many complexes are enhanced by irradiation of the low energy $d-d$ transitions, such as $t_{2g} \rightarrow e_g$ in octahedral coordination compounds (see 2). Quantum yields, Φ, defined as the ratios of moles of product formed (or reactant depleted) to the moles of photons absorbed, vary from very good, eg, chromium(III) ~ 0.5, to poor, eg, cobalt(III) (0.01, for ligand substitution (44). The substituted ligand is normally the stronger of the two on the axis with the weakest net pair of ligands as determined by spectrochemical relationships, ie, $CN^- > NO_2 > NH_3 > H_2O$, $F^- > Cl^-$. Exceptions do occur. Photochemical ligand dissociation is useful in the synthesis of multinuclear metal complexes such as diiron nonacarbonyl [15321-51-4] from iron pentacarbonyl [13463-40-6]

$$2[Fe(CO)_5] \xrightarrow{h\nu} [Fe_2(CO)_9] + CO$$

Or conversely, active radicals can be obtained by irradiating certain metal–metal bonded species:

$$[Mn_2(CO)_{10}] \xrightarrow{h\nu} 2[Mn(CO)_5]$$

Irradiation of coordination compounds in the charge-transfer spectral region can often enhance redox reactions. The quantum yields are variable.

The use of photochemical redox for practical energy transfer is being actively pursued (see HYDROGEN;HYDROGEN ENERGY). For example, sufficient energy is available in visible photons to split water into hydrogen and oxygen (45). Whereas derivatives of $[Ru(bpy)_3]^{2+}$ [15158-62-0], have been at the forefront of this area, a small amount of decomposition during the photochemical cycles had precluded commercialization of such systems. However, the encapsulated ruthenium(II) [116970-07-1] found in structure (**10**), is claimed to be 10^4 times more stable than $[Ru(bpy)_3]^{2+}$ (46).

(**10**)

Magnetism. Many coordination compounds are paramagnetic, ie, they have unpaired electrons. Combining several metal ions in one coordination compound can lead to a coupling of the individual magnetic moments. Their interaction depends on the distance between the magnetic ions and on the orientation of the magnetic moments. The magnetic moments can become aligned parallel to each other via exchange coupling below the Curie temperature, in which case one obtains ferromagnetism, or antiparallel to each other, leading to antiferromagnetism. Nonparallel alignment leads to ferrimagnetism. Magnetism of a compound can be measured with a Faraday or Gouy balance, with a superconducting quantum interference device (SQUID)-magnetometer or with nuclear magnetic resonance (NMR) methods for paramagnetic compounds, and the values obtained have to be corrected for the diamagnetic contribution.

For metal ions with 4, 5, 6, or 7 d-electrons in octahedral symmetry, two configurations are possible: low spin with the maximum of electrons paired, or high spin with a maximum of unpaired electrons. Considering, eg, a d^6 metal ion such as Co^{3+}, it can have, depending on the ligand field, no unpaired electrons, as in $[Co(H_2O)_6]^{3+}$ [15275-05-5] or four unpaired electrons, as in $[CoF_6]^{3-}$ [15318-87-3] per ion, corresponding to a low spin, respectively a high spin octahedral complex. If the two spin states are close in energy, they can be populated differently depending on the temperature. Having different magnetic properties, the high spin–low spin distribution can easily be monitored via the effective magnetic moment or color changes.

Spin frustration can occur upon antiferromagnetic coupling in confined cluster compounds of three or more metal ions, when the magnetic moments of the ions cannot be arranged antiparallel to all other neighbors. Then, a total magnetic moment is observed, and molecular magnets are obtained that are currently subject to research (see the Applications section). The interested reader can refer to O. Kahn's book on molecular magnetism (see General References) and its mathematical explanation via the Van Vleck equation.

6. Applications

6.1. Bactericides and Fungicides. Among marketed bactericides is the sodium salt of 2-pyridinethiol-1-oxide [3811-73-2], C_5H_5NOSNa. However, because this material is a strong skin irritant, the milder zinc chelate [13463-41-7] (**11**) is used in shampoos (see ANTIBACTERIAL AGENTS, HAIR PREPARATIONS) and also has been used in wood protection. The plant fungicides zineb [2122-67-7], $Zn(S_2CNHCH_2CH_2HCS_2)$, and maneb [12427-38-2], $Mn(S_2CNHCH_2CH_2NH CS_2)$, are polymeric chelates (see FUNGICIDES, AGRICULTURAL). A number of other coordination compounds, especially copper derivatives, are used as bactericides, fungicides, and disinfectants (47) (see DISINFECTANTS AND ANTISEPTICS). Some chelating agents used in the past, eg, ligands derived from 8-quinolinol, are no longer in use because of concern over their ability to facilitate unwanted metal-ion transport.

(**11**)

6.2. Catalysis. Hydrogenation reactions can be catalyzed by a wide variety of coordination compounds (48) (see CATALYSIS). For example, dicobalt octacarbonyl [10210-68-1], $Co_2(CO)_8$, has been suggested to be a suitable catalyst for the hydrogenation of olefins to alkanes, aldehydes to alcohols, acid anhydrides to acids and aldehydes, as well as for the selective hydrogenation of polyenes to monoenes, and for the hydrogenation and isomerization of unsaturated fats (see also OXO PROCESS). The list for chlorotris(triphenylphospine)rhodium(I) [14694-95-2], Wilkinson's catalyst is even longer (49). On the other hand, a number of chromium tricarbonyl arene derivatives such as benzenetricarbonyl-chromium [12082-08-5], $C_6H_6Cr(CO)_3$, appear to be quite selective for the hydrogenation of 1,3- and 1,4-dienes to form cis-monoenes by 1,4-addition. Other homogeneous hydrogenation catalysts include coordination compounds of Fe, Ru, Ir, Ni, Pd, Pt, Os, and Cu. Probably the most important homogeneous hydrogenation application has been the Monsanto asymmetric (or enantioselective) synthesis of L-dopa [59-92-7], $C_9H_{11}NO_4$, a chiral amino acid used to treat Parkinson's disease, using a rhodium(I) catalyst containing a chiral phosphine (50).

The palladium chloride process for oxidizing olefins to aldehydes in aqueous solution (Wacker process) apparently involves an intermediate anionic complex such as dichloro-(ethylene)hydroxopalladate(II) or else a neutral aqua complex $PdCl_2(CH_2=CH_2)(H_2O)$ [165824-16-8]. The coordinated $PdCl_2$ is reduced to a Pd(0) species during the olefin oxidation and is reoxidized by the cupric–cuprous chloride couple, which in turn is reoxidized by oxygen, and the net reaction for any olefin ($RCH=CH_2$) is then

$$2\,RCH{=\!=}CH_2 + O_2 \xrightarrow{\text{PdCl}_2-\text{CuCl}_2} 2\,RCH_2CHO$$

Whereas this reaction was used to oxidize ethylene to acetaldehyde, which in turn was oxidized to acetic acid, the direct carbonylation of methanol to acetic acid has largely replaced the Wacker process industrially (see ACETIC ACID AND DERIVATIVES). A large number of other oxidation reactions of hydrocarbons by oxygen involve coordination compounds as detailed elsewhere (51).

The oxo reaction or hydroformylation converts an olefin to an aldehyde containing one or more carbon atoms, using tetracarbonylhydrocobalt [16842-03-8], $CoH(CO)_4$. Phosphine substitution lowers the pressure required for the hydroformylation. Rhodium coordination species are also used for this purpose and give a higher normal to branched aldehyde ratio in the hydroformylation of alkenes, and both rhodium and platinum species appear useful for asymmetric hydroformylation (52). The carbonylation of methanol by CO to acetic acid is catalyzed by coordination compounds of all three Group 9 (VIII) metals (Co, Rh, and Ir); however, the Monsanto process using rhodium complexes allows the use of lower CO pressures and is used worldwide. Adiponitrile is prepared industrially by the hydrocyanation (addition to two moles of HCN) to butadiene using a $Ni[P(OC_6H_5)_3]_4$ [14221-00-2] catalyst.

Metal coordination compounds may also provide alternatives to the heterogeneous catalysts used for the water gas shift reaction. In fact, Ru, Rh, Ir, and Pt coordination compounds have all shown some promise (53).

The stereospecific polymerization of alkenes is catalyzed by coordination compounds such as Ziegler-Natta catalysts, which are heterogeneous $TiCl_3$–Al alkyl complexes. Cobalt carbonyl is a catalyst for the polymerizations of monoepoxides; several rhodium and iridium coordination compounds catalyze certain polymerizations; and nickel complexes can initiate living polymers. Cyclooligomerization is promoted by coordination compounds of nickel, as well. The products depend on the sites available for coordination; eg, complexes having four octahedral sites available for coordination by acetylene produce cyclooctatetraene, three facial sites yield benzene, etc. Other transition-metal species catalyze similar cyclizations to obtain benzene.

Olefin isomerization can be catalyzed by a number of catalysts such as molybdenum hexacarbonyl [13939-06-5], $Mo(CO)_6$. This compound has also been found to catalyze the photopolymerization of vinyl monomers, the cyclization of olefins, the epoxidation of alkenes and peroxo species, the conversion of isocyanates to carbodiimides, etc. Rhodium carbonylhydrotris(triphenylphosphine) [17185-29-4], $RhH(CO)[P(C_6H_5)_3]_3$, is a multifunctional catalyst that accelerates the isomerization and hydroformylation of alkenes.

These applications are mostly examples of homogeneous catalysis. Coordination catalysts that are attached to polymers via phosphine, siloxy, or other side chains have also shown promise. The catalytic specificity is often modified by such immobilization. Metal enzymes are, from this point of view, anchored coordination catalysts immobilized by the protein chains. Even multistep syntheses are possible using alternating catalysts along polymer chains. Other polynuclear coordination species, such as the homopoly and heteropoly ions, also have applications in reaction catalysis.

6.3. Coordination Polymers. In addition to catalysts, polymeric coordination compounds have applications in high temperature coatings and lubricants utilizing poly(metal phosphinates) LUBRICATION AND LUBRICANTS), and also in the chelated fiber called Enkatherm, which is the zinc chelate of (polyterephythaloyl oxalic-bis-amidrazone), PTO (**12**). Enkatherm has been said to resist temperatures of up to 1500°C, and to be superior to any other flame-resistant fiber (54) (see FLAME RETARDANTS FOR TEXTILES). Sensitivity to acid solutions, poor long-term stability, and photosensitivity have limited the use of such materials. On the other hand, there are a large number of potential uses for metal coordination polymers (55), including use as high energy lithographic resists (56).

(**12**)

Polyelectrolytes, charged polymer chains based on metal ion coordination, are not only of fundamental interest but exhibit unique properties due to their viscosity and high charge, so that they are used for the formation of functional coatings and micelles, modified membranes and structured surfaces, as well as support for functional groups and devices.

In supramolecular chemistry, metal−ion coordination is used in combination with other noncovalent bonding types for the construction of low dimensional functional devices. For example, in combination with hydrogen bonding or π stacking, nanoporous materials can be obtained. Depending on the metal ions employed, optical and magnetic properties can be introduced into the polymers.

6.4. Dyes and Pigments. Several thousand metric tons of metalated or metal coordinated phthalocyanine dyes (**13**) are sold annually in the United States. The partially oxidized metalated phthalocyanine dyes are good conductors and are called molecular metals (see SEMICONDUCTORS, ORGANIC COMPOUND SEMICONDUCTORS; PHTHALOCYANINE COMPOUNDS; COLORANTS FOR PLASTICS). Azo dyes (qv) are also often metalated. The basic unit for a 2,2(-azobisphenol dye is shown as structure (**14**). Sulfonic acid groups are used to provide solubility, and a wide variety of other substituents influence color and stability. Such complexes have also found applications as analytical indicators, pigments, ie, for cosmetics, and paint additives.

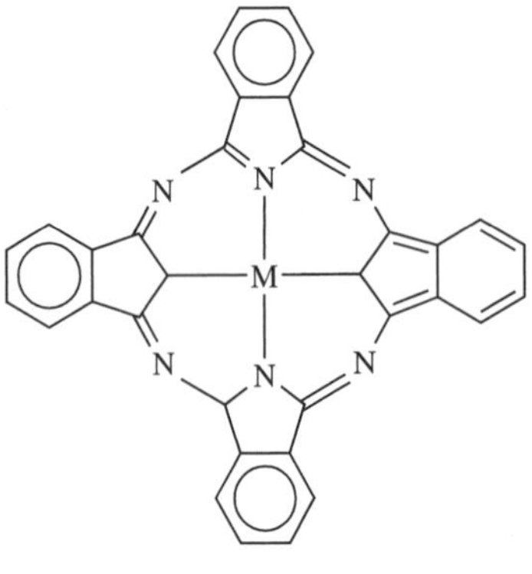

(13)

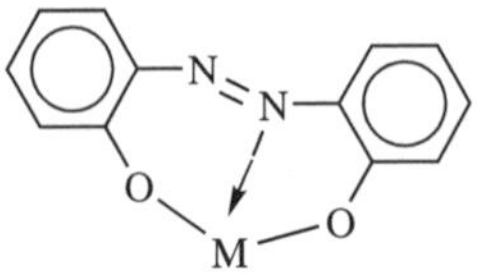

(14)

6.5. Photography. Coordination chemistry is important in classical black and white silver photographic systems, color photography, and in other special photographic areas (see PHOTOGRAPHY) (57). In order to have silver halide grains of appropriate size, silver halides are ripened by solutions containing a silver complexing agent. For example, $[AgX_2]^-$, where X is a halide, or $[Ag(NH_3)_2]^+$ [16972-61-5] can be used to partially dissolve and, through dilution, reprecipitate the silver halides in the gelatin emulsion. In fact, a light-insensitive emulsion consisting of $(NH_4)_2[AgBr_3]$ becomes light-sensitive AgBr when moistened with water or alcohol. Silver halides have increased light sensitivity when sensitized using 10^{-5} moles of a gold(I) or gold(III) coordination compound, such as ammonium dithiocyanatoaurate(I) [15066-31-6], $NH_4[Au(SCN)_2]$, or sodium tetrachloroaurate(III) [15189-51-2], $Na[AuCl_4]$, per mole of silver halide. Coordination compounds are used as emulsion stabilizers, developers, and are formed with the well-known thiosulfate fixers. Silver halide diffusion transfer processes and silver image stabilization also make use of coordination phenomena. A number of copper and chromium azo dyes have found use in diffusion transfer systems developed by Polaroid (see COLOR PHOTOGRAPHY, INSTANT). Coordination compounds are also important in a number of commercial photo-thermography and electrophotography applications as well as in the classic iron cyano blueprint images, a number of chromium systems, etc (58).

6.6. Electroplating. Aluminum can be electroplated by the electrolytic reduction of cryolite, which is trisodium aluminum hexafluoride [13775-53-6], Na_3AlF_6, containing alumina. Brass (see CAST COPPER ALLOYS) can be electroplated from aqueous cyanide solutions that contain cyano complexes of zinc(II) and copper(I). The soft CN^- stabilizes the copper as copper(I) and the two cyano complexes have comparable potentials. Without CN^- the potentials of aqueous zinc(II) and copper(I), as well as those of zinc(II) and copper(II), are >1V apart; thus only the copper plates out. Careful control of concentration and pH also enables brass to be deposited from solutions of citrate and tartrate. The noble metals are often plated from solutions in which coordination compounds help provide fine, even deposits (see ELECTROPLATING).

6.7. Fuel Additives. Antiknock fuel additives include lanthanide β-diketones, several mixed-ligand manganese carbonyl complexes, and tetraethyllead. Coordination compounds have been suggested as fuel oil additives. Metal deactivators for gasoline include Schiff-base ligands that minimize oxidation state changes in the traces of copper dissolving from fuel lines. For example, the planar copper(II) chelate [14522-52-2] (**15**) resists reduction to copper(I) in gasoline (see GASOLINE AND OTHER MOTOR FUELS).

(15)

6.8. Medical Applications. Calcium ethylenediaminetetraacetate [38620-52-9] is used to treat lead poisoning; the free ligand causes calcium loss (see LEAD COMPOUNDS, TOXICOLOGY). Many metal (main group as well as transition metal) coordination compounds show anticancer activity, among them Cp_2MCl_2 (M = Ti, V) as well as ionic species $[Cp_2MCl_2]^{n+}$ (M = Nb, Re, Mo), soluble ruthenium(II–IV) complexes used as antimetastatic agents, bis(β-diketonato)Ti(IV) compounds with aromatic substituents on the β-diketoiminato ligands active in the treatment of colon cancer, and the best known cisplatin, *cis*-dichlorodiammineplatinum(II) [15663-27-1], *cis*-$Pt(NH_3)_2Cl_2$, whose activity was discovered in 1971 (59,60) (see CHEMOTHERAPEUTICS, ANTICANCER). A number of other neutral coordination compounds of platinum(II) and (IV), and of other metals, are also under investigation. The improved cancer cure rates using combined chemotherapy and radiotherapy is leading to the development of Auger-emitting therapeutic drugs.

Radiopharmaceuticals based on a wide range of metal radionucleotides are used in diagnostics as well as therapy (61). Technetium-99m coordination compounds are used very widely as noninvasive imaging tools (62) (see IMAGING TECHNOLOGY; RADIOACTIVE TRACERS). Different coordination species concentrate in different organs. Several of the $[Tc^VO(chelate)_2]^n$ types have been used. In fact, the large majority of nuclear medicine scans in the United States are of technetium-99m complexes. Moreover, chiral transition-metal complexes have been used to probe nucleic acid structure (see NUCLEIC ACIDS). For example, the two chiral isomers of tris(1,10-phenanthroline)ruthenium(II) [24162-09-2] (**16**) interact differently with DNA. These compounds are enantioselective and provide an additional tool for DNA structural interpretation (63).

(16)

High throughput screening of immunoassays for the detection of antigens is available with water soluble and stable lanthanide complexes of Eu(III) and Tb(III) based on cryptates or multipodal ligands, which recognize antigens via specific binding sites and transfer energy after Ultraviolet radiation to a visible light emitting molecule.

Detection of abnormal tissue is realized by magnetic resonance imaging (mri), with contrast agents based on high spin Gd(III), Mn(II), and Fe(III) coordination compounds with a high number of unpaired electrons and long electron spin relaxation times. One of the best known is the coordination compound of gadolinium with DOTA [60239-18-1] (**17**), Na[Gd(DOTA)(H$_2$O)] used for the detection of blood–brain barrier abnormalities.

(17)

Anti infective, anti inflammatory and, anti arthritic agents are based mainly on transition metal complexes, whereas neurologically active compounds are based on lithium ion Gastrointestinal disorders are treated with Bi(III) coordination compounds. Gold complexes, such as (**18**) [34031-32-8], have been used to treat rheumatoid arthritis since 1935, and copper complexes may also be of some benefit as antiinflammatory drugs. There are many other examples available in the literature (64).

(18)

6.9. Chemosensors. Many coordination compounds are now constructed for their specific function arising from the combination of the metal ion with the ligand. The changes in the chemical and physical properties of complexes compared to their building blocks can be used in message transfers. For example, a lumophore can be attached to a metal-ion receptor unit via a spacer. Irradiating the lumophore without any metal ion fixed in the receptor, quenching by, ie, photoinduced electron transfer (PET) can occur, which means that the receptor is oxidized and the luminophore reduced. After complexation of a metal ion by the receptor, quenching is not possible any longer, and the lumophore will emit a characteristic luminescence on irradiation, indicating the presence of the specific metal ion (Fig. 5)(65–67).

Similar signaling devices have been developed for metal ions as well as anions. Anion receptors are, eg, metallacrown- or metallocenium-substituted macrocycles, and their binding can have an influence on the electrochemistry and the luminescence of the system. (68–70) Cooperativity is observed in systems that are able to recognize not only one specie, but also its counterion, eg, as it is the case for KI in a receptor cage formed by a Zn-porphyrine and a derived calix[4]arene (71). Luminescent lanthanide sensors can detect pH, pO_2, or selected anions (72).

6.10. Electronic Materials. Chemical vapor deposition (CVD) has become the method of choice in the growth of thin films of conductors, semi- or superconductors over a very large substrate area. The precursors for the preparation of such materials via CVD is mostly based on organometallic and inorganic chemistry. Especially the alkoxide coordination chemistry of Group 2 (II A) and transition metal ions yields soluble and volatile precursors suitable for CVD. For example, single source precursors for superconductors, containing Ba, Cu, and Y were in the focus of research for quite some time in order to obtain the high T_c–$YBa_2Cu_3O_7$–δ.

With the aim to miniaturize electronic circuits, rod like molecular wires of nanometric dimension (**19**) have been designed based on metal ion complexation

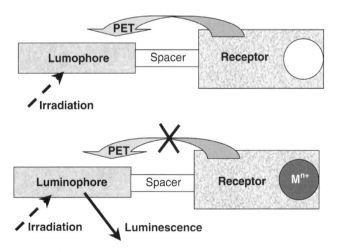

Fig. 5. Scheme showing the principle of a chemosensor based on blocking of luminescence via photo electron transfer when the receptor is empty and luminescence switched on in the presence of a metal ion in the receptor unit. [Based on A. P. de Silva and coworkers., *Coord. Chem. Rev.* **205**, 41–57 (2000)].

in order to transfer energy electronically (73).

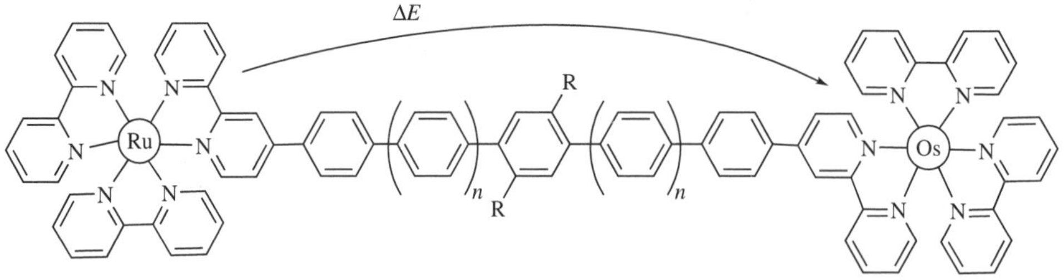

(19)

6.11. Magnetic Materials. In multinuclear coordination compounds, metal ions can communicate with each other magnetically via space (dipole–dipols) and/or orbitals of the ligands. In particular, metal complexes were investigated due to their physicochemical properties arising from the presence of dissimilar metal ions in close proximity. Unique magnetic properties are obtained by the combination of paramagnetic d and f metal ions within one ligand [212895-01-7] (**20**). The field of molecular magnets has galned enormous attention in the last few years, and coordination chemists are challenged by the synthesis of such potentially magnetic switches (74,75).

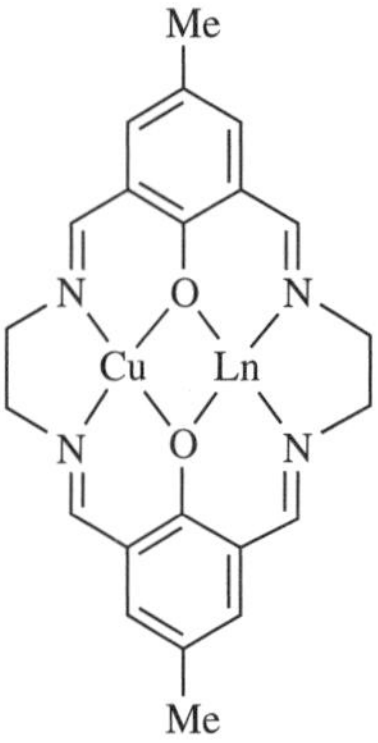

(20)

6.12. Mimetics of Biological Systems. Manganese is one of several first-row transition elements that are employed in biological systems to play a variety of metabolic and structural roles. One is the oxygen-evolving complex OEC, which oxidizes water to oxygen during photosynthesis, manganese catalase, manganese superoxide dismutase, arginase, and manganese ribonucleotide reductase. In order to model nature, numerous coordination compounds of manganese and other metal ions in various oxidation states have been synthesized. One example out of many is the dinuclear [Mn(2-OHsalpn)](2-OHsalpn $- N,N'$-(salicylidenimine)-1,3-diaminopropan-2-ol)$_2$, which is active in hydrogen peroxide disproportionation (76). The proposed mechanism for its activity is outlined in Figure 6.

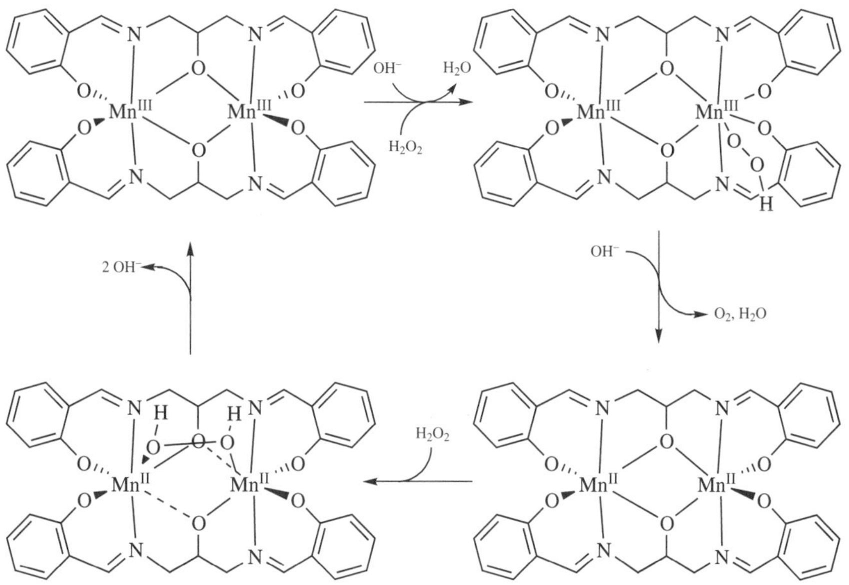

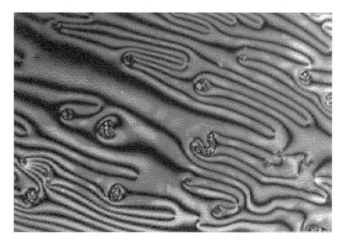

Fig. 6. Proposed mechanism of hydrogen peroxide disproportionation by the [Mn(2-OH-salpn)]₂ system.

6.13. Liquid Crystals. Metal-containing liquid crystals, metallomesogens, are used in the synthesis of flat screens with liquid crystal displays (LCDs). Due to their high anisotropy in structure, these compounds show birefringence (Fig. 7) and can be studied by polarized microscopy (77).

6.14. Miscellaneous. Numerous other applications of coordination compounds exist and may be found throughout this *series*. Tetrahedral cobalt(II) units are incorporated in cobalt blue glass (see COBALT AND COBALT ALLOYS; COLORANTS FOR CERAMICS; GLASS). Tetrahedral to octahedral coordination and the

Fig. 7. Nematic mesophase of an organomanganese complex.

concurrent blue to pink indicator color change is obtained from moisture on cobalt(II) chloride (see DESICCANTS; DRYING). Nickel cyano clathrates are used for organic isomer separation, and the closely related zeolites as catalysts (78) and as porous molecular sieves (see INCLUSION COMPOUNDS). A number of dioxygen complexes are being considered for artificial blood applications (see BLOOD, ARTIFICIAL); tetrakis(β-diketonato) anionic chelates of europium are employed in laser applications (see LASERS); the neutral tris(β-diketonato) lanthanide chelates are used as nmr shift reagents (see MAGNETIC SPIN RESONANCE); and coordination compound membranes are under consideration for ion-selective electrodes (see ELECTROANALYTICAL TECHNIQUES). The use of coordination compounds in extractive metallurgy includes complexes in leaching, solvent-extraction, ion-exchange, precipitation, and electrochemical processes (see METALLURGY SURVEY, EXTRACTIVE METALLURGY) (79). Similarly, nuclear fuel cell preparation and reprocessing use coordination compounds (80) (see NUCLEAR REACTORS CHEMICAL REPROCESSING NUCLEAR FUEL RESERVES).

BIBLIOGRAPHY

"Coordination Compounds" in *ECT* 1st ed., Vol. 4, pp. 379–391, by R. D. Johnson, The University of Pittsburgh; in *ECT* 2nd ed., Vol. 6, pp. 122–131, by N. Christian Nielsen, Central Methodist College; in *ECT* 3rd ed., Vol. 6, pp. 784–798 by R. D. Archer, University of Massachusetts; in *ECT* 4th ed., by R. D. Archer, University of Massachusetts; "Coordination Compounds" in *ECT* (online), posting date: December 4, 2000, by Ronald D. Archer, University of Massachusetts.

CITED PUBLICATIONS

1. R. G. Wilkins, *Kinetics and Mechanism of Reactions of Transition Metal Complexes*, VCH, Weinheim, Germany, 1991, Chapt.6.
2. G. J. Leigh, ed., *Nomenclature of Inorganic Chemistry, Recommendations 1990*, International Union of Pure and Applied Chemistry, Blackwell Scientific Publications, Oxford, UK, 1990.
3. B. P. Block, W. H. Powell, and W. C. Fernelius, *Inorganic Chemical Nomenclature, Principles and Practice*, American Chemical Society, Washington, D.C., 1990. pp. 144–147.
4. J.-M. Lehn, *Supramolecular Chemistry*, Wiley-VCH Weinheim, Germany, 1995;
5. M. Munakata, L. P. Wu, and T. Kuroda-Sowa, *Adv. Inorg. Chem.* **46**, 173–303 (1998).
6. *IUPAC Nomenclature of Inorganic Chemistry*, 2nd ed., Butterworths, London, 1971 and 1990.
7. *IUPAC Nomenclature of Inorganic Chemistry II*, 2000
8. G. J. Kubas, *Acc. Chem. Res.* **21**, 120–128 (1988).
9. A. Albinati and L. M. Venanzi, *Coord. Chem. Rev.* **200–202**, 687–715, (2000).
10. L. Stryer, *Biochemistry*, W. H. Freeman and Company, New York, 1988.
11. K. Itzod, *Adv. Inorg. Chem.* **50**, 33–107 (2000).
12. R. E. Mulvey, *Chem. Soc. Rev.* **27**(5), 339–346 (1998).
13. M. Driess, *Adv. Inorg. Chem.* **50**, 235–284 (2000).
14. L. Alderighi, P. Gans, S. Midollini, and A. Vacca, *Adv. Inorg. Chem.* **50**, 109–172 (2000).

15. H. Braunschweig and M. Colling, *Coord. Chem. Rev.* **223**, 1–51 (2001).

16. G. Linti and H. Schnöckel, *Coord. Chem. Rev.* **206–207**, 285–319 (2000).

17. A. H. H. Stephens and M. L. H. Green, *Adv. Inorg. Chem.* **44**, 1–43 (1997); R. M. Baum, *Chem. Eng. News* **70** (22), 25–27, 30–33 (June 1, 1992). Reviews advances in fullerene chemistry including metal coordination, but without a true bibliography. *Acc. Chem. Res.* **25**, 98–75 (Mar. 1992). A special issue on fullerenes, small 3D carbon cages such as C_{60}, although many other C_n are now known as well, including their metal derivatives.

18. G. G. Briand and N. Burford, *Adv. Inorg. Chem.* **50**, 285–357 (2000).

19. P. A. Gale, *Coord. Chem. Rev.* **199**, 181–233 (2000); P. A. Gale, *Coord. Chem. Rev.* **213**, 79–128 (2001).

20. J. H. Holloway and E. G. Hope, *Adv. Inorg. Chem.* **46**, 51–100 (1998).

21. N. Prokopuk and D. F. Shriver, *Adv. Inorg. Chem.* **46**, 1–49 (1998).

22. D. A. House, *Adv. Inorg. Chem.* **44**, 341–373 (1997).

23. J. Enemark and C. G. Young, *Adv. Inorg. Chem.* **40**, 1–88 (1993); D. C. Rees, M. K. Chan, and J. Kim *Adv. Inorg. Chem.* **40**, 89–119 (1993).

24. C. Bianchini and R. W. Zoellner, *Adv. Inorg. Chem.* **44**, 263–339 (1997).

25. K. Matsumoto and K. Sakai, *Adv. Inorg. Chem.* **49**, 375 (1999).

26. A. Messerschmidt, *Adv. Inorg. Chem.* **40**, 121–185 (1993).

27. G. Vicentini, L. B. Zinner, J. Zukerman-Schpector, and K. Zinner, *Coord. Chem. Rev.* **196**, 353–382 (2000).

28. P. Thuery, M. Nierlich, J. Harrowfield, and M. Ogden, *Calixarenes* **2001**, 561–582 (2001).

29. C. Piguet and J.-C. G. Buenzli, *Chem. Soc. Rev.* **28**(6), 347–358 (1999).

30. W. J. Evans, *Coord. Chem. Rev.* **206–207**, 263–283 (2000).

31. B. I. Kharisov and M. A. Mendez-Rojas, *Russ. Chem. Rev.* **70**(10), 865–884 (2001).

32. J. L. Sessler, A. E. Vivian, D. Seidel, A. K. Burell, M. Hoehner, T. D. Mody, A. Gebauer, S. J. Weghorn, and V. Lynch, *Coord. Chem. Rev.* **216–217**, 411–434 (2001).

33. D. L. Clark, *Los Alamos Sci.* **26**(2), 364–381 (2000); M. P. Neu, *Los Alamos Sci.* **26**(2), 416–417 (2000).

34. F. A. Cotton, *Chemical Applications of Group Theory*, 3rd ed., Wiley-Interscience, New York, 1990; S. F. A. Kettle, *Symmetry and Structure*, John Wiley & Sons, Chichester, UK, 1985.

35. G. Wilkinson, R. D. Gillard, and J. A. McCleverty, eds., *Comprehensive Coordination Chemistry*, Vol. 1, Pergamon Press, Oxford, UK, 1987.

36. R. DeKock and H. B. Gray, *Chemical Structure and Bonding*, Benjamin-Cummings, Menlo Park, Calif., 1980. F. Basolo and R. Johnson, *Coordination Chemistry*, Science Reviews, Northwood, UK, 1987. For more detailed treatments, see A. F. Williams, *A Theoretical Approach to Inorganic Chemistry*, Springer-Verlag, Berlin, 1979. J. S. Griffith, *The Theory of Transition Metal Ions*, Cambridge University Press, UK, 1961. M. Gerloch, *Magnetism and Ligand-Field Analysis*, Cambridge University Press, UK, 1983.

37. F. A. Cotton and G. Wilkinson, *Advanced Inorganic Chemistry*, 5th ed., John Wiley & Sons, Inc., New York, 1988. Chapt. 23.

38. Vol. 1, "Amino Acids," 1974; Vol. 2, "Amines," 1975. Vol. 3, "Other Organic Ligands," 1977. Vol. 4, "Inorganic Complexes," 1976. and Vol. 5, "Supplement," 1982. in A. E. Martell and R. M. Smith eds., *Critical Stability Constants*, Plenum Press, New York.

39. S. Ahrland, J. Chatt, and N. R. Davies, *Quart. Rev.* **11**, 265 (1958; S. Ahrland, *iStruct. Bonding (Berlin)* **5**, 118 (1968)).

40. R. G. Parr and R. G. Pearson, *J. Am. Chem. Soc.* **105**, 7512 (1983; R. G. Pearson, *J. Chem. Educ.* **64**, 561 (1987)).

41. H. Irving and R. J. P. Williams, *Nature (London)* **162**, 746 (1948); H. Irvingand R. J. P. Williams, *J. Chem. Soc.*, 3192 (1953).

42. H. Taube, *Electron Transfer Reactions of Complex Ions in Solution*, Academic Press, New York, 1970. A. Haim, *Acc. Chem. Res.* **8**, 265 (1975); see Ref. 1.

43. D. F. Shriver, P. W. Atkins, and C. H. Langford, *Inorganic Chemistry*, W. H. Freeman and Co., New York, 1990. Chapt. 17.

44. G. J. Ferraudi, *Elements of Inorganic Photochemistry*, John Wiley & Sons, Inc., New York, 1988; A. W. Adamson and P. D. Fleischauer, eds., *Concepts of Inorganic Photochemistry*, John Wiley & Sons, Inc., New York, 1975. 439 pp.; C. R. Bock and E. A. Koerner von Gustorf, in J. N. Pitts, Jr., G. S. Hammond, and K. Gollnick, eds., *Advances in Photochemistry*, Vol. 10, John Wiley & Sons, Inc., New York, 1977. pp. 221–310; V. Balzani and V. Carossiti, *Photochemistry of Coordination Compounds*, Academic Press, Ltd., London, 1970. 432 pp.

45. D. J. Cole-Hamilton and D. W. Bruce, in Ref. 35, Vol. 6., 1987. pp. 487–540.

46. L. De Cola, F. Barigelletti, V. Balzani, P. Belser, A. von Zelewsky, F. Voegtle, F. Ebermeyer, and S. Grammenudi, *J. Am. Chem. Soc.* **110**, 7210 (1988).

47. A. Spencer, in Ref. 35, Vol. 6, 1987. Chapt. 66.

48. A. Spencer, in Ref. 35, Vol. 6, 1987. pp. 229–316; J. Halpern, *Inorg. Chim. Acta* **50**, 11 (1981); Ref. 37, Chapt. 28; see Ref. 17.

49. M. Strem, *The Strem Catalog, No. 8*, Strem, Chemical, Newburyport, Mass., 1978. p. 115.

50. W. S. Knowles, *Acc. Chem. Res.* **16**, 106 (1983); J. Halpern, *Pure Appl. Chem.* **55**, 99 (1983).

51. H. Mimoun, in Ref. 35, Vol. 6. 1987. pp. 317–410.

52. A. Spencer, in Ref. 35, Vol. 6, 1987. pp. 258–266.

53. A. Spencer, Ref. 35, Vol. 6, 1987. pp. 269, 275, 278, 292; W. A. R. Slegeir, R. S. Sapienza, and B. Easterling, *ACS Symp. Ser.* **152**, 325 (1981); R. M. Laine, R. G. Rinker, and P. C. Ford, *J. Am. Chem. Soc.* **99**, 252 (1977); P. C. Ford and A. Rokicki, *Adv. Organometal Chem.* **28**, 139 (1988).

54. D. W. Van Krevelen, *Chem. Ind.* **49**, 1396 (1971); F. C. A. A. Van Berkel and H. Grotjahn, *Appl. Polym. Symp.* **21**, 67 (1973); *Text. Prog.* **8**, 124, 156 (1976).

55. C. E. Carraher, Jr., C. U. Pittman, Jr., J. E. Sheats, M. Zeldin, and B. Currell, eds., *Inorganic and Metal-Containing Polymeric Materials*, Plenum Publishing Co., New York, 1991; see Chapt. I and earlier reviews cited therein.

56. R. D. Archer, V. J. Tramontano, V. O. Ochaya, P. V. West, and W. G. Cumming, in Ref. 55, pp. 161–171.

57. D. D. Chapman and E. R. Schmittou, Ref. 35, Vol. 6, pp. 95–132; T. H. James, ed., *The Theory of the Photographic Process*, 4th ed., Macmillan, New York, 1977.

58. J. Kosar, *Light Sensitive Systems: Chemistry and Applications of Nonsilver Halide Photographic Processes*, John Wiley & Sons, Inc., New York, 1966; E. Brinckman, G. Delzenne, A. Poot, and J. Willems, *Unconventional Imaging Processes*, Focal Press, London, 1978.

59. S. J. Lippard, ed., *Platinum, Gold and Other Metal Chemotherapeutic Agents, ACS Symp. Series 209*, American Chemical Society, Washington, D.C., 1983. H. E. Howard-Lock and C. J. L. Lock, in Ref. 35, Vol. 6, 1987. pp. 755–778.

60. G. Natile and M. Coluccia, *Coord. Chem. Rev.* **216–217**, 383–410 (2001).

61. Z. Guo and P. J. Sadler, *Adv. Inorg. Chem.* **49**, 183–306 (2000).

62. C. J. Jones, in Ref. 35, Vol. 6, 1987. pp. 963–1009.

63. A. M. Pyle and J. K. Barton, *Progr. Inorg. Chem.* **38**, 413 (1987).

64. P. J. Sadler, *Adv. Inorg. Chem.* **36**, 1 (1991).

65. A. Prasanna de Silva, D. B. Fox, A. J. M. Huxley, T. S. Moody, *Coord. Chem. Rev.* **205**, 41–57 (2000).

66. L. Prodi, F. Bolletta, M. Montalti, and N. Zaccheroni, *Coord. Chem. Rev.* **205**, 59–83 (2000).
67. B. Valeur and I. Leray, *Coord. Chem. Rev.* **205**, 3–40 (2000).
68. J. J. Bodwin, A. D. Cutland, R. G. Malkani, and V. L. Pecoraro, *Coord. Chem. Rev.*, **216–217**, 489–512 (2001);
69. P. Beer and J. Cadman, *Coord. Chem. Rev.* **205**, 131–155 (2000).
70. M. H. Keefe, K. D. Benkstein, and J. T. Hunt, *Coord. Chem. Rev.* **205**, 201–228 (2000).
71. A. Robertson and S. Shinkai, *Coord. Chem. Rev.* **205**, 157–199 (2000).
72. D. Parker, *Coord. Chem. Rev.* **205**, 109–130 (2000).
73. B. Schlike, L. D. Cola, P. Belser, and V. Balzani, *Coord. Chem. Rev.* **208**, 267–275 (2000).
74. M. Sakamoto, K. Manseki, and H. Okawa, *Coord. Chem. Rev.* **219–221**, 379–414 (2001).
75. Magnetism: Molecules to Materials II, ed.: J. S. Miller and M. Drillon, Wiley-VCH Verlag GmbH, Weinheim, Germany, 2001.
76. N. A. Law, M. T. Claude, and V. L. Pecoraro, *Adv. Inorg. Chem.* **46**, 305–440 (1998).
77. J.-L. Serrano, ed., *Metallomesogens*, VCH, Weinheim, 1996.
78. G. J. Hutchings, *Chem. Br.*, 762 (1987); M. S. Spencer, *Crit. Rep. Appl. Chem.* **12**, 65 (1985).
79. M. J. Nicol, C. A. Fleming, and J. S. Preston, in Ref. 35, Vol. 6, 1987. pp. 779–842.
80. C. J. Jones, in Ref. 35, Vol. 6, 1987. pp. 881–963.

GENERAL REFERENCES

J. C. Bailar, Jr., ed., *Chemistry of Coordination Compounds, ACS Monograph 131*, Reinhold Publishing Corp., New York, 1956. Dated but very good on the historical and classical aspects of coordination chemistry.

F. Basolo and R. C. Johnson, *Coordination Chemistry*, 2nd ed., Sci. Revs., Northwood, UK, 1986. Easy-to-read paperback on coordination chemistry suitable for anyone who has studied basic chemistry.

F. Basolo and R. G. Pearson, *Mechanisms of Inorganic Reactions*, 2nd ed., John Wiley & Sons, Inc., New York, 1967. An excellent volume that stresses the reactions of complexes in solution; a background and a detailed theory section is included that is largely crystal field theory, but some advantages and disadvantages of molecular orbital theory are included.

B. D. Berezin and V. G. Vopian, trans., *Coordination Compounds of Porphyrins and Phthalocyanines*, John Wiley & Sons, Chichester, UK, 1981. The synthesis, structure, and properties of the macrocycles and their coordination compounds are included.

P. S. Braterman, *Reactions of Coordinated Ligands*, Plenum Press, New York; Vol. 1, 1986, includes over 1000 pages of carbon bonded ligand reactions, essentially organometallic and carbonyls; Vol. 2, 1989, discusses noncarbon bonded ligands.

F. A. Cotton and G. Wilkinson, *Advanced Inorganic Chemistry*, 5th ed., Interscience Publishers, a division of John Wiley & Sons, Inc., New York, 1988.

G. B. Kauffman, *Inorganic Coordination Compounds*, Heyden, London, 1981. An excellent volume on the history of coordination compounds prior to 1935.

S. Kawaguchi, *Variety in Coordination Modes of Ligands in Metal Complexes*, Springer-Verlag, Berlin, 1988.

L. F. Lindoy, *The Chemistry of Macrocyclic Ligand Complexes*, Cambridge University Press, Cambridge, UK, 1989.

W. A. Nugent and J. M. Mayer, *Metal-Ligand Multiple Bonds: The Chemistry of Transition Metal Complexes Containing Oxo, Nitrido, Imido, Alkylidene, or Alkylidyne*

Ligands, John Wiley & Sons, Inc., New York, 1988. Contains electronic and molecular structure, nmr, and ir spectroscopy, reactions, and catalysis.

G. W. Parshall, *Homogeneous Catalysis: The Applications and Chemistry of Catalysis by Soluble Transition Metal Complexes*, John Wiley & Sons, Inc., New York, 1980. 240 pp. An excellent treatment of catalysis by coordination compounds.

R. A. Sheldon and J. K. Kochi, *Metal-Catalyzed Oxidations of Organic Compounds*, Academic Press, Inc., New York, 1981. Mechanistic principles and synthetic methodology for homogeneous and heterogeneous metal-containing oxidation catalysts, many of which are coordination compounds, are discussed.

R. van Eldik, ed., *Inorganic High Pressure Chemistry*, Elsevier, Amsterdam, The Netherlands, 1986. High pressure coordination kinetics including solvent exchange, octahedral and four-coordinate substitution, electron transfer, photochemical, and bioinorganics are discussed.

R. G. Wilkins, *Kinetics and Mechanism of Reactions of Transition Metal Complexes*, 2nd ed., VCH. Weinheim, Germany, 1991. A critical and selected compilation of kinetics and mechanism data.

G. Wilkinson, R. D. Gillard, and J. A. McCleverty, eds., *Comprehensive Coordination Chemistry*, Pergamon Press, Oxford, UK, 1987, 7 Vols., Vol. 1, theory and background; Vol. 2, ligands; and Vol. 6, applications.

A. F. Williams, C. Floriani, and A. E. Merbach, *Perspectives in Coordination Chemistry*, VCH. Weinheim, Germany, 1992. Some state-of-the-art reviews included, July 1992.

J. J. Ziùlkowski, ed., *Coordination Chemistry and Catalysis*, World Scientific, Singapore, 1988. Compiles a very diverse collection of lectures by reknown leaders in this field.

E. C. Constable, *Coordination Chemistry of Macrocyclic Compounds*, Academic, San Diego, California, USA, 1999. Gives an overview on basic and recent macrocyclic coordination compounds.

C. E. Housecroft, *The Heavier d-Block Metals: Aspects in Inorganic and Coordination Chemistry*, Oxford Univ. Press, New York, 1999. Resumes the coordination chemistry of fourth and fifth row elements.

Chi-Ming Che, ed., *Advances in Transition Metal Coordination Chemistry*, JAI, Greenwich, Conn. 1996. Surveys some newer results on coordination compounds.

A. von Zelewsky, ed., *Stereochemistry of Coordination Compounds*, Wiley, UK, 1996. Gives an overview on the possibility of introducing chirality into coordination compounds.

J.-M. Lehn, *Supramolecular Chemistry: Concepts and Perspectives*, VCH, Germany, 1995. The introduction to supramolecular chemistry.

G. J. Leigh and N. Winterton, eds., *Modern Coordination Chemistry: The Legacy of Joseph Chatt*, Royal Society. Chemistry, UK, 2002. Overview over the last 25 years of coordination compounds.

B. F. G. Johnson and M. Schroeder, eds., *Advances in Coordination Chemistry*, Royal Chemical. Society, UK, 1996. Recent results on coordination chemistry.

O. Kahn, *Molecular Magnetism*, VCH Weinheim, 1993. Introduction into the principles of magnetism with comprehensive examples and understandable for a chemist.

Katharina M. Fromm
University of Geneva

COPOLYMERS

1. Introduction

Over the last 50 years, synthetic polymers have replaced traditional materials such as metals, ceramics, wood, and natural fibers in a large number of applications including automotive, construction, appliances, and clothing. In addition, complete new markets have been opened. This revolution has been possible because polymers have many interesting features including high strength/weight ratio, chemical inertness, and easy processability. Another important feature of the synthetic polymers is that the properties can be tuned to match the needs of a given application. This tuning process is often done by preparing new materials through copolymerization.

This article reviews the preparation, properties, use, and characterization of synthetic copolymers and discusses future trends.

2. Copolymer Structures

IUPAC (International Union of Pure and Applied Chemistry) defines a polymer (macromolecule) as: "A molecule of high relative molecular mass, the structure of which essentially comprises the multiple repetition of units derived, actually or conceptually, from molecules of low relative molecular mass". These molecules of low molecular mass are called monomers. Copolymers are macromolecules formed by polymerization of two or more different monomers (comonomers). Homopolymers are formed by polymerization of a single class of monomers.

Copolymers differ in the sequence arrangements of the monomer species in the copolymer chain. In terms of monomer sequence distribution different classes of copolymers can be distinguished (Table 1).

Statistical copolymers are copolymers in which the sequential distribution of the monomeric units obeys known statistical laws. Within this category, copolymers formed following a Markovian process of zero-order (Bernoullian distribution) are named *random copolymers* because the probability of finding a given monomeric unit at any given site of the chain is independent of the nature of the adjacent units. Nevertheless, the term of random copolymer is often used in a broader sense to refer to copolymers in which the comonomer units are evenly distributed along the polymer chain. This defination is the meaning used in this article.

An alternating copolymer is a copolymer comprising two species of monomeric units distributed in alternating sequence. A gradient copolymer is formed by polymer chains whose composition changes gradually along the chain.

Block copolymers are defined as polymers having a linear arrangement of blocks of different monomer composition. In other words, a block copolymer is a combination of two or more polymers joined end-on-end. Any of these polymers or blocks is comprised by monomeric units that should at least have one constitutional unit absent in the other blocks. The blocks forming the block copolymer can be different homopolymers, a combination of homopolymers and copolymers, or copolymers of different chemical composition in adjacent blocks.

Table 1. Classes of Copolymers in Terms of Monomer Sequence Distribution

Type of copolymer	Structure	Examples		
		Comonomers/reactants	Polymerization method	Name
statistical (random)	...ABBABABABAAABBABABAAB...	methyl methacrylate, butyl acrylate	free-radical polymerization	poly(methyl methacrylate-*stat*-butyl acrylate)
alternating	...ABABABABAB...	styrene, maleic anhydride	free radical polymerization	poly(styrene-*alt*-maleic anhydride)
periodic	...ABBABBABB... or —(ABB)$_n$	formaldehyde, ethylene oxide	free-radical polymerization	poly(formaldehyde-*per*-ethylene oxide-*per*-ethylene oxide)
gradient	...AAAABAAABAAABAAABBABBBBABBBBB...	styrene, butyl acrylate	controlled free-radical polymerization	polystyrene-*co*-butylacrylate
block	AAAAAABBBBBBBBBAAAAA	styrene, butadiene	ionic polymerization	Polystyrene-*block*-polybutadiene-*block*-polystyrene
graft	AAAAAAAAAA	styrene/acrylonitrile, polybutadiene	free-radical polymerization	polybutadiene-*graft*-poly(styrene-*stat*-acrylonitrile)

B
C
B
C
C
B

Table 2. **IUPAC Source-Based Copolymer Classification**

Polymer type	Connective	Example
unspecified or unknown	*-co-*	poly(A-*co*-B)
random (obeys Bernoullian distribution)	*-ran-*	poly(A-*ran*-B)
statistical (obeys known statistical law)	*-stat-*	poly(A-*stat*-B)
alternating (for two monomeric units)	*-alt-*	poly(A-*alt*-B)
periodic (ordered sequence for >2 monomeric units)	*-per-*	poly(A-*per*-B-*per*-C)
block (linear block arrangement)	*-block-*	polyA-*block*-polyB
graft (side chains connected to main chain)	*-graft-*	polyA-*graft*-polyB

A graft copolymer is a polymer comprising one or more blocks connected to the backbone as side chains, having constitutional or configurational features that make them different from the main chain.

IUPAC (1–4) proposed two naming methodologies for polymers. The first methodology is the *structure-based nomenclature system* that requires naming a polymer as poly(constitutional repeating unit), wherein the repeating unit is named as a bivalent organic radical according to the rules used for organic chemistry. This nomenclature can be difficult to apply if the structure is only partly known or unknown, unless assumptions are made. The second methodology is the *source-based nomenclature system* that requires naming the polymer by adding the prefix "poly" to the name of the actual or hypothetical monomers; an infix, called a connective, is placed between them to indicate the type of sequential arrangement of the constitutional units within the chains. This nomenclature is most often employed within the scientific community. Table 2 presents a summary of the IUPAC source-based copolymer classification. Some specific examples are given in Table 1.

Copolymers can also have widely different topologies. Table 3 presents the main classes as well as some examples with the corresponding IUPAC names.

3. Copolymerization Reactions

Table 4 summarizes the different types of polymerizations (6). Chain-growth polymerization involves polymer chain growth by reaction of an active polymer chain with single monomer molecules. In step-growth polymerization, polymer growth involves reactions between macromolecules. In addition, nonpolymeric by-products may be formed in both types of polymerizations. Condensative chain polymerization is very rare.

3.1. Chain-Growth Copolymerization. In chain-growth copolymerization, monomers can only join active chains. The activity of the chain is generated by either an initiator or a catalyst. The following classes of chain-growth copolymerizations can be distinguished according to the type of active center:

- Free radical polymerization (the active center is a radical).
- Anionic polymerization (the active center is an anion).
- Cationic polymerization (the active center is a cation).
- Catalytic polymerization (the active center is an active site of a catalyst).

Table 3. **Nomenclature of Some Common Copolymers**

Architecture	Example			
	Comonomers/reactants	Polymerization method	Name	Architecture
linear	styrene	free-radical polymerization	polystyrene	
branched	*n*-butyl acrylate	free-radical polymerization	*branch*-poly-(*n*-butyl acrylate)	
comb	poly (methylsiloxane, styrene)	atom-transfer radical polymerization (ATRP)	poly(methyl siloxane)-*comb*-polystyrene	
cross-linked/ network	butadiene	free-radical polymerization	*net*-polybutadiene	
star	methyl methacrylate	ATRP	4-*star*-poly (methyl methacrylate)	
hyper branched	4-(chloromethyl) styrene	ATRP	hyperbranched poly(chloromethyl styrene)[a]	

[a] IUPAC nomenclature for hyperbranched polymers is not available yet, in the meantime, IUPAC recommends following the rules in Ref. (5).

Table 4. **Classes of Polymerizations**

Chain-growth polymerization	Step-growth polymerization
$P_n + M \longrightarrow P_{n+1}$ (chain polymerization) polystyrene	$P_n + P_m \longrightarrow P_{n+m}$ (polyaddition) polyurethanes
$P_n + M \longrightarrow P_{n+1} + Z$ (condensative chain polymerization)	$P_n + P_m \longrightarrow P_{n+m} + Z$ (polycondensation) poly(ethylene terephthalate)

3.2. Step-Growth Copolymerization.   Monomer molecules consisting of at least two functional groups would undergo step-growth polymerization. These functional groups would be capable of reacting with each other, eg, a $-COOH$ group would react with $-OH$ and $-NH_2$ groups. The two reacting functional groups could be on the same monomer molecule, type A–B (eg, an aminoacid) or on two separate molecules, types A–A and B–B (eg, a diacid and a diol).

Step-growth polymerization proceeds by reaction of the functional groups of the reactants in a stepwise manner: monomers react to form dimer; monomer and dimer react to form trimer; dimer and trimer form pentamer, and in general:

$$n\text{-mer} + m\text{-mer} \longrightarrow (n + m)\text{-mer} \tag{1}$$

This reaction pathway requires to achieve very high conversions in order to produce high molecular weight polymers (>98–99%). A number of different chemical reactions may be used in step-growth polymerization including esterification, amidation, the formation of urethanes, aromatic substitution, and carbonate bond formation.

4. Free-Radical Copolymerization

Free-radical copolymerization can be divided into classical free-radical polymerization and controled radical polymerization.

4.1. Classical Free-Radical Copolymerization.   The characteristics of the materials produced by free-radical copolymerization are determined by the kinetics, rather than by the thermodynamics, of the chain-growth process. Therefore, it is important to adequately describe the copolymerization kinetics. There is considerable experimental evidence (7–9) showing that in many polymerization systems, propagation depends on the nature of the monomer and on the last two units of the growing chain. This is referred to as *penultimate model*. Nevertheless, copolymer composition can be well described by considering a model in which the reactivity of the propagation reaction is governed by the nature of the monomer and the terminal unit of the polymer radical (*terminal model*). In what follows, the penultimate model will be presented first and then the terminal model will be discussed.

Penultimate Model.   The features of the propagation reactions of the penultimate model for free-radical copolymerization are summarized in the following simplified reaction scheme:

$$R_{11} + M_1 \xrightarrow{k_{p111}} R_{11}$$

$$R_{21} + M_1 \xrightarrow{k_{p211}} R_{11}$$

$$R_{11} + M_2 \xrightarrow{k_{p112}} R_{12}$$

$$R_{21} + M_2 \xrightarrow{k_{p212}} R_{12}$$

$$R_{12} + M_1 \xrightarrow{k_{p121}} R_{21} \tag{2}$$

$$R_{22} + M_1 \xrightarrow{k_{p221}} R_{21}$$

$$R_{12} + M_2 \xrightarrow{k_{p122}} R_{22}$$

$$R_{22} + M_2 \xrightarrow{k_{p222}} R_{22}$$

where R_{ij} is a polymer radical whose last two units are of type i and j, respectively. From these reactions, four different monomer reactivity ratios (r_{ij}) and two radical reactivity ratios (s_i) can be defined as follows:

$$r_{11} = \frac{k_{p111}}{k_{p112}} \quad r_{21} = \frac{k_{p211}}{k_{p212}} \quad r_{12} = \frac{k_{p122}}{k_{p121}} \quad r_{22} = \frac{k_{p222}}{k_{p221}} \quad s_1 = \frac{k_{p211}}{k_{p111}} \quad s_2 = \frac{k_{p122}}{k_{p222}} \tag{3}$$

The instantaneous copolymer composition referred to monomer 1, Y_1, is determined by the relative rates of monomer consumption

$$
\begin{aligned}
Y_1 &= \frac{dM_1/dt}{dM_1/dt + dM_2/dt} \\[2mm]
&= \frac{1 + \dfrac{r_{21}\frac{[M_1]}{[M_2]}\left(r_{11}\frac{[M_1]}{[M_2]} + 1\right)}{r_{21}\frac{[M_1]}{[M_2]} + 1}}{2 + \dfrac{r_{21}\frac{[M_1]}{[M_2]}\left(r_{11}\frac{[M_1]}{[M_2]} + 1\right)}{r_{21}\frac{[M_1]}{[M_2]} + 1} + \dfrac{r_{12}\left(r_{22} + \frac{[M_1]}{[M_2]}\right)}{\frac{[M_1]}{[M_2]}\left(r_{12} + \frac{[M_1]}{[M_2]}\right)}}
\end{aligned} \tag{4}
$$

$$Y_2 = 1 - Y_1$$

where $[M_i]$ is the concentration of monomer i. The reactivity ratios determine not only the instantaneous copolymer composition, but also the comonomer sequence distribution in the copolymer chain. The probability distribution function for having a sequence of x units of monomer i in the copolymer chain is

$$F_i(x) = 1 - p_{jii} \qquad x = 1 \tag{5}$$

$$F_i(x) = p_{jii}p_{iii}^{x-2}(1 - p_{iii}) \qquad x > 1 \qquad i,j = 1,2 \tag{6}$$

where

$$p_{jii} = \frac{[M_i]}{[M_i] + [M_j]/r_{ji}} \tag{7}$$

$$p_{iii} = \frac{[M_i]}{[M_i] + [M_j]/r_{ii}} \tag{8}$$

The penultimate model can simultaneously account for the copolymerization rate, the copolymer composition, and the sequence distribution (9). However, its usefulness is limited by the fact that for a copolymerization of two monomers, the values of eight different propagation rate constants are needed, and the number of parameters rapidly increases as a multimonomer system is considered. Therefore, there is a strong interest in using simpler models. Fukuda and coworkers (7) proposed the *implicit penultimate model* in which $r_{ii} = r_{ji}$. Although there are theoretical reasons supporting the explicit penultimate model, its superiority over the implicit penultimate model seems to be marginal in most cases (9). Even in the case of the implicit penultimate model, the number of parameters is large, and further simplification of the model is desirable. This is achieved in the *terminal model* in which the reactivity of the propagation reaction is governed by the nature of the monomer and the terminal unit of the polymer radical.

Terminal Model.   According to the terminal model, the propagation reactions are as follows:

$$\begin{aligned}
R_1 + M_1 &\xrightarrow{k_{p11}} R_1 \\
R_1 + M_2 &\xrightarrow{k_{p12}} R_2 \\
R_2 + M_1 &\xrightarrow{k_{p21}} R_1 \\
R_2 + M_2 &\xrightarrow{k_{p22}} R_2
\end{aligned} \tag{9}$$

In this scheme, R_i represents a polymer radical with ultimate unit of type i.

The polymerization rate referred to monomer i, R_{pi}, is

$$R_{pi} = (k_{pli}p_1 + k_{p2i}p_2)[M_i][R^*] \qquad i = 1, 2 \quad (mol/Ls) \tag{10}$$

where kp_{ij} is the propagation rate constant of radicals of terminal unit i with monomer j, $[M_i]$ is the monomer concentration, $[R^*]$ is the concentration of radicals, and p_i is the time averaged probability of finding an active chain with ultimate unit of type i given by (10):

$$p_1 = \frac{k_{p21}[M_1]}{k_{p21}[M_1] + k_{p12}[M_2]} \qquad p_2 = 1 - p_1 \tag{11}$$

It is well documented that the terminal model is not adequate to describe the polymerization rate of an increasing number of free-radical copolymerizations

(7–9). However, it is widely used in practice because (a) for historical reasons, values of the terminal model reactivity ratios for many comonomer systems are available (11), and (b) in many cases, eg, emulsion polymerization, the uncertainties associated with [R*] are larger than the errors included by the terminal model.

The instantaneous copolymer composition is given by the Mayo-Lewis equation (12).

$$Y_1 = \frac{dM_1/dt}{dM_1/dt + dM_2/dt}$$

$$= \frac{1 + r_1([M_1]/[M_2])}{2 + r_1([M_1]/[M_2]) + r_2([M_2]/[M_1])} \tag{12}$$

$$Y_2 = 1 - Y_1$$

where $r_i = k_{pii}/k_{pij}$ are the monomer reactivity ratios.

It has been found that the Mayo-Lewis equation predicts well the evolution of copolymer composition (13–15). This equation shows that the instantaneous copolymer composition depends on the reactivity ratios, which in turn mostly depend on the chemical nature of the monomers involved, and on the ratio of monomer concentrations in the polymerization loci. Figure 1 shows the effect of the monomer molar ratio on the instantaneous copolymer composition for different reactivity ratios.

The reactivity ratios also determine the comonomer sequence distribution in the copolymer chain. The probability distribution function for having a

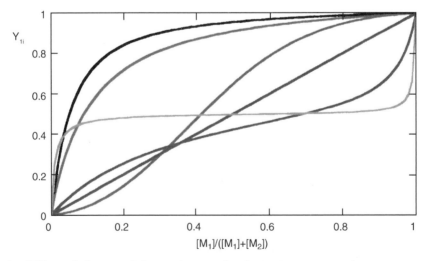

Fig. 1. Effect of the reactivity ratios on the instantaneous copolymer composition referred to monomer 1, Y_{1i}. Legend: (———) $r_1 = 22$ and $r_2 = 0.06$; (———) $r_1 = 10$ and $r = 0.1$; (———) $r_1 = 10$ and $r_2 = 5$; ——— $r_1 = r_2 = 1$; (———) $r_1 = 0.1$ and $r_2 = 0.5$; ——— $r_1 = 0.01$ and $r_2 = 0.02$.

sequence of x units of monomer i in the copolymer chain is

$$F_i(x) = p_{ii}^{x-1}(1 - p_{ii}) \qquad i = 1, 2 \tag{13}$$

where

$$p_{ii} = \frac{r_i[\mathrm{M_1}]/[\mathrm{M_2}]}{r_i[\mathrm{M_1}]/[\mathrm{M_2}] + 1} \tag{14}$$

The adequacy of the terminal model for describing the sequence distribution of copolymerization systems is controversial as it has been found inadequate for some systems such as α-methyl styrene–acrylonitrile (16), methyl methacrylate–acrylonitrile (17), and styrene–maleic anhydride (18), but adequate for vinyl acetate–butyl acrylate (13), and methyl methacrylate–methyl acrylate (14). Statistical methods to discriminate between the penultimate and the terminal models have been discussed (19). Certainly, until additional data are available, the terminal model seems to be a practical way for predicting copolymer composition and sequence distribution. Qualitatively, several cases may be distinguished:

- $r_1 \ll 1$ and $r_2 \ll 1$, both monomers add almost exclusively to the other monomer, and hence an alternating copolymer is obtained.

  ABABABABABABABABABABAB

- $r_1 < 1$ and $r_2 < 1$, both monomers add preferentially to the other monomer, and therefore a copolymer with some alternating character is formed. The alternating character decreases as the reactivity ratios approach unity.

  ABABAABABABBABABABABAABB

- $r_1 < 1$ and $r_2 > 1$, copolymer formed by a long sequences of monomer 2 separated by single monomer 1 units.

  BBBBABBBBBABBBBBABBBBBB

- $r_1 = r_2 = 1$, a random copolymer is formed.

  ABAABABBAAABABBBABAABB

- $r_1 > 1$ and $r_2 > 1$, the same monomer is added preferentially, and hence the copolymer is made of long sequences of each monomer.

  AAAABBBBAAAABBBBBAAAAABB

- $r_1 > 1$ and $r_2 < 1$, copolymer formed by a long sequences of monomer 1 separated by single monomer 2 units.

  AAAAAABAAAAABAAAAAABA

- $r_1 \gg 1$ and $r_2 \gg 1$, the same monomer is almost exclusively added and a mixture of homopolymer chains with small amounts of block copolymers is obtained

AAAAAAAAAAAAAAAAAAAAA

BBBBBBBBBBBBBBBBBBBBBBBB

BBBBBBBBBBBBBAAAAAAAAAA

Factors Affecting the Reactivity Ratios. *Chemical nature:* Reactivity ratios depend on both radical and monomer reactivities. Monomer reactivity increases with its ability to stabilize by resonance the radical formed from the monomer, which depends on the substituent X in $CH_2=CHX$, according to the sequence (20): $-Ph, -CH=CH_2 > -CN, -COR > -COOH, -COOR > -Cl > -OCOR, -R > -OR, -H$. Substituents composed of unsaturated linkages are most effective stabilizing the radicals because of the loosely held π-electrons. On the other hand, resonance stabilized radicals are the least reactive. The effect of resonance is more acute in radical than in monomer reactivity. This means that copolymerization between two monomers with stabilizing substituents or between two monomers without stabilizing substituents will be favored, whereas copolymerization between a monomer with a stabilizing substituent and a monomer without it will be difficult. Steric hindrance and polar effects also affect reactivity ratios.

Arguably, the most efficient method for producing graft copolymers is copolymerizing monomers with macromonomers. The reactivity of the macromonomer is primarily determined by the chemical nature of the polymerizable group in the macromonomer, but the degree of compatibility of the macromonomer with the propagating polymer chain may affect the macromonomer reactivity(21).

Polymerization conditions: Reactivity ratios are not substantially affected by the reaction medium used, provided that the system remains homogeneous. However, in heterogeneous systems, such as in emulsion polymerization, the apparent reactivity ratios are strongly affected by monomer partitioning between different phases. Thus, when monomers with widely different water solubilities are copolymerized, the monomer molar ratio in the polymer particles may be very different from the average ratio in the reactor, affecting the apparent reactivity ratios (22). Nevertheless, when the monomer partitioning is properly taken into account, the actual reactivity ratios should be used (23). In addition, the effect of the different water solubilities decreases as the solids content increases (24). The reactivity ratios are relatively insensitive to temperature (11).

Determination of the Reactivity Ratios. The reactivity ratios can be determined from experimental data and estimated from empirical correlations.

The determination of the reactivity ratios with small confidence intervals requires a careful experimental design, sensitive analytical techniques, and the use of a good method for parameter estimation. Unfortunately, these requirements are not always fulfilled in the determination of many of the binary reactivity ratio values published. Although, traditionally low conversion data have been used, higher precision is obtained running the experiments to high conversion (25). The error-in-variable estimation method, which allows accounting properly for all sources of error, is superior to classical linear and nonlinear least squares (26–28). The error-in-variable method also allows estimating reactivity ratios in terpolymerization (29), and from emulsion polymerization data by accounting for monomer partitioning (30).

The reactivity ratios can also be estimated from empirical equations. As discussed above, propagation constants depend on resonance stabilizations, polar, and steric effects. Price (31) and Alfrey and Price (32), proposed that the propagation rate constant for polymerization of radical 1 and monomer 2 be written as

$$k_{p12} = P_1 Q_2 \exp(-e_1 e_2) \tag{15}$$

where P_1 and Q_2 are associated to the resonance stabilization, and e_1 and e_2 depend on the polarity of the macroradical and monomer. By assuming that the same e values apply to the monomer and to the corresponding radical, the reactivity ratios are then expressed in terms of the Alfrey-Price Q,e scheme:

$$r_1 = (Q_1/Q_2) \exp[-e_1(e_1 - e_2)] \tag{16}$$
$$r_2 = (Q_2/Q_1) \exp[-e_2(e_2 - e_1)] \tag{17}$$
$$r_1 r_2 = \exp[-(e_1 - e_2)^2] \tag{18}$$

where each monomer has a Q resonance value and an e polarity value. The Q,e scheme requires that a monomer is chosen as a reference. Styrene was chosen as the reference monomer and a Q value of unity and an e value of -0.800 were assigned. Table 5 shows a selection of Q,e values and Table 6 presents a comparison between the values of the reactivity ratios predicted with this scheme and those estimated from experimental data.

Copolymerization with Depropagation.   Classical analysis of free-radical copolymerization considers that propagation occurs irreversibly. However, this is not the case for the polymerization of some monomers such as methyl methacrylate and α-methyl styrene at relatively high temperatures. The copolymerization of monomers that suffer depropagation cannot be accounted for by either the

Table 5. **Selection of Q and e Values**[a]

Monomer	Q	e
ethylene	0.016	0.05
vinyl chloride	0.056	0.16
methyl methacrylate	0.78	0.40
acrylamide	0.23	0.54
methyl acrylate	0.45	0.64
butyl acrylate	0.38	0.85
1-hexene	0.035	0.92
acrylonitrile	0.48	1.23
butadiene	1.70	−0.50
isoprene	1.99	−0.55
styrene	1.00	−0.80
vinyl acetate	0.026	−0.88
propylene	0.009	−1.69

[a] Ref. 13.

Table 6. **Comparison between the Values of the Reactivity Ratios Predicted by the *Q,e* Scheme and Those Estimated from Experimental Data**

Monomer system	Q,e scheme	Estimated from experimental data[a]
1. styrene	$r_1 = 0.70$	$r = 0.84$
2. butyl acrylate	$r_2 = 0.09$	$r_2 = 0.18$
1. methyl methacrylate	$r_1 = 1.91$	$r_1 = 2.15$
2. ethyl acrylate	$r_2 = 0.49$	$r_2 = 0.4$
1. styrene	$r_1 = 0.46$	$r = 0.6$
2. butadiene	$r_2 = 1.98$	$r_2 = 1.8$

[a] Ref. 11.

classical penultimate model or the classical terminal model. Lowry (33) developed models to predict the instantaneous copolymer composition when only one monomer tends to depropagate for three cases that considered different penultimate effects. Wittmer (34) and Krüger and co-workers (35) developed equations for the instantaneous copolymer composition when both monomers can depropagate. By considering the following terminal model scheme,

$$\mathrm{R}_{r,1} + \mathrm{M}_1 \underset{k_{pr11}}{\overset{k_{p11}}{\rightleftharpoons}} \mathrm{R}_{r+1,1}$$

$$\mathrm{R}_{r,1} + \mathrm{M}_2 \underset{k_{pr12}}{\overset{k_{p12}}{\rightleftharpoons}} \mathrm{R}_{r+1,2}$$

$$\mathrm{R}_{r,2} + \mathrm{M}_2 \underset{k_{pr22}}{\overset{k_{p22}}{\rightleftharpoons}} \mathrm{R}_{r+1,2}$$

$$\mathrm{R}_{r,2} + \mathrm{M}_1 \underset{k_{pr21}}{\overset{k_{p21}}{\rightleftharpoons}} \mathrm{R}_{r+1,1}$$

$$(19)$$

Krüger and co-workers (35) developed the following equation for the instantaneous molar ratio of the comonomers in the copolymer:

$$
\begin{aligned}
Y_1 &= \frac{d\mathrm{M}_1/dt}{d\mathrm{M}_1/dt + d\mathrm{M}_2/dt} \\
&= \frac{[\mathrm{M}_1]\{r_1([\mathrm{M}_1] + q_1 P_{21}) + [\mathrm{M}_2] - r_1 K_1 P_{11}\} - q_1 P_{21}(r_1 K_1 P_{11} + q_2 P_{12})}{[\mathrm{M}_1]\{r_1([\mathrm{M}_1] + q_1 P_{21}) + [\mathrm{M}_2] - r_1 K_1 P_{11}\} - q_1 P_{21}(r_1 K_1 P_{11} + q_2 P_{12})} \\
&\quad + [\mathrm{M}_2]\{r_2([\mathrm{M}_2] + q_2 P_{12}) + [\mathrm{M}_1] - r_2 K_2 P_{22}\} \\
&\quad - q_2 P_{12}(r_2 K_2 P_{22} + q_1 P_{21})
\end{aligned}
\tag{20}
$$

$$Y_2 = 1 - Y_1 \tag{21}$$

with

$$r_1 = \frac{k_{p11}}{k_{p12}} \qquad r_2 = \frac{k_{p22}}{k_{p21}} \qquad q_1 = \frac{k_{pr12}}{k_{p21}} \qquad q_2 = \frac{k_{pr21}}{k_{p12}} \qquad K_1 = \frac{k_{pr11}}{k_{p11}} \qquad K_2 = \frac{k_{pr22}}{k_{p22}}$$

$$(22)$$

the parameters P_{ij} are determined from a steady-state approximation on the radical species balance with the conditions that $P_{ii} + P_{ij} = 1$. By using a different analysis method, Wittmer (34) developed a different, although equivalent, equation. Palmer and co-workers (36) used the equations developed by Wittmer (34) and Kruger and co-workers (35) to analyze the copolymerization of methyl methacrylate and α-methyl styrene in a wide temperature range. It was reported that both approaches described the experimental data well with the Krüger equations being more stable and having better convergence properties. On the other hand, the data could not be fitted by the Mayo-Lewis equation (eq. 12).

Copolymer Composition Evolution. In classical free-radical polymerization, the number of polymer chains that are growing at the same time is rather small (10^{-8}–10^{-7} mol/L), and the time spent by a chain from initiation to termination is very short (typically 0.5–10 s). This means that the final copolymer is made of polymer chains formed at different moments in the polymerization process. This characteristic limits to random, alternating, and ill-defined graft the type of copolymers that are accessible using classical free-radical polymerization. This means that well-defined block and graft copolymers cannot be produced by means of classical free-radical copolymerization. On the other hand, the cumulative copolymer composition referred to monomer h, $\bar{Y}_h$, differs from the instantaneous one $\bar{Y}_{hi}$:

$$\bar{Y}_{hi} = \frac{1}{X_{Tf}} \int_0^{X_{Tf}} \mathbf{Y}_{hi} dX_T \tag{23}$$

where X_T is the overall molar conversion defined as the number of moles of monomer reacted divided by the total number of moles of monomer in the formulation. Figure 2 presents the effect of the reactivity ratio values on the evolution of the instantaneous copolymer composition in a batch reactor. It can be seen that a substantial composition drift occurred for monomers with different reactivity ratios, as well as for monomers of similar reactivity but using not equimolar feed compositions. This fact and the problems associated with heat removal led to the use of semicontinuous reactors for copolymer composition control (see the section Controlling Copolymer Microstructure).

Multicomponent Copolymerization. Multicomponent copolymerization involves three or more monomers. The polymerization rate of monomer i and the instantaneous copolymer composition are given by

$$\frac{d[\mathbf{M}_i]}{dt} = \left(\sum_{j=1}^n k_{pji} p_j \right) [\mathbf{M}_i][\mathbf{R}^*] \tag{24}$$

$$Y_i = \frac{\left(\sum_{j=1}^n k_{pji} p_j \right) f_i}{\sum_{i=1}^n \sum_{j=1}^n k_{pji} p_j f_i} \tag{25}$$

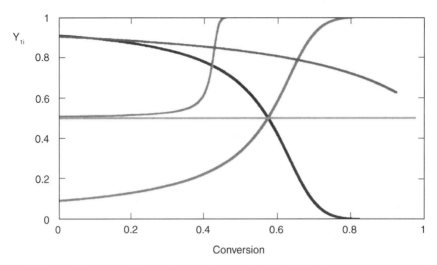

Fig. 2. Effect of the reactivity ratios on the evolution of the instantaneous copolymer composition Referred to monomer 1, in a batch reactor. For $[M_1]/([M_1]+[M_2]) = 0.5$: (——) $r_1 = 10$ and $r_2 = 0.1$; (——) $r_1 = 0.1$ and $r_2 = 10$; (——) $r_1 = r_2 = 0.01$; For $[M_1]/([M_1]+[M_2]) = 0.8$: (——) $r_1 = r_2 = 0.01$; (——) $r_1 = r_2 = 5$

where f_i is the mole fraction of unreacted monomer i at the polymerization locus.

4.2. Controlled Radical Polymerization. Free-radical copolymerization is attractive because of the huge number of monomers that can be copolymerized, the different media that can be used (both organic and aqueous), and the relative robustness of this technique to impurities. In the classical free-radical copolymerization considered above, only a few polymer chains are growing at the same time and the time spent in building a chain is very short (typically 0.5–10 s). This is convenient to produce random and alternating copolymers, but well-defined complex copolymer topologies (eg, gradient and block copolymers in Table 1) are not accessible through classical free-radical polymerization. In order to produce a well-defined block copolymer, all chains should start at the same time, and they should grow for some time in the presence of monomer 1 until the first block is formed. Then monomer 1 should be removed and a different monomer is added to produce the second block. In order to conduct such a process successfully, termination should be avoided during the polymerization.

In recent years, the development of controlled radical polymerization (CRP) methods has made it possible to conduct a free-radical polymerization minimizing the extent of termination. This allows preparing almost any kind of copolymer microstructure by means of a free-radical mechanism. All CRP methods have in common that a rapid dynamic equilibrium is established between a tiny amount of growing free-radicals and a large majority of dormant polymer chains. In these processes, each growing chain stays for a long time in the dormant state, then it is activated and adds a few monomer units before becoming dormant again. As the activation–polymerization–deactivation process is a random process, the molecular weight distribution of the growing chains becomes narrower as they grow longer. The composition of the polymer chain can be easily modified by controlling the monomer composition in the reactor. Termination between active radicals is minimized by simply maintaining its concentration at a low value. The CRP methods differ in the way in which these dormant species are formed. The most efficient CRP methods are stable free-radical polymerization best represented by nitroxide mediated polymerization (NMP), atom-transfer radical polymerization (ATRP), and reversible addition-fragmentation chain transfer (RAFT).

Nitroxide Mediated Polymerization. This field has been reviewed by recently Hawker and co-workers (37). The key aspect of this process is the reversible termination of the growing polymer chain with the nitroxide radical. In this way, the concentration of active chains is very low and the extent of the irreversible bimolecular termination is minimized (Fig. 3).

2,2,6,6-Tetramethylpiperidinyloxy (TEMPO) is one of the most commonly used nitroxide radicals (38), but it suffers from some serious drawbacks including the necessity of using high polymerization temperatures and its incompatibility with many important monomer families. Improved nitroxides based on phosphonate derivatives (39) and on the family of arenes (40) have been developed. With this second-generation nitroxides, a broader range of monomers including acrylates, acrylamides, 1,3-dienes and acrylonitrile, as well as monomers containing functional groups, such as amino, carboxylic acid and glycidyl, can be polymerized with accurate control of molecular weights and polydispersities. Good control of homopolymerization of methacrylates has not been achieved because these monomers are strongly affected by chain end degradation through hydrogen transfer (38). Nevertheless, random copolymers with up to 90% of methacrylate incorporation can be prepared in a controlled fashion (40); the reasons for this behavior are not clear currently. The reactivity ratios in NMP are the same as in conventional free-radical polymerization. This leads to one of the advantages of controlled radical polymerization compared to living anionic or cationic

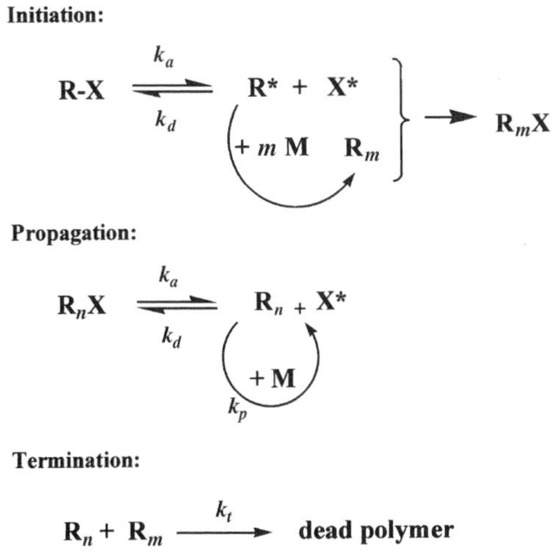

Initiation:

$$\text{R-X} \underset{k_d}{\overset{k_a}{\rightleftharpoons}} \left.\begin{array}{c} \text{R* + X*} \\ (+ m\,\text{M} \quad \text{R}_m) \end{array}\right\} \longrightarrow \text{R}_m\text{X}$$

Propagation:

$$\text{R}_n\text{X} \underset{k_d}{\overset{k_a}{\rightleftharpoons}} \text{R}_n + \text{X*} \atop \underset{k_p}{(+ \text{M})}$$

Termination:

$$\text{R}_n + \text{R}_m \xrightarrow{k_t} \text{dead polymer}$$

Fig. 3. Scheme of a NMP.

polymerizations, which is the ability of preparing well-defined random copolymers. In batch copolymerizations of monomers with different reactivity ratios, the chemical composition distribution of the chains formed by CRP considerably differs from that of the polymer chains produced in the classical free-radical polymerization (see the section Some Common Kinetic Features of the CRP Processes). Nitroxide mediated polymerizations allow us to produce narrow molecular weight distributions (MWD) up to a value of the number average molecular weight of ~200,000 g mol (40). At higher molecular weights, termination reactions become significant and this leads to a loss of control and the living character of the process.

ATRP.   An excellent review of this process can be found in (41). In the ATRP process, the reactor is charged with monomer and a certain amount of alkyl halides that act as initiator. The alkyl halides are dormant species that can be activated through a reversible redox process catalyzed by a transition-metal complex ($M_t^n - Y/$ ligand, where Y may be another ligand or the counterion). The active radical adds some monomer units as in classical free-radical polymerization until it suffers a deactivation reaction. (Fig. 4).

The main role of the initiator is to determine the number of polymer chains. Termination between active radicals is minimized by maintaining a low concentration of active chains. In a well-controlled process, <5% of the polymer chains suffer termination. The molecular weight is determined by the ratio of consumed monomer and the initiator. Well-defined polymers with molecular weights between 1000 and 150,000 g/mol have been obtained, but termination and chain transfer make it difficult to obtain longer polymers of well-defined microstructure. A challenge for the ATRP process is the removal and recycling of the catalyst, which has not been adequately achieved yet. Monomers with substituents that can be stabilized by resonance of the radical formed from the monomer [(eg, styrenes (42), (meth)acrylates (43,44), (meth)acrylamides (45), and acrylonitrile (46)] can be successfully polymerized by means of ATRP. Other monomers like olefins, halogenated alkanes, and vinyl acetate have not been polymerized yet using ATRP. On the other hand, acidic monomers have not been polymerized

Initiation:

$$R\text{-}X + M_t^n\text{-}Y / \text{Ligand} \underset{k_{\text{-add}}}{\overset{k_{\text{add}}}{\rightleftharpoons}} R\text{*} + X\text{-}M_t^{n+1}\text{-}Y / \text{Ligand}$$

$$\left(+ m\,M \qquad R_m \right) \Bigg\} \longrightarrow R_m X + M_t^n\text{-}Y / \text{Ligand}$$

Propagation:

$$R_n\text{-}X + M_t^n\text{-}Y / \text{Ligand} \underset{k_{\text{-add}}}{\overset{k_{\text{add}}}{\rightleftharpoons}} R_n + X\text{-}M_t^{n+1}\text{-}Y / \text{Ligand}$$

$$(+ M)$$

Termination:

$$R_n + R_m \xrightarrow{k_t} \text{dead polymer}$$

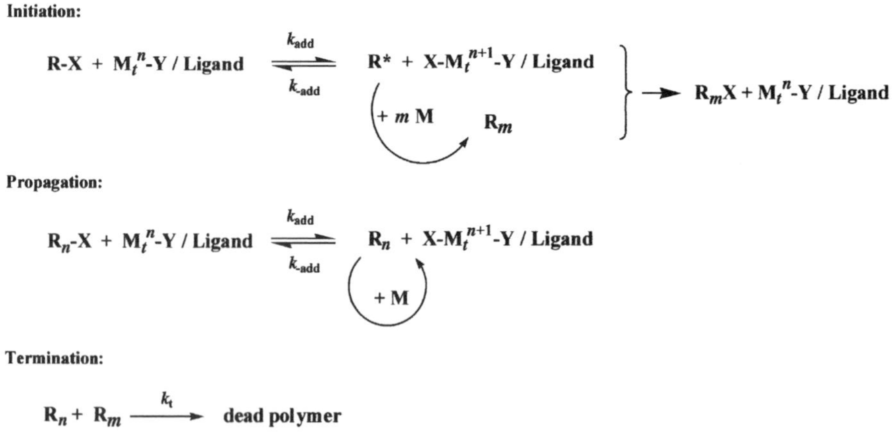

Fig. 4.   Scheme of a transition metal catalyzed ATRP.

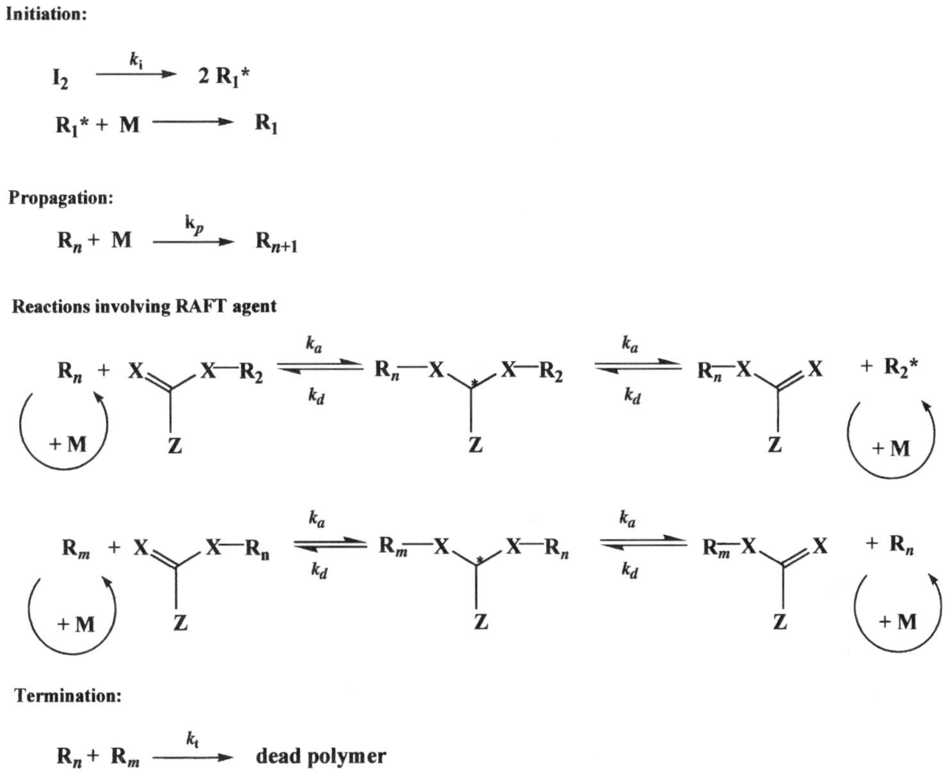

Initiation:

$$I_2 \xrightarrow{k_i} 2\,R_1^*$$

$$R_1^* + M \longrightarrow R_1$$

Propagation:

$$R_n + M \xrightarrow{k_p} R_{n+1}$$

Reactions involving RAFT agent

Termination:

$$R_n + R_m \xrightarrow{k_t} \text{dead polymer}$$

Fig. 5. Scheme of a RAFT process.

using this method because they can protonate ligands and form the corresponding salt. Nevertheless, the polymerization of the neutral salt of the acidic monomers is possible (47).

RAFT. This process is performed in the presence of certain dithio compounds (eg., dithiobenzoates), which act as highly efficient reversible addition–fragmentation chain transfers and provide the polymerization with living characteristics. The polymerization proceeds according to the scheme in Figure 5.

RAFT has a number of advantages as a synthetic method because it is applicable to a wide range of monomers (48), including many unreactive monomers, such as vinyl acetate (49). Since radicals are generated in a conventional way (eg, by means of a conventional initiator), it is tolerant of impurities. Polymerization conditions are typical of those used for classical free-radical polymerization and the process can be tailored to ambient temperatures (50). On the other hand, retardation may occur when the RAFT agent is used in high concentrations to produce low molecular weight polymers (51).

Some Common Kinetic Features of the CRP Processes. The key feature of the CRP processes is that molecular termination is minimized by maintaining the concentration of active radicals at a low value by establishing a rapid equilibrium between this small fraction of active radicals and a large majority of dormant polymer chains. The extent of bimolecular termination can be lowered by reducing the concentration of active chains, but at the expense of a longer process time. Therefore, there is a practical lower limit in the concentration of active chains. The relative influence of both bimolecular and monomolecular (eg, chain transfer to monomer) termination events on the MWD increases as longer polymer chains are produced. This represents a practical limit for the maximum molecular weight that can be achieved under controlled conditions, which currently is in the range of 150,000–200,000 g/mol (37,41).

Controlled radical polymerizations (NMP, ATRP, and RAFT) can be carried out in bulk, solution, and suspension polymerization. Mass transfer limitations of the radical trapping agent make difficult the implementation in conventional emulsion polymerization (52,53). This limitation has been overcome working in miniemulsion polymerization (51,54–60). In CRP, the reactivity ratios of the monomers are the same as in conventional free-radical polymerization (37,41). In addition, the reactivity ratios of the macromonomers are close to those of the monomers with the same polymerizable group, in contrast with classical free-radical polymerization where the reactivity of the macromonomers is substantially lower than that of the corresponding monomers. This may be due to the longer characteristic times for monomer and macroradical addition (61).

The chemical composition distribution of the chains formed by CRP considerably differs from that of the polymer chains produced in the classical free-radical polymerization. In a batch controlled process in which monomers with different reactivity ratios are polymerized, all polymer chains have a very similar chemical composition distribution (CCD) with a composition gradient along the chain. In the corresponding classical free-radical polymerization, the CCD of the chains formed at the beginning of the polymerization may greatly differ from that of the chains formed at the end of the process, and the composition along each chain is constant.

It is important noticing that the concepts of instantaneous and cumulative copolymer composition in CRP are different from those in classical free-radical polymerization. In CRP the instantaneous copolymer composition, which can be calculated with equations 4 or 12, refers to the composition of the part of the chains that is being formed in a given moment. In classical free-radical polymerization, the instantaneous copolymer composition is the composition of the chains being formed in a given moment. On the other hand, the cumulative copolymer composition in a CRP represents the average composition of the fraction of each chain formed up to the time considered, whereas in classical free-radical copolymerization, it is the average composition of all chains formed up to this time.

5. Anionic Copolymerization

Anionic polymerization is not spontaneous and requires the presence of initiators that provide the initiator anions. Anions can only attack those monomers whose electrons can be moved in such a way that a monomer anion results. Therefore, anionically polymerizable monomers should contain electron-accepting groups. These includes styrene, acrylic monomers, some aldehydes and ketones, and cyclic monomers such as ethylene oxide and other oxiranes, N-carboxy anhydrides, glycolide, lactams, and lactones that can be polymerized by ring-opening polymerization. The kinetic scheme of an anionic copolymerization may be summarized as follows:

Initiation:

$$RM \rightarrow R^- + M^+$$

$$R^- + M_1 \rightarrow R1_1^-$$

$$R^- + M_2 \rightarrow R2_1^-$$

Propagation:

$$R1_n^- + M_1 \xrightarrow{k_{p11}} 1_{n+1}^-$$

$$R1_n^- + M_2 \xrightarrow{k_{p12}} 2_{n+1}^-$$

$$R2_n^- + M_1 \xrightarrow{k_{p21}} 1_{n+1}^-$$

$$R2_n^- + M_2 \xrightarrow{k_{p22}} R2_{n+1}^-$$

$$(26)$$

This kinetic scheme does not include termination because in purified systems, most macroanions grow until all of the monomer present in the reactor is polymerized. Such polymerization is called living polymerization, because if additional monomer is added into the reactor the polymer chains undergo further growing. A characteristic of the living polymerization is that, provided that initiation is quick enough, all polymer chains grow to a similar extent yielding very narrow molecular weight distributions. In addition, block copolymers can be produced by adding a second monomer once the first one has completely reacted. Triblock and multiblock copolymers can be prepared by subsequent additions of different monomers. Also, graft, star, and hyperbranched polymers can be obtained by means of this technique by simply using suitable initiation systems (62).

The concept of reactivity ratios applies also to anionic copolymerization. However, the production of statistical copolymers in anionic polymerization is often difficult because the macroanions substantially differ in polarities yielding to widely different reactivity ratios, usually $r_1 \gg 1$ and $r_2 \ll 1$, which means that block copolymers are mainly produced. Typical reactivities encountered in anionic copolymerization are shown in Table 7 (63).

Reactivity ratios are affected by the solvent. Thus, Table 7 shows that styrene is in general less reactive than diene monomers except when tetrahydrofuran (THF) is used as a solvent, in which case the reactivity changes and styrene becomes more reactive. Statistical copolymerization of styrene and dienes with more polar monomers, which have much higher electronegativities, is unlikely to take place. For example, the copolymerization of styrene and methyl methacrylate only yields block copolymers. In general, only monomers with fairly small differences in electronegativities can be successfully copolymerized, and even in these cases the tendency is toward forming block copolymers.

Table 7. **Reactivity Ratios in Anionic Copolymerization**[a]

M_1	M_2	Solvent	Counterion	Temp°C	r_1	r_2
styrene	butadiene	hexane	Li$^+$	25	0.03	12.5
styrene	butadiene	benzene	Li$^+$	25	0.04	10.8
styrene	butadiene	THF	Li$^+$	−35	8	0.2
styrene	isoprene	benzene	Li$^+$	30	0.14	7.0
styrene	isoprene	cyclohexane	Li$^+$	40	0.046	16.6
styrene	isoprene	THF	Li$^+$	−35	40	0
isoprene	butadiene	hexane	Li$^+$	40	0.3	2.0
styrene	p-methylstyrene	benzene	Li$^+$	30	2.5	0.4
styrene	p-methylstyrene	THF	Li$^+$	0	1.3	0.9
styrene	p-methylstyrene	THF	Na$^+$	0	2	0.4
styrene	1,1-diphenylethylene	benzene	Li$^+$	30	0.7	0
isoprene	1,1-diphenylethylene	THF	Li$^+$, Na$^+$, K$^+$	0	0.1	0

[a] Ref. 63.

6. Cationic Polymerization

Cationic polymerization presents many similarities with anionic polymerization. Cationic initiators formed from carbocation salts, Brønsted acids, or Lewis acids, react with monomer to give monomer cations that upon addition of more monomer become macrocations. Both monomer cations and macrocations are fairly reactive and may react with the counterions instead of with monomer. Therefore, counterions should not be too nucleophilic. The nucleophilicity of the counterions depends on the solvent, and hence only a few solvents, including benzene, nitrobenzene, and methyl chloride, are suitable for cationic polymerization. Monomers suitable for cationic polymerization should have electron-donating groups: (a) olefins CH$_2$=CHR with electron-rich substituents, (b) compounds R_2C=Z with heteroatoms or hetero groups Z, and (c) cyclic molecules with heteroatoms as part of the ring structure. Although, there are many more cationically polymerizable monomers than anionically polymerizable ones, relatively few cationic polymerizations are performed industrially because macrocations are highly reactive and prone to suffer termination and chain-transfer reactions.

Thus, the cationic polymerization of alkenyl monomers suffers from the transfer reactions that affect the stability of the active intermediate species. Therefore, in order to minimize these transfer reactions and to produce high molecular weight polymers, very low temperatures must be used. When heterocyclic monomers are employed, higher temperatures can be used although care must be taken with the ceiling temperature that promotes depropagation in these cases.

A compilation of reactivity ratio data for styrene-like monomers are given in (62). The main trends are as follows: (a) copolymerization of styrene (A) with less basic monomers (B, obtained by adding an electron-withdrawing group, eg, p-halostyrene) results in r_A >1 and r_B; (b) copolymerization of styrene with more basic monomers (eg, α-methylstyrene and p-methylstyrene) results in r_A >1 and r_B >1. These trends can be altered by steric effects. When other nonstyrenic monomers are employed in the copolymerization it is difficult to establish clear trends. In addition, variables such as solvent polarity, temperature, and counterion effects may significantly affect the reactivity ratios, but it is not straightforward to determine the extent of the effect (64). In the copolymerization of heterocycle monomers, care should be taken when using the Mayo-Lewis equation to calculate reactivity ratios because this equation cannot be used if depropagation reactions are relatively important (see the section Copolymerization with Depropagation). Because of the living character of the polymerization, well-defined cationic polymerization is, together with anionic polymerization, one of the typical routes to prepare block and graft copolymers. In addition, it allows the preparation of prepolymers that can be further activated to generate branched and graft copolymers (65–69).

7. Ring-Opening Copolymerization

Ring-opening copolymerization (ROP) consists in opening a cyclic compound to produce a linear polymer as shown in the following scheme:

$$C_n X \rightarrow (C_n - X)_p \tag{27}$$

The functional groups denoted as X include $-CH=CH-$ and those that consist of one or two heteroatoms such as O, N, S, P, and Si. Ring-opening polymerization of heterocyclic monomers are classified depending on the nature of the propagating species and monomers (70). Ring-opening polymerization can proceed via (a) electrophilic propagating species, mainly cationic polymerization; (b) nucleophilic propagating species, mainly anionic polymerization; (c) zwitterion intermediates, in which the propagating chain and the monomer bear both cationic and anionic species. This type of polymerization is always catalyzed by an initiator (protonic acids including BF_3, $AlCl_3$, $AlBr_3$, and $SnCl_3$; tryalyloxonium salts; carbenium salts; and others) at very low temperatures to avoid depropagation.

Ionic ring-opening polymerization of heterocyclic compounds, such as ethylene oxide, THF, ethyleneimine, β-propiolactone, and caprolactam as well as the Ziegler-Natta ring opening of cyclic alkenes, such as cyclopentene and norbornene, are well known, but free-radical ROP is rather rare (71–73). An example of cyclic monomers that copolymerize with common monomers by radical ROP are the cyclic ketene acetals (74,75). The main deficiency of these monomers is their rather low reactivity to monomers such as styrene. Thus, for the copolymerization of 2-methylene-1,3-dioxepane and styrene in bulk at 120°C, the reactivity ratios are $r_{\text{ketene}} = 0.021$ and $r_{\text{st}} = 22.6$.

Regarding copolymerization, this technique is predominantly employed to produce block and also graft copolymers. In these cases, the concepts discussed for the copolymerization via cationic and anionic polymerization are also applicable here.

8. Catalytic Polymerization

Free-radical, anionic, and cationic polymerizations proceed by addition of monomer units to the active end of the growing polymer chain that in the course of polymerization separates from the bound initiator fragment. Catalytic polymerizations proceed by an insertion mechanism, in which the monomer units are inserted between the catalytic site and the growing polymer chain. Over 40% of the yearly polymer production is obtained by catalytic polymerization (76). This includes linear low density polyethylene (LLDPE) and high density polyethylene (HDPE), which are copolymers of ethylene and α-olefines (with decreasing amounts of α-olefines as a higher density is sought); polypropylene and high impact polypropylene; and EPDM elastomers (terpolymers of ethylene, propylene, and a nonconjugated diene, eg, 5-ethylidene-2-norbornene). Catalytic polymerizations are carried out using Ziegler-Natta, transition-metal and metallocene catalysts.

8.1. Ziegler-Natta Catalysts.   A typical commercial Ziegler-Natta catalyst for ethylene polymerization is produced by reacting $TiCl_4$ with a finely divided $MgCl_2$ stabilized with an electron donor (internal donor; eg, phthalate ester, diethers, and succinates) and activating the resulting system with trialkylaluminium compounds. In propylene polymerization, the addition of another electron donor (external donor, eg, alkoxysilanes) is usually needed to achieve high stereospecifity (76).

There is some debate about the mechanisms involved in the growth of polymer chains on the active center. The most widely accepted mechanism is that proposed by Cossee (77) in which propagation occurs at the transition-metal–alkyl bond (Fig. 6). The first step of this mechanism is the complexation of the olefin in a vacant coordination site followed by migration of the polymer chain and formation of a new metal–carbon bond. An exchange of the alkyl groups and the vacancy is needed to account for the formation of stereoregular polymers (78).

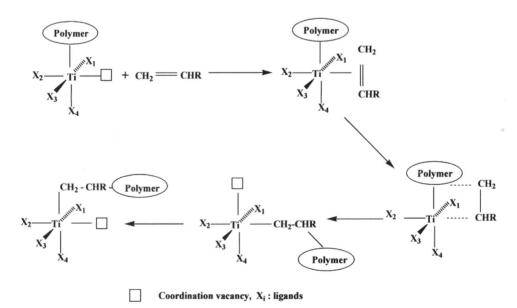

□   Coordination vacancy, X_i : ligands

Fig. 6.   Cossee's mechanism of monomer insertion.

Copolymerization adds complexity to the propagation step because four different reactions are involved

$$\text{catalyst}-M_1-\text{Polymer} + M_1 \xrightarrow{k_{p11}} \text{catalyst}-M_1-M_1-\text{Polymer}$$

$$\text{catalyst}-M_1-\text{Polymer} + M_2 \xrightarrow{k_{p12}} \text{catalyst}-M_2-M_1-\text{Polymer}$$

$$\text{catalyst}-M_2-\text{Polymer} + M_1 \xrightarrow{k_{p21}} \text{catalyst}-M_1-M_2-\text{Polymer}$$

$$\text{catalyst}-M_2-\text{Polymer} + M_2 \xrightarrow{k_{p22}} \text{catalyst}-M_2-M_2-\text{Polymer}$$

(28)

By considering this scheme, the instantaneous copolymer composition is (79)

$$Y_1 = \frac{dM_1/dt}{dM_1/dt + dM_2/dt} = \frac{1 + r_1([M_1]/[M_2])}{2 + r_1([M_1]/[M_2]) + r_2([M_2]/[M_1])}$$

$$Y_2 = 1 - Y_1$$

(29)

Note that the similarity of this equation with that giving the instantaneous copolymer composition for the terminal model of the classical free-radical copolymerization (eq. 12). However, in this case, $[M_i]$ is the concentration of monomer i at the active center, which due to mass transfer limitations may be different from that of the feed (80). On the other hand, the reactivity ratios, r_i, are a measure of the tendency for a comonomer to show preference for insertion into a growing chain in which the last inserted unit was the same, rather than the other comonomer. The effect of the values of the reactivity ratios on the chemical composition distribution is as in classical free-radical polymerization (the section Terminal Model).

Ziegler-Natta catalysts present multiplicity of active centers (81). Each centre may have a different activity, and hence different reactivity ratios. Therefore, the average copolymer composition of chains of different lengths is different, namely, there is a correlation between chain molecular weight and chain composition (Fig. 7). In addition, the catalytic activity changes with time. Consequently, the reaction scheme and the copolymerization equation may require modification to account for these features of the Ziegler-Natta catalysts. Reactivity ratios depend on operation conditions [eg, temperature (82), reactant concentrations (83,84), nature of the catalyst (85), and comonomer structure (86)].

8.2. Transition-Metal Catalysts. The Phillips catalyst discovered by Hogan and Banks (87) is the most common example of these catalysts. The Phillips catalysts encompass two families of supported chromium catalysts (88): (a) organochromium compounds, and (b) chromium oxide. Suitable supports include silica, aluminophosphates, and silica–titania. Polymerization of ethylene on these catalysts yields a linear polymer (HDPE). Copolymerization of ethylene and α-olefins yields linear polymers with short branches. The crystallinity of the copolymer, and consequently its density decreases as the copolymer content in α-olefins increases. Phillips catalysts are able to produce a wide range of polyethylenes (from HDPE to LLDPE). The chromium content of the catalyst is in the

Ziegler-Natta	Metallocene

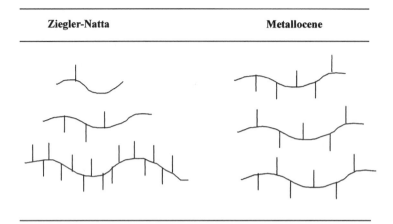

Fig. 7. Comparison between the polymers obtained using Ziegler-Natta catalysts and metallocenes.

range of 0.2–1 wt%, although only a small fraction of the chromium is active. The Phillips catalyst presents a distribution of actives sites leading to broad molecular weight distribution (89). The polymerization mechanism is still a matter of debate, although theoretical calculations (90) suggest that propagation may occur through the Cossee mechanism (77).

8.3. Metallocene Catalysts. Metallocenes are organometallic compounds that consist of a metallic atom coordinated with two cyclopentadienyl rings. Figure 8 shows the general structure of the homogeneous metallocene catalysts, where M is a transition metal, usually from Group 4 (IVB); B is a bridge formed by one or several atoms (usually C or Si); R is hydrogen, an alkyl group or any hydrocarbon group; and X is a halogen (usually Cl) or an alkyl group. Although metallocene catalyst are new in commercial applications, the first attempts to use metallocenes as catalysts for olefin polymerization were carried out in the 1950s. Natta and co-workers (91) and Breslow and Newburg (92) used metallocene activated with alkyl aluminium as catalysts in ethylene polymerization, but these systems showed a low activity and fell into oblivion until Sinn and Kaminsky (93) discovered that metallocenes activated with methyl aluminoxane (MAO) showed a high activity in olefin polymerization. Several reviews on metallocene catalysts are available (94–98). Metallocenes are single center catalysts and represent the most versatile way of synthesizing almost any kind of stereoregular polymer. Polymers with entirely new properties may be produced by varying the components (type of metal and its substituents, type and length of the bridge, and substituents of the rings) of the metallocene molecule (99,100). Metallocenes are single active center catalysts, and hence the composition of all copolymer chains is statistically the same, and independent of chain length (Fig. 7). This opens the possibility for a much tighter control of copolymer composition. In addition, metallocene catalysts exhibit higher reactivity for α-olefines than the conventional Ziegler-Natta catalysts (98). Figure 7 also illustrates that the molecular weight distribution of the polymer produced with metallocene catalysts is narrower than that obtained with Ziegler-Natta catalysts. For a given

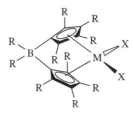

Fig. 8. General structure of homogeneous metallocene catalysts.

monomer system, the catalyst activity and the reactivity ratios strongly depend on the structure of the metallocene molecule (101), MAO/metallocene ratio (102) and temperature (102) among other variables.

The metallocene of Figure 8 is soluble in organic solvents, and hence can be directly used in solution copolymerization. However, soluble metallocenes present some serious drawbacks. First, high MAO/metallocene ratios are required to obtain adequate activity levels and stereochemical control. MAO is expensive and significantly increases production costs. In addition, most of polyolefins are produced by slurry and gas-phase processes, which need heterogeneous catalysts. Three main methods have been used to prepare supported metallocene catalysts (103): (1) direct impregnation of the metallocene on the support followed by treatment with MAO (104); (2) impregnation of the metallocene on a support previously treated with MAO (105); and (3) chemical anchoring of the metallocene on a modified support followed by treatment with MAO (106). In comparison with the homogeneous counterparts, the main advantages of the supported metallocenes are (103): (a) lower MAO/metallocene ratio; (b) higher molecular weights; (c) higher isotacticity for polypropylene, and (d) usable in slurry and gas processes. On the other hand, the supported metallocenes have a lower activity than the homogeneous ones and yield broader molecular weight distributions, likely because the interaction between the metallocene and the support leads to active centers with a distribution of catalytic activities.

Copolymerization of ethylene and α-olefins is used in the manufacture of the high tonnage LLDPE. In this process, metallocene catalysts exhibit higher reactivity for α-olefins than conventional Ziegler-Natta catalysts (98). The reactivity ratios of ethylene and α-olefins can be modified in a wide range by varying the structure of the metallocene catalyst. This allows tailoring the monomer sequence distribution in the copolymer and even alternating ethylene/α-olefin copolymers have been obtained for both long (1-octene) (107) and short (propylene) (108) α-olefins. Incorporation of styrene into ethylene/styrene copolymers is also easier with metallocene catalysts as compared with Ziegler-Natta catalysts (109). Highly syndiotactic poly(propylene-*co*-α-olefin) is one of the new polymers that can be synthesized by using metallocene catalysts (110).

There is some controversy about the mechanisms involved in polymer chain growth, and insertion mechanisms inspired in that proposed by Cossee (77) for Ziegler-Natta catalysts have been proposed (111,112). Production of EPDM by terpolymerization of ethylene/propylene/diene using metallocene catalysts has been reported (113).

One of the factors that is retarding the introduction of metallocene-based polyethylenes is the difficulties in processing caused by the narrow MWD (114). Processability is substantially improved in the case of polyethylene with long branches produced by means of the constraint-geometry catalyst (115,116). This catalyst has been used in the copolymerization of ethylene and styrene with a good incorporation of styrene (117).

9. Step-Growth Copolymerization

Step-growth polymerization proceeds by reaction of the functional groups of the reactants in a stepwise manner: monomer reacts to form dimer; monomer and dimer react to form trimer; dimer and trimer form pentamer, and in general

$$n\text{-mer} + m\text{-mer} \rightarrow (n+m)\text{-mer} \tag{30}$$

In this process, the molecular weight of the polymer continuously increases with time and the formation of polymer with sufficiently high molecular weight for practical applications requires very high conversions of the reactive groups (>98–99%). This requirement poses stringent conditions to the formation of poly-mers by polycondensation, such as the necessity for a favorable equilibrium and the absence of side reactions. In spite of these difficulties, a number of different chemical reactions may be used to synthesize materials by step polymerization. These include esterification, amidation, carbonate bond formation, the formation of urethanes, and aromatic ether and aromatic ketone bond formation among others.

All step polymerizations (including copolymerizations) fall into two groups depending on the type of monomer(s) employed. The first one implies the use of at least two bifunctional and/or polyfunctional monomers, each one possessing a single type of active group. In this context, a bifunctional monomer is a monomer with two functional groups per molecule and a polyfunctional monomer is a monomer with more than two functional groups. The monomers involved in this type of reaction are often represented as A–A and B–B, where A and B are the different reactive groups. An example of this reaction is the formation of polyesters from diols and diacids.

$$m\,\text{HOOC-R}_1\text{-COOH} + m\,\text{HO-R}_2\text{-OH} \rightleftharpoons$$
$$\text{HO}\text{+OC-R}_1\text{-COO-R}_2\text{-O}\text{+}_m\text{H} + 2(m{+}1)\,\text{H}_2\text{O} \tag{31}$$

In order to form a copolymer, at least three monomers are needed (eg, two diols and one diacid or one diol and two diacids). The second type of step polymeriza-tion involves the use of monomers with different functional groups in the same molecule, A-B type. To produce a copolymer by this step polymerization at least two A–B type monomers must be used (eg, two amino acids).

In comparison with homopolymers, the structure of the step copolymers contains two repeating units as it is shown in the following schemes:

$$-\text{CO}-\text{R}_1-\text{CONH}-\text{R}_2-\text{NHCO}-\text{R}_3-\text{CONH}-\text{R}_4-\text{NH}- \tag{32}$$

$$-\text{NH}-\text{R}_1-\text{CO}-\text{NH}-\text{R}_2-\text{CO}- \tag{33}$$

Structure 32 can be synthesized by using two diacids and two diols, and structure 33 by using two different aminoacids.

Different types of copolymers, as those discussed in chain-growth polymerization, are possible depending on the arrangement of the repeating unit on the chain. Thus, alternating, statistical (random) and block copolymers can be formed by step-growth copolymerization.

9.1. Kinetics of Step-Growth Copolymerization.

The kinetic analysis of the copolycondensations of two different diol diesters of dibasic diacids (118) can be used as an example of other copolycondensation kinetics. For this case and assuming that the hydroxyl chain ends attack the terminal ester groups of the chain through irreversible reactions (viz, it is assumed that the diols formed in the process are continuously removed), the following kinetic scheme can be written:

$$-COOR_1OH + HOR_1OOC- \xrightarrow{k_{11}} -COOR_1OOC- + HOR_1OH \tag{34}$$

$$-COOR_1OH + HOR_2OOC- \xrightarrow{k_{12}} -COOR_2OOC- + HOR_1OH \tag{35}$$

$$-COOR_2OH + HOR_1OOC- \xrightarrow{k_{21}} -COOR_1OOC- + HOR_2OH \tag{36}$$

$$-COOR_2OH + HOR_2OOC- \xrightarrow{k_{22}} -COOR_2OOC- + HOR_2OH \tag{37}$$

Reactions 35 and 36 are cross-reactions while reactions 34 and 37 are homoreactions. If $[OH]_1$ and $[COO]_1$ and $[OH]_2$ and $[COO]_2$ are the concentrations of hydroxyl and ester groups on chains 1 and 2, respectively, the formation rates of the diols (HOR_1OH and HOR_2OH) are given by

$$\frac{d[HOR_1OH]}{dt} = k_{11}[COO]_1[OH]_1 + k_{12}[COO]_1[OH]_2 \tag{38}$$

$$\frac{d[HOR_2OH]}{dt} = k_{21}[COO]_2[OH]_1 + k_{22}[COO]_2[OH]_2 \tag{39}$$

At the initial stage of the process, the following condition is satisfied

$$\frac{[monomer\ 1]}{[monomer\ 2]} = \frac{[OH]_1}{[OH]_2} = \frac{[COO]_1}{[COO]_2} = a \tag{40}$$

Dividing equations 38 and 39; rearranging the terms using the condition of equations 40 and considering $b = d[HOR_1OH]/d[HOR_2OH]$, the following equation is obtained

$$ak_{11} - \frac{b}{a}k_{22} = bk_{21} - k_{12} \tag{41}$$

The values of k_{11} and k_{22} can be obtained from homopolycondensation kinetics (118,119). If the left-hand term of equation 41 is plotted against b, the slope and intercept give the rate constants k_{21} and k_{12} of the cross-reactions. In a similar way, the rate of incorporation of the diols is

$$-\frac{d[-OR_1O-]}{dt} = k_{21}[COO]_2[OH]_1 + k_{11}[COO]_1[OH]_1 \tag{42}$$

$$-\frac{d[-OR_2O-]}{dt} = k_{12}[COO]_1[OH]_2 + k_{22}[COO]_2[OH]_2 \tag{43}$$

and setting diol ratios in the copolymer as c, dividing equation 42 by equation 43, and rearranging leads to

$$c = \frac{d[-OR_1O-]}{d[-OR_2O-]} = a \cdot \frac{k_{21} - ak_{11}}{ak_{12} + k_{22}} \tag{44}$$

Equation 44 allows us to built the copolymer composition diagram by calculating the ratio of diols in the polymer, c, for a given monomer ratio, a. This is equivalent to the analysis carried out in the section on the Terminal Model for the chain-growth polymerization that led to the Mayo-Lewis equation (eq. 12) (12). Han (118) demonstrated that the assumptions made to obtain equations 41 and 44 were in good agreement with the low conversion experimental data obtained in the copolycondensation of bis(2-hydroxy ethyl) terephthalate (HET) and bis(2-hydroxy-1-propyl) terephthalate (HDT) carried out in bulk at 160°C. This polycondensation obeys a second-order kinetics with respect to the concentration of ester and hydroxyl groups and the copolymer composition diagram shows an azeotrope composition at 29.5 mol% of HET.

Other systems that were accounted for by equations 41 and 44 are the polyamides synthesized by amide interchange reactions [N,N'-bis(2-aminoethyl)decandiamide (AES) and N,N,N', N'-bis(diethyleneiminodecandiamide) (DEIS)], and polyester amides synthesized by copolycondensation of a diamine amide of a diacid [AES and bis(2-hydroxyethyl) decandioate (HES)], although for those cases no azeotrope was obtained in the composition diagram (119,120). Table 8 shows the reactivity ratios obtained for these three systems.

In the works discussed above (118–120), the concept of equireactivity, namely, that all functional end-groups attached to the same residue have the

Table 8. Reactivity Ratios and Homoreaction Rate Constants of Different Copolycondensations Carried Out in Bulk[a]

Comonomer system	Temperature °C	Homocondensation rate constants kg/mol h	Reactivity ratios
HET (1)	160	1.69	$r_1 = 4.23$
HDT (2)		1.39	$r_2 = 1.70$
AES (1)	200	3.01	$r_1 = 1.30$
DEIS (2)		1.70	$r_2 = 0.59$
AES (1)	200	3.01	$r_1 = 10.03$
HES (2)		34.70	$r_2 = 0.19$

[a] Refs. (118–120).

same reactivity, was used. However, the penultimate effect can be important in some copolycondensation systems such as in phosgene, chloroformates, and phthalates (121).

Turska and co-workers (122–125) found that the composition of the copolymer varied during the solution copolycondensation of terephthaloyl chloride, 2,2-bis(4-hydroxyphenyl) propane and 2,2-bis(3,5-dichloro-4-hydroxyphenyl)propane in α-chloronaphthalene at 220°C. Theoretical considerations (123) demonstrated that it is possible to control the average composition of the copolymer by manipulating the reaction conditions for the case of three monomers. Moreover, these authors were able to show that if the homopolycondensations of monomer 1 with monomers 2 and 3 satisfy the Arrhenius dependence with temperature and have different activation energies, there is a temperature at which the rate constants are equal to each other. This temperature is called isokinetic temperature, and at this temperature, the composition of the copolymer is independent of the extent of the reaction. This was experimentally verified for the copolymerization of phenolphthalein, dichloro-3,3′-bisphenol A and terephthaloyl dichloride, which has a isokinetic temperature of 242°C (124). Boryniec (126) has developed kinetic equations for the copolycondensation of three bifunctional monomers when there is a change in reactivity of the intermonomer after reaction of the first functional group. The same author (127) has developed kinetic equations for the copolycondensation of a bifunctional monomer with two other monomers not reacting with each other considering invariable reactivity of the functional groups.

10. Random Copolymers

Random copolymers are by far the largest class of copolymers made and used today. They are mainly produced by catalytic polymerization and by classical free-radical polymerization. Some copolymers are also produced by ionic polymerization.

10.1. Catalytic Random Copolymers. Polyolefins produced by catalytic polymerization account for more than one-third of the yearly production of synthetic polymers (128). Polyolefins are made from a surprisingly short list of monomers, mainly ethylene and propylene, but also α-olefins (1-butene, 1-hexane, 1-octane), isobutylene and a few other monomers. These monomers are produced in large-scale petrochemical units, and hence they are available in large quantities at low cost. Polyolefins are extremely versatile materials with properties ranging from elastomers to thermoplastics to high strength fibers (129). One of the key characteristics of the polyolefins is crystallinity, which depends on the regularity of the chemical structure of the polymer chains. Highly regular polymers such as linear polyethylene and isotactic polypropylene are crystalline polymers, whereas some copolymers of ethylene–propylene and ethylene–α-olefin are amorphous materials.

Catalyst development and polymerization technology have made possible the continuous expansion of the properties of the polyolefins (130). The development of high mileage highly selective Ziegler-Natta catalysts allowed a control of the polymer microstructure (composition, CCD, MWD, tacticity, etc.) and polymer macrostructure (morphology, phase distribution, crystallinity) not achieved previously (76). This trend has been reinforced with the development of metallo-cene catalysts (131).

Polyolefins are mainly produced by means of particulate polymerization processes: gas-phase and slurry polymerizations. In these processes, the polymerization takes place inside each polymer particle and the properties of the final product strongly depend on the way in which the particle grows. The scientific understanding of the polymer growth mechanisms is critical to control both polymer microstructure and particle morphology, and to achieve the ultimate goal of improving and expanding the polymer properties envelope toward new specialty materials (132).

The main olefin copolymers are LLDPE, some grades of HDPE, ethylene–propylene elastomers, propylene copolymers, and high impact polypropylene (hiPP).

Copolymerization of ethylene and α-olefins lead to LLDPE. The crystallinity, and consequently the density of the polyethylene, is controlled by the amount and type of α-olefin incorporated to the backbone. Decreasing contents of α-olefin lead to higher densities. Thus, HDPE is produced at low or nil α-olefin content. On the other hand, ultralow density polyethylene is produced by copolymerizing ethylene and octane (eg, Attane, Dow Chemical). Comparing with low density polyethylene (LDPE) produced by free-radical polymerization, LLDPE exhibits a higher melting point (objects can be used at higher temperatures), is stiffer (thinner walls can be used), and presents higher tensile and impact strengths (more resistant films). However, the processability of the LLDPE is worse than that of the LDPE. A way of improving processability is to produce bimodal molecular weight distributions. This can be achieved by using dual-site catalysts such as a metallocene supported on a Ziegler-Natta catalyst (133), two different metallocenes on the same support (134), and a single metallocene activated with a mixture of two cocatalysts (135). Another alternative is to use two reactors in series producing different molecular weights in each reactor by using different hydrogen concentrations (hydrogen is an efficient chain-transfer agent) (136). In this case, the polymer particle consists of a blend of polymers of different molecular weights, and when there is a large difference in molecular weights, the effective mixing of the polymer fractions during melt processing may be limited. This limitation may be overcome by means of the "multizone circulating reactor" technology (137,138). This is a loop-like reactor in which different reaction conditions are maintained in each leg of the loop. Continuous circulation of the polymer particles allows the production of intimate blends of polymer chains having different molecular weights and/or compositions. It is claimed that this technology significantly expands the properties envelope (139).

LLDPE is mainly used in film applications with smaller markets in injection molding and wire-cable. Some important producers of LLDPE are Basell (Lupolex), Dow Chemical (Dowlex), ExxonMobil (LL grades), BP Solvay Polyethylene (Innovex), Chevron Phillips Chemicals (Marflex), Equistar Chemicals (Petrothene), Borealis (Borstar) and Atofina. Metallocene-based grades (mLLDPE) are also available in the market (Exceed, ExxonMobil; mPact, Chevron Phillips Chemicals; Boracene, Borealis; Evolue, Mitsubitsi Chemicals; Luflexen, Basell; Umerit, Ume Chemicals). Compared to conventional grades, mLLDPE provides a higher schock resistance (downgauged films), lower content of extractables (better organoleptic properties), and better optical properties (clarity and gloss). However, the processability of the mLLDPE is in general worse than that of the conventional LLDPE. Long branching mLLDPE produced with constraint geometry catalysts (115,116) exhibit LDPE-like processability with LLDPE-like performance (Elite, Dow Chemical).

Ethylene/propylene elastomers (EPM) are copolymers of ethylene and propylene with intermediate levels of each comonomer. These materials are completely amorphous and rapidly recover its shape after removal of a strain of at least 50%. In order to make these elastomers cross-linkable a nonconjugated diene is introduced in the formulation. These products are called EPDM for ethylene, propylene, diene monomer. Ethylidene norbornene and 1.4-hexadiene are common choices for the diene monomer. EPDMs are produced by using both Ziegler-Natta and metallocene catalysts. Some producers are DSM (Keltan), DuPont-Dow Elastomers [Nordel IP (metallocene)], ExxonMobil (Vistalon), and Enichem (Dutral). EPDMs are used in construction (roof sheeting, insulation sponge, seals, hoses-tubes, reservoir linings), automotive (sealing systems, hoses), impact modification of plastics, and wire and cable.

In propylene–ethylene copolymers with a low ethylene content, the insertion of ethylene in the propylene chain reduces the regularity of the chain leading to a semicrystalline polymer. The properties of these materials, called propylene random copolymers, depend not only on the total ethylene content, but also on the monomer sequence distribution (140). Other monomers such as 1-butene and 1,3-butadiene may be incorporated (141). Some producers are Basell (Moplen), Dow Chemical, Equistar (Petrothene), and ExxonMobil (PP9000 Series). The main uses are cast film and rigid and flexible packing.

High impact polypropylene is a multiphase material in which an ethylene–propylene elastomeric phase is finely dispersed in an isotactic polypropylene matrix. In principle, hiPP can be produced by blending iPP and the elastomeric copolymer, but it is simpler and more versatile in terms of product properties achievable to produce the heterophase material during polymerization. Thus, in the Spheripol process (Basell, 142) isotactic polypropylene particles are first produced in a slurry of liquid propylene, and then, the particles are transferred to a gas-phase fluidized bed reactor where the elastomeric material is produced by polymerizing an ethylene-propylene monomer mixture. Figure 9 illustrates the evolution of the particle morphology in this process. In the first reactor, the catalyst forced by the pressure exerted by the polymer that is being formed undergoes a rupture process leading to a porous particle composed of polymer microparticles containing catalyst fragments. In the second stage, because polymerization proceeds by an insertion mechanism, the newly formed elastomeric phase grows inside of the polypropylene microparticles becoming encapsulated by the polypropylene. The encapsulation may not be possible if the volume fraction of the elastomeric phase increases, and then the particle pores become filled with the ethylene–propylene elastomer. If the elastomeric phase reaches the surface of the particle, particle agglomeration may occur leading to reactor fouling and eventually to blocking. Therefore, the characteristics of the catalysts particle and the way in which the operation is conducted are critical for the success of the operation. In order to accommodate large amounts of elastomeric phase,

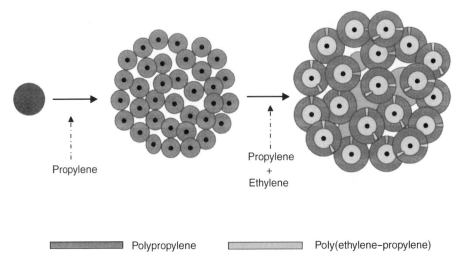

Propylene

Propylene
+
Ethylene

▬▬▬▬ Polypropylene   ▬▬▬▬ Poly(ethylene–propylene)

Fig. 9. The hiPP morphology development in a two-stage polymerization process.

particle porosity should be the maximum possible without compromising the particle mechanical integrity.

The ability of dissipating the impact energy, likely via both crazing and shear yielding, is the key property of hiPP. It has been established that the optimal balance of properties can be achieved when the compositions of the rubber fraction is in the range of 50–60 wt% of ethylene and the diameter of the elastomeric domains is ~1 μm (142). In addition, the crystallinity of the polypropylene matrix also plays an important role (143) and it is possible to improve the stiffness–impact performance by increasing crystallinity. This can be achieved by manipulating the molecular weight distribution of the polypropylene, because broad MWDs leads to higher stiffness, due to the effect of the MWD on the kinetics of crystallization (143).

The main applications of hiPP are automotive (interior and exterior trims, bumpers), rigid packaging, consumer goods, pails, and corrugated pipes. Grades containing ~70 wt% of amorphous, propylene-rich ethylene–propylene copolymers are supersoft polypropylene alloys suited for roofing and geomembranes. Some important producers are Basell, Atofina, Borealis, BP Amoco, ExxonMobil, Dow Chemical, and DSM.

10.2. Random Free-Radical Copolymers. Styrene–butadiene rubbers (SBR) is the largest volume synthetic rubber. It is produced in emulsion polymerization by free-radical polymerization and in solution by means of an anionic polymerization mechanisms. The large-tonnage grades of emulsion-polymerized SBR are produced in continuous stirred tank reactors in series. Conversion is maintained below 70% to avoid the formation of branches and gel that have a deleterious effect on rubber properties. There are two broad SBR types. The so-called "cold" grades that are produced at low polymerization temperatures (~5°C) and the "hot" grades produced at ~50°C. The tensile strength of SBR vulcanites increases with decreasing temperature (144) presumably because the molecular weight increases, the tendency for branching and cross-linking reactions decreases, and the ratio *trans*-1.4/*cis*-1.4 increases (145). The styrene/butadiene ratio for general purpose rubbers is in the region of 25/75, wt/wt. At higher contents of styrene, the elastic character of the material decreases and its plasticity increases. At the end of the polymerization, the latex is coagulated, washed, dried, and baled. Polymers with very high molecular weight that are difficult to process with ordinary equipment, can be processes upon addition of oils (oil extended SBR).

The main use of the cold SBR is for tires. Other applications include conveyor belts, rubber articles, and footwear. Hot polymerized SBR is used in sporting goods, shoe soles, and adhesives. Some important producers of SBR are Ameripol Sympol, Bayer, Dow Chemical, Enichem, Firestone Polymers, Goodyear Tire and Rubber, Japan Synthetic Rubber, Korea Kumho, Petroflex, Sinopec, and Zaklady.

Carboxylated styrene–butadiene latexes containing 50–60 wt% of styrene are directly used as dispersions for carpet backing (146) and paper-coating (147,148). Carboxylated styrene–butadiene latexes are produced by BASF, Dow Chemical, and Rhodia among others.

The production method of acrylonitrile–butadiene rubbers (NBR) is similar to that of the emulsion SBR. The properties of this material are strongly affected by its acrylonitrile content, which ranges from 10 to 50 wt%. As the acrylonitrile content increases the processability and oil resistance increase, but the low temperature flexibility decreases. The main use is in applications that require a rubber having good resistance to swelling in organic liquids. Aging and ozone resistance of NBR can be improved by selective hydrogenation of the double bonds of the butadiene units (hydrogenated nitrile rubber). The main producers are Bayer, Enichem, Goodyear, Japan Synthetic Rubber, Nitriflex, Uniroyal, and Zeon. Hydrogenated nitrile rubber is produced by Bayer and Zeon.

Styrene–acrylonitrile copolymers (SAN) are manufactured by emulsion, suspension, and bulk free-radical polymerization. Emulsion and bulk processes can be continuous. The properties of the SAN copolymers strongly depend on copolymer composition. Incorporation of acrylonitrile improves the chemical resistance, the barrier properties, the scratch resistance, and rigidity of polystyrene. Most SAN resins contain 15–30 wt% of acrylonitrile. SAN is used in applications that require better chemical resistance, toughness, and heat distortion temperature such as shower doors, cosmetic bottles, and toys. Most SAN is consumed in the production of acrylonitrile–butadiene–styrene (ABS resins) copolymers. Other applications are appliances, housewares, automotive, and packaging. Grades containing a high amount of acrylonitrile (60–80%) are used as barrier plastics. Some producers are Aiscondel (SAN 44), BASF (Lilerum), Daicel (Cevian), Dow Chemical (Tyril), and Mitsubitsi Chemicals (Saurex).

Ethylene–vinyl acetate copolymers are produced to practically cover the whole range of comonomer ratios from 98/2 (ethylene–vinyl acetate) to 4/96. Control of copolymer composition is easy because the reactivity ratios are close to 1.0. Bulk (mainly) and solution (in less extent) polymerization is used to manufacture copolymers containing 5–50% of vinyl acetate. Emulsion polymerization is used for copolymers with higher vinyl acetate content. Copolymers containing 3–25 wt% of vinyl acetate (EVA copolymers) are used in film applications (blow and cast films, shrink films, stretch film, disposable surgical gloves). Grades with higher vinyl acetate content are used as adhesives, diesel fuel additives, and asphalt modifiers. Dupont (Elvax), Equistar Chemical (Ultrathene), Polimeri (Greeflex), and Sumimoto (Evatate) manufacture EVA copolymers.

Vinyl acetate–ethylene dispersions are used in paint and coatings, adhesives, paper coating, and binders for nonwoven materials such as needlet felt carpet, cleaning cloths and sanitary products, moist wipes, bed covers, winter clothing, and tablewear. Air Products Polymers (Airflex) is the leading producer of these dispersions. Redispersable powder (Vinnapas, Wacker Polymer Systems) based on vinyl acetate–ethylene dispersions (in some cases including also vinyl esters of versatic acid (Veova monomers, Resolution Research) are used in construction (ceramic tile adhesives, mortars, plasters) and as powder paints. Redispersable powders for nonwoven are also available (Vinnex, Wacker Polymer Systems). Copolymers of vinyl acetate and branched vinyl esters are used for architectural paints. Marketed under the name of Veova 10 (Resolution Research) the most distinguishing feature imparted by this branched vinyl ester to the polymer is its resistance to hydrolysis. Some producers of these latexes are BASF, Clariant, Dow Chemical, Rhodia and Wacker Polymer Systems.

Dispersions of vinyl chloride and ethylene are used for paints and flame retardant systems (Airflex, Air Products Polymers). Redispersable powders based on vinyl chloride–ethylene copolymers are also available (Vinnol, Wacker Polymer Systems).

Acrylic copolymers are used for high temperature automotive coatings. Coating quality requires the use of solvent-borne systems, but the use of solvents is limited by environmental regulations. Therefore, there is a strong pressure to increase the solids content. In order to increase the solids content maintaining the viscosity in manageable values, low molecular weight ($\bar{M}_w$: 1000–10000 g/mol) is used. These low molecular weight copolymers are obtained copolymerizing a mixture of methacrylates and acrylates at high temperature (up to 150°C) (149). Monomers containing functional groups (eg, hydroxyl and carboxyl) are included in the formulation to allow cross-linking with suitable cross-linking agents (melamine and isocyanante for −OH, and epoxy for −COOH). A relatively large fraction of these functional monomers should be used to ensure that most of the short chains contain functional groups.

Acrylic dispersions are extensively used for coatings textiles and nonwoven fabrics, adhesives, floor care, caulks, and sealants. A typical formulation includes a hard (high T_g) monomer such as methyl methacrylate, a soft monomer (low glass transition temperature, T_g) such as butyl acrylate or 2-ethylhexylacrylate, and functional monomers to impart special characteristics (colloidal stability, adhesion, cross-linking) (147,150). The use of styrene as a hard monomer leads to styrene–acrylic latexes. Vinyl acetate is used in vinyl–acrylic latexes. The comonomer ratio is chosen to meet the required for the application. Thus, pressure sensitive adhesives must be soft and tacky ($T_g \cong -50$ to -25°C), architectural paints should form a continuous nonsticky film at room temperature (T_g from 5 to 30°C) and industrial coating and floor polishes should be hard and withstand exigent use (147,150). The T_g depends on copolymer composition and can be estimated using the Fox equation (151).

$$\frac{1}{T_g} = \sum_i \frac{w_i}{T_{g_i}} \tag{45}$$

where w_i is the weight fraction of monomer i in the copolymer and T_{g_i} the glass transition temperature of its homopolymer. Monomer composition also affect another important characteristics. For example, all acrylic paints show a good weatherability and durability but styrene–acrylic latexes with >20 wt% of styrene do not present a good exterior durability (150). Latex performance is also affected by copolymer microstructure (molecular weight distribution, branching, gel content, etc). Thus, in the case of pressure sensitive adhesives, high molecular weights provide resistance to shear, whereas resistance to peel is maximized for intermediate molecular weights, and tackiness requires low molecular weights (152). On the other hand, increasing the gel fraction from 1 to 32% leads to an increase of the shear resistance but further increase of the gel content to 55% results in a severe decrease of the shear resistance (153). Some important producers of acrylic latexes are Akzo-Nobel, Atofina, BASF, Dow, Rohm and Haas, and UCB.

Fluorinated polymers give excellent performance under highly demanding conditions requiring resistance to high temperature, chemical inertness, and low surface tension. Fluorinated homopolymers are often crystalline and copolymerization induces disorder of the polymer chain and consequently reduces crystallinity. Depending on the copolymer composition, fluorinated copolymers range from thermoplastic to elastomers. Thermoplastic fluoropolymers are produced by co(ter)polymerization of a main monomer with a relatively low amount of other monomer(s). Some thermoplastic fluoropolymers are

1. Tetrafluoroethylene (TFE)–hexafluoropropylene (HFP) copolymers (Teflon FEP, DuPont; Neoflon, Daikin).
2. TFE–perfluoropropyl vinyl ether (Teflon PFA, DuPont; Neoflon AP, Daikin; Hostaflon TFA, Ticona).
3. Vinylidenefluoride (VDF)–HFP (Kynarflex, Atofina).
4. VDF–TFE (Kynar SL, Atofina).
5. VDF–Chlorotrifluoroethylene (CTFE) (Foraflon, Atofina; Solef, Solvay).
6. VDF–TFE–HFP (Hostaflon TFB, Dyneon).

Fluoroelastomers require a high degree of disorder of the polymer chain, and hence 35–50% of the second comonomer is needed. These products are designed to maintain rubber-like elasticity under extremely severe conditions (high temperatures and in contact with chemicals). To prevent flow deformation under an imposed force, the fluoroelastomers are cross-linked. An excellent review on fluoroelastomers has been recently published (154). Some commercial products are

1. VDF–HFP (Daiel 801, Daikin; Fluorel, 3M/Dyneon, Viton A, DuPont).
2. VDF–CTFE (Kel F, Dyneon; Voltalef, Atofina).

3. TFE–perfluoromethyl vinyl ether (Kalrez, DuPont).

4. VDF–TFE–HFP (Tecnoflon, Ausimont; Viton B, DuPont).

Fluorinated copolymers are produced by free-radical polymerization in both aqueous and non-aqueous media. In aqueous media, perfluorocarbon surfactants are used and the polymer is separated and converted into various forms. Some products are sold as latexes. The product is recovered from nonaqueous polymerization by evaporating the fluorinated solvent.

Thermoplastic fluoropolymers are used in applications that require chemical inertness, excellent dielectric properties, nonaging characteristics, antistick properties, low coefficient of friction, and performance at extreme temperatures. These applications include electrical applications (wire-cable, molded electrical parts), lined pipes and fittings, heat exchangers, and conveyor belts.

The main uses of the fluorinated elastomers are O-rings and gaskets, shaft and oil seals, diaphgrams, hoses, and profiles.

10.3. Random Copolymers Produced by Ionic Polymerization. Butyl rubber is a copolymer of isobutylene and isoprene (1–3%) formed by cationic polymerization. The polymerization is carried out in a continuous slurry process with methyl chloride as the diluent. Isoprene is a strong chain-transfer agent and very low temperatures ($-90°C$) are used to produce high molecular weight polymer. The insaturared sites of the isoprene units enable vulcanization, and butyl rubber cures well although more slowly than polyiso-prene and polybutadiene. This causes problems when butyl rubber is blended with these polymers. Halogenation of some of the isoprene groups greatly increases the cure rate of butyl rubber. Substitution of isoprene by brominated p-methyl styrene leads to a copolymer more stable oxidatively, and with a good cur-ing reactivity. The applications of polyisobutylene copolymers take advantage of the low permeability of these copolymers. The main application is for the inner-liners of tubeless tires and inner tubes of tires. Other uses are adhesives, bin-ders, pipe wrap, caulking, and sealing compounds. ExxonMobil and Bayer are the main producers.

SBR can be also be produced by a solution process using organolithium initiators. These anionic systems show a marked sensitivity of the reactivity ratios to solvent type. Nonpolar solvents favor the incorporation of butadiene while polar solvents lead to styrene- rich copolymers (144). The main application is for tires. Firestone polymers, American synthetic rubber, Michelin, Repsol YPF, and Negromex are producers of this rubber. It is claimed that solution SBR gives low rolling resistance (less fuel consumption), improved wet grip, and good wear properties.

10.4. Controlling Copolymer Microstructure. The application prop-erties of copolymers are mostly determined by their microstructure (copolymer composition, monomer sequence distribution, molecular weight distribution, branching, cross-linking, etc.). Therefore, the final properties can be improved by controlling copolymer microstructure. Copolymers are "products-by-process" materials, and consequently their microstructure, and hence their final proper-ties, are determined by the process variables in the reactor (in a broad sense process variables include the type of polymerization process, catalysts, etc.). Competition and margin reduction are pushing polymer producers to achieve an efficient and consistent production of polymers with improved properties under safe and environmentally friendly conditions. Therefore, there is a strong interest in developing strategies for on-line control of polymerization reactors. Reviews discussing the field up to 1998 are available (155–160).

On-line control of copolymer microstructure is, in a large extent, associated to random copolymers produced by catalytic polymerization and classical free-radical polymerization. Catalytic polymerization is conducted in large scale continuous reactors, and hence grade transition is an important issue. On the other hand, due to proprietary reasons, data from those processes are not freely available to academic researchers, and this limits the number of works published in the open literature. Böhm and co-workers (161) developed a strategy to control the density (copolymer composition) and melt flow index (molecular weight) of a HDPE produced in a continuous slurry reactor. An on-line state estimator for the density and the melt flow index was developed based on measurements of the composition of the gas phase. A fuzzy-logic controller was used for grade transition. Control strategies for gas-phase production of LLDPE in a fluidized bed reactor have also been proposed (162–164). In these works, an extended Kalman filter was used for inferring the values of the density and melt flow index, and a model-based controller was employed to track the optimal trajectories in grade transition. Dynamic matrix control has been used to control product quality (terpolymer composition and Mooney viscosity) during changes in product specifications and in production rate and catalyst activity for an EPDM continuous stirred tank reactor (165).

Successful strategies for on-line control of copolymer (166–172) and terpolymer composition (173–175) of random free-radical copolymers have been implemented. On-line closed-loop control of the molecular weight distribution of linear polymers has been achieved (168,176–180). Open-loop strategies have also been proposed (181). The simultaneous closed-loop control of copolymer composition and MWD of linear polymers has been recently reported (182), as well as some open-loop strategies for this purpose (183,184). The application of the control strategies developed for linear polymers fail to control the polymer microstructure of nonlinear polymers (185) and although some success has been obtained in open-loop control based on complete mathematical models (186) the control of the microstructure of nonlinear polymers (MWD, branching, gel) still is an unresolved issue.

The implementation of strategies for on-line control of copolymer microstructure requires the availability of appropiate sensors. The development of on-line sensors for polymerization reactors have been difficult in comparison with other chemical processes because of the physical characteristics of the polymerization mixture that makes it difficult to use in-line sensors or circulation loops due to fouling and clogging of pipes, pumps, and the sensor itself (187,188). In the lab reactors, copolymer composition and conversion were monitored by gas chromatograph and densimeters (167,189), but these devices are not robust enough for application in industrial scale reactors. On-line reaction calorimetry and spectroscopic techniques [eg, Raman and near infrared (nir) spectroscopies] are more promising (190–192). Some success has been reported for the on-line measurement of the molecular weight distribution in solution polymerization processes (168), but in disperse systems, this is still an unresolved issue. Therefore, soft-sensors and/or state estimation techniques have been proposed to on-line infer the molecular weights (161,182).

11. Alternating Copolymers

There are several ways in which alternating copolymers can be produced. Thus, free-radical copolymerization of electron-donor monomers (such as styrene and α-olefins) and electron-acceptor monomers (such as maleic anhydride and fluorinated alkenes) yields alternating copolymers. These copolymerizations deviate from the predictions of the terminal model (193) and it occurs through the formation of a donor–acceptor charge-transfer complex between the two participating monomers, which enters into the propagation step as a single unit (194).

Alternating ethylene–α-olefin copolymers can be obtained using Ziegler-Natta catalysts at very low temperature (−70°C) (193) as well as at higher temperatures (0°C) using metallocene catalysts (107,108,195–197). Alternating CO–olefin copolymers can be produced in a methanol slurry phase using a catalyst system that includes a palladium salt of a carboxylic acid, a phosphorous or nitrogen bidendate base, an anion of an organic acid, and an oxidant such as 1,4-benzoquinone (198).

A limited number of alternating copolymers are commercially available. Alternating styrene–maleic anhydride copolymers (SMA) are produced by free-radical copolymerization. Low molecular weight SMA with a high maleic anhydride content (25–50 wt%) is alkali soluble and used in paper and textile sizing, floor polishes, printing ink, pigment dispersants, and coatings. High molecular weight SMA with low maleic anhydride content (<25 wt%) is used for molding and extrusion applications. Bayer (Cadon) and Nova Chemicals (Dyrlak) are producers of styrene–maleic anhydride copolymers.

Fluorinated alternating copolymers are produced by free-radical copolymerization of electron-withdrawing fluorinated alkenes with electron-donating monomers. Tetrafluoroethylene and ethylene are copolymerized in emulsion (using fluorinated surfactants) or in a non–aqueous media (using a fluorinated solvent) to yield an alternating copolymer that has a high tensile strength, moderate stiffness, outstanding impact strength, low dielectric constant, excellent resistivity, good thermal stability, and excellent chemical resistance. This copolymer can be cross-linked with electron beam and γ-ray radiation. The main uses of this copolymer are power and automotive wiring, injection-molded electrical components, pump impellers, and molding articles. Some producers are DuPont (Telzel), Asahi Glass (Aflon COP), Ausimont (Halon ET), and Daikin (Neoflon EP). Asashi Glass also produces TFE–propylene alternating copolymer (Aflon). Emulsion copolymerization of chlorotrifluorethylene (CTPE) with vinyl ethers leads to commercial alternating copolymers (Lumiflon, Asashi Glass; Zeffe, Daikin; Fluonate, Dainippon Ink; DX 2000, Atofina).

Alternating ethylene–CO copolymers are not of practical relevance because it is difficult to process without degradation (198). On the other hand, ethylene–propylene–CO terpolymers are perfectly alternating aliphatic polyketones that exhibit moduli, impact, and thermal characteristics of amorphous polymers with the chemical resistance of crystalline polymers while processing like a polyolefin. Shell commercialized these products under the trade name of Carilon, but the product was withdrawn from the market in February 2000.

12. Block Copolymers

12.1. Synthetic Methods. Block copolymers can be produced by means of several polymerization mechanisms including anionic polymerization, ring-opening polymerization, step-growth polymerization, catalytic polymerization, controlled free-radical polymerization, and combinations of some of these polymerization methods.

Anionic Polymerization. Anionic polymerization proceeds in the absence of termination and transfer reactions allowing the synthesis of block copolymers of the desired molecular weight, composition, and structure. The different synthetic methods are shown below for the case of styrenic thermoplastic elastomers (199)

(a) Succesive polymerization of monomers (200).

$$R^-Li^+ \xrightarrow{+nS} R\!-\!(S)_n^-Li^+ \xrightarrow{+mM} R\!-\!(S)_n\!-\!(M)_m^-Li^+ \xrightarrow{+nS}$$
$$R\!-\!(S)_n\!-\!(M)_m\!-\!(S)_n^-Li^+ \tag{46}$$

(b) Coupling diblock copolymers (200)

$$2(S)_m\!-\!(M)_m^-Li^+ + X\!-\!R\!-\!X \rightarrow$$
$$(S)_m\!-\!(M)_m\!-\!R\!-\!(M)_m\!-\!(S)_n + 2LiX \tag{47}$$

(c) Multifunctional initiation (201)

$$Li^{+-}R^-Li^+ \xrightarrow{+2nM} Li^{+-}(M)_n\!-\!R\!-\!(M)_n^-$$
$$Li^+ \xrightarrow{+2mS} Li^{+-}(S)_m\!-\!(M)_n\!-\!R\!-\!(M)_n\!-\!(S)_m^-Li^+ \tag{48}$$

where M is either butadiene of isoprene. In this case, the block copolymers are termed S–B–S and S–I–S, respectively. The polybutadiene and polyisoprene blocks contain double bonds that limit the stability of the products. Hydrogenation of S–B–S yields S–EB–S and hydrogenation of S–I–S gives S–EP–S.

Anionic polymerization requires ultrapure reagents and solvents, low temperature, and the absence of oxygen. An important limitation of anionic polymerization is that it can only be applied to a limited number of monomers (styrene, butadiene, isoprene, and cyclic monomers such as epoxides, anhydrides, lactones, and siloxanes). Nevertheless, polystyrene-*block*-polybutadiene-*block*-polymethylmethacrylate copolymers produced by anionic polymerization have been commercialized by Atofina (SBM).

Catalytic Polymerization. Polyolefin block copolymers are obtained by successive polymerization of different monomer mixtures on Ziegler-Natta catalyst under conditions in which chain termination by chain-transfer reactions is minimized (202–204) Isotactic–atactic stereoblock polypropylene have been synthesized by using a non–bridged metallocene catalyst in which the rotation of the indenyl ligands during chain growth allows the transition between blocks (205).

Controlled Free Radical Polymerization. The block copolymers produced by CRP are not as well defined as those produced by anionic polymerization, but have the advantage that almost any of the huge number of monomers that can be polymerized by free-radical polymerization can be used to synthesize block copolymers. Polystyrene-*block*-poly(*n*-butyl acrylate) (40), poly(butyl methacrylate)-*block*-poly(methyl methacrylate) (206), poly(methyl methacrylate)-*block*-poly(butyl methacrylate)-*block*-poly(methyl methacrylate) (207), poly(n-butyl acrylate)-*block*-poly(acrylic acid) (208), and polyHEMA-*block*-poly-MMA-*block*-polyHEMA (208) are examples of diblock and triblock copolymers produced by succesive addition of the monomers using monofunctional and difunctional CRP agents.

One of the interesting features of synthesizing block copolymers by means of CRP is that the initial block can be produced and stored before proceeding to the second block. In addition, CRP allows the incorporation of random copolymers as blocks (209).

Ring-Opening Polymerization. Some ring-opening polymerizations proceed by a living mechanism, which enables the preparation of block copolymers. The main difficulty being the rather limited choice of initiator catalysts that allow the living character of the polymerization to be met. Examples of block copolymer produced by anionic and cationic ring-opening polymerization have been reported in the literature (210–214): polyether-*block*-polyesther (213,214), polyester1-*block*-polyester2 (215), polyester-*block*-polycarbonate (216) polyvinyl-*block*-poly(ε-caprolactone) (217,218), polyvinyl-*block*-polycyclic amine (219), block copolymers of different caprolactones and ε-caprolactone and butadiene (219–222), and block copolymers of lactides and cyclic monomers (223).

Step-Growth Polymerization. Multiblock copolymers are produced by condensation reactions from polyurethanes, polyesters, and polyamides prepolymers. These copolymers are produced in a two-step process that for polyurethanes involves a first stage in which a long-chain diol is reacted with an excess of disocyanate yielding a soft prepolymer with isocyanate end groups. This prepolymer is reacted with a short-chain diol and additional diisocyanate, leading to a multiblock copolymer in which the ratio hard segments/soft segments is determined by the ratio short-chain diol/long-chain diol (224). Multiblock polyester copolymers are produced by a similar method using diacids (or diesters) and diols (225). Similarly, multiblock polyamide copolymers are prepared from diacids and diamines (226).

Combining Polymerization Techniques. Block copolymers can also be obtained by forming one block by means of a given polymerization method and the second one through a different polymerization mechanism. Some representative examples are given in Table 9.

12.2. Properties and Commercial Products. Block copolymers present unique structure–property relationships that are useful for a variety of applications including thermoplastic elastomers (TPE), elastomeric fibers, toughened thermoplastic resins, adhesives, compatibilizers for polymer blends,

Table 9. **Synthesis of Block Copolymers Combining Different Polymerization Techniques**

	Block 1		Block 2		
Monomer 1	Polymerization mechanism	Monomer 2	Polymerization mechanism	Reference
ε-caprolactone	anionic ring-opening polymerization	styrene	CRP-nitroxide	227
2,5-dioctyloxy-1,4-phenylene vinylidene	condensation	n-butyl acrylate	CRP-nitroxide	228
styrene	cationic	methyl acrylate	ATRP	229
THF	cationic ring opening	styrene	ATRP	230
MMA[a]	ATRP	ε-caprolactone	Anionic ring-opening polymerization	227
4-fluorophenyl sulfone	step growth	n-butyl acrylate	ATRP	231
isoprene	anionic	ethylene	Ziegler-Natta	232

[a] MMA = methyl methacrylate.

membranes, and surfactants. Table 10 presents some commercially available block copolymers.

Thermoplastic elastomers is the main application of block copolymers. These materials are composed of hard and soft segments, which form a processable melt at high temperatures and transform into a solid rubber-like object upon cooling. The transition between the strong elastic solid and the processable melt is reversible. Figure 10 illustrates this phase transition for an A–B–A block copolymer, where A is a short hard segment and B is a long soft segment. At low temperatures, the hard segments segregate forming a three-dimensional network with physical cross-links that act as the sulfur cross-links in vulcanized rubber. When temperature is increased above the T_g of the polymer forming the hard segments, the physical cross-links soften and a polymer melt is formed. Styrenic thermoplastic elastomers are commercially important A–B–A block copolymers. Polybutadiene and polyisoprene are common elastomeric midsegments. The presence of double bonds in the elastomeric part of these triblock copolymers limits the stability of the product. More stable polymers are produced by hydrogenation of S–B–S that yields polystyrene-*block*-poly(ethylene-butylene)-*block*-polystyrene. Similarly hydrogenation of S–I–S leads to polyisobutylene mid-segments.

Because polystyrene is an amorphous polymer soluble in many solvents, the styrenic thermoplastic elastomers present a poor oil and solvent resistance. This resistance can be improved by compounding and in some cases is an advantage because it allows the application from solution (199).

Polyethylene-*block*-poly(ethylenic-*stat*-α-olefin) copolymers produced by catalytic polymerization present a better resistance to oil and solvent because the hard segments (polyethylene) form crystalline aggregates. Multiblock copolymers formed by step-growth polymerization are high performance elastomers.

Table 10. **Commercially Available Block Copolymers**

Hard segments	Soft segments	Type	Trade name	Applications
styrene	butadiene or isoprene	A–B–A	Kraton D (Kraton polymers)	footwear, bitumen/asphalt modification, adhesives, sealants, household appliances, toys, tubing
styrene	ethylene–butylene isobutylene	A–B–A	Kraton G (Kraton polymers)	bitumen/asphalt modification, sealants, high performance adhesives, automotive, sports, medical equipment
styrene (high styrene content)	butadiene	triblock	Finaclear (ATOFINA)	health products, packaging
styrene	butadiene	linear/Star	Finaprene (ATOFINA)	footwear, compounds, bitumen modification, adhesives, plastic modification
polyamide	polyether	multiblock	Pebax (ATOFINA)	footwear, sporting goods, protective films, waterproof breathable materials
polyester	polyether	multiblock	Hytrel (DuPont)	automotive, fluid power, sporting goods, furniture and off-road transportation. Thin flexible membranes, tubing, hose jackets, wire and cable electrical connectors
polyamide	polyether	multiblock	Vestamid (Degussa-Hüls)	fuel lines, air brake tubing, hydraulic tubes, catheters, cable and wire, plastic-rubber components for the automotive industry, sport shoes
polyurethane	polyester/polyether/ poly carbonate	multiblock	Estane/Estagrip/Estaloc (Noveon)	automotive, power and handtools, wire and cable, footwear, consumer goods, health care
polystyrene/ poly-(methyl methacrylate)	polybutadiene	A–B–C	(SBM) ATOFINA	nanostructurated thermoplastic and thermoset materials, compatibilization of minerals and carbon black with polymers
polyester	polyester	multiblock	Arnitel V (DSM)	automotive, tubing, cable insulation, injective molding, films
poly(ethylene oxide) (hydrophilic block)	poly(propylene oxide) (hydrophobic block)	diblock	Synperonic (Uniquema)	surfactants for resin emulsification and emulsion polymerization

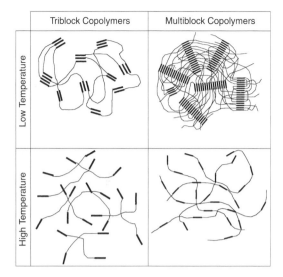

Fig. 10. Phase transition of thermoplastic elastomers.

Figure 10 illustrates the phase transition of a multiblock thermoplastic elastomer. In this case, the number of hard segments involved in each physical cross-link is much higher than for the A–B–A block copolymers. The hard segments of the multiblock copolymers of commercial importance are crystalline thermoplastics (polyurethanes, polyesters, and polyamides), therefore the cross-links are crystalline regions that provide a good oil and solvent resistance. Upon heating, the crystalline regions melt and a processable fluid is obtained. The soft segments are polyesters and polyethers.

The service temperature range of the thermoplastic elastomers spands from a temperature slightly above the T_g of the soft rubbery phase to a temperature slightly below the T_g or T_m of the hard segments. Values for the T_g and T_m of the different segments are given in Table 11.

The hardness of the thermoplastic elastomers depends on the ratio between soft and hard phases (199). The hardness range is very broad for styrenic triblock

Table 11. **Glass Transition and Melting Temperatures for Soft and Hard Segments**

Soft segment	T_g (°C)	Hard segment	T_g, T_m (°C)
butadiene	−90	styrene	95 (T_g)
isoprene	−60	polyurethane	190 (T_m)
ethylene-butylene	−60	polyester	180–220 (T_m)
isobutylene	−60	polyamide	220–275 (T_m)
polyethers	−40	polyethylene	70 (T_m)
polyesters	−60		
poly(ethylene-*stat*-α-olefin)	−50		

copolymers, in which at high styrene contents a clear flexible thermoplastic is obtained (eg, Finaclear from ATOFINA). Thermoplastic elastomers with crystallizable hard segments have limits on softness because a minimum length of the hard segment is required for crystallization.

Block copolymers, particularly styrenic triblock copolymers, are used as adhesives and sealants. The role of the chain architecture in the adhesion properties have been studied (233,234). Block copolymers are efficient compatibilizing agents for polymer blends. As little as 0.5–2% of diblock copolymers may be sufficient to achieve a good phase dispersion (235). Block copolymers are used in the surface modification of fillers (236) and as coatings for metal and glass surfaces (237). Amphiphilic block copolymers consist of hydrophilic polyoxyethylene segments and various hydrophobic parts (polypropylene oxide, polystyrene, etc) are efficient surfactants (238). New nanostructured materials can be prepared by assembly of block copolymers (239).

13. Graft Copolymers

High impact polystyrene (HIPS) and ABS resins are important graft copolymers produced by polymerizing vinyl monomers in the presence of polybutadiene through a free-radical mechanism. Grafting occurs by participation of the double bonds of the polybutadiene in the propagation reaction and by chain transfer to the polybutadiene followed by addition of monomer to the resulting allylic radical.

HIPS polymerization is carried out by first dissolving 4–12% by weight of polybutadiene rubber in a styrene monomer and adding a free-radical initiator. At the early stages of the polymerization, polybutadiene dissolved in styrene forms the continuous phase, whereas the newly formed polystyrene (swollen with styrene monomer) forms a separate phase stabilized with the grafted butadiene–styrene copolymer. As polymerization proceeds, the polystyrene phase becomes the major component and phase inversion leading to a multiphase material occurs. Figure 11 shows a transmission electron micrograph of a HIPS in which the rubber particles (dark areas) with polystyrene inclusions (clear areas) are dispersed in a continuous polystyrene matrix (clear background). Particle size severely affects the performance of HIPS. Small particles provide rigidity and gloss, whereas larger particles improve toughness (240). On the other hand, as the rubber content increases, impact strength increases while rigidity, heat distortion temperature, and clarity decrease. HIPS meets the application needs across a broad range of market segments including appliances, consumer electronics, packaging, housewares, disposables, and toys. Some important HIPS producers are BASF (Polystyrol), Dow Chemical (Styron, Aim), Nova Chemicals (Zylar), L.G. Chemical (Alphalac), and Hong Kong Petrochemical (SR and SRL grades).

ABS copolymers are produced in bulk polymerization by means of a process almost identical to that of the HIPS, but using styrene and acrylonitrile as polymerizing monomers. ABS polymers can also be produced in emulsion polymerization by polymerizing styrene and acrylonitrile on a polybutadiene seed. Commercial ABS is often a blend of this reactor produced ABS with SAN. The

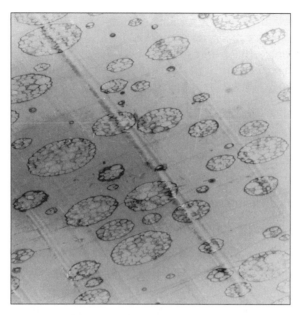

Fig. 11.　Structure of HIPS.

properties of the ABS copolymers can be modified by varying the relative amounts of the components, the degree of grafting, and the molecular weight. Thus, increasing the rubber content reduces tensile strength and increases impact strength. The effect of the rubber particle size is similar to that in HIPS, small particles lead to rigid and good surface aesthetics and large particles improve toughness. A combination of good properties is usually found with particles in the range of 0.3–0.5 μm (241). Table 12 summarizes some properties of ABS and HIPS. The ABS copolymers are used in electrical and electronic equipment, house and office appliances, and the automotive industry. Some trade names are Terluran (BASF), Magnum (Dow Chemical), Diastat (Mitsubishi Rayon), Cycolac (GE Plastics), Lustran (Bayer), and Estadine (Cossa Polimeri). Other producers and trade names are available at the web site http://www.ets-corp/tradenames/. Central Soft from Central Glass is a graft fluoropolymer (VDF/CTFE)-*graft*-VDF produced by free-radical polymerization.

　　　Polyolefin-based graft copolymers can be produced polymerizing a mixture of ethylene and α-olefin in the presence of two different metallocene catalysts (242,243). One of these catalysts has a low αolefin incorporating capability and produces primarily double-bond terminated polyethylene macromonomers. The

Table 12. **Properties of ABS and HIPS**

	ABS	HIPS
tensile strength (MPa)	27–55	16–28
izod impact (J/m)	106–640	58–150
vicat heat distortion (°C)	80–115	90–106

other catalyst has a higher affinity for α-olefins and also consumes macromonomers leading to a graft copolymer composed by a soft ethylene-α-olefin backbone with crystalline polyethylene branches. Polyolefin graft copolymers have also been produced via constraint geometry catalyst (242). Engage (DuPont-Dow Elastomers), Affinity (Dow Chemical), and Exact (ExxonMobil) are examples of commercial polyolefin-based graft copolymers.

Classical free-radical polymerization and catalytic polymerization only allow a limited control of the polymer architecture. Therefore, new and more selective grafting methodologies have been developed, which can be summarized by the three following main methods.

(1) "Grafting from" method. This method consists in initiating a polymerization of a monomer at some suitable reactive groups (Z) attached onto a prepolymer (PA) or from a macroinitiator with pendant functionality as indicate in the following scheme:

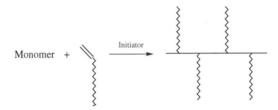

The control of the molecular weight of the grafted chains is not straightforward and can only be achieved if initiation is much faster than propagation reactions. Ionic grafting meets this requirement but the choice of monomers is limited (245,246). The grafting reaction can be done by radical polymerization [eg, using a polymer containing halogen with metal carbonyls such as molybdenum hexacarbonyl (247)], but care should be taken with the termination reaction that may produce cross-linked graft copolymer if it is by combination.

Controlled radical polymerization has opened a wide range of possibilities to graft a large number of monomers into prepolymers conveniently modified. Graft copolymers have been produced by means of a three-step nitroxide mediated radical polymerization. In the first step, a linear copolymer of styrene and p-chloromethylstyrene is obtained. In the second step, reaction of this copolymer with the sodium salt of a hydroxyl functionalized nitroxide yielded initiation sites for the third step that is a CRP grafting process (248). Graft copolymers can be easily produced by ATRP (41). An example of this method is the formation of graft copolymers by ATRP of vinyl monomers from pendant-functionalized poly(vinyl chloride) (PVC) (249)· In this case, polymerization was initiated by the chloroacetate moieties attached to the backbone. Monomers such as methyl acrylate (MA), n-butyl acrylate (n-BA), and styrene were grafted to PVC and graft copolymers of different T_g value than that of the PVC were produced. The decrease of T_g achieved for n-BA and MA grafts led to self-plastified PVC polymers.

In the CRP grafting processes, bimolecular termination of growing branches limits the number of grafts per chain. Thus, it has been found that for more than six initiating sites per backbone, bimolecular termination of adjacent growing chains is important (248).

(2) "Grafting onto" method. In this method, a prepolymer (or end-reactive oligomer–telechelic polymer), PB, containing a reactive functional group (G') reacts with the groups randomly attached (G) to a backbone of a second preformed polymer, PA, as shown in the following scheme:

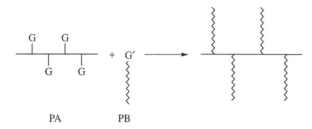

This method allows good control of the polymer microstructure because prepolymers with controlled structures can be employed. Both anionic and cationic techniques can be used to produce the backbone prepolymer with functionalized groups and also to couple the grafted polymer (produced by an ionic technique) onto the backbone. Examples of this method are grafting of living anionic polymers (eg, polystyril lithium) onto polymer backbones with reactive halogen or epoxide functionalities (250) and the grafting of living polytetrahydrofuran cations that can be grafted onto chlorohydroxylated polybutadiene and nitrile rubbers (251).

(3) "Grafting through" method. This is a popular approach because it can overcome some of the difficulties inherent to the other techniques. In this method, a macromonomer (viz, an oligomer bearing a polymerizable end-group) copolymerizes with a suitable monomer to form comb-like graft copolymers as shown in the following scheme:

Macromonomers can be prepared by ionic (252,253), free radical (254), and catalytic chain-transfer polymerization (255).

The interest of well-defined graft copolymers is due to the multiple properties that they may have in a single molecule. This feature opens the possibility of using this specialty and expensive polymers in applications that cannot be achieved with other polymers. Thus, graft copolymers have been used for improving processability, compatibility, dyeability, and water repellency (256–258). Because of their inherent surface activity, other uses for these copolymers are coatings, adhesives, fibers, films, and moldings (257). Recently, applications of graft copolymers for 100% solids *in situ* curing resins (by using macromonomers that cure under electron beam or ultra violet, uv, radiation) have been reported (259). The low cost and nonpolluting nature of the resulting coatings and the possibilities to improve film properties such as adhesion, tensile strength, and flexibility make them very attractive products.

Also, hydrogels were recently formed by grafting hydrophobic polymers in hydrophilic backbones. For example, graft copolymers of polystyrene and vinyl-pyrrolidone (NVP) were produced by using ATRP techniques (260), and also styryl-telechelic-polyisobutylene (PIB), and methacryloyl-telechelic-PIB have been copolymerized with NVP and 2-(dimethylamino)ethyl methacrylate to produce graft copolymers (261–263). These hydrogels have been employed for biomedical applications such as controlled release of drugs, enzyme immobilization, and contact lenses.

14. Star and Hyperbranched Copolymers

Star copolymers, from a wide range of monomers can be prepared by controlled radical polymerization. Thus, star copolymers can be synthesized by coupling alkoxyamine terminated linear copolymers with cross-linking agents (264). Heterogeneous star-block copolymers may be produced using linear polymers of different composition. Multifunctional small molecule initiators allow synthesizing star copolymers by using ATRP. Thus, six armed star-block copolymers of poly(methyl acrylate) and poly(isobornyl acrylate) (265) and eight armed star-block copolymers of PMMA and PBA have been produced (266). By using appropriate precursors containing multiple thiocarbonylthio groups, star polymers are also accessible by means of the RAFT process (267). Star copolymers were produced by Phillips Petroleum (268) by reacting a polystyrene-*block*-polybutadiene anion with SnCl$_4$. Some Kraton grades (Kraton polymers) are star copolymers produced by polymerizing divinylbenzene containing block copolymers to yield a cross-linked core from which block copolymer arms radiate outward. This method of producing the arms in the first place and then coupling them is called "arm first". Core first methods based on multifunctional initiators from which the arms are grown have been reported (269).

Hyperbranched polymers are commonly produced by (a) step-growth poly-condensation of AB$_x$ monomers, (b) self-condensing vinyl polymerization of AB* monomers, and (c) multibranching ring-opening polymerization of AB$_x$ mono-mers (270). Hyperbranched polymers can also be prepared by means of controlled radical polymerization. Thus, nitroxide mediated polymerization of self-conden-sing monomers that contain a double bond and an alkoxyamine initiating center yields to this type of topology (271). Similarly, hyperbranched polymers contain-ing a large amount of halogen end groups can be produced by ATRP of monomers containing a double bond and an initiator fragment (272). The terminal halogens may be replaced by other functionalities such as azido, amino, hydroxy, and epoxy by means of radical addition reactions (273).

Special attention must be paid to bimolecular termination when preparing stars, combs, dendritic, and hyperbranched topologies by means of controlled radical polymerization, because a relatively small fraction of terminated chains, which can be easily tolerated in the case of linear polymers, may have a catastrophic effect in complex architectures. Thus, for polymer chains growing in five directions, 5% of bimolecular termination will lead to 25% of chains linked together, and for chains growing in 20 directions this level of termination may lead to complete cross-linking and gelation (41). Hyperbranched polyethylene are produced by low pressure polymerization of ethylene on palladium and nickel catalysts (274,275).

Hyperbranched copolymers can, in general, be considered a subclass of dendritic polymers with the advantage that their production is less costly and more suitable for mass production. The unique properties of hyperbranched polymers are mainly due to their globular structure and to the large number of terminal functional groups. Hyperbranched polymers, in comparison to their counterpart linear polymers, offer better solubility in organic solvents, lower viscosities because of their spherical shape, and the choice of controlling the T_g by chemical modification of the end functional groups. Hyperbranched polymers cannot engage in chain entanglements, and hence their use in conventional structural applications is futile (276). Hyperbranched polymers with acrylate, vinyl ester, alkyl ether, epoxy, and OH functions are used as cross-linkers in coatings and thermosets (277). Very high functionality leads to a too fast cross-linking, and therefore an optimal number of reactive groups exists (278). Hyperbranched polymers provide an exceptional film hardeness that allows the use of low viscous–high solids–low molecular weight resins without compromising the coating performance, and also allows the use of entirely aliphatic monomers, which in turn resulted in excellent weatherability (279). Hyperbranched polymers are used as melt modifiers. Thus, strong reduction of the melt viscosity of linear polyamide has been observed when a small amount of hyperbranched polymer was added (280).

Hyperbranched polymers increase the toughness of glass and carbon-reinforced composites (281) and have also been used as dye carriers (282), for nonlinear optical materials (283), in molecular imprinting (284), and for the synthesis of nanoporous polymers with low dielectric constant (285).

Astramol and Hybrane from DSM are examples of the still limited number of commercially available hyperbranched polymers.

15. Characterization of Copolymers

Polymers are complex materials whose complete characterization requires the determination of a number of characteristics, which in turn determine the application performance. Polymer characteristics can be classified into four groups.

(a) Molecular structure. This characteristic refers to the arrangement and type of monomeric units in the polymer chain and includes copolymer composition, chemical composition distribution, and monomer sequence distribution. The techniques used to measure these characteristics are summarized in Table 13 and include solution fractionation methods, spectroscopic methods such as nuclear magnetic resonance (nmr), uv, infrared (ir) and Raman, pyrolysis GC, mass spectroscopy, and chemical tests.

Table 13. Experimental Techniques Used to Determine the Most Important Copolymer Characteristics

Characteristic	Property	Experimental technique
molecular structure	average composition, chemical composition distribution and monomer sequence distribution	infrared spectroscopy, ir nuclear magnetic resonance, nmr (^{13}C and ^{1}H nmr) raman spectroscopy cross-fractionation techniques (for CCD): SEC/TLC, HPLC/SEC, and SEC/SEC pyrolysis/GC (average composition) uv Spectroscopy X-ray Spectroscopy inverse gas chromatography, IGC size-exclusion chromatography (for CCD) uv and RI detectors MALDI–TOF mass spectrometry
molecular size	molecular weight and molecular weight distribution	size-exclusion chromatography (MWD) (refractive index, viscosity, and light scattering detectors are mostly used) light scattering (measurements in three different solvents are required for the apparent molecular weight) end-group determination ($\bar{M}_n$) (spectroscopic techniques and titration) colligative properties ($\bar{M}_n$) (membrane osmometry and vapor pressure osmometry are the most suitable ones) viscosity ($\bar{M}_v$) ultracentrifugation ($\bar{M}_w$) field-flow fractionation (MWD) neutron scattering field desorption mass spectroscopy (MWD, limit on 30.000 g/mol) liquid adsorption chromatography at the critical point conditions, LACCC (MWD of block copolymers) MALDI–TOF mass spectroscopy (averages and MWD of blocks)
molecular organization	crystallinity	inverse gas chromatography, IGC differential scanning calorimetry, DSC dynamic mechanical analysis, DMA crystaf, X-ray scattering, polarized optical microscopy

	Property	Method
	glass-transition temperature	inverse gas chromatography, IGC differential scanning calorimetry, DSC dynamic thermal analysis, DTA thermal mechanical analysis, TMA IGC
	cross-linking	raman spectroscopy X-ray diffraction
	branching	SEC/LLALS; SEC/MALLS; SEC/RI and viscosity and SEC/triple detector ^{13}C nmr and ^{1}H nmr (liquid and solid state) viscometry
	copolymer structure (block, graft...) and conformation	^{13}C nmr and ^{1}H nmr (dyads, triads, ... pentads), LACCC, static and dynamic light scattering ir, X-ray diffraction, neutron scattering, SAXS, WAXS Raman, TEM, SEM, MALDI–TOF MS pyrolysis/GC
	morphology	solid-state ^{13}C nmr, TEM, and SEM, neutron scattering, SAXS, DMA, Atomic force microscopy (AFM)
	purity and separation (block and graft copolymers)	thin-layer chromatography SEC/GPC, turbidimetry titration solvent extraction, ultracentrifugation, density gradient Centrifugation
mechanical, electrical and physical properties	stress–strain tests	Instron instrument
	crep tests	
	stress–relaxation tests	
	dynamical mechanical tests	dynamic and mechanical thermal analysis, DMTA
	impact tests	Izod and Charpy tests, falling-weight test bridge method, resonance method, time domain reflectometry
	dielectric relaxation	
	electrical breakdown	
	softening points	
	melt flow	
	melt viscosity	Brookfied and Money viscometers Bradender plasticgraph, capilary viscometers Inverse gas chromatography Wheather-Ometers
	solubility	
	permeability	
	stability	
	flammability	ignition tests, burning tests, oxigen redox tests

657

(b) Molecular size. The molecular size is defined by the average molecular weights, the MWD, degree of polymerization, hydrodynamic volume, radius of gyration, or other measurements relating to molecular dimension. There are a number of techniques that can be used to measure these characteristics such as: size-exclusion chromatography, membrane osmometry, vapor-pressure osmometry, end-group determination, light scattering, ultracentrifugation, and dilute-solution viscosity.

(c) Molecular organization. This group includes stereochemical configuration, isomerism, tacticity, branching, and cross-linking, which are of paramount importance for certain applications because they determine properties such as crystallinity and T_g, that definitely have a strong influence on mechanical properties. Molecular organization can be characterized by a number of techniques such as: thermal methods (differential scanning calorimetry, differential thermal analysis, thermogravimetry, etc), X-ray diffraction, solid-state nmr, ir and Raman spectroscopy, microscopy (optical and electron), inverse GC, neutron scattering, and others (see Table 13).

(d) Mechanical, electrical, and physical properties. These properties include crystallinity, T_g, and T_m, stress–strain, creep tests, impact tests, dynamic mechanical tests, dielectric relaxation, melt flow or viscosity, solubility, and others. The techniques used to determine these properties are summarized in Table 13. These properties are application specific and the experimental techniques are not always available commercially.

Table 13 presents a summary of the techniques used to characterize the copolymers. Additional information about these techniques can be obtained in (286–291). Nevertheless, the use of some of these techniques for copolymer characterization deserves some comments.

Size exclusion chromatography (SEC), also called gel permeation chromatography (GPC), is employed to determine molecular weight distributions and average molecular weights. This type of equipment separates the polymer molecules according to their molecular size, which means that at each elution time the polymer in the detection cell has the same molecular size. This is the basis of the so-called universal calibration that is a relationship between $[\eta]\,M$ and the retention time (where $[\eta]$ is the intrinsic viscosity and M is the molecular weight). The universal calibration assumes that a unique relationship between molecular weight and molecular size exists. However, the molecular size of a copolymer macromolecule depends not only on its size, but also on its composition, monomer sequence distribution, and topology (branched, grafted, etc). For example, the number of long-chain branches can be determined by SEC chromatograms if a combination of viscosity and light scattering detectors are used (287). However, this calculation usually requires us to assume the type of branching (star, bifunctional, tetrafunctional, etc.) (287), and hence special calibration curves should be developed for these copolymers (287,292).

SEC instruments can also be used to measure the copolymer composition distribution. In this case, a technique called orthogonal chromatography consisting in a set of two SEC systems is applied. The exit of the first one is connected to the injection valve of a second SEC system. The first SEC system is operated in a conventional way. The second one utilizes a HPLC-type separation mechanism by employing a weaker solvent mixture as the mobile phase (286,287). Other configurations of two separation techniques such as SEC–thin-layer chromatography (TLC), SEC–gradient high performance ligvid chromatography (HPLC) and HPLC–SEC have also been used to determine the chemical composition distribution (286–287). The main drawback of these techniques is that they are rather time consuming.

In the characterization of block and graft copolymers, the first concern is usually related to the separation of the block and graft copolymers from impurities such as homopolymers. Even if living anionic or controlled radical polymerization is used in the preparation of such copolymers, homopolymers are produced. The most useful techniques to carry out the fractionation are ultracentrifugation, density gradient centrifugation, TLC, GPC with dual or multiple detection systems, and turbidimetry. After fractionation according to the structure, the isolated block or graft copolymer is usually characterized by its overall composition and molecular weight. Composition can be determined by means of elemental analysis, uv, ir, nmr, and Raman spectroscopies; refractive index; pyrolysis–GC, and GC; X-ray fluorescence; and inverse GC. Determination of the molecular weight of block copolymers can be carried out by means of osmometry, viscometry, light scattering, and centrifugation, but the most widely employed technique is SEC (287) or more recently liquid adsorption chromatography under critical conditions (LACCC) (293,294). In the last few years, the structure and molecular weight distribution of the different segments composing block copolymers is being determined by matrix assisted laser desorption/ionization–time of flight (MALDI–TOF) mass spectroscopy. This technique is used alone or coupled with SEC instruments and also with LACCC chromatographic techniques (295–300). SEC techniques with refractive index (RI) and uv detectors for composition and low angle laser light scattering detector for branching have been used to determine the CCD and MWD of block copolymers (301).

16. Future Trends

Projections of future trends in polymers are highly speculative. Thus, in the prediction made in 1975 about the distribution of the market of plastics at the end of the twentieth century, the expectations were that most of the market would be taken by engineering and high performance plastics. However, the reality is that commodity plastics have extended their dominant position in the market with a share of ∼88% of the total market (302). Therefore, let this be a warning for this section.

For high tonnage copolymers, olefin-based copolymers will likely remain and even further strengthen their position in the market by entering other fields by out-performing other plastic materials or replacing more expensive and/or problematic conventional materials (132). The potential of the polyolefins is based on the easy availability of cheap monomers, the economics of the large-scale production, and the fact that they are both environmentally benign and extremely versatile in properties and applications. Developments in catalysts and polymerization processes have expanded the properties envelope making it possible for the polyolefins share of the global plastics market to have grown from 35 to 62% (132) in the last 25 years. This process will be reinforced with the development of new metallocene catalysts (131), which may be accelerated by the extensive application of high throughput screening techniques (303). Combinatorial materials research will have an important impact in the development of new polymeric materials. However, the development of new materials produced by new monomers is not expected unless these materials adapt to the existing production technologies and/or facilities. The development of water-resistant catalysts (304,305) will bring new products to the market.

Metallocene catalysts will also affect nonolefinic polymers because the highly ordered polymers obtained, eg, sindiotactic polystyrene (Questra, Dow Chemical; Xarec, Idemitsu) have properties not attained by traditional products.

Nanotechnology will likely have a critical influence on polymeric materials. Polymer nanocomposites will further expand the properties envelope bringing dramatic improvements in stiffness and gas-barrier properties (131).

In the specialty market, commercialization of controlled radical polymerization processes will bring a whole portfolio of new polymer materials. Block, graft, and hyperbranched copolymers of well-defined topology will have opportunities in markets such as coatings, adhesives, elastomers, sealants, lubricants, imaging materials, powder binders, dispersants, personal care products, detergents, photopatternable materials, and biological sensors. In order to fully exploit the potential of the controlled radical polymerization processes, more efficient, more selective, less expensive, and environmentally sound controlling agents are needed (41). In addition, these new materials will face strong competition from existing polymers to establish themselves in the market.

Polymer producers will continue to suffer strong competition and increasing social pressure to achieve a sustainable growth. Therefore, polymerization processes will be run more efficiently to achieve a consistent production of high performance polymers under safe and environmentally friendly conditions. This will require the development of robust and accurate on-line sensors for polymerization monitoring and efficient on-line optimization and control strategies.

BIBLIOGRAPHY

"Copolymers," in *ECT* 3rd ed., Vol. 6, pp. 798–818, by D. N. Schulz and D. P. Tate, Firestone Rubber & Tire Co.; in *ECT* 4th ed., Vol. 7, pp. 349–381, by Christine A. Costello and Donald N. Schulz, Exxon Research and Engineering Co.; "Copolymers" in *ECT* (online), posting date: December 4, 2000, by Christine A. Costello, Donald N. Schulz, Exxon Research and Engineering Company.

CITED PUBLICATIONS

1. IUPAC, *Pure Appl. Chem.* **57**, 1427 (1985).
2. IUPAC, *Pure Appl. Chem.* **66**, 2469 (1994).
3. IUPAC, *Pure Appl. Chem.* **69**, 2511 (1997).
4. IUPAC Recommendations on Macromolecular (Polymer) Nomenclature. *Guide for the Authors of Papers and Reports in Polymer Science and Technology*. Available at http://www.iupac.org/reports/IV/guide-for-authors.pdf.

5. G. R. Newkone, G. R. Baker, J. K. Young, and J. G. Traynham, *J. Polym. Sci. Polym. Chem.* **31**, 641 (1993).
6. H. G. Elias, *An Introduction to Polymer Science*, VCH, Weinheim, Germany, 1997, p. 50.
7. T. Fukuda, Y. D. Ma, and H. Inagaki, *Macromolecules* **18**, 17 (1985).
8. O. F. Olaj, I. Schnoll-Bitai, and P. Kreminger, *Eur. Polym. J.* **25**, 535 (1989).
9. M. L. Coote and T. P. Davis, *Prog. Polym. Sci.* **24**, 1217 (1999).
10. J. Forcada and J. M. Asua, *J. Polym. Sci. Polym. Chem. Ed.* **23**, 1955 (1985).
11. J. Brandrup, E. M. Immergut, and E. A. Grulke, eds., *Polymer Handbook 4th ed.*, Wiley-Interscience, New York, 1998.
12. F. R. Mayo and F. M. Lewis, *J. Am. Chem. Soc.* **66**, 1594 (1944).
13. A. S. Brar and S. Charan, *J. Polym. Sci. Part A: Polym. Chem.* **33**, 109 (1995).
14. M. C. López-González, M. Fernández-García, J. M. Barrales-Rienda, E. L. Madruga, and C. Arias, *Polymer* **34**, 3123 (1993).
15. A. Sáenz de Buruaga, J. R. Leiza, and J. M. Asua, *Polym React Eng.* **8**, 39 (2000).
16. J. Fleischhauer, G. Schmidtnaake, and D. Scheller, *Angew. Makromol. Chem.* **243**, 11 (1996).
17. K. Hatada, T. Kitayama, Y. Terawaki, H. Sato, R. Chujo, Y. Tanaka, R. Kitamaru, I. Ando, K. Hikichi, and F. Horii, *Polym. J.* **27**, 1104 (1995).
18. N. T. H. Ha and K. Fujimori, *Acta Polym.* **49**, 404 (1998).
19. A. L. Burke, T. A. Duever, and A. Penlidis, *Ind. Eng. Chem. Res.* **36**, 1016 (1997).
20. G. Odian, *Principles of Polymerization*. 2nd ed. Wiley-Interscience, New York, 1981, p. 455.
21. G. F. Meijs and E. Rizzardo, *JMS Rev. Macromol. Chem. Phys.* **C30** (384), 305 (1990).
22. J. Guillot and L. Rios-Guerrero, *Makromol. Chem.* **183**, 1979 (1982).
23. G. Arzamendi, J. C. de la Cal, and J. M. Asua, *Angew. Makromol. Chem.* **194**, 47 (1992).
24. L. M. Gugliotta, G. Arzamendi, and J. M. Asua, *J. Appl. Polym. Sci.* **55**, 1017 (1995).
25. F. L. M. Hautus, H. N. Lissen, and A. L. German, *J. Polym. Sci. Polym. Chem. Ed.* **22**, 3487 (1984).
26. K. K. Chee and S. C. Ng, *Macromolecules* **19**, 2779 (1986).
27. M. van den Brink, A. M. van Herk, and A. L. German, *J.Polym. Sci. Part A: Polym. Chem.* **37**, 3793 (1999).
28. A. L. Polic, T. A. Duever, and A. Penlidis, *J. Polym. Sci. Part A: Polym. Chem.* **36**, 813 (1998).
29. T. A. Duever, K. F. O'Driscoll, and P. M. Reilly, *J. Polym. Sci. Polym. Chem. Ed.* **21**, 2003 (1983).
30. J. C. de la Cal, J. R. Leiza, and J. M. Asua, *J. Polym. Sci. Part. A: Polym. Chem.* **29**, 155 (1991).
31. C. C. Price, *J. Polym. Sci.* **1**, 83 (1946).
32. T. Alfrey and C. C. Price, *J. Polym. Sci.* **2**, 101 (1947).
33. G. C. Lowry, *J. Polym. Sci.* **42**, 463 (1960).
34. P. Wittmer, *Adv. Chem.* **99**, 140 (1971).
35. H. Krüger, J. Bauer, and J. Rübner, *Makromol. Chem.* **188**, 2163 (1987).
36. D. E. Palmer, N. T. McManus, and A. Penlidis, *J. Polym. Sci., Part A: Polym. Chem.* **38**, 1981 (2000).
37. C. J. Hawker, A. W. Bosman, and E. Harth, *Chem. Rev.* **101**, 3661 (2001).
38. M. K. Georges, R. P. N. Veregin, P. M. Kazmaier, and G. K. Hamer, *Macromolecules* **26**, 2987 (1993).
39. D. Benoit, S. Grimaldi, S. Robin, J. P. Finet, P. Tordo, and Y. Gnanou, *J. Am. Chem. Soc.* **122**, 5929 (2000).

40. D. Benoit, V. Chaplinski, R. Braslau, and C. Hawker, *J. Am. Chem. Soc.* **121**, 3904 (1999).
41. K. Matyjaszewski and J. Xia, *Chem. Rev.* **101**, 2921 (2001).
42. WO Pat. 9,801,480, K. Matyjaszewski, S. Coca, S. G. Gayno, Y. Nakayama, and S. M. Jo.
43. J. S. Wang and K. Matyjaszewski, *J. Am. Chem. Soc.* **117**, 5614 (1995).
44. T. Grimaud and K. Matyjaszewski, *Macromolecules* **30**, 2216 (1997).
45. X. Huang and M. J. Wirth, *Macromolecules* **32**, 1694 (1999).
46. K. Matyjaszewski, S. M. Jo, H. J. Paik, and D. A. Shipp, *Macromolecules* **32**, 6431 (1999).
47. E. J. Ashford, V. Naldi, R. O'Dell, N. C. Billingham, and S. P. Armes, *Chem. Commun.* 1285 (1999).
48. J. Chiefari, Y. K. Chong, F. Ercole, J. Krstina, J. Jeffery, T. P. T. Le, R. T. A. Mayadunne, G. F. Meijs, C. L. Moad, G. Moad, E. Rizzardo, and S. H. Thang, *Macromolecules* **31**, 5559 (1998).
49. E. Rizzardo, Y. K. Chong, R. A. Evans, G. Moad, and S. H. Thang, *Macromol. Symp.* **111**, 1 (1996).
50. J. F. Quinn, T. P. Davis, and E. Rizzardo, *Chem. Commun.* 1044 (2001).
51. G. Moad, J. Chiefari, Y. K. Chong, J. Kristina, R. T. A. Mayadunne, A. Postma, E. Rizzardo, and S. H. Thang, *Polym. Int.* **49**, 993 (2000).
52. C. Marestin, C. Nöel, A. Guyot, and J. Claverie, *Macromolecules* **31**, 4041 (1998).
53. J. Qiu, B. Charleux, and K. Matyjaszewski, *Prog. Polym. Sci.* **26**, 2083 (2001).
54. T. Prodpran, V. Dimonie, E. D. Sudol, and M. S. El-Aasser, *Macromol. Symp.* **155**, 1 (2000).
55. P. J. Macleod, R. Barber, P. G. Odell, B. Keoshkerian, and M. K. Georges, *Macromol. Symp.* **155**, 31 (2000).
56. K. Matyjaszewski, J. Qiu, N. V. Tsarevsky, and B. Charleux, *J. Polym. Sci., Part A: Polym. Chem.* **38**, 4724 (2000).
57. K. Matyjaszewski, J. Qiu, D. A. Shipp, and S. G. Gaynor, *Macromol. Symp.* **155**, 15 (2000).
58. M. Lansalot, C. Farcet, B. Charleux, J. P. Vairon, and R. Pirri, *Macromolecules* **32**, 7354 (1999).
59. H. de Brouwer, J. G. Tsavalas, F. J. Schork, and M. J. Monteiro, *Macromolecules* **33**, 9239 (2000).
60. J. M. Asua, *Prog. Polym. Sci.* **27**, 283 (2002)
61. S. G. Roos, A. H. E. Muller, and K. Matyjaszewski, *ACS Symp. Ser.* **768**, 361 (2000).
62. R. F. Storey, B. J. Chrisholm, and K. R. Choate. *J. Macromol. Sci.: Pure Appl. Chem.* **A31**, 969 (1994).
63. S. Bywater, in H. F. Mark, N. M. Bikales, C. G. Overberger, and G. Menges, eds., *Encyclopedia of Polymer Science and Engineering*, Vol. 2, John Wiley & Sons, Inc., New York, 1987, p. 1.
64. J. P. Kennedy and E. Maréchal, *Carbocationic Polymerization*, Wiley-Interscience, New York, 1982.
65. J. M. Rooney, D. R. Squire, and V. T. Stannett, *J. Polym. Sci. Polym. Chem. Ed.* **14**, 1877 (1976).
66. M. Sawamoto, M. Miyamoto, and T. Higashimura, *Cationic Polymerization and Related Processes*, Academic Press, New York, 1984, p. 89.
67. S. Smith and A. J. Hubin, *J. Macrom. Sci. Chem.* **7**, 1399 (1973).
68. T. Saegusa, S. Mutsumoto, and Y. Hashimoto, *Macromolecules* **3**, 377 (1970).
69. E. J. Goethals, M. Van de Velde, and A. Munir, *Cationic Polymerization and Related Processes*, Academic Press, New York, 1984, p. 387.

70. Y. Chujo and T. Saegusa, in H. F. Mark, N. M. Bikales, C. G. Overberger, and G. Menges, eds., *Encyclopedia of Polymer Science and Engineering*, Vol. 14, John Wiley & Sons, New York, Inc., 1988, p. 622.

71. T. Endo and W. J. Bailey, *J. Polym. Sci. Polym. Lett. Ed.* **13**, 193 (1975).

72. T. Endo and W. J. Bailey, *J. Polym. Sci. Polym. Symp.* **64**, 17 (1978).

73. T. Endo and W. J. Bailey, *J. Polym. Sci. Polym. Lett. Ed.* **18**, 25 (1980).

74. L. N. Sidney, S. E. Shaffer, and W. J. Bailey, *Polym. Prepr. ACS Div. Polym. Chem.* **22**, 373 (1981).

75. W. J. Bailey, *ACS Symp. Ser.* **286**, 47 (1985).

76. G. Cecchin, G. Moroni, and A. Pelliconi, *Macromol. Symp.* **173**, 195 (2001).

77. P. Cossee, *Tetrahedron Lett.* **17**, 12 (1960).

78. P. J. T. Tait and N. D. Watkins, in G. Allen and J. C. Bevington, eds., *Monoalkene Polymerization: Mechanisms in Comprehensive Polymer Science*, Vol. 4, 1989, p. 535.

79. J. Boor, *Ziegler-Natta Catalyst and Polymerization*, Academic Press, New York, 1979, Chapt. 20.

80. T. F. McKenna, J. Dupuy, and R. Spitz, *J. Appl. Polym. Sci.* **63**, 315 (1997).

81. Y. V. Kissin, *J. Polym. Sci. Part A: Polym. Chem.* **39**, 1681 (2001).

82. L. L. Böhm, *Proc. IUPAC Macromol. Symp.* 245 (1982).

83. J. V. Seppälä and M. Auer, *Prog. Polym. Sci.* **15**, 147 (1990).

84. K. J. Chu, J. B. P. Soares, A. Penlidis, and S. K. Ihm, *Eur. Polym. J.* **36**, 3 (1999).

85. K. Y. Choi and W. H. Ray, *J.M.S. Rev. Macromol. Chem. Phys.* **C25**(1), 1 (1985).

86. K. Czaja and M. Bialek, *Polymer* **42**, 2289 (2000).

87. U.S. Pat 2,825,721, J. P. Hogan and R. L. Banks.

88. M. P. McDaniel, *Ind. Eng. Chem. Res.* **27**, 1559 (1988).

89. B. M. Wekhuysen and R. A. Schoonheydt, *Cat. Today* **51**, 215 (1999).

90. Ø. Espelid and K. J. Borve, *J. Catal.* **195**, 125 (2000).

91. G. Natta, P. Pino, G. Mazzani, and R. Lauzo, *Chim. Ind.* (Milan) **39**, 1032 (1957).

92. D. S. Breslow and N. R. Newburg, *J. Am. Chem. Soc.* **79**, 5072 (1957).

93. H. Sinn and W. Kaminsky, *Adv. Organomet. Chem.* **18**, 99 (1980).

94. J. B. P. Soares and A. E. Hamielec, *Polym. React. Eng.* **3**, 131 (1995).

95. S. S. Reddy and S. Sivaram, *Prog. Polym. Sci.* **20**, 309 (1995).

96. J. Huang and G. L. Rempel, *Prog. Polym. Sci.* **20**, 459 (1995).

97. K. Soga and T. Shino, *Prog. Polym. Sci.* **22**, 1503 (1997).

98. Y. Imanishi and N. Naga, *Prog. Polym. Sci.* **26**, 1147 (2001).

99. G. W. Coates and R. M. Waymouth, *Science* **267**, 217 (1995).

100. K. Koo and T. J. Marks, *J. Am. Chem. Soc.* **121**, 8791 (1999).

101. J. C. W. Chien and D. He, *J. Polym. Sci., Part A: Polym. Chem.* **29**, 1585 (1991).

102. J. C. W. Chien and D. He, *J. Polym. Sci., Part A: Polym. Chem.* **29**, 1595 (1991).

103. M. R. Ribeiro, A. Diffieux, and M. F. Portela, *Ind. Eng. Chem. Res.* **36**, 1224 (1997).

104. W. Kaminski and F. Renner, *Makromol. Chem. Rapid. Commun.* **14**, 239 (1993).

105. K. Soga and M. Kaminaka, *Makromol. Chem. Rapid. Commun.* **13**, 221 (1992).

106. K. Soga, H. J. Kim, and T. Shiono, *Makromol. Chem. Rapid. Commun.* **15**, 139 (1994).

107. T. Uozumi, K. Miyazawa, T. Sano, and K. Soga, *Macromol. Chem. Rapid. Commun.* **18**, 883 (1997).

108. M. K. Leclerc and R. M. Waymouth, *Angew. Chem. Int. Ed.* **37**, 922 (1998).

109. L. Oliva, L. Izzo, and P. Longo, *Macromol. Rapid. Commun.* **17**, 745 (1996).

110. N. Naga, K. Mizunuma, H. Sadatoshi, and M. Kakugo, *Macromolecules* **30**, 2197 (1997).

111. W. Kaminski and R. Steiger, *Polyhedron* **7**, 2375 (1988).

112. P. Corradini and G. Guerra, *Prog. Polym. Sci.* **16**, 239 (1991).

113. J. C. W. Chien, Z. Yu, M. M. Margues, J. C. Flores, and M. D. Ranch, *J. Polym. Sci., Part A: Polym. Chem.* **36**, 319 (1998).

114. R. D. Leaversuch, *Modern Plastics* 68–69, (April 1998).

115. J. A. M. Canich, U.S. Pat. 5,026,798 (1991) (to Exxon).

116. D. R. Neithamer and J. C. Stevens, Eur. Pat. Appl. 418,044 A2 (1991) (to Dow Chemicals).

117. G. Xu, *Macromolecules* **31**, 2395 (1998).

118. M. J. Han, *Macromolecules* **13**, 1009 (1980).

119. M. J. Han, *Macromolecules* **15**, 438 (1982).

120. M. J. Han, H. Ch. Kang, and K. B. Choi, *Macromolecules* **19**, 1649 (1986).

121. J. H. MacKey, V. A. Pattison, and J. A. Pawlat. *J. Polym. Sci. Polym. Chem. Ed.* **16**, 2849 (1978).

122. E. Turska, S. Boryniec, and L. Pietrzak, *J. Appl. Polym. Sci.* **18**, 667 (1974).

123. E. Turska, S. Boryniec, and A. Dems, *J. Appl. Polym. Sci.* **18**, 671 (1974).

124. E. Turska, S. Boryniec, and R. Jantas, *J. Appl. Polym. Sci.* **20**, 1849 (1976).

125. E. Turska, L. Pietrzak, and R. Jantas, *J. Appl. Polym. Sci.* **23**, 2409 (1979).

126. S. Boryniec, *Acta Polym.* **39**, 545 (1988).

127. S. Boryniec, *Acta Polym.* **40**, 622 (1989).

128. M. Balsam, P. Baum, and J. Engelmann, *Macromol. Mater. Eng.* **286**, A7 (2001).

129. D. L. Lohse, in C. D. Craver and C. E. Carraher, eds., *Applied Polymer Science in the 21st Century*, Elsevier, Amsterdam, The Netherlands, 2000, p. 73.

130. P. Galli, *J.M.S. Pure Appl. Chem.* **A36**, 1561 (1999).

131. K. B. Sinclair, *Macromol. Symp.* **173**, 237 (2001).

132. P. Galli and C. Vecellio, *Prog. Polym. Sci.* **26**, 1287 (2001).

133. U.S. Pat. 5,622,906 (1997) T. M. Pettijohm, (to Phillips Petroleum).

134. U.S. Pat. 5,719,241 (1994) G. Debras and A. Razavi, (to Fina).

135. Eur. Pat. 0,719,797 (1996) W. Michiels and A. Muñoz-Escalona, (to Repsol).

136. U.S. Pat. 5,326,835 (1994) A. Ahvenainen, K. Saranzila, H. Andtsjoe, J. Takakarhu, and A. Palmroos, (to Neste OY).

137. U.S. Pat. 5,698,642 (1997) G. Govoni, R. Rinaldi, M. Covezzi, and P. Galli, (to Montell).

138. WO Pat. 00/02929 (2000) G. Govoni and M. Covezzi, (to Montell).

139. A. de Vries and N. Izzo-Iammarrone, *Dechema Monographs* **137**, 43 (2001).

140. G. Guidetti, P. Busi, I. Giulianelli, and R. Zanetti, *Eur. Polym. J.* **19**, 757 (1983).

141. P. Galli, J. C. Haylock, T. Simomazzi, in J. Karger-Kocsis, ed., *Polypropylene Structure, Blends and Composites*, Vol. 2, Chapman & Hale, London, 1995.

142. U.S. Pat. 4,226,741 (1980) L. Luciani, N. Kashiva, P. C. Barbe, and A. Toyota, (to Montedison SPA).

143. C. Cecchin. *Macromol. Symp.* **78**, 213 (1994).

144. R. P. Quirk and M. Morton, in J. E. Mark, B. Erman, and F. R. Eirich, eds., *Science and Technology of Rubber*, Academic Press, New York, 1994, p. 23.

145. L. H. Howland, W. E. Messer, V. C. Neklutin, and V. S. Chambers, *Rubber Ag.* **64**, 459 (1949).

146. D. C. Blackley, in P. A. Lovell and M. S. El-Aasser, eds., *Emulsion Polymerization and Emulsion Polymers*, John Wiley, Chichester, 1997, p. 521.

147. A. J. De Fusco, K. C. Sehgal, and D. R. Bassett, in J. M. Asua, ed., *Polymerization Dispersions: Principles and Applications*, Kluwer, Dordrecht, The Netherlands, 1997, p. 379.

148. D. I. Lee, in J. M. Asua, ed., *Polymerization Dispersions. Principles and Applications*, Kluwer, Dordrecht, The Netherlands, 1997, p. 497.

149. M. Grady, 3rd IUPAC Sponsored International Symposium on Free Radical Polymerization: Kinetics and Mechanisms, Lucca, Italy, 2001.

150. P. M. Lesko and P. P. Sperry, in P. A. Lovell and M. S. El-Aasser, eds., *Emulsion Polymerization and Emulsion Polymers*, John Wiley & Sons, Chichester, U.K. 1997, p. 619.

151. T. G. Fox, *Bull. Am. Phys. Soc. Series* **1**, 2 (1950).

152. D. Satas, *Adhesive Age* **15**, 19 (1972).

153. C. Plessis, G. Arzamendi, J. R. Leiza, J. M. Alberdi, H. A. S. Schoobrood, D. Charmot, and J. M. Asua, *J. Polym. Sci. Part A: Polym. Chem.* **39**, 1106 (2001).

154. B. Améduri, B. Boutevin, and G. Kostow, *Prog. Polym. Sci.* **26**, 105 (2001).

155. J. F. MacGregor, A. Penlidis, and A. E. Hamielec, *Polym. Process Eng.* **2**, 179 (1984).

156. G. E. Eliçabe and G. R. Meira, *Polym. Eng. Sci.* **36**, 433 (1986).

157. M. Embirucu, E. Lima, and J. C. Pinto, *Polym. Eng. Sci.* **36**, 433 (1996).

158. M. A. Dubé, J. B. P. Soares, A. Penlidis, and A. E. Hamielec, *Ind. Eng. Chem. Res.* **36**, 966 (1997).

159. J. P. Congalidis and J. R. Richards, *Polym. React. Eng.* **6**, 71 (1998).

160. R. B. Mankar, D. N. Saraf, and S. K. Gupta. *J. Polym. Eng.* **18**, 371 (1998).

161. L. L. Böhm, P. Göbel, O. Lorenz, and T. Tauchnitz, *Dechema Monographs*, **127**, 257 (1992).

162. K. B. McAuley and J. F. MacGregor, *AIChE J.* **37**, 825 (1991).

163. K. B. McAuley and J. F. MacGregor, *AIChE J.* **38**, 1564 (1992).

164. K. B. McAuley and J. F. MacGregor, *AIChE J.* **39**, 855 (1993).

165. A. M. Meziou, P. B. Deshpande, C. Cozewith, N. I. Silverman, and W. G. Morrison, *Ind. Eng. Chem. Res.* **35**, 164 (1996).

166. J. Dimitratos, C. Georgakis, M. S. El-Aasser, A. Klein, in K. H. Reichert and W. Geister, eds., *Polymer Reaction Engineering*, VCH Verlag, Weinheim, Germany, 1989.

167. J. R. Leiza, J. C. de la Cal, G. R. Meira, and J. M. Asua, *Polym. React. Eng.* **1**, 461 (1992).

168. M. F. Ellis, T. W. Taylor, V. González, and K. F. Jonson, *AIChE J.* **40**, 445 (1994).

169. I. Sáenz de Buruaga, P. D. Armitage, J. R. Leiza, and J. M. Asua. *Ind. Eng. Chem. Res.* **36**, 4243 (1997).

170. I. Sáenz de Buruaga, A. Echeverría, P. D. Armitage, J. C. de la Cal, J. R. Leiza, and J. M. Asua, *AIChE J.* **43**, 1069 (1997).

171. P. Canu, S. Canegallo, G. Storti, and M. Morbidelli, *J. Appl. Polym. Sci.* **54**, 1899 (1994).

172. H. Hammouri, T. F. McKenna, and S. Othman, *Ind. Eng. Chem. Res.* **38**, 4815 (1999).

173. A. Urretabizkaia, J. R. Leiza, and J. M. Asua, *AIChE J.* **40**, 1850 (1994).

174. I. Sáenz de Buruaga, J. R. Leiza, and J. M. Asua, *Polym. React. Eng.* **8**, 39 (2000).

175. T. F. McKenna, G. Févotte, S. Othman, A. M. Santos, and H. Hammouri, *Polym. React. Eng.* **8**, 1 (2000).

176. P. Canu, S. Canegallo, M. Morbidelli, and G. Storti, *J. Appl. Polym. Sci.* **54**, 198 (1994).

177. A. Echeverría, J. R. Leiza, J. C. de la Cal, and J. M. Asua, *AIChE J.* **44**, 1667 (1998).

178. M. Vicente, S. Ben Amor, L. M. Gugliotta, J. R. Leiza, and J. M. Asua, *Ind. Eng. Chem. Res.* **40**, 218 (2001).

179. T. J. Crowley and K. Y. Choi, *Comput. Chem. Eng.* **25**, 1153 (1999).

180. T. J. Crowley and K. Y. Choi, *Ind. Eng. Chem. Res.* **36**, 3676 (1997).

181. A. Salazar, L. M. Gugliotta, J. R. Vega, and G. R. Meira, *Ind. Eng. Chem. Res.* **37**, 3582 (1998).

182. M. Vicente, J. R. Leiza, and J. M. Asua, *AIChE J.* **47**, 1594 (2001).

183. E. Saldivar and W. H. Ray, *AIChE J.* **43**, 2021 (1997).

184. C. Sayer, G. Arzamendi, J. M. Asua, E. L. Lima, and J. C. Pinto, *Comp. Chem. Eng.* **25**, 839 (2001).

185. M. Vicente, J. R. Leiza, and J. M. Asua, *ACS Symp. Ser.* **801**, 113 (2002).

186. M. Vicente, C. Sayer, G. Arzamendi, J. R. Leiza, J. C. Pinto, and J. M. Asua, *Chem. Eng. J.* **85**, 339 (2001).

187. D. H. C. Chien and A. Penlidis, *J.M.S. Macromol. Chem. Phys.* **C30**, 1, (1990).

188. O. Kammona, E. G. Ghatzi, and C. Kiparissides, *J.M.S. Rev. Macromol. Chem.* **57**, 39 (1999).

189. S. Canegallo, P. Canu, M. Morbidelli, and G. Storti, *J. Appl. Polym. Sci.* **47**, 961 (1993).

190. A. Urretabizkaia, E. D. Sudol, M. S. El-Aasser, and J. M. Asua, *J. Polym.Sci. Polym. Chem. Ed.* **31**, 2907 (1993).

191. P. D. Gossen, J. F. MacGregor, and R. M. Pelton, *Appl. Spect.* **47**, 1852 (1993).

192. M. Agnely, B. Amram, D. Charmot, S. Ben Amor, J. R. Leiza, J. M. Asua, C. Macron, J. P. Huvenne, and J. Sawatzki, International Symposium on Polymers in Dispersed Media, Lyon (France), 1999.

193. J. A. Seiner and M. Litt, *Macromolecules* **4**, 308 (1971).

194. K. Dodson and J. R. Ebdon, *Eur. Polym. J.* **13**, 791 (1977).

195. J. Furukawa, H. Amano, and R. Hirai, *J. Polym. Sci. Polym. Chem. Ed.* **10**, 681 (1972).

196. T. Uozumi, C. E. Ahu, M. Tomisaka, J. Jin, G. Tian, T. Sano, and K. Soga, *Macromol. Chem. Phys.* **201**, 1748 (2000).

197. W. Fan, M. F. Leclerc, and R. M. Waymouth, *J. Am. Chem. Soc.* **123**, 9555 (2001).

198. O. Sommazzi and F. Garbassi, *Prog. Polym. Sci.* **22**, 1547 (1997).

199. G. Holden, in C. D. Craver and C. E. Carraher, eds., *Applied Poymer Science, 21st Century*, Elsevier, 2000, p. 231.

200. H. L. Hsieh and R. P. Quirk, *Anionic Polymerization: Principles and Practical Applications.* Marcel Dekker, New York, (1993).

201. L. H. Tuang and G. Y. S. Lo, *Macromolecules* **27**, 2219 (1994).

202. P. Prabhu, A. Schindler, M. H. Theil, and R. D. Gilbert, *J. Polym. Sci. Part A: Polym. Chem. Ed.* **19**, 523 (1981).

203. WO Pat. 9,858,978 (2001) G. W. Verstrate, C. Cozewith, T. J. Pacansky, W. M. Davis, and P. Rangarajan, (to Exxon).

204. Eur. Pat. 1,108,733 (2001) K. Tanaka, Y. Sugita, M. Nakayama, and T. Nakamura, (to Idemitsu Petrochemical).

205. E. Hauptman, R. M. Waymouth, and J. W. Ziller, *J. Am. Chem Soc.* **117**, 11586 (1995).

206. C. Granel, P. Dubois, R. Jerome, and P. Teyssié, *Macromolecules* **29**, 8576 (1996).

207. Y. Kotani, M. Kato, M. Kamigaito, and M. Sawamoto, *Macromolecules* **29**, 6979 (1996).

208. B. Y. K. Chong, T. P. T. Le, G. Moad, E. Rizzardo, and S. H. Thang, *Macromolecules* **32**, 2071 (1999).

209. R. B. Grubbs, J. M. Dean, M. E. Broz, and F. S. Bales, *Macromolecules* **33**, 9522 (2000).

210. H. Hoecker, H. Kenl, S. Kuehling, and W. Hovestadt, *Macromol. Chem. Macromol. Symp.* **42/43**, 145 (1991).

211. C. X. Song and X. D. Feng, *Macromolecules* **17**, 2764 (1984).

212. D. W. Grijpma and A. J. Pennings, *Polym. Bull.* **25**(3), 335 (1991).

213. S. Inoue and T. Aida, in G. Eastmon, A. Ledwith, S. Russo, and P. Sigwalt, eds., *Comprehensive Polymer Science, The Synthesis, Characterization, Reactions & Applications of Polymers*, Vol. 3, Pergamon Press, New York, 1989, p. 553.

214. P. Dreyfuss and M. Dreyfuss, in G. Eastmon, A. Ledwith, S. Russo, and P. Sigwalt, eds., *Comprehensive Polymer Science, The Synthesis, Characterization, Reactions & Applications of Polymers*, Vol. 3, Pergamon Press, New York, 1989, p. 851.

215. T. Aida, K. Sanuki, and S. Inoue, *Macromolecules* **18**, 1049 (1985).
216. T. Aida, M. Ishikawa, S. Inoue, *Macromolecules* **19**, 8 (1986).
217. J. Henschen, R. Jerome, and P. Teyspié, *Macromolecules* **14**, 242 (1981).
218. R. P. Quirk, D. J. Kinning, and L. J. Fetters, in G. Eastmong, A. Ledwith, S. Russo, and P. Sigwalt, eds., *Comprehensive Polymer Science, The Synthesis, Characterization, Reactions & Applications of Polymers*, Vol. 7, Pergamon Press, New York, 1989. p. 1.
219. P. K. Bossaer, E. J. Goethals, P. J. MacKett, and D. C. Pepper, *Eur. Polym. J.* **13**, 489 (1977).
220. D. Tian, P. H. Orbois, and R. Jerome, *Macromolecules* **30**, 2575 (1997).
221. F. Stassin, Q. Hallenx, P. H. Dubois, C. H. Detrembleur, P. H. Lecounte, and R. Jerôme, *Macromol. Symp.* **153**, 27 (2000).
222. B. A. Rosenberg, Y. I. Estrin, and G. A. Estrina, *Macromol. Symp.* **153**, 209 (2000).
223. V. Simic, S. Rensec, and N. Spassky, *Macromol. Symp.* **153**, 109 (2000).
224. W. Mekel, W. Goyert, and W. Wieder, in G. Holden, N. R. Legge, R. P. Quirk, and H. E. Schroeder, eds., *Thermoplastic Elastomers-A Comprehensive Review*. Hanser & Hanser/Gardner, Munich, Germany, 1996, Chapt. 2.
225. R. K. Adams, G. K. Hoeschele, and W. K. Wisiepe, in G. Holden, N. R. Legge, R. P. Quirk, and H. E. Schroeder, eds., *Thermoplastic Elastomers-A Comprehensive Review*. Hanser & Hanser/Gardner, Munich, Germany, 1996, Chapt. 8.
226. R. G. Nelb and A. T. Chen, in G. Holden, N. R. Legge, R. P. Quirk, and H. E. Schroeder, eds., *Thermoplastic Elastomers-A Comprehensive Review*. Hanser & Hanser/Gardner, Munich, Germany, 1996, Chapt. 9.
227. C. J. Hawker, J. L. Hedrick, E. E. Malmstrom, M. Trollsas, D. Mecerreyes, P. H. Dubois, and R. Jerome, *Macromolecules* **31**, 213 (1998).
228. U. Stalmach, B. de Boer, A. D. Post, P. F. van Hutten, and G. Hadziioannon, *Angew. Chem. Int. Ed. Engl.* **40**, 428 (2001).
229. S. Coca and K. Matyjaszewski, *Macromolecules* **30**, 2808 (1997).
230. A. Kajiwara and K. Matyjaszewski, *Macromolecules* **31**, 3489 (1998).
231. S. G. Gaynor and K. Matyjaszewski, *Macromolecules* **30**, 4241 (1997).
232. P. Cohen, M. J. M. Abadie, F. Schue, and D. M. Richards, *Polymer* **22**, 1316 (1981).
233. L. Leger, *Macromol. Symp.* **149**, 197 (2000).
234. A. Falsafi, F. Bates, and M. Tyrrell, *Macromolecules* **34**, 1323 (2001).
235. R. Fayt, R. Jerome, and Ph. Teyssie, *Makromol. Chem.* **187**, 837 (1986).
236. Eur. Pat. 9937 (1980) D. D. Donermeyre and J. G. Martins, (to Monsanto).
237. Ger. Offen. 3,002,164 (1980) R. L. Vermillion and W. A. Bernett, (to 3M).
238. A. Laschewsky, *Adv. Polym. Sci.* **124**, 59 (1995).
239. K. L. Wooley, *J. Polym. Sci. Part A: Polym. Chem.* **38**, 1397 (2000).
240. Y. Okamoto, H. Miyagi, M. Kakugo, and K. Takahashi, *Macromolecules* **24**, 5639 (1991).
241. A. Echte, in C. K. Riew, ed., *Rubber-Toughened Plastics, Advances in Chemistry Series, No 222*, ACS, Washington D.C., 1989.
242. E. J. Markel, W. Weng, A. J. Peacock, and A. H. Dekmezian, *Macromolecules* **33**, 8541 (2000).
243. U.S. Pat. 6,114,457 (1998) E. J. Markel, C. H. De Gracia, and A. H. Dekmezian, (to ExxonMobil).
244. W.O. 9,712,919 (1997) H. Farah, M. Laugher, F. Hofmeister, T. H. Ho, M. Hughes, H. T. Phasu, S. P. Namhata, C. P. Bosnyak, R. T. Johnston, D. R. Parikh, R. M. Patel, C. L. Werling, and S. A. Ogol, (to Dow Chemical).
245. N. Hadjichristidis and J. E. L. Roovers, *J. Polym. Sci., Polym. Phys. Ed.* **16**, 851 (1978).

246. J. E. Puskas, P. Antony, Y. Kwon, C. Paulo, M. Kovar, P. R. Norton, G. Kaszas, and V. Altstädt, *Macromol. Mater. Eng.* **286**, 565 (2001).
247. G. C. Eastmond, *Pure Appl. Chem.* **53**, 657 (1981).
248. R. B. Grubbs, C. J. Hawker, J. Dao, and J. M. J. Frechet, *Angew. Chem. Int. Ed. Eng.* **36**, 270 (1997).
249. H. J. Paik, S. G. Gaynou, and K. Matyjaszewski, *Macromol. Rapid. Commun.* **19**, 47 (1998).
250. M. Takaki, R. Asami, H. Inukai, and T. Inenaga, *Macromolecules* **12**, 383 (1979).
251. G. C. Cameron and A. W. S. Duncan, *Makromol. Chem.* **184**, 1153 (1983).
252. G. O. Schulz and R. Milkovich, *J. Appl. Polym. Sci.* **27**, 4773 (1982).
253. R. Asami, M. Takaki, K. Kyuda, and E. Asakura, *Polymer J.* **15**, 139 (1983).
254. K. Ito, N. Usami, and Y. Yamashita, *Macromolecules* **13**, 216 (1980).
255. J. Chiefari, J. Jeffery, R. T. A. Mayadunne, G. Moad, E. Rizzardo, and S. H. Thang, *ACS Symp. Ser.* **768**, 297 (2000).
256. P. Dreyfuss and R. P. Quirk, in H. F. Mank, N. M. Bikales, C. G. Overberger, and G. Menges, eds., *Encyclopedia of Polymer Science and Engineering*, 2nd ed. Vol. 7, John Wiley & Sons, Inc., New York, 1987, p. 551.
257. R. Jerome, M. Henrioulle-Grauville, B. Bouterin, and J. J. Robin, *Prog. Polym. Sci.* **16**, 837 (1991).
258. Piirma, J. C. Chang, and M. Daneshuar, *Polym. Prepr., ACS Div. Polym. Sci.* **26**, 219 (1985).
259. S. J. Kubisen and G. S. Peacock, Technical Paper, Radcure Europe 85, FC 85 428 (1985).
260. K. Matyjaszewski, K. L. Beers, A. Hern, and S. G. Gaynor, *J. Polym. Sci. Part A: Polym. Chem.* **36**, 823 (1998).
261. J. P. Kennedy, *J. Polym. Sci. Polym. Symp. Ed.* **72**, 73 (1985).
262. B. Keszler and J. P. Kennedy, *J. Macromol. Sci. Chem.* **A21**, 319 (1984).
263. D. Chen, J. P. Kennedy, and A. J. Allen, *J. Macromol. Sci. Chem.* **A25**, 389 (1988).
264. D. Benoit, E. Harth, C. J. Hawker, and B. Helms, *Polym. Prepr.* **41**, 42 (2000).
265. K. Matyjaszewski, P. J. Miller, E. Fossum, and Y. Nakagawa, *Appl. Organomet. Chem.* **12**, 667 (1998).
266. J. Ueda, M. Kamigaito, and M. Sawamoto, *Macromolecules* **31**, 557 (1998).
267. Y. K. Chong, T. P. T Ley, G. Moad, E. Rizzardo, and S. H. Tang, *Macromolecules* **32**, 2071 (1999).
268. U.S. Pat. 4,136,137 (1979) H. L. Hsich and F. S. Naylor, (to Phillips Petroleum).
269. R. F. Storey, B. J. Chisholm, and K. R. Choate, *J. Macromol. Sci. Pure Appl. Chem.* **A31**, 969 (1994).
270. M. Jikei and M. Kakimoto, *Prog. Polym. Sci.* **26**, 1233 (2001).
271. C. J. Hawker, J. M. Frechet, R. B. Grubbs, and J. Dao, *J. Am. Chem. Soc.* **117**, 10763 (1995).
272. S. G. Gaynor, S. Edelman, and K. Matyjaszewski, *Macromolecules* **29**, 1079 (1996).
273. V. Coessens, Y. Nakagawa, S. G. Gaynor, and K. Matyjaszewski, *Macromol. Rapid. Commun.* **21**, 103 (2000).
274. L. K. Johnson, C. M. Killiam, and M. Brookhart, *J. Am. Chem. Soc.* **117**, 6414 (1995).
275. Z. Guan, P. M. Cotts, E. F. McLord, and S. J. McLain, *Science* **283**, 2059 (1999).
276. Y. H. Kim, *J. Polym. Sci. Part A: Polym. Chem.* **36**, 1685 (1998).
277. B. Voit, *J. Polym. Sci. Part A: Polym. Chem.* **38**, 2505 (2000).
278. D. Schmaljohann, B. Voit, J. F. G. A. Jansen, P. Hedriks, and J. A. Loontjens, *Macromol. Mater. Eng.* **31**, 275 (2000).
279. R. A. T. M. van Benthem, *Prog. Org. Coat.* **40**, 203 (2000).
280. T. Huber, F. Böhme, H. Komber, J. Kronek, J. Luston, D. Voigt, and B. Voit, *J. Macromol. Chem. Phys.* **200**, 126 (1999).

281. L. Boogh, B. Petterson, P. Kaiser, and J. A. Mason, *SAMPE J.* **33**, 45 (1997).
282. A. Sunder, M. Krämer, R. Hanselmann, R. Mülhaupt, and H. Frey, *Angew. Chem. Int. Ed. Engl.* **38**, 3552 (1999).
283. Y. Zhang, T. Wada, and H. Sasabe, *Polymer* **38**, 2893 (1997).
284. G. Maier and T. Griebel, *Polym. Prep.* **41**, 89 (2000).
285. C. J. Hawker, 13th Biennial Marvel Symposium, Tucson, Arif. March 1999.
286. J. I. Kroschimtz, ed., *Polymers: Polymer Characterization and Analysis*, Wiley-Interscience, New York, 1990.
287. H. G. Barth and J. W. Ways, eds., *Modern Methods of Characterization*, Wiley-Interscience, New York, 1991.
288. C. Booth and C. Price, eds., *Comprehensive Polymer Science. The Synthesis, Characterization, Reactions and Applications of Polymers*, Vol. 1, Pergamon Press, New York, 1989.
289. L. H. Tung, ed., *Fractionation of Synthetic Polymers—Principles and Practices*, Marcel Dekker, New York, (1977).
290. H. Inagaki and T. Tanaka, in J. W. Dawkins, ed., *Developments in Polymer Characterization*, Vol. 3, Applied Science Publishers, Elservier, New York, 1982.
291. C. D. Craver, ed., *Polymer Characterization. Spectroscopic Chromatographic, and Physical Instrumental Methods*. ACS, Washington D.C., 1983.
292. A. E. Hamielec and A. C. Ouano, *J. Liquid Chromaton.* **1**, 112 (1978).
293. J. Falkenhagen, H. Much, W. Stang, and A. H. E. Mueller, *Macromolecules* **338**, 3687 (2000).
294. H. Pasch, H. Much, and G. Schulz, *J. Appl. Polym. Sci.: Appl. Polym. Symp.* **52**, 79 (1993).
295. G. Wilezek-Vera, P. O. Danis, and A. Eisenberg, *Macromolecules* **29**, (1996).
296. G. Wilezek-Vera, Y. Yu, K. Waddell, P. O. Danis, and A. Eisenberg, *Macromolecules* **32**, 2180 (1999).
297. D. Yu, N. Vladimirow, and J. M. Frechet, *Polym. Mater. Sci. Eng.* **80**, 219 (1999), *Macromolecules* **32**, 5186 (1999).
298. E. Esser, C. Keil, D. Brawn, P. Montag, and M. Pasch, *Polymer* **41**, 4039 (2000).
299. H. Lee, T. Chang, D. Lee, M. S. Shim, H. Ji, W. K. Nomidez, and J. N. Mays, *Anal. Chem.* **73**, 1726 (2001).
300. H. J. Räder and N. Schrepp, *Acta Polymerica* **49**, 272 (1998).
301. E. Mechan, S. O'Donohue, and J. McConille, *Polym. Mat. Sci. Eng.* **69**, 269 (1993).
302. P. Barghoon, U. Stebani, and M. Balsam, *Acta Polymerica* **49**, 266 (1998).
303. A. Tuchbreiter, B. Kappler, J. Honerkamp, and R. Mülhaupt, *Dechema Monographs* **137**, 323 (2001).
304. R. Soula, C. Novat, A. Tomov, R. Spitz, J. Claverie, X. Drujon, J. Malinge, and T. Sandemont, *Macromolecules* **34**, 2022 (2001).
305. F. M. Bauers and S. Mecking, *Angew. Chem. Int. Ed. Engl.* **40**, 3020 (2001).

JOSÉ R. LEIZA
JOSÉ M. ASUA
The University of the Basque Country

COPPER

1. Introduction

Copper [7440-50-8], was one of the first metals to be used by early humans because, like gold and silver, it occurs in nature in the metallic state. The Chalcolythic Age marked the transition from man's use of stone implements to those made from copper. Humans first used copper for utilitarian objects. The "Ice Man" recently found in the Italian Alps carried a copper axe, an example of the copper tools believed to be prevalent during this time. The earliest recorded use of copper occurred in northern Iraq about 8500 B.C. Copper appeared elsewhere in Asia Minor and in Egypt around 7000 B.C. The oldest known copper smelting, dating to 3600 B.C., was conducted near the Timna mine in Elot, Israel, a site believed to be that of King Solomon's Mine (1). Indeed, copper is referenced several times in the Old Testament. Deposits on Cyprus were known as early as 3000 B.C. These mines became prized possessions of the empires that followed the Egyptians and became the chief source of copper for the Roman Empire. The metal was named *aes syprium* and subsequently *cuprum*, from which derives the English name, *copper*, and the symbol, *Cu*.

Copper is a soft metal, and although it was easy to work, it was not ideal for knives, axes, and other weapons. Thus, when humans learned how to alloy copper with tin to form bronze, about 3500 B.C., the harder and tougher alloy replaced copper in many of its uses. Because of its gold color, bronze quickly became a metal for decorative as well as utilitarian objects, giving rise to the Bronze Age. In time, humans learned how to smelt iron and to form steel, and since steel could cut through bronze, it largely replaced bronze in applications other than decorative items. Copper then fell into disuse until the advent of electricity in the late 1800s. Around 1850, with the invention of the dynamo, the electric motor and later, the telephone, copper became the preferred metal for electrical wire and other conductros. With electrification, copper rapidly came into its own, and its use has grown every year since. Maximum growth in the rate of production, about 6% per year, occurred during the period between the two world wars. Figure 1 tracks the production of copper in the Western world during the period 1810–1998 (2,3).

Agricola described the antecedents of today's copper smelting processes in *De Re Metallica*, published in 1556 (4). At that time, smelting was conducted at Mansfeld, Germany and Swansea, in Wales. Early copper production utilized so-called direct-smelting ores: ores with a grade in excess of 7% copper that could be smelted without first having to be concentrated. These ores were found in veins between rock masses, and early copper mines were therefore entirely located underground. Large-scale mining of low grade ores began early in the twentieth century, when Daniel Jackling developed the open-pit mining method at the Bingham Canyon Mine, located near Salt Lake City, Utah. Jackling's breakthrough, combined with development of the froth flotation process for concentrating low grade ores (by the Minerals Separation Company of London), made possible the exploitation of the massive porphyry copper ore bodies mined today in Arizona, New Mexico, Utah, and elsewhere (see section on occurrence).

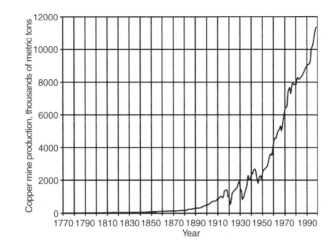

Fig. 1. Historical world copper production, 1770–1998 (2,3).

Today, the large porphyry copper ore bodies in Chile and Peru have made South America the world's leading copper-producing region. The United States ranks second, and it comprises the only region in which production and consumption are in balance. Central Africa, the Congo and Zambia, and the CIS follow the United States in production. Other important deposits are found in southern Oceania (Papua New Guniea, the Philippines, and Indonesia), Canada, Mexico and Poland. Since the United States is essentially self-sufficient in copper, copper-producing regions outside the United States supply Asia and Europe.

The copper industry is composed of *producers* and *fabricators*. Producers mine, smelt, and refine copper for sale in the form of cathodes or ingots, output that is collectively referred to as *refined copper*. Some producers also manufacture *wire rod*, selling it as a *refinery product*. Producers may additionally recover molybdenum disulfide, various precious metals, selenium, tellurium, copper sulfate, nickel sulfate, and sulfuric acid as byproducts. Fabricators manufacture useful shapes using combinations of refined copper, copper scrap, and alloying ingredients, as appropriate. The shapes include wire, rod and bar, tube, sheet, strip, plate, and foil and are known as *semifabricates*, or more commonly, *semis*. Fabricators known as *wire mills* manufacture electrical and mechanical wire. The broad term, *brass mills*, refers to fabricators that make tube and pipe, rod, and flat-rolled items, whether or not those products are made from brass. Other fabricators produce powered metal and speciality chemicals. Fabricators' products (semis) may be sold directly to manufacturers for the production of finished goods such as electrical and electronic equipment and automotive heat exchangers. Fabricators may also sell directly to end users in building construction and other industries. Smaller quantities of copper semis are sold to distributors.

2. Occurrence

Copper is one of the most abundant of the metallic elements in the earth's crust. Its average estimated concentration is 55 mg/kg (5), placing it below chromium

(200 mg/kg) and zinc (132 mg/kg) but above tin (40 mg/kg) and lead (24 mg/kg). It is estimated that there are 3×10^{18} tonnes of copper diffusely distributed in uppermost mile (1.6 km) of the continental crust (6). Only a small fraction of this copper is relatively concentrated, comprising an estimated 10^{10} tonnes concentrated in deposits with a grade of $\geq 0.25\%$ Cu (2). At the current global mine production rate, approximately 110 million tonnes per year, this fraction represents a million years' supply of copper theoretically available in the mineable portion of the earth's crust. There are also appreciable quantities of copper in deep-sea clays, in ocean manganese-based nodules scattered on the ocean floor, and in effluents from undersea fumaroles. None of these undersea sources have been economically exploited (see OCEAN RAW MATERIALS).

2.1. Minerals and Ores.   Almost all crustal copper is contained within igneous rock. Copper's strong chemical affinity for sulfur chiefly determines its manner of occurrence. Copper–iron sulfides have relatively low melting points and therefore crystallized last, between the other minerals that make up igneous rocks. Other sulfides were transported and deposited by hydrothermal processes, and are found in cracks and fissures. Copper ore minerals are classified as *primary, secondary, oxidized*, and *native* copper. Primary minerals were concentrated in ore bodies by secondary enrichment as a result of hydrothermal processes.

Secondary minerals formed when copper sulfides exposed at the surface became weathered, leached (dissolved) by groundwater, and precipitated near the water table (see METALLURGY, EXTRACTIVE). Copper, like gold and silver, can also occur as a primary mineral, in its metallic (native) form. A good example of the latter minerals are the lava-associated deposits of the Keweenaw Peninsula in Upper Michigan, which formed a significant portion of global copper production in the nineteenth century.

The five classifications of economically viable copper ore include porphyry deposits and vein replacement deposits; strata-bound deposits in sedimentary rocks; massive sulfide deposits in volcanic rocks; magmatic segregates associated with nickel in mafic intrusives, and native copper. A sixth type of deposit, one consisting of oxide minerals, is now recognized as a result of the development of leaching and solution purification technologies (see section on SXEW process).

Almost two-thirds of the world's copper resources are porphyry deposits. (The term *porphyry* is generally applied to a copper deposit that is hydrothermal in origin, and in which a large portion of the copper minerals is uniformly distributed as small particles in fractures and small veins.) Porphyry deposits usually contain $\leq 1\%$ copper. The most extensive of these deposits are located in western Canada, southwestern United States, Mexico, and the Andes Mountains of western South America. In addition to the porphyries, there are large bedded copper deposits in Germany, Poland, the CIS, Australia and central Africa. Most near-surface copper sulfide deposits have an oxidized cap of secondary mineralization. Since these oxide minerals are not treatable in a smelter, such deposits were previously not regarded as ores. Modern leaching technology now makes their recovery possible, and these minerals, together with oxidized portions of previously mined material, currently account for about 15–20% of primary-copper production.

3. Properties

Copper, with symbol Cu and atomic number 29, is one of the "noble" metals, and like gold and silver, it is a member of subgroup IB in the periodic chart of the elements. Its atomic electronic structure is described by the notation $1s^2 2s^2 p^6 3s^2 3p^6 3d^{10} 4s^1$, which depicts an argon core plus a filled $3d$ orbital and a single $4s$ electron. Copper owes its unique properties to this structure. For example, the filled $3d$ states limit compressibility and, consequently, scattering of conductance electrons due to thermal lattice vibrations. Lattice incompressibility and the loosely bound $4s$ and $3d$ electrons give the copper its high electrical and thermal conductivity (7). Copper's distinctive red color arises from absorption due to optical transitions between filled $3d$ states and empty conduction-band $4s$ states. Alloying alters this structure; additions of zinc, for example, produce progressively lighter golden yellows, while nickel additions yield pink to silver-white hues. (The cladding on U.S. coins other than the dollar contains 75% copper, 25% nickel.) The low ionization potential of the $4s$ electron, 7.724 eV, is responsible for the relative ease with which copper forms the Cu^+ ion, which is important to both industrial copper chemistry and the behavior of copper in the environment. The electron given up to form the Cu(II) state arises from the $3d$ orbital, whose ionization potential is only ~2 eV higher than that of the $4s$.

Copper has a face-centered cubic (fcc) crystal structure (space group: $A1$, $Fm3m$; $cF4$) at all temperatures below its melting point, 1358.03 K (1084.88°C). The lattice parameter at 298 K (25°C) is 0.361509 ± 0.000004 nm, the distance of closest approach (Burgers vector) is 0.255625 nm and the Goldschmidt atomic radius for 12-fold coordination is 0.1276 nm (8, 9). Copper's high ductility, an important commercial attribute is in large part due to the 12 (111)[110] slip systems in the fcc crystal structure. To a lesser extent, copper also deforms by twinning across {111} planes in the [112] direction.

Copper's atomic weight is 63.546; its nucleus contains 29 protons and 34–36 neutrons. There are two stable isotopes, ^{63}Cu (occurrence 69.09%) and ^{65}Cu (30.91%). Unstable isotopes range from ^{58}Cu to ^{68}Cu (9–11). All unstable isotopes are β emitters; however naturally occurring copper is not radioactive.

3.1. Physical Properties.
The density of pure, single-crystal copper at 278 K (20°C) is 8.95285 g/cm^3. As with all metals, density depends significantly on thermomechanical treatment, and densitities of 8.90526 g/cm^3 and lower have been reported for severely cold-drawn copper wire (12). The value 8.94 g/cm^3 at 298 K (25°C) is generally accepted in practice. A compilation of copper's physical properties can be found in Table 1.

From both technical and commercial standpoints, copper's most important physical property is its high electrical conductivity, highest among "engineering" metals and second only to silver. Copper's electrical properties are described by the International Annealed Copper Standard (IACS), which is defined as the volume conductivity of an annealed pure copper wire, one meter long, weighting one gram, having a density of 8.89 g/cm^3 at 298 K (25°C) and a resistance of exactly 0.15328 Ω. Designated "100% IACS" and corresponding to a volume resistivity of 16.70 Ω·m, this value is the standard against which all electrical copper products (wire, connector alloys, etc) are compared (13). Volume conductivity as high as 103.6% IACS has been reported for high purity copper (14), whereas

Table 1. **Physical Properties of Pure Copper**

Property	Value
atomic weight	63.546
atomic volume, cm^3/mol	7.11
mass numbers, stable isotopes	63 (69.1%), 65 (30.9%)
oxidation states	1, 2, 3
standard electrode potential, V	$Cu/Cu^+ = 0.520$
	$Cu^+/Cu^{2+} = 0.337$
electrochemical equivalent, mg/C[a]	0.3294 for Cu^{2+}
	0.6588 for Cu^+
electrolytic solution potential, V(SCE)[a]	0.158 ($Cu^{2+}+e^- = Cu^+$)
	0.3402 ($Cu^{2+}+2e^- = Cu$)
	0.522 ($Cu^+e^- = Cu$)
density, g/m^3	8.95285 (pure, single crystal)
	8.94 (nominal)
metallic (Goldschmidt) radius, nm	0.1276 (12-fold coordination)
ionic radius, M^+, nm[b]	0.096
covalent radius, nm[b]	0.138
crystal structure	fcc, *A1*, *Fm3m: cF4*
lattice parameter	0.0361509 ± 0.000004 nm (25°C)
electronegativity[b]	2.43
ionization energy, kJ/mol[b]	
first	745
second	1950
ionization potential, eV[a]	7.724 Cu(I)
	20.29 Cu(II)
	36.83 Cu(III)
Hall effect[a]	
Hall voltage, V at 0.30–0.8116 T	-5.24×10^{-4}
Hall coefficient, mV/(mA)(T)	-5.5
heat of atomization, kJ/mol[b]	339
thermal conductivity, W/(m)(K)	394^b, 398^a
electrical resistivity at 20°C, $n\Omega \cdot m$	16.70
temperature coefficient of electrical resistivitiy, 0–100°C[a]	0.0068
melting point	1358.03 K (1084.88°C)[a]
	1356 K (1083°C)[b]
heat of fusion, kJ/kg	205, 204.9, 206.8^a
	212^b
boiling point	2868 K (2,595°C), 2,840 (2,567°C)[a]
	2595 K (2,868°C)[a]
heat of vaporization, kJ/kg	7369^b
	4729, 4726, 4793^a
specific heat, kJ/(kg)(K)	0.255 (100 K)[a]
	0.384 (293 K)[b]
	0.386 (293 K)[a]
	0.494 (2000 K)[a]
coefficient of expansion, linear, μm/m	16.5
coefficient of expansion, volumetric, 10^{-6} K^{-1}	49.5
tensile strength, MPa	230 (annealed)[b]
	209 (annealed)[a]
	344 (cold drawn)[a]
elastic modulus, GPA	125 (tension, annealed)[a]
	102–120 (tension, hard-drawn)[b]
	128 (tension, cold drawn)[a]
	46.4 (shear, annealed)[a]
	140 (bulk)[a]

Table 1 (*Continued*)

Property	Value
magnetic susceptibility, 291 K, mks	-0.086×10^{-6b}
	-1.08×10^{-6a}
emissivity	0.03 (unoxidized metal, 100°C)[b]
	0.8 (heavily oxidized surface)[a]
spectral reflection coefficient, incandescent light	0.63[a]
nominal spectral emittance, $\lambda = 655$ nm, 800°C	0.15[a]
absorptivity, solar radiation	0.25[a]
viscosity, mPa · s (cP)	3.36 (1085°C)[a]
	3.33 (1100°C)[a]
	3.41 (1145°C)[b]
	3.22 (1150°C)[a]
	3.12 (1200°C)[a]
surface tension, mN/m (dyn/cm)	1300 (99.99999% Cu, 1084°C, vacuum)[a]
	1341 (99.999% Cu, N_2, 1150°C)[a]
	1104 (1145°C)[b]
	(see Ref. 15 for additional data)
coefficient of friction	4.0 (Cu on Cu) in H_2 or N_2[a]
	1.6 (Cu on Cu) in air or O_2[a]
	1.4 (clean)[a]
	0.8 (in paraffin oil)[a]
velocity of sound, m/s	4759 (longitudinal bulk waves)[a]
	3813 (irrotational rod waves)[a]
	2325 (shear waves)[a]
	2171 (Rayleigh waves)[a]

[a] Data from Ref. 12.
[b] Data from Ref. 3.

commercial oxygen-free electronic copper [Unified Numbering System (UNS) alloy C10100], oxygen-free copper (UNS C10200), and electrolytic tough pitch copper (UNS C11000) exhibit conductivities of at least 101% IACS. Alloying, even with minute additions of other metals, increases resistivity (decreases conductivity), as shown in Table 2 (9).

Copper's high thermal conductivity, 398 W(m) · (K) at 298 K (25°C), is exploited in applications ranging from ordinary heat exchangers to superconducting cables used in magnetic resonance imaging (MRI) equipment (although copper itself is not a superconductor). Maximum thermal conductivity, 19,600 W/(m) · (K), occurs near 10 K (-263°C). Conductivity decreases by only $\sim$7% between room temperature and the melting point, an important consideration in the design of heat exchagners. Alloying reduces thermal conductivity, but from an engineering standpoint higher strength and/or better corrosion resistance normally compensates for this loss. Copper's relatively high coefficient of thermal expansion, 16.7 m/(m) · (K) at 298 K, is one reason why bonds between copper and silicon semiconductor chips always include an transitional layer, usually of a metal, usually nickel, which has an intermediate expansion coefficient.

Copper is diamagnetic, although a number of ferromagnetic copper alloys are known. Ferromagnetism in such alloys arises not from copper itself but from the alloys' crystal structure or from precipitated ferromagnetic phases.

Table 2. **Increase in Resistivity of Copper Due to Small Solute Additions**

Solute	Room-temperature solubility, wt%	Copper, at%/cm	Observed range of copper, at%/cm
Ag	0.1	0.6	0.1–0.6
Al	9.4	0.95	0.8–1.1
As	6.5	6.7	6.6–6.8
Au	100.0	0.55	0.5–0.6
B	0.06	1.4	1.4–2.0
Be	0.2	0.65	0.6–0.7
Ca	<0.01	$(0.3)^a$	
Cd	<0.05	0.3	0.21–3.4
Co	0.2	6.9	6.7–7.0
Dr	<0.03	4.0	3.8–4.2
Fe	0.1	8.5	805–8.6
Ga	20.0	1.4	1.3–1.5
Ge	11.0	3.7	3.6–3.75
Hg		$(1.0)^a$	
In	3.0	1.1	1.0–1.2
Ir	1.5	$(6.1)^a$	
Li	<0.01	$(0.7)^a$	
Mg	1.0	$(0.8)^a$	
Mn	24.0	2.9	2.8–3.0
Ni	100.0	1.1	1.1–1.15
O	~0.0002	5.3	4.8–5.8
P	0.5	6.7	6.7–6.8
Pb	0.02	3.3	3.0–4.0
Pd	40.0	0.95	0.9–1.0
Ot	100.0	2.0	1.9–2.1
Rh	20.0	$(4.4)^a$	
S	~0.0003	9.2	8.7–9.7
Sb	2.0	5.5	5.4–5.6
Se	~0.0004	10.5	0.2–10.8
Sn	1.2	3.1	3.0–3.2
Te	~0.0005	8.4	2.8–3.5
Ti	0.4	$(16.0)^a$	
U	~0.1	$(10.0)^a$	
W		3.8	0.3
Zn	30.0	1.3	

^{a}Ref. 9.

Copper's low magnetic permeability avoids energy losses in electromagnetic devices such as motors and generators. Special heat treatments are employed to avoid traces of ferromagnetism in copper alloys in critical applications such as minesweepers.

Since pure copper does not undergo allotropic transformations, its mechanical properties depend entirely on grain size and on thermomechanical history, specifically, the degree of hot and cold working and/or annealing imparted during manufacture. Yield and tensile strengths (in MPa), respectively, of annealed oxygen-free (OF) copper between 4 K ($-269°C$) and 300 K ($27°C$) are described by the expressions (9)

$$\sigma_y = -8.60 - 0.0329\ T + 292\ d^{1/2} + 150\ I(SD = 9\ MPa)$$

$$\sigma_u = 419 - 1.19\ T + 0.00144\ T^2 + 156^{1/2}\ I\ (SD = 18\ MPa)$$

and for cold-worked OF copper by

$$\sigma_y = 124 - 0.241\ T + 14.1(CW) - 0.166(CW)^2 (SD = 32\ MPa)$$

$$\sigma_u = 412 - 0.664\ T + 2.73(CW) - 0.00695(CW)^2\ (SD = 32\ MPa)$$

where T is temperature in degrees K, d is grain size in μm, I is the impurity content in wt.%, CW is the degree of cold work expressed by the percent reduction in thickness and S.D. is the standard deviation. Uses of pure copper based on mechanical properties alone are quite limited, and most applications in which high strength, ductility or other such properties are required—usually often in combination with other attributes—are served by copper alloys, of which several hundred are now produced worldwide. Mechanical properties of representative wrought Nort American copper alloys are listed in Table 3 (15).

Alloying improves mechanical properties through solid-solution strengthening (α brasses, tin bronzes), precipitation hardening (beryllium coppers, chromium copper, zirconium copper), spinodal decomposition (copper–nickel–tin alloys), order–disorder transformations (β-brasses), diffusionless transformations (aluminum bronzes, aluminum brasses, nickel–aluminum bronzes), or combinations thereof. Alloys such as beryllium coppers and aluminum and manganese bronzes can attain strengths comparable with those of quenched and tempered steels. High strength, combined with the alloys' high corrosion resistance, enables copper-base alloys to compete against stainless steels and other advanced alloys for a large variety of industrial applications such as pump and valve components.

Many alloys are devised primarily to facilitate manufacturing processes. It can be argued that cast implements created in the Bronze Age exemplify this process. In the late twentieth century, copper alloy development centered on improvements in machining characteristics (lead-free brasses), corrosion resistance (marine and plumbing alloys), elevated-temperature resistance (electrical connector- and leadframe alloys), and, as in ancient times, castability.

3.2. Chemical Properties. Compared with alkali metals, copper's higher ionization energy and smaller ionic radius contribute to its forming oxides that are much less polar, less stable, and less basic than those of the alkali metals (16). The relative instability of copper's oxides is responsible for the occurrence of native, that is metallic, copper in nature (17). Cu(I) forms compounds with the anions of both strong and weak acids. Many of these compounds are stable and insoluble in water. Compounds and complexes of Cu(I) are almost colorless because the $3d$ orbital of the copper is completely filled. There is, however, a very strong tendency for Cu(I) to disproportionate in aqueous solutions into (Cu(II) and metallic copper.

$$2\ Cu^+ \rightarrow Cu^0 + Cu^{2+}$$

Whereas the cuprous (I) state is generally unstable in aqueous solutions, the cupric (II) state is quite stable. Ligands that form strong coordinate bonds bind Cu(II) readily to form complexes in which the copper has coordination numbers of 4 or 6. Formation of Cu(II) complexes in aqueous solution depends on the

Table 3. **Mechanical Properties of Selected Wrought Copper Alloys**[a]

Alloy [Unified Numbering System (UNS)]	Description of previous name	Form of specimen	Tensile strength, MPa	Yield strength, 0.5% extension (0.2% extension), MPa	Elongation in 50 mm, %	Shear strength, MPa	Fatigue strength, 10^6 cycles, MPa	Hardness, Rockwell B/F
C10200	oxygen-free	Rod	221–375	69–345	10–55	152–186	117	47–60/40/94
C11000	electrolytic tough pitch	Rod	221–379	39–345	16–55	152–200	177	45–60/40–87
C12200	phosphorus deoxidized	Tube	221–379	69–76	8–45	152–200	76–131	35–60/40–95
C14500	tellurium-bearing	Rod	221–296	76–338	10–50	152–200	NA	36–64/40–43
C15720		Rod	441–552	400–531	16–25	NA	NA	66–74/NA
C17200	beryllium copper	Rod	469–1379	(172–1227)	3–48	NA	NA	3–95/NA
C18200	chromium copper	Rod	310–593	97–531	5–40	NA	NA	65–83/NA
C19500		Plate	552–669	(448–655)	2–15	NA	NA	NA
C26000	cartridge brass, 70%	Rod	331–483	110–359	30–65	234–290	NA	60–80/65
C28000	muntz metal, 60%	Rod	359–496	138–345	25–52	269–310	NA	78/78–80
C36000	free-cutting brass	Rod	338–400	124–359	25–53	207–262	138	75–80/68
C37700	forging brass	Rod	359	138	45	NA	NA	NA/78
C50500	phosphor bronze 1.25% E	Plate	276–517	97–345	4–48	NA	221	64–79/68
C51000	phosphor bronze, 5% A	Rod	483–517	400–448	25	NA	NA	78/NA
C54400	phosphor bronze, B2	Rod	469–517	393–434	15–20	NA	NA	80–83/NA
C61300	aluminum bronze	Rod	552–586	331–400	35–40	276–331	NA	88–91/NA
C61800	aluminum bronze	Rod	552–586	269–293	23–28	296–324	179–193	88–89/NA
C63000	nickel–aluminum bronze	Rod	689–814	414–517	15–20	427–483	248–262	96–98/NA
C65500	high-silicon bronze A	Rod	400–745	152–414	13–60	296–427	NA	60–95/NA
C70600	copper–nickel, 10%	Tube	303–414	110–393	10–42	NA	NA	15–72/65–100
C71500	copper–nickel, 30%	Tube	372–414	172	45	NA	NA	35–45/77–80

[a]*Source:* Ref. 9.

678

ability of the ligands to compete with water for coordination sites. The stability and ease of formation of Cu(II) complexes are important factors influencing the bioavailability of copper in aqueous and marine environments. Until the late 1990s, copper's ecologic toxicity was grossly overestimated because of an inadequate awareness of the implication of this phenomenon, namely, that the bioavailability (and associated toxicity) of copper in aqueous environments can be several orders of magnitude lower than its concentration due to complexation with other ions.

3.3. Corrosion Resistance. Copper's third most important property is the collection of attributes generally termed its intrinsically high corrosion resistance. To a degree, copper exhibits the thermodynamic stability of silver and gold. Native copper nuggets found in Michigan, for example, are estimated to have remained stable in subterranean brines for as long as 10^9 years. Numerous copper and bronze archaeologic artifacts, some of them retrieved from the sea, have survivied for centuries. Although copper is thermodynamically stable in a few environments, its corrosion resistance is more often based on kinetics, that is the low rates of speed at which corrosion reactions proceed. Thus, in many potentially corrosive media (including, importantly, air, steam, and potable and marine waters) the metal tends to form adherent and protective surface films that, once established, effectively inhibit further attack unless the films are damaged or altered by changes in the environment (18). Copper's long and successful history of use for plumbing tube rests on this phenomenon, as does the formation of protective—and attractive—patina on copper roofs and bronze statuary. Copper roofing in certain rural atmospheres corrodes less than 0.4 mm in 200 years, whereas the copper skin of the Statue of Liberty has lost only 0.1 mm to corrosion in 100 years despite exposure to a marine/industrial atmosphere. Pure copper resists aerated alkaline solutions except in the presence of ammonia. Copper does not displace hydrogen from acid but dissolves readily in oxidizing acids such as nitric acid or sulfuric acid solutions containing an oxidizer such as ferric sulfate. Copper is also attacked by soft, low pH waters, ammonia solutions, amines, cyanide solutions, nitrates and nitrites, oxidizing heavy-metal salts, and certain sulfides. Aqueous solutions containing ammonia, amines, cyanide, nitrates, and nitrites attack copper and can, under certain conditions, generate stress corrosion cracking. Although copper is biostatic, the metal can exhibit microbially induced corrosion (MIC) (19–22). Chlorides, including seawater, do not seriously attack copper and its alloys, and this has led to many applications for which, for example, stainless steels are less well suited. On the other hand, titanium alloys have begun to replace copper alloys in naval and seawater-cooled power plant condensers. Despite the metal's known biostatic properties, it can exhibit microbially induced corrosion (MIC) (20,21).

4. Sources and Supplies

Copper enters into trade primarily in the form of concentrates, blister, anodes, refined copper ingots and cathodes, and copper semis. The former are provided by copper producers (miners, smelters, and refiners) the latter by fabricators, respectively.

4.1. Concentrates and Blister.   Concentrates, the product of mines and associated *concentrators*, contain 27–35% copper in the form of sulfide minerals. Most trade in concentrates is carried out by mines that do not have an associated smelter, or mines that produce more than their smelters can accommodate, in which case the surplus is sold. Some mines produce concentrates strictly for sale on world markets or for processing in foreign smelters. Concentrates also become available from time to time when the smelters that treat them are closed down for one reason or another. Blister copper is the first product from a smelter. It contains about 98% copper, some oxygen and iron, plus the noble metals and other impurities carried over from the ore. Blister copper may enter international trade when there is insufficient refinery capacity in the country in which it is produced, or as the result of contractual arrangements for the shipment of blister from the smelter for refining elsewhere. The price of concentrates and blister are negotiated on the basis of their copper content and the current price of copper on one of the commodity exchanges (see section on economic on aspects).

4.2. Refined Copper.   The term *refined copper* refers to metal with a copper content of ≥99.99%, although some special grades are as pure as 99.9999%. Refined copper is the product of either electrolytic refining of anode copper or electrowinning from hydrometallurgic or leach solutions. Commercial *refinery shapes* include electrorefined or electrowon copper cathodes and several types of continuously cast wire rod. Cathode and wire rod are now by far the most important forms of copper in international trade. There is a substantial international trade in refined copper since fabricators of copper and copper alloy semis often obtain their supplies from refineries in other countries.

4.3. Semis.   Semis (semifabricates) are fabricated product forms, such as rod, wire, sheet, plate, and tube. They are the products of wire mills, brass mills (which include rod mills and, tube mills and sheet and strip manufacturers), ingotmakers, and powdermakers. The price of semis is based on the commodity exchange price of copper; however, because their cost also includes alloying materials, energy, and labor input, the market price of semis is somewhat buffered and does not fluctuate to the extent that the price of refined copper does. End users often purchase semis directly from the fabricator when large quantities are involved; small quantities are purchased from distributors or supply houses.

4.4. Stockpiles.   The London Metal Exchange (LME) and the Commodity Exchange of New York (COMEX) (see Economic Aspects) maintain warehouses of stocked with refined copper. These warehouses are located at strategic locations close to major consuming centers of the world.

5. Recovery and Processing

Most copper is produced using a combination of mining, concentrating, smelting, and refining. About 15% of refined copper is produced through the leaching of mine and concentrator wastes, depleted tailings, and oxide ores with sulfuric acid derived from the smelting process's off-gases. The copper-containing leach solutions are purified and concentrated by *solvent extraction* (SX), and copper is recovered by *electrowinning* (EW). Whether from the traditional or SXEW paths, the final product is a high purity copper cathode.

5.1. Mining.   Copper ore is obtained from either open-pit or underground mines; the open-pit source is predominant today. The choice of mining method is a function of the ore grade, including its geologic setting (depth, shape, and amount of overburden) and economic considerations. Underground mining is generally more expensive and requires a higher ore grade (nominally >2% Cu) than does open-pit mining. In some instances a combination of open-pit and underground mining is used once a depth is reached in which open-pit mining is uneconomical. Conversely (e.g, San Manuel, Ariz.) open-pit mining has been used over areas that had previously been excavated underground. Open-pit mining can economically accommodate ore containing ≤1% copper and as lean as 0.2% copper where byproducts such as molybdenum and precious metals are present in sufficient concentration. Run-of-mine U.S. porphyry ores currently contain about 0.6% copper.

In underground mining, the ore body is accessed via tunnels or shafts and ore is excavated using *room and pillar* or *block caving* mining methods. Following excavation, the ore may be broken up in an underground crusher or hauled directly to the surface using hoists or specialized trucks. Underground copper mines dominated the U.S. industry in the nineteenth century but are now rare due to the widespread exploitation of low grade porphyry deposits. The world's largest underground copper, *El Teniente*, is located in Chile. In 1998, it produced more than 35 million metric tons of ore with an average grade of 1.16% Cu (>400,000 tons of copper contained).

In open-pit mining, nonmineralized overburden is first stripped away to reveal the orebody, which is then drilled, blasted, and excavated. Because the surface area of an open-pit mine continually increases with pit depth in order to maintain a safe pit wall angle, the overburden:ore *stripping ratio* is an important factor in a mine's economics. Most breaking of rock takes place during blasting, but in some cases primary crushers are located in the pit. Crushed ore is transported to the mill for concentration using trucks or conveyors, while overburden is hauled to waste dumps surrounding the mine.

Open-pit mining has become a highly efficient mining method, due to technological innovations in explosives, shovels, haulage trucks, computer modeling of the ore body for mine planning, and computerized control of the equipment. Electric shovels with a bucket capacity of 15–25 m^3 (500–900 ft^3)—sufficient to hold an average pickup truck comfortably—and haul trucks capable of hauling 100–350 short tons of ore per load are now common. Geosatellite positioning systems (GPSs) are used to control the position and dispatch of shovels and trucks. This system is so precise that one company has a completely robotic haul truck under test. Most of the open-pit mines in the southwestern United States move 30,000–250,000 tons of ore plus overburden per day. The world's largest copper mine is the *Escondida* open pit, located in Chile. Its sulfide and oxide operations currently produce more than 800,000 tons of copper contained per year.

5.2. Concentration.   *Concentrating* consists of crushing and grinding to liberate the copper mineralization from the host rock, followed by *flotation*, in which copper mineralization and waste materials, or *gangue* are separated. *Milling* refers to the crushing and grinding steps; however, the term *mill* is used interchangeably with the term *concentrator* to describe the facility in which the process takes place. Copper concentrators serving U.S. copper mines individually process 30,000–160,000 tonnes of low–grade ore per day.

Two or three stages of cone crushers are normally used for the first stage of size reduction. These crushers contain an eccentrically driven grinding cone that rotates in a fixed conical bowl, creating a moving wedge-shaped space. Ore is pinched and crushed as it drops into this space. Typical product sizes for three crushing stages in series are <20, <3, and <1 cm, respectively. Further size reduction is achieved by several stages of grinding in water to produce an ore slurry. In one process, grinding is conducted in large cylindrical vessels called *mills* into which ore, water, and steel rods or balls, called *grinding media*, are charged. These reduce the ore's particle size to <3 mm. Today, dry crushing, screening, and rodmilling is increasingly being replaced by semiautogenous (SAG) milling, in which large blocks of ore and hardened grinding balls are tumbled in very large mills. Impact of the ore against itself and the steel balls reduces the ore to the desired particle size. Effluent from the first stage of milling is passed to ball mills where the ore is further ground such that 75% of it is <0.25 mm in size. Wet cyclones classifiers recycle oversized ore particles between grinding stages.

Crushing and grinding are followed by *froth flotation* (see FLOTATION) where copper minerals are separated from the gangue and recovered. Copper concentrates commonly contain 25–35% copper; the upper limit is determined by the copper content of the mineral in question. Figure 2 shows a simplified flowchart for a typical copper concentrator. Flotation is conducted in large tanks, or *cells* into which air is injected or drawn, creating bubbles. Chemicals known as *collectors* are added to the cell, where they selectively coat and create a water-repelling film on exposed surfaces of sulfide minerals, causing the sulfides to cling to the rising bubbles. Other chemicals adjust pH, stabilize the froth, and depress unwanted minerals. The bubbles and their copper–mineral freight are skimmed from the surface froth. Gangue minerals, which are not attracted to the air bubbles, fall to the bottom of the cell, where they are removed. Common collectors include xanthates, dithiophosphates, or xanthate derivatives. Calcium or sodium cyanide and lime are used as depressants. Pine oil or long-chain alcohols are used as frothers. The pH is adjusted with lime, which optimizes the action of all of the reagents and depresses pyrite. Most copper flotation plants operate *rougher cells*, *cleaner cells*, and *scavenger cells* in series. Flow between these cells is illustrated in Figure 2. The froth from scavenger cells may be removed and reground in a special ball mill, from which it is reintroduced into rougher cells to further separate copper minerals from the gangue minerals.

The product, or *concentrate*, produced by the mill contains between 25 and 35% copper, depending on the copper content of the mineral(s) involved. The concentrate is dried by filtration and shipped to the smelter. Waste products, called *tailings*, are pumped to large *tailing ponds* where the water is decanted off and returned to the mill.

Molybdenite, MoS_2, occurs in many prophyry copper ore bodies. If the grade is sufficient, it can be recovered in the cocentrator as a byproduct of the copper mineral recovery. Molybdenite normally floats with the copper sulfides (see MOLYBDENUM AND MOLYBDENUM ALLOYS). Therefore, as shown in Figure 2, the copper

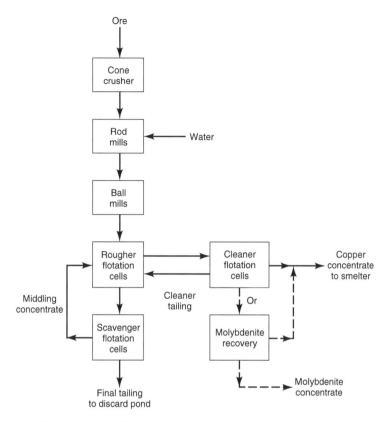

Fig. 2. Simplified flow diagram of a copper concentrator.

concentrate from the cleaner cells has to be separated from molybdenite in a separate flotation circuit.

5.3. Smelting. Copper concentrate usually (but not always) contains a mixture of copper and iron sulfides together with small amounts of gangue minerals. *Smelting* comprises the two operations needed to extract copper from concentrate using heat, flux, and oxidation. The first step, also called *smelting*, separates copper from the sulfur and iron and the gangue. In this step, much of the iron and a portion of the sulfur in the concentrate are oxidized, producing a mixture of molten copper and iron sulfides (known as *matte*), iron oxide (which is removed in the form of an iron silicon oxide slag and either discarded or processed further to remove additional copper), and sulfur dioxide gas.

Sulfur dioxide gas is recovered from the smelter and converted into sulfuric acid or other products. In a few installations, the gas is captured by lime and made into gypsum (crystallized calcium sulfate) for use in building materials. Sulfuric acid is normally sold, but in some cases it is used at the mine to leach copper oxide minerals in ores, waste dumps, and tailings. In 1998, sulfuric acid from copper (and other metal) smelting amounted to 4 million tonnes, or nearly 10% of U.S. production (23).

Figure 3 is a simplified flowchart for the smelting process. Material flows are given in metric tons per day. There are at least a dozen different versions

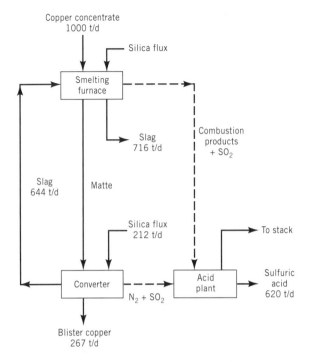

Fig. 3.   Flow diagram for smelting and converting.

of smelting furnaces; the most prevalent, in order of current tonnage processed worldwide, are Outokumpu (50%), reverberatory (25%), blast furnace (7%), El Teniente (7%), and Inco (5%). These furnaces differ mainly in the manner in which oxygen (in the form of air, oxygen-enriched air, or oxygen) is introduced into the reaction. The Outokumpu and Inco units are known as *flash furnaces* in that concentrates are blown into the furnace with oxygen or oxygen air.

Once in the furnace, the concentrates burn, or "flash." The oxidation reaction is *autogenous*, supplying nearly all the heat to sustain the furnace, operation; however, some natural gas is added to maintain proper heat balance. In the *reverberatory furnace*, copper concentrates and silica flux are introduced into a fuel-fired bed-shaped furnace hearth, where they are melted to form a low copper matte and slag. Fuel oil or natural gas provides the necessary energy and only a negligible amount of the heat value of the concentrate is utilized. Reverberatory and blast furnaces are being phased out because they are inefficient and the low sulfur dioxide concentration in their offgas renders the production of sulfuric acid difficult and costly.

5.4. Converting.   In a *converter* or *converting furnace*, matte from the smelting furnace is further oxidized to form copper and slag. The most common type of furnace, the Pierce–Smith converter—developed in the United States in 1906 and changed little since—consists of a simple cylindrical vessel with a large opening along one side and equipped to rotate about its longitudinal axis. In conventional converting, matte is charged into the furnace from open ladles, silica is added to flux iron in the matte, and air is blown into the charge through injectors, called *tuyeres*, located on the side of the converter. The sulfides are oxidized to sulfur dioxide, which is collected and sent to the sulfuric acid plant. Iron sulfide in the matte is oxidized by injecting air in a *slag blow*. The resulting iron oxide combines with silica to form an iron silica slag, which floats to the top of the melt and is poured off into ladles for either further processing or disposal. Following pouring, the now iron-free copper sulfide is oxidized by continued air injection in the *finish blow*. The product is *blister copper* (or simply *blister*) containing >98% Cu plus some sulfur, oxygen, and other impurities. This blister is then tapped into ladles for transport to the fire refining and anode casting operations.

Several continous (i.e., combined smelting–converting) technologies were introduced in the late 1990s. In the *Mitsubishi process*, oxygen or oxygen-enriched air is injected using lances, and process streams move continuously from one step to the next (24,25). In the *Kennecott–Outokumpu process* in which a flash furnace is used to smelt concentrates, matte is solidified, pulverized, and fed to a flash converter (26). Both processees avoid many of the air-quality problems associated with conventional smelting and converting operations. Most importantly, they avoid transfer of molten matte in large ladles, a procedure that generates ambient emissions of sulfur dioxide. The Mitsubishi and Kennecott–Outokumpu processes also produce concentrated sulfur dioxide gas, which is well suited for use in sulfuric acid plants.

5.5. Sulfur Recovery from Smelter Gases. The environmentally driven need to convert smelter gases to acid has been the most important force driving the development of new smelting technologies. Until the mid–Twentieth Century, sulfur dioxide gas was simply vented to the atmosphere via tall stacks. Today, it is normally captured and sent to a sulfuric acid plant, where it is catalytically converted to sulfur trioxide, then contacted with water to form sulfuric acid (see SULFURIC ACID AND SULFUR TRIOXIDE). The key to the operation is obtaining the sulfur dioxide in sufficient concentration (minimum $\sim$5%) to make acid efficiently. Today, the most modern smelters produce gases containing $\geq$35% sulfur dioxide. Since the maximum sulfur dioxide concentration that can be treated by most acid plants is $\sim$14%, the rich offgas from the smelter is mixed with weaker gas steams from the converters and from fugitive sources to prepare the gas feed to the acid plant.

Simple scrubbers, the devices first used to remove sulfur dioxide from offgases, are now employed primarily to treat gases with very low sulfur dioxide concentrations, such as those from dryers, anode furnaces, and building ventilation. Lime is the most commonly used absorbent; the resulting product is calcium sulfate. Scrubbers are used mainly in Japan, where the process has been taken one step further to the manufacture of *gypsum*, a crystallized form of calcium sulfate used in the manufacture of wallboard (see CALCIUM COMPOUNDS, CALCIUM SULFATE; SULFUR REMOVAL AND RECOVERY). Today, about 26% of the world's smelters are ranked "high capture" with regard to sulfur dioxide. The most efficient smelters exceed 99.5% SO_2 recovery.

5.6. Fire Refining. Fire refining adjusts the sulfur and oxygen levels in blister copper, removing impurities as slag or volatile products. The process is needed because excessive sulfur and oxygen levels result in excessive gas evolution during anode casting (the next step in the production of cathode), producing unacceptably rough anode surfaces, which, in turn, lead to low current efficiencies in the refining cell, uneven cathode deposits and excessive impurities. Fire-refined copper also is also marketed directly for fabrication into copper products.

Fire refining takes place in a reverberatory furnce or a rotary furnace that resembles a converter. Both types have capacities of 100–800 short tons (90–720 t). Most plants fire-refine and cast anodes within the smelter building to facilitate the supply of blister copper to the fire-refining furnace. Once charged, air is blown into the molten copper through tuyeres to complete the oxidation of some of the impurities and remove volatile impurities. Sodium carbonate flux may be added to remove arsenic and antimony. Blowing the blister copper raises the oxygen content to 0.6–1.0%, which is too high for casting. This oxygen content is reduced by a process known as *poling*, after the outdated method of feeding green tree trunks into the furnace as a reducing medium. Today, the oxygen content is reduced by feeding a reducing gas such as ammonia, re-formed gas, or a mix of natural gas and steam into the copper. The end product is cast directly into anodes for electrolytic refining or, rarely, into ingots for sale as fire-refined copper. Anode casting requires an oxygen content of 0.05–0.2% whereas copper to be used directly for fabrication into such things as bar stock, water tubing, or ingots for alloying (known as *tough pitch* copper) requires an oxygen content between 0.03 and 0.05%.

5.7. Electrorefining.   Copper intended for electrical uses requires further refining by electrolysis in order to raise its purity and electrical conductivity to the degree needed for electrical products. Total impurity content in the highest-grade copper is restricted to fewer than a few parts per million.

In electrofining, anodes produced from fire refined copper are dissolved electrolytically in acidic copper sulfate solutions and the copper concurrently electrodeposited (electroplated) onto copper *starting shets* to produce *cathodes*. Starting sheets are made beforehand on insoluble stainless steel or titanium cathodes called *mother blanks*. In some modern refineries, cathode copper is plated entirely onto stainless-steel blanks. Electrorefining is conducted in large tanks, and refineries are known as *tank houses*. A refinery having an annual production of 175,000 short tons might have as many as 1250 cells in the tank house, which may encompass several hectares. The volume of electrolyte in a modern tank house is typically 6000 m^3 for a copper production level of 500 short tons per day. Copper circulating in the electrolyte comprises approximately 10% of the refinery's annual production. Copper in the undissolved anodes represents an additional inventory. Cathodes are sold directly or are melted and cast into wire rod, ingots, and other forms. Some producers cast wire rod in their own refineries; others sell cathode to fabricators, where they are remelted and cast to rod or other intermediate products.

The basic process for electrorefining was developed around 1900. It remains essentially the same today, although several engineering improvements have been introduced. Periodic current reversal (PCR), for example, has enabled an increase in current density from the normal maximum of 240 A/m^3 to 300–350 A/m^3. Automation has led to significant cost savings. Figure 4 shows a simplified flowchart of a modern electrolytic refinery. There are four basic operations:

1. Anodes are dissolved and copper redeposited on starter sheets to form cathodes.

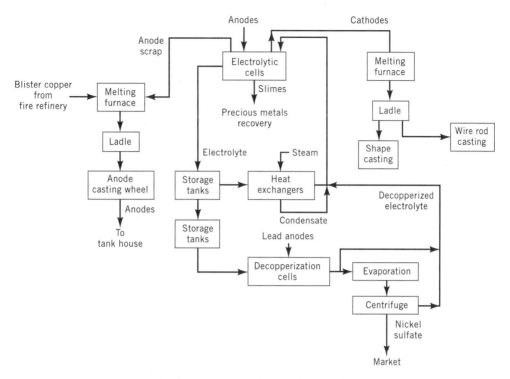

Fig. 4. Flow diagram for an electrorefinery.

2. Impurities (with the exception of silver) from the anode fall to the bottom of the cell and are collected as a sludge known as *anode slime*.

3. Electrolyte is treated to control the concentrations of copper and impurities circulating in it.

4. The refined copper cathodes are melted and cast into commercial shapes.

Anode slimes are shipped to a precious-metal refinery, where gold, silver, platinum, selenium, and tellurium are recovered. Sulfur, selenium, and tellurium in the slimes are combined with copper. The slimes may also contain lead and tin in the form of lead sulfate and tin sulfate.

Electrodeposition is the most important of the unit processes in electrolytic refining. It is performed in lead- or plastic-lined concrete cells, or more recently, in polymer–concrete cells. Composition, temperature, and flow rate of the electrolyte are of great importence to the quality of the cathode deposit. Changes in any one of these parameters can have a serious effect. Current density and voltage are two additional important variables. The trend is toward increased current density and cell voltage in order to change the ratio of copper produced to copper circulating in the electrolyte. In addition, some refineries have changed from the traditional 25–30-day anode cycle to a 9–14-day cycle using smaller anodes to reduce the copper inventory (27). This has reduced the amount of copper in inventory by ~60% (28). The majority of refineries operate at 175–240 A/m^3, copper concentrations of 30–50 gm/L, electrolyte temperatures of 55–65°C, and a circulation rate of 10–20 L/min.

The quality of the cathode deposit is extremely important since off-grade cathode cannot command established prices on metal exchanges. Dendrites, for example, can occlude electrolyte and insoluble particulate matter, thereby reducing quality. Surface quality is controlled by the addition of organic chemicals that influence nucleation through surface adsorption or complexation. The intent is to produce a cathode product that (1) is free from deep striations or fissures that would entrap electrolyte impurities, (2) does not develop nodular growths that might cause short circuits as well as entrap electrolyte impurities, and (3) is not so hard as to preclude straightening bent sheets.

Common additives include animal glue modified with lignin derivatives such as Orzan or Goulac, sulfonated oils (Avitrone), casein, and thiourea, as well as other proprietary reagents.

5.8. Hydrometallurigic Processes. Hydrometallurgical processes for copper can be categorized as:

1. Acid extraction of copper from oxide ore
2. Oxidation and solution of sulfides in waste rock from the mine, concentrator tailings (*dump leaching*), or slag piles
3. Dissolution of copper in concentrates to avoid conventional smelting
4. *In situ* leaching, otherwise known as *solution mining*

Copper ores have been leached in the United States since the 1920s, but hydrometallurgical processing did not become a major part of the copper industry until the availability of large amounts of sulfuric acid from the smelters. Hydrometallurgy was further enhanced by adoption of a solvent extraction and solution purification process developed in 1957 by General Mills (29). Combination of the solvent extraction process with electrowinning led to the development of the (*solvent extraction/electrowinning* (SXEW)) process. Previous leaching processes used *cementation*, the replacement of copper in solution for iron by contacting acidic copper-rich solutions with scrap iron. The product, known as *cement copper*, was fed to the smelter. The SXEW process enables the direct recovery of high purity copper, in cathode form.

An increasing portion of U.S. copper production is being obtained by the leaching of waste rock and tailings. There are currently over 30 installations of this type in the country; their production capacity is now about 850,000 tonnes of copper (30). Hydrometallurgy is also gaining importance in other parts of the world, and it has been predicted that by 2005, leaching and SXEW will account for 20% of the newly mined copper in the world (31).

Leaching of oxide ores is also becoming common; however, this practice is contrained by the availability of oxide mineralization at the mine site. Extraction of copper from oxide ore is accomplished by contacting the ore in a heap or pile, which is underlain by a plastic membrane. The ore is fragmented by blasting and is deposited in the pile in a manner so as to preserve porosity. Acid is dispersed over the top of the pile through a system of plastic garden sprinklers and allowed to percolate through the pile to be collected at the bottom as pregnant liquor. This liquor is piped to the solvent extraction copper recovery system, and spent acid is piped back to the pile for reuse. The acid is ultimately consumed by the limestone in the ore to form calcium sulfate—a very insolouble mineral—and by hydrated ferric sulfate. In fact, the amenability of an oxide ore for acid leaching is determined by the amount of limestone in the ore. Too much limestone consumes too much acid making the process uneconomical for that particular ore.

Copper sulfide minerals per se, cannot be leached by acid unless they are first oxidized. In practice oxidation is accomplished by long-term exposure to the atmosphere and by contact with naturally occuring *Thiobacillus thiooxidans and Thiobacillus ferroxidans* bacteria. Under the right conditions of moisture, temperature, and air, these bacteria have the ability to oxidize sulfide minerals to oxides and elemental sulfur. For example, in an environment maintained at a pH 1.5–4.0, *T. ferrooxidans*, combined with *T. thiooxidans* readily oxidizes the ferrous iron in solution to ferric iron. The ferric iron, in turn, has the capability of oxidizing cuprous sulfide to cupric oxide that is then leached by the addition of acid (32).

Heap leaching of waste rock and mill tailings is conducted in the same manner as leaching oxide ore. The exception is that attention must be paid to maintaining conditions in the heap that are conducive to bacterially induced oxidation and to the prevention or delay in the precipitation of hydrated ferric sulfate, which can lower the percolation rate of the heap.

Leaching of *copper concentrates* is not currently practiced except in a few special cases due to the lack of an efficient and cost-effective process to oxidize the copper minerals. A number of processes were developed for this purpose during the 1960s and 1970s, several using ferric chloride or sulfate as the oxidative agent; however, with the increased cost of energy beginning in 1974, all of these processes were passed over for commercial application. Work has been conducted using elevated pressure leaching following processes developed in the 1950s for the commercial leaching of nickel and cobalt sulfide ores (33,34). None of these processes have been reduced to commercial practice in view of the high cost of pressure leaching relative to the low value of copper.

In-situ leaching is defined as the in-place extraction of metals from ores in a mine or in dumps, heaps, slag piles or tailing piles. Unfortunately, many statistical compilations refer to copper recovered in this manner as *in-situ mining*. In fact, the term rightfully refers to *solution mining*, which involves the pumping of a leach solution underground where it contacts the ore body directly. Boreholes drilled into the ore body provide sites for injection and return of the leach solution. The orebody may be fractured *in situ* using explosives or hydro-fracture techniques developed by the petroleum industry (35).

Leach solutions are characteristically too dilute to be sent directly to an electrowinning cell. The solutions are therefore first concentrated by a process called solvent extraction. A flow sheet of the entire SXEW process is shown in Figure 5 (36). The process is conduced in three stages, each of which has two sub-stages; mixing and settling. The equipment used is therefore called a *mixer* / settler. In the first stage, copper is extracted from the pregnant leach solution by contacting it with chemicals called *extractants*, which are based on a family or organic chemicals known as *oximes*. Early extractants were members of this family called ketoximes. A variety of extractants based on aldoximes are now available that show performance superior to that of the ketoximes (37). The

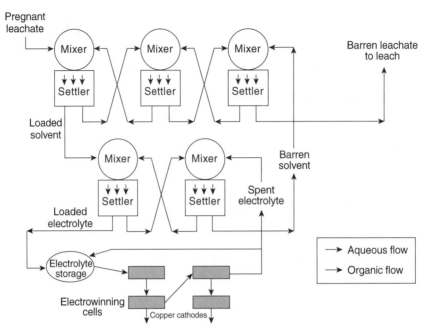

Fig. 5.   Flow diagram for the SXEW process (36).

extractants are dissolved in a kerosene-like diluent, forming what is called the *organic phase*. The extractants selectively remove copper from the weakly acidic leach solution and, in turn, increase its acidity. In the second stage, a process called *extraction*, the organic phase is separated from the depleted aqueous phase by settling. The copper is next recovered from the organic phase by contacting it with a strongly acidic aqueous solution. This operation is called *stripping*. Finally, the resulting copper-rich acidic aqueous solution is fed to an electrowinning cell, where copper is deposited onto cathodes much as in conventional electroplating. The stripped organic phase is returned to the loading stage, while the barren but now strongly acidified leach liquor is returned to the leaching operation.

Electrowinning is highly energy intensive in that only about 0.3 kg of copper per kilowatt–hour is deposited. This contrasts with about 3 kg of copper per kilowatt–hour deposited in electrorefining. Electrowinning cells must also operate above is a minimum potential of about 1.67 V, below which there is no deposition.

6. Fabrication

Copper and copper alloy shapes, known in the trade as *semifabricates*, or more commonly, *semis*, are made from refined copper, and/or copper and copper alloy scrap, with or without the addition of alloying ingredients. Fabricator's facilities are classified as *wire mills, brass mills, foundries*, and *powder plants*. Wire mills and foundries produce wire products and castings, respectively. The general term, *brass mills*, encompasses a variety of plants that produce, respectively, extruded rods and shapes, tube and fittings, and sheet and strip. Only some of these products are made from brass.

6.1. Wire. In order to make wire, refined copper cathodes are melted and cast into *wire rod*, an intermediate product for wire drawing. Wire rod production may take place in the refinery or at the wire mill. Common copper rod diameters and coil sizes are

- 8.0-mm (0.315-in.) or 11.7-mm (0.46-in.)-diameter rod in 113-kg (250-lb) coils
- 8.0-mm (0.315-in.)-diameter rod in 272 (600-lb) or 2268–2722-kg (5000–7000-lb) coil sizes
- Melting is generally conducted in an Asarco shaft furnace. Rod is continuously cast by one of four major processes: the Southwire process; the Properzi process; the Hazelett process, and the Hazelett–Contirod process. The General Electric dip-forming process and the Outokumpu upcast process are also used, but less frequently (38).

Wire rod is fed from coils into a drawing machine, or *drawbench*, which reduces the size of the wire by drawing it through a series of successively smaller-diameter dies. Dies used for drawing copper are made from diamond or tungsten carbide. Natural-diamond dies are used to draw wire diameters ranging from 0.01 to 4.06 mm (0.0004 to 0.16 in.). Tungsten carbide dies are used for sizes AWG (American Wire Gauge) 16 (1.291 mm, 0.05 in.) and larger. The dies are lubricated and cooled by water containing soaps and high fat emulsions. Wire from the drawing machine is taken up and packaged by one of four types of machines known as *spoolers, dead-block coilers, line-block coilers*, or a combination of spooler and line-block coiler known as a *Bundpacker* (39).

Most wire is insulated, standed, jacketed, and bundled into cable. Stranded copper wire and cable are made on machines known as *bunchers* or *stranders*. Various synthetic polymers are used for insulation depending on the service requirement. The most commonly used polymers are poly(vinyl chloride) (PVC), polyethylene, ethylenepropylene rubber (EPR), silicone rubber, polytetrafluoroethylene (PTFE), and fluorinated ethylene propylene (FEP). These insulation materials are formed on the wire by a coextrusion process. Fine magnet wire (otherwise known as *winding wire*, the type used in electric motors, transformers, and similar devices) is coated with a thin, flexible, heat-resisting enamel film. Enamels used include polyvinyl acetals, polyesters, and epoxy resins. Enameling is accomplished by multiple dipping or wiping with curing between stages, or by powder coating. The latter process requires no solvents and therefore simplifies environmental control measures. Oil-impregnated cellulose paper is used to insulate high voltage cable of the sort used in underground transmission lines. The paper is applied by winding tape around the individual conductors in a tight helix, with successive layers wrapped in alternating directions. The entire bundle is then pressure-impregnated with oil, assembled and encased in a sheath, which may be polymeric, metallic, or a combination of materials (40).

Stranded cable is made by paying individual filaments of wire off reels located alongside the stranding equipment. The wires are then fed over flyer arms that rotate about the takeup reel, giving the wires their twist. The bundles of twisted wires can themselves be twisted, bundled, and insulated using the same sort of stranding equipment (39).

High purity copper wire for electrical uses is by far the principal product manufactured by wire mills. Some mills do, however, also manufacture pure and alloyed copper wires for nonelectrical applications such as fasteners, springs, and cold-headed products (this is known as *mechanical wire*) and welding and brazing rods.

6.2. Alloying.   The principal products of brass mills are not limited to brass alone but include copper itself and many alloys. Alloys are prepared by the brass mills or by scrap recyclers and secondary refiners known in the trade as *ingotmakers*. Raw materials include refined copper cathodes, copper ingots, high quality copper or alloy scrap, and the various alloying ingredients required for each product. Alloying elements are usually added in the form of refined metal, although some clean scrap of known composition is also used. Melting and alloying take place in an electric melting furnace where some fire refining may also be performed in order to control oxygen content and other impurities. Once melted, the alloy is cast into a *billet* or *slab* for processing into the required shape.

6.3. Continuous Casting.   In the past, billets with large cross sections were cast statically: those for round products in the form of cylinders (or *logs*) of various dimensions, and those for flat products as thick rectangular slabs called *cakes*. Most mills now use continuous casting machines to produce these intermediate products. In the *vertical* continuous casting process (which is actually semicontinuous), molten metal is poured into the top of an open-ended mold cavity having water-cooled sides and a movable bottom. Hot metal is poured into the mold, where it freezes from the walls inward, forming a solid slug. As the metal solidifies, the bottom of the mold cavity is slowly withdrawn downward while more metal is poured into the top. The result is a continuous shape, up to about 7.6 m (25 ft) long and having the external geometry of the mold cavity. The cavity may be round, square, or rectangular. It may also be profiled for such things as gear blanks, or hollow, for sleeve-bearing blanks. The *horizontal* continuous casting process is similar, except that the mold's axis is oriented horizontally and molten metal is fed from one side. The opposite or exit side of the mold is initially plugged by a starter bar of the same diameter or cross-sectional shape as the finished casting. Grips or a set of pinch rolls slowly withdraw the starter bar followed by the billet that has solidified against it. Horizontal continuous casters generally produce solid rods, billets or flat slabs. Unlike vertical units, which are limited by the depth of the casting pit beneath the mold, horizontal casters can produce large cross-section products in any length. Cast rods used for wire drawing are normally <25 mm (<1 in.) in diameter. Billets for subsequent extrusion range up to about 305 mm (12 in.) in diameter, while cast slabs are normally between 9.5 and 16 mm (3/8 and 5/8 in.) thick. In either case, the casting is sawed into appropriate lengths for further processing in rolling mills, drawbenches, or extrusion presses, depending on the product from desired.

6.4. Rod, Bar, Shapes, Plate, Sheet, Strip, and Foil. In the terminology of copper products, *rod* designates round, hexagonal, or octagonal products supplied in straight lengths, that is, not coiled. *Bar* products are square or rectangular in cross section when sold in straight lengths. *Shapes* can have oval, half-round, geometric, or custom-ordered cross sections and are also sold in straight lengths. *Wire* can have any cross section; the term simply means that the product is sold in coils, reels, or spools. The remaining product designations are straightforward: *plate* is a thick, wide, flat product; *sheet* is a thin, wide, flat product; *strip* is a thin, narrow, flat product usually slit from coils of sheet; while *foil* is very thin sheet.

Rod, bar, and other shapes are normally made by extrusion, a process frequently likened to squeezing toothpaste from a tube. The billet is heated to temperatures approaching 815°C (1500°F) and placed in the extrusion press's cylindrical chamber. A hydraulically driven ram forces the softened metal through a steel die, which may be located at either end of the extrusion chamber. The die opening determines the profile of the finished extrusion. With *direct extrusion*, the extruded metal exits from the end opposite to the ram and in the same direction as the ram travels; with *indirect extrusion*, the die is contained in the face of a hollow ram and metal exits in a direction opposite to the ram's travel. Indirect extrusion produces products with better internal cleanliness. Some rod, bar and profiled wire products are also made by hot and/or cold rolling. Cold rolling hardens the metal, raising its *temper*.

Plate, sheet, strip, and foil are made by rolling continuously cast slabs of copper or copper alloys. For alloys that can be hot worked without cracking, the reduction process begins by heating the slabs, then hot-rolling them on very sturdy two-high roll stands. The rolling process is efficient since the hot metal is quite compliant. The rolls are reversible to permit rolling in forward and reverse directions, and several passes may be used to reduce the thickness to that required for the next stage of rolling. Horizontally oriented rolls reduce the strip's thickness while a pair of vertical edging rolls located in the same roll stand maintains the proper width. Progressively smaller working rolls are required as the strip gets thinner. Small-diameter rolls are normally supported by larger-diameter backup rolls in order to increase the stiffness across the width of the mill. Very thin gauges (such as foil) are rolled on Sendzimir, or "Z" mills, in which a cluster of rolls provides the necessary stiffness. The strip may be annealed between rolling passes; however, final dimensional tolerance is achieved by cold rolling. Cold rolling also imparts the temper required for the intended application. The rolling process must be carefully controlled since different alloys require different rolling conditions, and errors in hot-rolling or annealing temperatures or improper degrees of reduction can result in the production of expensive scrap.

6.5. Tube. Plumbing tube and commercial tube are made from refined (cathode) copper and/or high-grade copper scrap that has been fire-refined and deoxidized with phosphorous. Mills that operate reverberatory or induction (electric) furnaces can utilize scrap, which is usually cheaper than cathode, as delivered, since the furnaces can be used to fire refine the melt to the required purity. Such mills will only use cathode if scrap is in short supply or more costly because of high freight charges. Mills that operate simple shaft furnaces must use cathode as their raw material since no refining takes place in this type of furnace. On average, the North American copper tube industry uses about 65% scrap in its manufacturing process. Whether from cathode or scrap, the resulting metal must, when used for water tube, meet the composition requirements of UNS C12200 [deoxidized high residual phosphorus (DHP)] copper. This grade is quite pure (minimum 99.90% Cu); only phosphorus is added to remove oxygen, which makes the tube more formable and brazeable. Alloy tube or pipe may be brass or any other copper-based alloy.

Tubular products are produced by extrusion and drawing, by seam welding sheet that has been roll-formed into a tube, by continuous casting a shell and rolling the shell into a tube, and by the centrifugal casting process. For extrusion, a tube shell is first produced by continuous casting if a vertical casting machine is available. The tube shell is heated, placed in the extrusion press and extruded over a mandrel that extends through the shell and the die opening. If a horizontal casting machine is used, solid billets are pierced and extruded over a mandrel in a single operation. Extrusion produces a somewhat oversized, thick-walled tube, which is cold-drawn to final size. Drawing also reduces the tube's wall thickness. Copper or copper alloy tube is then drawn in multiple passes on draw blocks (*bull blocks*) or draw benches until it attains its finished size. Tube sold in coils is softened by annealing after drawing. Tube sold in straight lengths is straightened, cut to the appropriate length and shipped in the *hard-drawn* temper.

Plumbing tube has smooth surfaces inside and out. *Commercial tube* requires a more precisely controlled diameter but is otherwise quite similar. Some commercial tube is ridged or grooved inside and/or outside to improve heat conductivity. External grooving is accomplished using a roll-forming process; internal grooving is accomplished using an internal die. To produce inside-and-outside grooved tube a suitably formed die is rocked back and forth along the tube while a grooved and tapered mandrel controls the tube's internal diameter (41).

Extrusion and drawing lubricants used with annealed tube should not leave a carbonaceous film on the interior surface of the tube, since any remaining lubricant will be cracked to carbon during the annealing process. In some cases, the presence of carbon films on the interior surface of water tube may lead to pitting corrosion of the underlying copper (42). In North America, this problem is avoided by using a noncarbonizing lubricant. In Europe, special precautions are taken to either oxidize the carbonaceous material during annealing or to remove the internal residual film by a form of abrasive blasting (43).

Welded tube is made by drawing copper strips through a conical die such that the strip assumes a tubular shape. The mating edges of the strip are joined by induction or resistance seam welding as they exit the die. Excess metal (flashing) inside the tube is cleaned off by passing a broaching die through the tube. Large-diameter tube can be formed directly by this process; small-diameter tube is formed by drawing the as-welded tube to progressively smaller sizes. The welded tube process lends itself well to small-to-medium scale production (40).

The relatively new *cast and roll process* developed by Outokumpu OY is also suitable for modest production volumes, although the developer claims that it can compete with large-scale extrusion processes, as well. A tube shell is first continuously cast to yield a hollow cylinder with an external diameter of 87 mm (3.4 in.), a wall thickness between 18 and 20 mm (0.79 to 1.2 in.) and a length of 12.5 m (41 ft.). This shell is then reduced in size up to 95% by working it over an internal mandrel using planetary rollers. The size of the rolled tube can be reduced further by cold drawing (44).

Centrifugal and, occasionally, continuous casting processes are used to produce thick-walled pipe. For centrifugal casting, molten metal is poured into, and solidifies against, the walls of a spinning, hollow cylindrical steel or cast iron mold, forming a shell. As the metal solidifies, dross and impurities, which are lighter than the molten metal, segregate to the shell's inside diameter, where they are later removed by machining. The centrifugal casting process results in an exceptionally clean product with a fine, dense microstructure. Centrifugally cast copper alloy products can be used in such demanding applications as high pressure hydraulic cylinders.

6.6. Forging. Many copper and copper alloy products can be produced by forging (or *hot stamping*, as the process is called in the United Kingdom). Thick disks, called *slugs*, are sawed from continuous-cast and extruded billets and used as starting materials. Size and weight of the slugs are chosen such that the forged metal fills the die completely, producing some excess flashing—later trimmed off—at the die's mating surfaces. The cumulative deformation of extrusion and forging yields an extremely dense and fine-grained structure that makes forgings suitable for products such as high pressure valve and pump components. High metallurgical quality and the capability to produce near-net shapes enable the forging process to compete with permanent mold casting, investment casting, and machining processes, some of which are less costly. Copper and many copper alloys can be forged, but the overwhelming majority of copper-based forgings are made from *forging brass*, UNS C37700.

6.7. Cold Forming. Copper itself is quite ductile as a result its face-centered cubic (FCC) crystal structure (see section on properties). Many copper alloys, particularly those that share copper's crystal structure are also ductile and easy to form into useful products. Aside from pure copper for roofing products, most copper alloys produced in the form of sheet, plate, or strip are cold-formed in some manner. The large number of different sheet alloys derives in part from the need to optimize several properties (formability, electrical conductivity, corrosion resistance, color, etc) concurrently for various and diverse products. A detailed description of the many forming processes applied to copper alloys is beyond the scope of this article, but excellent literature on the subject is available (18).

Formability is influenced by alloy composition, temper (hardness or strength, often a function of grain size), thickness, and surface condition. From a commercial standpoint, the most important cold-forming alloys are brasses, the large variety of which range in zinc content from 5% to ~40%. As zinc content increases over this range, both strength and elongation increase, and formability increases accordingly. *Cartridge Brass*, UNS C26000 (70% Cu, 30% Zn), is among the most formable of all metals. Copper–tin alloys (known as *phosphor bronzes* for the deoxidizing phosphorus they contain) are also readily formable. *Nickel silvers*, which are brasses containing some nickel (and no silver), form well, although they tend to work-harden faster than do ordinary brasses and consequently require more frequent annealing. Lead, added to many brasses and other alloys to improve machinability impairs formability, but small amounts can be tolerated, a valuable feature in products that must be both drawn and machined (picture a threaded garden hose fitting). Grain size is controlled by a combination of deformation and annealing. Optimum formability for drawing and stretching operations is frequently that which provides maximum elongation. The situation is complex, however, and such factors as the influence on grain size of intermediate annealing operations must be taken into consideration. Grain sizes that are too coarse impair surface finish, a condition aptly called *orange peeling*. Grain orientation (actually crystal orientation, a phenomenon known as *texture*) must also be considered, since the drawing and bending properties of rolled sheet vary with respect to rolling direction, specifically, parallel or perpendicular to the length of the sheet.

6.8. Casting. Copper alloys are known for their excellent castability and, indeed, early cast artifacts were some of the first applications of copper. Statues, bells, and works or art are still cast in bronze, although such products now comprise a small fraction of the copper alloy casting market.

Casting is performed using any of the conventional methods: *sand, permanent mold, die, investment, continuous,* and *centrifugal*. Some alloys are better suited to one or more of these processes than are others, which is one of many factors that dictates process selection. In all cases, the melt is prepared in a furnace using scrap, prealloyed ingot, alloying elements, or combinations thereof. Refined cathode copper is rarely used in the foundry industry. Intermediate frequency induction furnaces are preferred for their fast melting rate and versatility. Low frequency induction furnaces, which must be kept energized and partially filled with molten metal, are used in foundries that primarily cast only one alloy.

The *sand casting process* continues to account for about 75% of copper-base castings in North America, largely because of its versatility in terms of casting size, weight, complexity, and number, and its relatively low capital requirements. It can be used with most copper alloys. Sand casting is less popular in Europe, where various permanent mold processes predominate. The North American situation is slowly changing, however, as environmental regulations, particularly those involving metal-containing offgases and the disposal of spent, lead-contaminated foundry sand have forced many sand foundries to choose between bearing the cost of pollution abatement and converting to a more environmentaly favorable process.

Large copper-base castings such as nickel-aluminum bronze marine propellers and bronze church bells are cast in molds made from sand bonded with resin or Portland cement and supported by a steel shell. Small castings are also made in resin-bonded sand, a process known as *shell molding*. The shell mold is formed by pressing resin-containing sand against a heated steel pattern. The resin sets under heat, forming a thin but strong crust. The shell molding process is capable of producing castings with fine surface detail and good dimensional fidelity.

Use of the *permanent mold casting process* (known as *gravity die casting* in the United Kingdom) for copper alloys is very popular in many parts of the world and is gradually gaining acceptance in North America. Plumbing hardware items, especially faucets, are now commonly produced by this casting method.

Here, split metallic molds, usually made from beryllium copper, beryllium nickel, die steel, or cast iron, are not truly "permanent" but are durable enough to last for tens of thousands of castings. Resin-bonded sand cores are inserted in the die cavity when the finished casting contains internal contours (as in a faucet). An automated version of the permanent mold casting method known as *low pressure die casting* utilizes a few inches of metallostatic head to force metal into the mold. In another version of the process, vacuum draws metal in the mold cavity.

The pore-free surface attainable with permanent mold casting is important to the plumbing fixture industry. Fine grain size, a result of rapid solidification, inputs strength, and an permanent-mold cast alloy cast will be significantly stronger than when cast in sand. This is advantageous in copper alloys applications competing against materials such as stainless steels. The casting process is best suited to castings weighing from one to several kilograms, although castings as heavy as 25 kg (55 lb) have been produced. Alloys such as yellow brass that narrow freezing ranges are best suited to the process. Aluminum bronzes, high-strength manganese bronzes and even pure copper are routinely cast in permanent molds.

Copper and its alloys can also be *die-cast* (*pressure die-cast* in the United Kingdom), although use of this process for copper alloys is uncommon in North America. Here, molten metal is forced into a water-cooled, split metallic mold under very high pressure. The process is fast, mechanized and labor-saving. Copper itself can be die-cast, but doing so requires special mold alloys that can resist thermal cycling to copper's high melting temperature. A large multiple-sponsor research project to enable the die-casting of electric induction motor rotors (almost all of which are currently made from aluminum) was launched in the late 1990s. If the project is successful and if die-cast copper motor rotors become widely accepted, they could increase worldwide demand by up to 400,000 tonnes of copper annually.

The *investment casting* or *lost wax process* dates to at least ancient Egypt. Here, wax patterns are dipped into a slurry of plaster or other refractory until a sufficient thickness is built up to form a mold. The wax is subsequently burned out of the mold. Numerous copper-based products are manufactured using investment casting, among them jewelry items (nickel silver), dental crowns (aluminum bronze), and mechanical components (yellow brass).

Copper alloys are routinely centrifugally cast to produce a large variety of industrial and military products. In addition to the hydraulic cylinders mentioned earlier, products include controllable-pitch marine propeller hubs, fittings, flanges, and a large variety of pump and valve housings. One relatively recent development is the ability to introduce two different alloys into the spinning mold sequentially, forming a composite structure. The alloys bond metallurgically, producing a casting with different properties on its inner and outer surfaces (45).

6.9. Machining. Machined products constitute an important copper market. Understandably, a large number of copper alloys are made in such a way as to provide optimum machining characteristics. The most popular machining alloy is *free-cutting brass*, UNS C36000, which contains nominally 60–63% copper, 2.5–3.7% lead, balance zinc. (Versions of the alloy can be found in all international specification systems.) Approximately 500,000 tonnes of free-cutting brass is consumed annually in the United States alone, all of it made entirely from scrap, about half of which is in the form of turnings recycled to the brass mills from machine shops. Free-cutting brass is widely accepted as having the best machining characteristics of all engineering metals. It is theoretically capable of attaining cutting rates nearly 5 times higher than those used with leaded free-machining steel, the alloy's major competition in the screw-machined products market.

High machinability in free-cutting brass and most other alloys derives from the addition of a few percent of lead. The lead breaks up machining chips (turnings), making the turnings easy to remove from the cutting zone. It also provides a measure of internal lubrication as it smears across the face of the cutting tool, enabling higher machining speeds. Leaded versions of numerous wrought copper alloys are commercially available. Many cast copper alloys contain lead to improve castability; enhanced machinability is a supplemental benefit. Leaded plumbing alloys, including free-cutting brass, have come under scrutiny by health authorities because of the possibility that small amounts of lead may leach into certain aggressive drinking waters. The U.S. Environmental Protection Agency's "Lead and Copper Rule" limits the level of such releases. All plumbing goods sold in the United States are now certified to meet these limitations. Plumbing product's whose shape or internal volume makes excessive leaching unavoidable can be cast in brasses containing bismuth and selenium (designated by the industry as *EnviroBrass*) in place of lead (45).

6.10. Powder Metallurgy (PM) and Composites. Products made from pressed and sintered metal powders constitute a small (1% of total consumption) but interesting market for copper. Copper and copper alloy powders used in such products are produced by gas or water atomization of molten metal, by electrolysis, and by solid-state reduction. Most copper-base powders are made from scrap. It is possible to achieve 100% of theoretical density in PM products, but items such as "oil-less" bearings—a high volume product—are intentionally left porous to retain lubricants. PM makes possible the creation of composite materials, which may or may not be entirely metallic and whose properties cannot be obtained by other means. Copper–graphite brushes for electric motors and copper-containing brake pads are examples of mixed structures. Modern examples of metal-matrix composites are the copper/refractory metal materials from which liquid-fueled rocket engine nozzles are made. Copper provides high thermal conductivity for efficient heat transfer, while the refractory metal (tungsten, molybdenum or niobium plus chromium) provides strength at elevated temperatures (46).

7. Economic Aspects

Copper is a world-traded commodity. Base prices of the several accepted grades of refined copper are established by buying and selling on the London Metal Exchange (LME), the Commodity Exchange of New York (COMEX), and the Chicago-based Mid-America Commodity Exchange (MACE). The main function of these commodity exchanges is to separate the manufacturing and speculating activities of the participants in the marketplace. The exchanges also provide the means by which natural market fluctuations can be mitigated to the benefit of both producers and consumers. While most refined copper is sold directly to customers, the prices that are negotiated between the buyer and seller are based generally on one of the major exchange prices at the time of delivery. The prices of the various grades of scrap are also generally derived as some fraction of the current LME or COMEX copper price.

Because copper is produced in many different parts of the world, there has never been a successful effort to control prices centrally such as in the case of as occurred with the Organization of Petroleum Exporting Countries (OPEC) vis-à-vis petroleum production. About 40% of global copper production was briefly controlled under a French organization, called Conseil Intergouvernemental des Pays Exportateurs de Cuivre (CIPEC), which represented eight producing countries, largely in Africa and Latin America. However, the cartel never successfully controlled prices, and it disbanded in 1992. The International Copper Study Group (ICSG) was formed in 1994 as an intergovernmental assemblage of copper producing and copper consuming countries in an attempt to stabilize prices by making the market more transparent. The United States and Canada are members of the ICSG.

The basis for copper pricing is simply the state of the world economy, which establishes the level of demand, counterbalanced against the installed capacity to produce and the unit costs of production, which establish the level of supply. Copper is an energy-intensive material; hence, the cost of energy-producing commodities, coal, oil and natural gas, have a direct effect on copper supply. Figure 6 (for which data were compiled from Metallgesellschaft, the American Bureau of Metal Statistics, and various industrial sources) shows a plot of the historic price of copper in both actual and constant 1994 dollars. It should be noted that ever since the 1974 "energy crisis," the constant dollar price of copper has trended downward. This occurred despite the fact that the average ore grade has also trended downward.

Measures taken by copper mining companies to reduce drastically both energy and labor costs enabled companies to absorb the lower prices for their

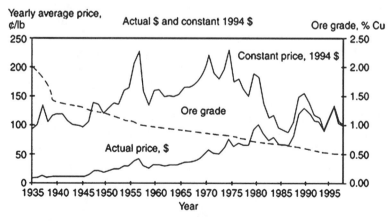

Fig. 6. Copper price versus ore grade. (*Data*: Matallgesellschaft, the American Bureau of Metal Statistics and various industrial sources.)

products. As indicated in Figure 6, which has been smoothed by using annual averages, the price of copper fluctuates. The reason for this is that copper world mining companies compete on the basis of lowest cost production. Thus, in times of surplus, the global copper price tends to follow production costs downward. As high cost producers temporarily close their mines and thus reduce production, or if copper consumption increases as a result of normal business cycle fluctuations, a copper shortage is created and the price of copper increases above the lowest cost producers' marginal price. Unfortunately, the inertia of the system caused by time delays in shutdown and start-up of production facilities causes a ratcheting of world copper prices.

8. Specifications, Standards, and Quality Control

Numerous national and international organizations specify standards for copper and copper alloy products, including the International Organization for Standardization (ISO), the British Standards Institution (BSI), the Deutsches Institut fur Normung (DIN), the Japanese Industrial Standards Committee (JISC), and the American Society for Testing and Materials (ASTM).

8.1. Refinery Products. ASTM B224 defines the standard classifications of copper according to the method of refining and characteristics determined by the method of casting or processing (47). The accepted basic standard for cathodes is given in ASTM B115 (48). ASTM B5 gives specifications for electrolytic tough pitch (ETP) copper in cakes, slabs, billets, ingots, ingot bars, and other refinery shapes. Electrolytic copper is one of the purest of all materials of commerce, and commercial ETP copper usually far surpasses the specifications of ASTM B5. In fact, ETP copper ranges from 99.94 to 99.96% Cu; even the highest level of impurities other than oxygen occur at concentrations of only 0.0015–0.003% (15–30 ppm). Oxygen, present as copper(I) oxide, is present up to 0.05% (500 ppm). In those coppers exceeding the specifications of ASTM B5, it is not uncommon to find conductivities of IACS of $\geq 101\%$. Oxygen–free (OF) copper is copper that has been specially processed to reduce the oxygen content in order to raise electrical conductivity and improve weldability. Oxygen-free copper C10200, for example, contains a maximum of 0.001% (10 ppm) oxygen. Oxygen-free electronic copper C10100, for electronic use, contains $\leq 0.005\%$ (5 ppm) oxygen and $< 0.01\%$ (100 ppm) total impurities.

8.2. Mill Products. Published specifications for copper semis deal with specific applications, such as building construction, and with certain copper alloys. ASTM B248M specifies the general requirements for wrought copper and copper alloy plate, sheet, strip, and rolled bar (metric); ASTM B249M contains requirements for wrought copper and copper alloy rod, bar, and other shapes (metric); ASTM B1 and B2 govern hard- and soft-drawn copper wire; ASTM B42 gives specifications for seamless copper pipe, while ASTM B68, B75, and B88 contain specifications for bright annealed copper tube, seamless copper tube, and copper water tube, respectively. In addition, ASTM B280 provides specifications for seamless copper tube used in air conditioning and refrigeration field service and ASTM B819, for seamless copper tube for medical gas systems.

8.3. Influence of Impurities. All *dissolved* impurities reduce the conductivity of copper (49) however, those impurities that form insoluble phases have little or no effect. For example, phosphorus, which is often used as a deoxidizer, sharply reduces electrical conductivity in OF copper, but it has little effect if the copper contains excess oxygen because insoluble phosphorus oxide then forms. Impurities also affect mechanical behavior; however, impurities in electrolytic copper are present at such low levels that they have little effect on the hot- or cold-working operations, including wire drawing and sheet rolling. The effect of impurities on annealing behavior is important in magnet wire. Enamels used to insulate the wire cure at a relatively low (baking) temperature. It is important that the copper also anneals at this temperature, so that the finished wire retains a minimum amount of spring-back after winding (50–52). Impurities such as antimony, sulfur, tellurium, and selenium raise the softening temperature of copper and thus raise its springback. Impurities such as bismuth, selenium, and sulfur have a strong effect in increasing springiness (53). Other impurities, such as silver, iron, nickel, cobalt and oxygen have no effect (54).

Impurities can affect the influence of thermal history on the mechanical properties of rolled copper products. For example, many impurities in commercial copper are present in metastable concentrations (above their solid solubility limits) at low (eg., 300°C) temperatures, but they can precipitate at somewhat elevated temperatures, thereby changing the metal's mechanical properties. Other impurities can revert from solution as oxides (in oxygen-bearing copper) as temperature is raised. In both cases, changes in temperature change the state of the impurities, and thereby the metal's hot-deformation characteristics. Since uniform deformation characteristics are obviously advantageous, efforts are normally made to ensure that most impurities are taken out of solid solution before deformation is attempted.

8.4. Quality Control. The spectrometer is the most suitable analytic instrument for quantifying the presence of most low–level residual impurities in copper. ASTM E414 specifies the standard method for the measurement of impurities in copper by the briquette dc arc technique (55).

Conductivity is measured using a 0.48-cm (12-gauge) wire, which is annealed at 500°C for 20 min, given a sulfuric acid pickle, quenched, and dried. The resistance of the prepared specimen is then compared with that of a standard wire of the same dimensions and structure.

Annealability, or the ease and rapidity with which copper softens on heating after cold working, is important in the drawing of wire products. Annealability is most often described by the half-hardness temperature, defined as the temperature at which annealing for one hour returns the metal's tensile or yield strength to the midpoint between that for the fully worked metal and that the fully annealed state. High purity copper, after cold working, reaches the half-hard stage when annealed for one hour at 140°C. Other tests for annealability include residual hardness after cold working and annealing and by spring elongation after work hardening and annealing.

9. Analytical Methods

The previous section described the analytic methods used in the quality control of refined copper. For general copper analyses, techniques such as atomic absorption analysis and X-ray fluorescence are used for control of operations and environmental measurements. Microquantities of copper are readily detected and identified by the formation of either the deep blue cupramine complex ion $[Cu(NH_3)_4]^{2+}$, a red-brown precipitate in the presence of potassium ferrocyanide, or a green precipitate with α-benzoin oxime. When substances containing copper are moistened with hydrochloric acid and heated on a platinum wire, a blue-green tinge to the flame results. Numerous other reagents, chiefly organic compounds applied in spot tests, have been employed for the detection of microquantities of copper.

10. Environmental Concerns

Copper production, and to some extent copper use and disposal practices, are regulated in the United States and elsewhere in the world by both state and federal regulations related to air and water quality and to waste disposal. The concerns regarding copper and the byproducts of its production in the natural environment can be divided into four categories:

- Natural flux to the atmosphere and the oceans
- Effluents from production
- Effluents from uses
- Lifecycle environmental impact

10.1. Natural Flux to Atmosphere and Oceans.   As noted earlier, copper is relatively abundant in the earth's crust. It can enter the natural environment in various forms and by a variety of mechanisms 56–59:

- In mists dispersed into the atmosphere from the sea where salt spray is generated in breaking waves
- As wind-blown powder and dust of geologic origin—mainly from deserts and other arid regions
- As wind-blown biogenic particles from agricultural areas, forests, and other vegetated areas
- As volcanic dust, ashes, and the like dispersed into the upper atmosphere
- As the result of volcanic and hydrologic activities at midocean ridges
- As biogenic refuse from aquatic flora and fauna
- By erosion of the seabed
- By erosion of riverbeds

It is estimated that as much copper is dispersed into the natural environment by natural forces as is by anthropogenic point sources—specifically, a total of ~20,000–51,000 tonnes/year (59).

10.2. Effluents from Production.   Most copper minerals are sulfides. Since many of these minerals are copper/iron sulfides, more sulfur is liberated from the ore by the smelting process than copper. During the smelting processes, these sulfides are oxidized to form gaseous sulfur dioxide. In the United States, under the Clean Air Act of 1970, stringent limitations are placed on the emission of sulfur dioxide from copper smelters. Similar regulations exist elsewhere in the world. As a result, measures have been taken by most smelters to restrict sulfur dioxide emissions to some extent. Today, two-thirds of the world's major smelters recover 95–99.5% of the sulfur in the ore. The majority of the remaining smelters recover at least 85% of the sulfur dioxide they generate. Sulfur dioxide is most often recovered in the form of sulfuric acid but also as elemental sulfur and calcium sulfate, an extremely insoluble and inert material that is similar in composition to the naturally occurring mineral, gypsum. Acid is either sold into the acid market or used to leach oxidized copper ores (see section on recovery and processing).

Copper mining operations have always been faced with a large solid-waste disposal problem. Waste consists of overburden (unmineralized rock and soil), tailings (copper mineral-depleted waste material from the mill), and slag (largely iron oxide, calcium oxide, and silica with low concentrations of other metals fused into a lavalike mass in the smelter) and depleted leach residues. All this material must be stored onsite. The wastes are regulated in the United States under the Solid Waste Disposal Act as amended in 1976 by the Resources Conservation and Recovery Act (RCRA). In addition, under the Emergency Planning Management and Community Right to Know Act of 1986 (EPCRA), otherwise known as the *Toxic Release Inventory* (TRI), the amount of any substance in the wastes that is listed among the 644 substances considered to be toxic under this act must be measured and reported to the USEPA annually. No risk to the public is implied, yet an operation as simple as moving unprocessed broken rock from one part of a mine to another must be reported as a "release to the land." Wastes naturally contain small concentrations of copper but may also contain other substances such as lead, manganese, and zinc, all of which are considered toxic under the Act.

Water effluents include process water and wastewater related to specific mining, milling, and smelter operations, plus storm water runoff that may come in contact with a facility's operations. Such effluents are monitored and regulated under the Clean Water Act of 1977. In general, in a copper mining/milling operation, most of the water used in the milling process is recycled back to the mill. In the U.S. Southwest as in other arid areas, the major water loss from a mining/milling site occurs by evaporation.

10.3. Effluents from Uses. A large portion of all of the copper in use is ultimately recycled for other uses (see section on recycling and disposal), although there is a finite amount of attrition into the environment. Copper contamination in the soil is not a major issue, since surface soils in the United States naturally contain, on the average, 18.2 ppm copper (range: 6.4–63.4 ppm) (60). In certain agricultural soils, where there is a copper deficiency, copper must be added as a mineral supplement to fertilizer. Surplus copper is remineralized into the soil.

The major concern over contamination due to copper uses is introduction of the metal to bodies of freshwater and saltwater where aquatic life may be adversely affected. (Because copper is an essential element, adverse effects only appear at copper intake levels higher than those required for metabolism.

Toxicity levels vary widely among species.) Most of copper's uses in aquatic applications utilize the metal's algicidal and fungicidal properties. In the United States, under the Clean Water Act of 1977, copper is regulated in freshwater at levels on the order of 6.5-34 ppb (61) and in saltwater at levels on the order of 2.9 ppb (data also from Ref. 63, but based on 1-h average concentration not to exceed once every 3 years). In both cases, the actual number is calculated under the law as a function of a number of variables, including water composition and pH. It should be noted that the restrictions on copper in wastewater discharge are significantly lower (measured in parts per billion) than the restrictions of copper in drinking water (in 1.0–1.3 ppm maximum; — see section on health and safety factors).

Copper enters bodies of water in rainwater runoff, through the use of copper algaecides in lakes and ponds, through the breakdown of copper-based antifouling paints, or in sewage effluent. Copper enters rainwater runoff from several sources: road traffic, through wear of brake linings and pads, tires, roadway surfaces; spillage of fuels, oils and antifreeze; chemicals, wood preservatives, antifouling paints, algicides; buildings, as corrosion products from copper roofs and facades and from plumbing systems; agriculture, by leaching of soils; and landfills, as seepage.

Road traffic–based sources are reportedly the main contributors of copper to the environment at large and are major contributors to aquatic systems, as well. Copper-based chemicals used as antifoulants, algicides, and similar products are the principal contributors to the aquatic environment and a major contributor to the total environment. These are followed in magnitude by landfills and agricultural sources. Plumbing systems, roofing and facades contribute only minor amounts of copper to the environment (62).

10.4. Lifecycle Assessment (LCA).   There is growing interest in quantifying the environmental impact of materials by performing "cradle to grave" analyses of the energy consumed in the materials' production and use, and by the materials' impact on the environment at each stage of their lifecycle. One objective of this LCA exercise is to compare the results obtained for various materials when making materials selection decisions for items ranging from consumer goods to sports arenas. LCA analysis potentially poses a threat to certain metals, especially ones whose production is particularly energy intensive. LCA is still in its infancy, and quantitative data are scarce for any material, however, copper should fare well in any LCA by virtue of the large amount of copper scrap that is recycled. Since scrap represents energy that has already been expended, its use in products significantly lowers the products' net energy content. For example, it is estimated that the mining, smelting, and refining of primary copper by traditional means requires about 62 MJ/kg. However, refined copper is now derived from smelting, leaching (SXEW) and scrap recycling. When the latter two energy-efficient sources are factored in, it is estimated that the resulting refined copper requires (in the United States) an average of only about 36 MJ/kg (63).

LCA-type comparisons can, for example, be applied to copper plumbing tube and chlorinated poly (vinyl chloride) (CPVC) pipe—products that compete in several global markets. Using the energy intensity for copper in the United States and recognizing that 65% of U.S. copper plumbing tube is currently manufactured from recycled scrap, it is estimated that 27 MJ/kg of energy are required to manufacture copper tube (64). It can further be shown that the energy required to manufacture the tube, transport it to an average-sized home in California, install it and to ultimately recycle it back to the manufacturer is 2118 MJ. The comparable amount of energy required for CPVC pipe is 2335 MJ. However, the estimated energy requirements for copper plumbing tube range from 760 MJ (if 100% scrap copper were used to make the tube) to 3900 MJ (100% primary copper). The corresponding range for CPVC is 1500 MJ to 3300 MJ, based on manufacturing data from a European Plastics manufacturers' association (65). From an environmental-impact viewpoint, it is significant to note that whereas copper is recycled for further use at the end of its useful life, CPVC must be disposed of in a landfill.

11. Recycling and Waste Disposal

Because of its high value, copper is among the most throughly recycled metals. A large and sophisticated industry gathers, grades, remelts, and returns copper-base metals to fabricators for reuse in commercial products. Recycled copper is known as *secondary copper* or simply *scrap*, of which there are three categories:

- *Prompt, home*, or *direct scrap*, which originates in smelting and/or fabrication operations. This scrap is recycled within the plant
- *New scrap*, which is metal returned to the fabricator by manufacturers as leftover material. Machine shop turnings (chips) are one example
- *Old scrap*, which includes all worn out, discarded, or salvaged products, such as telephone cable and plumbing tube from demolished buildings.

In North America, copper scrap is graded, according to standard definitions established by the Institute of Scrap Recycling Industries (ISRI). Definitions are based on form and the presence of contaminants (66):

No. 1 Scrap. Scrap that is of cathode quality and that requires only melting and casting. Examples include waste copper rod and bare wire.

No. 2 Scrap. Unalloyed copper scrap that is contaminated with other metals (such as might have been introduced by electroplating or soldering). It may be recycled to a primary smelter, where it is re-refined, reentering the market in the form of high grade cathode.

No. 3 Scrap. Low grade scrap of variable composition (10–88% copper) is processed in smelting furnaces, where it may be fire-refined to high purity. It may additionally be electrorefined to high grade cathode. Refining is performed by *primary* or *secondary smelters*. The latter specialize in scrap recover.

No. 4 Scrap. Alloyed scrap, consisting mainly of brasses, bronzes, and copper-nickels, is recycled to ingot makers or brass mills for the production of new alloys. The scrap is carefully sorted according to composition. The scrap is then melted and compositional adjustments made using air oxidation to

Table 4. **Average Lifetime of Recyclable Copper in Various Applications**[a]

Application	Estimated life in service, years
automobiles	8–15
small electric motors	10–12
electric cable	30–40
brass and other alloy parts	50–60
copper plumbing systems	60–80
roofing and facade sheet	>100

[a]Refs. 67–69.

remove aluminum, chromium, silicon, iron, and sometimes tin in the form of slag. Care is taken to avoid the loss of valuable metals such as zinc in brasses and tin in bronzes. High lead-content scrap (soldered copper–brass automobile radiators are a good example) is used to manufacture leaded brasses and lead-bearing casting alloys, providing a safe repository for lead that might otherwise enter the environment.

Scrap prices are established by the market and are posted daily in the trade press. Published prices generally fluctuate with the market price of refined copper, although supply–demand factors and supplier–client agreements often set the final prices.

Much like steel, copper is an industrial commodity that exists *within* the infrastructure—as electric motor windings, telecommunications cable, and components, where it may remain in service for 50, 80, or even 100 years before entering the scrap cycle. Table 4 lists estimated lifetimes for recyclable copper in various applications (67–69). As a result, the availability of old scrap for recycling is a function of the amount of copper that was used in the past. This situation differs from that for aluminum, whose main uses are short-term and for which metal returns to the scrap cycle within a matter of weeks or months. It has been estimated that the global copper reservoir in installed uses was about 33 million tonnes in 1940 and had grown to 190 million tonnes in 1991 (70). The latter figure corresponds to some 70–80% of all copper mined since 1900. The average growth rate of the installed copper reservoir was about 3 million tons/year over the 50-year period. In the United States, the growth rate of the installed copper reservoir was about 1.2 million tonnes/year during the period 1983–1991.

Because of the delay in the reappearance of scrap copper and the complex nature of scrap flow through the economy, statistical records of copper recycling are difficult to establish and it is impossible to determine precisely how much copper is actually recycled in any given year. The flow of scrap through the copper industry is shown in Figure 7, data for which were taken from, among others, the International Copper Study Group (71). Prompt scrap, which is internally recycled, does not appear in scrap recycling statistics. Available statistical data, however, do allow several important observations to be made:

1. In the United States, between 1978 and 1998, the fraction of the total consumption of copper derived from scrap varied between 35.5% (1998) and

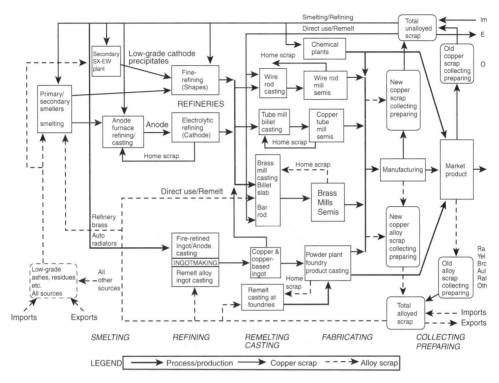

Fig. 7. Copper and copper alloy flowchart. [Data from ICSG (71)].

49.3% (1982). The average for the 21-year period was 43.3% (72). Of this, 68% was new scrap and 32%, old scrap.

2. Approximately 67% of all recovered copper scrap in Western countries does not undergo refining; specifically, it is melted and used directly by brass mills and other fabricators in the fabrication of alloy products. This is equivalent to 35% of all the copper used in the these countries (73). In the United States, where there is a domestic supply of refined copper, between 1978 and 1998 an average of 29% of all copper consumed was consumed directly by fabricators, mainly by brass mills (7). Of this scrap consumed, 52% was alloyed and the remainder unalloyed.

3. The functional distribution of copper uses suggests that approximately 92% of all copper consumed is recyclable. That is, the metal is used in forms and products that are not dissipative (74). Copper chemicals and powders, for example, are dissipative in their use, whereas, solid shapes, such as wire, plate, sheet, and bar, are not. Thus, very few copper or copper alloy products ever reach a landfill for final disposal. Only an estimated 0.5% of copper is lost from nondissipative forms, such as electronic circuitry (70). This factor is discussed in the section on environmental concerns.

4. Figure 8 shows that the amount of scrap used in any given year is a direct function of the price of refined copper (75). The higher the copper price, the greater the amount of scrap that is available for use. Copper scrap is itself a

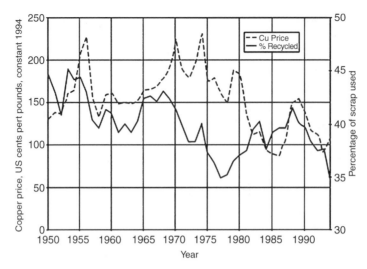

Fig. 8. World percentage of scrap used in refined copper versus copper price (61).

commodity, and scrap dealers tend to speculate by warehousing copper scrap when its price is depressed and offering it for sale when the price rises. As discussed in the environmental concerns section, the production of refined copper from scrap is significantly cheaper than that from ores and concentrates largely because of the reduced amount of energy required.

12. Health and Safety Factors

12.1. Human Metabolism. Copper is an essential trace element required for human metabolism. The total amount of copper in humans, based on individual tissue analyses, is estimated at 50–120 mg for adults and 14 mg for a full-term newborn infant. The concentration of copper in adults and infants is 0.7–1.71 milligrams per kilograms (mg/kg) of body weight and 4 mg/kg of body weight, respectively. The higher concentration of copper in newborn infants compensates for the lack of copper in mothers' milk. This additional copper is imparted to the fetus by the mother in the third trimester of pregnancy. The liver, brain, and heart have the largest concentrations of copper of all the organs in the body. The liver and the brain contain together one-third of the total copper content of the body. The liver plays a central role in copper metabolism and takes part in the storage, distribution, and disposal of copper from the body. All body tissues require copper in their metabolism, but some have greater metabolic needs than others. Copper is normally obtained from food and to a lesser extent, water intake; however, it is also provided in some vitamin and mineral supplements and it is added to baby formulae. According to the safe and adequate intake values established by the U.S. National Academy of Sciences, daily intake of copper should be 1.5–3 mg for adults and 0.7–2 mg for children (76).

Copper is a cofactor for many enzymes that act as catalysts for body functions. It plays a role in erythrocyte formation and development of bone, central nervous system (CNS), and connective tissue. It is necessary for the release of tissue iron and its movement to plasma. Copper is an antioxidant essential for cardiovascular health (77). It is also essential for the production of collagen, the fibrous component that binds heart muscles together, and elastin, a protein substance that makes heart and artery walls elastic. In addition to its metabolic role in the human body, copper acts as an antiinflammatory agent in rheumatoid arthritis and other diseases involving damaged tissue (78).

Copper deficiency is rare in Western countries since diets provide the necessary amount. However, the World Health Organization has expressed its opinion that copper deficiency is more of a potential problem than copper toxicity in humans (79). Children afflicted with chronic diarrhea or fed cow's milk exclusively are prone to copper deficiency (80). A deficiency of copper restricts the pickup of iron and thus inhibits the synthesis of hemoglobin leading to anemia. It can also lead to neutropenia, osteoporosis, arterial disease, and, in the case of severe deficiency, brain damage (81). The effects of copper deficiency are best observed in patients suffering from Menkes disease, a genetic disorder associated with a defect of copper transport and absorption in the body (82). The disease is characterized by rapidly progressive cerebral degeneration, bone lesions resembling those seen in scurvy, and elongation and tortuosity of the cerebral arteries. Death caused by severe, progressive neurodegeneration almost invariably occurs by the age of 3 years. There is no pharmacologic treatment available for Menke's disease.

12.2. Toxicity. Acute environmentally related copper toxicosis is not a major problem in humans as ingestion of excessive quantities of copper salts (25–50 mg of Cu) usually induces immediate expulsion by vomiting (83). In fact, 300 mg of copper sulfate in an 8-oz (0.24-l) glass of water was at one time prescribed extensively by the medical profession as an emetic (84). Death from the intake of huge excesses of copper sulfate has been reported in India, where it has been used as a vehicle for committing suicide (85). However, in at least one instance a patient survived the ingestion of 250 g of copper sulfate (86). Present regulations for the copper content of drinking water are based on the minimum level at which copper can be tasted in the water, about 2.6 mg/l, (87) and the level at which a nauseous feeling is felt by individuals exhibiting a sensitivity to the metal, 3-5 mg/l, (88). However, it has been observed in some communities with a copper content in drinking water of ≤8–10 mg/l that people become accustomed to the taste and do not exhibit nausea. In the United States, the regulatory level has been established at 1.0 mg/l and the action level, at 1.3 mg/l. The World Health Organization recommends a level of 2 mg/l based on acute effects (89). The WHO established no limits based on chronic effects.

Since copper is an essential nutrient, humans and other forms of life requiring copper in their metabolism have a variety of mechanisms that function to prevent overdose. Tolerance mechanisms (e.g., homeostasis) make copper generally innocuous (90). However, in the case of direct introduction of copper into the bloodstream, as has occurred in several cases of improperly constructed blood dialysis machines, death can result for copper overdose (88). A small proportion of the population, approximately 1 individual in 30,000, suffers from a genetic deficiency whereby copper is not efficiently removed from the system by the liver. This is known as *Wilson*'s disease (91). If not treated, patients can suffer severe neurologic and psychiatric disturbances and ultimately die of cirrhosis of the liver. The disease, however, can be very effectively treated pharmaceutically. Liver transplantation is the only permanent cure for the disease as it corrects the metabolic defect. In addition, it has been discovered that a very small population of infants suffer from what has been termed an *ecogenetic* disorder whereby affected infants, if exposed to excessive amounts of copper in their diets, succumb within the first year of life of cirrhosis of the liver (92). It appears that this disease is the same as what was previously called *Indian childhood cirrhosis*—a disease endemic to India where it has been customary to prepare infant food by boiling milk and cereal in copper or brass pots (93).

12.3. Medicinal and Antibacterial Qualities. Copper has been used in medicine for thousands of years. The first written record of its use is in the Smith *Papyrus*, an ancient medical record written between 2600 and 2200 B.C. (94). The *Papyrus* records the use of copper (probably in the form of the mineral malachite) to treat chest wounds and to sterilize drinking water. The ancient Egyptians used copper vessels for cooking and serving their food and bronze cups for drinking water (95). Modern research has shown certain copper compounds to be valuable in the treatment of inflammatory diseases, in some forms of cancer, and as an anticonvulsant agent (96). Likewise, copper plumbing tube has been found to be effective in restricting the growth of bacteria, including *Legionella pneumophila*, the cause of Legionnaires' disease (97). Copper–silver ionizers have been found to be effective in controlling *L. pneumophila* in hospital drinking water (98). Such systems have been widely used in the disinfecting of swimming pools (99). Further, research strongly suggests that the pathogen *Escherichia coli* O157 is killed within several hours' exposure to copper surfaces; *E. coli* O157 has caused countless outbreaks of foodborne and waterborne disease around the world and has caused tens of thousands of cases of bacterial gastroenteritis and numerous deaths among the very young and the elderly (100). The bacteriostatic properties of copper-bearing surfaces such as ordinary brass doorknobs and pushplates have also been credited for reducing infection in hospitals and other public buildings (101).

12.4. Worker Health and Safety. With the exception of vineyard sprayer's disease, no significant chronic effects of copper have been reported as a result of occupational exposure. In the case of vineyard sprayer's disease, 5 out of 15 patients who had used Bordeaux mixture, a mixture of copper sulfate and lime long used as a fungicide in agriculture, exhibited lung condensations varying from diffuse reticulonodular shadows to tumorlike opacities (102). A study of copper miners who had spent at least 20 years in copper mines showed no adverse effects (103). Metal fume fever, a 24–48-h illness characterized by dryness in the mouth and throat and headache has been reported in factory workers exposed to copper dust or fumes (104).

Copper mining activities fall under the aegis of the Federal Mine and Safety Act of 1977. The Act sets mandatory standards and requires training for new employees plus annual refresher training for all mine workers. It is administered by the Mine Safety and Health Administration (MSHA). The Occupational Safety and Health Act, which covers mills, smelters, refineries, and fabricators, is similar except that it is administered by the Occupational Safety and Health Administration (OSHA). Both MSHA and OSHA are part of the U.S. Department of Labor.

13. Uses

Figure 1 indicates that the demand for copper has increased continually since the first use of electrical power. Between 1994 and 1997, for example, the rate of growth in copper consumption was approximately 4.5% worldwide, one of the fastest-growth periods in history. Table 5 shows the consumption of copper in the western World from 1992 to 2000 (105).

Copper's useful properties and intrinsic attributes are exploited in the many commercial products in which the metal is used. In the United States, copper and copper alloy consumption, based on the metal's five most important attributes, is as follows (106):

Electrical conductivity 62%; corrosion resistance 20%; heat transfer 11%; structural 5%; aesthetic 2%.

Electrical and communications wire and cable together with plumbing products account for 85% of refined copper consumption and 63% of total consumption, including scrap. Electrical products are primarily manufactured from refined copper, whose degree of purity ensures high electrical conductivity. About 65% of the copper in plumbing products is derived from copper scrap. Most other applications, including most copper alloys, except those used in coinage, are also based mainly on scrap as a raw material.

Electrical uses constitute the largest single market for refined copper. In the United States, for example, wire and cable accounted for 67% of all refined copper consumed in 1998 (2). In rapidly emerging economies where power and communications infrastructures are being installed, the fraction of refined copper consumed as wire and cable can be as high as 80%. Likewise, in the United States, copper uses are found the markets described in the following subsections.

13.1. Building Construction.   Copper's use in homes and in commercial and industrial buildings is by far the metal's largest market, amounting to approximately 41% of total consumption. Electrical products, plumbing goods, and roofing sheet, respectively, make up the bulk of copper's building-construction applications. An average modern U.S. home, for example, contains approximately 439 lb (200 kg) of copper (see also BUILDING MATERIALS, SURVEY).

Electrical conductors in homes and commercial–industrial buildings are classified as *building wire and cable*. House wiring is almost exclusively copper; the main exception is service entrance cable (from the street to the primary electrical panel), which is often aluminum (see ELECTRICAL CONNECTORS). Penetration by aluminum of the interior house-wiring market, a factor in the 1960s, ebbed quickly after numerous fires were attributed to aluminum wire connections. Today little aluminum is used as interior house wire in the United States, and nearly none is used in Canada. Aluminum house wiring is not, however, uncommon in some developing countries.

Table 5. **Western World Consumption of Refined Copper ('000 Metric Tons)**

Region	1992	1993	1994	1995	1996	1997	1998	1999	2000 e
Western Europe[a]	3,268	3,097	3,341	3,388	3,345	3,534	3,728	3,763	3,920
United States	2,176	2,359	2,560	2,534	2,621	2,790	2,883	2,988	3,005
Japan	1,411	1,384	1,375	1,415	1,480	1,441	1,255	1,294	1,325
Other Asia	1,416	1,620	1,833	1,955	2,128	2,240	2,132	2,527	2,700
Latin America	450	469	503	511	619	714	874	894	984
Canada	156	186	199	190	218	225	246	267	280
Oceania	126	150	148	174	170	166	166	171	175
Africa	104	109	123	117	115	118	110	111	115
Total	*9,107*	*9,374*	*10,082*	*10,283*	*10,694*	*11,228*	*11,394*	*12,015*	*12,504*
Annual growth, (%)	*2*	*3*	*8*	*2*	*4*	*5*	*1*	*5*	*4*

Sources: World Bureau of Metal Statistics, International Copper Study Group and Codelco, Chile, e = estimated.
[a]Includes former German.

712

Plumbing tube, faucets, valves, and fittings for both potable water and heating use constitute copper's second largest use, accounting for approximately 14% of all copper consumed. Copper plumbing tube is overwhelmingly preferred in the U.S., Canadian, and British markets, and it is widely used in many other industrialized countries. Copper's reliability, based on its corrosion resistance in all but highly aggressive (acidic or exceedingly soft) waters, is the principal reason for the metal's continuing popularity. Ease of installation and ready availability are also important factors, as is, increasingly, appreciation of copper's ability to immobilize certain waterborne pathogens (see section on health and safety factors), thereby ensuring high water quality. Plastic plumbing products for potable water use made from poly (vinyl chloride) (PVC), chlorinated poly (vinyl chloride) (CPVC), crosslinked polyethylene (PEX), and at one time, poly-(butene-1) have encroached on copper tube markets in the United States. In Europe, plastics are challenging copper in the transition from galvanized steel pipe to alternative tubular products. In less developed countries, the plastic products' initial low cost and local availability has gained them market dominance. For energy efficiency reasons, copper tube has become the material of choice for most residential and commercial air-conditioning–refrigeration (ACR) systems. ACR applications represent about 9% of the market for copper and copper alloys in the United States. The desire for increased energy efficiency is driving a trend toward the increased use of thinner tubes, longer lengths, and enhanced-surface tubes.

Roofing sheet and related products such as gutters and downspouts are important (and highly visible) uses for copper. Much of the copper used in such applications is refined from scrap. In the past, copper roofing was relegated mainly to public buildings. This trend continues, although the 1990s saw significant growth of copper for architectural purposes in private homes, as well. In addition to roofing products, copper, brasses, and bronzes are used as wall and column sheathing, fascia, banisters and railings, architectural hardware, and decorative trim. Copper competes with many other building products in these applications, but the metal's wide spectrum of pleasing colors and textures and "natural" origin are seen by architects as unique advantages.

Copper, in the form of brass and bronze alloys, is widely used in builders' hardware, such as lock sets, doorknobs, hinges, and push- and kickplates. Coatings developed in the 1990s have ensured lifetime tarnish-resistant service in such products.

13.2. Electrical and Electronic Products. Copper's use in electrical and electronic products constitutes about 26% of the total market for copper in the United States (see also ELECTRONIC MATERIALS). Power utilities contribute to about 9% of total consumption. Included here are both electric-utility applications and equipment that connects to utility lines, such as switchgear and load centers. Copper cable competes well with aluminum in industrial applications, its position strengthened by the metal's higher electrical conductivity, which reduces the amount of energy otherwise lost as heat. Copper is not widely used for overhead transmission or distribution cables (except in parts of Japan and Africa) since its high density, compared with aluminum, requires more closely spaced towers. Many underground transmission and distribution cables are copper, however. Copper is preferred for ultra-high-voltage lines, 345 kV, because the metal's high electrical conductivity per unit volume and high thermal conductivity enable it to conduct large currents in narrow conduits. Copper's corrosion resistance is also advantageous here. There has been a trend, seen mainly in Europe, toward conversion of low to medium voltage (1–60-kV) overhead lines to underground cables.This shift has opened a new market for copper. Market drivers include underground cables' favorable lifecycle economics, increased safety and security, and aesthetics.

Copper magnet wire—that used in electric motors, transformers, and similar devices—continues to be an important copper product. About 60% of all electricity is used to drive electric motors, and a global need for better energy conservation has consequently led to a growing adoption of so-called high efficiency or premium-efficiency motors. For example, a standard 100-hp 74,57-kW AC motor, type D, having 73.5 lb (27.43 kg) of copper in its windings, exhibits an energy efficiency of 93.0%. A premium-efficiency motor of the same horsepower will have 125 lb (46.65 kg) of copper in its windings and will operate at 95.4% efficiency. Besides energy savings, there is a cost savings since the payback time for the premium motor is normally 2 years or less, depending on power cost, motor size, and duty cycle.Telecommunications represents about 9% of the market for copper in the United States. Copper ("twisted pair") communications cable, once seen as obsolescent, is actually an *increasing* market for copper. Fiber optic cable and to a lesser extent coaxial cable and microwave transmission have replaced copper in long lines. However, copper still dominates the subscriber loop (that part of the system located between the telephone central office and the customer), primarily in that portion of the loop between the digital loop carrier and the customer, where most of the copper used in telecommunications can be found. Market factors favorable to copper include its low cost relative to fiber, rapid growth in data communications, growth in Internet usage, recyclability, introduction of high-bandwidth products such as enhanced Category 5 cable and growing availability of digital subscriber line (DSL) telephone service. It is now possible, for example, for twisted-pair copper cables to accommodate data transfer at a rate greater than one gigabyte per second.

Approximately 4.5% of copper and copper alloy products are used in the electronics industry in the United States. Electrical contacts, connectors, and switch components are important outlets for copper and copper alloy strip. Pure copper foil, produced either by rolling, or more often by electrodeposition, is used in many printed circuits. Copper interconnects for integrated circuits, announced in the late 1990s, promise to increase processing speed several-fold over traditional circuitry, which rely on aluminum. While this application will account for only a few thousand pounds of copper annually, it clearly grants the metal standing as a "high technology" material.

Other important copper electrical products include busbar (which is made from plate or heavy tube), superconducting cable (in which copper acts as a thermal buffer), plus a countless variety of switch and contact components, connectors, leadframes, pole-line hardware, and high voltage switchgear products. Many of these products are made from brass, phosphor copper, and copper alloys that were specifically developed for their respective service conditions. Copper has no serious competitors here.

13.3. Transportation Equipment. Transportation uses (automobiles, trucks, buses, ships, aircraft, and aerospace) constitute about 12% of the market for copper and copper alloys in the United States. Automobile wiring harnesses represent the largest portion, about 8%, of this copper. The proliferation of electronic and electrical features in modern automobiles has increased the use of copper in wiring harnesses, motors, contacts, switches, and power supplies to the extent that these devices constitute most of the 25 kg (50 lb) of copper used in a modern American automobile. The remainder can be found in the cooling and brake systems and as an alloying ingredient in various aluminum alloy castings. The change from 12/14- to 36/42-V electrical systems will permit the use of thinner wires. However, the continuing proliferation of electrical and electronic devices will require larger number of wires, and the total amount of copper in wiring harness is therefore expected to remain about the same, if not increase.

Thin-gauge copper and brass strip for automotive radiators, once an important market, declined steadily after the introduction of aluminum radiators in the 1970s. However, radiators for trucks as well as those for the automobile after market continue to consume approximately 200,000 tonnes of copper annually, worldwide. Copper–brass radiators for new cars are also preferred in tropical countries where use of corrosion-inhibiting antifreeze compounds (necessary with aluminum but not copper) is uncommon. Many technical deficiencies in traditional copper–brass radiators were associated with the units' soldered assembly. An improved lightweight *brazed* copper–brass radiator developed by the International Copper Research Association and the International Copper Association, introduced in 1998, may regain some of copper's former original-equipment market share.

Copper alloys are widely used in marine applications ranging from propellers to seawater piping, pumps, heat exchangers, condensers, and other equipment. Alloys most commonly used include copper–nickels (condenser tubes), leaded brasses (tubesheets), aluminum bronzes (pipe, pump, and valve components), nickel–aluminum bronzes (propellers), aluminum brasses (tube), manganese bronzes (pump and valve components), and silicon bronzes (pump and valve components). Copper's traditionally large market share in these applications is being challenged by stainless steels and titanium, which offer properties comparable to copper alloys except for biofouling resistance. Copper–nickel sheathing applied to the legs of offshore platforms prevents buildup of algae, thereby increasing safety and reducing costs. Although technically viable, this market has been slow to develop. Ship hull sheathing, historically an important use for copper, has in modern times been limited to small craft. Copper, in the form of cuprous oxide and other specialty compounds is widely used in antifouling hull paints. Copper hull paints are environmentally superior to organotin coatings, which once threatened copper in this market but are now widely banned.

13.4. Industrial Machinery and Equipment. This market sector, which includes in-plant equipment, nonelectrical instruments, and off-highway vehicles, is the fourth largest for copper in the United States, representing approximately 11% of total consumption. A wide variety of copper alloys are used in industrial valves and fittings, sintered and cast bronze bearings, and heat exchanger (see also HEAT EXCHANGE TECHNOLOGY, HEAT PIPES) and condenser pipe and tube for both fresh water and seawater cooling.

13.5. Consumer and General Products. Approximately 9% of the total copper and copper alloys used in the United States are found in consumer and general products. Copper used in home appliances constitutes about 3% of this volume. Other uses are spread over an enormous variety, including consumer electronics, appliance cords, coinage, agricultural fungicides, wood preservatives, food additives, military and commercial ordnance, fasteners and closures, and utensils and cutlery.

Also included in this category are copper chemicals that, as a product form, represent about 1% of the copper market. Copper's unique chemical, toxicologic and nutritional properties enable it to serve a number of niche markets in both the agricultural and industrial sectors. Agriculturally, copper compounds are used as fungicides (55% of consumption) (see Fungicides, agricultural) livestock feeds (8%), crop nutrients (6%), and other uses. Among industrial uses, wood preservative (12%), antifouling paints (7%), mining and metallurgy (4%) chemicals and petroleum processing (4%), and textile and leather treatment (2%) are major uses (107). Copper sulfate, copper oxides and oxychlorides, and copper hydroxide are the dominant chemical product forms.

BIBLIOGRAPHY

"Copper" in *ECT* 1st ed., Vol. 4, pp. 391–431, by J. L. Bray, Purdue University, and H. Freiser, University of Pittsburgh; in *ECT* 2nd ed., Vol. 6, pp. 131–181, by H. Lanier, Kennecott Copper Corp.; in *ECT* 3rd ed., Vol. 6, pp. 819–869, by W. M. Tuddenham and P. A. Dougall, Kennecott Copper Corp.; "Copper" in *ECT* 4th ed., Vol. 7, pp. 381–428, by David B. George, Kennecott Coroporation; "Copper" in *ECT* (online), posting date: December 4, 2000, by David B. George, Kennecott Coroporation.

CITED PUBLICATIONS

1. R. Raymond, *Out of the Fiery Furnance—The Impact of Metals on the History of Mankind*, Macmillan, South Melbourne, Australia, 1984.
2. A. Sutulov, *Copper at the Crossroads*, Internmet Publications, Santiago, Chile, 1985, pp. 55–57.
3. WBMS, *World Metal Statistics, 49*, World Bureau of Metal Statistics, London, 1996.
4. H. C. Hoover and L. H. Hoover, transl. of G. Agricola, *De Re Metallica*, first Latin edition of 1556, Dover Publications, New York, 1950.
5. B. H. Manson, *Principles of Geochemistry*, J Wiley, New York, 1982.
6. J. D. Lowell, *Mining Eng.* **22**, 67–73 (1970).
7. J. M. Ziman, *Electrons and Holes; The Theory of Transport*, Oxford Univ. Press, London 1960.
8. R. C. Weast, ed., *Handbook of Chemistry and Physics*, 55th ed., CRC Press Boca Raton, FLa., 1974.
9. AIP, *American Institute of Physics Handbook*, 3rd ed., McGraw-Hill, New York, 1972.
10. N. J. Simon, E. S. Drexler and R. P. Reed, *Properties of Copper and Copper Alloys at Cryogenic Temperatures*, NIST Monograph, 177, National Institute of Standards and Technology, U.S. Dept. Commerce, Washington D.C. Feb. 1992.

11. Y. Wang, *Handbook of Radioactive Nuclides*, The Chemical Rubber Co., Cleveland, Ohio 1969.
12. R. E. Bolz and G. I. Tuve, eds., *Handbook of Tables of Applies Engineering Science*, 2nd ed., CRC Press, Boca Raton, Fla., 1974.
13. *Gmelins Handbook of Inorganic Chemistry* (Gmelins Handbuch der anorganischen Chemie), 8th ed., Verlag Chemie GmbH, 1955.
14. B. Landolt-Bornstein, III, *Technology, Classification, Value and Behavior of Metallic Materials* (Technik, Teil, Stoffwerte, und Verhalten von Metallischen Werkstoffen), Springer-Verlag, Berlin, 1964.
15. CDA, *Copper Rod Alloys for Machined Products*, Copper Development Association Inc., New York, 1996.
16. R. T. Sanderson, *Chemical Periodicity*, Reinhold, New York, 1960, p. 153.
17. Ref. 16 pp. 70–90.
18. J. H. Mendenhall *Understanding Copper Alloys*, Olin Brass, East Alton Ill, 1977, p. 108.
19. G. Joseph, K. J. A. Kunding, ed. *Copper-Its Trade, Manufacture, Use and Environmental Status*, ASM International, Materials Park, OH, 1999.
20. P. Angell, H. S. Campbell and A. H. L. Chamberlain, *Microbial Involvement in Corrosion of Copper in Fresh Water*, ICA Project 405, Interim Report, International Copper Association, Ltd., New York, Aug. 1990.
21. J. T. Walker, C. W. Keevil, P. J. Dennis, J. McEvoy, and J. S. Colbourne, *The Invluence of Water Chemistry and Environmental Conditions on the Microbial Colonization of Copper Tubing Leading to Pitting Corrosion, Especially in Institutional Buildings*, ICA Project 407, Final Report, International Copper Association, Ltd., New York, Aug. 1990.
22. ASM, *Metals Handbook*, 10th ed., Vol. **2**, ASM International, Materials Park, Ohio, 1990.
23. M. Mccoy, *Chem. Eng. News* 13 (Oct. 4, 1999).
24. *M. I. Process for Continuous Copper Smelting and Converting*, Mitsubishi Metal Corp., Tokyo, 1989.
25. P. Rutledge, *Eng. Min. J.* **176**(12), 88 (1975).
26. K. J. Richards and co-workers, eds., in *Advances in Sulfide Smelting*, Vol. **2**, TMS-AIME, Warrendale, Pa., 1983.
27. J. H. Acuma, paper presented at *AIME* Annual Meeting, Atlanta, Feb. 1977.
28. H. Ikeda and Y. Matsibara, in J. C. Yannopoulous and J. C. Agarwall, eds., *Extractive Metallurgy of Copper Int. Symp.* AIME, New York, 1976 **2**, 588.
29. D. W. Agers co-workers *"Copper Recovery from Acid Solutions Using Liquid Ion Exchange,"* paper presented at AIME Annual Meeting, New York, Feb. 1966.
30. *The Impact of SXEW Production on the Copper Industry*, Special Study, Brook Hunt & Associates, Ltd., London, Aug. 1994.
31. I. Knight, CRU International, 1999.
32. D. W. Duncan and P. C. Trussell, *Can. Met. Quart.* **3**, 43–45 (1964).
33. Anonymous, *Eng. Min. J.* **172**(2), 96–100 **172**(5), 88–94 (1971).
34. R. L. Braun, A. E. Lewis, and M. F. Wadsworth, *Trans. Met. Soc. AIME* **5**, 1717–1726 (1974).
35. D. H. Davidson, R. E. Weeks, and J. F. Edwards, *Generic In Situ Copper Mine Design Manual*, 5 vols, Bureau of Mines, U.S. Dept. Interior, Washington, D.C., 1988 (Available for National Technical Information Service, Springfield Va.).
36. A. K. Biswas and W. G. Davenport, *Extractive Metallurgy of Copper*, Pergamon Press, New York, 1980.
37. K. R. Suttil, *Eng. Min. J.* **42** 38 (Apr. 1990).
38. Ref. 19, p. 253.

39. Ref. 19, p. 256.
40. Ref. 19, pp. 255, 256.
41. ASTM, "Properties and Selection: Nonferrous Alloys and Pure Metals," in *Metals Handbook*, 9th ed., American Society of Metals, Materials Park, Ohio, 1979.
42. H. S. Campbell, *J. Inst. Met.* **77**, 345 (1950).
43. M. Edwards, J. F. Ferguson, and S. H. Reiber, *J. Am. Waterworks Assoc.* **86**(7), 74–90 (1995).
44. *Cast and Roll*, Outokumpu OY, Espoo, Finland, 1988.
45. CDA, *Copper Casting Alloys*, Copper Development Assoc. Inc., New York, 1994.
46. D. Ellis and H. M. Yun, "Progress Report on Development of a Cu-8 Cr-4 Nb Alloy Database for the Reusable Launch Vehicle (RLV)," *Proc. Copper*/Cobre 99, Vol. **1**, TMS, Warrendale, Pa., 1999.
47. ASTM, *Copper and Copper Alloys, Annual Book of ASTM Standards*, American Society for Testing and Materials, Philadelphia, Pt. 2.01; 1998.
48. ASTM, *Copper and Copper Alloys, Annual Book of ASTM Standards*, American Society for Testing and Materials, Philadelphia, Pt. 2.02; 1998.
49. J. S. Smart, Jr., "The Effect of Impurities in Copper," in A. Butts, ed., *Copper—The Science and Technology of the Metal, Its Alloys and Compounds*, ACS Monograph 122, Reinhold, New York, 1954, Chapt. 21.
50. J. S. Smart, Jr. and A. A. Smith, Jr., *Trans AIME* **166** (1946).
51. V. A. Phillips and A. Phillips, *J. Int. Met.* **81**, (1952).
52. S. Lundquist and S. Carlen, *Erzmetall* **9**, 145 (1956).
53. D. A. Reese and L. W. Condra, *Wire J.* **2**(7), 42 (1969).
54. K. E. Mackay and G. Armstrong-Smith, *J. Inst. Min. Metall.* **75C**, 269 (1996).
55. ASTM, "Analytical Atomic Spectroscopy; Surface Analysis," *Annual Book of ASTM Standards*, American Society for Testing and Materials, Philadelphia, Pt. 3.06.
56. P. E. Rasmussen, *Heavy Metals in the Environment: A Geoscience Prospective*, International Council on Metals and the Environment, Ottawa, Canada, 1996.
57. J. O. Nrigau, *Nature* **338**, 47–49 (1989).
58. J. O. Nriagu and J. M. Pacyna, *Nature* **333** 134–139 (1988).
59. J. O. Nriagu, *Environment* **37**, 7–33 (1990).
60. USGS, *Copper in Surface Soils of the United States*, Geological Survey, U.S. Department of the Interior, professional paper 1270.
61. EPA, *Ambient Water Quality Criteria for Copper—1984*, EPA-440/5-84-031, Office of Water Regulations and Standards, U.S. Environmental Protection Agency, Washington, D.C., 1985 (based on range in 4-day and 1 hour final chronic value for CaCo3 concentrations of 50–200 mg/L).
62. L. Lander and L. Lidestrom, *Copper in Society and the Environment*, 2nd ed., Swedish Environmental Research Group (MFG), Vasteras, Sweden, 1999, pp. 85–124.
63. W. H. Dresher, *Copper Availability and Production and Its Relationship to Plumbing Tube in the United States*, unpublished report for Copper Development Assoc. Inc., WHD Consulting, Tucson, Ariz. July 25, 1998, p. 14.
64. Ref. 65 p. 16.
65. R. Coulon, *Life Cycle Energy Profile of Copper Tube and CPVC Pipe for Supply of Potable Water in Residential Structures*, unpublished report for Copper Development Association Inc., Ecobalance, Inc., Rockville, Md. Aug. 12, 1998, p. 16.
66. ISRI, *Recyling Copper and Brass*, Institute of Scrap Recyling Industries, Washington, D.C., 1980.
67. R. Sundberg, Outokumpu OY Espoo, Finland, 1997.
68. G. E. Lagos and M. Andia "Copper Recyling and the Environment," Paper presented at Copper Recyling Int. Conf. (Brussels, March 3–4, 1997).
69. F. Arpaci and T. Vendura, *Metallstatistic* **47**(4), 340–345 (1993).

70. M. F. Henstock, "The Recyling of Non-ferrous Metals," *Int. Council on Metals in the Environment*, Ottawa, Canada, 1996, p. 124.

71. ICSG, International Copper Study Group, Vol. **1**, No. 1, Jan. 1997.

72. CDA, *Annual Data, 1999, Copper Supply and Consumption*, Copper Development Assoc. Inc., New York, 1999, pp. 13, 14.

73. CDA, *Annual Data, 1999, Copper Supply and Consumption*, Copper Development Assoc. Inc., New York, 1999, p. 121.

74. Ref. 72, p. 18.

75. Metallgesellschaft *Metallstatisik* 47 (1960); 58 (1971); 68 (1981); 78 (1991); 83 (1996).

76. EPA, *Summary Review of Health Effects Associates with Copper*, p. 59, U.S. Environmental Protection Agency, Washington, D.C., 1987.

77. K. Allen and L. Klevay, *Curr. Opinion Lipidol.* **5**, 22–28 (1994).

78. J. R. J. Sorenson, Copper complexes: A physiologic approach to the treatment of chronic disease, *Comp. Ther.* **11**, 49–69 (1985).

79. WHO/IPCS, *Environmental Health Criteria for Copper*, Vol. **200**, World Health Organization, Geneva, 1999.

80. B. Lonneral, J. G. Bell, and C. L. Keen, *Annu. Rev. Nutr.* **1**, 149 (1981).

81. D. M. Danks, "Copper Deficiency in Humans," in *Biological Roles of Copper* (Proc. CIBA Foundation Symp. 79), *Exerpa Medica*, Amsterdam, 1980.

82. D. M. Danks, in C. Scrive, A. L. Beaudet, W. S. Sly, and D. Vale, eds., *The Metabolic Basis of Inherited Diseases*, McGraw-Hill, New York, 1989, pp. 1411–1431.

83. L. H. Allen and N. W. Solomons, in *Absorption and Malabsorption of Mineral Nutrients*, Alan R. Liss, New York, 1984, pp. 199–229.

84. Anonymous, *Gastrointestinal Drugs*, in *Remington*'s Pharmaceutical Sciences, Mack Publishing Co., 1970, Chapt. 44, pp. 806–897.

85. H. K. Chuttani and co-workers *Am. J. Med.* **39**, 849–854 (1965).

86. W. Jantsch, K. Kulig, and B. H. Rumack, *Clin. Toxicol.* **22**, 585–588 (1985).

87. J. M. Cohen and co-workers *J. Am. Waterworks Assoc.* **52**, 660–670 (1960).

88. USEPA, *Drinking Water Criteria Document for Copper (Final Draft)*. U.S. Environmental Protection Agency, Environmental Criteria and Assessment Office, Cincinnati, EPA 600/X-84/190-1, 1985.

89. WHO, *Guidelines for Drinking-Water Quality*, 2nd ed., Vol. **1**, *Recommendations*, World Health Organization, Geneva, 1992.

90. D. R. Winge and R. K. Mehra *Intern. Rev. Exp. Pathol.* **31** 47–83 (1990).

91. J. C. Yarze and co-workers *Am. J. Med.* **92** 643–653 (1992).

92. T. Müller and co-workers *The Lancet* **347** 877–880 (1996).

93. M. S. Tanner, in S. R. Meadow, ed. *Recent Advances in Paediatrics* Churchill Livingston, Edinburgh, 1986, Vol. **8**, pp. 103–120.

94. H. H. A. Dollwet and J. R. J. Sorenson, *Trace Elem. Med.* **2**(2) 80–87 (1985).

95. Anonymous, *The Harvard Classics J.* **33** 23–24 (1910).

96. J. R. J. Sorenson and co-workers *Biol. Trace Elem. Res.* **5**, 257 (1983).

97. J. T. Walker and C. W. Keevil, *The Influence of Plumbing Tube Materials, Water, Chemistry and Temperature on Biofoulding of Plumbing Circuits with Particular Reference to the Colonization of Legionella pneumophila*, British Health Laboratory Services, Project 437, report of International Copper Assoc. Ltd., New York, 1994.

98. J. E. Stout and co-workers, *JAMA* **278**(17), 1404–1405 (1997).

99. M. T. Yahya and co-workers *J. Environ. Health* **51**(5), 282–285 (1989).

100. International Copper Association, Ltd., *New Research Indicates Copper Work Surfaces May Be Effective in Reducing the Health Threat from E. coli O157*, New York, 1999.

101. P. J. Kuhn, *Diagnos. Med.* (Nov./Dec. 1983).

102. T. C. Villar *Am. Rev. Resp. Dis.* **110**, 545–555 (1974).

103. I. H. Scheinberg, unpublished study for the FAO, 1967. reported in *Encyclopedia of Occupational Health*, Vol. **1**, International Labor Organization, Geneva, 1983, p. 567.
104. C. W. Armstrong and co-workers *J. Occup. Med.* **25**, 886–888 (1983).
105. E. Silva *Eng. Min. J.* **200**(3) 26–28 (1999).
106. CDA, *Annual Data, 1999, Copper Supply and Consumption*, Copper Development Assoc. Inc., New York, 1999.
107. CRU, *CRU Copper Stud.* **27**(7), (CRU International, Ltd., London) (1999).

GENERAL REFERENCES

A. K. Biswas and W. G. Davenport, *Extractive Metallurgy of Copper*, Pergamon Press, New York, 1980.

Eng. Min. J., Copper USA Supplement **191** (Jan. 1990)

I. B. Joralemon, *Copper: The Encompassing Story of Mankind*'s Fist Metal, Howell-North Books, Berkeley, Calif., 1973.

R. F. Mikesell, *The World Copper Industry: Structure and Economic Analysis*, Johns Hopkins Press, Baltimore, Md., 1979.

R. F. Prain, *Copper: The Anatomy of an Industry*, Mining Journal Books, London 1975.

S. D. Strauss, *Trouble in the Third Kingdom*, Mining Journal Books, London, 1986.

U.S. Bureau of Mines, annual *Minerals Yearbook*, Washington, D.C.

K. J. A. KUNDIG
Metallurgical Consultant

W. H. DRESHER
WHD Consulting

COPPER ALLOYS, WROUGHT

1. Introduction

Among the metals of commercial importance, copper [7440-50-8] (qv) and its alloys are surpassed only by iron (qv) and aluminum (see ALUMINUM AND ALUMINUM ALLOYS) in worldwide consumption. Typically, copper is alloyed with other elements to provide a broad range of mechanical, physical, and chemical properties that account for widespread use. The principal characteristics of copper alloys are moderate-to-high electrical and thermal conductivities combined with good corrosion resistance, good strength, good formability, unique decorative appearance, and moderate cost (see also CORROSION AND CORROSION CONTROL). Most copper alloys are readily hot and cold formed, joined (soldered, brazed, and welded), and plated (see also SOLDERS AND BRAZING FILLER METALS; WELDING).

Chief consumers of copper and copper alloys are the building construction industry for electrical wire, tubing, builder's hardware, plumbing, and sheathing (see BUILDING MATERIALS, SURVEY); electrical and electronic products for motors, connectors, printed circuit copper foil, and leadframes (see ELECTRICAL CONNECTORS; ELECTRONIC MATERIALS; INTEGRATED CIRCUITS); and the transportation (qv) sector for radiators and wiring harnesses. Other industries include ordnance, power utilities, coinage, decorative hardware, musical instruments and flatware. (see EXPLOSIVES AND PROPELLANTS; POWER GENERATION).

2. Alloy Designations

Copper is primarily alloyed to increase strength. Such alloying can, however, strongly affect other properties; eg electrical and thermal conductivities, corrosion resistance, formability, and color. Elements typically added to copper, both singly and in combination are zinc, tin, nickel, iron, aluminum, silicon, silver, chromium, titanium, and beryllium.

Copper and its alloys are classified in the United States by composition according to the Unified Numbering System (UNS) for metals and alloys. Wrought materials are assigned five-digit numerical designations which range from C10100 through C79999, but only the first three or sometimes four numerals are used for brevity. The designations of wrought copper alloys are given in Table 1. Designations that start with numeral 8 or 9 are reserved for cast alloys (see COPPER ALLOYS, CAST COPPER ALLOYS) (1).

Under the UNS designation system, coppers that contain no or very small amounts of intentional alloy additives (including Ag, P, Zr, Mg, Sn), ie, 99.3 wt% minimum copper, are distinguished from the dilute copper alloys. These latter are called the high coppers and contain a minimum of 96 wt% copper. Alloys of copper are grouped according to the principal elemental addition, such as zinc, tin, nickel, aluminum, silicon, or combinations of these. Important alloys within these groupings and the associated nominal compositions are listed in Table 2.

Table 1. **UNS Designation for Copper and Copper Alloy Families**

Alloy group	UNS designation	Principal alloy elements
coppers[a]	C10100–C15999	Ag, As, Mg, P, Zr
high coppers[b]	C16000–C19999	Cd, Be, Cr, Fe, Ni, P, Mg, Co
brasses	C20500–C28580	Zn
leaded brasses	C31200–C38590	Zn—Pb
tin brasses	C40400–C486	Sn, Zn
phosphor bronzes	C50100–C52400	Sn—P
leaded bronzes	C53200–C54800	Sn—P—Pb
phosphorus–silver	C55180–C55284	P, Ag—P
aluminum bronze	C60600–C64400	Al, Fe, Ni, Co, Si
silicon bronze	C64700–C66100	Si, Sn
modified brass	C662–C69950	Zn, Al, Si, Mn
cupronickels	C70100–C72950	Ni, Fe, Sn
nickel silvers	C73150–C77600	Ni—Zn
leaded nickel silvers	C78200–C79900	Ni—Zn—Pb

[a] Contains a minimum of 99.3 wt% copper.
[b] Contains a minimum of 96 wt% copper.

Table 2. **Nominal Compositions of Industrially Important Copper Alloys**

Alloy group	UNS designation	Elemental composition, wt%[a]
coppers (99.3 min Cu)	C101[b]	
	C107	0.085% min Ag
	C110[c]	
	C122	0.02 P
	C150,C151	0.05–0.2 Zr
	C155	0.11 Mg, 0.06 Ag, 0.06 P
	C1572	0.4 Al_2O_3
high coppers (96 min Cu)	C172	1.9 Be, 0.2 Co
	C17410	0.33 Be, 0.5 Co
	C17460	0.33 Be, 1.2 Ni
	C1751	0.3 Be, 1.7 Ni
	C182	0.9 Cr
	C190	1.1 Ni, 0.20 P
	C194	2.4 Fe, 0.03 P, 0.12 Zn
	C197	0.6 Fe, 0.2 P, 0.05 Mg
zinc brass	C260	30 Zn
leaded brass	C360	35 Zn, 3 Pb
tin brass	C425	9.5 Zn, 2.0 Sn
phosphor bronze	C510	5.0 Sn, 0.1 P
aluminum bronze	C638	2.8 Al, 1.8 Si
	C651	1.5 Si
silicon bronze	C654	3.0 Si, 1.5 Sn, 0.1 Cr
	C655	3.3 Si, 0.9 Mn
modified Cu—Zn	C688	22.7 Zn, 3.4 Al, 0.4 Co
cupronickel	C7025	3 Ni, 0.65 Si, 0.1 Mg
	C7026	2 Ni–0.5 Si
	C706	10 Ni, 1.4 Fe
	C725	9.5 Ni, 2.3 Sn
	C729	15 Ni, 8 Sn
nickel silver	C752	17 Zn, 18 Ni

[a] Remaining percentage is copper.
[b] Contains a maximum of 0.01 wt% impurities.
[c] Contains a maximum of 0.10 wt% impurities, excluding silver.

Most wrought alloys are provided in conditions that have been strengthened by various amounts of cold work or heat treatment. Cold worked tempers are the result of cold rolling or drawing by prescribed amounts of plastic deformation from the annealed condition. Alloys that respond to strengthening by heat treatment are referred to as precipitation or age hardenable. Cold worked conditions can also be thermally treated at relatively low temperatures (stress relief annealed) to produce a slight decrease in strength to benefit other properties, such as corrosion resistance, formability, and stress relaxation.

2.1. Temper. The system for designating material condition, whether the product form is strip, rod, or wire, is defined in ASTM Recommended Practice B601 (1). The ASTM system uses an alpha-numeric code for each of the standard temper designations. This system replaces the historical terminology of one-half hard, hard, spring, etc. Table 3 summarizes temper designations. Commercially, wrought tempers are specified by a narrow range of tensile properties for any given alloy.

Table 3. **Copper and Copper Alloy Temper Designations**

Condition	Historical	ASTM B601(1)
annealed	soft annealed	060
cold worked	1/4 hard	H01
	1/2 hard	H02
	3/4 hard	H03
	hard	H04
	extra hard	H06
	spring	H08
	extra spring	H10
	super spring	H14
cold worked and relief annealed	1/2 hard/RA	HR02
	hard/RA	HR04
	spring/RA	HR08
precipitation hardened alloys		
solution treated	A	TB00
solution treated + rolled	1/2 H	TD02
	H	TD04
	T002 + aged	TH04
mill hardened	AM,AT	TM00
	1/2 HM,1/2 HT	TM02
	HM, HT	TM04
	XHM	TM06
	XHMS	TM08

3. Product Forms and Processing

The output from brass mills in the United States is split nearly equally between copper and the alloys of copper. Copper and dilute copper alloy wrought products are made using electrically refined copper so as to maintain low impurity content. Copper alloys are commonly made from either refined copper plus elemental additions or from recycled alloy scrap. Copper alloys can be readily manufactured from remelted scrap while maintaining low levels of impurities. A greater proportion of the copper alloys used as engineering materials are made from recycled materials than most other commercial materials.

Generally, processing of wrought copper alloys will include melting, alloying, casting, homogenization, hot working, cold working, and annealing, with the latter two steps being repeated to achieve the desired product dimensions and property combinations. Wrought alloy product forms are varied and include plate, sheet, strip and foil, round and special cross-section bars, rod, and wire. Plate includes flat products that are >5 mm (0.2 in.) in thickness. The generally accepted difference between rod and wire is that the former is provided in straight lengths, as intended for screw machining of parts. Wire is usually a coiled product for redrawing to a smaller diameter or for cold heading. Diameter is not a distinguishing feature between rod and wire except for C110 electrical wire that is provided as 7.9 mm (0.312 in.) diameter rod intended for subsequent processing to wire.

3.1. Sheet and Strip. The manufacture of wrought copper materials starts with either semicontinuously cast slabs that are hot rolled, or cast plate that is thin enough, near 13 mm (0.5 in.), to be cold rolled directly. The surfaces of both hot rolled slabs and as-cast plate are milled to remove defects before proceeding to cold rolling and annealing operations. Spray casting and powder consolidation are alternative methods of producing feedstock for cold rolling and annealing.

Cold rolling is used to process to thinner gauges and to increase the strength of the finished product of copper and its alloys. A variety of mills, including four-high, tandem or reversing mills and cluster or Sendzimir, mills are used. On four-high mills, and especially on cluster mills, rolling forces on the work rolls are controlled to limit roll bending and thus ensure uniform thickness and flatness of the rolled product. Four-high mills are used for both in-process and finish rolling; cluster mills are most effectively used in the final rolling operation because of superior capability for maintaining tight control of thickness and shape. Cluster mills are also used for final rolling of foil as thin as 0.013 mm (0.0005 in.).

Strengthening by cold rolling is accompanied by decreased ductility. A softening heat treatment is needed when ductility is lowered to levels below that required for subsequent processing. Annealing, which is done at temperatures and for times sufficient to soften and recrystallize the cold worked material, can be either a batch or a continuous operations.

Batch (or bell) annealing is done within a sealed inner retort that contains a protective atmosphere. Continuous (strand or strip) annealing uses a furnace sealed at both ends to contain a protective atmosphere while permitting passage of the strip. Strand annealing can generally be carried out at strip thicknesses from 0.1 mm (0.004 in.) to 3.18 mm (0.125 in.).

Neutral or reducing atmospheres are used to protect copper alloys when being annealed. These atmospheres range from nitrogen to mixtures of nitrogen and hydrogen derived from cracked ammonia (qv) to which nitrogen is often added. Products of combustion from natural gas that has been partially burned so as to contain the reducing gases, carbon monoxide and hydrogen, may be used as well (see GAS, NATURAL). Strip is usually acid cleaned following annealing.

The final processing operation for strip may be tension leveling for the purpose of improving flatness across the width and removing curvature (coil set). Strip is passed through a set of several intermeshed rolls, in serpentine fashion, to mildly deform the material in bending while under tension. The amount of bending strain is varied locally across the width of sheet to balance more highly strained wavy edges or center-of-width buckles and thereby remove these undesirable features.

Most sheet is slit to various narrow widths before shipping for use in stamping presses and other applications. Slitting is done by opposing rotary disk knives that intersect the pass line of strip while it is moving at high line speed. The finished, slit product is shipped as flat pancake coils or is transversely wound on wide spools.

3.2. Rod and Tubular Products. Rod products having round and special cross-sections are hot extruded from cast billets. Seamless tubing is either hot extruded over a mandrel held in position within the die orifice, or is formed by rotary piercing of heated billets. Tubes of both round and other cross-sections can also be made at high line speed by high frequency induction welding of strip that is formed in-line before entering the welding head. Weld beads are scarfed from the internal and external surfaces, in-line with the welding operation. The tube is finally sized by cold drawing or tube reducing operations. These operations also develop a cold worked temper in the final product. Cold drawing can be done using either a fixed or floating mandrel. Alternatively, tube reducing uses semicircular grooved dies that oscillate along the tube over an internal mandrel. The latter method is capable of providing better tolerance control of diameter, wall thickness, and concentricity relative to cold drawing.

3.3. Wire. Most copper wire is cold drawn directly from continuously cast electrolytic cathode feedstock. Earlier, cast bars were hot worked to wire rods. By eliminating the hot working step, the continuous casting process offers higher quality, lower operating cost and longer product lengths. Alloy wires, on the other hand, are hot rolled from cast bars. Semifinished copper and copper alloy wire rod is subsequently processed through drawing, annealing, and acid cleaning operations to the finished condition. A common practice, especially for very fine wire products, is die shaving of rod at an intermediate processing step to facilitate processing and provide a high quality product. In addition to round wire, square, flat or shaped wire is also produced.

3.4. General Processing Behavior. Copper is readily strengthened by the cold working processes used to shape or form the material into the desired shape or thickness. Repeated cycles of working and annealing from the hot rolled or cast strip condition are typical for most copper alloys. Solid solution and dispersion strengthened alloys are usually annealed to their softest condition before the final working step. This fully annealed condition product can be used, without further cold work, to fabricate items which require the maximum ductility. More commonly, this product is cold worked to develop the desired balance of strength and ductility.

The effect of cold working the fully annealed condition by cold rolling of sheet on the yield strength, at 0.2% offset strain, and tensile strength of copper sheet is shown in Figure 1. As is typical for many alloys, the rolling curve exhibits an initial stage of rapid hardening followed by a lower hardening rate for continued reduction in thickness. The initial work hardening results from the generation and entanglement of dislocations in the initial, relatively defect-free, annealed grains. Dislocation generation and motion permit the initially equiaxed grains to elongate in the direction of the working operation.

Precipitation hardening is another means of increasing the strength of certain copper alloys, which involves heat treating the material to produce a fine array of very small particles in the alloy. These particles restrict dislocation motion and thus strengthen the material. Strengthening can also be accomplished via a combination of second phase precipitation, grain size, and cold work.

Figure 1 also shows the decrease in tensile elongation (a common measure of ductility) that accompanies the strength increase with cold working. Whereas these particular rolling curves include up to 70% reduction in thickness, pure copper is capable of being rolled much further without fracturing.

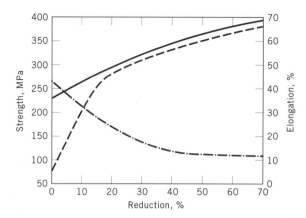

Fig. 1. The effect of cold rolling upon the tensile properties of unalloyed copper (C110): (—) represents tensile strength; (− − −), the 0.2% yield strength; and (−•−) the tensile elongation. To convert MPa to psi, multiply by 145.

Elements that can dissolve in copper, such as zinc, tin, and nickel, eg, increase annealed strength by varying amounts depending on the element and the quantity in solution. The effect of selected solution hardening elements on tensile properties of annealed copper alloys is illustrated by the data in Table 4, where the yield strength is the stress at 0.2% offset strain in a tensile test.

Solid solution hardened alloys also show different cold work hardening response relative to pure copper and each other as illustrated in Figure 2. These differences originate from the effects that the dissolved atom has on microplastic processes that occur during cold working. Finer grain sizes generally translate into stronger materials. Note that for alloy C260, finer grain size has higher strength at corresponding cold reduction. A higher quality, smooth surface that requires no buffing after forming operations is a further benefit from a fine grained microstructure.

The annealing response of cold worked copper is illustrated by Figure 3 in terms of changes in tensile properties determined at room temperature after an

Table 4. **Annealed Tensile Properties of Solution Strengthened Copper Alloys**

Alloy	Principal alloying element, %	0.2% Yield strength, MPa[a]	Tensile strength, MPa[a]	Elongation in 50 mm, %
C110		76	235	45
C220	10 Zn	83	260	45
C260	30 Zn	150	365	54
C425	10 Zn + 2 Sn	125	340	46
C510	5 Sn + 0.1 P	145	345	52
C521	8 Sn + 0.1 P	165	415	63
C706	10 Ni + 1.4 Fe	110	365	35
C752	18 Ni + 17 Zn	205	415	32

[a] To convert MPa to psi, multiply by 145.

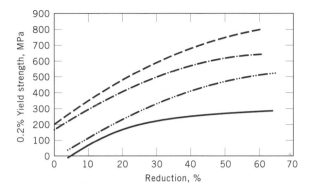

Fig. 2. Comparison of alloy cold rolling behaviors: (—) represents unalloyed copper (C110) having a grain size of 25 μm; (–••–) and (–•–), copper–30% zinc (C260) of 50- and 15-μm grain size, respectively; and (– – –), copper–8% tin (C521) having a 15-μm grain size. To convert MPa to psi, multiply by 145.

elevated temperature anneal for 1 h. Also important is the grain size obtained from each annealing temperature because this microstructural characteristic influences properties. In general, more uniform and finer grain size is promoted by annealing at a lower temperature, which is made possible by increasing the amount of cold work before annealing. Many products are offered in the fully soft or annealed condition to maximize available formability.

The annealing curves for several copper alloys are compared to copper in Figure 4. As for cold working, solute additions affect the annealing response by interaction with microplastic processes that cause softening via dislocation and subgrain elimination during the anneal.

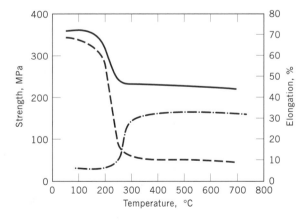

Fig. 3. Changes in tensile properties of cold rolled copper (C110), after 50% reduction in thickness, that attend annealing for one hour at each of the temperatures shown: (—) represents tensile strength; (– – –), the 0.2% yield strength; and (–•–), the tensile elongation. To convert MPa to psi, multiply by 145.

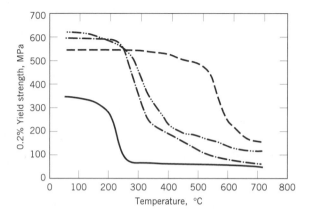

Fig. 4. Comparison of the annealing response in terms of yield strength of cold rolled (—) copper (C110) and cold rolled alloys; (—•—) copper–30% zinc (C260), (—••—) copper–5% tin (C510), and (– – –) copper–10% nickel (C706). The condition prior to annealing was after 50% cold reduction in thickness from the soft condition. The time of annealing at each of the temperatures is 1 h. To convert MPa to psi, multiply by 145.

4. Alloying for Strengthening

Elements added to pure copper can remain in solid solution in the copper or can form second phases separate from the copper, which forms the bulk of the alloy. Copper alloys where the principal elemental additions remain in solution comprise a group we will refer to as "solid solution alloys". Those copper alloys whose principal elemental additions cause discrete second phases to form during processing comprise groups referred to as "dispersed phase alloys" or "precipitation hardening alloys". Phase diagrams that show limits of solid solubility and equilibrium phases that form in binary and ternary combinations with copper are found in the literature (2,3).

4.1. Solid Solution Alloys.   Copper dissolves other elements to varying degrees to produce a single-phase alloy that is strengthened relative to unalloyed copper. The contribution to strengthening from an element depends on the amount in solution and by its particular physical characteristics such as atom size and valency. Tin, silicon, and aluminum show the highest strengthening efficiency of the common solute additives, whereas nickel and zinc are the least efficient (4,5). Many alloys are ternary combinations of these elements. The limiting factor in their alloy range is the extent to which these elements, either singly or in combination, remain dissolved in the copper parent during processing.

Alloys containing zinc, tin, aluminum, and nickel represent most of the commercial alloys, namely, the brasses, phosphor bronzes, cupronickels and nickel silvers (see Tables 1 and 2). The tensile properties of selected solid solution strengthened alloys, in the annealed condition, are listed in Table 4 (6). These alloys generally provide good formability, together with varying degrees of corrosion and oxidation resistance. Alloying for solution strengthening is accompanied by lowered electrical conductivity. Because of high alloy content, these solution strengthened alloys tend to have conductivities that are typically less than one-half that of unalloyed copper.

4.2. Dispersed Phase Alloys.   The presence of finely dispersed second-phase particles in copper alloys contributes to strength, by refining the grain size and increasing the amount of hardening due to cold working. These phases are developed in many commercially important copper alloys in several ways. A dispersion of fine particles can be incorporated into the alloy through thermomechanical processing where the alloy content exceeds the solid-state solubility limit, causing precipitation and coarsening of the excess solute as dispersed second-phase particles. These dispersed phases enhance the work hardening response relative to unalloyed copper to produce high strength while maintaining reasonably good conductivity.

One example of this kind of alloy is C194, where an anneal is used to form dispersions of iron particles. Solid solution alloys based on Cu–Zn and Cu–Al have also been modified by additions that react to form dispersions of intermetallic phases. Addition of aluminum and cobalt to a zinc-containing alloy (C688) and cobalt and silicon to an aluminum-containing alloy (C638) are two examples. Fine-grain sizes below ten μm and smaller amounts of cold work to develop strength combine to produce highly formable alloys that offer the highest strengths available from nonage-hardenable alloys.

Dispersed phases can also be produced by separation from the melt during casting. Examples include iron phosphides, as in C194, and cobalt or nickel beryllides, as in C172 and C17410, respectively.

Powder metallurgy is used to incorporate second phases into copper (see METALLURGY, POWDER). An example is the manufacture of oxide dispersed alloys like C15720. Aluminum oxide powder, that otherwise does not dissolve in copper, is incorporated in C15720 by mixing powders of copper, copper oxide particles, and a dilute copper–aluminum alloy. Hot extrusion is used to consolidate the mixture. Subsequent heat treatment reduces the copper oxide and the resultant oxygen diffuses into the alloy powder where it reacts with the dissolved aluminum to form uniformly dispersed sub micrometer-sized aluminum oxide particles. It is this ultrafine oxide that is principally responsible for the alloy's notable resistance to softening during subsequent high temperature exposure.

4.3. Precipitation (Age) Hardening Alloys.   Only a few copper alloy systems are capable of responding to precipitation or age hardening (7). Those that do have the constitutional characteristics of being single-phase (solid solution) at elevated temperatures and are able to develop into two or more phases at lower temperatures that are capable of resisting plastic deformation. The copper alloy systems of commercial importance are based on individual additions of Be, Cr, Ti, Zr, or Ni + X, where X = Al, Sn, Si, or P.

Processing involves heat treating the alloys at a sufficiently high temperature to put alloying elements into solution, followed by rapid cooling to near room temperature to retain this solid solution. A second lower temperature aging treatment to form the hardening phase particles is the final step. Cold work may be introduced between the two heat treatments to promote aging response and also add to final strength.

The preferred precipitation structure is one with uniformly dispersed particles within the grains of the alloy. Alloy composition and thermal treatments are chosen to achieve this structure through control of intermediate, metastable precipitate phases. The equilibrium, thermodynamically stable precipitate phases are generally coarse and provide little strengthening. When the latter are formed at grain boundaries, ductility of the material is significantly decreased. This grain boundary precipitation is avoided for strip that is intended to be formed into parts after having been previously hardened by the mill.

4.4. Special Addition Alloys. The most notable of the special additives to copper alloys are those added to enhance machinability. Lead, tellurium, selenium, and sulfur are within this group of additives. Because of increasing concern over lead toxicity, interest has centered on use of bismuth, which is nearly as effective as lead for improving machinability. The alloys that contain such additives are limited because of the difficulty they cause to hot and cold working. High zinc brasses such as C360, which undergo a change in crystal structure at elevated temperature, are favored because of the ability to accommodate lead and bismuth additions in processing. Whereas lead and bismuth are added to both copper and copper alloys, tellurium and selenium are added separately to copper (C145 and C147) to form copper-telluride and -selenide particles to enhance machinability.

Other special additions are used to deoxidize copper and prevent the formation of copper oxide. Such alloys may be preferred in applications where embrittlement by hydrogen through reaction with internally dispersed copper oxide particles is a concern, such as in C110. The most common deoxidized copper is C122, in which phosphorus reacts with copper oxide to form phosphorus pentoxide, which is removed as a slag while the copper is molten.

Many elemental additions to copper for strengthening and other properties also deoxidize the alloy. A side benefit of such additions is elimination of susceptibility to hydrogen embrittlement. Such deoxidizing additions include beryllium, aluminum, silicon, chromium, zirconium, and magnesium.

5. Properties

5.1. Strength. Table 5 illustrates the range of property combinations available in copper alloys by listing properties of selected commercially important alloys. The principal source of strengthening and the individual product forms in which each alloy is usually available are also identified.

Table 6 illustrates the increase in strength and the accompanying decrease in electrical conductivity that derives from alloying of copper. This trend is clearly apparent among solid solution strengthened alloys, namely those that contain zinc, tin, nickel, and zinc plus nickel as their principal alloying constituents. Notable exceptions to this trend are precipitation hardening alloys where the precipitating phases remove elements from solid solution leaving a leaner, higher conductivity matrix; eg, C1751, C182, and C7025. For the latter group, strength–conductivity combinations not possible with solid solution alloys are achieved.

Other properties are just as important as strength and conductivity for alloy selection. For example, the cupronickels have about the same strength as do copper–zinc brasses, and also have much lower conductivity. However, the

Table 5. **Properties of Copper Alloys**

UNS designation	Alloy type[a]	Product forms[b]	Electrical conductivity %IACS[c]	Modulus of elasticity, GPa[d]	Temper	Tensile strength, MPa[e]	Yield strength (0.2% offset), MPa[e]	Elongation, %
C110	pure	S, R, W, T	100	117	060	221	59	35
C151	D	S, R, W	95	117	H04	330	317	4
C1572	D	R, W	89	105	CW 97%	605	580	5
C172	P	S, R, W, T	20	128	TM04	980	860	12
C1751	P	S, R	45	130	AT	845	740	12
C182	P	S, R, W	80	130	TH04	460	405	14
C194	D	S	65	117	H08	503	482	2
C260	S	S, R, W, T	28	110	H08	662	600	2
C360	SA	S, R	26	97	H02	470	360	18
C425	S	S	28	110	H04	524	503	8
C510	S	S, R, W	15	110	H08	696	682	2
C638	D	S	10	117	H04	834	750	5
C688	D	S	18	117	H08	880	800	2
C655	S	S, R, W	7	105	H04	650	400	8
C706	S	S, R, T	9	124	H04	530	517	1
C7025	P	S	40	130	TM03	690	655	5
C725	S	S, R, W	11	138	H08	641	620	1
C729	P	S, W	7	127	TM04	860	790	15
C752	S	S, R, W	6	124	H08	655	641	1

[a] S is solution strengthened; D, dispersed phase; P, precipitation strengthened; SA, special addition.
[b] S is sheet and strip; R, rod and shapes; T, tube; and W, wire.
[c] IACS = International Annealed Copper Standard. Pure copper has a value of 100.
[d] To convert GPa to psi, multiply by 145,000.
[e] To convert MPa to psi, multiply by 145.

Table 6. Electrical Conductivity versus Strength Among Copper Alloys

UNS designation	Principal alloy element, %	Conductivity, annealed, %IACS	0.2% Yield strength hard temper, MPa[a]
C110		100	310
C210	5 Zn	56	365
C230	15 Zn	37	420
C260	30 Zn	28	495
C505	1.25 Sn	48	425
C510	5 Sn	15	560
C521	8 Sn	13	590
C706	10 Ni	9	520
C715	30 Ni	4	540
C1751	0.4 Be, 1.8 Ni	45	740
C182	1 Cr	80	400
C7025	3 Ni, 0.65 Si, 0.15 Mg	40	690

[a] To convert MPa to psi, multiply by 145.

corrosion resistance of the cupronickels far exceeds that of brass and is worth the higher cost if needed in the application. Similar trade-offs exist between these properties and formability, softening resistance, and other properties.

5.2. Electrical-Thermal Conductivities. Electrical conductivities of alloys (Tables 5 and 6) are often expressed as a percentage relative to an International Annealed Copper Standard (IACS), ie, units of % IACS, where the value of 100% IACS is assigned to pure copper having a measured resistivity value at room temperature of $17.24n \Omega \cdot m$. The measurement of resistivity and its conversion to % IACS is covered under ASTM B193 (8).

Copper has a high electrical conductivity that is second only to that of silver. The conductivity of silver in % IACS units is 108; gold, 73; aluminum, 64; and iron, 18. Wrought copper having a conductivity near 102% IACS is not uncommon because of improvements in refining practices since the standard was first established.

Electrical conductivity of copper is affected by temperature, alloy additions and impurities, and cold work (9–12). The electrical conductivity of annealed copper falls from 100% IACS at room temperature to 65% IACS at 150°C. Alloying invariably decreases conductivity. Cold work also decreases electrical conductivity as more and more dislocation and microstructural defects are incorporated into the annealed grains. These defects interfere with the passage of conduction electrons. Conductivity decreases by about 3–5% IACS for pure copper when cold worked 75% reduction in area. The conductivity of alloys is also affected to about the same degree by cold work.

Copper and its alloys also have relatively good thermal conductivity, which accounts for their application where heat removal is important, such as for heat sinks, heat exchanger tubes and electronic packaging (see HEAT EXCHANGE TECHNOLOGY, MICROELECTRONIC PACKAGING). Thermal conductivity and electrical conductivity depend similarly on composition primarily because the conduction electrons carry most of the thermal energy.

To a good approximation, thermal conductivity at room temperature is linearly related to electrical conductivity through the Wiedemann-Franz rule. This

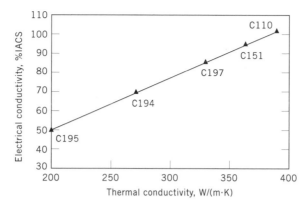

Fig. 5.   The Wiedemann-Franz relationship at 20°C between electrical and thermal conductivities of copper alloys having moderate to high conductivities.

relationship is dependent on temperature, however, because the temperature variations of the thermal and the electrical conductivities are not the same. At temperatures above room temperature, thermal conductivity of pure copper decreases more slowly than does electrical conductivity. For many copper alloys the thermal conductivity increases, whereas electrical conductivity decreases with temperature above ambient. The relationship at room temperature between thermal and electrical conductivity for moderate to high conductivity alloys is illustrated in Figure 5.

6. Formability

Copper and most of the wrought alloys are readily formed by bending, drawing, upset forging, stamping, and coining (13–15). The maximum formability condition is the fully soft or annealed condition. When additional strength or hardness is desired in the final part, the forming step is done starting with a cold worked temper or the part is not annealed between forming steps. Cold forming operations always work-harden alloys and in many cases sufficient strength for the application is produced in the formed part.

A variety of tests have been established to indicate whether a specific forming operation can be safely accomplished without failure. Three of the most commonly used tests are bend testing around a mandrel, limiting draw ratio testing, and bulge height formability testing. Temper, strip thickness, grain size, crystallographic texture, and alloy composition are important variables in these tests. An alloy in the annealed temper that can be formed into a particular shape may not be able to be similarly formed starting from a work hardened temper.

6.1. Bending.   The smallest radius over which strip of a particular alloy can be formed without failing is important in the selection of materials for a given application. The industry tests formability using samples cut from strip material to rank materials and thereby indicate whether a particular alloy/temper is suited for an application (15). Performance of the material during fabrication is, however, the best final judge of suitability.

In a bend test, a rectangular strip sample is formed around a die with a precisely machined edge of a known radius. The sample is formed by 90 or 180° about an axis in the plane of the sheet. The outer surface of the bend is inspected for cracking or unacceptably deep surface rumpling. The minimum bending radius (MBR) about which strip can be successfully formed depends on the strip thickness, t, and is reported with the test thickness specified or normalized with regard to thickness as MBR/t. Thus smaller MBR/t values indicate better formability, since tighter radii can be achieved for a given strip thickness.

Formability is not the same for all orientations of rolled temper alloy strip because of crystallographic texturing or mechanical fibering. The amount of directionality relative to the rolling direction depends on the amount of final cold reduction to temper, the intermediate processing history, and the specific alloy. For most alloys, bend formability is better around an axis that is perpendicular to the rolling direction relative to one that parallels the rolling direction. The industry has termed these goodway and badway bends, respectively. The terms longitudinal and transverse for goodway and badway bending, respectively, are also used. Longitudinal and transverse refers to the direction of material movement, relative to the rolling direction, during bending.

Examples of bend formability in alloys C510 and C725 are provided in Figure 6. These curves of bend formability illustrate several characteristics of most copper alloys. The minimum bending radius increases (formability decreases) with increasing strength (or temper). Also, the badway (transverse) bend formability degrades more rapidly with increasing strength than goodway formability, as shown for C510. The comparison alloy, C725, is more isotropic in formability for equivalent strength. This comparison illustrates that directional uniformity in properties is both alloy and process dependent. The minimum bend radii shown can be affected by special processing to produce better formability in some alloys, such as in phosphor bronze.

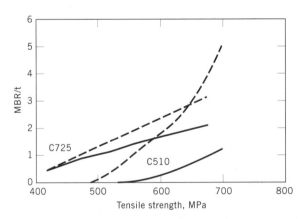

Fig. 6. The minimum radius over which alloy strip can be formed in bending without failure, normalized by thickness (MBR/t) for alloys C510 and C725 as affected by increasing strength (temper) where (——) represents orientation of bending relative to the rolling direction of strip, ie, goodway (longitudinal), and (– – –) badway (transverse). To convert MPa to psi, multiply by 145.

Precipitation strengthened alloys show the same general trend of degrading formability with increasing strength, albeit at high strength levels. The directionality of formability is often different for precipitation hardening alloys in the mill hardened temper with the badway bends being better than the good way bends. Bend formability of these alloys is often superior in the rolled temper condition; ie, before the aging treatment. Therefore, parts having the most demanding forming requirements can be first formed from rolled temper strip and then aged to obtain required strength. Disadvantages of this approach are possible distortion of formed parts during aging and the need for cleaning the parts after the aging anneal.

6.2. Drawing. Copper alloys are often deep drawn or cupped. A second drawing step is commonly applied to the more severe or "deeper drawn" shapes required for some applications, such as ballpoint pen cartridge cases.

The deep drawability of alloy sheet is often determined from its capability of being drawn into a cylindrical cup (13–18). In this test, a circular disk or blank is held in place with a hold down ring over the female die of a punch and die set. Lubricant used between the hold down ring and the workpiece allows the disk to slide under the high loads generated as the blank is forced by the punch through the die. The test is run using blanks of various radii until the largest radius is found that does not fracture during drawing; fracture usually occurs at the base of the cup. The results of this test are expressed in terms of a formability parameter called the limiting drawability ratio (LDR), which is defined as the ratio of the largest circular blank diameter to the punch diameter. This parameter is used to compare the relative drawability of various alloys or various conditions of the same alloy, with higher values indicating better performance.

This cupping test also provides a direct measure of anisotropy in the sheet resulting from crystallographic texture. Lobes, also called ears, that occur at the rim of formed cups are an indication of non-uniform thinning of the sheet. This effect is called planar anisotropy, the degree of which is measured from the average height of the ears. The location of earing with respect to the original rolling direction identifies the type of planar anisotropy. Earing is undesirable because it requires trimming and scrapping of metal from the rim area of the cup. Earing is directly related to the crystallographic texture and is reduced by control of texture through appropriate processing.

6.3. Plastic Strain Ratio. The plastic strain ratio(R) is the ratio of strains measured in the width and the thickness directions in tensile tests. This ratio characterizes the ability of materials to resist thinning during forming operations (13). In particular, high R-values indicate the ability of a sheet material to resist the thinning and failure at the base of a deep drawn cup. The plastic strain ratio is measured at 0, 45, and 90° relative to the rolling direction. These three plastic strain ratios R_0, R_{45}, and R_{90}, are combined to obtain the average strain ratio, called the R or the $\bar{R}$ value, and its variation in strain ratio, called ΔR:

$$\bar{R} = \frac{1}{4}(R_0 + 2\,R_{45} + R_{90})$$

$$\Delta R = \frac{1}{2}(R_0 - 2\,R_{45} + R_{90})$$

Much can be predicted from these parameters. Linear correlations have been established between the R value and the drawability parameter, LDR. Thus higher R values are associated with better deep drawability. ΔR, on the other hand, predicts the height and location of ears. For example, larger absolute values of ΔR predict higher ears during deep drawing cup shapes.

The sign of ΔR indicates the location of the individual ears with respect to the sheet rolling direction. For example, if ΔR is positive, the ears are located at 0 and 90° to the original rolling direction. This occurs when annealed copper has a pronounced cube crystallographic texture. When ΔR is negative, ears are at 45° to the rolling direction. This occurs when annealed fully recrystallized copper has a crystallographic texture that includes a significant component of the prior cold rolled texture. When annealed copper has no preferred crystallographic texture, called a random texture, ΔR is zero and there is no earing. But the random texture condition does not necessarily show the optimum R value, or drawability, that the material is capable of providing. The most desirable situation is for the ΔR parameter to be zero, predicting no plastic anisotropy, and the R value to be a maximum, predicting optimum drawability. In practice, it is difficult to achieve both situations with the same crystallographic texture. Over the years much work has been done with copper alloys and with brass in particular to develop mill processing schedules designed to provide sheet products having balanced textures that offer the best combination of high drawability and minimum earing.

6.4. Forming Limit Analysis. The ductility of sheet and strip can be predicted from an analysis that produces a forming limit diagram (FLD), which defines critical plastic strains at fracture over a range of forming conditions. The FLD encompasses the simpler, but limited measures of ductility represented by the percentage elongation from tensile tests and the minimum bend radius from bend tests.

The forming limit analysis involves deforming a clamped sheet specimen over a hemispherical punch into a dome shape until failure. At fracture, the dome height and the dimensional changes in a pregridded circle mesh pattern on the surface of a specimen are recorded, and the local strain pattern is calculated. From this strain pattern, the FLD is drawn, as shown in Figure 7 for C260 (14). Various deformation paths, ie, pure drawing (circles narrowed), pure bending (circles elongated), pure stretching (circles expanded), and in-between combinations modes, are produced by varying the test specimen width and the applied lubricant (13,14,17). The strains at the fracture, which are obtained from the distortion in the gridded circles closest to the fracture, are used to construct the forming limit diagram.

The region above the forming limit curve represents combinations of strain that lead to failure. Conversely, the region below the curve represents plastic deformation strains that the sheet material can withstand without failure. Pure drawing corresponds to a combination of positive and negative strain to the left side of the diagram. Pure biaxial stretching (both strains are positive), such as forming a bulge in a flat strip, is to the right side of the diagram. The case of bending is in between with strain equal to zero in one direction.

A second diagram, the limiting dome height (LDH) diagram, is also available from this analysis. The LDH diagram defines the limiting dome height

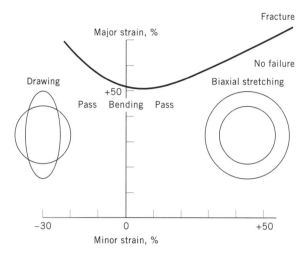

Fig. 7. The forming limit diagram (FLD) for C260–061 (annealed) showing limits to forming by various degrees of drawing (left of the origin), biaxial stretching (right of the origin), and pure bending (at the origin). Failure occurs at combinations of strains above the curve (14).

obtainable for these various forming modes and as such is a measure of the overall ductility (14).

6.5. Springback. During most kinds of cold-forming operations, regardless of material, elastic springback is encountered that can affect part dimensions (13). After metal is plastically deformed, elastic recovery occurs in the direction opposite to that of the plastic forming. The amount of springback can be predicted for simple bending, but prediction is more difficult for complicated shapes because plastic deformation is seldom homogeneous or uniform throughout formed parts.

A bending springback test is performed by bending a flat strip to a predetermined angle, A_0, which is the angle of the die around which the strip is bent. The strip is then released and the included angle, A_1, of the strip is measured. The springback function, A_0/A_1 shown in Figure 8, is a guideline measure of springback. The angle to which the strip was initially formed by the tooling is, A_0, may be 90° and thus the angle of bend that remains after forming, as A_1 in this index would be 90° or greater. The value of 1.0 represents no springback and increasing springback is indicated by decreasing values of this parameter. It is important to note that tests should be done at radii large enough to avoid cracking which would affect the amount of springback.

Manufacturers of copper alloy sheet and strip have developed springback data using this test as a guide for selecting alloys and tempers. The test is best used to compare springback among various tempers of an alloy or among different alloys. By allowing for springback, forming tools and dies can be designed to ensure that dimensional specifications are met in formed parts.

The degree of springback is higher for higher temper strip, for thinner strip, for larger radii of bending, and varies with the direction of bending (longitudinal or transverse). As illustrated by Figure 8 for C260, the amount of springback

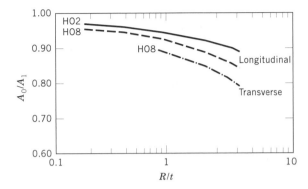

Fig. 8. Springback behavior for alloy C260, as defined by the ratio of the original forming angle, $90°$ or A_0, divided by the final relaxed angle, A_1. The effects of temper and orientation are shown as a function of bending radius R, normalized by the thickness of the strip t.

increases with temper and increases with applied bending radius. The stronger tempers exhibit more directional behavior.

6.6. Softening Resistance. The ability of being readily annealed or softened in a controlled manner to restore ductility is beneficial in mill processing, but resistance to softening of the wrought product is often preferred during fabrication and subsequent service (19). Joining operations such as welding, brazing, and soldering are prime examples where softening resistance is essential. Most alloying elements and impurities increase softening resistance. The amount by which softening resistance is increased is specific to the element being added and also depends on whether the element remains in solid solution or forms a second-phase particle.

Metallurgical mechanisms that impede the kinetics of thermally induced changes, such as recrystallization, may also enhance softening resistance. The lowering of diffusion rates so as to delay the onset of recrystallization results from nickel and manganese additions. Segregation of solute to dislocations, subboundaries, and grain boundaries to impede motion and thereby prevent recrystallization results from phosphorus and zirconium additions. The presence of fine precipitates to pin dislocations and internal boundaries effectively interferes with recrystallization; eg, age and dispersion hardened alloys. In general, finer and more homogeneous dispersions result in the most softening resistance.

In addition to compositional effects, softening resistance depends strongly on the amount of prior cold work; softening resistance is lowered with increasing amounts of cold work (Fig. 9). Thus the thermomechanical condition of the material must be known when applying softening data. Softening resistance data are best obtained using the pertinent service or process condition. Measurement of hardness or strength at room temperature after exposure for various times at a fixed temperature (isothermal data), or after exposure to various temperatures for a fixed time (isochronal data) provide the most useful softening information. For general comparisons between various alloys and conditions, the half-softening temperature is reported. This temperature is defined as that where hardness or strength (measured at room temperature) is lowered to the average

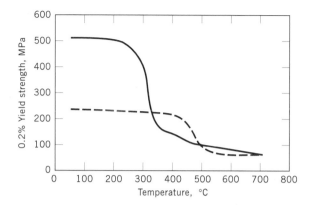

Fig. 9. The effect of prior cold work on softening resistance, as illustrated for C230 (copper−15% zinc) where (—) is CR 55%; (− − −) CR 10% in thickness. To convert MPa to psi, multiply by 145.

of the as-rolled and fully soft values after isochronal anneals (typically one hour). Half-softening temperatures of alloys shown in Figure 4, all having near 50% prior cold work are 220°C for alloy C110, 295°C for alloy C260, 340°C for alloy C510, and 525°C for alloy C706.

6.7. Stress Relaxation Resistance. Stress relaxation and creep both result from unwanted plastic deformation of metals under load at elevated temperature. Copper alloys are used extensively in applications where they are subjected to moderately elevated temperatures while under load. An important example is the spring member for contacts in electrical and electronic connectors. Critical to reliable performance is the maintenance of adequate contact force, or stability, while in service. Excessive decrease in this force to below a minimum threshold value because of losses in spring property can lead to premature open-circuit failure (see ELECTRICAL CONNECTORS). Such a loss is termed stress relaxation.

Stress relaxation (20−23) is the time-dependent decrease in load (stress) at the contact displacement resulting from connector mating. Creep, on the other hand, relates to time-dependent geometry change (strain or displacement) under fixed load, which is a condition that does not apply to connectors.

The amount of stress relaxation depends on the alloy, its temper, the temperature of exposure, and the duration of the exposure. Resistance to stress relaxation of copper is improved by alloying with solid solution elements, as well as by dispersion and precipitation strengthening. Changing temper to higher strength for a given alloy results in some loss in relaxation resistance, thus offsetting some of the initial strength gain of the higher temper by more loss due to stress relaxation. Relief annealing where yield strength is decreased slightly while causing little or no change in tensile strength is used to improve relaxation resistance.

Stress relaxation performance is usually determined in bending, by setting the initially imposed stress at the surface of the bent sample to be 50−100% of the tensile yield strength. Test details are presented in Ref. 23. Data is presented either as the percentage of the initial stress lost or the percentage that remains

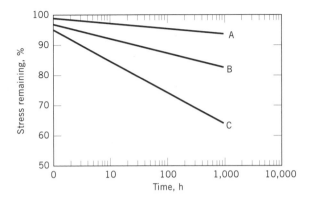

Fig. 10. Stress relaxation behavior at 105°C of the phosphor bronze alloy C521, in the A, relief annealed HR04 and rolled B, H04, tempers compared to C, rolled temper brass, C260–H04.

as a function of time at the exposure temperature. Often the data is plotted as percent stress remaining versus log time as shown in Figure 10, resulting in a plot where the data can be represented by one or two straight lines.

Individual effects of alloying and processing on stress relaxation behavior are illustrated by Figure 10. Alloying with zinc or tin improves stress relaxation relative to copper; C260 (copper–30% zinc) and C521 (copper–8% tin) are solid solution hardened alloys. The phosphor bronzes offer higher resistance to stress relaxation than brass. Relief annealing also improves resistance to stress relaxation as illustrated by comparison of C521 in a rolled temper (H04) with the same material after relief annealing (HR04). Acceptable stability is often limited to 70% stress remaining or higher, depending on the application. Thus brass in the H04 temper would not be acceptable for applications at this temperature where expected duration of exposure can exceed 1000 h.

The highest stress relaxation resistance, at both high strength and moderate conductivity, is available from precipitation hardened alloys. This effect is an important consideration for applications that have high exposure temperatures, as found in present-day automobile engine compartments where 125–150°C exposure temperatures are possible. Precipitation hardened alloys C172 and C7025 are compared to a relief annealed solid solution hardened alloy, C521, in Figure 11. It is important to note that the temperature of the alloy is comprised of the ambient temperature plus the temperature rise due to ohmic heating caused by the current flowing through the connector. This latter component is dependent on the electrical resistance of the alloy and a higher conductivity may assist in reducing relaxation by lowering the temperature of the alloy.

6.8. Fatigue Resistance.   Imposed cyclic stressing of metals may result in localized cracking that leads to fracture. The pattern of stressing can be in reversed bending, reversed torsion, and tension–compression, or half cycles of these such as bending in only one direction. The number of cycles of stressing that can be endured without fracture depends on the magnitude of the peak applied stress, the pattern of stressing, and the alloy's mechanical properties (20,24–29).

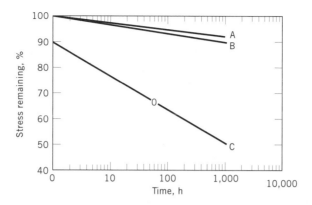

Fig. 11. Comparison between the stress relaxation resistance at 150°C, precipitation hardened alloys A, C172–TH02, and B, C7025–TM02, and C, a solution hardened alloy C521–HR02.

Fatigue properties are usually presented as graphs of applied stress, S, versus the logarithm of the number, N, of cycles to failure. The S–N curve for C510–H08 (spring) temper strip, tested in reversed bending, is shown in Figure 12 (25). Unlike steel, the S–N curve does not eventually become horizontal at high cycles so that an endurance limit does not exist. Thus fatigue resistance of copper and other face-centered cubic metals are reported as fatigue strength, which is defined as the applied stress at which failure occurs at 10^8 cycles.

Fatigue strengths for several copper alloys are listed in Table 7. Generally, fatigue strength increases with tensile strength of the material. This generality refers to the regime of high cycle fatigue where applied stress is a small fraction of its alloy's tensile strength (24). The rule-of-thumb is that the fatigue strength of a copper alloy is about one-third of its tensile strength.

Fatigue properties in bending are most appropriate for copper alloys as these are often used as spring contact components in bellows and electrical

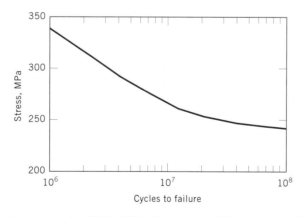

Fig. 12. Fatigue curve for C510–H08. To convert MPa to psi, multiply by 145.

Table 7. **Fatigue Strengths of Copper Alloys**[a]

Alloy	Average 0.2% yield strength, MPa[b]	Fatigue strength 10^8 cycles, MPa[b]
C172[c]	760	275
C194	480	150
C260	615	185
C510	690	235
C762	725	205

[a] All are H08, spring temper unless indicated.
[b] To convert MPa to psi, multiply by 145.
[c] TM02 (1/2 HM).

switches and connectors. These articles are usually designed for acceptable service lives at a moderate to high number of stress cycles.

Lead frames used in electronic packages have a different requirement for resistance to fracture such that the exposed leads must be capable of sustaining a few bending cycles without fracture but at high applied stress levels. Such conditions can be caused by handling damage prior to the packages being assembled onto a circuit board. The resistance to low cycle fatigue by a material is controlled by ductility rather than strength (26). The ability to straighten a damaged lead without its breaking is simulated in a lead bend fatigue test that is covered under Military Standard 883C (30,31). The test is done either on assembled packages or on appropriate samples prepared from alloy strip. Generally, lead materials are required to sustain at least three reversed 0–90° bends without fracture to ensure against scrapping packages.

6.9. Corrosion Resistance. Copper and selected copper alloys perform admirably in many hostile environments. Copper alloys with the appropriate corrosion resistance characteristics are recommended for atmospheric exposure (architectural and builder's hardware, cartridge cases), for use in fresh water supply (plumbing lines and fittings), in marine applications (desalination equipment and biofouling avoidance), for industrial and chemical plant equipment (heat exchangers such as condensers and radiators), and for electrical/electronic applications (connectors and semiconductor package lead-frames) (32) (see PACKAGING).

Atmospheric exposure, fresh and salt waters, and many types of soil can cause uniform corrosion of copper alloys. The relative ranking of alloys for resistance to general corrosion depends strongly on environment and is relatively independent of temper. Atmospheric corrosion, the least damaging of the various forms of corrosion, is generally predictable from weight loss data obtained from exposure to various environments (33) (see CORROSION AND CORROSION CONTROL).

Pitting corrosion may occur generally over an entire alloy surface or be localized in a specific area. The latter is the more serious circumstance. Such attack occurs usually at surfaces on which incomplete protective films exist or at external surface contaminants such as dirt. Potentially serious types of corrosion that have clearly defined causes include stress–corrosion cracking, dealloying, and corrosion fatigue (32–36).

Table 8. **Relative Susceptibility to Stress Corrosion and Dealloying of Commercial Copper Alloys**

UNS designation	Susceptibility Scale	
	Stress corrosion[a] (0–1000)	Dealloying[b] (0–10)
C260	1000	10
C230	200	4
C770	175	9
C422		4
C425	100	
C688	75	3
C510	20	
C521	10	
C706	0	
C194	0	0
C110	0	0

[a] 1000 is most susceptible and 0 is essentially immune under normal service conditions.

[b] 10 is most susceptible, 0 is least.

Stress corrosion is cracking that develops in sensitive alloys under tensile stress (either internally imposed or a residual stress after forming) in environments such as those containing amines and moist ammonia. The crack path can be either inter- or trans-granular, depending on alloy and environment. Not all alloys are susceptible to stress corrosion (33).

The relative susceptibility of several commercial alloys is presented in Table 8. The index used is a relative rating based on integrating performance in various environments. These environments include the harsh condition of exposure to moist ammonia, light-to-moderate industrial atmospheres, marine atmosphere, and an accelerated test in Mattsson's solution. The latter testing is described in ASTM G30 and G37 (37,38) and is intended to simulate industrial atmospheres. The index is linear. A rating of 1000 relates to the most susceptible and zero designates immunity to stress corrosion.

Dealloying refers to the selective removal of the more chemically active constituent of an alloy to leave a porous, weak deposit of the more noble constituent. Copper–zinc alloys containing >20% zinc are susceptible to dealloying, where it is called dezincification. High zinc brasses having zinc contents near 39% are particularly prone to this form of attack. In brass, dezincification is easily apparent to the unassisted eye as a reddish copper "blush" on the surface of the metal (32,36).

The relative susceptibility to dealloying of alloys that contain varying amounts of zinc is also summarized in Table 8. The index of susceptibility was derived from testing in a 3.4% sodium chloride solution at 40°C. A ranking of 10 relates to high susceptibility, zero relates to immunity. The relative rankings among the alloys may differ for different exposure time and environment.

Corrosion fatigue refers to the combination of chemical attack and cyclic stressing, where cracking propagates more rapidly than under static stressing.

Often, corrosion fatigue can be recognized by the presence of several cracks that initiate from stress raisers such as corrosion pits. The mode of fatigue failure is often transgranular.

Impingement or erosion attack can occur when liquids or gases impact metal surfaces at high velocity. The corrosion rate is high under such circumstances because any corrosion product films that can be protective if adherent are swept away as quickly as they are formed to leave exposed fresh surface.

6.10. Hydrogen Embrittlement. Copper alloys that contain cuprous oxide in their microstructures, as in C110, are potentially susceptible to embrittlement when heated in hydrogen-containing gases (39–41). Hydrogen that readily diffuses into the alloy and reduces the cuprous oxide to produce water vapor that subsequently collects under substantial internal pressure, to cause internal pores. This porosity is typically found along grain boundaries, leading to fractures along these boundaries (41). Accordingly, susceptible alloys are annealed during processing in nitrogen or very low hydrogen-containing gas, at low temperatures, and short times, to avoid embrittlement.

Oxygen-free or deoxidized copper and its alloys are not susceptible to hydrogen embrittlement. Phosphorus is a commonly used deoxidizer (as in C122– phosphorus deoxidized copper). Aluminum, beryllium, magnesium, phosphorus, silicon, and zirconium can react with oxygen and cuprous oxide to form more stable oxides that are not readily reduced by hydrogen. These elements are often added to alloys to improve mechanical properties with resistance to hydrogen embrittlement being an added benefit. Examples of the latter are C151, C172, C194, C197, C260, C510, C688, and C7025.

6.11. Solderability. Most copper alloys have good solderability (tinnability), meaning that they are wet by molten tin, tin–lead, and modifications of these to produce a continuous coating that has few to no pinhole sized nonwet areas. This characteristic of copper and its alloys accounts for the significant use of tin and solders (both lead containing and lead free) to provide corrosion resistance and in joining (see SOLDERS AND BRAZING FILLER METALS) (42).

The inherent solderability of copper alloys for coating and joining purposes is determined by a variety of methods. The most commonly used is the vertical dip test where a sample is first fluxed, then immersed for a specific time (~5 s) in tin or solder, removed, cooled and visually inspected. This test procedure is provided in Military-Standard (43,44) and ASTM Specifications. A solderability rating system is summarized in Table 9. The flux that is used must also be specified. Fluxes used by the electronics industry are very mild and include resin (Type R)

Table 9. **Solderability Rating System and Alloy Ratings**

Rating system class	Description	UNS designation[a]
I	uniform smooth coating, 100% wetting	C110, C194, C220, C510
II	>95% wetting	C195, C762
III	50–95% wetting	C260, C688
IV	<50% wetting	
V	no wetting	

[a] Ratings are for a mildly activated resin (Type RMA) flux.

and mildly activated resin (Type RMA). Joining operations, as in radiator manufacture and plumbing, generally use much more aggressive chloride or acid-type fluxes because these applications are more tolerant to residual flux than electronic equipment.

The inherent solderabilities of selected alloys are listed in Table 9. Class IV and V ratings with this particular flux indicate the presence of oxides or other surface contaminants that may be removable with more aggressive flux or acid pickling.

Tin and solder coatings are applied to copper and its alloys to provide the ability to be soldered after prolonged storage and to provide corrosion resistance (45–47). Such coatings are applied in several ways, each providing different coating characteristics. Coatings can be applied from a molten bath, wherein the strip surface is mechanically wiped as it exits the bath in order to produce a uniform coating thickness. Depending on the base alloy and process variables, the coating thickness is ~20–80 μm. Thicker coatings than those produced by mechanical wiping are made by directing strong air jets or "knives" at the surface. Reaction between the coating and the substrate copper alloy results in a 20–40 μm thick Cu–Sn intermetallic compound for both application methods. Finally, tin coatings can be applied by electrodeposition over a range of thicknesses of up to 300 μm. No intermetallic is formed during the plating process. However, rapid intermetallic growth subsequently occurs at room temperature, such that 20–30- μm thick intermetallic is present after 30 days. Further, electrotin coatings are often reflowed at just above the melting point by hot oil immersion or infrared radiation. This reflow operation results in a smooth, high quality surface. Additional intermetallic compound forms during the reflow operation and in storage or service.

Coating characteristics affect performance. For example, the intermetallic layer is easily oxidized during assembly and storage. If it is required that the material be soldered at some future time, it is necessary (but not always sufficient) for a residual layer of tin to be present. But this residual tin requirement is not necessary in many cases where the coating is intended to provide only tarnish resistance rather than extend shelf life solderability. Solder coatings are also degraded by diffusion of alloy constituents to the surface where they can react with the atmosphere to form sulfides, oxides, etc, that impede subsequent soldering. The quality of tin coatings is evident by several accelerated tests, including a 150°C bake test and a steam aging test (44).

6.12. Brazeability.   Brazing is, by definition, elevated temperature soldering, that is, soldering above the arbitrarily defined temperature of 425°C. Copper and its alloys are readily brazed and often are brazed in order to take advantage of the stronger and more stable brazed joint compared to the soldered joint. In addition, it is easy to match the properties of the filler metal to the copper alloy to be brazed because many of the brazing alloys are themselves copper-base alloys (42). As with soldering, brazing is done at a temperature above the melting point of the brazing (filler) alloy and below that of the alloy being joined. The filler or brazing metal is placed in or near the joining surfaces in the form of a strip, wire, or powder. With application of heat, the filler melts and by capillary action completely fills the gap and forms a metallurgical bond between the joining surfaces. For successful brazing it is necessary to set the proper clearance between the joining surfaces and to preclean the surfaces to be joined. In addition, oxidation protection must be provided via proper fluxing and atmosphere to ensure a sound braze and adequate flow characteristics of the filler metal.

Filler alloys for brazing of copper alloys are usually copper base. These include copper–zinc alloys (RBCuZn), copper–phosphorus alloys (BCuP-1), and copper–silver (Zn, Cd, or Li) alloys (BAg). The copper–zinc brazing alloys are low in cost, but they require a flux and are not recommended for pure copper because of the potential for corrosion. The phosphorus-containing filler alloys are self-fluxing and are reasonably low cost, but are to be avoided with high nickel cupronickels because of an embrittling reaction between nickel and phosphorus. The copper–silver base alloys (BAg) produce good sound joints for most copper alloys, but carry the increased cost burden of the silver component.

The heating required for the brazing operation is readily done using commonly available industrial equipment: muffle furnace, oxy-gas torch, and induction heating. To protect against oxidation and to ensure good filler metal fluidity, fluxing agents and appropriate atmospheres are specified for each heating method. The fluxing agents comprise various mixtures of borates, fluorides, boric acid, and wetting agents. These agents are usually applied in paste form. The atmospheres used in brazing furnaces are usually products of natural gas combustion, nitrogen, or dissociated ammonia, all having maintenance of low dew points. Whereas hydrogen can be used for many alloys, it must not be used for brazing electrolytic tough pitch copper, C11000, because of the embrittling steam reaction between hydrogen and the cuprous oxide particles. Moreover, the presence of hydrogen should be avoided during any brazing operation where cuprous oxide has the potential of forming.

Fluxes must be used when brazing the copper–zinc brasses. For copper alloys above ~20% zinc content, lower melting point brazes must be used to control zinc fuming and dezincification. Aluminum bronzes can be successfully brazed only when the proper flux (Type 4) is used to reduce the formation of alumina films. Various copper alloys can be successfully brazed only after they have received a stress relieving heat treatment and if they are heated slowly to the brazing temperature in order to avoid liquid metal embrittlement or fire cracking. Susceptibility to these latter problems is shown by the leaded alloys, the phosphor bronzes, the silicon bronzes, and the nickel silvers. When brazing precipitation hardening alloys, the brazing alloy and conditions can be chosen so that the alloy is solutionized during the brazing treatment. After a suitably rapid quench, the brazed alloy can be given an age hardening heat treatment.

6.13. Weldability.   Welding has long been an important method of fabricating copper alloy parts as these alloys are generally readily weldable (42). But, as is the case with all materials, successful welding requires the proper match of welding technique to the specific application and copper alloy. There are four primary concerns for successful welding of copper. The first and foremost is the high thermal conductivity of copper and its resulting ability to conduct heat away from the weld zone. The second concern is the chemical and in some cases toxic properties of the typical elements alloyed with copper. Adequate ventilation must be provided for the lead vapor, the zinc fumes, or the toxic compounds of Be, As, or Sb that can be emitted from the copper alloys containing these elements. The third concern is the ready solubility of oxygen in copper at elevated temperatures and the subsequent precipitation of cuprous oxide particles at grain boundaries during solidification and cooling, which if uncontrolled cause reduced strength and ductility in the weld zone. The fourth concern is that alloys containing Pb, Te, and S (eg, the free-machining alloys) are prone to hot shortness (hot cracking) during cooling.

Arc Welding.   has long been used to join copper alloys. The gas tungsten-arc welding (GTAW) and the inert gas metal arc welding (GMAW) methods are the preferred arc welding methods for copper alloys. The GTAW technique is appropriate for workpiece thicknesses up to ~3 mm (0.125 in.) because of its capability of delivering intense heat into a narrow area. The GMAW technique is used for workpiece thicknesses >12 mm (0.5 in.) because of its ability to provide rapid delivery of high heat. The rapid delivery of high heat is important for either technique to counter the high thermal conductivity of the high conductivity copper alloys. Besides heating efficiency, the importance of rapid delivery of the heat is to shorten the welding time to minimize oxygen pickup. The inert gases used are argon or mixtures of argon–helium . Although argon is cheaper, the helium content is often specified because of its higher thermal conductivity properties enabling higher heat input to the workpiece, especially for alloys with higher thermal conductivities.

In general, it is difficult to avoid porous welds of low ductility when arc welding oxygen-bearing electrolytic tough pitch copper, C11000. The situation is improved for arc welding of oxygen-free copper, but pickup of oxygen during welding must be avoided. Higher quality welds are more easily obtained when the workpiece and/or the filler metal is a deoxidized alloy or otherwise contains deoxidizing elements. For this reason the phosphorus deoxidized alloys, the silicon and aluminum bronzes, and the cupronickels are arc weldable.

In order to enable GTAW equipment to provide the required highly penetrating input of heat, a dc straight polarity current is applied. Using such equipment, copper alloy tubes are welded to copper alloy tube sheets. The ends of processed strip or wire are easily joined in this technique to create longer coils of strip, wire, or rod product. When age hardening alloys are welded, such as the beryllium or chromium coppers, they must be given a postwelding age hardening heat treatment. Such a treatment usually does not return the weld region to the same strength as the rest of the workpiece, but it is adequate for most applications. Arc welding is not recommended for the newer oxygen dispersion hardened alloys (C15720), because the desirable uniform array of fine oxide particles (alumina) is destroyed by melting of the alloy. Arc welding can be used for welding of copper to other metals, for which care must be taken to design for the differences in thermal expansion characteristics and during which the arc should be directed toward the higher conductivity alloy in the combination.

Resistance Welding.   has been successfully applied to copper alloys in all of its various spot, seam, or butt joining modes. Because the process depends on ohmic (I^2R) heating at the interface to be joined, the ability to resistance weld is inversely related to electrical conductivity of the alloys being welded. The current is applied via opposing electrodes on either side of the joint. The electrodes themselves are usually made of high conductivity, softening resistant copper alloys, such as chromium– or zirconium–copper. The electrodes may be coated with a refractory metal, such as molybdenum, to prevent electrode pickup or sticking to the work piece. Most copper alloys having strip thicknesses on the order of 0.025–3.2 mm can be welded by this method. The leaded copper alloys are susceptible to liquid metal embrittlement as well as lead bleed out. As is the case for arc welding, steps must be taken to control the oxide fuming from alloys containing high levels of zinc, beryllium, and lead. Projection welding can be used to prevent electrode sticking and to control the shape of the weld nugget in high conductivity strip thicker than 0.5 mm. In this method, point, linear, or annular projections (ridges) are formed or machined in the workpiece to concentrate the welding current where the heat is needed. Flatter electrodes can then be used, leading to less electrode pickup.

Induction Welding. also depends on I^2R heating. In this case, the electrical current is induced in the workpiece via an imposed high frequency magnetic field. The high frequency equipment is designed to concentrate the induced eddy currents near the surface at the edges of the workpiece to be joined. This method has been successfully applied to most copper alloys and, in particular, has been effectively used to make the longitudinal seam in the high speed manufacture of welded tube from copper alloy strip.

6.14. Machinability.   Copper and its alloys can be machined with differing degrees of ease. Special additives such as lead, bismuth, tellurium, selenium, and sulfur, are added, to enhance machinability (48), although other properties, such as formability and tensile ductility, normally suffer. These particular alloying elements form second-phase particles that promote chip fracture and the development of lubricating films at the tool-to-chip interface. Smaller, easier to handle chips and lower cutting forces leading to longer tool life result from use of these additives. Notable uses for special alloys having high machinability are rod, for high production rate screw machine items such as fasteners, connectors and plumbing components, and strip for keys.

Copper and its alloys are ranked according to a machinability rating index that is a percentage of the machinability of the most machinable copper alloy, C360, also known as free-cutting brass. The rating is based on relative tool

Table 10. **Machinability Ratings for Wrought Copper Alloys**

UNS designation	Rating[a]	Comment
C110	20	
C145	85	phosphorus deox, Te bearing (0.5%)
C147	85	sulfur bearing (0.45%)
C172	20	
C187	85	leaded copper
C260	30	
C353	90	high leaded brass
C360	100	free-cutting brass
C425	30	
C443	30	
C510	20	
C544	80	
C706	20	
C752	20	
C782	60	leaded nickel silver

[a] Rating is a relative scale based on C360 having a value of 100.

life, where C360 is assigned a rating of 100. The actual performance of an alloy is dependent on the operation (turning, drilling, threading, etc), tool design, the lubricant, and the precise criterion used to define machinability (tool life, cutting rate, surface finish). Machinability ratings for several wrought alloys are given in Table 10.

The characteristics of chips also follow the machinability rating. The most highly machinable alloys, (ie, those having ratings near 60 and greater) develop short, brittle chips that are easily cleared from tools to make them suited to high production rate machining. Somewhat longer but brittle helical chips are often characteristic of alloys having ratings near 30, such as C260. The least machinable alloys, such as copper, phosphor bronzes, and the copper-nickel alloys having ratings of 20, form long, continuous chips that are easily entangled and difficult to handle.

7. Alloy Specific Properties

The physical properties of pure copper are given in Table 11. The mechanical properties of pure copper are essentially the same as those for C101 and C110, discussed later, which are often described as copper.

Trace elements in copper exert a significant influence on electrical conductivity. Effects on conductivity vary because of inherent differences in effective atomic size, valency, and solubility. The decrease in conductivity produced by those elements appearing commonly in copper, at a fixed weight percent, rank as follows: Cd (least detrimental), Au, Zn, Ag, Ni, Sn, Mg, Al, Si, Fe, P, O, Ti(most detrimental). Table 12 summarizes these effects. In the absence of chemical or physical interactions, the increase in electrical resistivity is linear with amounts of each element, and the effect of multiatom additions is additive.

7.1. Coppers. The coppers represent a series of alloys ranging from the commercially pure copper, C101, to the dispersion hardened alloy C157. The difference within this series is the specification of small additions of phosphorus,

Table 11. **Physical Properties of Copper**[a]

Parameter	Value
mp, °C	1083
density, g/mL	8.94
electrical conductivity, % IACS min	101
electrical resistivity, n $\Omega \cdot$ m	17.1
thermal conductivity, W/(m $\cdot$ K)	391
coefficient of thermal expansion from 20–300°C, μm/(m $\cdot$ K)	17.7
specific heat, J/(kg $\cdot$ K)[b]	385
elastic modulus, GPa[c]	
tension	115
shear	44

[a] All properties at 20°C unless otherwise noted.
[b] To convert J to cal, divide by 4.184.
[c] To convert GPa to psi, multiply by 145,000.

Table 12. **Solubility Limits and Electrical Conductivity Effects of Elemental Additions to Copper**[a]

Element	Solubility at room temperature, wt%	Resistivity increase, $(\mu\Omega \cdot cm)/wt\%$	Electrical conductivity[b] at 0.5 wt %, % IACS
Ag	0.1	0.355	93
Al	9.4	2.22	62
As	6.5	5.67	38
Au	100.0	0.185	97
Be	0.2	4.57	44
Cd	0.5	0.172	98
Co	0.2	7.3	32
Cr	0.03	4.9	42
Fe	0.14	10.6	25
Mg	1.0	2.1	63
Mn	24.0	3.37	51
Ni	100.0	1.2	76
O	0.0002	21.0	14
P	0.5	14.3	20
Pb	0.02	1.02	79
Sb	2.0	2.9	55
Si	2.0	7.0	33
Sn	1.2	1.65	69
Ti	0.4	21.6	14
Zn	30	0.31	94
Zr	0.01	8.0	30

[a] Refs. (9–12).
[b] Assuming solubility at 0.5 wt%.

arsenic, cadmium, tellurium, sulfur, magnesium, zirconium, as well as oxygen. To be classified as one of the coppers, the alloy must contain at least 99.3% copper.

The mechanical properties of coppers having UNS designations between C10100 and C13000 are listed in Table 13. The coppers include high purity copper (C101, C102), electrolytic tough pitch (C110), phosphorus deoxidized (C120, C122), and silver-bearing copper (C115, C129, etc). The mechanical properties of

Table 13. **Tensile Properties of Copper (C10100–C13000)**

Temper	Tensile strength, MPa[b]	Yield strength,[a] MPa[b]	Elongation, %	Hardness, HRF[c]
OS 025[d]	235	76	45	45
H01	260	205	25	70
H02	290	250	14	84
H04	345	310	6	90
H08	380	345	4	94
H10	395	365	4	95

[a] 0.5% offset.
[b] To convert MPa to psi, multiply by 145.
[c] Hardness is on the Rockwell-F scale.
[d] Annealed to average grain size of 0.025 mm.

Table 14. **Tensile Properties of Dilute Alloy Coppers (C14300–C15720)**

Alloy	Temper	Tensile strength, MPa[a]	0.2% Yield strength, MPa[a]	Elongation, % 50 mm
C14300	H04	310	275	14
C14500[b]				
C14700[b]	H04	330	304[c]	20
C15000	TH04[d]	445	411[c]	18
C15100	H04	400	386	4
C15500	H04	425	394	5
C15720[b]	061	485	380	13

[a] To convert MPa to psi, multiply by 145.
[b] Rod of 25 mm diameter.
[c] 0.5% offset for rod products.
[d] Solution treated at 950°C, cold worked 90%, aged 1 h at 450°C.

these alloys are essentially the same. Other coppers, C142 through C157, offer higher strength at usually lower conductivity. The mechanical properties of alloys within the latter range are listed in Table 14.

Excellent resistance to saltwater corrosion and biofouling are notable attributes of copper and its dilute alloys. High resistance to atmospheric corrosion and stress corrosion cracking, combined with high conductivity, favor use in electrical–electronic applications.

7.2. C101 and C102. The oxygen-free coppers, C10100 and C10200, are manufactured by melting prime quality, electrolytically produced cathode copper under reducing conditions designed to restrict oxygen content to below 0.0005% (C101) and 0.001% (C102). C101 has a minimum copper content, which may also include a trace of silver, of 99.99%. Oxygen-free copper is immune to hydrogen embrittlement, and provides low outgassing tendency, along with very high (101% IACS min) conductivity at 20°C for annealed material.

7.3. C110. The most common commercial purity copper is C110. The principal difference between C110 and C102 is oxygen content, which is deliberately controlled at 0.04% in C110 to eliminate many impurities that are deleterious to conductivity. Oxygen is present as cuprous oxide particles, which do not significantly affect strength and ductility, but C110 is susceptible to hydrogen embrittlement. The properties of C110 are adequate for most applications and this alloy is less costly than higher purity copper.

7.4. Deoxidized Copper. Cuprous oxide can be readily removed from copper by small additions of phosphorus, calcium, and magnesium. Phosphorus is the most common deoxidizing element. Most phosphorus-deoxidized coppers contain residual phosphorus in concentrations of 0.01–0.04%. The most common alloy in this family is C122, containing 0.015–0.040% phosphorus. Because of residual dissolved phosphorus, the minimum electrical conductivity of C122 is 85% IACS, versus 100% IACS for C110. Applications involving brazing and welding in hydrogen-rich atmospheres are important uses for deoxidized copper where embrittlement by hydrogen has to be avoided.

7.5. Specialty Coppers. Additions are made to copper to satisfy specific needs. Tellurium at a nominal 0.5 wt% addition, sulfur at 0.35 wt%, and lead at 1 wt% enhance machinability. These alloys are identified as C145, C147, and

C187, respectively. The solubility limit for each element is <0.001%, so that the excess is present as second-phase particles that assist in fracture of chips and lubrication during machining.

Additions of 0.1 wt% cadmium to phosphorus deoxidized copper (C143), 0.05–0.2 wt% zirconium to copper (C150, C151), or Mg, Ag and P to copper (C155) significantly enhance softening resistance while maintaining high (85–95% IACS) conductivity. Because of environmental concerns, however, cadmium-containing alloys are being replaced by others that meet product and manufacturing requirements.

Extremely high softening resistance is available in alloy C15720 through dispersion strengthening by aluminum oxide. This alloy has been used successfully as a welding rod material and in other specific applications that can bear its cost.

7.6. High Copper Alloys. The high copper alloys contain a minimum of 96 wt% copper; most contain about 98.5 wt%. These alloys are found within the UNS classification of C162 to C199. As a group, they offer higher strength and better softening resistance than the specialty coppers. Like the coppers, the high copper alloys have excellent resistance to general and stress corrosion cracking. However, the high copper alloys have lower electrical conductivities than the coppers, they are more expensive, and special processing may be required to incorporate dispersoids or precipitates for optimal effectiveness. Nominal electrical conductivities and tensile properties of selected high coppers in hard rolled temper are summarized in Table 15.

7.7. Cadmium Copper (C162). Containing 1 wt% Cd, C162 has been used for many years in electrical connectors because of its excellent (90%

Table 15. **Conductivity and H04 (Hard) Temper Tensile Properties**

UNS	Electrical conductivity, % IACS	Tensile strength, MPa[a]	0.2% Yield strength, MPa[a]	Elongation in 50 mm, %
		High copper alloys		
C16200	90	414	310	5
C182[b]	80	460	405	14
C19000	60	579	531	3
C19200	65	448	414	7
C19400	65	462	434	4
C19500	50	593	572	2
C19700	80	448	414	7
		Cu—Zn brass alloys		
C210	56 (5% Zn)	379	365	5
C220	44 (10% Zn)	428	400	4
C230	37 (15% Zn)	469	421	7
C240	32 (20% Zn)	496	421	4
C260	28 (30% Zn)	524	496	10
C353	26 (34% Zn, 2% Pb)	503	462	12
C360[c]	26 (35% Zn, 3% Pb)	386	310	20

[a] To convert MPa to psi, multiply by 145.
[b] TH04 Temper material.
[c] H02 Temper material.

IACS) electrical conductivity and formability. Much of the cadmium is in solution; the remainder is dispersed as Cu_2Cd particles. Unique arc quenching ability, which reduces the surface degradation from electric-arc erosion during switching, is characteristic of this alloy. However, environmental concern because of cadmium's toxicity has all but eliminated its use, thus restricting its availability.

7.8. C190 to C197. The C190 to C197 alloys constitute a series of commercially useful copper alloys containing low levels of phosphorus; some contain Ni (C190–C191) and some contain Fe (C192–C197). They are relatively inexpensive and offer excellent combinations of strength, ductility, conductivity, and moderate softening resistance. As a group, these alloys are readily processed to provide a uniform array of iron and iron or nickel phosphide dispersoids for hardening and grain size control. Offering electrical conductivities of 50–80% IACS, they find extensive use in electronic products as leadframes and in electrical connectors. C194 (nominally, Cu–2.4 Fe–0.04 P–0.1 Zn) has the distinction of being the copper alloy that during the early 1970s replaced ASTM F30 (42 Ni— Fe alloy) as the dominant leadframe material in the electronics industry. Of increasing importance is the higher capacity for these alloys to dissipate heat from semiconductor packages and thus prolong integrated circuit (IC) life. Also of importance are their nonmagnetic properties permitting high clock speeds for IC devices (see INTEGRATED CIRCUITS; SEMICONDUCTOR TECHNOLOGY).

7.9. Copper–Zinc Brasses. copper–zinc alloys have been the most widely used copper alloys during the 1990s. It is no accident that the word brass is included in the name of many copper alloy manufacturers. The manufacture of brass buttons and other brass artifacts was the principal reason for the establishment of the U.S. copper alloy industry in Connecticut during the 1800s.

Brass alloys fall within the designations C205 to C280 and cover the entire solid solution range of up to 35 wt% zinc in the Cu–Zn alloy system. Zinc, traditionally less expensive than copper, does not too greatly impair conductivity and ductility as it solution hardens copper.

Brass alloys are highly formable, either hot or cold, and provide moderate strength and conductivity. Moreover, the alloys have a pleasing yellow "brass" color at zinc levels >20 wt%. The material is amenable to polishing, buffing, plating, and soldering. By far, the best known and most used composition is the 30 wt% zinc alloy, Cartridge Brass. Door knobs and bullet cartridges are the best known applications and illustrate the material's excellent formability and general utility. Properties of representative brass alloys in the hard temper are also summarized in Table 15.

The series of brasses, C312 to C385, contain from 0.25–5.0 wt% lead for the purpose of improving machinability. C360, having the composition Cu–35.4 Zn– 3.1 Pb, has become the industry standard for machinability performance. It is given the descriptive name of free-machining brass and a machinability rating value defined as 100 (Table 10).

Brasses are susceptible to dezincification in aqueous solutions when they contain >15wt% zinc. Stress corrosion cracking susceptibility is also significant >15 wt% zinc. Over the years, other elements have been added to the Cu—Zn base alloys to improve corrosion resistance. For example, a small addition of arsenic, antimony, phosphorus, or tin helps prevent dezincification to make brasses more useful in tubing applications.

Table 16. **Conductivity and Wrought Tensile Strength of Tin–Brasses Showing the Hardening Effect of Tin Additions**

UNS	Zinc, wt%	Tin, wt%	Electrical conductivity, % IACS	Tensile strength, MPa[a]	
				H04 Temper	H08 Temper
C411	10	0.5	32	448	552
C425	10	2.0	28	524	634
C220	10		44	428	524
C443	30	1.0	25	607	
C260	30		28	524	690

[a] To convert MPa to psi, multiply by 145.

7.10. Tin Brasses. The tin brass series of alloys consists of various copper–(2.5–35 wt%) zinc alloys to which up to ~4 wt% tin has been added. These are solid solution alloys that have their own classification as the C40000 series of alloys. Tin provides better general corrosion resistance and strength without greatly reducing electrical conductivity. As with all the brasses, these alloys are strengthened by cold work and are available in a wide range of tempers. These alloys offer the combined strength, formability, and corrosion resistance required by their principal applications, namely, fuse clips, weather stripping, electrical connectors, heat exchanger tubing, and ferrules. Several tin brasses have lead additions to enhance machinability. For example, the high leaded Naval Brass C485, containing Cu–37 Zn–0.7, Sn–1.8 Pb, has a machinability rating of 78.

Table 16 illustrates the property enhancements and tradeoffs seen when tin is added to a copper–zinc brass base composition. The most commonly used alloys for electrical connectors are the Cu–10 Zn–Sn brasses, such as C411, C422, and C425. These tin brasses offer improved corrosion and stress relaxation resistance relative to the base Cu–Zn alloy.

Admiralty Brass and Naval Brass are 30 and 40% zinc alloys, respectively, to which a 1% tin addition has been added. Resistance to dezincification of Cu–Zn alloys is increased by tin additions. Therefore, these alloys are important for their corrosion resistance in condenser tube applications. In these, as well as the other higher zinc compositions, it is common to use other alloying additives to enhance corrosion resistance. In particular, a small amount (0.02–0.10 wt%) of arsenic (C443), antimony (C444), or phosphorus (C445) is added to control dezincification. When any of these elements are used, the alloy is referred to as being "inhibited." For good stress corrosion resistance, it is recommended that these alloys be used in the fully annealed condition or in the cold worked plus stress relief annealed condition.

7.11. Tin Bronzes. Tin bronzes may be the most familiar of copper alloys with roots going back into ancient times. Whereas bronze is still used for statuary, these alloys are found in many modern applications, such as electrical connectors, bearings, bellows, and diaphragms. The wrought tin bronzes are also called phosphor bronzes because 0.03–0.35 wt% phosphorus is commonly added for deoxidation and improved melt fluidity during casting.

Table 17. **Conductivity and Wrought Tensile Strength of Tin Bronze Alloys**

UNS	Tin, wt%	Electrical conductivity, % IACS	Tensile strength, MPa[a]	
			H04 Temper	H08 Temper
C505	1.0	48	441	552
C511	4.0	20	545	696
C510	5.0	15	572	738
C521	8.0	13	634	800
C524	10.0	11	690	882
C544	4.0[b]	19	545	703

[a] To convert MPa to psi, multiply by 145.
[b] Also contain 4 wt % each of zinc and lead.

The several wrought alloys of commercial importance span the range of 1.0–10 wt% tin and are mostly used in work hardened tempers. These alloys are single phase and offer excellent cold working and forming characteristics. Strength, corrosion resistance, and stress relaxation resistance increase with tin content. The tin–phosphorus alloys are highly resistant to stress corrosion cracking and have good resistance to impingement corrosion attack. Unfortunately, conductivity decreases, cost increases (tin has been historically more costly than copper), and the capability of being hot processed is impaired as tin level is increased. Only compositions up to ~4.5 wt% tin show good hot processing characteristics. The richer compositions are cast in thinner cross-sections that can be cold rolled directly. Several cycles of working and annealing are employed to homogenize the cast microstructure.

One of the most highly used specialty tin bronzes is C544, containing Cu–4 Sn–4 Zn–4 Pb. Zinc provides increased strength to this tin bronze, whereas the lead addition provides good machinability (machinability rating of 80)and enhances wear resistance. C544 is used in many applications where these two latter properties are of the utmost benefit, namely, bearings, bushings, gears, pinions, washers, and valve parts. Table 17 indicates the range of properties available for the commonly used phosphor bronzes.

7.12. Aluminum Bronzes. Aluminum bronze alloys comprise a series of alloys (C606 to C644) based on the copper–aluminum (2–15 wt%) binary system, to which iron, nickel, and/or manganese are added to increase strength. The mechanical properties of commercially important alloys in this group are listed in Table 18.

Aluminum bronze alloys are used for their combined good strength, wear, and corrosion-resistance properties where high electrical and thermal conductivity are not required. Corrosion resistance results from the formation of an adherent aluminum oxide that protects the surface from further oxidation. Mechanical damage to the surface is readily healed by the redevelopment of this oxide. The aluminum bronzes are resistant to nonoxidizing mineral acids such as sulfuric or hydrochloric acids, but are not resistant to oxidizing acids such as nitric acid. However, these alloys must be properly heat treated to be resistant to dealloying and general corrosion.

The aluminum oxide surface layer that provides wear and corrosion resistance is not without drawbacks. This adherent film is difficult to remove during

Table 18. **Conductivity and H04 (Hard) Temper Tensile Properties of Aluminum Bronze Alloys**

UNS	Constituents, wt%			Electrical conductivity, % IACS	Tensile strength, MPa[a]	0.2% Yield strength, MPa[a]	Elongation in 50 mm, %
	Al	Fe	Ni				
C614	7	2.5		14	586	400	35
C615	8		2	13	862	620	5
C625	13	4.5		10	690	379	1
C630[b]	10	3.0	5	7	814	517	15

[a] To convert MPa to psi, multiply by 145.
[b] Also contains 1.5 wt% manganese.

industrial cleaning of the alloys. Cleaning techniques have been developed that fragment the tightly adherent aluminum oxide film before removing it by the nonoxidizing mineral acids commonly used in brass mill processing. Furthermore, excessive tool wear in stamping and shearing equipment is caused by the presence of this film. Soldering and brazing are also made difficult by this oxide film.

Two single-phase, binary alloys are used commercially: C606, containing 5 wt% Al, and C610, 8 wt% Al. Both alloys have a golden color and are used in rod or wire applications, such as for bolts, pump parts, and shafts. Most of the available aluminum bronzes contain additional alloy elements, however. For example, C608 is a commonly specified condenser or heat exchanger tube alloy that is essentially 5% aluminum to which 0.02–0.35 wt% arsenic has been added to improve corrosion resistance. Alloy C613, having 7 Al–2.5 Fe also has a 0.3 wt% tin addition for improved resistance to stress corrosion.

Most of the aluminum bronzes contain substantial iron, nickel, or manganese additions. These alloying elements increase strength by forming second phases during heat treatment. Iron, the most commonly added element in these bronzes, separates as an iron-rich particle that controls grain size while it enhances strength. Nickel also reacts with aluminum to form NiAl precipitates during heat treatment with the same result as the iron addition.

Several Cu—Al—Si alloys have commercial importance. For example, C636 containing 3.5 Al—1.0 Si, is a rod and wire alloy that is processed to around 590 MPa 85,000 (psi) tensile strength. C638, containing 2.8 Al–1.8 Si–0.4 Co, is a strip product that owes its good strength and formability to dispersion strengthening and a very fine grain size that results from precipitation of cobalt silicide particles. The latter alloy attains tensile strengths of 566 MPa (82,000 psi) and 883 MPa (128,000 psi) in the annealed and spring rolled (H08) tempers, respectively.

7.13. Silicon Bronzes.   Silicon bronzes have long been available for use in electrical connectors, heat exchanger tubes, and marine and pole line hardware because of their high solution hardened strength and resistance to general and stress corrosion. As a group, these alloys also have excellent hot and cold formability. Unlike the aluminum bronzes, the silicon bronzes have moderately good soldering and brazing characteristics. Their compositions are limited to below 4.0 wt% silicon because above this level, an extremely brittle phase

Table 19. **Conductivity and H04 (Hard) Temper Tensile Properties of Silicon Bronze Alloys**

UNS	Si, wt%	Electrical conductivity, % IACS	Tensile strength, MPa[a]	0.2% Yield strength, MPa[a]	Elongation in 50 mm,%
C651 wire	1.5	12	690	485	11
C654 flat[b]	3.0	7	790	700	6
C655 flat	3.0	7	650	400	8

[a] To convert MPa to psi, multiply by 145.
[b] Also contains 1.5 wt% tin.

(kappa) is developed that prevents cold processing. Electrical conductivities of silicon bronzes are low, however, because of the strong effect that silicon has in degrading this property. The three most popular of the commercially available alloys are listed in Table 19 with their typical properties.

7.14. Modified Copper–Zinc Alloys. The series of copper–zinc base alloys identified as C664 to C698 have been modified by additions of manganese (manganese brasses and the manganese bronzes), aluminum, silicon, nickel, and cobalt. Each of the modifying additions provides some property improvement to the already workable, formable, and inexpensive Cu—Zn brass base alloy. Aluminum and silicon additions improve strength and corrosion resistance. Nickel and cobalt form aluminide precipitates for dispersion strengthening and grain size control. The high zinc-containing alloys are formulated to facilitate hot processing by transforming to a highly formable (beta) phase at elevated temperature.

Representative properties of alloys from this group are summarized in Table 20. C664, C688, and C690 are strip products that are used in electrical connectors, springs, and switches. These alloys are designed to provide beneficial combinations of strength and formability in cold rolled tempers. All three utilize dispersion hardening through iron plus cobalt particles (C664), cobalt aluminides (C688), and nickel aluminides (C690).

Table 20. **Conductivity and Wrought Tensile Properties of Modified Copper–Zinc Alloys H04 (Hard) Temper**

UNS	Form	Composition, wt %	Electrical conductivity, % IACS	Tensile strength, MPa[a]	0.2% Yield strength, MPa[a]	Elongation in 50 mm, %
C664	flat	11.5 Zn–1.5 Fe–0.5 Co	30	605	585	5
C674	rod	36.5 Zn–2.8 Mn–1.2 Al–1.0 Si	23	634	379	20
C687	tube	20.5 Zn–2.0 Al–0.04 As	23	414	186	55
C688	flat	22.7 Zn–3.4 Al–0.4 Co	18	750	670	4
C690	flat	22.7 Zn–3.4 Al–0.6 Ni	18	780	700	4
C694	rod	14.5 Zn–4.0 Si	6	690	393	21

[a] To convert MPa to psi, multiply by 145.

C674 and C694 are commonly used in rod and wire forms. These alloys are used in valve stems, shafts, bushings, and gears. Dispersoids of second phases that come about from beta-phase decomposition products account for their higher strength. C687, inhibited by the arsenic addition, is primarily a tube alloy that is used in condensers, evaporators, heat exchangers, and ferrules.

7.15. Copper–Nickels. The copper–nickel alloy system is essentially single phase across its entire range. Alloys made from this system are easily fabricated by casting, forming, and welding. They are noted for excellent tarnishing and corrosion resistance. Commercial copper alloys extend from 5 to 40 wt% nickel.

Iron is added in small (usually 0.4–2.3 wt%) amounts to increase strength. More importantly, iron additions also enhance corrosion resistance, especially when precautions are taken to retain the iron in solution. Precipitation of the iron–nickel-rich phase does not result in strengthening and can cause degradation of corrosion resistance (49). A small (up to 1.0 wt%) amount of manganese is usually added to react with and remove both sulfur and oxygen from the melt. These copper alloys are most commonly applied where corrosion resistance is paramount, as in condenser tube or heat exchangers.

The common cupronickels are C704, which has 5 Ni–1.5 Fe; C706, 10 Ni–1.4 Fe; C710, 20 Ni; and C715, 30 Ni–0.7 Fe. Mechanical properties of these alloys are summarized in Table 21. The best resistance to aqueous corrosion is available from C715 containing 30 wt% nickel. Both C715 and the lower cost C706 are suited for condenser and heat exchanger tubes in power plants. This group of alloys is also highly resistant to stress corrosion and impingement corrosion.

C725, a 9 wt% nickel alloy that is further strengthened by 2 wt% tin, is used in electrical connectors and bellows. Properties are summarized in Table 21. The alloy has good resistance to stress relaxation at room and moderately elevated temperatures, which accounts for its use in electrical connectors and electrical circuit wire wrap pins.

Environmental factors must be considered when specifying copper–nickel alloys. The higher nickel compositions are less resistant to sulfides in seawater applications, for example. C713, containing 25 wt% nickel, is extensively used in U.S. coinage because of its good tarnish resistance. C722, containing Cu–17 Ni–0.8 Fe–0.5 Cr, is intended as a tube alloy having improved resistance

Table 21. **Conductivity and H04 (Hard) Temper Tensile Properties of Cupro–Nickel Alloys**

UNS	Alloy, wt %	Electrical conductivity, % IACS	Tensile strength, MPa[a]	0.2% Yield strength, MPa[a]	Elongation in 50 mm, %
C704	5.5 Ni–1.5 Fe	14.0	440	435	5
C706	10 Ni–1.4 Fe	9.0	518	500	5
C710	20 Ni	6.5	518	500	5
C715	30 Ni–0.5 Fe	4.6	580	545	3
C725	9 Ni–2 Sn	11.0	570	555	3

[a] To convert MPa to psi, multiply by 145.

Table 22. Conductivity and Hard Temper Tensile Properties of Cu—Ni—Zn Alloys

UNS	Nickel, wt %	Zinc, wt %	Electrical conductivity, % IACS	Tensile strength, MPa[a]	0.2% Yield strength, MPa	Elongation in 50 mm, %
C745	10	25	9	590	515	4
C752	18	17	6	585	510	3
C754	15	20	7	585	515	3
C757	12	23	8	585	515	4
C770	18	27	6	585	505	4

[a] To convert MPa to psi, multiply by 145.

to impingement corrosion in flowing water in condenser or heat exchanger applications.

7.16. Nickel–Silvers. Nickel–silver alloys, once called German silver, are a series of solid solution Cu–Ni–Zn alloys, the compositions of which encompass the ranges of 3–30 wt% Zn and 4–25 wt% Ni. This family of alloys falls within the UNS designation numbers C731 to C770. Leaded nickel–silvers that contain from 1.0 to 3.5 wt% lead for improved machining characteristics are designated as C782 to C799.

Nickel–silver alloys are not readily hot worked but have excellent cold fabricating characteristics. Because of the high nickel and zinc contents, these alloys exhibit good resistance to corrosion, good strength, and usually adequate formability, but have low electrical conductivity. Wire and strip are the dominant forms used for hardware, rivets, nameplates, hollowware, and optical parts. Table 22 lists common nickel–silvers and compares electrical conductivities and tensile strengths. Because of high nickel contents, these alloys are highly resistant to dezincification in spite of the high zinc content. Good resistance to corrosion in both fresh and seawater is another attribute.

Of the several leaded nickel–silvers, C782 (Cu–23Zn–10Ni–2Pb) is one of the most commonly used. This alloy is not hot processable and has limited cold processing capacity. It is usually available as strip for application as keystock, watch parts, and plates because of its excellent machinability, hardness and corrosion resistance.

7.17. Precipitation Hardening Alloys. Copper alloys that can be precipitation hardened to high strength are limited in number. In addition to the metallurgical requirement that the solubility of the added element(s) decreases with decreasing temperature, the precipitated phase that forms during aging must be distributed finely and have characteristics that act to resist plastic deformation.

Commercial precipitation hardening copper alloys are based on beryllium, chromium, titanium, and nickel, the latter in combination with silicon or tin. The principal attributes of these alloys are high strength in association with adequate formability. Electrical conductivity varies according to alloy and ranges from ∼20–80% IACS.

Usually, the aging treatment is done by the strip manufacturer, to produce "mill hardened" strip. The resulting strip properties are a practical compromise between good formability and less-than-peak strength. The advantage of these tempers is that the fabricator does not need to heat treat the strip, eliminating the need for post-form cleaning and the possibility of part distortion during the anneal. The highest strength can, however, be achieved by the fabricator performing the precipitation hardening thermal treatment following fabrication of parts from solution treated–cold rolled temper strip. This allows the fabricator to take advantage of the superior formability to make the most geometrically demanding parts and achieve high strength by aging after part fabrication.

7.18. C170 to C175 (Beryllium Copper). The addition of beryllium to copper results in a significant age hardening response, making these alloys one of the few nonferrous materials that can reach 1400-MPa (200,000 psi) tensile strength (7). These alloys are more precisely ternary alloys. Additions of cobalt or nickel control grain size through the formation of coarse beryllides that pin grain boundaries and promote the hardening precipitation reaction.

Table 23 illustrates yield strength–bend formability trade-offs associated with making parts from a mill hardened temper (TM02) versus forming rolled temper strip (H02) and then aging to produce a TH02 temper product. Parts that require the same formability can be made using both the rolled temper and mill hardened material, but it is possible to achieve higher strength by aging after forming. However, aging after forming can introduce dimensional changes that accompany the hardening reaction in beryllium coppers. Moreover, the parts must be chemically cleaned with aggressive reagents to remove beryllium oxide formed during the thermal treatment. For these reasons, mill hardened strip represents the majority of the beryllium copper that is used.

The trade-off between conductivity and formability, at comparable strength, is also shown in Table 23. More dilute compositions (C17410 and C1751) provide higher conductivity in the mill hardened condition, but at a sacrifice in formability relative to mill hardened C172.

Toxicity concerns regarding beryllium result principally from possible consequences from inhaling its oxide. Any operation that produces airborne particles or dust of the oxide must be carried out using proper precautions to personnel.

7.19. C180 to C182 (Chromium–Copper). Chromium–copper alloys have widespread use in those applications requiring good formability, conductivity,

Table 23. **Properties of Beryllium Copper Alloys**

UNS	Temper	Conductivity, % IACS[b]	Yield strength, MPa[c]	Formability, MBR/t[a]	
				GW	BW
C172	H02	15	586	0.5	1
	TH02[d]	20	1207	>5	>5
	TM02	18	758	0.5	1
C17410	HT	45	758	1.2	5.0
C17460	HT	50	792	1.5	1.5
C1751	TH04	50	758	2	2

[a] For 90° bend forming angle.
[b] Minimum values given.
[c] To convert MPa to psi, multiply by 145.
[d] H02 temper + aged at 315°C for 2 h.

Table 24. **Conductivity and Age Hardened Tensile Properties of Chromium–Copper Alloys**

UNS	Composition, wt %	Electrical conductivity, % IACS	Tensile strength, MPa[a]	0.2% Yield strength, MPa[a]	Elongation in 50 mm, %
C18080	0.5 Cr–0.1 Ag–0.1 Fe–0.06Ti–0.03Si	80	515	480	9
C181	0.8 Cr–0.15 Zr–0.04 Mg	80	496	455	10
C18135	0.4 Cr–0.4 Cd	92	469	421	12
C182	0.9 Cr	80	460	405	14

[a] To convert MPa to psi, multiply by 145.

stress relaxation resistance and strength. For example, both welding electrode tips and electrical connectors require moderate strength and relatively high conductivity, but the former also requires good softening resistance, whereas the latter also requires good formability. As much as 0.65 wt% chromium can be dissolved in copper to produce a uniform array of pure chromium precipitates during the age hardening treatment. The low room temperature solubility of chromium (<0.03%) ensures good (usually ~85% IACS) conductivity after the subsequent age hardening heat treatment. Moreover, compositions above the maximum solubility level are commercially made in order to take advantage of the dispersion of coarser chromium particles that are useful for control of grain size. Typical properties of chromium–copper alloys in age hardened temper are summarized in Table 24.

7.20. Cu–Ni–Si Alloys. This commercially important alloy family is precipitation hardened through the formation of Ni_2Si for optimum combinations of strength, conductivity, and formability. The composition and processing of C7025 also results in high temperature resistance to both stress relaxation and softening, required for part manufacturing and service of elevated temperature connectors. These alloys are notable also for good general corrosion and stress corrosion resistance that is equivalent to the highly resistant phosphor bronzes. High strength combined with moderate conductivity available from the alloy is important for applications where ohmic (I^2R) heating from high electrical currents are possible.

Cu—Ni—Si alloys are also available in more dilute versions such as C19010. Higher conductivity but lower strength is available from the latter alloy. Table 25 compares mechanical properties of the mill hardened condition of these alloys.

7.21. C726 to C729. Very high strength, nearly equal to that available from beryllium–copper, is possible with precipitation hardened Cu—Ni—Sn alloys. These alloys, referred to in the literature as spinodal alloys are hardened by a precipitation reaction known as spinodal decomposition. They encompass a range of compositions that includes Cu–7.5 Ni–5 Sn (C7265), Cu–9 Ni–6 Sn (C727) and Cu–15 Ni–8 Sn (C729). More dilute compositions such as Cu–4 Ni–4 Sn (C726) do not precipitation harden appreciably, as do C727 and C729. The latter two alloys are not easily hot worked and are usually made by strip casting followed directly by cold working, or by consolidation of alloy powder.

Table 25. **Properties of Precipitation Hardened Cu—Ni—Base Alloys**

UNS	Composition, wt %	Temper	Cond., % IACS	Tensile strength, MPa[a]	0.2% Yield strength, MPa[a]	Elongation in 50 mm, %
C180	2.5 Ni–0.6 Si–0.4 Cr	H50	48	690	525	12
C19010	1.3 Ni–0.25 Si–0.03P	Extra Hard	60	551	517	6
C7025	3 Ni–0.65 Si–0.15 Mg	TM00	40	690	530	12
		TM03	35	785	735	7
C7026	2 Ni–0.5 Si	TM04	50	670	620	6
C7265	7.5 Ni–5 Sn	TH04		1015	945	4
		TM04	15	860	755	12
C729	15 Ni–8 Sn	TH04		1240	1100	3
		TM04	8	860	790	15

[a] To convert MPa to psi, multiply by 145.

Like beryllium copper, C729 is available in the rolled temper condition, for subsequent aging after forming, or in the mill hardened condition. Typical properties of these alloys are shown in Table 25. In addition, these alloys exhibit notably excellent resistance to stress relaxation at high application temperatures, eg, 200°C, and in this respect outperform beryllium–copper. However, the 8% IACS electrical conductivity of the strongest Cu–Ni–Sn composition (C729) is lower than that of C172.

8. Economic Aspects

One of the factors affecting the relative pricing of commercial copper alloys is metal value. The metal value of copper was around $1.65/kg ($0.75/lb) at the end of 2002. Other factors affecting costs include metal value of the constituent alloy elements, order quantities, the supplier, and specified dimensional tolerances. Oxygen-free copper, C102, is sold as a premium product relative to copper. Brass, eg, C260, is less costly than copper because zinc is historically lower priced than copper. Tin-containing phosphor bronzes, such as C510, and nickel containing alloys such as the 700-series are more costly than brass and copper because of higher tin and nickel metal cost, respectively.

9. Applications

Alloy selection for a particular application generally involves consideration of a combination of physical and mechanical properties as well as cost. Heat exchangers, including automobile radiators and domestic heating systems, are prime examples. Alloys used for such applications require not only good corrosion resistance and reasonable thermal conductivity, but also the capabilities of being formed into a variety of shapes and joined easily by soldering or brazing. C110 and C122 coppers are therefore widely used in such applications. C260 brass has also been used extensively in automobile radiators because of the good formability it possesses for manufacture of headers and because of its strength. Copper–nickel alloys such as C706 and C715 find application in condensers used in power utilities and desalination units because of their good corrosion resistance and reasonable conductivity relative to alternative materials.

Electrical interconnections that range from home wall receptacles to miniature connectors used in electronic products all require good formability and other properties specific to their application. Automotive connectors are frequently situated in the harsh environment of the engine compartment, where resistance to stress relaxation is needed to provide stable contact force for reliable performance. Alloys like the beryllium coppers and C7025 are supplanting C260 brass and the phosphor bronzes in such elevated temperature applications. Interconnection systems, which carry large currents, require high electrical conductivity to minimize ohmic (I^2R) heating. Unacceptable loss in contact force due to stress relaxation can result in open-circuit failure of the connector if the operating temperature is too high for the alloy being used. The high copper alloys are therefore to be favored in such applications, provided that they also have adequate stress relaxation resistance. Alloys C151, C18080, C182, and C194 can more easily meet these property requirements. Computers, consumer electronics and telecommunications systems require connectors which carry only small, signal level, currents. The miniaturization trend in these applications, however, results in a requirement for copper alloys with yield strength >100 ksi, conductivity >45% IACS and good formability; eg, C7025 and C17460.

Strength, thermal conductivity and formability account for selection of particular alloys for use as leadframes in plastic encapsulated devices. Leads having high strength are needed to ensure against damage to the electronic package during handling before and during assembly to circuit boards. This requirement is particularly critical for high lead count packages where the leadframe material is thin (∼0.15 mm) and leads are extremely narrow (0.30 mm). The leadframe also provides a path for heat to be dissipated from the operating integrated circuit (chip) to prolong its service lifetime. Alloys that provide high strength near 690 MPa (100,000 psi) yield strength combined with high thermal conductivity (at least 175 W/(m·K), equivalent to around 40% IACS electrical conductivity) are preferred for such applications.

Other applications that take advantage of the high electrical and thermal conductivity of copper and its alloys include superconducting wires and RF/EMI shielding. Controlled electrical resistance is required for applications such as resistance heating as in heated car seats.

An additional virtue of many copper alloys is the capacity for deep drawing such as cartridge cases and flexible bellows. C260 and C510 are well known for their drawability. Be–Cu that is drawn and then aged is used for bellows that have excellent fatigue life.

Finally, coinage and architectural uses take advantage of the variety of distinctive colors inherent to copper and its alloys, from red, through gold-yellow and silver-white, as well as tarnish and corrosion resistance. Among the alloys used are C713, a white-appearing alloy used in U.S. coinage; gold-appearing aluminum–bronze (C6155) and brass (C260); and red-brown copper–tin bronze. Clad coinage having dissimilar copper alloys for the clad and core components have the further advantage of being difficult to counterfeit when used in vending machines. The worldwide trend has been to use coins to replace low denomination paper bills because the useful lifetime of coins is much longer.

Finally, it is important to note that the ability to be plated or coated with a wide range of materials including tin, solder, chromium, nickel, gold and palladium is an important attribute of copper alloys which contributes to their selection for many applications. For example automotive connectors may be tin or solder coated, whereas electronics connectors are nickel and gold plated with palladium used also. Decorative brass hardware may receive a chromium coating for cosmetic purposes.

BIBLIOGRAPHY

"Copper Alloys, Wrought" in *ECT* 1st ed., Vol. 4, pp. 431–458, by R. E. Ricksecker, Chase Brass & Copper Co.; "Copper Alloys (Wrought)" in *ECT* 2nd ed., Vol. 6, pp. 181–244, by R. E. Ricksecker, Chase Brass & Copper Co.; in *ECT* 3rd ed., Vol. 7, pp. 1–68, by R. E. Ricksecker, Chase Brass & Copper Co., Inc.; in *ECT* 4th ed., pp. 429–473; by John F. Breedis and Ronald N. Caron, Olin Corporation; "Copper Alloys, Wrought Copper and Wrought Copper Alloys" in *ECT* (online), posting date: December 4, 2000, by John F. Breedis and Ronald N. Caron, Olin Corporation.

CITED PUBLICATIONS

1. *Standard Practice for Temper Designations for Copper and Copper Alloys—Wrought and Cast*, ASTM B 601–92, American Society for Testing and Materials, Philadelphia, Pa., 1992.
2. "Alloy Phase Diagrams," Vol. 3, *Metals Handbook*, 10th ed., ASM International, Materials Park, Ohio, 1992.
3. T. B. Massalski, ed., *Binary Alloy Phase Diagrams*, Vol. 1, ASM International, Materials Park, Ohio, 1986.
4. M. M. Shea and N. S. Stoloff, *Met. Trans.* **5**, 755 (1974).
5. R. L. Fleischer, *Acta Met.* **11**, 203 (1963).
6. *Standards Handbook, Part—Alloy Data Wrought Copper and Copper Alloy Mill Products*, 8th ed., Copper Development Association, Greenwich, Conn., 1985.
7. *The Mechanical Properties of Copper–Beryllium Alloy Strip*, ASTM STP 367, American Society for Testing and Materials, Philadelphia, Pa., 1964.
8. *Standard Test Method for Resistivity of Electrical Conductor Materials*, ASTM B 193-87, American Society for Testing and Materials, Philadelphia, Pa., 1992.
9. F. Pawlek and K. Reichel, *Zeit. Metallk.* **47**, 347 (1956).
10. Y. T. Hsu and B. O'Reilly, *J. Met.* **29**, 21 (1977).
11. C. Drapier, *Rev. Metal.* **75**, 699 (1978).
12. J. W. Borough, *Trans. Met. Soc. AIME* **221**, 1274 (1961).
13. Forming and Forging, Vol. 14, *Metals Handbook*, ASM International, Materials Park, Ohio, 1988, pp. 575–590, 809–824, 877–899.
14. E. Shapiro and F. N. Mandigo, *Forming Limit Analysis for Enhanced Fabrication*, INCRA Monograph X, *The Metallurgy of Copper*, INCRA, New York, 1983.
15. K. Suzuki, K. Ueda, and M. Tsuji, *Trans. Jpn. Inst. Met.* **25**, 716 (1984).
16. D. V. Wilson, *J. Inst. Met.* **94**, 84 (1966).

17. W. A. Backofen, *Met. Trans.* **4**, 2679 (1973).
18. K. Laue and H. Stenger, *Extrusion: Processes, Machinery, Tooling*, ASM, Materials Park, Ohio, 1976.
19. W. L. Finlay, *Silver-Bearing Copper*, Corinthian Edition, Copper Range Co., New York, 1968.
20. J. H. Mendenhall, *Understanding Copper Alloys*, Winchester Press, New York, 1977.
21. A. Fox and E. O. Fuchs, *J. Testing Eval.* **6**, 211 (1978).
22. A. Fox, *J. Testing Eval.* **2**, 32 (1974).
23. *Standard Methods for Stress Relaxation Tests for Materials and Structures*, ASTM E 328-86, American Society for Testing and Materials, Philadelphia, Pa., 1992.
24. *Fatigue Crack Propagation*, ASTM STP 415, American Society for Testing and Materials, Philadelphia, Pa., 1967.
25. A. Fox, *J. Mater. JMLSA* **5**, 273 (1970).
26. C. E. Feltner and C. Laird, *Acta Met.* **15**, 1621 (1967).
27. J. P. Hickerson and R. W. Hertzberg, *Met. Trans.* **3**, 179 (1972).
28. P. O. Kettunen and U. F. Kocks, *Acta Met.* **20**, 95 (1972).
29. *Fatigue of Copper Alloys*, Vol. 19, *Metals Handbook*, Fatigue and Facture, pp. 869–873.
30. *Lead Integrity*, Method 2004, Standard 883C, Dept. of Defense, Washington, D.C., 1977.
31. *Factors Affecting Lead Bend Fatigue in P-DIPS*, D. Mahulikar and T. Hann, Microelectronics and Processing Engineers Conference, 1985.
32. *Corrosion*, Vol. 13. *Metals Handbook*, 9th ed., ASM International, Materials Park, Ohio, 1987.
33. J. M. Popplewell and T. V. Gearing, *Corrosion* **31**, 279 (1975).
34. A. Sterling, A. Atrens, and I. O. Smith, *Br. Corros. J.* **25**, 271 (1990).
35. K. D. Efird, *Corrosion* **33**, 3 (1977).
36. R. H. Heidersbach and E. D. Verink, *Corrosion* **28**, 397 (1972).
37. *Standard Test Methods for Use of Mattsson's Solution of pH 7. 2 to Evaluate the Stress Corrosion Cracking Susceptibility of Copper–Zinc Alloys*, ASTM G 37-85, American Society for Testing and Materials, Philadelphia, Pa., 1992.
38. *Standard Practice for Making and Using U-Bend Stress–Corrosion Test Specimens*, ASTM G 30-84, American Society for Testing and Materials, Philadelphia, Pa., 1992.
39. S. Harper, V. A. Callcut, D. W. Townsend, and R. Eborall, *J. Inst. Metals* **90**, 423 (1961).
40. M. Koiwa, A. Yamanaka, A. Arita, and H. Numakura, *Mater. Trans., Jpn. Inst. Metals* **30**, 991 (1989).
41. *Standard Methods of Test for Hydrogen Embrittlement of Copper*, ASTM B 577, American Society for Testing and Materials, Philadelphia, Pa., 1992.
42. *Welding, Brazing, and Soldering*, Vol. 6, *Metals Handbook*, 9th ed., ASM International, Materials Park, Ohio, 1983.
43. *Wetting Balance Solderability*, Method 2022.2, Military Standard-883C, Dept. of Defense, Washington, D.C., 1987.
44. *Solderability*, Method 2003.5, Military Standard –883C, Test Methods and Procedures for Microelectronics, Dept. of Defense, Washington, D.C., 1987.
45. S. P. Zarlingo and J. C. Fister, "Solderability Performance of Tin Coated Copper Strip for Connector Components," *Proceedings, 41st Electronic Components and Technology Conference*, Atlanta, Ga., IEEE, 1991, p. 229.
46. *Specification for Electrodeposited Coatings of Tin*, ATM B 545-83, American Society for Testing and Materials, Philadelphia, Pa., 1992.
47. *Specification for Electrodeposited Coatings of Tin –Lead Alloy (Solder Plate)*, ASTM B 579-82, American Society for Testing and Materials, Philadelphia, Pa., 1992.

48. *Machining*, Vol. 16, *Metals Handbook*, 9th ed., ASM International, Materials Park, Ohio, 1989, pp. 805–819.
49. E. W. Palmer and F. H. Wilson, *J. Metal.* **4**, 55 (1952).

GENERAL REFERENCES

References 13,17–20 are good general references.

Properties and Selection: Nonferrous Alloys and Special Purpose Materials, Vol. 2, *Metals Handbook*, 10th ed., ASM International, Materials Park, Ohio, 1990, pp. 215–345.

Materials Selection and Designs, Vol. 20, *Metals Handbook*, 10th ed., ASM International, Materials Park, Ohio, 1997.

Source Book on Copper and Copper Alloys, ASM International, Materials Park, Ohio, 1979.

Mechanical Testing, Vol. 8, *Metals Handbook*, 9th ed., ASM International, Materials Park, Ohio, 1985.

Powder Metallurgy, Vol. 7, *Metals Handbook*, 9th ed., ASM International, Materials Park, Ohio, 1984.

Copper '86 Copper Tomorrow, Conference Proceedings, Europa Metalli-LMI, Florence, Italy, 1986.

Copper '90 Refining, Fabrication, Markets, Conference Proceedings, Vasteras, Sweden, Institute of Metals, London, 1990.

Application Data Sheet—Standard Designations for Wrought and Cast Copper and Copper Alloys, Copper Development Association, Greenwich, Conn., 1992.

Copper and Copper-Base Powder Alloys, Metal Powder Industries Federation, Princeton, N.J., 1976.

E. A. Brandes and G. B. Brook, eds., *Smithells Metals Reference Book*, 7th ed., Butterworths, London, 1992.

Annual Book of ASTM Standards, Vol. 02. 01, *Copper and Copper Alloys*, American Society for Testing and Materials, Philadelphia, Pa., 1992.

S. H. Butt and J. M. Popplewell, "Corrosion Considerations on the Selection of Materials in Automotive Terminal Systems," S. A. E. Paper 700031, Automotive Engineering Congress, Detroit, Mich., Jan. 1970.

K. Dies, *Kupfer und Kupferlegierungen in der Technik*, Springer-Verlag, Berlin, 1967.

G. R. Gohn, J. P. Guerard, and H. S. Freynik, *The Mechanical Properties of Wrought Phosphor Bronze Alloys*, ASTM STP 183, American Society for Testing and Materials, Philadelphia, Pa., 1956.

Leadframe Materials, Electronic Materials Handbook, Vol. 1 Packaging, ASM International, 1989, pp. 483–492.

Materials Issues for Advanced Electronic and Optoelectronic Connectors, TMS Symposium, 1990.

RONALD N. CARON
Olin Corporation

PETER W. ROBINSON
Olin Corporation

COPPER COMPOUNDS

1. Introduction

Copper compounds, which represent only a small percentage of all copper production, play key roles in both industry and the biosphere. Copper [7440-50-8], mol wt = 63.546, $[Ar]3d^{10}4s^1$, is a member of the first transition series and much of its chemistry is associated with the copper(II) ion [15158-11-9], $[Ar]3d^9$. Copper forms compounds of commercial interest in the +1 and +2 oxidation states. The standard reduction potentials, E^0, for the reasonably attainable valence states of copper are

$$Cu^+ + e^- \longrightarrow Cu^\circ \quad E^0 = 0.52 \text{ V}$$

$$Cu^{2+} + 2e^- \longrightarrow Cu^\circ \quad E^0 = 0.34 \text{ V}$$

$$Cu^{2+} + e^- \longrightarrow Cu^+ \quad E^0 = 0.15 \text{ V}$$

$$Cu^{3+} + e^- \longrightarrow Cu^{2+} \quad E^0 = 1.80 \text{ V}$$

The copper(I) ion [17493-86-6] disproportionates spontaneously, $pK = -5.95$, in aqueous solution according to

$$2\ Cu^+ \longrightarrow Cu^\circ + Cu^{2+} \quad \Delta E^0 = +0.37 \text{ V}$$

The concentration of copper(I) ion remaining in solution is not appreciable. However, aqueous copper(I) ion can be stabilized by complex formation with various agents such as chloride, ammonia, cyanide, or acetonitrile.

The tri- or tetraamine complex of copper(I), prepared by reduction of the copper(II) tetraamine complex with copper metal, is quite stable in the absence of air. If the solution is acidified with a noncomplexing acid, the formation of copper metal, and copper(II) ion, is immediate. If hydrochloric acid is used for the neutralization of the ammonia, the insoluble cuprous chloride [7758-89-6], CuCl, is precipitated initially, followed by formation of the soluble ions $[CuCl_3]^{2-}$, $[CuCl_4]^{3-}$, and $[CuCl_5]^{4-}$ as acid is increased in the system.

The copper(I) ion, electronic structure $[Ar]3d^{10}$, is diamagnetic and colorless. Certain compounds such as cuprous oxide [1317-39-1] or cuprous sulfide [22205-45-4] are intensely colored, however, because of metal-to-ligand charge-transfer bands. Copper(I) is isoelectronic with zinc(II) and has similar stereochemistry. The preferred configuration is tetrahedral. Linear and trigonal planar structures are not uncommon, in part because the stereochemistry about the metal is determined by steric as well as electronic requirements of the ligands (see COORDINATION COMPOUNDS).

The stereochemical preference of the copper(II) ion is square planar or distorted octahedral because of the ligand field stabilization that arises from the $3d^9$ electronic configuration. This perturbation in an octahedral symmetry is known as Jahn-Teller distortion. Other configurations that occur for copper(II) include distorted tetrahedrons as well as a variety of five coordinate structures. Most copper(II) compounds are blue or green in color and exhibit a variety of magnetic phenomena (1). The majority of copper(II) compounds exhibit paramagnetic behavior as a result of the single unpaired $3d$ electron. There are, however, a significant number of polynuclear copper compounds that are sufficiently condensed to show spin–spin coupling of the unpaired electrons. This spin pairing may be so weak as to be observed only at near absolute zero temperatures or it may be strong enough to render the compound diamagnetic at room temperature or above. There have been reports of ferromagnetic polynuclear compounds as well. Probably the most significant, has been the high temperature superconductivity of copper oxide-containing materials (2,3).

2. Properties and Manufacture of Commercially Important Compounds

2.1. Copper(II) Carbonate Hydroxide.
Basic copper carbonate, also named copper(II) carbonate hydroxide [12069-69-1], occurs in nature as the green monoclinic mineral malachite. The approximate stoichiometry is $CuCO_3 \cdot Cu(OH)_2$. There are two grades available commercially, the light and the dense. The light grade is produced by adding a copper salt solution to a concentrated solution of sodium carbonate, usually at 45–65°C. The blue, voluminous azurite [12070-39-2], $C_2H_2Cu_3O_8$, forms initially and converts to the green malachite within two hours. The dense product can be produced by boiling an ammoniacal solution to copper(II) carbonate (4) or by addition of a copper salt solution to sodium bicarbonate at 45–65°C. A dense product can also be produced by simultaneous addition of copper(II) salt solutions and soda ash solutions at controlled pH. Pure $CuCO_3$ has not been isolated.

Basic copper carbonate is essentially insoluble in water, but dissolves in aqueous ammonia or alkali metal cyanide solutions. It dissolves readily in mineral acids and warm acetic acid to form the corresponding salt solution.

2.2. Copper Chloride.
Copper(I) chloride, CuCl, is a colorless or gray cubic crystal and occurs in nature as the mineral nantokite [14708-85-1]. The commercial product is white to gray to brown to green and of variable purity. Copper(I) chloride is usually produced at 450–900°C by direct combination of copper metal and chloride gas to yield a molten product (5–8). Once the reaction is initiated by heat it is self-sustaining and must be cooled. The molten product is variously cast, prilled, flaked, or ground depending on final use. Copper(I) chloride can be produced hydrometallurgically by reduction of copper(II) in the presence of chloride (9):

$$2\ CuCl_2 + \text{reducing agent} \longrightarrow 2\ CuCl + 2\ HCl + \text{oxidation product}$$

where the reducing agent can be sulfite, metallic copper, phosphorus acid, hydroxylamine, or zinc (10).

Copper(I) chloride is insoluble to slightly soluble in water. Solubility values between 0.001 and 0.1 g/L have been reported. Hot water hydrolyzes the material to copper(I) oxide. CuCl is insoluble in dilute sulfuric and nitric acids, but forms solutions of complex compounds with hydrochloric acid, ammonia, and alkali halide. Copper(I) chloride is fairly stable in air at relative humidities of less than 50%, but quickly decomposes in the presence of air and moisture.

Cupric chloride or copper(II) chloride [7447-39-4], $CuCl_2$, is usually prepared by dehydration of the dihydrate at 120°C. The anhydrous product is a deliquescent, monoclinic yellow crystal that forms the blue-green orthohombic, bipyramidal dihydrate in moist air. Both products are available commercially. The dihydrate can be prepared by reaction of copper carbonate, hydroxide, or oxide and hydrochloric acid followed by crystallization. The commercial preparation uses a tower packed with copper. An aqueous solution of copper(II) chloride is circulated through the tower and chlorine gas is sparged into the bottom of the tower to effect oxidation of the copper metal. Hydrochloric acid or hydrogen chloride is used to prevent hydrolysis of the copper(II) (11,12). Copper(II) chloride is very soluble in water and soluble in methanol, ethanol, and acetone.

Copper(II) oxychloride [1332-65-6], $Cu_2Cl(OH)_3$, is found in nature as the green hexagonal paratacamite [12186-00-4] or rhombic atacamite [1306-85-0]. It is usually precipitated by air oxidation of a concentrated sodium chloride solution of copper(I) chloride (13–15). Often the solution is circulated through a packed tower of copper metal, heated to 60–90°C, and aerated.

$$CuCl_2 + Cu \longrightarrow 2\,CuCl$$

$$6\,CuCl + 3\,H_2O + 3/2\,O_2 \longrightarrow CuCl_2 \cdot 3\,Cu(OH)_2 + 2\,CuCl_2$$

The mother liquor is separated from the product and returned to the tower. Copper(II) oxychloride is insoluble in water, but dissolves readily in mineral acids or warm acetic acid. The product dissolves in ammonia and alkali cyanide solution upon the formation of coordination complexes.

2.3. Copper(II) Fluorides. Copper(II) forms several stable fluorides, eg, cupric fluoride [7789-19-7], CuF_2, copper(II) fluoride dihydrate [13454-88-1], $CuF_2 \cdot 2H_2O$, and copper hydroxyfluoride [13867-72-6], CuOHF, all of which are interconvertible. When CuF_2 is exposed to moisture, it readily forms the dihydrate, and when the latter is heated in the absence of HF, CuOHF $\cdot$ H_2O results. The colorless crystals of anhydrous CuF_2 are triclinic in structure and are moisture sensitive, turning blue when exposed to moist air. Physical properties of CuF_2 are listed in Table 1. CuF_2 reacts with ammonia to form $CuF_2 \cdot 5NH_3$.

Several methods of synthesis for anhydrous CuF_2 have been reported, the most convenient and economical of which is the reaction of copper carbonate and anhydrous hydrogen fluoride to form the monohydrate, $CuF_2 \cdot H_2O$. Part of the water content from the monohydrate is removed by addition of excess HF. The excess HF is decanted and the remaining mass transferred to a Teflon-lined tray and dried under an atmosphere of hydrogen fluoride. The decanted material may also be dehydrated in a nickel or copper tray under an atmosphere of fluorine at 150–300°C. Both routes have successfully resulted in ultrapure (99.95%) white CuF_2 in good yields. The other method for the preparation of high purity anhydrous copper(II) fluoride is by the direct fluorination of commercially available CuOHF (16), or the action of a mixture of HF and BF_3 on $CuF_2 \cdot 2H_2O$ (17).

2.4. Copper Hydroxide. Copper(II) hydroxide [20427-59-2], $Cu(OH)_2$, produced by reaction of a copper salt solution and sodium hydroxide, is a blue,

Table 1. **Physical Properties of CuF$_2$**

Property	Value
molecular weight	101.54
melting point, °C	785 ± 10
boiling point, °C	1676
solubility, g/100 g	
water	4.75
anhydrous HF	0.01
aqueous 21.2% HF	12.1
density, g/cm^3	4.85
ΔH_f, kJ/mol^a	−539
ΔG_f, kJ/mol^a	−492
S, J/(mol · K)^a	77.45
C_p, J/(mol · K)^a	65.55

^{a} To convert from J to cal, divide by 4.184.

gelatinous, voluminous precipitate of limited stability. The thermodynamically unstable copper hydroxide can be kinetically stabilized by a suitable production method. Usually ammonia or phosphates are incorporated into the hydroxide to produce a color-stable product. The ammonia processed copper hydroxide (18–21) is almost stoichiometric and copper content as high as 64% is not uncommon. The phosphate produced material (22,23) is lower in copper (57–59%) and has a finer particle size and higher surface area than the ammonia processed hydroxide. Other methods of production generally rely on the formation of an insoluble copper precursor prior to the formation of the hydroxide (24–28).

Copper hydroxide is almost insoluble in water (3 µg/L) but readily dissolves in mineral acids and ammonia forming salt solutions or copper ammine complexes. The hydroxide is somewhat amphoteric dissolving in excess sodium hydroxide solution to form trihydroxycuprate [37830-77-6], [Cu(OH)$_3$]$^-$, and tetrahydroxycuprate [17949-75-6], [Cu(OH)$_4$]$^-$.

2.5. Copper Nitrates. The trihydrate [10031-43-3] crystallizes as blue rhombic plates. Copper(II) nitrate hexahydrate [13478-38-1], Cu(NO$_3$)$_2$ · 6H$_2$O, is produced by crystallization from solutions below the transition point of 26.4°C. A basic copper nitrate [12158-75-7], Cu$_2$(NO$_3$)(OH)$_3$, rather than the anhydrous product is produced on dehydration of the hydrated salts. The most common commercial forms for copper nitrate are the trihydrate and solutions containing about 14% copper. Copper nitrate can be prepared by dissolution of the carbonate, hydroxide, or oxides in nitric acid. Nitric acid vigorously attacks copper metal to give the nitrate and evolution of nitrogen oxides.

$$Cu + 4\ HNO_3 \longrightarrow Cu(NO_3)_2 + 2\ H_2O + 2\ NO_2$$

$$3\ Cu + 8\ HNO_3 \longrightarrow 3\ Cu(NO_3)_2 + 4\ H_2O + 2\ NO$$

The first reaction is favored at high temperatures and in the presence of concentrated acid.

The trihydrate is very soluble in water and ethanol. Decomposition begins around 80°C upon formation of the basic salt. At temperatures of 180°C the oxide is produced.

2.6. Copper Oxides.　Copper(I) oxide [1317-39-1] is a cubic or octahedral naturally occurring mineral known as cuprite [1308-76-5]. It is red or reddish brown in color. Commercially prepared copper(I) oxides vary in color from yellow to orange to red to purple as particle size increases. Usually copper(I) oxide is prepared by pyrometallurgical methods. It is prepared by heating copper powder in air above 1030°C or by blending copper(II) oxide with carbon and heating to 750°C in an inert atmosphere. A particularly air-stable copper(I) oxide is produced when a stoichiometric blend of copper(II) oxide and copper powder are heated to 800–900°C in the absence of oxygen. Lower temperatures can be used if ammonia is added to the gas stream (29–31).

Various hydrometallurgical processes can be used to prepare copper(I) oxide. Acidification with sulfuric acid of ammonia complexes of copper(I) precipitates a red product. Vacuum distillation (32) of $Cu_2(NH_3)_4CO_3$ produces a stable red copper(I) oxide. Addition of sodium hydroxide to the same carbonate initially gives a yellow material which, on heating, turns orange. A boiling slurry of basic copper sulfate and copper sulfate solution can be reduced using sulfur dioxide to give a red product (33). Solutions of sodium copper(I) chloride can be neutralized using sodium hydroxide to give products of various colors depending on conditions (Table 2). Electrolytic processes can produce copper(I) oxide using copper electrodes in brine. Yellow material is produced at room temperature; orange and red products are made as the temperature is increased.

Copper(I) oxide is stable in dry air, but reacts with oxygen to form copper(II) oxide in moist air. Cu_2O is insoluble in water, but dissolves in ammonia or hydrochloric acid. The product disproportionates to copper metal and copper(II) in dilute sulfuric or nitric acid.

Copper(II) oxide [1317-38-0], CuO, is found in nature as the black triclinic tenorite [1317-92-6] or the cubic or tetrahedral paramelaconite [71276-37-4]. Commercially available copper(II) oxide is generally black and dense although a brown material of low bulk density can be prepared by decomposition of the carbonate or hydroxide at around 300°C, or by the hydrolysis of hot copper salt solutions with sodium hydroxide. The black product of commerce is most often prepared by evaporation of $Cu(NH_3)_4CO_3$ solutions (37) or by precipitation of copper(II) oxide from hot ammonia solutions by addition of sodium hydroxide. An extremely fine (10–20 nm) copper(II) oxide has been prepared for use as a precursor in superconductors (38).

Table 2. **Copper(I) Oxides from the Reaction 2 NaCuCl₂ + NaOH ⟶ Cu₂O + H₂O + 4 NaCl**

pH	Temperature, °C	Product		
		Color	Particle size, μm	Reference
7.0		yellow	0.4	34
8.5	60	orange	1	35
alkaline	138	red	2.5	36
10.0[a]	55	purple	48	34

[a] Reactants added simultaneously.

Copper(II) oxide is less often prepared by pyrometallurgical means. Copper metal heated in air to 800°C produces the copper(II) oxide. Decomposition of nitrates, carbonates, and hydroxides at various temperatures also occurs.

Copper(II) oxide is insoluble in water, but readily dissolves in mineral acid or in hot formic or acetic acids. CuO slowly dissolves in ammonia solution, but alkaline ammonium carbonate solubilizes it quickly.

2.7. Copper(II) Sulfates. Copper(II) sulfate pentahydrate [7758-99-8], $CuSO_4 \cdot 5H_2O$, occurs in nature as the blue triclinic crystalline mineral chalcanthite [13817-21-5]. It is the most common commercial compound of copper. The pentahydrate slowly effloresces in low humidity or above 30.6°C. Above 88°C dehydration occurs rapidly.

$$CuSO_4 \cdot 5\,H_2O \xrightarrow{88°C} CuSO_4 \cdot 3\,H_2O \xrightarrow{114°C} CuSO_4 \cdot H_2O$$

$$\xrightarrow{245°C} CuSO_4 \xrightarrow{340°C} 3\,Cu(OH)_2 \cdot CuSO_4 \xrightarrow{600°C} CuO$$

The solubility and solution density as a function of temperature and acid concentration are given in Figure 1 (39). Crystallization of the pentahydrate can be effected both by concentrated and cooling or by addition of sulfuric acid. The former method is usually preferred because the crystals are of better structural integrity and thus more resistant to hard cake formation. Copper(II) sulfate can be prepared by dissolution of oxides, carbonates, or hydroxides in sulfuric acid solutions. Whereas copper metal does not displace hydrogen from acidic solution, aeration or oxygenation of hot dilute aqueous sulfuric acid in the presence of copper metal is a commonly used commercial method for $CuSO_4$ preparation (40). Solvent extraction is used to produce the pentahydrate from a variety of copper-containing liquors (4,39).

Copper(II) sulfate monohydrate [10257-54-2], $CuSO_4 \cdot H_2O$, which is almost white in color, is hygroscopic and packaging must contain moisture barriers. This product is produced by dehydration of the pentahydrate at 120–150°C.

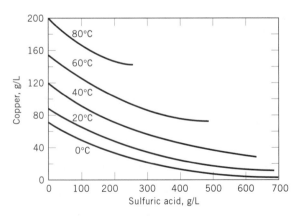

Fig. 1. Aqueous solubility of copper in copper sulfate as a function of sulfuric acid concentration at various temperatures. Reprinted with permission (39).

Trituration of stoichiometric quantities of copper(II) oxide and sulfuric acid can be used to prepare a material of limited purity. The advantages of the monohydrate as opposed to the pentahydrate are lowered freight cost and quickness of solubilization. However, these advantages are offset by the dustiness of the product and probably less than one percent of copper sulfate is used in the monohydrate form.

Anhydrous copper(II) sulfate [7758-98-7] is a gray to white rhombic crystal and occurs in nature as the mineral hydrocyanite. $CuSO_4$ is hygroscopic. It is produced by careful dehydration of the pentahydrate at 250°C. An impure product can also be produced from copper metal and hot sulfuric acid:

$$Cu + 2\ H_2SO_4 \longrightarrow CuSO_4 + SO_2 + 2\ H_2O$$

There are four basic sulfates that can be identified by potentiometric titration using sodium carbonate (41,42): langite [1318-78-1], $CuSO_4 \cdot 3Cu(OH)_2 \cdot H_2O$; brochantite [12068-81-4], $CuSO_4 \cdot 3Cu(OH)_2$; anterlite [12019-54-4], $CuSO_4 \cdot 2Cu(OH)_2$; and $CuSO_4 \cdot CuO \cdot 2Cu(OH)_2 \cdot xH_2O$. The basic copper(II) sulfate that is available commercially is known as the tribasic copper sulfate [12068-81-4], $CuSO_4 \cdot 3Cu(OH)_2$, which occurs as the green monoclinic mineral brochantite. This material is essentially insoluble in water, but dissolves readily in cold dilute mineral acids, warm acetic acid, and ammonia solutions.

Tribasic coppersulfate is usually prepared by reaction of sodium carbonate and copper sulfate. As the temperature of the reaction contents increases so does the size of the resulting particle. For use as a crop fungicide, intermediate (40–60°C) temperatures are used to obtain a fine particle. When lower temperatures are used to precipitate basic copper(II) sulfate, products high in sulfate and water of hydration are obtained.

3. Economic Aspects

Copper sulfate is the most important compound of copper. It is the starting material for the production of many other copper salts. However, 75% of production is used in agriculture, principally as a fungicide.

Today there are more than 100 manufacturers and the world's consumption is 200×10^3 t/yr (43).

Table 3 gives data on U.S. production of copper sulfate. Table 4 give U.S. export data and Table 5 gives U.S. import data for copper sulfate (44).

4. Analytical Methods

The specifications or typical analyses of selected copper compounds are given in Table 6 (45,46).

4.1. Separation. Preliminary separation of copper from a variety of interfering elements by extraction using diethyldithiocarbamate or dithizone has been practiced since the early 1940s (47–53) and is still commonly used (54,55). Excellent separations from a complex matrix can be effected using 2-hydroxyoximes (56), and precipitation of copper as the sulfide is commonplace.

Table 3. **U.S. Salient Copper Sulfate Statistics, t**[a,b]

Copper sulfate	1997	1998	1999	2000	2001
production	48,400	44,000	52,700	55,500	55,200
exports					
refined	92,900	86,200	25,200	93,600	22,500
unmanufactured[c]	628,000	412,000	395,000	650,000	556,000
imports					
refined	632,000	683,000	837,000	1,060,000	991,000
unmanufactured[c]	999,000	1,190,000	1,280,000	1,350,000	1,400,000

[a] From Ref. 44.
[b] Data are rounded to no more than three significant digits.
[c] Includes copper content of alloy scrap.

Hydrolysis to give basic salts of copper is quite often used, using collectors such as lanthanum (57) or iron(III) (58,59). Numerous ammonia-insoluble impurities in copper can be separated by coprecipitation with lanthanum using ammonia (60). Selective precipitation with trithiocarbonate ion (61), or concentration

Table 4. **U.S. Exports of Copper Sulfate by Country**[a,b]

Country	Quantity, t	Value, $\times 10^3$ \$
2000	10,300	25,100
2001:		
Australia	276	656
Brazil	49	134
Canada	843	1,620
Chile	302	620
China	693	1,510
Denmark	348	773
France	95	263
Germany	24	118
Hong Kong	226	420
India	1	3
Indonesia	392	828
Italy	197	480
Japan	992	2,370
Korea, Republic of	2,070	4,550
Mexico	110	246
Netherlands	116	253
New Zealand	169	392
Saudi Arabia	20	46
Singapore	559	1,300
Sweden	1,420	3,380
Taiwan	599	1,400
United Kingdom	1,250	3,110
Venezuela	8	22
other	239	600
Total	*11,000*	*25,100*

[a] From Ref. 44.
[b] Data are rounded to no more than three significant digits;
may not add to totals shown.

Source: U.S. Census Bureau.

Table 5. **U.S. Imports for Consumption of Copper Sulfate by Country**[a,b]

Country	Quantity, t	Value, $\times 10^3$ \$[c]
2000	2,550	9,910
2001:		
Australia	2,150	4,020
France	69	188
Germany	99	359
Hong Kong	195	220
Japan	5	70
Mexico	2,090	3,720
United Kingdom	6	43
other	35	68
Total	*4,650*	*8,680*

[a] From Ref. 44.
[b] Data are rounded to no more than three significant digits; may not add to totals shown.
[c] C.i.f. value at U.S. port.
Source: U.S. Census Bureau.

using ammonium pyrrolidine dithiocarbamate [5108-96-3] (APDC) or dithiooxamide (62) can be used. Ion exchange (qv) using Chelex 100 has been used to preconcentrate trace copper (63–66) Chelating ion exchangers have been developed for selective preconcentration (67) or are used in hydrometallurgical applications (68) (see CHELATING AGENTS; METALLURGY; TRACE AND RESIDUE ANALYSIS).

4.2. Determination. The most accurate (69) method for the determination of copper in its compounds is by electrogravimetry from a sulfuric and nitric acid solution (46). Pure copper compounds can be readily titrated using ethylene diamine tetracetic acid (EDTA) to a SNAZOXS or Murexide endpoint. Iodometric titration using sodium thiosulfate to a starch–iodide endpoint is one of the most common methods used industrially. This latter titration is quicker than electrolysis, almost as accurate, and much more tolerant of impurities than is the titration with EDTA. Gravimetry as the thiocyanate has also been used (69).

Table 6. **Analyses of Copper Compounds**

	Composition, wt%				
Assay	$Cu(OH)_2CuCO_3$	$Cu(NO_3)_2 \cdot 3\,H_2O$	CuO	$CuSO_4 \cdot 5\,H_2O$	$CuSO_4 \cdot 5\,H_2O$
grade	light	solution	black	technical	electronic
Cu	55.4	14.0	78.5	25.3	25.5
Fe	0.1	0.002	0.01	0.01	0.003
Pb	0.005	0.002	0.01	0.005	0.001
Zn	0.015	0.0005	0.05	0.01	0.0005
insolubles	0.05[a]		0.02[a]	0.05[b]	0.008[b]
ABD[c], kg/m^3	300	1450	1300	1000	1000

[a] Insoluble in HNO_3.
[b] Insoluble in H_2O.
[c] Apparent or free-falling bulk density.

Colorimetric procedures are often used to determine copper in trace amounts. Extraction of copper using diethyldithiocarbamate can be quite selective (61,63), but the method using dithizone is preferred because of its greater sensitivity and selectivity (51–53). Atomic absorption spectroscopy, atomic emission spectroscopy, x-ray fluorescence, and polargraphy are specific and sensitive methods for the determination of trace level copper.

5. Health and Safety Factors

Copper is one of the twenty-seven elements known to be essential to humans (70–73) (see MINERAL NUTRIENTS). The daily recommended requirement for humans is 2.5–5.0 mg (74). Copper is probably second only to iron as an oxidation catalyst and oxygen carrier in humans (75). It is present in many proteins, such as hemocyanin [9013-32-3], galactose oxidase [9028-79-9], ceruloplasmin [9031-37-2], dopamine β-hydroxylase, monoamine oxidase [9001-66-5], superoxide dismutase [9054-89-1], and phenolase (76,77). Copper aids in photosynthesis and other oxidative processes in plants.

Copper is toxic in exceedingly low concentrations to most fungi, algae, and certain bacteria and can be lethal to higher life forms in relatively high doses. The LD_{50}'s in mg/kg for rats are copper(II) acetate, 710; copper(II) chloride dihydrate, 163; anhydrous copper(II) nitrate, 940; and copper(II) sulfate pentahydrate, 300. For female rats, copper hydroxide has an LD_{50} of 2200 mg/kg. The acute oral toxicity in humans, LD_{LO}, is 100 mg/kg, but recovery from ingestion of 600 mg/kg has occurred. The symptoms of copper poisoning are nausea, vomiting, cramps, gastric disturbances, apathy, anemia, convulsions, coma, and death. Chronic toxicity from copper poisoning has not been definitely observed although there are numerous accounts of chronic poisoning resulting from the refining of copper. This poisoning is now thought to be a result of the common impurities in copper ores such as lead, selenium, and arsenic (78,79). Copper is known to accumulate in the liver, but it is uncertain whether pathological changes occur.

Inhalation of dusts can cause metal fume fever (80,81), and ulceration or perforation of the nasal septum. Mild discomfort has been noted with workplace concentrations as low as 0.08 mg/m^3. The workplace standard ACGIH (TLV) for copper dusts or mist is 1 mg/m^3 and 0.2 mg/m^3 for copper fume (82). OSHA PEL TWA for dust, mist is 1 mg/m^3 and for fumes and respirable particles is 0.1 mg/m^3 (83).

6. Uses

Examples of uses (43,84) of copper compounds are given in Table 7 which lists the materials of primary industrial importance. The majority of copper compounds are used as fungicides, nutritionals, and algicides.

6.1. Foliar Fungicides and Bactericides. Of the ~70,000 t/yr as copper in compounds used in agriculture, almost 75% is used in the control of fungi (see FUNGICIDES, AGRICULTURAL). The first reference to the use of copper as a fungicide dates to 1761 (95) where copper sulfate was used on wheat seed for the control of bunt. In 1807 (96) the discovery of copper as a fungicide was made

Table 7. **Uses of Copper Compounds**

Compound	CAS Registry number	Molecular formula	Uses
copper(I) acetate	[598-54-9]	$CuCH_3COO$	absorption of olefins
copper(II) acetate	[142-71-2]	$Cu(CH_3COO)_2$	fabrics, textiles, pigment, catalyst, paper treatment
copper(II) acetate monohydrate	[6046-93-1]	$Cu(CH_3COO)_2 \cdot H_2O$	manufacture of Paris Green fungicides, pigments, textiles
copper(II) acetate, basic	[52503-63-6]	$Cu(CH_3COO)_2 \cdot CuO \cdot 6H_2O$	insecticides, wood preserving, antifouling pigment
copper(II) arsenate	[7778-41-8]	$Cu_3(AsO_4) \cdot 4H_2O$	catalyst
copper(I) bromide	[7787-70-4]	$CuBr$	catalyst, brominating reagent, intensifier (photography)
copper(II) bromide	[7789-45-9]	$CuBr_2$	fungicides, animal feeds, catalyst, oil treatment
copper(II) carbonate, basic	[12069-69-1]	$CuCO_3 \cdot Cu(OH)_2$	catalyst, absorption of CO, fuel oil treatment
copper(I) chloride	[7758-89-6]	$CuCl$	catalyst, mordant, electroplating, pigment
copper(II) chloride	[7447-39-4]	$CuCl_2$	
copper(II) chloride dihydrate	[10125-13-0]	$CuCl_2 \cdot 2H_2O$	fungicides, pigment
copper(II) chloride hydroxide	[1332-65-6]	$CuCl_2 \cdot 3Cu(OH)_2$	wood preserving, textiles
copper(II) chromate(VI)	[13548-42-0]	$CuCrO_4$	hydrogenation catalyst
copper(II) chromate(III)	[12018-10-9]	$CuCr_2O_4$	electroplating, catalyst
copper(I) cyanide	[544-92-3]	$CuCN$	fluorinating agent, superconductors, cathode material, isomerization, herbicide, termite repellant, catalyst[a]
copper(II) fluoride	[7789-19-7]	CuF_2	
copper(II) formate	[544-19-4]	$Cu(HCO)_2$	mildewcide, bactericide
copper(II) fluoborate	[38465-60-0]	$Cu(BF_4)_2$	electroplating, electronics
copper(II) gluconate	[527-09-3]	$Cu(C_6H_{11}O_7)_2$	dietary supplement, breath freshener
copper(II) hydroxide	[20427-59-2]	$Cu(OH)_2$	fungicides, rayon manufacture, catalyst, antifouling pigment, electrolysis, electroplating
copper(I) iodide	[7681-65-4]	CuI	heat and light stabilizer in polymers, photographic emulsions, and light-sensitive paper, and in oil drilling to aid in corrosion inhibition in highly acid environments; used as feed additive, in cloud seeding, and as a double salt with mercury(II) iodide as a temperature indicator

Table 7 (*Continued*)

Compound	CAS Registry number	Molecular formula	Uses
copper(II) naphthenate	[1338-02-9]		fungicide, mildewcide in textiles, woods, and paints
copper(II) nitrate	[3251-23-8]	$Cu(NO_3)_2$	electrolysis and electroplating, electronics, fuel oil treatment, colorant, pyrotechnics, catalyst
Copper(II) nitrate trihydrate	[10031-43-3]	$Cu(NO_3)_2 \cdot 3H_2O$	fuel oil combustion improver, emulsifier, dispersant,
copper(II) oleate	[1120-44-1]	$C_{18}H_{34}O_2 \cdot 1/2Cu$	antifouling coating for fish nets and lines
copper(II) oxalate	[814-91-5]	CuC_2O_4	catalyst, stabilizer
copper(I) oxide	[1317-39-1]	Cu_2O	antifouling pigment, fungicide, pigment, catalyst, seed dressings
copper(II) oxide	[1317-38-0]	CuO	wood preserving, feed additive, pigment, catalyst, mineral supplements
copper(II) phosphate trihydrate	[7798-23-4]	$Cu_3(PO_4)_2 \cdot 3H_2O$	fungicide, corrosion inhibitor
copper(II) diphosphate hydrate	[10102-90-6]	$Cu_2P_2O_7 \cdot xH_2O$	electroplating plastics, aluminum, and zinc
copper(II) stearate	[660-60-6]	$C_{18}H_{36}O_2 \cdot 1/2Cu$	antifouling paints, wood and textile preservation
copper(II) sulfate	[7758-98-7]	$CuSO_4$	fungicides, algicides, antifouling paints, electrolysis and electroplating, electronics, fuel oil treatment, wood preserving
copper(II) sulfate pentahydrate	[7758-99-8]	$CuSO_4 \cdot 5H_2O$	fungicides
copper(II) sulfate, tribasic	[12068-81-4]	$CuSO_4 \cdot 3Cu(OH)_2$	
copper(I) sulfide	[22205-45-4]	Cu_2S	luminous paints, solar cells, semiconductors, lubricants
copper(II) sulfide	[1317-40-4]	CuS	antifouling pigment, manufacture of aniline black
copper(I) thiocyanate	[1111-67-7]	$CuSCN$	antifouling pigment

[a] From Refs. 85–94.

and the discovery of Bourdeaux mixture (copper sulfate plus lime) followed in 1882.

In order for copper to have any persistance as a foliar fungicide it must be rendered insoluble so that it is not washed off the leaf surface during rainfall. Copper compounds have been used successfully since the latter 1880s on fruit, vegetables, nut crops, and ornamentals. Copper is an effective broad-spectrum fungicide, although its action is more prophylactic in nature.

Whereas copper is not as effective as some organics against specific pathogens, pathogen resistance to copper has been minimal in the >100 year usage. Many organic fungicides are used less often then copper compounds because of residue tolerance, grower reluctance, or for regulatory reasons. Use of copper compounds as fungicides has seen a resurgence in the latter part of the twentieth century.

6.2. Plant and Animal Nutrient.   Copper is one of seven micronutrients that has been identified as essential to the proper growth of plants (97). Cereal crops are by far the most affected by copper deficiency (see WHEAT AND OTHER CEREAL GRAINS). Greenhouse studies have shown yield increases from 38% to over 500% for wheat, barley, and oats (98) using copper supplementation. A tenfold increase in the yield of oats was reported in France (99). Symptoms of copper deficiency vary depending on species, but often it is accompanied by withering or chlorosis in the leaves that is not ammenable to iron supplementation. In high concentrations, particularly in low pH soils, copper can be toxic to plants.

Copper compounds are used as feed additives in Europe and the United States primarily for chickens and swine (see FEEDS AND FEED ADDITIVES) (100,101). Copper increases the rate of gain and feed efficiencies of the animals. It is unclear whether this results from overcoming animal deficiencies or by enhancing preservation of feedstuffs.

6.3. Algicides.   Copper sulfate has been used to control algae in lakes and reservoirs since around 1905. In the United States it is used by every state either on a routine or an occasional basis (102). It provides an effective and ecologically sound method to control algal blooms. Algae cause water supplies to have odor and taste. The turbidity and scum that forms from algal blooms can blind filters and affect certain industries adversely. Significant algal blooms decrease recreational use and are a primary cause of oxygen depletion that results in fish kills.

Copper sulfate is by far the most common algicide. Other copper-containing algicides for use in domestic applications such as swimming pools are usually chelated to prevent hydrolysis and precipitation of the copper.

6.4. Other Uses.   Copper is one of the primary ingredients used in wood preservation. In combination with chromium and arsenic or zinc or borates it has largely replaced pentachlorophenol and creosote. As an antifouling pigment, copper(I) oxide has found renewed use because of regulatory problems of organotin compounds (see COATINGS, MARINE). Copper is much more amenable to the environment than the tin or lead alternatives. Copper phthalocyanines are excellent color-stable blue and green pigments for paints (see PHTHALOCYANINE COMPOUNDS). As an oxidation catalyst, copper with chrome is used extensively. Mixed oxides of chromium and copper show promise as a replacement for platinum group metals in emission control devices. Copper used with zinc or with zinc and chromium is employed as a low temperature shift catalyst (LTS) for the synthesis of fuels and methanol from coal gases (see COAL CONVERSION PROCESSES; FUELS, SYNTHETIC). Electroplating uses copper sulfate baths extensively and electroless copper baths are used to produce circuit boards (see ELECTROLESS PLATING). Copper oxide is used in air-bag technology, and oxides and sulfides are used as solar collectors (see SOLAR ENERGY). Copper compounds are also used as fuel additives to minimize sulfide and carbon monoxide emissions.

BIBLIOGRAPHY

"Copper Compounds" in *ECT* 1st ed., Vol. 4, pp. 467–479, by L. G. Utter and S. B. Tuwiner, Phelps Dodge Corp.; in *ECT* 2nd ed, Vol. 6, pp. 265–280, by E. A. Winter, M. J. Montesinas, and J. E. Singley, in *ECT* 3rd ed., Vol. 7, pp. 97–109, by R. N. Kust, Kennecott Copper Corp.; in *ECT* 4th ed., Vol. 7, pp. 505–520, by H. Wayne Richardson, CP Chemicals, Inc.; "Copper Compounds" in *ECT* (online), posting date: December 4, 2000, by H. Wayne Richardson, CP Chemicals, Inc.; "Copper Compounds" under "Fluorine Compounds, Inorganic" in *ECT* 1st ed., Vol. 6, p. 693, by F. D. Loomis; "Copper" under "Fluorine Compounds, Inorganic" in *ECT* 2nd ed., Vol. 9, pp. 583–584, by W. E. White; in *ECT* 3rd ed., Vol. 10, pp. 719–720, by D. T. Meshri, Advance Research Chemicals Inc.; in *ECT* 4th ed.; Vol. 11, pp. 338–340, by Dayal T. Meshri, Advance Research Chemicals, Inc.; "Fluorine Compounds, Inorganic, Copper" in *ECT* (online), posting date: December 4, 2000, by Dayal T. Meshri, Advance Research Chemicals, Inc.

CITED PUBLICATIONS

1. W. Hatfield and R. Whyman, in R. Carlin, ed., *Transition Metal Chemistry*, Vol. 5, Marcel Dekker, New York, 1969, pp. 47–179.
2. C. P. Poole, Jr., T. Datta, and H. A. Farach, with M. M. Rigney and C. R. Sanders, *Copper Oxide Superconductors*, John Wiley & Sons, Inc., New York, 1988.
3. W. E. Hatfield and J. H. Miller, Jr., eds., *High-Temperature Superconducting Materials, Preparations, Properties, and Processing*, Marcel Dekker, Inc., New York and Basel, 1988. Original paper: J. G. Bedmorz and K. A. Muller, *Z. Phys. B-Condensed Matter* **64**, 189 (1986).
4. C. Merigold, D. Agers, and J. House, *International Solvent Extraction Conference*, London, 1971.
5. Ger. Pat. 1000361 (1955), F. Bittner (to Degussa).
6. Fr. Pat. 2009852 (1969), (to Degussa).
7. Ger. Pat. 1813891 (1958), (to Norddeutsche Affinerie).
8. U.S. Pat. 3,679,359 (1972), E. Haberland and W. Perkow (to Norddeutsche Affinerie).
9. R. Keller and H. Wycoff, *Inorg. Synth.* **2**, 1–4 (1946).
10. Ger. Pat. 3305545 (1983), E. Mack and L. Witzke (to Goldschmidt).
11. Ger. Pat. 1080088 (1958), H. Niemann and K. Herrmann (to Schering Corp.).
12. U.S. Pat. 2,367,153 (1945), C. Swinehart (to Harshaw).
13. U.S. Pat. 3,202,478 (1965), A. Hindel, S. Raval, S. Damani, H. Damani, and K. Damani (to Sudhir Chem. Co.); Brit. Pat. 912,125 (1962).
14. Ger. Pat. 1159914 (1963), E. Podschus (to Bayer AG).
15. PL 55953 (1968),(to Instytut Przemyslu Org.).
16. J. R. Lundquist, *Final Report Pacific Northwest Laboratories*, Seattle, Wash. NASA CR-72571, June 12, 1969; U.S. Pat. 3,607,015 (Sept. 21, 1971), J. R. Lundquist, R. Wash, and R. B. King (to NASA).

17. U.S. Pat. 2,782,099 (Feb. 19, 1957), D. A. McCaulay (to Standard Oil of Indiana).
18. U.S. Pat. 1,800,828 (1931), W. Furness (to Cellocilk Co.).
19. U.S. Pat. 2,104,754 (1938), D. Marsh and B. Marsh.
20. U.S. Pat. 2,536,096 (1951), P. Rowe (to Mountain Copper).
21. U.S. Pat. 25,225,242 (1950) P. Rowe (to Lake Chem. Co.).
22. U.S. Pat. 24,324 (1957), W. Furness (to Copper Research).
23. Ger. Pat. 1592441 (1965), W. Furness (to Kennecott Copper); U.S. Pat. 3,194,749 (1965), (to Kennecott Copper).
24. U.S. Pat. 3,231,464 (1966), E. Dettwiler and J. Filliettaz (to Rohm and Haas).
25. U.S. Pat. 4,418,056 (1983), M. Gonzalez (to Cuproquim S.A.).
26. Brit. Pat. 83/01912 (1983), J. Giulini and A. Meyer (to Giulini Adolfomer Ind.).
27. U.S. Pat. 4,490,337 (1984), H. W. Richardson (to Kocide Chemical Corp.).
28. U.S. Pat. 4,808,406 (1989), N. C. Brinkman (to Kocide Chemical Corp.).
29. U.S. Pat. 3,466,143 (1969), H. Day (to Calumet and Hecla).
30. U.S. Pat. 2,758,014 (1956), J. Drapeau and P. Johnson (to Glidden).
31. U.S. Pat. 2,891,842 (1959), J. Drapeau and P. Johnson (to Glidden).
32. U.S. Pat. 3,492,115 (1970) (to S. Mahalla).
33. U.S. Pat. 2,665,192 (1954) P. Rowe (to Mountain Copper).
34. Jpn. Kokai 78/133775 (1978), (to Nippon Chem.).
35. Brit. Pat. 936,922 (1963), A. Campbell and A. Taylor (to ICI).
36. Jpn. Kokai 80/71629 (1980), (to Nippon Chem.).
37. W. Kunda, H. Veltman, and D. Evans, *Copper Met. Proc. Extr. Met. Div. Symp.*, 27–69 (1970).
38. "Dowa Mining," *Adv. Ceram. Rep. Ang.*(1990).
39. H. Moyer, *AIME Annual Meeting*, New Orleans, La., Feb. 18–22, 1979.
40. U.S. Pat. 2,533,245 (1950), G. Harike (to Tennessee Copper).
41. L. Markov and K. Balarev, *Izv. Khim* **15**, 472–481 (1982).
42. H. Weiser, W. Milligan, and E. Cook, *J. Am. Chem. Soc.* **64**, 503–508(1942).
43. "Uses of Copper Compounds," Copper Development Asssociation, Inc., www.copper.org/compounds, accessed May 2003.
44. D. L. Edelstein, "Copper," *Minerals Year Book*,U.S. Geological Survey, Reston, Va., 2001.
45. Z. Marczenko, *Separation and Spectrophotometric Determination of Elements*, Ellis Horwood Ltd., West Sussex, UK, 1986; E. B. Sandell, *Colorimetric Determination of Traces of Metals*, Interscience Publishers, New York, 1959.
46. C. Freedenthal, *Copper Compounds, Encyclopedia of Industrial Chemical Analysis*, John Wiley & Sons, Inc., New York, 1970, 651–680.
47. L. I. Butler and H. O. Allen, *J. Assoc. Office. Agr. Chem.* **25**, 567 (1942).
48. L. Gerber, R. I. Claassen, and C. S. Boruff, *Ind. Eng. Chem., Anal. Ed.* **14**, 364 (1942).
49. I. Stone, *Ind. Eng. Chem., Anal. Ed.* **14**, 479(1942).
50. T. C. J. Ovenston and C. A. Parker, *Anal. Chim. Acta* **4**, 135 (1950).
51. G. H. Bendix and D. Grabenstetter, *Ind. Eng. Chem., Anal. Ed.* **15**, 649 (1943).
52. S. L. Morrisson and H. L. Paige, *Ind. Eng. Chem., Anal. Ed.* **18**, 211 (1946).
53. P. M. Heertjes, *Chem. Weekblad* **42**, 91 (1946).
54. P. L. Schuller and L. E. Coles, *Pure Appl. Chem.* **51**, 385 (1979).
55. E. A. Allen, P. K. N. Bartlett, and G. Ingram, *Analyst* **109**, 1075 (1984).
56. R. Dubczynski and H. Maleszewska, *Chem. Anal.* **32**, 619 (1987).
57. Z. Marczenko and K. Kasiura, *Chem. Anal. (Warsaw)* **10**, 449 (1965).
58. N. A. Rudnev, G. I. Malofeeva, N. P. Andreeva, and T. V. Tikhonova, *Zh. Analit. Khim.* **26**, 697 (1971).
59. A. I. Novikov, A. A. Shaffert, and E. K. Schekoturova, *Zh. Analit. Khim.* **32**, 1108 (1977).

60. ASTM, *Chemical Analysis of Metals and Metal Bearing Ores*, VO3.05, P195, Philadelphia, Pa., 1987.
61. K. Singh, R. Gupta, and P. Bhatia, *Acta Chim. Hung.* **113**, 3 (1983).
62. R. Santelli, M. Gallego, and M. Valcarcel, *Anal. Chem.* **61**, 1427 (1989).
63. W. Van Berkel, A. Overbosch, G. Feenstra, and F. Maessen, *J. Anal. At. Spectrom.* **3**, 249 (1988).
64. S. Pai, T. Chen, and G. Wong, *Anal Chem.* **62**, 774 (1990).
65. J. P. Riley and D. Taylor, *Anal. Chim. Acta* **40**, 479 (1968).
66. H. Watanabe and H. Ohmori, *Talanta* **28**, 774 (1981).
67. A. Warshawsky, in M. Streat and D. Naden, eds., *Ion Exchange and Sorption Processes in Hydrometallurgy*, Society of Chem Industry, Chichester, UK, 1987, p. 166.
68. J. Melling and D. West, in D. Naden and M. Streat, eds., *Ion Exchange Technology*, Ellis Horwood Ltd., UK, p. 724.
69. I. Sarudi, *Z. Anal. Chem.* **130**, 301 (1950).
70. H. E. Stokinger, *Patty* **2A**, 1620–1630 (1981).
71. S. Cohen, *JOM, J. Occup. Med.* **16**, 621–624 (1974).
72. W. Dreichmann and H. Gerarde: *Toxicology of Drugs and Chemicals*, Academic Press, Inc., New York, 1969.
73. W. Mertz, *Nutr. Today* **18**, 26 (1983).
74. *A Critical Review of Copper in Medicine Report No. 234*, International Copper Research Association, Inc., Goteborg-New York, 1975.
75. D. Kertesz, R. Zito, and F. Ghiretti, in O. Hayaishi, ed., *Oxygenases*, Academic Press, Inc., London, 1962.
76. *Biological Roles of Copper*, no. 79, Ciba Foundation Symposium, Excerpta Medica and Elsevier-North Holland, Amsterdam, The Netherlands, 1980, p. 343.
77. H. Siegel, ed., *Metal Ions in Biological Systems*, Vol. 12, *Properties of Copper*, Marcel Dekker, New York, 1981, p. 384.
78. E. Browning, *Toxicity of Industrial Metals*, 2nd ed., Butterworths, London, 1969.
79. R. Fabre and R. Truhaut, *Precis de Toxicologie*, Centre Doc. University, Paris, 1960.
80. A. Askergren and M. Mellgren, *Scand. J. Work, Environ. Health* **1**, 45–49 (1975).
81. R. Gleason, *Am. Ind. Hyg. Assoc. J.* **29**, 375–376 (1968).
82. *American Conference of Governmental Industrial Hygienists* (ACGIH), Cincinnati, Ohio, 1991, p. 121.
83. R. J. Lewis, Sr., *Sax's Dangerous Properties of Industrial Materials*, 10th ed., Vol. 2, John Wiley & Sons, Inc., New York, 2000.
84. H. W. Richardson, ed., *Copper Compounds Application Handbook*, Marcel Dekker, New York, 1992.
85. Jpn. Kokai Tokkyo Koho, 02, 302,311(Dec. 14, 1990), I. Harada, M. Aritsuka, and A. Yoshikawa (to Mitsui Tiatsu Chemicals).
86. B. Leng and J. H. Moss, *J. Flourine Chem.* **8**, 165 (1976).
87. J. H. Moss, R. Ottie, and J. B. Wilford, *J. Fluorine Chem.* **3** 317 (1973).
88. Jpn. Kokai Tokkyo Koho 01, 133,921 (Nov. 18, 1987), S. Aoki and co-workers(to Fujikura Ltd.).
89. Jpn. Kokai Tokkyo Koho, 63,313,426 (Dec. 21, 1988), Y. Tanaka, T. Shibata, and N. Uno (to Furukawa Electric Co. Ltd.).
90. Jpn. Kokai Tokkyo Koho, 63,288,943 (Nov. 25, 1988), T. Kyodo, S. Hirai, and K. Takahashi (to Sumitomo Electric Industries Ltd.).
91. I. G. Ryss, *The Chemistry of Fluorine and its Inorganic Compounds*, State Publishing House for Scientific and Chemical Literature, Moscow, 1956, Eng. Trans. ACE-Tr-3927, Vol. II, Office of Technical Services, U.S. Department of Commerce, Washington, D.C., 1960, p. 643.
92. G. N. Wolcott, *P. R. Agri. Exp. Stu. Bull.* **73** (1947).

93. H. Martin, R. L. Wain, and E. H. Wilkinson, *Ann. Appl. Biol.* **29**, 412 (1942).

94. Jpn. Kokai Tokkyo Koho, 63, 49,255 (Mar. 2, 1988), Y. Kawasaki (to Matsushita Electric Industrial Co. Ltd.).

95. H. Shulthess, *Abhandl. Naturf. Gesell. Zurich*,**1**, 498 (1761).

96. *Prevost, Phytopath. Classic* **6**, 1 (1807).

97. U. Gupta, "Copper in Agricultural Crops," in J. O. Nriagu, ed., *Copper in the Environment*, Part 1, John Wiley & Sons, Inc., New York, 1979.

98. U. Gupta and L. B. McLeod, *Can. J. Soil. Sci.* **50**, 373–378 (1970).

99. L. Duval, *C. R. Acad. Agric. Fr.* **49**, 1216–1220 (1963).

100. G. Cromwell, T. Stahly, and W. Williams, *Feedstuffs* **53**, 30, 32, 35, 36 (1982).

101. L. DeGoey, R. Wahlstrom, and R. Emerick, *J. Anim. Sci.* **33**, 52–57(1971).

102. D. M. McKnight, S. W. Chrisholm, F. M. M. Morel, INCRA Project No. 252, *Copper Sulfate Treatment of Lakes and Reservoirs: Chemical and Biological Considerations*, Ralph M. Parsons Lab, Division of Water Resources & Hydrodynamics, Massachusetts Institute of Technology, Cambridge, Mass., 1981.

H. Wayne Richardson
CP Chemicals, Inc.
Dayal T. Meshri
Advance Research Chemicals, Inc.

COPYRIGHTS

1. Introduction

Copyright has been grouped with other forms of legal protection under the general term "intellectual property". It is a means of protecting that particular form of creativity that has been variously referred to as originality of authorship, expression of ideas, or writings of an author. It is distinct from other forms of intellectual property that do not protect original expression of authorship, eg, patents, that protect novel inventions or discoveries; trademarks that protect terms and symbols identifying the source or origin of goods and services; and trade secrets that protect confidential, proprietary information.

2. Historical Background

Although there is considerable scholarly speculation on ancient precursors of copyright, the fact is that no significant copyright protection existed in theory or practice until the technological development of a means of mass duplication of creative works—the printing press. The law did not need a coherent protective structure for written and graphic works when the only means of copying them, by hand, was so labor-intensive as to require an investment potentially far in excess of the worth of the copy.

The invention of the printing press meant that written and graphic works could be duplicated and disseminated in vast numbers. Hence, the necessity of protecting the right to copy—the "copyright"—arose. Interestingly, copyright protection in England emerged in the fifteenth and sixteenth centuries as a means of censorship and control, rather than as a means of encouragement of authorship. The Crown feared the dissemination of treasonous writings; the Church feared the spread of heretical writings. And so, no work could be published other than through the Stationers' Company—the printers' trade guild, which was under the control of the Crown. Those who were allowed to publish through the Stationers' Company gained copyright protection for their works.

That was the case until the late seventeenth century, when the exclusive grant of copyright through the Stationers' Company expired. In 1710, Parliament enacted the first true copyright statute, the Statute of Anne. This law was noteworthy in two respects. First, and perhaps most significantly, it granted copyright protection to the author, as creator of the work, rather than to the printer, as exploiter of the work. Second, it limited the duration of copyright protection to a fixed term of years, after which the work would go into the public domain, and be free for all to use.

In the United States, immediately after independence, 12 of the 13 original states enacted their own copyright statutes (in part as a result of lobbying and efforts by Noah Webster and James Madison). But, as was the case in many other areas, the need for a uniform, national copyright law became apparent, as copyright easily transcends state borders—it is a simple matter to bring a book from one state to another and copy it. This ease of piracy, coupled with the lack of effective enforcement and the inhospitability of state courts to out-of-state copyright owners were obvious problems.

Thus, the Constitutional Convention, with little debate, included the power to enact a national copyright (as well as patent) law among the enumerated powers of Congress:

The Congress shall have the Power...to promote the Progress of Science and useful Arts, by securing for limited times to Authors and Inventors the exclusive Right to their respective Writings and Discoveries (1).

The purpose of copyright, then, is to "promote progress" in learning (the eighteenth century sense of the word "science"), for the good of all society. The way to do so is to grant economic property rights to creators, as the incentive to pursue their vocation and so enrich society. But there are limitations on those property rights: they may only endure for "limited times," and may only be granted for "writings" of "authors".

The First Congress after the Constitution was adopted enacted a national copyright statute in 1790. Thereafter, the development of federal copyright law followed a pattern that endures to this day. As developments in technology and forms of mass entertainment, communications, scholarship, and the arts occurred, the copyright law would be amended to accommodate them. Periodically, a total revision of the copyright statute would become necessary. Such complete revisions occurred in 1831, 1870, 1909, and 1976.

The 1976 Copyright Act is the basic law governing copyright today (2). But the principles and, to a degree, provisions of the 1909 Act are still of importance, eg, in the provisions regarding duration. The process continues: There have been many significant amendments to the 1976 Act since it went into effect on January 1, 1978.

3. Copyrightability

Under United States law, a work is either protected (copyrighted), or unprotected and free for all to use (in the public domain). But what sorts of works may be protected—are "copyrightable"?

The Copyright Act specifies that copyright extends to "original works of authorship fixed in any tangible medium of expression, now known or later developed, from which they can be perceived, reproduced or otherwise communicated, either directly or with the aid of a machine or device" (3). Many of the requirements for copyrightability may be gleaned from this provision:

- The work must be an "original work of authorship". Thus, unlike patent rights, originality, and not novelty, constitutes the touchstone of protection. Courts have defined what makes a work "original" both positively (the "spark of creativity," or "something...which is one man's alone") and negatively (as constituting that which has not been copied from another, even if not unique or novel) (4,5).

- The work must be the product of an "author". At base, this means that a human being, at some point, has created the work, even if at the behest or for the ownership of a corporate entity, and even if the creative process uses a machine or device (such as a camera or a computer program) as a tool in the creative process.

- The work must be "fixed in a tangible medium of expression". To a very limited extent, there are some works that are not so fixed, such as purely improvised and unrecorded pieces of music or choreography, extemporaneous speeches, or live, unrecorded and ephemeral broadcasts. Unfixed works are protected, but by state common law copyright, and not the federal statute. All works that are "fixed" are governed exclusively by the federal statute.

3.1. The Subject Matter of Copyright. The law goes on to specify, by way of example, the types of works that are covered: literary works, musical works (including lyrics), dramatic works (including accompanying music), pantomimes and choreographic works, pictorial, graphic and sculptural works, motion pictures and other audiovisual works, sound recordings, and architectural works. This list is nonexhaustive (3).

The test for copyrightability is nonsubjective. It matters not whether the work is "good" or "bad" art, or even obscene. Such considerations are not relevant to copyrightability.

The law also specifies that works of the United States Government are not subject to copyright protection (6), which is not to say that the United States may not own copyrights, but only that works created by Government employees are common property, and so are not copyrightable. However, works created pursuant to Government grants or using Government funds may be copyrightable, and owned by those outside the Government who receive the funds, depending on the regulations of the particular Government agency making the grant. Indeed, if both parties so agree, the copyright in those works may be transferred to, and owned by, the Government.

3.2. The Idea/Expression Dichotomy.   Copyright protects the expression of ideas, but not ideas themselves. Thus, copyright will not protect any procedure, process, system, method of operation, concept, principle, or discovery, regardless of the form in which it is described, explained, illustrated, or embodied (7). From this principle many others are derived.

- Copyright does not extend to titles, phrases, or, as a general rule, forms. Such items do not have the requisite originality of expression to distinguish them from the ideas they represent.

- Copyright does not extend to facts or news, but only to the particular form of expression of those facts. Thus, eg, the fact that a particular chemical causes a particular reaction may not be protected by copyright. But the text of an article describing that reaction and describing the experimental procedure to elicit it will be protectable, for those textual descriptions will constitute the expression of the fact, and not the fact itself. That copyright protection will not prevent anyone from recreating the experiment, but only from copying the article about it.

- Similarly, copyright does not protect research. As the Supreme Court has said, copyright protects creativity, not effort, no matter now significant that effort is (4).

- There are circumstances where the idea and expression are not distinguishable. It has been held, eg, that such "merger" of idea and expression occurred in a jewelry pin made in the exact form of a honeybee, or in simple sweepstakes rules. There was no other way (or only a very limited number of ways) to depict a bee in gold, or to express those contest rules. In such cases, copyright will not protect the work, for that would protect not only the expression, but the idea itself.

3.3. Utilitarian Works.   Utilitarian works may be copyrightable, but only to the extent that they contain copyrightable subject matter (8). The copyrightable subject matter must be physically or conceptually separable from the purely utilitarian object. Thus, eg, a common straight-backed chair will not be copyrightable, for there is nothing about it that is physically or conceptually separable from its "chair-ness", its purely utilitarian function. But if that chair contains a carved lion's head on its back, the lion's head is physically or conceptually separable and, therefore, copyrightable.

3.4. Compilations.   Copyright extends not only to works that can exist on their own, but to compilations of such works or even of public domain material. The Copyright law imposes a three-step test for such copyrightable compilations. They must first constitute the collection and assembling of preexisting data or materials. Second, those materials must be selected, coordinated or arranged in a particular fashion. And, third, that selection, coordination or arrangement must itself possess sufficient originality and creativity to constitute an original work of authorship. Thus, eg, the alphabetical listing of all subscribers to a telephone company's service, as in an ordinary "white pages" telephone directory, does not constitute a copyrightable compilation—no selection was made (all subscribers were listed) and no arrangement or coordination rose to the level of original expression (the listings were merely ordered alphabetically). On the other hand, the anthologizing of articles on a particular subject such as in an encyclopedia does constitute a copyrightable compilation. (Compilations of materials which can each stand on their own as copyrightable works, such as an encyclopedia, journal, or newspaper, are called "collective works.")

In no event will copyright in a compilation extend to, affect, or enlarge the protection of the underlying preexisting materials. Rather, it is only the original expression contributed by the author of the compilation—such as the selection of articles in an encyclopedia—to which the compilation copyright extends.

4. Copyright Ownership

Copyright is a property right. Although it differs from most other forms of property in that it is intangible, it nevertheless has the essential elements of property, and is governed by the principles of property ownership.

At the outset, the intangible nature of copyright requires a distinction between the intangible property of the copyright (called a "work") and the material object in which the copyrighted work is, quite literally, embodied (termed a "copy" or "phonorecord," terms that include such diverse media as paper-and-ink, computer disks, and audiotapes). Ownership of the copyrighted work does not constitute ownership of the material object in which it is embodied, and vice versa. Copyright ownership vests initially in the author or authors of the work.

4.1. Joint Authorship.   When more than one author has created a work, the work is said to be a "joint work". Under the law, such a joint work is one prepared by two or more authors with the intention that their contributions be merged into inseparable or interdependent parts of a unitary whole. Thus, joint authorship can occur when a composer and a lyricist collaborate on a song: Even though their contributions, the music and lyrics, can exist independently of each other (the music as an instrumental, the lyrics as a poem), they were created as interdependent parts of a unitary whole (the song). Two scientists collaborating on an article for a scientific journal are similarly joint authors: their contributions cannot be "teased out" of the article they have written, and so are inseparable parts of a unitary whole. But note that the test of joint authorship is intention: The creation must be made with the intention that the contributions be merged into one work, and that intention cannot be imputed after the fact of creation if it was not there to begin with. Case law has held that each contribution on its own must constitute copyrightable subject matter.

4.2. Joint Ownership.   Joint ownership of copyright occurs when there is joint authorship. But it may also occur in other ways—eg, by transfer of a copyright to two or more individuals (such as when an author bequeaths his copyright to two children).

As is the case with other forms of property, joint ownership of copyright is legally termed a tenancy-in-common: Each joint owner is presumed to own an undivided proportional interest in the entire work. For example, if there are three joint owners, each is presumed to own one-third of the entire work. The presumption may be defeated by an express agreement of the parties.

4.3. Works-Made-for-Hire.   There are many instances when, although a person has created a work, that creation has been made at the behest of another. In many such circumstances, common sense tells us that the person doing the creation should not own the copyright. As an easy example, consider a company that manufactures an appliance, and has one of its employees write an instruction manual for the appliance. Logically, the company, and not the employee, should own the copyright in the instruction manual.

Such situations are governed by the work-made-for-hire doctrine of the Copyright law. Remember that, under the Copyright law, copyright ownership vests initially in the author of a work. In cases of works made for hire, the law specifies that the employer or other person for whom the work is prepared is deemed to be the author (9). Thus, in our example, the appliance company would be deemed to be the author, and hence the initial copyright owner, of the copyrighted instruction manual.

The law very specifically defines what is, and therefore what is not, a work made for hire, in allowing for two, and only two, possibilities.

First, a work made for hire is a work prepared by an employee within the scope of his/her employment. As the law does not define "employee" or "employment", there were differences of opinion over the meaning of these terms until the Supreme Court resolved the matter. Employment, the court ruled, has the same meaning as is commonly understood under the law of agency (10). Thus, while many factors determine "employment", such as the method of payment for services, whether taxes and the like are withheld, where the work is done, who supplies the tools and instrumentalities for the work, the duration of the engagement, and so on, it is safe to say that a relatively formal and commonly understood employment relationship is required, as distinguished from a situation of special commission or independent contractor. It is also worth noting that, in a work-made-for-hire situation, the parties may nevertheless agree in writing that the work is not a work made for hire.

Second, a work may be a work made for hire if it is specially ordered or commissioned, but only if it meets both of the following requirements: It must be a work that falls into one of these nine categories: (1) a contribution to a collective work; (2) part of a motion picture or other audiovisual work; (3) a translation; (4) a supplementary work; (5) a compilation; (6) an instructional text; (7) a test; (8) answer material for a test; or (9) an atlas. And, the parties must expressly agree in a written instrument signed by them that the work is to be considered a work made for hire.

Thus, the circumstances under which an independently commissioned work is considered a work made for hire are limited indeed. As a result, the commissioning party is likely to seek a transfer of copyright, in the form of an assignment or license.

5. Transfers and Licenses of Copyright

Copyright, as we have seen, is a form of property. Like other forms of property, it may be freely transferred. There are, however, certain special rules for the transfer of copyrights, and certain aspects of the law concerning transfer of property are of special importance to copyright.

Ownership of the copyright in a work is distinct from ownership of the material object (the copy or phonorecord) in which the copyrighted work is embodied. So, too, the transfer of one does not constitute transfer of the other. For example, if a painter sells her painting (ie, the material object, such as canvas and oils), she does not automatically transfer the copyright in it. And sale of that copyright (eg, so as to allow reproduction of the oil painting in printed posters) does not transfer the material object.

It has often been said that copyright is a bundle of many different rights. As we shall see, there are many different ways in which a particular copyright may be exploited. The law allows copyright ownership to be virtually infinitely divisible. That is, each of those rights, in any subdivision conceivable, may be sold separately. The owner of any particular exclusive right is deemed to be the owner of copyright for that right. Thus, eg, the author of an article may sell the exclusive right of first publication of that article, but nothing else, to another, and that will result in two "owners of copyright" in that article—the purchaser of the right of first publication, who will own only that right, and the author, who will own all other rights.

As a practical matter, an important distinction must be made between two methods by which copyrights are exploited. On the one hand are assignments or transfers of ownership of the copyright, either in whole or in part. In such case, the purchaser becomes the outright owner of the copyright or the particular right at issue. On the other hand are licenses of copyright, either in whole or, more likely, in part. A license is merely the permission to use the copyrighted work in the particular manner specified. While exclusive licenses constitute transfers of copyright ownership for the particular rights involved, nonexclusive licenses do not. The distinction is of importance because the Copyright Act requires that transfers of copyright ownership must be in writing to be valid, whereas nonexclusive licenses need not be reduced to writing.

It will also be remembered that, in the case of joint ownership of works, the joint owners were treated as tenants-in-common. For purposes of transfer of the copyright, this means that each coowner may only transfer his own interest in the copyright, and not his coowner's interest. Thus, a coowner may not grant an exclusive license (which constitutes a transfer of copyright ownership) without his coowner's permission. But any coowner may grant a nonexclusive license to use the copyright without his coowner's permission. If he does so, however, he is subject to a duty to account to his coowners for their proportional shares of the profits realized by the nonexclusive license.

It was thought that, due to unequal bargaining power, authors would not be able to realize the true ultimate value of their works in initial transfers of copyright. Accordingly the law provides to authors or, if they are dead, their surviving spouses and children (or, if they have none, their executors or administrators), a "termination right" (11). Any transfer of copyright made after January 1, 1978 by an author may be terminated between 35 and 40 years after the transfer is made, and the copyright "recaptured". The technical formalities concerning such terminations are intricate.

6. Copyright Formalities

Changes to the Copyright law, starting with the 1976 Copyright Act and continuing with the Berne Convention Implementation Act of 1988, have radically changed United States Copyright law regarding copyright formalities. It is safe to say that many formalities that previously were of paramount importance have now been eased or entirely eliminated.

6.1. Copyright Notice. In the past, the law contained an absolute requirement that each copy of a published work bear a proper copyright notice. This notice formality was a major trap for unwary copyright owners. Failure to comply with the technicalities of the law's notice provisions resulted in the unintentional loss of protection for many works.

However, an amendment to the law abolished the notice requirement for all works first published on or after March 1, 1989. For such works, no copyright notice is required. Notice, however, is still required on all copies of works publicly distributed before that date. And, notice will still be widely used even when it is not required, so as to inform the world of the copyright status of the work. Notice consists of three elements: (1) the symbol © (the letter "c" in a circle) or the word "Copyright" or the abbreviation "Copr."; (2) the name of the copyright owner; and (3) the year date of first publication.

Although copyright notice is no longer a prerequisite to copyright protection, it is still valuable as a nonlegal matter. It serves to warn off infringers, and to identify the copyright owner to those seeking a license.

6.2. Copyright Deposit. The law requires that copies of every published work be submitted to the United States Copyright Office, which is a branch of the Library of Congress. The purpose of this requirement is to stock the shelves of the Library. This requirement is usually satisfied as part of the registration process; failure to make deposit may ultimately lead to a fine, but will not affect the existence of the copyright.

6.3. Copyright Registration. Although one hears about "copyrighting a work", the term is usually inexactly used. Although the speaker is referring to registering the work with the Copyright Office, the fact is that copyright registration is not required for copyright protection. To the contrary, federal copyright protection exists from the moment a work is created, that is, fixed in a tangible medium of expression, even if it is never registered.

Although copyright protection is not dependent on registration, registration does have important advantages. First, in the case of works of United States authors, no lawsuit for copyright infringement may be brought until a work is registered. Second, many important remedies in a lawsuit, such as recovery of statutory damages and attorneys' fees, are not available to a copyright owner unless registration has preceded the infringement (there is a 3-month grace period from the publication date for published works). Third, the certificate of copyright registration that the Copyright Office provides is *prima facie* evidence of the facts it contains, and shifts the burden of proof concerning those facts from the copyright owner to the defendant in a lawsuit.

Copyright registration is easily accomplished, even by non-attorneys. The copyright claimant completes a relatively simple form, and returns it to the Copyright Office with a nominal fee and deposit copies of the work. (Special provisions allow for nondisclosure of trade secrets or full computer programs and the like in the deposit.)

6.4. Other Formalities. As we have noted, transfers of copyright ownership must be in writing to be valid. The Copyright Office will record any documents pertaining to copyrights, including transfers. Such recordation sometimes has important consequences, as in the perfection of security interests in copyrights.

7. Copyright Duration

Two different regimes of copyright duration apply in the United States: one for works first created, published or registered for copyright on or after January 1, 1978 (new law works), and one for works published or registered before that date (old law works) (12). In all cases, copyright terms run through December 31 of their anniversary year. (The terms given below embody a 20-year extension of all existing copyrights which became effective October 27, 1998.)

7.1. New Law Works. For new law works, the basic copyright term is the life of the author and 70 years after the author's death. In the case of joint authors, the "life" in question is that of the longest surviving joint author.

For works where the duration of the author's life is not known—anonymous and pseudonymous works, and works made for hire—the term is 95 years from publication or 120 years from creation, whichever expires first.

7.2. Old Law Works. Protection for pre-1978 registered or published works endures under a system of dual terms. There is an initial term of 28 years from the earlier of publication or registration, followed by a renewal term of an additional 67 years, for a total of 95 years of protection. For works first published or copyrighted before 1964, renewal required registration in the Copyright Office in the last year of the initial term. If renewal was not made, the work fell into the public domain. For works first published or copyrighted from 1964 to 1977, renewal is automatic, but, in the last year of that initial term, an application for renewal of copyright may be filed in the Copyright Office, which will provide certain benefits to the renewal claimant.

The law contains a complicated provision, which case law has further elaborated, concerning ownership of renewal rights. As a general matter, renewal rights do not vest until the last year of the initial term, and then vest in the following individuals: (1) the author; (2) if the author is dead, the author's surviving spouse and children, as a class; (3) if there are no surviving spouse or children, the author's executor (ie, for the beneficiaries under the author's will); and (4) if the author did not leave a will, the author's next of kin under applicable state law.

Further, as a general rule, while these renewal rights may be assigned away in advance, such advance assignments are only binding if those making them survive to the time when renewal rights vest (13,14). For example, if an author assigned their renewal term rights in advance of renewal, but died during the initial term of copyright leaving a surviving spouse at the time of renewal, the surviving spouse owns the renewal rights and the assignment of the renewal term rights is ineffective.

8. Copyright Rights

The Copyright Act grants copyright owners six exclusive rights (15). These rights include not only the right to do the specified actions, but also to authorize them.

8.1. The Right to Reproduce in Copies.   The most basic copyright right, of course, is the right to reproduce the copyrighted work in copies or pho-norecords. This is the right to prevent unauthorized duplication of the work—eg, the printing of an article without the copyright owner's consent.

8.2. The Right to Prepare Derivative Works.   Many copyrighted works serve as the basis for derivative works, in which the underlying work is recast, transformed or adapted. Examples would be translations, motion pictures made from novels, and musical arrangements. This can be a major source of income for copyright owners. Of course, permission is necessary to make a derivative work. The copyright in a derivative work does not affect the copyright status of the underlying work, and the copyright in the derivative work extends only to the material contributed by the author of that work, as distinguished from the preexisting material.

8.3. The Right of Public Distribution.   Obviously, the exclusive right to reproduce the copyrighted work also entails public distribution of copies, by sale or other transfer of ownership. This right, too, is the copyright owner's.

8.4. The Right of Public Performance.   Certain types of works—eg, musical compositions, plays, or choreographic works—are meant to be performed. Public performance of those works is the copyright owner's exclusive right. This general right is not applicable to sound recordings, as noted below.

8.5. The Right of Public Display.   Other types of works, notably pictorial, graphic, and sculptural works, are meant to be displayed. Again, their public display is the copyright owner's exclusive right.

8.6. The Performance Right in Sound Recordings.   The right publicly to perform sound recordings is limited to digital subscription transmissions, which are defined in detail. Note that this limitation does not apply to the musical compositions embodied in the sound recording, which are governed by the general performing right discussed above.

9. Moral Rights

In addition to copyright rights, the copyright law was amended effective June 1, 1991 to grant very limited additional rights to authors of certain types of works, even if they have parted with copyright ownership (16). These "moral rights" are applicable only to works of visual art that exist in single copies or multiples of up to 200. Even within this limited category of works, there are many exceptions—eg, moral rights do not apply to works made for hire. The moral rights are those of attribution (the right to have the author's name attached to or deleted from the work) and integrity (the right to prevent mutilation or distortion of the work which would prejudice the author's honor or reputation).

10. Limitations and Exemptions

The law contains several limitations on copyright rights and exemptions for certain uses. We will here touch only on the most important.

10.1. Fair Use. The best known exemption is the fair use doctrine. Certain uses of copyrighted works that would otherwise be infringements are excused from liability because they are "fair". The law gives examples in a non-exhaustive list: uses for purposes such as criticism, comment, news reporting, teaching, scholarship, or research. But, even within those examples, each case must be judged on its particular merits and facts, and no particular use will be presumed to be a fair use.

The Copyright Act requires the courts to consider at least four factors in each case, as follows: (*1*) The purpose and character of the use (including whether it is of a commercial or nonprofit educational nature); even some commercial uses may be fair uses (eg, legitimate parodies which do not take too much of the copyrighted work). Courts frequently focus on the "transformative" nature of the use. (*2*) The nature of the copyrighted work; it has been held, eg, that use of an unpublished work is less likely to be a fair use than use of a published work, and the use of a scholarly or scientific work more likely to be a fair use than the use of works of pure entertainment. (*3*) The amount and substantiality of the portion used in relation to the copyrighted work as a whole; the less used, and the less significant the portion used, the more likely that fair use will be found. (*4*) The effect of the use upon the potential market for or value of the copyrighted work; this has sometimes been said to be the most significant fair use factor, because any real harm to the copyrighted work or its exploitation (via market substitution, rather than criticism or comment) will defeat the purpose of copyright protection (17).

10.2. First Sale Doctrine. Although the copyright owner has the exclusive right to distribute copies to the public, the bona fide possessor of a particular copy may, in most circumstances, further dispose of that copy without the copyright owner's consent. Thus, eg, the purchaser of a book may freely resell the copy she purchased (hence, used book stores do not violate Copyright law). Similarly, a copy of a work legitimately owned may be displayed publicly, as in the case of a picture hung in a museum. The Copyright Act has been amended to prohibit the rental of sound recordings or computer software, even though that rental would have been permitted by the first sale doctrine. Frequently, computer software will be sold under a so-called "shrinkwrap" license, which purports to bind the purchaser to limitations on the use and transfer of the software beyond the requirements of the law.

11. Infringement

Anyone who violates the exclusive rights of a copyright owner is liable for infringement, in a lawsuit brought in federal court. There is a 3-year statute of limitations on copyright infringement actions.

11.1. The Test for Infringement.   It is rare that actual evidence of copying exists. Thus, proof of copying is usually circumstantial, and is shown by a two-part test. First, the alleged infringer must be shown to have had access to the copyrighted work. Second, the two works must appear to their hypothetical intended audience to be substantially similar, and that substantial similarity must be of protected expression, not unprotected ideas or concepts. The two parts of the test may be seen to be in balance: while both must be present, the greater the evidence of substantial similarity, the less evidence of access is necessary, and vice versa.

Infringement of rights other than that of reproduction (such as the rights of public performance or display) are more easily proven directly by evidence of the infringing acts (for example, by a tape recording of the infringing public performance).

11.2. Remedies.   A copyright owner successfully proving infringement has three types of remedies available: recovery of monetary damages, injunctive relief, and recovery of costs including attorneys' fees.

Damages may be recovered in two alternative measures, at the choice of the copyright owner. First, the copyright owner is entitled to his actual damages and the infringer's profits that result from the infringement. These measures of damage are often difficult to prove, and so the law allows for statutory damages in the alternative (provided copyright registration has been timely made). Statutory damages are assessed by the court, in its discretion, between $750 and 30,000 for each work infringed (and not for each act of infringement). The limits may be lowered to $200 for truly innocent infringement, or raised to $150,000 for willful infringement.

Injunctive relief—making the infringer stop infringing—is often more important to the copyright owner than recovering damages. The court may craft appropriate injunctive relief.

Within the court's discretion, and again subject to timely registration, the prevailing party in an infringement suit may be awarded the costs of the litigation, including a reasonable attorney's fee.

11.3. The Internet.   The Digital Millennium Copyright Act, enacted on October 28, 1998, affected the liability of service providers when their services are used to distribute copyrighted works on the Internet. The DMCA protects against anti-circumvention technologies, which are designed to frustrate the electronic protection of copyrighted works, and also prohibits modification of copyright management information, which identifies copyrighted works electronically. It also clarifies the liability of on-line service providers and Internet access providers, by limiting the remedies against certain such providers, and providing "notice and takedown" remedies for copyright owners.

12. International Copyright

Because copyright easily transcends national boundaries, several international copyright conventions have been developed to protect copyrights internationally. The best known and most widely effective conventions are the Berne Convention for the Protection of Literacy and Artistic Works and the Universal Copyright

Convention; the United States is a signatory to both. More recently, two treaties dealing with the use of copyrighted works in the digital environment have been created, the WIPO Copyright Treaty and the WIPO Performances and Phonograms Treaty. In varying degrees, the treaties specify minimum standards which each member country's copyright law must meet. Even with adoption of those minimum standards, national copyright laws vary significantly from country to country.

International copyright treaties generally follow the principle of national treatment. Each member country treats nationals of other countries at least as well as it treats its own nationals.

BIBLIOGRAPHY

"Trademarks and Copyrights," in *ECT* 3rd ed., Vol. 23, pp. 348–374, by J. A. Baumgarten and C. H. Lieb, Paskus, Gordon & Hyman.; "Copyrights and Trademarks," in *ECT* 4th ed., Vol. 7, pp. 521–547, by L. H. Deinard, N. J. Stathis, and M. E. Rosenberger, White & Case; "Copyrights and Trademarks, Copyrights," in *ECT* (online), Posting date: December 4, 2000, by I. Fred Koenigsberg, White & Case.

CITED PUBLICATIONS

1. U.S. Const., Art. I, Sec. 8, Cl. 8.
2. 17 U.S.C. §§101 et seq. (1988).
3. 17 U.S.C. §102(a) (1988).
4. *Feist Publications, Inc.* v. *Rural Telephone Serv. Co.*, 499 U.S. 340 (1991).
5. *Bleistein* v. *Donaldson Lithographing Co.*, 188 U.S. 239 (1903).
6. 17 U.S.C. §105 (1988).
7. 17 U.S.C. §102(b) (1988).
8. *Mazer v. Stein*, 347 U.S. 201 (1954).
9. 17 U.S.C. §201(b) (1988).
10. *Community for Creative Non-Violence* v. *Reid*, 490 U.S. 730 (1989).
11. 17 U.S.C. §203 (1988).
12. 17 U.S.C. §§302–305 (1991).
13. *Fred Fisher Music Co.* v. *M. Witmark & Sons*, 318 U.S. 643 (1943).
14. *Stewart* v. *Abend*, 495 U.S. 207 (1990).
15. 17 U.S.C. §106 (1988).
16. 17 U.S.C. §106A (1991).
17. 17 U.S.C. §107 (1988).

I. Fred Koenigsberg
White & Case LLP

CORROSION AND CORROSION CONTROL

1. Introduction

Corrosion is the natural degradation of materials in the environment through electrochemical or chemical reaction. Traditionally, the definition of corrosion refers to the degradation of metals and has not included the degradation of non-metals such as wood (rotting) or plastics (swelling or crazing), but increasingly, natural degradation of any engineering material is being regarded as corrosion. The vast majority of the technologically significant corrosion involves the deterioration of metallic materials, and only the corrosion of metallic materials is discussed here.

2. Electrochemical Nature of Corrosion

Ores are mined and are then refined in an energy intensive process to produce pure metals, which in turn are combined to make alloys (see METALLURGY; MINERALS RECOVERY AND PROCESSING). Corrosion occurs because of the tendency of these refined materials to return to a more thermodynamically stable state (1–3). The key reaction in corrosion is the oxidation or anodic dissolution of the metal to produce metal ions and electrons

$$\mathrm{M} \xrightarrow{k_1} \mathrm{M}^{n+} + n\, e^- \tag{1}$$

The ions, M^{n+}, formed by this reaction at a rate, k_1, may be carried into a bulk solution in contact with the metal, or may form insoluble salts or oxides. In order for this anodic reaction to proceed, a second reaction which uses the electrons produced, ie, a reduction reaction, must take place. This second reaction, the cathodic reaction, occurs at the same rate because the electrons produced by the anodic reaction must be consumed by the cathodic reaction to maintain electroneutrality. Therefore, $I_c = I_a$, where I_c and I_a are the cathodic and anodic currents, respectively. The cathodic reaction, in most cases, is hydrogen evolution or oxygen reduction.

The four elements necessary for corrosion are an aggressive environment, an anodic and a cathodic reaction, and an electron conducting path between the anode and the cathode. Other factors such as a mechanical stress also play a role. The thermodynamic and kinetic aspects of corrosion determine, respectively, if corrosion can occur, and the rate at which it does occur.

3. Manifestations of Corrosion

The most common form of corrosion is uniform corrosion, in which the entire metal surface degrades at a near uniform rate (1–4). Often the surface is covered by the corrosion products. The rusting of iron (qv) in a humid atmosphere or the tarnishing of copper (qv) or silver alloys in sulfur-containing environments are examples (see also SILVER AND SILVER ALLOYS). High temperature, or dry, oxidation, is also usually uniform in character. Uniform corrosion, the most visible form of corrosion, is the least insidious because the weight lost by metal dissolution can be monitored and predicted.

An especially insidious type of corrosion is localized corrosion (1–3,5,6), which occurs at distinct sites on the surface of a metal while the remainder of the metal is either not attacked or attacked much more slowly. Localized corrosion is usually seen on metals that are passivated, ie, protected from corrosion by oxide films, and occurs as a result of the breakdown of the oxide film. Generally the oxide film breakdown requires the presence of an aggressive anion, the most common of which is chloride. Localized corrosion can cause considerable damage to a metal structure without the metal exhibiting any appreciable loss in weight. Localized corrosion occurs on a number of technologically important materials such as stainless steels, nickel-base alloys, aluminum, titanium, and copper (see ALUMINUM AND ALUMINUM ALLOYS; NICKEL AND NICKEL ALLOYS; STEEL; and TITANIUM AND TITANIUM ALLOYS).

Two types of localized corrosion are pitting and crevice corrosion. Pitting corrosion occurs on exposed metal surfaces, whereas crevice corrosion occurs within occluded areas on the surfaces of metals such as the areas under rivets or gaskets, or beneath silt or dirt deposits. Crevice corrosion is usually associated with stagnant conditions within the crevices. A common example of pitting corrosion is evident on household storm window frames made from aluminum alloys.

Another type of corrosion is dealloying which has also been called parting or selective leaching. Dealloying (1–4) is the preferential removal of one of the alloying elements from an alloy resulting in the enrichment of the other alloying element(s). Common examples are the loss of zinc from brasses (dezincification) (see COPPER ALLOYS, CAST ALLOYS) and the loss of iron from cast irons (graphitization).

Corrosion may also appear in the form of intergranular attack, ie, preferential attack of the boundaries between the crystals (grains) in metals and alloys (1–5). Intergranular attack generally occurs because the grain boundary and the grain have different corrosion tendencies, ie, different potentials. Intergranular corrosion often leads to a loss in strength or ductility of the metal.

Corrosion also occurs as a result of the conjoint action of physical processes and chemical or electrochemical reactions (1–5). The specific manifestation of corrosion is determined by the physical processes involved. Environmentally induced cracking (EIC) is the failure of a metal in a corrosive environment and under a mechanical stress. The observed cracking and subsequent failure would not occur from either the mechanical stress or the corrosive environment alone. Specific chemical agents cause particular metals to undergo EIC, and mechanical failure occurs below the normal strength (yield stress) of the metal. Examples are the failure of brasses in ammonia environments and stainless steels in chloride or caustic environments.

When a stress is cyclic rather than constant, the failure is termed corrosion fatigue. Fretting corrosion results from the relative motion of two bodies in contact, one or both being a metal. The motion is small such as a vibration. Erosion corrosion results from the action of a high velocity fluid impinging on a metal surface. Metals and alloys can also experience cracking in liquid metal environments. This form of corrosion is referred to as liquid metal cracking (LMC) (1,2,7). For example, mercury (qv) promotes cracking in highly stressed copper alloys and high strength aluminum alloys, and liquid copper promotes cracking of steels and stainless steels. Titanium and nickel alloys are also susceptible to LMC in specific environments.

Galvanic corrosion (1–4) occurs as a result of the electrical contact of different metals in an aggressive environment. The driving force is the electrode potential difference between the two metals. One metal acts principally as a cathode and the other metal as the anode. Galvanic corrosion can result from the presence of a second phase in a metal. An example is manganese sulfide [18820-29-6], MnS, inclusions in steels. Galvanic corrosion is also an important consideration for the environmental stability of metal matrix composites (qv) such as graphite reinforced aluminum. Galvanic corrosion can accelerate many of the other types of corrosion.

Microbiologically influenced corrosion results from the interaction of microorganisms and a metal (3,8–11). The action of microorganisms is at least one of the reasons why natural seawater is more corrosive than either artificial seawater or sodium chloride solutions. Microorganisms attach to the surfaces of metals and can, for example, act as diffusion barriers; produce metabolites that enhance or initiate corrosion; act as sinks or sources for species involved in cathodic reactions, such as oxygen and hydrogen; increase the pH at the surface as a result of photosynthesis; or decrease the pH by production of acid metabolites. A more detailed discussion of the various forms of corrosion may be found in the literature (1–4).

4. Origin of Corrosion

In the presence of oxygen and water the oxides of most metals are more thermodynamically stable than the elemental form of the metal. Therefore, with the exception of gold, the only metal that is thermodynamically stable in the presence of oxygen, there is always a thermodynamic driving force for corrosion of metals. Most metals, however, exhibit some tendency to passivate, ie, to form a protective oxide film on the surface that retards further corrosion. Were it not for the passivation by oxide films, most metals and alloys would corrode rapidly.

The thermodynamics of electrochemical reactions can be understood by considering the standard electrode potential, E^0, the potential of a reaction under standard conditions of temperature and pressure where all reactants and products are at unit activity. Table 1 lists a variety of standard electrode potentials. The standard potential is expressed relative to the standard hydrogen reference electrode potential in units of volts. A given reaction tends to proceed in the anodic direction, ie, toward the oxidation reaction, if the potential of the reaction is positive with respect to the standard potential. Conversely, a movement of the potential in the negative direction away from the standard potential encourages a cathodic or reduction reaction.

Table 1. **Standard Electrode (Reduction) Potentials, 25°C**[a]

Electrode	E^0, V[b]
$Li^+ + e^- \rightarrow Li$	−3.045
$Zn^{2+} + 2e^- \rightarrow Zn$	−0.763
$Fe^{2+} + 2e^- \rightarrow Fe$	−0.440
$Cd^{2+} + 2e^- \rightarrow Cd$	−0.403
$Co^{2+} + 2e^- \rightarrow Co$	−0.277
$Ni^{2+} + 2e^- \rightarrow Ni$	−0.250
$Sn^{2+} + 2e^- \rightarrow Sn$	−0.140
$Pb^{2+} + 2e^- \rightarrow Pb$	−0.126
$Sn^{4+} + 2e^- \rightarrow Sn^{2+}$	0.15
$Cu^{2+} + e^- \rightarrow Cu^+$	0.158
$Cu^{2+} + 2e^- \rightarrow Cu$	0.337
$Fe^{3+} + e^- \rightarrow Fe^{2+}$	0.771
$Ag^+ + e^- \rightarrow Ag$	0.799
$Br_2(l) + 2e^- \rightarrow 2Br^-$	1.065
$Cl_2(g) + 2e^- \rightarrow 2Cl^-$	1.358

[a]Ref. 3.
[b]Against standard hydrogen electrode (SHE) set at $E^0 = 0.000V$

A piece of zinc placed in an aqueous $1\,M\,Zn^{2+}$ solution attains a potential of approximately − 0.763 V vs standard hydrogen. If Cu^{2+} ions are then introduced into the solution, the zinc begins to corrode because the zinc surface exists at a potential which encourages the cathodic reaction for copper, ie, the standard potential of the zinc reaction is negative to that of the copper and the Cu ions electroplate onto the zinc via the reaction $Cu^{2+} + 2\,e^- \rightarrow Cu$. This coupling causes the zinc potential to shift, ie, to be polarized, in the positive direction and the reaction $Zn \rightarrow Zn^{2+} + 2\,e^-$ then proceeds. The potentials of the two reactions are polarized toward each other, away from the respective standard potentials. The rates of these two reactions must be the same to preserve electroneutrality, and additionally, some of the zinc surface must be exposed to the solution. That is, if a continuous coating of copper were to be electroplated on the zinc surface such that none of the zinc were exposed, further oxidation would be prevented.

Corrosion occurs even if the two reactants involved are not at standard conditions. In this case, the nonstandard equilibrium potential for each reaction, often referred to as the reversible potential, can be calculated from the Nernst equation. Additional information on thermodynamic aspects of corrosion can be found in the literature (1–4,12,13).

The larger the potential difference between the equilibrium potentials of the anodic and cathodic reactions involved in corrosion, the greater the driving force which is more traditionally represented as the change in Gibbs free energy, ΔG. The free-energy change for a reaction is related to the difference in the equilibrium potentials of the two reactions, cathodic minus anodic by

$$\Delta G = -nFE \qquad (2)$$

where n is the number of electrons involved in the overall reaction, F is the Faraday constant, and E, the cell voltage or the difference in the equilibrium potential values of the cathode and the anode, respectively, given in volts.

The standard reduction potential of oxygen (taken at pH = 0)

$$O_2 + 4\,H^+ + 4\,e^- \longrightarrow 2\,H_2O \tag{3}$$

is 1.229 V, a much more positive number than the metal reactions given in Table 1. Therefore, the free-energy change for a coupling of any of the metal reactions, considered as the anode, and the oxygen reaction, considered as the cathode, would be a large negative number; ie, the reaction is thermodynamically highly favored to proceed. Reduction of oxygen is the most common cathodic reaction in the corrosion of metals in neutral and alkaline media. In acidic media the common cathodic reaction is hydrogen evolution

$$2\,H^+ + 2\,e^- \longrightarrow H_2 \tag{4}$$

The standard potential for this reaction is 0.000 V. Most of the metal reactions have standard potentials below this value.

5. Electrochemical Equilibrium Diagrams

The thermodynamic data pertinent to the corrosion of metals in aqueous media have been systematically assembled in a form that has become known as Pourbaix diagrams (13). The data include the potential and pH dependence of metal, metal oxide, and metal hydroxide reactions and, in some cases, complex ions. The potential and pH dependence of the hydrogen and oxygen reactions are also supplied because these are the common corrosion cathodic reactions as discussed above. The Pourbaix diagram for the iron–water system is given as Figure 1.

If the potential of a metal surface is moved below line A, the hydrogen reaction line, cathodic hydrogen evolution is favored on the surface. Similarly a potential below line B, the oxygen reaction line, favors the cathodic oxygen reduction reaction. A potential above the oxygen reaction line favors oxygen evolution by the anodic oxidation of water. In between these two lines is the region where water is thermodynamically stable.

The Pourbaix diagram is constructed from data for a variety of reactions. The expressions for these reactions can be (1) potential dependent and pH independent, (2) pH dependent and potential independent, (3) both pH and potential dependent, or (4) both pH and potential independent. The dependence is determined by whether the reaction of interest is electrochemical or chemical and whether or not it involves the components of water. From this diagram it is possible to determine if there is a driving force for corrosion at any given potential–pH combination and what the corresponding cathodic reaction is likely to be. Regions where insoluble corrosion products form are indicated. These are regions where passivity is possible. Passivity generally results from the formation of an oxide film on a metal surface. The film can dramatically reduce further corrosion of the metal. Regions of immunity, where there is no thermodynamic driving

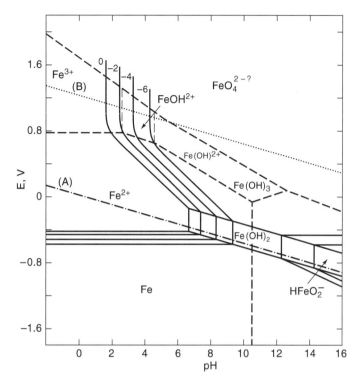

Fig. 1. Pourbaix diagram for the iron–water system at 25°C, considering as solid substances only Fe, Fe(OH)$_2$, and Fe(OH)$_3$, where (−·−·), line A, represents the hydrogen reaction line; (···), line B, the oxygen line(13). The values 0,−2, −4, −6 are molar concentrations of the ions involved in each reaction.

force for the metal to corrode, can also be determined. Figure 2 presents a simplified picture of the Pourbaix diagram of Figure 1 showing the regions of immunity, possible corrosion, and possible passivation. A complete collection of electrochemical equilibrium diagrams can be ordered from NACE International (13).

Pourbaix diagrams are only thermodynamic predictions and yield no information about the kinetics of the reactions involved nor are the influences of other ionic species which may be present in the solution included. Complexing ions, particularly halides, can interfere with passivation and can influence the position of the lines in a Pourbaix diagram. Further, in real corrosion situations the local chemistry in pits, crevices, and cracks differs from that of the bulk concentration and influences the actual electrochemical reactions at those sites. More comprehensive information on Pourbaix diagrams and the use of them in corrosion is available (1–3,12,13).

6. Kinetics of Electrochemical Reactions

Even in uniform corrosion, a corroding metal surface has numerous local anodes and cathodes. The sites of these local reactions may be fixed by microstructural

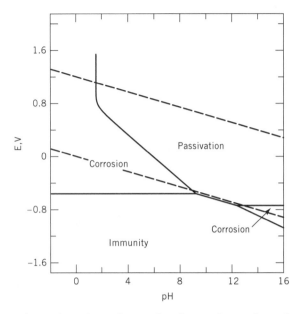

Fig. 2. The thermodynamic regions of corrosion, immunity, and passivation of iron in an iron–water system assuming passivation by a film of Fe_2O_3(13).

features or may change as corrosion proceeds. The oxidation reaction at anodic sites on the metal surface can be represented as in equation 1. A corresponding reduction reaction must be occurring at cathodic sites. In acidic solutions this would likely be hydrogen evolution as shown in equation 4. The potentials of these two reactions would be moved toward each other, away from the respective equilibrium potentials, and the metal surface would assume an overall uniform potential. The dotted lines in Figure 3 illustrate the relationships of the individual electrochemical reaction kinetics as influenced by electrochemical potential. These lines also illustrate the interaction of the two reactions on the metal surface. Diagrams of this sort are referred to as Evans or mixed potential diagrams. The terms are used interchangeable in the corrosion community.

The two dashed lines in the upper left hand corner of the Evans diagram represent the electrochemical potential vs electrochemical reaction rate (expressed as current density, ie, current divided by surface area) for the oxidation and the reduction forms of the hydrogen reaction. At point A the two are equal, ie, at equilibrium, and the potential is therefore the equilibrium potential, E^0, for the hydrogen reaction for the specific conditions involved. Note that the reaction kinetics are linear on these axes. The change in potential for each decade of log current density is referred to as the Tafel slope (14). The Tafel slope is related to the activation energy for the reaction. A more detailed treatment of Tafel slopes can be found elsewhere (1–4,14,15).

The value of the current density where the oxidation and reduction forms of a reaction cross, ie, at point A or point B in Figure 3, is referred to as the exchange current density, i_0. For any equilibrium point, the exchange current density is the rate of the two forms of the reaction, ie, cathodic and anodic, occur-

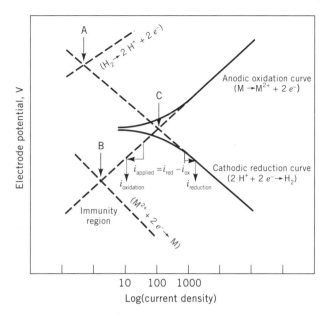

Fig. 3. Hypothetical Evans diagram and polarization curve for a metal corroding in an acidic solution, where point A represents the exchange current density, i_0, for the hydrogen electrode at equilibrium; point B, the exchange current density at the reversible or equilibrium potential, E_{rev}, for $M \rightleftharpoons M^{2+} + 2e^-$; and point C, the corrosion current density, i_{corr}, at the open-circuit corrosion potential, E_{corr}. See also discussion in text.

ring at equilibrium. The greater the exchange current density the more kinetically favored is the reaction. The exchange current density for the reduction of hydrogen on platinized platinum is 10^{-3}A/cm^2, whereas on mercury it is 10^{-13}A/cm^2, indicating that platinum is a good catalyst for hydrogen evolution. The exchange current densities for most electrochemical reactions fall in between these two values. Exchange current density and Tafel slope values for a variety of other electrochemical reactions of interest to corrosion can be found in the literature (2,3,16). The example in Figure 3 involves hydrogen evolution on a corroding surface and the exchange current density value for the hydrogen reaction on that surface must then be applied.

In the lower left part of the Evans diagram, the oxidation and reduction forms of metal dissolution reactions are given. Assuming the data of this diagram are accurate, a prediction of corrosion rate and the potential of the corroding metal surface can be made. The potentials of the hydrogen and the metal reactions are pulled toward one another following the dashed lines, or Tafel slope, appropriate for the particular polarization; ie, the anodic line is followed for the metal dissolution reaction, the cathodic line followed for the hydrogen reaction. Point C where the two cross is the only potential where the rates of the anodic and the cathodic reactions are equal. This defines the corrosion potential, E_{corr}, and the corrosion current density, i_{corr}, from which the corrosion rate can be calculated by Faraday's law. The other dotted lines were not considered because the rates at the corrosion potential are so low as to be insignificant.

The solid line in Figure 3 represents the potential vs the measured (or the applied) current density. Measured or applied current is the current actually measured in an external circuit; ie, the amount of external current that must be applied to the electrode in order to move the potential to each desired point. The corrosion potential and corrosion current density can also be determined from the potential versus measured current behavior, which is referred to as polarization curve rather than an Evans' diagram, by extrapolation of either or both the anodic or cathodic portion of the curve. This latter procedure does not require specific knowledge of the equilibrium potentials, exchange current densities, and Tafel slope values of the specific reactions involved. Thus Evans diagrams, constructed from information contained in the literature, and polarization curves, generated by experimentation, can be used to predict and analyze uniform and other forms of corrosion. Further treatment of these subjects can be found elsewhere (1–4,17).

7. Galvanic Corrosion and Cathodic Protection

Galvanic corrosion results when two or more dissimilar metals or alloys immersed in the same electrolyte are in electrical contact (1–4). This form of corrosion is one of the most common and most preventable. The metal or alloy in a galvanic couple having the lower corrosion potential has its potential pulled in the positive direction by the coupling with the metal or alloy that has the higher corrosion potential, generally causing the one with the lower potential to experience accelerated corrosion. Conversely, the metal with the higher corrosion potential experiences a negative shift in potential as a result of the coupling causing it to corrode less and to support additional cathodic reaction. Figure 4 is a galvanic series for a variety of metals and alloys listing the corrosion potentials of these materials in seawater. Free corrosion potentials, not coupled or equilibrium potentials, are given and these allow a determination of which metal has the lower potential in a galvanic couple. Generally, the larger the potential difference between components in a galvanic couple the faster will be the corrosion of the one having the lower potential. However, the kinetics of the particular reactions involved must be evaluated for an accurate determination of corrosion rate.

Galvanic corrosion can be used to a corrosion advantage. If, eg, a metal such as zinc or magnesium, having a low corrosion potential in most environments, is coupled with a steel component, the zinc or magnesium pulls the potential of the steel down causing the steel to corrode less. When a sacrificial metal or alloy, called a sacrificial anode, is attached to a structure having a higher corrosion potential to intentionally pull the potential of the higher potential metal down and thus decrease the corrosion rate, it is called cathodic protection. This method of corrosion mitigation is common for underground pipelines (qv), residential hot water heaters, and hulls and tanks (qv) of ships. The lowering of potential can also be achieved by the application of external electrical current, ie, impressed current cathodic protection. Electrons are pumped into the structure to lower the potential and thus discourage anodic (corrosion) reactions. A second electrode (anode) is needed for this procedure and this electrode is often a highly inert

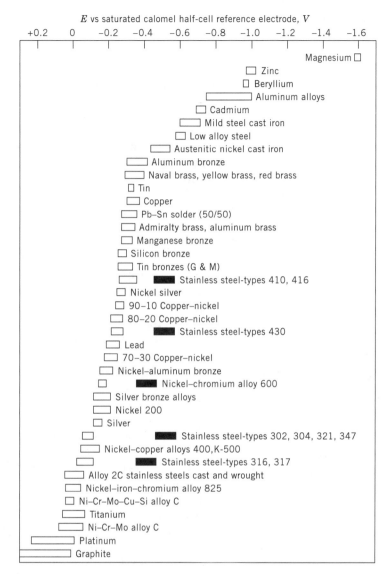

Fig. 4. Galvanic series in flowing seawater. Certain alloys may become more active in low velocity or poorly aerated seawater and the potentials exhibited under these conditions are indicated as darkened potential ranges (1).

material such as platinum to prevent it from being corroded away. However, scrap iron or steel is sometimes used as the impressed current anode. This approach remains functional until the scrap metal anode corrodes away.

The lower potential metal in a galvanic couple does not always have its corrosion rate accelerated. For metals that form a passive film, coupling with another metal of higher potential can cause the film-forming metal's potential to shift from a value at which it corrodes to one at which it passivates and therefore corrodes less. When this is done intentionally the procedure is referred to as anodic protection, ie, achieving protection by intentionally shifting the potential in the positive direction. Anodic protection is generally achieved by adding oxidizers to the electrolyte or by an external electrical circuit.

The higher potential metal in a galvanic couple is not always rendered less corrodible. A passive metal can sometimes be pulled out of its passive region and into a more corrosive region. A metal can also have its potential pulled down to the point where hydrogen reaction is possible and this can lead to one of the forms of EIC called hydrogen induced cracking (HIC). Another example involves organic coatings (qv) on metal structures. These coatings can be damaged by having the potential of the structure shifted too far in the negative direction. This is one of the consequences of excessive cathodic protection.

8. Environmental Effects

The environment plays several roles in corrosion. It acts to complete the electrical circuit, ie, supplies the ionic conduction path; provides reactants for the cathodic process; removes soluble reaction products from the metal surface; and/or destabilizes or breaks down protective reaction products such as oxide films that are formed on the metal. Some important environmental factors include: the oxygen concentration; the pH of the electrolyte; the temperature; and the concentration of anions.

Reduction of oxygen is one of the predominant cathodic reactions contributing to corrosion. Awareness of the importance of the role of oxygen was developed in the 1920s (18). In classical liquid drop experiments, the corrosion of iron or steel by drops of electrolytes was shown to depend on electrochemical action between the central relatively unaerated area, which becomes anodic and suffers attack, and the peripheral aerated portion, which becomes cathodic and remains unattacked. In 1945, the linear relationship between rate of iron corrosion and oxygen pressure from 0 to 2.5 MPa (0–25 atm) was shown (19).

The concentration dependence of iron corrosion in potassium chloride [7447-40-7], sodium chloride [7647-14-5], andlithium chloride [7447-44-8] solutions is shown in Figure 5 (20). In all three cases there is a maximum in corrosion rate. For NaCl this maximum is at $\sim$0.5 N ($\sim$3 wt%). Oxygen solubility decreases with increasing salt concentration, thus the lower corrosion rate at higher salt concentrations. The initial increase in the iron corrosion rate is related to the action of the chloride ion in concert with oxygen. The corrosion rate of iron reaches a maximum at $\sim$70°C. As for salt concentration, the increased rate of chemical reaction achieved with increased temperature is balanced by a decrease in oxygen solubility.

The corrosion rate of iron in aerated water is also a function of pH and generally follows the pattern in Figure 6 (21). At pH 4–10 , the rate is controlled by the availability of oxygen. In more acidic solutions (lower pH) the corrosion rate is accelerated, and the reduction of hydrogen ion replaces the reduction of oxygen as the rate controlling cathodic reaction. Many metals follow approximately the same behavior with the exception of those metals that dissolve to form amphoteric ions. Zinc forms the zincate ion, ZnO_2^{2-}, which causes zinc to corrode excessively above a pH 12 (see Zinc compounds); whereas Al forms the aluminate ion

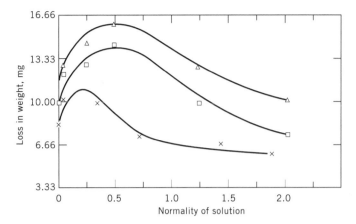

Fig. 5. Corrosion–concentration curves for alkali chlorides where $\times$ denotes LiCl; $\square$, NaCl; and $\triangle$, KCl (20).

[11098-82-1],, AlO_2^-, which increases the dissolution rate above a pH of ~ 8 (see ALUMINUM COMPOUNDS, ALUMINATES).

Chlorides, which are ubiquitous in nature, play an important role in the corrosion of metals. Chlorides and other anions also play an important role in localized corrosion, ie, the breakdown of the insoluble protective reaction product films, eg, passive films, that prevent corrosion of the underlying metal. A variety of mechanisms attempting to explain the role of chloride in general and in localized corrosion have been proposed (6,22–25).

Very often the environment is reflected in the composition of corrosion products, eg, the composition of the green patina formed on copper roofs over a

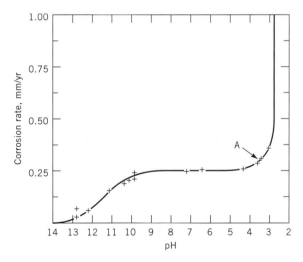

Fig. 6. Effect of pH on corrosion of iron in aerated water at room temperature (21). Point A is where H_2 evolution begins. To convert mm/yr to mils per year (mpy), multiply by 39.37.

Table 2. **Composition of Green Patina on Copper from Different Locations**[a]

Location of structure	Age of structure, year	Composition of green patina, %			
		$CuCO_3$	$CuCl_2$	$Cu(OH)_2$	$CuSO_4$
urban	30	14.6		9.6	49.8
rural	300	1.4		58.5	25.6
marine	13	12.8	26.7	52.5	2.5
urban–marine	38		4.6	61.5	29.7

[a]Refs. 26 and 27.

period of years. The determination of the chemical composition of this green patina was one of the first systematic corrosion studies ever made (see COPPER). The composition varied considerably depending on the location of the structure as shown in Table 2 (26,27).

9. Alloy Composition and Metallurgical Factors

A primary factor in determining corrosion behavior of metals and alloys is their chemical composition. Alloys having varying degrees of corrosion resistance have been developed in response to environmental needs. A good example of how corrosion resistance can be successfully changed by altering the composition can be seen in the alloying of steels. At the lower end of the alloying scale are the less costly and less corrosion resistant low alloy steels. These are iron-base alloys containing from 0.5–3.0 wt% Ni, Cr, Mo, or Cu and generally small amounts of P, N, and S. At the higher end of the alloying scale are the more costly and significantly more corrosion resistant stainless steels. These latter alloys contain a minimum of 10.5 weight percent Cr and, depending on the grade, other alloying elements such as Ni, Mo, and N. The corrosion resistance of these alloys is based on the protective nature of the surface film, which in turn is based on the physical and chemical properties of the oxide film.

In addition to alloying there are other metallurgical factors, such as crystallography, grain size and shape, grain heterogeneity, second phases, impurity inclusions, and residual stress, that influence corrosion. The technologically important structural materials are polycrystalline aggregates. Each individual crystal is referred to as a grain. Grain orientation can affect corrosion resistance as evidenced by metallographic etching rates and pitting behavior (22). Grain shape may likewise vary greatly depending on the alloy and processing history. Alloys, particularly in the as-cast condition, generally exhibit chemical inhomogeneity such that there is segregation of alloying elements and impurities to the grain boundary regions. These heterogeneities, which can also develop during subsequent processing such as welding (qv) or heat treatment, can produce different electrochemical characteristics at the grain boundary relative to the grain interior and can lead to intergranular corrosion. This problem can be of great practical importance, especially to wrought stainless steels and nickel alloys. Second phases, such as ferrite grains in an otherwise austenitic stainless steel and beta grains in an otherwise alpha brass, can be of considerable importance in some alloy systems and some forms of corrosion. Residual stresses from cold-working or other sources can lead to increased corrosion rates (22,24,28), and are also important in stress–corrosion cracking.

The following sections briefly discusses some of the alloying and metallurgical factors in a three technologically important alloy systems. A more complete discussion of the corrosion behavior of these and other alloy systems and the influence of metallurgical factors on each is available (4).

9.1. Stainless Steels. The stainless steels, by virtue of the alloying additions and processing, can be categorized as ferritic (body-centered cubic structure), austenitic (face-centered cubic structure), duplex (a combination of ferritic and austenitic), martensitic (body-centered tetragonal or cubic structure), and precipitation hardened. There are some 180 different alloys that can be recognized as stainless steels. These alloys contain a minimum of 10.5 wt% Cr and, depending on the grade, other alloying elements such as Ni, Mo, and N (5,29). The concentration of particular alloying elements are based on the desired properties of the stainless steel. Figure 7 shows the compositional and property linkages for the stainless steels (29).

Figure 7 Compositional and property linkages for the stainless steels (29).

The corrosion resistance of stainless steels results from the formation of a passive film and, for this reason these materials are susceptible to pitting corrosion and to crevice corrosion. The resistance to pitting or crevice corrosion is improved by additional alloying with Cr, Mo, and N for the austenitic and duplex

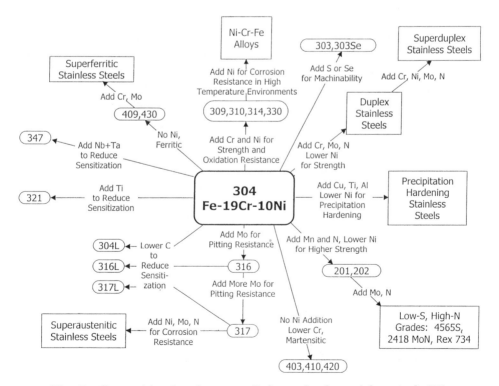

Fig. 7. Compositional and property linkages for the stainless steels (29).

stainless steels and alloying the ferritic stainless steels with additional Cr and Mo. The manufacturers of stainless steels have developed indexes for resistance to pitting and crevice corrosion based on the composition of the alloy. The index is known as the pitting resistance equivalent number (PREN). PREN is given by PREN = %Cr + 3.3 (%Mo) + 16 (%N) + 1.65 (%W). In general, the higher the value of PREN, the greater the resistance to crevice and pitting corrosion. The major problem with using a PREN is that it is based solely on composition and ignores the detrimental effects of microstructure, such as Cr depleted zones and alloy segregation (29).

The most common and serious metallurgical factor affecting the corrosion resistance of stainless steels is termed sensitization. This condition is caused by the precipitation of chromium-rich carbides (qv) at the grain boundaries, giving rise to chromium depleted grain boundary areas. For example, heating a type 304 stainless steel containing 0.039% carbon for 10 h at 700°C will reduce the chromium level from 19% to <13% in the region next to the carbide precipitate (29). The depleted areas are anodic to the grain interior and tend to dissolve thereby causing intergranular corrosion. Sensitization can also make stainless steel more prone to stress–corrosion cracking (SCC). There are several measures available to mitigate sensitization. Low carbon grades such as AISI 304L and 316L are available that have much less tendency toward sensitization. Also, alloying additions of titanium or niobium and tantalum can be used to tie up carbon. Another area of concern with stainless steel alloys is EIC and this topic is treated in more detail later in this article for stainless steels as well as for other selected alloys. A more detailed discussion of stainless steels may be found in the literature (5,29).

9.2. Copper Alloys. Copper and its alloys have good corrosion resistance in many nonoxidizing aqueous environments and atmospheric conditions. Several copper alloys are exceptionally resistant to certain atmospheres. The copper–nickel alloys are the most corrosion resistant of the commercial copper alloys (3). The two most common Cu–Ni alloys are 90% Cu–10% Ni and 70% Cu–30% Ni. They have good resistance to corrosion in both fresh and salt water (4).

The most known of the copper alloys is brass, which is produced by the addition of zinc. The zinc improves the mechanical properties by solid solution strengthening. However, the zinc reduces the corrosion resistance. A variety of names are associated with brasses with a particular zinc content: 40% Zn–Cu alloys are referred to as Muntz metal; 30% Zn–Cu alloys are referred to as yellow brass; and 15% Zn–Cu alloys are termed red brass (2). Brasses are susceptible to dealloying in the form of dezincification, ie, the preferential loss of zinc from the alloy. Brasses having zinc concentrations of 15% or greater are prone to dezincification and dezincification is generally more severe in brasses that have two metallurgical phases. For the two-phase alloys, greater than ~38 wt% Zn, dezincification usually starts in the high zinc content β-phase and is followed by dezincification of the lower zinc α-phase (4). Naval brass and admiralty brass are similar in composition to Muntz metal and yellow brass, respectively, but have 1 wt% tin added for improved resistance to dezincification. Red brass is relatively resistant to dezincification but is more susceptible to impingement attack than, for example, yellow brass (2).

Conditions that favor dezincification include stagnant solutions, especially acidic ones, high temperatures, and porous scale formation (2). Tin tends to inhibit dealloying especially in cast alloys (4). Additions of small amounts of arsenic, antimony, or phosphorus can increase the resistance to dezincification. These elements are, however, not entirely effective in preventing the dezincification of the two-phase (α-β) brasses because dezincification of the β-phase is not prevented (4). Another area of corrosion concern involves applied or residual stresses from fabrication that can lead to EIC of brasses in the form of SCC.

9.3. Aluminum Alloys. The corrosion resistance of aluminum, like stainless steels, is provided by a passive film that protects the surface from degradation by the environment. However, in environments that contain aggressive anions such as chloride, the passive film can be degraded locally causing film breakdown and, depending on the geometry of the sample, pitting or crevice corrosion.

Copper, silicon, magnesium , zinc, and manganese are some common alloying additions to aluminum. Most of the alloying elements are added to aluminum to produce alloys having improved mechanical properties. However, the strengthening phases that result from the alloying can disrupt the passive oxide layer on aluminum and lead to localized corrosion. Also, the second phase constituents can produce local galvanic cells as discussed below. Common alloying elements that are added to steels and many other metals to improve corrosion resistance, such as Cr and Mo, have a very low solubility in aluminum and result in second phase precipitates. The presence of the second phase constituents, in turn, lowers the corrosion resistance. A number of surface modification techniques that produce single phase, metastable aluminum alloys have been shown to increase the localized corrosion resistance as compared to pure aluminum (30–34). These techniques include ion implantation, laser alloying, and magnatron sputtering. However, due to capital costs and the sizes of the sample that can be treated, these techniques have not gained widespread use.

Aluminum alloys are susceptible to intergranular corrosion when precipitates form. Intergranular corrosion in aluminum alloys is caused by a potential difference, ie, galvanic couple. The formation of the anodic and cathodic areas in the couple varies depending on the alloying additions (3). Reference (35) is a compilation of potentials reported for intermetalic phases in aluminum alloys. Stress–corrosion cracking is characteristically intergranular for aluminum alloys. Wrought high strength aluminum alloys, whether the products are rolled, forged, or extruded, tend to be highly textured because of the manner in which the secondary phases or inclusions are strung out. Exfoliation corrosion, a type of intergranular corrosion, proceeds along subsurface paths. The corrosion products have a greater volume than the alloy and produce a layer type attack. Selection of alloys with proper tempers can provide resistance to exfoliation corrosion (4). The texture is also is important to the SCC of high strength aluminum alloys (see ALUMINUM AND ALUMINUM ALLOYS).

10. Environmentally Induced Cracking

Environmentally induced cracking is a brittle fracture process caused by the conjoint action of a mechanical stress and a corrosive environment. There are several different types of EIC: stress–corrosion cracking (SCC), corrosion fatigue cracking (CFC), and hydrogen induced cracking (HIC). In SCC and HIC the stress may be normal stress experienced in operation or residual stress from welding, heat treatment, and/or cold work. CFC results from cyclic stresses. In any of these, the corrosion rates are generally very low and the stress level required for the cracking is generally well below the normal yield stress of the particular alloy involved. The concentration of the chemical species responsible for the cracking need not be high. EIC failures are often catastrophic such as the Pt. Pleasant, West Virginia, bridge collapse that killed 46 people in 1967 (3). Cracking in steam pipes, underground pipelines, and aircraft have also been attributed to EIC.

The chemical species that can lead to EIC in each alloy system are fairly well known although the exact mechanisms of crack initiation and propagation are not thoroughly understood. The species that promote EIC in one alloy system do not necessarily promote EIC in others. A discussion of EIC and the chemical species that promote it can be found in the literature (3).

The cracking in EIC can proceed either intergranularly, ie, between the grains, or transgranularly, ie, across the grains. Small changes in the environment can cause a shift from one type to the other as can relatively small changes in the composition of the alloy. Likewise the crack may be single and unbranched or have multiple origins and also be branched. Thus neither intergranular or transgranular cracking is indicative of or excludes EIC as the cause of a service crack. More detailed information regarding EIC is available (1,3,4,36).

10.1. Copper Alloys.   Copper alloys under an applied or residual stress are susceptible to SCC in environments containing ammonia (qv) or ammonium compounds (qv). In addition to ammonia, water and oxygen, or another cathodic reactant, are required to cause SCC. Ammonia and ammonium compounds are sometimes found in the atmosphere, in cleaning compounds, in fertilizers (qv), or in chemicals used for water treatment. Brasses containing <15% zinc are highly resistant to SCC, whereas brasses containing 20–40% zinc are highly susceptible and this susceptibility increases only slightly as zinc concentration increases from 20 to 40% (4). SCC is usually intergranular but can occur transgranular under certain conditions (2,4).

Trace quantities of nitrogen oxides may also cause SCC. These failures are probably the result of the nitrogen oxides being converted to ammonium salts on the brass surface. The premature failure of yellow brass brackets in the humidifier chamber of an air-conditioning system was traced to this latter cause (2). In another case, SCC of 12% Ni–23% Zn–Cu alloy, known as nickel brass, parts of the Central Office Telephone equipment in Los Angeles occurred within 2 years of installation for similar reasons. The source of nitrogen was the air in the Los Angeles area which has high concentrations of nitrogen oxides and suspended nitrates (see AIR POLLUTION). The nitrates settle as dust on the brass parts (2). The susceptibility to SCC can be minimized by (1) proper alloy selection; (2) thermal stress relief; (3) avoiding contact with ammonia and ammonia compounds; and (4) using an inhibitor (2,4).

10.2. Aluminum Alloys.   Both the 2000 and the 7000 series aluminum alloys have experienced significant SCC in service. These high strength alloys in the wrought form are highly textured, the grains are flattened and elongated into specific crystallographic directions due to the processing. SCC behavior differs greatly according to the direction of tensile stress with respect to the texture directions. The alloys are most vulnerable to SCC if stressed parallel to the short transverse grain direction, ie, parallel to the thinnest dimension of the grain and most resistant if stressed only parallel to the longest grain dimension. In practice, it is the short transverse-direction stresses that cause SCC problems. Hence prudent practice is to avoid designs in which high sustained stresses are imposed across the short transverse direction. For example, one avoids an interference-fit fastener, or a taper pin fastener oriented to stress a part across the vulnerable texture unless the alloy is inherently of low susceptibility to SCC. Table 3 summarizes alloys and tempers in various categories of susceptibility as judged from specimens having short transverse, ie, most vulnerable, orientations.

In addition to the possibility of selecting alloys having minimum SCC susceptibility while retaining other properties as needed, there are other steps possible to reduce the SCC probability: (1) avoid designs that permit water to accumulate; (2) avoid conditions in which salts, especially chlorides, can concentrate; and (3) where available and otherwise acceptable, use a clad alloy (2) (see METAL SURFACE TREATMENTS).

10.3. High Strength Steels. Steels that owe their strength to heat treatment, whether martensitic, precipitation hardened, or maraging, and whether stainless or not, are susceptible to SCC in aqueous environments, including

Table 3. **Susceptibility Classification of Commercial Wrought Aluminum Alloys in Plate Form—Short Transverse Orientation**[a]

Aqueous susceptibility	Aluminum alloy	Temper[b]
very low	1100	all
	3003, 3004, 3005	all
	5000, 5050, 5052, 5154	all
	5454, 6063	
	5086	O, H32, H34
	6061, 6262	O, T6
	Alclad: 2014, 2219, 6061, 7075	all
low	2219	T6, T8
	5086	H36
	5083, 5456	controlled
	6061	T4
	6161, 5351	all
	6066, 6070, 6071	T6
	2021	T8
	7049, 7050, 7075	T73
moderate	2024, 2124	T8
	7050, 7175	T736
	7049, 7075, 7178	T6
appreciable	2024, 2219	T3, T4
	2014, 7075, 7079, 7178	T6
	5083, 5086, 5456	sensitized
	7005, 7039	T5, T6

[a]Ref. 37.
[b]Standard mill designation.

water vapor. The primary factor in determining the degree of SCC susceptibility of a given steel is its strength. There is no sharply defined threshold strength that defines a threshold susceptibility, but above 1200 MPa (174,000 psi) yield strength the problem becomes of increasing concern, until at 1400 MPa (~200,000 psi) susceptibility is generally very high. This does not mean that steels cannot be used at such strength levels or even higher, but for steels to be used at these strengths, great care must be exercised in design such that any tensile or bending stresses are small, or that moisture is excluded from the surface of the steel.

High strength steels are also susceptible to HIC. Cadmium electroplating is a useful protection measure for steels, but hydrogen either must not be codeposited with cadmium, or if it is codeposited, as in the cyanide plating bath, hydrogen must be safely redistributed by thermal treatment before the high strength steel component is stressed. Otherwise the steel may experience HIC.

10.4. Stainless Steels. Austenitic stainless steels undergo SCC when stressed in hot aqueous environments containing chloride ion. The oxygen level of the environment can be important, probably through its effect in establishing the electrode potential of the steel. The total matrix of combinations of stress level, chloride ion and oxygen concentrations, and temperature that cause SCC has not been worked out. At about the yield strength stress and ca 290°C, 1 ppm chloride and 1 ppm dissolved oxygen are about the minimum levels that initiate SCC. At room temperature, SCC by chloride seldom occurs in austenitic stainless steels except when the steels are heavily sensitized. When these steels are not heavily sensitized, there is not usually SCC except at elevated temperatures. Whereas there is no well-defined threshold temperature for vulnerability to SCC in stainless steels, above ~60°C, it becomes of increasing concern.

The relative susceptibilities of the common grades of austenitic stainless steel to SCC in chloride do not differ greatly, although the steels that can and do become sensitized are decidedly inferior in this condition. The high purity ferritic grades of stainless steel offer appreciable improvement in SCC resistance as compared to the austenitic grades, but they are not immune to cracking. Care must also be exercised to avoid the ductile-to-brittle transition that most steels undergo as the temperature is decreased and potential embrittlement by σ-phase formation in elevated temperature service. The standard methods for avoiding chloride SCC in austenitic stainless steels include: avoidance of fabrication stresses; minimizing chloride ion level; and minimizing oxygen concentration in the environment. Additionally, where feasible, it is possible to mitigate against SCC in these alloys by cathodic protection, but excessive cathodic protection can lead to HIC.

11. Inhibitors

Corrosion inhibitors are substances that slow or prevent corrosion when added to an environment in which a metal usually corrodes. Corrosion inhibitors are usually added to a system in small amounts either continuously or intermittently. The effectiveness of corrosion inhibitors is partly dependent on the metals or alloys to be protected as well as the severity of the environment. The main factors, which must be considered before application of a corrosion inhibitor to an aqueous system, are the compatibility of the inhibitor and the metal(s), the salt concentration, the pH, the dissolved oxygen concentration, and the concentration of interfering species such as chlorides or metal cations. In addition, many inhibitors, most notably chromates, are toxic and environmental regulations limit use. (38).

Inhibitors act and are classified in a variety of ways (1,3,38,39). The classifications used herein closely follow the discussion in (38). Types of inhibitors discussed below include (1) anodic, (2) cathodic, (3) organic, (4) precipitation, and (5) vapor-phase inhibitors.

11.1. Anodic Inhibitors. Passivating or anodic inhibitors produce a large positive shift in the corrosion potential of a metal. There are two classes of anodic inhibitors which are used for metals and alloys where the anodic shift in potential promotes passivation, ie, anodic protection. The first class includes oxidizing anions that can passivate a metal in the absence of oxygen. Chromate is a classical example of an oxidizing anodic inhibitor for the passivation of steels. The second class of anodic inhibitors contains ions which need oxygen to passivate a metal. Tungstate and molybdate, eg, require the presence of oxygen to passivate a steel. The concentration of the anodic inhibitor is critical for corrosion protection. Insufficient concentrations can lead to pitting corrosion or an increase in the corrosion rate. The use of anodic inhibitors is more difficult at higher salt concentrations, higher temperatures, lower pH values, and in some cases, at lower oxygen concentrations (38).

11.2. Cathodic Inhibitors. Cathodic inhibitors act to retard or poison the cathodic reaction or selectively precipitate onto cathodic areas producing diffusion barriers to cathodic reactants, thereby reducing the rate of the cathodic reaction. There are three types of cathodic inhibitors: (1) hydrogen poisons, (2) oxygen scavengers, and (3) cathodic precipitates. Hydrogen poisons are chemical species such as arsenic or antimony that retard the hydrogen reduction reaction. Because the hydrogen poison slows the cathodic reaction, and because the cathodic and anodic reactions must proceed at the same rate, the whole of the corrosion process is slowed. A potentially serious drawback upon use of hydrogen poisons is that the hydrogen on the surface can be more easily absorbed into the metal or alloy and can lead to HIC in susceptible materials. Oxygen scavengers prevent corrosion by tying up the oxygen in solution, thereby making it unavailable for the cathodic reaction. The most common oxygen scavengers used in water at ambient temperatures are sulfur dioxide [7446-09-5] and sodium sulfate [7757-82-6]. Cathodic precipitate inhibitors such as calcium carbonate [471-34-1] or magnesium carbonate [546-93-0] precipitate onto the cathodic areas producing a film that reduces the cathodic activity (38).

11.3. Organic Inhibitors. Generally, organic inhibitors adsorb on the entire metal surface and impede corrosion reactions (38). Organic inhibitors consist of broad classes of organic compounds. For example, aliphatic organic amines (qv) adsorb by the surface-active NH_2 group that forms a chemisorptive bond with the metal surface. The hydrocarbon tails orient away from the interface toward the solution, so that further protection is provided by the formation of a hydrophobic network that excludes water and aggressive ions from the metal surface (39). Organic inhibitors influence both anodic and cathodic reactions to varying degrees depending on the potential, the chemical structure of the inhibitor, and the size of the inhibitor molecule. Soluble organic inhibitors produce a protective layer that is only a few molecules thick, whereas insoluble inhibitors added as dispersions can build a film to a thickness of several hundredths of a centimeter (38).

11.4. Precipitation and Vapor-Phase Inhibitors. Precipitation inhibitors are film-forming compounds that produce barrier films over the entire surface. Phosphates and silicates, which are the most common, do not provide the degree of protection afforded by chromate inhibitors, but are useful in situations where nontoxic additives are required. Two main drawbacks to the use of phosphates and silicates are the dependence on the water composition and the control required to achieve maximum inhibition (38,39).

Vapor-phase inhibitors are volatile compounds that adsorb onto metal surfaces, and retard or prevent corrosion by a variety of mechanisms (38). Inhibitors such as dicyclohexamine nitrate [3882-06-02] can protect a variety of metals including steel, aluminum, and tinplate. A number of vapor-phase inhibitors are commercially available as powders or tablets. However, vapor-phase inhibitors attack nonferrous metals to varying degrees, thus the manufacturers' recommendations should be checked before application. The system to be protected must be closed to maintain the volatile compound, but objects as large as the interior of an ocean-going tanker have been treated by this technique.

11.5. Environmentally Compatible Inhibitors. The recent and expected environmental regulations concerning the release of hazardous materials and the growing list of materials considered as hazardous are impacting the use of many effective inhibitors such as chromate. Thus, new research and development on environmentally acceptable inhibitors is proceeding. This is a great challenge as inhibitor and coating users and manufacturers have to replace commercial products that have decades of acceptable field performance in regard to corrosion (40,41).

12. Coatings for Corrosion Prevention

Coatings (qv) are applied to metal substrates to prevent corrosion. Generally, the coating protects the metal by imposing a physical barrier between the metal substrate and the environment. However, the coating can also act to provide cathodic protection or by serving as holding reservoirs for inhibitors. Coatings may be divided into organic, inorganic, and metallic coatings (1–3,42). For a more in-depth coverage of this topic see (1–3) (see also COATINGS, MARINE; METAL SURFACE TREATMENTS).

12.1. Coating Types. Organic coatings afford protection by providing a physical barrier and are often used as holding reservoirs for corrosion inhibitors. Organic coatings include paints, resins, lacquers, and varnishes (see also BARRIER POLYMERS; PAINT). These coatings usually comprise four basic constituents: binder, pigments and fillers, additives and solvents. The properties of the coating strongly depend on the formulation of the coating (43). The effective application of an organic coating requires (*1*) proper surface preparation of the metal substrate, (*2*) selection of the proper primer or priming coat, and (*3*) selection of the appropriate top coat(s). Poor performance of organic coatings most often results from improper surface preparation or poor application (1). Maintenance programs involving periodic inspection and repair are necessary for reliable corrosion protection by organic coatings. Organic coatings probably protect more metal on a tonnage basis than any other means of corrosion protection. As a rule, however, organic coatings should not be used in environments that would rapidly corrode the metal if the coating were compromised. For example, paint would not be used to protect the inside of a tank car used for shipping hydrochloric acid because one small coating defect would result in the rapid perforation of the tank car wall.

Inorganic coatings are also used to provide a barrier between the environment and the metal (1,2,42). Inorganic coatings include chemical conversion coatings, glass (qv) linings, enamels (see ENAMELS, PORCELAIN OR VITREOUS), and cement (qv). Chemical conversion coatings are produced by intentionally corroding the metal surface in a controlled manner. This is done so as to produce an adherent corrosion product that protects the metal from further corrosion. Anodization of aluminum, one of the more commonly used conversion coatings techniques, produces a protective aluminum oxide film on the aluminum metal. Another example of a chemical conversion coating is phosphatizing for the protection of automobile bodies. Porcelain enamel coatings, that are inert in water and resistant to most weather are routinely applied to steel, cast iron, and aluminum (4). They are commonly seen on appliances and plumbing fixtures. Glass-lined metals are used in process industries where there is concern over corrosion or contamination of the product. For example, glass-lined metals are used in the pharmaceutical industry (4). Portland cement coatings have been used to protect steel and cast-iron water pipes (2).

Often metallic coatings, in addition to providing a barrier between the metal substrate and the environment, provide cathodic protection when the coating is compromised (2,3). Metallic coatings (qv) and other inorganic coatings are produced using a variety of techniques including hot dipping, electroplating, cladding, thermal spray techniques, chemical vapor deposition, or by surface modification using directed energy (laser or ion) beams. Two classic metallic coating techniques and a relatively recent technique are described below.

13. Coating Methods

Hot dipping is one of the oldest methods used for coating metals. In hot dipping, coatings are applied by dipping the metal in a molten bath commonly of zinc, tin, lead, or aluminum (2,4). The best known hot dipping procedure is that of coating steel with zinc to produce galvanized steel. In addition to providing a barrier between the steel substrate and the environment, the zinc acts as a sacrificial anode and provides cathodic protection to the underlying steel when the coating is breached.

Electroplating (qv) consists of immersing a metal in a plating bath and electrochemically plating the solution species onto the metal substrate (1,4). Electroplating is used to provide coatings for a variety of reasons including corrosion protection, wear resistance, decoration, and to build up the dimensions of a substrate. Additives to the plating bath can improve coating properties such as grain size, strength, uniformity, and brightness. The electroplate can be a single metal, an alloy, or a sequence of layers of different composition. Coating thicknesses can range from on the order of thousandths of a mil to 20 mils (1). Zinc, nickel, tin, and cadmium are the most commonly plated materials on a tonnage basis. Electroplated tin is used as a protective coating for food cans (see FOOD PACKAGING). The electroplated tin provides a physical barrier and galvanic protection (in most food products) if the coating is compromised. However, the tin coating cannot provide galvanic protection in the presence of dissolved oxygen so that food should not remain in tin plated cans after opening (3).

The ability to modify metal surfaces using directed energy beams is a new approach being used to improve the corrosion resistance of metals (31,44). Techniques such as ion implantation (qv), ion beam mixing, and ion beam assisted deposition (IBAD), as well as laser-surface alloying and processing have been shown to improve the corrosion behavior of metals (see LASERS). These techniques are versatile and have been used to modify the cathodic or anodic reactions, improve the nature of passive films, and produce barrier coatings (31). Ion implantation has been used to improve the corrosion resistance of titanium in hot acids and aluminum and steels in aqueous chloride environments (44). However, due to capital costs and the sizes of the sample that can be treated, these techniques have not yet gained wide spread use.

BIBLIOGRAPHY

"Corrosion Inhibitors", in *ECT* 2nd ed., Vol. 6, pp. 317–346, by C. C. Nathan; "Corrosion and Corrosion Inhibitors", in *ECT* 3rd ed., Vol. 7, pp. 113–142, by R. T. Foley and B. F. Brown, The American University; in *ECT* 4th ed., Vol. 7, pp. 548–572, by Patrick & Moran, United States Naval Academy, and Faul M. Natishan, Naval Research Laboratory; "Corrosion and Corrosion Control" in *ECT* (online), posting date: December 4, 2000, by Patrick J. Moran, United States Naval Academy and Paul M. Natishan, Naval Research Laboratory.

CITED PUBLICATIONS

1. M. G. Fontana, *Corrosion Engineering*, McGraw-Hill Book Co., Inc., New York, 1986.
2. H. H. Uhlig, and R. W. Revie, *Corrosion and Corrosion Control*, John Wiley & Sons, Inc., New York, 1985.
3. D. A. Jones, *Principles and Prevention of Corrosion*, Macmillan Publishing, New York, 1992.
4. *ASM Handbook*, ASM International, Materials Park, 1987.
5. A. J. Sedriks, *Corrosion of Stainless Steels*, John Wiley & Sons, Inc., New York, 1979.
6. Z. Szklarska-Smialowska, *Pitting Corrosion of Metals*, National Association of Corrosion Engineers, Houston, Tex., 1986.
7. H. L. Logan, *Corrosion Basics: An Introduction*, L. S. V. Delinder, ed., National Association of Corrosion Engineers, Houston, Tex., 1984.
8. S. C. Dexter, *Bull. Electrochem.* **12**, 1 (1996).
9. S. C. Dexter, and P. Chandrasekaran, *Biofouling* **15**, 313 (2000).
10. B. Little, P. Wagner, and F. Mansfeld, *Int. Mater. Rev.* **36**, 253 (1991).
11. R. Pope, B. Little, and R. Ray, *Biofouling* **16**, 83 (2000).

12. M. Pourbaix, *Lectures on Electrochemical Corrosion*, Plenum Press, New York-London, 1973.
13. M. Pourbaix, *Atlas of Electrochemical Equilibria in Aqueous Solutions*, NACE, Houston, Tex., 1974.
14. J. Tafel, *Z. Physik. Chem.* **50**, 641 (1905).
15. J. O. M. Bockris and A. K. N. Reddy, *Modern Electrochemistry*, Plenum Press, New York, 1970.
16. H. Kita, *J. Electrochem. Soc.* **113**, 1095 (1966).
17. R. Baboian, *Electrochemical Techniques in Corrosion Engineering*, NACE, Houston, Tex., 1986.
18. U. R. Evans, *J. Soc. Chem. Ind.* **43**, 315 (1924).
19. W. H. J. Vernon, *J. Sci. Inst.* **22**, 226 (1945).
20. C. W. Borgmann, *Ind. Eng. Chem.* **29**, 814 (1937).
21. W. Whitman, R. Russell, and V. Atteri, *Ind. Eng. Chem.* **16**, 665 (1924).
22. J. Kruger, *Int. Mater. Rev.* **33**, 113 (1988).
23. H. H. Strehblow, *Werkstoffe Korros.* **35**, 437 (1984).
24. Z. Szklarska-Smialowska, *Corrosion* **27**, 223 (1971).
25. G. S. Frankel, *J. Electrochem. Soc.* **145**, 2970 (1998).
26. W. H. J. Vernon, and L. Whitby, *J. Inst. Met.* **44**, 389 (1930).
27. W. H. J. Vernon, *J. Inst. Met.* **49**, 153 (1932).
28. *Corrosion in Action*, The International Nickel Co, New York, 1977.
29. A. J. Sedriks, *ASM Handbook*, ASM International, Materials Park, 2002.
30. V. Ashworth, W. A. Grant, and R. O. M. Procter, *Corr. Sci.* **16**, 661 (1976).
31. E. McCafferty, G. K. Hubler, P. M. Natishan, P. G. Moore, R. A. Kant, and B. D. Sartwell, *J. Mater. Sci. Eng.* **86**, 1 (1987).
32. W. C. Moshier, and G. D. Davis, *J. Electrochem. Soc.* C145 (1988).
33. P. M. Natishan, E. McCafferty, and G. K. Hubler, *J. Electrochem. Soc.* **135**, 321 (1988).
34. A. R. Srivatsa, C. R. Clayton, and J. K. Hirvonen, *Advances in Coating Technology for Corrosion and Wear Resistant Coatings*, TMS, Warrendale, 1995.
35. R. G. Buchheit, *J. Electrochem. Soc.* **142**, 3994 (1995).
36. L. S. V. Delinder, *Corrosion Basics: An Introduction*, National Association of Corrosion Engineers, Houston, Tex., 1984.
37. E. H. Spuhler, and C. L. Burton, in *ALCOA Green Letter*, Aluminum Company of America, Pittsburgh, 1970.
38. N. Hackerman, and E. S. Snavelyin, *Corrosion Basics: An Introduction*, L. S. V. Delinder, ed., NACE, Houston, Tex., 1984.
39. McCaffertyE., in J. H. Delinder, ed., *Corrosion Control by Coatings*, Science Press, Princeton, N. J. 1979, p. 279.
40. S. R. Taylor, H. S. Isaacs, and E. W. Brooman, *Environmentally Acceptable Inhibitors and Coatings*, The Electrochemical Society, Inc., Pennington, 1997.
41. D. C. Hansen, and E. McCafferty, *J. Electrochem. Soc.* **143**, 114 (1996).
42. N. E. Hamner, *Corrosion Basics: An Introduction*, L. S. V. Delinder, ed., NACE, Houston, Tex., 1984.
43. J. H. W. D. Wit, in P. Marcus and J. Oudar, eds., *Corrosion Mechanisms in Theory and Practice*, Marcel Dekker, Inc., New York, 1995, p. 581.
44. McCaffertyE., P. M. Natishan, and G. K. Hubler, *Nucl. Instr. Meth.* **B56**, 639 (1991).

PAUL NATISHAN
Naval Research Laboratory

PATRICK MORAN
U.S. Naval Academy

COSMETICS

1. Introduction

Cosmetics are products created by the cosmetic industry and marketed directly to consumers. The cosmetic industry is dominated by manufacturers of finished products but also includes manufacturers who sell products to distributors as well as suppliers of raw and packaging materials. Cosmetics represent a large group of consumer products designed to improve the health, cleanliness, and physical appearance of the human exterior and to protect a body part against damage from the environment. Cosmetics are promoted to the public and are available without prescription.

The difference between a cosmetic and a drug is often confusing. In the United States the inclusion of a drug constituent, as defined by the Food and Drug Administration (FDA), in a cosmetic product may make the product a drug; whenever there is a claim for pharmacological activity of one of a product's constituents, the product is a drug. Some products are identified as quasi or over-the-counter (OTC) drugs according to each country's regulations. The composition, claim structure, and distribution of OTC products may be more tightly regulated than those of pharmacologically inactive cosmetics. The difference between an ordinary cosmetic and a quasi or OTC drug may not be readily apparent; it is based on statutory regulations. Certain types of products, such as hair-growth products and skin rejuvenators, are not cosmetics, and OTC claims for hair growth or skin rejuvenation are not allowed in the United States.

Cosmetics, regardless of form, can be grouped by product use into the following seven categories: (1) skin care and maintenance, including products that soften (emollients and lubricants), hydrate (moisturizers), tone (astringents), protect (sunscreens), etc, and repair (antichapping, antiwrinkling, antiacne agents); (2) cleansing, including soap, bath preparations, shampoos, and dentifrices (qv); (3) odor improvement by use of fragrance, deodorants, and antiperspirants; (4) hair removal, aided by shaving preparations, and depilatories; (5) hair care and maintenance, including waving, straightening, antidandruff, styling and setting, conditioning, and coloring products (see HAIR PREPARATIONS); (6) care and maintenance of mucous membranes by use of mouthwashes, intimate care products, and lip antichapping products; and (7) decorative cosmetics, used to beautify eyes, lips, skin, and nails.

2. History

Cosmetic preparations and usage are rooted in antiquity, when suspensions of natural pigments in lipids were evidently used to enhance appearance, and fragrant plant concoctions were widely traded. The use of cosmetics for adornment is recorded in biblical writings, and the use of soap (qv), probably a hydrolysate of animal lipids by wood ashes, was encouraged for cleanliness. The benefits of bathing were fully known to the ancients, who built elaborate bathhouses. Bathing became less popular in Western cultures during the Middle Ages but again became accepted during the eighteenth and nineteenth centuries.

The use of fragrant substances has been continuous, and the use of lipids or emollients for anointing is fully documented in historical writings. However, it is probably not justifiable to identify the recipes passed on from antiquity as cosmetics. The compositions based on folklore and mysticism were replaced by more scientifically acceptable products beginning about 1875. The first edition of a handbook of cosmetic chemistry published in 1920 included a foreword noting that scientific cosmetic chemistry did not exist prior to that publication (1). A few years later, texts on cosmetic chemistry and other formularies became available (2).

The Society of Cosmetic Chemists, with individual memberships, was founded in the United States after World War II, based on the belief that scientific expertise and exchange were the foundations for future expansion of the cosmetic industry. Prior to that time, knowledge of cosmetic formulation was jealously guarded. Related scientific societies emerged in other countries and have since joined to form the International Federation of Societies of Cosmetic Chemists.

3. Economic Aspects

Economic summaries of the cosmetic industry, commonly documented by sales volume, are sometimes based on unit sales, sometimes on manufacturers' sales in monetary units, and sometimes on consumer spending. Figures normally include contributions by private labeling operations but do not necessarily reflect the value of the industry service sector, which includes suppliers of raw materials, beauticians, testing laboratories, and other specialists. Moreover, product categories cannot be rigidly defined. For example, the differentiation between a deodorant (a cosmetic) and an antiperspirant (an OTC drug) is often obscured by its trade name.

A summary by broad categories is given in Table 1. A more detailed breakdown of U.S. cosmetic sales is provided in Table 2. The U.S. Commerce Department reports a modest (8%) increase in the value of cosmetic industry shipments between 1989 and 1991, despite a 13% decrease in total industry employment.

Numerous cosmetic trade organizations exist. Foremost among them are the Cosmetic, Toiletry, and Fragrance Association (CTFA), formerly the Toilet Goods Association; the European Cosmetics Industries Federation (COLIPA);

Table 1. **U.S. Manufacturers' Sales of Cosmetics and Toiletries**[a]

Product line	Annual sales, $ \times 10^6$		Average annual growth, %
	1985	1990	
hair care	3,630	4,760	5.5
skin care	2,500	3,510	7.0
color cosmetics	2,540	3,380	5.9
fragrances	2,370	2,840	3.7
other toiletries	4,160	5,270	4.8
Total	15,200	19,760	5.4

[a]Courtesy of Kline & Company, Fairfield, N.J.

Table 2. **1991 U.S. Cosmetic Sales**[a]

Product line	Sales, $\$ \times 10^9$
hair care	5.10
fragrances	3.90
makeup (color)	3.00
skin care	1.80
antiperspirant and deodorants	1.60
dentifrices	1.20
mouthwashes	0.58
shaving	0.58
hair coloring	0.57
sun care	0.40
nail care	0.37
Total	20.00

[a]Based on U.S. Department of Commerce *Economic Industry Reports* and estimates by the staff of *Drug and Cosmetic Industry*.

and the Japanese Cosmetic Industry Association (JCIA). These organizations provide member companies with regulatory and technical information and supply documentation on the industry's practices to governments and consumers. The cosmetic industry supports a number of trade journals. Comprehensive annual listings of companies, individuals, and products are available (3–5).

4. Regulation of the Cosmetic Industry

In the United States, the 1938 revision of the Federal Food and Drug Act regulates cosmetic products and identifies these materials as:

(*1*) articles intended to be rubbed, poured, sprinkled, or sprayed on, introduced into, or otherwise applied to the human body or any part thereof for cleansing, beautifying, promoting attractiveness, or altering the appearance, and (*2*) articles intended for use as a component of any such articles, except that such term shall not include soap.

This definition establishes the legal difference between a drug and a cosmetic. It is clearly the purpose of, or the claims for, the product, not necessarily its performance, that legally classifies it as a drug or a cosmetic in the United States. For example, a skin-care product intended to beautify by removing wrinkles may be viewed as a cosmetic because it alters the appearance and a drug because it affects a body structure.

The FDA is responsible for enforcing the 1939 act as well as the Fair Packaging and Labeling Act. In light of the difficulty of differentiating between cosmetics and drugs, the FDA has in recent years implemented its regulatory power by concluding that certain topically applied products should be identified as OTC drugs. As a group, these OTC drugs were originally considered cosmetics and remain among the products distributed by cosmetic companies. They include acne, antidandruff, antiperspirant, astringent, oral-care, skin-protectant, and sunscreen products.

The use or presence of poisonous or deleterious substances in cosmetics and drugs is prohibited. The presence of such materials makes the product "adulterated" or "misbranded" and in violation of good manufacturing practices (GMP), which are applicable to drugs and, with minor changes, to cosmetics (6).

In contrast to prescription drugs, OTC drugs and cosmetics are not subject to preclearance in the United States. However, the rules covering OTC drugs preclude introduction of untested drugs or new combinations. A "new chemical entity" that appears suitable for OTC drug use requires work-up via the new drug application (NDA) process. In contrast, the use of ingredients in cosmetics is essentially unrestricted and may include less well known substances.

4.1. Color Additives. The FDA has created a unique classification and strict limitations on color additives (see also COLORANTS FOR FOOD, DRUGS, COSMETICS, AND MEDICAL DEVICES). Certified color additives are synthetic organic dyes that are described in an approved color additive petition. Each manufactured lot of a certified dye must be analyzed and certified by the FDA prior to usage. Color lakes are pigments (qv) that consist of an insoluble metallic salt of a certified color additive deposited on an inert substrate. Lakes are subject to the color additive regulations of the FDA and must be certified by FDA prior to use. Noncertified color additives require an approved color additive petition, but individual batches need not be FDA certified prior to use.

Hair colorants, the fourth class of color additives, may be used only to color scalp hair and may not be used in the area of the eye. Use of these colorants is exempt, that is, coal-tar hair dyes may be sold with cautionary labeling, directions for preliminary (patch) testing, and restrictions against use in or near the eye. The FDA diligently enforces the rules governing color additives and limits the use of, or even delists colorants deemed unsafe. The list of substances specifically prohibited for use in cosmetics is short.

Under the Fair Packaging and Labeling Act, the FDA has instituted regulations for identifying components of cosmetics on product labels. To avoid confusion, the CTFA has established standardized names for about 6000 cosmetic ingredients (7). Rigid U.S. labeling requirements mandate that ingredients be listed in order of descending concentration. Similar regulations are expected for European cosmetics within the next few years (8).

4.2. European Regulations. Regulations for cosmetics differ from country to country but, in general, are similar to or patterned after U.S. regulation. Thus, the identification of a cosmetic in the European Community differs only marginally from that in the United States. A 1991 European Economic Community (EEC) directive defines a cosmetic as:

any substance or preparation intended for placing in contact with the various external parts of the human body (epidermis, hair system, nails, lips and external genital organs) or with the teeth and the mucous membranes of the oral cavity with a view to cleaning them, perfuming them, protecting them, keeping them in good condition, changing their appearance and/or correcting body odours.

EEC proposals also assert that cosmetic products must not damage human health when applied under normal or reasonably foreseeable conditions of use.

Finally, the EEC directive states that the label of a cosmetic should include a list of ingredients in descending order of weight at the time of manufacture.

Although the EEC is still in the process of completing cosmetic regulation, the final directive is expected to require member states to ban marketing of cosmetics that contain prohibited ingredients, an amount of a substance in excess of that proscribed, coloring agents, preservatives, or uv filters not specifically allowed.

4.3. Japanese Regulation.   Japanese regulations of cosmetics are similar to those already discussed. The safety and quality of cosmetic products are regulated under the Pharmaceutical Affairs Law with detailed requirements for approval and licensing of manufacturing and import, for labeling and advertising standards, and for reporting safety data to the Ministry of Health and Welfare.

Cosmetics in Japan are defined as externally used articles for cleaning, beautifying, promoting attractiveness, and altering the appearance of the human body and for keeping the skin and hair healthy, provided that the action of the article on the human body is mild. Articles intended for use in diagnosis, treatment, or prevention of disease and those intended to affect the structure or any function of the body are identified as quasi drugs and are excluded. Japanese law identifies the following as quasi drugs: products for the prevention of foul breath or body odor; products for the prevention of prickly heat; products for the prevention of hair loss, promotion of hair growth, or removal of hair; hair dyes; agents for permanent waving of hair; and agents combining cosmetic effects with the purpose of preventing acne, chapping, itchy skin rashes, chilblains, or disinfection of the skin or mouth.

The Japanese regulations include both a list of substances that may not be used in cosmetics and a list of ingredients that may be used but which must conform to the specifications of the Japanese Standards of Cosmetic Ingredients. Some ingredients are allowed only in some types of cosmetic preparations. The use of certain hormones is controlled, and concentration limits exist for another group of ingredients. Japanese regulations differ significantly from U.S. regulations with regard to the formal recognition of an allowed, or positive, list.

Regulatory changes and discussions of the impact of regulations on the manufacture and import of cosmetic products are available in manuals published by the CTFA (9).

5. Product Requirements

5.1. Safety.   Cosmetic products must meet acceptable standards of safety during use, must be produced under sanitary conditions, and must exhibit stability during storage, shipment, and use. Cosmetics are not lifesaving or life-prolonging drugs, and the requirements for innocuousness are absolute. In the United States the manufacturer bears the responsibility for not using injurious or questionable ingredients. The safety of each ingredient used in each finished cosmetic product must be adequately substantiated prior to marketing. In countries that have positive lists of ingredients that may be used in cosmetics, the burden for testing each finished cosmetic products is reduced. Positive listing assumes, without requiring evidence, that no adverse effects result from the use of a mixture of safe ingredients.

For many years the safety of cosmetic ingredients has been established using a variety of animal safety tests. More recently animal welfare organizations have urged that this type of safety testing be abandoned. Despite widespread use of cosmetics without professional supervision, the incidence of injury from cosmetic products is rare. In part, this is the result of extensive animal safety testing of components as well as of finished products. Such animal testing was considered mandatory from about 1945 to about 1985. Since the mid-1980s animal testing has been significantly reduced.

In vitro safety testing technology is becoming more common. Validation of these methods is based on comparisons with early animal safety data. Whether these *in vitro* tests can ensure the safety of all products that reach the consumer cannot be predicted (10). As of this writing, the principle that *in vitro* testing may be substituted for *in vivo* testing for complete safety substantiation has not been accepted by regulatory agencies. In the United States, the CTFA created the Cosmetic Ingredient Review (CIR) for the purpose of evaluating existing *in vitro* and *in vivo* data and reviewing the safety of the ingredients used in cosmetics. The review of ingredients is prioritized based on frequency of use, concentration used, the area of use, and consumer complaints. The CIR conclusions are available from the CTFA.

In addition to the CIR process the cosmetic industry has instituted a second, important, self-regulatory procedure: the voluntary reporting of adverse reactions, which is intended to provide data on the type and incidence of adverse reactions noted by consumers or by their medical advisors. This reporting procedure creates early awareness of problems handled outside hospital emergency facilities or centers for acute poisoning.

Safety testing of a finished cosmetic product should be sufficient to ensure that the product does not cause irritation when used in accordance with directions, neither elicits sensitization nor includes a sensitizer, and does not cause photoallergic responses. Some of the methods for determining animal or human responses to cosmetics are noted in Table 3.

A particularly critical test for establishing the safety of cosmetics is the exaggerated-use test, in which panelists, often under medical supervision, use a product at frequencies that exceed the normally expected usage. Any adverse reactions, including subjective reports of burning or itching without clinical symptoms, suggest that the product should be examined further. This test also can be used to elicit comments concerning product acceptability.

Repeated usage of certain common cosmetic ingredients can elicit a response within the sebaceous gland apparatus that generates comedos. The cause of this phenomenon is not entirely clear, but an animal (rabbit ear) test purportedly measures the comedogenic potential of cosmetic ingredients or finished products (24). Controversy surrounds the identity of comedogenic substances and the concentration required to elicit the response. Thus use of cosmetic ingredients that have been suspected of causing comedogenicity are generally avoided.

5.2. Production Facilities. The manufacture of acceptable cosmetic products requires not only safe ingredients but also facilities that maintain

Table 3. **Cosmetic Safety Tests**

Determination	Test vehicle	
	Animals	Humans

primary irritation	acute
	albino rabbit with or
	    without scarification[a]
		chronic (21 days)
		occlusive; chamber
		    scarification[b]
		Duhring soap chamber[c]

contact allergy	guinea pig maximization
	    (with or without
	    Freund's omplement adjuvant)[d]
		repeat patch test
		    (open and closed);
		    maximization test[e]

phototoxicity	hairless mouse and other species plus uvA irradiation[f]	human test[f,g]
photoallergy	guinea pig plus uvA/uvB irradiation[h]	photomaximization test[i]
eye irritation	Draize rabbit eye test[j]	eye instillation using diluted product
comedogenicity sting test	rabbit ear[k]	in-use test face[l,m]

[a] Ref. 11.   [b] Ref. 12.   [c] Ref. 13.   [d] Ref. 14.   [e] Ref. 15.   [f] Ref. 16.   [g] Ref. 17.
[h] Ref. 18.   [i] Ref. 19.   [j] Ref. 20.   [k] Ref. 21.   [l] Ref. 22.   [m] Ref. 23.

high standards of quality and cleanliness. Most countries have established regulations intended to assure that no substandard product or batch is distributed to consumers. Good Manufacturing Practices (GMP) represent workable standards that cover every aspect of drug manufacture, from building construction to distribution of finished products. GMPs in the United States that have been established for drug manufacture are commonly used in cosmetic production (6,25).

5.3. Contamination.   Manufacturers of cosmetics must be careful to guard against chemical and microbial contamination. Chemical contamination, which may result from the presence of undesirable impurities in raw materials, is avoidable by adhering to rigid specifications for raw materials. Compendial specifications and publications by the CTFA and other professional societies form the basis of most intracompany raw material specifications. Moreover, all packaging components must meet not only physical and design specifications but also such chemical requirements as extractables and absence of dust and similar contaminants (see PACKAGING, COSMETICS AND PHARMACEUTICALS).

Chemical contamination arising from overheating or other decomposition reactions during processing or from improper storage of incoming supplies must also be avoided. For these reasons, adherence to documented production processes and periodic reassays of stored supplies are required. Additionally, final chemical or physical examinations of the finished and filled products are required to ascertain that no inadvertent chemical contamination has occurred during manufacture and that no undesirable ingredients are present.

An entirely different type of contamination arises from the presence of microbiota in a product. As in the case of chemical contamination, compendial requirements for microbiological purity exists. Pharmacopoeial standards vary from country to country, and manufacturers must use the specifications and kill times that meet local requirements. As of this writing, the criteria in the

British Pharmacopoeia are more stringent than those established by the CTFA, which are stricter than those in the *United States Pharmacopoeia*. In order to meet commonly accepted standards of microbial purity, manufacturing facilities must be periodically cleaned and all products that can support microbial growth must contain an effective preservative (6).

5.4. Stability. An additional mandatory requirement for cosmetic products is chemical and physical stability. Interactions between ingredients that lead to new chemical entities or decomposition products are unacceptable. Stability testing becomes particularly critical if the product includes an active or drug constituent for which a specific performance claim is made. In the absence of an expiration date, a cosmetic product or an OTC drug should be stable for 60 months at ambient temperature. This temperature is a function of climatic zones. Therefore, controlled temperature storage, sometimes at controlled relative humidity, is universally recognized as ideal despite its attendant cost. In order to demonstrate long-term chemical stability on the basis of short- or intermediate-term studies, formulations are stored routinely at elevated temperatures, normally 37, 45, or 50°C. Changes are extrapolated to ambient temperatures using the Arrhenius equation for reaction rates.

Another type of chemical change is initiated by light, which may trigger autolytic, that is, free radical (Type I) or singlet oxygen (Type II) reactions. These changes are routinely classified as oxidation. Rancidity in cosmetics, especially those containing unsaturated lipids, is commonly prevented by use of antioxidants (qv).

Requirements for physical stability in cosmetics are not as rigid as those for chemical stability. As a rule, minor changes in viscosity or appearance are acceptable to users. More drastic changes, resulting from separation of an emulsion because of creaming or oiling, are not acceptable. Short-term physical, or viscosity, changes cannot be extrapolated to long-term performance. Changes observed during static viscosity tests have little predictive value for long-term viscosity or emulsion stability. Short-term dynamic viscosity tests also do not allow prediction of long-term viscosity changes, but these can sometimes be used to predict changes in the nature of emulsions. Zeta potential and particle size determination can provide predictive information on emulsion behavior.

5.5. Performance. Consumer acceptance is a criterion on which cosmetic marketers cannot compromise. Whereas the likes and dislikes of consumers are in a state of constant flux, some product features are critical. A deodorant that does not deodorize or a hair coloring that fades in sunlight is unacceptable. Performance is tested by *in vitro* techniques during formulation, but the ultimate test of a product's performance requires in-use experience with consumers and critical assessment by trained observers. Performance tests can sometimes be combined with in-use safety tests, and protocols for such programs have been developed.

6. Ingredients

Manufacturers of cosmetics employ a surprisingly large number of raw materials. Some of these ingredients are active constituents that have purported beneficial effects on the skin, hair, or nails, for example, acting as moisturizers or conditioners. These substances are generally used in limited quantities. Other ingredients are used to formulate or create the vehicle. These are bulk chemicals used in comparatively large amounts. The resulting combination of various substances affects the nature (viscosity, oiliness, etc) of the finished cosmetic. As a rule numerous combinations and permutations are tested to optimize textural characteristics and to match these to consumers' preferences. Finally, cosmetics may include substances added primarily to appeal to consumers. These ingredients need not contribute appreciably to product performance.

About 6000 different cosmetic ingredients have been identified (7). These can be divided into smaller groups according to chemical similarity or functionality. Table 4 represents a breakdown by functionality on the skin or in the product. The chemical identity of only one ingredient that performs the desired function is given. In most cases, other equally effective substances exist. The diversity of functions required in cosmetics is evident, and cosmetic ingredients may perform more than one function or belong to more than one chemical class. A typical example is sodium DL-2-pyrrolidinone-5-carboxylate (sodium PCA) [28874-51-3], $NaC_5H_7NO_3$. Chemically, this compound may be viewed as an amide, a heterocyclic compound, or an organic salt; functionally, it is a humectant and skin-conditioning agent.

Ingredients exhibiting certain functions are required in many types of cosmetic products. Antioxidants and preservatives are especially critical for product shelf life and quality during usage. Shelf life is defined herein as that period of time during which a product in an unopened package maintains its quality and performance and shows no physical or chemical instability. Antioxidants and preservatives do not contribute to physical stability but are included in cosmetic products to ensure oxidative stability and to control microbial contamination. Once a package has been opened, oxidative processes may cause the product to deteriorate, and microbial species may gain access to the product. These additives are expected to impart some protection even under these circumstances.

6.1. Antioxidants. Some antioxidants useful in cosmetics are listed in Table 5. The operant mechanisms are interference with radical propagation reactions, reaction with oxygen, or reduction of active oxygen species. Antioxidants are intended to protect the product but not the skin against oxidative damage resulting from ultraviolet radiation or singlet oxygen formation.

6.2. Preservatives. Several microorganisms can survive and propagate on unpreserved cosmetic products. Preservatives are routinely added to all preparations that can support microbial growth. The choice of a preservative for a given product is difficult. Anhydrous preparations and products containing high levels of ethanol or i-propanol may not require the addition of preservatives.

Contamination during manufacture is common, even when microbially clean ingredients are used. Water, which is almost ubiquitous in cosmetic products, is especially troublesome and must be free from contaminating microorganisms. All other ingredients should be screened for the presence of microbial species and batches of raw materials of dubious purity may have to be rejected. Cleanliness during manufacture, processing, and filling must be strictly maintained. Despite these precautions, microbial integrity of products may require the presence of one or more preservatives that are compatible with the product's

Table 4. **Cosmetic Functions and Representative Ingredients**[a]

Function	Ingredient[b]	Molecular formula	CAS Registry number
Biologically active agents			
antiacne	salicylic acid	$C_7H_6O_3$	[69-72-7]
anticaries	monosodium fluorophosphate	Na_2HPO_3F	[10163-15-2]
antidandruff	zinc pyrithione	$C_{10}H_8N_2O_2S_2Zn$	[13463-41-7]
antimicrobial	benzalkonium chloride		[8001-54-5]
antiperspirant	aluminum chlorohydrate	$Al_2ClH_5O_5$	[12042-91-0]
biocides	triclosan	$C_{12}H_7Cl_3O_2$	[3380-34-5]
sunscreen	octyl methoxycinnamate	$C_{18}H_{26}O_3$	[5466-77-3]
skin protectant	dimethicone	$(C_2H_6OSi)_nC_4H_{12}Si$	[9006-65-9]
			[63148-62-9]
		$(C_2H_6OSi)_n$	[9016-00-6]
external analgesic	methyl salicylate	$C_8H_8O_3$	[119-36-8]
Nonbiologically active agents			
abrasive			
skin	oatmeal		
teeth	dicalcium phosphate	$Ca_2(HPO_4)_2$	[7757-93-9]
antifoam	simethicone		[8050-81-5]
antioxidant	ascorbic acid	$C_6H_8O_6$	[50-81-7]
antistatic agent	dimethylditallow alkylammonium chlorides		[68783-78-8]
binder	hydroxypropylcellulose		[9004-64-2]
chelator	hydroxyethyl ethylenediamine triacetic acid (HEDTA)	$C_{10}H_{18}N_2O_7$	[150-39-0]
colorant			
pigment	ultramarine	$Na_7Al_6Si_6O_{24}S_2$	[1317-97-11]; [1345-00-2]; [12769-96-9]
dye	FD&C Red No. 4	$C_{18}H_{16}N_2O_7S_2 \cdot 2Na$	[4548-53-2]
emulsion stabilizer	xanthan gum		[11138-66-2]
film former	PVP	$(C_6H_9NO)_x$	[9003-39-8]
hair colorant	*p*-phenylenediamine	$C_6H_8N_2$	[106-50-3]
hair conditioner	sodium lauroamphoacetate	$Na_2C_{18}H_{35}N_2O_3 \cdot HO$	[14350-96-0]
humectant	glycerol	$C_3H_8O_3$	[56-81-5]
deodorant			
mouth	zinc chloride	$ZnCl_2$	[7646-85-7]
external	cetylpyridinium chloride	$C_{21}H_{38}ClN$	[123-03-5]
preservative	propylparaben	$C_{10}H_{12}O_3$	[94-13-3]
emollient	octyl stearate	$C_{26}H_{52}O_2$	[22047-49-0]
skin-conditioning agent			
general	pyrrolidinone carboxylic acid (PCA)	$C_5H_7NO_3$	[98-79-3]
occlusive	petrolatum	C_nH_{2n+2}	[8009-03-8]
film forming	hyaluronic acid		[9004-61-9]
solvent	ethanol	C_2H_6O	[64-17-5]
cleansing agent	sodium lauryl sulfate	$C_{12}H_{25}NaO_4S$	[151-21-3]
emulsifying agent	polysorbate 65		[9005-71-4]
foam booster	cocamide DEA		[68140-00-1]
suspending agent	sodium lignosulfonate		[8061-51-6]
hydrotrope	sodium toluenesulfonate		[12068-03-0]
viscosity-controlling agent			
decrease	propylene glycol	$C_3H_8O_2$	[57-55-6]
increase	hydroxypropylmethyl-cellulose		[9004-65-3]

[a]Additional functions may be found in Ref. 26.
[b]CTFA adopted names are used; this notation is used for cosmetic labeling.

Table 5. **Free-Radical-Inhibiting Antioxidants or Reductants Useful in Cosmetics**[a,b]

Antioxidant	CAS Registry number	Molecular formula
ascorbic acid	[50-81-7]	$C_6H_8O_6$
ascorbyl palmitate	[137-66-6]	$C_{22}H_{38}O_7$
butylated hydroxyanisole (BHA)	[25013-16-5]	$C_{11}H_{16}O_2$
butylated hydroxytoluene (BHT)	[128-37-0]	$C_{15}H_{24}O$
t-butyl hydroquinone	[1948-33-0]	$C_{10}H_{14}O_2$
cysteine	[52-90-4]	$C_3H_7NO_2S$
dilauryl thiodipropionate	[123-28-4]	$C_{30}H_{58}O_4S$
dodecyl gallate	[1166-52-5]	$C_{19}H_{30}O_5$
ellagic acid	[476-66-4]	$C_{14}H_6O_8$
erythorbic acid	[98-65-6]	$C_6H_8O_6$
kaempferol	[520-18-3]	$C_{15}H_{10}O_6$
nordihydroguaiaretic acid	[500-38-9]	$C_{18}H_{22}O_4$
propyl gallate	[121-79-9]	$C_{10}H_{12}O_5$
quercetin	[117-39-5]	$C_{15}H_{10}O_7$
sodium ascorbate	[134-03-2]	$C_6H_7NaO_6$
sodium sulfite	[7757-83-7]	Na_2SO_3
thioglycolic acid	[68-11-1]	$C_2H_4O_2S$
tocopherol	[59-02-9]; [1406-18-4]	$C_{28}H_{48}O_2$

[a]Ref. 26 includes a more comprehensive listing.
[b]Use levels are normally about 0.1% and rarely exceed 0.2%.

ingredients. Products should not support the growth or viability of any microbial species that may have been accidentally introduced. Preservatives are also required to reduce contamination by consumers during normal use. Powerful preservative action to create self-sterilizing products is required. Whereas production of sterile cosmetics may be practicable, maintenance of sterility during use is problematical, because fingers and cosmetic applicators are not sterile.

Pharmacopoeias and CTFA publications provide guidelines for challenge test procedures and limits on microbial counts (25). The compendial requirements for kill of microorganisms vary significantly, and alternative test methods may be required (27). As a general rule, pathogenic organisms should be absent (28). Table 6 lists a number of antimicrobial preservatives used in cosmetic products. Experience has shown that some of the most commonly used preservatives are inactivated by a variety of surfactants. For example, the parabens (esters of p-hydroxybenzoic acid) are exceptionally sensitive to the presence of nonionic surfactants, presumably as a result of micellization of the antimicrobial by the surfactant. Over the years, preservation problems have resulted in the introduction into cosmetics of unusual substances that exhibit suitable antimicrobial spectra. However, some of these ingredients reportedly are irritants or sensitizers. Controversies in the scientific literature over the use of these substances are aggravated by regulatory acceptance or prohibition, which may differ from country to country. Table 6 includes preservatives that may be barred in certain countries.

Local restrictions concerning the inclusion of preservatives and other constituents are dependent on the cosmetic product's method of use. Products that are allowed to remain on the skin are differentiated from those that are meant to

Table 6. **Antimicrobial Preservatives Useful in Cosmetics**[a,b]

Name	CAS Registry number	Molecular formula
benzoic acid[c]	[65-85-0]	$C_7H_6O_2$
benzyl alcohol	[100-51-6]	C_7H_8O
5-bromo-5-nitro-1,3-dioxane	[30007-47-7]	$C_4H_6BrNO_4$
2-bromo-2-nitropropane-1,3-diol	[52-51-7]	$C_3H_6BrNO_4$
butylparaben	[94-26-8]	$C_{11}H_{14}O_3$
calcium propionate	[4075-81-4]	$CaC_6H_{10}O_4$
chlorobutanol	[57-15-8]	$C_4H_7Cl_3O$
m-cresol	[108-39-4]	C_7H_8O
o-cresol	[95-48-7]	C_7H_8O
p-cresol	[106-44-5]	C_7H_8O
DEDM hydantoin	[26850-24-8]	$C_9H_{16}N_2O_4$
dehydroacetic acid	[520-45-6]	$C_8H_8O_4$
diazolidinyl urea	[278-92-2]	$C_{11}H_8O_2$
dimethyl oxazolidine	[51200-87-4]	$C_5H_{11}NO$
DMDM hydantoin	[6440-58-0]	$C_7H_{12}N_2O_4$
7-ethylbicyclooxazolidine	[7747-35-5]	$C_7H_{13}NO_2$
ethylparaben	[120-47-8]	$C_9H_{10}O_3$
formaldehyde	[50-00-0]	CH_2O
glutaral	[111-30-8]	$C_5H_8O_2$
glyoxal	[107-22-2]	$C_2H_2O_2$
imidazolidinyl urea	[39236-46-9]	$C_{11}H_{16}N_8O_8$
iodopropynyl butylcarbamate	[55406-53-6]	$C_8H_{12}INO_2$
isobutylparaben	[4247-02-3]	$C_{11}H_{14}O_3$
isopropylparaben	[4191-73-5]	$C_{10}H_{12}O_3$
MDM hydantoin	[116-25-6]	$C_6H_{10}N_2O_3$
methylchloroisothiazolinone	[26172-55-4]	C_4H_4ClNOS
methyldibromoglutaronitrile	[35691-65-7]	$C_6H_6Br_2N_2$
methylisothiazolinone	[2682-20-4]	C_4H_5NOS
methylparaben	[99-76-3]	$C_8H_8O_3$
phenethyl alcohol	[200-456-2]	$C_8H_{10}O$
phenol	[108-95-2]	C_6H_6O
phenoxyethanol	[122-99-6]	$C_8H_{10}O_2$
phenylmercuric acetate	[62-38-4]	$HgC_8H_8O_2$
phenylmercuric benzoate	[94-43-9]	$HgC_{13}H_{10}O_2$
phenylmercuric borate	[102-98-7]	$HgC_6H_7BO_3$
o-phenylphenol	[90-43-7]	$C_{12}H_{10}O$
propylparaben	[94-13-3]	$C_{10}H_{12}O_3$
Quaternium-14	[27479-28-3]	$C_{23}H_{42}N \cdot Cl$
Quaternium-15	[51229-78-8]	$C_9H_{16}ClN_4 \cdot Cl$
sodium dehydroacetate	[4418-26-2]	$NaC_8H_7O_4$
sodium phenolsulfonate	[1300-51-2]	$NaC_6H_5O_4S$
sodium phenoxide	[139-02-6]	NaC_6H_5O
sodium pyrithione	[3811-73-2]	NaC_5H_5NOS
sorbic acid[c]	[110-44-1]	$C_6H_8O_2$
thimerosal	[54-64-8]	$NaHgC_9H_9O_3S$
triclocarban	[101-20-2]	$C_{13}H_9Cl_3N_2O$
triclosan	[3380-34-5]	$C_{12}H_7Cl_3O_2$
zinc pyrithione	[13463-41-7]	$ZnC_{10}H_8N_2O_2S_2$

[a]Ref. 26 includes a more comprehensive listing.

[b]Use levels are product dependent but generally do not exceed 0.25%.

[c]The acid salts are also used.

be rinsed off. Components of products left on the skin can be expected to penetrate the viable epidermis and to be systematically absorbed. Products that are rinsed off shortly after skin contact, such as shampoos, can, if properly labeled, contain preservatives that might elicit adverse reactions if left on the skin. Typical examples of such preservatives are formaldehyde, formaldehyde releasers such as Quaternium 15 or MDM hydantoin, and the blend of methylchloroisothiazolinone and methylisothiazolinone.

Decorative eye cosmetic products have been reported to be subject to pathogenic microbial contamination. Regulatory agencies in several countries, therefore, permit the use of mercury-containing preservatives in eye makeups. The infections reported were to a large extent caused by contamination during use, and the introduction of self-sterilizing preparations seems warranted.

6.3. Lipids.   Natural and synthetic lipids are used in almost all cosmetic products. Lipids serve as emollients or occlusive agents, lubricants, binders for creating compressed powders, adhesives to hold makeup in place, and hardeners in such products as lipsticks. In addition, lipids are used as gloss-imparting agents in hair-care products. The primary requirements for lipids in cosmetics are absence of excessive greasiness and ease of spreading on skin. Oily lipids, principal constituents of emulsions (creams and lotions), are well suited for inclusion in massage products, oils used to treat the skin (bath oils), ointments, suntan oils, and the like. Selection for a specific application is made on the basis of chemical inertness and physical properties. Petrolatum, mineral oils, polymeric silicones, polybutenes, and related substances are ingredients used for skin and hair conditioning. Conditioning is cosmetic jargon for describing a substance's beneficial effect on the substrate. For example, quaternary compounds are substantive to skin and hair proteins and thus can produce conditioning effects. Similarly, lipidic compounds without substantive functional groups, for example, tricaprin, condition skin merely by their presence on the surface. A selected listing of cosmetically useful lipids is provided in Table 7.

6.4. Solvents.   Solvents can be added to cosmetics to help dissolve components used in cosmetic preparations. Water is the most common solvent and is the continuous phase in most suspensions and water/oil (w/o) emulsions. Organic solvents are required in the preparation of colognes, hair fixatives, and nail lacquers. Selected solvents are used to remove soil, sebum, and makeup from skin. Solvents used in cosmetics include acetone, denatured alcohol, butoxyethanol (ethyleneglycol monobutylether), diethylene glycol, dimethyl isosorbide, ethyl acetate, heptane, isopropyl alcohol, mineral spirits (boiling range 110–155°C), polyethylene glycol (mol wt from 200 up to 15,000), propylene glycol, toluene, and tricaprin (glyceryl tri-n-decanoate). A comprehensive listing may be found in Reference 26. The selection of solvents for use in cosmetics is a complex task because of odor as well as topical and inhalation toxicities.

6.5. Surfactants.   Substances commonly classified as surfactants (qv) or surface active agents are required in a wide variety of cosmetics. These are often categorized on the basis of ionic character but are grouped in Table 8, which includes at least one member from each of the various chemical types of surfactants, on the basis of utility in cosmetics. Prolonged contact with anionic surfactants can cause some swelling of the skin. Although this is a temporary phenomenon, skin in this swollen condition allows permeation of externally

Table 7. **Cosmetically Useful Lipids**[a]

Material	CAS Registry number	Molecular formula
Emollients		
butyl oleate	[142-77-8]	$C_{22}H_{42}O_2$
caprylic/capric glycerides	[65381-09-1]	
cetyl lactate	[35274-05-6]	$C_{19}H_{38}O_3$
dibutyl sebacate	[109-43-3]	$C_{18}H_{34}O_4$
diisobutyl adipate	[141-04-8]	$C_{14}H_{26}O_4$
ethyl linoleate	[544-35-4]	$C_{20}H_{26}O_2$
glyceryl isostearate	[32057-14-0]	$C_{21}H_{42}O_4$
hydrogenated palm kernel glycerides[b]		
isodecyl myristate	[17670-91-6]	$C_{24}H_{48}O_2$
isopropyl stearate	[112-10-7]	$C_{21}H_{42}O_2$
lauryl lactate	[6283-92-7]	$C_{15}H_{30}O_3$
mineral oil	[8012-95-1]	C_nH_{2n}
myristyl myristate	[3234-85-3]	$C_{24}H_{56}O_2$
oleyl oleate	[3687-45-4]	$C_{36}H_{68}O_2$
PPG-10 cetyl ether	[9035-85-2]	$(C_3H_3O)_2C_{16}H_{34}O$
propylene glycol dicaprylate	[7384-97-6]	$C_{15}H_{19}NOS \cdot HCl$
squalene	[111-02-4]	$C_{30}H_{50}$
wheat germ glycerides	[58990-07-8]	
Occlusive agents		
acetylated lanolin	[61788-48-5]	
butyl stearate	[123-95-5]	$C_{22}H_{44}O_2$
caprylic/capric triglyceride	[65381-09-1]	
dimethicone	[9006-65-9]	$(C_2H_6OSi)_nC_4H_{12}Si$
hydrogenated rice bran wax[c]		
lauryl stearate	[5303-25-3]	$C_{30}H_{60}O_2$
paraffin	[8002-74-2]	C_nH_{2n+2}
pentarerythritol tetrastearate	[115-83-3]	$C_{77}H_{148}O_8$
petrolatum	[8009-03-8]	C_nH_{2n+2}
propylene glycol dipelargonate	[225-350-9]	$C_{21}H_{40}O_4$
stearyl erucate[d]		$C_{40}H_{78}O_2$
trilinolein	[537-40-6]	$C_{57}H_{98}O_6$
Natural lipids		
apricot kernel oil	[72869-69-3]	
beeswax	[8006-40-4]	
carnauba	[8015-86-9]	
castor oil	[8001-79-4]	
coconut oil	[8001-31-8]	
japan wax	[8001-39-6]	
jojoba wax	[66625-78-3]	
lanolin	[8006-54-0]	
mink oil[e]		
olive oil	[8001-25-0]	
ozokerite	[8021-55-4]	
rice bran oil	[68553-81-1]; [84696-37-7]	
sesame oil	[8008-74-0]	
sunflower seed oil	[8001-21-6]	
vegetable oil	[68956-68-3]	
walnut oil	[8024-09-7]	

[a]Ref. 26 includes a more comprehensive listing.
[b]This is a hydrogenated mixture of mono-, di-, and triglycerides derived from palm kernel oil.
[c]Prepared by partial hydrogenation of rice bran wax.
[d]Erucic acid, *n*-octadecanol ester.
[e]Oil obtained from subdermal fatty tissue of genus *Mustela*.

Table 8. **Cosmetic Surfactants**^{a}

Material^b	CAS Registry number	Molecular formula
Cleansing agents		
ammonium laureth sulfate^{c,d}	[32612-48-9]	$(C_2H_4O)_nC_{12}H_{26}O_4S \cdot H_3N$
cetalkonium chloride	[122-18-9]	$C_{25}H_{46}N \cdot Cl$
DEA myristate	[53404-39-0]	$C_{14}H_{28}O_2 \cdot C_4H_{11}NO_2$
decyl polyglucose^{d,e}		
dioctyl sodium sulfosuccinate^{c,d}	[577-11-7]	$C_{20}H_{38}O_7S \cdot Na$
disodium cocoamphodiacetate^d	[68650-39-5]	
disodium laurimino dipropionate^d	[3655-00-3]	$C_{18}H_{35}NO_4 \cdot 2Na$
lauryl betaine^{c,d}	[683-10-3]	$C_{16}H_{33}NO_2$
lauryl pyrrolidone^d	[2687-96-9]	$C_{16}H_{31}NO$
nonoxynol-12	[9016-45-9]	$(C_2H_4O)_nC_{15}H_{24}O$
myristamine oxide^{c,d}	[3332-27-2]	$C_{16}H_{35}NO$
PEG-50 stearate	[9004-99-3]	$(C_2H_4O)_nC_{18}H_{36}O_2$
potassium dodecylbenzenesulfonate^d	[27177-77-1]	$KC_{18}H_{30}O_3S$
potassium oleate	[143-18-0]	$KC_{18}H_{34}O_2$
sodium cocoyl glutamate^d	[68187-32-6]	
sodium C_{14-16} olefin sulfonate^d	[68439-57-6]	
sodium laureth phosphate^{c,d}	[42612-52-2]	
sodium lauryl sulfate^{c,d}	[151-21-3]	$NaC_{12}H_{26}O_4S$
sodium methyl oleoyl taurate^c	[137-20-2]	$NaC_{21}H_{41}NO_4S$
sodium nonoxynol-25 sulfate	[9014-90-8]	$(C_2H_4O)_nC_{15}H_{24}O_4S \cdot Na$
sodium oleoyl isethionate^d	[142-15-4]	$NaC_{20}H_{38}O_5S$
sodium stearate^c	[822-16-2]	$NaC_{18}H_{36}O_2$
TEA-abietoyl hydrolyzed collagen^d	[68918-77-4]	
TEA-lauryl sulfate^d	[139-96-8]	$C_{12}H_{26}O_4S \cdot C_6H_{15}NO_3$
TEA-oleoyl sarcosinate^c	[17736-08-2]	$C_{21}H_{39}NO_3 \cdot C_6H_{15}NO_3$
Emulsifying agents		
ceteareth-10	[68439-49-6]	
cetrimonium bromide	[57-09-0]	$C_{19}H_{42}N \cdot Br$
laneth-5	[3055-95-6]	$C_{22}H_{46}O_6$
lecithin	[8002-43-5]	
nonoxynol-9	[14409-72-4]	$C_{33}H_{60}O_{10}$
PEG-20 dilaurate	[9005-02-1]	$(C_2H_4O)_nC_{24}H_{46}O_3$
PEG-8 oleate	[9004-96-0]	$(C_2H_4O)_nC_{18}H_{34}O_2$
poloxamer 407	[9003-11-6]	$(C_3H_6O \cdot C_2H_4O)_x$
polyglyceryl-8 oleate	[9007-48-1]	
polysorbate 60	[9005-67-8]	
sorbitan sequioleate	[8007-43-0]	
sucrose stearate	[25168-73-4]	$C_{30}H_{56}O_{12}$
Foam boosters		
cocamine oxide	[61788-90-7]	
lauramide DEA	[120-40-1]	$C_{16}H_{33}NO_3$
myristamide MIPA	[10525-14-1]	$C_{17}H_{35}NO_2$
myristaminopropionic acid	[14960-08-8]	$C_{17}H_{35}NO_2$
Hydrotropes		
ammonium xylenesulfonate	[26447-10-9]	$C_8H_{10}O_3S \cdot H_3N$
potassium toluenesulfonate	[16106-44-8]	$C_7H_8O_3S \cdot K$
sodium methyl naphthalene sulfonate	[26264-58-4]	$C_{11}H_{10}O_3S \cdot Na$
Solubilizing agents		
cetareth-40	[68439-49-6]	
oleth-44	[9004-98-2]	$(C_2H_4O)_nC_{18}H_{36}O$
PEG-40 stearate	[9004-99-3]	$(C_2H_4O)_2C_{18}H_{36}O_2$

Table 8 (*Continued*)

Material[b]	CAS Registry number	Molecular formula
Suspending agents		
behentrimonium chloride	[17301-53-0]	$C_{25}H_{54}N \cdot Cl$
benzethonium chloride	[121-54-0]	$C_{27}H_{42}NO_2 \cdot Cl$
sodium lignosulfonate	[8061-51-6]	
sodium polystyrene sulfonate	[9003-59-2]	$(C_8H_8O_3S \cdot Na)_x$

[a]Ref. 26 includes a comprehensive listing.
[b]CFTA names are used.
[c]Belongs to a chemical class especially useful in facial and body washes.
[d]Belongs to a chemical class especially useful in shampoos.
[e]Decylether of a glucose oligomer.

applied substances. Nonionic surfactants as a group are generally believed to be mild even under exaggerated conditions. The more hydrophobic nonionics, those that are water dispersible (not water-soluble), can enhance transdermal passage. Amphoteric surfactants as a group exhibit a favorable safety profile. Finally, cationic surfactants are commonly rated as more irritating than the anionics, but the evidence for generalized conclusions is insufficient.

6.6. Colorants. Color (qv) is used in cosmetic products for several reasons: the addition of color to a product makes it more attractive and enhances consumer acceptance; tinting helps hide discoloration resulting from use of a particular ingredient or from age; and finally, decorative cosmetics owe their existence to color.

Organic Colorants. The importance of coal-tar colorants cannot be over-emphasized. The cosmetic industry, in cooperation with the FDA, has spent a great deal of time and money in efforts to establish the safety of these dyes (see COLORANTS FOR FOOD, DRUGS, COSMETICS, AND MEDICAL DEVICES). Contamination, especially by heavy metals, and other impurities arising from the synthesis of permitted dyes are strictly controlled. Despite this effort, the number of usable organic dyes and of pigments derived from them has been drastically curtailed by regulatory action.

In addition to the U.S. certified coal-tar colorants, some noncertified naturally occurring plant and animal colorants, such as alkanet, annatto [1393-63-1], carotene [36-88-4], $C_{40}H_{56}$, chlorophyll [1406-65-1], cochineal [1260-17-9], saffron [138-55-6], and henna [83-72-7], can be used in cosmetics. In the United States, however, natural food colors, such as beet extract or powder, turmeric, and saffron, are not allowed as cosmetic colorants.

The terms FD&C, D&C, and External D&C (Ext. D&C), which are part of the name of colorants, reflect the FDA's colorant certification. FD&C dyes may be used for foods, drugs, and cosmetics; D&C dyes are allowed in drugs and cosmetics; and Ext. D&C dyes are permitted only in topical products. Straight colorants include both the organic dyes and corresponding lakes, made by extending the colorant on a substrate such as aluminum hydroxide or barium sulfate. The pure dye content of these lakes varies from 2 to 80%; the organic dyes contain over 80% pure dye. Colorants certified for cosmetic use may not contain more

Table 9. **Inorganic Pigments Useful in Makeups**

Material	Molecular formula	Color
titanium dioxide	TiO_2	white
zinc oxide	ZnO	white
talc	steatite	whitish
barium sulfate	$BaSO_4$	white
mica		glossy, colorless
titanium dioxide–ferric oxide coated mica	[a]	glossy, nacreous, multicolored
		multicolored
guanine[b]	$C_5H_5N_5O$	nacreous
bismuth oxychloride	$BiOCl$	white, nacreous
iron oxides	85% Fe_2O_3	yellow to orange
umber	Fe_2O_3/Fe_3O_4	brown
sienna	Fe_2O_3 (ignited)	red
	Fe_3O_4	black
chrome hydroxide green	$Cr_2O(OH)_4$	bluish green
chrome oxide greens	Cr_2O_3	green
ferric ammonium ferrocyanide	$Fe(NH_4)[Fe(CN)_6]$	blue
ferric ferrocyanide	$Fe_4[Fe(CN)_6]_3$	blue
manganese violet	$Mn(NH_4)P_2O_7$	violet
ultramarines[c]		blue, violet, red, pink, green

[a]Material is a mixture.
[b]Guanine [73-40-5], an organic dye, is also known as CI 75170 [73-40-5].
[c]Materials are fusion mixtures.

than 0.002% of lead, not more than 0.0002% of arsenic, and not more than 0.003% of heavy metals other than lead and arsenic.

Inorganic Colorants. In addition to various white pigments, other inorganic colorants such as those listed in Table 9 are used in a number of cosmetic products. These usually exhibit excellent lightfastness and are completely insoluble in solvents and water.

Naturally occurring colored minerals that contain oxides of iron are known by such names as ochre [1309-37-1], umber [12713-03-0], sienna [1309-37-1], etc. These show greater variation in color and tinting power than the synthetic equivalents, and the nature and amount of impurities in the national products is also variable. Most of the pigments identified in Table 9 are, therefore, manufactured synthetically. They are primarily used in skin-makeup products and in eye-area colorants.

Nacreous Pigments. For many years nacreous pigments were limited to guanine (from fish scales) and bismuth oxychloride. Mica, gold, copper, and silver, in flake form, can also provide some interesting glossy effects in products and on the face. Guanine is relatively costly, and bismuth oxychloride darkens on exposure to light and is difficult to suspend because of its high specific gravity. An entirely new set of colored, iridescent, inorganic pigments, which may be described as mixtures of mica and titanium dioxide (sometimes with iron oxides), has been created by coating mica flakes with titanium dioxide. The wavelengths of light reflected from these compositions can produce a complete range of colored interference patterns. The particle size of the mica must be controlled and may not exceed 150 μm, at least in the United States. Additional color effects can be created by sandwiching the mica, TiO_2, and Fe_2O_3.

7. Specialized Cosmetic Technologies

Several specialized technologies have been perfected for cosmetic products. Among these, emulsification, stick technology, and powder blending are prominent.

8. Emulsification

Emulsification is essential for the development of all types of skin- and hair-care preparations and a variety of makeup products. Emulsions (qv) are fine dispersions of one liquid or semisolid in a second liquid (the continuous phase) with which the first substance is not miscible. Generally, one of the phases is water and the other phase is an oily substance: oil-in-water emulsions are identified as o/w; water-in-oil emulsions as w/o. When oil and water are mixed by shaking or stirring in the absence of a surface-active agent, the two phases separate rapidly to minimize the interfacial energy. Maintenance of the dispersion of small droplets of the internal phase, a requirement for emulsification, is practical only by including at least one surface-active emulsifier in the oil-and-water blend.

The addition of emulsifiers (see Table 8) lowers the energy of the large interfacial area created by forming a huge number of small droplets from a single large drop. In practical emulsification technology, this thermodynamic emulsion stabilization is augmented by two other features. One is the formation of a rigid interfacial film on the surface of the droplets of the internal phase (29). This film, sometimes exhibiting the optical characteristics of a liquid crystal, acts as a mechanical barrier to the coalescence of the droplets of the internal phase. Finally, the droplets may be stabilized by the formation of an electric double layer, which favors the electrical repulsion between charged particles. The latter requires the presence of an electrolyte or an ionized emulsifier.

The coalescence of internal phase droplets can be further decreased by raising the viscosity of the external continuous phase through addition of gums or synthetic polymers, for example, cellulosic gums such as hydroxypropyl methyl-cellulose [9004-65-3], fermentation gums such as xanthan gum [11138-66-2], or cross-linked carboxyvinyl polymers such as carbomer [39007-16-3]. The increased viscosity also counteracts changes in the emulsion resulting from differences in the specific gravity of the two phases as mandated by Stokes' law. An advance in cosmetic emulsification technology has resulted from the development of cross-linked carboxyvinyl polymers, in which some of the carboxylic acid residues are esterified with various fatty alcohols. These polymers possess the ability to act as primary emulsifiers and thicken the system when some of the remaining carboxylic groups are neutralized with alkali (see CARBOXYLIC ACIDS).

The selection of emulsifiers, auxiliary emulsifiers, gums, and other components is complicated and largely empirical. Despite the lack of a rigid theoretical basis, the hydrophile/lipophile balance (HLB) is the most useful approach for the selection of nonionic emulsifiers (30). The inclusion of ionic emulsifiers was not contemplated in the original formulation of the HLB system. The HLB system also does not account for the effect of low HLB viscosity increasing ingredients, such as cetyl alcohol or glyceryl monostearate. The precise selection of the desired blend from commercial nonionics for emulsification is often frustrating (31). Methods for selecting suitable blends of emulsifiers and stabilizers (32,33), for preparing emulsions (34,35), and for studying stability (36) have been published. The technical literature also includes publications dealing with the theory of emulsification and the structure of emulsions and of microemulsions (37,38).

Conventional cosmetic emulsions (macroemulsions) normally contain about 70% or more of the external phase, which may be a mixture of components. The internal phase is routinely introduced into the external phase at an elevated temperature with vigorous agitation. The emulsifiers are distributed according to their solubility between the two phases. The level of emulsifiers (rarely more than about 10%) is kept low since excessive amounts may destabilize emulsions or form a clear solubilizate. Auxiliary emulsifiers and other components are included in the phases in which they are soluble.

The term multiple emulsion describes a w/o emulsion in an o/w emulsion. For example, when a w/o emulsion is added to water, no dispersion is expected unless the aqueous phase is fortified with a suitable emulsifier. The resulting dispersion may then be a blend of a w/o and an o/w emulsion, or it may be a multiple emulsion of the w/o/w type. In this latter case, the initial w/o emulsion becomes the internal phase of the final product. Generally, these preparations are not very stable unless they are produced under rigidly controlled conditions (32,39,40).

Microemulsions or solubilized or transparent systems are very important in the marketing of cosmetic products to enhance consumer appeal (32,41). As a rule, large quantities of hydrophilic surfactants are required to effect solubilization. Alternatively, a combination of a solvent and a surfactant can provide a practical solution. In modern clear mouthwash preparations, for example, the flavoring oils are solubilized in part by the solvent (alcohol) and in part by the surfactants. The nature of solubilized systems is not clear. Under normal circumstances, microemulsions are stable and form spontaneously. Formation of a microemulsion requires little or no agitation. Microemulsions may become cloudy on heating or cooling, but clarity at intermediate temperatures is restored automatically.

Formation of liposomal vesicles under controlled conditions of emulsification of lipids with phospholipids has achieved prominence in the development of drugs and cosmetics (42). Such vesicles are formed not only by phospholipids but also by certain nonionic emulsifying agents. Formation is further enhanced by use of specialized agitation equipment known as microfluidizers. The almost spontaneous formation of liposomal vesicles arises from the self-assembly concepts of surfactant molecules (43). Vesicles of this type are unusual sustained-release disperse systems that have been widely promoted in the drug and cosmetic industries.

8.1. Stick Technology.   Cosmetic sticks can be divided into three categories: sticks molded in the container; sticks molded separately and then encased; and sticks formed by compression.

Container Molding. Antiperspirant, deodorant, sunscreen, and antiacne sticks are container molded. The amount of dispersed ingredients makes them brittle and difficult to handle mechanically.

The required solids are suspended in a wax–emollient blend at about 60 to 80°C and milled. The liquid suspension is then cooled to about 55°C, and the more volatile ingredients are added. The mass is placed into containers, which are commonly provided with a threaded shaft for raising or lowering the product.

Antiperspirant sticks based on this molding technique have become more popular since volatile low mol wt cyclomethicones [69430-24-6] have been used successfully as the lipids and fatty alcohols as the waxes. This type of product delivers the active antiperspirant to the site as a clinging powder without excessive oiliness.

Deodorant and cologne sticks are formed by allowing sodium stearate to gel in a suitable organic solvent, usually ethanol or propylene glycol. The soap and the solvent are heated under reflux until the soap is dissolved. The solution is cooled to about 60°C; fragrance, color, and the like are added; and the mass is placed into suitable containers.

Stick Molding. Various types of lipsticks and eye-shadow sticks are stick molded. A wax-containing lipid mass is milled with the pigment at elevated (about 75°C) temperatures until it is uniform. The lipid mass, at a temperature about 10°C above its melting point, is then poured into metallic molds. Deaeration is essential to prevent unsightly depressions on the molded sticks. To avoid sudden congealing, the molds are customarily heated to a temperature above room temperature before filling; after filling, they are chilled to temperatures well below room temperature. After unmolding, the sticks may be inserted into various types of containers (swivel, metal, or plastic). In order to formulate acceptable molded sticks, slowly developing surface anomalies or defects must be avoided. Foremost are the excrescence of solid fatty substances (also called bloom) and the exudation of liquid substances (also called sweating). Both of these defects are attributed to polymorphic transformation. The selection of the proper blend of lipids to create an acceptable makeup stick is complex. Some of the lipids used in such products and their primary characteristics are listed in Table 10. Other types of sticks, for example, eyebrow pencils and lip liners, are molded similarly but may be inserted into wooden pencil stock, trimmed, and appropriately finished.

The criteria for a good cosmetic makeup stick are manifold: the sticks must pay off, that is, deliver the desired amount when used at or near room temperature; sticks must withstand exposure to moderately elevated temperatures; they should not break during normal use; stick components must not elicit unpleasant, for example, oily or warming, sensations after application; pigments must be uniformly distributed and must not react photochemically with the remaining stick components to cause rancidity or photoirritation; and finally, the film produced on the body site must be resistant to rubbing off or transferring to eating utensils or clothing.

Compressed-Powder Sticks. Compression of a blend of solids using a suitable binder or by extruding a water containing magma results in compressed-powder sticks.

Table 10. **Lipids Used in Cosmetic Molded Sticks**[a]

Lipid	CAS Registry number	Characteristics
	Oils	
castor oil	[8001-79-4]	high gloss; high viscosity
mineral oil	[8012-95-1]	high gloss
	Esters / Alcohols	
butyl stearate	[123-95-5]	rapid wetting of pigments; controls sweating at elevated temperatures
guerbet alcohols		
decyl tetradecanol	[58670-89-6]	satiny gloss
octyl dodecanol	[5333-42-6]	satiny gloss
isocetyl alcohol	[36311-34-9]	similar to castor oil
isopropyl myristate	[110-27-0]	same as butyl stearate
isopropyl palmitate	[142-91-6]	same as butyl stearate
oleyl alcohol	[143-28-2]	high gloss; turns rancid
	Fats	
cocoa butter	[8002-31-1]	tendency to bloom
glyceryl monostearate	[31566-31-1]	high melting
hydrogenated cocoglycerides[b]		variable properties
hydrogenated vegetable oil	[68334-28-1]	greasy
lanolin	[8006-54-0]	wets pigments
	Waxes	
beeswax	[8006-40-4]	low gloss
candelilla	[8006-44-8]	low melting
carnauba	[8015-86-9]	very hard
ozokerite	[8021-55-4]	thermally stable
	Dye solvents	
ethoxydiglycol acetate	[112-15-2]	bitter
polyethylene glycols	[25322-68-3]	
propylene glycol	[57-55-6]	

[a]Ref. 26 includes comprehensive listings.
[b]These are blends of mono-, di-, and triglycerides of hydrogenated coconut oil.

8.2. Powder Blending.

8.2. Powder Blending.   Cosmetic powders serve two primary functions. One group, commonly called body powders or talcs, is applied to the skin to provide lubricity and to absorb excessive moisture. The second group, commonly referred to as face powders, exists in both loose and compressed forms and is used to impart some color to the skin and to dull excessive oiliness.

Most powders, including medicated powders, depend on talc to provide lubricity and a matte finish on the skin. Talc is generally blended with other constituents, such as those listed in Table 11. The plate-like nature of mined talc makes this hydrous magnesium silicate (steatite) an important skin-care constituent. Loose body and makeup powders utilize additional bulk ingredients. The products can also include antimicrobial agents, dyes, and pigments. The selection of a fragrance must be made with great care; some bulk ingredients are alkaline, and the perfume oil on the surface of particles is subject to oxidation, especially if pigmented ingredients are included.

The basic manufacturing process involves thorough blending of the components, especially the pigments, and comminution with the aid of a variety of

Table 11. **Powder Ingredients Used for Cosmetics**

Ingredient	Chemical identity	CAS Registry number	Comment
chalk		[13397-25-6]	opaque, alkaline
kaolin (clay)	aluminum silicate	[1332-58-7]	
	attapulgite	[1337-76-4]	
	fuller's earth	[8031-18-3]	opaque, low gloss
	hectorite	[12173-47-6]	
	montmorillonite	[1318-93-0]	
magnesium carbonate		[546-93-0]	absorbent
metallic soaps	magnesium stearate	[557-704-0]	hydrophobic, lubricant
	zinc ricinoleate	[13040-19-2]	
silica	fumed	[7631-86-9]	absorbent
	xerogel	[112945-52-5]	
starch	corn starch	[9005-25-6]	hygroscopic
	rice starch	[9005-25-8]	
talc	hydrated magnesium silicate (steatite)	[14807-96-6]	opaque lubricant
titanium dioxide		[13463-67-7]	opaque, white
zinc oxide		[1314-13-2]	opaque, adherent
zirconium silicate		[10101-52-7]	opaque, white

mills to reduce the particle size. Loose powders are filled without additional processing.

If compression is required to provide a stick or pan-type of product, the bulk components must be held together with a binder. Common binders are various lipids, polymers, polysaccharides, and waxes. Some binder compositions include water, which is removed by drying the compact. The amount of binder must be carefully controlled to yield a solid, nonfragile compact that is soft enough to pay off. Excessive amounts of or improperly compounded binders glaze during use because of transfer of skin lipids to the compact.

When the bulk containing the binder is uniform, it is compressed on pneumatic, hydraulic, or ram-type presses. Compression can be carried out in presses provided with suitably designed cavities or in metallic pans. The pans are filled with the powder mass, and a plunger with a cross-sectional shape similar to that of the pan is used to compress the tablet. The resulting tablets are commonly used with powder puffs or cosmetic brushes.

9. Skin Preparation Products

Products for use on the skin are designed to improve skin quality, to maintain (or restore) skin's youthful appearance, and to aid in alleviating the symptoms of minor diseases of the skin. Many of these products are subject to different regulations in different countries. Skin products are generally formulated for a specific consumer purpose.

9.1. Skin-Care Products.   Preparations are generally classified by body part and purpose. For example:

Product	Purpose
baby preparations	
oils and lotions	cleansing, soothing
diaper-rash products	prevention, cure
powders	drying
foot preparations	
antifungals	anti-infective
emollients	soothing, crack-prevention
powders	drying
facial preparations	
lotions and creams	smoothing, protecting, rejuvenating
body preparations	
lotions and creams	smoothing, protecting
oils	smoothing, protecting
powders	drying
hand preparations	
lotions and creams	smoothing, protecting
gels	antichapping

The smoothing or emollient properties of creams and lotions are critical for making these emulsions the preferred vehicles for facial skin moisturizers, skin protectants, and rejuvenating products. On the body, emollients provide smoothness and tend to reduce the sensation of tightness commonly associated with dryness and loss of lipids from the skin. Although a wide variety of plant and animal extracts have been claimed to impart skin benefits, valid scientific evidence for efficacy has been provided only rarely.

Emulsion components enter the stratum corneum and other epidermal layers at different rates. Most of the water evaporates, and a residue of emulsifiers, lipids, and other nonvolatile constituents remains on the skin. Some of these materials and other product ingredients may permeate the skin; others remain on the surface. If the blend of nonvolatiles materially reduces the evaporative loss of water from the skin, known as the transepidermal water loss (TEWL), the film is identified as occlusive. Application of a layer of petrolatum to normal skin can reduce the TEWL, which is normally about $4–8$ g/(m^2 · h), by as much as 50 to 75% for several hours. The evaporated water is to a large extent trapped under the occlusive layer hydrating or moisturizing the dead cells of the stratum corneum. The flexibility of isolated stratum corneum is dependent on the presence of water: dry stratum corneum is brittle and difficult to stretch or bend. Thus, any increase in the water content of skin is believed to improve the skin quality.

The ability to moisturize the stratum corneum has also been claimed for the presence of certain hydrophilic polymers, for example, guar hydroxypropyl trimonium chloride [65497-29-2], on the skin. By far the most popular way to moisturize skin is with humectants, some of which are listed in Table 12. It is

Table 12. **Skin Conditioners and Moisturizers**[a]

Material	CAS Registry number	Molecular formula
glycerol	[56-81-5]	$C_3H_8O_3$
2-pyrrolidinone-5-carboxylic acid (PCA)	[98-79-3]	$C_5H_7NO_3$
sodium lactate	[72-17-3]	$C_3H_5NaO_3$
urea	[57-13-6]	CH_4N_2O
cholesterol	[57-88-5]	$C_{27}H_{46}O$
hydrolyzed glycosaminoglycans[b]		
hydrolyzed soy protein	[68607-88-5]	
linoleic acid	[60-33-3]	$C_{18}H_{32}O_2$
tocopheryl acetate	[7695-91-2]	$C_{31}H_{52}O_3$
witch hazel distillate[c]		
sodium hyaluronate[d]		
myristyl betaine	[2601-33-4]	$C_{18}H_{37}NO_2$

[a]Ref. 26 includes a more comprehensive listing.
[b]Mixed polysaccharides from animal connective tissue.
[c]Nonalcoholic steam distillate of parts of *Hamamelis virginiana*.
[d]Sodium salt of hyaluronic acid [9004-61-9].

claimed that humectants attract water from the environment and thereby provide moisture to the skin.

Studies of the interactions between water and the lipid constituents of the stratum corneum suggest that the supply of water per se is not responsible for skin quality and condition. Water vapor from lower layers provides a constant supply of moisture to the epidermis. Instead, the ability of the skin to retain the moisture is critical, and this ability depends on the lipid lamellar bilayers that occupy the spaces between the cells of the stratum corneum (44–46).

In the United States, products claimed to reverse or alleviate the stigmata of facial skin aging are considered drugs. Claims for improvement of fine wrinkling, mottled hyperpigmentation, and roughness associated with photodamage, on the part of products containing all *trans*-retinoic acid [302-79-4], have received some favorable comments from regulatory advisory panels. Other approaches for anti-aging products are based on desquamation by α-hydroxyacids, for example, lactic acid [50-21-5]. Finally, a number of substances, such as hyaluronic acid [9004-61-9] and collagen [9007-34-5], have been claimed to improve the appearance of wrinkled skin (see also ANTIAGING AGENTS).

The amounts and types of lipids used in skin-care products control their application properties. Methods for assessing these characteristics using expert panelists have been described (47).

The ability of skin-care products to supply moisture to the skin remains in question. In the United States, however, the OTC panel has sanctioned the use of skin-protectant ingredients such as glycerin, which may play roles in the skin's water ecology. Products for the care of body skin are similar to preparations formulated for the care of facial skin. Products for overall body care should leave a dry, satinlike finish even though relatively high levels of unctuous lipids are used. Facial night creams may leave the skin somewhat oily, whereas facial day creams must provide a dry finish.

Hand-care products are designed to reduce chapping and cracking, especially prevalent during cold, dry, winter weather. Hand-care products are commonly fortified with various humectants, and products for the elbows and feet may include abrasives. Bath powders impart lubricity to body skin, absorb moisture, and provide some fragrance. These are formulated without pigments to preclude the staining of clothing. Skin-care formulations have been published for skin protectants (48), face creams (49), hand creams (50), and body creams (51).

9.2. Antiacne Preparations.   Antiacne products are designed to alleviate the unsightly appearance and underlying cause of juvenile acne. Generally, acne is a mild disease of the follicular duct in which sebaceous secretion is not readily allowed to pass to the surface of the skin because of a hyperkeratotic restriction in the duct. The retained sebum may undergo chemical changes or be altered by microbial species, with consequent inflammatory responses. In the past, cosmetic preparations were designed to remove sebum from the skin surface with solvents or cleansers and work against microorganisms with antibacterials. In addition, acne was cosmetically treated with abrasives in the hope that scrubbing would relieve the ductal blockage.

As of 1991 in the United States, OTC antiacne preparations may contain only a few active drugs, for example, sulfur [7704-34-9], resorcinol acetate [102-29-4], resorcinol [108-46-3], salicylic acid [69-72-7], and some combinations (52). OTC anti-acne constituents may be included in a variety of conventional cosmetic preparations, which then become OTC drugs. These include lotions, creams, solutions, facial makeups, facial cleansers (including abrasive cleansers), and astringents. Products must contain the specified drugs at the designated concentrations. Compositions of antiacne products have been published (53).

9.3. Sunscreens.   Radiation that reaches the earth's surface from the sun is limited to wavelengths above about 285 nm because shorter wavelengths are absorbed by ozone (qv). Investigations of the impact of ultraviolet (uv) light on human skin have identified the range from 285 to about 320 nm as uvB and that between 330 nm and visible light as uvA. Both uvA and uvB have the potential to burn skin, resulting in an acute sunburn that is painful and can have damaging long-term effects, such as wrinkling, actinic keratoses, or carcinomas. The flux of uvA and uvB that reaches human skin depends on altitude and latitude. Equatorial regions receive maximum flux. Clouds, dust, and reflected uv light from the ground or water also affect flux and may provide some wavelength specificity to the radiation. The flux required to produce a barely observable erythema, the so-called minimal erythemal dose (MED), depends on the energy of the uv radiation. Thus, much higher fluxes of uvA than of uvB are required for the production of one MED. The estimation of the MED (a rapid skin response) is not a quantitative measure of long-term skin damage, especially because the lower energy uvA penetrates much deeper into the skin than does the shorter wavelength uvB.

The use of uv light absorbing substances is accepted worldwide as a means of protecting skin and body against damage and trauma from uv radiation. These colorless organic substances are raised to higher energy levels upon absorption of uv light, but little is known about mechanisms for the disposal of this energy.

These substances can be classified by the wavelengths at which absorbance is maximal. Absorption throughout the incident uv range (285 to about 400 nm) affords the best protection against erythema.

It is also possible to deflect uv radiation by physically blocking the radiation using an opaque makeup product. A low particle size titanium dioxide can reflect uv light without the undesirable whitening effect on the skin that often results from products containing, for example, zinc oxide or regular grades of titanium dioxide.

In vitro absorption-spectrophotometry techniques are available to assess a sunscreen's efficacy, but the preferred methods are *in vivo* procedures in which a small body site is irradiated with the desired wavelengths for different periods in the presence or absence of a uv protectant. Procedures vary from country to country to determine the incremental timing of the exposure that ultimately allows quantification via sun protective factor (SPF). In the United States, sunscreen preparations are considered OTC drug products, and the SPF must be specified (54). Even in countries that do not identify these products as drugs, SPF labeling has become customary.

The SPF is defined as the ratio of the time required to produce a perceptible erythema on a site protected by a specified dose of the uv protectant product to the time required for minimal erythema development in the unprotected skin. An SPF of 8 indicates that the product allows a subject to expose the protected skin 8 times as long as the unprotected skin to produce the minimum erythemal response. The measurement can be quite subjective unless skin color and the history of reactions to sun exposure of the test subjects are taken into account. The MED range for Caucasians at 300 nm averages 34 mJ/cm^2. The range is 14–80 mJ/cm^2. Perspiration or the use of artificial irradiation devices can create additional problems.

Because perspiration and bathing are commonly associated with sun exposure, the need to determine the SPF after bathing or long after application to the body site is important. In use, the quantity of screen applied and its uniform distribution over the exposed area control the achieved SPF. Methods for assessing the water-resistant or waterproof qualities of sunscreen products have been established by the FDA.

A list of uv absorbing substances found useful in protective sunscreen products is provided in Table 13. Some information on the levels permitted in products in both the United States and the EEC is included. Descriptions and specifications of sunscreens have been published (55).

In principle, emulsified sunscreen products are similar to emollient skin-care products in which some of the emollient lipids are replaced by uv absorbers. The formulation of an effective sunscreen product generally requires combination of a uvB and a uvA absorber if an SPF above about 12 is desired. Two or more of the sunscreens listed in Table 13 normally constitute about one-half of the nonvolatiles found in sunscreen lotions. The other half consists of an emollient (solvent) and emulsifying and bodying agents. If water-resistant qualities are desired, polymeric film formers, for example, acrylates–octylacrylamide copolymers [9002-93-1], or water-repellent lipids, for example, dimethicone [9006-65-9], are included.

Table 13. **Cosmetic Uv Absorbers**

Ingredient	CAS Registry number	Quantity approved, %	
		U.S.[a]	EEC[b]
uvA Absorbers			
benzophenone-8	[131-53-3]	3	
menthyl anthranilate	[134-09-6]	3.5–5	
benzophenone-4	[4065-45-6]	5–10	5
benzophenone-3	[113-57-7]	2–6	10
uvB Absorbers			
p-aminobenzoic acid (PABA)	[150-13-0]	5–15	5
pentyl dimethyl PABA	[14779-78-3]	1–5	5
cinoxate	[104-28-9]	1–3	5
DEA *p*-methoxcinnamate		8–10	8
digalloyl trioleate	[17048-39-4]	2–5	4
ethyl dihydroxypropyl PABA	[5882-17-0]	1–5	5
octocrylene	[6187-30-4]	7–10	
octyl methoxycinnamate	[5460-77-3]	2–7.5	10
octyl salicylate	[118-60-5]	3–5	5
glyceryl PABA	[136-44-7]	2–3	5
homosalate	[118-56-9]	4–15	10
lawsone (0.25%)	[83-72-7]		
plus dihydroxyacetone (33%)	[96-26-4]		
octyl dimethyl PABA	[21245-02-3]	1.4–8	8
2-phenylbenzimidazole-5-sulfonic acid	[27503-81-7]	1–4	8
TEA salicylate	[2174-16-7]	5–12	2
sulfomethyl benzylidene bornanone	[90457-82-2]		10
urocanic acid (and esters)	[104-98-3]		2
Physical barriers			
red petrolatum		30–100	
titanium dioxide	[13463-67-7]	2–25	

[a]Ref. 55.
[b]Tentative.

More recently anhydrous sunscreens have become popular. Products of this type are based on blends of emollient lipids and acceptable uv absorbers. Formulations of sunscreen products have been published (56).

9.4. Facial Makeup. This classification applies to all products intended to impart a satinlike tinted finish to facial skin and includes liquid makeups, tinted loose or compressed powders, rouges, and blushers.

In modern liquid makeups and rouges, the required pigments are extended and ground in a blend of suitable cosmetic lipids. This magma is then emulsified, commonly as o/w, in a water base. Soaps, monostearates, conditioning lipids, and viscosity-increasing clays are primary components. Nonionic emulsifiers can replace part or all of the soap. Pigment levels in these emulsions are about 15% but can range as high as 60% if the stabilizing clays are included. The viscosity of these types of preparations varies from fairly thin fluids to thixotropic viscous lotions to firm creams. These products must not dry too rapidly to permit spreading on the skin and feathering of the edges for proper shading.

Tinted dry powders form the second type of facial makeup. Commonly, the blended solids are compressed into compacts. The finished products, sold as compressed powders, rouges, or blushers, are applied to the face with the aid of powder puffs, brushes, or similar devices. Facial makeup compositions have been published for rouge (57), powder (58), and makeup (59).

9.5. Skin Coloring and Bleaching Preparations. Products designed to simulate a tan, to lighten skin color in general, or to decolorize small hyperpigmented areas such as age spots either impart to or remove color from the skin. Skin stains are intended to create the appearance of tanned skin without exposure to the sun. The most widely used ingredient is dihydroxyacetone [96-26-4] (2–5% at pH 4 to 6) which reacts with protein amino groups in the stratum corneum to produce yellowish brown Maillard products. Lawsone [83-72-7] and juglone [481-39-0] are known to stain skin directly. Stimulation of melanin formation is another approach to artificial tanning. Commercialization, which is limited, depends primarily on topical application of products containing tyrosine or a tyrosine recursor.

The number of cosmetically acceptable bleaching ingredients is very small, and products for this purpose are considered drugs in the United States. The most popular ingredient is hydroquinone [123-31-9] at 1–5%; the addition of uv light absorbers and antioxidants reportedly helps to reduce color recurrence. Effective bleaching requires repeated localized applications of a cream or ointment type of preparation. Formulations for skin lighteners (60) and skin darkeners (61) have been published.

10. Astringents

Astringents are designed to dry the skin, denature skin proteins, and tighten or reduce the size of pore openings on the skin surface. These products can have antimicrobial effects and are frequently buffered to lower the pH of skin. They are perfumed, hydro-alcoholic solutions of weak acids, such as tannic acid or potassium alum, and various plant extracts, such as birch leaf extract. The alcohol is not only a suitable solvent but also helps remove excess sebum and soil from the skin. After-shave lotions generally function as astringents.

In the United States, some astringents, depending on product claims, are considered OTC drugs (62). Only three ingredients, aluminum acetate [139-12-8], aluminum sulfate [10043-01-3], and hamamelis [84696-19-5], are considered safe and effective.

10.1. Antiperspirants and Deodorants. There are many forms of antiperspirants and deodorants: liquids, powders, creams, and sticks. Deodorants do not interfere with the delivery of eccrine or apocrine secretions to the skin surface but control odor by reodorization or antibacterial action. Deodorant products, regardless of form, are antimicrobial fragrance products. An important antimicrobial or cosmetic biocide used in many products is triclosan [3380-34-5]. Other active agents include zinc phenolsulfonate [127-82-2], p-chloro-m-xylenol [88-04-0], and cetrimonium bromide [57-09-0]. There have been claims that ion-exchange polymers and complexing agents provide protection against

Table 14. **Antiperspirant Ingredients**

Name	CAS Registry number	Molecular formula
Aluminum chlorohydrates[a]		
aluminum chlorohydrate	[1327-41-9]; [12042-91-0]	$Al_2(OH)_5Cl \cdot nH_2O$
aluminum sesquichlorohydrate	[11097-68-0]	$Al_2(OH)_{4.5}Cl_{1.5} \cdot nH_2O$
aluminum dichlorohydrate	[1327-41-9]; [12042-91-0]	$Al_2(OH)_4Cl_2 \cdot nH_2O$
Aluminum zirconium chlorohydrates[a,b]		
aluminum zirconium octachlorohydrate		$Al_8Zr(OH)_{20}Cl_8 \cdot nH_2O$
aluminum zirconium pentachlorohydrate		$Al_8Zr(OH)_{23}Cl_5 \cdot nH_2O$
aluminum zirconium tetrachlorohydrate	[57158-29-9]	$Al_4Zr(OH)_{12}Cl_4 \cdot nH_2O$
aluminum zirconium trichlorohydrate		$Al_4Zr(OH)_{13}Cl_3 \cdot nH_2O$
Aluminum salts		
aluminum chloride	[7446-70-0]	$AlCl_3$
buffered aluminum sulfate	[10043-01-3]; [18917-91-4]	$Al_2(SO_4)_3$ and $Al(CH_3CHOHCOO)_3$

[a]Partially dehydrated derivatives complexed with polyethylene glycol or propylene glycol exist. In the United States, derivatives in which some of the water of hydration has been replaced by glycine are particularly popular. Aluminum zirconium tetrachlorhydrex gly and related derivatives can be used in OTC antiperspirant products.
[b]The Al/Zr ratio is variable.

unpleasant body odors. In addition, delayed-release, that is, liposomal or encapsulated, substances of diverse activity have been employed.

The mechanism of antiperspirant action has not been fully established but probably is associated with blockage of ducts leading to the surface by protein denaturation by aluminum salts. The FDA has mandated that an antiperspirant product must reduce perspiration by at least 20% and has provided some guidelines for testing finished products. Some antiperspirant chemicals are listed in Table 14 (63).

Clear solutions of antiperspirants have been on the market for about 100 years. Cream and lotion types are o/w emulsions commonly formulated using nonionic emulsifiers to avoid aluminum salt formation, especially by carboxylic acids. Cream antiperspirants are generally distributed in jars, whereas lotions are dispensed from roll-on types of containers.

Antiperspirant aerosols (qv) can be wet or dry. In the wet type, the antiperspirant chemical is dissolved in a suitable solvent, such as water–ethanol, combined with emollients and so on, and dispensed after pressurization using an acceptable propellant. Dry aerosols may be based on a finely milled antiperspirant component suspended with emollients and suspending agents in a volatile liquid that is lost after dispensing.

Most antiperspirant sticks are molded. Sticks dominate in the U.S. market, whereas lotion and cream antiperspirants are preferred in Europe. Stick antiperspirant products may include suspending agents, coupling agents to wet the antiperspirant chemical (about 20–25%), and emollients. The blend is prepared at about 65°C and poured at about 55°C. Antiperspirant (64) and deodorant (65) compositions have been published.

11. Cleansing Preparations

Cleansing preparations are products, based on surfactants or abrasives, that are designed to remove unwanted soil and debris from skin, hair, and the oral cavity. Soaps (qv) are the best-known cleansers but are not considered cosmetic products unless they are formulated with agents that prevent skin damage or contain antimicrobial agents. Soaps are the least costly and most popular skin cleansers available. Use as hair cleansers is limited, however, by the tendency to form insoluble alkaline earth soaps, which leave a dulling film on hair, and the taste of soaps generally precludes use in oral-care preparations. Soaps dominate the skin cleanser market, although they have been shown in closed patch skin tests to cause some irritation owing to their alkalinity. Modern skin cleansers (liquids and bars) for sensitive skin employ various synthetic detergents (see Table 8).

11.1. Skin Cleansers. Their mildness, foaming qualities, water solubility, and tolerance of slightly acid conditions (pH 5–6) make many of the surfactants listed in Table 8 attractive for use in formulating facial and body cleansers. Irritant qualities of preparations based on one or more of these surfactants can be further modified by the addition of lipids or agents that lower the defatting (drying) tendencies of the finished product.

The solubility characteristics of sodium acyl isethionates allow them to be used in synthetic detergent (syndet) bars. Complex blends of an isethionate and various soaps, free fatty acids, and small amounts of other surfactants reportedly are essentially nonirritant skin cleansers (66). As a rule, the more detersive surfactants, for example alkyl sulfates, α-olefin sulfonates, and alkylaryl sulfonates, are used in limited amounts in skin cleansers. Most skin cleansers are compounded to leave an emollient residue on the skin after rinsing with water. Free fatty acids, alkyl betaines, and some compatible cationic or quaternary compounds have been found to be especially useful. A mildly acidic environment on the skin helps control the growth of resident microbial species. Detergent-based skin cleansers can be formulated with abrasives to remove scaly or hard-to-remove materials from the skin.

Foaming bath and shower preparations are based on blends of surfactants and various conditioning agents, many of which are derived from plants, and may contain a relatively high percentage of fragrance. Surfactants are those identified in Table 8 as being beneficial in facial and body washes. Foam boosters are used to create the billowing foam desired in many of these products; sequestering agents are added to prevent the formation of alkali metal salts, especially when soap is used.

Cream-type skin cleansers have been used for many years, particularly on the face. The classical cold cream consists of mineral oil (50–60%), beeswax ($\simeq$15%), borax ($\simeq$1%), and water (30–40%). Neutralization of the cerotic acid in beeswax by borax yields the emulsifier to form the cream. When applied to the face, the mineral oil acts as a solvent for sebum, soil, and makeup. The remains are tissued off, leaving the skin clean with a lubricating, oily finish. This very simple composition can be modified with additional emulsifiers, thickening agents, and cosmetic additives. There are numerous cosmetic wipe-off cleansers, although the popularity of these oil-based products has declined. When still more emulsifiers, especially water-soluble nonionics, are included, the original wipe-off products are converted into the more popular rinse-off types. The principal cleansing agents in these products are the oily components; in contrast with the detergent-based cleansers, the surfactant is included only to aid in removal of the product during rinsing, not for detergency.

11.2. Hair Cleansers. Except for a few specialty preparations, hair cleansers, or shampoos, are based on aqueous surfactants. The detergents of choice in shampoos are listed in Table 8. The most popular surfactants in shampoos are alkyl sulfates and alkylether sulfates, commonly used at about 10–15% active. These ingredients by themselves do not provide the dense, copious foam desired by consumers, and additives are required, especially for use on oily hair or scalps. The foam boosters usually found in finished shampoos are fatty acid alkanolamides, fatty alcohols, and amine ocides. In the so-called superamide foam boosters the ratio of alkanolamine to fatty acid is 1:1. In addition to about 1–2% of an amide, most commonly the diethanolamide of lauric acid [120-40-1] or of coconut fatty acids [61791-31-9], almost all shampoos contain a hair-conditioning agent. The detergent removes lipids from the hair's surface, leaving hair with limited gloss and often difficult to comb. Hair that has been defatted by detergents has a tendency to retain a static electric charge. The resulting fly-away hair can be avoided by making the hair more conductive by means of moisture or electrolytes or by lubricating the hair with a lipid or conditioning agent. The preferred conditioning agents are materials substantive to hair, that is, not readily rinsed off by water.

Excessive degreasing by shampoos can be overcome by treatment with an after-shampoo (cream) rinse or a hair dressing. It is considered desirable to control the effects of excessive surface degreasing at the time of shampooing. A wide variety of hair-conditioning additives has been recommended and tested. Only a few have gained wide acceptance, for example:

Name	CAS Registry number
dialkyl (C_{12}–C_{18}) dimethylammonium chloride,	eg, [53401-74-9]
hydrolyzed collagen	[9015-54-7]
polymeric quaternary derivatives	eg, [26590-05-6]
	or [53568-66-4]
potassium cocoyl hydrolyzed collagen	[68920-65-0]
sodium cocoamphoacetate	eg, [68334-21-4]
sodium lauroyl glutamate	[22923-31-7]
stearamidopropyl betaine	[68920-65-0]

Modern shampoos containing one or more of these conditioners have been designated as two-in-one products. They provide good cleaning and leave hair conditioned without the need for a second treatment. Eye stinging and irritation caused by shampoos can be reduced by including nonionic surfactants (with 10–45 polyoxyethylene groups) or by adding an amphocarboxylate.

Dandruff, a benign scaling skin disease of the scalp, is commonly viewed as a hair problem. The etiology and therapy of dandruff are similar to those of seborrheic dermatitis (67). Antidandruff shampoos are formulated using antimicrobial or desquamating agents to reduce the lipophilic yeasts (qv) widely believed to be the cause of scalp flaking. In the United States, shampoos for which antidandruff claims are made are OTC drugs. The choice of active agents is limited to coal tar [8007-45-2], zinc pyrithione [13463-41-7], salicylic acid [69-72-7], selenium sulfide [7488-56-4], and sulfur [7704-34-9], which can be added to shampoos or other scalp preparations. Some of the fungicidal azoles and piroctone olamine (1-hydroxy-4-methyl-6-(2,4,4-tri-methylpentyl)-2-(1*H*) pyridinone, ethanolamine salt [68890-66-4] are active against these causative organisms but are not recognized in United States OTC regulations.

11.3. Oral Cleansing Products. Toothpastes and mouthwashes are considered cosmetic oral cleansers as long as claims about them are restricted to cleaning or deodorization. Because deodorization may depend on reduction of microbiota in the mouth, several antimicrobial agents, either quaternaries, such as benzethonium chloride [121-54-0], or phenolics, such as triclosan [3380-35-5], are permitted. Products that include anticaries or antigingivitis agents or claim to provide such treatment are considered drugs.

Mouthwashes are hydro-alcoholic preparations in which flavorants, essential oils (see OILS, ESSENTIAL), and other agents are combined to provide long-term breath deodorization. Palatability can be improved by including a polyhydric alcohol such as glycerin or sorbitol (see ALCOHOLS, POLYHYDRIC). Occasionally, anionic and nonionic surfactants are used to help solubilize flavorants and to help remove debris and bacteria from the mouth.

Dentifrices (qv), or toothpastes, depend on abrasives to clean and polish teeth. The principal ingredients in toothpastes or powders are 20–50% polishing agents, such as calcium carbonate, di- or tricalcium phosphate, insoluble sodium metaphosphate, silica, and alumina; 0.5–1.0% detergents, for example, soap or anionic surface-active agents; 0.3–10% binders (gums); 20–60% humectants, such as glycerol, propylene glycol, and sorbitol; sweeteners (saccharin, sorbitol); preservatives, such as benzoic acid or *p*-hydroxybenzoates; flavors, for example, essential oils; and water. The most widely used surfactant is sodium lauryl sulfate, which is available with high purity. It produces the desired foam during brushing, acts as a cleansing agent, and has some bactericidal activity.

Transparent dentifrices can be prepared from certain xerogel silicas through use of high levels of polyhydric alcohols. Clarity depends on matching the refractive indexes of the silica and the liquid base. Compositions for liquid facial cleansers (68), shampoos (69), conditioning shampoos (70), dandruff shampoos (71), surfactant bars (72), toothpastes (73), and mouthwashes (74) have been published.

12. Shaving Products

Cosmetic shaving products are preparations for use before, during, or after shaving.

12.1. Preshaves. Preshave products are used primarily for dry (electric) shaving. Solid preshaves are usually compressed-powder sticks based on lubricating solids, such as talc or zinc or glyceryl stearate. Liquid preshaves are intended to remove perspiration residues and tighten and lubricate the skin. The alcohol content is relatively high (50–80%) to accelerate drying. The remaining ingredients may be polymeric lubricants, such as 1–2% polyvinylpyrrolidinone (PVP) [9003-39-8], emollients, such as 1–5% diisopropyl adipate [6938-94-9], and up to about 5% propylene glycol.

12.2. Shaving Creams. Despite the replacement of soap by synthetic detergents, products for wet shaving continue to be based on soap. Shaving creams and soaps are available as solids, that is, bars; creams, generally in tubes; or aerosols. Solids are essentially pure soaps applied to the face as foams with a brush. Shaving creams may be nonlathering (emulsion) and rarely consist entirely of lubricating lipids.

The principal ingredients of shaving creams and aerosols are liquid soaps, usually a blend of potassium, amine, and sodium salts of fatty acids, formulated to create a foam with the desired consistency and rinsing qualities. The soap blend may include synthetic surfactants, skin-conditioning agents, and other components. Modern nonaerosol shaving creams may contain 20–30% soap (potassium or triethanolamine (TEA)), up to about 10% glycerine, emollients, and foam stabilizers. Aerosol shaving creams are dilute forms of the cream types and are dispensed from the container with the aid of hydrocarbon propellants (up to about 10%). Aerosol shaving creams may also include some emulsifiers to ensure uniform emulsification of the propellant during the short shaking period before dispensation.

The objectives of shaving creams include protecting the face from cuts by cushioning the razor. Beard-softening qualities are attributable almost exclusively to hair hydration, which also depends on pH. Lubrication must be provided, primarily between the blade and the hair fibers through which it passes. Blade technology based on polymeric coatings reduces the need for this type of lubrication by shaving creams.

12.3. After-Shaves. After-shave preparations serve the same function as and are formulated similarly to skin astringents. After-shave balms are hydro-alcoholic or alcohol-free emulsions that supply soothing ingredients, for example, witch hazel, and emollients, for example, decyl oleate [3687-46-5], to the skin. Menthol, which provides a cooling sensation, is a common constituent of after-shaves.

Compositions for preshaves (75), shaving creams (76), and after-shaves (77) have been published.

13. Nail-Care Products

Over the years the cosmetic industry has created a wide variety of products for nail care. Some of these, such as cuticle removers and nail hardeners, are functional; others, such as nail lacquers, lacquer removers, and nail elongators, are decorative.

13.1. Functional Nail-Care Products. Cuticle removers are solutions of dilute alkalies that facilitate removal, or at least softening, of the cuticle. Formulations containing as much as 5% potassium hydroxide have been reported. Such preparations may contain about 10% glycerine to reduce drying, and thickeners, such as clays, to reduce runoff. Lipids and other conditioners are included to reduce damage to tissues other than the cuticle.

Nail hardeners have been based on various protein cross-linking agents. Only formaldehyde is widely used commercially. Contact with skin and inhalation must be avoided to preclude sensitization and other adverse reactions. The popularity of products of this type is decreasing because the polymers used in nail elongators can be used to coat nails to increase the mechanical strength.

13.2. Decorative Nail-Care Products. Nail lacquers, or nail polishes, consist of resin, plasticizer, pigments, and solvents. The most commonly used resin is nitrocellulose, prepared by esterification of celluloses with nitric acid, with a degree of substitution between 1.8 and 2.3 nitrate groupings per anhydroglucose unit. Ethylacetate, butylacetate, and toluene are typical solvents. Toluenesulfonamide–formaldehyde resin [25035-71-6] and similar polymers, for example, the terpolymer of 2,2,4-trimethyl-1,3-pentanediol, isophthalic acid, and trimellitic anhydride, are the resins of choice as secondary film formers for optimal nail adhesion. Other resins, such as alkyds, acrylates, and polyamides, can also serve as secondary film formers.

Camphor, dibutyl phthalate [84-74-2], and other lipidic solvents are common plasticizers. Nail lacquers require the presence of a suspending agent because pigments have a tendency to settle. Most tinted lacquers contain a suitable flocculating agent, such as stearalkonium hectorite, a reaction product of hectorite [12173-46-6] and stearalkonium chloride [122-19-0].

The blend of pigments used to create a particular shade must conform to regulations covering pigments and dyes in cosmetics. Regulations vary among countries and undergo frequent changes. The selected pigments may not stain the nails, and any organic dye or pigment that might exhibit solubility in the mixture of lacquer solvents is avoided. Typical organic dyes include monoazo dyes, such as D&C Red No. 6 Barium Lake and D&C Red No. 34 Calcium Lake, and pyrazole dyes, such as FD&C Yellow No. 5 Aluminum Lake (see AZO DYES). Inorganic pigments, such as iron oxides and titanium dioxide, can be incorporated. Colored pigments do not usually exhibit the opacity and reflective brilliance demanded of modern nail enamels. Bismuth oxychloride, mica, and guanine were extensively used in the past to provide nacreous reflections, whereas titanium dioxide provided opacity. These substances have been replaced by the highly reflective and nacreous synthetic mica–titanium dioxide pigments. Generally, nacreous pigments are supplied as suspensions in nitrocellulose-containing solvent blends, whereas other pigments are mixed with nitrocellulose and plasticizers and processed through a roller mill. Nitrocellulose processing must be done with extreme caution in an explosion-proof environment. The level of pigment in nail enamels generally does not exceed about 5%. The solvent level is about 60–70%, however, and the dried lacquer may contain as much as 10–15% pigment.

Nail lacquer removers are simply acetone or blends of solvents similar to those used in nail lacquers. It is commonly accepted that solvents have a drying effect on nails, and nail lacquer removers are often fortified with various lipids such as castor oil [8991-79-4] or cetyl palmitate [540-10-3].

Nail elongators are products intended to lengthen nails. These have become extremely popular. In earlier compositions, polymerization was conducted by mixing monomers, oligomers, and catalysts on the nail (78). More recently, nail elongation is achieved by adhering a piece of non-woven nylon fabric (referred to as nail wrap) to the nail with a colorless lacquer. This process may be repeated until the desired nail thickness has been reached. After shaping, the artificial nail is further decorated. Compositions for nail lacquers (79) and nail hardeners and conditioners (80) have been published.

14. Hair Products

Cosmetics for hair care fall into several categories: cleansers or shampoos, conditioners, fixatives, coloring products, waving and straightening products, and hair removers (see HAIR PREPARATIONS).

14.1. Hair Conditioners.   Hair conditioners are designed to repair chemical and environmental damage, replace natural lipids removed by shampooing, and facilitate managing and styling hair. The classical hair-conditioning products were based on lipids, which were deposited on hair either directly, with oils or pomades, or from emulsions. Liquid and semisolid brilliantines are formulated from mineral oil or vegetable or animal fats thickened with waxes (ozokerite), fatty alcohols (cetyl alcohol), or polymers (for example, polyethylene), and are normally dispensed from jars or tubes. Emulsion products are commonly based on an oil phase that consists of mineral oil, lanolin, and synthetic or vegetable-derived lipids. The emulsifiers vary widely and may include anionics (soap), nonionics (alkyl polyoxyethylene ethers), or cationics (eg, PEG-2 stearmonium chloride [606087-87-8]). These o/w type emulsions may be thickened with various gums and may contain plant extracts, antimicrobial agents (quaternaries), uv screens, and hair-fixative polymers such as PVP.

Microemulsions, temporary emulsions, that is, two-layer hair dressings, and clear solutions of nonvolatile lubricants are on the market. Hair tonics, usually hydro-alcoholic, achieve similar effects by including lipid substances or synthetic emollients, such as the mono butyl ethers of polypropylene oxides [9003-13-8] (10–50 mol). The primary benefits of these lipid-based products are lubrication and improvements in hair gloss and hair-holding (dressing) qualities. Hair holding and manageability result from the tendency of the lipid components to make the fibers adhere to each other laterally, not from the coating of individual hairs.

An entirely different, and in the 1990s more popular, type of hair conditioning is achieved by treating hair with substantive quaternary compounds or quaternary polymers. Quaternaries are sorbed by hair, retained despite rinsing with water, and removed only by shampooing. Quaternaries that are not easily removed cannot be used, because they tend to build up on the hair, making it overconditioned and limp. The most widely used quaternary is stearalkonium chloride [122-19-0], which has been used at 3–5% concentration in cream rinses for many years. More recently, many useful quaternaries have become available; some are listed in Table 15. Table 15 also includes some quaternary polymers that are not only substantive to hair but also possess hair-fixative properties. Despite the commercial success of many conditioning quaternaries, efforts to synthesize better performing derivatives continue.

Table 15. **Hair-Conditioning and Polymeric Fixative Compounds**[a]

Material	CAS Registry number
Hair conditioners	
disoyadimonium chloride	[61788-92-9]
hydroxyethyl cetyldimonium chloride	[24625-03-4]
stearalkonium chloride	[122-19-0]
quaternium 22	[51812-80-7]
quaternium 79 hydrolyzed milk protein[b]	
Hair conditioners with fixative properties	
polyquaternium 4[c]	
polyquaternium 6	[26062-79-3]
polyquaternium 7	[26590-05-6]
polyquaternium 10	[53568-66-4]
polyquaternium 11[d]	
polyquaternium 22	[53694-17-0]
Hair-fixative polymers	
polyvinylpyrrolidinone (PVP)	[9003-39-8]
shellac	[9000-59-3]
vinyl acetate–crotonic acid–vinyl neodecanoate copolymer	[55353-21-4]

[a] Ref. 26 includes a more comprehensive listing.
[b] Material is the reaction product of a fatty acid amide of *N,N*-dimethylpropylene-diamine and epichlorohydrin and hydrolyzed milk protein.
[c] Dialkyldimethyl ammonium chloride–hydroxyethylcellulose copolymer.
[d] Vinylpyrrolidinone–dimethylaminoethyl methacrylate copolymer, dimethyl sulfate reaction product.

A third type of hair-conditioning product relies on the use of proteins, amino acids (qv), botanicals, and amphoterics. Many ingredients have been identified as hair conditioners. Some of them are claimed to be substantive to the hair, whereas others are claimed to penetrate into the hair and repair previously incurred damage. Some of these hair-conditioning substances have been incorporated into newer delivery systems, such as mousses. These water-based products are dispensed as foams, which are rubbed into the hair and may then be rinsed off with water or allowed to remain on the hair for conditioning and styling benefits. All types of hydrolyzed proteins, such askeratin, soy, yeast, and wheat, chemically modified and free amino acids, such as cystine, aspartic acid, and lauroyl glutamate, and biological additives, such as casein, beer, eggs, nettle extract, and horse chestnut extract, have been formulated into products containing amphoteric and other more conventional cosmetic ingredients. Reference 26 includes an extensive list of chemicals used in hair conditioners and related products.

14.2. Hair Fixatives. These products are designed to assist in hair styling and in maintaining the style for a period of time. In contrast with hair dressings, hair fixatives do not leave an oily residue on the hair but tend to coat the hair with film-forming residues after drying. As in the case of hair dressings, style-holding qualities depend primarily on fiber–fiber adhesion and to a minor extent on fiber coating. The products may be conveniently divided into two groups: those that are applied to damp or wet hair, hair-setting products, and those that are applied to hair after styling, hair sprays. Styling requires that hair be formed into and retain the desired configuration. Curlers of various designs provide a wavy style, whereas hot combing results in essentially straight hair.

Wave-setting products can be applied to wet hair and should not interfere with or delay drying. Such products commonly aid in wet styling of hair. After drying, these products are claimed to help retain the style, regardless of frequent combing or exposure to high humidity. Wave sets can be formulated with water-soluble polymeric substances or with polymers that show solubility only in hydro-alcoholic media. Some of the preferred hair-fixative polymers (see Table 15) are combined with lubricants or emollients and other excipients. The viscosity of these products can vary from that of a water-thin fluid to a rather firm gel. The set-holding polymer constitutes about 1–3% of the product, and the viscosity-increasing substance is commonly a cross-linked polyacrylate, for example, carbomer [9007-16-3].

Hair sprays are applied from aerosol cans or pumps to dried and styled hair. Hair-spray products containing little or no water are preferred, because the presence of significant levels of water tends to soften a preexisting style. In the past, hair sprays were alcoholic solutions of polymers that were propelled with fluorinated or chlorinated highly volatile solvents. The alcohol concentration was kept as low as possible to reduce excessive wetting. As a result, the fixative resins had to exhibit good solubility in the propellant blends. Environmental regulations today preclude the use of these propellant solvents. Thus higher levels of alcohols are now used and the propellants of choice are low concentrations of hydrocarbons.

14.3. Hair Colorants.   Hair colorants are commonly divided into temporary, semipermanent, and permanent types. Decolorizing (bleaching) represents a fourth type of hair coloring (81) (see BLEACHING AGENTS).

Hair bleaching removes the pigment melanin from the hair shaft by oxidative destruction. Alkaline hydrogen peroxide is the agent of choice. Because hydrogen peroxide is unstable at elevated pH, it is frequently supplied in pure form (6–10%) and is combined at the time of use with an ammonia or an amine-containing product to provide approximately 3–6% H_2O_2 at a pH of about 8.5 to 9.5. Thickening is required in order to retain the blended oxidizing mixture on the hair. Thickeners and conditioning agents, for example, fatty alcohols and protein derivatives, can be formulated into the alkalizing component or, occasionally, into the hydrogen peroxide. Surfactants are required to assure that every hair fiber is thoroughly wet by the blended mixture. Bleaching by hydrogen peroxide is enhanced by the presence of a peroxydisulfate, such as potassium persulfate [7727-21-1]. Bleaching damages the hair by converting some cystine to cysteic acid. In addition, the high pH induces swelling and cuticular damage. These adverse effects are counteracted by conditioning after treatment or by including some protectants in the hair-bleach product.

Temporary hair colorants are removed from the hair by a single shampoo. Temporary hair colorants usually employ certified dyes that have little affinity for hair (see Table 9). They are incorporated into aqueous solutions, shampoos, or hair-setting products.

Semipermanent hair colorants employ dyes that are absorbed directly by the hair. These dyes add color to the preexisting (natural) hair color and are

Table 16. **Semipermanent Hair Dyes**[a]

Name	CAS Registry number	CI Number	Chemical type
Pigment Violet 19	[1047-16-1]	46500	quinacridone
Pigment Yellow 13	[5102-83-0]	21100	diazo
Basic Violet 3	[548-62-9]	42555	triarylmethane
Basic Red 76	[68391-30-0]	12245	monoazo
Direct Red 80	[2610-10-8]	35780	tetraazo
Disperse Blue 1	[2475-45-8]	64500	anthraquinone
HC Blue 2	[33229-34-4]		nitro-p-phenylenediamine
HC Yellow 4	[52551-67-4]		nitroaniline

[a]Ref. 26 includes a comprehensive listing.

useful primarily for blending in gray fibers. These dyes may fade significantly owing to exposure to sunlight and also are gradually removed by shampooing. Dyes selected for this purpose should not stain the scalp or skin during application. Typically, temporary hair colorings are distributed as pourable lotions. Formulations may include alkanolamides, polymeric substances, fatty alcohols, thickeners, and conditioners commonly employed in all hair cosmetics. The chemical nature of the dyes is highly diverse and varies among manufacturers. Solvents, carriers, or complex solubilizers may be required when pigments are used. The CTFA lists temporary hair dyes with other substances as "Color Additives—Hair Colorants" (26). A few key chemical types are identified in Table 16.

Development of the desired shade depends to a large degree on the tendency of the individual dyes in the mixture to adhere to the hair in proper proportions. Performance evaluation on many different types of hair, for example, natural, bleached, and permanently waved, is required.

Permanent hair colorants, frequently identified as oxidation dyes, show much greater resistance to fading and shampoo loss than do semipermanent hair colorants. As a rule, these dyes remain on the hair; it is common practice to dye only that portion of the hair shaft that has emerged from the scalp since the last application. In permanent dyeing, a lotion containing developers and couplers is blended with hydrogen peroxide and then applied to the hair. The objective is uniform penetration of the various components into the hair, oxidation of the developer to a reactive intermediate, and formation of a colored dye stuff with the coupler. The dyes are synthesized within the fiber and migrate outward only slowly because of their size. In addition, the reactions occur in alkaline media, and some of the peroxide bleaches the hair. Thus it is possible to generate colored hair lighter than the original shade. Dye formation is complex because shading is achieved by employing several developers and several couplers in the same dye bath. The process is illustrated by p-phenylenediamine, which is oxidized by the peroxide to a quinone diimine. This short-lived intermediate can react, for example, with resorcinol to yield a brownish indoaniline. Table 17 provides some insight into the many interactions that exist from just a few components. Further shading is possible by including semipermanent colorants (see Table 16), especially nitroaniline derivatives.

In hair coloring a light ash blond shade may require as little as 0.5–1% of intermediates, whereas a true black may require up to about 5%. In principle,

Table 17. **Intermediates Used in Oxidation Hair Dyes**^{a}

Material	CAS Registry number	Molecular formula
Developers		
4-amino-3-nitrophenol	[119-34-6]	$C_6H_6N_2O_3$
p-aminophenol	[123-30-8]	C_6H_7NO
2-methoxy-*p*-phenylenediamine sulfate	[42909-29-5]	$C_7H_{10}N_2O \cdot xH_2SO_4$
p-methylaminophenol	[150-75-4]	C_7H_9NO
p-phenylenediamine	[106-50-3]	$C_6H_8N_2$
N-phenyl-*p*-phenylenediamine	[101-54-2]	$C_{12}H_{12}N_2$
phloroglucinol	[106-73-6]	$C_9H_6O_3$
toluene-2,5-diamine	[95-70-5]	$C_7H_{10}N_2$
Couplers		
o-aminophenol	[95-55-6]	C_6H_7NO
2,4-diaminophenoxyethanol HCl	[66422-95-5]	$C_8H_{12}N_2O_2 \cdot 2HCl$
2,6-diaminopyridine	[141-86-6]	$C_5H_7N_3$
hydroquinone	[123-31-9]	$C_6H_6O_2$
1,6-naphthalenediol	[83-56-7]	$C_{10}H_8O_2$
m-phenylenediamine	[106-45-2]	$C_6H_8N_2$
pyrocatechol	[120-80-9]	$C_6H_6O_2$
pyrogallol	[87-66-1]	$C_6H_6O_3$
resorcinol	[108-46-3]	$C_6H_6O_2$

^{a}Ref. 26 includes a comprehensive listing.

the formulator blends precursors that yield red, blue, and yellow dyes. The base in which the components are dissolved or suspended is similar to that used in simple bleaches and may include alkanolamides, various types of surfactants, thickening agents, and solvents. Removal of undesirable dyes is achieved by treating the discolored hair with a powerful reductant of the sulfite family.

Permanent coloration can also be achieved by exposing hair to certain metals: copper, silver, and especially lead salts. Preparations containing aqueous solutions of lead acetate may include a source of sulfur, usually thiosulfate, which may react with cystine in the hair to produce some cysteine or may react directly with the metal ion to form dark metallic sulfides. Preparations of this type, which darken hair gradually, are not universally considered safe.

14.4. Hair Waving and Straightening Products. The development of hair-waving and hair-straightening products requires a careful balance between product performance and hair damage. The hair-waving process essentially depends on converting some cystine cross-links in keratin to cysteine residues, which are reoxidized after the configuration of the hair has been changed. Sometimes hair straightening can also be achieved by a similar, relatively innocuous chemical change in the hair. As a rule, however, much more chemical destruction is required to achieve rapid and permanent straightening than to achieve permanent waving.

Permanent waving depends on the metathesis of a mercaptan and the cystine in hair while the hair is held in a curly pattern on a suitable device (curler). The most commonly used mercaptan is thioglycolic acid [68-11-1], although some other nonvolatile mercaptans can be employed. The active species is the mercaptide anion. Thus, adjustment to a pH between about 8.8 and 9.5 using amines or especially ammonia is required. A typical hair-waving product may consist of a 0.5–0.75 N solution of thioglycolic acid adjusted to a pH of about 9.1 with ammonia. The product generally includes a nonionic surfactant, to ensure thorough wetting of the wound hair tress, and a fragrance. Opaque lotion products can be created by adding the actives to mineral oil or other lipid-containing emulsions. The thioglycolate lotion is allowed to remain on the hair for about 10–30 min; then the hair is rinsed with water. Next, an oxidizing solution consisting of a dilute (1.5–3%) acidified hydrogen peroxide solution or of a potassium or sodium bromate is applied to the hair. This so-called neutralizing solution oxidizes the cysteine residues to cystine (without bleaching) in a new configuration within about 5–10 min. The neutralizer may contain a variety of hair conditioners and is removed from the hair by thorough rinsing with water after unwinding.

Alternatively, the metathesis can be effected by sulfites or bisulfites that convert cystine into one cysteine residue and one thiosulfate (Bunte salt) residue. Hair waving based on sulfites is slower than that based on mercaptans and is more likely to cause changes in hair color.

Acidic waving systems based on mercaptans have recently achieved some popularity. The preferred mercaptan is glyceryl thioglycolate [30618-84-9], which is relatively odorless and provides sufficient active anionic mercaptide species at a neutral pH.

Hair straightening is more difficult than hair waving. Kinky hair has a tight crimp that cannot be straightened by winding over a rod or curler. Two processes for straightening exist. One, based on thioglycolates, effects the same chemical change as that occurring during permanent waving. The other, more aggressive, process is based on (1–8%) sodium hydroxide (or guanidine). The exact concentration depends on the temperature at which the process is carried out. In order to hold the hair straight, hair-straightening products are viscous. The hair is combed repeatedly during the process, which has the effect of reconfiguring the hair but can lead to serious hair damage from excessive pulling. The chemical reactions with sodium hydroxide involve formation of cysteine and dehydroalanine residues in the hair with some loss of sulfur. The cysteine and dehydroalanine can subsequently react to form the thioether, lanthionine [922-55-4], which helps repair the mechanical strength of the fiber to some extent. Similar chemical reactions occur when steam is allowed to interact with hair, such as during hot pressing, which was an earlier technique for straightening hair.

Conditioners, lipids, acid rinses, and related cosmetics have been developed to minimize hair damage from these rather destructive processes.

14.5. Hair Removers. Hair removers are designed to remove hair from the skin surface without cutting in order to avoid undesirable stubble. Cosmetic products have been developed for chemical destruction of hair, that is, depilation, and for facilitating mechanical hair removal, that is, epilation.

Depilatories epitomize the chemical destruction of hair and allow hair removal by scraping with a blunt instrument or by rubbing with terry cloth. Chemical depilatories are based on 5–6% calcium thioglycolate in a cream base (to avoid runoff) at a pH of about 12. The pH is maintained with calcium or strontium hydroxide. Hair destruction is rapid, requiring not more than about 10 min.

Treatment with a depilatory is followed by careful rinsing with water and various conditioning products intended to restore the skin's pH to normal. This type of treatment does not destroy the dermal papilla, and the hair grows back.

Epilation is required for permanent hair removal. The most effective epilation process is electrolysis or a similar procedure. Epilation can also be achieved by pulling the fibers out of the skin. For this purpose, wax mixtures (rosin and beeswax) are blended with lipids, for example, oleyl oleate, which melt at a suitable temperature (about 50–55°C). The mixture is applied to the site (a cloth tape may be melted into the mass) and after cooling is rapidly pulled off the skin. A similar process can be carried out with a tape impregnated with an aggressive adhesive.

Compositions have been published for cream rinses (82), hair conditioners, dressings, and mousses (83), hair-styling products (84), hair sprays (85), hair colorants (86), hair-waving products (87), hair-straightening products (88), and depilatories (89).

15. Decorative Cosmetics

Decorative cosmetics are products intended to enhance appearance by adding color or by hiding or deemphasizing physical defects. In Western cultures, most decorative cosmetics are for use on the face. Products in this category are various types of powders, facial makeups, and lip- and eye-coloring products. Regardless of the site of application or the type of product, all decorative cosmetics must meet certain critical performance criteria as outlined in Table 18. The dyes and pigments used in these products must be stable in the finished

Table 18. **Performance Criteria for Decorative Cosmetics**

Characteristic	Typical components	Comment
covering power	titanium dioxide	hides defects
	zinc oxide	
	magnesium oxide	
	zirconium silicate	
slip	talc	easy application
	zinc or magnesium stearate	smooth sensation
	starch	
	emollient lipids	
absorbency	chalk	absorbs skin secretions
	silica	without color change
	starch	
	synthetic polymers	
adherence	emollient lipids	clings to skin
	volatile solvents	limits ruboff
	polymeric substances	dries to hard film
	gums	
pigmentation	certified pigments	provides color
	noncertified pigments	

preparation and must not fade or discolor as a result of exposure to the variable environment of the skin.

15.1. Lip Makeups. Intensely pigmented coloring products have been used for many years to accentuate and modify the appearance of the lips. These products are marketed in soft stick forms (lipstick), as pastes (tinted lip gloss), and as hard sticks (lip liner). Lipsticks are manufactured via the molding process. The brilliant colors required for lipsticks are produced primarily by a limited number of available organic dyes and lakes. Formulation of acceptable shades is difficult and subject to the vagaries of fashion. For many years, lipsticks that caused a permanent stain on the lips were popular. Such staining is no longer desirable, and skin-staining dyes such as eosin (D&C Red No. 21) are rarely used. Traces of an inoffensive fragrance and of an antioxidant are commonly included in the lipid base (see Table 10). Sophisticated lipsticks may also contain moisturizers and uv light screens.

High gloss lipsticks use castor oil or 2-octyldodecanol. The more wear-resistant fat-based sticks are generally somewhat duller. There is a large number of possible ingredients, but the performance of most sticks is comparable. Feathering of the lipstick film, that is, creeping of color into crevices surrounding the lip tissue, is avoided by controlling the rheological properties of the applied stick mass.

Tinted and untinted soft lipstick masses are distributed in pans as lip glosses. These are applied with the fingers or with lipstick brushes. Hard, tinted, pencil-type sticks have been marketed as lip liners. Chap sticks are unpigmented lipsticks intended to alleviate scaling and to prevent cracked lips. Compositions of lipsticks, lip glosses, and chap sticks have been published (90).

15.2. Eye Makeup. Since antiquity, eye makeup preparations have been used to beautify the area surrounding the eye. Various forms of eye shadows color the eyelid; mascaras color and lengthen the eyelashes; eyeliners delineate the portion of the eyelid from which the eyelashes emerge. The appearance of eyebrows can be altered with various types of makeup, and, finally, false eyelashes and eyebrows have been marketed for years.

In the United States the use of coal-tar dyes in eye makeup is generally prohibited. The use of permanent and temporary hair colorants (Tables 16 and 17) and of organic dyes and their lakes is precluded. As a result, only insoluble inorganic pigments can be used (Table 9). The sensitivity of the eye mandates that coarse or irritating particulate matter not be used. All eye preparations must be properly preserved and, preferably, self-sterilizing to prevent accidental introduction of pathogens into the eye.

Eye shadows may be molded or compressed. The technology used for molded sticks resembles that used in lipsticks except that the choice of pigments is restricted. Glossy lipids are generally avoided. If gloss is desired, it is achieved by the inclusion of a nacreous pigment. The basic technology for producing compressed eye-shadow sticks or powder compacts is that of other powder products.

Mascaras are available in three basic forms: cakes; creams; and mascaramatics, narrow containers provided with brushes. Cakes are commonly prepared by milling pigments into (sodium) soap chips and compressing this blend into metal or plastic pans. A small brush wet with water is used to transfer the mascara to the eyelashes. Cake mascaras can be modified with ingredients that improve adhesion of the mascara to the eyelashes, for example beeswax [8006-40-4] and dihydroabietyl alcohol [26216-77-31]; resist washing off or smearing by tearing, for example, aluminum mono-, di-, or tri-stearates; and lengthen the hairs (polymeric fibrous filaments).

Cream mascaras are pigmented, viscous, o/w emulsions; soap emulsions are common. Viscosity is increased with glyceryl monostearate to help suspend the pigment. Cream mascaras are distributed in small jars or narrow-orifice tubes and are applied with brushes.

Mascaramatic mascaras have the largest share of the market. Emulsion mascaramatics are cream-type mascaras dispensed from containers that include a closure provided with a wand ending in a small brush. In solvent mascaramatics, mascara masses are pigment suspensions in thickened hydrocarbon solvents such as isoparaffins and petroleum distillates. The thickeners include waxes (microcrystalline [63231-60-7], carnauba [8015-86-9], or ouricury [68917-70-4]), polymers (hydrogenated polyisobutene [61693-08-1]), and esters (propylene glycol distearate [6182-11-2] or trilaurin [538-24-9]).

Mascara pigmentation is usually black or brown–black. Mascaras during and after application are extremely close to the cornea, and any potential irritant must be rigidly excluded. The use of lash-elongating synthetics, such as rayon, nylon, and the like, has not resulted in significant problems.

The use of false eyelashes is rare. These are prepared from natural or synthetic fibers attached to a tinted lash strip, which can be glued onto the lid with the aid of an adhesive.

Eyeliners are available in two popular forms. One of these is a deeply pigmented emulsion that is applied with a fine brush. The emulsion must be viscous to avoid running and should dry to a waterproof film. The emulsion can be patterned after the emulsions used in mascaras. Glossy eyeliners require the use of nacreous pigments suspended in polymeric film formers, for example, acrylic acid copolymers.

The second type of eyeliner is a soft crayon, in pencil form, that delivers mass with minimal pressure. Products of this type may contain as much as 70% talc, 5% pigment, and 5% aluminum stearate. The lipid portion may include squalene [111-02-4] and alkanolamides.

Eyebrow pencils are used to outline the brow, especially after plucking of undesirable fibers. They are commonly prepared by stick molding. Eyebrow pencils are generally harder than eyeliner pencils. Like other pencils, the extruded mass is normally encased in wood. Some eyebrow pencils may be quite soft, approaching the texture of lipsticks.

Formulations of eye shadows (91), mascaras (92), and eyeliners (93) have been published.

BIBLIOGRAPHY

"Cosmetics" in *ECT* 1st ed., pp. 545–562 by F. E. Wall, Consulting Chemist; in *ECT* 2nd ed., Vol. 6, pp. 346–375 by H. Isacoff, International Flavors and Fragrances (U.S.); in *ECT* 3rd ed., Vol. 7, pp. 143–176, by H. Isacoff, International Flavors and Fragrances, Inc.; in *ECT* 4th ed., Vol. 8, pp. 572–619, by Martin M. Rieger, M & A Rieger Associates.

CITED PUBLICATIONS

1. H. Truttwin, *Handbuch der Kosmetischen Chemie*, 2nd ed., J. A. Barth, Leipzig, Germany, 1924.
2. M. G. de Navarre, *The Chemistry and Manufacture of Cosmetics*, D. Van Nostrand Company, Inc., New York, 1941; F. Chilson, *Modern Cosmetics*, The Drug and Cosmetic Industry, New York, 1934 (formulation only).
3. *CTFA Buyer's Guide*, CTFA, Washington, D.C.
4. *Cosmetics &* Toiletries, Who's Who, Allured Publishing Co., Wheaton, Ill., Jan., 1993.
5. *Soap Cosmetics Chemical Specialties, Blue Book*, PTN Publishing Co., Melville, N.Y., Apr.
6. *CFR 21*, Part 211, U.S. Government Printing Office, Washington, D.C. 20402.
7. *CTFA International Cosmetic Ingredient Dictionary*, 4th ed., CTFA, Washington, D.C., 1991.
8. A. C. deGroot and I. R. White, *Contact Dermatitis* **25**, 273 (1991).
9. *CTFA International Resource Manual* and *CTFA International Color Handbook*, CTFA, Washington, D.C., 1992.
10. P. M. Silber, *Cosm. Toil.* **106**(IV), 56 (1991).
11. J. H. Draize and co-workers, *J. Pharmacol., Exp. Ther.* **82**, 377 (1944); *CFR* 16, 1500.42.
12. H. P. Frosch and A. M. Kligman, *Contact Dermatitis* **2**, 314 (1976).
13. H. P. Frosch, in H. P. Frosch, ed., *Principles of Cosmetics for Dermatologists*, Mosby, St. Louis, Mo., 1982.
14. B. Magnusson and A. M. Kligman, *J. Invest. Dermatol.* **52**, 268 (1969).
15. A. M. Kligman, *J. Invest. Dermatol* **47**, 393 (1966).
16. P. D. Forbes and co-workers, *Food Cosm. Toxicol.* **15**, 55 (1977).
17. T. Maurer, *Contact and Photocontact Allergens*, Marcel Dekker, New York, 1983.
18. L. C. Harber and A. R. Shalita, in F. N. Marzulli and H. I. Maibach, eds., *Dermatotoxicology and Pharmacology*, Hemisphere, Washington, D.C., 1977.
19. K. H. Kaidbey and A. M. Kligman, *Contact Dermatitis* **6**, 161 (1980).
20. J. H. Draize and E. A. Kelley, *Proc. Sci. Sect. Toilet Goods Assoc.* **17**, 1 (1952); *CFR* 16, 1500.41.
21. A. M. Kligman and O. H. Mills, Jr., *Arch. Dermatol* **106**, 843 (1972).
22. H. P. Frosch and A. M. Kligman, *J. Soc. Cosm. Chem.* **28**, 197 (1977).
23. K. Lammintausta and co-workers, *Dermatosen* **36**, 45 (1988).
24. M. Lanzet, *Cosm. Toil.* **101**(II), 63 (1986); *Cosm. Toil.* **105**(X), 26 (1990).
25. *CTFA Microbiology, Quality Assurance, and Occupational Safety Guidelines*, CTFA, Washington, D.C.; *Cosm. Toil.* **105**(III), 79 (1990).
26. *CTFA International Cosmetic Ingredient Handbook*, 2nd ed., CTFA, Washington, D.C., 1992.
27. D. S. Orth, *Cosm. Toil.* **106**(III), 45 (1991).
28. W. J. L. Smith, in N. Estrin, ed., *The Cosmetic Industry*, Marcel Dekker, New York, 1984, Chapt. 21.
29. S. Friberg, *J. Soc. Cosm. Chem.* **30**, 309 (1979).
30. P. Becher, *J. Disp. Sci. Technol.* **5**, 81 (1984).
31. Y. Saiato and co-workers, *JAOCS* **67**, 145–148 (1990).
32. B. Idson, in H. A. Lieberman and co-eds., *Pharmaceutical Dosage Forms, Disperse Systems*, Vol. I, Marcel Dekker, New York, 1988, Chapt. 6.
33. M. Rosoff, in Ref. 32, Chapt. 7.
34. K. Shinoda and H. Kunieda, in P. Becher, ed., *Encyclopedia of Emulsion Technology*, Vol. 1, M. Dekker, New York, 1983.

35. M. M. Rieger, in L. Lachman and co-eds., *The Theory and Practice of Industrial Pharmacy*, 3rd ed., Lea and Febiger, Philadelphia, 1986.

36. M. M. Rieger, *Cosm. Toil.* **106**(V), 59–69 (1991).

37. H. E. Junginger, *Pharm. Weekblad* **6**, 141 (1984).

38. G. M. Eccleston, *J. Soc. Cosm. Chem.* **41**, 1–22 (1990).

39. C. Prybilski and co-workers, *Cosm. Toil.* **106**(XI), 97 (1991).

40. A. Bevacqua and co-workers, *Cosm. Toil* **106**(V), 53 (1991).

41. F. Comalles and co-workers, *Int. J. Cosm. Sci.* **11**, 5 (1989).

42. M. Riaz and co-workers, in H. H. Lieberman, ed., *Pharmaceutical Dosage Forms: Disperse Systems*, Vol. 2, Marcel Dekker, New York, 1989, Chapt. 16; N. Wiener, *Antimicr. Agents Chemother.* **34** (10), 107 (1990).

43. J. N. Israelachvili, *Intermolecular and Surface Forces*, Academic Press, New York, 1985.

44. A. J. I. Ward and C. duReau, *Int. J. Pharm.* **74**, 137 (1991).

45. R. O. Potts and co-workers, *J. Invest. Dermatol.* **96**, 495, 580 (1991).

46. G. Imokawa and co-workers, *J. Invest. Dermatol.* **96**, 529 (1991).

47. G. V. Civille and C. A. Dus, *Cosm. Toil.* **106**(V), 83 (1991).

48. *Cosm. Toil.* **107**(III), 96, 97 (1992); **106**(XII), 71 (1991).

49. *Cosm. Toil.* **106**(XII), 101–104 (1991); **107**(VII), 101–108 (1992).

50. *Cosm. Toil.* **106**(XII), 105, 106 (1991).

51. *Cosm. Toil.* **106**(XII), 97–100 (1991); **107**(VII), 110–114 (1992).

52. *Fed. Reg.* **56**, 41008–41020 (Aug. 16, 1991).

53. *Cosm. Toil.* **107**(III), 96 (1992).

54. *Fed. Reg.* **43**(166), 38206–38267 (Aug. 25, 1978).

55. N. A. Shaath, *Cosm. Toil.* **101**(III), 55 (1986); **102**(III), 21 (1987); **102**(III), 69 (1987); idem, in N. J. Lowe and N. A. Shaath, eds., *Development, Evaluation and Regulatory Aspects*, Marcel Dekker, New York, 1990; *Fed. Reg.* **43**(166); 38206–38269 (Aug. 25, 1978).

56. *Cosm. Toil.* **105**(XII), 92, 122–130, 133–134 (1990); **106**(XII), 107, 108 (1991).

57. *Cosm. Toil.* **104**(VII), 83–84 (1989).

58. *Cosm. Toil.* **104**(VII), 85 (1989).

59. *Cosm. Toil.* **104**(VII), 86–94 (1989); **106**(XII), 109–110 (1991).

60. *Cosm. Toil.* **107**(V), 24 (1992); **106**(IX), 32 (1991).

61. *Cosm. Toil.* **106**(VI), 33, 34 (1991); **106**(XII), 107, 108 (1991).

62. *Fed. Reg.* **84**, 13440–13499 (Apr. 3, 1989).

63. *Fed. Reg.* **47** (162), 36492–36505 (1982).

64. *Cosm Toil.* **102**(X), 54 (1987); **105**(IV), 76–80(1990); **106**(XII), 78, 110 (1991).

65. *Cosm. Toil.* **102**(X), 53, 119 (1987); **105**(IV), 75, 76 (1990); **106**(XII), 78 (1991).

66. U.S. Pat. 4,954,282 (Sept. 4, 1990), K. J. Rys and co-workers (to Lever Brothers Co.).

67. *Fed. Reg.* **56**, 63554–63569 (Dec. 4, 1991).

68. *Cosm. Toil.* **102**(X), 112, 113, 116–117 (1987); **106**(VI), 34 (1991).

69. *Cosm. Toil.* **102**(X), 115 (1987); **106**(IV), 89–93 (1991).

70. *Cosm. Toil.* **106**(IV), 83–86 (1991); **107**(II), 29 (1992).

71. *Cosm. Toil.* **106**(IV), 91–92 (1991).

72. *Cosm. Toil.* **107**(VII), 116, (1992).

73. *Cosm. Toil.* **102**(X), 87 (1987); **107**(III), 93–95 (1992).

74. *Cosm. Toil.* **107**(III), 95 (1992).

75. *Cosm. Toil.* **105**(IV), 81 (1990).

76. *Cosm. Toil.* **105**(IV), 81–84 (1990).

77. *Cosm. Toil.* **105**(IV), 84–87 (1990); **106**(XII), 111 (1991).

78. L. J. Viola, in E. Sagarin and co-eds., *Cosmetics: Science and Technology*, 2nd ed., Vol. 2, John Wiley & Sons, Inc., New York, 1972, p. 543.

79. *Cosm. Toil* **106**(VI), 36 (1991).
80. *Cosm. Toil.* **104**(VII), 97, 98 (1989).
81. J. F. Corbett, *Cosm. Toil* **106**(VII), 53 (1991).
82. *Cosm. Toil.* **106**(IV), 94–98 (1991).
83. *Cosm. Toil.* **102**(X), 107–109, 110–111 (1987); **106**(VII), 72, 73, 75, 76, 83 (1991).
84. *Cosm. Toil.* **106**(VII), 78–80 (1991).
85. *Cosm. Toil.* **106**(VII), 29, 80, 82 (1991).
86. *Cosm. Toil.* **106**(VII), 71 (1991).
87. *Cosm. Toil.* **102**(X), 109–110 (1987); 106 (VII), 77 (1991).
88. *Cosm. Toil.* **106**(VII), 49, 77, (1991).
89. *Cosm. Toil.* **105**, 87 (1990).
90. *Cosm. Toil.* **104**(VII), 67–76 (1989).
91. *Cosm. Toil.* **104**(VII), 76–78 (1989).
92. *Cosm. Toil.* **104**(VII), 79–81 (1989); **106**(IX), 33 (1991).
93. *Cosm. Toil.* **104**(VII), 78–79 (1989); **106**(VI), 36 (1991).

GENERAL REFERENCES

E. Jungermann, ed., *Cosmetic Science and Technology Series*, Marcel Dekker, New York, continuing series.

Kosmetikjahrbuch, H. Ziolkowsky KG, Augsburg, Germany.

F. N. Marzulli and H. I. Maibach, eds., *Dermatoxicology*, 4th ed., Hemisphere Publishing Co., Washington, D.C., 1991.

A. Nowak, *Cosmetic Preparations*, Micelle Press, London, 1991.

M. Rieger, *Surfactants in Cosmetics*, Marcel Dekker, New York, 1985.

K. Schrader, *Grundlagen und Rezepturen der Kosmetika*, Hüthig Verlag, Heidelberg, Germany, 1989.

W. C. Waggoner, ed., *Chemical Safety and Efficacy Testing of Cosmetics*, Marcel Dekker, New York, 1990.

Aerztliche Kosmetologie, G. Braun, Karlsruhe, Germany.

International Journal Cosmetic Science, Blackwell Scientific Publications, Oxford, UK.

Japanese Cosmetic Science Society Journal, Tokyo, Japan.

Journal of the Society of Cosmetic Chemists, New York.

Cosmetics & Toiletries, Allured Publishing Co., Wheaton, Ill.

Drug & Cosmetics Industry, Edgell Communications, Inc., Cleveland, Ohio.

Fett-Wissenschaft Technologie, Konradin-Industrieverlag, Leinfelden-Echterdingen, Germany.

HAPPI, Rodman Publishing Corp., Ramsey, N.Y.

Manufacturing Chemist, Morgan-Grampian, London.

Parfümerie und Kosmetik, A. Hüthig, Heidelberg, Germany.

Parfums, Cosmetiques, Aromes, Paris.

Seifen, Oele, Fette, Wachse, A. Ziolkowsky KG, Augsburg, Germany.

Soap Cosmetics Chemical Specialties, PTN Publishing Co., Melville, N.Y.

Soap, Perfumery and Cosmetics, United Trade Press, Ltd., London.

Martin M. Rieger
M & A Rieger, Associates